THE
ENCYCLOPEDIA
OF THE
ENVIRONMENT

THE
ENCYCLOPEDIA
OF THE
Environment

The René Dubos Center for Human Environments, Inc.

Ruth A. Eblen and William R. Eblen

EDITORS

HOUGHTON MIFFLIN COMPANY

Boston • New York

Copyright © 1994 by Houghton Mifflin Company

For information about permission to reproduce selections from
this book, write to Permissions, Houghton Mifflin Company,
215 Park Avenue South, New York, New York 10003.

Library of Congress Cataloging-in-Publication Data
The Encyclopedia of the environment / Ruth A. Eblen and
William R. Eblen, editors.
p. cm.
Includes index.
ISBN 0-395-55041-6
I. Environmental sciences — Encyclopedias. I. Eblen,
Ruth A. II. Eblen, William R.
GE10.E53 1994 94-13669
363.7'003 — dc20 CIP

Printed in the United States of America

Book design by Robert Overholtzer

DOW 10 9 8 7 6 5 4 3 2 1

Contents

Contributors

John F. Ahearne
Sigma Xi,
The Scientific Research Society

Bruce N. Ames
University of California
at Berkeley

Frank M. Andrews
Institute for Social Research
University of Michigan

Robin M. Andrews
Virginia Polytechnic Institute
and State University

Ronald M. Atlas
University of Louisville

R. U. Ayres
The European Institute of
Business Administration

Victor R. Baker
University of Arizona

John M. Bancroft
University of Arizona

Robert S. Banks
Robert S. Banks Associates, Inc.

Douglas S. Barasch
New York, New York

Brent Barker
Electric Power Research
Institute

Donald G. Barnes
Alexandria, Virginia

Jonathan Barnett
City College of New York

Bradley W. Bateman
Grinnell College

Sandra S. Batie
Virginia Polytechnic Institute
and State University

Dinah Bear
President's Council
on Environmental Quality

Alfred M. Beeton
Great Lakes Environmental
Research Laboratories

Norman A. Berg
American Farmland Trust

Joseph A. Berry
Carnegie Institution
of Washington

Niko Besnier
Yale University

Peter E. Black
State University of New York
College of Environmental
Science and Forestry

John P. Bongaarts
The Population Council

Daniel B. Botkin
Center for the Study of
the Environment and
George Mason University

Douglas E. Bowers
U.S. Department of Agriculture

Thadis W. Box
New Mexico State University

Lynn Brady
University of Washington

Patrick L. Brezonik
University of Minnesota

Shirley A. Briggs
Rachel Carson Council, Inc.

Lydia Bronte
Phelps Stokes Institute

Janet Brown
Buckminster Fuller Institute

Peter G. Brown
University of Maryland

Patricia Buffler
University of California
at Berkeley

Leon F. Burmeister
University of Iowa

Philip J. Burton
University of British Columbia

Dallas Burtraw
Resources for the Future

Warren T. Byrd, Jr.
University of Virginia

John Cairns, Jr.
Virginia Polytechnic Institute
and State University

Lynton Keith Caldwell
Indiana University

Ross Capon
National Association of
Railroad Passengers

John Carey
Greystone Communications

S. R. Carpenter
University of Wisconsin

Ralph Cavanagh
Natural Resources Defense
Council

Alok K. Chakrabarti
New Jersey Institute
of Technology

Jason W. Clay
Rights & Resources

Joshua T. Cohen
Harvard School of Public Health

David E. Cole
University of Michigan

Rob Coppock
World Resources Institute

Malcolm Cormack
Virginia Museum of Fine Arts

Max Costa
New York University
Medical Center

Ken Cottrill
Downington, Pennsylvania

Kathleen Courrier
World Resources Institute

Vincent T. Covello
Columbia University

Pam J. Crabtree
New York University

James Craig
Virginia Polytechnic Institute
and State University

Craig Cramer
Editor, *The New Farm*

Ronald G. Cummings
Georgia State University

Kenneth W. Cummins
University of Pittsburgh

Dale Curtis
Greenwire

Herman E. Daly
World Bank

Joel Darmstadter
Resources for the Future

Terry Davies
Resources for the Future

Paul F. Deisler, Jr.
University of Texas
School of Public Health

C. C. Delwiche
University of California at Davis

Edward F. Denison
The Brookings Institution

Kenneth L. Dickson
University of North Texas

Terry M. Dinan
Congressional Budget Office

Peter B. Doeringer
Boston University

Robert Dorfman
Harvard University

Robert D. Doyle
University of North Texas

Elisabeth M. Drake
Massachusetts Institute
of Technology

James Drake
University of Tennessee,
Knoxville

Faye Duchin
Institute for Economic Analysis,
New York University

Norman E. Duncan
NED & Associates, Inc.

Riley E. Dunlap
Washington State University

Donald N. Duvick
Iowa State University

Dana Duxbury
Dana Duxbury and Associates

Ruth A. Eblen
The René Dubos Center
for Human Environments, Inc.

William R. Eblen
The René Dubos Center
for Human Environments, Inc.

R. S. Eckaus
Massachusetts Institute
of Technology

W. T. Edmondson
University of Washington

Jake Ehlers
Garrison, New York

John R. Ehrenfeld
Massachusetts Institute
of Technology

Gary B. Ellis
Silver Spring, Maryland

Hugh W. Ellsaesser
Lawrence Livermore National
Laboratory (retired)

Jerry Evensky
Syracuse University

William E. Fassett, Ph.D., R.Ph.
University of Washington

Kenneth E. Feith
Potomac, Maryland

Joseph Fiksel
Decision Focus Incorporated

John Firor
National Center for
Atmospheric Research

Anthony C. Fisher
University of California
at Berkeley

Lianne Sorkin Fisher
Global Forum

Susan Fisher
National Environmental
Stewardship Commission of
the Episcopal Church

Robert Fishman
Rutgers University

Peter Flack
Flack and Kurtz

Harvey K. Flad
Vassar College

Ronald Lee Fleming
The Townscape Institute

Michael S. Flynn
University of Michigan

Rodney Fort
Washington State University

Kennith E. Foster
University of Arizona

Jerry F. Franklin
University of Washington

Marjorie A. Franklin
Franklin Associates, Ltd.

Luisa M. Freeman
Applied Energy Group, Inc.

Nicholas Freudenberg
Hunter College

William R. Freudenburg
University of Wisconsin –
Madison

Ralph I. Freudenthal
West Palm Beach, Florida

Jack J. Fritz
The World Bank

Charlotte Frola
Solid Waste Association
of North America

Don Fullerton
Carnegie Mellon University

Connell B. Gallagher
University of Vermont

Daniel L. Gallagher
Virginia Polytechnic Institute
and State University

Francis H. Geer
St. Philip's Church in
the Highlands

Aharon Gibor
University of California
at Santa Barbara

Amihai Glazer
University of California
at Irvine

Alberto Goetzl
Seneca Economics and Trade

Joseph L. Goldman
International Center for
the Solution of Environmental
Problems

Thomas J. Goliber
The Futures Group

Michael A. Gollin
Keck, Mahin & Cate

Gordon Goodman
Stockholm Environment Institute

William G. Gordon
Marine fisheries consultant,
Fairplay, Colorado

Thomas J. Goreau
Global Coral Reef Alliance

Robert A. Goyer, M.D.
Chapel Hill, North Carolina

John D. Graham
Harvard School of Public Health

Robert Gramling
University of Southwestern
Louisiana

Eric Grant
Denver, Colorado

Thomas P. Grumbly
U.S. Department of Energy

Caroline Clarke Guarnizo
National Academy of Sciences

Duane J. Gubler
Centers for Disease Control
and Prevention

Elias P. Gyftopoulos
Massachusetts Institute
of Technology

Susan G. Hadden
University of Texas at Austin

Jay D. Hair
National Wildlife Federation

Charles A. S. Hall
State University of New York
College of Environmental Science
and Forestry

F. Kenneth Hare
Oakville, Ontario

Jack R. Harlan
University of Illinois

Glenn W. Harrison
University of South Carolina

Carl Haub
Population Reference Bureau

Thomas A. Heberlein
University of Wisconsin

J. Douglas Helms
U.S.D.A. Soil Conservation
Service

H. Lanier Hickman, Jr.
Solid Waste Association
of North America

P. J. Hill
Political Economy Research Center
and Wheaton College

Edward Hoagland
Bennington College

Briavel Holcomb
Rutgers University

John P. Holdren
University of California
at Berkeley

Roger LeB. Hooke
University of Minnesota

Elwood Hopkins
The Mega Cities Project, Inc.

David A. Hounshell
Carnegie Mellon University

Howard Hu
Harvard University Medical
School

Charles Hutchinson
University of Arizona

John B. Jackson
Santa Fe, New Mexico

Harold K. Jacobson
University of Michigan

Daniel H. Janzen
University of Pennsylvania

James M. Jasper
New York University

Kerry M. Joels
Alexandria, Virginia

Marcha Johnson
Brooklyn, New York

Thomas B. Johnson
National Coal Association

Carl F. Jordan
University of Georgia

Joseph Kastner
Grandview, New York

Robert W. Kates
Trenton, Maine

Charles B. Keely
Georgetown University

Nathan Keyfitz
Harvard Center for Population
and Development Studies

Fazlun Khalid
World Islamic Federation
for the Environment

James W. Kirchner
University of California
at Berkeley

Lawrence R. Klein
University of Pennsylvania

Charles D. Kolstad
University of Illinois, Champaign-
Urbana and University of
California at Santa Barbara

William Kornblum
The City University of New York

Melvin Kranzberg
Georgia Institute of Technology

Lester B. Lave
Carnegie Mellon University

Luna B. Leopold
University of California
at Berkeley

Howard Levenson
California Integrated Waste
Management Board

Charles Levenstein
University of Massachusetts,
Lowell

Jeffrey R. Levine
Cambridge, Massachusetts

Mark A. Levine
Sive, Paget & Riesel, P.C.

Herbert S. Levinson
Transportation Consultant
New Haven, Connecticut

Michael H. Levy
Environmental Strategies
& Solutions, Inc.

Henry T. Lewis
University of Alberta

Laurence A. Lewis
Clark University

Gene E. Likens
Institute of Ecosystem Studies

Paul J. Lioy
University of Medicine and
Dentistry of New Jersey
and The Environmental and
Occupational Health Sciences
Institute

Geoffrey H. Lipman
World Travel and Tourism Council

Morton Lippman
New York University

Charles E. Little
Kensington, Maryland

Craig G. Lorimer
University of Wisconsin

Walter Loss
Brookhaven National Laboratory

Thomas E. Lovejoy
Smithsonian Institution

Jean Lythcott
Teachers College,
Columbia University

Jeffrey K. MacKie-Mason
University of Michigan

Richard S. Magee
New Jersey Institute
of Technology

Peter Maille
Arlington, Virginia

Christopher Manes
Cathedral City, California

Jonathan Marks
Yale University

Daniel Martin
International Coordinating
Committee on Religion and
the Earth

Philip L. Martin
University of California at Davis

G. R. Marzolf
Louisville, Colorado

James F. Mathis
Summit, New Jersey

W. Parker Mauldin
The Rockefeller Foundation

Russell C. Maulitz
Presbyterian Medical Center
of Philadelphia

Ernst Mayr
Harvard University

K. Jill McAfee
Advaced Sciences, Inc.

Roger O. McClellan, D.V.M.
Chemical Industry Institute
of Toxicology

Joseph McFalls, Jr.
Villanova University

Samuel J. McNaughton
Syracuse University

Eleanor M. Mc Peck
Radcliffe College

Curt Meine
International Crane Foundation

Martin V. Melosi
University of Houston

James Merchant
University of Iowa

Angela G. Mertig
Washington State University

Paulette Middleton
ASRC/State University of
New York at Albany

Diann J. Miele
Rhode Island Department
of Health

Alan S. Miller
The Center for Global Change

Richard B. Milner
American Museum of Natural
History

Jill Minick
Case Western Reserve University

Robert Cameron Mitchell
Clark University

Abdelsamie Moet
Case Western Reserve University

Gary Moll
American Forests

David W. Moody
U.S. Geological Survey

William R. Moomaw
Tufts University

Barbara Scott Murdock
St. Paul, Minnesota

Norman Myers
Oxford University

J. Arly Nelson
University of Texas

Robert H. Nelson
School of Public Affairs
University of Maryland

Barbara R. Niederlehner
Virginia Polytechnic Institute
and State University

William A. Nierenberg
Scripps Institution
of Oceanography

William D. Nordhaus
Yale University

John T. Novak
Virginia Polytechnic Institute
and State University

James Nybakken
Moss Landing Marine Laboratory
California State University

Rice Odell
Washington, D.C.

Eugene P. Odum
Institute of Ecology
University of Georgia

H. T. Odum
University of Florida

Gilbert S. Omenn, M.D.
University of Washington

Amy Ong Tsui
University of North Carolina
at Chapel Hill

Elinor Ostrom
Indiana University

Wilfred Owen
Arlington, Virginia

Talbot Page
Brown University

Martin Palmer
International Consultancy on
Religion, Education and Culture
The Manchester Metropolitan
University

Richard Pardo
Chemonics International

Bruce C. Parker
Virginia Polytechnic Institute
and State University

Bentham Paulos
Abt Associates Inc.

Stephen C. Peck
Electric Power Research Institute

Janice E. Perlman
The Mega Cities Project, Inc.

Joseph M. Petulla
University of San Francisco

David Pimentel
Cornell University

Stuart L. Pimm
University of Tennessee,
Knoxville

Henry C. Pitot, M.D.
University of Wisconsin –
Madison

Robert E. Pollack
Columbia University

Frank J. Popper
Rutgers University

Jerry Powell
Resource Recycling

Ranchor Prime
International Consultancy on
Religion, Education and Culture
The Manchester Metropolitan
University

Paul C. Pritchard
National Parks and Conservation
Association

Norville S. Prosser
Spotsylvania, Virginia

Susan W. Putnam
Harvard School of Public Health

E. L. Quarantelli
University of Delaware

George Rainer
Flack and Kurtz

David P. Rall, M.D.
Washington, D.C.

Sir Shridath Ramphal
Guyana

Stephen Rattien
National Academy of Sciences

Jack Raymond
New York, New York

Randall R. Reeves
Okapi Wildlife Associates

William C. Reffalt
Herndon, Virginia

William A. Reiners
University of Wyoming

Paul Relis
California Integrated Waste
Management Board

Stephen J. Reynolds
University of Iowa

Laurie L. Richardson
Florida International University

Joseph V. Rodricks
ENVIRON International Corp.

Haia K. Roffman
AWD Technologies, Inc.

Bernard Roizman
University of Chicago

Victor A. Roque
Orange and Rockland Utilities, Inc.

Barbara Gutmann Rosenkrantz
Harvard University

David Rothenberg
New Jersey Institute of
Technology

Wolfgang Rüdig
University of Strathclyde

David Runnalls
Institute for Research on
Public Policy
Ottawa, Ontario

Charles M. Salter
Charles M. Salter Associates, Inc.

Milton R. J. Salton
New York University School
of Medicine

Hal Salwasser
University of Montana

V. Alaric Sample
American Forests

R. Neil Sampson
American Forests

Sheldon W. Samuels
Ramazzini Institute for
Occupational and Environmental
Health Research

Peter M. Sandman
Newton Center, Massachusetts

Hernan Sanhueza
International Planned Parenthood
Federation

Lynn Scarlett
Reason Foundation

Harry E. Schwarz
Clark University

Vincent Scully
Yale University

Sheldon J. Segal
The Population Council

A. M. C. Şengör
İstanbul Teknik Üniversitesi

Linda K. Shaw
Oak Ridge National Laboratory

Renee A. Shaw
Oak Ridge National Laboratory

Paul Shepard
Bondurant, Wyoming

Warren K. Sinclair
National Council on Radiation
Protection and Measurements

Steven W. Sinding
The Rockefeller Foundation

David Sive
Sive, Paget & Riesel, P.C.

Lisa Skumatz
Synergic Resources Corp.

John Slade, M.D.
University of Medicine and
Dentistry of New Jersey

Paul Slovic
Decision Research, Inc.
and University of Oregon

David A. Somers
University of Minnesota

Frank Spencer
Queens College of the
City University of New York

Charles A. Starner
Jerry Kugler Associates

Chauncey Starr
Electric Power Research Institute

Harold K. Steen
Forest History Society

Frederick Steiner
Arizona State University

Andrew Stoeckle
Abt Associates Inc.

Thomas B. Stoel, Jr.
Washington, D.C.

Arnold Strassenburg
State University of New York
at Stony Brook

Karen B. Strier
University of Wisconsin –
Madison

Maurice F. Strong
Chairman, Ontario Hydro and
Chairman, Earth Council

Richard L. Stroup
Montana State University
and Political Economy Research
Center

Roger B. Swain
Horticulture

Lee M. Talbot
Lee Talbot Associates
International

Joel A. Tarr
Carnegie Mellon University

Jefferson W. Tester
Massachusetts Institute
of Technology

Peter S. Thacher
Stonington, Connecticut

William Trager
The Rockefeller University

Tom Turner
Sierra Club Legal Defense Fund

Robert Twombly
City College of New York

Arthur C. Upton, M.D.
New York University Medical
Center

Mark J. Utell, M.D.
University of Rochester

Etienne van de Walle
University of Pennsylvania

Carlton S. Van Doren
Texas A & M University

Jan van Schilfgaarde
U.S. Department of Agriculture

Willem van Vliet –
University of Colorado, Boulder

N. C. Vasuki
Delaware Solid Waste Authority

Hassan Virji
International START Secretariat

Vukan R. Vuchic
University of Pennsylvania

Mark R. Walbridge
George Mason University

H. Jesse Walker
Louisiana State University

James C. G. Walker
University of Michigan

William Waller
University of North Texas

Barry Walden Walsh
Garrett Park, Maryland

Brent Walth
Portland, Oregon

Herbert Han-pu Wang
Abt Associates Inc.

George W. Ware
University of Arizona

Karen J. Warren
Macalester College

Donald Watson
Rensselaer Polytechnic Institute

David H. Wegman, M.D.
University of Massachusetts,
Lowell

Bernard Weiss
University of Rochester School
of Medicine

Richard P. Wells
Abt Associates Inc.

Patricia Werner
University of Florida

Arthur H. Westing
Westing Associates

Gilbert F. White
University of Colorado

Robin K. White
Oak Ridge National Laboratory

John Wiens
Colorado State University

David L. Wigston
Northern Territory University,
Australia

Chris F. Wilkinson
Technology Sciences Group, Inc.

Delta Willis
New York, New York

Jennifer R. Wolch
University of Southern California

M. Gordon Wolman
Johns Hopkins University

Daniel Woltering
ENVIRON Corporation

Julia Wormser
Abt Associates Inc.

Lisa M. Wormser
Surface Transportation Policy
Project

Lynn L. Wright
Oak Ridge National Laboratory

Mark Wright
Washington, D.C.

Gerald L. Young
Washington State University

Ervin H. Zube
University of Arizona

Jeffrey M. Zupan
Chestnut Ridge, New York

Contributors of Unsigned Entries

Orestes Anastasia
Susan M. Barlow
Kate Bickert
Adam Button
David Case
Michael Casey
Matthew D'Amore
Scott Gentile

Carol B. Goldburg
Wendy Horng
Eunice Kang
Julie A. Kenyon
Robert Lebeau
Terzah Lewis
Kimberly A. Lindblade
Leland Malkus

Andre McCloskey
Charles Mendler
James Brett Moreland
Jon Moskowitz
Joanne Newman
Eduardo Paguaga
Michael Paque
Anne H. Peracca

Sara Peracca
M. James Riordan
Julie Ruttenberg
Scott Savaiano
Alisa Shen
Patrick Stern
K. Gordon Young
Mary E. Yurlina

Preface

We cannot escape from the past, but neither can we avoid inventing the future. With our knowledge and a sense of responsibility for the welfare of humankind and the Earth, we can create new environments that are ecologically sound, aesthetically satisfying, economically rewarding, and favorable to the continued growth of civilization.

— René Dubos

Humans and their environments—not as separate, static entities, but as interacting components of complex dynamic systems—are the focus of *The Encyclopedia of the Environment*. The very process of living transforms the total environment profoundly and lastingly, and this is particularly true of human activities. From the humanistic point of view, the environment is the place where the natural and human orders undergo some kind of fusion.

The Encyclopedia of the Environment is based on the legacy of René Dubos, the world renowned scientist and humanist. His early understanding of the need to study bacteria in the soil, where they live, rather than in petri dishes in the laboratory, led to his teaching the world the principles of finding and producing antibiotics. The same ecological insight shaped his contributions to improving environmental quality and changing public policy and prompted *The New York Times* to dub him "the philosopher of the Earth."

René Dubos founded The Center for Human Environments that bears his name in response to the need for a humanistic approach to environmental problems. The Center emerged early in 1975 from a collaboration between René Dubos, then Professor Emeritus at The Rockefeller University, and Total Education in the Total Environment—a nonprofit organization with extensive experience in environmental education, founded in 1964 by William R. Eblen. As chairman, René Dubos formulated the Center's socio-environmental philosophy, guided its policies, and together with his wife Jean participated in all of its activities until his death in 1982. Dubos bequeathed his environmental library and archives from the previous twenty-five years to the Center. The Center's motto is the Dubos admonition "Think Globally, Act Locally"™.

The goals of the Center are to help develop creative policies for resolving environmental problems and to create new environmental values. In order to achieve these goals the Center is involved in separate but complementary programs designed to acquire, organize, and disseminate knowledge about the interdependent technical and humanistic aspects of socio-environmental problems.

People expect more from the environments in which they live than conditions suitable for their health, resources to run the economic machine, and whatever is meant by good ecological conditions. Dubos said that people want to experience the sensory, emotional, and spiritual satisfactions that can be obtained only from an intimate interplay, indeed from an identification with the places in which they live. This interplay and identification constitutes an individual's sense of place.

Entries in *The Encyclopedia of the Environment* reflect this interplay between human beings and environments from two complementary points of view. On the one hand, human beings always transform the environments in which they live and function—practically all inhabited environments are artificial in the sense that they have been profoundly altered by human cultures. On the other hand, human beings are shaped by the environments in which they develop, each culture reflecting the influence of the environment in which it has been created and has evolved.

This leads to the following tenets on which *The Encyclopedia of the Environment* and the programs and activities of The René Dubos Center are based:

- Human life implies interventions into nature. Properly managed, these interventions can be not only ecologically sound but also create new environmental values. In many parts of the world human interventions have resulted in "humanized" environments that are ecologically stable, economically profitable, esthetically pleasurable, and favorable to the continued growth of civilization.

- We now have the knowledge that makes it possible to take advantage of the resiliency of nature, provided there is the social will to act. Admittedly, human interventions into nature have often been destructive and are now responsible for the ecological problems which threaten the whole planet. There is overwhelming evidence, however, that many forms of environmental damage can be repaired. When properly managed, damaged eco-

xvi *Preface*

systems have recovered more rapidly and at lower cost than had been anticipated.

- Any change implies risk. There cannot be any progress without social or technological changes and, therefore, without risk. Our societies have to develop better methods of anticipating risks associated with changes and for evaluating the willingness of people to accept these risks. The acceptability of risk is probably always conditioned by the desirability of certain social values.

- Changes inevitably have consequences that cannot be predicted and that differ in their effects on various social systems. The usual methods for dealing with these difficulties are (a) increasing governmental regulation to improve safety and fairness, and (b) confrontations between opposing factions. Regulation and confrontation are costly and tend to paralyze initiatives essential for progress. Opportunities that enable scholars to discuss the socio-environmental problems arising from new developments with scientists, technologists, and other decision-makers are necessary; the discussion of these problems is best carried out not in an adversarial atmosphere but in a cooperative spirit.

- The general public is increasingly involved in all important decisions concerning environmental problems and technological developments. While such participatory democracy is desirable, it can be useless and even dangerous unless based on knowledge. The knowledge needed by the general public in this regard is not technical but rather focused on costs, benefits, and long-range socio-cultural consequences.

An additional dimension for selecting entries for *The Encyclopedia of the Environment* was provided by Dubos' "Five E's of Environmental Management": Ecology, Economics, Energetics, Esthetics, and Ethics.

The following key questions also helped delineate the selection of entries: What do we need to know as individuals, consumers, citizens, and voters to make informed choices and take specific actions on environmental problems? What do we need to know about the roles of business, industry, government, and special interest groups in socio-environmental problems? And what do we need to know about legislative, regulatory, and institutional changes implemented by government and supported by different constituencies?

World authorities—specialists with many different viewpoints—responded to these questions by providing the best available nontechnical information in a single volume reference work for the general reader. Entries were selected to create a common ground for

understanding the basic terms and concepts involved in complex environmental issues. They emphasize the broad range and marked diversity encompassed by the concept of the total environment—from *acid rain* to *zoning laws*. Because of space limitations biographical entries exclude living leaders but include selected representatives from various fields to give the reader a sense of the breadth of the relationships between humans and their environments. Many other leaders—living and deceased—are cited throughout.

The Encyclopedia of the Environment is a milestone in The Decade of Environmental Literacy launched by the Dubos Center in 1990 in cooperation with the United Nations Environment Programme (UNEP). This reference book is by its very nature a work-in-progress. We hope it provides a conceptual framework that will encourage readers to collaborate with us in improving and expanding it.

Acknowledgments

The wellspring for *The Encyclopedia of the Environment* is The René Dubos Center's forum program. Since 1977 the Center has conducted seventeen forums and related activities, involving over 2,000 authorities and more than 650 organizations representing many different vantage points. We remain very grateful to Nancy Wright, a theologian and editor, for proposing Ruth Eblen's idea for *The Encyclopedia of the Environment* to Paul Barnabeo, formerly an editor at Houghton Mifflin Company. His enthusiasm and scholarly interest in Dubosian philosophy and creative efforts in the early stages of planning, got us off to a good start.

The Encyclopedia has truly been a collaborative effort, but we would be remiss in not citing those individuals who have contributed so very much in special ways, above and beyond the call. First and foremost is Jonathan Latimer, formerly Reference Publisher at Houghton Mifflin Company. Every aspect of the Encyclopedia has benefited from his knowledge, experience, and technical expertise, and his vital editorial guidance and encouragement made this work a reality. In addition to contributing to its early development, Peggy Tsukahira, editor of Dubos' *Celebrations of Life*, brought her special knowledge of Dubos' philosophy and her sensitivity to international and multicultural issues to the editing of the Encyclopedia.

The Executive and Advisory Boards for the Encyclopedia have given generously of their time and expertise to all stages of its development, especially in recommending expert authors. Many wrote key articles and helped edit entries. In addition, the following helped organize groups of entries devoted to their fields: Norman Berg (agriculture); John Cairns (natural environment); Riley Dunlap (organizations); John Keith (settlements); Lester Lave (economics); Thomas Merrick (population); Gilbert Omenn (health and medicine); Neil Sampson (land); Lynn Scarlett (waste); Chauncey Starr (energy); David Sive (law and public policy); and Arthur Upton (health and medicine). We are also grateful to the following contributors who provided recommendations and scholarly advice: Antonia Abbey, Richard K. Bambach, Jack Cranford, Paul Deisler, John Ehrenfeld, Susan Fisher, Roger Kasperson, Barbara Murdock, Gerard Piel, Jonathan Piel, Jack Raymond, George Simmons, and Ervin Zube.

The staff of The René Dubos Center made important contributions as well, including Michael Casey, who organized the entries and established the data bank; Sara Peracca, who acted as editorial assistant; Erica Mohan, who provided administrative assistance; and Ann McCooey, who gave general support.

And finally, we are very grateful to the authors of the Encyclopedia. They represent the best and the brightest in their areas of expertise, and we are very proud to have had this opportunity to bring their collective wisdom to *The Encyclopedia of the Environment*.

Ruth A. Eblen, Executive Director,
and William R. Eblen, President,
The René Dubos Center for Human
Environments, Inc.

A

ACID RAIN

Acid rain is the deposition of airborne acids, not just in rain, but also in snow, fog, and dry acidic particles. The primary source of acid rain is burning of coal and oil in electrical powerplants, industrial boilers, and internal combustion engines. Fossil fuel burning produces sulfur dioxide (SO_2) and oxides of nitrogen (NO_x). In the atmosphere, SO_2 and NO_x react chemically with ozone and other compounds to form sulfuric and nitric acids—the acids in acid rain.

Before SO_2 and NO_x fall to earth as acids, however, they can be carried by wind for hundreds of miles. Thus, acid rain may originate in one state, region, or nation, and fall in another. Roughly half of Canada's acid rain comes from the U.S., and 80% of Scandinavia's acid rain comes from elsewhere in Europe. Natural, unpolluted rain can contain small amounts of organic acids or windblown alkali dust, and thus typically has a pH ranging from 4.9 to 6.5. However, downwind of major industrial regions (such as the American midwest) average rainfall is often 10–100 times more acidic (pH 4.5 to 4). Individual acid storms can be over five times more acidic (0.7 pH unit lower) than these long-term averages.

There is clear evidence that acid rain causes biological damage in acid-sensitive lakes and streams. Small changes in the balance between acids and bases in stream and lake water can cause large changes in pH. Many aquatic organisms lose the ability to absorb and retain sodium in low-pH waters. Acid waters also often have high concentrations of dissolved aluminum, which disrupts organisms' salt balances and interferes with gill function. Often, reproductive failure occurs before direct damage to adult fish. Less conspicuous organisms, notably amphibians and aquatic invertebrates, often succumb first to acidification. Therefore, although disappearance of large sport fish is one of the most visible effects of acidification, it is not a good early warning indicator.

Small changes in acid loading, and thus lake acidity, may make lakes uninhabitable for many species. Even in waters that are not chronically acidic, organisms may be threatened by short-lived acid pulses that occur when spring snowmelt or intense rainstorms flush acids through the watershed faster than they can be neutralized. Other lakes may be naturally acidic, be-cause they have high concentrations of organic acids. Organisms living in such waters are adapted to their chemistry; in contrast, organisms living in acid-sensitive, but unacidified, waters may not survive the chemical changes that result from acidification.

Lakes and streams receiving the same acid rain may be too acidic for fish, or may be habitable, depending on how well their watersheds neutralize acids. Lakes and streams in watersheds with sedimentary bedrock will usually not be acidic, because most sedimentary rocks contain carbonate minerals, which dissolve rapidly to buffer (neutralize) acids. In contrast, watersheds underlain by igneous and metamorphic rocks (which contain little or no carbonate minerals) can be moderately or highly acid-sensitive. The acid sensitivity of these watersheds is determined by how much buffering capacity is stored in their soils. This, in turn, depends on how rapidly alkaline chemicals are released by mineral weathering. Acid rain may damage some lakes, while having little effect on others in the same region, due to local geological differences.

A direct link has not been established between acid rain and forest dieback in the U.S., except for forests on mountainsides that are frequently bathed in highly acidic clouds and fogs. In some European forests, acid rain may be contributing to forest dieback by leaching nutrients such as calcium, magnesium, and potassium from soils. Forest dieback probably results from a combination of stresses, including climate, drought, acid rain, ozone, and other air pollutants. A simple, direct relationship between acid rain and forest health is therefore difficult to prove.

Acid rain itself probably has little direct effect on human health, but it may indirectly create health risks for some people. For example, as water becomes more acidic, it becomes more corrosive; thus acidification may accelerate leaching of lead from pipes and solder, leading to higher concentrations of lead in drinking water (although most community water supplies add alkaline chemicals to drinking water to control corrosion). Similarly, coal combustion—which leads to acid rain—also releases mercury, and lake acidification converts mercury to its toxic methyl form. Thus, some people who routinely eat fish from acidified lakes could be exposed to high levels of methyl mercury. Furthermore, the same combustion processes that cause acid rain also generate ozone, airborne particulates, and acidic aerosols, all of which are potentially hazardous when inhaled. Therefore, measures to control acid rain may benefit human health, even if acid rain itself does not directly make people sick.

Emissions of SO_2 and NO_x, the building blocks of acid rain, can be reduced by switching to low-sulfur coal and oil supplies, chemically or physically cleaning sulfur from coal and oil before combustion, mixing powdered limestone with the fuel during combustion, or using "scrubbers" to remove SO_2 and NO_x from flue gases after combustion. NO_x emissions can also be controlled by lowering the combustion temperature or restricting the supply of oxygen. SO_2 and NO_x emissions can also be cut through energy conservation. Energy-efficient light bulbs and refrigerators, for example, reduce electricity demand and thus reduce acid rain, usually at lower cost than other technical solutions. Where acid rain cannot be adequately controlled, individual acid lakes can be neutralized by repeated additions of crushed limestone. This usually permits some fish to return, but does not restore the lakes' original pre-acidification chemistry or biology.

Now that scientists generally agree on the causes and effects of acid rain, the political debate has shifted from whether a problem exists at all, to whether it is a problem worth solving and, if so, who should pay the cost. The environmental costs of acid rain (and thus the benefits expected from controlling it) are potentially large but highly uncertain, while the costs of emission control are both substantial and highly certain. Some people still maintain that we should not control acid rain unless we are sure that it is economically worthwhile to do so. Others contend that emission control is cheap insurance against potentially widespread, insidious, and irreversible environmental damage. This viewpoint has apparently been persuasive. In the mid-1980s, Canada and seventeen European countries jointly agreed to reduce their SO_2 emissions by 30%, in an ultimately successful attempt to pressure Britain and the U.S. to make similar reductions. In the U.S., the 1990 Clean Air Act amendments signaled the end of a decade of political indecision, mandating 40% cuts in SO_2 emissions from 1980 levels.

Now that commitments have been made to reduce the emissions that cause acid rain, there is a growing sense of optimism that the acid rain problem will be solved. This complacency may be premature. Drastic reductions in acid deposition can reverse the acidification of lakes, streams, and forest soils, but there is growing evidence that the proposed reductions will be too small for many ecosystems to recover. Restoring fish habitat in some Norwegian streams, for example, may require cutting emissions by 80% or 90%, far beyond any reductions now planned.

Dealing with acid rain will require ongoing sci-entific work and political commitment, long after its novelty as an environmental issue has faded. Although field data show that rainfall acidity is now declining in many regions, it remains—and will remain for the foreseeable future—far higher than natural levels. If it remains high enough to leach away buffering capacity from acid-sensitive watersheds faster than geochemical processes can resupply it, long-term progressive acidification of some lakes, streams, and soils will continue even as rainfall acidity declines. There is little indication that acid rain, or its effects, will simply go away.

JAMES W. KIRCHNER

For Further Reading: National Acid Precipitation Assessment Program, *Acidic Deposition: State of Science and Technology* (1990); James L. Regens and Robert W. Rycroft, *The Acid Rain Controversy* (1988); David W. Schindler, "Effects of acid rain on freshwater ecosystems," *Science* (1988).

ACOUSTIC ENVIRONMENT

Sound in the environment occurs across four dimensions: time plus the three dimensions of space. Wavelength frequencies occur in the horizontal dimension; sound energy, in pressure per square centimeter, occurs in the vertical dimension; and reverberation period occurs in the longitudinal dimension (depth). Many sounds coexist in the same space at the same time.

Sound energy is generated by the vibration of a body such as vocal cords. The movement of the vibrating surface compresses the air molecules directly adjacent to it, causing a pressure wave. The pressure wave travels away from the source of the sound at a constant speed, 1,130 feet per second in air. When the pressure wave reaches the organ of perception, it is converted to audible sound. In humans, this occurs when sound vibrations reach the inner ear and stimulate the auditory nerve.

Millions of years ago the ability to hear sounds made by dangerous animals was necessary to allow our ancestors to protect themselves in their environment. The human hearing mechanism evolved to be so sensitive in hearing sound that no microphone or noise measurement system has ever been built that can measure the low sound levels audible to humans.

The intensity, or loudness, of sound is measured in decibels (dB). The decibel scale is a logarithmic scale that expresses power ratios. An increase of 10 dB requires a ten-fold increase in power; an increase of 20 dB requires a twenty-fold increase in power. Humans can detect a change in sound intensity as small as 1 dB. The sound levels human beings are typically exposed to range from 0 to 120 dB. (For a figure showing decibel ranges for common sounds, see "Pollution, Noise." Measurements of sound levels expressed in decibels are used by the U.S. Occupational Safety and Health Administration in setting regulations to limit the risk of hearing damage by workers, by communities establishing noise ordinances, and for describing speech interference, sleep interference, and annoyance.

It has been recognized since ancient times that hearing loss occurs among workers in noisy industries. A popular description of loud noise is "deafening." Noise that is not damaging can be annoying. Two thousand years ago a ruler in Rome proclaimed that horse-drawn carts would not be allowed within the city limits after dark because the noise disturbed the populace and interfered with sleep.

To combat the deleterious effects of noise in the modern environment, the U.S. Environmental Protection Agency, the Canadian National Research Board, and other organizations throughout the world have researched the effects of noise on people. Standards have been promulgated to protect the hearing of workers in factories and to limit the exposure of people in their homes and offices to noise generated by aircraft, mechanical equipment in buildings, and vehicles.

CHARLES M. SALTER

ACQUIRED IMMUNE DEFICIENCY SYNDROME

See AIDS.

AEROBE

See Bacteria.

AEROSOLS

See Atmosphere; Smog-Tropospheric Ozone.

AESTHETICS

See Esthetics.

AGE COMPOSITION

The age composition of a population represents the number of people of a given age and sex in the population. It can be depicted by a population pyramid, which shows the proportion of the population in each

age group. The sum of the proportions in all age groups equals 100% of the population.

There are three general types of so-called population pyramids: rapid, slow, and near zero growth. A rapid-growth population pyramid is actually shaped like a pyramid, since each age group is larger than the one before it. This shape results primarily from high fertility. If couples in one generation average six children, as they do in Kenya, for example, their children's generation will be about three times larger than their own. Thus, the pyramid would be about three times as wide at the base as in the middle. Rapid-growth population pyramids can also acquire their distinctive shape because of high mortality in the past, meaning that the older age groups have fewer surviving members and thus occupy a relatively small section. The base can also be broadened if mortality, particularly infant mortality, has been declining in recent years, resulting in younger age groups having more survivors and occupying a relatively large section of the pyramid. Because the vast majority of people in rapid-growth populations are young and capable of childbearing, there is tremendous momentum for future population growth. Even if these people only have four-child families, which is the average for less developed countries, their children's generation would be twice the size of their own.

The shape of a slow-growth population pyramid indicates the process of change from a rapid-growth to a nearly stationary population, in response to changes in fertility and mortality. The United States is typical of these "middle age," slow-growth societies.

Nearly stationary populations, such as that of Denmark, are better represented as rectangles than as pyramids. All age groups are about the same size, since the birth rate and the death rate have been low and relatively constant for a long time. This means that each age group is about the same size at birth and, since relatively few people die before old age, the groups remain close in size until late in life, when mortality rates rise. Because a relatively high proportion of people in near-stationary populations are elderly, these populations are called "old" societies. They are at or near zero population growth.

Population pyramids can also be shaped by migration. Since migration is age-selective, it alters the shape of both the place-of-origin and place-of-destination pyramids. In general, since migrants tend to be disproportionately young, the place-of-origin pyramid grows "older" and the place-of-destination pyramid grows "younger."

The age composition of a society has a profound influence on its demographic and social character. For example, age structure affects population growth, and it is especially relevant to social problems that usually have age-dependent causes or consequences. The affected individuals are often disproportionately from specific age groups. For example, the chronically ill are disproportionately elderly. As the proportion of elderly people in a population increases, so does the proportion of chronically ill people. Thus, changes in age composition can alter the severity of a social problem, even if there is no change in the underlying causes.

JOSEPH MCFALLS, JR.

AGENT ORANGE

See Herbicides; War and Military Activities: Environmental Effects.

AGRICULTURE

The development of agriculture probably did more to alter the face of the Earth than any other human intervention in the environment before large cities. It is generally believed that agriculture did not originate in any one place but was discovered gradually in many different parts of the world between 5000 and 9000 B.C. The earliest humans practiced hunting and gathering, harvesting what edible foods grew naturally and following the migration of animals. After thousands of years humans realized that food production could be increased if seeds taken from wild plants were planted near desirable places for permanent settlements and animals confined or raised within the same area. Early domestications such as sheep (9000 B.C. in Iraq), cattle (6000 B.C. in Greece), barley (7000 B.C. in Iran), squash (5000 B.C. in Mexico), potatoes (2500 B.C. in South America), and rice (3000 B.C. in India) show that the trend to domesticate promising sources of food was worldwide.

As soon as people started to plant for harvest, they began to alter the ecology around them. To plant seeds, native grasses had to be uprooted, trees and brush removed. Plows and hoes were developed to turn soil and keep out weeds. Agriculture raised the level of nutrition and permitted the development of occupations unrelated to food, making cities possible. The most dramatic changes occurred in dry areas where people turned to irrigation. In addition to transforming deserts into gardens, the complex social systems necessary to harness rivers and divert water to fields resulted in some of the great early civilizations. The Sumerians, in the delta of the Tigris and Euphrates rivers, used the bounty from their irrigated fields to support a highly specialized urban society. Egyptians, who enjoyed high yields from deposits left by the annual

flooding of the Nile, developed by 3000 B.C. a strongly bureaucratic society around the control of those waters. This self-sustaining system supported a stable rural economy for thousands of years.

With the beginning of agriculture came two problems which have plagued farmers ever since—soil depletion from continually planting the same crop in the same place and soil erosion caused by the runoff from cultivated fields. In addition, some irrigated lands, such as those in Iraq, eventually had to be abandoned due to salt buildups in the soil—a problem that has also been seen in modern irrigation systems.

Early farmers not only altered the land, they altered plants and animals. By continually selecting seeds from plants with the most desirable characteristics, they produced varieties that were much better suited to cultivation than those growing in the wild. For example, wild wheat dropped its seeds as soon as they ripened. By saving seeds from mutants that held the grains longer, farmers were able to develop a wheat that was more practical to harvest. Likewise certain animals were chosen for domestication. Between 10,000 and 5000 B.C. most of the plants and animals currently used in agriculture were first domesticated and improved.

Trade between civilizations spread plants and animals far from their native habitat. But each part of the world developed its own food preferences based on climate and growing conditions. Chinese agriculture was shaped by the heavy pressure of population in a country where only 10% of the land was arable. By the 4th millennium B.C. millet was grown in north China and rice in south China. By 2200 B.C. the extensive Chinese system of irrigation had begun. Meat (mainly pork) formed a small part of the Chinese diet; human wastes were needed to supplement animal manure as fertilizer. Before the end of the first millennium B.C., population growth had forced intensive cultivation and a search for more productive varieties. About A.D. 1000 Champa rice, a drought-resistant, early maturing variety, was introduced. Farmers expanded rice growing into hilly areas, transforming steep slopes into terraced and irrigated rice fields. The 16th-century introduction of such New World crops as corn and potatoes permitted further expansion onto poorer soils.

On the Indian subcontinent a sophisticated urban civilization developed in the 3rd millennium B.C. in Mohenjo-Daro, a culture based on highly productive flooded lands along the Indus River. As in the case of Egypt, farmers worked with the natural cycle of floods without any need for fertilizer. By 1000 B.C. agriculture had spread over much of the subcontinent. Wheat, barley, millet, and many fruits and vegetables were widely grown; rice was planted on land that could be irrigated. Agricultural technology changed slowly in India. Land was more abundant than in China and virgin soil was still available in the 17th century.

In the Americas, corn, potatoes, squash, beans, and other native crops were found to be highly productive. North American Indian men continued to hunt while women practiced agriculture. Peruvians constructed an elaborate system of terraces. African agriculture was established by 3000 B.C. below the Sahara. Africans domesticated sorghum and raised cattle and some wheat. The arrival of Asian crops such as yams and bananas about 2000 years ago permitted agriculture to expand south into humid areas.

The Western world came under the domination of Rome in the two centuries B.C. Rome's conquered provinces usually preserved much of their old style of agriculture and land tenure arrangements, ranging from a continuation of the forced labor on state-owned lands that had characterized the Nile area to the many small holdings in Roman Britain. In Italy itself, competition from other provinces and erosion caused a decline in grain production in favor of wine, olive oil, fruit trees, and other crops. Large-scale farms called latifundia operated by slave labor replaced peasant farmers in many areas. A two-field system with half lying fallow each year became common. So did the use of rotations that included legumes and root crops for animal feed. By the Roman era, most hand tools resembled their modern counterparts. Plows, however, still dug too shallowly to turn a furrow.

During the 1000 years that constituted the Middle Ages, Western agriculture made slow progress. The centuries following the barbarian invasions witnessed the growth of the feudal manorial system under which a lord or cleric commanded the services of the peasants who farmed their lands. Concentrations of free peasants who owned small holdings remained in Germany and England. New inventions such as the wheeled German plow, which turned a deep furrow, and the horse collar, likely a Chinese idea which permitted horses to pull much greater loads, permitted more intensive cultivation. The open field system characterized medieval agriculture. Under it, peasants on a manor were assigned a number of long, narrow strips to farm scattered throughout the manor's lands. This assured some equity in dividing good and poor land but made it difficult for individual farmers to improve their acreage. Peasants were also required to do a certain amount of work on the lord's land. The two-field rotation of the Romans persisted for centuries although in some places it was supplanted by a three-field system in which two different grain crops alternated with fallow, thus increasing production. The custom of keeping livestock on common ground and woodland and allowing it to forage the fields at certain times discouraged attempts to improve breeds or increase the amount of available manure. In most things, the manor was self-sufficient.

Beginning about the 10th century, much new land was brought under cultivation. The pressure of growing population pushed agriculture into marginal waste lands and forests. Marshes were drained in the Netherlands; German settlers opened farmland in the Baltics. Another trend, in the later Middle Ages, was to convert the services peasants owed their lord to money rents. Along with this was a movement by landlords to evict tenants and enclose fields in order to specialize in profitable market crops such as wool. This was a period of peasant unrest. While evicted tenants often went to cities, other peasants were able to become freeholders. On the other hand, in the eastern German states and Russia, serfdom was just taking hold in the 16th century.

Medieval agriculture, from an ecological point of view, was a fragile system that could barely sustain its low level of productivity. Manure and legumes were used inefficiently. Even pasture and woodlands tended to be overgrazed by the few animals on a typical manor. Europe, however, was on the verge of a sustained increase in production. Part of this change came from new crops. Columbus' voyage to the New World in 1492 had a dramatic impact on European agriculture. Previously Europeans favored cereals such as wheat, which produced scarcely three or four grains for every one sown. Local famines were common in Europe when poor weather ruined a crop. But two American plants entering Europe in the 1500s, corn and potatoes, offered dramatically improved yields for a minimum of labor. Corn, used mostly for animal feed, made an increase in livestock (and therefore manure) feasible. Potatoes, adopted more slowly, improved the diets of peasants wherever they were grown, especially in the German states and Ireland. During the Renaissance, agricultural manuals became popular with landowners seeking to improve their productivity.

By experimenting with new methods, 18th-century British landowners achieved a great improvement in their country's agriculture, one that would have a lasting impact on agriculture around the world. As landowners gained control over their estates through enclosure, they made two critical advances. One was to break the cycle of low productivity that had so long characterized European agriculture. Progressive British farmers began using the so-called Norfolk system, a four-course rotation that eliminated fallow land by offsetting grains with legumes and root crops, such as turnips. The roots and legumes provided fodder for more animals, which added more manure to the soil. The Norfolk system, in other words, put the requirements of plants and animals in balance, to their mutual benefit. Many farmers further improved their lands by adding marl and draining heavy soils. A second advance was the careful breeding of animals; over the course of the century the weight of livestock going to market doubled. By the late 18th century, new farm machines such as threshers and seed drills were also helping to improve productivity. Both advances were bolstered by scientific inquiries into biology and agriculture.

The success of these techniques depended on environmental factors. The lighter soils of south and west England benefited most, making these areas even more agricultural than before. The heavy clay soils of northern England were less adaptable. Here, cottage industries and urban factories became increasingly important, especially where water power was available. As agriculture advanced, regional specialization came to characterize many parts of Europe. The British continued to lead the way in the 19th century. By 1870 only 14% of the labor force was engaged in agriculture yet it produced 80% of the country's food.

While British agriculture was improving, a very different sort of agriculture was taking shape in Britain's North American colonies. Whereas British land was in short supply and labor cheap, American colonists found a vast amount of cheap land but little labor. British practices of intensive cultivation and careful fertilizing gave way in America to an extensive agriculture that mined the soil. American farmers, who produced crops for sale whenever they could reach a market, found it made economic sense to clear new land, plant it until the soil wore out, and then abandon it. This was true even among Southern farmers, many of whom solved their labor problem by buying slaves. In the tobacco areas of the Chesapeake, for example, a common plan was to kill the forest by girdling trees and plant tobacco, corn, and wheat without fertilizer for a couple of years each until the land would yield no more. After this the farmer, who had since cleared other fields, would leave the land fallow for 20–30 years and then start over or else move to fresh Western land. This practice exhausted the soil, encouraged erosion, and resulted in much faster, thinner settlement of the West and larger farms than would otherwise have occurred.

Not all American farmers followed these wasteful practices. Some ethnic or religious groups, notably the Germans, retained European customs, balancing animal husbandry with plant cultivation so the soil's fertility could be maintained. In the 18th century, gentleman farmers such as George Washington followed British developments closely and worked hard to restore worn-out land. By the 1820s agricultural societies and fairs were encouraging farmers to use fertilizers, stop erosion, and carefully rotate their crops. The difficulty of finding labor beyond one's immediate

family made Americans inventors and early adopters of new machinery. In 1837 Cyrus McCormick designed the first practical reaper. The steel plow, perfected in the 1830s and 1840s, permitted rich prairie soils to be farmed. Threshing machines, seed planters, hay rakes, and other machines were also improved. In 1862 the U.S. Department of Agriculture was established and soon began doing significant research on new plant varieties, animal diseases, soil chemistry, and other areas.

The rapid expansion of agriculture led to overproduction and lower prices. In the 1860s farmers entered politics with demands for lower railroad rates and better credit. In the 1880s farmers began plowing the dry lands of the High Plains, which had previously been used for grazing. Dry land farming, an American development, permitted the planting of wheat and some other crops on land with less than 20 inches of rainfall annually by special cultural practices including a fallow period after each crop. These dry land areas, however, were in danger of severe wind erosion when the rains failed. In the most marginal areas, optimistic farmers settled during wet periods, only to abandon their farms in times of drought. In the South cotton monoculture intensified with sharecropping on former plantations. With the closing of the frontier by the late 19th century, though, many American farmers were adopting the more careful farming methods that researchers had been recommending.

Farming in the U.S. entered a new phase following World War I. Surplus production and low prices became a persistent problem. Poor conservation practices prevented many farmers from keeping their land productive. With the onset of a general depression in 1929, the U.S. government became one of the first to support farm prices by purchases of surplus commodities and by taking land out of production. This latter program was soon linked to improving conservation on good farmland and converting marginal lands back to forest, grass, or other conserving uses. The Dust Bowl of 1934–35 dramatized the dangers of attempting to plant dry lands. With assistance from the Soil Conservation Service, farmers planted shelterbelts of trees to break the wind, plowed hills crosswise rather than up and down, and alternated strips of cropland with grasses or legumes.

Following World War II a technological revolution occurred which transformed farming in the U.S. and many other parts of the world. Until the 1940s American farmers used fertilizers only sparingly, and mostly natural ones like manure. Heavy pesticide applications were uncommon. After the war, cheap fertilizers and effective pesticides such as DDT became widely available. At the same time, farm machinery based on the internal combustion engine, after a half century of development, was widely adopted in the U.S. and other developed nations. Research also made great progress in crop yields, especially through the use of hybrids. Animal breeding yielded equally impressive results.

When combined, these advances had an extraordinary impact. In the U.S., yields between 1945 and 1965 rose by 56% for wheat, 74% for cotton, more than double for corn and rice and more than triple for grain sorghum. Milk per cow increased by 74%. Comparable increases were recorded for many other crops and livestock. Insects and diseases which had plagued farmers for years seemingly disappeared overnight. After 1945 American farms became more highly capitalized and larger, leading to a one-third drop in the number of farms by 1960. The general farm with its variety of crops and livestock nearly vanished in favor of specialized operations producing one or two products for market. Food prices dropped to the lowest percentage of family expenditures of any country in the world.

This agricultural revolution came at a price, however. Environmental problems with some of the new chemicals began appearing shortly after farmers started using them in large quantities. A case in point was the fire ant eradication program in the South during the 1950s. Massive aerial pesticide applications failed to eliminate the ants but killed much wildlife and may have even accelerated the spread of fire ants. More ominous was the discovery of pesticide residues in food and in people who had applied them. Moreover, surviving insects often developed into resistant strains. Publication of Rachel Carson's *Silent Spring* in 1962 helped inspire a movement to regulate and test pesticides, develop less persistent chemicals, and find alternate ways of restraining pests, such as biological controls.

The new agricultural methods had other problems, as well. Chemical runoff from fertilizers and pesticides polluted rivers, bays, and groundwater. Some farmers tore out shelterbelts or eliminated other conservation practices in order to make large fields suitable for machinery. The new agriculture also increased farmers' dependence on the outside world. The expense of equipment, chemical inputs, and seeds along with rising land prices added to farm debt and made it harder for new farmers to enter the business. The use of internal combustion engines added energy to the list of input costs. Finally, the great productivity of agriculture increased the importance of exports and government price and income programs, making farmers more vulnerable to policy changes and international market swings.

These developments in the U.S. and, less universally, Europe, influenced agriculture around the world. The Green Revolution of the 1960s, combining better strains of wheat and rice with pesticides and chemical fertilizers, brought great gains in production to countries such as India. Communist governments tried to modernize agriculture with large cooperative farms but faced numerous reverses. China made impressive gains in production when it moved toward free market principles in the 1980s. Plantation agriculture continued to be important in many developing countries for the production of export crops such as coffee, tea, bananas, and rubber. But most farms in those countries remained poor subsistence farms that could not easily afford the improvements required to take part in the Green Revolution. Moreover, environmental complications have spread with the technology. For example, some pesticides once used in the U.S. but now banned there are in use overseas and show up in foods imported into the U.S.

Since the 1970s much attention has been focused on solving agriculture's environmental problems. Research has emphasized safer pesticides and integrated pest management, an approach that combines chemical, biological, and cultural tools. Farmers have shown an interest in sustainable agriculture by which environmentally friendly methods are used in an economically viable way. Government programs in the 1980s put more emphasis on sound conservation procedures. The Food Security Act of 1985 set up a 45 million acre conservation reserve to encourage farmers to remove environmentally fragile land from production. This is the largest long-term reserve ever established in the U.S. Penalties were also imposed for plowing up marginal land or wetlands. The 1990 farm bill required that farmers practice proper conservation to qualify for price support programs. Environmental provisions have become a regular part of farm legislation and are expected to continue to be a major factor in farm bills for the foreseeable future.

DOUGLAS E. BOWERS

For Further Reading: J. D. Chambers and G. E. Mingay, *The Agricultural Revolution, 1750–1880* (1966); Gilbert C. Fite, *American Farmers: The New Minority* (1981); Douglas Helms and Douglas E. Bowers, eds., "History of Agriculture and the Environment," in *Agricultural History* (1992).
See also Animals, Domesticated; Anthropology; Biotechnology, Agricultural; Desertification; Erosion; Farming; Fertilizers; Green Revolution; Hazards for Farm Workers; Integrated Pest Management; Irrigation; Land Use; Pastoralism; Pesticides; Plants, Domesticated; Pollution, Non-Point Source; Ranching; Soil Conservation.

AGROFORESTRY

Agroforestry is a land use system that combines, on one piece of land, trees with crops, livestock, or both.

The trees, crops, and livestock may be managed either at the same time, or one after the other. For example, in dry land in Africa scattered acacia trees often grow in the midst of cultivated peanut fields. In more humid zones, farmers may practice shifting cultivation, where they clear the forest, cultivate it for a number of years, and then let the forest return.

There are three types of agroforestry systems. An agrosilvopastoral system includes crops, trees, and livestock. An example would be opening the above mentioned peanut field to livestock during the off season. An agrosilvocultural system with, for example, shifting cultivation would consist wholly of agricultural crops and trees. When farmers combine trees and livestock on the same land, it is called silvopastoralism. Examples of silvopastoralism are the grazing of cattle on underbrush in tree plantations, and trees growing in pasture.

These systems have provided for the livelihood of people worldwide. Agroforestry was a tradition in Europe until the middle ages. Today, farmers in Asia, Africa, and tropical America continue these age-old agroforestry systems. Indeed, there are farmers in the U.S. that currently practice agroforestry.

The benefits of agroforestry, especially for low-income farmers on marginal land, include increased productivity, security, and sustainability. For example, some palms produce durable wood for construction, leaves useful as thatching for roofs, and woody leaf stalks that can be fashioned into furniture. These trees are also tall and small-crowned, and they use water sparingly. Thus, they have only a small negative effect on crops while producing valuable products. By intercropping these "multipurpose trees" with crops, farmers render their fields more productive.

"Homegardens" are a form of agroforestry practiced by subsistence farmers worldwide. They are a diverse mix of fruit trees, food crops, and fodder for livestock. The plants are arranged to form a multilayered canopy that uses the sun, water, and soil very efficiently. The diversity of homegardens buffers the farmer from the risk involved in subsistence agriculture, and thus increases his or her security.

Other agroforestry systems employ trees or shrubs that can trap atmospheric nitrogen and make it available to crops. The fertilizing effect of these trees can increase crop yields, lengthen periods of cultivation, and shorten fallow periods. All of these factors have a critical impact on agricultural sustainability.

Traditionally, farmers in developing countries have used trial and error to derive appropriate agroforestry systems. While this method has produced amazingly effective results, it is a slow process. With population mounting and timber extraction continuing, traditional agroforestry systems are being disrupted. The result is often increased poverty, environmental degradation, and consequently, still more disruption to the system. These circumstances have prompted research stations around the world to react. Scientists and farmers are now collaborating to adapt agroforestry practices to the conditions existing today, so that farmers will once again be able to manage their land productively, securely, and sustainably.

PETER MAILLE

AIDS

About 1980, physicians, initially in the U.S. but within a year or two in Europe and Africa as well, noted a strange illness that seemed to bear little resemblance to any disease previously recognized. Marked by weight loss, frequent bouts of pneumonia, swollen lymph glands, and most conspicuously by the youth of many victims, this newly recognized illness was soon discovered to stem from a malfunction of the immune system. Between late 1982 and the end of 1983 investigators worldwide sought to identify a cause, whether infectious or toxic, for the mysterious syndrome. These investigators, concentrated especially in France (at the Pasteur Institute, Paris) and the U.S. (in New York, San Francisco, and at the National Institutes of Health) soon noted the signs of a virus that seemed a likely candidate.

Ultimately designated the human immunodeficiency virus—HIV-1, often abbreviated simply to HIV—this agent caused a slow and apparently inexorable erosion of the patient's defenses. Once patients' cellular immunity was impaired, the path was clear for invasion by certain forms of cancer (notably lymphoma and Kaposi's sarcoma), as well as by opportunistic infections (OIS). In different combinations, diarrhea, pneumonia, and brain infections all ensued.

In some ways this "piggybacking" of infections on top of deficient immune function resembled certain congenital immunodeficiency states that had long been noted in some children. But investigators soon discerned that the vicious downward spiral of this "new" disease was triggered through infection from sexual or blood-borne contact. The latter usually resulted from addicts' shared needles or from blood transfusions. The disease was ultimately dubbed the *Acquired Immuno-Deficiency Syndrome*, or AIDS.

It soon became clear that the fullest possible understanding of AIDS was possible only within an ecological conceptual framework. Such a framework, already being applied in some public health and medico-historical circles, helped revolutionize the theory of medicine. It was one of the very few positive advances in the disaster of AIDS.

In many ways the progress of AIDS at the biological level was and remains the most controversial part of the story. It was clear that this "disease," AIDS, was a final pathway for a much larger number of individuals infected with the HIV virus. But how does this virus work? It seemed to have the ability to elude the usual defenses of humans and other organisms through its inner molecular workings. The virus had evolved two complementary capabilities: (1) the ability to home in on the very effector cells of the human host that might develop immunity against both HIV and the OIS that followed it, and (2) a cancer-like ability to mutate, to vary its genetic makeup and hence to escape from conventional modes of treatment.

Many considered this ability to mutate part of the secret of the next higher level of the eco-biology of AIDS: its epidemiology. By definition, epidemics are ecological catastrophes. They are significant, measurable perturbations in the equilibrium among key parts of the environment: disease agents, host populations that harbor those agents, and technology. Such perturbations are deviations from the pre-existing "background intensity" of such disease, which raises the question of how many, if any, cases of AIDS occurred before 1980.

Most investigators consider it quite likely that AIDS already existed at least by the late 1950s. Evidence of this has come from epidemiology, clinical medicine, and molecular virology.

Some scientists postulated a causative role for early lots of polio vaccine in spreading the virus beginning in 1959. Such speculations, though unproven, underscored the role of technology at least in amplifying and spreading the disease if not in creating it. As for the ultimate origin of AIDS, some theoreticians as well as empirical AIDS investigators, in studies remarkably parallel to human evolutionary theory, postulated that simian immunodeficiency viruses in African monkeys may have mutated into forms infectious to humans.

Although AIDS in Africa had afflicted patients throughout the general population, in Europe and the U.S. the disease initially spread fastest among gay men, intravenous drug users, transfusion recipients, and their heterosexual partners; and, geographically, within large cities. The disease and its victims met hostility and often moral outrage from many people. According to some observers, such as Randy Shilts (*And the Band Played On*), this caused delays in recognition, treatment, and research funding. By the 1990s it was abundantly clear, however, that HIV disease had spread beyond the first "high risk" groups into the general population.

Everywhere, the human environment affected the spread of AIDS: in France, lax oversight of blood banking practices; in the U.S., moralistic shunning of hon-

est discussion about risky behaviors; in Africa, re-use of needles designed to be disposable; in Romania, shameless experimentation in orphan children with blood "mini-transfusions." These practices, and many others, represented examples of the social ecology of disease and society's failure to manage it.

Based on late-1980s definitions of the stages of HIV disease, there were approximately 200,000 Americans living with AIDS in 1992, with many hundreds of thousands more displaying some evidence of HIV infection. Although other illnesses, such as heart disease and breast cancer, continue to cause more deaths and morbidity than AIDS, HIV disease, more than any other human ill, clearly underscores the ability of humankind to impact and indeed undermine the human environment. It highlights humans' dependence on the relationships between their environments and their society, and on one another. It drives home the precariousness of those relationships, and the need, sensed early on by some (particularly gay) victims, to seize control of and alter them. Thus AIDS, much like syphilis and tuberculosis in earlier times, has become an emblematic disease of the late 20th century.

RUSSELL C. MAULITZ

For Further Reading: Robert J. Blendon, Karen Donelan, Richard A. Knox, "Public opinion and AIDS: lessons for the second decade," *Journal of the American Medical Association* (February 19, 1992); Elizabeth Fee and Daniel M. Fox, eds., *AIDS: The Burdens of History* (1988); Mirko Grmek, *History of AIDS* (1990).

AIR

See Atmosphere; Clean Air Act; Pollution, Air.

AIR TRAVEL

See Transportation, Air.

ALASKAN PIPELINE

See Trans-Alaskan Pipeline.

ALCOHOL

See Fuel, Future.

ALGAE

Algae are a large, diverse group of plants. Most are aquatic, and are ubiquitous throughout marine and fresh waters. Ranging in size from micrometers to the giant kelps of coastal temperate waters, they exhibit a highly variable morphology. Microscopic forms can be single cells, branched or unbranched filaments, or arranged in sheets. Individuals may be simple or highly elaborate with intricate walls made of substances such as plates of calcium carbonate, scales of organic matter, or overlapping walls of glass. Many planktonic microalgae ("phytoplankton") have ornate spines and projections, which maintain buoyancy to ensure a photic (lighted) habitat. An even more extensive range of morphology is seen in macroalgae, including seaweeds. Here types range from calcareous coralline algae that can be as important as coral in building coral reefs, to spongy, higher plant-like forms, to the 150 foot tall kelp "forests." Algae are different from other plants in that they are non-vascular and do not form gametes in the process of reproduction.

All algae are photosynthetic, and possess green chlorophyll *a* as the main photoreactive pigment. Algae exhibit a wide range of light-harvesting accessory pigments, which capture light energy and transfer it to chlorophyll *a*. The dominant accessory pigment is the basis for the common name of most major algal groups. Red algae have the red pigment phycoerythrin; green algae have green chlorophyll *b*; brown algae have yellow, orange, or brown carotenoids; and blue-green algae (which are actually photosynthetic bacteria) have the blue-green pigment phycocyanin.

Algae are important ecological constituents of all aquatic environments, in terms of both biology and water chemistry. Biologically, as primary producers they form the base of the aquatic food chain, a role played by photosynthetic organisms in all photic environments. Production of oxygen via algal photosynthesis is the main source of dissolved oxygen, required to support aquatic animal life. Macroalgae provide a habitat for small invertebrates and aquatic animals, and are particularly important in providing a refuge for larval fish. Symbiotic algae, such as those found in the tissues of tropical corals and terrestrial lichens, provide photosynthetic and nutrient byproducts to their hosts.

Algae are also very important in terms of aquatic chemistry. Besides production of oxygen, algal photosynthesis affects water chemistry by changing pH levels. Natural water systems have a natural pH buffering system based on dissolved carbon dioxide and related chemical equilibrium products (such as carbonic acid and bicarbonate). As the photosynthetic process consumes dissolved carbon dioxide, pH increases. This, in turn, can affect chemical (including nutrient) reaction rates, normally in highly balanced equilibria in healthy aquatic ecosystems.

Algae are also important in mediating aquatic biogeochemical cycles: the complex suite of interrelated processes by which elements are cycled between biological, geological, and chemical systems. The glass walls of microscopic diatoms are composed of silica. Uptake of dissolved silica by diatoms controls the concentration of silicon in the world's oceans. Blue-green algae are particularly important in the biological aspects of the nitrogen cycle of aquatic systems; they "fix" atmospheric nitrogen to biologically available forms. Typically, nitrogen is one of the two main limiting nutrients for aquatic productivity (the other being phosphorus). Blue-green algal nitrogen fixation can be the main source of this nutrient to the aquatic biota, especially in nutrient-poor areas such as the Sargasso Sea. On an even larger scale, it is postulated that algae are important in regulating the Earth's climate. Part of the Gaia hypothesis suggests that a sulfur gas (dimethyl sulfide, or DMS) produced by marine phytoplankton may be the most important source of cloud condensation nuclei seeding clouds in remote marine environments. It is postulated that as the Earth's temperature increases due to changes in the planet's atmosphere (the greenhouse effect), marine phytoplankton will produce more DMS which will produce more clouds, thereby raising the albedo of the Earth and decreasing mean global temperature.

Certain algae are of direct importance economically. Agar, produced by red algae, is widely used in all areas of microbiology, including medical microbiology, and is a common food additive. Carrageenan, also produced by red algae, is an important constituent of ice cream and cosmetics. Blue-green algae are a recognized source of high-quality protein, usable by humans.

Despite the many benefits of algae, at times they can be detrimental. The most common algal problem is a "bloom," a massive overgrowth of algae. This usually accompanies eutrophication, or nutrient enrichment, caused by the input of limiting nutrients into aquatic water bodies via sewage or agricultural runoff. Enrichment of phosphate can cause a bloom of nitrogen-fixing blue-green algae. Enrichment of phosphate and fixed nitrogen together can cause blooms of all algal types. A severe bloom can suffocate fish and aquatic animals. During the night, in the absence of oxygen-producing photosynthesis, algae consume oxygen for respiration. In shallow lakes under bloom conditions the massive amount of algae present can effectively utilize all available dissolved oxygen, resulting in anoxia.

An even more detrimental situation is that of toxic algal blooms. Some freshwater blue-green algae produce toxins that can kill livestock. Red tides are blooms of microscopic algae (dinoflagellates), so concentrated that the water turns red. Some of these dinoflagellates produce neurotoxins (such as saxitoxin), which can accumulate in mussels and clams rendering them lethal for consumption, thus impacting coastal economy.

It is becoming increasingly evident that the study of algae (called phycology) is of utmost importance in understanding the interconnectedness and functioning of aquatic systems, under both healthy and perturbed conditions. One of the newest areas of environmental research is remote sensing: the use of satellite and aircraft image data to study ecosystems on a regional scale. Aquatic remote sensing is based on water color, which is very strongly influenced by algal pigments. Detection of algal chlorophyll *a* by satellites has allowed the study of oceanic primary production on a global scale. New research on the detection of algal accessory pigments is allowing determination of the type of alga present, which gives much information on nutrient status, biogeochemistry, and the overall health of large areas of marine and fresh waters.

LAURIE L. RICHARDSON

For Further Reading: Harold C. Bold and Michael J. Wynne, *Introduction to the Algae: Structure and Reproduction,* 2nd ed. (1985); W. Marshall Darley, *Algal Biology: A Physiological Approach* (1982); Philip Sze, *The Biology of the Algae,* 2nd ed. (1993).
See also Coral Reefs; Greenhouse Effect; Nitrogen Cycle; Oxygen Cycle.

AMAZON RIVER BASIN

See Tropical River Basins.

AMES TEST

The Ames test, properly termed the Salmonella/Mammalian-Microsome Mutagenicity Test, was introduced by Bruce N. Ames in 1973. The test is frequently used as an indicator of carcinogenic risk by assessing the capacity of chemicals to cause mutations, or genetic changes, in these bacterial cells.

The test uses mutant strains of the bacterial species *Salmonella typhimurium,* which require the amino acid histidine to grow. As a result of the mutagenic effect of a test chemical, new mutations are detected as revertants, which are mutants that regain the condition of not requiring histidine for growth. Thus, when spread on the surface of an agar medium containing all the necessary growth requirements except

histidine, the revertant bacterial strains will grow and form countable, visible colonies on the agar medium. In the absence of mutations, growth will not occur on the medium — a negative test result.

The chemical being tested may be combined with homogenates of rat or human liver cells, to provide the biotransformation enzymes that convert the chemical to an active form. A wide variety of procarcinogens require metabolic activation by these microsomal enzymes before becoming detectable as mutagens. However, direct-acting carcinogens do not require such mammalian cell activation to produce mutations. Once a compound is tested at several concentrations, a dose-response curve can be drawn to show the positive relationship between dose and response and to compare the potency of various chemicals.

The Ames test is the most convenient test available for mutagenicity. It is relatively inexpensive and may take only a few days to test many chemicals, while live animal studies or human epidemiological studies may take years and cost millions of dollars per chemical.

Since about half of the chemicals that produce cancers in rats or mice test positive in the Ames test, it misses up to 50% of the carcinogens that test positive at maximal doses in rodents. Many of the "missed carcinogens" may not present hazards for humans. Also, some chemicals that are mutagenic do not give carcinogenic results in test animals. Despite a huge literature much more needs to be learned to more reliably predict carcinogenic risk for people.

GILBERT S. OMENN

ANAEROBE

See Bacteria.

ANALYSIS, BENEFIT-COST

See Benefit-Cost Analysis.

ANALYSIS, RISK-BENEFIT

See Risk-Benefit Analysis.

ANIMAL HUSBANDRY

See Animals, Domesticated.

ANIMAL RIGHTS

Controversy over humans' treatment of non-human species has been growing gradually in the European world for several hundred years. With urbanization and industrialization, fewer people had daily contact with animals as resources, and more had experience with animals as pets, or "companion animals" in the parlance of animal rights activists today. Urban dwellers in the 19th century recognized many traits that other animals share with humans, notably the ability to feel pain and the capacity for emotions such as love and loyalty. The first animal protection legislation was passed by the British Parliament in 1822, and a Society for the Protection of Animals was founded two years later. Similar societies were founded in large numbers in the United States later in the century, and modest efforts were made in several other European countries.

Although the term *animal rights*, as well as the underlying sentiment that animals should live free from human interference, was developed by British philosopher Henry Salt in 1892, it did not gain prominence until the 1980s. Whereas traditional humane societies were concerned that animals not be subjected to unnecessary pain, especially that inflicted by brutal and poorly educated owners, the claim that animals have rights implies that humans should not use them for our own purposes in any way: not by eating them, hunting them, riding them, wearing items made from them, experimenting upon them, or entertaining ourselves with them.

Movements promoting animal rights, currently active in most industrial nations but especially in Britain and the United States, date from the late 1970s and early 1980s. They synthesized the compassion found in traditional humane organizations; a critique of science, technology, and unconstrained development articulated by the environmental movement; and motifs from many human rights causes. In the English-speaking world, several works by professional philosophers found an audience among activists who founded hundreds of groups in the early 1980s. Membership in these groups soared, especially in the late 1980s. In the United States, as many as one million people joined explicit animal rights groups, in addition to the millions supporting traditional humane organizations.

The new animal rights organizations pursue diverse strategies to help animals. The Animal Liberation Front (ALF), which began in Britain but was soon imported to the United States, favors sabotage and laboratory break-ins, which have not only "liberated" animals but also yielded notorious videotapes showing experimenters in an unfavorable light. Even activists within the movement debate whether such tactics help or hurt the movement's public image. People for the Ethical Treatment of Animals (PETA), the United States' largest animal rights group with several hundred thousand paying members, deploys many tactics, especially those involving media attention. Older organizations, such as the Humane Society of the United States, founded in 1954, have also grown more radical under the influence of new groups and ideas, although they continue their traditional practices of lobbying and education.

Several distinct philosophical positions underlie this movement, for which "animal rights" is primarily a political and ideological label. As a philosophical term, animal rights implies that non-human species have absolute moral rights to be treated as ends, and not as means for some human purpose, much as we grant humans the right not to be experimented upon without informed consent (if then). This position has been best elaborated by Tom Regan in *The Case for Animal Rights* (1983). Its proponents have not specified a particular characteristic that endows an animal with rights, for example, language, consciousness, ability to plan its life, or a central nervous system. They have also not determined whether animals such as leopards have rights, while others such as lobsters do not.

A second philosophical basis for "animal liberation," not based on rights, was presented by Peter Singer in his 1975 book *Animal Liberation*, which was a major inspiration for the movement that followed, partly because it popularized the epithet *speciesism* as a parallel to *racism* and *sexism*. Singer's utilitarian argument is that many non-human species feel pleasure and pain as vividly as humans do, and therefore their experiences should enter into a proper accounting of a society's costs and benefits. Singer concludes that we should drastically reduce the pain we inflict on other species, although he provides no argument against taking the life of an animal if this could be done painlessly. The standard challenge to utilitarian arguments such as this is whether it is possible to compare the pleasures and pains of different individuals, not to mention different species. Singer assumes, but cannot prove, that the pleasures of a pig in living its life are greater than the pleasure of a ham-lover in eating that pig. Still, even Singer doubts the ability of oysters to experience pleasure and pain.

British philosopher Mary Midgley provides a third argument for animal protection. In *Animals and Why They Matter* (1983) she returns to a traditional appeal to compassion, on the grounds that humans and domesticated animals have formed "mixed communities" for thousands of years, forming a variety of intricate bonds with each other. Dachshunds and commercial breeds of turkeys, for example, could hardly survive without humans, who in turn feel an attachment and loyalty to their animals. One feature of Midgley's framework is its implication that domestic and wild animals be treated according to very different rules: nurturing compassion for domestic animals, non-interference (except to protect endangered habitats) for wild animals. She avoids the criticism that many environmentalists have leveled against the animal rights movement: that activists' concern for individual animals rather than entire species makes little sense when applied to wild animals.

Animal rights activists, especially in the United States and the United Kingdom, have occasionally resorted to radical tactics, including damage to research laboratories, costing millions of dollars. Their concerns have broadened as well, to include unpopular species such as snakes, bats, and even insects. Many animal rights advocates believe that keeping pets is a form of oppression, even though it is precisely this experience with animals that makes many people concerned with protecting them.

Some animal protection measures seem to have wide public support, others very little. The fur industry has been sharply curtailed by the animal rights movement in several European countries as well as, increasingly, in the United States. Cosmetic companies have changed their testing practices, often reducing the number of live animals used by 90% and in some cases eliminating them altogether. The animal rights movement has had little effect on eating habits, on the other hand, even though the food industry uses by far the most animals (five billion annually in the United States alone). It has also stopped very few medical and scientific experiments using animals, since the public and most policymakers continue to believe that such experiments save human lives. Activists have nonetheless inspired greater sensitivity among scientists as well as legislation for better care of lab animals. Public sentiment in the 1990s seems to favor increased compassion for non-human animals, but not the granting of full moral rights.

JAMES M. JASPER

For Further Reading: Robert Garner, *Animals, Politics and Morality* (1993); James M. Jasper and Dorothy Nelkin, *The Animal Rights Crusade: The Growth of a Moral Protest* (1992); James Turner, *Reckoning with the Beast* (1980).

ANIMALS, DOMESTICATED

A population of animals is said to be domesticated when it has been selectively bred over many generations to fit human purposes and habitats. The first domesticated animal for which there is archeological evidence is the dog, probably bred by prehistoric hunt-

ing peoples to aid in the hunt. The first domesticated dogs we have recovered so far, based on skeletal remains, are from Iran c. 12,000 B.C. and Idaho c. 11,000 B.C. Early selections were for adaptation to the human household and for hunting various kinds of animals in various ways. Breeds were developed with long, rangy bodies and legs to run down gazelles, deer, rabbits, etc.; hounds were developed with keen sense of smell to follow game and hold it until the arrival of the hunter; pointers, setters, and retrievers were bred for fowling; sturdy agile breeds for bear hunting or baiting; low slung breeds for hunting badgers and ferrets; terriers for rat killing; and so on. Other breeds were developed for work: huskies for pulling sleds; large heavy breeds for pack animals; others for sheep and cattle herding; and still others as watch dogs, fierce dogs bred and trained to protect flocks or people. Some breeds were selected for food; the Mesoamerican Indians developed a dog that could not bark and could be corn-fed. Maize is not a good diet for carnivores, however, and the ceramic effigies that come down to us show dogs with nutritional deformities. Some dogs were bred for fighting so that people could wager on them, as in cock fighting. Finally, there are breeds developed as pets and companions with no other utilitarian purpose in mind.

The diversity among domesticated dogs is enormous, and the wild gene pool is large as well. Several strains of wolves, coyotes, and jackals have probably contributed directly or indirectly. The red wolf of Southeastern U.S. is near extinction, not from overhunting, but from interbreeding with both coyotes and domesticated dogs.

Most mammals are easily tamed by putting them under human care at an early age — the earlier the better. If milking animals are already available, even newborn offspring can be raised and adopted into human company. Otherwise, adoption must be delayed until near weaning time, although there are records of women suckling dogs, pigs, cats, and other animals. But taming is only a step toward domestication, a process that requires changing the genetic structure of the population. Wild animal populations carry a substantial genetic load, i.e., a large number of recessive alleles that are not expressed or are rarely expressed because they are covered by dominant alleles. As soon as humans select for any trait whatever and mate like with like, there is some inbreeding and the genetic load begins to unravel. Unexpected traits appear in subsequent generations. The route to domestication is almost automatic when humans control the breeding system.

Carefully thought out integration of crop plants and farm animals provides the most sustainable agricultural systems we know. Forage crops are important in soil maintenance or soil building, as are the animal manures properly returned to the soil. Organic matter is the key to good tilth and lasting fertility of soils, and animals can contribute substantially to its maintenance. Unfortunately, integrated systems are not as common as they should be. Nomadic and seminomadic systems make little use of animal manures, and in many parts of the world, dung must be burned for fuel because there is no other available.

In the Near East, the first domesticates associated with agriculture were sheep and goats. It is difficult or often impossible to distinguish between them from the kinds of bones likely to be found in archeological sites. Such bone fragments are often reported as sheep/goat or simply caprids. The first hint at some kind of husbandry is found in several sites, c. 9000 B.C., where there is a shift with time from a more or less random representation of age and sex to a definite bias toward younger animals and male sex. This does not necessarily mean domestication because similar trends have been found in gazelle and deer remains, animals that were not domesticated, but it does suggest some form of game management and selective killing if not herding. In later sites, clear evidence of domestication appears in archaeological contexts.

Under domestication, female sheep have reduced horns or none at all. Wild sheep have no wool—the hair is rather deer-like—and their bodies are taller, rangier, and less compact. Wool, long thin tails, and long fat tails are traits that evolved under domestication. Hair sheep are still raised in Africa and South India, but most breeds now have wool. The wild sheep of the Near East prefer rolling terrain and move high into the mountains in summer and range to lower elevations in winter. Both sheep and goats were domesticated by 7000 B.C.

The wild goat prefers rugged cliffs and rocky outcrops. It has backward curving scimitar-shaped horns with a pronounced keel facing forward; domesticated goats have twisted horns often in a more or less horizontal position. Stature is also reduced. The long hair of the Angora goat is a late development, but welcome to the nomads who weave their black tents of mohair. Sheep and goats together with wheat and barley formed the core of the agricultural complex that evolved in the Near East.

Wild cattle once roamed over a huge range in Eurasia and North Africa with different races characteristic of different regions. They attracted the attention of people everywhere, eliciting awe, reverence, respect, and appreciation. The great hall of the bulls in Lascaux cave in France suggests a religious concern going back to the ice age. The site of Çatal Hüyük in Turkey, dated to c. 6000 B.C., contains some 50 shrines featuring bull skulls and figures of bulls on the walls. Blood of wild cattle has been identified on an altar and sac-

rificial knife in the site of Çayönü, Turkey, c. 7000 B.C. A number of sites in North Africa indicate that some hunting people specialized in killing wild cattle.

Remnants of a special regard for cattle can be found today from Iberia where bulls are publicly and ceremoniously killed in the bull ring, usually on Sundays; to India where holy men stage riots in favor of antislaughter laws to protect the sacred cattle; and in Assam and Burma where the Mithan race of cattle is reared for sacrifice. Cattle were sacrificed by the Romans and were sacred to ancient Egyptians. The Biblical book of Numbers 28 and 29 indicates a yearly sacrificial slaughter of 113 bulls, 37 rams, 1093 lambs, and 30 goats for the Hebrew temple, not including sin offerings, guilt offerings, or free-will sacrifices of the people. Several present-day tribes in Africa are almost completely dependent on the blood and milk of cattle for sustenance and feel a deep bond with the animals that feed them.

The Tibetan yak and the water buffalo are other bovids brought into the domestic fold. Yak hair is spun and woven into coarse cloth or pounded into felt to cover the portable yurts of the nomads. Water buffalo, including the carabao race, are well suited to hot climates provided water is available for wallowing. They are also sacrificed in religious rites.

All of the bovines can furnish work, milk, meat, hides, and manure for fuel or fertilizer. They pull plows or carts, lift water for irrigation, trample out grain, or pull a threshing sledge. They may be used as pack animals or to turn rollers to crush sugar cane or millstones to grind flour, etc.

Of the equids, the donkey and the horse have provided the most service. The onager, a wild ass of Asia, may have been tamed and used to some extent, but never served humanity as much as the horse and donkey. The donkey, probably domesticated in Northeast Africa, is remarkably strong for its size and a survivor under abuse. The horse is larger, more delicate, and more popular. It may have been domesticated in Central Asia, but the first remains show up in the Ukraine, c. 3500 B.C. Many breeds have been developed by selective breeding: Arabians for speed and endurance; the Thoroughbred for racing; the quarter horse for handling cattle; the Belgian, Clydesdale, and others for heavy work; Shetland ponies as pets and children's mounts; and trotters for racing. Criollos and mustangs have evolved without much human selection. Horse flesh is commonly consumed in Western Europe. A mule is the offspring of a male donkey and female horse; a hinny is the reciprocal. Both are usually sterile, but very hardy.

Of the camelids, the dromedary or one-humped camel was probably domesticated in Arabia, where archeological remains were found c. 3000 B.C. The bactrian or two-humped camel is native to the cold deserts of central Asia. The earliest remains were found in Iran and Turkmenistan and dated c. 2500 B.C. Both are used for work and transportation, and the dromedary for racing. They can be hybridized; the offspring has one hump and males are sterile but females fertile. The American camelids are the llama, used for transport and meat, and the alpaca for wool and meat. The alpaca can be crossed with the wild vicuña to produce extraordinarily fine wool. Both were domesticated in the Andes by 4000 B.C.

The wild progenitor of the domesticated pig once ranged over much of Eurasia and parts of North Africa. Remnants persist in swamps and forest reserves. Young animals are easily tamed, and it is likely that there was more than one domestication. There was a decrease in stature, changes in the shape of the skull, and the reduction of tusks in domestic breeds. The animal is efficient in converting waste vegetable material into meat and also efficient on high energy rations. It has found a useful place in many agricultures of the world and is especially well integrated into Chinese food production systems. The earliest archeological evidence of domesticated pigs is from the site of Çayönü in Turkey, dated c. 7000 B.C. Bones and teeth are found in early Neolithic sites in China c. 6000 B.C. and swine were introduced to New Guinea about 4000 B.C. The taboo against pork now observed by Hebrews and Muslims may have come into effect c. 2400 B.C. At least there is a marked decline in abundance of pig remains about this time, although there are alternative explanations such as destruction of oak forest for agriculture, etc.

While deer are easily tamed, the only domesticate of the family of consequence is the reindeer, herded mainly by Lapps in the far north of Scandinavia and Siberia. It has been introduced to Alaska and South Georgia Island, but commercial herds are confined to the northern rim of Eurasia. The animals are used for meat, work, and hides. They are ridden and used as pack animals.

The European rabbit was domesticated and a number of breeds developed. It is primarily a meat animal; the fur is soft but fragile. Hair is plucked, usually twice a year, from the Angora breed.

The guinea pig or cavy was domesticated in South America and was once more widespread than now. It is still important in the diet of Andean Indians, who keep it in the household, usually in the kitchen where it feeds on scraps. It is prolific and an inexpensive source of meat. Other rodents are kept by peoples in tropical America for food, but not truly domesticated.

The house cat has long been a popular pet and companion, and has served the household as a killer of mice and rats. The wild race was once distributed over most of Europe, including the British and Mediterranean Islands. The ancient Egyptians were probably important in the domestication process because they deified the cat. Male cats were sacred to the sun god Ra and females to the goddess Bast. More than fifty breeds have been developed since, mostly for cat shows. The Angora (from Angora-Ankara) is of note in having one blue and one brown eye, in being deaf, and in having long hair. The cheetah, a very long-legged, fast-running cat, has been reared for hunting gazelle and other antelopes.

Mammals domesticated for the fur trade include silver fox, mink, and chinchilla. The hamster and gerbil were taken into the household for pets and are used in biomedical research.

The chicken is the most widely exploited of domesticated birds. It is thought to be derived from the red jungle fowl of southeast Asia and apparently moved northward to China and then westward. Archeological signs of it have been found in several sites in China c. 6000 B.C. and bones turn up in the Harappan civilization of the Indus Valley, 2500–2000 B.C. Diversity among domestic breeds is enormous. Selections have been made for egg production, meat production, crowing, plumage, cock fighting, and a variety of unusual traits. Reading entrails for divination is widespread.

The Peking duck is perhaps the best known of domestic ducks, both as a breed and a culinary delight. It is probably descended from the common mallard, which has a very wide range and was probably domesticated in Europe as well as in China. Remains in China date to 2500 B.C. The mallard is an excellent flier, but the domesticated race has lost the ability. In the Americas, the muscovy duck was domesticated around the Caribbean Sea and adjacent Central American isthmus. It is quite different in appearance from European and Asian ducks and is now common as a feral bird in city parks and ponds around the Gulf of Mexico.

The common goose was probably domesticated in southeast Europe from the wild greylag around 3000 B.C. The Romans plucked geese twice a year for down and force fed them to make fatty livers for paté. Both practices continue today. The common pigeon that infests city parks and nests under eaves of public buildings is descended from the rock dove. Our city buildings substitute nicely for the cliffs and rocky outcrops of its native habitat. The wild races range from the British Isles across southern Europe and North Africa to the Near East. It was probably domesticated in Iraq c. 4500 B.C. The park pigeon is a weedy race, but true domesticates have been bred by pigeon fanciers. Pigeon production is a sizeable industry in the Near East, both for food and for the guano used as garden fertilizer. Homing pigeons have been used to send messages, but more often compete in races for wagers.

Native Americans both kept and domesticated the native wild turkey. The bird is easily tamed, and kept animals served not only as food and a source of feathers, but are excellent watch animals, setting up a loud alarm on the approach of strangers, snakes, or predators. Recent breeding has produced breeds with wider breasts and more white meat and less dark meat. In the United States, it is especially popular for feast days. Turkey bones were reported from Tehuacán, Puebla, Mexico c. 200 B.C.–A.D. 700, but the bird could well have been domesticated before that.

The guinea fowl is a raucous bird of Africa appreciated for its gamy flavor and dark meat. It was well known to the classical ancients, but we have no earlier evidence.

Cage birds have been domesticated for pets and companions and include finches such as the canary and zebra finch, parrots such as the budgerigar (budgie), parakeet, cockatiel, macaw, green parrot, cockatoo, Myna birds, nightingales, and others. The canary is descended from a wild finch of the Canary Islands and has been bred for singing. It has been used to detect lethal gases in coal mines, to which it is more sensitive than are humans.

Falcons and cormorants are captured in the wild, tamed, trained, and used for hunting but are not truly domesticated.

Goldfish, a colorful race of carp, are a popular domesticate among fanciers who have bred strange and bizarre races with fancy fins, bubble eyes, and other deformations, and also among commercial fish producers for the fish market. It was referred to in Chinese literature 640 B.C.–A.D. 508, but probably not domesticated until the Sung Dynasty A.D. 960–1278. Plain carp have been bred in Europe and the Near East for meat production by aquaculture. Catfish and trout are other meat fish raised artificially. Even hatchery-raised salmon are different from wild populations, and these are becoming the majority of salmon in the Northwest and Alaska. Eels are raised in tanks, but do not reproduce in captivity. A number of tropical fish are bred for pets and the aquarium trade.

The European honey bee supports a substantial industry, producing honey and wax. Bees are commonly moved from pasture to pasture by truck to exploit seasonal flowers. They are in demand by orchardists and some seed producers as pollinators. Other species of bee are often better pollinators, but are more difficult to raise in quantity. The Aztecs domesticated a stingless bee, which would surely be appreciated by bee keepers, but its honey production is very low compared to that of the European bee.

Silk production traces back to the Chinese Neolithic period. The commercial silk worm no longer exists in the wild, so the insect is fully domesticated. The worms eat only mulberry leaves which are supplied by special nurseries. The mulberries are planted together and pruned to a short stub from which a number of leafy switches grow. The leaves are stripped off at a convenient height, and there is no need to climb trees to obtain silk worm fodder. At the proper time, racks of curled bamboo strips are provided for pupation. A cocoon may contain up to 3 kilometers of monofilament, strong for its diameter and with a sheen that is much appreciated. Most of the finest silks still come from China and Japan, but industries are established in France, Italy, Turkey, Iran, and elsewhere.

JACK R. HARLAN

For Further Reading: E. Isaac, *Geography of Domestication* (1970); C. A. Reed, ed., *Origins of Agriculture* (1977); R. J. Wenke, *Patterns in Prehistory*, 2nd ed. (1984).

ANIMISM

Animism is the belief that all animate and inanimate objects are endowed with spirits or souls, or that an immaterial force animates the universe. Examples of animistic beliefs can be seen in a wide variety of religious traditions ranging from Shinto in Japan to the various Scandinavian myths surrounding the cosmic tree Yggdrasill.

The object of adoration can be an animal, the sky, water, thunder, or even the entire earth. The Konde of Africa worshiped Mbamba, a sky divinity to whom prayers were offered in times of drought. Among certain North American Indians, such as the Pawnee and Hopi, the earth and sky are considered partners in providing the means for life. The earth was personified in Greek myth as the goddess Gaia, who still endures in expressions like "Mother Earth" and "Mother Nature." The animating or attribution of a soullike quality to natural phenomena is often an effort to communicate with spirits in order to secure food, to treat illness, or to avert disaster. For example, in South America, Pachamama (Mother Earth) is worshiped at various times throughout the year by people dwelling in the Andes. Crops are planted at the times when Pachamama is most fertile.

Animism originally referred to a theory on the origin of religion presented by English anthropologist E. B. Tylor (1832–1917) in his 1871 work, *Primitive Culture*. Tylor defined animism as "the belief in Spiritual Beings" and concluded that it was the earliest form of religion. In his study of "primitive" cultures, whom he believed resembled early humankind, Tylor concluded that the experiences of death and dreaming led early humans to invent a spiritual aspect to the human physical experience. Their belief—that they were actually seeing dead kin or visiting strange places while dreaming—led early humans to conclude that a spirit was functioning outside the body. According to Tylor, the "primitives" soon attributed a spirit to all things in nature, both animate and inanimate, thereby forming the first religion.

Tylor's theory on the origin of religion met with criticism and debate over a possible earlier form of religion. Scholars such as R. R. Marett (1866–1943), Emile Durkheim (1858–1917), and Sir James G. Frazer (1845–1941) offered alternate theories regarding the origin of religion. Although Tylor's theory is no longer accepted by anthropologists or historians, the term *animism* endures today in the vocabularies of both disciplines. Today, animism does not describe a specific religion, but rather the practices or elements of various religious traditions. Anthropologists doubt that genuine worship of a natural object or force exists in its own right. Nevertheless, animistic tendencies in various religious and cultural traditions over the ages have provided the framework for a harmonious relationship between human beings and their environments.

ANTARCTICA

The fate of Antarctica, considered Earth's last pristine wilderness, rests upon unanimous agreements of numerous nations. For two centuries of increasingly complex multinational activities involving exploration, science ranging from beyond the upper atmosphere to the depths of Antarctic oceans and ice sheets, tourism, territorial claims, and political maneuvering, the world has edged closer to an ultimate goal: creation of an international preserve for science and other peaceful purposes. The history of Antarctica, especially the Antarctic Treaty (1959), indicates progress has been made toward this goal, albeit with compromises and setbacks. One impetus for this quest has been Antarctica's unspoiled natural environments and communities of living organisms.

The only continent where human beings are not indigenous, Antarctica is geographically isolated and even more so by its vast seasonal pack ice belt reaching 1500 kilometers from the South Pole. The highest, coldest, windiest, and driest continent, it also is the most lifeless although surrounded by a rich, biologically productive ocean. Antarctica is one and a half times the size of the continental U.S. and 98% is covered with permanent snow and ice often more than 3000 meters thick. The remaining 2% ice-free land occurs mostly along the Peninsula, in the mountains, and ringing the continent. Within the latter are a few larger expanses of snow- and ice-free ground called dry valleys. These areas contain rocks, gravels, sands, primitive soils, lakes or ponds, and a sparse biota.

A unique coastline feature is the occurrence of vast floating ice shelves that advance seaward, break off, and produce the large tabular icebergs so distinctive of the Southern Ocean. In the Southern Ocean between 50° and 60°S, the Antarctic Convergence occurs, where colder Antarctic surface waters sink beneath warmer water masses. However, circumpolar deep waters flow southward bringing warmer, more saline, nutrient-rich waters from lower latitudes where they surface at the Antarctic Divergence and stimulate pronounced biological productivity.

Discovery and early exploration of the Antarctic Peninsula began in the 1820s by Nathaniel Palmer and James Weddell, fur sealers from the U.S. and the U.K., respectively. The Antarctic Continent was discovered and named in 1841 by Charles Wilkes of the U.S. Exploring Expedition, who set a party ashore at the Shackleton Ice Shelf to collect rocks and biological specimens. The Norwegian explorer C. E. Borchgrevink made the first winter-over stay on the continent (1898–1900). There followed numerous other explorations, among them those of Robert Scott (1901–04, 1910–13) and Ernest Shackleton (1907–09, 1914–16) of the U.K., and Roald Amundsen (1910–12) of Norway, who reached the South Geographic Pole. From 1928 to the 1950s many U.S. expeditions were dominated by Richard E. Byrd of the U.S. Navy, who pioneered in the introduction of modern technology (planes and ground vehicles) and propounded the importance of Antarctica for scientific studies. A number of other nations became involved in scientific, commercial, and political pursuits during the first century of Antarctic exploration.

Significant progress was made when, through the United Nations, an International Geophysical Year (1957–58) was organized. This IGY involved worldwide, cooperative, scientific efforts of 66 countries and more than 10,000 scientists and technicians, working at 2,500 stations, 50 of which were located in Antarctica under the supervision of the 12 nations that in 1959 generated and signed the Antarctic Treaty. Additions to the Antarctic Treaty have been made continually since its implementation and have largely proved beneficial in preserving the use of Antarctica for peaceful purposes.

Antarctic ecology links the ocean with land. The ocean surrounding Antarctica has a greater biomass density and perhaps four times the productivity per unit area than any other part of the world's oceans. High photosynthetic production by plankton algae supports a larger standing crop of zooplankton, mostly krill and copepods, upon which depend higher levels of the marine food web—fish, squid, whales, seals, and birds. The more than 30 million Antarctic seals (95% crabeater plus some Weddell, leopard, Ross, etc.), more than 40 million Adelie penguins, and 18 other bird species breed on land or ice and link the ocean—their food source—with the land.

Ecosystems restricted to land, in contrast to the ocean, are simple with few species and low productivity, yet scientifically among the more interesting on Earth. Most organisms are bacteria, algae, fungi, lichens, mosses, liverworts, microarthropods, nematodes, and tardigrades. Life is largely restricted to the vicinity of ice-free areas both in or on snow (algae), rocks (lichens), primitive soils, and widely varying aquatic habitats. Planktonic communities and attached algal mats in freshwater-hypersaline lakes, ponds, glacial or snow bank meltstreams, and seasonal ponds atop glaciers all characterize simple and attenuated food webs. Productivity and community structure are regulated by light, temperature, water, and nutrients, as elsewhere, but the limitations in Antarctica can be more extreme, albeit in some locations birds and seals provide nutrients. Terrestrial and aquatic communities in Antarctica overall are vulnerable to disturbances and recover slowly.

Fur seals were the first resource over-exploited, and by the 1830s were reduced to nonharvestable levels. Miraculously, fur seal populations had recovered by 1900 and were no longer of economic interest. Whaling in Antarctic waters began in the early 1900s with annual harvests mounting until about 1930. By 1950, population declines became obvious—first the great blue whale, then progressively smaller species as factory ships harvested species more frequently encountered. Finally in 1979, all Antarctic whaling except for the minke whale was stopped. It is not known whether the over-exploited whale species will recover to their pre-1900 numbers as did the faster-breeding fur seals. Krill harvesting with some Antarctic fishes and squid began in 1972 and continues to expand without constraint. Over-exploitation and depletion of krill stocks—upon which nearly all marine fish, squid, birds, and mammals depend—could bring disasters heretofore unknown in Antarctic history.

Mineral resources have been discovered and/or predicted mainly on land and continental shelves. These fall under stricter regulations imposed by the Antarctic Treaty and appended agreements. To date no mineral exploitation has occurred. Tourism by ships and planes has been a major Antarctic industry since the 1970s. The numerous visits have been brief and no hotels or resorts have been built so far. Mineral exploitation, tourism, and large scientific ventures are the potential activities on land most likely to endanger the more sensitive Antarctic communities.

In 1992 the Antarctic Treaty's umbrella, originally signed by 12 nations, reached 39. The 27 additional nations agreeing to the Treaty's regulations include 14 conducting substantial scientific research in Antarctica and hence having consultative status (voting rights) and 13 presently non-active (non-voting). Of the original 12 nations, Argentina, Australia, Chile, France, New Zealand, Norway, and the United Kingdom have made territorial claims, whereas Belgium, Japan, South Africa, the former U.S.S.R., and the U.S. have not. The chronology of nations laying claims to Antarctic territory, efforts to settle conflicts, and steps toward conservation and wise use of Antarctica's resources reveal that progress has been slow, yet remarkable in view of the growing numbers of people, enterprises, and nations becoming interested in Antarctica.

Few have experienced firsthand the unique beauty of Antarctica, such as the tabular icebergs, wind-sculpted mountain peaks and rocky crags, and the motion and sounds of birds amidst fog and pack ice at the Divergence. These and much more have been extolled in prose, poetry, symphonic music, art, and photography, upon which no price tag can be placed.

BRUCE C. PARKER

For Further Reading: *Antarctic Journal of the United States* (1966–present); Robert K. Headland, *Chronological List of Antarctic Expeditions and Related Historical Events* (1989); Bruce C. Parker, ed., *Conservation Problems in Antarctica* (1972).
See also Glaciers; International Geophysical Year; Law, International Environmental; Oceanography; Ozone Layer; Plankton; Whaling Industry.

ANTHROPOCENTRISM

Anthropocentrism places humankind at the center of importance above all universal phenomena. The anthropocentric position assumes that human beings exist completely separate from and independently of the natural physical world. The assumption leads to the conceptualization of the natural world as an object that "surrounds" humankind, the conscious subject. This conceptualization promotes an instrumental view of the natural world as primarily knowable in terms of its usefulness to humankind.

Philosophical discourse that occurred with the humanist movement of the Renaissance during the fourteenth century marks the rise of anthropocentric thought in modern history. Humanism is a loosely defined term referring to a variety of disciplines and beliefs that focus on the human realm of existence. Humanism was concerned with the development of individual virtue and mind; a human intelligence capable of self-inquiry and critical analysis was deemed a free intelligence. This concern with the attainment of autonomy and the emphasis of the idea of the individual led many scholars, including Petrarch and Mirandola, to assert the unique potentiality of humankind and its resulting earthly preeminence. The human character was regarded as having no fixed limits, with the potential to develop to levels unattainable by other animals, and being able to determine its own future.

Currently, anthropocentrism is discussed in the context that most global ecological problems can be attributed to humankind's tendency, especially in western modern-industrial cultures, to view the natural world as an instrument that exists solely to benefit or serve humankind. Recent attempts to abandon anthropocentrism and adopt a wider view placing humankind within the context of other living organisms and the nonliving physical world have led to the "deep ecology" perspective and to other movements based on "holism."

ANTHROPOLOGY

Anthropology is the study of the human species. It encompasses the study of cultural and biological diversity across time and space, and spans a range of intellectual approaches from the humanistic and interpretative to the scientific. The scientific aspect of anthropology is comparative in nature, and consequently the subjects of anthropological study are not only other human societies, but other primate species as well.

There are generally recognized to be two basic tenets of modern anthropology. The first is the principle of natural selection, initially formulated by Charles Darwin in 1859. Natural selection explains the apparent fit between an organism and its environment as a consequence of genetics and history: Long term trends in reproduction within a diverse population of organisms alter the genetic composition of the population in such a way as to make the average member of a descendant population better adapted. This replaced a view that considered adaptation to be ahistorical — the result of a direct creative act.

The second principle is cultural relativism, popularized by Franz Boas. This gives us the basic framework for comparing different cultures. Cultural relativism holds that diverse human societies are neither "better" nor "worse" than others, for they can only be compared in narrow ways — in much the same way that an elephant cannot be judged to be better or worse than a kangaroo. Cultures with simple technologies may have very complex social systems and languages, and neither "simple" nor "complex" merits a value judgment equivalent to "good" or "bad." Even aspects of morality, including the premium placed on human life

and welfare, egalitarianism, tolerance of behavioral diversity, selflessness, pacifism, and piety, often vary in inconsistent ways, thus precluding again the ranking of cultures in any but the most limited of senses. Cultural relativism replaced a view known as ethnocentrism, wherein entire cultural systems were considered to be superior or inferior to others.

Together, natural selection and cultural relativism give anthropologists theoretical tools for the comparative study of humans and other primates, and their interactions with their environments. Environments enter into anthropological research in three ways. First, in primate ecology, as the relationships between our closest relatives and their surroundings. Second, in the relationships between our ancestors and their surroundings. And third, in the study of contemporary societies.

Primate Ecology

The human species, *Homo sapiens*, is one member of a diverse taxonomic group, the Order Primates. Because of the common history implied by that fact, comparisons of humans to nonhuman primates can provide insights into the complex events that have shaped the course of becoming human. Through the 1960s, studies of nonhuman primates in their natural habitats focused on two groups: the great apes (chimpanzees, gorillas, and orang-utans) and the savanna-dwelling macaques and baboons. Apes hold a unique position in anthropological studies because they are our closest living relatives. Indeed, chimpanzees and gorillas share an especially intimate relationship to us, being distinct from our species for so short a time that their genes are less than 2% different from our own. Baboons and macaques have also received special attention, because the ecological pressures these semi-terrestrial Old World monkeys confront are probably similar to those that the first ground-living human ancestors faced.

Nevertheless, the vast majority of primates are arboreal forest-dwellers, restricted to the tropical forests of South America, Africa, and Asia, and studying them can illuminate general ecological and behavioral principles applicable to all primates. Today, nonhuman primates and their tropical forest habitats are under increasing pressure from human economic expansion, and consequently many are threatened with extinction. Primate ecology is thus valuable not only for a comparative perspective to understanding the human condition, but as well for the preservation of the very biological diversity that holds the clues to our origins.

Ecological studies of nonhuman primates focus on their diets, habitat use and ranging behavior, grouping patterns, social organization, and reproduction. Differences in feeding behavior also correlate with differences in anatomy, especially in the structure of the teeth and jaws. Studying these regularities may someday help us to understand the dietary adaptations that affected the form of our own dentition.

The distribution and seasonal availability of primate foods play a major role in determining group size and social structure, but small species may also form groups as a defensive strategy against predators. Members of a few primate species, such as the nocturnal prosimians and the diurnal orang-utan, lead semi-solitary lives, but the vast majority live in complex social networks. These vary from monogamous pairs of animals with their dependent young, to single-male/multi-female polygynous troops, to multi-male/multi-female polygamous groups. Even within the same form of society, the particulars of group life may vary. For example, the Brazilian muriqui, the Kenyan baboon, and the Tanzanian chimpanzee all live in multi-male, multi-female groups. Yet the stable core of baboon society lies in the relations among the adult *females*, while that of the muriqui and chimpanzee lies in the relations among the *males*. Chimpanzees also have highly fluid patterns of association, while baboons are strongly cohesive, and muriquis appear intermediate.

Sexual dimorphism of body size and canine teeth is found only within polygynous or polygamous primates; in monogamous primates and a few polygamous ones, such as the muriqui, males and females are about the same size and have similar sized canine teeth. This is a practical starting point for interpreting the evolution of human social behavior from fossils. In the canine teeth, the course of human evolution has involved an apparent reduction in sexual dimorphism. Modern humans, however, are quite sexually dimorphic in facial and body hair, and in the distribution of subcutaneous fat, traits which cannot be detected in fossils, and appear to have little or no homolog in our closest relatives. It will require many more years of physiological and ecological studies to understand the ways in which humans, for all their fundamentally primate biology, are also unique.

Human Evolution

In spite of the extraordinary genetic similarity of humans to chimpanzees and gorillas (closer, it has been estimated, than virtually indistinguishable species of fruitflies), humans have diverged radically from the apes in a short period of time.

The principal distinctions of the human species that can be approached in the fossil or material record are bipedal walking, small canine teeth, and large brains (and their correlates or consequences, tools). Other differences are not detectable in the prehistoric record, such as the loss of body hair, articulate speech, and socio-sexual behavior.

The attribute that seems to be most directly responsible for the evolutionary success of the human species, however, is one most inaccessible to direct historical analysis: the development of culture. To anthropologists, culture is a cumulative historical property of a society, communicated through language. One aspect of culture is technology, the ways in which groups of people solve problems posed by their environments. Though other species are known to use tools, in humans the social and linguistic aspects of culture permit technologies to be spread and refined by other people and other societies, thus giving culture in essence an evolutionary history of its own.

The initial phase of human evolution appears to have been the adoption of bipedal walking as a means of going from place to place, as opposed to swinging in trees, like the other apes. This occurred sometime in the latest part of the Miocene epoch, between about 8 and 5.5 million years ago. Though there are relatively few fossil sites known from this time, it appears as though the African apes of that age were living in drier, more open habitats than their ancestors, as apparently tropical forests were giving way to temperate woodlands. The global forces that culminated in the drying-out of the Mediterranean about 5.6 million years ago created new local climates, and led to interchanges of animals and new competitors. It was in this general framework that our ancestors began to be differentiated from the other apes.

We designate the bipedal apes, those on our own line, as hominids, members of the family Hominidae. They include, over the last few million years, three genera: *Australopithecus*, *Paranthropus* (often subsumed within *Australopithecus*), and our own genus, *Homo*. The earliest of these was *Australopithecus*, whose fossils from a site called Tabarin are older than 4 million years. The earliest tools, however, are from only about 2.4 million years ago, which is roughly the same time that our own genus first appeared.

It is very likely that even the earliest hominids used tools, probably not unlike those now known to be used by modern-day chimpanzees. These tools, however (which include using chewed-up leaves as sponges, twigs for collecting termites, and stones to smash open nuts), would not be preserved or recognized in the archaeological record. What we call "the earliest tools" are really the earliest *recognizable* tools—chipped stones.

The earliest stone tools were most likely used in processing wild animal foods, such as skinning and slicing carcasses, as wild plant foods would not require sharp-edged stones. Since chimpanzees and baboons are known to hunt occasionally, it is likely that early hominids did as well. In those non-hominid primates, however, meat contributes a much smaller proportion of the overall diet than in most contemporary human societies.

The inclusion of significant quantities of meat in the human diet had three important consequences. First, the diet would be augmented with a rich nutritional source, permitting a move out of the tropics, into temperate climates where vegetable foods would be seasonally scarce. Second, the social coordination and cooperation that accompanies meat acquisition and consumption even in nonhuman primates appears to have been an important basis for the evolution of sharing and for the emergence of social networks of reciprocity. In all primates, the collection of vegetable foods tends to be an individualized activity, and in humans these foods are shared exclusively within the family. Meat, by contrast, is a communal food, involving several family groups in the hunt and in the distribution of the kill. And third, the establishment of a sexual division of labor, with women specializing in the collection and processing of spatially and temporally predictable vegetable foods, which comprised the majority of the diet; and men in the acquisition of rich, less predictable and riskier, and generally more highly valued, animal foods.

Shifts in subsistence among hominid ancestors reflect the beginnings of a major change in human beings from the largely autonomous, nonhuman patterns of subsistence to a reliance on cooperative social groups. Perhaps the most important advantage of such cooperative networks was that they, along with improved technology, helped to insure a stable supply of food. Another uniquely and universally human cultural development along these lines was the use and control of fire, which also broadened the spectrum of edible foods through cooking. Certain plant foods could be detoxified and made edible, and meat could be smoked and saved for later consumption. With these fairly simple origins, humans ultimately were able to adapt to far more varied circumstances, including the periodic "ice ages" of prehistoric EurAsia, than could any other primate.

Technology, which is the only aspect of culture that is reliably preserved in the record of early hominids, has obviously been paramount in the ecological success of the human species, but may also have had a hand in reducing the biological diversity of the hominids. Two million years ago there were two or three genera of hominids, half-a-million years ago there was one species of hominid, and today there is but one recognized subspecies. We find in the archeological record a general "improvement" in tools, suggesting that the hominids possessing better tools survived at the expense of those that lacked them. It appears as though technologically simpler cultures have been almost invariably driven to extinction (either directly or indirectly) by more technologically sophisticated cultures. Today this trend continues as "cultural extinction," as Western industrialized culture supplants other systems of belief and behavior, irretrievably diminishing the rich and diverse modes of human existence. The development of ethical codes to preserve other lifeways, and indeed other lives, is an outgrowth of liberal thought of the 20th century.

Historical and Contemporary Human Societies

A second turning point with major consequences for the relationships of humans to their environments, but with even more profound consequences for the relationships of humans to other societies, was relatively quite recent in the human species, and centered on the origins of food production. Hunter-gatherers, with relatively simple technologies, live at fairly low population densities, moving often, and with few material possessions. The spread of food production (which occurred independently on several continents) carried with it different environmental relationships.

Crop growers had to remain in the same place for extended periods of time, to maintain (and if necessary, to defend) one's fields. There was also a necessity for concentrated labor at certain times of the year, which created a desire for larger families and extended kin ties. The advantages of agriculture involved a far more stable and regular supply of food than any animals had ever experienced. The disadvantages involved a nutritional risk due to a reliance on fewer kinds of foods in the diet, and a greater threat from infectious disease epidemics, which can occur at high population densities, detailed by anthropologist Mark Nathan Cohen in *Health and the Rise of Civilization* (1989). Other consequences seem to have included the origins of social inequalities, both of gender and class, as the storage and control of key resources altered relationships of power among individuals and groups. More subtle and long-term were the consequences of a new relationship with environments.

Early agriculturists experienced the earliest known population explosion, one that has continued up to the present day. Anthropologists are not agreed on whether an expanding population made agriculture a necessary and irreversible choice of human societies, or whether expanding population was itself a simple consequence of the agricultural way of life. It is clear, however, that in the history of our species, food production and population expansion have been intimately and inextricably linked to each other.

Agriculture also brought over-exploitation, as fields that had been sown and harvested for millennia with the same crops reduced their yields under various stresses, notably that of salinization. Faced with a constant feedback loop of increasing population and degrading resources, agriculturists consistently had to expand to work more land, at the expense of the technologically simpler hunters and gatherers. The hunter-gatherers that remained by the beginning of this century had been forced into the most marginal environments on earth.

The paradox is that the hunter-gatherer way of life, living in an equilibrium with environments, avoiding over-exploitation of local resources, and maintaining low population densities, had been the "normal" human way of life for literally hundreds of thousands of years, since the origin of our species. The "family values" which generally subsume the removal of women from productive to exclusively reproductive and domestic roles are fairly ephemeral developments in the span of human history.

A more recent change in the relationship between people and their surroundings came with the industrial revolution, which occurred only once (in northern Europe, a fairly late recipient of agriculture) and spread to other parts of the world. This economic transformation is still occurring, and has major consequences for all human populations—consequences that for all our species' vaunted intelligence and foresight, are often approached on a short-term, case-by-case basis.

Humans are probably the most successful species on earth at colonizing new areas (though the species that accompany them, such as rats and roaches, have proven to be hardy colonists as well). Our success is due to technology, and the possibilities it affords for controlling and exploiting environments. This same technology is also successful at degrading environments on a large scale, in ways that are evident from the earliest salinization of the Fertile Crescent to the latest smog in southern California.

Cultural changes made possible the global population boom, and they also afford us the means to control reproduction: that is, to make conscious decisions about reproduction at the individual and social levels. The desire to do so stems from an ethical concern to maximize the quality of life for the existing human population. And yet our ethics also respect the right of individuals to control their bodies, either by choosing to procreate, or not to procreate. At the level of the individual, it is ethically difficult to coerce someone into making either reproductive decision contrary to their wishes. When considered as groups, however, people make very consistent decisions about reproduction: modernization, education, and upward mobility across social classes lead directly to a reduction in the growth rate of populations.

A few generations ago, European social theorists were apprehensive about the fact that the middle and upper classes reproduced less than the lower classes. As they considered class to be a signal of a person's

genetic value, they interpreted the population growth of poorer people as a harbinger of genetic decline for the nation. This view has been out of favor since the end of World War II. It is now widely appreciated that a broad spectrum of genetic potentialities exists in all classes and races, and an effective social policy must involve the cultivation of those potentialities independently of the group from which the individual is derived. Thus, the rate of reproduction of the lower classes can be altered by removing barriers to their upward mobility, and the nation can run smoothly as long as excellence is widely sought and nurtured.

Though culture has been the primary means by which humans have been able to cope with their environments, culture also has provided the source of new problems for humans to cope with. For example, patterns of disease and mortality have been radically altered. While most of the diseases faced by hunters and gatherers are endemic (e.g., yellow fever or hookworms), populations with improved sanitation and medicine but higher population densities suffer more from epidemics (such as bubonic plague and influenza).

With medical knowledge and technology increasing lifespans, the integration of elderly people into society, including specialized care when needed, becomes a more pressing problem. Children are physically maturing earlier than ever before (women having their first menstrual period and men reaching their maximum height), probably due to calorie-rich diets. However, these diets high in sugar and fat promote heart disease and diabetes at later ages. And the nature of large urban centers dictates the co-existence of a great amount of behavioral and ideological diversity, again a recent problem in the span of human history, with which societies have dealt in ways ranging from extreme tolerance to intolerance.

Even more fundamentally, the subsistence modes of hunter-gatherer societies involve direct foraging; with the rise of agriculture-based economic systems, most people subsist by working for someone else. The nature of this work created a number of different social relationships previously unknown: employment, subsidy, exploitation, colonialism, patronage, and alienation. Humans all over the world have been forced to adjust psychologically to very new social environments that result from transformations of their economic systems.

How any particular society solves its problems is, of course, dependent upon the available technology, the motivations of its individual members, its social institutions, and power relations. Effective solutions to environmental problems are therefore obliged to compromise among the interests and needs of the total environment, the economic and political institutions, and the immediate needs of local people. Many times an environmental policy conceived by one segment of society is doomed because it is shaped by a genuine concern for the ecosystem, but not for the welfare of the local people, who perceive the policy as oppressive, insensitive, and ethnocentric. A recent celebrated case involved a ban on seal hunting and its effect on the Inuit people of Canada, documented by anthropologist George Wenzel in *Animal Rights, Human Rights* (1991). All people believe their lives are ultimately more valuable than those of other species, so expecting them to abandon their lifeways, or to accept a decline in the quality of their lives, is unrealistic unless solutions ultimately involve them, rather than being imposed upon them.

In the global society of the 21st century, the most industrialized nations are responsible both for causing and ultimately for solving the problems we perceive. The development and implementation of a global strategy is ultimately an anthropological issue, different in its specifics, but ultimately similar in kind to the problems faced by all human societies at all times.

JONATHAN MARKS
KAREN B. STRIER

For Further Reading: R. G. Klein, *The Human Career: Human Biological and Cultural Origins* (1989); A. F. Richard, *Primates in Nature* (1985); I. Tattersall, E. Delson, and J. Van Couvering, eds., *Encyclopedia of Human Evolution and Prehistory* (1988).

ANTIBIOTICS

Antibiotic(s) is the term coined by Selman Waksman in 1947 to discriminate the antimicrobial products produced principally by bacteria and fungi from the many organic and inorganic compounds (e.g., dyes, detergents, soaps, alcohols, phenol, heavy metals) exhibiting antibacterial activity, generally at substantial concentrations. Implicit in Waksman's usage and definition are (1) the selective action of antibiotics, (2) their ability to inhibit growth at very low concentrations or high dilutions (thereby distinguishing them from non-specific, antimicrobial agents active at high concentrations), and (3) their biological origins principally from microorganisms. This definition is still universally accepted.

The definition of antibiotics and the implications of the concepts of antibiotic therapy were preceded by many important, key, pioneering investigations and observations leading to the emergence of the antibiotic era. Microbial antagonism, a phenomenon whereby one organism prevents the growth of, or indeed kills another microorganism, had been recognized by the

founders of microbiology including Pasteur and Koch, and it provided the rationale for the search for antibacterial substances even before the turn of this century. The dramatic, almost chance observations by Fleming in 1929 of the inhibitory effects of a *Penicillium* mold on the growth of *Staphylococcus aureus* triggered a new era in the search for novel antimicrobial agents produced by other microorganisms. Although a number of chemically synthesized compounds (e.g., dyes, prontosil, sulfonamides) had been found to be very useful antibacterial agents, Rene Dubos was one of the first investigators to initiate the quest for specific agents by a rational, microbiological approach in his 1939 studies of a bactericidal substance extracted from a soil bacillus isolate. His highly perceptive, landmark achievement established the feasibility of isolating antagonistic bacteria producing a bactericidal (i.e., one capable of killing bacteria—in this case the pneumococcus) substance which could be purified and its activity defined. Moreover, he clearly demonstrated the protective action of the agent in preventing the lethality of infected mice with various type-specific pneumococci. This pioneering "first" in the antibiotic era resulted in the recognition and crystallization of the two antibacterial agents tyrocidine and gramicidin. Dubos's achievement served to stimulate and encourage the Oxford group of Florey and Chain to embark on the stabilization, purification, and demonstration of the dramatic therapeutic potentials of penicillin. Thus, the new and exciting era of antibiotics had been ushered in.

In common with other antimicrobial agents, antibiotics may be either bactericidal or bacteriostatic in their action. Bactericidal substances are irreversible, lethal or killing agents. Bacteriostatic compounds inhibit growth and can, under appropriate conditions, be reversed, permitting the continuation of growth of the surviving bacteria. Tyrocidine is bactericidal, whereas chloramphenicol is bacteriostatic. Expansion of our knowledge of the specific biochemical targets of antibiotics has permitted a more precise analysis as to whether an antibiotic is bactericidal or bacteriostatic. At one time it was believed that penicillin was bacteriostatic, but with the discovery that the enzymes responsible for the construction of the bacterial cell wall are sensitive to penicillin, it became apparent that these antibiotics were bactericidal for actively growing, sensitive bacteria. Moreover, it is now known that the β-lactam antibiotics such as penicillin interact with the target enzymes (penicillin-binding proteins) forming a chemical bond that is not easily broken.

Recognition of the specific targets and their location in the microbial cell has greatly enhanced our understanding of the mechanisms of antibiotic action and clarified whether the interactions between target and antibiotic are bactericidal or bacteriostatic. This has enabled us to place the receptor-drug concepts of the father of chemotherapy, Paul Ehrlich, on a more sophisticated, biochemical-molecular basis and paved the way for innovative chemical and biological modifications of many antibiotic substances. In addition, knowledge of the specific targets and their sites in microbial cells has permitted the establishment of antibiotic categories exhibiting selective activity for particular targets or structures in cells. The antibiotics now known can thus be classified on the basis of their target sites as well as on their chemical structures.

In some instances there are strong relationships between target specificity and chemical structure. Because of their chemical structures, polypeptide antibiotics such as polymyxin, tyrocidine, and gramicidin disrupt the essential barrier functions of cell membranes. The cells become leaky and can no longer keep essential substances inside. The β-lactam antibiotics block the formation of the bacterial cell wall, which is essential for its survival. Aminoglycoside antibiotics bind to parts (subunits) of the cells' machinery responsible for the formation of proteins, which are essential for cell division and survival. The polyene antibiotics such as amphotericin B and nystatin attach to the sterols in the fungal membrane barriers, disrupt them, and cause leakage of vital cellular constituents. Antibiotics may thus interact with their targets leading to reversible or irreversible impairment of their function.

An enormous number and variety of microorganisms have been screened for antibiotic production. Many different fungi and bacteria produce antibiotic substances, and the ability to produce a particular class of antibiotic is not restricted to a single microbial group (e.g., β-lactams can be produced by both fungi and bacteria). Bacteria of the *Streptomyces* group are particularly prominent as producers of many antibiotic compounds, a substantial number of which are highly active, therapeutic agents.

The diversity of the microbial groups producing antibiotics is more than matched by the tremendous chemical diversity of these compounds. The variety of chemical structures of the large number of antibiotics isolated and chemically characterized is too large to list comprehensively. Few fall neatly into a single chemical category and most are complex structures. As the chemical knowledge of the structure of antibiotics advances with the growth in understanding the interactions with their targets, it becomes possible to pinpoint specific types of reactive groups in the antibiotic. This kind of information is particularly helpful in classifying these agents.

The overall impact of the discovery of antibiotics and emergence of the antibiotic era has been enormous. Antibiotics have played a major role in reducing human pain and suffering, reduced the length of stays in hospitals, and controlled infectious diseases that were almost invariably fatal in the pre-antibiotic era. Antibiotics have reduced the threat of other serious bacterial infections including pneumococcal pneumonia, streptococcal infections, meningitis, and tuberculosis. Most life-threatening bacterial infections and a number of fungal infections are amenable to antibiotic therapy.

The widespread use of antibiotics has not been without its setbacks, for the emergence of resistant bacteria has created new problems for the control of hospital and community acquired infections. This has led to the emergence of strategies to combat resistant organisms and the development of alternate antibiotics and chemotherapeutic agents to treat infectious diseases. It has become evident that as microorganisms acquire genetic resistance to "older" antibiotics, new agents will be required, thus extending the antibiotic era into the foreseeable future.

In addition to their value in the fight against microbial infectious disease, it has been suggested that antibiotics have contributed to the increased longevity of human beings. Although this suggestion is not uncontroversial, because the introduction of antibiotics coincided with advances in hygiene, there is no question as to the tremendous benefits and impact they have conferred on the human race. Assessment of the impact of antibiotics on our environment is much more difficult to gauge, except perhaps as contaminants in the food chain due to their usage in domesticated animals thus calling for constant vigilance and adequate monitoring.

MILTON R. J. SALTON

For Further Reading: E. F. Gale, E. Cundliffe, P. E. Reynolds, M. H. Richmond, and M. J. Waring, *The Molecular Basis of Antibiotic Action,* 2nd ed. (1981); William R. Scheibel, *Antibiotics* (1990); Selman A. Waksman, *Microbial Antagonisms and Antibiotic Substances* (1947).

APPROPRIATE TECHNOLOGY

See Industrialization; Technology; Technology and the Environment.

AQUACULTURE

Aquaculture is the science, art, and business of cultivating freshwater and marine plants and animals under controlled conditions. It is the water-based equivalent of agriculture and follows the same basic principles. Nutrition, water temperature, and other factors must be controlled to provide optimal growth rates and allow harvesting at opportune seasons of the year.

Aquaculture is practiced throughout the world for both subsistence and commercial production. The historic antecedents go back to ancient Chinese and Egyptian times, or nearly 4,000 years. However, the main increase in worldwide production occurred after World War II, particularly in Asia, the United States, and Europe. Growth in the aquaculture industry has been concurrent with wider government recognition of it as a legitimate and important enterprise. As a result, United States commercial production of catfish, trout, salmon, and penaeid shrimp increased by 241 percent between 1980 and 1987 and is growing steadily. By 1989 worldwide aquaculture production had reached 13.4 million pounds.

Seaweed cultivation is another growing component of aquaculture, especially in countries such as China and Japan where most of the demand exists. The demand for byproducts of seaweed or algae, used in medicine, livestock feed, and industrial chemicals, is increasing in non-Asian countries. Seaweed colloids such as carrageenan are routinely used in ice cream, cosmetics, vitamins, laxatives, and industrial lubricants.

The scale of aquaculture production ranges from individual and family farms to cooperatives and highly intensive large-scale businesses. Likewise, the level of technology varies, based on the degree of access to financial and physical resources, the needs of the aquacultural species, the farmer's management capabilities, and cultural and historical factors. In industrialized countries such as the United States aquaculture, like agriculture, is becoming increasingly intensive. As a result, dense populations of fish are now raised in more confined spaces in an effort to achieve higher levels of efficiency. This type of modern aquaculture is highly science-based and requires intensive managerial skills on the part of the farmer.

In contrast, in developing countries aquaculture is often associated with natural ecosystems, such as local embayments, allowing the farmer less direct control of the elements. In addition, farmers producing under open, estuarine conditions face other constraints, such as water pollution. In its efforts to produce low-cost, high protein products for the commercial market, fish farming in developing countries often must compete with agriculture for skilled labor and capital investment.

Regardless of the method of approach, successful commercial fish farming depends on improvements in life-cycle biology, with attendant advances in nutrition, genetics, pathology, parasitology, and other disciplines. Research and development have been relatively latent in modern aquaculture. As the prestige of the business increases, however, government and institutional support of such programs will help improve aquaculture methods just as it helped agriculture.

AQUIFER

An aquifer is a subsurface formation of permeable sediment or rock containing an exploitable supply of water. Underlying an aquifer is a confining layer, usually rock. When that layer is composed of semiimpermeable material, such as shale, which retards downward movement of water, it is called an *aquitard*. When that layer is composed of impermeable material, such as unfractured bedrock or clay, which prevents further downward migration of water, it is called an *aquiclude*.

An aquifer may be confined or unconfined. A confined aquifer is bounded above and below by aquitards or aquicludes. An unconfined aquifer is overlaid by a zone of aeration, a region of porous soil or rock adjacent to the earth's surface through which water passes en route to the aquifer. As water accumulates above the confining layer, it saturates the small, often microscopic air spaces in the soil or rock, called interstices. When the region above the confining layer is saturated to the point where it can yield usable quantities of water, it is called the zone of saturation, a term often considered synonymous with aquifer, although an aquifer encompasses the zone of saturation and its fringes in the adjacent strata. Only the water available within the zone of saturation is termed groundwater.

The process by which aquifers are replenished is called recharging. Unconfined aquifers may be recharged by rain or snowfall across their entire overlying surface, provided the ground is sufficiently permeable. However, the primary sources of replenishment for unconfined aquifers are recharge zones, low-lying areas such as wetlands, swamps, and stream banks where large amounts of precipitation gather before flowing into a surface body of water such as a stream or lake. Sometimes artificial recharge areas are created, for example to allow urban runoff to recharge aquifers used for municipal water supply.

In a confined aquifer, also known as an *artesian aquifer*, groundwater is confined under pressure because of two factors. First, recharge of a confined aquifer usually occurs where the confining layers tilt upwards and rise to the surface. As the aquifer is recharged, pressure builds because the water entering the aquifer weighs down upon the water that preceded it. Second, additional pressure results from the weight of the adjacent rock on the aquifer.

The upper surface of the water in an unconfined aquifer is called the water table. The equivalent for confined aquifers is the potentiometric surface, the imaginary level to which the confined groundwater would rise if it were unconfined. If the potentiometric surface of an unconfined aquifer is above the ground, penetration of the aquifer by a well will result in an artesian well, one that flows above ground under its own pressure.

Although it is convenient to portray aquifer systems in uniform terms, in nature their configuration, boundaries, and composition are always unique. For example, the water table above an aquifer may be portrayed graphically as a level plane, which suggests that it always occurs at the same depth. Actually the depth of the water table varies greatly throughout the area of the aquifer and fluctuates continuously. The configuration of the water table is controlled by diverse factors such as soil composition, local groundwater use, and seasonal precipitation. Also, the composition and configuration of the aquifer itself is affected by geologic processes ranging from the gradual erosion of subsurface materials by water to sudden events such as earthquakes and volcanic eruptions.

ARAL SEA

The Aral Sea, located in the semiarid region of South Central Asia, was once the fourth largest inland body of water in the world and supported a thriving fishing industry. As a result of a large-scale irrigation project begun in 1960, the ecological balance of the region has been disrupted. This project caused the Aral's shores to recede drastically and led to a host of environmental problems that have rendered the land and sea incapable of supporting the region's growing population.

Drawing water from two rivers that feed the Aral Sea, the Amu Darya and Syr Darya, the irrigation project was initiated to increase the agricultural productivity of the region. At the outset, the project was successful and agricultural output increased. As the irrigation project continued, however, it led to increased salinity of the land, which in turn adversely affected agricultural productivity and the living standards of people in the region.

In the last thirty years, the Aral Sea's level has dropped by more than forty feet, reducing its volume by 66 percent and its area by 40 percent. As the sea diminishes, it leaves salt-covered plains in its wake. The salt is spread by windstorms and deposited widely, increasing the salinity of the soil. This problem is aggravated by inefficient irrigation practices, which have created waterlogged soil in some areas. As the water level rises, the water itself is brought closer to the surface. Then it evaporates, leaving behind a layer of salt.

This soil chemistry proves toxic to most plants. In addition, the increased salinity of the sea has caused a decrease in the fish population and changed wetland habitats, contributing to a dying ecosystem and reducing the biodiversity of the region. As the Aral's size has diminished so has its moderating effect on regional climate, causing colder winters and hotter summers and thereby reducing or changing the growing seasons.

Cotton is the primary crop grown in the region. The cultivation of cotton requires large amounts of water and often pesticide and herbicide use as well. Since the Aral Sea basin is a closed water system, these chemicals accumulate in the underground water supply and contaminate drinking water. This contamination has contributed to a regional health crisis, as birth defects and cancer have increased dramatically. According to the Worldwatch Institute and others, the future of the Aral Sea is not hopeless if some suggested methods are instituted. Farmers could switch to less water-intensive crops and apply water-saving techniques such as lining irrigation canals to avoid seepage, improving drainage systems, and using drip irrigation instead of furrow or flood systems. The current demand for water could be relieved somewhat by using existing underground water; this would help lower water tables and result in lower salinity levels. Charging for water use could also give an increased incentive to save. Family planning could be encouraged as well, since the area no longer supports the population growth that resulted from the brief period of economic prosperity after 1960. Together, these tactics would help reverse the decline of the Aral Sea region, restore enough of its waters to support the fishing industry, and lead to a new era of sustainable agricultural development.

ARCHEOLOGY

Archeologists use material remains to reconstruct past human behavior. Most archeological studies have centered on objects that are made and used by humans and recovered through archeological excavations: both artifacts (portable objects) and features (non-portable remains). These include house foundations, stone tools, and pottery shards that can be used to study prehistoric settlement patterns, technology, and trading practices. In addition, archeological remains termed *ecofacts* can be used to reconstruct past environments and the ways in which these environments were exploited by prehistoric human populations. For example, pollen samples extracted from prehistoric sediments can be used to identify past vegetation patterns. Samples of prehistoric land snails, insects, and rodent remains recovered from archeological deposits can also aid in the reconstruction of past environments when the habitat preferences of these species are well known. The identification and analysis of plant remains and animal bones recovered from archeological sites can also reveal prehistoric hunting and gathering practices, agricultural and animal husbandry patterns, and diet. This information is crucial for the reconstruction of past economies and lifeways.

Archeologists view the relationship between humans and their environment as a dynamic one. Past human behavior was not simply determined by the environment; past foraging and farming practices also had a measurable impact on the prehistoric environment. For example, millennia of irrigation agriculture in southern Mesopotamia irreversibly altered the landscape, eventually making it unsuitable for agriculture in any form. The evaporation of flood waters used in irrigation left salt in the soil, initially favoring the use of salt-tolerant crops such as barley. Over time, the soils became so saline, however, that they could no longer support grain agriculture.

The question of the origins of agriculture in the Middle East represents a classic case study in the relationship between archeology and the environment. From the appearance of the earliest human ancestors, some four to five million years ago, up until the end of the Pleistocene Ice Age, about 10,000 years ago, all humans made their living by foraging, gathering wild plants, and hunting and/or scavenging wild animals. At the end of the Ice Age, beginning around 8000 B.C., the first farming communities appeared in the Near East. Within 2,000 years, farming communities based on wheat and barley agriculture and on the herding of sheep and goats were widespread across the Near East from Greece to Pakistan. This pattern of mixed farming provided the economic basis for the great civilizations of the Old World, including Sumer, Egypt, Greece, and Rome.

Archeologists have long been concerned with the relationship between the origins of agriculture and the climatic changes that occurred at the end of the Pleistocene. In the first half of this century, archeologists such as V. Gordon Childe argued that the end of the Ice Age led to a widespread desiccation of much of the Middle East. This led to the concentration of humans, plants, and animals into well-watered oases, where cereal crops and herd animals were initially domesticated. Recent pollen and geological studies indicate that the pattern of climatic change in the Middle East at the end of the Pleistocene was more complex. Pollen analyses suggest that the period of worldwide warming that began about 13,000 B.C. was accompanied by increasing moisture during the growing season in the Middle East. This climatic amelioration would have increased both the tree cover and the availability of food plants such as wild wheat and barley. This warm-

ing trend was interrupted by a colder, drier phase between about 9000 and 8000 B.C. At the end of this colder, drier period, beginning about 8000 B.C., permanent farming villages whose inhabitants cultivated barley and wheat appeared in the well-watered areas of the Near East, such as Jericho, for the first time.

Some archeologists have argued that this cold dry episode, termed the Younger Dryas, may have been a major catalyst for the adoption of agriculture in the Near East. Others have taken a more broadly ecological viewpoint, arguing that the late Pleistocene climatic changes when combined with changes in human behavior, including the establishment of permanent village communities beginning around 10,500 B.C., may have created the conditions that favored the adoption of cereal agriculture. These archeologists argue that the climatic changes that occurred at the end of the Pleistocene in the Near East would have favored the spread of the annual cereals and legumes that are the wild ancestors of crop plants such as wheat, barley, and lentils. With the establishment of increasingly permanent communities during the late Pleistocene, human activities such as using fire to clear underbrush, cutting firewood, and even trampling would have further encouraged the growth of these annuals. Thus human behaviors acted in concert with late Pleistocene climatic changes to create the conditions that favored the propagation of the wild ancestors of the earliest crop plants.

The effect of late Pleistocene climatic changes on human populations in temperate Europe was quite different. The worldwide rise in temperature that began in the late Pleistocene led to an expansion of the forest cover in much of Europe north of the Alps. Animals such as reindeer and wild horses, which are adapted to open steppe and tundra environments, were replaced by forest-dwelling forms such as deer and wild pigs. Human populations adapted to these dramatic environmental changes by adopting a broad-based foraging strategy, including the hunting of deer and wild pigs, fishing, and fowling. Animal bones and artifacts such as bird darts, fish traps, and bows and arrows reflect this broad-based hunting strategy. The activities of these European hunters also affected the environment of early post-glacial Europe. Pollen evidence indicates that these foragers used fire to clear the heavily forested European environment. Burning significantly increased the productivity of these environments by favoring the growth of berries and bushy plants such as hazel, which was used as a source of food and raw materials for basketry.

The evidence from environmental archeology indicates that even in periods of marked climatic change, such as the end of the Ice Age, not only did the environment affect human behavior, but human subsistence activities had a pronounced effect on prehistoric environments.

PAM J. CRABTREE

For Further Reading: Karl W. Butzer, *Archaeology as Human Ecology* (1982); John G. Evans, *An Introduction to Environmental Archaeology* (1978); Donald O. Henry, *From Foraging to Agriculture: The Levant at the End of the Ice Age* (1989).

ARCHITECTURE

Definitions of Architecture

In its root definition, *architecture* means the "mastery of building." Its practice is as old as the history of building and it developed in skill and sophistication with the rise of the world civilizations and city building throughout the globe. Architecture as a recognized profession could be said to date at least from 2680 B.C. when the name of the architect Imhoptep was inscribed on the stepped pyramid of Zoser. The great works of architecture of the ancient civilizations of India, China, Egypt, Greece, and Rome continue to inspire us by their grandeur and technological expertise.

Architecture is conventionally defined as the application of art and science to the design and construction of buildings. Architecture is usually taken to comprise the building and its setting and thus includes landscape design and urban design, each recognized as a professional specialization. The masterworks of architecture have reflected and at times created the vision for the highest aspirations of culture. John Ruskin wrote in 1848 that "Architecture is the art which so disposes and adorns the edifices raised by man . . . that the sight of them contributes to his mental health, power, and pleasure." Architecture can be appreciated as both timeless and timely, for its enduring and universal aesthetic qualities as well as for its unique context in time and place, and for the details of why, where, and how it was built.

The Temple of Athena on the Acropolis in Athens is recognized as a great work of Classical Greek architecture with universal qualities that inspire us across many millennia and many civilizations. Its function as a religious monument or its stone engineering and construction can be part of our appreciation, but the aesthetic lessons of its classic design—composition, proportion, and scale—can be applied to other designs, evidenced by the classic architectural tradition that has been replicated in other places and times. It is these aesthetic principles that distinguish the architecture of Ancient Greece, setting it apart from "mere" building. But consider a "mere" building, a farmhouse or barn, normally built without an architect or pretense to architectural style. Nevertheless, the design and construction of the rural barn can also inspire an aesthetic response of delight as an example of straightforward functionality, ingenious use of materials, and suitability to a particular landscape and climate. From this standpoint, the distinction between temple and barn—between "architecture" and "mere building"—is not so easily made. The barns and farmhouses attest to the aesthetic qualities of building art and craft, which can be appreciated fully by viewing the building as part of the context of its materials, climate, and landscape.

Distinctions between art and craft, architecture and building, appearance and practicality characterize a continuing debate: Is architecture best defined as "a part of" or as "apart from" its physical and environmental context? Should building and its environmental setting be a functional and aesthetic whole? The answers determine the extent to which the concerns of climate and environmental responsiveness should concern architecture and architects. The environmental view—in which architecture is seen as part of its local place, culture, and climate—is a more enlarged definition of architecture that has developed out of the emerging concern and understanding of the impact of building and architecture upon the global environment.

Until the 20th century, it could be said that we designed houses, buildings, and cities that respond to the natural climate. They were built as protection from the extremes of weather and used the sun's heating effect and the wind's cooling effect to create buildings that work with nature. But that no longer appears to be the general case. The design of our buildings, infrastructure, and cities now has an evident impact upon both the local and global climate. In the United States, approximately 35% of energy use is devoted to heating, cooling, and lighting buildings; much is powered by fossil fuels, which contribute to air and water pollution. In our metropolitan areas, vast portions of land have been built upon, eliminating the role of the natural landscape in cleansing air and water, the role that Barry Commoner terms the "environmental sponge." With the increased scope and size of modern building practices, the way that we design buildings and cities threatens to change the natural climate and environment. Rather than building designs that respond to conditions of climate and environment, we now see that climate and environment respond to conditions of building design and construction.

The recognition of the connection between the environmental impact of individual buildings and the quality of the global environment has developed in the past twenty years, but has only recently become part of public awareness and the architectural profession's expressed concern.

Architecture, Climate, and Environment: The Evolution of an Idea

Prior to the modern era, architects and builders had little other than local materials, natural resources, and their own ingenuity to provide for human comfort and health in buildings situated in a wide range of demanding climates. By understanding local climate, devising ways to use the sun for warmth and to create shade and breeze for cooling, the familiar elements of regional architectural styles were devised: the central fireplace and hearth, sun rooms, porches, balconies, courtyards, shading trellises, and shutters. The history of architecture is in part testimony to our increasing control of climate through building design and technology.

The anonymous beginnings of primitive buildings and the great architectural traditions that arose throughout the world show the ingenuity of climate-responsive design. The traditional Swiss farmhouse was oriented for southern sun, with interior zones that could be closed down for winter heat conservation, and with roof lines and materials used to gain added insulation from the snow. The indigenous building of the Himalayan cultures, a hemisphere away but with a climate similar to the Swiss Alps, developed remarkably similar design techniques. The Islamic and Moorish architectural tradition developed out of hot arid conditions. The garden courtyard, shaded and often graced with a cooling water fountain, became the architectural version of "paradise on Earth," exemplified by the interior courtyards of the 14th-century Palace of the Alhambra, Granada, Spain. In colonial southern United States, the humble "shot-gun" house (two separate rooms connected by a covered passageway) and the "high style" plantation manors both exemplified a response to the hot humid climate by utilizing shaded porticos, cross-ventilation, and ventilated roofs.

The ancient Roman Forum Baths at Ostia combined south-facing openings for direct winter solar heating and underfloor heating (warmed by running oven chimney flues under the floors). This may be the first example in the Western world of radiant heating although similar designs developed independently in Japan and Korea. From the earliest time, buildings were oriented to benefit from the sun's winter heat, if only through small doors and windows. Beginning with the development of manufactured glass, effective methods of trapping the sun's heat arose to directly heat buildings through large sun-facing windows, skylights, and glass conservatories. Through the use of glass-enclosed gardens, greenhouse designers demonstrated that it was possible to "change the climate," in effect creating springtime conditions for plant propagation even in

harsh northern winter conditions. In the early 19th century, J. C. Loudon created a remarkable series of greenhouse designs that combined sun, thermal storage, sun-shading, and underground radiant heating, the identical elements promulgated as "passive solar heating" a century and a half later. From these 19th-century experiments in indoor and outdoor gardening grew the modern science of building climatology, which gave the term *microclimate* to define the boundary layer near the ground where temperature and humidity are significantly affected by specific vegetation, soil and land contours, and in turn by buildings.

Beginning in the early 1930s, Chicago architects Fred and William Keck began a decade-long investigation of south-facing windows in residences, the first in the United States to be called "solar houses." In the late 1930s, two masters of modern architecture, Walter Gropius and Marcel Breuer, studied the local climate as major determinants in designs of their own homes in Lincoln, Massachusetts, evidenced by generous south-facing and properly shaded windows. Frank Lloyd Wright, in his Usonian house designs in Wisconsin and in Taliesin West in Arizona, built in the late 1930s, utilized and expressed the local climate and indigenous materials and landscaping, giving ample testimony that an understanding of climate underlies a mastery of architecture, regardless of style.

In the 1950s, climate-responsive design was promoted in a series of articles published in *House Beautiful* magazine and the *Bulletin of the American Institute of Architects* (1949–52) under the editorial direction of James Marston Fitch and Paul Siple. These articles featured ways to utilize climate design techniques, such as white reflective roofs for warm climates and earth-sheltering and solar orientation for cool climates. However, by the late 1950s, at the very moment when climate-responsive design was being widely promulgated in professional and popular housing literature, advances in heat-resistant glass and in compact mechanical heating and air conditioning were developed. This allowed architects to ignore local climate, although at the cost of energy efficiency. The era of air conditioning had begun.

The Equitable Savings and Loan Building, built in Portland, Oregon, in 1948, was the first fully sealed and air-conditioned modern office building in the United States. In 1952, the Lever House was the first modern skyscraper in New York City to use newly developed heat-absorbing tinted glass as a means to reduce undesired solar gain. Shortly thereafter, the same glass technology was specified for the east and west curtain-wall windows of the United Nations Building, rather than using exterior sun-shades that had been earlier proposed for the design by architects Le Corbusier and Oscar Niemeyer. These designs established the style of all entirely glass-faced modern office buildings that was to dominate contemporary architecture for the ensuing decades. They were copied throughout the globe regardless of climate and orientation, far from the ideal of climate-responsive design and at a cost to comfort, energy efficiency, and environmental quality that was only later to become fully evident.

In 1973, the OPEC oil embargo curtailed imported oil shipments to the United States and abruptly brought the issues of oil dependence, energy availability, and cost to the public consciousness. Energy conservation in buildings became a part of national energy policies throughout the industrialized world. One response to oil shortages in the United States was to promote solar heating for houses, revitalizing the 1930s work of the Keck brothers. At the time of the oil embargo there were fewer than a dozen solar houses in the United States. A decade later, solar heating for houses was widespread, demonstrating that solar heating was a simple, straightforward approach that could reduce fossil fuel use and heating costs without changing accepted architectural styles and building practices.

The integration of climate and human comfort— "the bioclimatic approach to architectural regionalism"—had been proposed in the mid-1950s by Victor and Alydar Olgyay. This highlighted the fact that architectural design begins with the physiological requirements of human comfort and seeks ways to use climatic elements to provide these requirements naturally and efficiently. In the mid-1970s, the term *passive solar design* was used to describe similar applications to provide winter solar heating and summer sun-shading and natural cooling ventilation. Applied in an approach that balanced comfort and energy efficiency appropriate for all seasons, these climatic design concepts are now a recognized part of the principles of environmentally responsive architecture.

Following the development of residential-scale solar design in the 1970s, attention moved to designing energy-efficient approaches for larger buildings. These included the technique known as daylighting, a practical application of solar energy and a cost-effective replacement for electric energy, used prior to the widespread availability of electric lighting in 19th-century factory and office buildings. The 1984 Tennessee Valley Authority Headquarters in Chattanooga, Tennessee, employs natural lighting for a large modern office complex, using light reflectors and light-wells to bring daylight into the building interior.

The environmentally responsive approach to large buildings combines passive heating and cooling, daylighting, and the creation of beneficial microclimates in and around buildings by the configuration of building design and landscape. The approach begins with analysis of the "end use" comfort and energy requirements: the temperature, humidity, air flow, and lighting levels required for human comfort and, as appropriate, for healthy gardens and landscapes. Energy efficiency is accomplished by matching the requirements for comfort to ambient energy sources, such as air and ground temperature, wind, sun, and daylighting. Design elements include using the sun-facing side of the building for direct winter solar heating; using light-wells or atriums within the building for natural lighting, solar heating, ventilation, and gardens; using the roof area for skylights and possibly roof gardens; and using the basement or below-grade portions of the building for efficient storage of heat or for ice-making for cooling. The 1983 New Canaan, Connecticut, Nature Center illustrates this in a combined visitors center and solar greenhouse designed for plant propagation and garden design. The building design itself is conceived as a natural energy system that restores environmental quality to its site.

Architecture as an Expression of the Environmental Ideal

The environmental approach extends climate-responsive design to all of the energy and environmental impacts of a building, including design of its systems of air quality, water supply, waste recovery, and recycling. The 1992 National Audubon Headquarters Office in New York City illustrates design responses to these concerns by demonstrating roof gardening, natural lighting, efficient and individualized heating and cooling controls, and a recycling system within the building, itself a remodeling of an existing structure and example of adaptive reuse.

In all of these examples, the concern is not only for the health of people using a building, but includes improvement of environmental quality through site restoration, landscaping, and community infrastructure, inspired by an environmental and biological systems view that buildings and infrastructure should create the conditions of environmental sustainability. This definition of architecture extends beyond consideration of the building and its setting alone to include the building's impact upon air and water quality, and the protection of land to provide for agricultural sustainability and biological diversity viewed from a local, regional, and global perspective.

The concern for environment—the emerging view that all human activity and enterprise has global impact, ultimately affecting the quality of biological and ecological health on Earth—has resulted in an increased concern of the architectural profession to address not only the health, safety, and welfare of building occupants but also the health of environmental and ecological systems of which all building design and construction form a part. They suggest that a "first principle" of all architecture and environmental design could be posed as the architectural equivalent to the Hippocratic Oath: to design buildings that "do no harm" and instead implement and express an ideal "pattern of the world as we would have it," to create and sustain healthy environments for all of life that graces our planet.

DONALD WATSON

For Further Reading: Victor Olgyay, *Design With Climate: Bioclimatic Approach to Architectural Regionalism* (1953, 1992); Sim Van der Ryn and Peter Calthorpe, *Sustainable Communities: A New Design Synthesis for Cities, Suburbs and Towns* (1986); Donald Watson and Kenneth Labs, *Climatic Design* (1983, 1993).

ARCHITECTURE: THE NATURAL AND THE MANMADE

Architecture is one of the major strategies whereby human societies mediate between the individual and nature's laws. Architecture mitigates nature's horrors for the individual and does so in various ways, mostly by trying to make the natural world make sense in human terms. Every culture has attempted this through its architecture, in one way or another.

Most cultures have made use of some variation on two great stratagems. The first, which is pre-Greek and non-Greek and generally worldwide, is one in which human beings attempt to imitate natural forms in their own ritual buildings, and thereby to draw nature's power down to the human community. The other is the opposite, in which human beings attempt to introduce into nature a shining image of a peculiarly human power. This aggressive stratagem was fundamentally invented by the Greeks, and we've been involved in it one way or another ever since.

We're very fortunate that on this continent, and not so long ago, human beings built great works of the first type, in which they basically tried to imitate nature's forms and thereby to draw them to themselves. The greatest ceremonial site on the whole continent is Teotihuacan. There the Avenue of the Dead runs directly to the base of the Temple of the Moon, behind which rises the mountain that is called Tenan ("Our Lady of Stone"). That mountain, running with springs, is basically pyramidal in shape and notched in the center. And the temple imitates the mountain's shape, intensifies it, clarifies it, geometricizes it, and therefore makes it more potent, as if to draw the water down from the mountain to the fields below. In order to accomplish that better, it introduces strong horizontal lines of fracture into its own mass, as if it is settling and squeezing and thereby forcing the water out of the mountain itself.

Off the axis of the Avenue of the Dead is the great Temple of the Sun, which is oriented toward the sunset of a summer day, setting it in the celestial order, and backed by a mountain mass, setting it in the terrestrial order. Thus, the site is locked into the forms of the sky as well as those of the earth.

The same is true in those wonderful cultures of the American Southwest, which are still alive, and where, unlike Teotihuacan, the original rituals are still in active force among the people. Those cultures produced the great pueblos of New Mexico and Arizona, of which Taos, in northern New Mexico, is probably still the most traditional in terms of its forms. Its main house, North House, is pyramidal. Unlike the Mexican temple, though, it does not have a perfect geometric shape. It conceals its fundamental pyramidal quality in the asymmetry of its step-backs. But that makes it no less the manmade mountain, echoing the shape of Taos Mountain beyond it, and in fact helping that mountain to release its water in the form of a never-dying spring that runs winter and summer from the great horned peaks that cradle the sacred Blue Lake. Like the pyramid base at Teotihuacan, it abstracts the mountain's shape in manmade drumbeat rhythm. But its inner rooms are called by the same word that denotes a cavern in the mountain, and its step-backs are like the shoulder of the mountain, so that the building is one with the mountain but, again, intensifies it. Seen from the west, North House looks like a skyscraper. It goes up five stories high and then descends in the steps of a typical Pueblo sky-altar. Like the structures of Mexico, it abstracts nature into sharply geometric forms; here the clouds are translated into a clear-cut, cubistic human rhythm.

A phenomenon of the manmade form fundamentally imitating yet intensifying the natural one is common all through Mesoamerica, and it shapes the great site of Tenochtitlan itself. However, when we go south to the classic Maya areas—especially, say, to Tikal in Guatemala—at first it looks as if another principle must be in action. First of all, there are no conspicuous sacred mountains to be seen. Second, the base of Temple 1 at Tikal is very, very high. And to our minds, instead of suggesting a mountain, it tends—because of its verticality and symmetry, and because it has a sort of a body, and a head and a mouth, and a crown—to suggest a human image. It tends, as a matter of fact, to suggest the image of the great king Ah Cacao, who is buried inside it.

That impression is intensified by the fact that directly across the plaza of that acropolis at Tikal stands the temple of Ah Cacao's queen, Lady Twelve Macaw. We know from the carving on the lintel of the temple that she was short and broad and stocky, and the temple is exactly that shape. It's as if this is the exemplary royal couple: the queen heavy, solemn, and enthroned; the king tall, heroic, and dominating the space for a great distance.

The buildings at Tikal are the first great skyscrapers on the American continent. They rise above the buildings of the town and above the rain forest. They rise up from the earth and they touch — indeed, they scrape — the clouds. The sense of an organic line in the roof comb makes Maya temple architecture very different from that of Mexico or the Pueblos; the Maya roof traces the clouds much more tremulously and completely in its forms.

The ziggurats of Mesopotamia are intended to connect earth with heaven — it seems that the objective is to create something that looks like a formidable mountain mass, like the great mountains of Lebanon that Gilgamesh, the hero-king of Uruk, scaled to find the cedars for the door of his temple. The basic mass of the ziggurat, with the heavy brick deep inside it, was also of great significance. After Gilgamesh lost immortality, he comforted himself with the thought that his temple — even its core — was built with well-fired brick. So his immortality was in his work in the city, which would endure beyond his time.

The Egyptian pyramid was taking shape at the same time. Although in the beginning, as in the case of the stepped pyramid of King Zoser at Saqqara, the pyramid was influenced by the Mesopotamian ziggurat, nothing could be more different. Whereas everything in the ziggurat expresses a forward-and-back play of masses, making us feel the mountainous bulk of the whole thing, everything in the Egyptian pyramid seems to have no three-dimensional mass at all. Big as it is, there's no sense of one plane behind the other in blocks; the planes seem to recede into space like a staircase. The pyramid is like a vision of something taking off from the ground, totally dematerialized, without three dimensions at all, and climbing into the sky. And when the Egyptians eventually perfected their pyramid type, that effect was even stronger. Everything about the great pyramids at Gizeh denies their corporeal mass. That effect, of course, was even stronger originally, when the pyramids were all sheathed in blinding white limestone. They were drawing the rays of Ra, the sun god, down to the earth; they were aimed at the sun so that, magically, the pharaoh would eventually find himself in the boat of the sun god. So, in the pyramids, the sacred mountain dematerializes into the rays of the sun.

The Greeks change all that. They change it with their temple, which introduces into the landscape, for the first time, a building that is not basically reflective of natural forms but is in every way manmade and sug-

gestive of human form. The old sacred shapes remain sacred — the cones and the horns of the mountains — but columns suggestive of the human force take on a new importance. In Paestum, the two temples of Hera are oriented exactly toward a cone-shaped hill reminiscent of the sacred conical hill of Hera at Argos, in mainland Greece. This Hera, however, is embodied in a form in which the columns suggest the vertically standing bodies of human beings — very muscular ones. The whole organization of the temple suggests that other great creation of the Archaic period, the phalanx, in which the hoplites are like columns in the sense that they are free-standing, self-sufficient geometric forms. It brings human wish, the perfection of a special human cultural order, into the old order of natural things, and sets up a dialogue between nature's will and the human wish — out of which the whole luminous structure of Greek classic tragic thought takes form. And it completely changes the old way in which manmade abstraction is applied to a topographical shape. Now it is applied to shapes that will suggest to us, empathetically, the bodies of human beings.

Every temple has to be much like all others, since absolute differences of character can only be read between beings of the same species. So every temple is like all temples, yet in the Archaic period, each one is also absolutely different from all others, as indeed human beings are. At Paestum the columns of the Temple of Athena are slender compared with the heavy columns of Hera. Also, their point of entasis — of maximum swelling — is high, whereas it is quite low in Hera's columns. So we regard the temples of Hera as weights, heavy on the earth; but Athena's columns, like the whole tight body of her temple itself, we read as lifting up rather than pressing down.

In the Classic period of the fifth century B.C., the Greeks looked in two directions. First, they looked backward toward the old way of the kings and the heroes. Olympia, the site of the sanctuary of Zeus, is marked by a gentle conical hill — the Hill of Kronos, the dead father of Zeus. It is the quietest of all the great sacred sites. All the sculptures there are concerned with the law and the breaking of the law and the justice of Zeus and the civilization of the world by Herakles. In the end it is all deeply calm and profoundly conservative, touching some deep core in all human beings.

Then, during the very next decade — and the same great school of sculptors does it — Athens breaks free, breaks all the laws. It is the perfect image of democracy breaking out of the old law, the old limit, the old conservative way. Everything about the Parthenon has too many elements in it, like the whole democracy of the city crowded around it. She has a row of eight columns across the front — really too many for the eye to take in easily all at once. She has another row of columns behind that, so she's partly Ionic. She is a condensation of Greece, of what Athens is, and of the Athenian Empire all the way to the other side of the Aegean. Regardless of where you stand, the Parthenon always seems to be moving, to be turning toward you. It's what was impossible, apparently, before this new dynamic unit known as democracy, with all its self-destructive aspects and all its power.

The Romans were very different. In one sense, compared with the Greeks, they were very conservative always. But in a basic sense, the topography of Italy is very different from that of Greece. In Rome, there are very few sacred mountains; it's the slope that's sacred. And the greatest site of all is the site of Fortuna Primigenia, at Praeneste. The goddess Fortune, at once the nurse and the offspring of Jupiter, nurses him with her waters, which are led down the mountain slope through the body of her temple, distributed into pools and underground tunnels, and released to the fields below.

For the Romans, the word *templum* means a sacred place, not a building. If the Greek temple is like the Greek phalanx, which won its battles by marching in step as a solid body and smashing the enemy, the Roman sanctuary at Praeneste is like the Roman legion, in which the legionaries were trained to fight in open order, like the columns of the sanctuary, and so could open out and envelop the enemy. When you arrive high up at the hemicycle at Praeneste and look back toward the Tyrrhenian Sea, you feel like the commander of the vista; the whole great structure opens its intervals around you and extends its wings like the legion deploying to enclose the world.

The sanctuary at Praeneste illustrates the Roman instinct to enclose, just as the empire encloses the Mediterranean. Roman architecture came to enclose space completely, to develop the interior as a controlled universe of its own. In a very Roman way, by concentrating on the interior, you can make the world more perfect than its outside. Therefore, the Pantheon is a great planetarium, with the planets standing around the side and the sun marking them out. It's like the Roman Empire itself—a web of provinces around a central sea, with the sun above. But it's more than that. It really is the whole of universal space shaped. When you get in and walk into the shaft of sunlight, all the rest goes black; you are in the blackness of some vast space where the confines are unknown. Even though the coffers and the columns are heavy and classical, they disappear, and you are brought into a whole universe perfectly shaped by the imagination of Rome.

Eventually, the Romans combined the vertical, domed space with the long, horizontal shape in Hagia Sophia, at Constantinople. Most of all, they combined,

as the builders of the Pantheon had not, the square with the circle. Between the entrance and the altar at the far end, a circle rises in what seems a magical way from the square below it. All mass dematerializes; the big pillars are screened, the upper walls seem to float. The result is a pure, transcendental world—dedicated, like the Parthenon, to a goddess who embodies wisdom, but totally different in the sense that the natural world is changed by human art into something more perfect, more exact. Built on the elemental geometric laws of the circle and the square, Hagia Sophia can symbolize for us all the great interior spaces of Medieval Europe, like those of Gothic architecture, leaping up from the square to circle into the harmony of the spheres and connecting the kingdom of France with a cosmic order—all of that is part of a movement initiated by Rome.

Human beings came back outside again and began to want to see a kind of divinity in nature itself during the Renaissance. Praeneste became the model for the great new gardens, of which the first was Bramante's Court of Belvedere in the Vatican, where the architect brought the old pagan sacred mountain right into the papal enclosure to balance the dome of the basilica that he was building at the other end of the court.

The Italian garden is the direct ancestor of the English Romantic garden in its various forms. Out of it too comes the French classic garden, but that finally becomes a totally different thing. The whole tradition of the sacred mountain, which is essential in the Renaissance Italian garden, disappears in it. It was originally there—for instance, at St. Germain-en-Laye, one of the first French gardens based on those of Italy. There we have the height, with the old royal palace, and a view back toward Paris, across the Ile-de-France; it was laid out on the slope, like the Praeneste.

But by the time that Nicholas Fouquet, in 1656, began to lay out his great garden at Vaux-le-Vicomte and brought together Le Notre the landscape architect, Lebrun the painter, and Le Vau the architect, a totally different concept had taken form. First of all, Fouquet was not a noble; his purchase of Vaux-le-Vicomte brought the title with it. Fouquet set out to make a noble setting for himself by moving away the three villages on the site and planting a forest. The forest was the place, from the time of Charlemagne and before, where the French nobility found virtue by hunting. You had to have a forest to be noble. The garden was designed to look as if it were hollowed out of the woods. The thin planes of the *parterres* were made to look as if they were pushing the forest back. They were designed to look as thin as paper, with the water sliding directly underneath them. The mass of the chateau was placed behind the moat, so that it would not compromise the weightlessness of the rest of the earth's surface. We are drawn into the chateau, led directly through and released to the view of the garden.

It is that great leap across the landscape that counts, that wonderful leap in which the separate *parterres* of Italy are merged into one long horizontal plane. In Italy the evergreen hedges are thick, so the eye is caught and trapped in the labyrinth of the planting; at Vaux the planting is razor-thin, so the eye skims across it like a high-velocity weapon. And in the distance is all that's left of the sacred mountain: a figure of Hercules, the Farnese Hercules, old, heavy, and tired.

Here, as in almost all of Le Notre's gardens, there is a burst of energy up the middle and various degrees of expansion on both sides. The subtlety of the garden at Vaux is that it isn't the same on one side as it is on the other. The forest comes close to the garden on the right but gives way to the left, so that we feel a wonderful lateral expansion in that direction. But it's the great burst up the middle that seizes our eyes, in the high-velocity view toward Hercules. The impression is one of enormous release, as if all of a sudden we are opened to a new sense of the vastness of the earth and of our own importance as individuals who can control it with our perspective view.

The curious use of the term *pourtraiture* by the French writers of landscape treatises in the seventeenth century referred to the process whereby geometric shapes are transferred in scale to the ground. The word *pourtraiture* surely has at least two meanings in this context: a portrait of the client and a portrait of the earth. The Italian architects felt that they could embody the character of the client better in the garden than in the villa because the garden was more flexible.

Fouquet eventually pushed Louis XIV too far, and Louis imprisoned him—and then took Le Notre, Le Vau, and Lebrun and made a portrait of the new France, a portrait of himself, at Versailles. The avenues of the town come, as it were, from the borders of France to the person of the king in the center of the palace, beyond which the view opens to the Bassin, which reflects the sky and extends indefinitely into space. Indeed, the avenues go on, by implication, to the continental borders of the new France, the first modern nation at continental scale. At Versailles, a human absolutism rules the world. The sacred mountain disappears; indeed, even the chateau as a mass disappears to become simply a fence. The absence of a mansard roof allows the building to look one-dimensional; it resembles a palisade, defining the sweep of water and sand before it. This is the new continental France of the Sun King. And it claimed for the French monarchy much of the same kind of link with cosmic forces that Gothic architecture had embodied not so many centuries before.

Versailles became the model for capitals in such emerging nations as the United States. Washington, which was designed by L'Enfant, basically on the pattern of Versailles, took a long time to grow up to the scale of the Grand Bassin. Not until the Potomac was partly filled in during the early twentieth century did the Grand Bassin become the reflecting pool that would lead to the Lincoln Memorial.

The ultimate garden is, of course, Paris, where in the nineteenth century Haussmann's *etoiles* came into existence. The avenue from the Tuileries that was imagined by Le Notre became the Champs-Elysees. It and the other avenues create a great *etoile*, cutting through the solid clumps of mansarded buildings, which recall the trees of Versailles. And the boulevards and the other streets are narrow enough so that the scale, like that of the *allees* of Versailles, is kept alive at a pedestrian level, creating a totally new landscape—a manmade landscape that in the nineteenth century became the modern garden, the consummate work of modern art. Here modern painting took form, in love with this new kind of landscape, this urban garden.

All things that are great seem to carry the seeds of their own destruction within them. Le Corbusier's Ideal City of 1922, though he specifically denies that it has any connection with the planning of Louis XIV, is nonetheless clearly based on it. There is the basic grid, as in Washington; then the radiating avenues go out from several centers of power and, as in Washington, a mall, which Le Corbusier calls a *jardin anglais*. But progressively, from the outer perimeter inward, he destroys the definition of the grid and the relationship of the buildings to the street. While he uses more-or-less traditional buildings with courtyards at the outer edge, he moves inward to a building type that is a bit like Versailles. It is a long slab-like fence, stepping in and out, *a redents*, running through a long *jardin anglais*, that obliterates the grid of streets. Finally, in the center, he adds cross-axial skyscrapers built within enormous superblocks. A super-highway, not a boulevard, runs down the center. The old definition of the city is destroyed; the automobile takes over. The image of what a good deal of American redevelopment did to our cities through Le Corbusier's imagery is already there in his drawings of 1922 and 1925.

When Louis pushed out to the Rhine, the Pyrenees, and the Alps, those frontiers were fortified by Sebastien le Prestre de Vauban. And they were fortified in echelon, defending what every French schoolchild is taught is the rather miraculously perfect geometric shape of France within its natural frontiers. The arts of fortification and of landscape architecture were virtually the same in the seventeenth century. Fortifications were designed to produce the same kind of landscape art as the gardens themselves and were called by the same name, *etoiles*. At Chantilly, for example, Vauban collaborated with Le Notre, the great landscape gardener. The courtyard in front of the chateau looks like a citadel with bastions at each corner. Each bastion has a face and a flank. In the flank is at least one gun that was intended to fire across the curtain wall of the citadel and the opposite face of the bastion across the way. The bastion is "eared" to protect the gun and to shield its flank. The forms of the fortification take shape according to the trajectory of missile weapons. In the fortifications, as in the gardens, a series of intersecting diagonals radiate "indefinitely" across the landscape.

This connection between the two arts extended throughout French history. In the 1840s, when all at once the center of power moved to Paris and the railroads were pushed to the frontiers, all of France became one citadel, or one garden, with the radiating avenues running from Paris to Vauban's fortifications on the frontiers. Hence the classic garden symbolized the image of the new, centralized France and in part created it, while the fortifications, using some of the same forms, defended that France in depth behind the new frontiers. That is why France, from that day to this, stands as the first continental state of its scale to be integrated both culturally and in relation to the landscape. It is all one great act of civilization and control, running from the center and radiating out to the mountains and the sea that surround it.

It is probably a fact that the major act of identification that most human beings are still able to make late in the twentieth century is with the nation-state, of which France remains almost the archetypal symbol. That, of course, has drawbacks. For instance, there's no need to romanticize the original inhabitants of North America as being preservationists; they were not. But they were people who believed that all life was the same—that is to say, that human structure is the same as the structure of the mountain; that the life of mankind is the eagle life and the snake life and the cloud life. Existence is shared equally by all. So at Taos—as at Paestum, for that matter—the view is from the manmade to the mountain, which is equally meaningful and real. But in the new state of human domination it is simply the human perspective, setting all objects into *its* order and extended beyond the visible horizon.

Whatever the case, it's clear that France bears some of the most striking monuments to the manifestation of the worst aspect of the modern state, which is modern war. One such monument is on the heights of the Somme, where the English and the French fought all through the summer of 1916—fundamentally for nothing, so far as military advantage was concerned. There Sir Edwin Lutyens built a great memorial to the hundreds of thousands who were killed. When we come to it across the rolling fields from Amiens, we begin to feel very unprotected. We can imagine the machine-gun fire sweeping across the fields, so we look toward the little folds in the earth that open to the left and right of the road and seem to offer refuge. Immediately we become aware that the infantry had done the same thing before us, because there's a little graveyard on almost every one of the gullies, where the troops were caught by the mortars and the artillery. When we get to the height—the objective—the memorial of Thiepval looms over us, stepping mountainously up and back in brick and white trim like one of the American skyscrapers of the 1920s. Again, classical architecture seems to take on the character of a face: the tondi (the wreaths) become eyes; the great arch becomes a scream; the whole monster as we approach it, becomes one great empty image of agony and terror and horror, engulfing us. It is the ultimate "portrait" of landscape art that rises up to consume us all. It stands behind a carpet of grass. There is no path for us, the living. We have to violate the grass to approach it. Closer, we are enveloped by the creature's great gorge. One sarcophagus, like a palate, lies within it, under the arch. We must go left or right, up diagonal stairways, to climb toward it. The white stone panels are covered with the names of the dead—many thousands of names.

Lutyen's Thiepval ferociously guards the dead and menaces the living. But Maya Lin's Vietnam Veterans Memorial in Washington, D.C., is enormously gentle. The ground opens for all of us. We're drawn into it, touching the cool face of death with our hand. We commune with the dead. In that sense they have a country still. This is why this memorial so broke the hearts of the veterans of that war, who thought that their country had cast them out forever. The ages crowd in with us and them, remembering the surface of the earth and the cut into it, the Classic gesture to the horizon and to the temple and the sun. The impulse remains to respect the integrity of the earth, to find a truth in it, and beyond dying, to shape the community with it for the common good.

VINCENT SCULLY

For Further Reading: Vincent Scully, *The Earth, The Temple, and the Gods: Greek Sacred Architecture* (1962); Vincent Scully, *Pueblo: Mountain, Village, Dance* (1975); Vincent Scully, *Architecture: The Natural and the Manmade* (1991). *See also* Architecture; Esthetics; Landscape; Landscape Architecture; Urban Planning and Design; Ziggurats.

ARCTIC RIVER BASINS

The world's oceans are fed by rivers that drain a land area that is only 40% as large. However, in the case of

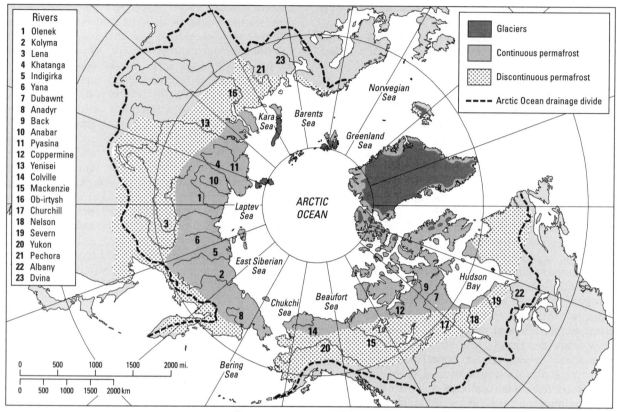

Figure 1. Arctic river basins.

the Arctic Ocean, just the opposite situation prevails—the Arctic Ocean is only 40% the size of the land area that drains into it. Therefore it is not surprising that four of the world's ten longest rivers drain into the Arctic Ocean and that the effect of river discharge on it is greater than on any of the other oceans of the world.

The landmass that drains into the Arctic Ocean ranges in latitude from 48° to 82° N and the vegetation ranges from steppe grasses and forests (especially taiga) to tundra and ice-covered deserts. Variations in climate are equally great. Many of the Arctic's rivers are exotic in that they pass through different climatic and vegetation zones on their way to the sea. One of the major environmental factors affecting them, as well as the human activity developed along them, is permafrost.

Permafrost, a condition in which the ground remains below 0°C for more than one year, is present over about one-fifth of the Earth's land surface. It is especially well developed at high altitudes and latitudes and therefore is found throughout most of the Arctic's drainage basins. Permafrost may be continuous or discontinuous and favors the development of many small surface forms such as ice wedges and ice-wedge poly-

gons. The presence of permafrost, when combined with long periods of below-freezing temperatures and the presence of a complete snow and ice cover, affects drastically the discharge characteristics of rivers.

During the season with such conditions (up to nine months over the northern parts of Alaska, Canada, and Siberia) there is no surface or subsurface (groundwater) flow except in those few locations where streams are fed by springs. Some of these rivers (such as the Colville in Alaska) have no discharge during the winter season. In contrast, rivers originating outside the zone of permafrost flow throughout winter, but along most of their courses the flow is beneath an ice cover. In the discontinuous and non-permafrost areas groundwater can contribute to the river's volume throughout the year.

With snow melt, discharge increases rapidly with flooding beginning in the headwater areas. Snow melt and river ice breakup progresses downstream. In the most northerly parts, breakup may not occur until June. In the zone of continuous permafrost, all of the snow-melt water either evaporates (at a low rate compared to other areas) or flows into the drainage channels because it cannot penetrate downward past the permafrost table.

Much of the Arctic is low-lying, has gently sloping terrain, and possesses wide floodplains. Many of the rivers are multi-channeled and braided. These conditions, combined with low evaporation rates and the lack of percolation, insure that large areas of floodplains and deltas are very watery environments during the short summer.

Under such harsh environmental conditions it is not surprising that the drainage basins of rivers that flow into the Arctic Ocean have been little altered by humans, especially if compared with other drainage basins such as those formed by the Nile, Mississippi, Indus, Rhine, and Yangtze Rivers.

Although such development is minimal and widely spaced, it is varied and includes many of the activities encountered elsewhere. Agriculture, animal husbandry, fisheries, forestry, mining, and the development of hydroelectric power plants, transportation facilities, and defense installations are all represented.

The coastal zones, and especially deltaic areas, have been the location of much of the development. Because overland transportation is very difficult, transport by water has been emphasized. Supply ships moving along the Arctic coastlines make contact with coastal and riverine villages during the ice-free periods of the year. The development of the northeast passage has proceeded quite rapidly and is now a year-round operation through the use of nuclear-powered transport ships, which are sometimes aided by nuclear-powered icebreakers. The Russian Arctic has developed much more rapidly than that of North America.

Because many of the rivers that flow into the Arctic Ocean are navigable for long distances—the Yenisei is navigable along nearly its entire length—settlements have been developed along them and much of the development was expeditious. For example, during the early 1940s settlements such as Indigirskiy and Marshalskiy were established as staging locations for planes being ferried from the U.S. to the Soviet Union. Further, prior to and subsequent to World War II, early warning stations were established along the Arctic coast. In the process of constructing such facilities, much was learned about how to effectively meet the environmental challenges of the Arctic.

Many locations have been affected by human activities. The construction of buildings, roads, bridges, dams, and airfields, in addition to farming, animal grazing, mining, and lumbering, if improperly done, cause thawing of the permafrost and aggravate erosion. The simple act of running a tractor over the tundra during summer can initiate severe thaw and erosion, so that in many arctic areas mechanical transport that is likely to damage soil and vegetation is prohibited.

Although such direct impacts have and will result in environmental changes, those of an indirect nature are becoming increasingly significant. Atmospheric circulation patterns can transport pollutants to the Arctic as occurred during the Chernobyl disaster of 1986 when the biotic portion (both flora and fauna) of northern Scandinavia suffered drastically. Other arctic processes—such as the release of methane that occurs during permafrost degradation—can also play major roles in global change.

The former Soviet Union originally drew up plans to transfer northward flowing water from such rivers as the Divina, Pechora, and Ob to the south. Although these plans have been shelved, if such diversions were to occur, the environmental changes on the Russian drainage basins would be much greater than any other human modifications that have occurred to date. Such diversions would flood and destroy forest land and agriculture fields, alter fisheries, and impair navigation. Equally drastic might well be the changes in the Arctic Ocean and in climate that would accompany discharge reductions.

H. JESSE WALKER

For Further Reading: Gail Osherenko and Oran R. Young, *The Age of the Arctic* (1989); Sergei P. Suslov, *Physical Geography of Asiatic Russia* (1961); H. Jesse Walker, *The Morphology of the North Slope* (1973).
See also Floodplains; Oceans and Seas; Rivers and Streams.

ARMY CORPS OF ENGINEERS

See United States Government.

ARTIFICIAL INTELLIGENCE

Artificial intelligence (AI) technology is a branch of computer science that originated in the early 1960s and came into widespread commercial usage during the 1980s. It consists of various methods for developing computer programs (often called *knowledge-based* systems or *expert* systems) that are capable of logical reasoning for interpreting complex information and seeking solutions to problems. Although most current applications of AI are in the areas of industrial and financial decision-making, a number of organizations have introduced AI methods into the field of health, safety, and environmental management.

Applications of AI usually involve the computer-aided representation of human knowledge and reasoning in symbolic form. Therefore, the design and development of knowledge-based systems has traditionally required the specialized expertise of *knowledge engineers*. However, recent advances are making it possible for business and scientific professionals with little or no background in computer programming to develop simple expert systems. While applications vary widely, such systems are generally used for providing expert advice, solving problems, or performing other "intelligent" tasks. Just as spreadsheet programs enable professionals to create models that represent their quantitative knowledge, expert systems enable them to create logical, deductive models that represent their qualitative knowledge.

There have been many environmental applications of expert systems; for example, advising emergency response teams about how to deal with industrial accidents such as chemical spills. In the regulatory arena, the Environmental Protection Agency is applying expert systems to assist in granting hazardous waste site permits, in modeling water quality, and in a number of other environmental engineering applications. Many large manufacturing companies have established internal AI groups that are building expert systems to advise on diagnosing and repairing equipment failures, safety testing, and a host of similar tasks. This rapid expansion of expert system applications has been made possible by the commercial availability of low-priced general-purpose development tools, called *shells*, which have shortened the typical system development time from years to months.

Recent innovations in AI enable the construction of *intelligent assistants*, which allow specialized knowledge to be captured and applied to a variety of environmental management activities. With this approach, system users can simultaneously access qualitative knowledge bases, quantitative data bases, conventional analytic models, and expert advisory systems. Intelligent assistants are less powerful but more versatile than expert systems. Whereas an expert system for chemical emergency response might actually recommend mitigation strategies, an intelligent assistant might simply identify relevant information about the chemicals involved and then suggest alternative strategies to consider.

Applications to Environmental Management

There are several types of human knowledge that may contribute to environmental management applications of AI: empirical knowledge about organisms and their environment; situational knowledge about local environmental conditions; judgmental knowledge about human beliefs and priorities; theoretical knowledge about biological phenomena; and normative knowledge about policies and acceptance criteria. A knowledge-based system that logically introduces these types of knowledge can be useful in the *environmental management process*, which typically consists of the following stages:

Hazard identification is the first step of the process,

and involves applying screening criteria and logical deduction to the product or activity being considered. If no possible adverse outcomes are detected in the hazard identification step, then there may be no need for further screening. If certain classes of potential risk are identified, then the knowledge-based system can invoke more detailed screening criteria that are relevant to each class. Specific risks (e.g., increased cancer rates) might be identified and assigned a sufficient level of concern to warrant further assessment.

Risk assessment is the next step in the process, and involves developing quantitative estimates of hazard. The knowledge-based system might suggest the use of certain scientific models, such as biological dose-response models, that can produce theoretical estimates of the degree of potential hazard. Naturally, such estimates are based on numerous assumptions and may have large associated uncertainties. If the data are insufficient to support quantitative analysis, the knowledge-based system might instead use qualitative models to establish the nature, if not the degree, of possible risks. This is often accomplished through *rule-based reasoning* involving a chain of logical inferences.

Risk evaluation is the next step in the process. Once potential risks have been assessed, it is possible to introduce value judgments regarding the degree of concern about a specified hypothesized endpoint. The system can assist in this evaluation; for example by providing comparisons with the risks associated with other environmental hazards.

Intervention decision-making is the final step in the process, and requires normative judgments about appropriate methods for controlling or reducing risks. Choice of these methods may be based upon logical rules, or may require more formal risk/benefit balancing methods. Knowledge-based systems can support the selection and application of such methods, as well as interpretation of the results.

Categories of AI Techniques

It is useful to distinguish three different categories of generalized AI techniques:

Data interpretation techniques involve screening data to detect patterns, to identify potential problems or opportunities, or to discover similarities between current and past situations. Examples of applications are ranking waste sites and analyzing geological samples.

Problem diagnosis techniques involve investigating known problems to recognize characteristic symptoms, to develop and confirm hypotheses about possible causes, and to suggest strategies for repair or recovery. Examples of applications are diagnosing medical diseases and troubleshooting equipment failures.

Decision support techniques involve evaluating alternative choices to explore their possible consequences, to compare their relative costs and benefits, and to recommend appropriate action plans. Examples of applications include planning emergency responses and selecting remedial actions.

In many cases two techniques can be combined; for example, data interpretation may be used to identify problems, after which a decision is made among several possible solutions.

In summary, artificial intelligence, or knowledge system technology, provides a number of useful capabilities to support environmental management. Because of the limitations of quantitative methods, risk assessment often requires judgmental reasoning, and knowledge-based systems can provide consistent support to humans in evaluating the various exposure and risk factors relevant to environmental concerns. Specifically, knowledge-based systems can help to preserve and disseminate specialized knowledge regarding characterization of environmental hazards. They can also support the practice of risk assessment methodology including risk identification and trade-off analysis, interpret available data about substance/organism/ecosystem combinations to suggest risk considerations, and support decisions regarding monitoring, controlling, or mitigating environmental impacts.

JOSEPH FIKSEL

For Further Reading: J. J. Cohrssen and V. T. Covello, *Risk Analysis: A Guide to Principles and Methods for Analyzing Health and Environmental Risks* (1989); E. Feigenbaum, P. McCorduck, and H. P. Nii, *The Rise of the Expert Company* (1988); P. Klahr and D. A. Waterman, *Expert Systems: Techniques, Tools and Applications* (1986).

ASBESTOS

Asbestos refers to a group of naturally occurring magnesium silicate minerals that are mined from the earth, as are iron, lead, and copper. While these other minerals break into dust particles, asbestos divides into millions of fibers, some of which may be so fine that they are undetectable to the naked eye.

Four types of asbestos have wide commercial use. Chrysotile accounts for 95% of the world's asbestos production. It is a member of the serpentine group, a rock type consisting primarily of magnesium. The remaining 5% includes crocidolite, amosite, and anthophyllite, which come from the amphibole group, an igneous or metamorphic type of rock consisting of as many as eighteen different minerals.

Chrysotile is made up of magnesium, silicon, and water and consists of rolled up sheets formed from two

layers. The walls of the asbestos fibers are composed of a number of these individual sheets bent into scrolls.

All varieties of asbestos have substantial resistance to heat and chemical deterioration. Thus, they have found their way into more than 3,000 commercial and industrial applications since the asbestos industry began in the 1870s, when mining for chrysotile began in Quebec, Canada.

In the United States, asbestos had been used for insulation from 1870 to 1900, yet it was not until World War II that it appeared widely in shipyards and other construction in forms such as acoustic and decorative tiles and sprayed-on fire proofing. It is estimated that more than half of all buildings constructed between 1950 and 1970 contain some form of asbestos, particularly multistory buildings and schools. Literally millions of homes, schools, state and federal office buildings, and commercial and industrial structures have benefited from its fire-resistant properties.

However, exposure to asbestos fibers has resulted in disease and death. This effect has come to be known as "The Asbestos Problem." When asbestos fibers are inhaled, they elicit a defense mechanism in which white blood cells migrate to the air sacs and bronchial tubes in an attempt to engulf and destroy the fibers. This walling off of the fibers thickens the air sacs with a protein deposit. If enough of the air sacs are affected, normal respiration is impeded and shortness of breath occurs with even light exercise.

The three major respiratory diseases associated with asbestos are asbestosis, a fibrotic scarring of the lungs, lung cancer, and mesothelioma, which is cancer of the lining of the lungs. The development of respiratory disease due to asbestos exposure may occur long after the exposure. For asbestosis, it may take five to fifteen years to appear; for lung cancer, fifteen to twenty years; and for mesothelioma, twenty to forty years.

ASH

Ash is the general term for the residues of combustion or incineration. Ash is generated mainly by coal-burning power plants, municipal waste combustors, and various incinerators (e.g., hazardous waste, waste sludge, and hospital waste). Municipal waste combustion ash (MWC ash) is also referred to as "incinerator ash," "waste combustor residues," "waste-to-energy residues," and similar terms.

The main objective of solid waste consumption is to reduce the volume of municipal solid waste while producing energy. This is accomplished by burning the combustible matter in the waste, which can reduce the volume by 80% to 90% and the weight by 70% to 80%. The 20% to 30% that remains is MWC ash.

MWC ash consists of two main components: bottom ash and fly ash. Bottom ash is the coarse, dense, non-combustible material that remains on the furnace grate after combustion and is usually collected by conveyor and cooled by water quenching. Fly ash is the lighter, finer material that is suspended in the flue gas and collected in air pollution control (APC) devices such as cyclones, electrostatic precipitators (ESP), fabric filters, and scrubbers. In scrubbers the fly ash is commingled with wet or dry lime acid gas control residues, which contain unspent lime or lime reaction products. The fly ash can vary between 5% and 20%, by weight, of the total incinerator residue, depending on the heating value of the municipal solid waste, the type of furnace, the combustion efficiency, and the type of APC equipment.

Typically MWC ash is physically and chemically varied because the wastes that enter the combustors vary. The consistency of bottom ash is similar to gravel. Combined ash has almost an equal gravel and sand content, and when scrubber residue is present, it has properties like volcanic ash. Municipal solid waste contains many different components that incorporate a tremendous variety of chemicals. They contain salts, inorganics, and many organic compounds, which are major human health concerns. When municipal solid waste is combusted, the organic compounds in the waste are largely destroyed. The inorganic metal constituents become concentrated as the volume of the waste is reduced. Products of incomplete combustion, such as dioxins, mainly adhere to small fly ash particles. Fly ash also typically contains higher concentrations of the more volatile metals, such as lead and cadmium. The less volatile metals, such as manganese, silicon, and aluminum, stay mostly in the bottom ash.

In most situations in the U.S., the fly ash and the bottom ash are joined in the facility to form combined ash, which is safer to handle and less likely to pollute groundwater. The change in pH reduces the likelihood that heavy metals from the ash will enter the groundwater. The combined ash is disposed of in monofills (ash-only landfills) or together with garbage in municipal solid waste landfills (co-disposal). The trend for the future is increased reliance on monofill disposal.

HAIA K. ROFFMAN

ASWAN HIGH DAM

The Aswan High Dam is located on the Nile River in the Nubia region of southern Egypt. Work began on the project in 1960 and was completed in 1970 at the cost of 1 billion dollars. The crest of the dam stands 111 meters above the riverbed and is 3,820 meters long.

The dam is 980 meters wide at its base and 40 meters wide at its crest.

The dam was constructed to improve navigation, protect the population centers and agriculture of Lower Egypt from the yearly flood cycle of the Nile, and generate electricity. A much smaller dam six kilometers north of the High Dam, completed in 1902, allowed Egyptian authorities limited control over the Nile. The High Dam doubled the amount of land under cultivation. Without the dam Egypt would have probably continued to suffer from extensive flooding, persistent devastating drought, and energy shortages. However, the dam has led to some unforeseen ecological and social consequences.

Despite the increase in agricultural production, the gains envisaged in terms of employment and agricultural development were eliminated by the growth in population between 1960 and 1970. Since the High Dam was completed, Egypt has experienced a drastic cut in per capita cultivated area. The rise in groundwater level and the resulting increase in soil salinity have caused 50 percent of the irrigated land to deteriorate to medium or poor quality soil. The High Dam stopped the downstream flow of the Nile's fertilizing silt, causing riverbed degradation, erosion along the river's banks, the close of a third of the brick-making factories downstream, and destruction of the Nile sardine industry. Farmers have been forced to replace the silt with chemical fertilizers, further degrading the water quality and bringing other health hazards. For instance, since the building of the High Dam there has been a reported increase in the cases of schistosomiasis, a debilitating parasitic disease caused by a tiny blood fluke. The Nile Delta region, weakened by the disappearance of reinforcing silt, is experiencing serious erosion and is in danger of inundation by the waters of the Mediterranean.

Because the land surrounding the river was submerged, Egyptian and Sudanese authorities were forced to relocate approximately 90,000 Nubians and desert nomads. These people have been placed in new settlements near the lake region in Egypt and the Sudan. While community services in the new settlements are much improved according to modern standards, the trauma of dislocation and fundamental changes in living conditions have caused damage to traditional Nubian and nomad societies.

The impoundment of the Nile created Lake Nasser, a freshwater lake 480 kilometers long and with a total surface area of 6,276 square kilometers, which stretches over Egypt's southern border with Sudan. The lake supports a fishing industry and several brick-making factories. The flooding caused by the creation of the lake necessitated the relocation of 23 ancient Egyptian and Nubian temples, most notably Abu Simbel. A large tourist industry was stimulated as a result of improved access to Aswan City via the Nile and by a new airport. The growth of Aswan City can also be attributed to the power station located at the dam site. It has the capacity to produce 2.1 million kilowatts, allowing for the industrialization of Aswan City and the implementation of a rural electrification program in southern Egypt.

ATMOSPHERE

The atmosphere is the gaseous envelope surrounding Earth and retained by its gravitational field. In a biological sense the atmosphere serves both as the integument or skin and the exterior circulatory system of our planet in that it protects the planet surface from bombardment by energetic particles and rays and the numerous small meteorites and it also redistributes heat, moisture, dust, and biospheric nutrients. Despite these important functions, the atmosphere has changed over time. Its composition relative to the solar abundance of the elements makes it clear that the primordial gases were swept away by the solar wind as the solar nebular material forming planetary earth coalesced and that our present atmosphere has evolved

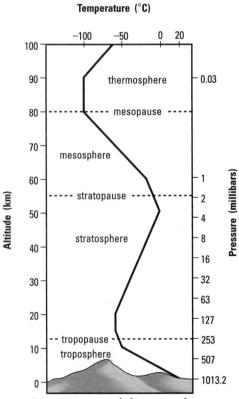

Figure 1. Layers of the atmosphere.

from outgassing from the material of the primordial planet.

Layers of the Atmosphere

The atmosphere also changes in many ways with altitude (Figure 1). In an unconfined gas such as the atmosphere, pressure is determined by the weight of the overlying gas. The mean pressure at sea level is 1013.2 millibars or 14.7 pounds per square foot. With altitude the overlying column decreases so the pressure falls — by about half for each 6 kilometers (km) or 18,000 feet (ft).

As the pressure on a gas is reduced, the gas expands and as it expands it loses energy (cools) because it is doing work on the surroundings. The adiabatic (without addition or removal of energy) rate of cooling for an ascending air parcel is 10°C/km (5.5°F/1000 ft). However, if the air is saturated with water vapor, cooling causes some of the vapor to condense, releasing its latent heat of condensation, which then reduces the amount of cooling of the rising air. The rate of cooling for saturated or moist adiabatic ascent of air parcels varies with water content, which is determined by the temperature and pressure of the parcel. For surface air at 30°C (86°F) it can be as low as 3.5°C/km. By the time the parcel has reached about 12 km (40,000 ft), so much of the water vapor has been condensed out that the parcel cools at essentially the dry adiabatic rate. The average rate is about 5.5°C/km (3°F/1000 ft) and this is near the rate of decrease of air temperature normally observed in ascending through the lower atmosphere.

The decrease of air temperature with altitude ends at the tropopause, the boundary between the troposphere below — where most of what is called weather occurs — and the stratosphere above. This boundary is found near 17 km (55,000 ft) in the tropics and descends to about 10 km (30,000 ft) at the poles. However, the slope is not uniform. There is a sharp or even discontinuous decrease in altitude at the so-called tropopause breaks — near 40° latitude in summer and 25–30° latitude in winter.

In the stratosphere air temperature rises with altitude to the stratopause or upper boundary of the stratosphere near 55 km (33 mi), where the air temperature is near or above that at the surface. The warmth of the stratosphere is due to the absorption of solar ultraviolet by ozone. Above the stratopause is the mesosphere where temperature again decreases with altitude at near the dry adiabatic rate to the mesopause near 80 km (48 mi) altitude. Above the mesopause is the thermosphere where temperature again increases with altitude to the exosphere where the atmosphere is so thin that it no longer behaves as a gas. Temperatures here can reach 700° to 1500°C. The thermosphere is the region in which the most energetic short wave part of the solar spectrum is absorbed by nitrogen and oxygen.

In the upper thermosphere gases exist primarily in atomic and ionized states because of the energy of the solar flux to which they are exposed. The term ionosphere is sometimes used to denote the region where ions are most dense. However, ionization tends to occur in layers at different altitudes—denoted D, E, and F layers with increasing altitude. The lowest or D layer normally occurs near the mesopause. These layers rise and fall diurnally and with the seasons and strengthen with solar activity. They are important to radio communication because they tend to reflect radio waves.

Composition of the Atmosphere

The present composition of the atmosphere is given in Table 1. The major constituents are nitrogen, oxygen, argon, carbon dioxide, and neon. Water vapor, which is highly variable and ranges up to no more than 1% by volume, is what in many respects distinguishes Earth from the other planets in the solar system. The many remaining constituents are generally referred to as trace constituents since most account for less than one part per million (ppm) by volume and are variable to highly variable. A non-gaseous, but nevertheless, important constituent of the atmosphere is aerosol or airborne particulates.

Oxygen is probably the most important constituent of our atmosphere because it is not only essential for life but also is required to form many of the constit-

Table 1
Composition of the atmosphere by molecular or volume ratios.

Substance	Formula	Abundance	Comment
nitrogen	N_2	78.1%	long lived
oxygen	O_2	20.1%	long lived
water	H_2O	0.1-1.0%	variable
argon	Ar	0.0934%	permanent
carbon dioxide	CO_2	0.037%	variable, increasing
neon	Ne	1.8 ppm	permanent
helium	He	5.2 ppm	escaping
methane	CH_4	1.6 ppm	variable, increasing
krypton	Kr	1.0 ppm	permanent
hydrogen	H_2	0.5 ppm	variable
nitrous oxide	N_2O	0.3 ppm	increasing
carbon monoxide	CO	0.1 ppm	variable, increasing
sulfur dioxide	SO_2	< 0.1 ppm	highly variable
ozone	O_3	< 0.1 ppm	highly variable
xenon	Xe	0.001 ppm	permanent
ammonia	NH_4	< few ppb	highly variable
nitric oxide	NO	< few ppb	highly variable
nitrogen dioxide	NO_2	< few ppb	highly variable

uents that are important for life, such as water, ozone, carbon dioxide, and the gaseous forms of sulfur and nitrogen required for airborne replenishment of these elements as they are leached from the soil.

Since oxygen is too active chemically to have remained free, it had to have a source other than outgassing. It is currently believed that free oxygen was originally produced from water vapor ascending high enough to encounter ultraviolet light energetic enough to decompose the water molecules into hydrogen and oxygen, and the hydrogen then escaped because its kinetic energy exceeded the gravitational attraction of the planet. At first oxygen so released was removed by oxidation of surface materials such as iron. Once these exposed materials were oxidized, the surface environment was transformed from the original reducing environment into an oxidizing one and oxygen began to accumulate. When oxygen pressure reached about 50 billionths of PAL (present atmosphere level), oxygen-based photosynthesis became possible, leading to a new source of free oxygen. Such levels may originally have occurred only in local oxygen oases in the ocean rather than in the atmosphere as a whole.

Chlorophyll, which gives plants their green color, made it possible for plants to use the energy of sunlight to convert carbon dioxide and water to sugars and biomass with a release of free oxygen. With the most common plants this chemical process is partially reversed in the absence of sunlight in the process of respiration. Oxygen-based photosynthesis is believed to have started by at least 2.2 BYBP (billion years before present) and perhaps as early as 3.5 BYBP. After that point oxygen could have accumulated in the atmosphere more rapidly. Present estimates are that oxygen rose to 0.1 PAL about 650 MYBP (million years before present) and reached 1.0 PAL by 400 MYBP. With the development of an oxygen atmosphere, ultraviolet from the sun would have split some of the molecules into atoms which would have recombined with other oxygen molecules forming ozone in the usual manner. Current estimates are that oxygen at 0.1 PAL and its accompanying ozone would have absorbed the solar ultraviolet that is capable of denaturing protein and thus would have allowed life forms to move out of the ocean onto land.

Atmospheric Composition, Minor Constituents

It should be kept in mind, however, that trace and variable constituents can still be very important ones. Some of the trace constituents are of interest because of their special properties. For example, krypton, radon, tritium, carbon-14, and other gases that are radioactive and thus easy to measure are useful as tracers for studying atmospheric motion. The CFCs (chlorofluorocarbons such as freons) and other organic halogen compounds (such as halons and methyl bromide) are of interest because they can survive to rise into the stratosphere where they are photolytically decomposed, providing sources of chlorine and bromine to the stratosphere where they can destroy ozone.

Water vapor is responsible for most of what is thought of as weather: clouds, fog, rain, hail, snow. It is the circulation of water vapor through our atmosphere forming the precipitation that waters plants and fills the streams, lakes, and ultimately the oceans that makes life in our biosphere possible. Water vapor also plays an extremely important role in distributing absorbed solar energy to make the surface temperature or climate more uniform by tempering it at the extremes of hot and cold. Most of the solar energy received in low latitudes is used to evaporate water rather than to raise local temperatures. The water vapor that evaporates from the tropical oceans, with its latent heat of vaporization, serves primarily to drive the tropical convection of the Intertropical Convergence Zones, monsoons, and tropical cyclones. Because these intense convective storms are zonally distributed in the summer hemisphere they assist in driving the global-scale circulation with updrafts in low latitudes of the summer hemisphere and primary downdrafts in the subtropics of the winter hemisphere. Part of the solar-driven tropical updraft extends up through the tropical tropopause and induces circuit-completing currents in the lower stratosphere downward over the winter pole. The released latent heat in the ascending convective currents primarily supplies the buoyancy that keeps the convective plume rising rather than warming the surrounding atmosphere.

It is in the compensating downward currents in the subtropics and over the winter pole, which are heated due to adiabatic descent, that heating is detected. This represents an almost instantaneous transport of heat from the subsolar surface of the Earth to the subtropics and to the winter pole. The subsiding subtropical branches of this circulation stabilize and suppress precipitation in the subtropics, producing the typical aridity and deserts of these latitudes. The updraft regions with their copious precipitation in the Intertropical Convergence Zones, monsoons, and tropical cyclones are responsible for the rain forests and lush vegetation of the moist tropics.

Similarly, ozone, which is present from hundredths of a ppm at the surface to about ten ppm near 30 km altitude, absorbs most of the energetic and harmful part of the ultraviolet radiation from the sun which is not absorbed by nitrogen and oxygen.

Because of both the variation in ozone column depth and the absorbing path length due to zenith angle of the sun, ultraviolet flux at the surface increases about 50-fold (5,000%) from the poles to the equator on an annual mean basis. That's an increase of 1.2% of the polar value per mile. It is more useful to regard ultraviolet as having approximately six doublings in 6,000 miles or roughly a 1% increase of the local value for each 10-mile displacement toward the equator. Since a 1% decrease in the depth of the ozone layer is considered to cause a 2% increase in surface ultraviolet, it is equivalent to moving approximately 20 miles closer to the equator. While the Antarctic ozone hole allows the ultraviolet flux to increase there in the spring to values typical for summer, they are still only about one-tenth of those experienced near the equator all year round.

Ozone and the even more active and less plentiful hydroxyl radical (OH) are very important scavengers of the atmosphere. By chemical attack they are constantly removing many of the constituents being continually injected into the atmosphere by volcanoes, fires, the living biosphere, and even more the dead and decaying organisms of the biosphere; keeping our atmosphere clean, transparent, and odorless, i.e., unpolluted. The ultraviolet radiation that does penetrate the atmosphere aids directly in this process and also provides the energy for reformation of ozone and the hydroxyl radical. This is a self-stimulatory process in that higher levels of hydrocarbons and oxides of nitrogen lead to higher levels of ozone which can lead to plant damage, and some believe, to health effects in urban areas. This is what creates the Los Angeles smog type of air pollution. While it is clear that plant-damaging levels of ozone can arise from natural emissions of the biosphere, it is less clear whether natural emissions alone can produce the health effects levels of ozone as established in present pollution control regulations or whether such levels require the addition of man-induced emissions.

Carbon dioxide plays key roles for the biosphere both in serving as the main food for plants and in maintaining a moderate climate by serving as a greenhouse gas. While nitrogen, oxygen, and ozone absorb essentially all of the most damaging ultraviolet, most of the rest of the solar radiation penetrates to the surface where it can be used by plants for photosynthesis. On the other hand the outgoing infrared radiation is absorbed and reradiated by the so-called greenhouse gases such as water vapor, carbon dioxide, methane, ozone, nitrous oxide, and the freons. Thus the Earth's surface, due to the greenhouse effect, receives not only the radiant energy from the sun but also the so-called back radiation from clouds, aerosols, and greenhouse or infrared absorbing gases in the atmosphere. This makes the planet's surface 33°C (60°F) warmer than it would be without an atmosphere. This is the basis for speaking of the greenhouse effect of the atmosphere. Increases in greenhouse gases will enhance this effect and is presumed to lead to climate warming.

The primeval source of carbon dioxide is outgassing, mainly from volcanoes. It is removed from the atmosphere by fixation in the organic material of plants and animals. The organic material of dead organisms accumulates and remains for long periods as humus in soils and peats. As these were buried under sediments and compressed, they formed the coal, petroleum, natural gas, lignites, and oil shales which we now recover and use as fossil fuels. Carbon dioxide also dissolves in water, so the oceans contain about 60 times as much as the atmosphere. While the carbon dioxide of the ocean is also fixed into organic material through photosynthesis by phytoplankton, a small part of this material eventually sinks to the bottom and is buried in the sediments. An even larger amount is fixed into the shells of nanoplankton of the oceans as calcium carbonate which accumulates in the sediments, sometimes forming deep layers like the white cliffs of Dover.

Since these competing processes have not always been in balance, the level of carbon dioxide in the atmosphere has fluctuated. Present research indicates that the level of carbon dioxide in the atmosphere 500 MYBP may have been about 15 times the present level. In the Cretaceous, the time of the dinosaurs 135–65 MYBP, it is estimated to have been 5 to 10 times the present level. This is also the period of warmest climate so far documented for the Earth; the global annual mean surface air temperature is believed to have been about 10°C (18°F) warmer than the present 15°C (59°F). However, tropical temperatures appear to have been very little different from those at present with the bulk of the warming in polar regions. Similarly, carbon dioxide fluctuated with temperature during the glacial/interglacial cycles of the Pleistocene of the last million years, dropping to about 200 ppm during the glacials and rising to about 300 ppm during the interglacials. This has been one of the strongest arguments for greenhouse warming due to the rise in carbon dioxide from burning of fossil fuels. However, the available data tend more to indicate that the changes in carbon dioxide lagged behind those of temperature rather than vice versa. If the changes in carbon dioxide had preceded those of temperature, we would be hard pressed to explain what had caused carbon dioxide to fluctuate on a 100,000-year cycle without the biospheric expansions resulting from the interglacial warmings available as the driving mechanism.

Carbonyl sulfide, carbon disulfide, hydrogen sulfide, sulfur dioxide, and dimethyl sulfide gases provide sulfur, which is largely oxidized to sulfuric acid and sulfate. These are important sources of aerosol particles and acidity of rain. These gaseous forms of sulfur are produced mainly by soil bacteria and various organisms that live in the ocean. Sulfur dioxide from the burning of fossil fuels and smoke from coal and wood fires were responsible for the first recognized forms of air pollution. Carbonyl sulfide is longer lived and aside from explosive volcanic eruptions is believed to be the main source of sulfur to the stratosphere. It is oxidized there to sulfuric acid which collects water vapor and forms the particles of the so-called Junge layer of aerosol in the lower stratosphere near 20 km altitude. The return of sulfur dioxide, sulfuric acid, and sulfates to the surface represents an important source of fertilization for the biosphere since sulfur is required for plant growth and, being soluble, is quickly leached from surface soils.

The principal forms of fixed nitrogen to the atmosphere are ammonia, nitrous and nitric oxide, and nitrogen dioxide. Aside from nitrous oxide, these are chemically active and within hours to days are oxidized to nitric acid and nitrates, which also make important contributions to aerosols and rain acidity. The sources of these fixed nitrogen compounds are believed to be almost equally natural and anthropogenic. The natural sources are soil bacteria, lightning, and downward transport from the ionosphere. The return of nitric acid and nitrates to the surface again is an important source of fertilization for the biosphere. Fixed nitrogen is generally the nutrient that is most limiting to plant growth due to its unavailability both on land and in the ocean. Only certain plants such as the legumes are able to enter into symbiotic relationships with soil bacteria to fix gaseous or molecular nitrogen.

While sulfates and nitrates are key constituents of aerosols, there are also windblown dust and meteoric debris. But the largest contribution is from organic material from the biosphere. Hydrocarbon emissions, including aromatics like terpenes, pinenes, and isoprenes and longer-chain hydrocarbons such as waxes, have been estimated to be as large as a billion metric tons per year. It is these that account primarily for the blue hazes painted in many preindustrial landscape paintings and the many "blue ridges" and "smoky" and "disappearing" mountains that acquired their names also before the industrial era. Aerosols are important not only because of their effects on visibility; they also serve as condensation nuclei for the formation of cloud, fog, and rain droplets. Since they generally reduce the sunlight reaching the surface and absorb and emit like black bodies, they can be important in determining climate, at least locally.

HUGH W. ELLSAESSER

For Further Reading: Heinrich D. Holland, *The Chemical Evolution of the Atmosphere and Oceans* (1984); John S. Lewis and Ronald G. Prinn, *Planets and Their Atmospheres: Origin and Evolution* (1984); J. William Schopf, ed., *Earth's Earliest Biosphere: Its Origin and Evolution* (1983).

See also Aerosols; Carbon Cycle; Clean Air Act; Climate Change; Energy, Wind; Meteorology; Mineral Cycles; Nitrogen; Nitrogen Cycle; Oxygen; Oxygen Cycle; Ozone Layer; Photosynthesis; Pollution, Air; Radiation; Smog; Stratosphere; Troposphere; Ultraviolet Radiation; Water Cycle.

ATOMIC ENERGY COMMISSION

See Nuclear Power; United States Government.

AUTECOLOGY AND SYNECOLOGY

Autecology is the study of the relationships of individual species to their environments. In contrast, synecology is concerned with the interactions of groups of organisms within communities. The myriad of peculiar and often amazing adaptations displayed by Earth's species is a result of the interactions between organisms and the living (biotic) and nonliving (abiotic) factors with which they strive to survive. Autecology focuses on single species or single environmental factors to determine how they influence each other. Because environmental physiology (or physiological ecology) often requires complex equipment and facilities, many investigations are done in a laboratory, rather than in the field, so that single factors can then be controlled.

The ability of humans to use plants and animals for food depends on an understanding of the autecological requirements of the individual species involved. Autecological research is also important to efforts in conservation and restoration of species. Unless its physiological requirements and tolerances are known, a rare or endangered species facing diminishing numbers or a loss of habitat cannot be saved. The success of a species in a reconstructed habitat depends on what it is adapted to in its native environment. Exactly what these individual components are, or how forgiving, or plastic, the species is, may or may not be readily perceptible. Either way, these autecological issues need to be addressed before sound species-management decisions can be made.

Synecology, also known as biocoenology or community ecology, examines the distribution of communities around the globe and the responses of groups of organisms within communities to each other and to their environments. It is the study of community dynamics—the interaction and influence of one community member upon another and changes in species composition over time. Historically, it developed from an area of plant ecology that involved sampling, mapping, and describing vegetation types or communities

across the landscape. Environmental issues involve synecological questions, such as how introduced organisms will interact with natural species and food webs. Community level interactions such as interspecific competition, predation, or herbivory, and pollinator or seed dispersal factors are many of the potential focal points in the study of synecology.

AUTOMOBILE INDUSTRY

The automobile has profoundly affected the environment, particularly in modern industrial societies, where the unprecedented mobility it provides has dramatically altered human life styles and expanded the geographical zone of our activities. In their early days at the beginning of this century, motor vehicles contributed to environmental quality by using gasoline, a byproduct of producing kerosene, and eliminating the solid waste problems horses created in cities. However, over recent decades, automobiles have faced increasing public scrutiny for their own environmental pollution and energy use.

Automotive Production

The automotive industry—composed of vehicle manufacturers and their numerous suppliers of materials, parts, and components—is a significant consumer of energy and source of manufacturing emissions. In the U.S. alone, there are 20 engine plants, over 40 assembly plants, hundreds of metal stamping facilities, and thousands of supplier facilities, ranging from small plants to huge industrial complexes that produce everything from microchips and electric motors to tires and steel. These activities employ more than one million people, and automotive production consumes significant fractions of many basic materials.

Supplier manufacturing and automotive assembly consume significant quantities of all forms of energy. Natural gas heats plants, fuels paint drying ovens, and is an ingredient in some manufacturing processes. All industrial facilities consume electric power, which is produced using some combination of coal, liquid petroleum, natural gas, hydropower, and nuclear power. Electric power accounts for about 75% of the industry's power consumption expenditures. Other important fuels are natural gas, about 15% of energy expenditures, and coal and coke, at over 8%. The industry also relies on steam, oil, and propane. The industry's total energy bill in 1990 was nearly $3 billion, and may have been even higher because of the difficulty in tracking energy use throughout the entire automotive supply structure.

Because of regulatory, economic, and social forces,

practically all industry production processes have substantially reduced their energy consumption in the past decade or so. The automotive industry's use of fuels other than electricity declined almost 17% from 1984 to 1990, due partly to conservation efforts and partly to conversion to electricity.

Automotive plants have released sulfuric acid and other "smokestack emissions" into the atmosphere, solvents and other chemical pollutants into ground and water, and plant wastes into the solid waste stream. However, it is not clear that the industry polluted disproportionately to other manufacturing industries, and its significance as a source of pollution largely reflected the size and complexity of its operations. In recent decades the automotive industry has been a leader in the U.S. and much of Europe in minimizing industrial pollution, often achieving regulatory goals and targets well ahead of schedule.

The Automobile

At the present time, there are over 550 million motor vehicles in the world, with nearly 190 million in the U.S. alone, of which over 140 million are passenger cars. The vehicles that make up this worldwide stock have vastly differing environmental effects, reflecting differences in design, manufacturing, maintenance, and use. U.S. passenger cars made since 1980 are far more fuel efficient than those made earlier, and newer cars emit far lower levels of many environmental pollutants than their older counterparts. While the U.S. and Japan have imposed increasingly restrictive emission control requirements since the 1960s, Europe has only recently adopted extensive emission standards, and many developing nations have no emission standards at all. Furthermore, vehicle use patterns differ widely, and total use affects fuel consumption and emissions levels, as does average trip length, with short trips decreasing fuel efficiency and increasing emissions per mile.

Resource Conservation

Petroleum meets 41% of all U.S. energy needs, about half of which is used by highway vehicles, so motor vehicles account for over 20% of total U.S. energy consumption. This makes the automobile a logical target for energy and petroleum conservation. At present it is probably unrealistic to propose the elimination or even serious curtailment of automobile travel, since the automobile is so tightly integrated into our modern world economy and society, and alternative forms of transportation cannot yet meet many of our transportation needs. As any energy conversion process results in the creation of some form of waste, we cannot totally eliminate it, but rather must minimize it by including conservation of scarce resources and waste minimization in our decisions.

Significant national and international efforts are underway to find alternatives to petroleum and to develop modifications to present fuels. Gasoline still has some clear advantages. Economically available liquid petroleum will last 40 years or more with current technology and substantially more if economical extraction technologies for tar sands, oil shale, and other resources are developed. Gasoline has an extremely high ratio of energy to weight, which is an advantage for mobile consumption. Because the refining and distribution systems already exist, gasoline has an important economic advantage over most alternative fuels.

Nevertheless, several alternative fuels do possess significant advantages from an environmental standpoint. Presently under consideration are compressed natural gas (CNG), liquid petroleum gas (LPG), methanol, ethanol, and hydrogen, as well as reformulated gasoline, a modified blend of gasoline with alcohol or ether. The spark-ignited engine is inherently adaptable to practically all seriously considered alternative fuels. In most cases, the challenge presented by these alternative fuels is not so much compatibility with the engine, but issues such as the economics of fuel supply, storage on board the vehicle, and vehicle range. For example, CNG has a high octane number and thus good resistance to engine knock, and potentially low emissions. However, it is difficult to provide onboard storage for CNG sufficient to ensure adequate vehicle range, and it requires very high pressure storage. There are ready supplies of both CNG and LPG, although LPG is more dependent on the supply of liquid petroleum. These fuels generally burn cleaner than gasoline, particularly during the cold start or early stages of the operating cycle of the engine.

The alcohol fuels possess less certain environmental benefits, particularly because of the higher levels of some pollutants they yield, such as various aldehydes. A fundamental defect of methanol is that natural gas would probably be the source for it, and that conversion would consume about 30% of the energy available in the gas prior to its use to fuel the vehicle. Ethanol can be derived from several sources, although biomass currently seems most likely. At this point, economic ethanol production is highly dependent on tax incentives.

Hydrogen is a promising alternative fuel because the lack of carbon in the combustion process ensures that no carbon monoxide, carbon dioxide, or unburned hydrocarbons are produced. However, hydrogen combustion does yield nitrogen oxides. Hydrogen is difficult to produce and also poses problems in on-board storage and probable limits on vehicle range.

Economic considerations in the early 1990s seem to

favor natural gas, LPG, and reformulated gasoline for motor vehicle fuel. However, for the next 10 to 20 years, gasoline derived from liquid petroleum will remain our primary motor vehicle fuel. Reformulated gasoline is already available in many regions and will probably grow in use.

Automobile Emissions

Since the 1960s, the automobile has been the subject of intense social concern as a factor in atmospheric pollution. Most automotive emissions result from the combustion process; however, loss of vapor from the fuel system prior to combustion can be an important source of hydrocarbons. Current U.S. vehicles all employ some evaporative emission control technology to reduce this emission source.

Of the combustion emissions, some are currently regulated, such as unburned hydrocarbons (HC), carbon monoxide (CO), and nitrogen oxides (NO_x), but there are also unregulated compounds of potential concern, such as carbon dioxide (CO_2) and aldehydes (partially burned hydrocarbons). Considerable disagreement and uncertainty remains about the environmental and health effects of these and other emissions, but we are making progress in understanding them, even as we are making progress in the reduction of regulated emissions. The exact role of CO_2 in global warming is not completely understood, but since CO_2 is a byproduct of the combustion of any compound containing carbon, this is an issue of serious concern.

The gasoline-burning engine has numerous operating characteristics, and optimization of some often yields sub-optimal performance on others. For example, "lean-burn" engines confer advantages for fuel efficiency, but exacerbate problems with NO_x emissions.

Related but different processes produce the emissions of primary concern today, CO, HC, and NO_x. Fuel combustion with insufficient air increases carbon monoxide levels. However, even with lean fuel-air ratios (excess oxygen), some CO is present since this intermediate product of combustion does not entirely disappear even under the best conditions. The primary oxide of nitrogen produced by engines is nitric oxide, NO. Levels of NO depend on three primary variables: temperature, time, and composition. The higher the maximum cylinder temperature and the longer the period of time held at these high temperatures, the greater the NO formation.

The process of hydrocarbon formation is complex and not completely understood. The combustion process in a transient, spark ignition engine is very difficult to analyze as well. It appears that hydrocarbons come from incomplete combustion of the fuel, including fuel lost through leakage, misfiring, and becoming trapped in cylinder crevices, and from quenching at relatively cool combustion chamber surfaces.

Emission control efforts were initially stimulated by California because of its early concern with photochemical smog. Its first emission standards issued in 1959 required relatively modest emission reduction. Since then, a series of major California and federal government regulatory initiatives have reduced carbon monoxide and unburned hydrocarbons by over 95% and nitrogen oxides by 75%.

Early emission control technologies were quite simple, involving little more than recycling crankcase breather gases back to the inlet manifold for return to the combustion chamber. Control efforts in the late 1960s and early 1970s focused on optimizing internal engine processes such as ignition timing, fuel-air mixture ratio, combustion chamber design, and compression ratio for reduced emissions. Later came exhaust gas recirculation for nitrogen oxide control. However, with ever-tightening standards, engine optimization strictly for emissions began to result in significant deterioration of fuel economy, performance, and driveability. It was difficult for control technology to keep pace with escalating regulatory requirements. Finally in the 1970s, two critical technologies were developed that achieved truly effective emission control: the exhaust catalytic reactor or converter and electronic engine control.

The exhaust catalytic converter is an after-engine chemical reactor that facilitates the near elimination of pollutants. Its development allowed designers to optimize the engine for fuel economy, driveability, and performance, as the catalytic converter takes on much of the burden for emission control. During the late 1970s, electronic microprocessor-based engine control became commercially feasible, which improved both various engine variables and the functioning of the catalytic converter.

Very restrictive requirements for carbon monoxide (CO), unburned hydrocarbons (HC), and nitrogen oxides (NO_x) drove catalytic reactor technology considerably beyond the early oxidation reactors. The three-way catalyst achieved not only oxidation of CO and HC, but also the chemical reduction of NO_x. However, the proper operation of this system requires very precise control of the engine fuel-air ratio quite near the chemically correct, or stoichiometric, value. This precision was achieved with the exhaust gas oxygen sensor, coupled with electronic control. The next step toward further reductions in vehicle emissions will probably require even more sophisticated control technology, such as preheated catalysts or catalytic reactors closely coupled to the engine, which will help reduce the CO and HC emissions during the initial—or "cold start"—and warmup phases of engine operation, by hastening warmup of the catalyst.

In addition to the requirements of the 1991 Clean Air Act, California has passed a stringent set of its own new standards to be phased in during the 1990s. These include specific requirements for the introduction of so-called low emission (LEV), transitional low emission (TLEV), and ultra-low emission vehicles (ULEVs). Beginning in 1998, California will require 2% of all vehicles sold to achieve an essentially zero emission level (ZEV). These California requirements will undoubtedly require a multiple-system approach, probably including the increased use of advanced control technology, reformulated gasoline and alternative fuels, and electric vehicles.

Recycling and Solid Waste Disposal

As of the early 1990s, about 75% of the typical scrapped vehicle in the U.S. is recycled, ranging from parts and components to bulk metals, primarily due to technical developments that permit separating and shredding its metal content, and the value of scrap steel in manufacturing new steel. Many key materials used in present-day vehicles are currently recycled, and initial efforts to recycle virtually all others are under way. Still, a significant fraction of the content of scrapped vehicles—called "fluff"—ends up in increasingly scarce landfills, including various fluids and trace metals that could be considered hazardous waste.

Energy conservation, improved air quality, and solid waste reductions are complex goals, and responses in one arena often create problems in another. Simply comparing the energy requirements of producing a ton of aluminum or a ton of steel, steel is the more energy-conserving choice. However, specific automotive applications often require fewer pounds of aluminum because of its lower density. Aluminum's reduced weight yields lower vehicle fuel consumption and exhaust emissions, and recycling aluminum does not incur the energy penalties incurred in its original production.

Environmental goals are not always congruent. Thus the electric car promises the elimination of automotive emissions, but increases power plant emissions, and the batteries it requires could pose a major solid waste problem. Growing pressures to reduce motor vehicle fuel consumption and exhaust emissions will almost surely result in lighter weight vehicles in the future. In part, this will be achieved by reducing vehicle size and improved engineering, but the industry will almost surely rely heavily on the substitution of lighter materials like plastics for steel, risking increased solid waste problems, as the vehicle may become less recyclable. Thus a full life cycle systems analysis must be performed if we are to make truly optimal energy conservation and pollution reduction decisions.

DAVID E. COLE
MICHAEL S. FLYNN

For Further Reading: John B. Heywood, *Internal Combustion Engine Fundamentals* (1988); National Research Council, *Fuels to Drive Our Future* (1990); D. J. Patterson and N. A. Henein, *Emission From Combustions and Their Control* (1972).
See also Clean Air Act; Energy; Fuel, Fossil; Fuel, Future; Fuel, Liquid; Fuel, Synthetic; Pollution, Air; Transportation; Vehicles, Alternative Energy.

B

BACTERIA

Bacterial Biodiversity

The bacteria are organisms with cells lacking a nucleus (prokaryotic). Most bacteria are unicellular, but some form multicellular associations. There are two major groups of bacteria, eubacteria and archaebacteria (Archaea), that are distinguished by differences in ribosomal RNA and by morphological and physiological characteristics.

Many of the archaebacteria have unusual physiological properties that enable them to occupy niches in extreme environments. Some grow at high temperatures, such as in hot springs. Others grow at high salt concentrations, such as in salt lakes. The methanogens are a highly specialized physiological group of archaebacteria that live where there is no atmospheric oxygen (anaerobic) and form methane from carbon dioxide.

The eubacteria, which are usually simply called bacteria, exhibit extreme diversity. Several of the major groups are distinguished primarily on cell shapes (rods, cocci [spherical], curved, or filament-forming); spore production; staining reactions (color after staining according to the Gram stain procedure—pink-red for Gram negative or blue-purple for Gram positive); and motility (without hairlike organelles of locomotion, called flagella; with flagella surrounding the cell, called peritrichous flagella; with flagella projecting from the end of the cell, called polar flagella). Other major bacterial groups are defined based upon their metabolism, in particular how they generate energy in the form of ATP (photosynthetic using light energy; chemolithotrophic using inorganic compounds; heterotrophic using organic compounds); and whether the metabolism is anaerobic, aerobic (using molecular oxygen), or facultatively anaerobic (capable of both aerobic and anaerobic metabolism).

Cyanobacteria, or blue-green bacteria, carry out a type of photosynthesis that resembles that of higher plants. It results in the production of molecular oxygen. Other photosynthetic bacteria carry out photosynthesis without the production of oxygen. Such anaerobic photosynthetic bacteria include the purple nonsulfur bacteria, purple sulfur bacteria, green sulfur bacteria, and green flexibacteria. These anaerobic photosynthetic bacteria use light of longer wavelengths than algae and cyanobacteria and, hence, are able to grow at greater depths within lakes.

Chemolithotrophic bacteria use inorganic compounds to generate ATP. Some chemolithotrophic bacteria, called nitrifying bacteria, convert ammonia (NH_4^+) to nitrate (NO_3^-). Others, such as *Thiobacillus thiooxidans*, convert reduced sulfur compounds to sulfate. These metabolic transformations of inorganic compounds by chemolithotrophic bacteria cause global-scale cycling of various elements between the air, water, and soil.

The Gram-positive cocci include the genus *Staphylococcus*, which typically form grape-like clusters, and the genus *Streptococcus*, which occur in pairs or chains. Species of *Staphylococcus* commonly occur on skin surfaces where they live without causing disease. *Staphylococcus aureus*, however, is a potential human pathogen, infecting wounds and also causing food poisoning. Some members of *Streptococcus* are also human pathogens. For example, rheumatic fever is caused by *Streptococcus pyogenes*. Several *Streptococcus* species are responsible for the formation of dental cavities, tooth decay.

Two most important genera of bacteria, *Bacillus* and *Clostridium*, are both Gram-positive rods. *Bacillus* species can grow in the presence of air, whereas *Clostridium* species are strictly anaerobic. Food spoilage by *Bacillus* and *Clostridium* species is of great economic importance. Several *Clostridium* species are important human pathogens. For example, *Clostridium botulinum* is the causative agent of botulism, *Clostridium tetani* causes tetanus, and *Clostridium perfringens* causes gas gangrene.

The Gram-negative facultatively anaerobic rods include intestinal (enteric) bacteria, such as *Escherichia coli*, which are motile by means of peritrichous flagella and often live in the human intestinal tract. Much of what we know about bacterial metabolism and bacterial genetics has been elucidated in studies using *E. coli*. Additionally, *E. coli* is employed as an indicator of fecal contamination in environmental microbiology. The genera *Salmonella* and *Shigella* contain many species, many of which are important human pathogens. In particular, typhoid fever and various gastrointestinal upsets are caused by *Salmonella* species; bacterial dysentery is caused by *Shigella*.

The Gram-negative aerobic rods encompass a metabolically diverse group of bacteria. Many *Pseudomonas* species are nutritionally versatile and are capable of degrading numerous natural and synthetic organic compounds. *Azotobacter, Rhizobium,* and *Bradyrhizobium* are capable of fixing atmospheric nitrogen. *Rhizobium* and *Bradyrhizobium* species infect leguminous plant roots where they cause the formation of tumorous growths called nodules within which they live in a mutually beneficial relationship. *Agrobacterium* produces tumorous growths on infected plants, known as galls. *Agrobacterium tumefaciens* causes galls of many different plants and is an extremely important plant pathogen, causing large economic losses in agriculture. *Photobacterium* is a luminescent bacterium that is sometimes associated with fish that rely upon light from bioluminescence for their food finding and other behavioral activities.

The spiral and curved bacteria group are helically curved rods that may have less than one complete turn (comma-shaped) to many turns (helical). *Campylobacter fetus* is a curved bacterium that frequently is the cause of gastrointestinal infections in infants. *Bdellovibrio,* which also is curved, has the unique characteristic of being able to penetrate and reproduce within the cells of other bacteria. Spirochetes are helically coiled rods, with cells wound around one or more central axial fibrils. Many spirochetes are human pathogens. *Treponema pallidum* causes syphilis. *Borrelia burgdorferi* causes Lyme disease.

The budding and/or appendaged bacteria are grouped together because they produce cell appendages. Many of the appendaged bacteria grow well at low nutrient concentrations. *Caulobacter,* for example, is able to grow in very dilute concentrations of organic matter in lakes and even distilled water.

The sheathed bacteria are those bacteria whose cells occur within a filamentous structure. The formation of a sheath enables these bacteria to attach themselves to solid surfaces. It also affords protection against predators and parasites. *Sphaerotilus natans,* a sheathed bacterium that is often referred to as the sewage fungus, normally occurs in polluted flowing waters, such as sewage effluents, where it may be present in high concentrations just below sewage outfalls.

Some bacteria exhibit gliding motility on solid surfaces. These bacteria lack the specialized structures that other bacteria use to propel themselves. Myxobacteria are gliding bacteria that aggregate to form fruiting bodies. The fruiting bodies of myxobacteria occur on decaying plant material, on the bark of living trees, or on animal dung, appearing as highly colored slimy growths that may extend above the surface of the substrate.

Actinomycetes are bacteria that resemble fungi in appearance because they form filamentous growths. The production of antibiotics by actinomycetes, such as *Streptomyces griseus,* is extremely important in the pharmaceutical industry. Actinomycetes often grow slowly in soils where they are important in decompositional processes.

The rickettsias are intracellular parasites, that is, they can only reproduce within living host cells. Many rickettsias are carried by insects and cause diseases in humans. For example, *Rickettsia rickettsii* is transmitted by ticks and causes Rocky Mountain spotted fever. The chlamydias also grow only within animal cells. In birds they cause respiratory diseases and generalized infections. For example, the disease psittacosis, parrot fever, is caused by *Chlamydia psittaci.*

Bacterial Biogeochemical Cycling Activities

The activities of bacteria within the biosphere have a direct impact on the quality of human life. Because of their ubiquitous distribution and diverse enzymatic activities, bacteria play a major role in biogeochemical cycling (the movement of materials via biochemical reactions through the global biosphere). The major elements of living biomass, carbon, hydrogen, oxygen, nitrogen, phosphorus, and sulfur, are most intensively cycled by bacteria; other elements are generally cycled to a lesser extent. Higher forms of life, including humans, depend upon the essential biogeochemical cycling activities of bacteria. The biogeochemical cycling activities of bacteria are essential for the survival of plant and animal populations and determine, in large part, the potential productivity level of a given habitat.

The biogeochemical cycling of carbon primarily involves the transfer of carbon dioxide and organic carbon between the atmosphere, where carbon occurs principally as inorganic CO_2, and the hydrosphere and lithosphere, which contain varying concentrations of organic and inorganic carbon compounds. The metabolism of photosynthetic and chemolithotrophic bacteria that manufacture their own food contributes to primary production, the conversion of inorganic carbon dioxide to organic carbon. Some ecosystems, notably deep sea thermal vents, are totally dependent on chemolithotrophic bacteria as the source of organic carbon. Once carbon is fixed (reduced) into organic compounds, it can be transferred from population to population within the biological community, supporting the growth of a wide variety of heterotrophic organisms.

Bacterial decomposition of dead plants and animals and partially digested organic matter is largely responsible for the conversion of organic matter to carbon dioxide and the reinjection of inorganic CO_2 into the atmosphere. Some natural organic compounds, such as lignin, cellulose, and humic acids, are relatively resistant to attack and decay only slowly. Various synthetic compounds, such as DDT, may be recalcitrant, that is, completely resistant to enzymatic degradation. Many modern problems relating to the accumulation of environmental pollutants, such as plastics, reflect the inability of bacteria to degrade rapidly enough the concentrated wastes of industrialized societies.

Biological nitrogen fixation, the conversion of N_2 to ammonia or organic nitrogen compounds, is restricted to a limited number of bacterial species. The fixation of atmospheric nitrogen is carried out by free-living bacteria and by bacteria living in symbiotic (mutually dependent) association with plants that produce nitrogenase. Symbiotic nitrogen fixation by *Rhizobium* or *Bradyrhizobium* is most important in agricultural fields, where these bacteria live in association with leguminous crop plants. In aquatic habitats, cyanobacteria, such as *Anabaena* and *Nostoc*, are very important in determining the rates of conversion of atmospheric nitrogen to fixed forms of nitrogen.

The two steps of nitrification, the formation of nitrite from ammonium, and the formation of nitrate from nitrite, are carried out by chemolithotrophic bacteria. In soils, *Nitrosomonas* is the dominant bacterial genus involved in the oxidation of ammonia to nitrite, and *Nitrobacter* is the dominant genus involved in the oxidation of nitrite to nitrate. Nitrification is very important in soil habitats because the transformation of ammonium ions to nitrite and nitrate ions results in a change from a cation to an anion. Positively charged cations are bound by negatively charged soil clay particles and, thus, are retained in soils, but negatively charged anions, such as nitrate, are not absorbed by soil particles and are readily leached from the soil.

Denitrification, the conversion of fixed forms of nitrogen to molecular nitrogen, is another important process in the biogeochemical cycling of nitrogen mediated by bacteria. Some aerobic bacteria can use nitrate in place of oxygen, reducing nitrate as a result of anaerobic respiration. Some bacteria, such as *E. coli*, are only able to reduce nitrate to nitrite, but a variety of other bacteria are able to carry out the two subsequent anaerobic respirations by which nitrite ion is reduced to nitrous oxide gas (N_2O) and subsequently to molecular nitrogen (N_2).

Sulfur can exist in a variety of oxidation states within organic and inorganic compounds, and oxidation-reduction reactions—mediated by bacteria—change the oxidation states of sulfur within various compounds, establishing the sulfur cycle. Hydrogen sulfide is formed by sulfate-reducing bacteria that use sulfate during anaerobic respiration. Hydrogen sulfide can accumulate in toxic concentrations in areas of rapid protein decomposition; it is highly reactive, and is very toxic to most biological systems. It can react with metals to form insoluble metallic sulfides. The predominant source of hydrogen sulfide in different habitats varies. In organically rich soils, most of the hydrogen sulfide is generated from the decomposition of organic sulfur-containing compounds. In anaerobic sulfate-rich marine sediments, most of it is generated from the reduction of sulfate by sulfate-reducing bacteria, such as members of the genus *Desulfovibrio*. Anaerobic sulfate reduction is important in corrosion processes and in the biogeochemical cycling of sulfur.

Although hydrogen sulfide is toxic to many organisms, photosynthetic sulfur bacteria use it during their metabolism. Anaerobic photosynthetic bacteria often occur on the surface of sediments, where there is light to support their activities and a supply of hydrogen sulfide from sulfate reduction and anaerobic degradation of organic sulfur-containing compounds. Some photosynthetic bacteria deposit elemental sulfur as an oxidation product, whereas others form sulfate.

Acid mine drainage is a consequence of the metabolism of sulfur and iron-oxidizing bacteria. Coal in geological deposits is often associated with pyrite (FeS_2), and when coal mining activities expose pyrite ores to atmospheric oxygen, the combination produces large amounts of sulfuric acid. When pyrites are mined as part of an ore recovery operation, oxidation may produce large amounts of acid. The acid draining from mines kills aquatic life and renders the water it contaminates unsuitable for drinking or recreational use. Strip mining is a particular problem with respect to acid mine drainage because this method of coal recovery removes the overlying soil and rock, leaving a porous rubble of tailings exposed to oxygen and percolating water.

Bacterial Degradation of Wastes and Pollutants

Human activities create vast amounts of wastes and pollutants. The release of these materials into the environment sometimes causes serious health problems and may preclude desirable usage of our land and water resources. The use of rivers, for example, as a habitat for fish, as a source of irrigation and drinking water, and for the disposal of sewage depends on the careful management of the amounts of wastes entering the ecosystem and the levels of pathogenic bacteria associated with their release. Proper treatment of wastes, employing microbial biodegradation and disinfection of potable water supplies in order to kill contaminating pathogenic bacteria, can greatly improve the safety and quality of water supplies and the status of human health.

One consequence of urbanization is the need to remove sewage and other organic wastes from population centers. Waterways that are normally used for waste removal under the premise that "the solution to pollution is dilution" can be overwhelmed by such concentrated inputs of organic matter. A high BOD (biochemical oxygen demand) generally indicates the presence of excessive amounts of organic carbon. The dissolved oxygen in natural waters seldom exceeds 8 milligrams per liter because of its low solubility, and it is often considerably lower because of heterotrophic microbial activity, making oxygen depletion a likely consequence of adding wastes with high BOD values to aquatic ecosystems. Exhaustion of the dissolved oxygen content is the principal result of a sewage overload on natural waters. Oxygen deprivation kills strictly aerobic organisms, including fish, and the decomposition of dead organisms within the water body creates an additional oxygen demand. Fermentation products and the reduction of the removal of nitrate and sulfate give rise to noxious odors, tastes, and colors, making the water putrid and septic.

The treatment of domestic sewage reduces the BOD, due to suspended or dissolved organics, and the number of enteric pathogens so that the discharged sewage effluent will not cause unacceptable deterioration of environmental quality. Primary treatments rely on physical separation procedures to lower the BOD; secondary treatments rely on microbial biodegradation to further reduce the concentration of organic compounds in the effluent; and tertiary treatments use chemical methods to remove inorganic compounds and pathogenic bacteria. Municipal sewage treatment facilities are designed to handle organic wastes but are normally incapable of dealing with industrial wastes containing toxic chemicals, such as heavy metals. Industrial facilities frequently must operate their own treatment plants to deal with waste materials.

The importance to public health of clean drinking water requires objective test methods to establish high standards of water safety and to evaluate the effectiveness of treatment procedures. To monitor water routinely for the detection of actual enteropathogens, such as *Salmonella* and *Shigella*, would be a difficult and uncertain undertaking. Instead, bacteriological tests of drinking water establish the degree of fecal contamination of a water sample by demonstrating the presence of indicator organisms. The most frequently used indicator organism is the normally nonpathogenic coliform bacterium *Escherichia coli*. Positive tests for *E. coli* do not prove the presence of enteropathogenic organisms but do establish this possibility. Because *E. coli* is more numerous and easier to grow than enteropathogens, the test has a built-in safety factor for detecting potentially dangerous fecal contamination. *E. coli* meets many of the criteria for an ideal indicator organism, but there are limitations to its use as such, and various other species have been proposed as additional or replacement indicators of water safety.

Human exploitation of fossil fuel reserves and the production of many novel synthetic compounds (xenobiotics) in the 20th century have introduced into environments many polluting compounds that bacteria normally do not encounter and, thus, are not prepared to biodegrade. Many of these compounds are toxic to living systems, and their presence in aquatic and terrestrial habitats often has serious ecological consequences, including major kills of indigenous biota. The disposal or accidental spillage of these compounds has created serious modern environmental pollution problems, particularly when microbial biodegradation activities fail to remove these pollutants quickly enough to prevent environmental damage. Sewage treatment and water purification systems are usually incapable of removing these substances if they enter municipal water supplies, where they pose a potential human health hazard.

We rely on bacteria to biodegrade our waste materials and have come to assume that anything thrown into environments will disappear; and to an incredibly large extent, this is true. Bacteria have a vast capacity for rapidly degrading organic materials and, thus, can be relied upon to act as biological incinerators. However, the bacteria associated with waste materials that are released into terrestrial and aquatic ecosystems include pathogens that can cause serious outbreaks of human disease. The multiple usage of water resources as sources of drinking and irrigation water and for recreational purposes mandates careful management of the release of wastes and pollutants into these aquatic ecosystems.

When chemicals are released directly into environments as a result of careless or accidental spillages, we must generally rely on the metabolic activities of bacteria to remove the polluting substances. When xenobiotics are resistant and bacteria fail in their role as biological incinerators, environmental pollutants accumulate. The proper chemical design of synthetic compounds, which makes them susceptible to microbial attack, is essential for maintaining environmental quality. Bacterial degradation of pollutants is used in bioremediation to remove polluting substances and to restore environmental quality. In some cases environmental conditions are modified to stimulate indigenous bacterial populations, for example, through aeration or the addition of nitrogen-containing fertilizers. In other cases, seed cultures are added that can degrade the specific polluting substance. Using recombinant DNA technology, it is possible to create bacteria capable of degrading complex mixtures of toxic environmental pollutants, and the development of such bacteria may prove useful in treating the myriad chemicals in industrial wastes. The human race depends on bacteria to make the world a better place in which to live and, in the end, we remain dependent on microbial biodegradation for recycling our wastes and maintaining the environmental quality of the biosphere.

RONALD M. ATLAS

For Further Reading: Ronald M. Atlas and Richard Bartha, *Microbial Ecology: Fundamentals and Applications* (1993); J.C. Fry, ed., *Microbial Control of Pollution* (1992); R. Mitchell, ed., *Environmental Microbiology* (1992).
See also Biodegradable; Biotechnology, Environmental; Carbon Cycle; Irrigation; Mineral Cycles; Mining Industry; Nitrogen Cycle; Oxygen Cycle; Phosphorus; Photosynthesis; Sulfur; Wastewater Treatment, Municipal; Water; Xenobiotic.

BAIKAL, LAKE

Lake Baikal (also spelled Baykal) is a unique ecosystem and the world's oldest lake. Located in southeastern Siberia about 50 miles north of the Mongolian border, it is in terms of area the eighth largest lake in the world, occupying 12,000 square miles and stretching 395 miles in length averaging 30 miles in width. It has the largest volume of surface fresh water of all lakes and contains one fifth of the fresh water on the earth's surface and four fifths of the fresh water in Russia. Its maximum depth of 5,314 feet makes it the world's deepest lake. It is revered by local people as the "Holy Sea" and has long played a role in Russian literature and culture.

Although water enters the lake from 336 rivers and streams and from precipitation and subterranean sources, the only outlet is the Angara River. The water level varies from two to three feet during the year, rising to its highest level during August and September and falling to its lowest during March and April. Lake Baikal's water is clear to a depth of 130 feet, as it contains few minerals and has a low level of salinity.

Because of its immense size, the lake has a modifying effect on the climate in the immediate area, where the summers are cooler than in the surrounding areas and the winters are warmer. The lake is frozen from January until May.

Over 1,500 different species of animals are supported at different depths, 600 species of plants live on or near the surface, and 326 species of birds are also supported by the lake. At least 1,300 of these species are unique to the lake. These include the Baikal seal, a freshwater shrimp (*Acanthogammarus victori*), the galomyanka fish, which gives birth to live young, and the fish *Comechorus baicalensis*, which lives at depths below 3,300 feet (1,000 meters). Although most lakes are nearly lifeless below 1,000 feet (300 meters),

the circulation in Baikal carries oxygenated water to three times this depth. The lake is a unique resource for scientific study. In 1991 scientists from the United States Geological Survey joined Soviet researchers in collecting sediment cores from the lake floor.

Industries around Lake Baikal include mica and marble mining, cellulose and paper plants, ship building, fisheries, and timber. As a result of the large quantities of industrial effluent entering the lake every day, conservationists have raised concerns about possible environmental hazards. In 1971 the Soviet government issued a decree that called for protection and rational use of resources and prevention of emissions of pollutants into the air and water from industrial plants on the shore of the lake.

BAMBOO

Bamboo, a grass with a hard, woody stem, has been used by humans throughout history, playing an integral role in the cultural and natural ecosystems of tropical and subtropical areas. Commonly referred to as the "poor man's timber," bamboo's versatility makes it very important in the rural areas of the developing world. According to the properties of different species, it has supplied the principal or preferred material for commercial as well as domestic products, such as barrels, baskets, bridges, brooms, cages, carts, chairs, chopsticks, cradles, cupboards, fences, fishing poles, flutes, houses, ladders, lanterns, medicines, paper, rayon, and xylophones.

Bamboos are found on all continents except Europe and Antarctica. They grow at temperatures ranging from −40°F to 180°F, at latitudes from 46° N to 47° S, and at altitudes from sea level to 13,000 feet. In the Western Hemisphere the natural distribution of bamboos extends from the southern part of the United States southward to Argentina and Chile. Gaps in this distribution occur principally in relatively arid regions or in places where agriculture has destroyed the natural forest cover.

There are approximately 1,500 species of bamboo. They are distinguished from other grasses by characteristics such as woody stems and stalked leaf blades. Bamboos comprise a highly varied array of plants that range in size and habit from miniature ornamentals a few inches high to long, slender climbers to giants two feet in diameter and more than 100 feet tall. Due to their fast growth rate, some species are extremely valuable as an agricultural crop. These species have the capacity to grow over 3 feet in 24 hours, reaching full maturity within a year.

Regardless of geographic location or climatic condition, plants of the same bamboo species flower at the same time—an unusual phenomenon. Some species of bamboo flower only once every 120 years. After a bamboo has flowered the stem withers and dies. The bamboo must regenerate either from seed or from the surviving rhizomes (underground stems that send up shoots).

Animals also depend on bamboos. For instance, the giant panda eats bamboos exclusively. To support its size, it must eat as much as 40 pounds of bamboo leaves and stalks every day. This narrow diet makes pandas extremely vulnerable to fluctuations in the availability of bamboo. In the 1970s huge acreages of bamboo flowered and died; consequently, many pandas starved.

Current research focuses on determining an individual species' suitability for specific uses such as construction, handicrafts, or tools as well as discovering ways to intensify bamboo production for food. In addition to this research, people in the developing world are being trained in the basic techniques of building and weaving with bamboo. The resultant cottage industries provide employment for the surrounding communities, developing the local economies.

Bamboo is said to have influenced human evolution over the last 100 to 200 million years. As the Chinese philosopher Pou-Son-Tung stated, "A meal should have meat, but a house must have bamboo. Without bamboo, we lose serenity and culture itself."

BAT

See Pest Control.

BENEDICT OF NURSIA

(ca. 628–689/90), Saint Benedict, Benedictine monk, author of *The Rule of Benedict*. The work of Benedict of Nursia was one of the most important formative elements in the Christian monastic movement. As such, he can be seen as the founding father of the ethic of humanistic environmental management that has become a defining feature of that movement.

When Benedict lived there had already been almost 200 years of development in the monastic tradition. The earliest monks were urban Christians who embraced a life of prayer, poverty, and celibacy. They continued to live and serve in their communities. A second wave of monks embraced the "desert experience" and withdrew from their communities into the wilderness. The first "desert monks" were individual hermits who sought out a solitary and isolated life. Later, groups of these solitary monks would band together

under a set of agreed-upon rules and form a monastic community in the wilderness.

Benedict did not set out to form an order. The son of rural gentry, he was drawn to a contemplative spiritual life. He withdrew into the wilderness and lived for three years by himself in a cave. Soon his fame as a spiritual guide spread and he was drawn out of his solitary life. Benedict was chosen by a group of local monks to become their leader, or abbot. Tradition has it that Benedict set about to reform their easy-going ways. They were so disturbed by their new abbot's zeal that they attempted to poison him. He left and resumed his solitary life, but was soon called to lead other, more responsible communities. In the later stages of his life, Benedict was the spiritual leader of a whole chain of separate monasteries, and noble families from around Italy sent their sons to him for spiritual training.

Benedict's spirituality and monastic method was communicated to his followers by a written rule that he prepared for the monasteries that turned to him for leadership. The rule described how to go about choosing a just and able abbot. Further, the rule instructed this abbot in developing a balanced monastic life that combined equal elements of prayer, study, and manual labor. It is this third element of monastic life that allowed Benedict to make his unique contribution to Christianity and to environmental management. Benedict felt that manual work was essential to the spiritual health of his monks. Monasteries under the rule of Benedict not only managed their daily needs of food and shelter by communal effort, but also set about to undertake major projects of restoration and development. One group of monks would develop beautiful orchards or vineyards on previously arid land. Another group would drain a swampland and turn it into productive farms. A third would open an uninhabitable stretch of forest to human settlers. Because of the effectiveness and loose structure of the rule of Benedict, monasteries following his teaching sprang up across Europe and, as they undertook their developmental work, they changed the landscape of the continent forever.

In the Christian ethic of nature, there are two distinct directions. The first is an ethic of conservation that is best represented by Francis of Assisi and his followers. They see all nature as brothers or sisters and try to conserve and embrace its wildness and variety. The second strain of Christian ethic is best represented by Benedict and those who live by his rule. It is a strain of environmental management that teaches that humanity can develop and enhance the wonders of nature through careful management and hard work. In the Judeo-Christian tradition, when it is properly applied, these two ethics complement each other.

In Christian teaching it is clearly acknowledged that human beings differ in their spiritual needs and aspirations.

It is also clear that the quality of our relationship as human beings to the world we inhabit can be different. For some, conservation and the celebration of the natural world are most important. For others, conservation is complemented by a desire to enhance nature.

It is essential that environmental work in the tradition of Benedict of Nursia continues. In many monasteries around the world, the rule of Benedict is still followed by monks who are developing and enhancing their homes, but the true followers of Benedict in the modern world are the many scientists and environmentalists who advocate sound environmental management along with conservation.

Benedict of Nursia can be regarded as a patron saint of those who believe that true conservation means not only protecting nature against degradation, but also developing human activities that favor a creative, harmonious relationship between human beings and nature.

FRANCIS H. GEER

For Further Reading: Rene Dubos, "Franciscan Conservation vs. Benedictine Stewardship," in *The World of Rene Dubos* (1990); David Hugh Farmer, ed., *Benedict's Disciples* (1980); David Parry, ed., *Household of God,* an English translation of the Rule of Benedict with a short commentary (1980).
See also Christianity; Francis, Saint; Religion.

BENEFIT-COST ANALYSIS

Benefit-cost analysis (B-CA) is a technique for evaluating the social contributions, or benefits, and the impairments, or costs, of proposed or actual projects undertaken, usually by governments, in the public interest. It is closely related to business profit-loss calculations, but differs significantly because it deals with undertakings whose results are not measurable in dollars and cents and with purposes much more complicated than those of business firms.

Benefit-cost analysis has become widely used, beginning with the Water Resources Act of 1936 in which the Congress required the Corps of Engineers not to undertake any project unless "the benefits to whomsoever they may accrue are in excess of the estimated costs." More recently President Reagan required benefit-cost analyses for all federally financed projects in his Executive Order 12291 (1981). It has come to be a routine part of designing and deciding about investment projects in most governments throughout the world as well as in international agencies such as The World Bank.

B-CA rests on the philosophic postulate, sometimes called radical individualism, that holds that the satisfaction of each member of society is a separate goal and that each adult member is the authoritative judge of the satisfaction she or he would derive from any governmental undertaking. This principle gives rise to the criterion of "Pareto efficiency," originally formulated by Vilfredo Pareto (1848–1923), an Italian engineer, economist, and sociologist, who laid the foundations of modern welfare economics. The Pareto criterion holds that no project should be adopted if there is an alternative that some members of society would prefer and that none would regard as inferior. It does not take into account the amounts of satisfaction that the society's members would derive from a project, largely because of the difficulty of measuring different people's amounts of satisfaction on a uniform scale.

The Pareto criterion offers only incomplete guidance to project choices. There are frequently many Pareto-efficient alternatives to choose from, because any project qualifies unless there is some particular alternative to it that the members of the society unanimously agree is as good or better; such unanimity is rare. In addition, as a practical matter, it is not feasible to consult the preferences of every member of a society with respect to all conceivable alternatives to a proposed project.

The "Hicks-Kaldor compensation principle" provides a practical alternative. According to this principle, a Project A should be judged socially preferable to a Project B if—and only if—the advocates of A would find it worthwhile to offer the advocates of B enough compensation so that the latter would prefer Project A with the compensation to Project B without it. This formulation measures the amount of satisfaction each person would derive from a project by the amount she or he is willing to pay to have that project instead of an alternative, and gives equal weight to the satisfactions of all members of the society, rich or poor. It can be shown that a project that passes the Hicks-Kaldor test when compared with all available alternatives would also satisfy the Pareto criterion if the offered compensation were actually collected and distributed to supporters of rejected projects.

When a choice is to be made between two or more alternatives, the compensation principle thus recommends that the one for which the people as a whole are willing to pay the most is the one that should be chosen. Although it is virtually never practical to collect or pay the compensations, the proposal for which people are willing to pay the most can be regarded plausibly as the one that the society, on balance, prefers. Benefit-cost analysis provides the data needed for identifying this preferred project. It has therefore become an important ingredient of the process of designing governmental undertakings and choosing among them.

There are two broad approaches to estimating aggregate willingness to pay (wtp) for a specific project. One is generally called "contingent valuation," though "hypothetical valuation" would be more descriptive. Its basis is, simply: If you want to know how much people would be willing to pay for a particular project, ask them. Accordingly, in a contingent valuation study, a properly randomized and stratified sample of the people affected is drawn, the project and its expected results are explained to each, and each is asked how much she or he would be willing to pay to have the project instead of the status quo or some other basis of comparison. (Opponents of the project are permitted to indicate negative wtp.) The community's net wtp for the project is estimated by scaling up the total reported by the sample. Many economists doubt the reliability of this method because it depends on unsubstantiated verbal assertions.

The alternative to contingent valuation is "synthetic valuation." It consists in tracing a project's consequences and those of its significant alternatives in detail, estimating the social value or cost of each consequence in terms of wtp, and totaling the values. Synthetic valuation is laborious, expensive, and subject to substantial error. It requires estimates of the following data:

- The cost to the government of administering and enforcing the program over the period it will be in operation.
- The probable responses of the people and organizations affected. The desired response cannot be taken for granted. For example, it is unlikely that all drivers will obey a speed limit.
- The costs to people and organizations of complying. These costs usually include out-of-pocket expenses, disruptions such as loss or change of employment, inconveniences such as driving slowly, and so forth.
- The consequent reductions in discharges of pollutants or other abuses of the environment and the effect of those reductions on the state of the environment (e.g., the amount that nitrogen oxide discharges will be reduced and the effect of the reduction on the frequency of smog episodes).
- The social value of each of the improvements in environmental conditions.

Sometimes the social value of a particular kind of benefit is estimated by a public opinion survey, as in the contingent valuation approach. To avoid relying on such unverifiable oral testimony, analysts often use the amount people pay for a marketed commodity that confers a similar benefit, or an estimate of the amount that well-informed and thoughtful people would be willing to pay for the type of benefit in question, or some other indirectly derived estimate.

Finally, all the expected benefits and costs are added up to obtain the overall social value of the project.

In many instances, all available methods for estimating the public's willingness to pay for specific benefits or to avoid specific costs invoke unacceptable assumptions or are otherwise inapplicable. In those cases, the B-CA cannot be completed; the best that can be done is to estimate the net value of the benefits for which monetary equivalents can be found and to note the magnitudes of the remaining consequences in the most meaningful units available. Several devices can then be used to assist in reaching practical judgments.

One is "cost-effectiveness" analysis. This method avoids the need to assign monetary values to all benefits and costs by specifying "required" levels of attainment of beneficial results and then seeking the project that satisfies the requirements at lowest net cost. Thus, one might seek the most economical or "cost-effective" way to reduce the concentration of sulfur dioxide in a city's atmosphere to a specified level.

A major shortcoming of cost-effectiveness analysis is that the stipulated levels of attainment may prove uneconomically stringent or lax when the costs of meeting them have been computed. Thus, a "sensitivity analysis" is often employed to compare the costs of attaining a range of levels of environmental quality. A subjective judgment is then required to select the socially desirable balance between environmental quality and cost of attainment.

Three technical questions arise in virtually all benefit-cost analyses: how to allow for uncertainty, how to incorporate events that occur at different times, and how to indicate the effect of the project on the different groups that comprise the community.

The preceding discussion makes it amply clear that a B-CA consists of error-prone estimates of an uncertain future. Nevertheless, the results of benefit-cost analyses are generally presented as point estimates, as if they were known with certainty. At the very least, a responsible analysis should indicate ranges of uncertainty that bracket reasonable estimates of benefits and costs. In many instances, the plausible ranges of costs and benefits will overlap, making it difficult to decide whether the project is worthwhile; concealing the ranges of uncertainty is especially harmful in such instances. More sophisticated treatments of uncertainty depend on estimating the probability distributions of benefits, costs, and related data, and merge into the specialized discipline of "decision analysis."

Time enters in an essential way into virtually every

B-CA. Thus it is necessary to compare costs incurred at one date with benefits that accrue perhaps many years later. A device called discounting, borrowed from commercial arithmetic, is generally used to compare consequences that occur at different dates. The idea is to select a base date, usually the date at which the project is begun, and for each of the project's consequences compute a consequence that occurs at the base date and is deemed to have the same social value as the consequence being evaluated. For instance, a cost of $100 to be incurred five years from now might be judged equivalent to a cost of $100/(1+r)^5$ payable at once if the applicable rate of discount is r percent per year. The discounted values are generally called "present values," and the total present values of the benefits and costs of a project are simply the sums of the present values of its consequences at all dates.

Discounting, though used nearly universally, has given rise to a number of questions and a large literature. The most searching questions concern the morality of giving less weight to future consequences than to immediate ones. The moral qualms should not be dismissed lightly, but there are justifications, related to the justification for considering interest payments as legitimate personal and social costs. In fact, the most reasonable rate of interest to use in computing present values is the marginal productivity of capital, which measures the social advantages of having capital available sooner rather than later.

Decisions entail varying effects across different groups within a community and such distributional consequences of governmental undertakings must be addressed. The original mandate to compare benefits and costs without regard to "whomsoever they may accrue" to was a trap. No political decision can afford to ignore the fact that, generally, some people enjoy a project's benefits while others bear the costs. Benefit-cost analyses should therefore be disaggregated enough to show how the benefits and burdens of a project impinge on the major population segments affected.

Despite all its difficulties and limitations, B-CA has proved to be an indispensable tool for social decision-making. It cannot incorporate all the subtle (and sometimes not-so-subtle) considerations that enter into social decisions—for example, the moral propriety or social equity implicit in social decisions—and so cannot supplant human hunch and judgment in social decision-making. But it is an unexcelled procedure for organizing relevant information. It calls attention to pertinent facts in the public and political discussions that lead to a decision, and at the same time constrains the introduction of irrelevant and misleading issues and claims. It also promotes openness and consistency in judging the benefits and costs of diverse projects. But the final arbiter in decisions about governmental, and therefore political, projects must always be unquantifiable human judgment.

ROBERT DORFMAN

For Further Reading: John A. Dixon and Maynard M. Hufschmidt, eds., *Economic Valuation Techniques for the Environment* (1986); Edward M. Gramlich, *Benefit-Cost Analysis of Government Programs* (1981).

BENNETT, HUGH H.

(1881–1960), American soil scientist, soil conservation leader, author. A native of Anson County, North Carolina, Bennett graduated from the University of North Carolina at Chapel Hill in 1903, and then joined the Bureau of Soils in the U.S. Department of Agriculture. While making soil surveys in the southern United States, Bennett became convinced of the threat soil erosion posed to the country's future agricultural productivity. His numerous speeches and articles soon earned him a reputation as the nation's leading advocate of soil conservation, and he was selected in September 1933 to head a temporary New Deal agency, the Soil Erosion Service in the Department of the Interior. On April 27, 1935, President Franklin D. Roosevelt signed the Soil Conservation Act, which created the Soil Conservation Service (SCS) in the Department of Agriculture. Bennett set the course of the nation's soil and water conservation programs as the first chief of SCS, a position he held until November 13, 1951.

Bennett came to be regarded as the "father of soil conservation." He was significant in elevating concern about soil erosion from the level of a few disparate voices to a national movement of awareness and commitment. Soil conservation joined forestry and protection of scenic areas as national conservation concerns. His successes are evident in federal laws for soil conservation, professional organizations, public interest organizations committed to soil and water conservation, and increased emphasis on soil conservation in university curricula.

Bennett accomplished this task at a time when a few dedicated scientists in the federal government became advocates for their respective causes, promoted federal legislation, and then served as heads of federal agencies they had virtually created. Gifford Pinchot's advocacy of forest conservation and Harvey W. Wiley's fight for pure food and drug legislation parallel Bennett's vision.

Before becoming the first head of the Soil Conservation Service, Bennett had already had a 30-year career as a soil scientist, involving extensive periods in the field observing the effects of soil erosion domestically and in several foreign countries. Gullies were obvious to the casual observer, but Bennett publicized the danger of sheet erosion, a process in which an almost imperceptible layer of soil is removed from the field. Thus, Bennett had scientific credentials and credibility to reach a national audience.

As a scientist Bennett wrote for professional journals, but after commencing his crusade for soil conservation, he wrote for magazines with a wider and sometimes more influential audience. If not as eloquent as some of the naturalist writers, he wrote clearly and with commitment about his cause. While he recognized the need to reach the general public through the popular press, his best known article was a government publication, *Soil Erosion A National Menace*, USDA Circular 33. Co-authored with William R. Chapline, this piece provided a general survey of erosion conditions which was used in securing legislative support for a national program of soil conservation.

Bennett had obvious political skills and was a master at seizing the opportune moment. He successfully lobbied for funds in 1929 for a series of soil erosion experiment stations and then supervised their work. When it became obvious that funds would be available for soil conservation, he pushed his ideas and his candidacy to head up the work. His sense of the dramatic was on display during the Senate Public Lands Committee hearings on the Soil Conservation Act in April 1935. Realizing that a great dust storm from the Great Plains was blowing eastward, he used its sky-darkening arrival to dramatize the cause of soil conservation and win approval for the legislation creating the Soil Conservation Service.

The most valuable element of Bennett's character was his passion for his crusade. As a long-time colleague remarked, he loved to carry the message. He spoke with a fervor that impressed politicians on Capitol Hill, scientists at the Cosmos Club, or farmers on the courthouse square alike.

After elevating soil to a national concern and securing legislation for a permanent commitment to its conservation, Bennett made several decisions that influenced national soil conservation programs, especially the Soil Conservation Service. He recognized the complex causes of soil erosion and insisted that numerous disciplines be involved in devising solutions. Bennett did not believe in panaceas, but thought that the solution to a complex problem should rely on analytical contributions from several physical and biological sciences including agronomy, biology, forestry, engineering, range management, and soil science. SCS recruited from all these fields and then devised courses to give the field staff broader training in a variety of disciplines. Bennett also insisted that SCS should work directly with farmers on conservation measures rather than simply disseminate information. Plans for

conservation work on a farm should be designed specifically for that farm and be based on the capability of the land. Personal contact made programs more effective and created political support for conservation programs.

The viability of soil and water conservation as national concerns was further assured by the creation of the Soil Conservation Society of America (now the Soil and Water Conservation Society) and The Friends of the Land. Bennett was an influential founding member of both groups. The former group, made up largely of people personally involved in the field of soil conservation, publishes the *Journal of Soil and Water Conservation*. The Friends of the Land drew members from diverse backgrounds who were concerned with conservation issues. It published a well-written, at times eloquent magazine, *The Land*, whose authors came from diverse fields in business, science, literature, and other areas.

Hugh Hammond Bennett is buried in Arlington National Cemetery, Arlington, Virginia.

J. DOUGLAS HELMS

For Further Reading: Hugh Hammond Bennett, *Elements of Soil Conservation* (1947); Wellington Brink, *Big Hugh: The Father of Soil Conservation* (1951); D. Harper Simms, *The Soil Conservation Service* (1970).

BENZENE

See Pollution, Air; Pollution, Water: Processes.

BHOPAL

The Bhopal tragedy occurred in the early morning hours of December 3, 1984. Methyl isocyanate, a toxic gas used in the manufacture of pesticides, leaked from a Union Carbide plant in Bhopal, India, and drifted over the city. Approximately 3,800 people died and more than 200,000 were injured (about 2,720 permanently).

The toxic gas escaped from a storage tank due to a runaway chemical reaction. A hose was mistakenly attached to a pressure gauge opening in the tank, allowing the introduction of large amounts of water (approximately 120 to 240 gallons). In addition, there was an abnormally high level of chloroform present as a result of the distilling process. The catalyst for the reaction was iron in the stainless steel tank walls.

The methyl isocyanate gas was not contained due to a variety of human and mechanical errors. The refrigeration system had not been operating since June 1984. Two pieces of safety equipment that destroy contaminated methyl isocyanate and vent gases, the vent gas scrubber and flare, were also not fully in service. In addition, the water-spray system could not contain the escaping gas. Because the plant had one of the best safety records in India, local management may have become complacent. The surrounding community had no emergency plan of action.

The accident heightened world concern over the problems of environmental safety in and around industrial plants, especially in Africa, Asia, and Latin America. Minimum standards for safety at the Bhopal plant had never been mandated by the state government of Madhya Pradesh. Shortly before the accident an Indian government study found that only sixteen of the fifty pesticide-manufacturing plants in the country had significant pollution-control systems. The government, however, lacked the technical and institutional capacity to implement toxic chemical-control laws.

Indian and Indian government shareholders of Union Carbide India owned 49.1 percent of the Bhopal plant, and government policy had required that the plant be staffed entirely by Indians. Although Union Carbide Corporation was held responsible for the accident, multinational corporations in general were not condemned. The tragedy was viewed as a single incident that could have happened to any company. In response to the disaster the Indian government installed new housing units, provided clinics, and instituted job programs, supported in part by Union Carbide, to retrain some of the disabled people in leatherwork and tailoring. The corporation also provided millions of dollars in relief efforts and settled all claims by a payment of $470 million.

The most positive response to the accident by the chemical industry has been the creation of the Responsible Care Program, started in Canada in 1985 and now being implemented worldwide. Its guiding principles deal directly with improving community awareness and emergency response, employee health and safety, pollution prevention, process safety, distribution, and product stewardship.

BIOASSAY

Bioassays test the carcinogenic potential of environmental pollutants or commercial substances on living tissue samples or on nonhuman organisms. The underlying assumption of these tests is that substances that cause cancer in laboratory tests are very likely to cause cancer in humans.

Bioassays can be divided into two groups: short-term and long-term. Short-term bioassays use animal tissue or cells to test a potentially carcinogenic substance.

Long-term bioassays employ live animals exposed to a test substance.

Short-term bioassays are easy to perform, inexpensive, and quick: a researcher can predict the carcinogenicity of a substance in two or three days. Short-term tests have the advantage of providing information about the mechanism of effect, especially whether or not a chemical is damaging DNA and causing mutations. Unfortunately, certain types of carcinogens may slip past a short-term test because standard assays check only for specific types of carcinogenic effects. Therefore, researchers tend to use short-term bioassay data in conjunction with other data. Although long-term bioassays take much longer to perform and cost a great deal more, they are able to test for a wider range of effects. Lifetime bioassays in rats and mice are a highly standardized protocol now, yet they are limited by differences between rodents and humans (and even between rats and mice) in metabolism and in the effect of certain chemicals. Also, there is an inherent statistical limitation in testing groups of 50 rodents and extrapolating to risks of one in one million from the observed incidence of cancers or of other effects.

Because of the disparity in cost, time, and required space, experimenters usually assess carcinogenicity with short-term assays before performing long-term tests. If a substance fails short-term assaying, it can be regulated in the environment or removed from the market without going through long-term tests. If a substance passes short-term testing but fails long-term testing, the substance will be put through more rigorous testing. Unless human epidemiological evidence shows carcinogenicity, substances that pass both tests are allowed on the market.

Although the results of assays correlate strongly with what scientists know about cancer, there are a few discrepancies. Certain substances test as carcinogenic even when human epidemiological evidence makes the claim that they are safe. Researchers call this kind of test a false positive. More controversial and disturbing are substances that test negative for carcinogenicity but are actually carcinogenic, referred to as false negatives. False negatives have led many people to question the safety of certain substances and the value of bioassay testing.

Another area of controversy comes out of the statistical evaluation of test data. Because cancer researchers lack a full understanding of the complexities underlying the disease's processes, they lean toward conservative or "worst-case" estimates of risk. This practice leads to the regulation of potentially carcinogenic substances far below the level where they present danger. Overregulation can waste time and money. Such regulations are often tested in court, creating legal battles that can take years to resolve. In a growing number of instances, specific chemicals have been shown by studies of mechanisms of action to act differently in rodents than in humans. These findings are an exciting development in the emerging science of chemical risk assessment.

Bioassays have been an integral part of risk assessment for carcinogenicity. The same kinds of short-term and long-term strategies are now being applied to detect other adverse effects besides carcinogenicity, namely teratogenicity (ability to cause birth defects), mutations, and damage to specific organ systems.

BIOCHEMICAL OXYGEN DEMAND

See Pollution, Water: Processes.

BIODEGRADABLE/PHOTODEGRADABLE

See Bioremediation; Composting; Plastics; Waste Treatment.

BIODIVERSITY

Biodiversity is a term intended to encompass the variety of life on Earth. Most easily thought of in terms of species and subspecies (e.g., the diversity of birds comprises approximately 9600 species) it also applies to genetic variation within species, differences between populations, and variation at higher levels of organization such as the community or even variation within a biome.

Biological diversity has essentially existed since the beginning of life on Earth. There has been a general tendency for biodiversity to increase although there have been a few moments of biotic crisis when biodiversity has been greatly reduced, such as at the end of the Permian or Cretaceous periods.

Currently biodiversity represents the collective biological resource base in support of humans. Originally mostly food, fiber, and materials for shelter, biodiversity has come to encompass a wide and growing set of resources such as paper, pulp and its derivatives, pharmaceuticals, latexes, resins, and essential oils. Entire government departments and agencies are oriented around various aspects of the biological resource base: agriculture, health, fisheries, forests, etc.

In recent years the ability to extract benefit from natural variety has advanced with science and technology such that it is possible to generate wealth at the level of the molecule. One of the outstanding examples of this is the polymerase chain reaction that uses a heat resistant enzyme to cause a chain reaction replication of genetic material. This is now a fundamental part of diagnostic medicine in the industrialized world as well as a critical part of much of biotechnology. This was not practical until a heat resistant enzyme was identified from a bacterium *Thermus aquaticus* in a hot spring in Yellowstone National Park. The value of this single molecule through its role in diagnostic medicine and the resultant healthier and thus more productive population must be several billions of dollars annually.

Biodiversity in its natural aggregations and activities is responsible for an array of public services generally treated as free by economic systems. These services include the major global cycles of energy (in particular the fixing of solar energy through photosynthesis), water, and elements such as carbon, nitrogen, and phosphorus. They also include regional processes such as the hydrological cycle in the Amazon basin, through which a significant portion of Amazonian rainfall is generated internally in the basin, as well as more local processes such as watershed function. For example, the Panama Canal, which requires 52 million gallons of fresh water for each ship that transits the canal, is dependent on the tropical forest watershed surrounding Gatun Lake both for the water supply and the protection of the canal from serious sedimentation.

The relation between these processes and biological diversity is not clear-cut. Clearly less diverse biological systems are capable of performing many of these ecological functions (e.g., a plantation forest until harvest will provide significant watershed protection). The rare species that characterize almost all ecosystems simply are unlikely by their rarity to play a critical role in ecosystem function. Yet it is important to note that the presence of these species may reflect recurring conditions during which they may be more abundant and play a vital role. In addition species are known in some instances to play major roles in ecosystem structure. The area of biodiversity and ecosystem function is ripe for research.

Biodiversity also has enormous aesthetic significance. As a biological entity, humans find almost all aspects of life beautiful, or at least interesting. That is why tens of millions of people pursue hobbies such as bird watching. Nature tourism, often called ecotourism, is one of the most significant sources of income in some countries such as Kenya and Costa Rica. Some of the same fascination lies behind ornamental gardening or attendance at zoos and aquaria, which is greater in the U.S. than that at all major sports events combined.

A much undervalued aspect of biodiversity is its intellectual value. Each and every species, unique by definition, has something important to contribute to our understanding of how living systems work. Important concepts that have emerged from the study of biodiversity, whether deliberate or accidental, include vaccination and antibiotics—fundamental building blocks in modern medicine. Only with the discovery of the thermal vents in the ocean bottom did scientists realize life could exist at temperatures in excess of the boiling point of water. This library function of biodiversity may in the end be the most powerful contribution it has to make to human society. An additional ethical perspective is based on a fundamental respect for other forms of life.

The human impact on biodiversity has in some instances promoted greater variety in strains of crops and domestic animals. The human impact has been largely negative because the state of biodiversity reflects all other environmental problems. In the beginning, human impact was mostly through direct effects on particular species by overharvest or in some instances deliberate elimination of "undesirable" species. Most of the extinction generated by humans has been indirect, through the destruction of habitat and the often unintentional introduction of diseases and predators, leaving us with disastrous consequences on island biotas (e.g., the flora of St. Helena, or the dodo and solitaire in the Mascarene islands).

Today habitat destruction advances on a continental scale. No element of this is more important than tropical deforestation, which occurs at about 100 acres per minute. In coastal areas the resulting siltation leads to the smothering of coral reefs. To habitat destruction must be added, however, the problems of toxic substances and regional pollution problems such as acid rain causing forest death in eastern Europe. The greatest threat of all to biodiversity is the prospect of global climate change. Projected rates of change are greater than any species have been known to accommodate during previous periods of natural change. In addition, biodiversity is increasingly confined to isolated areas surrounded by human-modified landscapes creating an obstacle course for any species that might otherwise be able to track its requisite climatic conditions.

The field of conservation biology and conservation per se are actively addressing these threats and challenges but so far with insufficient resources. It will take far more than increases in science and more traditional conservation. In addition, human activity, and in particular, economic development have to change in character so as to accommodate, indeed promote, conservation of biological diversity. That can only come to pass with serious value changes in human society.

THOMAS E. LOVEJOY

For Further Reading: Ruth Patrick, ed., *Diversity* (1983); Edward O. Wilson, *The Diversity of Life* (1992); Edward O. Wilson and Frances M. Peters, eds., *Biodiversity* (1988).

See also Biome; Biotechnology, Environmental; Climate Change; Deforestation; Diversity; Ecological Stability; Ecology as a Science; Ecosystems, Natural; Enzyme; Ethics; Human Ecology; Natural Habitats; Photosynthesis; Resource Management; Water Cycle.

BIOENERGY

Bioenergy is energy derived from plant matter, or "biomass." Green plants capture solar energy and store it as chemical energy in the form of cell walls in the plants' stalks, stems, and leaves and as oils or starch in the seeds, fruits, or roots. Both plants and the waste materials derived from them (such as sawdust, wood wastes, and agricultural wastes) are referred to as biomass. Biomass can be used directly as a solid fuel to produce heat, or it can be converted to other bioenergy carriers such as liquid and gaseous fuels.

Bioenergy encompasses a variety of renewable-energy technologies that are locally available and used the world over. Wood and dung have been burned for millennia in fireplaces and stoves for heating and cooking. In several industrialized countries, some cities burn trash to generate electricity. Industries around the globe burn wood wastes (bark and branches) to generate heat and electricity for running mills. Charcoal provides the energy for steelmaking in Brazil. And 50% of the transportation fuel in that country is ethanol, derived from the fermentation and distillation of sugar from sugar cane. Biogas digesters, which rely on bacteria to decompose biomass, are widely used in China and India to convert animal and plant wastes into the gas methane, which is used for heating, cooking, and electricity. Biomass used in these and other ways currently accounts for about 15% of the world's total energy use and 38% of the energy use in developing countries.

Though bioenergy contributes substantially to meeting global energy requirements, traditional bioenergy systems do have some economic and environmental disadvantages. First, solid biomass fuels (wood, straw, trash) have a fairly low energy content per unit of weight compared with that of fossil fuels. Second, biomass fuels often must be collected over a wide area to obtain sufficient amounts to fulfill the demand. These characteristics make transportation-and-handling expenses a large component of biomass fuel costs and limit the size of bioenergy systems. A further disadvantage of some biomass conversion systems is their low efficiency in converting biomass to usable energy. Inefficient bioenergy systems require more fuel than fossil systems and produce more waste products such as particulate emissions.

Technology development is overcoming the disadvantages, however. One important development is the design and implementation of dedicated feedstock supply systems to furnish the biomass fuel. These systems consist of energy crops (fast-growing trees and grasses) developed to produce high yields on a limited land area, and specialized harvest and transportation methods designed to reduce biomass fuel costs. Energy crops, which are produced with standard agricultural techniques, could become an important alternative crop. Their presence could add to the landscape diversity and aesthetics of a region. Research indicates that it would not be economically or environmentally beneficial to replace natural forest systems with energy crops. Because dedicated feedstock supply systems must be adapted to local conditions, they will differ among regions and countries. Dedicated feedstock supply systems could probably be implemented at many locations in the world today, but when fossil fuel prices are low, biomass fuels are economically competitive at only a few locations.

Many developments in conversion technology are also occurring. Some are very simple, such as developing low-cost means of reducing the moisture content of biomass before it is combusted or gasified. Many other advances require complex mechanical and chemical engineering. One example is the linking of jet turbines with small gasifiers for the generation of electricity. Though preliminary trials show great promise, better methods of gas cleanup must be found before the technique can be marketed. Another example is the use of gasifiers to produce methanol, a liquid transportation fuel, by employing a combination of chemical catalysts, heat, and pressure to convert the gases to liquids. This technology has been demonstrated, but researchers continue to search for better gas cleanup methods and catalysts. Pyrolysis, a process that generates chemical changes as a result of the action of heat, is traditionally used to make charcoal from wood. New pyrolysis systems are being developed to produce biocrude oils that can substitute for petroleum products.

Nearly all the new processes being investigated to increase the efficiency and reduce the emissions of fossil fuels can also be fueled by biomass. Fuel cells, for example, represent a new technology without any moving parts that cleanly, quietly, and efficiently converts the stored chemical energy of a fuel into electrical energy. The hydrogen needed by a fuel cell can be derived from coal, oil, natural gas, or biomass. Small fuel cell development units (200 kW size) are currently being field tested and much larger units are being developed for the production of electricity at power plants. Someday small fuel cells may also directly power cars and trucks.

Vehicles run on bioenergy when they use the fuel

ethanol made from the sugars and starches in sugar cane, sugar beets, or corn. Ethanol from plant stalks and stems or biomass wastes (composed largely of cellulose) will be available in the future as a result of advances in the genetic engineering of yeasts and bacteria for converting cellulose into sugars. The first demonstration units are already under construction, and full commercialization is anticipated by the 2010s.

New bioenergy technology development promises significant social, economic, and environmental advantages through the wise use and deployment of bioenergy systems. Biomass fuels, for instance, have an inherent environmental advantage over fossil fuels because they contain very little sulfur, a major contributor to acid rain. Bioenergy systems further exhibit the potential for no net release of carbon dioxide to the atmosphere when the systems rely on biomass from energy crops. The release of carbon dioxide in the burning of the biomass is offset by the carbon dioxide used by the energy crops for photosynthesis and growth. Also, when perennial trees and grasses are grown for energy, little erosion occurs; therefore erosion-prone cropland that might otherwise provide no income to farmers can be used in an environmentally sound manner to grow crops. Also the waste products of bioenergy conversion processes, ashes and sludges, contain valuable nutrients that can replenish the soil. Perhaps the greatest benefit is that any country with arable land and biomass wastes can implement bioenergy systems.

LYNN L. WRIGHT

LINDA K. SHAW

For Further Reading: M. Brower, *Cool Energy: Renewable Solutions to Environmental Problems* (1992); T. B. Johansson, H. Kelly, A. K. Reddy, and R. H. Williams, eds., *Renewable Energy: Sources for Fuels and Electricity* (1993); C. J. Weinberg and R. H. Williams, "Energy from the Sun," *Scientific American* (1990).

BIOLOGICAL CONTROL

Biological control makes use of natural enemies to control pests, mimicking the complex processes of a food chain involving competition between species for nutrients. This competition—in the form of predation and parasitism—occurs from the microbial level to progressively higher species, with the ultimate beneficiary at the end of the food chain being the human population.

The practice of biological control involves the planned manipulation of living organisms (microbes, insects, birds, and other fauna) by humans for the purpose of controlling pest populations in agroecosystems. Through the encouragement and active cultivation of a biologically diverse system—including food crops, auxiliary plants, and weeds—beneficial predators and parasites may be attracted and flourish at the expense of unwanted pests and pathogens. With the maintenance of a reliable habitat for beneficial organisms, that population may act as a buffer between desired crops and a pest population, with minimal deleterious effects on food crops. Should the balance change, the potential exists for those same beneficial organisms to become pests. Therefore, successful biological control requires the maintenance of a suitable environment for beneficial organisms, involving intensive managerial control and knowledge of the agroecosystem, with a minimum of chemical inputs.

Biological control is an integral part of integrated pest management (IPM). Like IPM, the practice is significantly more intensive in terms of labor and management than chemical control, since it requires an intimate knowledge of different pest populations, life cycles, plant resistances, and a complex of symbiotic and conflicting biological relationships occurring within a given agroecosystem. For example, certain wasps parasitize the grubs of moths and butterflies by laying their eggs directly inside the host, ladybugs prey on aphids and mealybugs, and green lacewings attack leafhoppers and thrips. Larger fauna—such as snakes, birds, and poultry—can also be incorporated as living population controls for garden pests. When pollen-bearing trees, flowers, and weeds are provided as shelter and as an alternate food source for both humans and beneficial organisms in combination with careful monitoring of competition, biological control may naturally reduce the chance of pest attacks on desired crops.

Beneficial populations include two general classes: (1) predators and (2) parasites and parasitoids. Predators are usually, but not always, carnivorous and typically generalists in their diet, consuming the pest population that is most abundant at a given time. Predators are generally well adapted to dietary changes, as one prey population recedes and another emerges. Parasites and parasitoids may be insects or microbes (for example, viruses, bacteria, fungi, nematodes, protozoans, and rickettsias), and are more specialized as to their hosts, since they depend on the biological functions of the latter to flourish. Whereas parasites do not kill their hosts to achieve sustenance, parasitoids do eventually kill their hosts. At the point that the host is weakened or killed, the parasitoid is dominant, effectively disrupting the life cycle of the host. When beneficial organisms are introduced at strategic points in the reproductive cycles of pests, that pest can be controlled for an entire growing season. Beneficial organisms may occur naturally in the field with their re-

spective hosts, or they can be purchased from commercial suppliers for release into farming systems.

The commercial success of biological control has not been fully established, as its effectiveness is highly variable and dependent upon a range of factors, including: cultural practices, climate, regional pest populations and management at a larger geographical level, and unanticipated pest infestations.

BIOMES

Biomes are climate-mediated, regional units of the earth's biota. They are the sum of plant and animal communities coexisting in the same region, or climatically equivalent geographic regions. They are usually named by one or a few dominant life-forms. Life-forms of the biota, not individual species, are the biological units used for the analysis, description, and classification of biomes. Included in any particular biome type are all successional stages of the constituent plant and animal communities as well as landscapes influenced by human activity (anthropogenic ecosystems), both past and present.

Understanding the nature of a biome means understanding the accommodation of its constituent biota to the abiotic environment. This is different from the ecosystem concept, wherein abiotic components, such as local (micro) climate and soil, are considered an integral part of the system with the biotic components, plants, animals, and microbes. Ecosystem processes are analyzed as interactions (flows of energy and matter) between abiotic and biotic components, and between the biota themselves. By contrast, at the geographic scale of a biome, the abiotic environment, particularly macroclimate, is the context or primary determinant set of the system, rather than an interacting component. Biomes are not simply large ecosystems.

Terrestrial biomes are circumscribed by area. At a landscape scale they are defined by their physiognomic-dominant plant life-forms, which also provide the names of most terrestrial biomes—either directly, as in deciduous forest, or indirectly, as in tundra, savanna, or desert. At a regional scale terrestrial biomes are equivalent to the "formation" of plant ecologists.

Marine and freshwater biomes are circumscribed by volume, and related to the properties of the water body, such as salinity, water movement, and the shape and structure of the water basin. They may be named after dominant animal life-forms, such as coral reefs, or part of the water body, such as benthic or pelagic biomes. They may also be defined in terms of the presence or absence of primary production, as photic or nonphotic, or by the size or movement of the water body, such as pond, lake, or stream.

Intertidal and wetland biomes are usually defined by substrate, as in rocky or sandy shores, but also by geomorphic processes, such as erosion or deposition, or by geomorphology, such as floodplain, riparian plain, or estuary.

Regional variation in the names of terrestrial biomes reflects the fact that all human cultures have traditionally recognized major plant formations. For example, steppe is the name of the winter cold/late-summer drought grasslands of Eurasia, which have regional equivalents in the Great Basin of North America. They are represented in the Southern Hemisphere by the pampa of eastern Argentina, and the tussock grasslands of Otago on the South Island of New Zealand. Steppe is also an example of a climate regime and biome type corresponding to a great-soil group, in this case the Chernozems (Mollisols, in the USDA soil taxonomy). Similarly, the shrub formation of California called chaparral corresponds to the Mediterranean maqui. The Russian name taiga is used commonly for the boreal coniferous forest biome found on all Northern Hemisphere continents.

The savanna biome is an example of where a more precise scientific definition has been agreed upon by scientists, due to increasing knowledge of the responses of biota to the main determinants: all tropical and subtropical ecosystems characterized by a continuous cover of warm-season grasses that show seasonality related to water, and in which woody species are significant but do not form a closed canopy or continuous cover. Nevertheless, local names reflect the high degree of variation within the biome types. In the case of savanna, there is considerable variation in the length of the dry season, types of animal grazing, and ratios of woody species to grass species. In South America, the dry thorn forest and scrub savannas of Argentina and Brazil are called chaco and pantanal respectively; the open savanna grasslands of Venezuela are called llanos, and those of Brazil are campos.

Regional variation in traditional names of biome types may also reflect the impact of humankind on the ecosystems making up the biome. The Mediterranean maqui results from the cutting down of the dominant trees every twenty years; the trees regenerate as shoots, forming human-height bushes. In areas where the shoots are cut every six to eight years, and regularly burned and grazed, an open vegetation is formed, called garrigue; regional equivalents are tomillares in Spain and phrygana in Greece. These biome-type, physiognomic variants result from many centuries of human management.

Primary Determinants of Biomes

Using macroclimate as the context for classifying biomes usually results in temperature and rainfall (precipitation) being considered as primary determinants. Temperature regimes result in biota accommodating to (and hence groups of biomes classified as) cold, cool, temperate, sub-tropical, and tropical regimes. Rainfall variation results in biota accommodating to extremes of water availability, from very low (desert) to high

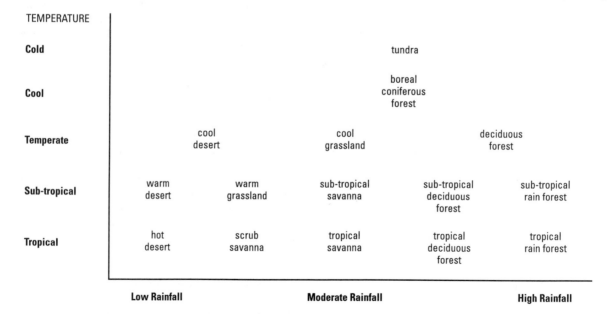

Figure 1. Classification of biomes by macroclimate.

(constant rain). This gives the broad classification of terrestrial biomes shown in Figure 1.

This simple classification illustrates that at tropical latitudes there are a variety of biomes, associated with variation in the primary determinant, water availability. Toward the poles, temperature becomes the primary determinant, and the number of biomes at each latitudinal climatic belt decreases. In the tundra biome, which occupies a substantial area of northern North America and Eurasia, low temperature can determine low water availability; permafrost soils effectively "freeze" availability of both soil moisture and available nutrients.

As the gradients of temperature and rainfall decrease from the equator to the poles, so does the physiognomic and structural complexity of the plant formations of the biomes. This variation in the life-forms of the dominants has been used as the basis of another approach to biome classification, such as that adopted by UNESCO (Table 1).

However, neither of these classifications takes account of seasonal variation in macroclimatic factors, which can be an important determinant of some biomes. For example, a key determinant of the savanna biome is the pattern of strong seasonality in available soil moisture, which sets a limit on the maximum plant productivity, and a constraint on the type of plants that can survive the alternating periods of drought and favorable water relations. This emphasizes that the "rainfall" determinant of biomes is really a combination of precipitation and evapotranspiration—water gained from the atmosphere, and lost by evaporation from the soil surface and transpiration by plants. Thus, this variable is best termed plant available moisture (PAM), and has been recognized as the primary determinant for savanna biomes, which experience a dry season annually.

In contrast, tropical and subtropical rain forest are marked by virtual absence of seasonality, with relatively constant temperatures and year-round rainfall. Deciduous forest of the temperate zone can still be termed rain forest if rainfall is high, but seasonality of temperature variation will occur. Moisture gradients across the temperate and cool-temperate belts of North America result in the coastal and montane evergreen rain forests of the northern Pacific coast, and the deciduous summer forest and mixed deciduous forest of the east. There are cool-temperate rain forests in southern South America on the Pacific coasts of Chile and Tierra Del Fuego, on the Atlantic coasts of the British Isles in Europe, and in Tasmania.

Secondary Determinants of Biomes

In addition to the primary biome climatic variable determinants of temperature and plant available moisture (PAM) there are three major secondary determinants of biomes. The major abiotic secondary determinant is nutrient availability, or plant available nutrients (PAN). In the ecosystem concept nutrients are considered as constituents of interactive processes through biogeochemical cycles, and individual nutrients or groups of nutrients may be regarded as rate limiting for plant growth. At the geographic scale of a biome, PAN is in the context of the abiotic environment determining the system. Within a biome type determined by the climatic variables of temperature and PAM, variations will occur due to nutrient availability, ranging from nutrient-rich (eutrophic) to nutrient-poor (dystrophic) variants.

Recent classifications of savanna biomes have utilized the concept of the PAM/PAN plane, where individual savanna locations can be plotted as points along gradients of available moisture (PAM) and nutrients (PAN). Thus in Africa, a general association of infertile soils (ultisols and oxisols, USDA soil taxonomy) with high rainfall areas, and more fertile soils (vertisols and alfisols) with aridity, has led to the concept of two variants of the African savanna biome: moist/dystrophic and arid/eutrophic.

Another important concept of nutrient availability as a biome determinant is the nature of the main reservoir of nutrients. In temperate and summer deciduous rain forests, the major nutrient reservoir is the soil, whereas in many tropical rain forests it is the plant biomass, so that PAN relies on a high turnover of nutrients within the ecosystems that comprise the biome. This has consequences for human impact on biomes, such as forest clearance.

Animals are also secondary determinants of biomes. Although terrestrial biomes are characterized by their dominant plant life-forms, they are secondarily affected by animal activities, especially those of herbivores. At the ecosystem level, detailed plant-animal interac-

Table 1
Classification of biomes by dominant life-form.

Biome type	Life form of dominants
Closed forest	complete tree canopy, evergreen or deciduous
Woodland (open forest)	tree canopy covers > 40% of area, but crowns mostly, not touching one another
Savanna	grass or other herbs plus < 40% tree cover
Scrub	woody dominants 0.05–5 m tall
Dwarf scrub (heath)	woody dominants < 0.05 m tall
Grassland	grass and other herbs, few if any shrubs or trees
Swamp (marsh, bog)	submerged and/or emergent herbaceous hydrophytes; sometimes with shrubs or trees

tions are an important aspect of ecosystem function, e.g., herbivores and detritivores are important for nutrient turnover. At the biome scale, animal determinants of plant biomass are distinguished, not by individual species, but by the nature and degree of impact.

Variations in grazing pressure may cause a shift in the life-form composition of areas within biomes. The Palouse prairie of the northwest U.S., and the Pacific prairie of California were, prior to European colonization, dominated by bunch grasses. Much of this region is now cropland. Overgrazing in the drier and less fertile areas, unsuitable for cultivation, has resulted in invasion and dominance by sagebrush. Similarly, the natural vegetation of southern Texas and northern Mexico is described as savanna, but much of the region is now dominated by thorn woodland; this has changed over the last 200 years, with the development of ranching in the region. Grazing animals act as a vector for the dispersal of the thorn scrub, particularly in dry years. Thus the mesquite biome savanna variant in this region is a recent, herbivory/anthropogenic determined, physiognomic conversion from grasslands and true tree/grass savanna to thorn scrub.

Whereas herbivory is an important secondary determinant of savannas, the herbivores vary from one geographic region to another. In Australian savannas insects are the major grazing native herbivores; ecological equivalents of the large grazing ungulates of Africa are missing in the native fauna. Introduced grazing ungulates have greater carrying capacities in northern Australia than they can achieve in the same biome type of their native continent, most likely due to release from natural predators and pathogens.

Fire is also an important secondary determinant of biomes, being a system-attribute that, like animals, can "fashion" the physiognomy of vegetation of constituent ecosystems. Fire as a natural phenomenon is started mainly by lightning. Its impact upon vegetation depends on its frequency and intensity, which in turn is the result of a complex interaction of flammability and fuel load (amount of plant biomass). Almost all terrestrial biomes experience natural fire, but those most shaped by its effects are in seasonally dry climates.

Fire mediates biogeochemical cycles by releasing nutrients from the burned biomass. The newly available nutrients can cause a "flush" of vegetation regeneration after fire. Manipulation of this nutrient flush by human cultures is exemplified by slash-and-burn agriculture, which has occurred in forests ranging from cool-temperate to tropical latitudes. This appears to be an ancient human manipulation of forest biomes, with regional variants, such as in North America (e.g., Iroquois in northeast U.S.), Europe (Neolithic "ladnam" clearances), and much of southeast Asia and Australia,

where it is still practiced (e.g., Malaysia, Papua New Guinea).

Individual plant species may show adaptations to fire. For example, a variety of woody taxa have developed lignotubers or burls that enable them to regenerate after fire, and epicormic buds in several woody species allow shoot regeneration after fire. Such adaptations coincidentally allow sustained response to human management, such as coppicing, pollarding, and shredding. These techniques of management occur historically in most regions of summer deciduous forest and Mediterranean biomes. At the biome scale there are physiognomic expressions of adaptations to fire. The maqui and garrigue expressions of the Mediterranean biome are examples of vegetation resulting from human management of physiognomic adaptations to an abiotic determinant—in this case fire.

Grassland may respond to fire primarily by regeneration from protected meristems or from seeds. Examples are the tall-grass prairies of the Great Plains of North America, with mainly perennial species which regrow after frequent fires, and northern Australian savannas where an annual grass, *Sorghum intrans*, sets self-burying seeds that germinate after annual dry-season fires. The widespread tropical and subtropical grass, *Imperata cyclindrica*, regenerates after fire from its root system, and then sets seed. In New Guinea repeated anthropogenic fires encourage *Imperata*-dominated vegetation at the expense of forest regeneration, resulting in a fire-disclimax biome variant, locally known as kunai.

Biomes and Global Change

For many thousands of years, human cultures have caused shifts in the balance of three secondary determinants of biomes. For example, (1) nutrient availability has been changed by vegetation clearance and fire, as in shifting cultivation (slash-and-burn agriculture); (2) herbivory has been changed by grazing-species introductions as a result of colonization, by transposition of systems of animal husbandry, and by population explosions of insects from non-specific use of pesticides; (3) fire, either intentionally or accidentally, has been used as a vegetation management tool for at least 40,000 years, and this has changed the ratio of fire-sensitive/fire-tolerant biota within biomes, which has resulted in shifts in physiognomic structure. Sustained harvesting of ecosystems has led to change in their physiognomic composition, and hence of the biomes of which they are constituents.

Change in the primary determinants of biomes—climatic variables of temperature and precipitation regimes—has occurred throughout Earth history, but cultural evolution has created unprecedented human-kind-induced rates of change in these primary determinants, particularly within the last 400 years. Within the time-scales of Earth geological history, the biota of biomes have accommodated to climatic change by geographical displacement of physiognomic dominants and by evolution. Species reflect these slower changes in their eco-physiological tolerances.

However, at the time-scale of recent human-induced global change, the responses of individual species and ecosystems are uncertain. Responses are being studied experimentally and by mathematical modeling in order to understand the descriptive and prescriptive expression of these responses in various biomes. Indeed, global biospheric change itself is recognizable only at the scale of the biome.

PATRICIA WERNER
DAVID L. WIGSTON

For Further Reading: Michael G. Barbour and William Dwight Billings, eds., *North American Terrestrial Vegetation* (1988); V. E. Shelford, *The Ecology of North America* (1978); H. Walter, *Vegetation of the Earth in Relation to Climate and the Eco-Physiological Conditions,* trans. J. Wiesler (1973). *See also* Biosphere; Chaparral; Climate Change; Coral Reefs; Deserts; Ecosystems, Natural; Fire; Floods; Forests, Deciduous; Grasslands; Lakes and Ponds; Tropical Rain Forests; Tundra; Wetlands.

BIOREMEDIATION

The process of bioremediation uses microorganisms to degrade and detoxify chemically contaminated soil or water. It employs the same principles as do the biological wastewater treatment plants used in many municipalities. Most chemical wastes will eventually biodegrade naturally. The goal of bioremediation is to speed up that process by providing the microorganisms, either naturally present or introduced, with the required pH, temperature, moisture, and sources of carbon and oxygen. A contaminated area such as an illegal dump for waste oil may thus be "remediated" by the appetites of living bacteria enhanced and encouraged by the proper environmental conditions.

A number of bioremediation techniques have been developed. The easiest to design and control are wastewater treatment reactors. These include aerated lagoon systems and ponds that have turbines to mix oxygen and fertilizer media into the sludge. Another way to provide agitation and enhance the contact of bacteria with wastes uses bacteria-impregnated rotating disks. A slightly drier method of bioremediation is land farming. Contaminated soils are treated aerobically in polyethylene-lined beds that can be tilled, drained, and even sown with plants to control moisture levels. Composting, in which contaminated soil is mixed with wood chips and periodically turned over and aerated, is another bioremediation technique.

If it is too costly to excavate waste matter or if its removal is deemed undesirable, bioremediation at the site is an alternative. The contaminated site is sprayed or injected with bacteria and nutrients. Compounds that provide oxygen for reactions, such as hydrogen peroxide, can be added as well. The injection techniques used for below-ground wastes can be slow and require the right geology and hydrology in order to be successful.

Anaerobic degradation is yet another method that can be used for wastes that are converted into acids. Anaerobic bacteria convert acids into methane gas, which is then collected. Some hazardous organic molecules are too large to be absorbed by a bacterial cell. In such cases, a white rot fungus may be used to crack these oversized compounds into molecules small enough for bacterial degradation. The choice of any of these techniques involves many factors, including cost, volume to be treated, and public health and safety.

Bioremediation has been embraced by industry and public environmental agencies as a promising solution to the proliferation of hazardous waste sites across the United States. Bacterial digestion of hazardous chemicals has yielded exciting results in chemical-specific laboratory trials as well as in numerous site clean-ups. The United States Environmental Protection Agency considers bioremediation as a potential component of Superfund clean-up strategies.

Some major constraints on successful implementation of bioremediation include the chemical composition of waste materials and external environmental conditions. There are many chemicals in the environment, with new ones released every year. Bioremediation data exist for only a small fraction of these chemicals. In addition, unknown contaminants at a waste site may yield results that differ from laboratory trials.

BIOTA

Biota refers to the species of a region, including single-celled organisms, fungi, and bacteria, as well as the more conspicuous plants and animals. The particular species that make up a biota reflect many factors. History of the region is important. The unique marsupial fauna of Australia is the result of the isolation of this large island from the rest of the world at a critical time in mammalian evolution. The physical environment plays a large role in shaping biotas. First, some environments favor particular kinds of organisms such as large mammals in the high Arctic and reptiles in deserts. Second, the physical environment acts as a selective agent for adaptations to particular physical conditions. The white fur of the polar bear and the glossy scales and streamlined bodies of sand-swimming lizards are splendid examples. The biota is also shaped by interactions among species. The visual and chemical lures used by plants to attract specific pollinators represent millions of years of shared existence. Local biotas are affected by disturbance; a landslide bares the ground and creates opportunities for colonization by plants that can only germinate in full sunlight. Thus, one biota is replaced, for a short time at least, by another.

Human activities are having an increasing influence on local and global biotas. The common plants and animals of a region are often invaders that accompanied human activity. Still others are intentionally introduced species that have become part of the "natural" landscape. Rene Dubos described how the hedgerows of Europe that today support a diverse biotic community are themselves the product of deliberate cultivation by people. As land use and settlement patterns change, so too do biotic communities.

The interactions of abiotic, biotic, and cultural factors point to the resilience and vitality of the Earth. Rock doves, originally from the cliffs of the Middle East, now find a suitable habitat on the ledges of highrise buildings. The diversity and integrity of the biotic communities of the world represent a storehouse of biological knowledge and successful adaptation.

ROBIN M. ANDREWS

BIOTECHNOLOGY, AGRICULTURAL

Agricultural biotechnology may be broadly defined as technologies that involve genetic modification and utilization of living organisms to improve production and quality of food, feed, and fiber produced in agricultural systems. Improved crop varieties, more productive animal breeds, and selection of microorganisms to improve production of fermented food products are historical examples of biotechnology using this broad definition. Of the numerous biotechnologies applied to agricultural systems, the most potentially useful but controversial is genetic engineering.

Genetic engineering, also referred to as transformation, is the use of molecular genetic techniques to alter and transfer genes isolated from any organism to another organism to add new and useful properties to the recipient. Genes are pieces of chromosomes and are composed of DNA. The DNA sequence of a gene encodes the information that controls specific metabolic function and the developmental traits of organisms. The genetic code that converts the DNA sequence into

the properties of an organism is universal, enabling genetically engineered genes (transgenes) to function properly in unrelated organisms. The scientific foundation underlying genetic engineering originated with the demonstration by O. T. Avery and colleagues at the Rockefeller Institute for Medical Research in 1944 that DNA was the key genetic molecule, and the discovery by James Watson and Francis Crick in 1952 of the molecular structure of DNA. Further important developments were (1) elucidation of gene structure and the regulation of gene expression; (2) discovery and characterization of restriction enzymes that can be used to precisely cut and reconstruct transgene DNA molecules; and (3) development of methods to introduce transgenes into the chromosomes of recipient organisms to produce genetically engineered organisms (GEOs), also referred to as transgenic or recombinant organisms.

Genetic engineering involves (1) isolation and modification of the gene to be transferred, using molecular genetics techniques to insure appropriate transgene expression levels in the recipient organism; (2) introduction and stable integration of the transgene into the chromosomes of the recipient organism; and (3) selection of GEOs that exhibit desired transgene expression levels and inheritance. At the present level of technology, one or more transgenes are usually transferred so that for the most part the GEO differs from its normal counterpart in only one or a few properties. Methods vary widely for different recipient organisms, but genetic engineering is now possible in most bacteria, plants, and animals used in food processing and agricultural systems. Examples that illustrate the genetic distance between organisms that have been genetically engineered include expression of a cow gene encoding growth hormone in a common laboratory strain of bacteria for cheaper production of that hormone, and expression of a bacterial gene that confers insect resistance in corn plants. These examples illustrate that genetic engineering differs from all other genetic manipulations because species barriers are overcome, enabling transfer of genetic information between organisms that would not occur in nature.

The goal of genetic engineering is to produce GEOs that possess novel and useful traits. Applications in agriculture involve three different strategies. One strategy—also used to produce human pharmaceuticals and enzymes used in industrial processes—is to use transgenic bacteria to produce peptide hormones and recombinant vaccines that are used in animal agriculture or recombinant enzymes that are used in food processing. Another strategy, biopharming, involves using transgenic animals or plants to produce human pharmaceutical products more cheaply compared to conventional methods. The application that has the

Table 1

Examples of genetically engineered organisms being tested for use in agricultural systems.

Recipient organism	Donor, transgene or genetic manipulation	Property
Bacteria		
Pseudomonas syringae	deleted *ice+* nucleation gene	reduced ice nucleation in plants during frosts
Clavibacterxyli subsp. *Cynodontis*	*Bacillus thuringiensis* delta-endotoxin gene	resistance to European corn borer in corn
Plants		
tomato	delayed fruit softening during ripening	improved post-harvest quality
tomato	*B. thuringiensis* delta-endotoxin gene	resistance to lepidopteran insects (tobacco hornworm)
corn	*Streptomyces hygroscopicus bar* gene	herbicide tolerance
alfalfa	Alfalfa mosaic virus coat protein gene	virus resistance
Animals		
pigs	extra copies of somatotropin genes	leaner meat
fish	extra copies of somatotropin genes	increased growth rate

greatest potential for benefit involves genetic engineering to improve the properties of an organism followed by planned introduction of the GEO into large-scale agricultural systems. Examples of transgenic bacteria, plants, and animals being tested for agricultural applications are listed in Table 1.

Potential benefits and risks of applications of genetically engineered organisms in agricultural and food processing systems have been and will be extensively debated by scientists, citizens, and policy makers. Utilization of specific GEOs has the potential to improve efficiency and profitability of agricultural systems. Development of GEOs that resist insects, diseases, and environmental stresses will reduce the need for applications of pesticides and ensure crop productivity. Reducing the use of these inputs will reduce environmental pollution associated with fertilizer and pesticide applications in cropping systems.

The potential risk associated with GEOs arises in part from the fact that the genetic transfer used is unlikely to occur in nature. Therefore, genetic engineering may impart the GEO with unpredictable properties that may pose risks to human or animal health, or have unforeseen negative impacts on managed or natural ecosystems. The safety of genetically engineered food products and how these products should be tested and regulated for consumer protection is being debated rigorously. The level of risk associated with a GEO is related to the properties of the recipient organism and the transgenic traits. Generally speaking, if a well-characterized recipient organism (e.g., tomato) is genetically engineered so that a well-characterized gene is turned off to enhance product quality, then this product is unlikely to pose a risk to the consumer. On the other hand, if an organism is genetically engineered with a gene that produces a protein that may cause allergic reactions in some humans upon consumption, then obviously this GEO poses a risk to human health.

Unforeseen negative impacts on ecological systems include direct effects of the GEO and the possibility of transgene transfer to related nondomesticated species. The possibility of uncontrolled spread of GEOs has been likened to invasions of nonindigenous pests such as the zebra mussel or kudzu plant, which have caused major ecological disruptions and economic loss. The probability of this occurrence is related to the properties of the recipient organism and whether the transgenic trait provides a selective advantage compared to indigenous organisms. For example, genetically engineered bacteria would appear to be difficult to contain and may have the potential to transfer their transgene to other bacterial species. On the other hand, transgenic farm animals exhibit low probability of ecological damage because they are grown in contained facilities. Crop plants vary in their competitiveness and weediness and therefore differ in risk potential.

In accordance with concerns for the risk, federal and, in some cases, state regulations govern genetic engineering research and containment, and planned introduction of GEOs. Research is conducted in contained laboratories and facilities to avoid inadvertent release of experimental GEOs. Federal and, in some cases, state permits are required for field-testing GEOs; over 300 have already been conducted in the U.S. No unforeseen negative impacts have been associated with these small-scale contained tests. Commercialization of the first GEOs is planned for the mid 1990s.

DAVID A. SOMERS

For Further Reading: Bernard D. Davis, ed., *The Genetic Revolution* (1991); Larry Gonick and Mark Wheels, *The Cartoon Guide to Genetics* (1991); Jean L. Marx, ed., *A Revolution in Biotechnology* (1989).

See also Agriculture; Animals, Domesticated; Bacteria; Biotechnology, Environmental; Biotechnology, Medical; DNA; Genetics; Pest Control; Plants, Domesticated; Viruses.

BIOTECHNOLOGY, ENVIRONMENTAL

The term *environmental biotechnology* was introduced in the first of a series of meetings in Seattle (1984 and 1987) and in Braunschweig, Germany (1990). It refers to technologies that involve genetic modification and use of microorganisms in such applications as the bioconversion of organic waste, the clean-up of oil spills, and separating metals from ores. Biotechnology has a growing place in the remediation of hazardous waste sites, in new industrial processes, in recycling of wastes and byproducts, in biopesticides, and in design of biodegradable materials, including plastics.

Microorganisms can use many chemicals as food and energy sources. With complete aerobic metabolism, using oxygen as final electron acceptor, the end products are carbon dioxide and water. With some chemicals, especially man-made chemicals, oxidation may be incomplete or may be better carried out by anaerobic organisms, which use nitrate or sulfate as final electron acceptor. Some organisms can biotransform chemicals but obtain no energy yield; such organisms need additional substrate to grow, a co-metabolism scheme.

With existing organisms selected from nature, bioremediation is highly effective in soils for removing petroleum hydrocarbons—including alkanes, aromatics (benzene and its relatives), and chloroaromatics (polychlorophenols)—and organophosphate pesticides. In groundwater, microorganisms can be highly effective for dealing with gasoline hydrocarbons (benzene, toluene, xylene) and chlorinated solvents (trichloroethylene). Microorganisms are not effective, however, in degrading PCBs, heavy metals, and chlorinated pesticides.

Rational biotreatment requires knowledge in three disparate areas: (1) the physiology, biochemistry, and genetics of the detoxification pathways of potentially useful microorganisms; (2) an appreciation of the microenvironments in which the treatment may be performed and the structure and function of indigenous and inoculated microbial communities; and (3) field site engineering, so as to design and implement optimal growth conditions and interface with physical and chemical methods of containment and destruction of the chemical contaminants.

Recombinant DNA technologies are being employed to modify microorganisms so that genes for biodegradation steps that are missing can be introduced to enable the organism to complete the metabolism of target compounds to harmless final products (preferably carbon dioxide and water). Also, organisms may be modified to make them resistant to inhibition by chemicals commonly present in the contaminated sites or formed by the metabolism of important target chemicals. It may be feasible to increase the breadth of substrate specificity, enabling the organism to degrade more chemicals. It is best and safest to use organisms, now containing one or more inserted genes, that are well adapted to the environments into which they are to be introduced. Techniques exist to insert marker genes whose products make it feasible to track the introduced organism and determine its distribution and fate and, thereby, its safety.

Natural or genetically engineered microorganisms may be employed either by direct injection into groundwater or onto soils, or by incorporation into bioreactors. Bioreactors provide a practical means of assuring a controlled environment for biodegradation, avoiding questions about release of the organisms to the general environment.

There likely will be intense international competition in all areas of biotechnology. In Japan the Bio-Industry Development Corporation has an expansive mandate and strong funding support from the key government agencies. Germany has invested in a major new research institute in Braunschweig. Several U.S. biotech start-up companies are focused on this opportunity. Environmental biotechnology has the potential to contribute to pollution prevention as well as remediation of contaminated sites on site.

GILBERT S. OMENN

For Further Reading: Bernard D. Davis, ed., *The Genetic Revolution* (1991); G. S. Omenn, *Environmental Biotechnology* (1988); G. S. Omenn, A. Hollander, eds., *Genetic Control of Environmental Pollutants* (1984).
See also Bioremediation; Biotechnology, Agricultural; Biotechnology, Medical.

BIOTECHNOLOGY, MEDICAL

Medical biotechnology is the application of bioengineering principles to medical research and treatment. In 1992, approximately 1,231 U.S. companies employing 79,000 workers were engaged in some aspect of biotechnology. Two-thirds of the companies were engaged in developing either human diagnostic or therapeutic products; more than 600 diagnostic products and at least 18 human therapeutic drugs and vaccines were being marketed. Diagnostics include tests to detect the presence of a disease-related gene, such as the mutation causing cystic fibrosis. Drugs and vaccines include the human growth hormone, hepatitis B vaccine, alpha-interferon, and erythropoietin.

Two technical breakthroughs of the 1970s led directly to today's successes in medical biotechnology. In 1972 Herbert Boyer of the University of California and

Stanley Cohen of Stanford University invented the process known as recombinant DNA technology or gene splicing. In this way, genes and genetic traits could be transferred from one organism to another in the laboratory. In 1975 Cesar Milstein and George Kohler at Cambridge University fused antibody-producing cells with malignant cells to produce hybridomas. These hybrid cells produced previously unavailable amounts of identical antibodies, known as monoclonal antibodies. The twin tools of recombinant DNA technology and monoclonal antibodies are at the core of medical biotechnology. Dramatic advances in instrumentation were critical, also, including DNA and protein synthesizer machines.

Advances in medical biotechnology often offer enormous benefits at great cost. In patients with kidney disease, for example, the kidney's impaired ability to produce the hormone erythropoietin causes anemia, a deficiency in red blood cell production. Recombinant human erythropoietin can restore a normal red blood cell count, eliminate the need for blood transfusions, and improve the quality of life for patients. Treatment is expensive, on the order of $10,000 per year per patient, in addition to costs for kidney dialysis.

The use of recombinant human growth hormone is also illustrative. Growth is ordinarily regulated by human growth hormone secreted by the pituitary gland. Insufficient growth hormone secretion may limit a child's adult height to no more than four feet. Such children can be treated with growth hormone. Until 1985 only a severely limited supply of growth hormone was available from cadavers. At the same time that growth hormone from this source was found to carry the threat of viral contamination, a recombinant product was marketed, making pure human growth hormone available in abundant supply. An estimated 15,000 American children may be candidates for recombinant growth hormone therapy. Therapy costs $20,000 to $30,000 per year for up to five years and entails three injections per week.

With the need to balance benefit to the patient, benefit to society, cost to the patient, and cost to society, medical biotechnology increasingly involves ethical analyses. As formidable technical barriers continue to fall, purely scientific and medical questions that begin with "Could we . . . ?" often evolve into ethically tinged questions that begin with "Should we . . . ?"

This evolution in the direction of ethics and societal values was foreshadowed in the infancy of biotechnology in the early 1970s. Twice, in 1973 and again in 1975, biologists gathered at the Asilomar Conference Center in California to examine the ways in which recombinant DNA research should be carried out. Agreements were reached on safety and security procedures—procedures that, in time, proved unnecessarily strict and were substantively relaxed in 1978. The legacy of this self-reflective process endures to the present, however, and has been crucial to the conduct of medical biotechnology. That legacy is the Recombinant DNA Advisory Committee of the National Institutes of Health, conceived in the wake of Asilomar, and, in 1993, in its nineteenth year of continuous existence.

The Recombinant DNA Advisory Committee stands as the governor of the next dramatic phase of medical biotechnology: human gene therapy. This procedure involves the actual replacement or repair of defective genes in living human cells, not just replacing their missing actions. Gene therapy was undertaken for the first time in 1990 in a girl with a fatal immune deficiency disorder and then in adults with malignant melanoma. Several other protocols are underway. Such somatic-cell gene therapy is used to correct genetic defects in an individual patient and poses potential risks to only that patient. Germ-line gene therapy, which has not yet been undertaken, would affect succeeding generations.

The Human Genome Project, a 15-year joint effort of the National Institutes of Health and the U.S. Department of Energy, began to take shape in the mid 1980s and surfaced as a line item in Federal Government budgets in 1988. It is intended to equip biomedical researchers with the tools they need to search for genes quickly and cheaply. Once genes have been pinpointed and isolated, researchers can then begin to understand the genetic errors that result in disease. This information will constitute the engine of medical biotechnology well into the 21st century.

Already the mapping of the human genome is yielding dividends in the detection and combatting of important genetic diseases. Cystic fibrosis is the most common, life-shortening, recessive disorder affecting Caucasians of European descent. Between 1,700 and 2,000 babies with cystic fibrosis are born annually in the United States. In 1989 scientists identified the most common genetic mutation that causes cystic fibrosis. Quickly thereafter, scientists developed tests to detect mutations in the segment of DNA—the cystic fibrosis gene—that is responsible for the disease. These DNA tests can screen and identify cystic fibrosis carriers before they have a child with cystic fibrosis. Beyond the approximately 30,000 Americans who have cystic fibrosis, as many as 8 million Americans could be cystic fibrosis carriers. These millions of cystic fibrosis carriers are, today, largely unidentified. The tools of medical biotechnology now permit their identification, should they desire to be identified.

An increasing number of predictive genetic tests will become available in the next decade. Increasingly individuals will have the opportunity to avail themselves of knowledge about their genetic endowments and the likely medical prospects for themselves and their offspring. Issues to be confronted include genetic services delivery, public education, professional capacity, financing, stigmatization, discrimination, quality assurance, cost-effectiveness, and ethics in resource allocation.

GARY B. ELLIS

For Further Reading: Robert M. Cook-Deegan, *The Gene Wars: Science, Politics, and the Human Genome* (1993); Kathi E. Hanna, ed., *Biomedical Politics* (1991); U.S. Congress, Office of Technology Assessment, *Cystic Fibrosis and DNA Tests: Implications of Carrier Screening* (1992).
See also Biotechnology, Agricultural; Biotechnology, Environmental; Ethics; Genetics; Health and Disease.

BIRTH DEFECTS

See Genetics.

BOD

See Pollution, Water: Processes.

BOG

See Wetlands.

BOULDING, KENNETH E.

(1910–1993), economist, philosopher, and poet. In several unique ways Kenneth Boulding helped place the examination of environmental problems in intellectual frameworks embracing both social and natural sciences beginning in the 1950s. With a cheerful disregard for conventional academic boundaries, he forged new structures of thought, and interpreted scientific findings in poetic and often amusing modes that broadened understanding of issues facing the human race.

The son of a Liverpool plumber, he began his studies in chemistry at Oxford, took a first in economics, never bothered with a Ph.D., and ended his career at the University of Colorado with 35 honorary degrees from universities around the world. He soon established his credentials as an economist with a text, *Economic Analysis*, that went through numerous editions (1941, 1948, 1955, 1966), but he enjoyed pioneering new fields of investigation that reached beyond classical economics.

In collaboration with colleagues from other disciplines he shaped the analysis of the grants economy as an alternative to traditional market economy. He was a key contributor to the framing of general systems theory. He laid the groundwork for interdisciplinary approaches to peace research and to studies of conflict resolution. His concern for the meaning of human betterment was important in outlining futurist research. The interaction of natural and human systems had a significant part in all of those efforts. This helped correct the then-prevailing tendency to view environmental questions within the tradition of a single discipline. Several fresh concepts emerged.

With Barbara Ward and Buckminster Fuller, he was influential in popularizing the notion of Spaceship Earth as a vehicle with the human family as a crew heading for an uncertain goal. He inspired a shift in economic analysis towards evolutionary or ecological economics that sought to take full account of the character of natural systems. He examined the evolution of social organizations in a framework he called "ecodynamics." His book *The Image* had a profound effect upon understanding of the factors involved in perception of the environment.

Boulding's ideas commanded attention because of their originality, the positions he held, and his felicity in expressing them. As president of the American Economic Association in 1968 and of the American Association for the Advancement of Science in 1979 he received recognition from some scientific groups that were looking skeptically at emerging environmental issues. He was more influential as a participant in interdisciplinary groups addressing major national and international policies.

One early gathering on environmental questions was the 1955 Princeton conference on *Man's Role in Changing the Face of the Earth.* That sharpened understanding of the diversity of views toward human impact on physical and biological systems. In the Airlie House meeting in 1968 on *The Careless Technology* some of the more potent ideas about needed national policy, such as environmental impact statements, took shape. During the 1970s Boulding was especially active in promoting the re-thinking of national energy policy, and was a member of the National Research Council's committee on Nuclear and Alternative Energy Systems. This and related activities led to extensive commentary on social aspects of government choices of strategies to promote research or operating policies affecting energy conservation and sources. As a regular contributor to *Technology Review* during 1974–1982 he reached a large audience of technologists with his lively and innovative thinking.

The quality of expression that attracted a serious hearing for his ideas in a variety of places is illustrated by the type of verse with which he often summed up a long, erudite discussion. On a few occasions it stirred up opposition. One instance was when informal circulation of verse he had written on the topic of a proposed report on water development in California persuaded the State legislature it should abolish the committee of which he was a member before the formal document could see the light of day. Generally, however, Boulding's light treatment of weighty topics caught their essence and commended them to careful consideration. The following is from the proceedings of the Princeton gathering in 1955.

THE CONSERVATIONIST'S LAMENT*

The world is finite, resources are scarce,
Things are bad and will be worse.
Coal is burned and gas exploded,
Forests cut and soils eroded.
Wells are dry and air's polluted,
Dust is blowing, trees uprooted.
Oil is going, ores depleted,
Drains receive what is excreted.
Land is sinking, seas are rising,
Man is far too enterprising.
Fire will rage with Man to fan it,
Soon we'll have a plundered planet.
People breed like fertile rabbits,
People have disgusting habits.
Moral: The evolutionary plan
          Went astray by evolving Man.

THE TECHNOLOGIST'S REPLY*

Man's potential is quite terrific,
You can't go back to the Neolithic.
The cream is there for us to skim it,
Knowledge is power, and the sky's the limit.
Every mouth has hands to feed it,
Food is found when people need it.
All we need is found in granite,
Once we have the men to plan it.
Yeast and algae give us meat,
Soil is almost obsolete.
Men can grow to pastures greener,
Till all the earth is Pasadena.
Moral: Man's a nuisance, Man's a crackpot,
          But only Man can hit the jackpot.

In jovial as well as sober vein, Boulding gave the environmental movement a greater and more humane sense of coherence over four decades.

GILBERT F. WHITE

For Further Reading: Kenneth E. Boulding Collected Papers (1971–1985); Kenneth Boulding, *The Image: Knowledge in*

Life and Society (1956); Kenneth E. Boulding, *Towards a New Economics: Critical Essays on Ecology, Distribution and Other Themes* (1992).

BUDDHISM

To understand Buddhism it is necessary to understand the role of karma, the guidance of dharma, and the insights of compassion. Buddhism holds that, as the Buddha said, "When that exists, this comes to be; on the arising of that, this arises. When that does not exist, this does not come to be; on the cessation of that, this ceases."

At the time of the Buddha (ca. 500 B.C.), the Vedic religions of India taught that humanity was part of a constant cycle of birth, death, and rebirth. The "soul" or atman passed from one body to another, for instance, from animal to tree or from insect to animal. The human form was seen as one of the most advanced. The reasons for this were as stated above. Whatever actions one did, led to further consequences, the cycle of karma. The Buddha—the term means the enlightened one—broke into this concept with a radical new understanding: the dharma (teaching) of the Buddha. He taught that there is no "soul," no atman. All that exists is the buildup, or sum total of karma. Thus if those actions which create karma can cease, the karma will cease and without karma, there is nothing to continue the cycle of birth, death, and rebirth.

To achieve the extinction of karma, it is necessary to rid oneself of any idea of the self. It is to realize that all things are no-thing, empty except for their results of believing they are some-thing. Thus the Buddhist looks at the whole of life and sees that underlying it all, human, animal, or vegetable, is the same cause—karma—and the same answer, the dharma of the Buddha.

In relation to the environment, this teaching is of vital importance. An action which is undertaken for personal, selfish, or greedy motives will cause the further creation of bad karma, of hurtful actions. However, an action undertaken in the guidance or spirit of the Eightfold Path with its dharma about right livelihood and right conduct will create contentment and happiness. This positive human attitude is, in the final analysis, rooted in genuine and un-self-ish compassion and loving kindness that seeks to bring light and happiness for all sentient beings. The consequences of this for the natural environment are obvious.

The radical new understanding of karma taught by the Buddha never denied a basic insight of karma, namely that all beings are interrelated. The following verses from the Avatamsaka Sutra express this sense of interrelatedness of all existence:

---

*By Kenneth Boulding, from *Man's Role in Changing the Face of the Earth*, William L. Thomas, ed., University of Chicago Press, 1956, p. 1087. Reprinted by permission of University of Chicago Press.

All lands are my body
And so are the Buddhas living there;
Watch my pores,
And I will show you the Buddha's realm.
Just as the nature of Earth is one
While beings each live separately,
And the Earth has no thought of oneness or
    difference
So is the truth of the Buddha.

This doctrine is known as that of mutual interpenetration and infusion of all phenomena. It guides Buddhist thinking about any and all aspects of the natural world, for no part is separate from the rest. To end the hold of karma on any one being is to wish to end the hold of karma on all beings. This is the compassion which lies at the very heart of Buddhism and which is also central to Buddhist understandings of the environment. In Mahayana Buddhism (the Buddhism of Tibet, China, and much of Japan) this principle of compassion is expressed in the Bodhisatva. This being is one who has extinguished all effects of karma and who could cease existence, but who holds back in order to help rescue other sentient beings still trapped in their karma. The compassion of the Bodhisatva is expressed in the following text from the Vajradhvaja Sutra:

> A Bodhisatva resolves: I take upon myself the burden of all sufferings. . . . All beings I must set free. The whole world of living beings I must rescue, from the terrors of birth, of old age, of sickness, of death and rebirth, of all moral offence, of all states of woe of the whole cycle of birth and death.

The very way of life of Buddhism is to be increasingly aware of and in tune with the patterns of nature and to see through the confusion and chaos to the truth of the interpenetration of all. Meditation often starts with breathing awareness and then moves out from there to a heightened awareness of your surroundings. This has led traditions such as the Thai Forest monks to seek sanctuary for meditation in the forests. The Japanese Zen gardens are an articulation of the Buddhist understanding of pattern within all life; the gardens are designed as places of meditation and reflection upon the true nature of nature. In all such exercises, the penetration through the falsehood of self to the realization of no-self, of no-thing, is central.

Over the centuries, Buddhism has developed in both agricultural and urban settings. It has always sought to show compassion for all living beings. Today, it is often in the compounds of Buddhist temples, on holy Buddhist mountains, and in monasteries that nature still flourishes, while often all around the environment has been overused or destroyed. This accords with Buddhist teaching that the physical world deteriorates as the morality of the people degenerates. Where the dharma is taught and followed, nature still flourishes. Where greed, selfishness, and corruption thrive, the physical world collapses. For centuries, Buddhism has borne witness to this truth. In recent years, spurred on by the scale of the degeneration of traditional Buddhist societies through Western influence and by the stark evidence of environmental destruction, Buddhists have begun to actively promote Buddhist compassion for the environment. For instance in Thailand, Forest monks have taken a stand against the destruction of the forests and have established tree nurseries and reforestation schemes based on Buddhist principles of compassion. These engaged Buddhists are part of a growing worldwide network of such Buddhists who believe that the spiritual life must include practical expressions of compassion. They are living out the dharma by expressing practical compassion in the midst of great corruption and greed. Increasingly Buddhists worldwide are realizing that the world needs, perhaps as never before, the wisdom and the compassion of the Buddha.

MARTIN PALMER

For Further Reading: Martine Batchelor and Kerry Brown, eds., *Buddhism and Ecology* (1992); Susan J. Clark, *Celebrating Earth's Holy Days* (1992); Lewis G. Regenstein, *Replenish the Earth* (1991).

BUILT ENVIRONMENT

See Architecture.

BUREAU OF LAND MANAGEMENT

See United States Government.

C

CANALS

The channelization of water in canals has been undertaken by human societies throughout recorded history. Canals have been built to move water from places of surplus to those of deficit, to drain wetlands, or as waterways for transportation. These activities have been significant in the development of civilizations, and have also resulted in changes to their environments.

The locations of hearths of most early cultures are in or near major river systems. The growth of urban centers is associated with the expansion of agriculture and the availability of food surpluses. One of the most important technological developments in this first agricultural revolution was the use of canals to manage water. This collective effort occurred more than 4,000 years ago along the Tigris and Euphrates rivers of Mesopotamia and the Nile in Egypt, as well as in the Indus Valley and in China; and later among the Maya, Inca, and Aztec of the western hemisphere. Many irrigation systems were developed to expand agriculture into semiarid or arid regions. Twentieth-century examples include the Gazira project in the Sudan, as well as projects in California and Arizona.

Environmental effects of irrigation canal systems include salinization, dehydration, and human health problems. Excess salts have occurred in the soils of many irrigated areas (especially where there are high evapotranspiration rates). The divergence of water from its natural flow system may also lead to the dehydration of wetlands with ecological consequences to the native flora and fauna, as in Florida's Everglades. Water-borne diseases, such as schistosomiasis (Bilharzia), have increased and affected human populations in Africa.

Rivers are transportation routes, and canals have been dug as adjuncts to the natural system, or to allow ships to bypass such obstacles as waterfalls or rapids. In the late 18th century English canals were the major transport arteries of the Industrial Revolution, and were responsible for the growth of many towns. In North America canals were laid out in great number during the early and mid-19th century. In 1825 the Erie Canal was built and forged the main link between the coastal port of New York City and its hinterland. Although railroads later captured most of the transport of bulk materials, canals remain an important feature of shipping transportation systems, with the Kiel Canal in Germany the world's busiest.

Not only barges and ships but also various species of fauna and flora can travel these canals. Along the St. Lawrence Seaway, the deepening of the Welland Canal allowed the movement of lampreys from Lake Ontario to the western Great Lakes where they quickly attached themselves to gamefish and affected the fishing industry.

Perhaps the best-known canals are those that have been constructed to shorten the route between two large bodies of water. The sea-level Suez Canal was built between 1859 and 1869 to enable ships to bypass a much longer route around the Cape of Good Hope. The Panama Canal, which links the Atlantic with the Pacific, was considered one of the world's most impressive feats of engineering when it was built between 1904 and 1914 because it required the construction of locks to raise and lower ships through the higher central region of the isthmus.

Such interoceanic canals may cause ecological changes by the diffusion of species from one ecosystem to another. The deepening of the Suez Canal in 1956 has resulted in the migration of over 100 new invertebrates from the Red Sea to the Mediterranean.

Recently, regional environments degraded by canal systems have undergone restoration efforts. The Soviet Union abandoned a complex canal scheme in an effort to halt the drying up of the Caspian Sea; and the Kissimmee River in Florida has been dechannelized to reflood its banks. Aesthetic as well as ecological concerns will be as important as economic in future water transfers.

HARVEY FLAD

CANCER

Cancer is a disease distinguished by any of various malignant neoplasms (tumors) characterized by the proliferation (rapid growth) of anaplastic (less differentiated) cells that tend to invade surrounding tissue and metastasize (spread) to new body sites. It is the most feared disease in industrialized countries—borne out by the statistics of the trends in incidence, mortality, and survival of patients. Even with age adjustment of the data, the incidence of breast, lung, melanoma, prostate, and lymphoid cancers is increasing. Similarly, the overall mortality continues to increase, owing both to the increasing incidence of cancers and the increased longevity of the general population. More than two-thirds of cases occur in individuals over 50. On the other hand, cancer is a major cause of death in children during the first decade of life—second only to accidents. Survival for children and young adults following treatment has risen dramatically in a number of types, such as leukemia; kidney, testicular, and adrenal cancers as well as cancer of the lymphatic system, especially Hodgkin's disease, the latter even in older adults. Unfortunately, survival in the major cancers—lung, breast, colon, and prostate—has not increased so markedly and in some instances (e.g., lung) has remained essentially unchanged over the past two decades.

Cancer is one form of a general class of diseases known as neoplasia (literally, new growth). Neoplasia is defined as a heritably altered, relatively autonomous growth of tissue(s). Cellular changes characteristic of neoplasia are transmitted in a heritable manner from one cell to another. These changes are primarily in the growth or replication of the cell, but even more important is the autonomy assumed by the cancerous cell in failing to obey the biological mechanisms that govern the growth and metabolism of individual cells and the interactions of cells with the rest of the organism. Neoplasms occur only in tissues (cell populations), making this disease unique to multicellular organisms.

Neoplasms may be classified as either benign or malignant. Malignant neoplasms successfully form metastases, i.e., secondary growths separate from the primary neoplasm, and almost all of them exhibit abnormalities in chromosome structure. The term *cancer* is synonymous with malignant neoplasm, whereas the term *tumor* indicates a readily defined mass of tissue distinguishable from normal living tissue. Many tumors, such as a scar, an abscess, or a healing bone callus, are not neoplastic.

Neoplasms are named according to the tissue from which they arise, plus a suffix. Names of benign neoplasms usually carry the suffix *-oma*; malignant neoplasms are indicated by the suffix *-sarcoma* for cancers arising from supportive tissues such as muscle, bone, and fat, or the suffix or term *carcinoma* referring to cancers arising from epithelial tissues such as skin, stomach, bowel, bladder, or breast. Exceptions to this kind of nomenclature include thymomas, which are either malignant or benign neoplasms of the thymus gland, and such descriptive terms as dermoid, a benign neoplasm of the ovary. Teratomas consist of malignant tissues of both epithelial and mesenchymal origin. Leukemias are malignant neoplasms that usually arise in the bone marrow or lymphatic tissues of the body and have a major portion of their cells circulating in the bloodstream.

The Causes of Cancer

Carcinogens are agents that cause cancer and may be chemical, physical, biological, or genetic in nature. A variety of different chemicals in our environment cause cancer in a variety of species. The structures of chemical carcinogens vary. Polycyclic hydrocarbons occur in smoke and other air pollutants; aromatic amines occur naturally or as a result of synthetic processes in industry; and nitrosamines occur naturally (including within the human body) in minute amounts. Examples of chemical carcinogens found in industry or as environmental pollutants include benzene, arsenicals, benzidines, bis(chloromethyl)ether, chromium and nickel compounds, and vinyl chloride. Several drugs (including alkylating agents used to treat cancer) are also carcinogenic. Estrogens and androgens—both synthetic and naturally occurring male and female hormones—are carcinogenic to humans and animals under certain conditions. Foodstuffs may also be contaminated with potent carcinogens such as aflatoxin B_1.

Most chemical carcinogens do not induce their effects directly, but rather must be metabolized to a more active form. Normal cells are equipped with many mechanisms capable of such "activation" and the formation of "ultimate" forms of chemical carcinogens. Such ultimate forms are mutagenic because of their ability to act directly with DNA, resulting in an alteration of the structure of genes and other components of the genome.

Ultraviolet and ionizing radiation are carcinogenic for both human and animal tissues. Epidemiologic studies, as well as basic investigations, have demonstrated that exposure to ultraviolet light from the sun or other sources is carcinogenic for tissues of the skin. Ionizing radiation, as evidenced by the atomic bomb casualties in Japan at the end of World War II, can induce leukemia, as well as cancer of the thyroid, breast, stomach, uterus, and bone. Medical, diagnostic, and therapeutic applications of ionizing and ultraviolet radiation are thus not without a carcinogenic potential. Carcinogenesis from physical objects foreign to the organism, such as plastic films or disks or metal implants, results in sarcomas in lower animals, but such relationships in humans have as yet been inconclusive. On the other hand, asbestos—especially the crocidolite form—has been causally related to the development of both bronchogenic carcinoma and malignant mesothelioma in humans. In some instances, such as sarcomas induced by plastic films or disks, destruction of the crystalline or physical structure of the foreign body results in the loss of carcinogenic action.

A variety of parasites have been associated in a causal manner with a number of animal and plant cancers. Bacteria and worms have been shown by epidemiologic and direct demonstration to be the cause of the development of specific neoplasms in both plants and mammals. However, the best studied and probably the most prevalent biological agents that cause cancer are viruses. A relatively large number of viruses have been shown to be carcinogenic in lower vertebrates and several have been closely linked to the development of certain human neoplasms. The Epstein-Barr virus causes Burkitt's lymphoma, prevalent in Africa. The same virus is closely associated with a cancer of the throat (nasopharyngeal carcinoma), common in certain parts of China but occurring throughout the world. Some carcinomas of the female genital tract, anus, esophagus, and skin are associated with the presence of DNA from specific strains of the human papillomavirus. The hepatitis B virus and/or components of its genetic material (DNA) are found in the vast majority of hepatocellular carcinomas in the human, identifying this virus as a causative agent for this disease. A close link has also been established between the development of human T-cell leukemias/lymphomas and an RNA oncogenic (carcinogenic) virus termed HTLV-1. This latter association appears most frequently in certain parts of Japan, the West Indies, and to a lesser extent the southeastern United States.

Another disease caused by an RNA virus similar in structure to HTLV-1 is the Acquired Immune Deficiency Syndrome (AIDS). Viruses which cause this disease, termed Human Immunodeficiency Viruses (HIV), exert their effects by destroying cells critical to the ability of the immune system to respond to foreign invaders, bacteria, viruses, etc., as well as in the host immune response to cancer. This latter effect is seen by a high incidence of several types of cancers in patients with AIDS, notably lymphomas and Kaposi's sarcoma.

Viruses are classified by the type of nucleic acid (DNA or RNA) that makes up their genome. The Epstein-Barr, papilloma, and hepatitis B viruses are DNA viruses, whereas HTLV-1 is an RNA virus. All or part of the DNA sequences in oncogenic (associated with tumors) DNA viruses are incorporated into the host cell DNA during infection. This process is a major factor resulting in the neoplastic transformation. Oncogene RNA viruses, however, must have their RNA genome copied into DNA by an enzyme called reverse transcriptase, the information for which is supplied by the virus. The DNA copy may then be incorporated into the host cell genome, as with DNA sequences of DNA viruses.

In general, oncogenic viruses contain within their genome one or more genes that are essential for the process of converting a normal to a neoplastic cell. These genes are termed oncogenes. Oncogenes of many oncogenic RNA viruses have evolved from normal genes, with the same function occurring in the genome of target cells. The cellular counterpart of a viral oncogene is termed a proto-oncogene. Other viral oncogenes, especially those of DNA viruses, do not appear to have normal cellular counterparts.

A number of types of human cancer may be the result of genetic predisposition. Although many types occur, the total contribution of direct hereditary cancers to the total cancer incidence is relatively small (less than 5%). The inherited predisposition to cancer may be the result of mutations in specific genes termed tumor suppressor genes resulting in a dramatically high incidence of specific cancer types. Some cancers may also be associated with certain structural chromosomal abnormalities within the neoplastic cell. Such changes, such as those that cause chronic myelogenous (originating in the bone marrow) leukemia, are the result of changes within somatic (body) rather than germ cells.

The Development of Cancer (Pathogenesis)

There is always a latent period of time between the initial exposure to the carcinogenic agent and the appearance of the neoplasm. The first stage of carcinogenesis, *initiation*, appears to result in one or more mutational changes in the initiated cell, giving the cell a growth advantage in the presence of a promoting agent. The promoting agent, when administered continuously, induces the second stage of carcinogenesis, *promotion*, inducing the selective expansion of initiated cells over their normal counterparts. Removal of the promoting agent after expansion of initiated cell populations results in a loss of many cells in the stage of promotion, indicating the reversible effect of the promoting agent on initiated cells. The third stage is *progression*, in which initiated cells or their progeny in the stage of promotion may enter the stage spontaneously as a result of an error in the mitotic cycle, or by the direct action of "progressor" agents, most if not all of which are capable of breaking chromosomes. Cells in the stage of progression exhibit the evolution of changes in the number and structure of cellular chromosomes and possess the potential for biological malignancy or cancer itself. In the third stage, biologic malignancy implies the spread of the neoplasm by metastatic growth of cells that originated from the primary growth and spread through the bloodstream or lymphatic system, by direct implantation within body cavities or by surgical intervention in which inadvertent transplantation of cancer cells occurs. Rarely cells in the stage of progression remain latent and nonreplicating in the host for years. In contrast, cells of some cancers may actually stop growing, differentiate, then remain quiescent for the rest of the lifespan of the host.

Cancer Prevention

Of the known causes of cancer, dietary factors, smoking, and reproductive mores are important in the development of human cancer. Cigarette smoking and other forms of tobacco abuse are now felt to be the primary causative agent in the development of approximately one-third of human cancers in western civilizations. These include lung, bladder, esophageal, mouth, and uterine cancers. Excessive intake of alcoholic beverages acts as a promoting agent in the development of several human cancers including those of the pharynx, liver, and esophagus. Late menopause and early menarche as well as the lack of childbearing or bearing the first child after age 30 increase the risk of developing breast cancer; women having multiple sexual partners have a greater risk of developing cancer of the uterine cervix. "Passive" prevention of cancer by regulation or voluntary control of exposure to real and potential carcinogens is the goal of many programs in this country; active prevention is the self-administration of preventive agents such as retinoids (vitamin A-like chemicals) or dietary factors exhibiting the ability to prevent oxidation within the living cell. Carcinogenic potential can be identified by tests for mutagenicity (the Ames test) and clastogenicity (chromosome breakage), and by whether they induce neoplastic transformation either *in vivo* or *in vitro* (within or outside the living organism). Unfortunately, vaccines against carcinogenic viruses, such as Epstein-Barr and papillomaviruses, have not yet been developed for general use, although an effective vaccine for hepatitis B has been marketed.

U.S. governmental action to regulate the exposure of humans to chemical carcinogens in industry include the Delaney clause, safeguarding foodstuffs from the addition of known carcinogens, and the Toxic Substances Control Act (TSCA), which seeks to prevent the entrance of carcinogenic agents into the human environment from industrial chemicals, pesticides, and related agents. U.S. industrial workers are now well protected by the Occupational Safety and Health Act, and similar legislation is now in effect in many industrialized nations.

Diagnosis, Treatment, and Survival

The successful treatment of cancer usually depends on early diagnosis. Several methods have been developed. One of the most widely used and successful is the "Pap" test developed by Dr. George N. Papanicolaou. In this test cells obtained from readily available areas of the body, such as the mouth, breast milk, urine, uterine cervix, and biopsy material, are examined under the microscope for their staining and structural characteristics. Since such characteristics of neoplastic cells are dramatically different from normal cells, it is possible in most instances to make at least a tentative diagnosis on the basis of the test alone. Chemicals circulating in the blood, such as the prostate-specific antigen (PSA), have been used for the early diagnosis of cancers arising in specific tissues. Radiographic screening, especially of the breast by mammography, has also been widely used.

For appropriate therapy, the diagnosis must be accurately determined, usually by microscopic examination of the tissue by the pathologist. In the ideal circumstance surgery can remove all neoplastic cells from the body. When metastatic lesions are present, surgery is useful in removing at least most of the cancerous tissues. Surgery can also alleviate problems caused by the neoplasm, such as blockage of blood circulation, obstruction of the bowel, or severe pain.

Radiation therapy involves the use of both radiant and particle-ionizing energy. Such therapy is most successful with certain types of neoplasms such as lymphomas, leukemias, and many carcinomas. Toxic chemotherapy, which seeks to kill cancer cells, is utilized primarily in the therapy of neoplasms that have spread beyond the primary site. Within the last two decades, chemotherapy has been responsible for a dramatic improvement in the cure of certain types of cancer, namely, Hodgkin's disease, acute lymphoblastic leukemia in children, testicular cancer, and Wilms' tumor of the kidney. All such neoplasms exhibit a rapid growth rate, making them more susceptible to chemotherapeutic agents that are most effective in killing dividing cells. However, various drugs exhibit their therapeutic effectiveness against cancer cells in different ways; thus, the utilization of a combination of three to six drugs has been found to be most effective in therapy and also in preventing the occurrence of resistance to drugs by the neoplasm.

Treatment of cancer cells with hormones that alter their growth and/or differentiation is also effective, particularly in neoplasms of endocrine organs or those serving as targets for hormones. Cancers of the breast and prostate are frequently treated in this way. In addition, stimulation of the immune defenses of the host or treatment utilizing antibodies specific for components of the neoplasm have been used in attempts to eliminate cancer cells that remain following other types of therapy.

Cancer Research — Past, Present, and Future

Cancer research has focused on the identification of agents that cause the disease and their mechanism of action, especially chemicals. Within the last two decades, knowledge of the neoplastic process itself has mushroomed. The identification and characterization of specific genes critical in the development of neoplasms occupy a major component of cancer research. The mechanisms involved in the three stages of cancer development are also being widely studied. Vaccination against the major viruses causing human cancer is a realistic possibility for people throughout the world. There are numerous educational programs that encourage the modification of lifestyles as a method to prevent many types of human cancer. The detailed mechanisms of cell replication are being uncovered. Such knowledge will lay the foundation for the control of cancer cell replication. Therapy may also involve causing the neoplastic cells to differentiate, thereby preventing their continued replication and making them potentially responsive to controlling factors within the body. An understanding of the mechanisms underlying the continuing evolution of chromosomal changes evident in the stage of progression may lead to the successful therapy of many neoplasms in the human that cannot be controlled by known methods.

HENRY C. PITOT, M.D.

For Further Reading: Devra L. Davis and David Hoel, eds., *Trends in Cancer Mortality in Industrial Countries* (1990); Elizabeth C. Miller, "Some Current Perspectives in Chemical Carcinogenesis in Humans and Experimental Animals: Presidential Address," in *Cancer Research* (1978); Henry C. Pitot, *Fundamentals of Oncology* (1986).
See also AIDS; Asbestos; Carcinogens, Natural; DNA; Epidemiology; Health and Disease; Ionizing Radiation; Life-Style Factors; Mutagen; Occupational Safety and Health; Parasitism; Polycyclic Aromatic Hydrocarbons; Radiation; Tobacco; Toxic Chemicals; Toxic Substances Control Act; Ultraviolet Radiation; Viruses.

CAPITAL

Capital is defined by three attributes: (1) it is produced by labor, (2) it increases the productivity of labor, and (3) it lasts a long time (longer than one year). Examples of physical capital include buildings, refineries, delivery trucks, and carpenters' tools. Examples of nonphysical capital include human capital—the education and training that make workers more productive—and discoveries and new technology.

People invest their labor to create physical capital, such as a boat constructed by a fisherman, that will make future production more efficient. A fisherman with a boat and a net is many times more productive

than one without this capital. While physical capital is important, the most productive investments in this century have been investments in education and research and development. The resulting capital and scientific and technological knowledge are the foundation of modern economies. Such capital is vital for the technologies of transport—including automobiles, trains, aircraft, and spaceships—and for the ability to fulfill human needs with maximum efficiency. For example, personal computers are much more economical in terms of energy and materials than their less powerful mainframe "grandparents." The pace of technological change is so rapid that much physical capital is obsolete before it is worn out; thus depreciation is governed more by obsolescence than by physical wear and tear.

Economists and accountants estimate depreciation of a firm's capital stock in order to calculate the profit being earned by the firm. A similar notion has been applied to a nation's nonrenewable resources and environment. For example, France's National Patrimony Accounts are an estimate of the extent to which current production and consumption have reduced the quality of the environment and the stock of nonrenewable resources available to future generations. Such assessments reveal that increases in economic activity and income are at the expense of available resources and environmental quality.

CARBON

Carbon is a nonmetallic element that is chemically inert at ordinary temperatures. It exists in three forms: an amorphous form that is a dull black, typified by soot; a shiny black soft crystalline form called graphite; and the transparent crystalline form, diamond, which though commonly colorless also exists with tinges of other colors. Naturally occurring carbon is 98.89% the C-12 isotope. Its atomic weight is 12.011; its atomic number is 6.

There are more known compounds of carbon than there are of all other elements combined. This unique ability of carbon derives from its capacity to bond with itself to make long chains, branched chains, and rings that have bonding sites left for the attachment of a variety of other elements and groups. Because carbon has this ability, it is the very basis of life.

Carbon is virtually insoluble in most solvents. It is difficult to assign melting and boiling points because graphite sublimes at 3500°C; the melting point is about 3550°C and the boiling point about 4825°C. The density of carbon increases from amorphous, through graphite to diamond, the hardest substance known. At elevated temperatures all forms of carbon burn to form carbon dioxide.

Carbon forms two kinds of compounds, inorganic and organic. Common inorganic compounds include carbon monoxide and carbon dioxide, carbonates and bicarbonates, cyanides and carbides. Organic compounds fundamentally involve carbon in combination with hydrogen, oxygen, and some with nitrogen.

Carbon can be extracted from organic compounds readily; it is the black substance that appears when any organic material is charred. In the laboratory carbon can be released from table sugar by concentrated sulfuric acid. The destructive distillation of wood also yields carbon. A pure form of carbon is obtained industrially from the incomplete combustion of natural gas, oil, or acetylene—their hydrogen produces water and their carbon is isolated in a fine particle form called carbon black.

Known since antiquity in its relationship to fire, carbon's discovery as a substance cannot be associated with one person. Its chemical identity as a nonmetallic element, however, was asserted by Antoine Lavoisier in the 1780s.

Carbon is widely distributed as an element in stars (including the Sun), comets, meteorites, and planets. It is found on Earth in its pure form as graphite, diamond, and soot, but it is found most commonly in its compounds. Carbon dioxide gas forms an average of 0.036% of Earth's dry air by volume. By contrast the atmosphere of Mars is 96.02% carbon dioxide. Carbon dioxide is dissolved in all the waters of Earth. In the rocky crust, deposits of carbonates of calcium (limestone), magnesium, and iron, and deposits of coal, oil, and natural gas all contain carbon. All life forms are made of carbon as the essential element.

Without carbon's chemical behavior there would be no carbohydrates, fats, proteins, or nucleic acids—in short, there would be no life. The molecules required to serve as building blocks for life, to provide the immense diversity among living things, and for the processes that are required to sustain life must be highly stable, large, and although very similar, have small but key variations. The skeletons of these molecules are long chains, branched chains, and rings of up to thousands of carbon atoms bonded to each other. Carbon is unique in this ability. The carbon skeletons are the source of the similarity among the variety of molecules, for example, proteins. The skeleton is stable because carbon atoms bond to each other with single, double, or triple covalent (electron-sharing) bonds. Each carbon atom is involved in four bonds. When carbon atoms are chained with single bonds, each atom has two bonds involving the carbon atoms on either side of it in the chain, leaving two other bonding sites. Atoms, or groups of atoms, usually oxygen, hydrogen,

nitrogen, or another carbon, attach to the chain at these other sites. These are also strong covalent bonds and add stability to the molecule. The diversity within molecules arises in part from the nature and placement of these groups. The four atoms that bond with a carbon atom are arranged in a tetrahedron around that carbon atom; so three-dimensional spatial arrangement is also a source of the diversity among molecules that are otherwise highly similar.

Carbon enters the living world from the abiotic environment in the form of carbon dioxide gas. Organic molecules are made from carbon dioxide and water by green plants in photosynthesis. Carbon returns from the biotic environment to the nonliving world also as carbon dioxide gas. Respiration, putrefaction, and the burning of organic fuels such as coal, natural gas, peat, oil, wood, and charcoal all return carbon dioxide to the air.

Carbon is a crucial element in the nuclear fusion of hydrogen to helium in stars. The first step in the process involves the capture of a proton—a hydrogen nucleus—by the nucleus of a C-12 atom. The C-12 nucleus cycles through isotopes of nitrogen and carbon and back to the C-12 nucleus. In this sense carbon catalyzes the nuclear fusion reaction.

Industrial use of carbon includes cutting and drilling with diamond, because it is so hard that it will cut anything, and the use of graphite as pencil lead and as a lubricant. Rubber is made abrasion-resistant by the addition of carbon, so 95% of carbon black produced is used in the tire industry. Carbon black is also used in typewriter ribbons, carbon paper, inks, and paints. Carbon is used to extract metals from their ores in three ways: it serves as the electrodes in the electrolytic extraction of aluminum from its oxide, it reduces the oxides of iron, zinc, and lead in industrial furnaces, and it reduces the impurities in copper ore so they can be separated from the copper. Industrial use of carbon compounds includes plastics, pharmaceuticals, anesthetics, and fertilizers. The major use, however, of carbon compounds is as fuels. The burning of fuels releases the potential energy in the carbon to carbon bonds and converts it to heat, mechanical, and electrical energy. This energy conversion is about 40% efficient.

Carbon dating enables us to accurately date artifacts made from plant material. The method relies on a comparison of the ratio of the amount of the isotope C-14 to that of the isotopes C-12 and C-13 in the artifact and in the carbon dioxide of the atmosphere.

JEAN LYTHCOTT

For Further Reading: Isaac Asimov, *The World of Carbon* (1958); F. W. Gibbs, *Organic Chemistry Today* (1961); Herman F. Mark and the Editors of LIFE, *Giant Molecules* (1966);

C. H. Snyder, *The Extraordinary Chemistry of Ordinary Things* (1992).
See also Atmosphere; Carbon Cycle; Carbon Monoxide; Fuel, Fossil; Fusion; Photosynthesis.

CARBON CYCLE

The global carbon cycle concerns the storage and especially the movements of carbon on and near the surface of the Earth. Carbon (C), the basic element of life, can take many forms: Carbon dioxide (CO_2) is a gas that can be in the atmosphere or dissolved in water; limestone ($CaCO_3$) can be a rock or can dissolve in water; fats, carbohydrates, proteins, and fossil fuels are carbon-containing biological compounds. Huge reservoirs of carbon are stored in the deep ocean, ocean sediments, and in certain sedimentary rocks of the Earth's mantle. Small reservoirs are contained in fossil fuels—carbonaceous materials in which the ratio of carbon to other materials is great enough to be mined and burned with an energy and economic gain. In addition, small quantities of carbon are stored in the atmosphere, the surface of the ocean, and the biota, especially the terrestrial phytomass (plants) and soil.

Carbon can move (cycle) easily among these forms. For example, CO_2 diffuses across the surface of the ocean, or into the leaves of plants where it is transformed into wood and other materials. Rivers dissolve limestone rock and carry the carbon to the sea. People burn fossil fuels, turning their carbon into CO_2, which enters the atmosphere.

The smaller reservoirs of carbon—the fossil fuels—are the focus of the current debate concerning the global carbon cycle and human activity. The largest reservoirs are sealed off from the atmosphere and therefore have little or no effect on atmospheric concentration of carbon dioxide. Most of the deep ocean is sealed off by a permanent vertical thermocline (steep temperature difference that prevents advective mixing). This exists from about 50 degrees North to 50 degrees South. The ocean sediments and uneconomic fossil carbon do not react with the atmosphere.

Compared to ancient civilizations, the modern citizens of industrialized nations are enormously wealthy in terms of the amount of energy we have doing our bidding. In one day, the average American uses the energy equivalent of about seven gallons of oil or the labor of 80 people. Most of this work is supplied from fossil fuels made of carbon. For a dollar spent, about a half liter of oil—or its equivalent in other fuels—is burned somewhere to provide that good or service, and about a kilogram of carbon dioxide is released into the atmosphere.

Oil, natural gas, and coal provide most of the energy that powers the economies of modern nations. Fossil fuels are remnants of plant growth in ancient ecosystems, and they all contain the same basic elements, principally carbon and hydrogen, with smaller but variable amounts of oxygen, sulfur, and other trace materials. Because of these similarities, combustion of any fossil fuel produces more or less similar byproducts and hence environmental impacts.

The Release of Carbon Dioxide

No method is known by which most of the energy contained in the chemical bonds of fossil fuels can be released without producing large volumes of carbon dioxide and water. Although carbon dioxide technically can be removed from stack gases, the dollar cost is astronomical and the energy cost would be a very large proportion of the energy produced by the original combustion, negating most of the effects of carbon dioxide removal. About half of the CO_2 produced from fossil fuels remains in the atmosphere for decades—even centuries. Even if the net flux of CO_2 to the atmosphere were stopped immediately, the existing CO_2 would be reabsorbed only very slowly.

The reason that human activities have such a large impact on atmospheric CO_2 is that the Earth's atmosphere contains only trace quantities of this gas, about 0.036% or 360 parts per million (ppm). It is believed that the present concentration of carbon dioxide is about 20% to 50% higher than it was a century ago, although there are no direct measurements of CO_2 in the atmosphere before 1957. Some new data on CO_2 concentrations of ancient glaciers, presumed to be in equilibrium with ancient atmospheres and stable since then, indicate that the atmosphere may have varied from 190 to 300 ppm CO_2 over the past several thousands of years. These data also show that ancient fluctuations of CO_2 were correlated closely with temperatures. Thus it is believed that ancient climates were determined largely by contemporary atmospheric CO_2. (It is possible, however, that global temperatures changed for other reasons and high temperatures *caused* a release of CO_2 from the oceans or biosphere.)

Since 1957 the concentration of carbon dioxide has been monitored daily at Mauna Loa, Hawaii, a high mountain where well-mixed air thought reasonably characteristic of the mid Northern Hemisphere passes by and where an average parcel of air circles the Earth about once every two weeks. The data show annual fluctuations that are principally a result of net photosynthesis removing CO_2 from the atmosphere during spring and summer and net respiration returning CO_2 to the atmosphere during fall and winter. The record also shows a clear and unequivocal increase from each year to the next, an increase that appears unquestionably a result of human activity.

An important question, unresolved despite 30 years of intense scientific effort, is the imbalance of the global carbon cycle—that is, of the 5.5 billion metric tons of carbon released into the atmosphere from fossil fuel burning each year during the 1980s, less than 3 billion metric tons can be found as increases in atmospheric carbon and only about another 2 billion metric tons are thought to be found in the sea. Where is the missing carbon?

Early analysts of the problem thought that the terrestrial phytomass was a net sink for carbon dioxide, fixing it through the "beta effect" of increased photosynthetic rates due to enhanced concentrations of atmospheric carbon dioxide, a plant nutrient. This increased growth would, in theory, absorb the missing carbon. Unfortunately it has been difficult to verify this theory, for experiments show that even when photosynthesis is enhanced by increased CO_2, there is rarely a substantial increase in the carbon accumulated by the plant. Other scientists have shown clearly that changes in terrestrial phytomass, especially tropical forests, instead *contribute* an additional one half to two billion metric tons of carbon per year to the atmosphere from changing patterns of human land use, principally the expansion of agriculture and pastures. These changes are especially important in the tropics because these forests often contain more carbon per hectare than temperate ecosystems and the forests are being converted to uses that store much less carbon per unit of land. For example, changing tropical forests in Central America to pastureland reduces the amount of carbon stored from about 100 to 200 metric tons per hectare to about 5 metric tons per hectare. On the other hand, many forests in industrialized temperate areas, such as New England and Finland, are recovering their carbon content, for where human populations are relatively low, food can be grown on less land due to higher yields from increased fertilizer use, and because food is imported from other regions.

The principal cause of the changing patterns of land use and subsequent carbon release in the tropics is human population growth, sometimes combined with the various social structures that encourage and maintain ownership of much of the most fertile land in the hands of a very small percentage of the population.

More recent research has offered a number of resolutions to the "missing carbon" problem. In fact, many natural forested ecosystems appear to act as carbon sinks relative to the atmosphere, removing small but significant quantities of carbon in soils, sediments, and, via other discharge, the ocean. This sequestering appears to be enhanced by the increased anthropogenic mobilization of nitrogen and other nutrients, as well as the continuing fertilization effect of the increasing

CO_2, which together would increase the photosynthesis of plants, both on land and in the sea. Thus it is possible that natural vegetation operates as a carbon sink whether or not the beta effect is significant.

The Impact of Increasing Carbon Dioxide

At or near current concentrations, carbon dioxide has no known deleterious effects on the physiology of humans or other animals. But relatively small changes can affect plants because CO_2 is a plant nutrient. Carbon dioxide concentrations increase the photosynthesis of most plants even at moderately enhanced levels, so that the levels of atmospheric carbon dioxide that are likely to be present in coming decades may stimulate the growth of individual plants and indeed the entire terrestrial biota.

The temperature of the Earth is a result of the balance between incoming short-wave solar radiation and the long-wave radiation that is reflected back to space. Minute quantities of carbon dioxide and water vapor in the Earth's atmosphere are relatively transparent to incoming short-wave radiation, but absorb and retain a small proportion of the outbound long-wave radiation, converting it to sensible heat. As the concentration of carbon dioxide in the atmosphere increases, more long-wave radiation is absorbed and converted to heat, leading at least in theory to a global warming. This has been labeled the greenhouse effect.

Most scientists who study the problem believe that increased carbon dioxide concentrations will increase temperature significantly—perhaps one to four degrees Celsius by the middle of the next century, and that these changes may be just beginning to be distinguishable from normal variations. These climate changes also could shift rainfall distributions significantly, changing global agricultural patterns for better or worse, depending on the location. One recent analysis suggests an increase in the frequency of drought in the tropics and subtropics, something that many farmers in Southern Africa and Northern Mexico believe is already happening. Additionally, a rise in temperature could cause the ocean's water to expand and to melt parts of the polar ice caps, which, by increasing the sea level, would cause enormous economic damage by flooding coastal cities, including, initially, their expensive subways.

It is important to note that the projected temperature changes are based on models that are complex, quite controversial, and still essentially unverified. In particular, there is a great deal of uncertainty pertaining to vertical heat transfer in water vapor and to the role of clouds. Some scientists believe that climate warming may be far less than predicted by the models most often used, and that the increased crop productivity that should result from the increased CO_2 will be essential for feeding future human population increases. On the other hand, other research suggests that other residuals of industrial society, including methane, nitrous oxide, and chlorofluorocarbons (CFCs) could double the warming resulting from increased CO_2. But the very large projected increase in CO_2 means that eventually the climatic impacts are likely to be very large and potentially irreversible at least for thousands of years.

For this reason, most of the scientific community believe it is wise to limit the anthropogenic production of CO_2 as much as possible. Efforts underway include conservation of various kinds to reduce the use of fossil fuels; increasing availability of nuclear and solar power; and curbing substantially the growth rate of the human population.

CHARLES A. S. HALL

For Further Reading: R. Paul Detwiler and Charles A. S. Hall, "Land Use Change and Carbon Exchange in the Tropics," in *Science* (1988); Charles A. S. Hall, Cutler J. Cleveland, and Robert Kaufmann, *Energy and Resource Quality: The Ecology of the Economic Process* (1986); P. Peter Tans, Inez Y. Fung, and Taro Takahashi, "Observational Constraints on the Global Atmospheric CO_2 Budget," in *Science* (1990).
See also Atmosphere; Carbon; CFCs; Climate Change; Energy; Fuel, Fossil; Greenhouse Effect; Meteorology and Climatology; Oceanography; Population; Tropical Rain Forests.

CARBON DIOXIDE

See Atmosphere.

CARBON MONOXIDE

Carbon monoxide (CO) is a tasteless, odorless, colorless, toxic gas that, because of its many sources, is a ubiquitous air pollutant. When inhaled, CO rapidly diffuses from the lungs into the blood, where it reversibly binds to hemoglobin to form carboxyhemoglobin. Hemoglobin has an affinity for CO about 220 times greater than that for oxygen. Thus, when CO occupies available binding sites in hemoglobin for oxygen, it reduces the capacity of the blood to carry oxygen. Additionally, carboxyhemoglobin formation reduces the ability of tissues to extract oxygen from hemoglobin. The tissues that are most sensitive to diminished oxygen—the brain and the heart—are particularly vulnerable to hypoxia (a deficiency in the amount of oxygen).

The primary source of carbon monoxide is from the combustion of carbonaceous materials, including gasoline, natural gas, oil, coal, wood, and tobacco. The

principal source of CO in outdoor air is motor vehicle emissions; ambient concentrations vary from 1 part per million (ppm) in rural areas to 140 ppm in urban areas with heavy traffic. Because cigarette smoke may contain CO at 400 ppm, cigarette smokers often have blood carboxyhemoglobin levels in the range of 9% to 15%.

Carbon monoxide causes toxicity from oxygen deprivation to tissue, and it is not directly toxic to the lungs. At carboxyhemoglobin levels between 10% and 30%, headaches, lightheadedness, decreased visual acuity, and impaired cognitive function may result. In the northern U.S., improved residential insulation to conserve energy has increased the risk of both acute and chronic CO intoxication in the home from improperly vented coal- or wood-burning stoves. During the winter, the nonspecific symptoms of headache, lethargy, and nausea resulting from CO toxicity have sometimes been mistakenly attributed to viral infections. At carboxyhemoglobin levels greater than 50%, syncope, seizures, and coma may occur. Each year several thousand people in the U.S. die from accidental CO poisoning as the result of sustained exposure to high concentrations of CO from improperly vented heaters or automobile exhaust.

Initial treatment of CO poisoning involves removal from the source and institution of basic life-support measures. High levels of oxygen are administered to facilitate dissociation of CO from the hemoglobin molecule.

Standards promulgated by the Environmental Protection Agency are designed to protect the public by keeping blood carboxyhemoglobin levels in nonsmokers at levels below 2%. To attain this goal, no more than a one-hour level of 35 ppm of CO, or an eighthour level of 9 ppm should be reached. Chronic exposure to 9 ppm CO in sedentary, healthy subjects would result in a carboxyhemoglobin level of 1.8%. The Occupational Safety and Health Administration guidelines limit workers' exposure to CO to no more than 50 ppm averaged over eight hours.

MARK J. UTELL, M.D.

CARBON TAX

The ineffectiveness and inefficiency of centralized command and control regulation have led regulators and political leaders to experiment with market-based incentives. Rather than command that an action be taken, the approach is to provide an economic incentive. Proponents argue that economic incentives: (1) will cause polluters to act more rapidly, (2) give more flexibility to polluters to achieve abatement at the least cost, and (3) end many of the current bottlenecks preventing regulators from acting quickly.

Citizens in many nations have expressed concern about human activities that may induce global climate change. Numerous proposals have been made to curtail emissions of greenhouse gases, of which carbon dioxide is the most important. A proposal using market forces would place a tax on the carbon dioxide emissions from each activity to motivate individuals and businesses to lower their emissions, either through fuel switching or by increasing energy efficiency. The largest anthropogenic source of carbon dioxide emissions is the burning of fossil fuels. Thus, these fuels have been the focus of carbon tax proposals.

Per unit of energy produced, the burning of coal results in almost twice the amount of carbon dioxide as does natural gas, with carbon dioxide from burning oil falling in between the two. A carbon tax would thus fall most heavily on coal. A significant tax would do much to discourage coal production and use. It would also stimulate other energy sources, such as hydroelectric, nuclear, and solar, and encourage development of more fuel-efficient vehicles and curtailment of the number of miles driven each year.

In addition to lowering carbon dioxide emissions, a carbon tax could raise a substantial amount of revenue. As with any tax, this revenue could be used to lower the government deficit, increase spending on public programs, or lower other taxes. At the same time, it would raise the cost of producing some goods and services, thus decreasing their sale. A tax proposed by the European Commission (EC) was estimated to have a small but significant effect on EC production and competitiveness internationally. The EC proposed to implement this tax only if its international competitors did so as well, which the United States refused to do.

The United States has contended that the scientific uncertainty of the anthropogenic contribution to global climate change is too great to warrant expensive abatement. While there is disagreement as to what the desirable reduction of carbon dioxide emissions should be, there is agreement among nations that a market incentive approach, such as a carbon tax, is the desirable way of accomplishing emissions reduction.

CARCINOGEN

See Cancer.

CARCINOGENS AND TOXINS, NATURAL

Natural toxins are chemicals that are produced by organisms for the purpose of harming other organisms,

often in defense against predators. Natural toxins are ubiquitous in the environment and abundant in human diets. All plants, for example, produce toxins to protect themselves against fungi, insects, and animal predators such as humans. Tens of thousands of these natural pesticides have been discovered, and every species of plant contains its own set of different toxins, usually a few dozen in each species. When plants are stressed or damaged, e.g., during a pest attack, they increase their levels of natural pesticides manyfold. Many plant toxins, even in low doses, are acutely toxic to humans, which is why only a small subset of plants is edible for humans. Bacteria and fungi also produce toxins, some quite potent in humans. A wide variety of insects, reptiles, and fish produce venom to deter predators and/or capture prey. Virtually any tissue can be a target of at least some natural toxins.

In laboratory animal cancer tests, rats or mice are given the maximum tolerated dose of a chemical (often hundreds of thousands of times above the weight-adjusted dose that humans usually receive). This high dose is used so that a carcinogenic effect can be detected without using a large number of animals. Half of all chemicals tested—natural and synthetic alike—turn out to be carcinogenic at these high doses. The reason for this high level of positive results may be that the high dose itself is commonly toxic enough to induce cells to divide, and cell division is a risk factor for carcinogenesis. The same chemical may pose very little risk when used at low doses (i.e., realistic exposures), where cell division is not induced.

Examples of Natural Human Carcinogens

Many natural plant toxins have been implicated in human cancer. For example, the mold toxin aflatoxin, present in many grains and nuts, is a potent liver carcinogen. The toxins of betel nuts, which are chewed with tobacco in various countries of the world, are implicated as a cause of oral cancer. The phorbol esters present in some folk remedies and herb teas are thought to cause nasopharyngeal cancer in China and esophageal cancer in Curaçao. The pyrrolizidine toxins found in comfrey tea, various herbal medicines, and some foods cause liver cancer in rats and possibly in humans. Many of the common elements present in food at low doses (e.g., salts of lead, cadmium, beryllium, nickel, chromium, selenium, and arsenic) are rodent or human carcinogens at high doses. Certain natural and necessary hormones are also carcinogenic because they induce chronic cell division; high levels of estrogen, for example, are associated with increased risk of cancers of the breast, uterus, and ovaries.

Ames et al. have estimated that Americans eat approximately 1,500 milligrams (mg) per person per day of natural toxins, equivalent in weight to about 5 standard aspirin. By contrast, the Food and Drug Administration (FDA) in 1988 assayed food for residues of the 200 synthetic chemicals thought to be of greatest importance, including most synthetic pesticides and a few industrial chemicals; the results show that the human intake of these residues averages only about 0.09 mg per person per day. In addition, Americans ingest about 5,000 to 10,000 different natural pesticides and their breakdown products annually. Roughly half of these natural plant pesticides (27 out of 52 tested) are carcinogenic in rodents at high doses. It is probable that almost every plant product in the supermarket contains naturally occurring rodent carcinogens.

Nevertheless, the consumption of more fruits and vegetables is the most effective way to lower one's cancer risk, other than refraining from smoking, because these foods also contain anticarcinogenic vitamins and antioxidants. Studies have shown that people who eat a large number of fruits and vegetables (at least five pieces per day) are at significantly lower risk for developing most types of cancer, including cancers of the lung, larynx, oral cavity, esophagus, stomach, colon and rectum, bladder, pancreas, cervix, and ovary.

The cooking of food is also a major dietary source of potential carcinogens. Cooking produces about 2,000 mg per person per day of burnt material that contains many known rodent carcinogens. Roasted coffee, for example, is known to contain 826 volatile chemicals and several hundred nonvolatile chemicals. Of these, 26 have been tested so far, and 19 are rodent carcinogens. Nevertheless, there is no clear evidence that coffee poses a significant risk to humans.

Defenses Against Toxins

A significant proportion of an animal's resources is directed toward defense against toxins. Most vertebrates have elaborate mechanisms to detect, avoid, detoxify, and eliminate toxins. Such defense mechanisms include olfactory and gustatory chemoreceptors for detecting toxicity; nausea, vomiting, and subsequent food/odor aversions for avoiding or expelling toxins; the continuous shedding of those cells most extensively exposed to toxins (the cells of the surface layers of the mouth, esophagus, stomach, intestine, colon, skin, and lungs are discarded every few days); and the induction of a wide variety of general detoxifying enzymes in the liver and various other organs. The co-evolutionary race between toxin-secreting organisms and their predators has led to a huge spectrum of different toxins and a variety of anti-toxin strategies.

Natural and synthetic chemicals share the same general toxicological and biochemical properties. Animal defenses that evolved to counter toxins are mostly of a general type, as might be expected, since the number of potentially toxic natural chemicals is so large. Humans are thus generally well buffered against exogenous toxins, whether natural or synthetic.

BRUCE N. AMES

For Further Reading: B. N. Ames, M. Profet, and L. S. Gold, "Nature's chemicals and synthetic chemicals: Comparative toxicology," in *Proceedings of the National Academy of Sciences, USA* (1990); R. F. Keeler and A. T. Tu, eds., *Handbook of Natural Toxins: Plant and Fungal Toxins*, vol. 1 (1983); G. A. Rosenthal and D. H. Janzen, eds., *Herbivores: Their Interaction with Secondary Plant Metabolites* (1979).

CARRYING CAPACITY

Carrying capacity is the number of organisms of a particular species—whether human, animal, or plant—that an area (or the Earth) can support without irreversibly reducing its capacity to support them in the future at the desired level of living. From an ecological point of view, carrying capacity is a powerful constraint to population growth, if not an absolute determinant. From an economics point of view, carrying capacity for humans is subject to endless expansion through technology and policy interventions.

Carrying capacity is thus a function of many interacting factors including food and energy supplies and ecosystem services (such as provision of fresh water and recycling of nutrients). For humans, additional factors include lifestyles, social institutions, political structures, and cultural constraints. Carrying capacity is ultimately determined by the component that yields the lowest capacity.

The Earth's human carrying capacity is dependent on the degree of sustainable development achieved. There is evidence that human numbers with their consumption of resources, plus the technologies deployed to supply that consumption, are often exceeding carrying capacity already. The World Hunger Project has calculated that the planetary ecosystem could, with present agrotechnologies and assuming equal distribution of food supplies, sustainably support no more than 5.5 billion people even on a vegetarian diet. The global population reached 5.5 billion in 1992. If humans derived 15% of their calories from animal products, as do many people in South America, the sustainable population would decline to 3.7 billion. If they gained 25% of their calories from animal protein, as is the case with most people in North America, the Earth could support only 2.8 billion people.

It is possible that new breakthroughs in food-production technologies could increase these numbers. Present technology brought about a "Green Revolution" between 1950 and 1984. During that time there was a

2.6-fold increase in world grain output, representing an average increase of almost 3% per year, and raising per-capita production by more than one third. But since 1985 there has been far less annual increase, even though the period has seen the world's farmers investing billions of dollars to increase output, supported by the incentive of rising grain prices and the restoration to production of idled U.S. cropland. Crop yields have plateaued; it appears that plant breeders and agronomists have, at least temporarily, exhausted the scope for technological innovation. The 1992 harvest was little higher than that of 1985, but there were an additional 625 million people to feed. While world population has increased by almost 13%, grain output per person has declined by nearly 9%.

An example of reaching the limits of carrying capacity on a national scale is Kenya, whose present population of 25 million people is projected to expand to 113 million by the time zero growth is attained in the 22nd century. Even if the nation were to employ Western Europe's high-technology agriculture, it could not support more than 52 million people off its land resources, and even if it were to achieve a birth rate of two children per family forthwith, the population would still double because of demographic momentum (52% of Kenyans are age 15 or under, meaning that proportionately large numbers of potential parents already exist). So Kenya will have to depend on steadily increasing amounts of food from outside to support itself. But in part because of its high population growth rate, 3.7%, its per-capita economic growth remains low at only 1.9%. Moreover Kenya's terms of trade have been declining, meaning the country faces the prospect of diminishing financial reserves to purchase food abroad. Its export economy will have to achieve as-yet-unattained high levels if the nation is to be able to buy enough food to meet its ever-growing needs.

Kenya shows many signs, then, of already exceeding its carrying capacity. Preventive measures are most effective early, when population starts to grow rapidly. For Kenya today, an immediate and vigorous effort to slow its population growth to achieve the two-child family in 2010 instead of the projected 2035 could hold its population to 72 million, 41 million less than its present trend indicates.

NORMAN MYERS

For Further Reading: Robert S. Chen, W. H. Bender, R. W. Kates, E. Messer, and S. R. Millman, *The Hunger Report: 1990* (1990); Paul R. Ehrlich, G. C. Daily, A. H. Ehrlich, P. Matson, and P. Vitousek, *Global Change and Carrying Capacity: Implications for Life on Earth* (1989); Robert S. McNamara, *A Global Population Policy to Advance Human Development in the 21st Century* (1991).

See also Appropriate Technology; Biotechnology, Agricultural; Ecological Stability; Ecology as a Science; Family Planning Programs; Green Revolution; Population; Sustainable Development.

CARSON, RACHEL LOUISE

(1907–1964), American scientist and writer. Rachel Carson's role as "mother of the environmental movement" is based mainly on the impact of her last and most famous book, *Silent Spring* (1962), in which she changed course from her earlier books on the sea to one touching the general public in everyday life. After teaching readers about the natural world as an amazingly intricate and awesome habitat, she showed them the vulnerability of the closely interwoven ecology of the Earth through the global impact of reckless pesticide use.

Rachel Carson was in the older scientific tradition of the naturalist, seeing the whole through understanding the details as they form the larger picture. The desire to bring this comprehension to others—from fellow scientists to the general public—requires a kind of writing talent inherent in literature rather than technical papers. Her education prepared her for both. Her undergraduate studies were in writing and zoology, and her graduate focus was marine biology. Her years in the U.S. Fish and Wildlife Service (1935–52) honed her skill in transforming research results into clear reports significant to scientists and understandable by the public. By her own writing and editing of other studies, as Editor in Chief of the Service, she educated a generation of colleagues in scrupulous accuracy and literate presentation.

Rachel Carson chose substantial and difficult subjects for her unofficial writing. *Under the Sea Wind* (1941) and *The Sea Around Us* (1951) brought the latest knowledge of the oceans into syntheses both all-encompassing and enthralling. The extraordinary worldwide success of *The Sea Around Us* permitted her to resign from the government to devote herself to writing. *The Edge of the Sea* (1955) further assured her high standing with fellow scientists, writers, and the general public.

Those with Rachel Carson's ability to be alert to the frontiers of scientific thought while seeing the need to put this understanding into universal language include some of her colleagues and mentors such as William Beebe, Aldo Leopold, Raymond Dasmann, Archie Carr, and Marston Bates. Though they each had an appreciative audience, her remarkable capacity to reach a worldwide following gave a far greater range to the view she shared with them of humanity's place in nature, and people's responsibility to their fellow creatures.

Early recognition of the adverse effects of the wide use of new chemical pesticides at the end of World War II came from research on fish and wildlife. This was well known to Rachel Carson. As evidence mounted, use skyrocketed, up to some 638 million pounds of active ingredients annually in the United States by 1960. The urgency of the situation was clear to those close to the problems. Trying to counter these fast-growing hazards meant opposing strong commercial interests in this country and internationally. The public had been taught that this new large-scale use of poisons was the key to greater agricultural production and to curtailing other pest problems. To make an effective portion of the public realize the grave dangers, someone of known reputation for scientific integrity was essential. From her previous writings she had the standing to assure respectful attention, and the independence to withstand the inevitable attacks. Seeing the situation clearly, and unable to enlist anyone else for the task, she put aside other plans. The evidence was scattered and complex, and her explanation had to be lucid and compelling. First it was necessary to give the public a firm grasp of the *ecology* of the Earth (making the word and concept generally known for the first time), and then to show how these new assaults damaged essential elements in the web of life.

The book galvanized both the public and their governments, bringing a new impetus to conservation (soon called the environmental movement) with innovative new laws and agencies created. Many private groups were organized. The influence of *Silent Spring* has continued to grow, with an ever-widening circle of public understanding and vastly increased research. It sold better in 1992 than in any year since that of its first publication. *Silent Spring* also gave a basis for confronting problems of similar world contaminants: other untested chemicals and pollutants and nuclear hazards. Carson questioned attitudes and practices of modern industrial society in relation to the natural world, providing the basis for a sounder ethical and economic course for the future.

As the importance of Rachel Carson's work is increasingly realized, many find inspiration in her example of unstinting, difficult work in the face of both public attacks and personal burdens. Her determination to finish *Silent Spring*, an unpredictable task in uncharted ways, brought her through extraordinary trials. She finished the book and saw its first impact before she died of the cancer she had battled in the last years.

SHIRLEY A. BRIGGS

For Further Reading: Paul Brooks, *The House of Life: Rachel Carson at Work* (1972); Frank Graham, *Since Silent Spring* (1970); Douglas H. Strong, *Dreamers and Defenders: American Conservationists* (1988).

CERCLA

See Superfund.

CFCS

See Ozone Layer; Pollution.

CHANGING COURSE

A substantial part of the book *Changing Course: A Global Business Perspective on Development and the Environment*, by Stephen Schmidheiny, Chairman of the Swiss conglomerate UNOTEC, and the Business Council for Sustainable Development (originally 50 business leaders of diverse nationality), is devoted to a collection of 38 case studies chosen to illustrate seven themes: management leadership, industrial partnerships, stakeholder cooperation, finance, cleaner production, cleaner products, and sustainable resource use. The cases are taken from around the world and from companies of all sizes and are offered as a "snapshot of current best practice," not as an ideal. The cumulative picture that results is less a detailed blueprint than an inspirational statement of possibilities, consistent with limited experience and few hard and fast rules. It is a reflection of the existing lack of precise definitions and goals—*why* corporate environmental practices have to change rather than *what* changes are needed and *how* they can be accomplished.

While described as success stories, the case studies illustrate some of the difficulties created by attempting to put sustainable development theory into corporate practice. In one case, an environmentally minded furniture company incurred large costs in attempting to identify a reliable supplier of sustainably harvested teak and was forced to drop several product lines. This example is nevertheless described as a success because the company avoided a consumer backlash and received an environmental award. In a similar vein, a credit company is included because of its development of a detailed system for protecting itself against investments that might bring exposure to hazardous waste liability.

Valuable examples from developing countries illustrate that these themes are not unique to the industrialized countries. A company producing rayon in India, for example, was forced to become more efficient because of increases in the cost of wood pulp—a key ingredient in rayon—due to deforestation. In six years, a series of internally generated ideas for improving efficiency and reducing costs allowed a 20% increase in production simultaneous with significant reductions in energy use, chemical consumption, and emissions. The economics were so favorable that a senior corporate executive referred to the need for pollution control measures as "a blessing in disguise." Despite its efforts, the company was subjected to more stringent government regulation after local residents brought a court case challenging odors and colored wastewaters, demonstrating that the best of environmental intentions do not necessarily insulate firms from public pressures.

Significantly, many of the case studies indicate the potential for reducing costs as well as pollution. The result is "a potent source of competitive advantage." Indeed, the nexus between environmental concerns and corporate planning verges on the inevitable; adherents will succeed, while "Companies that fail to change can expect to become obsolete."

A strength of the book is that it focuses squarely on business roles and responsibilities. The authors repeatedly point to internal factors as the greatest obstacles to innovative pollution prevention. A successful pollution prevention program, they assert, is not solely or even primarily a matter of technology, and "ultimately comes down to desire." "Eco-efficiency" requires "profound changes in the goals and assumptions that drive corporate activities, and change in the daily practices and tools used to reach them . . . a break with business-as-usual mentalities and conventional wisdom that sidelines environmental and human concerns."

The authors touch on many of the themes and strategies developed by the World Resources Institute and other experts working on means for implementing sustainable development, from full cost pricing to the revision of national income accounts to reflect environmental degradation. As members of the business community, it is not surprising that they emphasize the importance of free markets, the elimination of government subsidies, and the use of more flexible regulatory instruments. They avoid some potentially divisive issues, such as the possibility that resource constraints might eventually require moving beyond simply reducing packaging to avoiding consumption of some goods and services altogether.

Prior to the Earth Summit (as the authors note) the "prevailing view of the links between business and the environment was that environmental protection and profitability are natural opposites." Perhaps due partly to *Changing Course*, the evidence for this proposition is no longer so convincing. From the pages of leading business journals to the World Bank's 1992 *World Development Report*, the fundamental importance of sustainable development to successful economies is being increasingly trumpeted. The debate is shifting to even more difficult questions about what we mean by sustainable development and its implications for how business should be conducted.

Some companies have begun to experiment with concepts such as "design for environment" that consider the environmental impacts of production and use as integral features of product design. Sun Oil had to modify the Valdez principles—standards for corporate practice advocated by environmentalists—which, read literally, could be construed to preclude oil production, before accepting them. Other companies have found that making life-cycle comparisons of the impact of alternative products is far from an established science. For example, even a meaningful comparison of the environmental consequences of disposable and cloth diapers is extremely complicated. Numerous assumptions must be made about the energy used to transport and wash cloth diapers and about the disposal method used for plastic diapers.

The bold vision outlined in *Changing Course* is a future in which business leads rather than follows environmental regulation and market forces naturally drive production toward a truly sustainable global economy. If we have truly begun to move in this direction, the authors of *Changing Course* should be given credit for taking some of the first steps.

ALAN S. MILLER

For Further Reading: Frances Cairncross, *Costing the Earth* (1991); Donald Huisingh and Institute for Local Self-Reliance, *Proven Profits From Pollution Prevention: Case Studies in Resource Conservation and Waste Reduction* (1986); Stephen Schmidheiny with the Business Council for Sustainable Development, *Changing Course: A Global Business Perspective on Development and the Environment* (1992).

CHAPARRAL

The term *chaparral* refers both to a specific kind of vegetation and to the biome characterized by that vegetation, with its related abiotic and biotic factors. Chaparral vegetation consists of small-leaved evergreen shrubs, bushes, and small trees. Because of the harsh climate conditions the plants rarely exceed 2.5 meters in height, although dense thickets tend to form.

In this biome the hot, dry summers permit limited or no growth in vegetation; the cooler, wet winters are the primary growth season. The few changes in vegetation that occur during the course of the year include very bright, colorful springs in which plants flower and produce their fruit before the onset of summer. Leaves also change color from a dull green to a brighter green at the beginning of winter. The dry summers make the vegetation prone to fire, which can sweep across large areas of chaparral. Fires occur about once every ten to forty years. Most chaparral plants regenerate from surviving roots, and many species require fire for germination. Chaparral can return to normal density and growth after about ten years. More frequent burning, which is a practice in some methods of agriculture, can convert chaparral into grasslands.

Except for occasional appearances in the wet season, medium-sized and large browsing mammals (such as deer) do not find chaparral hospitable because of the many unpalatable herbs and shrubs and the difficulty of passage through the thick vegetation. Small animals, such as lizards, rabbits, chipmunks, and insect-eating birds, are year-round residents, although conditions are difficult to tolerate.

California chaparral generally occurs on the wet slopes of the coastal mountains, below an altitude of 1,000 meters and wherever the average yearly rainfall is between 250 and 500 millimeters. The most common shrubs in California chaparral are sages, mountain lilacs, desert buckwheat, and scrub oak. At higher altitudes, precipitation is greater and the vegetation generally changes into fir and pine forest. Chamiso, or ribbonwood, grows most commonly where the soil is drier.

Agriculture first began near the Mediterranean chaparral, or maquis, about 10,000 years ago. Though human activities have removed a large amount of this chaparral, much of the vegetation still remains. When left alone, chaparral is generally succeeded by trees and turns into forest. The Australian chaparral, or mallee, contains the greatest diversity of vegetation, as compared with other scrublands; it has about 1,000 shrub species and large eucalyptus forests. Chaparral is also found in central South America and the southern tip of Africa.

CHEMICAL INDUSTRY

The substances and properties of nature have been used to create new materials since early humans discovered they could use fire to alter certain characteristics, e.g., harden pottery and refine metal ores. Their elementary experimentation with chemical reactions created many new products. Later, Egyptian and Roman alchemists developed chemical processes for preparing medications and potions. These early "chemists" learned to make glass and began a leather tanning industry. By the late Roman period, soapmaking, natural pigment preparation, the production of gunpowder, and other relatively simple industrial processes were underway.

By the 18th century, chemical manufacturing was still carried out on a relatively small scale. It consisted primarily of the purification of inorganic substances such as salt, sulfur, mercury, and arsenic; the manufacture of sulfuric and hydrochloric acids; and the preparation of ethyl alcohol by fermentation. Sulfuric acid was produced in far larger amounts than all of the other substances combined. In the early 1800s, large-scale production of alkali (also called soda ash or anhydrous sodium carbonate) commenced; this is regarded by many as the beginning of the chemical industry. Alkali was used to manufacture soap, glass, and other chemicals. The operation of alkali manufacturing plants in England, France, and Belgium in the 1800s produced sizable amounts of alkali and also resulted in the formation of wastes, which were readily dumped nearby.

In almost every chemical manufacturing process, when two or more starting materials are mixed under carefully controlled conditions to form a desired product, undesired waste materials also form. The small volume of wastes gathered during the industry's early growth coupled with a lack of concern for worker health resulted in unrestricted effluent flow to the environment.

An example of hazardous workplace exposure and environmental contamination is the early process for manufacturing alkali. Initial alkali production required mixing salt (sodium chloride) and sulfuric acid to form sodium bisulfate. This created a waste, hydrochloric acid, which was continuously released into smokestacks and into the air. Exposure to this gas rotted workers' teeth and frequently caused bronchitis. Hydrochloric acid emissions seriously damaged local crops and caused illness to livestock. Another step in the process created tons of black ash. Some production sites had almost one million tons of this sulfide-smelling ash piled on-site. Noxious fumes were released with little concern for adverse health consequences. Many years later, when the value of hydrochloric acid was discovered, it was collected as a marketable by-product and its release into the environment significantly decreased. The process for manufacturing alkali was changed to make it more efficient and noxious releases from the plants subsided.

Soon sulfuric acid production grew, as did the manufacture of other inorganic products. In the early 1800s there was very limited understanding of organic chemistry. (Organic chemicals are carbon-containing chemicals.) The only organic chemicals manufactured on a commercial scale were ethyl alcohol, acetic acid, and charcoal. The textile industry was still using expensive natural pigments because synthetic colors were unknown.

In most large cities lighting was achieved by using illuminating gas made from coal tar. In 1815 benzene was prepared from coal tar; this discovery created the organic chemical industry. In 1828 urea, another organic chemical, was made from two inorganic substances. Soon after, aniline was produced synthetically as was the aniline dye mauve (also called aniline purple). By the early 1900s German chemists had developed manufacturing processes for several dyes based on aniline chemistry. With these new processes came new chemical products and new process wastes. Apparently no effort was made to assess the toxicity of these products or wastes to humans or the environment. Decades later, several aniline-based dyes and azo dyes were found to cause cancer and other diseases.

While chemical manufacturing in Britain and Germany rapidly expanded, the American chemical industry was slow to develop. In the early 1800s Du Pont was established to make gunpowder. By the late 1800s several American coal tar dye companies were founded. Dow Chemical began in 1892 to make chlorine bleach. In 1901 Monsanto was established to produce saccharin from coal tar. Other companies followed in the U.S. and Europe.

World War I created a need for large amounts of many chemicals and this triggered the explosive growth of the chemical industry that found increased production demands to be of paramount importance. Little thought was given to the disposition of chemical wastes and plant effluents. There was little evidence that synthetic chemicals were hazardous to human health or to the environment.

By the end of World War I, electric lights had become popular and the production of illuminating gas, and its byproduct coal tar, dropped dramatically. A byproduct from the steel industry, coke (partially burned coal from steel ovens) became the replacement for coal tar and provided a valuable source of other organic chemicals (e.g., toluene, cresols). The many tons of waste from processing coke were buried in abandoned mines, dumped in rivers, and hauled to the ocean for disposal. Many volatile process effluents were collected as products while others were released to the atmosphere as byproducts.

In 1917 Union Carbide was created by the merger of several companies to examine the possibility of obtaining organic chemicals from natural gas and petroleum. This was the start of the petrochemical industry. In the 1930s the development of catalytic cracking of natural gas and petroleum, a process used to separate chemicals by temperature and pressure, provided a means for isolating many new organic chemicals.

Refined petrochemicals include gasoline, oil, waxes, antifreeze, solvents, and plastics.

The discovery of synthetic fibers created other major products and processes within the chemical industry. Initially, cellulose was converted to rayon and acetate. This was followed by the synthesis of nylon. Other fibers, including polypropylene, spandex, polyester, and acrylics soon followed. Since almost all of the raw materials used to manufacture synthetic fibers are derived from petroleum, the petrochemical industry continued to expand to meet these ever-growing needs. Many of these synthetic organic chemicals developed during the early 20th century readily degrade in the environment.

In the mid-1900s a new class of chemicals was commercialized. Many of these halogenated chemicals (e.g., chlorides, bromides) accumulate in animal tissues and persist in the environment for many years. At the time of their development, their benefits were thought to clearly outweigh any potential adverse effects. Highly chlorinated pesticides such as chlordecone, DDT, and chlordane were developed and widely used. Many years later these chlorinated pesticides were found to bioaccumulate in food-chain animals and in humans, and were shown to be toxic to certain species. Their continued use was then banned.

The polychlorinated byphenyls (PCBs) were developed as flame retardants, plasticizers, and insulating fluids for electrical transformers. Their major attribute, chemical stability even under severe conditions, has resulted in bioaccumulation in humans and other animal species, and in widespread environmental contamination. Several industries dumped thousands of tons of PCBs for many years into the Great Lakes and major rivers, resulting in widespread contamination that has caused local fishing industries to close, water supplies in many towns to become hazardous, and apparent health problems to develop.

The introduction of new chemicals has occasionally resulted in new and unexpected occupational illnesses. For example, polyvinyl chloride (PVC), a plastic polymer made by binding together vinyl chloride monomer molecules, is used to manufacture pipes, shower curtains, floor tiles, and many other products. When PVC was first manufactured, many workers developed mesothelioma, a rare liver cancer. A lengthy investigation showed its cause was exposure to vinyl chloride monomer. Now workers are protected during the manufacture of PVC. Other occupational diseases occurred as the chemical industry grew. Dye workers frequently developed bladder cancer; an investigation found a correlation between occupational exposure to 2-naphthylamine and to benzidine, and the development of bladder cancer. Even though these correlations were shown in the 1930s, worker exposure to these carcinogens continued into the 1960s. The 1974 OSHA Standard established procedures for the handling of cancer-causing chemicals and mandated worker protection, forcing the dye and pigment industry to change their manufacturing processes to minimize worker exposure to these and other recognized human carcinogens.

Some chemical manufacturers recognized the need to determine the toxicity and potential environmental effects of their products and began examining these issues in the 1950s. However, product safety assessments were not commonplace until the 1960s. The largest chemical manufacturers established toxicology facilities to accurately assess the hazards associated with the manufacture, use, and disposal of their products. Even with the availability of company laboratories and external contract testing facilities, companies primarily tested only those chemical products for which safety assessments were required (i.e., pesticides, food additives, pharmaceuticals).

As recently as the 1960s, it was common and permissible to bury chemical wastes on-site or in regional dumps. Little thought was given to the possibility that buried chemicals might leach through soil and contaminate aquifers from which towns derived their drinking water, or that volatile emissions from chemical dumps could make neighborhoods unsafe. Regulations were slow to develop and the chemical industry gave environmental concerns a low priority. As the public became aware of the hazards associated with chemicals and their byproducts, the industry lost credibility. The image of a large industry polluting the environment, threatening human health, and selling unsafe products went unchallenged for many years.

Several efforts were undertaken to enhance the industry's image. Almost all chemical manufacturers in the U.S. belong to the Chemical Manufacturers Association (CMA). The CMA made several efforts to regain public confidence through articles, advertisements, and other means. Voluntary cleanups and pollution control efforts were not evident to the public and the industry's image remained poor. When the U.S. EPA was created in 1970, laws were passed to protect the environment and the EPA became the primary government agency for enforcement. The Clean Air Act and Occupational Safety and Health Act of 1970, The Federal Water Pollution Control Act of 1972, and the Toxic Substances Control Act of 1976 gave EPA the authority to regulate the environment and the industry responsible for its contamination. The EPA identified and publicized many buried waste sites existing throughout the country.

In 1985 the Canadian chemical industry implemented a program to regain public trust. It was adopted by the U.S. chemical industry in 1988 and administered by the CMA. Called Responsible Care, the program

had been accepted by all 175 CMA members. Through Responsible Care, the chemical industry has made a firm commitment to significantly improve its interactions with the environment and the public. All CMA member companies must adhere to ten principles of conduct. These include recognizing and responding to community concerns about chemicals; manufacturing chemicals that can be produced, transported, used, and disposed of safely; making health, safety, and environmental considerations a high priority; and operating manufacturing facilities in a way that protects the environment, facility employees, and the public. The full implementation of the program will afford the public an opportunity to reexamine the chemical industry, its actions, goals, and accomplishments, and decide whether greater trust is warranted.

As needs arise for new chemicals, the potential impact of these substances on the environment should be a major consideration. The chemical industry recently committed to substantially cut current effluent levels, improve processes so as to create less effluent, identify methods for safely destroying existing chemical wastes, and to more effectively communicate with local officials and townspeople.

In terms of output, the U.S. chemical industry is the largest in the world ($1,245 billion in 1991), followed by Japan, West Germany, France, and the U.K. According to reports issued in 1991 by the Organization for Economic Cooperation and Development, the chemical industry in industrial countries more than doubled its output since 1970 while its energy consumption per unit of production fell by 57%. In terms of employees, the U.S. chemical industry is by far the largest (1,072,000 in 1991), followed by West Germany, Japan, and the U.K. Chemical industry employment was relatively stable during the 1980s and early 1990s. Between 1981 and 1990 the U.S. share of world chemical exports declined by nearly 5% while other developed nations maintained their export shares. Export share lost by the U.S. has gone to newly industrialized countries (Hong Kong, Taiwan, South Korea, and Singapore) and developing Asian nations. According to the U.S. Council on Environmental Quality, the U.S. chemical industry in 1991 had capital expenditures totaling $2,068 million and operating costs totaling $4,047 million for pollution abatement (air, water, and solid waste).

RALPH I. FREUDENTHAL

For Further Reading: Chemecology (a periodical); Chemical Manufacturers Association, *U.S. Chemical Industry Statistical Handbook* (1992); Allen J. Lenz, *The U.S. Chemical Industry Performance in 1992 and Outlook* (1993); Stephan Schmidheiny with the Business Council for Sustainable Development, *Changing Course* (1992).

See also Bhopal; Clean Air Act; Clean Water Act; Environmental Protection Agency, U.S.; Hazardous Waste; Labeling; Love Canal; Occupational Safety and Health; PCBs; Pesticides; Petroleum Industry; Plastics; Resource Conservation and Recovery Act; Source Reduction; Technology; Toxic Chemicals; Toxicology.

CHEMICALS, TOXIC

See Toxic Chemicals.

CHEMOTHERAPY

The term *chemotherapy* was introduced by the German physician-scientist Paul Ehrlich at the turn of the century to refer to the "use of drugs to injure an invading organism without injury to the host." Ehrlich proceeded to develop several chemotherapeutic agents, or compounds of known chemical composition (drugs) that shared the ability of quinine to injure invading organisms more than the host. Romanovsky had observed in 1891 that quinine selectively damaged the malarial parasite in the blood of treated patients, leading him to conclude that the chemical harmed the parasite more than the host. Previously, the role of microorganisms in causing infections, plagues, and epidemics had been identified by Robert Koch in the late 1800s. His four postulates clearly established a relationship between the host (human) and its parasite (infectious agent), which set the stage for chemotherapy.

Implicit in the definition of chemotherapy is the fact that chemicals can demonstrate *selective toxicity*, which refers to the degree to which a chemical selectively harms a parasite without harming the host. Selective toxicity was defined by Ehrlich in quantitative terms as the *chemotherapeutic index*. This index compares the dose-response relationship for the desired effect of a drug, such as injury to the parasite, with that of its undesired effect, such as toxicity to the host. The difference between the desired effective dose and the dose producing undesirable effects in the host defines the safety and degree of selectivity for chemotherapeutic agents in particular, and for all drugs in general. Virtually all drugs are potentially toxic at some elevated dose. Therefore, the utility of a drug is a function of the degree to which it is selective. When the host and its parasite are markedly different biologically, such as humans and bacteria, the development of drugs with high selectivity is easier, as evidenced by the penicillins. This situation is in sharp contrast to instances in which the "parasite" resembles or is actually derived from host tissues, such as in most cancers. One of the mechanisms by which drugs produce their selective effects, also predicted by Ehrlich, is due to the existence of specific sites, or receptors, on parasites or on tissue cells that possess a high affinity for the drug molecules.

J. ARLY NELSON

CHERNOBYL

On April 26, 1986, a nuclear accident occurred at the Chernobyl Nuclear Power Station in the Ukraine in which roughly 7 tons of radioactive fuel were blown out of the building. The accident occurred when operators who were evaluating the reactor's response to a loss of electrical power reduced the reactor's output from 3,200 to 1,600 thermal megawatts. After 12 hours, the power was reduced and then suddenly fell to 30 megawatts. Although at least 30 control rods should have remained in the core to ensure proper control, to increase power 204 of the 211 control rods were withdrawn, leaving only seven control rods to prevent a runaway chain reaction.

The impact was catastrophic and almost immediate. Forty seconds after the control rods were removed, an uncontrolled power surge overheated the reactor. Due to steam vapor lock in the fuel channels, adequate cooling was not available. The excess heat caused the fuel rods and cooling tubes to shatter, allowing the uranium fuel to react with the water coolant. The high-pressure steam released blew off the entire top of the reactor, destroying the upper part of the building. The exposure of graphite to the air led to a continuous graphite fire, creating a thermal column of smoke over the reactor.

Scientists immediately began searching for the missing uranium fuel. At extreme risk to their health, they probed the rooms below the reactor and discovered large, highly radioactive masses of sand and fuel. From these and other discoveries over the ensuing three years, scientists determined that the explosion had dislodged sand packed around the reactor and had caused the reactor base to drop four meters. The nuclear fuel combined with the sand, burned through the concrete floor, and spilled into the rooms and corridors below the reactor—a meltdown had occurred. When the mixtures cooled, the fuel stabilized. The damaged reactor, which will be radioactive for at least 100,000 years, is now enclosed by steel-reinforced concrete—the "sarcophagus."

Thirty-one people are officially reported to have died from direct effects of radiation due to the explosion at Chernobyl. Because of possible exposure to high levels of radiation, 135,000 residents of the surrounding countryside were evacuated following the accident.

The 600,000 emergency workers involved in the immediate cleanup, containment, and reconstruction effort were exposed to extreme radiation levels, in some cases ten times the recommended maximum dose. Since the accident, more than 200 cases of radiation sickness have been reported. There has been an increased incidence of thyroid cancer among children who ingested radioactive milk immediately after the accident.

Over twenty countries were exposed to radioactive particles from the accident. Because of prevailing weather patterns, Scandinavia, Eastern Europe, and Southern Europe suffered the greatest exposure outside of the former Soviet Union. The radioactive materials that were released included radioactive cesium and iodine, both of which are carcinogenic at high doses, capable of causing leukemia, cancer of soft tissue, and thyroid cancer. Due to the millions of acres of heavily contaminated land, the contaminated Kiev Reservoir, and consumption of contaminated produce, low-level exposure to radiation from the accident still occurs. In addition, the 800 sites of buried radioactive debris are leaching radioactive materials into the groundwater. However, since scientists disagree about the long-term effects of exposure to low doses of radiation, estimates of the total number of cancers that will result worldwide from the Chernobyl accident vary from a few thousand to more than a million.

CHINA SYNDROME

See Nuclear Reactor.

CHLORINATION

Chlorination is a process used to treat water to make it safe from waterborne microbial diseases. This process is used to treat both drinking water and sewage by adding various forms of chlorine to water. The widespread use of chlorine for disinfection of municipal water supplies in the early 1900s closely parallels the decrease in microbial diseases such as typhoid and cholera in the United States. The decrease in typhoid deaths in Philadelphia following installation of chlorination facilities in 1913, and the subsequent documentation of the correlation between the incidence of waterborne disease and inadequate chlorination by Abel Wolman in the 1930s, provided dramatic evidence of the link between protection of public health and chlorination. Today, disinfection of sewage prior to discharge into streams and lakes is used to reduce the potential for disease transmission due to whole-body contact.

Chlorine is a strong oxidizing agent and is therefore very reactive. As a result, chlorine is dangerous to workers and must be handled carefully. Chlorine is most economically applied to water as chlorine gas, where it reacts to form hypochlorous acid (HOCl). In this form, it is a potent disinfectant, typically reducing waterborne bacterial and viral concentrations by several orders of magnitude at concentrations of less than 1 part per million. Hypochlorous acid also reacts with ammonia to form chloramine, a less potent but much more stable disinfectant, which therefore persists much longer than hypochlorous acid.

The use of chlorine is not without controversy, because it reacts with organic matter in water to create a series of chlorinated byproducts that are potential human carcinogens. At present, the trihalogenated methanes (THMs), primarily chloroform, are regulated, but soon other disinfection byproducts (DBPs) will also be limited. This regulation is expected to result in changes in disinfection practice, including the use of alternate disinfectants, such as ozone or chlorine dioxide. However, these disinfectants also have health-related concerns. Although the use of chlorinated ammonia (chloramines) greatly reduces the formation of chlorinated organic byproducts, it is much less effective at killing pathogenic microorganisms in water.

Chlorination—which has for 75 years provided developed countries with safe drinking water—is now being reevaluated, and changes in the practice of water disinfection will continue to occur. However, because of its proven record of preventing waterborne disease, it is unlikely that an acceptable substitute will be found that will be as widely used as chlorine.

JOHN T. NOVAK

CHLORINE

See Pollution, Air; Water Supply, Municipal.

CHLOROFLUOROCARBONS

See Ozone Layer; Pollution.

CHLOROMETHANE

See Pollution, Air.

CHRISTIANITY

Christianity is firmly based upon belief in a creator God who has created all life for a purpose, not by chance. Within Christianity, the purpose of God's creation is differently understood. In the Early Church, which came into existence after the death and resurrection of Jesus Christ ca. 30 A.D., the world was not of great importance, because they expected Jesus to return at any moment and the old world to pass away. Within a generation of Christ's death, this hope had faded and the churches began to come to terms with the world as it was. Many of these churches, formed in the late first to mid third century A.D., were fundamentally infuenced in their understanding of Christianity by Greek philosophical thought in that they saw God as a remote creator who had made himself accessible through Jesus Christ and present in the world through the Holy Spirit. Ideas in the Old Testament of God's compassion, love, and celebration of the rest of creation were largely ignored. The human's relationship with God was central, and the role of the rest of creation was at best peripheral.

With the collapse of the old Roman Empire and the rise of the Church as a religious and secular power, attitudes in Europe began to change. St. Augustine (354–430), who was deeply influenced by the dualism of the gnostics and Manichaeans, saw the spiritual world as infinitely superior to the material world. This dualism—the spiritual equaling good and the material equaling bad—was not an inherent dimension of Christianity. Indeed, Christianity, with its belief in the incarnation of God in human form in Jesus Christ, has at its center a material-world affirming vision. But through the gnostic influences of St. Augustine, anti-material ideas gained credibility and led to the development of a dualistic approach within Christianity. At the same time, other approaches were being developed. The rise of monasticism in the fifth century integrated concern for the spiritual and the material. The siting of monasteries away from urban centers and the reliance of the monks and nuns on agriculture soon produced a form of Christianity that integrated the spiritual pursuits of Christianity with a deep knowledge of and love for nature. This had two very important manifestations during the sixth and seventh centuries whose influence is being rediscovered today.

Starting in Italy but soon spreading across France, Britain, and other areas of Western Europe, the Benedictines fused prayer and work into a powerful tool for revitalizing the environment. Much of Western Roman Europe had been left environmentally exhausted by the demands made upon it by the agrobusiness of the late Roman Empire. Into this earliest of Europe's environmental crises came the Benedictines. By a careful process of reforestation, landclearing, canals, creation of lakes, and husbanding of appropriate animals, the Benedictines restored large areas of degraded land to proper agriculture and began to stabilize the natural environment. The agriculture and way of life established by the Benedictines became the norm for traditional agriculture in Europe, the loss of which we are now mourning. For at the heart of Benedict's vision was the belief that humanity had been put on Earth to cultivate and care for it, derived from Genesis 2:15.

In Ireland, Scotland, Northern England, and Wales, the Celtic Church fused together the nature veneration and sensitivity of the former Celtic faiths such as Druidism, with a poetic and deeply creation-centered spirituality from Christian monasticism. Celtic Christianity flourished for over 200 years before being subdued by the Roman Church, though the influence of the Celtic vision of Christianity continued in many ways for centuries afterwards. In Celtic spirituality, the human being is part of the greater purpose of creation which exists to give praise to God and to be the object of God's love.

In other parts of the world, beyond the influence of state religion Roman Christianity, the Churches developed a creation spirituality which reflected the different areas they inhabited. For instance, in Russia, hermits and monks sought refuge in the vast forests and wilderness areas and brought a sense of the sacred back to nature, often in association with older beliefs that they absorbed from the pre-Christian faiths and cultures.

However, in much of Europe, the Church as a state force sought to eradicate other or earlier forms of belief. "Paganism" in particular was vilified and with this vilification came a great fear of the natural world as a place of evil forces and spirits. Nevertheless, this same church created people such as St. Hildegard of Bingen with her creation spirituality and St. Francis of Assisi with his powerful message that all of creation, including human beings, were part of one great family under God. This enabled Francis to speak sincerely of the animals and birds, the elements and the planets as his brothers and sisters, and to address the Earth as Mother Earth.

The next major development came with the Protestant Revolution. Fused with rising nationalism and capitalism, and absorbing the strongly anthropocentric teachings of the Renaissance, Christianity came to see nature as a foe against which humanity was set. The utilitarian understanding of nature that began to emerge from both religious and secular thinkers in the 17th century led to the development of mass exploitation of nature. While it is an overstatement to say Christianity created this attitude—it did not do so in any Orthodox countries, for instance—Christianity in its Protestant forms, with a few notable exceptions, did nothing to prevent the abuse of nature caused by European expansion and the rise of industry.

In recent years, Christianity has had to come to

terms with the environmental crisis. For some this has meant looking back to earlier models such as the Benedictines, Celtic Church, or St. Francis. For others it has meant a willingness to completely reexamine the beliefs of Christianity in order to tell a renewed story of our purpose on Earth. Broadly speaking three responses have emerged. First is dominion. Some Christians believe God has given humans dominion over nature, but argue that this dominion has to be used sensibly. Second is stewardship. We are here to use but also to care for the planet. We should never abuse. Third is blessing or priesthood. We should be a servant, a priest or a blessing to creation. Our role is to be partners both with God and with the rest of creation in continuing life on Earth. It is out of these differing visions that Christianity today is seeking to respond to the environmental challenge.

MARTIN PALMER

For Further Reading: Elizabeth Breuilly and Martin Palmer, eds., *Christianity and Ecology* (1992); Philip N. Joranson and Ken Butigan, eds., *Cry of the Environment* (1984); World Council of Churches, *Searching for the New Heavens and the New Earth: An Ecumenical Response to UNCED* (1992).

CITIZEN SUITS

Citizen suits are legal suits that a plaintiff may bring against a defendant alleged to violate requirements of a law. The alleged violator can include the federal government or any federal agency. Citizen suit provisions generally contain the same principles under most statutes.

The first citizen suit provision in the United States was included in the 1970 amendment to the Clean Air Act. Congress included the provision to augment and prompt federal and state enforcement of statutory violations. In 1972, Congress passed the Clean Water Act, which contained a similar citizen suit provision. Since that time every major federal environmental statute enacted, except for the Federal Insecticide, Fungicide, and Rodenticide Act, has included a citizen suit provision.

Two general conditions must be met before a civil action may be initiated by a citizen. First, a notice of intent to sue of a period of days (depending on the statute) must be given to the Environmental Protection Agency, the state, and the alleged violator. Second, government agencies must not be diligently prosecuting the alleged violator for the same actions. Both of these requirements are meant to prompt and allow for federal enforcement and to avoid redundant litigation.

After a violation of the statute has been proven, the court may award costs of litigation—which usually include reasonable attorney's and expert witness's fees—to any prevailing or substantially prevailing party. This suggests that the defendant may in some cases be awarded costs of litigation. However, this rarely is the case because of the nature of citizen suits under environmental statutes, and of the imbalance of resources, which usually favors the defendants.

Most citizen suits are settled in the form of negotiated consent decrees. These decrees are agreements under the jurisdiction of the court that typically contain injunctive relief, penalties, and attorney's fees as agreed to by both parties. Citizen suit provisions change as environmental laws are amended and reworked by the Congress of the United States.

CLEAN AIR ACT

The Clean Air Act of 1963 was the first national legislation in the United States aimed at air pollution control. The act required the United States Public Health Service to research the sources and effects of air pollution. Federal grants were provided to states whereby they were to set and enforce air quality regulations. In 1965 amendments were added setting national standards for automobile exhausts of carbon monoxide and hydrocarbons. In 1970 an extensive set of amendments were added, ostensibly rewriting the Clean Air Act. The Act—one of the longest and potentially most far-reaching regulatory programs ever to be enacted—was again amended in 1977 and 1990.

The 1970 amendments directed the Environmental Protection Agency (EPA) to establish two types of national ambient air quality standards for seven major pollutants found to be harmful to human health or the environment: sulfur dioxide, carbon monoxide, hydrocarbons, lead, nitrogen oxides, ozone, and particulates. Primary standards were set to protect human health, and an added margin of safety was provided for vulnerable segments of the population, such as children and the elderly. Secondary standards were to maintain visual clarity of the air and to prevent damage to crops, buildings, water, and materials. The deadline for the nation to meet the primary air quality standards was 1982. However, as of 1990, almost all major urban areas in the United States had failed to meet them.

The act also directed the EPA to set maximum emission limits, or new source performance standards, for new or expanding factories and plants on an industry-by-industry basis, taking into account the costs, energy requirements, and environmental effects of the standards. Among the materials regulated by emission standards are asbestos, benzene, beryllium, mercury, polychlorinated biphenyls, and vinyl chloride—most of which have no natural source in the environment. For existing plants, guidelines were to be issued informing the states of pollution control equipment, which could be economically retrofitted to existing technology. The lack of enforcement of the original act is apparent in the 1990 amendments, which call for guidance documents for retrofit controls on existing sources of pollution to be issued within two years.

An implementation plan is required of each state that specifies how federal air standards will be achieved. Federal standards can be used as a guideline for states to develop specific emission limitations for individual plants, but the state limits must match or be stronger than federal standards. The 1990 amendments require that complete state implementation plans be submitted by 1993.

The 1990 amendments sought to close the gap between the goal and the actuality of clean air in the United States. The legislation addressed the following areas of concern: acid rain, urban smog, toxic air pollutants, ozone protection, marketing pollution rights, and workers' compensation.

CLEAN WATER ACT

The Federal Water Pollution Control Act of 1970 has been amended four times: in 1972, 1977, 1981, and 1987. Together with its amendments, this comprehensive and technically rigorous piece of legislation, known as the Clean Water Act, seeks to protect the nation's vast water resources from pollution.

The 1972 amendments comprehensively readdressed water pollution control laws dating back to 1899 and greatly changed the nation's water pollution control policy by, for the first time, placing limitations on effluents and setting water quality standards for surface water. The act's goal was to make the nation's surface waters, such as lakes, rivers, and streams, safe for fish, shellfish, wildlife, and people by 1983, and to eliminate the discharge of pollutants into all navigable waters by 1985.

To accomplish this ambitious goal, the act required that industries discharging pollutants directly into surface water use the "best practicable" pollution control equipment by 1977. More advanced equipment, the "best available technology economically achievable," was to be in place by 1983. Direct dischargers also had to assure that water quality was maintained and that the current use of a body of water (such as for recreation, fishing, or waste disposal) was not diminished by their discharges. Indirect dischargers, or industries that released pollutants into city wastewater systems instead of directly into surface water, were directed

to pretreat their discharge before it was released.

In addition to effluent and discharge limitations, the act required that municipal treatment plants have secondary treatment levels in place by 1977, and that they use the "best practicable treatment technology" by 1983. The act authorized $18 billion in federal construction grants for the states to build wastewater treatment plants that were 75% federally financed, and $24.6 billion to clean up currently polluted waters. A state-run pollution discharge permit program was to be established, and the EPA created specific standards for discharges considered to be toxic.

The 1972 amendments, with their strict requirements for advanced treatment technologies to be in use by certain deadlines, were found to be too rigorous. The 1977 amendments addressed this issue by extending many of the deadlines and allowing for "innovative and alternative" technologies to be used in meeting the effluent standards.

In 1987, the act was amended and strengthened as state management programs for nonpoint sources of pollution were added. This type of pollution, which is diffuse and cannot be traced to a pipe or other specific source, comes from various human activities such as agriculture, mining, urban runoff, construction, and individual sewage disposal. Issues awaiting further legislative action include protection of ground water and wetlands and promotion of water conservation.

CLEAR-CUTTING

Clear-cutting is a method of harvesting timber in which all trees within a given area regardless of size or age are felled at one time. In temperate forests this method is often used to promote the regeneration of tree species that require direct sunlight for seed germination and early growth. These "shade-intolerant" species represent the early stages of forest succession, and are often eventually replaced by other late-successional tree species that germinate in the shade of the established forest. Many of the early-successional tree species are fast-growing and of high commercial value relative to the late-successional species, and thus are favored in the management of forests where wood production is the primary objective.

The practice of clear-cutting has been criticized for its ecological and aesthetic impacts. If used improperly, clear-cutting can result in the loss of tree cover for an extended period of time. High soil temperatures from direct exposure to sunlight can accelerate the decomposition of soil organic matter, and reduce the soil moisture, nutrients, and microorganisms important to the rooting of new trees. In extreme situations, this process can result in soil erosion, impoverishing the soil and increasing sedimentation of streams and fish habitat. This has been a significant problem in moist tropical forests where clear-cutting has been practiced on a large scale, often on highly erodible soils and in tree species poorly suited to regeneration following clear-cutting. Where such clear-cutting has been followed by extensive burning or livestock grazing, large areas have remained deforested for many years.

Concern for the protection of biological diversity has brought additional criticism of clear-cutting because it replaces the existing forest with an early-successional stage forest. To maintain biological diversity, it is important to maintain a variety of habitat conditions in forests of differing successional stages. Over time, however, timber harvesting has resulted in a shrinking proportion of late-successional forests, limiting habitat for plant and animal species found in those particular ecosystems. Scientists are currently experimenting with new approaches to timber harvesting whereby most of the trees can be harvested while still maintaining the essential ecological characteristics of late-successional forest ecosystems.

For many forest areas, there are alternative methods of timber harvesting that are ecologically and aesthetically preferable to clear-cutting. In other areas, particularly those dominated by shade-intolerant tree species, clear-cutting is a viable method of timber harvesting that can be accomplished with little negative effect. Clear-cutting represents one end of a spectrum of timber harvest alternatives, going from the removal of all trees at once to the "selective harvest" of only a few mature trees at a time from within the forest. The best choice of harvest method depends upon the existing and desired tree species, site characteristics (soil type, slope, rainfall), and other ecosystem management considerations, including an awareness of forest conditions in adjacent areas and their ecological relationships to one another.

V. ALARIC SAMPLE

CLIMATE CHANGE

The word *climate* is used to summarize the weather of a particular geographic area over a specified time period. Climatic change is the difference between climatic conditions of two locations in the same season, or over different times at the same place. Changes in the Earth's climate usually compare the yearly average conditions over the whole globe for two periods of time several decades apart.

The climate of a particular location, or of the whole Earth, results from many physical actions in the atmosphere, the ocean, and on land. On the time scale of days, weeks, and seasons, weather is affected mostly by things that happen in the atmosphere. The slower changes in climate, however, also are affected by events and changes in the ocean, ice sheets, and snow fields, and green plants on land.

The atmosphere is the most obvious part of the climate system. The most apparent components of the atmosphere that affect climate are the direction and strength of wind, air pressure, temperature, and humidity. Additional parts of what can be called the atmospheric climate are the distribution of clouds and precipitation, and of shortwave and longwave radiation. The amount of energy received from the Sun in the form of sunlight must be balanced over time by the energy emitted from the Earth into space. Otherwise, the planet would slowly warm up (if less is given off than absorbed) or cool down (if more is emitted than absorbed). Finally, the presence of pollution from human activity also affects the Earth's climate by altering the composition of atmospheric trace gases that affect absorption of solar and infrared energy.

On a global scale the distribution of many atmospheric properties is closely related to the surface properties and configuration of the ocean. The principal oceanic climate components include the distribution of ocean temperature, salinity, and current. Dissolved gases and suspended matter are also important, especially in terms of the absorption and storage of carbon in the deep ocean. Carbon dioxide, an atmospheric trace gas with important influences on the radiative balance of the planet, is constantly exchanged between the oceans and the atmosphere. The ocean thus affects climate in several direct and indirect ways.

In addition to the atmosphere and ocean, the global distribution of ice and snow is an important part of the climate system. The extent and thickness of ice and snow fields, and their temperature and motion, affect the other parts of the climate system. Sea and lake ice and snow cover react fairly rapidly to changes in atmospheric and oceanic conditions. Continental ice sheets and mountain glaciers change much more slowly (see Glaciers). Snow and ice not only respond to climate change, but can also affect climatic conditions. The larger the proportion of the Earth's surface covered by snow and ice, the more energy will be reflected into space instead of absorbed at the surface. Thus changes in snow and ice cover affect climate by altering the absorption of energy from the sun thereby altering climate.

Changes in land surface operate on many time scales. The seasonal growth and decline of grasses and other plants have two important short-term impacts on weather and climatic conditions. First, the seasonal changes in plant cover alter the ability of the Earth's surface to reflect sunlight. This affects the amount of energy absorbed in a particular location, which directly changes soil temperature, soil moisture, and other local conditions. Second, the changing metabolism of green plants as they absorb carbon dioxide and give off oxygen affects the atmospheric concentration of carbon dioxide. The measurements of atmospheric carbon dioxide taken on the mountaintop station at Mauna Loa in Hawaii since 1959 clearly show this "respiration" for the northern hemisphere. During the spring and summer months the atmospheric concentration of carbon dioxide decreases slightly; during fall and winter it increases.

Modifying patterns of land use can alter climate over longer time scales. Replacing forest with grassland changes the surface properties that are relevant for climatic processes with effects as described above that persist over decades. The expansion of deserts, with the virtual elimination of vegetation, has the same or stronger effects (see Desertification). If changes in land use are geographically large enough, they may even change global climate. Local and regional climatic consequences due to land use changes are probably most pronounced in tropical rain forests. Much of the moisture for precipitation in tropical rain forests is water vapor evaporated from the surface or given off by growing plants. Extensive deforestation in these areas can directly affect most of the components of local climate. The rates of deforestation in the wet tropics in recent years may also be enough to affect global climatic conditions. The data on the extent of global deforestation as well as the calculations of the implications of deforestation for global climate change, however, are controversial. Although many scientists are convinced that current rates of deforestation contribute to climate change, the exact amount of the effect is uncertain.

Changes in cities and urban areas can also affect local climate. Dark asphalt streets absorb much more energy from sunlight than lighter-colored materials. This is called the "urban heat island effect." It is one reason the daytime temperature in most cities is a few degrees higher than the surrounding areas. As urban areas expand, their influence on local climate increases.

Both agricultural and industrial activities can also affect climate directly. All human activities, from subsistence agriculture to the most advanced industries, emit various greenhouse gases. The most important for climate change is carbon dioxide, which is the principal greenhouse gas generated by human activity. Some industrial pollutants, such as sulfur dioxide, cool the surface since they reflect incoming solar radiation before it can warm the Earth. Temperature records show some cooling near heavily industrialized areas, but it is difficult to assess the overall effect of pollu-

tion. Our best estimate is that the global average temperature has risen about 0.5°C (about 1°F) over the last 100 years.

The single most important factor in the Earth's climate is radiation from the Sun. Sunlight warms the atmosphere, land, and ocean, and provides the energy that drives the Earth's weather systems. Most scientists think that changes in the amount of light received from the Sun, however, are so small that they do not affect the Earth's climate over a few hundred years. Over thousands and millions of years, changes in the Earth's distance from the Sun due to its orbit are more important than variation in amount of light given off by the Sun. These orbital variations are caused primarily by the gravitational attraction of Jupiter and Saturn. They cause changes in the seasonal and latitudinal distributions of solar energy even though the global annual solar radiation changes very little. The regional redistributions of solar energy are thought to be the principal causes of the ice ages and warm periods between them. The glaciers of the most recent ice age covered the greatest area about 18,000 years ago. All of human civilization has developed and flourished in the warm period since the last glacial period. The Earth's orbit should lead to another cool period in about 3,000 years.

Volcanic eruptions also can affect climatic processes by altering the amount of sunlight reaching the Earth's surface. Volcanoes can inject sulfur dioxide and other particles high into the atmosphere, and these particles can reflect sunlight and prevent it warming the Earth's surface. The strength of the effect depends on the gases emitted during eruption and how high into the atmosphere they are injected. El Chichon, a volcano that erupted in Mexico in 1982, had little effect on the global climate because it did not get much material into the high atmosphere. In contrast, Mt. Pinatubo, which erupted in the Philippines in 1991, is expected to have the effect of lowering global average temperatures by as much as 0.5°C (about 1°F) for two or three years. However, the effects of individual volcanoes seldom last as long as a decade.

More important for long-term alteration of the climate are changes in the atmospheric concentrations of "greenhouse gases" like carbon dioxide, halocarbons (of which chlorofluorocarbons, or CFCs, are the most important), and methane. When present in the atmosphere, these gases trap energy that would otherwise be radiated from the Earth's surface into space, thereby warming the planet. Carbon dioxide and methane are both products of natural biological processes and would be present in the atmosphere even if there were no people at all. Their natural concentrations in the atmosphere help make the planet livable by raising surface temperatures above what they would otherwise be. Atmospheric concentrations of greenhouse gases remained essentially unchanged for about 10,000 years prior to the industrial revolution. Over about the last 200 years, however, human activity dramatically increased the concentrations of these gases in the atmosphere. During the same time period, the global average temperature has risen about 0.5°C (1°F). Because of the complex interactions between clouds, oceans, ice cover, and other components of the climate system, however, it is not possible to prove that this temperature increase is due only to the increased concentrations of greenhouse gases.

Our knowledge of past climate depends heavily on what are called "proxy data." Reliable instrument measurements of temperature, for example, are only available for about the last 100 years. To estimate climate further into the past, we must rely on other artifacts that relate to climatic conditions. Tree rings can be used to estimate conditions in a particular location. The width of the ring produced as a tree grows in a given year can be measured to indicate the seasonal conditions the tree experienced. Measurements of pollen deposited in lake and bog sediments provide information from which past patterns of vegetation can be reconstructed. These are less precise than tree ring data because many plants are tolerant to different conditions and thus blur the record of climate change. Since pollen data indicate broad changes in patterns of vegetation cover, which occur relatively slowly, they cannot be dated as to precise years. Glaciers typically respond to changes in climatic conditions, so they can be used as indicators of climate change. Evidence from studies of ice cores and glacial deposits can be used to estimate both the direction and magnitude of climatic changes. Of these proxy data, glaciological evidence is the least precise.

Studies of past climate data identify two important changes in the past 1,500 years: the Medieval Warm Epoch and the Little Ice Age. The Medieval Warm Epoch lasted from about 1000 to 1400 A.D., although there are differences in timing from place to place. The Medieval Warm Epoch was primarily confined to Europe and North America. Even within these areas there are considerable variations. There is evidence, for example, that Alpine glaciers were advanced during the 12th century as far as at any time in the last 3,000 years.

From 1200 to 1600 A.D. the climate of those parts of the globe that were warm seems to have cooled significantly. By the 17th century, it appears that conditions colder than today were found throughout those regions. Warming did not begin until the 19th century, and the period from 1400 to 1800 A.D. is referred to as the Little Ice Age. There were also complex patterns of differential warming within this period. For example,

there appears to have been warming in Europe around 1500 A.D.

An important aspect of both the Medieval Warm Epoch and the Little Ice Age is that the climate was not continuously warm or cold. Rather it continued to experience the year-to-year fluctuations we see in climate today, and there also were differences from place to place. It appears that Britain experienced a series of hot summers during the 17th century, which was also a time of frequent severe winters. Changes in the variability of climate may be as important in determining the impact of climate on humans as changes in the averages.

Determining the exact effect of climatic changes on various peoples is complex. The results of unusual climatic conditions for harvests, for example, depend on both the crop and the regional climate. In semi-arid or semi-humid areas moist and cool summers produce excellent harvests, while droughts are most feared. In humid western and central areas, droughts only rarely cause harvest failures. In these regions, bad harvests are mostly (but not exclusively) related to prolonged cold spells in spring and to cool and wet summers. In assessing the human consequences of climate change it is necessary to pay careful attention to these regional differences.

Climate change will not mean worse conditions for all people everywhere. Milder winters and longer growing seasons should improve the agricultural productivity of parts of the northern latitudes in North America and Russia. A large country with many climatic zones, the United States should be able to adjust to gradual climate changes, although such adjustments may be expensive.

People in different parts of the world are not equally vulnerable to changes in climatic conditions. For example, a storm of the same wind strength and duration today would usually cause many more deaths along the coast of Bangladesh than along the coast of Louisiana or Florida. People with more financial resources and better weather forecasting are usually able to protect themselves more efficiently against weather extremes. Countries with more financial and human resources should be able to respond to climate change more appropriately.

ROB COPPOCK

For Further Reading: National Academy of Sciences, *Policy Implications of Greenhouse Warming* (1991); National Research Council, *Toward an Understanding of Global Change: Initial Priorities for U.S. Contributions to the International Geosphere-Biosphere Program* (1988); National Research Council, *Global Environmental Change: Understanding the Human Dimension* (1992).
See also Tropical Rain Forests.

COAL

A black or brownish-black burnable sedimentary rock of organic origin, coal is a major energy resource that has played a significant role in the industrial and economic development of many countries.

Coal is called a fossil fuel because it is derived from plants that grew in vast swamps 300 million years ago. Over vast periods of time, the plant matter was compressed and altered by geological processes, including pressure and temperature, and transformed into coal. How the plant matter was altered determined the type or rank of coal that was formed. Although not actually a form of coal, peat is the first stage in the coal formation process. Lignite, the lowest grade of coal, is basically peat that was subjected to increased pressure and heat. Still more pressure and layers of overlying strata resulted in the formation of the higher coal ranks—subbituminous, bituminous, and anthracite or hard coal.

Coal's diverse and complex chemical makeup (chiefly carbon, hydrogen, and oxygen, with smaller amounts of sulfur, nitrogen, and trace elements) contributes to its usefulness as an energy resource. From an environmental perspective, some of these same elements pose potential problems as they undergo changes during the normal combustion process. Consequently, the issue of air quality, along with the physical impact of mining on land and water quality, have been the chief environmental concerns traditionally associated with the use of coal.

Significance as an Energy Resource

By any measurement criteria, coal ranks among the world's major energy resources. According to World Energy Council data, "recoverable" reserves—coal that can be mined using existing technology—totals 1.79 trillion tons worldwide. Although coal is widely distributed around the globe, three nations account for more than half of recoverable reserves—China, 11%; and 23% each in the U.S. and the area encompassed by the former Soviet Union.

Using reserves/production (R/P) ratios, it is estimated world reserves of coal will last twice as long as the combined reserves of oil and natural gas—238 years vs. 43.4 years and 58.2 years, respectively.

The Energy Information Administration (EIA) places U.S. recoverable reserves at 285 billion tons, enough to last nearly 250 years at current consumption rates. According to the Department of Energy (DOE), coal comprises 80% of total U.S. fossil energy reserves, compared with less than 3% for oil and 4% for natural gas.

In the U.S. coal is widely distributed geographically, with 38 states having significant reserves. The largest

quantities are in Montana (120 billion tons), Illinois (78 billion tons), Wyoming (69 billion tons), West Virginia (37 billion tons), and Pennsylvania (29 billion tons).

Coal Exploration, Mining, and Distribution

Prior to mining, coal deposits must be explored and evaluated. Once a particular deposit has been selected as a potential mining site, several years of exploration, planning, and development take place before production begins. From a planning perspective, this includes addressing such issues as the market for the coal, land ownership, mineral rights, coal quality and quantity, environmental protection, and mining and transportation methods. In the U.S. it can take as long as a decade, with an investment of several million dollars, before these issues are resolved and mining permits are secured from federal and local authorities.

Coal exploration is a complex technology directed at finding commercially viable seams (high quality and energy content) that can be mined in an efficient and environmentally responsible manner. This is an undertaking that utilizes numerous technical and professional experts, along with modern technology—including aerial and satellite photography, electronic distance measuring devices, test borings, laboratory evaluation, and seismic analysis.

The mining of coal itself has traditionally had environmental implications, mostly of a negative nature. Modern mining operations, however, are far removed from the practices of bygone eras, when removing the coal from the ground in the least expensive and fastest way possible was the only concern. Coal mining has become a highly regulated activity, with specific laws and regulations dealing with extraction, miner health and safety, and land reclamation.

How a particular deposit is mined usually depends on several factors, including the depth of the coal bed from the surface and the geologic character of the terrain. In general, coal that is 200 feet or more from the surface is usually mined by "underground" or "deep mining" techniques. Shallower deposits are extracted by surface mining methods.

Underground mining, which has become increasingly mechanized, involves digging an underground portal or shaft to the coal, constructing a series of "rooms" into the coalbed, and leaving "pillars" of coal to help support the mine roof. Most underground coal (88%) is mined by continuous or longwall mining machines.

Surface mining is a large-scale earth moving operation involving taking away the covering layer of rock and soil; extracting the coal; backfilling with soil; and restoring the site to its original vegetation and appearance. This restoration process, called reclamation, is required by both federal and state law, and is incorporated into all phases of normal operations at modern U.S. surface coal mines.

Since passage of the federal Surface Mining Control and Reclamation Act in 1977, the coal industry has reclaimed about 2 million acres of mined lands, an area larger than the state of Delaware. In addition, more than 100,000 additional acres of abandoned mines, remnants of neglect from long ago, have also been reclaimed through money paid by today's coal producers into a national trust fund.

After mining, coal is subjected to various cleaning and washing procedures, which remove rock and most impurities, including some pyritic sulfur that is not chemically bound to the coal. The cleaned coal is classified, or sized, by passing over a series of screens, and then blended in mixing conveyors before being delivered into a rail hopper, barge, or truck for shipment. Much of this operation is highly automated and computerized.

In the U.S. about 60% of coal transported to market goes via railroad. Another 15% is carried by barges on inland waterways, with the remainder hauled by truck, conveyor, or slurry pipeline. About 10% of U.S. coal production is exported by ship to overseas customers each year.

Coal Use and the Environment

Although there are many secondary uses and byproducts produced from coal, its major value is as a fuel for electric utility power plants. Nearly eight out of every ten tons of coal used in the U.S. goes for this purpose.

In recent years, much of the growth in energy consumption has been in the form of electricity, 55% of which is generated by coal-fired power plants nationwide. Between 1980 and 1992, total U.S. coal consumption increased significantly, from 703 million tons to 928 million tons. The greater reliance on domestically produced coal electricity has played an important role in U.S. economic growth and in reducing dependence on imported petroleum.

The basic method of using coal to produce electricity at a power plant has remained unchanged for most of the 20th century—coal is burned in a boiler that contains water, which is converted to steam that turns a giant turbine. The turning motion of the turbine blades causes an electric current to be created in a generator, which is sent out of the power plant as electricity to homes and businesses.

The combustion process causes changes in coal's chemical components—especially carbon, sulfur, and nitrogen—which can pose significant air quality problems if released into the atmosphere in large quantities. Coal's popularity as a primary fuel of transportation and industry earlier in the century, combined with a general lack of awareness of environmental impacts, resulted in the lingering image of coal as a dirty, polluting fuel.

Coal's renaissance since the Clean Air Act of 1970, fueled by increasing use by electric utilities, has been accompanied by a heightened environmental consciousness, and the end result has been noticeable and dramatic. According to EPA National Ambient Air Quality data, despite a doubling of coal consumption in the past 20 years, the pollutants most closely associated with coal burning have declined significantly: 26% for sulfur dioxide (SO_2) and 61% for particulates.

These improvements are the result of several initiatives, but at their root they reflect the expenditure of millions of dollars by coal producers and utilities to comply with the world's strictest national air quality law—the Clean Air Act. Additional and even more stringent amendments to the Clean Air Act were passed by Congress in 1990.

Particulate control is accomplished primarily through technology. Boiler debris is periodically removed from the furnace, but fly ash, which rises up the stack with flue gases, is collected by specially designed equipment. This includes a baghouse, which uses fabric bags to filter out particulate-laden gases; and the electrostatic precipitator, which uses high-voltage discharge electrodes to give the dust particles an electric charge so they can be collected. Both methods of collection are highly effective, with removal capacities usually in the 99.5% range.

The removed ash, once considered a nuisance waste material, is today recycled as a multi-use product. Fly ash, which accounts for about 80% of coal ash byproducts, is used as an additive in concrete and grout. Bottom ash is useful in highway paving applications and as a construction aggregate. Ash is also used as an additive in roofing materials, mineral wool insulation, bricks, concrete blocks, paints, and plastics, as well as fill in some surface mine reclamation projects. Research is continuing to find other useful applications for ash, such as in reinforcing artificial ocean fish reefs, and also recovering ash's inherent mineral resources, including carbon, magnetite, alumina, and iron oxide.

When coal is burned, the sulfur present in the fuel combines with oxygen at the rate of one-to-two to form sulfur dioxide. If emitted into the atmosphere, SO_2 may bond to particles of dust and smoke and be transported long distances. In addition, SO_2 combined with water vapor can produce sulfuric acid, a component in the formation of acid rain.

Control of SO_2 emissions has been the focus of much research and study. Four basic approaches have been responsible for the steady decline in coal-based SO_2 emissions since 1970:

- *Pre-Combustion Removal*—Also known as beneficiation, this involves crushing coal to separate impurities and processing in a liquid medium to wash away the sulfur. The result is a reduction of 30% or more in pyritic, or non-chemically bound, sulfur.

- *Post-Combustion Removal*—Usually involves the chemical "cleaning" of exhaust gases at power plants, using a flue gas desulfurization (FGD) system, more commonly known as scrubbers. In this process, the power plant exhaust gases are sprayed with a slurry mixture of water and limestone or other reagent. A reaction occurs with the SO_2, resulting in the formation of a calcium sulfite and/or calcium sulfate sludge, which has the consistency of toothpaste. This sludge is normally buried in a disposal area somewhere near the power plant. Although expensive to install and operate (costing anywhere from $100 million to $400 million each), FGD systems have proven effective in removing up to 90% of SO_2 from coal combustion gases at new facilities.

- *Low Sulfur Coal*—Some coals are lower in sulfur content than others. In some cases, utilities can more easily meet Clean Air Act emissions standards by burning a low sulfur coal.

- *Coal Blending*—It is also possible in some cases to meet emissions requirements by mixing or "blending" coals of varying sulfur content prior to burning in a utility boiler.

Clean Coal Technologies

As the 20th century comes to a close, the basic technology of simply burning coal to release its energy is undergoing radical and revolutionary change. Spurred by environmental considerations and an increasing need for more abundant, efficient, and cost-effective energy supplies, innovative new processes to more cleanly and completely extract power from coal are being deployed for large-scale commercial use.

These "clean coal technologies" are a diverse group of non-traditional power generating and pollution control concepts resulting from years of research and development in hundreds of government, academic, and private industry laboratories.

Clean coal technologies are divided into four basic categories:

- *Pre-Combustion*—These are technologies applied before coal is burned to produce energy. They improve the effectiveness of physical coal cleaning or washing, or employ chemical, biological, or other techniques to remove high percentages of sulfur and ash.

- *Combustion*—Technologies that remove pollutants inside the combustor or boiler while the coal burns. The best-known and most effective of these processes include atmospheric (AFBC) and pressurized (PFBC) fluidized-bed combustion, advanced combustors, and limestone injection multistage burners (LIMB). All offer varying degrees of sulfur dioxide and/or nitrogen oxide emissions reductions while improving operating efficiency.

- *Post-Combustion*—Technologies that clean flue gases emitted from coal burning and that are generally located in the ductwork leading to the smokestack, or in advanced versions of present-day flue gas desulfurization systems (scrubbers). Emerging alternatives to the conventional scrubber systems include in-duct sorbent injection and advanced flue gas desulfurization systems.

- *Conversion*—Technologies that change coal into a gas or liquid that can be cleaned and used as a fuel. Basic coal gasification and liquefaction technologies have been known for many years. The techniques involve heating and squeezing coal under high pressure until it breaks down into a gas or liquid. When the coal particles break apart, sulfur and other impurities can be removed effectively, and the resulting liquids and gases can often be used in equipment designed to burn petroleum without making major alterations. The most promising conversion technologies include integrated gasification combined cycle (IGCC) and coal oil coprocessing.

Clean coal technologies can be "retrofit" (or applied to existing plants to reduce emissions); "repowering" (which replace a significant portion of an original plant and increase capacity while reducing emissions); or "greenfield" (incorporated in a new construction project application). Virtually all offer enhanced reductions of sulfur and nitrogen oxides and particulates. In addition, their higher efficiencies produce less carbon dioxide per unit of fuel consumed, which has implications for the debate over reducing the impact of human activity on atmospheric greenhouse gas concentrations. For example, processes such as PFBC and IGCC, which boost generating efficiencies from the current 33% to the 40%–45% range, result in carbon dioxide emissions reductions of 10%–15%.

The $5 billion joint industry/government Clean Coal Technology Program was created by Congress in 1986 and implemented through the U.S. Department of Energy. Its goal has been to foster and accelerate the commercial development and deployment of the most promising clean coal technologies. Some of the processes developed under this program are already in the market place or near commercial deployment, and others are expected to follow in the early years of the 21st century.

Coal's Future

Since the 1970s, the U.S. economy has become increasingly dependent on electricity to meet its energy needs. In general, each percentage increase in Gross National Product over the past 20 years has resulted in just over a 1% rise in the demand for electricity.

Continuing the trend of recent years, utilities are expected to rely greatly on coal in the future as their primary baseload fuel for electricity generation. According to the Department of Energy, America will need more electricity in the future for new homes and factories, as well as for innovative new applications, such as electric cars. Electricity use will be more important than ever to the health and well-being of the nation's economy.

DOE forecasts that 280,000 megawatts of new electric capacity should be built by utilities before 2010 to meet an expected rapid rise in new electricity demand. Although some of these plants will use other forms of energy, abundant and less-costly coal is the likely choice to provide most of this electricity, particularly through the use of cleaner and more efficient clean coal technologies. DOE's National Energy Strategy says coal could account for up to 75% of the nation's electricity generation by 2030, under a high growth scenario.

U.S. coal and clean coal technologies may also play a significant role in meeting the needs of developing nations for providing energy supplies in an environmentally acceptable manner.

THOMAS B. JOHNSON

For Further Reading: National Coal Association, *Facts About Coal* (1993); Office of Surface Mining Reclamation and Enforcement, *Surface Coal Mining Reclamation: 15 Years of Progress, 1977–1992, Part I* (1992); U.S. Energy Information Administration, *Coal Data: A Reference* (1991).
See also Acid Rain; Clean Air Act; Electric Utility Industry; Energy; Mining Industry; Pollution, Air; Restoration Ecology; Sulfur.

COASTLINES AND ARTIFICIAL STRUCTURES

The most conspicuous boundary on Earth is that which joins land and sea. With a length of approximately one million kilometers, the zone it delineates has served as a source of livelihood, enjoyment, and inspiration for humans throughout much of their history. It is a zone whose resources have been exploited heavily and whose physical form has been modified drastically. Nonetheless, within the last few decades humans have begun to realize that its potential is finite, that it is being altered at ever-increasing rates, and that wise management would be prudent.

Earth's natural shoreline is highly varied. Segments are represented physically by cliffs (about 50%); dunes, barriers, and beaches (about 20%); mudflats and deltas (about 2%); lagoons; and estuaries. Biologically, it is occupied by both terrestrial and marine organisms such as mangroves, marsh grasses, seaweeds, corals, algae, and shellfish. Glacial ice is present along about 5% of the coastlines of the world, with permafrost dominating along most of the coast facing the Arctic Ocean.

The shoreline itself is one of the most dynamic environments in that it is regularly impacted by waves, tides, and currents and occasionally by such forces as typhoons, tsunamis, and sea-level changes. Chemical and biological processes are important along some segments of the shoreline, especially in the tropics. Not only do these active processes result in frequent changes in the natural forms and materials of the coastal zone, but they also impact many of the human constructs that line the shore.

The earliest significant changes to the shoreline began with the development of agriculture, domestication of animals, and deforestation. These first changes were unintentional and indirect. Flow regimes and sediment loads changed and they in turn altered the growth rates of deltas. Intensification of these early endeavors, especially agriculture, led to the construction of permanent settlements (eventually cities), occupational specialization, political organization, commercial development, and trade, all of which became involved, either directly or indirectly, with shoreline modification.

Shorelines, like river courses, tend to localize human activities. Two types of coastal locations singled out were those that could be reclaimed for agriculture, aquaculture, and solar salt production; and those that provided protection for ships used in trade and war. Reclamation early involved the building of dikes, levees, and drainage ditches. As ships became larger and more numerous and trade expanded, coastal engineers constructed breakwaters to protect harbors and even drained swamps to create them.

Although such endeavors had rather humble beginnings, it did not take long for major changes to occur. Alexandria, an important port on the Mediterranean Sea, was the world's largest city in 300 B.C. and Marco Polo labeled Hangchow, China, the world's busiest port in the 13th century. Subsequent to the period of exploration, during which most of the coastlines of the world became known around the world, the Industrial, Commercial, and Technological Revolutions led to a rapid expansion in shoreline settlement and artificialization. For example, in 1500 only 25% of the world's largest cities were coastal; today it is 70%. At the present time, in many countries more than half of the population lives in the coastal zone, a percentage that is increasing. Japan has more than 4,000 harbors, one for each eight kilometers of shoreline; 51% of its shoreline is bordered by artificial structures. Other examples include England with 38%, South Carolina with 39%, and Belgium with 85%.

The types of structure built along a coast are affected by economics, politics, esthetics, law, and technology as well as the availability of materials and the nature of the problem. Most structures are built to control erosion, to improve navigation, or for reclamation. Structures designed to serve as an interface between the land and the sea include seawalls, revetments, bulkheads, dikes, and levees. Seawalls may be massive (some over 12 meters high) when used for protection of typhoon- or tsunami-prone coasts; they represent the most common anti-erosion type structure. Other structures, built at various angles to the shoreline, include groins, jetties, piers, and even drainage pipes. Yet another group includes those built offshore to serve as detached breakwaters. They, along with groins, are especially used to protect and restore beaches, 70% of which are suffering erosion in the 1990s.

The materials used vary greatly. Boulders (riprap), timber, and steel are time-honored materials, although concrete is usually used for seawalls. In recent decades, fabricated armor units such as tetrapods (which may weigh as much as 50 tons), have become popular. Presently more than 40 different designs for these wave de-energizing armor units have been developed.

Although most structures around the world have served the purpose for which they were built, many have not and, indeed, many have aggravated the problems they were intended to solve. Groins and detached breakwaters, the objective of which is to trap sand, usually cause losses (and therefore increase erosion) downdrift. Breakwaters change the shoreface (both form and material) and can lead to undermining at the base of the structure and increased wave action.

In the course of history, much of the shoreline has been modified to suit human purposes. In the process, negative impacts, such as sinking, have often intensified the need for the use of protective structures. Tokyo, a prime example, has over a million people living below the zero meter line, and thus protected from flooding by seawalls. However, such cases may pale in comparison to what will be needed if the predicted sea-level rise, associated with global warming, occurs. Whereas retreat from the coast may be an option in some instances it generally would not be acceptable, nor would be the conversion of cities to a Venice-like condition. The major options appear to be either to use

fill to build up the base of cities or to increase the height and strength of seawalls. The cost of such endeavors will be great although most of them can be done progressively. Such may not be the case for locations like the low-lying islands of the Pacific and Indian Oceans. At the Earth Summit conference held in Rio de Janeiro in June 1992, the President of the Maldives stated that to build a seawall that would protect his country would cost 1.5 billion dollars.

H. JESSE WALKER

For Further Reading: P. Bruun, *Design and Construction of Mounds for Breakwaters and Coastal Protection* (1985); CERC (Coastal Engineering Research Center), *Shore Protection Manual* (1977); H. J. Walker, ed., *Artificial Structures and Shorelines* (1988).

COGENERATION OF ELECTRICITY AND HEAT

Cogeneration is the concurrent generation of electricity (motive power) and process heat (steam). It saves fuel—typically 10% to 30%—because either waste energy from a heating process is used for the generation of electricity, or waste energy from a power plant is used for heating applications.

Cogeneration affords one of the largest opportunities for saving fuel because many common processes have sizable waste energies suitable for this technology. It encompasses many different energy recovery and energy conversion devices. Some of the energy conversion devices, such as steam turbines and reciprocating diesel and spark-ignition engines, have been in common use for decades. Others, such as turbines with an organic material as a working fluid and thermionic converters, are just now being commercialized or are still undergoing testing.

Small-scale cogeneration facilities save capital because the equipment is built in a manufacturing plant rather than at the site of the facility, and in a much shorter time than that required for a large central electric power station. This latter feature is a valuable tool for electric utility planners who may have to predict electricity demands a decade before a new large power plant would finally come into service. In addition, cogeneration facilities are beneficial to the total environment because they satisfy specified energy needs with less primary energy.

Power devices for cogeneration fall into two classes: topping units and bottoming units. Topping units take advantage of the fact that many low-temperature direct-fired processes such as drying, baking, space heating, and washing are inefficient because they consume directly the high-quality (high-temperature) energy of combustion products for tasks that actually require only low-quality (low-temperature) energy. The effectiveness of fuel use in such processes can be increased substantially by first using the high-quality energy of fuel combustion in a diesel engine, gas turbine, or steam turbine to drive an electric generator, and then recovering the exhaust energy of the unit to perform heating tasks needing temperatures of only 150°-600°F.

Bottoming units are applicable to high-temperature processes such as the production of metals and ceramics in furnaces and kilns operating at 1,000°F and above. Waste energy from the process is directed to a power conversion device driving an electrical generator. In a typical application, furnace exhaust gas, still containing a large quantity of high-quality energy, is directed to a boiler where steam is generated. The steam drives a turbine-generator engine and produces electricity. The combined system uses about 30% less energy than when the furnace heat and electricity are produced separately. Cogeneration by means of waste energy recovery with a bottoming engine is particularly attractive because it produces electricity with no incremental consumption of fuel and often can be installed in existing facilities.

The major energy conversion techniques used in cogeneration are:

Steam Turbines: Steam turbines have been used for both cogeneration and conventional power generation throughout much of this century. In a paper mill, for example, a high-pressure topping turbine extracts part of the energy from a high-pressure steam flow. The remaining energy in the exhaust steam is used to operate paper mill machinery such as digesters, blenders, and dryers. A typical electrical output would be about 50 kWh (kilowatt-hours) per million Btu (British thermal units) of steam energy delivered to the mill machinery.

In a district heating installation, waste energy from a power plant is fed, in the form of low-pressure steam or hot water, to a network that supplies the heating needs of a city or a residential and commercial complex of buildings.

Low-pressure steam turbines are used as bottoming units. They recover waste energy from relatively high-temperature exhaust gases of a process by means of a waste heat boiler, or from the spent steam of intermediate-temperature industrial processes.

Steam topping and bottoming turbines are feasible from about 2,000 kW (kilowatts) up to 50,000 kW with presently available hardware. They have reasonable capital and installation costs. For district heating applications, the power plant can be much larger, and the capital and installation costs are dictated by the type of plant under consideration and the costs of the district heating network.

Diesel Engines: Diesel engines are applicable as topping units when a high ratio of electrical output to process heat is required—up to 400 kWh per million Btu of heat delivered to the process. Process steam and hot water are produced by recovery boilers coupled to the exhaust stack and to the cooling water of the engine. Systems from 100 kW to several thousand kilowatts can be built. However, these systems use medium-speed and high-speed diesel engines, the type generally used in trucks, construction equipment, and rail locomotives. Such engines burn high-grade distillate petroleum, which is likely to be expensive and often in short supply in years to come.

A more versatile diesel engine for topping large systems, up to about 30,000 kW, is the large slow-speed, two-stroke diesel engine. This engine, often used for propulsion of large ships, is capable of burning very-low-grade fuels such as high-sulfur crude or heavy residual oil. Recent experiments have shown that it may even be capable of burning a powdered coal-water slurry. System costs, including heat recovery boilers, are affordable.

Combustion Gas Turbines: Combustion gas turbines are well suited as topping units for large systems, particularly where natural gas or clean-burning by-product fuels such as refinery gas are available. Gas turbine systems offer low capital cost, particularly in large systems of 10,000-150,000 kW. Also, the high exhaust gas temperature of gas turbines permits their integration with a great variety of industrial processes.

Spark-Ignition Engines: Spark-ignition engines that burn natural gas can also be used as topping units. Very low capital cost is achieved by converting high power automobile engines for use in small cogeneration modules of about 60 kW output. Each module produces about 500,000 Btu per hour of process heat in the form of low-pressure steam and hot water. One to ten modules could be used in applications such as shopping centers, hospitals, apartment buildings, and light industrial sites, to supply all on-site electrical and process heat needs.

Organic Rankine Turbines: An organic Rankine turbine is an advanced type of bottoming unit. It uses an organic material as a working fluid and is capable of recovering efficiently the energy from low-temperature (300°-600°F) waste streams. It can be built in a wide range of sizes, from 50 kW to 30,000 kW or more. Output per unit of waste energy input will generally be 20% to 30% greater than that obtainable with steam-turbine bottoming units.

The technologies described above provide the basis for virtually all cogeneration systems. Other technologies now in the research and development stage, such as thermionic converters and Stirling cycle engines, may also play a role in future cogeneration systems.

In its most elementary form a thermionic converter consists of one electrode connected to a high-temperature energy source (about 3,000°F), a second electrode connected to a low-temperature energy sink (about 1,000°F) and separated from the first by an intervening evacuated space, and leads connecting the two electrodes to an electrical load. Electrons boil off the hot electrode by the process of thermionic emission, condense on the colder electrode, and return to the hot electrode via the load. Thermionic converters may eventually be used as topping units for gas turbines and high-temperature industrial furnaces.

Cogeneration plants of various sizes are installed throughout the industrialized world. Their advantages become more and more valued as nonrenewable energy sources become scarcer and energy prices increase.

ELIAS P. GYFTOPOULOS

For Further Reading: Ezekail L. Clark, "Cogeneration— Efficient Energy Source," in *Annual Review of Energy* (1986); Elias P. Gyftopoulos, Lazaros J. Lazaridis, and Thomas F. Widmer, *Potential Fuel Effectiveness in Industry* (1974); Joan M. Ogden, Robert H. Williams, and Mark E. Fulmer, "Cogeneration Applications of Biomass Gasifier/Gas Turbine Technologies in the Cane Sugar and Alcohol Industries," in *Energy and the Environment in the 21st Century* (1991).
See also Electric Utility Industry; Energy, Electric.

COLLECTIVE BARGAINING

Collective bargaining is the process of negotiation and dialogue between a union, representing employees, and an employer. Both parties agree by mutual obligation and by law to discuss topics ranging from wages and hours to health benefits to health and safety protocols. The obligation does not compel either party to agree to a proposal or require a concession; however, a written contract may be requested. The collective bargaining process is extensively regulated by United States laws administered by the National Labor Relations Board such as the National Labor Relations Act and the Labor Management Relations Act of 1947. Most collective bargaining agreements include provisions that outline union and employer rights, wages and benefits, work time and time off, seniority rights and grievance procedures, methods for assuring worker participation in decisions affecting the workplace, and administration of the agreement.

In some workplaces, such as large manufacturing plants, different groups of employees may have different interests. These groups may form separate bargaining units to negotiate with the employer for those individual interests. The constituents of specific bargaining units must be approved by the regional National Labor Relations Board, and each bargaining unit may be represented by different unions. Efforts by employers to play one unit against the other and gain concessions are known as *whipsawing*.

Collective bargaining is playing a growing role in the environmental arena. Union activity has been crucial in enforcing and supplementing Occupational Safety and Health Act regulations. Concerns over the health effects of secondhand smoke are working their way into collective bargaining agreements. For example, provisions are being made for smokers to have a separate lounge or area where they can smoke without affecting others.

At times workers and environmentalists are on opposing sides, due to the belief that environmental protection inevitably leads to a loss of jobs. Collective bargaining offers opportunities to protect jobs and the natural environment simultaneously. As energy technologies change, bargaining agreements can provide programs to assist workers in transition and encourage a spirit of cooperation to help industry adapt successfully to increasing regulation. Disputes among citizens and environmental organizations, government agencies, corporations, labor unions, and other groups over a number of issues—for instance, environmental quality, biotechnology, natural resources, agriculture, food and nutrition, and technology—are increasingly being addressed by conflict management organizations.

Such collaboration concerns international issues as well. The North American Free Trade Agreement, for example, has been opposed by both unions and environmentalists. Environmentalists and United States unions fear that industries will relocate in Mexico to avoid United States laws that maintain labor and environmental standards.

COMMENSALISM

See Symbiosis.

COMMUNICATION

Communication is a process whereby people exchange meaning with others. The meaning that is communicated may be an explicit message as in the sentence "Give me the book," or it may be an implicit signal about an emotional state as in a sigh. The ability to communicate is a principal characteristic of the human species. As Colin Cherry notes, "Man is a communicating animal; communication is one of his oldest activities." Communication is also a principal tool in the development and preservation of cultures. It is the primary means by which humans transmit ideas across geographical space and through time.

There has been much debate about the communication capabilities of humans versus animals. Traditionally human communication has been distinguished from animal communication by the ability of people to encode meanings in symbolic forms such as words or pictures and manipulate those symbolic forms to create new meanings. For example, a person who knows the English language can manipulate words to create new and unique sentences. In addition, humans pass these skills along to their children and the communication process continues across generations. While animals can communicate about emotional states such as hunger, perceived danger, and sexual desire, it has been presumed that they cannot use symbols such as words to capture and communicate abstract ideas.

The accepted belief that the manipulation of symbolic forms is uniquely human has been challenged in the past few decades. Several chimpanzees and gorillas have been taught sign language or a set of symbols on a keyboard. Also there have been a few documented cases where animals combined symbols in new and unique ways. However, these cases have been challenged by most scientists, who note that instances where animals have combined symbols to (apparently) create new meanings have been very rare. They may even be random events. The use of symbolic forms to communicate has not passed beyond animal and trainer. The apes have not been able to communicate symbolically with others in their species, nor have they passed the skill along to other generations. Nonetheless, there is a growing recognition that the communication capabilities of animals are greater than had been accepted. Research about the boundaries between human and animal communication is certain to continue.

Origins Of Human Communication

No one knows for certain when humans began to communicate or what the earliest forms of communication were like. However, it is inferred that when humans were able to make tools and transfer this knowledge across geographic space as well as across generations, communication skills must have been in place. This implies that communication was acquired by humans a few million years ago. The earliest forms of communication probably involved vocal signals, e.g., shouts, grunts, and primitive words, that were pre-linguistic. That is, they did not form a complete language as we use that term today. In addition, early forms of communication almost certainly included a system of gestures.

Human communication is rooted in biology but shaped by culture. In order for humans to acquire speech, biological changes had to occur through evolution. Early humans did not have fully developed vocal tracts necessary to speak. Studies of early human anatomies suggest that speech existed between 90,000 and 40,000 years ago, and language was in common use by 35,000 B.C.

What did early humans communicate about? Here too, much of our knowledge about early communications is based upon inferences. We infer that humans communicated about making tools and that much communication surrounded vital activities such as hunting and farming. It is also accepted that the earliest preserved symbols created by humans, such as cave paintings and stone monuments, were not art as we use that term today but forms of communication. Early humans created them to send messages across generations about their lineage, work, and possessions as well as to communicate with deities they worshipped. In this sense, early communication was about vital, functional tasks and important social or religious values.

The Communication Process

In order to understand the communication process, it is useful to distinguish two categories of communication. The first includes the general process of person-to-person communication. The second includes the process of sending messages.

Person-to-person communication is a continuous, multi-channel process. When humans interact with each other, there is a rich and continuous flow of communication between them, even when people are silent. Communication flows through multiple channels such as spoken words, tone of voice, gestures, facial expressions, touch, body orientation, and spatial relations. While these are discrete channels of communication, they function in concert with each other. For example, tone of voice can modify the meaning of words, and facial expressions such as a wink can tell a listener that a spoken utterance should not be taken seriously. Since the flow of communication is so rich, there are many opportunities for errors to occur. Errors can include a signal that is not noticed (e.g., a wink that is not seen) or a signal that is misinterpreted. For this reason, many of the signals in a communication exchange between individuals are redundant—they reinforce other signals as well as modify them.

Person-to-person communication always occurs in a social context, e.g., buying groceries in a supermarket, talking in a classroom setting, or eating dinner in a family setting. The social context for communication is a crucial component in the communication exchange. The social context establishes the code or set of rules for communication. For example, people interact in different ways depending upon whether they are in a classroom or a bowling alley. People adapt their communication behavior based upon their understanding of the code or rules for the setting and interpret the communication behavior of others against the code or rules. Here too, there are many ways in which errors can occur, e.g., when one person's understanding of the rules for communicating in a given setting differs from another person's understanding of the rules.

These characteristics of person-to-person communication demonstrate that the meaning exchanged in a communication exchange is fluid and dynamic. This is why two people interacting with a third person may have very different interpretations of what happened. Indeed, explicit meaning is often an elusive concept when dealing with person-to-person communication. Meaning can change if the social context changes and meaning can change based upon how a person perceives and interprets the many signals in the multichannel communication exchange.

The second general category of communication involves sending messages. This may be part of a person-to-person communication exchange as described above, or one person or a group may create communications for a mass audience, e.g., writing a book, creating a film, or writing a warning label for a pesticide container. In all of these cases, a person or group seeks to create a message, transmit it to others, and have them understand the message as it was intended.

Scholars who have studied these forms of communication distinguish the sender of the message, the receiver, the content of the message, the channel or means of transmission (as well as noise that may interfere with the transmission), and the impact or effects of the message. These are components in the communication process. The process is captured in a question: Who, Says What, To Whom, In What Channel, With What Effect?

The context in which messages are created and interpreted is also important. Consider a photograph of a factory smokestack. During the Great Depression of the 1930s, such a photograph would likely be interpreted to mean "work" or "prosperity." However, in the context of the 1990s when people are concerned about the environment such a photograph is likely to mean "pollution." Public information campaigns by commercial advertisers as well as public interest groups very often involve a struggle to affect the context in which people will interpret messages they receive from others.

In focusing upon mass media, many researchers have been concerned about the power of media to shape messages as well as transmit them. This is the basis for Marshall McLuhan's famous aphorism, "The medium is the message." For example, television fosters the use of powerful visual images. Those who create news and other content are driven to seek powerful visual images in order to win and hold the attention of audiences. However, in doing so, they may change or distort the meaning of an event.

Other researchers are concerned about the power of mass media to create culture and overpower existing cultural values. Media shape our understanding about society through the messages they convey to everyone. These commonly experienced messages help to form our expectations about how the world works and how we should behave. Other messages that shape cultural values, e.g., those from local institutions such as schools, government, churches, and social organizations, are often not presented in as powerful a form and they are not commonly experienced in the way millions of people commonly experience a television program. A child's attitude toward the environment may be influenced more by an adventure show than by public service announcements about recycling and fire safety. The power of media to shape culture, and through cultural values to guide behavior, is problematic when so many of the messages carried by mass media deal with violence, exploitation of groups within society, and disdain for existing social norms.

Communication Technology: Origins

Communication technologies are tools used in the communication process. One of the earliest uses of communication technology involved pigments, sticks, and other instruments to create cave paintings more than 25,000 years ago in Europe and North Africa. Writing as a system of communication and the accompanying technologies used to create and preserve writing emerged in the 4th century B.C. The Egyptians developed a writing system of pictographs and hieroglyphs, and the Sumerians developed cuneiforms as a writing system. Other writing systems were independently developed in China and by the Mayans in Mexico and Central America.

The emergence of writing illustrates the promise, the obstacles, and the potential pitfalls associated with many communication technologies. Writing liberated humans from the onerous task of preserving knowledge through memory and transmitting it through spoken words. Prior to the development of writing, knowledge was preserved in stories and other oral forms that individuals memorized and later passed on to subsequent generations. Writing also allowed knowledge to be spread geographically and preserved accurately over time. However, as in the case of many communication technologies that were to follow writing, there were a number of problems and obstacles to overcome. For example, the first generation of writing materials—stone and clay tablets—was not very portable. This provided an incentive to improve existing technology by developing lighter materials such as papyrus, leather, and parchment. This pattern of upgrading technology after it has been introduced has continued with each new communication technology.

Writing technology also foreshadowed many subsequent technologies in that it was resisted by some groups who felt they would lose power under the new technology environment. For example, Plato complained about the resistance to writing by the education establishment in ancient Greece. Although writing was used in government and business for approximately 100 years before Plato's day, the bureaucrats who maintained an oral system of education resisted the adoption of writing in education.

Writing's impact on society was not entirely beneficial. It fostered a new class of elite scribes who could read and write and who had access to written documents. These written documents became instruments of power and many people were denied access to them. Further, the average citizen who could hear and understand the earlier system of oral knowledge now lacked the skill (i.e., literacy) to learn from a system of written knowledge. Access to technology, skill in using it and, in many cases, ability to pay are potential obstacles to broad adoption of all technologies.

The Emergence of Modern Communication Technologies

The pace of technological change was slow following the invention of writing. The next major advance was the development of a printing press that used movable type in Mainz, Germany, by Johannes Gutenberg in the mid-15th century. The printing press changed Western society dramatically by supporting the mass production of written documents. The capability to produce and distribute printed materials broadly also led to the spread of literacy.

By the 19th century the pace of technological change had accelerated. Samuel Morse invented the telegraph in 1835 and Alexander Graham Bell developed the telephone in 1876. The principal effect of these two technologies was to speed up the flow of information. As information moved more quickly, the pace of change in society accelerated and communities as well as countries became linked in new ways. This allowed many businesses to expand. Satellite offices could be managed from a central office with links by telegraph and telephone. It also helped to create a truly international economy. At the same time, the telephone and telegraph increased the pace of colonization. Colonies in Africa, Asia, the Middle East, and Latin America could be more easily controlled through communication links to a parent state.

The beginning of the 20th century marked the introduction of several important communication technologies, including radio, the phonograph, motion pictures, and color photography. These technologies created mass media—communication services that would eventually enter nearly every household. They also became powerful vehicles for shaping political and social attitudes as well as purchasing behavior. In the 20th century the major institutions of society included not only government, religion, and educational organizations but also RCA, CBS, General Electric, and other organizations that controlled the mass media.

In the second half of the 20th century, communication technologies have become entwined with virtually every aspect of business, government, and home life. People in Western society depend on television, computers, and advanced telecommunication technologies to work and socialize. Telecommuting, working at home with a computer and a modem or fax machine, helps reduce energy use and pollution. These technologies were made possible by the construction of new communication networks that reach throughout individual countries and the world, e.g., satellites, cable television, and new telecommunication networks. They were also aided by reduced costs for equipment, miniaturization of component parts such as transistors and microchips, and portability of many communication technologies.

As we approach a new century, the pace of technological change is accelerating even further. This is made possible by advances in microchip technology and the digital compression of signals as well as the implementation of fiber optics in communication networks and the launching of more powerful communication satellites. These technologies will support a broad range of futuristic services, including cable systems with hundreds of channels, video telephones, movies delivered on demand over cable and telephone lines, interactive television programs, shopping in virtual stores that people can walk through via a communication network, and multi-media instruction that links classrooms across a state or the country. At the same time, there is much uncertainty surrounding these technologies and the services that will be carried on them. Little is known about the economic viability of the new networks, consumer and business demand for new services, and the social impacts that will follow. In this sense, communication technology may be outpacing our ability to use it or to understand the consequences of implementing it.

Communication and Culture

There is a more profound change underway than the acceleration of technological developments in communication technology. It involves the convergence of the social and the technological in forming a culture. Television, computers, and communication networks are not just instruments to transmit information. They are vital components in the fabric of social, education, and business life. They create an environment in which we live just as air and water create an environment for human life. Further, they shape cultural values as well as individual attitudes and behavior. As such, there is a need to treat them as we do the natural ecosystems that support human life and culture.

JOHN CAREY

For Further Reading: Gregory Bateson, *Steps To An Ecology of Mind* (1972); Colin Cherry, *On Human Communication* (1966); W. Russell Neuman, *The Future of the Mass Audience* (1991).
See also Fiber Optics; Labeling; Language; Risk Communication; Technology.

COMMUNITIES, LOW AND MODERATE INCOME

Throughout most of the industrial regions of the world people with limited economic resources and low social status usually bear the added burden of living under dangerous environmental conditions. Residents of poor communities suffer inordinately from industrial accidents and the effects of ecocide. Built precariously on flood plains, steep mine faces, and other unstable landscapes, communities of the poor and the more marginal working classes are frequent victims of avoidable ecological disasters—dam ruptures, mud slides, flash floods, and other results of a disregard for sound environmental planning.

In the late stages of urban industrial civilization, marked by the explosive growth of metropolitan regions with enormous energy budgets, the hazards of poor environmental planning and control are shared somewhat more democratically. Residents of moderate-income communities located near nuclear energy installations, urban landfills and toxic waste sites, or vacation colonies crowded onto unstable ecological zones like barrier islands, are also increasingly harmed by what would have been avoidable ecological disasters. Other moderate-income communities, located near airports or chemical sluiceways (the Love Canal community in northern New York State being one of the most notable examples), suffer the more insidious and initially hidden effects of pollution.

On the fringes of cities in the industrializing world, and inside the older cities of North America, segregated communities host populations of poor and moderate-income people who suffer from excessive mortality due to interpersonal violence (associated especially with illegal drug markets) and from the adverse effects of poor housing and lack of pollution controls. The in-

cidence of respiratory diseases (virulent TB, asthma, and emphysema, for example) is far higher in central city ghetto communities than anywhere else in the U.S. With the exception of the social democratic nations of Europe and a few examples in Oceania (e.g., New Zealand), poor communities—be they central city ghettos or other communities in relative poverty—suffer from the effects of failure to provide adequate low-cost housing, health and educational resources, and parks and open space. France, for example, provides over 25% percent of its citizens with affordable "social housing," built by one or another nonprofit agency with some public subsidy. In the United States less than 5% of the population live in subsidized housing, and in much of the existing federally subsidized public housing and other forms of nonprofit, low-rent shelter, rates of occupancy exceed 120%.

Overcrowding in low-income housing and the destruction and abandonment of subsidized housing in some communities (e.g., Chicago and Newark) help explain the rapid rise of homelessness since the late 1970s. Deindustrialization in the older manufacturing regions of Europe and North America, in favor of industrialization in low-wage regions with less stringent environmental controls, is producing additional increments of poverty and homelessness in low-income communities and is leaving large areas of environmentally degraded land and water for possible but unspecified future restoration.

Examples of poor and moderate-income communities that have unsuccessfully adapted to their environments abound. Many of these examples are drawn from case studies of communities in the early stages of rural and urban industrialization. Much of the literature on communities in Appalachia, for example, is about problematic adaptations to adverse environmental and social conditions. The classic work on Appalachia by Harry Caudill (1963) documents the development of a hill country sub-culture which sustains families and communities over generations. Too often, Caudill argued, this sub-culture fosters fatalism toward the powerlessness of the poor when confronted by the policies of distant corporate owners of the mines and forests. Kai Erikson's study of the Buffalo Creek Disaster (1976) is another modern classic about a mining community faced with industrial and environmental disaster. Hundreds of poor Appalachian families were wiped out when the haphazardly constructed Buffalo Creek Dam, a project of the regional coal company, burst and wiped out the hamlets along the creek below. People in the villages knew the company's dam was poorly designed and built but thought there was little they could do about it until it was too late to protest. A point which emerges from all the literature on communities in rapidly industrializing but impoverished regions is that local organizing in favor of environmental control and quality of community life can successfully confront fatalism and prevent disasters.

The literature also carries the warning, however, that outward appearances of failure need careful assessment before viable communities are destroyed. Janice Perlman was one of the first social scientists to point out in 1976 that the shantytown communities in Brazil and elsewhere in Latin America often appear to be poor adaptations, especially from a purely environmental standpoint. Socially, however, they are often quite cohesive and stable human environments and become centers of creative entrepreneurism. Perlman and others have shown that the fervor with which state planners raze the favellas or shantytowns often leads to the destruction of the communities and their replacement with sterile, ill-conceived poverty tracts located even further from the residents' places of employment and exhibiting their own severe environmental problems. These experiences point to the urgent need for more democratic planning and less hasty destruction of what seem to be environmentally unsound communities.

Residents of low-income and blue collar communities are often said to accept the tradeoff between jobs and environmental stress. "Smokestacks mean healthy paychecks" was the slogan voiced in communities built adjacent to heavy industrial plants in the industrialized nations earlier in the century. Accidents such as those at Three Mile Island, Bhopal, Chernobyl, and hundreds of less well-known episodes which affect the lives of low-income community residents even more immediately, have led them to reject their earlier fatalism. They increasingly demand more environmentally sound economic development. Where they feel it is appropriate, significant segments of their populations support strict environmental control and sound ecological adapations, a trend which the resistance to deforestation shown by communities in the Amazon suggests is likely to develop fitfully throughout the world.

WILLIAM KORNBLUM

For Further Reading: Cynthia M. Duncan, ed., *Rural Poverty in America* (1992); William Kornblum, *Sociology in a Changing World,* 3rd ed. (1994); Alejandro Portes and Richard Schauffler, "Competing Perspectives on the Latin American Informal Sector," *Population and Development Review* (March 1993).
See also Bhopal; Chernobyl; Equity; Homelessness: Causes; Love Canal; Rural Communities; Three Mile Island; Urban Challenges and Opportunities; Urban Renewal.

COMMUNITY

An ecological community is an assemblage of species that occur together in space and time. Often, there are interactions among these species, although not all species present in an area will necessarily interact. Some ecologists use the term *association* to distinguish sets of species that simply co-occur from sets of interacting species (a *community*).

The community represents an organizational level above the population or species in an ecological hierarchy. Communities may be considered at many scales, from assemblages of organisms found in a square centimeter of forest floor or a square meter of intertidal shoreline to those occurring over square kilometers. Ecologists usually specify communities in terms of kinds of organisms (e.g., bird communities, lizard communities, microbial communities) or environments (e.g., rocky intertidal communities, forest communities, old-field communities). The determination of community scale and boundaries is arbitrary, and depends on the questions being asked or the management mission at hand.

One major thrust of research in community ecology has focused on describing the structure and distribution of communities over the landscape. During the 1920s two distinctly different views of community organization developed. One held that communities were highly organized by virtue of interdependencies among the species, and that communities with a particular species composition would therefore occur repeatedly wherever environmental conditions were similar. Some proponents of this view likened the community to a "supraorganism," and compared the apparently orderly progression of community types in ecological succession to the development of an organism and the final "climax" community to the mature adult. This view fostered the description and classification of community "types" in a manner similar to that applied to species by systematists.

Other ecologists doubted that communities exhibited such a tight organization. At its extreme, this "individualistic" view held that species responded to environmental conditions completely independently of one another, occurring where conditions met their needs. The community present in a location was no more than a set of species that happened to have similar ecological requirements. Succession, by this view, involved the disappearance and appearance of species at a location as environmental conditions changed through time.

Both approaches have been transformed by modern computer technology, which permits information on the species composition at a large number of locations to be analyzed in a variety of ways. Ecologists viewing communities as discrete types have used clustering procedures to classify the communities quantitatively, whereas those adopting a less structured view have employed multivariate techniques to describe how community composition changes along environmental gradients. These analyses show that many species are indeed distributed independently of one another on environmental gradients. However, because many environmental features do not vary continuously in nature, there are discontinuities in environmental gradients, and this leads to groupings of species that can be classified as community "types." As with most controversies, there are elements of truth in both views.

A second major thrust of community research has focused on interactions among co-occurring species. These interactions, such as competition or predation, are important to understanding why particular sets of species do or do not co-occur and why some communities contain more species (have greater *biodiversity*) than others. Many ecologists have held that, because populations grow until their numbers are limited by the availability of resources such as food or breeding habitat, species that have similar resource requirements will frequently compete. If the species overlap extensively, the competitively dominant species may exclude the other species from the area, and thus from the community. If competition is widespread among sets of ecologically similar species (ecological *guilds*), it may set limits to the diversity and species composition of the community. Alternatively, competition may force species to adjust their ecological requirements so as to reduce overlap between them. Such adjustments may be either short-term (behavioral or physiological) or long-term (evolutionary). In either case, they may permit the formerly competing species to coexist, enhancing community diversity. Many communities in low-latitude, tropical regions are more diverse than those in high-latitude locations. The greater availability in the tropics of resources that have less seasonal fluctuation, combined with finer competitive adjustments among species, may at least partly explain this pattern.

Competition is not the only way in which species may interact. For competition to play a major role in determining community composition, species populations must be limited by resource availability. Predation may prevent populations of competitors from reaching levels at which one can exclude the other. Parasitism may have similar effects. Unless they lead to the local eradication of species, both predation and parasitism may act to enhance community diversity. Other forms of environmental disturbance, such as disease outbreaks, habitat alteration by fire or human development, or abnormally harsh weather, may also depress populations of competitors below resource-

limitation levels, permitting them to coexist. These effects are collectively contained in the *intermediate disturbance hypothesis*, which holds that community diversity will be greatest when disruptions occur at intermediate frequencies and strengths. When such disturbances are weak or infrequent, competitive interactions may reduce diversity, while frequent and severe disturbances may lead to the disappearance of some intolerant species.

Competition is therefore likely to determine community composition only when there is a general equilibrium between resource supplies in the environment and resource demands by the species. But natural environments may not often be stable enough to permit such equilibrium conditions to develop. Other interactions among species may vary in occurrence and intensity, and weather conditions often vary substantially between years.

The complexity of interaction and disturbance effects on communities and the nonequilibrium nature of natural ecosystems make it difficult to develop predictions about communities, especially about the precise species composition of communities in particular locations. Thus, even though species occur as members of communities, most management of natural, agricultural, or urban environments is focused on particular species of interest rather than on communities. The webs of interactions among species in communities, however, act to spread the effects of management actions or unplanned human interventions in unforeseen directions. Eradication of predators to enhance game populations, for example, may lead to increased competition among the game species and the eventual exclusion of some species. Introduction of a predator to control populations of a pest species may allow a competitive subordinate of that species to reach outbreak levels. Many of our management failures or surprises stem from an inadequate consideration of the community matrix in which species are embedded. The development of approaches to community management is in its infancy, but it should clearly be a high priority.

JOHN A. WIENS

For Further Reading: J. Kikkawa and D. J. Anderson, eds., *Community Ecology: Pattern and Process* (1986); P. A. Keddy, *Competition* (1989); J. Diamond and T. J. Case, eds., *Community Ecology* (1986).

COMPOSTING

Composting is a method of recycling organic residues and wastes for use as fertilizer. Compost is the product of the decomposition of materials by microbial organisms, which break down the waste material into a stable fertilizer with a carbon/nitrogen ratio of ten to one. Composting supplies a large quantity of nutrients—nitrogen, phosphorus, and potassium—which helps to improve soil texture and health, in turn yielding healthier plants.

The composting process mirrors the natural decomposition processes in soils. In natural soils various materials are metabolized, excreted, and broken down by bacteria and microorganisms to form humus, a stable source of nutrients readily available for plants. Because gardening and agriculture remove so much of the humus layers from the natural soil, the aim of composting is to replace that nutrient source and reincorporate it back into the soils. Therefore, composting is not only a method of producing fertilizer for gardens, but also for maintaining a balance within the local ecosystem from which it derives.

There are two basic approaches to making compost: aerobic and anaerobic. The aerobic technique involves piling various organic materials in layers and turning the pile regularly to give the active bacteria and other microorganisms enough oxygen to continue consuming, breaking down material, and reproducing. As the microorganisms flourish in the compost pile, the temperature rises to between 120° and 150° F. At this point the pile is turned, the temperature drops, and the process begins anew. With each turning, new stages of organisms take over, until activity virtually ceases. After three months the end result is a stable, viable, nutrient-rich soil.

The anaerobic method is a slower process. It also involves piling organic material in layers, but the material is allowed to break down without turning. The temperature in the pile remains low, and there is insufficient oxygen for rapid decomposition. Theoretically, the end result should be the same. Although the aerobic method is more labor-intensive than the anaerobic method, the production time from start to finish is at least 50% less. In addition, the high temperatures in the aerobic pile kill off weed seeds and unwanted pathogens—many of which would survive in the anaerobic compost.

COMPREHENSIVE ENVIRONMENTAL RESPONSE, COMPENSATION, AND LIABILITY ACT

See Superfund.

CONSERVATION

See Environmental Movement.

CONSERVATION EASEMENTS AND OTHER COVENANTS

In setting aside land for the preservation of natural habitats and wildlife, many environmental organizations and private citizens have sought the use of conservation easements and other covenants—private legal agreements between two or more parties that determine how a specific piece of land will be used. As the value of land resources increases, individuals turn more frequently to private agreements to resolve present and potential conflicts over their use.

An easement is a grant of an interest in land by the possessor of the land to another. The owner of the easement is then entitled to limited use or enjoyment of the land and protection from interference by a third party in such use or enjoyment. Once the easement is conveyed, the use of the land is not subject to the will of the landowner and cannot be terminated at will. The term of the easement can be for a certain number of years, for life, or in fee. An easement that has the potential of enduring forever is called an easement in fee simple. A conservation easement may convey to the easement owner the right to maintain the land as a wilderness area. Once this is established, the owner of the easement can legally compel the owner of the land to refrain from acts that go against the easement.

The purpose of these easements is to form an agreement that will be binding on future generations of owners of the property involved. In order for the agreement to be binding on these future generations, the successors (heirs and assigns) must have notice of it. For example, Mrs. Smith may have granted to XYZ Conservation Organization an easement to maintain her land as a wildlife preserve for the next one hundred years. When Mrs. Smith's heirs become owners of that land, they also must respect the easement and cannot subject the land to uses that go against the easement, such as commercial development or mineral excavation.

A covenant differs from an easement in that it is not a property interest in the covenantor's land but a contractual limitation upon the covenantor's estate. A covenant is a promise to do or refrain from doing a certain thing pertaining to the use of the land by the owner or possessor thereof. The covenantee has no rights in the land itself. The following is an example of a promissory covenant: "Mrs. Smith conveys property to Mr. Jones and his heirs, and Mr. Jones promises on behalf of himself, his heirs and assigns, to use the property for conservation purposes." Covenants can also place a durational limit on the estate conveyed. In this case, Mrs. Smith may convey property to Mr. Jones for as long as Mr. Jones and his heirs use the property for conservation purposes.

CONSERVATION MOVEMENT IN THE U.S.

The word *conservation* refers to the act of saving or preserving an item from being wasted, of guarding and protecting something of value from being lost. Since the beginning of the 20th century it has taken on a more specific meaning that includes the maintenance and supervisory protection of forests or other natural resources under private or public control.

Medieval Activities and Conservation

Until the modern industrial age, the ordinary routines of most people called for activities that depended on land and forests. When human populations were small and challenges to staying alive great, few people thought about the long-range consequences of their actions until conditions of survival were threatened.

Beginning in ancient times and extending through the Middle Ages, hundreds of tribes and ethnic groups worked their way through Europe from Asia, and later from feudal Europe back toward the East. They cleared forests for agriculture, irrigated their fields, diked and built canals, drove back wild animals, made polders to hold back the sea, grazed sheep or cattle in mountain meadows, killed animals that threatened their farm produce, and did whatever else they thought was necessary to survive. The centuries were marked by the retreat of forests, loss of heath and marshes, the effects of mining and quarrying, and the creation of new towns with larger farms and flocks to provide for them.

This march of civilization took many centuries. In Judaeo-Christian cultures, humans' use of nature was accepted as sanctioned by God and the feudal king. Charlemagne encouraged forest clearing "in order to improve" his possessions for himself and his successors. But for the same reasons Charlemagne also protected many of his forests for hunting the "royal game," and exacted a tax for the use of the forest.

During the Middle Ages in Europe there was widespread concern, expressed in royal decrees, about quickly rising human populations, the rapid increase of industries that required large amounts of wood, and the practice of forest clearing by fire. But consciousness of the need for conservation did little to stop rapid clearing of forests worldwide.

When the industrial age arrived in Europe in the late 17th and 18th centuries, themes of conservation were less prominent because of the prevailing optimistic views of the Enlightenment that had taken hold. All nature could be brought under the control of reason, and technologies of the new age, leaders believed, would abolish poverty and usher in a new era of comfort in democratic societies.

American Settlers and Conservation

The first European settlers in America found dense woods and abundant wildlife. They systematically cleared forests, planted fields, and pushed back wildlife. By the late 1700s, entrepreneurs found they could make a great deal of money from wood, fur, gold, iron ore, and coal. They extracted what they wanted from the land, and when nearby resources were depleted, they moved to more lucrative areas.

During the 19th century, as logging and mining became more efficient, and as growing towns were tied together by railroads, the landscape changed dramatically. Public lands on which forests grew were pillaged for fuel and railroad ties. Laws were made for quick settlement, not for conservation. By the end of the 1880s, when the frontier closed, the federal government had lost control over its best land, and much public land had gone to private owners who did what they wanted.

There was also enormous waste in the search for other natural resources. Strip mining for coal left Appalachia in ruins. Petroleum was allowed to drain into streams, catch fire, or evaporate because there seemed to be plenty of it. Hundreds of species of wildlife were decimated or depleted. Hundreds of thousands of bison were deliberately exterminated in order to make room for cattle and bring native Americans under control. Beavers, seals, polar bears, walruses, whales, and other animals brought such an attractive price that their populations were quickly brought to near extinction.

It was inevitable that a reaction should set in because of the despoilment of nature. European travelers to North America during the 1700s complained about American waste of soil and land. George Washington was embarrassed by the way his fellow farmers treated the land, but he admitted that land was cheap and labor expensive. It took extra work and money to fertilize fields, rotate crops, protect against erosion, and otherwise conserve the soil.

The voices of protest arose from people and groups embodying a number of diverse values and interests. In the 19th century, agricultural reformers preached conservation of soil. Scientists and sportsmen promoted forest, wildlife, and natural resource conservation. By the turn of the twentieth century their ideas entered the political agenda, primarily through the conservation program of Theodore Roosevelt, who loved the outdoors for its own sake, but who was even more concerned about the economic necessity of conservation of natural resources for future generations.

Conservation values in the U.S. can be classified into the biocentric, the ecologic, and the economic. Those who look at the natural environment from the biocentric point of view concentrate value in nature for and in itself, apart from the human uses of it. They uphold rights for beings of the natural world as equal to rights of human beings. The biocentric tradition stems from deep religious feelings which were expressed in ancient times through the religions of India, China, Japan, and aboriginal cultures. These religions—Taoism, Buddhism, Shintoism, and animism—have expressed reverence for nature because of the divine principle that creates and sustains material and animal life together with human beings in the world.

The ecologic emphasis is derived from a scientific understanding of interrelationships and interdependence among the parts of natural communities. The important ecological concept for this group is a model of a stable community made up of plants and animals, preferably rich and diverse, and a traceable flow of energy that can be disrupted by natural disasters or, more commonly, by human activity and interference.

The economic perspective, sometimes more broadly called the utilitarian approach to conservation, focuses on the optimal use of natural resources for the longest period of time. This value orientation hates to waste anything, especially natural resources. The efficiency interest of economists, agency officials, or decision makers has often placed them into an active defense of the natural environment within the economic wing of the conservation movement.

The Biocentric Perspective

The biocentric viewpoint in America was first expressed in the transcendentalism of Ralph Waldo Emerson and his disciple Henry David Thoreau. Emerson saw America and its treasures of wilderness as ready to be overcome by a "material interest." Thoreau's *Walden* is also an essay against materialistic commercialism. Emerson and Thoreau perceived the value of nature or wilderness in itself; John Muir experienced the fact, and his newspaper accounts, journal articles, essays, and letters are filled with a passionate expression of that experience. He founded the Sierra Club to protect the beauties of Nature and lived a life that has inspired countless exponents of the wilderness mystique.

Although Emerson, Thoreau, and Muir can be described as the early pillars of the American biocentric perspective, their own values were shaped by dozens of earlier well-known 19th-century figures. For example, John James Audubon's *Birds of America* fed a growing appetite for natural history. He lamented the "destruction of the forest" and consequent loss of thousands of wild birds and animals. Joining him in his criticism were James Fenimore Cooper (especially in *The Prairie*), painter Thomas Cole, writer Washington Irving, poet William Cullen Bryant, essayist Charles Lanham, traveler-historian Francis Parkman,

and artist George Catlin, who first proposed the idea of establishing national parks.

During the latter part of the 19th century and into the 20th century there was a growing movement to preserve the wilderness for the "spiritual benefit" of Americans. In 1882 a former taxidermist, William Temple Hornaday, became the first director of the New York Zoological Park and became famous as a zealous spokesman for wildlife protection. He was an adamant foe of sport hunting, worked tirelessly for laws against the sale of wild game and the importation of wild bird plumage, and wrote close to 20 books on wildlife preservation.

By 1920 the biocentric perspective was summed up in Albert Schweitzer's ethical maxim of "reverence for all life, plants and animals as well as one's fellow man." By the middle of the 20th century the tradition was the driving force of the new environmental movement. The Endangered Species Act of 1973 institutionalized the perspective, and new theories of Deep Ecology and Gaia made it into an explicit philosophical religion. The new environmental movement's most influential spokesman, David Brower, summed it up: "I believe in wilderness for itself alone. I believe in the rights of creatures other than man."

The Ecologic Perspective

The ecologic view rests on scientific understanding of and respect for natural systems rather than valuing single components of the ecosystems, such as individual animals or trees in themselves.

George Perkins Marsh in his 1864 book *Man and Nature* sought to understand the biological laws which explain successional processes and balance in nature. He was fascinated by the damage humans could do to the earth by overcutting, overgrazing, and thoughtless agricultural practices. The scientific presumption in Marsh's writings is that nature enjoys a self-regulating balance, even under the stress of great natural catastrophes, which enables it to restore itself to a kind of primordial harmony as long as human activity does not interfere with the process irreversibly. Marsh's formulation of a scientific viewpoint was influential in the first U.S. conservationist political movement toward the end of the century. His ideas influenced the policies of Franklin Hough, the first head of the Division of Forestry, and Charles S. Sargent, an active member of the American Forestry Congress, as well as Gifford Pinchot, the famous conservationist head of the Forest Service under Theodore Roosevelt.

Other early American ecological scientists were Frederic E. Clements and Henry Cowles, who worked on midwestern grasslands and sand dunes at the turn of the century. In 1915 enough ecologists were doing research to establish the Ecological Society of America.

Through the efforts and writing of Aldo Leopold, whose *Game Management* was published in 1933, ecology began to be applied to biological systems management. Many sportsmen's clubs joined efforts to manage wildlife through ecological studies because game managers were beginning to recognize hunting as a way to reestablish a balance in some ecosystems. For example, Leopold saw that the Kaibab deer in the American Southwest no longer had natural predators to control their numbers because ranchers had wiped out wolves, coyotes, and pumas. State laws were changed to permit regulated hunting to thin out the populations of some wildlife.

Recreational hunters had been calling for regulation of game animals for hunting for about 100 years before these efforts. The New York Sportsmen's Club was founded in 1844 for the purpose. Many writers like Henry William Herbert (writing under the pen name of Frank Forester), George Bird Grinnell, an editor of *Field and Stream* magazine in 1876, and others campaigned to protect game animals. They increased their political strength during the presidency of Theodore Roosevelt. In 1922 the Izaak Walton League of America was established to work for game protection and clean water.

In 1948 two more writers, natural scientists Fairfield Osborn, in *Our Plundered Planet*, and William Vogt, in *Road to Survival*, added the human factor to the ecological perspective. They noted the lack of ecological balance between human populations and the natural resources on which mankind depends for survival and reminded their fellow citizens of the devastating impact humanity was having on the land. These writers are the links to biologists such as Paul Ehrlich and Garrett Hardin, who have been major influences in the modern environmental movement since the first Earth Day in 1970.

The Economic Perspective

Those who espouse the "wisest," most efficient use of natural resources over the longest period of time have generally been considered to belong to the "utilitarian" wing of the conservation movement. In the past this group has been associated with the federal government and more recently with a new breed of environmental economists.

In 1877 the secretary of the interior under Rutherford Hayes, Carl Schurz, emphasized that forests should be preserved and managed for the long-term use of the American people. Not until 1891, however, did the president receive the power from Congress to set aside forest reservations. By the end of Theodore Roosevelt's administration, thanks to a growing public awareness, almost 170 million acres of land had been set aside as national forests and parks. The person who

was largely responsible for the forest reserve clause, Bernhard Fernow, was trained in forest management in Germany. He argued for government responsibility in limiting rights of private parties over natural resources and for the need to manage them for the long-term public good of society.

Gifford Pinchot has been recognized as the embodiment of the values of early economic conservationism. Steward of Theodore Roosevelt's Progressive conservation program, Pinchot wrote, "The first great fact about conservation is that it stands for development . . . (not just) husbanding of resources for future generations . . . but the use of natural resources now existing on this continent for the benefit of the people who live here and now. . . . In the second place conservation stands for the prevention of waste. . . . The third principle is this: The natural resources must be developed and preserved for the benefit of many, and not merely for the few. . . . Conservation means the greatest good for the greatest number for the longest time."

These notions remained strong into the Great Depression of the 1930s. President Franklin D. Roosevelt took up the theme of the need for government intervention to control the "havoc" that had historically characterized U.S. natural resource use and allocation. Roosevelt had a long-time interest in conservation and practiced it on his own estates. The president saw severe drought, over-cut forests, massive soil erosion, and other environmental problems as symbols as powerful as the 15 million unemployed Americans during his tenure. He selected a number of well-known conservationists to help him develop policy—Secretary of Interior Harold Ickes; agricultural experts Henry Wallace and Rexford Tugwell; Henry Morgenthau; and the "father of soil conservation," Hugh Hammond Bennett. Roosevelt's central idea during the Depression was the vision of a new start for the country with wise use of natural resources over the longest possible time for its citizens.

Sustainable Development

The central ideas of the conservation movement have most recently been formulated in terms of "sustainability." In 1987 the World Commission on Environment and Development published *Our Common Future*, based on three years of discussions with experts and people from all over the world. The document calls for sustainable development as "development which meets the needs of the present without compromising the ability of future generations."

The notion refers to the need to protect our world's natural resources and the maintenance and improvement of living standards, as traditional conservationists of the economic perspective have maintained. But ecologists insist that the term includes the preservation of ecological systems. And biocentric advocates believe that sustainability is not possible without reverence for nature in and for itself.

During the 1992 U.N. Conference on Environment and Development, all three perspectives of conservation were discussed under such issues as intergenerational fairness, large-scale damage of tropical forests, the decline of species diversity, global warming, air and water degradation, and global carrying capacity. The reaction to these threats has come from strong traditions of conservation in movements spreading around the world.

JOSEPH M. PETULLA

For Further Reading: Roderick Nash, *Wilderness and the American Mind* (1973); Joseph M. Petulla, *American Environmental History*, 2nd ed., (1988); Joseph M. Petulla, *American Environmentalism: Values, Tactics, Priorities* (1980).
See also Deep Ecology; Environmental Movement; Environmental Organizations; Gaia Hypothesis; *Our Common Future*; Religion; Soil Conservation; Transcendentalism.

CONSUMER WASTE

See Recycling.

CONSUMPTION

Consumption is the total amount of goods and services that are obtained and used by individuals. Goods and services encompass a wide range, some supplied by the market (such as food or housing) and some supplied by governments or nature; they include both tangible commodities (e.g., meat and potatoes) as well as intangible services such as the beauty of a flower garden. Some goods and services are produced by humans, e.g., housing, some are produced by nature, e.g., scenic resources that we all enjoy, and some are "given" to humankind, such as our endowment of minerals in the ground.

The process of consuming is not a straightforward matter of physically ingesting or wearing out a commodity as it is enjoyed. Consumption can also involve simply viewing a beautiful scene or benefiting from national defense—a passive form of consumption. Another twist is that consumption need not always contribute to one's well-being. People consume pollution by simply living in a polluted area. Pollution is a "bad" rather than a "good," but in all other respects it is consumed and thus is a part of consumption. Nearly everything that impinges on a person's well-being can be considered consumption.

Consumption is not confined to the standard process whereby a good is destroyed in the process of enjoyment (as would be the case when a hamburger is consumed). Many environmental goods are totally unaffected by the process of consumption. People do pick flowers, but they can also consume a flower garden by looking at it: the flower garden remains unaffected by having been viewed. Short of people treading on each others' toes, consuming a scenic vista, television program, or library book need not reduce the prospects for others to enjoy that same commodity.

The ultimate form of passive consumption involves existence value. Many people gain pleasure from simply knowing that elephants exist in Africa, even though they have no plans to visit Africa. They are consuming the existence of African elephants. This is a somewhat controversial concept, largely because existence value is difficult to observe and difficult to separate from other ways in which we view the consumption of commodities.

A subtlety arises with goods that have a multi-year lifetime and provide services over that lifetime. An example would be a boat that will last ten years and provide recreation services over that period. The boat is not being consumed, but rather the daily or annual services provided by the boat as long as it exists.

The concept of consumption of conventional goods and services is straightforward; when the concept is expanded to environmental goods and services, things become a little more cloudy. Environmental consumption can include the ingestion of polluted air (or, viewed positively, clean air), the viewing of a scenic vista, the enjoyment of a wilderness experience, the pleasure from experiencing nature, the displeasure from the killing of dolphins, and a wide variety of other environmental effects that influence individual well-being.

CHARLES D. KOLSTAD

CONTINGENT VALUATION

The Contingent Valuation Method (CVM) attempts to estimate the dollar value of changes in environmental quality. It estimates values for goods and services that do not have market prices. Values estimated with this method are intended to provide analysts and policy makers with the information that is required for comparisons of benefits and costs associated with proposed programs for environmental improvements. Important tradeoffs exist between improved environmental quality, jobs, and prices for goods and services.

Estimates for costs associated with environmental programs may often be straightforward, for example, the cost of control equipment for smokestacks, emission controls on automobiles, or restricted lumbering activity to preserve the habitat of an endangered species.

The estimation of benefits associated with such programs, however, involves difficult and subjective questions. What is the dollar value of cleaner air or cleaner water? What value does society place on the preservation of a potentially extinct species? What values accrue to society from reduced risks to public health and safety that might result from more rigid regulation of disposal sites for toxic materials? CVM is presently the only technique for estimating these values that can be used for many environmental issues.

In CVM a sample of the population responds to a questionnaire that is administered either by telephone, by mail, or by in-person interviews. The questionnaire generally describes a proposed environmental improvement that could result from a specific set of programs. Secondly, a "provision rule" states the conditions under which payments reported in the CVM survey would result in the realization of the proposed environmental improvement. Finally, the survey participants are asked the maximum amount of money that they would be willing to pay (often in the form of higher taxes), or alternatively, whether they would be willing to pay a specific sum, *contingent upon* a realization of the posited environmental improvement. Thus the valuation process used in the CVM involves an environmental improvement and a "payment" that are hypothetical.

The CVM is an effort to stimulate decision making processes in ways that allow the researcher to infer values of individuals for an environmental improvement from people's behavior in a controlled setting. In this regard, most CVM studies attempt to simulate either market processes or voting/referendum processes.

There exists no theory that relates to individual valuation behavior in markets or referendums under conditions in which the good being purchased or the issue on which people are to vote is hypothetical and implied economic commitments are hypothetical. Therefore, as a theoretical basis for applications of the CVM one must presume that the received economic theory of individual behavior in markets where real economic commitments are made, or the majority rule principle derived in social choice theory, is relevant for the hypothetical context of the CVM. The consistency of people's valuation behavior in the CVM with that assumed in value theory or the majority rule principle is, of course, an empirical question. Unfortunately, there does not currently exist a body of empirical evidence that might establish this consistency in any compelling way. Thus there exists no basis for drawing unequivocal conclusions as to the theoretical substance of values derived with the CVM.

Empirical evidence on the extent to which CVM values represent real economic commitments is mixed. A small number of studies have directly compared CVM values for an item with values for the same item obtained under real economic conditions. Most suggest that CVM values may substantially overestimate real economic commitments.

The accuracy of a CVM survey may well depend on the investigator's success in structuring the questionnaire in a way that appropriately communicates a real valuation context. A great deal of contemporary research is focused on this issue, making use of verbal protocols with subjects in focus groups as a means for exploring how a subject (as opposed to the investigator) interprets the context for valuation questions. Results from these ongoing efforts are promising in terms of the potential for improving the accuracy of CVM values. It is premature, however, to assess the extent to which this promise might be realized.

CVM research over the last two decades has substantively expanded and enhanced our understanding of "nonuse values"—values held by individuals who do not use the environmental resource in question. For example, a proposed program for improving the water quality of a river in New Mexico may be valued by a resident of South Carolina, even though he may never have visited the river for any recreational purpose or plan to visit it in the future. He is clearly a nonuser of the river, yet may still be prepared to pay some amount of money to see the proposed program implemented. A number of CVM studies have substantiated the existence of this class of nonuse value. The quantitative magnitude of these values remains an open question.

CVM research has begun to identify the range of environmental issues that are seemingly of concern to people and the *potential* range of values that people may attach to them. It has become clear that the value people assign to a particular project must also be influenced by their perception of other environmental programs they may wish to support and by the limitations of their ability to actually pay. For example, CVM studies have demonstrated the public's hypothetical willingness to pay for improved air quality in cities such as Denver and Los Angeles as well as in recreation areas such as Grand Canyon National Park; improved water quality in specific lakes and streams such as the Monongahela River of West Virginia and Pennsylvania; the protection of endangered species of wildlife; reductions in environmental risk associated with the siting and transportation of toxic materials; reductions in the frequency and seriousness of oil spills; and the preservation of wilderness areas.

Average values estimated by CVM range from $15 to as much as $100 or more per household per year for each program. Obviously actual dollar support for the entire range of programs cannot be assumed. Further, when responding to a CVM questionnaire on, for example, improved air quality in a specific national park, the subject may interpret the question as including improved air quality in several parks, all parks, improved air quality throughout the U.S., or possibly environmental improvements of *all* kinds. The CVM must be further refined to elicit the subject's valuation for one specific environmental issue with *other* potential environmental (or, more broadly, all publicly provided) goods being appropriately taken into account.

RONALD G. CUMMINGS
GLENN W. HARRISON

For Further Reading: J. B. Braden and C. K. Kolstad, eds., *Measuring the Demand for Environmental Quality* (1991); R. G. Cummings, D. S. Brookshire, and W. D. Schulze, *Valuing Environmental Goods: An Assessment of the Contingent Valuation Method* (1986); Robert C. Mitchell and Richard T. Carson, *Using Surveys To Value Public Goods: The Contingent Valuation Method* (1989).
See also Benefit-Cost Analysis; Life Cycle Analysis; Risk-Benefit Analysis.

CONTRACEPTION

The ability to control fertility is a recent development for human beings. If the entire era of humankind is considered as a 24-hour day, the 10,000 years of the historical epoch becomes but a second or two, and the period during which women have been able to control their fertility is a scant millisecond. Except for this minuscule fraction of the human past, the usual pattern of life for surviving females was to achieve the menarche and, from then on, to experience alternate periods of pregnancy and lactation until early death. It was only rarely that women reached the menopause. When the risk of dying during pregnancy or childbirth ended with secondary infertility, the affliction may have been welcomed, provided the social customs of the time did not call for harsh punishment for young women who could no longer bear children. From such cases the survival value of reduced fertility for both mothers and their children must have been evident, for methods to prevent pregnancy have been pursued throughout recorded history.

Early attempts were usually based on preventing sperm from making their way to the arena of fertilization. Methods used ranged from withdrawal (*coitus interruptus*) to the introduction into the vagina of sub-

stances intended to kill entering sperm. Camel dung, prairie grass, and other herbal products are among the ingredients recorded in the folklore of ancient cultures. In more recent times, sperm have been confronted with various forms of vulcanized barriers or plunged into lethal pools of jelly, cream, foam, or effervescent fluids. For all these methods, ancient or modern, the scientific basis has been the realization that the ejaculate contains the male factor responsible for starting pregnancy, an observation recorded in Sanskrit thousands of years ago in the Indian vedas.

It was not until 1932, when Kyusaku Ogino proposed the rhythm method of periodic abstinence, that medical science focused attention upon the woman's ovulatory cycle as a key event for controlling fertility. Soon afterwards, endocrinologists began to marshal the knowledge necessary to prevent ovulation; when they succeeded, fertility control was revolutionized. The era of hormonal contraception was launched and with it have come "the pill," contraceptive injections, and, more recently, the first long-acting contraceptive implant.

The pill (oral contraceptive) was developed by Gregory Pincus and co-workers at the Worcester Foundation for Experimental Biology, Worcester, Massachusetts, and introduced in the U.S. in 1960. By 1990 it had become the country's most popular method. More than one out of four American couples who use a contraceptive choose one of the 47 oral contraceptive products on sale as prescription drugs (Table 1). Most products are a combination of two female hormones, estrogen and progestin. Some use progestin only. Oral contraceptives work primarily by preventing ovulation. They can also thicken the cervical mucus, making it less penetrable by sperm. Epidemiological studies have shown that while oral contraceptive use is associated with increased risks of several serious conditions, including heart attack, stroke, and thromboembolism, the risk to healthy nonsmokers is very

Table 1

Methods of fertility control used by American couples (1990 National Survey of Family Growth)

Method	Percent of couples using contraception who use method
Oral Contraceptives	27.7
Tubal Sterilization	24.8
Condom	13.1
Vasectomy	10.5
Diaphragm	5.2
Rhythm	2.1
Withdrawal	2.0
IUD	1.8
Spermicides	1.7
Sponge	1.0

small. There is no proven increase in risk of cancer among oral contraceptive users, and for ovarian and uterine cancers there appears to be a protective effect.

One of the hormones in oral contraceptives, progestin, can also be used in an injectable or implant form. Injectable preparations can last from one to three months and are used by about six million women in 90 countries. After many years of controversy, the first injectable contraceptive was approved for use in the U.S. in 1992. In 1990 the Food and Drug Administration approved the world's first implant contraceptive that can give virtually certain protection against pregnancy for five years. Marketed as NORPLANT^R, this method is spreading rapidly throughout the world. It consists of six small, flexible tubes containing the contraceptive hormone that are placed in the superficial layer under the skin by a minor surgical procedure. They can be removed at any time to restore normal fertility.

Tubal sterilization is popular among American women. It is chosen by six out of ten sexually active women who do not intend to have a child in the future. Male sterilization is less popular. About 10% of American couples use vasectomy to control their fertility. It involves cutting or blocking the vas deferens (the narrow tube that carries sperm from the testis) to prevent sperm from mixing with seminal fluid prior to ejaculation. Neither surgical procedure can be considered reversible, because success rates of the reversal surgery are low for both men and women.

Intrauterine contraceptive devices (IUDs) are the most widely used, reversible method of fertility control, worldwide. They are not, however, used extensively in the U.S., where fewer than 2% of contracepting couples use them. IUDs account for more than half the contraceptive use in China and about 30% in the Scandinavian countries. The most advanced IUDs (the Copper T380, for example) are among the most effective reversible methods available. They are not associated with serious health risks and provide greater satisfaction to their users than other methods, according to surveys carried out in the U.S.

Among barrier methods used for contraception, the condom is the most popular. In the U.S., it is used by 13% of all couples who are using contraceptives. In addition to protecting against pregnancy, condom use can help prevent the spread of sexually transmitted disease, including HIV/AIDS. Other barrier methods in use are diaphragms, vaginal sponges, cervical caps, and spermicides. As a whole, barrier methods are not extremely effective. On the average, pregnancy rates with these methods exceed 15%. The major problems that lead to failure appear to be noncompliance or unavailability of the method at the time of coitus.

Worldwide, there has been a dramatic increase in the use of methods to regulate fertility, due in part to the initiation of national family planning programs in developing countries in the 1960s. At that time less than 10% of eligible couples were using some form of fertility control. By 1990 the figure exceeded 50%, a total of over 500 million couples.

SHELDON J. SEGAL

For Further Reading: Susan Harlap, Kathryn Kost, and Jacqueline D. Forrest, *Preventing Pregnancy, Protecting Health: A New Look at Birth Control Choices in the United States* (1991); W. Parker Mauldin and John A. Ross, "Family Planning Programs: Efforts and Results, 1982–1989," in *Studies in Family Planning* (1991); Sheldon J. Segal, Amy O. Tsui, and Susan M. Rogers, eds., *Demographic and Programmatic Consequences of Contraceptive Innovations* (1989).
See also Family Planning Programs; Human Reproduction; Population.

CORAL REEFS

Coral Reefs: Rain Forests of the Ocean

Coral reefs are often called "the rain forests of the oceans" because they are among the richest marine ecosystems in species, productivity, biomass, structural complexity, and beauty. Like rain forests, reefs provide evolutionary lessons from intricate interactions between organisms. Both rely on structural frameworks built by a single group of living organisms: trees in forests and corals in reefs. Like rain forests, coral reefs thrive in nutrient-poor habitats by containing many species whose complex food chains recycle essential nutrients with great efficiency, making reefs especially sensitive to any process that disrupts recycling.

Almost every group of marine organisms reaches its greatest species diversity in coral reefs. Over a quarter of all marine fish are found in reefs, and estimates of fish productivity suggest that around 10% to 15% of the total worldwide catch comes from there. Reefs occupy only around 600,000 square kilometers, less than 0.2% of the ocean surface, making their biodiversity and productivity per unit area many times greater than other marine ecosystems. Most reefs form long, narrow strips along the edge between shallow and deep water, rather than occupying large areas, like rain forests; oceans cover more than two thirds of the Earth, yet tropical rain forests cover over ten times the area of reefs. Because biodiversity and productivity of marine ecosystems are so much more concentrated, comparing areas of damaged reefs to deforested areas on land seriously understates their great vulnerability.

Corals: Architects of the Reef

Corals are simple, bottom-dwelling animals whose fundamental unit is the polyp, which has a common opening to take up food and excrete wastes, surrounded by a ring of tentacles. Each polyp sits in its own cup in a limestone skeleton, which the coral constantly builds as it grows upwards. Polyps use weak stinging cells in their tentacles to capture small animal plankton from the water. Reef-building corals live in large colonies made by repeated divisions of genetically identical polyps. These colonies come in an astonishing variety of branching, leafy, or massive forms, which may grow continuously for thousands of years. They are related to jellyfish, sea anemones, and a variety of soft corals that lack massive limestone skeletons.

The cells of reef-building corals contain symbiotic algae that release most of the organic matter they make from photosynthesis to their coral hosts. Corals consequently look much like plants and grow over each other in competition for light. They are also able to take up dissolved and particulate organic matter from seawater. This wide range of potential food sources places corals at many levels of the food chain simultaneously, acting like a producer, herbivore, carnivore, and decomposer. The algae remove carbon dioxide and excreted nutrients, while supplying food and oxygen, and greatly enhancing the rate at which corals deposit their skeleton. Virtually all coral skeletons are white, but they are hidden beneath tissues with a wide range of colors, derived from pigments in the symbiotic algae.

Reef Structure and Natural Stresses

Coral reefs—dynamic wave-resistant structures built by the skeletons of living organisms—absorb the energy of breaking waves that transport food and nutrients to corals and clean their surfaces of sediment and waste matter. Because the most active growth is in the wave-breaking zone, reefs form linear structures facing the waves, often parallel to shorelines, protecting them from erosion. Only corals are able to build these structures. Other associated species live in nooks and crannies in the reef, in surrounding beds of sand made up of the remains of skeletons of corals and other reef organisms, and in associated protected ecosystems such as seagrass beds and mangroves. Fish and turtles from adjacent ocean waters use reefs as feeding or breeding sites.

Coral reefs are periodically devastated by storms or earthquakes, and take decades to recover. They are also subject to a variety of natural stresses, such as population explosions of the coral-eating crown-of-thorns starfish in the Indian and Pacific Oceans. Corals and other reef organisms are also subject to mortality from a variety of diseases whose impacts are spreading worldwide. It is uncertain whether these diseases are natural events or caused by pollution.

As a result of their dependence on symbiotic algae, coral reefs can grow only in conditions suitable for the algae. Coral symbiosis requires warm, bright, clean marine waters, confining reefs to shallow well-lit tropical waters, free from excessive turbidity (muddiness) and pollution. Extremes of temperature, salinity, or light can cause corals to expel their algae, losing most of their food supply and capacity for rapid growth. This phenomenon is called bleaching because the corals lose color. Thus corals and coral reefs are extremely sensitive to environmental change and habitat degradation.

Corals and Geological History

Reefs have been built over the last 500 million years by a variety of marine organisms, including corals, sponges, algae, clams, and worms. Modern corals probably build the largest and fastest-growing reef structures in history because of the remarkable enhancement of skeleton growth caused by symbiotic algae. Many ancient reef builders are thought not to have had this advantage, but 65 million years ago, reefs were constructed by massive clams that did. After a period without reefs, modern corals and reefs spread. Over the last several million years oscillating ice-age and non-ice-age conditions have caused ocean levels to swing up and down by 100 to 150 meters, forcing corals to migrate up and down in response.

Around 130,000 years ago global climate was one to two degrees Celsius warmer than today, conditions similar to those projected in coming decades from global warming. Fossil reefs around the world show that sea levels were then around five to eight meters higher than today's levels. On a finer scale, annual growth bands in many corals contain information about tropical climate change over the past centuries that is not obtainable by any other means.

Reefs also record changes in geological processes. Areas of rising crust are marked by reefs killed by being uplifted above the water. If the crust is subsiding or sea level rising slowly enough, reefs are able to grow upwards. Corals typically grow around a centimeter a year, but much of this material is eroded by waves or by boring organisms such as sponges, clams, and worms, so the reef framework itself accumulates at a few millimeters per year. Volcanoes forming near mid-ocean spreading centers gradually subside as the crust on which they stand cools and moves away. Reefs around their edges grow upward as volcanic rock sinks, leaving behind circular reef structures, or atolls. Charles Darwin's speculation that these were remains of drowned volcanoes was confirmed nearly a century later by drilling through atoll limestone. Atolls are the majority of reefs in the central Pacific and Indian Oceans.

Human Stresses to Reefs

Easy access by the rapidly expanding human populations of adjacent tropical coastal lowlands is causing mounting stress to reefs. Corals can be directly damaged by boats, anchors, handling by divers, dredging, reef mining for limestone, oil spills, and leakage of toxic chemicals from land sources or passing ships. Spearguns, trawls, explosives, and poisons are often used to collect fish or reef organisms for food, aquarium specimens, jewelry, or curios. In most regions favored reef fish are severely depleted, and fish are smaller and less diverse. Changes in fish populations may remove the species that control the abundance of other reef organisms (such as seaweeds) thus allowing the spread of "weedy" species.

Deforestation of watersheds and coastal habitats greatly accelerates erosion, allowing soil washed into the sea to smother reefs. The reef structure ceases growth and is gradually eroded by organisms boring into and weakening it. Drainage of canals in coastal wetlands damages reefs by freshwater and sediment discharges. If sewage generated by coastal populations does not undergo tertiary treatment to remove excess nitrogen and phosphorus, these nutrients stimulate prolific growth of seaweed, which overgrows corals. Expanding areas of dead or dying reef are seen around many coastal towns, resort areas, and populated areas in the tropics.

Over 100 countries have reefs, and they are the most valued marine natural resource in most. Almost all have documented serious deterioration of reefs close to population centers in the past decade. Coral reef countries obtain the major part of their fish from reefs. Their tourism is largely based on diving, fishing, and boating in the reef waters, or on the sand produced by the death of reef organisms. If reefs die, sand replenishment is cut off and beaches are destroyed by erosion. Deterioration of shore protection, best provided by healthy reefs, can cause tremendous losses along coasts during storms. The economic value of reefs is often recognized only after it is impaired. Many reefs are being damaged by poorly planned coastal development that was intended to capitalize on the beauty of reef-related habitats.

Starting in the 1980s, massive bleaching of coral reefs was reported in the Pacific, Caribbean, and Indian Oceans. In extreme events almost all corals in a reef may bleach white, and many die. The survivors slowly recover, taking up to 10 months to regain normal color and growth. During this period they are starving, unable to grow or reproduce. These events appear to be due to unusually high water temperatures. They follow periods when water temperatures were one degree Celsius above long-term averages in the warmest seasons of the year. It is still controversial whether these are linked to the record global temperatures of that period or to local weather extremes. Nevertheless there are clear signs that many reefs around the world have been under thermal stress. This suggests that coral reefs may be among the most climate-sensitive ecosystems.

Before the mass bleaching of the 1980s, coral reefs were under increasing stress worldwide due to local causes. If bleaching continues, it may prove impossible for reefs to adapt to rising temperatures and sea levels from future global climate change. Hurricane strength could also increase with global warming, placing further stress on many reefs. Reef recovery requires abatement of external stresses, limiting change to rates within the capacity of the reef to adapt. This includes treating sewage, reforesting watersheds, controlling overfishing, stopping destructive utilization of reef organisms and materials, and preventing global warming. Unless such actions are taken together, many reefs may succumb in coming decades. Alternative technologies will have to be developed to restore reefs and provide shore protection. Serious ecological damage to reefs and economic damage to many tropical countries could result from climate change and environmental mismanagement, and the lowest-lying island nations on mid-ocean atolls could vanish entirely if sea level, temperature, and storm strength rise.

THOMAS J. GOREAU

For Further Reading: T. F. Goreau, N. I. Goreau, and T. J. Goreau, "Corals and Coral Reefs," *Scientific American* (1979); B. Salvat, "Coral Reefs, a Challenging Ecosystem for Human Societies," *Global Environmental Change* (1992); United Nations Environment Programme and International Union for the Conservation of Nature, *Coral Reefs of the World* (1988).

COST EFFECTIVENESS

Cost-effectiveness analysis is a tool for attaining a specified goal at least cost. It is most useful when the goal is well established, such as a specific target for a measure of ambient air quality. It may also be very useful when current knowledge does not permit a thorough examination of what should be the goal. If a full cost-benefit analysis cannot be undertaken, and the target is chosen somewhat arbitrarily, a good cost-effectiveness study can still decide how best to achieve that target.

For example, consider the social benefits of achieving an ozone standard in the range 0.08 to 0.30 parts per million in southern California. These benefits cannot be estimated with confidence. In contrast, the cost of each method for achieving a given level of ambient air quality is possible (though difficult) to measure. Since the estimates of benefits are highly uncertain, little may be gained by a detailed comparison of estimated benefits and costs. Instead, any standard might be reasonable, within the range where estimated incremental benefits appear to be comparable with estimated incremental costs.

For whatever standard is selected, cost-effectiveness analysis provides a tool for finding the least-cost combination of methods to achieve the standard. Thus, an attractive two-step approach is to find a "reasonable" standard and then minimize the cost of attaining it.

As in the southern California example, there are many ways to achieve the standard: Half the population could be moved out of the area, all industry could be banished, all motor vehicles could be banished, or stringent emissions controls could be applied to automobiles and factories. These different approaches have quite different social costs. It is important to choose the one or the combination that has the lowest social cost, while still cleaning the air.

Despite its virtues, it is important to recognize that cost-effectiveness analysis does not tell us how to achieve the greatest net social benefit; it only tells us how to minimize the cost of achieving the specified target. But this tool has been able to suggest considerable cost savings. For example, requiring a 50% abatement of sulfur dioxide at all sources is much more expensive than a program that achieves the same abatement at least cost (by allowing some sources to over-comply and others to do nothing).

This analysis also can be used to compare regulatory policies to price incentives such as effluent fees or marketable permits. For a regulatory "command and control" policy to minimize the cost of attaining the target, the regulator would have to know exactly the right combination of abatement methods. If a polluting firm must pay a fee or buy a permit for every unit of emissions, in contrast, the firm itself can be expected to find the cheapest ways to reduce pollution.

DON FULLERTON

DAMS AND RESERVOIRS

Reservoirs are used to provide reliable water supply, provide for navigation, hydroelectric power generation, and flood control. Dams provide the control that enables society to develop water resources for agricultural, urban, and industrial uses. In the U.S., about 2,500 reservoirs with capacities of 5,000 acre feet or more provide about 480 million acre feet of storage, thus, about one quarter of the annual runoff can be stored. Storage capacity is dominated by large reservoirs; the 574 largest reservoirs store almost 90% of the total.

The dam building and diversion phase of water resource development is virtually complete in the U.S. for reasons of politics, economics, and site availability. Furthermore, there is increasing recognition that impoundment has unintended effects on riverine systems, many of which are not well understood and some of which are damaging to human uses.

Characteristics of Reservoirs

Reservoirs are lake-like (lacustrine) segments of river systems and differ from natural lakes. The drainage basin of a natural lake is typically about 10 times the area of the lake itself, but the drainage basin of most reservoirs is relatively larger; 500 times the area of the reservoir would not be uncommon. This results from the fact that dams are built on larger rivers, rather than on headwater streams. Additional aspects of this feature are that (1) reservoir shape tends to be elongate (drowned river valleys), (2) sediment loads deposit in the upstream ends of reservoirs forming characteristic "deltas," (3) longitudinal gradients of suspended and dissolved constituents of water are common, and (4) reservoir conditions tend to reflect the land uses in the drainage area.

Natural and Induced Phenomena in Reservoirs

Impoundment changes natural patterns in rivers and initiates new patterns of change. There is more reason to expect continued change than to expect equilibrium.

Sediment Accumulation.

Sediment transport by rivers to the sea is interrupt-ed by impoundment. Sediment deposits reduce reservoir storage capacity and thus flood protection capability. Perhaps a more significant and current issue is that, because materials adsorb to silts and clays, sediment deposits represent "interim sinks" for nutrients and contaminants that, prior to impoundment, were transported to the sea. Materials such as organic pesticides, toxic byproducts of industry, trace metals and salts from irrigation return flow, and agricultural and domestic fertilizers are of increasing concern to many water users.

Eutrophication.

Increased plankton algal growth results as nutrients accumulate in lakes. This response is known as eutrophication. It occurs naturally as lakes "age" but human influence hastens the process and has been implicated in a wide range of problems that pose public health threats to water supplies and recreational resources.

The transport of nutrients by rivers has not caused widespread eutrophication problems in rivers because planktonic biota are not characteristic of free flowing rivers. With impoundment, however, plankton develop rapidly and respond immediately to nutrient and contaminant inflows. Eutrophication is more an immediate response to impoundment than an accelerated process.

Extinction and Invasion.

The effects of impoundment on the biota are not always separable from other aspects of water quality deterioration. The recent extinction of fishes is well documented. The fate of other aquatic organisms in the face of environmental change is less well known. There is little doubt, however, that as flowing reaches diminish, as flow regimes change, and as water qualities in tailwaters are altered, dam construction is significantly implicated as a cause in changing the environment in which aquatic organisms evolved or to which they are adapted. Such habitat alteration also creates environments suitable for invaders. Furthermore, as more flowing water is made available for navigation through construction of locks and dams and as commercial traffic increases, the opportunity for inadvertent transfer of species from native environments to new ones has increased. These may be innocuous events but in some instances exotic species have caused major water resource problems.

Most often problems arise when invading species that have competitive advantages, or are released from natural control mechanisms, respond with explosive population growth and have a life history feature that is damaging to system integrity or to human use of water. Recent examples include the sea lamprey that entered the Laurentian Great Lakes when the Welland canal was opened. It was thought to have many damaging effects on the commercial fishery. The Asian clam (*Corbicula*) spread through the U.S. in the 1960s, and more recently the zebra mussel (*Dreissena*) was introduced into the Great Lakes near Detroit and is spreading rapidly. Purposefully introduced organisms, such as the common carp, often respond similarly.

Interactions of Rivers and Reservoirs: Opportunities for Management

The hydrologic regime of the impounded river often dominates the character of its reservoirs. At low river flows the downstream end of the reservoir is lacustrine because water is retained for longer periods of time, yet the upstream end remains riverine. When the river is flooding, the riverine zone moves farther into the reservoir. Under these conditions operators of dams may increase the discharge and riverine features of the reservoir are experienced further downstream. Such reservoirs are said to have short retention times and are called "run-of-the-river" reservoirs.

Lock and dam systems used for navigation on the Ohio River and upper Mississippi, and several large reservoirs that control water for hydroelectric power production on the Columbia River, are examples of run-of-the-river operation. Water storage and flood control operations as at Hoover Dam and Glen Canyon Dam on the Colorado River yield longer water retention times in Lake Mead and Lake Powell. Such deep reservoirs, with large volumes relative to the discharge of the river, have more lacustrine characteristics.

Dam operations and management decisions surrounding the uses of water resources define an unequivocal difference between reservoirs and natural lakes. Once the decision to build a dam is made, decisions about its operation must follow. These include choosing the magnitude, scheduling, duration, and routing of discharges (through turbines or through bypasses, locks, or spillways). The depth from which water is withdrawn and the magnitude of the drawdown may be controlled. Many of these decisions fall easily within the normal range of options open to operations engineers.

The full range of operational options, while incorporated into the design specifications for the dam, are, however, almost never used. Learning how to achieve water quality management goals through dam operations and design are distinct challenges for the future.

G.R. MARZOLF

For Further Reading: Geoffry E. Petts, *Impounded Rivers, Perspectives for Ecological Management* (1984); Loren D. Potter and C.L. Drake, *Lake Powell: Virgin Flow to Dynamo* (1989); Kent W. Thornton, B.L. Kimmel, and F.E. Payne, *Reservoir Limnology* (1990).
See also Floods.

DARWIN, CHARLES

(1809–1882), English naturalist. The father of the greatest intellectual revolution of our time, Charles Darwin, was born on February 12, 1809, at Shrewsbury, England, the second son of Dr. Robert Darwin, a successful physician. His grandfather, Erasmus, likewise a successful physician but also a poet and thinker, had already expressed an interest in evolution in his *Zoonomia*. From his childhood on, Charles Darwin showed, as he himself later said, that he was "born a naturalist." He collected beetles and was insatiable in his quest for knowledge about nature. When 22 years old, he had the great luck to be invited to join H.M.S. *Beagle* as naturalist and companion of Captain Robert Fitzroy. Fitzroy had been commissioned to survey the coasts of South America to provide information for making better charts. The voyage lasted five years, Darwin returning to England in October 1836. In an eminently readable travelogue (*Journal of Researches*), Darwin tells us about all the places he visited—volcanic and coral islands, tropical forests in Brazil, the vast pampas of Patagonia, a crossing of the Andes from Chile to Tucuman in Argentina, the Galapagos, coral islands in the Pacific, and much more.

This rich experience served Darwin as the background of his studies for the rest of his life. Being independently wealthy, he could afford to devote himself entirely to scientific studies. In 1839, he married his cousin Emma Wedgwood and in 1842 the young couple moved from London to Downe (Kent), about 17 miles south of London, where Darwin died on April 19, 1882, never again having left England.

In March 1837 Darwin learned from the ornithologist John Gould that the mockingbirds he had collected on three islands in the Galapagos were not three varieties, as he had thought, but three species. This gave him the idea that new species can evolve by geographic isolation; since these Galapagos mockingbirds were clearly derived from a South American mainland species, it also gave him the theory of common descent. The mechanism by which evolutionary change is effected, natural selection, occurred to him in September 1838 after reading Malthus' *Essay on the Principle of Population.* Yet it took Darwin another 20 years of intensive study before he was ready to publish his epoch-making volume *On the Origin of Species* (November 24, 1859). During these years Darwin published the results of the geological researches he had made on the *Beagle* voyage, and books on the origin of coral reefs, on volcanic islands, and on the geology of the mainland of South America. He also devoted eight years to a study of the living and fossil *Cirripedia* (barnacles), the resulting four volumes remaining the authoritative monograph of these animals almost up to our time.

In June 1858, Darwin received a manuscript written in the Moluccas by the naturalist Alfred Russel Wallace proposing a theory of evolution by natural selection that was essentially the same as the one Darwin had formulated almost 20 years earlier but never published. Short accounts of the two theories were published side-by-side in July 1858 and Darwin succeeded in writing in 16 months his great work *On the Origin of Species by Means of Natural Selection or the Preservation of Favoured Races in the Struggle for Life.*

Five major theses were proposed in this work. First, species are not constant and the living world continues to evolve. Second, all organisms, including humans, are descended from common ancestors, perhaps ultimately from a single origin of life. Third, there is an ongoing process leading to a multiplication of species and an extinction of old species. Fourth, all evolutionary change is gradual; there are no sudden saltations. And fifth, all evolutionary change is due to the production of almost unlimited variation and subsequent selection.

The fact of evolution and the principle of common descent were quickly accepted by most knowledgeable biologists. The principle of natural selection, however, was vigorously opposed, not becoming the consensus of biologists until the so-called evolutionary synthesis of the 1930s and '40s. The modern concept of evolution held by biologists is remarkably close to Darwin's original proposal.

After 1859 Darwin remained as active as ever, not only preparing six editions of the *Origin*, but also publishing numerous other books. What puzzled him most was the nature and the origin of variation, and to this he devoted in 1868 a two-volume work, *The Variation of Animals and Plants under Domestication.* Without a knowledge of genetics, a science that was not founded until 1900, Darwin was unable to solve this problem. In 1871 he published *The Descent of Man, and Selection in Relating to Sex*, a two-volume work in which he not only firmly placed Man into the evolutionary stream, but also elaborated his theory of sexual selection which, after long neglect, is now one of the most active areas of evolutionary biology. According to this theory, success in leaving offspring may be due not only to better adaption to the environment but also to a capacity to be a more successful reproducer.

In the last twenty years of his life, Darwin published a remarkable series of botanical treatises beginning with a wonderful book on orchids and their fertilization by insects (1862), on insectivorous plants, on cross- and self-fertilization in plants, and on the power of movements of plants. These works made him one of the leading botanists of his era. His writings were full of brilliant ideas but were not properly followed up by other botanists until almost a hundred years later. Darwin also published a pioneering book on behavior, *The Expression of the Emotions in Man and Animals* (1872), again a volume far ahead of his time. His last work was *The Formation of Vegetable Mold through the Action of Worms, with Observations on Their Habits* (1881). Here Darwin showed the important role of earthworms in the formation of the topsoil, a truly important ecological study. Actually, however, all of Darwin's ideas have had a decisive impact on ecology, and a whole field, evolutionary ecology, has developed in recent years dealing with the evolutionary aspects of the environment.

ERNST MAYR

For Further Reading: Charles Darwin, *The Autobiography of Charles Darwin* (N. Barlow, ed.; 1958); Ernst Mayr, *Evolution and the Diversity of Life* (1976); Ernst Mayr, *The Growth of Biological Thought* (1982).
See also Evolution; Genetics; Mendel, Gregor; Wallace, Alfred Russel.

DDT

DDT (dichloro-diphenyl-trichloro-ethane) is a synthetic, fat-soluble, chlorinated hydrocarbon insecticide that was introduced in the 1930s and became the most widely used—and best studied—pesticide in the world. It and other chlorinated hydrocarbons were inexpensive, highly effective, broad-spectrum pesticides, and for a number of years were used indiscriminately until certain adverse effects became apparent.

DDT is stable and remains active long after initial application. It was this stability, its solubility in fat, and its distribution in the total environment that led to its role as a widespread environmental pollutant. Once applied or dispersed it was transported in an uncontrollable way through environments. Residues were found in estuarine and even open ocean water. Once in the food chain, it resisted ultimate metabolism and bioconcentrated in predators. Among those predators were humans, and during the 1960s evidence accrued that residue levels of DDT or its metabolites were detectable with very high prevalence in human fat tissue (as a consequence of their fat solubility) and human milk (as a consequence of its fat content).

Once registered for use on 334 crops and agricultural commodities, it was banned in the U.S. in 1972, and the uses of many other chlorinated hydrocarbons were similarly either banned or severely restricted during the next decade.

The highest reported levels in 1983 were from countries with active DDT-spraying programs at that time or just previously stopped, such as India, Mexico, China, and Guatemala; Scandinavian countries had lower levels; and the U.S. levels were intermediate.

DDT is not very toxic for adult human beings. Feeding studies in volunteers carried out at relatively high doses for up to 18 months did not result in appreciable illness. Similar results were found in studies in workers with intensive occupational exposure. At very high doses it is an excitatory neurotoxin, and causes hyper-irritability and seizures. There are no documented cases of death due to DDT ingestion, since at the amounts involved, the toxicity of the diluent, typically kerosene, becomes important. The accumulated DDT, which encompasses both metabolites and DDT, is manifest in human tissue primarily as DDE (dichlorodiphenyl dichloroethane). DDE is the biomarker for DDT. Human beings do not further metabolize DDE nor excrete it much except in breast milk.

There are few reports of illness occurring in the general population from DDT, despite the exposure of everyone born since 1940. Few studies have been carried out in newborns, or on subtle toxicities like enzyme induction or disturbed estrogen metabolism. There was an exposure epidemic in 1981. Triana, Alabama, is a town of about 600 people downstream from a defunct DDT manufacturing facility. Many residents regularly ate fish caught locally that were heavily contaminated with DDT residues. This resulted in the highest serum DDT yet recorded, 76.2 ng/ml (nanograms per milliliter), as compared to the national mean of 15.0 ng/ml. A number of values were also comparable to those seen in heavily exposed workers. No directly attributable illness has been reported.

It was the combination of its physical, chemical, and biological properties that made DDT and other chlorinated hydrocarbons such a threat to life-support systems on Earth. Environmental effects included the prevalence of abnormally thin, calcium-deficient eggshells in birds of prey, which led to a decrease in population.

Unfortunately, the substitution of nonpersistent pesticides has not only failed to solve all of the original problems associated with DDT but also introduced new problems. For example, some degradation products of the nonpersistent pesticides have at least as much potential for nontarget damage as DDT. In addition, the increase in the use of organophosphate (nonpersistent) insecticides has been accompanied by a large increase in worker poisonings. DDT has saved—and continues to save—millions of human lives, especially in developing countries, where so many deadly diseases are transmitted by insects and arachnids.

DEBT-FOR-NATURE SWAPS

A debt-for-nature swap is a financial arrangement that provides funding for the maintenance of already existing nature preserves or parks. Its primary purposes are to provide this funding and to help build indigenous institutions designed to protect the preserves. Debt-for-nature swaps are an often misunderstood form of triage designed to help preserve biodiversity. The largest misunderstanding about the swaps is the misconception that they involve a third world country ceding title (or sovereignty) to large tracts of wilderness in return for the cancellation of their debts to the first world.

Swaps were first arranged in the late 1980s as a creative response to the third world debt crisis. At the time, many third world countries owed more money to first world lenders than they could reasonably hope to pay back, so the original lenders were left with billions of dollars of loans that were likely to go into default. As is typical in such cases, the lenders began to sell the loans to third parties for less than their face value. For instance, if Bolivia owed a New York bank $1 million but there was doubt about Bolivia's ability to pay, the New York bank might sell the loan to a Boston bank for less than $1 million. This transaction would both help the New York bank to recover some of the money it originally loaned and give the Boston bank a chance to make $1 million for an investment of much less money. If the Boston bank bought the loan for 50 cents on the dollar, it would stand to make $1 million on a $500,000 investment, provided Bolivia eventually repaid the loan. The availability of these deeply discounted loans created the possibility for making debt-for-nature swaps.

Every swap involves at least three parties: the government of a third world country, the bank (or government) in the first world that originally loaned them the money, and a nongovernmental organization (NGO) such as the World Wildlife Fund. The NGO buys the loan from the bank at a small proportion of its face value and arranges for the third world government to pay a large sum to an indigenous group (or groups) working to protect a nature preserve. Everyone benefits in the transaction. The bank gets some of its original loan back and takes a bad loan off its books. The third world government is relieved of the obligation of paying the loan back to the bank; it also gets to pay off the loan in local currency rather than in expensive foreign currency. The NGO gets a major environmental improvement. Finally, there is the creation of stronger indigenous institutions working for the protection of preserves. By 1992 over 20 swaps had been arranged, for a total of more than $100 million.

BRADLEY W. BATEMAN

DECOMMISSIONING

Decommissioning is the process of dismantling and decontaminating nuclear power plants, plutonium production plants, and nuclear submarines; it also includes management of the resultant waste products. After approximately thirty years of use, nuclear plant pipes and vessels become brittle and radioactive from continuous neutron bombardment. The radioactivity increases each year the plants operate.

Three methods for decommissioning nuclear power plants have been proposed: (1) to decontaminate and dismantle the facility immediately after shutdown; (2) to do a preliminary cleanup of the radioactive waste and keep the structure intact for 50–100 years, eventually to be dismantled; or (3) to permanently entomb the facility, which requires covering the reactor with reinforced concrete and erecting barriers. In all three methods, workers must wear protective clothing and breathing apparatus. Due to the high levels of radioactivity, they must limit the time they spend in contaminated environments. In the first and second methods, high-pressure water jets and chemical solvents are used to decontaminate materials, such as pipes and machinery, that contacted radioactive materials during the life of the reactor. In the process, the solvents themselves become radioactive and must also be properly disposed of. Since spills may occur, any exposed soil should also be considered radioactive waste.

Radioactive material that cannot be decontaminated by the first method is shipped to safe waste burial facilities; decontaminated material may be recycled or buried in a landfill. In the second method, after initial attempts at decontamination the remaining structure is left standing so the radioactive materials may decay naturally. This results in lower radioactive levels when the plant is totally dismantled 50–100 years later. To keep the reactor intact, whether for decades as in the second method or permanently as in the third method, on-site surveillance is required. This surveillance incurs costs, since there is a need for a maintenance crew and for scientists to monitor levels of radioactivity.

The estimated cost of decommissioning nuclear power plants in the United States ranges from $50 million to $3 billion, or approximately 40 percent of the cost of the initial investment for each plant. Waste disposal is approximately 40 percent of the total cost of decommissioning. Since reactor designs have changed rapidly over the years, there is no standard decommissioning system. Of the 20 power reactors shut down in the United States between 1957 and 1985, there were 9 different designs. As of 1992 only a few plants in the world have been decommissioned.

Financial planning is needed to ensure that plants have sufficient funds to pay for decommissioning. Currently, nuclear power customers in the United States are charged a monthly fee that is put into a decommissioning account. Some utilities, however, have used this money for general purposes, including the construction of new plants. Twenty-nine percent of United States nuclear power companies do not expect that their current funding method will cover decommissioning costs. In contrast, reactor operators in Sweden must pay annual fees to their government, which invests them in decommissioning accounts. Germany and Switzerland have similar programs.

DEEP ECOLOGY

See Ecology, Deep.

DEEP WELL INJECTION

Deep well injection is a technically sound and costly process that involves injecting selected industrial and treated municipal wastes into a deep underground formation (injection zone) using a well specifically located, designed, and constructed for that purpose. Typically, injection zones are one-half to two miles below the surface and are separated from sources of drinking water by several impermeable overlying rock formations (confining zone) that are many hundreds of feet thick.

Deep well disposal of liquids began in the petroleum industry in the 1930s when it became common practice to dispose of saltwater (brines) that usually accompanies oil and gas production by injection into underground formations. Deep well disposal of other wastes began in the 1950s. Disposal wells are used by municipalities for sewage disposal and by the petrochemical, pharmaceutical, paper, mining, automotive, and food processing industries for waste disposal.

Injection zones characteristically contain saltwater

and no potentially usable resources such as drinking water. The wastes are pumped into, and occupy, pore space within the zone. Properly located, constructed, operated, and monitored deep wells are designed to permanently isolate and contain injected waste within this zone.

Protection of drinking water is the primary concern in the construction of a disposal well and is achieved by assuring permanent containment of the waste. A well is constructed in stages. First a borehole is drilled to a depth below all drinkable water. A steel surface casing or pipe is installed the full length of this borehole and cement is placed outside the casing from the bottom to the top to seal the casing into the hole. This provides a barrier of steel and cement to protect drinking water zones.

The next step is to continue drilling below the surface casing into the injection zone. Samples and instrument measurements are routinely taken to evaluate the type and thickness of the rock formations. Another protective casing string is installed from the surface to the injection zone and again cemented the entire depth to seal the space outside the casing, which can be constructed from steel or corrosion-resistant materials such as fiberglass or titanium. This provides two cement and two casing barriers to protect the drinking water zone.

A smaller pipe, called injection tubing, is installed inside this protective casing and waste is injected through this inside tubing. The space between the tubing and the casing (the annulus) is secured and sealed at the top with a wellhead and sealed at the bottom with a packer. The annulus is filled with a noncorrosive fluid under pressure. The fluid is maintained at a different pressure from the waste within the injection tubing. The annulus and injection tubing pressures are continuously monitored and recorded. Any leak that occurs either in the injection tubing or protective casing will result in a change in annulus pressure and the well is immediately shut down so the cause of the pressure change can be determined and rectified. This insures that all wastes reach and are retained in the injection zone. Thus, this well-within-a-well provides multiple layers of protection.

When a well is retired from service, the well borehole must be securely plugged to prevent any movement of the waste. A properly sealed well, using cement and other materials, permanently confines the waste within the injection zone.

DEFORESTATION

Deforestation is defined as the conversion of forest land for non-forest uses, such as pasture or cropland, or for dams, highways, airports, and residential or industrial development. Logging is not generally considered deforestation even if the standing forest is cleared, as long as the land continues to be devoted to forest use following timber harvesting. Unmanaged or uncontrolled timber or fuelwood harvesting that exceeds the ability of the forest to maintain itself is generally described as forest degradation. Severe forest degradation often leads to deforestation, especially in arid and semiarid regions such as sub-Saharan Africa.

The building of roads into previously untouched forest areas can make these areas more accessible to poor rural families in search of land for subsistence farming, and thereby lead to deforestation for crop or pasture. Where soils are poor, such lands are often abandoned after a few years, reverting to native grasses or scrub growth.

The primary causes of deforestation in the United States are the clearing of forests to build roads; airports; dams; and residential, commercial and industrial developments. The U.S. Forest Service estimates that the U.S. will lose about 28 million acres of forest by the year 2040, primarily in the South and Pacific Coast states, mainly to roads and urban development.

In the developing countries of the world the primary causes of deforestation are clearing for agriculture and over-harvesting for fuelwood. Tropical deforestation has become a matter of particular concern because of the potential loss of plant and animal species that grow or live in the humid tropical forests of the world. Although tropical moist forests cover only 7% of the earth, they contain at least half of the world's plant and animal species, many of which have yet to be studied or even identified.

Large-scale deforestation also contributes anywhere from 10% to 30% of the annual human-caused emissions of carbon dioxide to the earth's atmosphere. Carbon dioxide is one of the primary greenhouse gases involved in a potential global warming effect. Growing forests, on the other hand, act as carbon sinks by removing carbon dioxide from the atmosphere and storing it in trees and soil. For example, the vast forest areas of Siberia, which cover an area the size of the U.S., are thought to contain up to half as much stored carbon as do the Amazon forests. Increasing utilization of the Siberian forests is sure to be closely watched by the global environmental community.

In 1989, the Food and Agriculture Organization of the United Nations (FAO) began an inventory and assessment of world forest resources in order to update a similar survey conducted in 1980. A primary focus of the FAO assessment has been the amount and rate of deforestation in the world's tropical forests. Its estimates for 87 tropical countries showed that the annual rate of tropical deforestation between 1981 and 1990 was 17 million hectares (42 million acres). This is a rate of about 1% of the world's tropical forest as of 1980. Africa accounted for 5.1 million hectares, Asia for 3.5 million hectares, and Latin America for 8.4 million hectares of the annual total.

In Africa the major causes of tropical deforestation included bush fires, fuelwood harvesting, and conversion to agricultural uses. In Asia, FAO found that the rate of deforestation had more than doubled since 1980, from 2 million hectares a year to 4.7 million. High population density and rural poverty were the main causes of deforestation in that region, with almost 75% of the deforestation coming from agricultural clearing. Unmanaged forest exploitation was also a major contributor. In Latin America, FAO cited the failure of government land-use policies as a direct cause of deforestation during the 1980s. For example, in the Amazon basin alone, 4.1 million hectares had been cleared annually for agricultural use, even though 94% of the area's soils are unsuitable for sustainable agriculture. Similar situations prevailed in a number of other South and Central American tropical countries, including Colombia.

The underlying causes of deforestation are social and economic. In poorer tropical countries landless rural families clear the forest to obtain land for food. Typically, forest trees are cut and burned to enrich the soil with ashes. The land is then farmed for several years before the soil becomes too poor to grow food crops. The farmers then move on and repeat the process. This is known as shifting cultivation, a traditional method becoming destructive by overuse from growing rural populations. The abandoned land usually reverts to native grasses or scrub growth, or it may be used as pasture until the soil is totally exhausted, and then abandoned.

Deforestation is also carried out by governments wishing to sell the trees for timber and put the land into export crops such as coffee, tea, rubber, palm oil, or sugar in order to obtain foreign exchange or reduce heavy foreign debt. Total indebtedness of the developing countries doubled during the 1980s, from $600 billion to $1.2 trillion, resulting in an outflow of capital from developing countries to industrialized countries in 1988 of $32 billion. Under these circumstances most developing countries lack funds to invest in conservation or sustainable forestry. Some governments also use forest areas for resettlement programs, designed to help landless people become small farmers. If not well planned, such programs can needlessly destroy thousands of hectares of forest.

The consequences of uncontrolled deforestation, especially in mountainous areas such as the Peruvian Andes or the hill country of Nepal, can lead to severe soil erosion and deeper poverty for the mountain people and flooding of homes and farms of their lowland neighbors. Siltation of reservoirs and irrigation systems in the lowlands is another costly consequence of destructive mountain land use.

World population continues to grow, and with it the need in most countries to grow more food. This means that deforestation will not be halted in the foreseeable future. There is hope, however, that through better information gathering and analysis, land use and population planning, and improved methods of farming, deforestation can be limited to cases of necessity, rather than of mere convenience or expediency. It is especially important for every country to identify the social, environmental, and economic benefits of its forests and set policies to maintain and protect essential forest areas for soil and water protection, flood control, food and fodder production, forest products of all kinds, and cultural and traditional values. Until all of these benefits are properly weighed in the balance, forests will continue to shrink.

Solutions to deforestation include bringing natural forests under sustainable management, harvesting only as much as the forest grows; conserving the plants and animals of the tropical forest through habitat conservation that will allow local people to earn income from the forest (otherwise they will clear it to survive); investment in reforestation to create new wood resources and to rehabilitate degraded forest lands, and investments in improved agriculture systems for small farmers, to reduce shifting cultivation.

To fund these efforts, many countries threatened by deforestation require international financial and technical assistance. Through such agencies as FAO, The United Nations Environment Program (UNEP), and through individual country assistance programs such as the U.S. Agency for International Development (USAID), industrialized countries can assist developing countries to improve tropical country land and forest management. Many countries, including the U.S., consider tropical forestry conservation among their highest development assistance priorities. A possible global agreement on forest conservation is under discussion at the international level. However, even more must be done to reduce debt and poverty in developing countries if global forest conservation is ever to become a reality. The cost of slowing deforestation is enormous, but the costs of not controlling it will be even greater.

RICHARD PARDO

For Further Reading: Jessica Mathews, ed., *Preserving the Global Environment: the Challenge of Shared Leadership* (1991); Duncan Poore, et al., *No Timber Without Trees: Sustainability in the Tropical Forests* (1989); Narendra P. Sharma, ed., *Managing the World's Forests: Looking For Balance Between Conservation and Development* (1992).

DELANEY CLAUSE

The Delaney Clause prohibits the use of food additives that are found to induce cancer. The clause was drafted by Rep. James Delaney and was enacted as part of the 1958 Food Additives Amendment. It is currently codified as part of the Food, Drug, and Cosmetic Act and is part of the mandate of the Food and Drug Administration (FDA) to assure food safety.

The act states the general rule that the FDA cannot approve a substance for use in food unless there are data establishing it as "safe." The Delaney Clause provides:

> No additive shall be deemed to be safe if it is found to induce cancer when ingested by man or animal, or if it is found, after tests which are appropriate for the evaluation of the safety of food additives, to induce cancer in man or animal.

Accordingly, if an additive is "found to induce cancer," it cannot be used in food.

When the Environmental Protection Agency (EPA) was established in 1970, authority to set tolerances for pesticides in food was transferred from the FDA to the EPA, which therefore shares responsibility for implementing and interpreting the clause.

The terms in the key phrase of the Delaney clause — "found to induce cancer" — are not defined in the act. The legislative history of the clause suggests that Congress intended the FDA to exercise its scientific judgment as to the carcinogenicity of particular food additives.

To do so, the regulators must evaluate biological, biochemical, toxicological, and statistical data such as the animal model, the number of animals used, the route of administration, the identity of the substance tested, the dose levels, the duration of exposure, age of the animals, alternative explanations for cancers that occur, and theories as to the mechanism of causation of such cancers. Some carcinogens operate independently. Some promote the effects of other carcinogens. Some must be present in a concentration that exceeds a biological threshold before they will have any effect. The decision of whether a substance induces cancer, therefore, requires the exercise of agency discretion.

Although regulatory agency decisions applying the Delaney Clause are rarely reversed by the courts, there has been some controversy since the clause was enacted over how the FDA and EPA should interpret scientific studies relating to cancer. Initially, the Delaney Clause was construed as requiring a complete ban on all compounds associated with cancer in man or animals. Under this "zero-tolerance" approach, the benefits of using the additive and the magnitude of the risk involved were seen as irrelevant.

However, the number of known carcinogens has increased from a handful in 1958 to a substantial proportion of all known chemicals today. Accordingly, interpretation of the Delaney Clause has shifted to a "negligible-risk" approach. Typically, under this *de minimis* standard, a carcinogen can be used if there is less than a one in one million chance that an individual will experience cancer from daily exposure over a lifetime.

In 1985 the FDA approved the use of methylene chloride to decaffeinate coffee using the negligible-risk approach, finding that the risk of cancer was one in 12 million. In contrast, in the 1987 case of *Public Citizen v. Young*, the FDA was precluded from using the negligible-risk approach for color additives. As compared with preservatives, sweeteners, and flavoring agents, color additives lack nutritive value and have no beneficial effect on the health and safety of the food supply and so are subject to stricter regulation.

MICHAEL A. GOLLIN

DEMAND

In economic terms, *demand* is the desire for a product or service combined with the ability to pay for it. The demand for a product depends on preferences and incomes, the possible competitive market prices of the product, and the prices of any of the product's substitutes and complements. The consumer's income plays a primary role in demand. In economic terms the poor demand little, although they might desire a great deal.

An economic analysis of demand focuses on the principal relationship between prices and quantities. At the several prices that could prevail in a market during some period of time, different quantities of a product or service would be bought. Demand, then, is considered to be a catalog of prices and quantities, with one quantity bought for each possible price of a particular good or service. The graphical relationship between price and quantity is called a *demand curve*. In a demand curve, price is plotted on the vertical axis and quantity on the horizontal axis. The demand curve shows that an inverse relation exists between price and quantity. The curve falls from left to right, indicating that smaller quantities are bought at higher prices and larger quantities are bought at lower prices.

Exceptions to this rule may occur if consumers value a product because of its prestige and will therefore buy more of it at a higher price. The demand curve may shift due to a stronger consumer desire for the product (perhaps due to advertising), an increase in consumers' incomes, a rise in the prices of substitutes for the product, or a fall in the prices of complements to the product.

DEMOGRAPHIC TRANSITION

During the development process that transforms an agricultural society into a modern industrial one, fundamental demographic changes take place. The term *demographic transition* refers to the general pattern of changes in birth, death, and population growth rates that accompanies development. Population size is nearly stable before and after the transition, but rapid growth occurs in the intervening period.

Five transitional stages can be identified:

- *Pretransition.* Birth and death rates are both high and population growth is near zero. This situation prevailed in most of the world before the onset of the industrial age.
- *Early transition.* The death rate declines as a result of a lower incidence of epidemics and famines and improvements in standards of living, nutrition, and public health measures. Since fertility remains high, population growth accelerates.
- *Mid-transition.* As the cost of children to parents rises (e.g., for education) and as their economic value declines (e.g., for agricultural labor) couples begin to adopt contraception to limit family size, causing a drop in the birth rate. Population growth reaches its maximum.
- *Late transition.* The death rate reaches a minimum and with a continued decline in fertility, the rate of population growth, while still positive, slows.
- *Posttransition.* A new equilibrium between births and deaths is established resulting in near zero growth in population size.

This broad description of the demographic transition is consistent with a large variety of actual patterns of change over time, although it does not take migration into account. The timing of the onset of the transition, the duration of the different phases, and levels of birth and death rates differ from region to region and from country to country.

The populations of the developed world have now essentially completed their transition. Fairly reliable historical records indicate that a sustained decline in mortality began in Western European populations in the second half of the 18th century. In North America, Europe, Japan, Australia, and New Zealand, the onset of the decline in the birth rate occurred in the late 19th and early 20th centuries. This decline since then has not been steady, however, as an unusually low fertility rate during the depression of the 1930s was followed by the postwar baby boom. Today the average rate of population growth of developed countries is only a fraction of one percent per year, and in a few countries such as Germany, Sweden, and Hungary population size is actually declining.

In the other countries of Asia and in those of Africa and Latin America, a sustained reduction in mortality was first observed in the first half of the 20th century. Low cost public health measures have yielded very rapid mortality reductions, especially in the decades since World War II. By the late 1960s the average death rate had dropped to 15 per 1000 which, together with a still largely unchanged birth rate of 40 per 1000, yielded an annual growth rate of 25 per 1000 or 2.5%. This rate of natural increase is substantially above those observed historically in European populations in mid-transition, a difference that is in part attributable to relatively higher levels of fertility in contemporary developing countries. Countries in the developing world today are at widely differing stages of the transition. During the 1970s and 1980s birth rates declined substantially except in sub-Saharan Africa, which in 1990 was still in the early transitional stage. The steepest reductions occurred in East and Southeast Asian countries where socioeconomic development and family planning programs have been conducive to rapid changes in reproductive behavior. Hong Kong and Singapore have nearly completed their transitions.

Large increases in population occur over the course of the demographic transition. The population of the industrialized world in 1992 stands at 1.2 billion with very slow additional growth projected to bring the total to 1.3 billion by the end of the 21st century. This is nearly a seven-fold increase over the estimated population of about 190 million in 1750. Much larger increases are expected in the developing world as it completes its transition in the 21st century. Between 1750 and 1990 the population of the developing world grew from approximately 0.57 billion to 4.1 billion. Current projections indicate substantial further growth with the population total reaching 10 billion in 2100, a more than seventeen-fold increase over the course of the transition. Providing food, housing, and a reasonable standard of living for the additional 6 billion people expected in the next century could put enormous stress on the world's economic and environmental resources.

JOHN BONGAARTS

For Further Reading: R. Bulatao, et al., *World Population Projections, 1989–90 Edition* (1990); P. Demeny, "The World Demographic Situation," in *World Population and U.S. Policy: The Choices Ahead* (J. Menken ed.) (1986); T. W. Merrick, "World Population in Transition," *Population Bulletin*, vol. 41, no. 2 (1989).

DEMOGRAPHY

Demography is the study of human populations, including their size, composition, and distribution, as well as the causes and consequences of changes in these factors. Populations are never static. Their size, composition, and distribution change through the interplay of three demographic processes: fertility, mortality, and migration.

Fertility refers to the number of births that actually occur in a population. In 1990, fertility rates of national populations ranged from an average of 8.1 children per woman in Rwanda to 1.3 children per woman in Italy. The average for the United States was 2.1, and for the world, 3.6. Fertility must be distinguished from its sister term, fecundity, which refers to the physiological ability to have children.

The second cause of population change is mortality. The death rate, which measures the proportion of a population that dies each year, usually is expressed as the number of deaths per 1,000 persons in a given year. Worldwide, death rates range from only about 4 in Costa Rica and Jordan, to the low 20s in Ethiopia, Mali, and a few other African countries. In 1990, the rate for both the United States and the world was 9 deaths per 1,000 persons.

The third cause of population change is migration, the movement of people into or out of a specific geographic area for the purpose of changing residences. Migration can add to or subtract from an area's population, depending upon whether more people move in or out. International migration involves movement across a national border, while internal migration includes moves within a country. Migration can occur in great waves in response to world events—such as the mass exodus from East to West Germany in 1990—or as a slow trickle.

Population size and growth are determined by the net effects of the three demographic processes specified above. Population growth can be positive, as it is worldwide and in most national populations; negative, as it was for the city of Detroit in the 1980s; or zero, as it currently is in Denmark.

People have many characteristics that have demographic dimensions including their sex, age, race, and ethnicity. A population has characteristics that correspond to the personal traits of its individual members. The age composition of a population, for example, is determined from the collective ages of all its members. The age, sex, race, and ethnic compositions of a population are fashioned solely by the prime demographic forces of fertility, mortality, and migration.

The distribution of a population is how the people and their population-related actions, such as births and deaths, are spatially dispersed. Demographers keep tabs on the distribution of populations by world region, by country, by province or state within countries, by urban and rural area, and by segments of metropolitan areas (for example, central city, suburb, and neighborhood). Like the compositions of a population, the geographic distribution of a population is determined solely by fertility, mortality, and migration.

JOSEPH MCFALLS, JR.

DEOXYRIBONUCLEIC ACID

See Genetics.

DEPARTMENT OF AGRICULTURE

See United States Government.

DEPARTMENT OF DEFENSE

See United States Government.

DEPARTMENT OF INTERIOR

See United States Government.

DESALINATION

Desalination is the process of reducing the salt content of water to make it suitable for domestic or industrial use. Although there are many economic and technical problems associated with the desalination of sea water, there is considerable interest in doing so to supplement limited fresh water supplies, especially during periods of drought and in the use of brackish groundwater near oceans, where salt levels are high. The processes most frequently used for desalination are: flash distillation and reverse osmosis.

Flash distillation is most frequently used for desalination of salt water because it operates virtually independently of the concentration of influent solids. In this process, the salt water (brine) is heated under pressure and then discharged through a series of stages, each at a lower pressure. Because of the drop in pressure, each stage will rapidly form vapor (flash) to reach equilibrium conditions for that stage. This vapor, which leaves behind the salts, is condensed to form a low-salt distillate. The cost of "fresh" water produced by flash distillation depends on the design and manufacture of the process. Because flash distillation requires a substantial input of energy, its use has been limited to situations where other drinking water sources are unavailable and where sea water is the major water source.

The desalination process used most frequently for less salty water is reverse osmosis. This process uses a simple concept that has been widely observed in nature. That is, if a salt water solution is separated from a fresh water solution by a semipermeable membrane that allows the passage of water but not salts, water will flow from the fresh water side of the membrane to the salty side in an effort to equalize the salt concentration and pressure. In the reverse osmosis process, pressure is applied to the salty side, resulting in the "reversed" flow of water across the membrane to the fresh water side. The process depends on the types of membranes used and the character of the water being treated. Water with a higher salt concentration requires greater pressure, so the process has found widespread use for moderately salty or brackish waters such as those in coastal areas of Florida.

While other desalination processes have been developed, their cost and effectiveness are not competitive with flash distillation or reverse osmosis. Because of the high costs involved, widespread use of desalination will continue to be limited to those areas where fresh water is unavailable. Since much of the cost depends on energy availability, it is unlikely that technological breakthroughs can increase the use of desalination.

JOHN T. NOVAK

DESERTIFICATION

As defined by the United Nations Environment Programme, desertification is land degradation in arid, semiarid, and dry subhumid areas resulting mainly from adverse human impact. The fluctuation in size of deserts due to their variable climate is not desertification. Generally, desertification encompasses three broad types of processes: (1) vegetation degradation, (2) soil degradation, and (3) salinization and waterlogging.

The word *desertification* was first used in 1949 by the French forester Aubréville to describe deforestation in tropical and subtropical Africa. It came into general use in the 1970s to describe the severe environmental degradation that accompanied droughts in the semiarid Sahel region of Africa. These conditions led to the landmark United Nations Conference on Desertification (UNCOD) in 1977, which established desertification as an issue and attempted to mobilize global resources to address it.

About half of the world's dryland area is in economically productive use as rangeland, or for rainfed or irrigated agriculture. Of this subarea, about 60% is estimated to be degraded, or desertified, resulting in annual economic losses of about $42 billion in lost income worldwide. Models of global climate change suggest that drylands will expand with increases in area of up to 17%.

Grasslands and shrub and tree savannas are dominant vegetation types in semiarid and subhumid drylands. Grasses provide forage for grazing animals such as cattle, while trees and shrubs meet some of the needs of browsing goats and camels. The availability of these resources mirrors the variability of climate. Thus, at any one time, there are limits to the number of animals that might be supported (carrying capacity). Traditional subsistence pastoral systems maintained relatively small and mobile herds with a mix of grazers and browsers.

Economic and political factors often exert pressures that are inconsistent with sustained, long-term viability of these sytems. Carrying capacity may be exceeded inadvertently when animal prices are low, or intentionally for short term gain. Worldwide preference for beef encourages herds of cattle only, placing inordinate pressure on grass resources. In Africa, the demise of colonial power and establishment of new states led to administrative boundaries that limited the movement of nomads and their herds. In attempts to sedentarize and assist nomads across the Sahel, governments developed wells to water animals. Impacts that formerly were spread over large areas became concentrated around wells. Finally, there are few investment opportunities in much of North Africa. Animals are a traditional form of wealth, so many Africans invest in animals rather than put savings in banks.

Overuse of rangelands results in a reduction in dense, shallow-rooted grass species and an increase in less desirable, deep-rooted, woody shrubs. If use pressure is sustained, even shrubs may be removed, either physically or through soil degradation. In developed countries, there is a reduction in carrying capacity, at least for cattle, and an increase in erosion, which threatens water resources. In developing countries, herd compositions change somewhat, favoring browsers that exploit remaining woody species.

Woody shrubs also have value for fuel or other products. Fuelwood is a major element of the economy in most developing countries. In times of economic stress, wood provides an opportunity to generate income through the production of charcoal. In developed countries, there has been a resurgence in the demand for fuelwood for home heating and special purpose cooking (e.g., mesquite-broiled meats).

Removal of vegetation may lead to soil degradation by increasing the effectiveness of raindrops in loosening soil particles for removal by water and wind.

Trampling by animals also can compact soil, reduce infiltration rates, and thus further enhance erosion. Reduced vegetation cover and rainfall infiltration increase surface temperatures, retarding the accumulation of nitrogen. All these factors adversely affect shallow-rooted plants, such as grass, and favor deep-rooted shrubs. Removal of topsoil and reduction in infiltration capacity threatens all plants, and damage can become permanent if left unabated. Expansion of rainfed farming into marginal areas also is regionally important. High wheat prices, inexpensive land, and relatively favorable climatic conditions led to conversion of rangeland to more risky dry farms on the Colorado high plains during the 1970s, threatening parts of the region with dust bowl conditions.

Soil particles create other problems once mobilized. Moving dunes of sand and silt are a universal symbol of desertification, although they are important only locally. Dunes pose threats to agricultural fields, towns, roads, railroads, canals, and other structures. Blowing sand also kills crops directly through abrasion. Eroded soil carried in suspension decreases water quality and can collect in canals and reservoirs, reducing their effective lives.

Irrigation occupies a relatively small area but is the most capital intensive and economically important form of agriculture in drylands. Thus, perhaps the most costly form of desertification is salinization and waterlogging of irrigated lands. In humid regions, salts (soluble materials found in rock) are dissolved and carried away in streams. Drylands have little free water, drainage basins often lack an outlet, and salts cannot be readily removed. When rains do occur, salts are redissolved and concentrated where runoff collects. Thus, the dissolved load in both surface and ground water in drylands is often quite high. Evaporative rates are high because of high temperatures and low humidity.

Large volumes of water, often with high levels of dissolved salts, must be applied to dryland crops. Too little water leads to deposition of salts near the soil surface (salinization). If too much water is applied and there is no drainage, soil becomes saturated or waterlogged. Either condition reduces yields and ultimately destroys the productive capacity of the soil. Prevention or repair requires careful management and substantial investment to deliver and control irrigation water and to carry away drainage water. Salinization and waterlogging are found in virtually all irrigated drylands, from the San Joaquin Valley of California to the Indus of Pakistan.

The cost of salinization and waterlogging is high— about $250 per hectare of annual income lost from moderately degraded land—and reclamation is costly. The size of the existing investment and the economic and political influence of irrigated regions usually provoke action by national and local governments.

The financial impact of vegetation and soil degradation outside irrigated areas is less, with about $38 per hectare being lost annually in moderately degraded rainfed agriculture, and about $7 per hectare on moderately degraded rangeland. Attempts to mitigate vegetation and soil degradation in drylands have met with little success for several reasons. The value of the resource is small and attention usually is directed elsewhere (e.g., irrigated agriculture). The number of people affected is comparatively small, as is their political and economic power. As a consequence, the economic and population pressures that lead to vegetation degradation in drylands have proved irresistible, particularly in developing countries.

In the United States, vegetation and soil degradation were recognized early as a consequence of the dust bowl of the 1930s. Soil conservation research and extension programs were initiated, and policies to protect vulnerable lands were enacted that are still in force. However, a combination of climate and economics can still result in degradation. The general improvement in rangelands in the western United States probably owes as much to unfavorable changes in the economics of cattle ranching as to good management policies.

Few international attempts to combat desertification have been successful. Once degraded, drylands are difficult and expensive to repair. The necessary resources and political will may be mustered for irrigated lands, but it is less likely for rainfed farmland and even less so for rangeland. If the United States is used as a model for successfully combatting desertification, the solution lies in development of a strong integrated economy that is not dependent on the drylands themselves.

CHARLES F. HUTCHINSON

For Further Reading: W. H. Schlesinger, et al., "Biological Feedbacks in Global Desertification," *Science* (1990); United Nations Environment Programme, *World Atlas of Desertification* (1992); M. M. Verstraete and S. A. Schwartz, "Desertification and Global Change," *Vegetatio* (1991).
See also Deserts.

DESERTS

Deserts as defined by climate are regions where a combination of high temperatures and low rainfall cause evaporation to exceed precipitation. It is actually more accurate to speak of **arid lands** than of deserts.

The world's arid lands are divided into three classes of relative aridity on the basis of a formula: P/ETP, with the first term standing for an area's mean annual precipitation and the second for mean annual potential evapotranspiration. Dividing P by ETP yields a ratio expressed as a decimal fraction, and this number classifies a dry place as **hyperarid** (less than 0.03), **arid** (0.04 to 0.20), or **semiarid** (0.21-0.50). This formula yields values that are comparable across a variety of conditions, unlike rainfall figures alone, because rainfall evaporates more slowly in the temperate zones than in the tropical zones.

Arid lands generally have extreme diurnal temperature fluctuations, as the temperature of dry air drops precipitously when the sun goes down. The precipitation pattern is usually drought-and-deluge; it may not rain for weeks, months, or years on end, but when it rains it often pours. Most streams are ephemeral; they carry water only after a heavy rain or as a result of snowmelt at adjacent higher elevations. By contrast, perennially flowing arid lands rivers like the Nile, Murray, Euphrates, Rio Grande, and Indus are allogenic, meaning that they rise in areas of heavy rainfall and survive a desert passage to empty into the sea.

Geologically, the arid lands are underlain principally by three types of structures: (1) half or more of the dry lands lie atop sedimentary structures containing aquifers and deposits of such fossil fuels as petroleum and natural gas; (2) slightly more than a third of them are dominated by remnants of the Gondwana Shields, which date back 1.5 billion years and may contain deposits of precious and semiprecious metals; (3) the smallest portion of these lands are Hercynian structures dotted by volcanic structures and sometimes containing coal deposits.

Topographically, arid landforms range from Alpine mountain ranges (e.g., the Rocky Mountains of North America), to basin-and-range topography (e.g., the Great Basin of North America), to broad tablelands and plateaus (e.g., the low tablelands of eastern Arabia, the central Asian steppes, and the Gondwana Shield surfaces of Australia and southern Africa), to extensive sedimentary plains (e.g., the Patagonian Plains of South America).

Biologically, all plants and animals native to the arid lands have developed special strategies to cope with an environment in which drought is the rule rather than the exception. Arid lands plants and animals can be divided into strategists of four basic kinds: (1) drought escapers, such as the desert wildflowers that "escape" into their tough seeds until conditions favorable to germination and growth return, or insects that "escape" into the egg or larval stage until wet weather returns and then run through their entire life cycles in a matter of days; (2) drought evaders, such as those perennial plants (saltbush is one) with extremely efficient, deep, and widespread root systems, or those animals (includ-

ing snakes and lizards) that burrow into the ground or hollow out pockets in loose sand in which to ride out the heat of the daylight hours; (3) drought resistors, such as the succulents and cacti that store water in roots and trunks, or the camel that minimizes water loss by limiting its panting and sweating and by reducing the water content of its urine and excrement; (4) drought endurers, such as those perennial woody shrubs and trees that go dormant during droughts (including the entire genus of acacia shrubs) or that have no leaves at all, or animals such as lungfish and frogs that estivate during seasons of drought. (Estivation is a special form of hibernation in which animals adjust their bodily functions to achieve minimum water loss over prolonged periods.) Members of the human species fall into a fifth category, one we might label technological manipulators, since our kind could not tolerate arid conditions without such ancient and modern artifice as constructed shelter, special clothing, assisted transportation, well drilling, and air conditioning. The last item, especially, has contributed mightily to the dramatic increase in arid lands population in the current century.

In addition to these shared characteristics, each subregion has its own defining qualities. The **hyperarid** regions support almost no perennial vegetation, but annuals may appear in wet years; agriculture and livestock grazing generally are impossible; and precipitation can vary from year to year by as much as 100%. The **arid** regions support scattered vegetation, including bushes, small woody shrubs, and succulent, thorny, or leafless shrubs; very light grazing is possible but rainfed agriculture is not; annual precipitation varies from 80-150 millimeters (mm) to 200-350 mm, and can vary interannually by 50% to 100%. The **semiarid** regions include steppe lands with some savannas and tropical scrub vegetation; good grazing areas are scattered through such regions; rainfed agriculture is possible but unpredictable; annual rainfall varies from 300-400 mm to as much as 800 mm in summer-dominant rainfall areas and from 200-250 mm to 450-500 mm in winter-dominant rainfall areas at Mediterranean and tropical latitudes, and interannual precipitation varies by 25% to 50%.

The hyperarid, arid, and semiarid regions together account for 18,932,258 square miles (an area equivalent to that of North America and South America combined), or 32.9% of Earth's total land area. Most of these arid lands are found in latitudes of 15° to 30° in both the Northern and Southern Hemispheres, a pair of globe-girdling belts known jointly as the **arid zone**. While the American continents contain more than half of all arid lands, nearly half the countries of the world include arid lands. And although few people inhabit the hyperarid regions (the "true deserts," where human population densities average as few as two person per square mile), an estimated 728 million people (14% of the planet's 1990 human population) live in the arid and semiarid regions; of those, more than 524 million (72%) inhabit the latter.

None of these figures is static, however. Some modern deserts once were underwater, some were better watered than they are now and so supported a more luxuriant flora and a much less specialized fauna, and some deserts were smaller while others were larger. Six thousand to eight thousand years ago, for example, it appears that the Sahara was one-third its present size and that rainfed vegetation supported not only herds of domestic livestock but also wild populations of giraffes, elephants, and lions. In our own time the deserts are expanding again, in a process called desertification. A consensus now is building among researchers that desertification is an artifact largely of human activity, both in the form of direct stress on the ecosystems of arid and semiarid regions, leading to the transformation of climatically marginal lands into biological deserts, and of indirect stress, leading to the greenhouse effect and global warming.

As arid lands researcher Harold E. Dregne observed in 1985, "The history of arid land development is one of wholesale removal of trees and shrubs, overgrazing, cultivation of climatically marginal lands, waterlogging and salinization of low-lying irrigated lands, sedimentation of irrigation canals and reservoirs, formation of dunes and hummocks where the vegetative cover has been disturbed, and blinding dust storms." This process continues. Massive tree and shrub cutting are epidemic in countries that lack alternatives to fuelwood. Overgrazing plagues both developing and industrialized nations. In the industrialized nations, particularly, "the desert blooms" but at a heavy price: inadequate drainage of irrigated farmland leads to salt-poisoned fields (salinization) and to increasingly saline water for all subsequent users downstream. Some desert soils contain minerals that react with the minerals in water to produce a ground surface as hard as concrete, for example, in many parts of the southwestern U.S. In addition, heavily irrigated lands are often subjected to high doses of chemical fertilizers and pesticides, a practice that can lead to a buildup of toxics well beyond the directly affected agroecosystem.

Today, development of highly integrated urban-industrial complexes, as in the desert and steppe regions of the U.S., Australia, and Asia, are major causes of human population and economic growth in arid lands. This is nothing new—the early civilizations of Egypt and Mesopotamia concentrated large populations at oases in the midst of a region dominated by hunting-and-gathering tribes and dryland farmers—but the reach of modern cities exceeds that of their predeces-

sors: to feed, clothe, and shelter the urban multitudes, greater and greater demands are being levied on the arid hinterlands.

An alternative to the traditional exploitation of arid lands may lie in the concept of **sustainable development**, defined by M. A. Toman in 1992 as "development that meets the (human) needs of the present without compromising the ability of future (human) generations to meet their own needs." Inherent in this concept are the hope of preservation and of nurturing over time, of stewardship and of self-imposed limits, of intergenerational fairness and of awareness of local and global environmental carrying capacity.

That an alternative approach to development is needed is obvious in a quick review of the current prominence of the eight traditional uses of arid lands: (1) nomadic hunting and gathering economies, whose demands on the land are relatively light, now are limited primarily to the Kalahari Bushmen of Africa and to a declining handful of Aboriginals in Australia; (2) nomadic pastoralism has been much reduced in favor of (3) sedentary ranching, but both rely on grazing, an inefficient method of converting plant foods to human use and one that often degenerates into overgrazing; (4) rainfed (dryland) farming, which depends more heavily on the weather than on technology, continues to be widespread where practical, but (5) irrigated farming now is practiced on an enormous scale worldwide, leading to the problems already discussed and to drawdowns of aquifers that rainfall cannot possibly replenish in less than the thousands of years required to accumulate these underground reservoirs in the first place; (6) mining (of metals, fuels, and groundwater) continues to be a major economic activity in arid lands; (7) recreation and tourism are on the rise in arid lands worldwide, which discourages some destructive uses of the land (such as strip mines) but increases the direct stress on delicately balanced ecosystems through such disruptive activities as off-road vehicle use; (8) urban development, as we have seen, is the driving force behind both population expansion in arid lands and growing demand in that quarter for food, fiber, minerals, and fuels.

All of these traditional uses of arid lands are consumptive, and making efforts toward sustainable development is vital. A few examples of this approach already in use, at least experimentally, in arid lands worldwide include: afforestation and reforestation projects to provide shade and fodder for livestock, fuelwood for human use, erosion control, land stabilization, and habitat for wildlife; developing low water-use field crops and landscape plants (see Xeriscape); implementing enlightened grazing policies that recognize the limited carrying capacity of arid environments; replacing chemical fertilizers and pesticides with organic soil amendments and natural pest controls; employing the techniques of bioremediation, as in the use of constructed wetlands to treat municipal wastewater, agricultural runoff, and acid mine drainage; designing cities and towns with energy efficiency and water conservation as paramount goals, and reducing the incentive for environmental plunder by ensuring economic stability for the world's rapidly growing human population.

JOHN M. BANCROFT
KENNITH E. FOSTER

For Further Reading: R. L. Heathcote, *The Arid Lands: Their Use and Abuse* (1983); Map of the World Distribution of Arid Regions, *Man and the Biosphere (MAB) Technical Notes 7* (1979); National Geographic Society, *The Desert Realm: Lands of Majesty and Mystery* (1982).

DETERGENTS

A detergent is a substance that is particularly effective in dislodging dirt and foreign particles from soiled surfaces. Soap, which is made from natural substances, was the first detergent. The problems with soap are that is reacts with certain dissolved minerals in hard water to form an unsightly white dust that remains on clothing after washing, and a great deal needs to be used. The first synthetic detergents were produced by the Germans during World War I and were made available to the general public after World War II. An advantage of these synthetic detergents was that they worked well in both hard and soft water. The term *detergent* now refers primarily to synthetic household detergents used for dishwashing and laundry. Detergents are also used in industry for textile scouring, dyeing, bleaching, and finishing; for metal cleaning; for cleaning and sterilizing food-processing equipment; and for making cosmetics.

Detergents work through various processes to remove dirt and oil. A detergent is a type of surfactant—an agent that aids in the emulsification of oil and dirt particles. Once emulsified, the suspensions of oil and dirt are prevented by the detergent from being redeposited on the soiled surface. Detergents also contain water softeners, bleaches, brighteners, enzymes, and builders. The builders in laundry detergents—most commonly sodium silicate, sodium carbonate, or various phosphates—increase the emulsifying power of the detergent. Detergents that lack builders are enhanced by special enzymes, which effectively speed up cleaning reactions while themselves remaining unchanged.

During the 1950s and 1960s detergent foam contaminated waterways throughout the nation and around the world. It was discovered that the bacteria that digest soap molecules could not biodegrade synthetic detergent molecules. Detergents were reformulated with a changed molecular structure that resulted in a biodegradable detergent. Biodegradable detergents have replaced most, if not all, nonbiodegradable detergents.

Another problem associated with detergents is cultural eutrophication, or the accelerated aging of a lake or pond. As part of the natural aging process of a lake or pond, mineral and nutrient levels increase, resulting in an increase in plant life. This increase causes a corresponding decrease in the amount of dissolved oxygen needed by other organisms. Accelerated eutrophication due to human activity is harmful to the aquatic ecosystem. Phosphate builders from laundry detergents, for example, provide an excess of nutrients, resulting in a proliferation of algae (algal bloom). When the algae die, the bacterial decomposition depletes the lake or pond of oxygen. Cultural eutrophication can result in massive fish kills, or even in a dead lake. The effects can reach coastal waters via the rivers that empty the lakes.

The process of cultural eutrophication is reversible and can be stopped if excess nutrient and organic matter supplies are cut off. Programs were enacted during the 1970s and 1980s to halt the eutrophication of the Great Lakes and the Chesapeake Bay area. In many areas across the country, phosphate-containing detergents are banned, and land-use management plans to protect aquatic ecosystems have been proposed.

DEWATERING

Dewatering is a process commonly used in wastewater treatment to remove water from residuals (sludges) generated during the purification process. Sludges produced by the removal of organic and inorganic solids from wastewater contain anywhere from 95% to 99% water. In order for these sludges to be treated and disposed of in a safe and cost-effective manner, excess water must be removed. In the early days of sewage treatment, dewatering was usually accomplished by applying the sludge to open sand beds where the sludge would drain and then air dry over a period of several months. In order to accelerate the process and reduce the unpleasant conditions associated with open sludge beds, mechanical dewatering processes in enclosed buildings are now most often used. These processes usually reduce the moisture content of municipal sewage sludges to 70% to 85%, although moisture contents as low as 50% are possible with some high-pressure processes.

The most commonly used dewatering processes are centrifuges and belt filter presses. If a drier sludge is required, a recessed plate filter press may be used. In general, the selection of a dewatering process depends on the type of sludge to be dewatered and the eventual disposal method. If a sludge is to be incinerated, the driest possible sludge cake is needed, which requires a high-pressure device. If the sludge is to be used as a fertilizer and soil conditioner, a lower solids content may be acceptable, depending on the hauling distances involved. Because dewatering results in a substantial reduction in the volume of sludge, longer hauling distances usually require a more thoroughly dewatered sludge.

In order for dewatering processes to work more efficiently, chemical agents are usually added to the sludge. These chemical agents, most often high-molecular-weight synthetic organic polymers, cause sludge particles to flocculate, thereby speeding the release of water. Because the theory behind sludge conditioning is not well understood, selection of the proper sludge conditioning chemical often requires trial and error testing. The cost of chemicals may be a significant part of the overall operating cost of waste treatment operations. Therefore, the selection of appropriate chemicals and minimization of their use are important priorities in a wastewater utility.

Dewatering also is being used increasingly at plants that treat water to be used for drinking. While many of the same considerations are involved in the handling of water treatment residuals, there are important differences that affect the selection of a dewatering process. Because sludges generated by the treatment of drinking water contain few fertilizing chemicals, such as nitrogen and phosphorus, and little organic matter, they are seldom applied to the land. Rather, they are frequently disposed of in landfills. This option tends to favor mechanical devices—usually centrifuges and belt filter presses—which dewater sludges to a state of dryness acceptable to landfill operators. As with wastewater sludges, chemical conditioning is usually used in conjunction with mechanical dewatering.

JOHN T. NOVAK

DICHLORODIPHENYLTRICHLOROETHANE

See DDT.

DIOXINS

Dioxins are a family of aromatic chemical compounds known as polychlorinated dibenzo-*para*-dioxin, or PCDD. Dioxins are not manufactured but are formed as impurities in the process of chlorinating phenols used in producing herbicides. They are also formed when polychlorinated biphenyls (PCBs) are heated or burned. The dioxin family of compounds have as a nucleus a triple-ring structure consisting of two benzene rings connected through a pair of oxygen atoms.

The term *dioxin* is most commonly used to refer to the compound 2,3,7,8-tetrachlorodibenzo-*para*-dioxin, also known as TCDD or tetrachlorodioxin. This type of dioxin is thought to be the most toxic of the 75 theoretically possible chlorinated dioxins. It is usually produced in trace amounts during the production of the herbicide 2,4,5-T (2,4,5-trichlorophenoxy acetic acid). As such it is a contaminant or byproduct found in Agent Orange, a 50:50 combination of 2,4,5-T and 2,4-D (2,4-dichlorophenoxy acetic acid).

There have been numerous studies of workers and civilians in an attempt to discern the effects of exposure to dioxin. Many of the studies have centered around workers exposed to dioxin during manufacturing accidents. Workers involved in the manufacture of 2,4,5-T or its precursor trichlorophenol may be exposed to dioxin when explosions or runaway reactions occur during the formulation process.

The first recorded accident during the manufacture of 2,4,5-T occurred in the United States in 1949. Other accidents or high-exposure incidents have since occurred in different parts of the world. Health effects resulting from exposure to chlorinated organic chemicals such as dioxin include a skin disease known as chloracne. Although acute effects of dioxin exposure vary, studies of exposed workers have thus far showed no measurable increases in mortality. However, serious effects of dioxin on small animals have been well documented.

In addition to workers involved in the manufacturing processes mentioned, people who spray 2,4,5-T, Agent Orange, or other dioxin-containing pesticides or compounds risk exposure to dioxin. There is also a risk of exposure for people in or near areas sprayed with these chemicals.

Much of what is known about the effects of environmental exposure to dioxin is the result of the Seveso experience. On July 10, 1976, dioxin was accidently released from a chemical plant in Seveso, Italy. A cloud of vapor spread over an area of about 880 acres, containing almost 40,000 people. The more serious contamination occurred directly south of the plant, covering about 215 acres. The town was evacuated when children began developing rashes and both domestic and wild animals in the area began dying. Among the general population, however, scientists found no significant general health problems.

DISCOUNT RATE

A discount rate allows the comparison of costs paid and benefits received at different times. It allows the conversion of future—or past—values into present value.

A product today is different from the same product delivered a year from now. In a similar way, an American dollar is different from a French franc. To know the relative value of the two currencies, we need to know the rate of exchange. To know the value of a product delivered in the future relative to the same product delivered now, we need to know the discount rate. Just as the exchange rate allows the conversion of the value of one currency into the value of the other, the discount rate allows the conversion of the value of a product given or received in the future into its value today. Recognizing the present value of a future benefit or cost is crucial in evaluating alternative projects when some of the alternatives have benefits or costs that occur at different points in time.

Calculating Present Value

The discount rate is often used interchangeably with the interest rate commonly used to project the growth over time in the value of a monetary sum. If the interest rate is 5%, a bond purchased for $1,000 today will be worth $1,050 one year from now. That is, the investor can put aside $1,000 today and get back $1,000 plus 5% of that amount one year later.

To find the value of a given sum invested today after one year's time, we multiply the sum today ($1,000 in this case) by 1.05. The future value one year later of $1,000 now is $1,050. In the general case the future value (FV) is the product of the present value (PV) and one plus the interest rate $(1 + i)$: $FV = PV(1 + i)$. Alternatively, using the same logic and rearranging the formula, we could calculate the present value of the future amount by using: $PV = FV/(1 + i)$.

The latter calculation illustrates the discounting process. We divide the amount to be delivered one year from now by the factor $(1 + i)$, which in effect is the exchange rate between an amount delivered today and an amount of the same present value delivered in one year. This converts the future value into an equivalent value delivered today. PV is the (discounted) present value of FV, delivered one year from now.

Similar procedures can be used to calculate the present value of future amounts received (or paid) two or more years in the future. Starting as before, the $1,050 received in one year could be reinvested after one year, to grow to $1,050(1.05) or $1,102.50 after two years, or $1,000(1.05)(1.05). That is, $1,102.50 is the future value, two years later, of $1,000 invested now; or $1,000

is the present discounted value of $1,102.50 received in two years.

Using the same logic and procedure, we can evaluate alternatives that involve expenditures and receipts at differing points in time. Receipts add to the present value of a project, while expenditures or costs reduce the present value. The sum of discounted costs and receipts (or benefits) is the net present value of the alternative.

Consider a simple example. An owner of a woodlot plans to use the site to build a new house next year. He must clear a number of maple trees to build the house. A contractor has offered to harvest the trees this week, and pay the owner $1,000 immediately, or to wait one year before harvesting them and pay the owner $1,100 then. No other costs or benefits are involved. Will the owner wait?

If he calculates using a discount rate of less than 10%, waiting will be the better option. A payment of $1,100 one year from now is worth $1,047.62 now if discounted at 5%, more than the value of harvesting now. Using a discount rate of 10% would make the present value of waiting a year just $1,000, making the owner indifferent between harvesting now and harvesting in one year. In other words, a lower discount rate increases the present value of the benefits to be received in the future, making waiting for them more attractive. (In a parallel way, a lower discount rate increases the present value of future costs, making acceptance of them less attractive.) A higher discount rate, as the name implies, places a larger discount on events in the future, making current expenditures and receipts more important.

Determining the Discount Rate

Discount rates used by individuals to compare costs and benefits across time are nearly always positive. That is, delivery today is valued more than delivery later. There are two underlying reasons for this: (1) individuals are impatient, having what economists call a positive time preference—wanting availability sooner rather than later; and (2) earlier returns allow other projects to be undertaken. Many productive investment projects can get more from the same resources if more time can be taken in the project's completion. To build 500 newly designed airplanes, to produce an apple orchard, or to restore a burned-over forest to its natural state, all take time as well as the use of resources. To some degree, each can be speeded up, at a cost. But typically when more time is available, each of these productive projects can be accomplished at lower cost.

Another factor that can influence the discount rate used by individuals is the expected rate of inflation. If an investor can purchase a bond for $1,000 now that will pay $1,050 in one year, and there is no inflation, then she will just be willing to buy it if her discount rate is 5%. But if she expects 4% inflation, then the $1,050 one year from now will, she expects, buy only as much as $1,010 will buy now. If she discounts the future money payment at 5% in a noninflationary environment, she will discount it at approximately 5% + 4% = 9% when she expects 4% inflation. But if she is using the market interest rate as her cost of investing in a project, inflation will already be included in the market rate. Market interest rates are determined by exchanges among lenders and borrowers who have inflationary expectations, and add them to their own noninflationary discount rates.

A third factor that may be used by an individual to adjust his discount rate is the risk that an expected payment will not be received. For international investments, the risk of political instability can be important. Years ago, when international oil companies invested in the development of oil fields in exchange for contracted rights to pump the oil at specified prices, some companies found that after the oil fields were developed and production had begun, the host governments voided the agreements and simply confiscated the fixed capital and equipment. Some of the expected returns to the investors did not materialize.

Currently, a firm or organization that would like to purchase rights to rain forests and other valuable habitat in nations around the world, either as a speculative venture or simply as a preservation project, must contend with the possibility that the rights it purchases may be unilaterally voided or confiscated. The future benefits, such as access to and control of the purchased land, might not be available as expected. Taxes on net benefits will also bring much the same result, though generally with more predictability. Decision-makers can either reduce the expected benefits, or use a higher discount rate in evaluating the benefits of potential projects subject to such risk and uncertainty.

Use of the Discount Rate in Benefit/Cost Analysis

An investment project involves a stream of costs (required sacrifices) and benefits (which may or may not include money payments) over time. The project will probably be undertaken if the present value of the benefits exceeds the present value of the costs, or, said differently, if the ratio of discounted benefits to discounted costs is greater than unity.

The choice of a discount rate is often critical to this calculation. A small change in the rate can change the ratio of benefits to costs from positive to negative, or vice versa. If the decision-maker is evaluating a project in the private sector, the market rate at which the

funds could be borrowed or invested in another project is a strong guide in choosing a discount rate.

For a simple project with no uncertainty, the use of a discount rate based on the interest rate will reveal the market's evaluation of the project. If there is uncertainty, the decision-maker can turn to the capital markets for information on rates of return, or discount rates for projects with similar risk.

For private firms and individuals, taxes are another consideration. Taxes reduce the portion of net project benefits that accrue to the investor. A person willing to invest money at 10% interest will require 13.3% when the returns are taxed at 25%. Multiple taxes are frequently present, and each is used to reduce the calculated benefits, or to raise the discount rate used to evaluate the project. If the firm is a corporation, for example, and if there is a corporate income tax on profits gained, as well as a personal income tax on the project's net proceeds when distributed as dividends, both taxes will reduce the benefits to investors of undertaking the project.

When the government undertakes a project, what discount rate should the government agency use in evaluating projects? No taxes are paid on the net benefits to the public but the costs of the project must be paid by citizens, in the form of taxes now or taxes later to fund bond repayments. Taxes now will reduce consumption, or savings, or some of each. Bond payments must be made with market interest added. What is sacrificed, if current consumption is reduced by taxpayers? Since each individual is either borrowing to finance consumption or is reducing savings, each is paying (or failing to receive) the market interest. If future taxpayers will repay bonds to finance a government project, they will pay the market interest. If investors pay a standard rate of income tax on the proceeds they have saved and invested, the market rate will reflect that tax rate, exceeding the untaxed rate of return required to bring forth the needed savings. Nevertheless, the fact that any government project requires the sacrifice of private projects valued at least at the market rate, argues that the market rate should be used as a discount rate in evaluating possible government projects. Indeed, to the extent that corporate projects, yielding a higher return but facing multiple taxes on payments to investors, are displaced by the required taxation, an even higher rate might be justified.

Are future generations cheated by the discounting procedure? Some people argue that discounting benefits and costs reduces the welfare of future generations by making future costs and benefits seem less important by discounting them. If project benefits and costs were not discounted, as some have suggested, then projects with very slow repayments would look as attractive as those which repay costs very quickly. Yet projects that repay quickly provide the opportunity to reinvest the same resources, providing more net benefits over time than the former.

For example, suppose that projects A and B, of equal $1 million cost (all paid today) each return $2 million in benefits, but that A returned the benefits at the end of 5 years, while B returned the same benefits at the end of 25 years. With no discount rate, the two projects both have a benefit/cost ratio of 2. But the benefits to society are quite different. If the proceeds of A are reinvested so that for both projects, the initial costs are the same and the benefits are taken in year 25, project A can be undertaken once in the first 5 years, twice in the second 5 years (as the $2 million in benefits are reinvested), four times in the third, 16 times in the fourth, and 32 times in the fifth 5-year period. With the same initial sacrifice of $1 million, and over the same 25-year period, projects such as A can return $64 million compared to only $2 million for projects such as B. Use of a discount rate brings into perspective the productiveness of time, making more attractive those projects which tie up resources for a shorter time.

Discount Rates and Shortsighted Individuals

It is easy to imagine the possibility of individual decision-makers who are personally shortsighted. A large landowner who is elderly and in poor health, with no heirs to care about, may in fact not personally value even large payoffs accruing to his property in the future. Such a person might be tempted to act with a very high discount rate, and to adopt very shortsighted policies such as overgrazing land, harming the land but increasing profits in the short run. However, for even this individual, a better option to gain short term benefits would be not to mismanage the land, but simply to sell all or part of it to an investor not suffering from a shortsighted view.

Whoever has a low discount rate by virtue of a low rate of time preference will value a long-lived asset most. Thus in a market setting, long-lived assets tend to be owned by individuals with low discount rates and long time horizons, or by those who expect to sell to others when they want quick cash. The ability of owners to buy and sell encourages even short-sighted individuals such as the elderly landowner above to manage assets as if they cared about the future. In effect, ready buyers are representing future generations, just as traders in Florida represent New Yorkers who want Florida oranges but will never visit Florida. They do so by being willing to hold assets that must provide only normal rates of return, either in current benefits paid or in rising asset values as their payoff time approaches or their scarcity value increases. In effect, markets move long-lived assets from those with high discount rates to those with normal discount rates, much as traders move oranges from Floridians who personally value added oranges very little (and who may not care at all about New Yorkers) to those in New York who will pay good prices in New York markets.

In contrast to market owners, elected politicians cannot survive by going against the wishes of current voters in order to sell into a market serving the interests of future generations. Only owners of property whose value is currently affected by the government decision have an incentive to lobby for farsighted policies. Elected officials and their appointees placed in such a position may be pressured to act as if they have high discount rates—that is, as if they value the present far more than the future.

RICHARD L. STROUP

For Further Reading: James D. Gwartney and Richard L. Stroup, "Capital, Interest, and Profit," in *Economics: Private and Public Choice*, 6th ed. (1992); President's Council of Economic Advisers, "Economic Growth and Future Generations," in *Economic Report of the President: 1993* (1993); Warren Scoville, "Did Colonial Farmers Waste Our Land?" in *Southern Economic Journal* (April 1953).

See also Benefit-Cost Analysis; Debt-for-Nature Swaps; Economics of Renewable and Nonrenewable Resources.

DISEASE

See Health and Disease.

DISPOSAL CHARGES

Disposal charges offer a method of motivating producers and consumers to consider the cost of disposing of products when deciding how to produce or to buy them, and whether to recycle them. Most households currently pay for waste disposal services either through their local property taxes or by a fixed fee to a private collector. Under this pricing system, households do not have a monetary incentive to reduce the amount of waste that they dispose of by changing their consumption behavior or by increasing their recycling efforts. Likewise, producers do not have an incentive to make less wasteful goods, such as goods with less packaging, or to use recycled materials to produce the goods that they sell. Three types of disposal charges encourage producers and consumers to take disposal costs into account.

1. *Household Charges.* Some communities have begun to charge households according to each bag or can of trash they dispose of. Under these "unit-based

pricing" programs, households can save money by buying goods with less packaging or by recycling and composting their waste. Studies of unit-based pricing programs indicate that they may significantly decrease the amount of waste sent to the landfill or incinerator, but that they also present some potential problems. Particularly worrisome is the fact that unit-based pricing provides households with an incentive to dispose of their waste illegally.

2. *Combination Disposal Tax and Reuse Subsidy.* Under a combination disposal tax and reuse subsidy policy, producers and importers would be taxed according to the cost of disposing of the goods that they produce. This would encourage them to reduce the amount or toxicity of waste associated with their products. In addition, all firms that use recycled materials would receive a subsidy (reflecting avoided disposal costs), thereby encouraging increased recycling. An advantage of this policy is that it would not provide an incentive for illegal disposal. It would, however, not be feasible to administer a set of taxes and subsidies on all consumer products. A combination tax and subsidy policy might target items that have potential for increased recycling, or that are particularly damaging, such as old car batteries, tires, and used oil.

3. *Deposit Refund System.* Under a deposit refund system, consumers would pay for the cost of disposing of a product in the form of a deposit at the time of purchase. If the product is returned, all or part of the deposit would be refunded. The most familiar form of a deposit refund system is the bottle bill, which exists in ten states. Deposit refund systems may be used to encourage the return of products that can be recycled, such as bottles and cans, or products that pose potential health hazards when disposed of in incinerators or landfills, such as car batteries or used oil. Like the combined disposal charge and reuse subsidy, deposit refund systems do not provide an incentive for illegal disposal.

TERRY DINAN

DISSOLVED OXYGEN

All fish and higher forms of aquatic life require oxygen that is dissolved in water to survive. The amount of oxygen dissolved is usually measured as milligrams of oxygen per liter of water (mg/l). Oxygen is not very soluble in water. The maximum amount that will dissolve is between 8 and 11 mg/l. Higher temperatures and dissolved salts decrease this saturation level. Most fish require at least 2 mg/l, and many game fish require up to 4 mg/l. If the oxygen content drops below these values, fish kills can result. In addition, waters with very low levels of oxygen are usually associated with offensive odors.

There are two primary sources of dissolved oxygen: the atmosphere and plant photosynthesis. When dissolved oxygen concentration drops below its saturation value, oxygen transfers from the atmosphere to the water. The rate of transfer depends on the amount of surface mixing and the difference between the actual dissolved concentration and the maximum saturation concentration that the water can hold.

Another source of dissolved oxygen is aquatic plants and algae. During daylight hours, the plants photosynthesize and produce oxygen. Plants also respire; they consume oxygen to combine with organics and produce energy. Thus, at night, there is a net consumption of oxygen by plants. The combined effect is a diurnal curve in which the dissolved oxygen concentrations vary in a cyclic process—rising after dawn, increasing until late afternoon, and falling during the evening hours before rising again the next morning.

The most important mechanism for the removal of dissolved oxygen is the breakdown of organic matter. When an organic waste such as sewage is discharged to a water body, bacteria in the water use the organics as food. If dissolved oxygen is present, the bacteria break down the waste by combining the organic molecules with oxygen yielding energy for growth. In a water body, this bacterial respiration can deplete oxygen from the water.

This bacterial respiration is so important that a standard measure of the strength of an organic waste is the amount of oxygen that bacteria will consume in breaking down the waste. This measure is called the biochemical oxygen demand (BOD) of a waste. To protect the aquatic life in a stream and ensure minimum dissolved oxygen levels, limits are set on the amount of BOD that a sewage plant can discharge to a water body. These limits vary by the size of the water body, the amount of dilution that takes place, how fast the particular waste degrades, and the temperature of the water. Recall that warmer waters hold less oxygen. In addition, bacteria degrade wastes more rapidly at higher temperatures. Thus for most streams, the critical period for dissolved oxygen is during the summer when flows are low, temperatures are high, and little dilution occurs. Dissolved oxygen concentrations are likely to be lowest at these times, and fish kills are more likely to occur.

Another process that can remove oxygen is eutrophication. When nutrients such as nitrogen and phosphorus enter a water body, excessive growth of algae can occur. The nutrients frequently originate from agricultural runoff and from sewage discharge. When the algae die, they settle to the bottom and are consumed by the bacteria there. This reduces the oxygen in the water.

DANIEL L. GALLAGHER

DISTRIBUTION

Economic distribution can be divided into two separate concepts, the first of which deals with the movement of goods from manufacturers to consumers. This includes the role that mechanisms such as market prices play in the allocation of the total flow of goods and services throughout an economy, from producer to wholesalers to consumers. The second concept deals with the way in which economic wealth is shared in any society. Information about the distribution of income and wealth among various groups is used to evaluate notions of equity and distributive justice.

Economic theory assumes that goods and services are produced for the most part because producers believe that investors will profit from their activity; that is, the costs incurred will be less than the price received. Producers compare the current total cost of input and distribution to the price at which they think they will be able to sell their outputs (see Demand). In order to produce the correct quantity of a product, a producer must predict consumer preferences. When producers do not accurately predict consumer preferences, the market supply of a particular good will be greater or less than the quantity demanded at the expected price. The price of the good will then fall or rise in order to clear the market. This economic theory works well for products for which storage is not possible, such as cut flowers, but less well for manufactured goods.

Individual income is closely related to education and training—the large portion of the labor force with little education receives low wages and a small proportion of national income. Education and training are keys to raising the income of individuals and groups who earn low wages and suffer high unemployment.

Many people believe that social equity is satisfied if individuals who earn low wages have access to education and better jobs but choose not to take advantage of the opportunities or cannot compete. Even in these cases, society provides a safety net to ensure that individuals have at least a minimum standard of living.

DIVERSITY

Diversity is defined as a state or instance of difference, unlikeness, multiformity, or variety. The concept of diversity can be applied to discussions of living organ-

isms as they function in the natural environment and to human beings and the variety of cultures and ethnicities around the globe.

The true extent of the biological diversity, or biodiversity, of living organisms on Earth is unknown. Although approximately 1.4 million species have been formally described, estimates of the total number of species on Earth range from 3 to 30 million.

Species diversity is measured using several different parameters. Species richness refers to the total number of species in a given area or ecosystem. Species evenness is a measure of the relative abundance and distribution of the various species in a given area. Generally, biological diversity increases as latitude decreases. Tropical rain forests and coral reefs near the equator are the most diverse ecosystems on Earth.

Species loss occurs naturally as particular groups within populations become extinct. Human activity has greatly accelerated the rate of species loss. Extinctions of plant and animal species around the world are currently occurring at a rate one thousand times faster than at any time during the last 65 million years.

Diversity is also relevant to agriculture. Throughout history humankind has used approximately 7,000 different species of plant for food. Today people rely on approximately 20 species, three of which (corn, wheat, and rice) constitute 65% of the world's food supply. Plant breeders obtain genetic material from wild plant varieties to improve crops—for instance, to make a species more resistant to disease. Many of today's crops share genetic material, making them equally vulnerable to pests and disease.

Biodiversity is also an important source of medicine and industrial products. Approximately 40% of the prescription drugs sold in the United States contain materials originally derived from wild plant species. Wild plant species also represent a potential source of material for producing new drugs to combat cancer and HIV. Industrial products, such as rubber, petroleum substitutes, and timber, have been extracted or produced from plants, while many other such valuable products have yet to be developed.

Just as biological diversity facilitates Darwinian evolution, so is cultural diversity essential for social progress. All over the world ethnic and regional groups are asserting their identity and recapturing autonomy. This trend increases not only cultural diversity but also the rate of social change.

DNA

DNA (deoxyribonucleic acid), found in all nucleated cells, comprises the genetic material passed on from generation to generation. Specifically, it is the fundamental component of genes and chromosomes, determining every function of the cell and giving each type of cell distinctive features. A unique quality of the DNA molecule is its ability to self-replicate with the aid of certain enzymes (substances that alter the speed of chemical reactions without being permanently changed themselves).

The DNA molecule was first observed in cell nuclei in 1869 and named *nuclein*. In 1944 the Oswald Avery group at The Rockefeller Institute found DNA to be the carrier of genetic information. James Watson and Francis Crick (1962 Nobel laureates) described the actual structure of DNA in 1953 as a "double helix" consisting of two parallel strands, with sugar-phosphate backbones attached to nitrogen-containing, or nitrogenous bases. These strands are linked by the nitrogen base crosspieces arranged in pairs of nucleotides. The nucleotides are made up of a phosphate group, a deoxyribose sugar, and either a purine (adenine or guanine, A or G) or a pyrimidine (cytosine or thymine, C or T) nitrogenous base. The overall appearance of the double helix is that of a twisted ladder, with nucleotide pairs meeting to form hydrogen-bonded rungs, and the sugar and phosphate groups forming the sides. In the double helix model, the order of nitrogen bases along one strand determines the matching bases on the other, complementary strand. That is, every adenine (A) must be joined to thymine (T), and every guanine (G) must be joined to cytosine (C). No other pairings are possible. Thus, each strand of the double helix is the complement of each other, according to the A-T and G-C base pairing rule. The order of the nitrogenous base pairs specifies different "messages" for the cell.

The structure of DNA assures reliable self-replication, in which the double helix is "unzippered" in a precisely controlled enzymatic process that brings complementary nitrogen bases together to create new pairs. This process essentially builds two new DNA "ladders" from the original two "half-ladders." In this way, DNA replicates and provides full copies to daughter cells when parent cells divide.

DNA is found in the nuclei of all animal and plant cells and also exists outside the nucleus within mitochondria (structures producing most of the energy in cells) and chloroplasts (the parts of the plant cell necessary for photosynthesis). DNA's main function is to provide instructions for protein synthesis. In a manner similar to the self-replication process, DNA opens its double helix to allow a segment of one strand to be "copied" into a ribonucleic acid messenger molecule (mRNA), which consists of a string of nucleotides that complements the DNA base sequence. This mRNA then travels to structures in the cell called ribosomes, which "read" the sequence of nucleotides of the

mRNA in groups of three, or triplets. Each triplet specifies a particular amino acid. In this manner, by using the coded information originating from a DNA molecule, ribosomes can connect a series of amino acids to create many copies of a specific protein molecule.

DOMESTICATION

Domestication derives from the Latin word *domus*, meaning house or household. Domestication is an evolutionary process in which the genetic constitution of populations is altered to confer greater and greater fitness for the man-made habitats provided. When populations are extracted from the wild and introduced into habitats altered by man, a combination of natural and artificial selection pressures modify the genetic constitution to confer increasing fitness for the new habitats. Since domestication is an evolutionary process, all degrees of intermediate states and conditions may be found between the extremes of fully wild races and fully domesticated races. Fully domesticated races cannot survive without human intervention.

In domesticated plant races, the mechanisms for seed dispersal are usually altered or suppressed, making them dependent on humans for survival. In wild cereals, the inflorescence fragments when ripe and seeds fall (shatter); in domesticated cereals, fragmentation is delayed until after harvest or suppressed entirely. In other crops, pods and capsules that split open in wild races do not do so in domesticated races. Some races of plants propagated vegetatively become sterile or even lose the capacity to flower. Other modifications include loss or decrease in seed dormancy; larger seeds, fruits, or tubers; more uniform ripening; fewer but larger inflorescences; reduction in toxic compounds; enlarged storage organs; greater yield and palatability; etc.

Some intermediate races of plants can escape into wild habitats and thrive, others may thrive in man-made habitats without intentional care and become weeds. Weed races are typical byproducts of the domestication process in plants. There are weed races of some kinds of wheat and of rice, maize, barley, oats, rye, sorghum, millets, potato, tomato, cabbage, carrot, beet, radish, sunflower, and so on. Even some perennial crops have weed races, e.g., orange, guava, grape, pear, pomegranate, asparagus, etc. Since domestication very rarely results in new species, the wild, weed, and domesticated races are genetically compatible and tend to interbreed wherever they occur together. Genetic traits may be transferred in all directions among the three classes. In some cases, it is difficult or even impossible to distinguish between wild and weedy races, habitat being the primary criterion. Some wild plants are adapted to naturally disturbed or unstable habitats such as talus slopes, river banks, dunes, blow-downs in forests, etc., and some of these have adapted readily to habitats disturbed by man. Some crops do not have weed races and, of course, many weeds are not closely related to crops.

For animals, the primary criterion that renders races dependent on human aid is loss of defense against predation. Most domesticated animals (in appropriate habitats) can survive or even thrive without the aid of humans where there is little or no predation. Cats and dogs that must live by predation in the wild can become feral and do well, but there are breeds of both that could not adapt. The weedy habits of some plants are echoed in animals by species that have insinuated themselves into man-made habitats, e.g., the house sparrow, starling, rock dove (pigeon), brown rat, house mouse, house fly, and fruit fly. There are domesticated races of the pigeon, rat, mouse, goldfish and fruit fly that cannot survive without human aid.

The term "cultivated plant" is often used interchangeably with "domesticated plant" in the literature, but there is a difference. Cultivation refers to activities that tend or care for the plant; true domestication requires genetic changes that confer changes in adaptation. It is quite possible to cultivate wild plants. Wild flowers are often brought into the garden for their charm and beauty. Wild plants may be used in landscaping, as rootstock for grafting, or as pollinators; they may be enhanced by fire, tillage, pruning, fertilization, or irrigation. Such activities are included in "cultivation."

The terms tame and domesticated are also often confounded. It is quite possible to tame wild animals without domesticating them. In such cases, tame refers to docility and accommodation to human presence. Domesticated animals have been altered genetically from their wild ancestors by human selection over generations.

JACK R. HARLAN

For Further Reading: J. R. Harlan, *Crops and Man*, 2nd ed. (1992); D. Rindos, *The Origins of Agriculture: An Evolutionary Perspective* (1984); P. J. Ucko and G. W. Dimbleby, eds., *The Domestication and Exploitation of Plants and Animals* (1969).

DOSE RESPONSE RELATIONSHIP

A deliberate, specified, controlled exposure, as in a planned experiment, is called a *dose*, and a detected effect is called a *response*. The level or degree of response is generally affected by the level or intensity of the dose, the degree of response increasing with increasing dose. This kind of direct relationship between dose and response is what is meant by the term *dose response*, a term most often applied to results from studies using test animals.

Exposure can be to chemicals or mixtures of chemicals; to physical agents such as heat, sound, or radiation; or to more than one of these agents at once. Responses can be adverse, beneficial, both, or of no known consequence. In studies of dose response, the particular response to be studied, called the *endpoint*, needs to be carefully defined so as to make its detection as clear as possible.

Dose, too, needs to be carefully defined. The degree of response for a given endpoint can also depend on the route by which the dose is administered (for animal exposure to chemical substances, for example, by inhalation, skin contact, or oral ingestion); whether the dose is administered at a constant rate, intermittently, or all at once; the total time period during which doses are administered; and the period in the organism's life during which dosing occurs. Dose is generally defined as the amount of chemical or physical agent to which an organism is exposed. However, emphasis is increasingly being placed on methods for measuring or estimating the dose that is delivered internally to specific sites, utilizing biomarkers of exposure. Doses of chemical agents in animal testing are generally classified as acute (short-term administration of high doses), chronic (long-term administration of low doses), and subacute (intermediate-term administration of intermediate doses).

The subjects under study and the conditions of study also need careful definition. In animal studies, for example, species, strain, gender, age, type and method of supplying feed, general health, and many other possibly confounding factors can influence test results profoundly. Even with the greatest of care in defining the conditions of study in either animal or human studies, the measurements of response are still subject to experimental and other types of error and uncertainty. Dose response thus has special value: if dose response is clearly observed, the effect observed is probably a real one; if dose response is questionable, there may be no actual effect, or something may be wrong with the study. Thus, dose response offers a valuable test of the value of experimental findings.

PAUL F. DEISLER, JR.

DOXIADES, CONSTANTINOS APOSTOLOS

(1913–1975), architect, urban planner and writer.

With his apt middle name, Constantinos Apostolos

Doxiades was a zealous and exceptionally articulate apostle of long range city planning. He was called "an oracle among planners" and "busy re-modeler of the world." He was a highly successful businessman with clients around the world.

The problem, as Doxiades saw it, was that "the expanding city is eliminating nature, and when it does not eliminate it, it contributes to its degradation, crushing man in the process. . . . We are heading toward much larger cities which will be born out of the merger of the existing ones into a continuous system which will cover our countries from one end to the other and over national boundaries to form a universal city. . . . We try to avoid this course toward the great city with its inhuman dimensions, but this is unrealistic. Our civilization makes such a city an imperative necessity. . . . Our only realistic hope is to try to transform the inhuman nature of this city, to create human conditions within the inhuman frame."

Doxiades gained early recognition in the post-World War II reconstruction of Greece. In the mid-1960s, at the height of his career, he was consultant to more than 100 national governments, city administrations, regional associations, and private groups. Ada Louise Huxtable, the leading architecture journalist at the time, said he was "influencing a good part of the environmental destiny of the world."

By then he and his staff had designed Islamabad, the new capital of Pakistan; provided housing for a half million people in Karachi; and had contracts with Spain, France, Iraq, Jordan, and Ghana. They were working on a characteristically expansive project in South America to design new cities and towns with coordinated resources, transportation, and communications in geographically related sectors of Brazil, Argentina, Bolivia, Paraguay, and Uruguay; and on a plan for a trans-Asian highway. In the United States, he was engaged in Louisville, Detroit, Cincinnati, Philadelphia, and Washington, D.C. He hesitated to offer any ideas about New York (or Paris and London), but he came up with a plan, sponsored by the non-profit Great Lakes Megalopolis Development Corporation, for an area between Buffalo and Milwaukee that would control water pollution and land use.

Doxiades had a gift for imagery and Greek-rooted catchwords. He coined the term *ekistics*, from the Greek word for house, to denote the science of human settlements. He said most urban development, responding to unrestrained population growth, was creating overlapping *megalopolises*. They soon would constitute a world city, which he called *Ecumenopolis*. Ecumenopolis would either be a disaster or be liveable, depending upon wise assessment, planning, and management, i.e., good ekistics. When he received the annual award of the Aspen Institute for Humanistic

Studies in 1966, he discussed *anthropocosmos*, his word for the human world. He spoke dramatically of yet another new science, *anthropics*, based on the wholeness of humanity. His own ideal city of the future, he called *Dynapolis*.

When Doxiades spoke of architecture, he was not speaking of single monumental structures, but of everything that went into the human habitat. It was not enough, he felt, to create a skyscraper or a Taj Mahal or even a square in Florence. Architecture was a medium for providing for populations, especially housing for the masses, answering "the real question: how we want to live." He described himself as a "master builder."

Most of Doxiades's plans called for change from existing patterns in which cities expanded in widening concentric circles while the outdated cores withered and died. He proposed forced growth along linear axes for business and industrial development. He placed residential areas, including parks and neighborhood services, away from but within easy reach of the axes. He provided roads for local vehicular traffic, but laid out main traffic arteries away from the central axes. A key concept was the removal of large-scale auto traffic from the city, especially the surfaces. He said it was cheaper to maintain tunnels than to keep re-surfacing streets damaged by excessive auto traffic.

Doxiades was born in Stenimachos, Bulgaria. His father, a prominent pediatrician, fled with his family to Athens in 1914 at the outbreak of World War I. His parents assisted war refugees and his father became Greek Minister of Refugees, Social Welfare, and Public Health. Young Doxiades helped in finding housing for the homeless. Thus, Doxiades said, he learned in his early youth the practical and emotional implications of human habitation.

Doxiades obtained his architecture degree at Athens Technical University and received an engineering doctorate at Berlin-Charlottenburg Technical University. He later attributed his love of learning and aesthetics to his humanistic teachers in Greece and his Bauhaus-oriented mentors in Germany. After World War II he served in the Greek delegation at the founding of the United Nations in San Francisco. In 1948 he chaired the United Nations Working Group on Housing Policies. In 1951, hospitalized by overwork and ulcers, he went to Australia with his wife and children. He returned to Athens in 1953 to found Doxiades Associates and built up a 700-person staff. The offices overlooked the Parthenon. Once a year, with a mercantile instinct for networking and publicity, Doxiades invited prominent philosophers, architects, environmentalists, businessmen, journalists, and government officials onto his yacht for seminar-like discussions and a cruise of the Greek islands.

For all of the Greek-rooted labels, his "science" of ekistics embraced a somewhat traditional procedure: first, the collection of data on population patterns, commercial needs, environmental studies, and so forth. Next, existing trends and plans were analyzed. Finally, the analysis was applied to achieving specific goals. What Doxiades contributed was his undoubted professional skills, practical experience, and attention-getting language.

When he died at the age of 62 at his home in Athens, a colleague estimated he had affected some 60 million lives all over the globe. It was ruefully noted that "many of his firm's costly and voluminous studies— backed with charts and computer printouts—came to grace a shelf in someone's office." On the other hand, he left a ringing "creed" (his term) for the environmentally conscious world in general and for his profession in particular: "Architecture," he wrote, "is the discipline not of designing houses or buildings, much less of designing monuments, but of building the human habitat."

JACK RAYMOND

For Further Reading: C. A. Doxiades, *Architecture in Transition* (1963); C. A. Doxiades and a symposium, with Rene Dubos, *Anthropopolis: City for Human Development* (1974); Christopher Rand, "The Ekistic World," *The New Yorker* (May 11, 1963).

DRUGS

A drug is any substance used as a medicine or in making medicines. Most drugs are chemical substances that, upon use, elicit a beneficial response, such as the cure or mitigation of a disease or a pathological condition. Recreational use of amphetamines, ethanol, cocaine, and opiates compromises the beneficial criterion.

Historical Development

Drugs have always been important to humans. Shortly after our early ancestors satisfied their hunger, they probably looked for something to relieve aches and pains. The uses of medicinal plants were among the earliest information to be recorded, and plant drugs were assigned sufficient value to be among the earliest items of commerce. There was a gradual refinement in the use of drugs; this progressive development accelerated parallel with scientific advancements. Plants were the primary source of drugs until the mid-20th century when there was a major shift to the use of synthetic chemicals.

Increased medicinal research following World War II resulted in significantly improved separation tech-

Table 1
Chronology of landmark drugs

1630s Cinchona bark (fever or Jesuit's bark) was introduced into Spain from the New World. It became the accepted drug for treatment of malaria. Quinine, the active ingredient, was isolated in 1820. Resistance to synthetic antimalarial drugs had become common, but quinine is still effective.

1780s A physician recognized that digitalis tea, a folk remedy, was an effective treatment for dropsy (congestive heart failure). Digitoxin and digoxin were isolated from digitalis in the 1930s. Digoxin is still among the most frequently dispensed drugs in the U.S.

1806 Morphine was isolated from opium. It was the first natural drug constituent to be isolated. Many physicians still consider it the drug standard for pain relief. In the search for more useful synthetic analogs, separation of analgesic property from addictive effect has been a problem; heroin was an early failure.

1899 Aspirin (acetylsalicylic acid), an early synthetic drug, was introduced. It became a popular nonprescription analgesic and anti-inflammatory drug. Willow bark, a folk remedy, contains salicin, the progenitor of aspirin.

1924 Ephedrine was introduced into occidental medicine.

1940s Penicillin initiated the antibiotic era. These "miracle" drugs effectively control many pathogens that previously caused prolonged illness or death (*see* Antibiotics).

1950s Modern chemical synthesis produced thiazide diuretics for treating hypertension, phenothiazines for schizophrenia, and corticosteroids for inflammatory conditions.

1990s Interferon and drugs that modulate the immune response, which show promise for control of previously refractory diseases and tumors, were produced by a combination of molecular and genetic manipulation (*see* Biotechnology, Medical).

niques and the recognition that the best therapeutic control for most potent drugs is attained when the drug is used as a pure, single active ingredient in a suitable dosage form. Thus, we now have the ubiquitous tablet. Other modern developments involve using natural drug constituents as chemical models and synthesizing analogs that are selected for enhanced potency or fewer adverse effects; cocaine from coca leaves was the progenitor of benzocaine and other more useful local anesthetics, and ephedrine from the Chinese herb Ma-huang gave rise to a long list of antihistamines. Adjusting the dose of a drug may produce distinct therapeutic responses, such as cough suppression with low doses of codeine and pain relief with higher doses. One of the current frontiers is the increasingly sophisticated multi-drug dosage regimens that are yielding promising results in the treatment of difficult pathologies such as some types of cancers. Table 1 shows a chronology of a few landmark drugs to illustrate the development of our modern therapeutic arsenal.

Use and Control

Laws controlling drug use in the U.S. are based on two assumptions: first, that drugs are inherently risky and their use is justified only when the medical benefits outweigh the risks, and second, that better appreciation of risks and benefits can be obtained only through scientific evaluation in controlled clinical trials. Drugs marketed in the U.S. must be proved safe and effective in animal tests and then in well-controlled clinical trials with humans. The drugs are then said to have approved medical uses (indications), and are also said to be licit drugs. Drugs without approved medical uses cannot be marketed legally, and their use is considered illicit. Licit drugs may also be subject to inappropriate use.

Licit Uses

Drugs that have at least one approved use may be sold without prescription if adequate directions can be developed for safe and appropriate use by lay persons without supervision by a health professional. These nonprescription drugs are also commonly called "over-the-counter" or OTC products. Approximately 700 active ingredients are currently approved for OTC sale, and these compounds are combined to make over 300,000 individual nonprescription products marketed in the U.S.

Drugs that require a prescription are those for which diagnosis and safe use cannot be assured without supervision of a health practitioner. These drugs are called legend drugs in the U.S. because their label must bear the legend, "Caution: Federal law prohibits dispensing without prescription," a requirement of the Food, Drug and Cosmetic Act (FDCA). Drugs that are *adulterated* (contaminated, spoiled, or improperly manufactured or stored) or *misbranded* (inappropriately labeled) can be seized by the FDA and removed from the market.

Manufacturers of licit drugs may not promote them for uses that are not part of their approved labeling. However, physicians may prescribe legend drugs for nonapproved uses. In general, prescribers rely on the scientific and medical literature for evidence of the safety and effectiveness of these nonapproved uses.

In recent years, the FDA has approved conversion from prescription-only to OTC status for a number of drugs, a trend that is expected to continue. These drugs include ibuprofen (e.g., Advil®, Nuprin®) for pain and inflammation, loperamide (e.g., Imodium AD®) for diarrhea, and miconazole and clotrimazole (e.g., Gyne-Lotrimin® and Monistat®) creams for vag-

initis. These products enable patients to self-treat conditions that formerly required a physician's visit.

Inappropriate Uses

The inappropriate use of licit drugs (particularly legend drugs) is a growing concern for public health officials and managers of health care programs. Categories of inappropriate use include: unnecessary treatment (e.g., using an antibiotic for treating a common cold); treatment for an excessive period of time (e.g., appetite suppressants beyond a few weeks); simultaneously using two or more drugs that interact adversely with one another; using drugs in patients who are known to be intolerant of them; and use of newer, costly drugs, when older, less expensive products would be equally effective. Of particular concern is the use of drugs when non-drug therapy (e.g., diet, exercise, physical therapy, counseling) would be more effective or less costly.

Illicit Uses

Most illicit drugs alter perception or other mental processes in ways that persons find pleasurable. Societies have tended to restrict or prohibit use of drugs for purely recreational purposes. Drugs restricted to use by adults only, and often restricted to use in specific locations, include nicotine (in tobacco) and alcohol. Their sale to or use by minors is considered illicit, as is their use by adults in ways that pose danger to others. Other frequently abused drugs, which are generally prohibited except for legitimate medical use, comprise six major classes: stimulants (e.g., cocaine and Ritalin®); barbiturates (e.g., Nembutal®); benzodiazepines (e.g., Valium® and Xanax®); narcotics (e.g., codeine and Demerol®); marijuana; and psychotropics (e.g., LSD, peyote, and PCP).

Control of distribution and use of substances, licit and illicit, with a high potential for abuse is the objective of the Controlled Substances Act (CSA). Enacted in 1970, the CSA imposes criminal sanctions for distribution or possession of certain drugs other than for legitimate medical purposes. Health practitioners, manufacturers, and distributors must register with the Drug Enforcement Agency (DEA) and maintain records of all receipts and distributions of the five categories of drugs subject to the act, which are known as controlled substances.

Environmental Impact

Drugs may have variable ecological influences, direct and indirect. Proper drug use provides a direct beneficial influence on the targeted individual. This can be considered a positive ecological impact of limited scope. The use of drugs in agriculture (e.g., antibiotics and hormones in livestock feed) has been associated with concerns about human exposure to drugs through treated animals. However, since biologic responses to drugs normally follow a characteristic dose-response profile (no response with subthreshold dosage levels, more pronounced responses with increasing dosage levels), the low levels of inadvertent drug exposure of nontargeted individuals to properly used drugs are virtually without ecological significance.

There are risks associated with the use of most drugs, and adverse effects do occur. Broad-spectrum antibiotics sometimes cause a significant alteration in the ecological balance of the intestinal microflora. When many of the dominant intestinal microorganisms are killed or inhibited, unaffected minor components of the normal microflora proliferate without the usual ecological constraints. This gives rise to such situations as overgrowth by *Candida* (a yeast) in the intestine or pseudomembranous colitis (caused by *Clostridium difficile*, spore-forming anaerobic bacteria). The problems (nausea, abdominal pain or tenderness, and diarrhea) caused by such ecological changes in the intestinal microenvironment can be managed by discontinuing use of the antibiotics or by initiating secondary therapy.

Specialized agronomy and silviculture practices have been developed for only a few of the plants used for drug purposes. Most medicinal plants are merely collected from native flora. A rapid increase in demand for a slow-growing or uncommon species may cause the species to become endangered or possibly even pushed to the point of extinction. A recent case for concern involved taxol, a plant constituent that provides promising treatment of ovarian cancer. Taxol occurs in low concentrations in the bark and needles of the Pacific yew. The native stands of this rare and slow-growing tree cannot meet a sustained demand for the drug. Until alternate approaches are fully developed for the production of taxol, the Pacific yew tree may become endangered.

Solvents and chemicals used to isolate or synthesize drug substances have the potential for causing a negative environmental impact. However, good manufacturing practices can ameliorate ecological damage (see Pharmaceutical Industry). In addition, the United States Environmental Protection Agency has recently become interested in potential environmental dangers arising from large scale disposal of discontinued, recalled, damaged, or outdated antibiotics and chemotherapeutic agents, especially in landfill sites or into water.

Societal Impact

On balance, legitimate medical drug use has been considered a positive factor in American society, a view validated in part by the steadily increasing life span

and quality of life. One societal phenomenon associated in part with drugs, therefore, is the "graying of America."

Concerns have been expressed about possible negative effects of certain classes of drugs, particularly psychoactive agents. One concern is the "medicalization" of social problems. For example, minor tranquilizers and alcohol are frequently used for treating the strains and stresses of everyday life. Critics of these uses suggest that American society must learn to use non-drug approaches to stress, such as biofeedback, counseling, or self-discipline.

Our largely Protestant heritage may account for the attitude that "if a drug makes you feel good, it must be bad." On this view, drug treatment of anxiety and depression is not as socially accepted as is treatment of other diseases, such as high blood pressure. At its extreme, this view equates drug and alcohol addiction with moral laxity, and encourages punitive approaches to controlling substance abuse.

An opposing view labels drug use a normal human behavior, largely a matter of personal choice, and believes education and medical management of drug abuse are preferable to criminal sanctions. Its adherents argue for "decriminalization" of substance abuse, if not outright legalization. They consider the high costs associated with misuse as largely those of society's response (arising from criminal sanctions), rather than direct effects of the misuse. The treatment of substance abuse in some settings (particularly the workplace and especially for alcoholism) now predominantly follows the medical, rather than criminal, approach. However, many substance abusers have little access to medical forms of treatment and are more likely to enter the criminal justice system.

Drugs as tools of civilization are, on any account, swords that cut both ways. A society's success in using drugs to promote individual and collective health while avoiding the consequences of inappropriate use is surely one measure of its cultural advancement.

LYNN R. BRADY
WILLIAM E. FASSETT

For Further Reading: Oakley Stern Ray and Charles Ksir, *Drugs, Society & Human Behavior* (1990); Mickey C. Smith, *Small Comfort: A History of the Minor Tranquilizers* (1985); Weldon L. Witters, Peter J. Venturelli, and Glen Hanson, *Drugs and Society*, 3rd Ed. (1992).
See also Cancer; Health and Disease.

DUBOS, RENÉ JULES

(1901–1982), French-born American scientist and humanist. A free lance of science, René Dubos was one of

René Dubos receiving at Dartmouth College in 1960 one of his more than forty honorary doctorates. Courtesy of The Rockefeller Archive Center.

Left: René Dubos, at work in The Rockefeller University Laboratory on the first antibiotics, 1932. Courtesy of The Rockefeller Archive Center.

René Jules Dubos, Professor Emeritus, The Rockefeller University.

René Dubos receiving the Arches of Science Award in 1966, with Glenn Seaborg, Nobel laureate (*left*), and Dael Wolfle, executive officer, AAAS (*right*). Source: W. H. Houlton.

René Dubos with Barbara Ward, his co-author on *Only One Earth*, at Stockholm in 1972. Courtesy of The Rockefeller Archive Center.

René Dubos managing his land in upstate New York. Courtesy of The Rockefeller Archive Center.

René Dubos with William and Ruth Eblen at the René Dubos Center's reception in honor of the publication of his last book, *Celebrations of Life*, in 1981.

the most influential thinkers of the 20th century. Professor Emeritus at The Rockefeller University, he was an internationally known microbiologist and experimental pathologist who first demonstrated the feasibility of obtaining germ fighting drugs from microbes. He is also known around the world as an educator, Pulitzer Prize–winning author, and lecturer. He was the recipient of numerous awards for his scientific achievements and his books, and holds honorary degrees from more than forty colleges and universities in the United States, Europe, and South America. These degrees were all the more remarkable for their diversity since they included three honorary M.D.'s and twenty in fields other than science.

Dubos's ecological approach led to his discovery of gramicidin—the first antibiotic to be produced commercially and used clinically—which became the cornerstone of the antibiotic arsenal. Other scientific achievements included his ground-breaking work on tuberculosis, pneumonia, the mechanisms of acquired immunity, natural susceptibility, and resistance to infection. He was intensely concerned with the effects that environmental forces—physiochemical, biological and social—exert on human life.

Beginning in the 1950s, Dubos addressed these concerns in over two dozen books, including *Mirage of Health; The Dreams of Reason; Man Adapting; So Human An Animal; A God Within*, and with Jean-Paul Escande in *Quest: Reflections on Medicine, Science, and Humanity*. In *Louis Pasteur* and *Louis Pasteur and Modern Medicine* Dubos recognized that Pasteur, with whom he strongly identified, approached problems from an ecological point of view. Dubos's last book, *Celebrations of Life*, prompted *The Washington Post* to say: "Dubos is the foe of all sorts of fashionable simple-mindedness, especially of the scientific-determinist variety, and we shall have to complicate our minds to get his message straight." *The Wooing of Earth* is a philosophical study of the interplay between human beings and the earth that led the *New York Times* to dub him the "Philosopher of Earth."

Dubos was appointed chairman of the committee of experts for the first U.N. Conference on the Human Environment held in Stockholm in 1972. And he co-authored with Barbara Ward, the British economist, the report that would become the conceptual guidelines for that now historic conference: *Only One Earth: The Care and Maintenance of A Small Planet*. This landmark book reveals how extensive Dubos's participation was in all phases of the environmental crusade, starting with his examination of the effects of a few environmental factors on the course of microbial infections and reaching into the widest aspects of global and even cosmic ecology. Dubos's concept of the human environment was also at the heart of the second

U.N. Conference (the "Earth Summit") held in 1992 in Rio de Janeiro.

Through his interest in the total environment he became increasingly involved in the theoretical and practical problems bearing on environmental quality and on the future of technological societies. In 1975, Dubos founded The Center for Human Environments that now bears his name in collaboration with Total Education in The Total Environment, Inc. (T.E.T.E.), a nonprofit organization with extensive experience in environmental education. The center's mission, he said, would be to help the general public and decision-makers formulate policies for the resolution of environmental conflicts and for the creation of new environmental values. To this end, Dubos formulated the center's philosophy, developed its guidelines, and took an active part as chairman in all of its activities until his death in 1982. This is the legacy, along with his environmental library and archives, that serves as the basis for all of the programs and related activities of The René Dubos Center for Human Environments.

René Dubos was born in St. Brice-sous-Foret in the farming country of the Ile de France, north of Paris, a region that had been profoundly transformed by human beings since the late Stone Age. The region's historical development conditioned Dubos's ecological philosophy and convinced him that human beings can profoundly alter the surface of the earth without desecrating it, and indeed, humans can create new and lasting ecological values by working in collaboration with nature.

Dubos's extraordinary successes in everything he undertook were achieved at the cost of immense labor and against tremendous odds. He had a younger sister and brother and his parents operated a small butcher shop, first where he was born and later in Henonville where they lived until he was about thirteen. He attended a one-room schoolhouse in the small village where he won all of the local academic prizes and, by the age of seven, he helped teach the younger children. He loved to run and race but after winning a bicycle race he arrived home exhausted only to learn that he was severely ill with rheumatic fever. As a consequence, he developed a heart lesion and was bedridden for almost two years. This not only prevented him from involvement in sports but also conditioned the rest of his life. He read a good deal and said he recalled looking forward to receiving a series of pamphlets that his parents subscribed to for him that arrived every week until he was eleven or twelve. They contained American stories in French about Buffalo Bill, Nick Carter, and Nat Pinkerton, and they made him decide to visit America some day.

Dubos would say later that his first real awareness of *joie de vivre* came when he was eight years old and partially recovered from his illness. His mother allowed him to go out with her for the first time on a beautiful, sunny day to take a short walk to the village to buy milk. That walk, he would say, was one of the most important events of his life. The road was at best ordinary and dull, but this day it seemed to stretch out in front of him like an enchanted world. The few people they met along the way made him feel that contact with human beings was an immensely exciting experience. He remained in love with the world, despite his disability, for the rest of his life.

When he was thirteen years old his father sold the butcher shop and bought another one in Paris and the family moved there. With the outbreak of World War I, his father was called to the Army and served throughout the war. His mother kept the shop open and Dubos would help her when he was not in school. Despite the time he spent in the butcher shop, he was very successful in school, and around 1915 he won a special competitive fellowship to one of the lycees in Paris—College Chaptal, where he studied science and modern languages. While there, he read the works of the historian Hippolyte Taine, who dealt with the French fablist, La Fontaine. Taine proposed that the character of La Fontaine can be reconstructed from the characteristics of the Ile de France country in which he lived. Taine's concept of the molding force of the environment upon historical events and personalities influenced much of Dubos's subsequent scientific work.

Returning from the Army in 1918, Dubos's father was stricken by what may have been a brain abscess and died early in 1919. Dubos decided he would quit school altogether and work to help his mother, but he soon realized that abandoning school would be far worse for her. As he put it, they were full of illusions and hopes about the future. Fortunately, he received a fellowship which permitted him to complete his secondary schooling. His mother sold the butcher shop and they moved to another suburb of Paris and rented a small grocery store. Through an artist friend of his mother, Dubos learned to paint and draw, which he enjoyed very much, and he acquired an interest in art history that lasted throughout his life. He contemplated going to L'Ecole de Physique et Chemie, but he was prevented by a recurrence of his heart condition. By pure chance, the only school left for which he could compete was the National Institute of Agronomy where he studied for two years. He later admitted that he was "outright bad" in some of the technical courses. He received the lowest grade in bacteriology—the science in which he would achieve such success—and also in chemistry which he did not like. He told his mother after his final exam that it was the last time he would set foot in a laboratory because he had made up his mind that it was not the life for him.

In 1922 he took a job as an associate editor for the *Journal of International Agricultural Intelligence* in Rome. His duties were to read technical and semitechnical papers and prepare abstracts of them for publication in developing countries. He came across a paper by Sergei Winogradsky, a Russian soil bacteriologist who had worked for the Pasteur Institute. Winogradsky argued that to understand bacteria it was essential to find a technique for studying microbes in their own environment, in competition with other bacteria, and using whatever resources the soil provided, instead of isolating a bacterium from the soil. Dubos was so excited by the article he immediately decided he would become a bacteriologist.

Although he enjoyed Rome very much, Dubos was getting restless and wanted to go to America. He attended the University of Rome to learn the techniques of bacteriology and worked for eighteen months translating technical books to make enough money to go. He also worked as a technical assistant to an International Congress of Soil Science where he met Selman Waksman from Rutgers University and Jacob Lipman, Director of the New Jersey Agricultural Experiment Station. With a vague offer of work from Lipman, Dubos left for America in 1924, and went to Rutgers the evening he arrived in New York. Lipman had in fact forgotten him, but offered him a room in exchange for babysitting his twin sons. Through Waksman, Dubos received a small fellowship of about twenty-five dollars a month. He spent the first year there taking courses in chemistry and biochemistry and studying bacteriology under Waksman at the agricultural experiment station.

Dubos insisted that the most important part of his training was the reading he did on his own. He especially credited an English journal *The Biochemical Journal* for teaching him biochemistry—by reading every paper and then going back to the textbooks to develop an understanding of those parts where he lacked theoretical training.

For his doctoral thesis at Rutgers University, he decided to study microorganisms in soil that decompose cellulose. That study introduced concepts to him that proved to be very useful in his later life. He realized that the performance of microorganisms changed with variations in the soil environment and that they could be manipulated. And his first paper, imbued with this viewpoint that was not then commonly held, was published in the journal *Ecology*.

After finishing his thesis Dubos became so interested in problems of dealing with the most fundamental biochemical phenomena of life that he sought further scientific training. Eventually he met Oswald T. Avery, one of the most famous bacteriologists at the then Rockefeller Institute—a meeting that he would refer to as the most important event in his life. Many years later Dubos would write a biography of Dr. Avery, *The Professor, The Institute, and DNA* that documents his enormous admiration and respect for him.

Avery had a timid, almost shy manner, and after listening to Dubos he said his laboratory was very interested in a substance that is a sort of semi-cellulose — the complex polysaccharide that surrounds and envelops the virulent pneumococcus. "If we could ever find an enzyme that would decompose that substance . . . it would open all sorts of possibilities." Dubos immediately replied, to his own surprise, that "if there were no enzyme that could . . . it would accumulate . . . and there would be mountains of it now" and "since I have nothing to do this summer, I'd like to come and work on it."

Dr. Avery offered him a fellowship at $1,800 a year to work in his laboratory and he accepted immediately. Within an incredibly short period of time, three or four months, Dubos succeeded in finding a soil microbe that could destroy the carbohydrate capsule that protected the pneumococcus bacterium from the human body's defenses. He did so by forcing a microorganism to switch its feeding regimen to a carbohydrate, when it was the only sustenance available. Dubos discovered that a microbe's adaptive enzyme was put into use only when it was needed and that an array of potentialities are implanted in any organism's genome that can be switched on as required. That was in 1929 and two decades later, after molecular biology had been opened up, Jacob, Lwoff and Monod followed this line of investigation and were awarded Nobel Prizes for their work.

The discovery of a soil microbe that could produce an enzyme that would dissolve the heretofore impervious armor of the pneumonia bacterium led Dubos to a daring hypothesis, one that would change our world:

"This encouraged me to postulate that one could find in nature other kinds of microbes capable of producing substances having anti-infectious activities. To this end, I developed techniques for the discovery in soil, water, sewage and other materials of microbes that could act on microbial agents of disease. In 1939, I demonstrated in fact that the soil microbe *Bacillus brevis* produced two different antibacterial substances one of which — a polypeptide that I called Gramicidin — was highly active against various agents of disease in vivo as well as in vitro."

Gramicidin was used extensively in the early 1940s for fighting infections in wounds, but it proved to be too toxic for widespread application to humans. But the fundamental value of his historic paper "Bactericidal Effect of an Extract of a Soil Bacillus on Gram Positive Cocci" was that it revealed the possibility of

putting fantastic new weapons in the arsenal against infectious diseases. At Dubos's suggestion Florey and Chain went back to the unappreciated and ignored effects of a bread mold found by Fleming and produced the greatest panacea in human history — penicillin, for which Chain, Fleming, and Florey were awarded Nobel Prizes. Selman Waksman began his hunt for an antibiotic and came up with streptomycin, which earned him a Nobel Prize as well.

Although Dubos received international acclaim and offers by large pharmaceutical companies of lucrative positions in antibiotic research, he chose to stay in academic life. He reasoned that antibiotic research would no longer be intellectually challenging for him. More importantly, he had become increasingly interested in the mechanisms of disease and in the influence of environmental forces and forms of stress on the susceptibility and resistance of animals and human beings to infection.

A tragic event in Dubos's personal life convinced him that the onset and outcome of disease were determined in most cases less by the virulence of the microbe than by the effect of the total environment on the response of the patient. His first wife, Marie Louise Bonnet, developed acute tuberculosis in 1940 and entered a sanatorium. She returned to New York City in 1942 but her tuberculosis soon became reactivated and she died shortly thereafter. Based on his own investigations, Dubos concluded that his wife had contracted tuberculosis as a child, but recovered spontaneously. Born and raised in Limoge, her father had died of silico-tuberculosis — a disease that was extremely common among porcelain workers. Although the Duboses lived comfortably, she had been profoundly disturbed emotionally by World War II disasters in France and by a more recent tragedy in her own family. Dubos hypothesized that these influences had activated her latent tuberculosis. This personal tragedy motivated a profound change in his scientific and social life.

He left The Rockefeller University for the first and only time to accept an appointment to a dual chair at Harvard University Medical School as Professor of Comparative Pathology and Tropical Medicine. Although he found his two years at Harvard intellectually stimulating, he mourned his wife and he agreed to return to The Rockefeller University provided he could organize a department of environmental biomedicine devoted to research on tuberculosis. He returned in 1944 to chair his new department where he would remain until 1971 when he became Professor Emeritus. In 1946 he married his second wife, Jean Porter, a laboratory assistant. Together they would write *The White Plague*, one of a series of his extremely influential books on health and disease.

René Dubos would become known as "The Despairing Optimist"—the title he gave his column of ten years in *The American Scholar*. He believed that "Optimism is a creative philosophical attitude, because it encourages taking advantage of personal and social crises for the development of novel and more sensible ways of life." He said that optimism is not only essential for action but also constitutes the only attitude compatible with sanity.

RUTH A. EBLEN

For Further Reading: René Dubos, *The Professor, The Institute, and DNA* (1976); René Dubos, *The Torch of Life* (1962); Gerard Piel and Osborn Segerberg, Jr., eds., *The World of René Dubos* (1990).

See also Environment; Five E's of Environmental Management; Gaia Hypothesis; Historical Roots of Our Ecological Crisis; Human Adaptation; Humanized Environments; Invariants of Humankind; Nature vs. Nurture Controversy; Pasteur, Louis; Shelters; Technological Fix; A Theology of the Earth; Think Globally, Act Locally; Wilderness Experience.

DUNE STABILIZATION

See Coastlines and Artificial Structures.

DUST STORMS

See Desertification.

E

EARTH DAY

Facilitated by the widespread social activism of the 1960s, the first Earth Day was held on April 22, 1970. The idea of setting aside a specific day to focus national attention on environmental problems was conceived by Senator Gaylord Nelson (D-Wis.). The goal was to increase public awareness and mobilize citizen participation. Senator Nelson and Representative Paul McCloskey (R-Calif.) sponsored the group Environmental Teach-In, which organized Earth Day activities that featured public education teach-ins, speeches, and demonstrations about environmental issues.

Reflecting public concern about environmental protection, Congress adjourned for the day and nearly 40 senators and representatives spoke at local gatherings. René Dubos, Barry Commoner, Paul Ehrlich, and Ralph Nader gave speeches marking the importance and magnitude of the occasion. A quarter of a million people rallied in Washington, D.C., and 100,000 marched down Fifth Avenue in New York City to show their support. At least 2,000 colleges, 10,000 high schools and elementary schools, and 2,000 communities participated in the event with special classes and community cleanups. Following its initial success, Earth Day was extended to Earth Week by presidential proclamations in 1971 and 1972.

Although Earth Day has been observed annually since 1970, it wasn't until the twentieth anniversary of the original event that it received worldwide attention. In 1990 Earth Day was celebrated around the world. From Hong Kong to Kenya, festivals, tree plantings, theater productions, and concerts took place. In Kenya 1.5 million trees were planted and on Vancouver Island, Canada, Boy Scouts replanted trees from an old forest scheduled to be cut down. Chinese Premier Li Peng spoke on April 22 about the need for environmental protection. Environmentalists gathered in Munich's main square and released 10,000 balloons carrying environmental messages in the hope that neighboring countries would take note. Other festivities around the world included a sunrise ceremony in Halifax, Nova Scotia, an environmentally focused drama competition by the Sacred Earth Trust in London, and the unveiling of a 2½-meter garbage sculpture called *Monument to the Unknown Refuse* in Toulouse, France.

Public concern for environmental quality peaked with the first Earth Day and the 1972 UN Conference on the Human Environment held in Stockholm and then declined during the 1970s and 1980s. It soared to new heights worldwide with the twentieth Earth Day and the 1992 Earth Summit in Rio de Janeiro, which commemorated the twentieth anniversary of that now historic Stockholm conference.

EARTH FIRST!

See Environmental Movement, Radical.

EARTHQUAKES

See Natural Disasters.

ECOFEMINISM

In 1974 Françoise d'Eaubonne coined the term *ecoféminisme* to bring attention to women's potential for bringing about an ecological revolution. Since then, ecofeminism has gained international recognition as a grassroots, women-initiated, non-violent activist movement designed to bring about local, regional, and international change in the maltreatment of both women and nonhuman nature. It includes homemakers organizing to eliminate toxics from their homes and communities (e.g., Mothers of East Los Angeles), activists hugging trees to prevent bulldozers from felling them (e.g., the Chipko movement in India), protesters making peace encampments (e.g., the Women's Pentagon Actions and Seneca Peace Encampment), artists' depictions of women and nature in their photographs, paintings, music, stories, and dance (e.g., ecofeminist "land art" and "performance art"), and academics teaching courses on "feminism and the environment." As a form of political activism, ecofeminism has drawn from and contributed to the convergence of the women's, civil rights, peace, and environmental movements.

Ecofeminism also is a name for a variety of theories concerning the causes, nature, and solutions to social and environmental domination. These theories grow out of and reflect a diversity of positions primarily from ecology, history, feminism, literature, theology, and philosophy. From ecology, ecofeminism draws on the notions of both the connectedness and disconnectedness of various life forms, and the importance of ecosystemic well-being. From history, ecofeminism draws on the insights of intellectual historians, anthropologists, sociologists, and others who have articulated positions on the rise of patriarchal institutions and the domination of women and nonhuman nature. From feminism, ecofeminism draws on the analyses of the domination of women from a variety of feminisms (e.g., liberal, Marxist, radical, socialist, and Third World); it extends these analyses to a critique of the interconnections among social systems of domination (e.g., sexism, racism, classism) and the domination of nature (naturism). From literature, ecofeminism draws on women nature writers, feminist utopias, and other genres to explore the experiences, representations, and symbolic associations between women and nature and to develop an ecofeminist literary theory.

From theology, ecofeminism draws on pre-patriarchal religions (e.g., paganism), Earth-based spiritualities (e.g., Goddess worship), feminist-inspired interpretations of the Gaia hypothesis, and feminist theologies to develop ecofeminist theologies. From philosophy, ecofeminism has pointed to the conceptual dichotomies of mainstream Western philosophy (e.g., mind-body, reason-emotion, culture-nature, public-private, man-woman, human-animal), value-hierarchical ("Up-Down") thinking, mainstay philosophical notions (e.g., of rationality, knowledge, objectivity, the self) and fields (e.g., ethics, epistemology, philosophy of science) as sources of male-gender (androcentric) bias; it has prompted the development of ecofeminist philosophies aimed at overcoming both these gender and anthropocentric biases in environmental ethics and philosophy.

Nearly all ecofeminist theories draw attention to the historical language used to feminize or sexualize nature (e.g., Mother Nature, rape of nature, virgin timber) and to naturalize women (e.g., women as cows, foxes, chicks, bitches, beavers, old bats, old hens). In patriarchal contexts where what is female- or nature-identified is already perceived to be inferior, of lower status, or of less value than what is male- or culture-identified, such language serves only to reinforce harmful domination metaphors linking women and nature.

Despite important differences among them, all ecofeminists agree that there are significant interconnections among social systems of domination and environmental exploitation, an understanding of which is necessary for any adequate environmentalism, environmental policy, or environmental philosophy.

Broadly conceived, then, ecofeminism is a movement and theory committed to exposing and eliminating the nature and causes of the patriarchal domination of women and nonhuman nature.

The relationship between ecofeminism and other kinds of environmentalism is both clear and unclear. What is clear is that ecofeminist activism is a grassroots, community-based, coalition-building movement that has enormous appeal to local women, poor women, women of color, women of the South (i.e., from the Southern hemisphere). Ecofeminist positions span the spectrum of positions in environmentalism: from traditional conservation and preservation emphases, to reformist animal rights and human stewardship concerns, to newer bioregionalist land ethics, and, ultimately, to indigenous, multicultural, pluralistic approaches to environmental problems.

What is not yet clear is how compatible the various expressions of ecofeminist activism and philosophy are with other—not explicitly feminist—environmentalisms (e.g., land stewardship projects, deep ecology, alternative energy). It also is not clear how compatible ecofeminist positions are with each other (e.g., Must ecofeminists be animal rights advocates and vegetarians? Is spirituality an essential ingredient of ecofeminist philosophy?).

This lack of clarity is to be expected. Ecofeminism is not the name of one position, one movement, one philosophy. Consequently, there is not one ecofeminist environmentalism. Nonetheless, ecofeminist environmentalism tends to be structurally pluralistic, reflecting the historical and material realities of women cross-culturally and emphasizing a "many" rather than "one" approach to pressing environmental problems. For example, while solar energy might be an "appropriate technology" for the West, supplying solar stoves for women in the African bush who cook before dawn and after dusk may not be. Or, while the consumption of animals raised under "factory farming" conditions may be morally prohibitive for many, the consumption of animals raised under humane conditions or of "wild animals" may not be morally prohibitive for those for whom there are no alternatives (e.g., the Inuit). Or, while toxic wastes are an environmental problem for all life—human and nonhuman—they are an especially crucial problem for Americans of color, who bear an alarmingly disproportionate burden in the location of commercial hazardous waste facilities and uncontrolled toxic waste sites in the United States.

In many respects, then, the relation of ecofeminism to other kinds of environmentalism has yet to be seen. But this much is certain: Any environmentalism that promotes practices and policies that contribute to the subordination of women, however unintentionally, unconsciously, or nonmaliciously, is inadequate from an ecofeminist point of view.

KAREN J. WARREN

For Further Reading: Carolyn Merchant, *The Death of Nature: Women, Ecology and the Scientific Revolution* (1980); Judith Plant, ed., *Healing the Wounds: The Promise of Ecofeminism* (1989); Karen J. Warren, ed., *Hypatia: A Journal of Feminist Philosophy* (1991).

ECOLABELING

Ecolabeling, or Green Labeling, is the advertising of a product's environmental benefits on the product or its package. Many polls show that consumers prefer products they consider environmentally sound. In a 1992 survey, 57% of Americans said they avoided products that harm the environment; in Canada, over three quarters said so.

One of the most common ecolabels is the "chasing arrows" symbol for recycling. This label means that some part of the product or package is recyclable or recycled. However, the label does not always mean that the product or package can be recycled locally. Other examples of ecolabeling include packages marked "biodegradable," and detergents labeled as containing no phosphorus.

The practice of ecolabeling became widespread after Earth Day 1990. A survey conducted by Marketing Intelligence Service in 1990 found that between 1985 and 1990, the fraction of products introduced in the U.S. with environmental claims rose from 0.5% to 9.2%.

As ecolabeling grew more popular, so did what some saw as abuses of the practice. For example, many products are labeled "degradable," although in a normal landfill they will not degrade. In 1991 Procter and Gamble was forced by ten states to drop claims that its disposable diapers were easily compostable; very few facilities will compost disposable diapers.

The practice of issuing false or misleading claims in an ecolabel is known as "greenwashing." Concern over greenwashing has given rise to programs all over the world designed to certify manufacturers' ecolabels. Many industrialized nations have government programs that certify products as having met certain environmental specifications.

In the late 1970s, the West German government established the "Blue Angel" program, which awards a seal of approval to products that the government decides are less harmful to the environment than others in the same product type.

Canada began its "Environmental Choice" green certification program in 1988. Under this program, the government evaluates products for their use of recycled material and their use of water and energy. It issues an EcoLogo—three doves entwined in the shape of a maple leaf—to those that pass its specifications.

The Japanese, British, and Indian governments have also begun or announced programs of green certification. In addition, the European Community has announced a planned certification program for its member countries.

In the U.S., several states have passed laws restricting the use of certain ecolabels. California has restricted the use of the term "recyclable" on products and packages to those sold in areas of the state where recycling facilities that accept the product or package exist. Rhode Island has banned the use of the "chasing arrows" recycling symbol on products and packages because it could be easily misinterpreted. New York is planning a 75-day application process for the use of the "chasing arrows" symbol.

Many manufacturers are concerned that differing state laws regarding ecolabels may make it difficult for them to market products nationwide. Some companies have dropped ecolabels from their products because of state laws that they complained were too confusing or conflicting.

There have been some efforts to set up more standardized laws governing ecolabeling in the U.S. In 1990 a group of 11 state attorneys general formed a task force to design proposals for national standards. They produced the "Green Reports," which listed voluntary guidelines for states to follow in regulating environmental claims until national standards could be set up.

In July 1992, following several successful lawsuits against companies making environmental claims about their products, the Federal Trade Commission (FTC) announced voluntary guidelines for the use of ecolabels. These guidelines included standards for the use of terms such as "recyclable," "biodegradable," and "compostable." According to the FTC, any company observing these guidelines in its ecolabels will be safe from their prosecution. Some members of Congress have supported the introduction of guidelines for ecolabeling in the Resource Conservation and Recovery Act.

In the private sector, two major independent certification organizations have also formed in the U.S. in the last few years: Green Seal and Scientific Certification Systems.

Green Seal was founded by Denis Hayes, chair of Earth Day 1990. The non-profit organization sets standards for a particular product line, and allows manufacturers whose products and packaging meet those standards to use the Green Seal label. For example, Green Seal standards for bathroom and facial tissue require that the product be made entirely of recovered paper, with at least 10% post-consumer recycled content.

Green Seal has also issued standards for tissue and re-refined engine oil. It is developing standards for compact fluorescent light bulbs, household cleaners, coffee filters, and house paint.

Scientific Certification Systems (SCS), a California company that has applied for non-profit status, also verifies manufacturers' environmental claims for their products. However, they verify on a case-by-case basis, rather than across a whole product type. If the company's claims, most of which involve recycled content, are verified by SCS, the company is allowed to display the SCS "Green Cross" logo along with its environmental claims.

SCS has also begun to examine products with an eye toward their overall effect on the environment. The company then produces an "Environmental Report Card" that lists the resources going into a product in comparison to those going into a "typical" product in that line. The first "Environmental Report Card," issued for a brand of plastic trash bags made of recycled material, included the information that the recycled bags required about one-fourth of the electricity needed to produce "normal" trash bags. SCS does not follow the environmental impact of a product after it is made.

Some experts have pointed out that one cannot determine which products are less harmful to the environment than others without looking at the full impact of the product, from creation to destruction. This "cradle-to-grave" look at a product is known as "life cycle assessment."

Most scientists feel that life cycle assessment cannot be accurately done until some scientific debates are resolved. For example, while many experts argue that paper packaging is better for the environment than plastic, others say plastic actually does less damage to the environment, especially if it is recycled.

JEFFREY R. LEVINE

For Further Reading: Consumer Reports, "Selling Green" (October 1991); Bradley Johnson and Christy Fisher, "Seals Slow to Sprout," *Advertising Age* (April 20, 1992); Alex Pham, "It's Not Easy Being Green: FTC Issues Some Guidelines," *Washington Post* (July 29, 1992).

ECOLOGICAL ECONOMICS

A new field of study, *ecological economics*, considers the unified systems of the environment and the human economy. Policies for management are sought which can make the economy and the environment

symbiotic. Ecological economics began among intellectuals in the mid-18th and the 19th centuries who considered energetics and land as the basis for value. Measures of resource value were sought in estimates of nature's work. But in the 20th century when economic growth based on abundant fossil fuels was accelerating, there were only sporadic contributions and concerns with the environmental basis for the economy. When environmental life support was taken for granted, ideas of unlimited human creativity were prevalent. Human willingness to pay became the main concept of value.

Following the oil shortage crisis of 1973 the energy basis of economic wealth was recognized as fundamental by a few economists. People of many backgrounds were drawn to the task of revising economics to include the unpaid part of the system of humanity and nature. A sustainable economy requires human society to adapt to the environment following principles derived from the study of ecological systems. Sustainability requires recycling materials, adapting to the pulsing of earth processes, designing with nature's hierarchy of many scales in organizing the landscape, preserving essential information and genetic diversity, and optimizing efforts in order to increase productivity.

Among many new approaches are five subject areas for study and application:

(1) Understanding the combined system of the environment and the economy (for example, unifying landscape ecology and geographical economics): Sustainable prosperity depends on mutual reinforcement between nature's systems of resource production and the human economic processes. Each stimulates the other. By contributing to the environmental production processes, the human economy causes more resources to flow into the human prosperity. One of the main ideas in ecological economics, Lotka's 1924 "maximum power principle," is:

> In the process of self organization, those system designs and policies are reinforced and sustained which draw in more resources and develop more efficiency in their use.

In the long run, fitting the human economy to help the environmental resource production will make the human economy more sustainable and prosperous. Rather than a competition between human jobs and environmental protection, it is a symbiosis that maximizes environmental and human welfare together. Finding economic and environmental management policies that will develop the mutual reinforcement is one of the main concerns of the field.

(2) Identifying principles and designs common to both ecological and economic systems: Although of-

ten discovered independently, many principles in ecology and economics are similar, such as equations and models for production, limiting factors, recycling of materials, utilization of by-products, dependence on energy laws, and input-output. There is now active reorientation underway to learn and share concepts.

(3) Understanding and managing the interface between environmental and economic systems: Forestry, fisheries, agriculture, and tourism are examples of ecologic-economic interface (Figure 1). The interface shows contributions from the environment without payment, as money is only paid to people for their work and the contributions of the assets they "own," including environmental assets. Unless special arrangements are made to reinforce the environmental production system (reinforcing pathway in Figure 1), the free market tends to drain stocks and reduce environmental production that is useful to the economy.

One of the ways the reinforcement can occur is with recycling of materials. Rather than allowing the by-products of the human economy to accumulate in great dumps (landfills) that divert land from useful purpose and often leak toxic substances into ground waters, the materials can be returned to environmental processes in places and concentrations where they assist environmental productivity. For example, return of nutrient-rich waters from treated sewage to wetlands was tested in many kinds of ecosystems in Florida and is now becoming a worldwide practice in fitting cities to their environments. This was practiced on a smaller scale in older cultures.

(4) Development of appropriate measures of environmental value and understanding their relationship to market values: Figure 1 shows nature's work on the left generating, for example, a contribution of wood to the economy. Nature receives no flow of money (the dashed lines). As the wood is brought into the econo-

my, human services are contributed and these do receive money. For example, the dashed line in the middle of the diagram brings the money from sales to the landowner and forester. They send the money back to the economy (to the right again) buying goods and services for investment and consumption. Environmental valuation measures the work of nature in generating the real wealth (the wood). Energy measures of this work do not change and are not affected by the use or non-use of the wood, in contrast to economic valuation.

Market values are what people and businesses are willing to pay. Market prices make the human part of the system efficient. Market values respond inversely to the environmental resource, prices being least when the environmental system is contributing most. Market value of restoring natural capital (resource reserves in nature) has been suggested as a resource value (dashed lines in Figure 1), but such values don't include the work of nature. If only market value is used for decisions about the environment, the environmental resource system tends to be used up so that its further production is lost. Weedy systems take over. Examples are the many failed fisheries of the world where free markets caused the resource to be overfished. Other examples are the scrubby vegetation that replaces over-harvested tropical forest. In order to have sustainable environmental production by forests and fisheries, work from the human economy must be put back into nature to reinforce the desired system, as in Figure 1.

Market value and environmental values are very different and should not be confused. People and businesses have to use market value to guide their buying and selling. But for deciding how to manage the environment, to judge pollution impact, or decide what land use is best for the economy overall, environmental valuation is necessary.

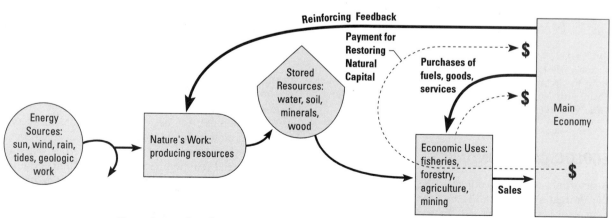

Figure 1. Interface between environmental production and economic use.

Much of the research in ecological economics concerns new measures of environmental contribution. Some evaluate the "energy embodied" in the previous work of nature in generating a product or service (solid lines in Figure 1). One method assigns input energies to pathways of a network according to the input-output data on some circulating quantity such as money.

Another measure, Emergy—spelled with an "m"—calculates environmental work and human work on a common basis. It is the available potential energy, expressed in energy units (emjoules) used directly and indirectly to make a product or service. The maximum power principle implies that public policies to make a vital economy should be those that maximize the emergy production and use for the coupled system of the economy and the environment working together.

(5) Learning the way global international policies on trade and finance should drive the utilization of natural resources: Because market prices don't recognize the real contribution of resources, inequities of capital and foreign trade can cause unsustainable stripping of minerals, forests, soils, and fishery stocks in undeveloped countries to support overdevelopments in other countries. Currencies of rural undeveloped countries have high emergy/money ratios (8 to 48 solar emjoules per 1993 dollars) compared to the urban, developed nations' emergy/money ratios (0.5 to 3 solar emjoules per 1993 dollars). Large differences in this ratio cause inequity in foreign trade (2 to 30 times more

EMERGY in raw resources traded than in the buying power of the money paid). When less developed countries borrow from developed countries they may pay back 5 to 10 times more EMERGY. Trade and borrowing would lead to mutual prosperity if price were based on emdollars (emergy-evaluated currency).

HOWARD T. ODUM

For Further Reading: L. C. Braat and W. F. J. Van Lierop, eds., *Economic-Ecological Modeling* (1987); R. Costanza, *Ecological Economics* (1991); J. Martinez-Alier, *Ecological Economics* (1987).

ECOLOGICAL PYRAMIDS

An ecological pyramid is a graph depicting (usually) the number of individuals of species at different trophic levels. For example, in Figure 1a, the energy which plants capture from the sun during photosynthesis may end up in the tissues of a hawk. It gets there via the birds the hawk has eaten, the insects eaten by the birds, and the plants on which the insects fed. The plant-insect-bird-hawk system is the food chain, and each stage, a trophic level. More generally the trophic levels are called producers (plants), herbivores or primary consumers (the insects), carnivores or secondary consumers (the bird) and top-carnivores or tertiary consumers (the hawk). The numbers of individuals at each level often drop dramatically. There are more

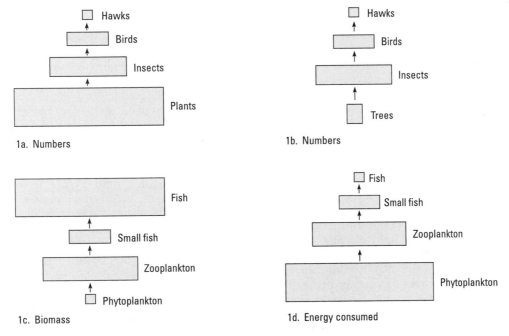

Figure 1. Ecological pyramids.

plants than insects, more insects than birds, and top predators are the rarest of them all. Because of this pattern, the graph of numbers is shaped like a pyramid, hence the name.

This simple idea is appealing, but we must be careful how we document it. As described, it is a pyramid of numbers. It would not resemble a pyramid if the plants were trees (Figure 1b). A few individual trees might support millions of insects, then fewer birds, then fewer hawks. A more sensible approach would be to use biomass. There is much more plant biomass because each tree is so massive.

Such a pyramid of biomasses need not to be based on living plants. In most systems there is a detrital food chain based on the decaying tissues of plants and animals. This may be the only food chain in systems which are sinks into which organic material flows, such as the bottom of an ocean or lake and the leaves on the forest floor. The pyramid of biomasses would then involve the dead plant material, the detritivores that feed on it, their predators, and so on.

Ecological pyramids reflect underlying energetic laws. The first law of thermodynamics tells us that when we convert energy from one form to another the amount of energy remains constant. An approximate way of stating the second law is that the amount of useful energy decreases at each conversion. When the insects eat the plant they convert the energy locked up in the plant tissues into insect tissue. Yet they also pay an energetic cost in doing so. Mammals and birds pay an even greater cost. Only about 1% of the energy of the food mammals and birds consume goes to produce new tissues. The rest is "lost" as heat, both to keep warm and as a byproduct of the conversion process. There are fewer birds than insects because of these energetic losses. We measure this energy in calories in the same way that we count the energy content of our diet in calories. (In common usage, each calorie is actually one thousand times the scientific calorie. The scientific calorie is the amount of energy that it would take to raise the temperature of one gram of water 1° C. Ecologists often report values in units of grams of carbon, which are easier to measure and which can be converted roughly into calories.)

Even if we measure biomass, the resulting values do not always form a pyramid. Familiar examples include oceanic ecosystems. We think of the ocean as blue, but it can be green, in which case there is a high biomass of phytoplankton. When the ocean is blue, it has a low biomass of these microscopic plants, while the biomass of zooplankton and even fish may be higher (Figure 1c). There are terrestrial systems too where animal biomass appears to exceed plant biomass, such as among the large numbers of mammals grazing the vegetation in some East African parks.

Suppose the biomass of the mammals exceeds that of the grasses. The plants may be grazed down to resemble well-manicured lawns. What matters is not the instantaneous masses of the grass and mammals (standing crops) but the flow of energy through each. Each day, each square meter of a productive grassland may produce 50,000 calories of new tissue. This is the primary productivity. The herbivores may consume all of this new growth. In doing so, only about 1% of these calories (500 calories) would go toward producing baby mammals; the rest would be lost in the process of respiration and keeping warm. These 500 calories may eventually be consumed by lions and other predators, which might get 5 calories for growth. This is the pyramid of energy. By the second law of thermodynamics it has to be a pyramid (Figure 1d).

The reason biomasses need not form a pyramid is because standing crops and energy flows need not be related. Typically, large biomasses produce large amounts of new tissue, small biomasses, small amounts. But in the case of grasslands and ocean phytoplankton, the grazers keep the plants rare, but they are still very productive. The grazers may be 100-fold less productive, but still have a higher standing crop.

Ecologists are just beginning to ask how the biomasses at different trophic levels should be related to the energy entering the food chain. In other words, they ask: What is the shape of the energy pyramids? Some commonly used mathematical models suggest that with one species per trophic level, the top species will always increase in biomass with increasing plant productivity. Thus, plants will increase if there are only plants. Herbivores should increase if there are only plants and herbivores. Plants and carnivores, but not herbivores, increase if there are three levels. The patterns alternate depending on whether the number of levels is odd or even. It has been suggested that over large gradients in productivity, there should be stepwise increases in plant, herbivore, and carnivore biomass as systems change from functionally one to three trophic levels.

Terrestrial plant productivity ranges from about 10 to 1,000 grams of carbon per square meter per year. Aquatic values span a similar range. In the least productive ecosystems, herbivore biomass appears to increase much faster than plant productivity. As the two quantities change, so does the shape of the biomass pyramid. At higher productivities, the ratio of the two quantities appears to be approximately constant. Quite how the biomasses of herbivores change is the subject of debate, however.

STUART L. PIMM

For Further Reading: Samuel J. McNaughton, M. Oesterheld, D. A. Frank, and K. J. Williams, "Ecosystem-level patterns of

primary productivity and herbivory in terrestrial habitats," in *Nature* (1989); J. Moen and L. Oksansen, "Ecosystem trends," in *Nature* (1991); L. Oksanen, D. Fretwell, J. Arruda, and P. Niemelä, "Exploitation ecosystems in gradients of primary productivity," in *The American Naturalist* (1981); Stuart L. Pimm, *Food Webs* (1982).
See also Carbon Cycle; Energy; Food Chains; Thermodynamics, Laws of.

ECOLOGICAL RISK

An ecological risk encompasses the nature, magnitude, and probability of harm to systems of living things (ecologies). Examples of ecological risk are the potential harm to marine life from an oil spill, habitat loss and possible species extinction resulting from the cutting of timber, and probable impact to multiple and interlocking ecological systems from global warming due to the emission of greenhouse gases. In addition to these anthropogenic (human-induced) changes, ecosystems experience naturally occurring environmental changes and cycles. The risk of species extinction from a large meteorite is an example of such an ecological risk, and the extinction of the dinosaurs 65 million years ago is a possible instance of this risk being realized.

The study of ecological risks grew rapidly in the 1980s and continues into the 1990s in response to increased scientific activity, U.S. legislative and regulatory mandates (e.g., National Environmental Policy Act, Toxic Substances Control Act, Clean Water and Air Acts, and Superfund) and overall public awareness. Much of the applied work is for the purposes of prevention and control and for apportioning responsibility and costs when cleanups are required.

Ecological risk assessment is the process of evaluating scientific information on the adverse effects of stressors on the environment. It is similar in approach to human health risk assessment, but its focus extends beyond impacts on individual organisms (e.g., an endangered animal) to include populations, communities, key resources, biodiversity, and ecosystem recovery. The process of performing an ecological risk assessment can be viewed as having five main steps which culminate in a specific decision or action, often described as risk management. The process begins with a clear articulation of the scope of the problem and the goals of performing the assessment. The remaining four steps can be performed concurrently or in differing order than listed below.

First is the description of the impact(s) of some initiating action or stressor on an ecological system. This step requires some understanding, and possibly simulation modeling, of the biological system and its environment in order to prepare a baseline characterization (pre-impact), compared with the probable characteristics of the system after encountering the stressor. In addition, the inherent hazard of the stressor along with the magnitude and spatial and temporal patterns of exposure must be known or estimated. For example, an oil spill involves a specific ecosystem with certain habitats and biota (e.g., a coastal estuary), exposed under particular environmental conditions (e.g., time of year and weather) and a stressor with known hazard and exposure (e.g., chemical toxicity profile for a given oil type, the amount released, and transport and fate estimates). Prior knowledge and experience are used to describe the ecosystem in the pre-impact condition, and estimates, measures, and professional judgment are used to characterize the system post-impact.

Second is an evaluation of the likelihood of occurrence of various possible outcomes or impacts. Most ecological assessments involve only a qualitative analysis of the likelihood and form of the impact, although the trend is toward more quantitative analysis. For a given stressor, adverse effect thresholds are compared to measured or estimated exposures to establish a first approximation of the likelihood that an impact will occur. Some assessments identify the "most likely impact," some identify several alternative impacts (including the "reasonable worst case"), and some assess probabilities of "credibly possible cases" using quantitative techniques including simulations and Monte Carlo studies. For the oil spill example, the process of determining what impacts are most likely to occur will reflect the sum of knowledge from prior similar situations, the availability of data specific to the present situation, and the interpretation skills employed.

Third is an assessment of the level of confidence or uncertainty attached to the analysis itself. This part of the assessment is often done qualitatively, in discussions of why a particular data set or model was used, how realistic are the assumptions of the model, and how far its conclusions can be extrapolated and its implications trusted. This more general level of uncertainty assessment corresponds, in physical experiments, to the analysis of systematic error (having to do with errors in the models and concepts), in contrast with the more direct level of uncertainty assessment, error analysis (having to do with instrument and data error, and which is also applicable to assessments of ecological risk). This very important step is often the least developed.

Even small-scale ecological systems are complicated; data are often limited and expensive to gather, with pervasive uncertainties. Studies are often limited as a consequence. One form of partial ecological risk analysis is to focus on a single "indicator" species that is rare or endangered, is considered particularly sensitive or at least is well studied, or is economically important. Another form of partial analysis is "worst case analysis." Feasible worst case assumptions are chosen for both the exposure conditions and effects thresholds. If, under these extreme assumptions, the ecological risk is within the acceptable range, the assessment is often accepted without requiring further analysis of the uncertainties. Otherwise, further study is needed to provide more realistic data and assumptions in order to reduce the level of uncertainty and complete the assessment. An ecological assessment for the oil spill example may focus on an important fishery or on a nursery habitat in the estuary, although some analysis of indirect impacts through species interactions (e.g., loss of food supply for the fish) would probably be warranted. Worst case assumptions might include: no recovery of the spilled oil, and mixing of the oil with the water and sediment so there is maximum exposure to aquatic biota.

Fourth is a valuation of the possible impacts. It is particularly difficult to know how to value the decline of a population, the loss of a species, the reduction of a habitat, or a change in ecosystem stability. There is much discussion of how to improve this process, and ultimately valuation is a societal judgment. "Society's view" ranges from "a thing is right when it tends to preserve the integrity, stability, and beauty of the biotic community" to the more frequently used cost-benefit analysis. In practice today, many of the valuations are defined by the processes of legislation, regulation, and court interpretation. An example that would apply in the oil spill case is the Natural Resource Damage Assessment provision of the Superfund Act, which lays out both scientific and economic methods to assess injuries and damages to natural resources and the framework for a risk management decision.

General understanding of ecological risks and the development of assessment techniques and strategies to deal with them are improving in response to the overall increase in environmental awareness.

DANIEL WOLTERING
TALBOT PAGE

For Further Reading: John Cairns, Jr., et al., *Predicting Ecosystem Risk* (1992); Robert Lacy, "Loss of Genetic Diversity from Managed Populations: Interacting Effects of Drift, Mutation, Immigration, Selection and Population Subdivision," in *Conservation Biology* (1987); Aldo Leopold, *A Sand County Almanac* (1966).
See also Risk; Superfund.

ECOLOGICAL STABILITY

The prescientific belief in the balance of nature can be found in the writings of Greek and Roman philosophers and has persisted in philosophical and religious discussions since then. According to the idea of the balance of nature, nature undisturbed by human action achieves a permanence of form and structure that would persist indefinitely. Nature is believed to exist in a perfect balance. Often there is a religious basis to this belief: if the world were made by a perfect, omniscient, and omnipotent being, then it must be perfect in structure and form. If it is perfect now, then any change will make it no longer perfect. Therefore it must be able to persist in a single undisturbed condition. Furthermore, the ability of nature to persist in a constant condition is taken as one of the characteristics of its perfection.

In these prescientific discussions the question invariably is raised: if nature must be perfect and constant by design, why is it not found to be so? Traditionally, two answers have been given, both placing the blame on people: either nature was perfect without people and anything they do to the environment destroys its perfection, or the human purpose on Earth is to carry out the final actions to create the perfect balance, and our failure to accomplish these tasks leads to the observed instability in nature. In the first case, nature is seen as separate from human beings—people are outside nature and have only negative effects on it. In the second case, people are part of nature, but fail to act properly. Echoes of these beliefs can be found in contemporary discussions of environmental degradation.

Throughout the history of the science of ecology, the idea of ecological stability has often been vague and implicit. Where it has been defined explicitly, it is almost invariably equivalent to, and usually consciously borrowed from, the physical concept of the stability of a mechanical system, as in engineering and physics. This kind of stability will be referred to here as "classic static stability." A system with classic static stability has two characteristics: an equilibrium or rest condition and the ability to return to that rest condition when disturbed. The problem with this definition is simply that ecological systems fail to meet these two criteria. Ecological systems, in general, do not have fixed equilibria. Thus this classic definition is not suitable to real ecological systems.

In rejecting classic static stability for populations and ecosystems, one is also rejecting the classic concept of the balance of nature, and one is also rejecting long-held and deeply felt cultural and religious beliefs. This change in a traditional culturally dominant belief comes slowly, and many of the contemporary approaches to environmental issues continue to assume, either explicitly or implicitly, that as long as human influences are removed, the natural condition will be one of constancy.

Classic static stability has been used commonly, even when it was not recognized as such, in ecology and in environmental sciences. Two examples will illustrate this use, one from the analysis of individual populations and the other from the analysis of ecosystems.

The concept of classic static stability is illustrated by the logistic growth curve, the classic curve used to describe the growth of a single population. A differential equation describes this curve in terms of the size of a population, the intrinsic rate of increase (birth rate minus death rate), and the carrying capacity. This equation results in a population that grows to a value called the carrying capacity. At this value the population remains the same, so the carrying capacity is an equilibrium population. The logistic also returns to its equilibrium condition when it is disturbed from it. If the size of a population is decreased below the carrying capacity, or forced to a size greater than the carrying capacity, the growth rate changes so that the population returns to the carrying capacity.

The logistic has been used widely throughout the 20th century to describe population growth. It has been one basis for projections of the ultimate size of human populations. It is still used in the management of fisheries. It has been demonstrated to hold for certain populations in a laboratory, such as microbial organisms grown in a test tube. However, no documented case demonstrates that the logistic growth curve holds for a wild population under field conditions. Observations of animal populations under field conditions indicate that change rather than constancy is the rule, even for situations that have been believed to be long undisturbed by human influences.

The logistic fails as a realistic model of population growth for several reasons. It assumes a constant environment, ignoring the environment outside the population. In reality, the environment is always changing. The logistic also assumes that the factor known as the intrinsic rate of increase (defined as the difference between an intrinsic birth rate and intrinsic death rate) is a constant, uninfluenced by changes in gene frequencies or any other factor. For these and other reasons, the logistic is a poor model of population growth, and does not reflect the actual stability characteristics of real populations embedded in ecological communities of many species and within ecosystems subject to environmental and internal changes.

Classical theory of ecological succession provides another important example of the implicit acceptance of classic static stability. According to the classic theory of ecological succession, a cleared land area passes through a series of vegetation stages that ends in a permanent condition known as the "climax." This climax was supposed to persist indefinitely, without change. In addition, a natural ecosystem was also supposed to be capable of returning to its original climax condition if it were disturbed. Although this concept of ecological succession was predominant throughout the first half of the 20th century, few contemporary ecologists agree with it, and most reject it. However, many analyses of environmental issues have an implicit assumption that the classic climax condition is the actual condition of undisturbed natural systems.

Contemporary ecological research demonstrates that ecosystems are always changing, both in response to external environmental factors such as climate, and in response to internal processes, such as chemical cycling within ecosystems. Fossil pollen deposits in lakes and wetlands of North America and Europe provide evidence of the changes in vegetation over thousands of years, in most cases for the Holocene (the past 11,000 years) and in many cases for an even longer period. These records show that vegetation has changed continually in geographic distribution.

As a case in point, consider the Boundary Waters Canoe Area of Minnesota, an area designated by U.S. Federal Law as a "wilderness," and therefore an area which is supposed to be "natural" and within which the influence of human beings is supposed to be unnoticeable. According to classic ecological theory, the forests of the Boundary Waters should be dominated by species that occur in old-age, undisturbed boreal forests, such as white and black spruce and balsam fir. These species should have dominated that area, and areas subject to similar climates, indefinitely. However, this area was covered by the continental ice sheet as recently as 10,000 years ago. Pollen records indicate that the first vegetation to characterize the area after the continental ice sheet melted was tundra, and the tundra was then replaced by a series of ecological communities of differing composition, with major alterations in composition and species dominance occurring at intervals of approximately 1,000 years.

Over shorter time periods the communities also experienced change. Vegetation communities are characterized by external climatic disturbances such as storms, and internal disturbances including fire, insects, and disease. A study of the Boundary Waters Canoe Area shows that, on average, any location within the wilderness has burned approximately once a century. Since many species of woody plants have an average lifetime of 100 years or more, these systems would undergo major changes before they had a chance to reach an equilibrium condition. Many of the woody plants found in this and other ecosystems have evolved with such disturbances, are adapted to them, and require them for their persistence.

Internal processes also can lead to changes over time. Recent evidence suggests that, in long-undisturbed forests, water gradually leaches nutrients below the rooting depths of the large trees. Without some physical disturbance to turn over the soil, the nutrients become unavailable, and the stage that has maximum biomass and diversity is replaced by a stage of much lower biomass and diversity. No stage persists indefinitely.

Contemporary formal theory in ecology is beginning to support the idea that natural ecological systems are characterized by changes, not by constancy. For example, models of forest growth that involve random processes, internal chemical cycling, and external climatic changes predict that long-undisturbed forests undergo non-periodic changes.

In spite of the growing evidence that rejects the classic concept of static stability, this concept still plays a role in management of renewable biological resources and in the discussion of environmental issues. When a system that changes naturally is managed or conserved as if it were constant, failure to achieve the management or conservation goals is typically the result. To illustrate this point, two more examples will be presented. As before, the first example concerns an individual population and the second example concerns an ecological community and an ecosystem.

The first example concerns the management of fisheries. The logistic growth equation has been and continues to be the primary basis for setting harvest quotas of fish. The traditional goal of fisheries management has been to maximize production and therefore to maximize yield. A population growing according to the logistic growth curve has its maximum growth rate when the population size is half the carrying capacity. This population size has become known as the size of "maximum sustainable yield." A hypothetical logistic population allowed to reach this level and grow during the next time period can be harvested back to the one-half carrying capacity level. If the situation is static, then the population will grow exactly the same amount in the next time period and can be harvested by exactly the same amount again. The harvest could be repeated indefinitely at the maximum level as long as the conditions of the logistic were met.

Real populations of fish are not constant over time, nor are their habitats and environments. For example, there is evidence that salmon on the west coast of Oregon are influenced by changes in oceanic upwellings. In years of strong upwellings, there is a high abundance of fish returning to freshwater streams; in years of weak upwellings, the returns are low. A harvest plan that assumes a constant environment and a logistically growing population will overestimate the harvest that the population can sustain in many years; this will lead to the population being driven to lower and lower levels from which its recovery, even in years of strong upwellings, will be below that predicted by the logistic.

A solution attempted to this dilemma has been to suggest that the goal of fisheries management should be changed from maximizing yield to optimizing yield. However, in practice, the definition of an optimum yield has been simply to reduce the harvest by some set percentage below that calculated as the maximum from the logistic growth equation, as a way of providing a somewhat more conservative estimate of the possible harvest. In this case, the logistic growth equation, and therefore the concept of classic static stability, still lies behind the approach to management.

The second example concerns the conservation of an endangered species that lives in a forest ecosystem. Consider an old-age forest that experiences fire under natural conditions and contains species adapted to fire, and is set aside in a nature preserve within which fire is suppressed. With the natural disturbances removed, the system undergoes alterations, and the ecosystem achieves conditions that were not known before. The Kirtland's warbler is an endangered species that nests only in jack pine stands on sandy soils in Michigan. This warbler began to decline rapidly in abundance in the 1960s. It was recognized that fire suppression in its habitat was one of the major problems. Jack pine has cones that open only after they are heated by fire. Jack pine is intolerant of the shade of other trees, and small trees of this species will not grow or survive in an established forest. Thus when fire is suppressed, jack pine declines. As a result of fire suppression, the habitat of the Kirtland's warbler was disappearing. A program that included setting aside land for the warbler, within which control burns are set, has helped to conserve this species. This solution accepted the changing, dynamic character of natural ecological systems.

Some conclude that if classic static stability does not exist, then little can be predicted about the changes in states of natural populations and ecosystems. This is not the case. To say that a system lacks a fixed equilibrium is not to say that it cannot have well-characterized dynamics. The solution to the decline in the Kirtland's warbler is a case in point. Although jack pine forests undergo frequent disturbances, their responses are well characterized and the changes in the state of these forests can be projected with considerable realism.

Once the idea that populations and ecosystems undergo change is accepted, concepts to replace classic static stability can be formulated. A number of concepts have been suggested, but this remains a topic of development in the science of ecology.

One approach is to recognize that natural populations and ecosystems pass through a finite series of states and that the conditions that one finds are within well-specified bounds. Understanding this, it is possible to talk about whether a human-induced environmental impact alters the set of states that occur, or alters the bounds (the range) of conditions that can occur. It is possible to ask whether the way some variable changes over time is altered by the human effects.

When a system passes through a series of states, some of the states are revisited. It is possible therefore to ask how often a specific, desirable state will recur, and what might be done to increase the frequency of the occurrence of that state and decrease the average time between recurrences. For example, people who hope to harvest fish at a high level over a long time period can ask how to affect the population and its habitat so that the desired population levels that can sustain a high harvest will recur frequently.

In summary, natural ecological systems fail to have the characteristics required by a traditional concept of stability—they lack fixed equilibria. New concepts enable us to characterize the behavior of these systems over time, and therefore to replace the old, static idea of stability with new concepts that accept the dynamic qualities of ecological systems.

DANIEL B. BOTKIN

For Further Reading: Daniel B. Botkin, *Discordant Harmonies: A New Ecology for the 21st Century* (1990); F. N. Egerton, "Changing Concepts of the Balance of Nature" in *Quarterly Review of Biology*, Vol. 48 (1973); Clarence J. Glacken, *Traces on the Rhodian Shore* (1967).

ECOLOGY, DEEP

The term *deep ecology* was coined by Norwegian philosopher Arne Naess in 1972. Following a long career in philosophy of science, semantics, and studies of Spinoza and Gandhi, Naess turned his analytic skill to the articulation of two directions the burgeoning environmental movement might take: a shallow ecology, in which pollution and resource depletion would be countered by quick-fix solutions that abate the problems while covering up their causes, and a deep ecology, which would look for the fundamental facets of our culture that lead to degradation of our habitat. Deep ecology demands a change in the basic ideas underlying civilization so that nature will be respected as valuable in itself and also as part of human identity. If we come to think of the natural world in this way, it will be impossible to pave it over with our own purposes. The deep ecology position suggests that only such a renovated conception will prevent further devastation of our surrounding world.

The distinction between shallow and deep ecology began to find favor among those seeking to evaluate whether proposed solutions to problems arising from human interaction with nature took into consideration the severity of the problems they were designed to solve. Supporters of deep ecology have taken the term from this point in two directions: (1) developing a philosophy of deep ecology, and (2) using it as a basis for a movement that sees fundamental social and ethical change as the only solution to the environmental crisis.

The philosophy of deep ecology demands not only new moral choices but a new way of conceiving the way things are. Relations, not entities, are considered to be primary. Thus the link between humanity and nature comes before any sense of what humanity and nature are in isolation. Natural systems have intrinsic worth, value in themselves apart from how humans might appreciate or spurn them. This has been called ecocentrism (thinking of the environment first) as opposed to anthropocentrism (thinking of humanity first). There is also an imperative for people to connect to the world as far as possible; Naess calls this Self-realization, with a capital *S*, to indicate that it is the Great Self of nature (like the Hindu *atman*) which we strive to reach, without losing the limits of our own egos (self with small *s*.) Self-realization can be increased by identifying with as much of the natural world as possible, by empathizing with the struggles and triumphs of even those organisms most distant from the human sphere. The more we identify with, the larger we become. The more we respect the Earth, the less we will be able to conceive of harming it.

There is some debate as to what constitutes the deep ecology movement. Some say it includes all those who believe that fundamental social changes are necessary to remedy environmental conflicts, while others say it should include only those who explicit take up the banner of deep ecology. The latter position is often taken by radical environmental groups such as Earth First! who ask for "no compromise in defense of Mother Earth," demanding wilderness areas that would be free of any human trespassing and unconditional support for the rights of any endangered species over any individual human intrusion. Others use the term *deep ecology* to refer to any personal ecological philosophy that admits a spiritual, reverential attitude, intuitively sensing the intrinsic value of nature. Naess has said that for such an intuition to become deep ecology, it must lead from wisdom to action, demanding that we change our lives to follow our beliefs. The main change is toward appreciating quality of life, rather than standard of living, defining a sense of quality that can apply to all cultures across the globe. Naess summarizes this desired way of life as "simple in means, rich in ends."

As a philosophy, deep ecology has encouraged many to stand up for their intuitive feeling of being one with nature or about the absolute importance of complexity, diversity, and richness of life on Earth. The discipline of conservation biology, for example, was founded by scientists who wanted to connect their moral belief that diversity of species should be preserved for its own sake with biological research on how this preservation can best be accomplished. Science founded on such an ethical premise is an example of deep ecology in practice.

The term *deep ecology* is widely used as a label for the diverse series of perspectives on the environment that urge basic reconsideration and change. It continues to gain adherents well beyond the limits of those versed in its special terminology. Its call for systematic description of 'ecosophies'—worldviews that depict an affirmation of the connection between humanity and nature—is beginning to be answered. The message of deep ecology is more a command for further understanding than an already completed perspective.

Deep ecology has been challenged for being too critical of those working on immediate, pragmatic solutions to specific environmental issues. Specific actions may always appear to be shallow, because they cannot change everything, while deep ecological perpectives seem distant from immediate situations, a bias shared by many philosophical approaches to real-world problems. Deep ecology deepens itself when it speaks less in opposition to 'shallow' immediate answers and concentrates instead on how society might be reorganized, based upon respect for the integrity of a nature which has room for tempered human purposes. If the personal aspect of intuitive identity with nature is over-emphasized to the detriment of the political self, deep ecology will accomplish little as philosophy or movement. The biggest challenge for adherents of deep ecology will be to design substantive social reforms based on the premise of human humility within nature.

DAVID ROTHENBERG

For Further Reading: Bill Devall and George Sessions, *Deep Ecology: Living as if Nature Mattered* (1985); Naess, Arne, *Ecology, Community and Lifestyle,* tr. and ed. David Rothenberg, (1989); Peter Reed and David Rothenberg, eds., *Wisdom in the Open Air: The Norwegian Roots of Deep Ecology* (1992).

ECOLOGY, HUMAN

See Human Ecology.

ECOLOGY, MARINE

Marine ecology is the science that treats the complete spectrum of interrelationships existing between marine organisms and their environments and among groups of these organisms. It is sometimes considered synonymous with oceanography. It is more restrictive, however, since the broader discipline of oceanography encompasses all physical, chemical, geological, and biological aspects of oceans. An important central theme of marine ecology is that organisms do not exist in isolation, but as groups that interact with each other, with other groups, and with the surrounding physical and chemical environments. In increasing complexity, organisms exist in populations, communities, and ecosystems. They may be distributed as free floating organisms (the plankton), as free swimmers (the nekton), or as bottom dwellers (the benthos).

Marine ecology differs significantly from terrestrial ecology. Because water absorbs light, the production of organic material through photosynthesis is limited to about the upper 100 meters: the photic zone. Since the average depth of the oceans is 3,800 meters, most of the ocean is without light and thus without any primary production; it is dependent upon food materials either falling from the lighted zone or transported in from the land. The productivity of the upper layers of the oceans is only a fraction of that of the land because the necessary nutrients (nitrates and phosphates) are either rare or inaccessible. The dominant plants in the oceans are microscopic and are free-floating as phytoplankton. They are consumed by tiny animals—the zooplankton—and they in turn are consumed by various planktonic and nektonic carnivores. Therefore, in the seas, most large animals are carnivores, and food chains are longer than on land.

Because they are buoyed up by the dense water, marine organisms invest less energy in structural materials. They tend to be dominated by protein, whereas terrestrial organisms are dominated by carbohydrate; a change in biochemistry that makes marine organisms shorter-lived than land organisms. Because the ocean waters contain large numbers of organisms permanently suspended as plankton, a whole new group of organisms not represented in terrestrial ecology is found, namely the filter feeders. They strain food from the passing water.

Oxygen is much less abundant in water than in air and its concentration also varies depending upon temperature and salinity. However, with few exceptions, there is sufficient oxygen at all depths in the oceans to sustain life.

The oceans are profoundly three-dimensional systems. The greatest density and variety of organisms are in the upper layers, where the light permits primary production. Organisms are stratified by depth, and as one progresses deeper the kinds of organisms change, and their numbers decrease due to lack of food.

Within each community in the seas the biological mechanisms for regulation and perpetuation of the component organisms are the same as those in terrestrial ecology, namely competition, predation, grazing, and recruitment.

JAMES NYBAKKEN

ECOLOGY, RESTORATION

See Restoration Ecology.

ECOLOGY AS A PERSPECTIVE

Ecology is sometimes characterized as the study of a natural "web of life." It would follow that humans are in the web or that they manipulate its strands. But the image of a web is too meager and simple for the reality. A web is flat and finished and has the mortal frailty of the individual spider. Although elastic, it has insufficient depth. However solid to the touch of the spider, it fails to denote the *oikos* – the habitation – and to suggest the enduring integration of the primitive Greek domicile with its sacred hearth, bonding the Earth to all aspects of society.

Ecology deals with organisms in an environment and with the processes that link organism and place. But ecology as such cannot be studied; only organism, Earth, air, and sea can be studied. Although it is the subject of scientific research, it is also a way of thinking and understanding that is very old. It is not a single discipline: there is no one body of thought and technique that frames an ecology of humans. As a scope or way of seeing, such a *perspective* on the human situation has been part of philosophy and art for thousands of years. It badly needs attention and revival.

Ecology raises questions about reciprocity. What do humans do in nature and what does nature do in them? What is the nature of the transaction? Biology states that the relationship is circular, a mutual feedback. Human ecology may not be limited strictly to biological concepts, but it cannot ignore them or even transcend them. It emerges from biological reality and grows from the fact of interconnection as a general principle of life. It must take a long view of human life and nature as they form a mesh or pattern going beyond historical time and beyond the conceptual bounds of other humane studies.

Individuals have their particular integrity. In one aspect the self is an arrangement of organs, feelings, and thoughts – a "me" – surrounded by various boundaries: skin, clothes, and insular habits. This idea is conferred by the whole history of civilization, and is verified by self-consciousness. The ecological view is the self as a center of organization, drawing on and influencing the surroundings, in which the personal perimeter and behavior contact the world instead of excluding it. Both views are real and their reciprocity significant. Both are needed to have a healthy social and human maturity.

The second view – that of the self as defined by relationships – has been given short shrift in societies emphasizing autonomy, individuality and competition. Attitudes do not change easily. The conventional image of a person has outlines stylized to fit the fixed curves of people's vision, hidden from themselves by habits of perception. Because they learn to talk at the same time they learn to think, language encourages them to see themselves – or a plant or animal – as an isolated sack, a thing, a contained self. Ecological thinking, on the other hand, requires vision across boundaries. The epidermis of the skin is ecologically like a pond surface or a forest soil, not a shell so much as a delicate interpenetration. It reveals the self ennobled and extended rather than threatened, because the beauty and complexity of nature are continuous with humans.

Ecology as applied to people faces the task of renewing a balanced view where now there is human-centeredness. It implies that individuals must find room in "their" world for all plants and animals, even for their otherness and their opposition. It further implies exploration and openness across an inner boundary – an ego boundary – and appreciative understanding of the animal in themselves, which their heritage of Platonism, Christian morbidity, duality, and mechanism have long held repellent and degrading. The older countercurrents – relics of pagan myth, the universal application of Christian compassion, philosophical naturalism, nature romanticism, and pantheism – have been swept away.

How simple people's relationship to nature would be if they only had to choose between protecting their natural home and destroying it. Most of their efforts to provide for the natural in their philosophy have failed – run aground on their own determination to work out a peace at arm's length. Their harsh reaction against the peaceable kingdom of sentimental romanticism was evoked partly by the tone of its dulcet facade but also by the disillusion to which it led. Natural dependence and contingency suggest emotional surrender to mass behavior and other lowest common denominators. In this light ecology can make us feel threatened. Our historical disappointment in the nature of nature has created a cold climate for ecologists who assert once again that we are limited and obligated. Somehow ecological thought must manage in spite of the chill to reach the centers of humanism and technology, to convey there a sense of our place in a universal vascular system without depriving us of our self-esteem and confidence.

Ecologists' message is not truly bad news. Our natural affiliations define and illumine freedom instead of denying it. Being more enduring than individuals, ecological patterns—spatial distributions, symbioses, the streams of energy and matter and communication—create among individuals the tensions and polarities so different from dichotomy and separateness. The responses, or what theologians call "the sensibilities" of creatures (including ourselves) to such arrangements grow in part from a healthy union of the two kinds of self already mentioned, one emphasizing integrity, the other relatedness. But it goes beyond that to something better known to 13th century Europeans or Paleolithic hunters than to ourselves. If nature is not a prison and Earth a shoddy way-station, we must find the faith and force to affirm its metabolism as our own—or rather, our own as part of it. To do so means nothing less than a shift in our whole frame of reference and our attitude toward life itself, a wider perception of the landscape as a creative, harmonious being where relationships of things are as real as the things. Without losing our sense of a great human destiny and without intellectual surrender, we must affirm that the world is a being, a part of our own body.

Such a being may be called an ecosystem or simply a forest or landscape. Its members are engaged in a choreography of materials and energy and information, the creation of order and organization. The pond is an example. Its ecology includes all events: the conversion of sunlight to food and the food chains within and around it, people drinking, bathing, fishing, plowing the slopes of the watershed, drawing a picture of it, and formulating theories about the world based on what they see in the pond. They and all the other organisms at and in the pond act upon one another and are linked to other ponds by a network of connections.

The elegance of such systems and delicacy of equilibrium are the outcome of a long evolution of interdependence. Even society, mind, and culture are parts of that evolution. There is an essential relationship between them and the natural habitat. Internal complexity, as in the human mind, is an extension of natural complexity, measured by the variety of plants and animals and the variety of nerve cells.

The creation of order, of which human beings are an example, is realized also in the number of species and habitats, an abundance of landscapes lush and poor. Even deserts and tundras increase the planetary opulence. To convert all "wastes"—all deserts, ice-fields, and marshes—into cultivated fields and cities would impoverish rather than enrich life esthetically as well as ecologically. Nature is a fundamental "resource" to be sustained for our own well-being. But it loses in the translation into usable energy and commodities. Ecology may testify as often against our uses of the world, even against conservation techniques of control and management for sustained yield, as it does for them. Although ecology may be treated as a science, its greater and overriding wisdom is universal.

That wisdom can be approached mathematically, chemically, or it can be danced or told as a myth. It has been embodied in widely scattered economically different cultures. It is manifest, for example, among pre-Classical Greeks, in Navajo religion and social orientation, in Romantic poetry of the 18th and 19th centuries, in Chinese landscape painting of the 11th century, in Whiteheadian philosophy, in Zen Buddhism, in the world view of the cult of the Neolithic Great Mother, in the ceremonials of Bushman hunters, and in the medieval Christian metaphysics of light. What is common among all of them is a deep sense of engagement with the landscape, with profound connections to surroundings and to natural processes central to all life.

It is difficult in our language even to describe that sense. English becomes imprecise or mystical—and therefore suspicious—as it struggles with "process" thought. It belongs to an idiom of social hierarchy in which all nature is made to mimic humans. The living world is perceived in that idiom as an upright ladder, a "great chain of being," an image that seems at first ecological but is basically rigid, linear, condescending, lacking humility and love of otherness.

The sardonic phrase, "the place of nature in people's world," offers, tongue-in-cheek, a clever footing for confronting a world made in humans' image and conforming to words. It satirizes the prevailing philosophy of anti-nature and human omniscience. For such a philosophy nothing in nature has inherent merit. As one professor put it, "The only reason anything is done on this Earth is for people. Did the rivers, winds, animals, rocks, or dust ever consider my wishes or needs? Surely, we do all our acts in an earthly environment, but I have never had a tree, valley, mountain, or flower thank me for preserving it." This view carries great force, epitomized in history by Bacon, Descartes, Hegel, Hobbes, and Marx.

Some other post-Renaissance thinkers are wrongly accused of undermining our assurance of natural order. The theories of the heliocentric solar system, of biological evolution, and of the unconscious mind are held to have deprived the universe of the beneficence and purpose to which humans were special heirs and to have evoked feelings of separation, of antipathy towards a meaningless existence in a neutral cosmos. Modern despair, pathological individualism, and predatory socialism were not, however, the products of Copernicus, Darwin, and Freud. Each showed the interpenetration of human life and the universe to be richer and more mysterious than had been thought.

Darwin's theory of evolution has been crucial to ecology. It might have helped rather than aggravated the growing sense of human alienation had its interpreters emphasized predation and competition less. Its bases of universal kinship and common bonds of function, experience, and value among organisms were obscured by pre-existing ideas of animal depravity. Evolutionary theory was exploited to justify the worst in people and was misused in defense of social and economic injustice. Humanitarians opposed the degradation of individuals in the service of industrial progress, the slaughter of American Indians, and child labor, because each treated humans "like animals." The temper of social reform was to find good only in attributes separating humans from animals.

It is not surprising to find substantial ecological insight in art. There are poems and dances as there are prayers and laws attending to ecology. Essays on nature are an element of a feedback system influencing people's reactions to their environments. The essay is as real a part of the community as are the songs of choirs or crickets.

Truly ecological thinking need not be incompatible with place and time. It has an element of humility that moves individuals to silent wonder and glad affirmation. It offers an essential factor, like a necessary vitamin, to all engineering and social planning, to poetry and understanding. There is only one ecology, not a human ecology on one hand and another for the subhuman. No one school or theory or project or agency controls it. It means seeing the world mosaic from the human vantage without being human-fanatic. It must be used to confront the great philosophical problems of people — transience, meaning, and limitation — without fear. Affirmation of its own organic essence will be the ultimate test of the human mind.

PAUL SHEPARD

For Further Reading: Calvin Luther Martin, *In the Spirit of the Earth* (1992); Paul Shepard, *Man in the Landscape* (1991); Alan Watts, *The Book: On the Taboo Against Knowing Who You Are* (1966).

See also Darwin, Charles; Ecology as a Science; Gaia Hypothesis; Religion.

ECOLOGY AS A SCIENCE

Prior to the 1970s ecology was thought of as a branch of biology, and defined as "the study of organisms in relation to environment." But now, ecology has emerged from its roots in biology to become a separate discipline that integrates organisms, the physical environment, and humans. This emergence is in line with the Greek root of the word *ecology* which is *oikos* meaning "household." Accordingly, *ecology* translates as the "study of the environmental house, including all its inhabitants, in which we live and in which we place our human-made structures and domesticated plants and animals." Or, alternatively, extending the older definition: "study of organisms in relation to environment and environment in relation to organisms."

Because the discipline is so broad and increasingly global in scope, both holistic and reductionist approaches must be involved. At the present time ecologists tend to identify themselves as either ecosystem ecologists who focus on large units (e.g., lakes, forests, watersheds, whole landscapes) as systems, or population ecologists who focus on species or communities of species as evolutionary units. Such a dichotomy between "top-down" and "bottom-up" approaches is, of course, arbitrary, since the overall goal of ecological research is the understanding of how both wholes and parts function and how they interact in time and space.

Again, because the discipline is so broad it tends to fragment into numerous subdivisions, or interfaces, with their own organizations, meetings, symposia volumes, and journals. Examples are conservation ecology, population genetics, ecological economics, agroecology, restoration ecology, systems ecology, and many more.

Environmental science differs from ecology in that in addition to ecological principles it focuses on the resource base (air, water, soil, food, minerals), climate and earth science, and on the technology of assessment and control of pollution and other human perturbations. Environmental studies, as taught in many schools and colleges, deals with ecological concepts in relation to economic, legal, ethical, historical, social, and demographic considerations; in other words, the interface between environmental sciences and the humanities.

Despite the fragmentations and overlaps, unifying principles are gradually emerging. With so youthful a discipline some of the concepts proposed as basic principles at this point are tentative. The major principles of ecological science can be summarized under the following nine headings:

Ecological Energetics

An ecosystem is a thermodynamically open, far from equilibrium system that requires a continuing inflow and outflow of energy. Energy flow is one-way in that energy cannot be reused, in contrast to materials that can be used over and over again without loss of utility. Accordingly, input and output environments are an essential part of the concept. For example, in considering a forest tract as an ecosystem, what is coming in and going out (energy, materials, and organisms) is as important as what is inside the tract. The same holds for a city viewed as an ecosystem. It is not a self-contained unit ecologically or economically; its well-being depends not only on what goes on within city limits, but even more so on the external life-support environment that (1) provides air, water, food, fuel, and minerals and (2) processes outgoing wastes generated by the intense consumption of resources imported from outside.

Because of the openness and interdependence of ecosystems, "source-sink" situations often develop in which a productive area or population (the source) may export to an adjacent less productive area or population (the sink). For example, a productive salt marsh may export organic matter or organisms to the less productive coastal waters. Or a species in one area may have a higher rate of reproduction than needed to support the population, and surplus individuals may provide recruitment for an adjacent area of low reproduction.

Material Cycling

Vital materials, such as water or nitrogen, pass back and forth between organisms and environment and between various components of the environment. Ecologists call these more or less circular pathways biogeochemical cycles. Energy of some kind is always required to drive such cycles. For example, solar energy drives the water cycle (evaporation, clouds, rain) while mineral nutrients in the soil are recycled by microbes that use the energy in organic matter. Recycling of human-made materials such as paper or metals requires human and fuel energy. Such recycling in human affairs becomes appropriate when supplies or depositories (landfills, etc.) become limiting, or when the value of the recycled product equals or exceeds the cost of collection and remanufacture. In nature, vital elements such as phosphorus tend to be hoarded and recycled wherever available forms are scarce.

From the standpoint of the biosphere as a whole, biogeochemical cycles fall into two groups: gaseous types with a large reservoir in the atmosphere (e.g., nitrogen), and sedimentary types with a reservoir in the soils and sediments of the Earth's crust (e.g., phosphorus). Microorganisms, such as the nitrogen fixers and the sulfide reducers, play major roles in the cycling of nearly all of the biogenic (vital to life) mineral nutrients. When these "friendly" microbes are stressed by pollution, the functional well-being of the whole ecosystem is at risk.

Hierarchical Organization

To understand, study, and deal with this complex world it is helpful to think in terms of levels of organization hierarchies (graded series or rank orders). The ecological levels of organization are commonly arranged as follows: organism, population, biotic community, ecosystem, landscape, biome, biogeographic region, biosphere. The ecosystem is the first level in this hierarchy that is complete in that it includes both organisms and the physical environment interacting to function as a system.

Interactions at the population level (e.g., between a parasite and its host) that tend to be non-equilibrium or cyclic are very often constrained by the slower interactions that characterize larger systems such as large forests, oceans, or the atmosphere. Accordingly, higher levels of organization tend to be more homeostatic (kept in equilibrium) than the lower-level components. This is an important principle because so often what seems to be the case at one level may not hold for another level. For example, flooding on a river floodplain can be a bad thing for an animal caught in the rising water, or for a person who has been unwise enough to build a house on the floodplain, but flooding is necessary for the long-term survival of the flood-adapted vegetation and associated fauna. In this case, flooding may be detrimental for some organisms but it has an overall positive value at the ecosystem level.

An important consequence of hierarchical organization is that as components, or subsets, are combined to produce larger functional wholes, new properties emerge that were not present or not evident at the level below. Accordingly, an *emergent property* of an ecological level or unit is one that results from the functional interaction of the components, and therefore is a property that cannot be predicted from a study of components that are isolated or decoupled from the whole unit. The interaction of algae and coelenterate animals to produce an efficient nutrient cycling mechanism in a coral reef is an example. This principle is a more formal statement of the old adage that "the whole is more than the sum of the parts."

Scales, both spatial and temporal, are also important considerations. Ecological relations as observed on a small scale or on the short term may be quite different from what is seen on larger scales and longer times. The complex interaction of ecosystems in time and space is little understood and is currently the focus of both experimental and modeling research.

One thing is certain: If we are serious about sustainability in the use of our environment we must raise our focus in management and planning to large landscapes and beyond. To deal with floods, for example, we must understand and manage the whole watershed on the long term, not just "quick-fix" the river (build a dam, for example) for a short-term advantage. Our future depends on the maintenance of environmental quality at the higher levels of organization.

Ecosystem Development or Ecological Succession

Paralleling the youth-to-maturity growth in the individual, and perhaps also in society, biotic communities and ecosystems undergo development from pioneer to mature or climax stages in a process known as ecological succession. In the early successional stages much of the available energy is directed toward growth, but as the mature stages are approached more and more primary production (sun energy converted to organic matter) goes to maintenance of the complex system that has been produced so that growth in size (biomass) slows and stops. During this development sequence, species composition changes with time as opportunistic species with high reproductive rates (known as r-strategists) that colonize the early stages are replaced by species adapted to the crowded conditions of the mature stages (known as k-strategists).

A somewhat similar shift in energy partitioning occurs during the development of cities. As they grow larger, more and more available energy (and money) has to be devoted to maintenance and repair (i.e., pumping out the disorder inherent in large, complex systems). Then taxes have to be increased to cover these increased costs. Ultimately, businesses find it difficult to grow because of increasing costs, and accordingly move to less developed areas.

When ecosystems are perturbed either by natural events (such as storms) or by human abuses of the environment (such as severe erosion or pollution) the biotic community is often set back to an earlier stage of development. Succession then becomes nature's repair process. Unfortunately, an increasing amount of the total environment worldwide has become so degraded as a result of long-term misuse that natural succession fails to repair even when the abuse stops. Then humans have to devote energy, effort, and money to artificially rehabilitate the ecosystem. Accordingly, environmental engineering and restoration ecology are becoming big business as we recognize that human survival will depend more and more on repairing and maintaining our life-supporting biosphere.

On the positive side, regular pulses such as tides or periodic fires or annual harvesting and replanting as in agriculture, can "pulse stabilize" a system in an early productive successional state, provided that there are organisms and communities that are adapted to the pulsing environment. In many cases (as in tidal marshes) the adapted community is able to use the pulse as an energy subsidy. Such pulse-stabilized ecosystems add productivity and diversity to the landscape and the biosphere.

Food Chains and Food Webs

As energy passes along a food chain (plant-herbivore-predator, for example) it decreases in quantity but increases in quality (concentration) at each successive transfer. Thus, predators are less numerous than herbivores but per capita (or gram for gram) they have a greater influence on the function and species composition of the ecosystem. The same decrease in quantity and increase in concentration occurs in fossil fuel formation since coal and oil have a far greater utility quality than the original plant material from which they were derived.

In terms of primary energy sources (available plant production) there are three basic food chains: the grazing food chain beginning with living plant tissue, and two detritus food chains beginning with non-living plant material, namely, the food chain beginning with particulate organic matter (POM) such as dead leaves, and the food chain beginning with dissolved organic material (DOM) exuded or extracted from cells or vascular systems. Food webs develop as networks as a result of cross feeding between these pathways creating a diversity of interactions between producers and consumers.

Food webs are not just a matter of "who eats who"; there are many direct and indirect positive interactions as well. For example, the saliva of grazing animals contains growth substances that stimulate the regrowth of plants on which the grazers depend. Another example: root fungi called mycorrhizae feed on the photosynthate they extract from the root, but in return provide the plant with minerals they are able to extract from the soil that are not available to the roots alone. "Reward feedback" is the term recently proposed for a situation where an organism "downstream" in the food chain benefits an "upstream" organism on which it depends. Such mutual aid increases the chances of survival for both the herbivore and the plant, the predator and its prey, and the parasite and its host.

Growth Forms and Carrying Capacity

The two extreme ways in which populations grow when opportunity for growth presents itself (e.g., beginning of the growing season, or availability of new resources) are: (1) exponential growth in which the population doubles in size at regular time intervals, and (2) logistic or self-controlled growth in which the growth rate decreases as density increases and levels off at or before some limiting or saturation level is reached. Exponential growth creates such a momentum that the population tends to overshoot limits, resulting in boom and bust cycles. These two growth forms are often designated as J-shaped and S-shaped or sigmoid growth, respectively, from the shape of the graph curves obtained when population size (density) is plotted against time. In the real world most growth is intermediate between these very fast and very slow patterns.

Carrying capacity is the saturation level in the logistic growth formula, and is usually defined as the population size that can be sustained by a given environment (or resource) without degrading that environment's ability to support that level of use. In reality, it is a two-dimensional concept involving not only density but the intensity of per capita use. These two characteristics track in a reciprocal manner, that is, as the intensity of per capita use goes up, the number of individuals that can be supported goes down. For example, for a given resource base more people can be supported in a developing country than in a developed one because the per capita consumption is so much less in the former. The term *cultural carrying capacity* is often used for the theoretical sustainable level for a given quality of life (in terms of per capita consumption).

Biodiversity

Biodiversity may be defined as the variety of life forms, the ecological roles they play, and the genetic diversity they contain. Diversity of life forms is considered important not only because of the direct or indirect importance of individual life forms (species, for example) to society, but because of the redundancy and stability that diversity of ecological roles contributes to the ecosystem. Concern for the loss of biodiversity must extend both below and above the species level. Loss of genetic diversity, for example, can result in rapid extinction, while the variety of life forms in any region depends on the size, variety, and dynamics of patches (ecosystems) and corridors that interconnect habitats. Accordingly, efforts to preserve biodiversity must focus on the landscape level of organization.

Evolutionary Ecology

From the ecological viewpoint there are two kinds of natural selection, or two aspects of the struggle for existence: organism vs. organism, which leads to competition, predation, and other negative interactions; and organism vs. environment, which leads to mutualism and other cooperative interactions. To survive, an organism must not only compete with other organisms for energy and space but it must also adapt to or modify its environment and its community in a cooperative manner. Mutualism (interaction of unrelated species for mutual benefit) has special survival value when resources become scarce, or where soil or water is nutrient poor (as in some coral reefs or rain forests). Accordingly, although natural selection is most evident at the genetic level it also occurs at higher levels.

The Gaia Hypothesis and Global Concerns

Since the beginning of life on earth, organisms have not only adapted to physical conditions but have modified the total environment in ways that have proven beneficial to life, as, for example, putting oxygen into the atmosphere (by green plants) and reducing carbon dioxide (by limestone-depositing marine organisms). The concept that organisms and the physical environment have co-evolved over geological time so that the biosphere is a holistic system, or a sort of self-regulating super-ecosystem, is known as the Gaia hypothesis (Gaia is the Greek goddess of the Earth). While this hypothesis is controversial, there is no doubt that humans are now modifying the Earth in ways that are not beneficial. Humanity could be considered a parasite on the biosphere for life support; we need to learn how to be a prudent parasite that does not destroy its host. This means reducing wastes and wanton destruction of resources, promoting the sustainability of renewable resources, and investing more in Earth care.

EUGENE P. ODUM

For Further Reading: Paul Colinvaux, *Ecology -2* (1993); Eugene P. Odum, *Ecology and Our Endangered Life-Support Systems*, 2nd ed. (1993); Leslie A. Real and James H. Brown, *Foundations of Ecology: Classic Papers with Commentaries* (1991).

ECONOMIC CONSERVATION

Economic conservation is the conservation of resources based on the criterion of economic efficiency, which requires that resources be allocated to their highest valued use (over an indefinite time period). Thus, resources should be conserved if their economic value is greater when they are allocated for future as opposed to present use. Consequently, the "highest valued use" can be maintaining resources in a wilderness or pristine condition indefinitely, if resource maintenance outweighs the benefits of resource consumption.

Value originates at the individual level and is measured in terms of the tradeoffs or choices that individuals would make in the event of resource scarcity. Benefits and costs represent an appropriate aggregation of value across individuals. These concepts are meaningful only in marginal terms, that is, the value ascribed to an incremental change in resource use or economic activity. Economic conservation requires a framework that facilitates the comparison of marginal benefits and marginal costs at different times. Critical to this evaluation within a formal context is the role of a discount rate.

A modern definition of benefits and costs is inclusive of any goods or services to which people attach value, including those stemming from passive use or nonuse of the resource. Hence, the visual amenities from a beautiful panorama or the vicarious pleasure derived from merely knowing of its existence should be included in an economic assessment. Furthermore, the reallocation of resource use can generate different patterns of consumption and savings over time, with potential ramifications that should be considered. For instance, the present consumptive use of a resource might lead to the accumulation of wealth that can facilitate economic development or technological advancement, thereby providing a higher standard of living in the future.

One historical distinction between alternative approaches to conservation characterized the preservationist and conservationist movements in the late 19th and early 20th centuries. Preservationists such as John Muir argued for preserving wilderness areas in a relatively natural and undeveloped state for their own intrinsic value. Conservationists such as Gifford Pinchot, natural resource advisor to President Theodore Roosevelt, were allied with the preservationists in their concern about the rapid exploitation and depletion of natural resources in North America. However, the conservationists argued for "rational resource use" from a utilitarian perspective.

Economic conservation would provide some comfort to the preservationist view, due to the significant sources of value now recognized to stem from nonuse or passive use. However, it differs in one fundamental area, since, from the economic perspective, intrinsic value has no meaning except for the subjective value ascribed to a resource by individuals.

A contemporary application of economic conservation stems from debates about energy use. Oil price shocks in the 1970s reinforced the impression that resource availability is subject to absolute constraints. Energy conservation became an idea that had intrinsic value for many, and pollution externalities further reinforced this view. In contrast, the principles of economic conservation suggest that, while absolute limits to energy resources exist in some sense, they are no less relevant than the dynamic process of economic change and technological advancement that can lead to the discovery of new resources and more efficient production processes.

DALLAS BURTRAW

ECONOMIC DEVELOPMENT

Economic development implies advancement in many dimensions: political modernization, cultural and social change, better education, and improved health conditions, as well as improvement in the quality of economic life through greater availability of private and public consumption goods. All of this depends on increases in the production of goods and services through enhanced productivity of labor and other resources. That, in turn, requires improvements in labor skills, increases in the amount of capital equipment with which labor works, and augmentation of human effort with other sources of energy.

By far the largest proportion of this energy must come from fossil fuels: coal, petroleum products, and natural gas. There are presently no large-scale, cost-effective alternatives, except for nuclear power, which has its own environmental hazards, and relatively modest amounts of hydroelectric power. The production and use of fossil fuels generate environmental threats at all phases of the fuel cycles: disruption of local ecosystems in the extraction processes, wastes that can pollute ground water, and emissions of air pollutants (such as sulfur dioxide, nitrous oxides, and ozone) as well as greenhouse gases when the fuels are burned.

Expanded production also requires more intensive use of chemicals, often derived from fossil fuels, and mineral products, such as steel. Producing and using the chemicals and metals generates mining, industrial, and agricultural wastes. In addition, as income grows and the use of industrial goods increases, growing personal consumption creates more environmental hazards, related largely to the disposal of waste products ranging from plastic goods to automobile exhausts.

The rapid urbanization that accompanies development accentuates pollution problems. High rates of migration from the countryside are seldom matched by the construction of new sewage and water treatment facilities. As a result these infrastructure services become overtaxed and pollution spreads.

Yet developing countries were never environmentally pristine before beginning their development processes, as is obvious from the prevalence in these countries of diseases transmitted through polluted waters. So, while the processes of development inevitably add to environmental stresses, the same processes, through improvements in waste disposal, in working conditions and living standards, can reduce the environmental hazards associated with poverty.

Development will also, necessarily, result in growth in greenhouse gas emissions from increased use of fossil fuels. Increases in paddy rice cultivation and animal stocks will also generate larger amounts of methane. The clearing and burning of forests and use of wood for fuel not only generate greenhouse gases, but also reduce the potential for absorbing carbon dioxide. In addition, the clearing of forests increases land erosion and reduces the natural habitat of local fauna and, thus, contributes to the reduction of biodiversity.

Improved efficiency in the use of fossil fuels can reduce both local and greenhouse gas emissions per unit of output. Unless there are major new technical innovations, however, industrialization and growth will swamp this effect in developing countries.

International negotations on trade and the control of global warming are generating growing pressures on developing countries to improve their environmental controls. The developing countries will have to balance these pressures with their desires for improvement in other material conditions of life.

R. S. ECKAUS

For Further Reading: Partha Dasgupta, *The Control of Resources* (1982); P. C. Stern, O. R. Young, and D. Druckman, eds., *Global Environmental Change: Understanding the Human Dimensions* (1992); World Bank, *World Development Report, 1992* (1992).
See also Economics; Energy; Greenhouse Effect; Urbanization; Waste.

ECONOMIC EFFICIENCY

Economic efficiency requires that actions (e.g., policies or decisions) be chosen to maximize the total resultant benefits. In the context of environmental issues, this typically requires that actions strike an appropriate balance between the cost of environmental protection and the environmental benefit of this protection. The appropriate balance rarely involves preserving nature at any cost, nor does it involve increasing human consumption without regard for the environment.

This appropriate balance occurs at the abatement level where the incremental costs of control are equal to the incremental benefits of control. If abatement were increased by a unit from this level, additional costs would exceed additional benefits and vice versa. When this condition is met, total benefits cannot be increased by a change in emissions, and the correct balance between costs and benefits is achieved.

Finding this balance between the costs of human activities and their environmental impacts is usually difficult to achieve in practice. In order to balance incremental control costs against incremental environmental benefits, both need to be valued in a common metric. Typically, incremental control costs are valued using the market prices of goods and services needed to achieve the control. Unfortunately environmental benefits are rarely traded and consequently there are no market prices.

There are three empirical approaches to measuring the value of environmental attributes. The household production function approach is based on observed expenditures for goods and services that are used in conjunction with an environmental attribute; for example, travel costs to recreation sites may be used to derive a measure of willingness to pay for use of the site. The hedonic price approach estimates implicit prices for individual environmental attributes, such as the presence of clean air, using observed prices of market-traded commodities (e.g., houses) by comparing the prices of houses located in areas with clean vs. polluted air. Direct elicitation of preferences relies on questionnaires (e.g., regarding visibility and entrance fee tradeoffs for a national park) or on experiments to elicit preferences directly. However, each of these approaches has some theoretical and/or practical difficulties.

Further practical problems arise when environmental effects are long-lived or even irreversible. That is, even if the annual incremental benefit of reducing a particular environmental harm is known, cumulating this marginal benefit over time raises issues of whether and to what extent future harm is less costly than current harm. In the parlance of economists, the question is what "discount rate" should be applied in converting future environmental harm to equivalent present harm. Here there are two views. One is that environmental goods are substitutes for other goods and that the same discount rate should be used for preservation of environmental attributes as is used for other intertemporal tradeoffs such as savings and investment decisions. A second view is that a lower "social" rate of discount should be used for the preservation of environmental goods.

STEPHEN C. PECK

ECONOMIC INCENTIVES

Economic incentives such as taxes, subsidies, and tradable permits work by making the polluter directly face the cost created by pollution. Using economic incentives is supported by a remarkable and surprisingly uncontroversial result in economic theory known as the Second Welfare Theorem: under certain idealized conditions, a competitive free-market economy will achieve Pareto optimality. Pareto optimality is a widely accepted, weak condition for a well-functioning economy: there are no changes that can improve one person's welfare without harming at least one other person. All undisputable welfare improvements have been made. However, the conditions required for competitive markets to have this property are far from the conditions of real economies. The central tasks of environmental economics are to identify *market failures* that prevent Pareto optimality, and to design remedies for these failures.

Externalities are the most pervasive type of market failure. An externality is present when the activity of one person has an inadvertent impact on the well-being of another person. Environmental externalities are common because the use of many resources imposes costs on society, but the user of the resource is not charged a price equal to the cost imposed. For example, the smoke emitted by a factory unintentionally worsens air quality for nearby residents. The factory owner pays for the steel and oil used in production, but not for the use of clean air. Pollution is not the only type of environmental externality, of course; global warming and the loss of wilderness are two other examples.

Externalities prevent the achievment of Pareto optimality because the polluter does not equate his or her incremental benefit of generating another increment of pollution to the incremental social cost of that pollution. To see this, suppose a factory can lower its production cost by buying much dirtier but slightly cheaper coal. The benefit to the factory is a small cost savings. The cost to society is the greater air pollution. If the factory owner keeps the cost savings and doesn't bear the cost of dirty air, even the smallest cost savings will be pursued, no matter how dirty the coal. Suppose the cost savings is $1, but the social cost of the additional pollution (say, from additional lung disease) is $10. An improvement is possible by stopping the use of the dirtier coal, while transferring $1 from the neighbors to the factory owner. The factory owner still gains $1, while the neighbors pay only $1 to gain $10 in clean air benefits.

When costly side effects can be ignored by a polluter, there will be too much pollution. Several policies can give polluters an *economic incentive* to consider side effects when deciding how much pollution to generate.

When the noxious activity costs the victim more than it costs to prevent, the problem may be resolved through negotiation and side payments. Although negotiation may work well for disputes between neighbors, most environmental problems involve too many people for negotiations to be feasible. For instance, one factory's smoke may harm millions of residents over thousands of square miles. In general, correcting environmental externalities will require some form of government intervention. The three most important economic interventions are taxes, subsidies, and tradable permits. These all work by *internalizing the externality*, that is by making the polluter directly face the cost created by the pollution.

Taxes: Since 1920 economists have recommended imposing a tax on polluting activity equal to the incremental social cost imposed by that activity. Then as long as the social cost of the pollution is greater than the cost of prevention or cleanup, the polluter will want to reduce the pollution, to the point at which the social benefits of further reductions are not sufficient to warrant the costs of obtaining the reductions.

Subsidies: Rather than impose a tax to discourage pollution the government can offer an equal subsidy per unit of reduction. A subsidy per pound of gunk *reduced* creates the same incentive as a tax per pound *produced:* each pound of gunk eliminated raises profits by the amount of the subsidy or tax. The main difference between taxes and subsidies is distributional: the cost of control can be paid by taxpayers (through a subsidy), or by some combination of the factory's owner, workers, and customers (through a tax).

Tradable permits: A very different approach to using economic incentives for environmental problems is to create a market in which the polluter must pay a price for the use of the formerly unpriced input (e.g., clean air). The usual method is for the government to issue permits for a fixed amount of gunk and to allow individuals and firms to buy and sell the permits. The fewer the permits the higher will be their market price. By controlling the quantity of permits the government can control the permit price so that the polluter has to pay the same amount per pound of gunk as it would under a tax or subsidy. Thus, all three methods can solve equivalently the problem of equating the costs and benefits of externalities, while using the polluter's self-interest to obtain the socially desirable level of control.

One advantage of using these economic incentives is that a given level of pollution control can be attained at the least cost. For example, with tradable permits, the permits will be most valuable to polluters with the highest control costs; they will purchase the permits, while those with lower control costs sell their permits and reduce their pollution. This result contrasts with the use of emission standards: all polluters must control to a given level, even though it will likely be cheaper to have some polluters control a bit more while others control an equal amount less.

As simple and effective as economic incentives seem for environmental remedies, there are a number of complicating problems. For example, when there is uncertainty about the benefits or costs of control, taxes and tradable permits are no longer equivalent. Which method is more effective depends on the shapes of the incremental cost and benefit functions.

Another problem is that the effectiveness of economic incentives depends on the competitiveness of the markets in which polluters operate. Suppose a polluting firm is a monopolist. Polluting monopolists impose two costs on society: the pollution generated, and a reduction in production in order to raise price and profits. Imposing a pollution tax on a monopolist reduces the pollution but simultaneously exacerbates the problem of too little production. Thus a tax may actually make society worse off. More generally, any time there are other market imperfections aside from the pollution externality of concern, economic incentives are not guaranteed to raise social welfare. The incentive policy may reduce pollution appropriately but interact unfavorably with the other market imperfections, causing some social loss to offset the gains from pollution control.

A third serious difficulty is that severe external harms can lead to problems for policy design known as non-convexities, and corrective taxes are no longer guaranteed to achieve a Pareto optimum. Indeed, the usual incentives policies may make society worse off. In particular, it is difficult to design effective economic incentives policies when an environmental problem is likely to cause severe and irreversible damage.

Despite the problems with economic incentives policies, most economists view incentives approaches to be more effective than alternatives. However, there has been substantial recent attention to various practical implementation problems that throw further doubt on the universal preferability of incentives policies. These problems include the possibility that other, non-economic objectives are important (such as equity and administrative simplicity), political constraints, high monitoring costs, and other technological limitations.

Research on the possibilities and limitations of economic incentives for environmental remedies is an active area. Meanwhile, many recent policies have been based on economic incentives. These include the use of tradable permits for leaded gasoline, chlorofluorocarbons, and sulfur dioxide emissions; deposit-refund systems for beverage containers and lead batteries; and taxes on toxic solvent wastes and carbon dioxide.

JEFFREY K. MACKIE-MASON

For Further Reading: William Baumol and Wallace Oates, *The Theory of Environmental Policy* (1988); Anthony Fisher, *Resource and Environmental Economics* (1981); Thomas Schelling, "Prices as Regulatory Instruments," in Thomas Schelling, ed., *Incentives for Environmental Protection* (1983).

See also Economics; Marketable Discharge Permits; Pollution Trading; Tradable Permits.

ECONOMICS

Economics refers both to a theory on how to allocate scarce resources and a body of material on the economy. This duality can be confusing, especially because the theory is a general one that helps to understand a wide array of issues, including why the environment is polluted and what approaches are likely to be most productive in improving environmental quality. At the same time, most of what is written about the economy and business has little or nothing to do with economic theory. Thus, some economists make important contributions to theory and practice without mentioning unemployment, tax policy, or stock prices. Most business leaders, business reporters, and business economists have little or no knowledge of formal economic theory.

An Introduction to Economics

An economy answers the questions: (1) What is to be produced? (2) How is each good and service to be produced? (3) How much of each good and service should be produced? (4) Who will work to produce this output? (5) Who will consume the goods and services that are produced? In a market economy, consumers have the most influence on what and how much of each good and service is produced: each consumer decides how much of each good and service to purchase. Thus consumers answer questions (1) and (3). Producers answer question (2) within society's general framework of laws and regulations. A market economy answers question (4) when the producers offer wages to hire workers, buy raw materials, and offer profits to attract investment in order to acquire capital. Markets work by providing incentives to consumers, workers, producers, and property owners: if there are too few computer programmers, the wages of programmers will rise, attracting others to become programmers. Question (5) is answered through the process of providing income to consumers and revenue to companies and government. The higher the income to consumers and the higher the revenue to companies or government, the greater their purchasing power.

During the 20th century economies in the developed world have grown rapidly, increasing per capita income. Even economists marvel that an economy as complicated as that of the U.S. is able to function under such a decentralized organization. No one gives orders to ensure that milk and bread are produced and delivered to stores; no one orders steel mills to produce raw material for building cars and bridges. Somehow, the free market's distribution system is able to offer millions of goods and services every day with few cases where too much or too little is produced. Of course the economy does not function perfectly, as evidenced by regional pockets of unemployment, business cycles with their periodic increases in unemployment, and occasional shortages of goods, which lead to rapid price increases. Nonetheless, the U.S. economy does a remarkable job in providing highly desired goods and services, employment, investment to businesses, and savings to provide investment.

No government intervention is required to ensure that people throughout the U.S. are offered sufficient food, clothing, housing, recreation, and other services. There are trouble spots: regions with persistently high unemployment, misallocated long-term investment (particularly investment in education and other "human capital"), and gaps between government and private sector responsibilities. However, the most persistent problems come when government tries to override market forces, forgetting the inherent contradiction when government wants to change specific market outcomes, but not affect the rest of the economy.

The complexity of a market economy can be seen in the attempts of the former communist economies to transform themselves into market economies. Communist firms never had to ask themselves what they should produce or whom they should hire. Consumers never had to decide whether to invest in firms and if so, which to invest in. Workers and companies looked to the government to solve problems, rather than working on the solutions themselves.

Economic Terminology

In a market economy the demand for a product is the total quantity that all consumers would buy at a particular price, for example, the total quantity of sirloin steak that Manhattan consumers want to purchase at $4 per pound. At $5 per pound some consumers desire to purchase less sirloin, finding pork, chicken, or fish

a more attractive buy. The responsiveness of demand to price is termed the "price or demand elasticity." The demand for "normal" or "superior" goods rises as income increases; for example, environmental concern is income elastic. Supply is the total quantity that will be offered on the market at a particular price, for example, the total quantity of soybeans offered at $5 per bushel. The higher the price, the greater the quantity that producers are likely to offer to sell. The intersection of supply and demand schedules defines the price and quantity for a market equilibrium.

The fundamental problem of economic theory is how an individual can achieve her goals when her resources are limited. Almost all goods are scarce in the sense that less are available than consumers desire. For example, consumers are offered a vast array of attractive goods and services but are able to purchase only a small proportion of what is offered. Economists model the situation by assuming that consumers are seeking to maximize their satisfaction (or happiness or utility) in choosing among the offered goods and services. Each consumer is assumed to have a "utility function" that indicates how much satisfaction can be expected from purchasing any of the goods or services. For example, this utility function indicates how much a particular consumer would prefer steak to fish for dinner.

In formal terms, economic theory shows how to maximize this utility function (or get the most satisfaction) by choosing which of the available goods and services to purchase with a fixed amount of income. The consumer is assumed to know the prices and other attributes of each good. The theory advises a consumer to stop purchasing additional units of a good (or additional units of quality) when the incremental satisfaction from the last dollar spent on quantity or quality is equal across all goods and services that are purchased. For example, even though the first taste of a rich chocolate dessert gives enormous satisfaction, additional bites give successively lower levels of satisfaction. The theory advises you to stop purchasing additional amounts of the dessert when the last dollar spent on dessert gives the same level of satisfaction as the last dollar spent on housing, clothes, and vacations.

Economic Theory and the Environment

Economic theory has been developed far beyond the simple principles stated above. Economists are able to detail the advantages of a competitive market economy and why taxes on income are generally better than sales taxes. Economists have investigated many issues involving environmental protection and remediation.

Pollution is generated not because some company or

individual desires to despoil environments. Rather, the pollution is created because all activities create undesired residuals. Just living results in exhaling carbon dioxide (a greenhouse gas) and excreting residuals from digesting food. Making a product or service similarly creates undesired residuals. As long as people lived in small groups and engaged in simple production, the discharges created at worst small, temporary problems. However, the combination of population growth and high per capita income results in discharge levels that can pollute rivers and lakes or even destroy the ozone layer in the stratosphere or change the Earth's climate.

Littering is a moral issue; a tiny amount of personal convenience is the only reason to use the world as a garbage can and lessen other people's enjoyment of the total environment. In contrast, discharging undesired residuals is not a moral issue; the socially desired level of discharge is not zero. Animals must discharge residuals, as must the production of goods and services. Materials are not created or destroyed in economic activity; this means the materials coming into a process must all emerge (in the form of desired goods and services and undesired residuals). However, some activities create large quantities of noxious pollution. If no satisfactory way can be found to control the discharge, such activities are banned, for example, producing coke with beehive coke ovens.

Generally there is a choice concerning both the chemical composition of the residuals and the place they are discharged. For example, burning coal to produce electricity creates large quantities of sulfur dioxide. This gas can be emitted from the smoke stack, later causing acid rain, or transformed into a gypsum sludge that is discharged into holding ponds. Sulfur dioxide can also be collected and used to make sulfuric acid, a valuable product.

Changing the chemical form and location of the discharge can be expensive. The cheapest way to dispose of these residuals is simply to dump them into the local environment. If the polluter bore the full social cost of disposal, he could make informed decisions. However, only a tiny proportion of the social cost of polluting is borne by the individual who dumps. Polluting is a "tragedy of the commons" in that each person is motivated to overuse the common property (in this case, the atmosphere), creating a polluted environment that no one desires. In economic jargon, environmental pollution is the result of a "market failure." The public interest requires reducing discharges to levels that will preserve the desired level of environmental quality without needlessly curtailing production and consumption activities.

Beginning with the Clean Air Act of 1970, the U.S. legislated a host of statutes to protect and enhance en-vironmental quality. In an economy as large and complicated as that of the U.S. (or any of the other large, developed nations), a centralized command-and-control system for pollution control cannot be effective or efficient. The millions of workplaces and products and tens of millions of residences mean that it is literally impossible for a regulatory agency to give detailed orders to each polluter or regulate each polluting factory or product. Instead, a decentralized system is needed to preserve the basic decentralized structure of a market economy.

The decentralized incentives might take the form of green taxes (effluent fees) or tradable discharge allowances. A green tax is an adjustment in price to account for the difference between social and private harm done by each discharge. For example, the U.S. currently levies a tax on each pound of CFCs (compounds that destroy stratospheric ozone); another example is the proposed "carbon tax" intended to encourage limiting carbon dioxide emissions. In 1975 Congress enacted legislation to increase the corporate average fuel economy (CAFE) in automobiles; an economic approach to accomplish the same goal is a Btu tax, since it discourages energy use. The decentralized alternative to adjusting price is to control quantity. Under this approach, the Environmental Protection Agency (EPA) specifies the quantity of pollution that may be discharged, creating an "allowance" for each unit of discharge. The allowances can be bought or sold. For example, under the 1990 Clean Air Act each coal-burning electricity generation plant must reduce its sulfur dioxide emissions by about half. A plant that reduces these emissions from 100 to 30 would be able to sell 20 allowances to another company. Allowance (emissions) trading gives electric utilities freedom to choose a control strategy that minimizes control costs; the cost of controlling sulfur dioxide is estimated to be reduced by $1 billion to $2 billion, a 25% to 50% savings.

In principle, all resources are either renewable or depletable. For example, the sunshine so essential to agriculture and the water used for irrigation are renewable resources; reasonable current use does not diminish the availability of these resources to future generations. The iron ore and petroleum used today are depletable resources; the resources used now will not be available to future generations. One major problem is the difficulty of acquiring a property right in a wild animal; property rights can be acquired by killing or capturing an animal. Deciding not to fish for salmon this year so that more will breed does no good unless all fishermen agree. Plants and animals are a renewable resource—until they are pushed to extinction. Preserving species diversity has become a major goal of environmentalists.

Economics of Public Policy

Economists have helped to develop more systematic ways of evaluating alternative public policies. The approach is a scientific examination of each alternative that assesses its costs and potential for accomplishing social goals. The best known policy analysis tool is benefit-cost analysis. In this framework, each of the inputs and outputs of a program are quantified and valued in dollar terms to find the best social choice. Two principal difficulties in conducting a benefit-cost analysis are valuing nonmarket services (as through determining people's willingness to pay for quality improvements) such as environmental quality, and finding a way to make effects commensurate over time. A less demanding alternative to benefit-cost analysis is cost-effectiveness analysis.

Industrial development in Europe and the U.S., some farming practices, and much urban development have devastated some local environments. The culprit is not population or economic activity per se. Rather environmental damage resulted from thoughtless actions or trying to dump residuals cheaply without consideration of the long-term consequences. Attaining a high standard of living, having a large population, and living in large cities need not harm environments. Technology can be made much cleaner, if environmental quality is a high priority. For example, environmental quality in the U.S. has improved since the 1970 Clean Air Act, at the same time that economic activity almost doubled and population increased almost 25%.

Developing countries seek to attain high living standards through industrialization. Industrialization is more difficult than assumed, since it requires development of social and economic institutions.

For the U.S., environmental regulation currently costs about $120 billion per year, about 2% of the GNP. By using these resources more effectively, and perhaps with some increase in environmental expenditures, the quality of air and water could be improved, and general environmental quality increased.

Sustainable development is economic development that attends to environmental quality. Many people are concerned about intergenerational equity, about this generation using up resources that should be left for future generations; these concerns are addressed in seeking sustainable development. With care for environmental quality, the use of renewable and depletable resources, and investment in efficient ways to produce goods and services, environmental quality can improve as living standards grow. However, sustainable development will not occur by chance. Rather, sustainable development will result only from hard work, scientific progress, and agreement on the common goal of improving environmental quality as we improve living standards.

LESTER B. LAVE

For Further Reading: Maureen L. Cropper and Wallace E. Oates, "Environmental Economics: A Survey," *Journal of Economic Literature* (1992); Paul R. Portney, *Public Policies for Environmental Protection* (1990); Thomas H. Tietenberg, *Environmental and Natural Resource Economics* (1986).

See also Benefit-Cost Analysis; Capital; Carbon Tax; Consumption; Contingent Valuation; Cost Effectiveness; Demand; Discount Rate; Distribution; Ecological Economics; Economic Conservation; Economic Development; Economic Efficiency; Economic Incentives; Economics of Greenhouse Effect; Economics of Renewable and Nonrenewable Resources; Economics of Technical Change; Economy, Primary/Secondary/Tertiary/Quarternary; Externalities; Free Market; Free-Rider; Human Resources; Industrial Metabolism; Labor; Limits to Growth; Materials Balance; Opportunity Cost; Price System; Productivity; Property Rights; Purchasing Power; Scarcity; Social and Economic Infrastructure; Standard of Living; Supply; Transaction Costs.

ECONOMICS OF RENEWABLE AND NONRENEWABLE RESOURCES

Producing goods and services requires natural resources, such as petroleum and lumber. Petroleum is a nonrenewable (depletable) resource because there are a fixed number of barrels underground; extracting oil depletes the underground reserve by the amount extracted. Lumber is a renewable resource if the trees that are cut are replanted, providing trees to be harvested in the future.

Since using a nonrenewable resource deprives future producers of the use of that resource, the owners of the resource must consider both the current and future prices of the resource, as well as the current and future extraction costs. It is senseless to extract a barrel of oil if the price is less than the extraction cost. Similarly, a barrel of oil should not be extracted today if the net price (price minus extraction cost) is less than the present value of net price in the future. For example, if the discount rate is 10% per year and the net price of oil is $15 today and is expected to be $20 next year, the owner would do better to delay production for one year, since the present value of the oil next year is $18.

In a world where extraction costs remained unchanged over time and producers had confidence that they knew the future prices of a nonrenewable resource, the net price of the resource would be expected to increase by the discount rate. Thus, the price of petroleum or iron or copper would be expected to rise by the discount rate each year. However, extraction costs have decreased over time, new reserves have been discovered, and there is great uncertainty about future price; the real prices (after accounting for inflation) of nonrenewable resources have tended to decline over time.

For renewable resources, the discount rate also plays a crucial role. A stand of trees represents an investment. If the trees are harvested today, they bring a certain amount of money, say $100. If the trees are not harvested until next year, two things happen. First the trees grow, and so there is more lumber to sell; suppose that the trees would sell for $120 next year. Second, however, if the trees were sold today, the $100 could be invested to earn a return. Suppose that $100 invested today would yield $110 in one year. In this example, the owner is better off by letting the trees grow for another year.

Newly planted trees grow rapidly, fully mature trees stop growing entirely. As trees mature, the growth rate slows. At some point in the growth cycle of a tree, the additional growth provides precisely 10% growth in revenue. The following year, the growth would be smaller. The owner maximizes profit by harvesting the trees when the growth rate has declined to the interest rate (which measures what could be earned by investing the money in other projects). Actually, since trees grow rapidly when they are young, the trees should be cut somewhat sooner and replanted.

This analysis assumes that the resources provide benefits only when extracted. Often this is not so. A standing forest, for example, provides habitat for many species of plants and animals, some, perhaps, like the spotted owl in the Pacific Northwest region of the U.S., considered endangered. Oil in the ground may not provide benefits, but extracting and consuming it ordinarily carries environmental costs, in the form of air and water pollution, and emission of the greenhouse gas carbon dioxide. In principle, these external benefits and costs can be folded into the theory of optimal use. For the forest, if the habitat benefit is positively related to the age (and size) of the standing trees, then the optimal harvest will be pushed into the future. In some cases it may be advantageous to defer the harvest indefinitely, that is, to preserve the forest. Similarly, the environmental costs of oil production from a given tract of oil-bearing land may outweigh the benefits of the oil, in which case it will be most advantageous—optimal—not to produce from the tract. In principle, this is clear. The difficulty comes in translating principle into practice, since this involves estimation of the environmental costs and benefits, a challenging task that lies at the frontier of environmental economics.

ANTHONY C. FISHER

For Further Reading: Anthony C. Fisher, *Resource and Environmental Economics* (1981); Robert M. Solow, "The Economics of Resources or Resources of Economics," *American Economic Review* (May 1974); Thomas Tietenberg, *Environmental and Natural Resource Economics* (1988).
See also Benefit-Cost Analysis; Discount Rate; Economics; Resources, Nonrenewable.

ECONOMICS OF RISK

By the year 2000, the U.S. Environmental Protection Agency projects that public and private expenditures on environmental protection will exceed $150 billion per year. However, even these vast resources are inadequate to eliminate all environmental threats. The "economics of risk" refers to the problem of optimally allocating resources to address these threats, asking specifically:

- Are we targeting the "biggest" risks?
- How much should we spend to reduce various risks?
- How can we achieve risk reduction goals at a minimum cost to society?

Setting Priorities

Environmental protection advances social goals such as "clean" air and drinking water, "safe" food, and a "pristine" environment. Achieving these goals consumes resources in the form of capital and labor—resources which could otherwise be used to advance other social goals such as education, nutrition, and national defense, or even other environmental goals. The fact that resources are scarce means that any decision to expend resources on an environmental goal places a relative value on that goal even though that value may be implicit and unscrutinized.

Since resource allocation is inevitable, it makes sense to identify those problems which are the most important. This means first quantifying the size of various risks to human health and the environment and then devoting resoures to reducing large risks. For example, if the resources needed to eliminate each of two carcinogens from the environment are equal, it is preferable to expend resources first on the more dangerous carcinogen.

In addition to making the need to choose among competing goals explicit, economics also helps compare risks that are qualitatively different. For example, society must compare risks to the environment to human health risks. Within each of these domains, further comparisons are required, e.g., the health risks to the elderly and to children; or the relative likelihood of cancer, neurotoxicity, and reproductive health threats.

Economists can help shed light on these delicate issues through scientific methods that gather information about the choices people actually make. These "revealed preference" techniques study the value tradeoffs people make when they invest their own resources to reduce health and ecological risks. For example, consumer willingness to pay higher costs for safer products helps identify how much they value improved health. Willingness to pay more to live in cleaner areas of the Los Angeles basin reflects not only the value people place on community characteristics such as improved school systems, but also the value they place on a cleaner environment. While these methodologies help evaluate the relative importance of competing goals, they do not provide a precise solution to this problem. Ultimately, society as a whole, through the actions of politically accountable officials, must use judgment to make these decisions.

How Safe Is Safe Enough?

Just as scarce resources make it necessary to rank environmental problems, the fact that resources are finite also makes it prudent to decide how much various risks should be reduced. Often, greater societal value can be achieved by using less resources to reduce a risk rather than using more resources in an attempt to completely eliminate a risk.

The case of environmental lead contamination illustrates this point. Lead is a well-known neurotoxin to which children are especially vulnerable. Lead pervades the environment, contaminating air, soil, house dust, food, and drinking water. Because it contaminates so many different media, complete elimination of lead from the environment is impractical. Nonetheless, during the late 1970s and through the 1980s, average blood lead levels among American children decreased by two thirds as leaded gasoline and the use of lead solder in food cans were phased out. Economists have shown that improvements were accomplished at very little net cost to the public.

Making further gains in this area would be very costly. Residential lead paint, as well as soil and house dust contaminated by decaying lead paint, are the most important remaining lead contamination sources. Complete remediation of these sources can cost on the order of $10,000 per house. Since there are 57 million houses in the country with lead paint, the total cost of removing lead from homes could run into the hundreds of billons of dollars, or more than the United States spends on all environmental risks in any one year. These resources might be better used to address other pressing issues such as education and health care.

How to Best Reduce Risks

Some approaches to reducing environmental risks are more costly than others. Technological standards, which prescribe the technology to be used to address a specific environmental risk, often fail to identify the least expensive alternative.

Economists advocate regulatory schemes that prescribe performance goals but allow regulated parties to choose the least expensive technology. For example, the regulation of sulfur dioxide emissions by the Clean Air Act Amendments of 1990 is accomplished by allocating a limited number of marketable permits issued by the Environmental Protection Agency, each of which licenses a firm to generate a certain amount of pollution. Firms that do not need all the permits they hold can sell them. This gives them an incentive to invest in pollution control and induces industry to focus pollution control resources at those facilities for which improvements are least expensive. This flexibility can achieve large economic savings since the costs of emission reduction often vary enormously from source to source.

The disadvantage of the permit scheme is that its added flexibility makes its environmental results difficult to predict. Certain geographical areas may end up with excessive pollution if a firm uses a large number of permits in that area. Modifications in the program to ameliorate this and other problems make it less flexible and more complex, compromising some of the cost savings, but it is apparent that economic incentives will play an increasingly important role in environmental policy.

JOSHUA T. COHEN
JOHN D. GRAHAM

For Further Reading: John Graham, Laura Green, and Marc Roberts, *In Search of Safety* (1988); Robert Hahn and Robert Stavins, "Incentive-Based Environmental Regulation: A New Era from an Old Idea," *Ecology Law Quarterly* (1991); Paul Portney (ed.), *Public Policies for Environmental Protection* (1990).
See also Best Available Technology; Lead.

ECONOMICS OF TECHNICAL CHANGE

There have been several studies on productivity and industrial competitiveness. All have concluded that the competitive supremacy of the U.S. has declined in the international marketplace. A decline in growth of productivity has been identified as a major cause of economic problems for the U.S.

Paul R. Lawrence identified five critical problems for the manufacturing sector that go beyond the productivity problem: (a) changes in pattern of consumer demand with changes in income levels, (b) slow growth in demand for manufactured goods due to sluggish overall economic growth in the global economy, (c) long-term decline in employment in the manufacturing sector due to technological changes, (d) shifts in the pattern of U.S. international specialization due to changes in comparative advantage resulting from changes in relative factor endowments and productive capabilities associated with foreign economic growth and policies, and (e) short-term changes in exchange rates and cyclical conditions.

In its study of the manufacturing sector in the U.S., the MIT Commission noted that manufacturing productivity in the U.S. grew in the 1980s, but at the cost of domestic jobs. The commission proposed a broader measure of economic vitality by focusing on *productive performance* to capture the influence of factors such as quality, timeliness of service, flexibility, speed of innovation, and command of strategic technology.

Using U.S. patents as a surrogate measure of technological position, Chakrabarti showed that Japan has steadily increased its technological strength in high-technology product areas, particularly equipment and intermediate products, since 1980. Germany has also shown growth, but the U.S. has remained stagnant. These data confirm the findings of other studies that the U.S. position in high technology has weakened in recent years with Japan emerging as a dominant force.

Economists have found a close relationship between basic research performed in industry and productivity growth. Japanese and German firms finance a greater proportion of total research than their American counterparts. In 1983, 49% of research in the U.S. was financed by business; the comparable figures for Japan and Germany were 65.3% and 58.1%

Japan has been consistent in maintaining a steady level of productivity growth although during the 1980s it slowed down somewhat. Productivity growth rates in the U.S. have not been impressive, but did improve during the 1980s. The U.S. compares favorably with Germany in productivity but lags behind Japan in machinery, electrical engineering, and transport equipment. In other industrial sectors, the U.S. still leads in productivity.

The importance of manufacturing has decreased in the industrial nations. The service sector accounted for 63% of jobs in the U.S. in 1987. The comparable figures in Japan and Germany are 58% and 57%. In 1990 in international trade the U.S. maintained a $64 billion net surplus in the service sector while it incurred a net negative balance of $77 billion in the trading of manufactured goods. Productivity in the service sector has become an important issue in the U.S.

Technical innovation depends on an organized commitment to basic and applied research, as well as on organization and management of information. How well industry can make use of discoveries depends on many factors such as market conditions, regulations, and availability of necessary technology. Other economic factors such as demand and relative prices have influence on the technological paradigms, rate of technical progress, and precise trajectory of technological advance. To the extent some of the economic factors are influenced by public policy, technical change can be induced.

Productivity and technical change are the engines of economic growth. The race for technological development is pursued by all rival nations. In determining the course of development, all sectors—corporate, government, and organized labor—have important roles to play.

ALOK K. CHAKRABARTI

For Further Reading: William J. Abernathy, Kim B. Clark, and Alan M. Kantrow, *Industrial Renaissance: Producing a Competitive Future for America* (1985); Martin N. Baily and Alok K. Chakrabarti, *Innovation and the Productivity Crisis* (1988); Michael L. Dertouzos, Richard K. Lester, Robert M. Solow, and the MIT Commission on Industrial Productivity, *Made in America* (1989); John W. Kendrick, *Understanding Productivity: An Introduction to the Dynamics of Productivity Change* (1977); Robert M. Solow, "Technical Change and Aggregate Production Function," in *Review of Economics and Statistics* (Vol. 39, 1957).

ECONOMY, PRIMARY/SECONDARY/ TERTIARY/QUATERNARY

Economic activity is not exclusively limited to materials extraction, transformation, and disposal. However, such activities, including energy production and transportation, are a major part of the real economy. It is difficult to imagine an economic system that could function without any dependence on physical materials or external energy sources. It is therefore helpful to conceptualize the relationship between materials use, economic activity, and environmental impacts by introducing a simple taxonomy of economic activities.

Primary sectors are concerned with extraction of raw materials, (mining, quarrying, agriculture, logging), physical processing to concentrate crude materials and dispose of wastes (beneficiation of ores, washing of coal, distillation of crude petroleum, debarking of logs, threshing of grain, peeling and coring of fruits, etc.), and thermal or chemical processing (roasting, smelting, calcining, electrolytic or oxygen refining, alloying, dehydrogenation, refining, pulping of wood, cooking or baking of food products, etc.). Production of finished fuels and electricity is part of the primary sector. The outputs of the primary sectors are metals, wood and paper products, food products, building materials, commodity chemicals, finished fuels, and electricity.

Secondary sectors are concerned with the conversion (shaping, forming, joining, weaving, assembly, coating, painting or dyeing, printing, etc.) of finished materials into manufactured products, including structures. The so-called fine chemicals and pharmaceuticals, cosmetics, and other such products are regarded as secondary.

Tertiary activities generate final services, rather than products. Tertiary services are produced by means of specialized material products or infrastructure. Examples of tertiary services include transportation, communications, retail and wholesale sales, and real estate. Services provided by consumer goods can be included in this category.

Quaternary activities are essentially "pure" services, for example, entertainment, education, child care, or governance. They may be produced and delivered either by other persons or by organizations.

It is clear that primary production activities are intrinsically much "dirtier" than secondary activities. This is because primary activities are, by definition, concerned with the physical separation of potentially useful materials from the nonuseful materials among which they are found in natural form. The unwanted materials are, by definition, wastes and pollutants. The most important examples of wastes associated with primary activities are mine wastes, smelter wastes, and agricultural wastes. For example, for each ton of steel and aluminum produced at least a ton of mine waste is generated. For copper, a ton of virgin metal implies at least 170 tons of mine waste. In the case of rare metals like gold, platinum, and uranium, the amount of mine waste generated per ton of product is much larger.

Smelter wastes are unwanted fractions of mineral ores. They are emitted as gases, such as carbon dioxide, carbon monoxide (which can be recycled as a low-grade fuel), and sulfur dioxide (which can be "scrubbed" by converting it into a solid waste) and solid wastes, like slag, which carry significant contamination by toxic heavy metals. In the case of zinc, a ton of sulfur is produced per ton of metal; for copper and lead the sulfur emissions are smaller, but still far from negligible. Agricultural wastes are unwanted fractions of the harvested plant material such as tree bark, stems, inedible leaves, inedible roots, chaff, corn husks, etc. Some of this can, in principle, be used as animal feed, and some can be burned for energy. Neverthless, much of it is discarded and ends up in waterways.

The notion of "clean production," which has become popular in recent years, has strict limits of applicability in the primary sector. Finding economic uses for mine wastes, smelter wastes, and food processing or pulp or paper wastes is a real challenge.

The secondary sector generates far less waste than the primary sector, and much more of it can be eliminated by clever engineering. Much secondary sector waste consists of intermediate materials, such as solvents, lubricants, catalysts, paints and pigments, cleaning agents, and so forth. The notion of clean production is far more applicable in the secondary sector, where the wastes tend to be intermediate materials (such as the examples noted above) used dissipatively or finished materials used inefficiently. Examples of the latter are metal cuttings from the engineering sector or fabric cuttings from the clothing industry. There is a strong economic incentive to minimize dissipative losses and to use finished materials with maximum efficiency.

Wastes from the tertiary sector are primarily consumption-related. Food products and drugs are actually consumed; other items, such as packaging materials, cleaning agents, garden supplies, lubricants, and fuels, are dissipated or chemically transformed into wastes. Durable goods and structures are worn out rather than being consumed. When their useful lives are over they must be repaired, reconditioned, recycled, or disposed of in some way. Opportunities for waste minimization exist, but in most cases they will require institutional innovations rather than engineering innovations to be effective. By contrast, quaternary activities are comparatively (though not absolutely) waste-free.

R. U. AYRES

ECOSYSTEMS

Ecosystems (short for ecological systems) are functional units that result from the interactions of abiotic, biotic, and cultural (anthropogenic) components. Like all systems they are a combination of interacting, interrelated parts that form a unitary whole. All ecosystems are "open" systems in the sense that energy and matter are transferred in and out. The Earth considered as a single ecosystem has many attributes of a living, evolving organism. It constantly converts solar energy into myriad organic products; it has increased in biological complexity over time.

Natural ecosystems, made up of abiotic factors (air, water, rocks, energy) and biotic factors (plants, animals, and microorganisms), evolved on the Earth and have functioned for more than three billion years. The Earth's biosphere, including the atmosphere (air), hydrosphere (water), and lithosphere (land), constitutes a feedback or cybernetic system that reflects what René Dubos referred to as "a co-evolutionary process" between living things and their physical and chemical environments. The largest ecosystem (the universe or total environment) is made up of many smaller ecosystems interlocked through cycles of energy and chemical elements. The flow of energy and matter through ecosystems, therefore, is regulated by the complex interactions of the energy, water, carbon, oxygen, nitrogen, phosphorus, sulfur, and other cycles that are essential to the functioning of the biosphere. The food chains (grazing and detritus) usually overlap and their interwoven complexes make up the intricate

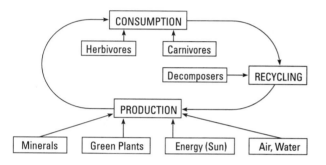

Figure 1. Natural ecosystem. Source: William and Ruth Eblen, *Experiencing the Total Environment*. New York: Scholastic Book Services, 1977.

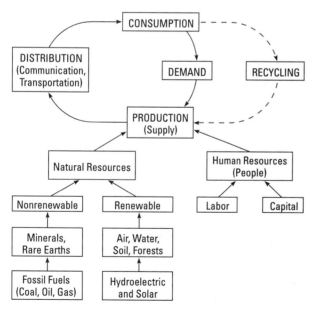

Figure 2. Cultural ecosystem. Source: William and Ruth Eblen, *Experiencing the Total Environment*. New York: Scholastic Book Services, 1977.

food web of ecosystems that depend directly on the other cycles in the total system.

Cultural ecosystems, made up of abiotic, biotic, and cultural factors, evolved when humans were added to natural ecosystems. Not all changes are due to the influence of random natural events; rather, some are side effects of human activities, and now many are the result of deliberate social choices. The reciprocal interplay between natural and cultural components in cultural ecosystems is illustrated by people humanizing their environments and in turn the humanized environments influencing the evolution of human societies.

Comparisons between natural and cultural ecosystems emphasize some of the basic similarities and differences. For example, economic production in the cultural ecosystem requires human resources (capital and labor) that natural production does not. However, energy, air, water, and minerals are essential inputs for both natural and economic (cultural) production. Consumer demand stimulates production in the economic cycle; natural production is provided by photosynthesis in green plants, and herbivores (primary) and carnivores (secondary) are consumers.

WILLIAM R. EBLEN

ECOTAGE

See Environmental Movement, Radical.

ECOTERRORISM

See Environmental Movement, Radical.

EFFLUENT

See Pollution.

ELECTRIC UTILITY INDUSTRY

In 1882 the first central power station in the U.S. began operation in Manhattan, supplying electricity for lamps. Within 50 years, hundreds of power stations were delivering electricity across a network of power lines. So crucial had electricity become that when national leaders suggested shutting off all current for one minute to commemorate Thomas Edison's funeral in 1931, the plan had to be abandoned. A nationwide shutdown would have paralyzed the country. In 1974 electricity represented 28% of all the energy used in the U.S. By 1990 that figure had climbed to 36%. Experts predict that by 2010 it will reach 45%.

Electric power has dramatically transformed human life and society. The development of electric motors revolutionized industrial processes during the early 1900s, bringing about rapid advances in productivity. More recently, electricity has powered the information revolution. Today, virtually everything we do—from illuminating our homes to powering our factories—depends on electricity.

But electricity has played another, equally important role. Electric power, unlike almost all other energy sources, is clean at the point of use. Centralized electricity generating plants have made it possible to isolate emissions associated with electricity generation to a limited number of sites. That in turn has made pollution easier to control. Historically, the shift from millions of individual polluting sources to several thousand electric power plants has played a significant role in enabling society to reduce the environmental consequences of energy use.

The Electric Power Industry

The role of electric utilities is to provide electric power safely, reliably, at a reasonable cost, and with as little impact on the environment as possible. America's electric utility industry is large and diverse, consisting of more than 3,000 individual utilities. They include investor-owned electric companies and rural electric cooperatives, as well as systems owned by federal, state, district, county, and municipal governments. Investor-owned utilities generate more than three-quarters of the electricity used nationwide, making them one of the country's largest businesses. In 1990 investor-owned utilities received revenues of more than $139 billion and spent approximately $117 billion on fuel, taxes, and operations and maintenance costs.

Electric power is measured in units of power called watts—or, more commonly, in kilowatts, the equivalent of 1,000 watts. One kilowatt-hour is enough to illuminate a 100-watt light bulb for 10 hours. In 1990, the nation's electric utilities supplied nearly 2.5 trillion kilowatt-hours of electricity. To generate that power, utilities used more than 600,000 short tons of coal, 186,000 42-gallon barrels of oil, and 2.4 trillion cubic feet of gas. Nuclear power contributed about 20% of the nation's electricity. Hydroelectric power plants generated about 5%. A smaller but growing percentage of electricity is generated from wind, solar, and geothermal sources.

Privately-owned utilities operate as a business monopoly that is controlled and regulated. State public utility commissions regulate rates, accounting records, service standards, and service area conditions and boundaries. The Federal Energy Regulatory Commission regulates interstate electric rates and services.

The Nuclear Regulatory Commission regulates the nuclear power industry.

Because the generation of electricity has a wide impact on natural resources and the environment, electric utilities are also closely regulated by federal, state, and local commissions that oversee environmental protection. At the federal level, the Environmental Protection Agency (EPA) grants permits for nearly all generating facilities. Increasing concern over the effects of energy use on the environment has led to regulations that impact virtually all aspects of electric utility operation.

Electricity Generation

The greatest impact of electricity on the environment occurs during the generation of electric power. Energy comes in many forms, from the chemical energy contained in coal or natural gas to the kinetic energy possessed by flowing water. Electricity is a refined energy that can be generated from a wide variety of primary energy sources. Many utilities rely on a variety of ways to generate electricity. In choosing their generating mix, utilities weigh several important considerations: cost, availability, reliability, land use, fuel supply, transmission and distribution to customers, and the environmental impact.

Fossil Fuel

Three-quarters of the electricity used in the U.S. is generated from fossil fuels, including coal, oil, and natural gas. Coal is the nation's most abundant fossil fuel resource and the leading fuel for electricity generation—the source of about 55% of the electricity consumed in the U.S. Electricity generated from coal is relatively inexpensive, averaging about 6¢ per kilowatt-hour.

There are drawbacks to fossil fuels. Athough supplies remain abundant, they are not unlimited, and most of the oil the U.S. uses is imported. The burning of fossil fuels releases a variety of potentially harmful gases. These include sulfur oxides and nitrogen oxides, which have been linked to acid rain, as well as carbon dioxide, which is implicated in atmospheric changes that could potentially lead to global climate changes.

Nuclear Power

Nuclear power plants currently produce roughly 20% of the electricity used in the U.S. In many European countries, nuclear power contributes a greater percentage of electric power. Unlike fossil fuels, nuclear power does not emit substances into the air, and is virtually unlimited. But nuclear generation does produce high- and low-level radioactive solid wastes that must be disposed of in secure depositories—an issue that remains highly controversial. Because some of these wastes remain radioactive for thousands of years, they must be stored so that they will remain intact for many generations. Many local communities have been reluctant to agree to store reactor wastes in their own regions.

Another significant concern has been safety. Stringent regulations have dramatically increased the cost of constructing nuclear facilities. Many experts believe nuclear power plants can be built and operated safely. A new generation of nuclear power plants is being designed that would be safer and far less expensive to construct than current plants. Whether these generating facilities are built will ultimately depend on how society weighs the advantages of nuclear power against its potential risks.

Hydropower

By the 1930s, hydroelectric plants tapped the power of moving water to generate roughly 40% of the nation's electricity. Hydroelectric facilities still generate about 12% of the electricity we use, by far the largest contribution of any of the renewable energy resources. Worldwide, about 15% of electric power is produced from hydropower. In many regions, hydroelectric power remains the cheapest, cleanest, and most flexible source of electricity. Although hydroelectric facilities do not emit potentially dangerous substances into air or water, they do affect the local environment. When rivers are dammed, thousands of acres upstream are submerged and the water flow downstream is altered. The dam becomes an obstacle to fish that must swim upstream to spawn. Other aquatic life may be affected by changes in water temperature and oxygen content. To ensure that hydropower remains a viable resource, utilities are developing methods to mitigate the effects, including ways to divert spawning fish populations around generating turbines.

Wind Power

Wind power is one of the most promising forms of renewable energy. Winds of sufficient velocity to supply extensive clusters of wind turbines, or wind farms, are found in many locations across the U.S., particularly the Great Plains and mountainous regions on the east and west coasts. Already, more than 17,000 wind turbines generate about 1,500 megawatts of electricity, mostly in California and Hawaii, where strong winds average 15 to 20 miles per hour. The electricity produced by all of these wind farms would meet the power needs of 330,000 households. Electricity currently produced by wind farms displaces the energy equivalent of 3.8 million barrels of oil and avoids about 1 million tons of carbon emissions. Using wind power instead of fossil fuels avoids other emissions, as well,

such as sulfur oxides, nitrogen oxides, and particulates.

Wind power is expected to contribute an increasing portion of the nation's power. Wind-generated electricity currently costs about 7¢ to 9¢ per kilowatt-hour. Technological advances will lower that cost to as little as 5¢ in some regions, making wind power competitive with coal-based power in areas where high winds exist. Wind turbines that operate efficiently at lower wind speeds will also expand the regions in which wind power can be generated cost-effectively.

Solar Power

Solar power represents another important renewable energy for electricity generation. Since no fuel is consumed, virtually no pollutants are emitted. Since the 1970s, the cost of generating electricity from the sun has dropped dramatically. Solar power still remains significantly more expensive than electricity generated from fossil fuels. The use of large arrays of solar modules to generate significant amounts of electricity, for example, is still five to six times more expensive than coal-based power. As photovoltaic technologies improve, however, the cost could drop to between 6¢ and 8¢ by 2010, making photovoltaic-generated power economically competitive. Already solar power is cost-effective when electricity is needed in remote areas.

Biomass

Biomass, or vegetation, could provide another enormous source of energy. Electric utilities are currently exploring the potential of dedicated biomass power plants, generating facilities that would literally "grow electricity" by harvesting trees specifically for electric power production. Estimates suggest that a 100-megawatt power plant can be fueled entirely by a plantation of trees extending in a six-mile radius around the plant—producing enough electricity for 100,000 people.

Transmission and Distribution

Once electricity is generated, it is delivered to customers by more than 4.8 million kilometers of transmission and distribution lines and over 300,000 kilometers of underground cables. The U.S. power delivery system represents the largest interconnection of electrical transmission, distribution, and control equipment in the world.

Fundamental changes in power delivery are underway. The widespread use of microprocessors has increased the demand for higher quality power. Engineers are employing a range of new technologies, including the application of versatile, high-power semiconductors, computer processors, advanced automation techniques, and diagnostic systems to improve power quality, cut losses, and enhance the ability of utilities to control the flow of electricity across the nation's power grid.

In general, electric power delivery poses few environmental concerns apart from the visual impact of transmission towers on the landscape. In recent years, concern has been raised over the potentially harmful health effects of the magnetic fields created by electric power lines and electric appliances and components. Some studies have suggested a weak link between magnetic field exposure and childhood leukemia. The nature and extent of any possible risk, however, and the mechanisms that may be at work, are still unknown. The Electric Power Research Institute (EPRI), which conducts research and development on behalf of many of the nation's electric utilities, is conducting far-ranging research into field effects in order to establish a scientific basis for understanding. Meanwhile, a variety of transmission and distribution line configurations are being designed that dramatically decrease magnetic fields without reducing the ability to carry current.

The Benefits of Electrification

Electricity has improved the quality of life in ways that are almost beyond measure — from lighting homes and offices to providing the power for life-saving medical technologies. Electric power will play an important role in feeding and clothing the world's burgeoning population. In many applications, the use of electricity in place of primary energy sources such as gas or oil can save primary energy, extending the contribution of limited natural resources. Equally important, electricity can be generated from a variety of renewable energy resources, making it an essential part of a sustainable energy future. Significant progress is being made in limiting the environmental impact of electricity generation and use and in improving the efficiency of electric applications. Even so, electricity use will continue to affect the environment in a variety of ways, making it necessary for society to continue to weigh the risks of electricity generation and use against its benefits.

BRENT BARKER

For Further Reading: Leslie Lamarre, "New Push for Energy Efficiency," *EPRI Journal* (April/May 1990); Leslie Lamarre, "A Growth Market in Wind Power," *EPRI Journal* (December 1992); Taylor Moore "In Search of a National Energy Strategy," *EPRI Journal* (January/February 1991); Taylor Moore, "High Hopes for High-Power Solar," *EPRI Journal* (December 1992).

See also Bioenergy; Clean Air Act; Coal; Electromagnetic Fields; Energy, Electric; Fuel, Fossil; Hydropower; Law, Energy; Nuclear Power; Petroleum; Pollution, Air; Solar and Other Renewable Energy Sources.

ELECTROMAGNETIC FIELDS

Electromagnetic fields, EMF, are the power-frequency (60-Hz) electric and/or magnetic fields produced by the transmission, distribution, and utilization of electric energy. This definition, however, is not entirely satisfactory from a scientific point of view. For example, at 60 Hz, electric and magnetic fields can be considered to be decoupled. They are two distinct physical agents with different properties, a notion not implied by the EMF terminology. Further, discussion of EMF exposure has considered only the strength of the EMF field without consideration of other characteristics of electromagnetic waves, such as spatial orientation.

This ambiguous terminology notwithstanding, the past two decades have seen mounting public, media, legislative, regulatory, and scientific concern that EMF exposure may have carcinogenic and other adverse health effects. This is largely a consequence of the more than ninety epidemiologic studies published over the past fifteen years that have examined possible associations between EMF exposure and various health endpoints, with increasing attention to possible associations between various cancers and exposures in both occupational and community settings. Occupational studies have examined workers in the electric utility industry and in other "electrical occupations" with presumed exposure to elevated EMF levels. Community studies have examined children and adults living in homes in which fields from nearby electric power facilities are the presumed risk factor.

Although high-voltage transmission lines are the most visually imposing source of EMF exposure, 60-Hz electric and magnetic fields are ubiquitous in the indoor environment. They are produced by a variety of sources in residences and workplaces, including low-voltage distribution lines, service drops, plumbing system-grounded neutrals, household appliances, computer and other office equipment, and a wide variety of industrial equipment (such as arc welders). The present exposure assessment and epidemiologic research focuses primarily on two areas: (1) background residential 60-Hz magnetic field exposure from sources outside the home (e.g., primary and secondary distribution lines, service drops, and grounding systems) and its association with childhood cancer; and (2) occupational exposures of electrical workers to 60-Hz electric and magnetic fields and their association with adult leukemia, brain tumors, and (possibly) male breast cancer.

At present, neither body of research is sufficiently persuasive to justify a causal interpretation. Most importantly in this regard, a coherent mechanism by which 60-Hz electric and magnetic fields can interact with biological tissue has not yet been elucidated and accepted by the bioelectromagnetic research community, nor has a comprehensive body of toxicologic research been conducted. Thus, although effects have been reported from both laboratory and epidemiologic research, there is an insufficient basis to determine whether 60-Hz electric and/or magnetic fields pose even a potential human health hazard, and, if so, the extent to which routine exposures actually constitute a public and/or occupational health risk.

ROBERT S. BANKS

ELECTROMAGNETIC SPECTRUM

The electromagnetic spectrum is the term used to refer to all waves caused by the acceleration of electric charge. This spectrum includes radio waves, visible light, and x rays.

Figure 1 shows a water wave traveling along the surface of a lake. Features which characterize the waveform are the amplitude, the maximum vertical displacement of a water droplet from the smooth surface of the lake, and the wavelength (the horizontal distance that separates two adjacent crests). The motion of the waveform through time is characterized by the wavespeed (the horizontal distance any crest moves in one second) and the frequency (the number of crests that pass a stationary observer in one second). Three of these quantities are related by a simple equation:

$$\text{wavespeed} = \text{wavelength} \times \text{frequency}$$

If an electric charge oscillates at a low frequency, the electromagnetic wave that spreads out in all directions from this source will have a large wavelength; charges that oscillate at a high frequency produce a wave with a small wavelength. What oscillates in the case of an electromagnetic wave is electric and magnetic fields. These fields are not tangible matter, as water is, but they carry energy, and they influence the behavior of the electric charges that are constituent of all matter.

The wavespeed of electromagnetic waves traveling through vaccum is 3×10^8 meters per second (186,000 miles per second). The transmitter at a radio station might include a circuit that causes electric charges to move up and down in the antenna 1,000,000 times per second (1 megahertz or 1000 kilohertz). This antenna would transmit radio waves with a wavelength of 300 meters, in the middle of the AM broadcast band. FM frequencies are much higher. The middle of the FM band is about 100 megahertz. The wavelength of a wave at this frequency is 3 meters (see Figure 2.)

AM and FM waves travel long distances through air without much loss of amplitude (and thus power). One limitation to the range of a transmitter is the fact that electromagnetic waves travel in straight lines; they do not follow the curvature of the Earth. However, radio waves can travel from a transmitter at one place on Earth to a receiver thousands of miles away provided that the wavelength is short enough. The frequency of such transmitters vary from 1.5×10^{10} hertz to 10^{12} hertz; the wavelengths are from 2 centimeters to 0.3 millimeters. Short radio waves travel in straight lines but they are reflected from a layer of ions (charged particles in the upper atmosphere) and return to the surface where they can be detected by shortwave radio receivers.

Radar and microwaves also employ wavelengths in the one-centimeter range, and are generated by electrical circuits that oscillate at a frequency near 3×10^{10} hertz. Storm clouds contain ions (charged particles) created by the rapid motion of the saturated

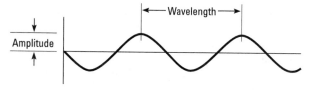

Figure 1. Water characteristics.

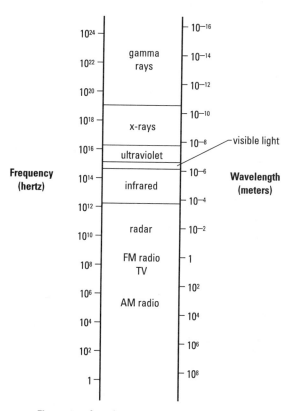

Figure 2. The electromagnetic spectrum.

clouds relative to the surrounding air, and these ions reflect radar waves just as the ionization layer in the stratosphere reflects short radio waves. The thin metal plate in the window of a microwave oven contains holes that are a few millimeters in diameter. This plate is opaque to microwaves, yet it allows visible light, which is an electromagnetic wave of shorter wavelength, to pass through the holes.

It is not possible to build electric circuits that produce electromagnetic waves with still shorter wavelengths. However, the electron oscillations in the atoms of an incandescent solid do radiate such waves. This region of the electromagnetic spectrum is called heat radiation or infrared rays. The frequencies of these rays range from 10^{12} to 3.9×10^{14} hertz, and the wavelengths 3×10^{-4} to 7.7×10^{-7} meter (about the thickness of a razor blade edge). Detectors of infrared radiation include all objects that absorb heat radiation, including human skin and solar collectors.

An atom has a discrete set of allowed energy levels. If a collection of atoms, e.g., a gas such as hydrogen, absorbs energy, thereby putting some atoms into higher energy levels (excited states), individual atoms will spontaneously return to the lowest possible level (the ground state). Each such transition is equivalent to a brief electrical oscillation, and collectively the atoms emit waves. Some of these oscillations correspond to frequencies between 3.9×10^{14} and 7.5×10^{14} hertz, with wavelengths 7.7×10^{-7} to 4×10^{-7} meter. This narrow range is detectable by the human eye and is called visible light. Chlorophyll (in green plants) and photographic film use visible light energy.

The same kind of atomic transitions that produce visible light produce ultraviolet radiation when the frequencies exceed 7.5×10^{14} hertz. The wavelength band for the ultraviolet ranges from 4×10^{-7} to 2×10^{-8} meter. Ultraviolet light can cause damage to human cells, and thus skin cancer. Cancer would be even more prevalent on Earth if it were not for ozone in the upper atmosphere that absorbs much of the ultraviolet radiation in sunlight.

Any acceleration of a charge may produce some radiation. Electrons that are moving rapidly as a result of acceleration through a large voltage will emit a burst of x rays when they are suddenly stopped by a metal target. The wavelengths of such x rays range between 2×10^{-8} and 4×10^{-11} meter. X rays are very penetrating, though they are absorbed more by dense matter, such as bone, than by soft tissue, which is mostly water. The fact that x rays fog photographic film was strongly involved in their discovery, and it is still the most common method of detection.

The energy that holds protons and neutrons together in a nucleus is about a million times greater than the energy that binds electrons in an atom. When nuclei decay from an excited energy state to a lower state, this transition corresponds to a charged particle oscillation with a frequency a million times greater than atomic transition frequencies, about 5×10^{20} hertz. The wavelength is 6×10^{-13} meter. Radiation in this region of the electromagnetic spectrum is called gamma radiation. Gamma rays that emerge from fundamental particle interactions can have energies much larger, and wavelengths even smaller.

Although it is useful to describe electromagnetic radiation in the language of waves, it often manifests itself as localized quanta of energy called photons. At long wavelengths, the wave character of the radiation dominates while at very short wavelengths the particle characteristics cannot be ignored. The accepted model for electromagnetic radiation views it as having a dual wave-particle nature.

The boundaries in Figure 2 that separate radiowaves, infrared, visible light, ultraviolet, x rays, and gamma rays from one another are not rigid. Near these boundaries the names used should be dictated by how the radiation is produced, not by its frequency or wavelength. What is truly remarkable is not that the various kinds of electromagnetic radiation are so different in the ways they are produced and detected, nor even the dual nature of the radiation, but rather that many different phenomena can be described in terms of a single model involving only a small number of parameters.

ARNOLD STRASSENBURG

For Further Reading: Gordon L. Hester, "Public Policy Implications of Possible Adverse Health Effects from Electric and Magnetic Fields," in *Management of Hazardous Agents*, D. G. LeVine and A. C. Upton, eds., 1992.
See also Atmosphere; Energy; Greenhouse Effect; Ionizing Radiation; Laser; Microwaves; Ozone Layer; Photosynthesis; Radiation; Ultraviolet Radiation.

EL NIÑO

El Niño is a large ocean current of warm water off the coast of Peru and Ecuador, flowing southwards against the coastal and offshore currents. The easterly tradewinds cause the warm surface water to move away from the continent, resulting in an upwelling of nutrient-rich cold water. The current is called "El Niño," which is Spanish for "the boy." The term refers to the Christ Child, because the effects of a pronounced El Niño current are usually experienced just before Christmas. Episodes of intensification of El Niño—when the easterly winds weaken—occur at irregular intervals, although usually twice every decade. The episodes last for about twelve months. Due to their connection to intensifications of the El Niño current, changes in Pacific Ocean currents and global atmospheric conditions are now referred to as El Niño.

During an intensification of the current, the warm waters of El Niño spread, preventing deep nutrient-rich cold water from rising, and causing it to occupy a smaller area on the ocean surface. During these episodes, marine and bird populations dependent on the cold water nutrients disappear almost completely. Before 1972 Peru supplied 38% of the world's fishmeal, which is used as animal feed. In 1972, a severe El Niño year, Peru's harvest of anchovies, a main ingredient in fishmeal, decreased to almost nothing. As sea temperatures rise, atmospheric pressure above the eastern tropical Pacific drops, causing large amounts of rain to fall along the normally arid Peruvian coast. El Niño is an important element in the global climate system. The warm sea surface temperatures associated with El Niño are linked to large-scale changes in the distribution of atmospheric pressure over the Pacific and Indian ocean region, and are often associated with weather anomalies around the world.

During the severe 1982 to 1983 episode of El Niño, extreme weather changes occurred around the globe. Indonesia experienced its worst drought in a decade, causing a sharp decline in rice production and the advent of fires, and taking a heavy toll on the nation's rainforests. Drought also struck Australia, South Africa, Ethiopia, and agricultural regions of the former Soviet Union. Due to the failure of the annual monsoon rains, rice production on the Indian subcontinent fell significantly. Floods caused extensive damage to Kenyan agriculture. Uncharacteristic snow in the spring of 1983 caused damage in Armenia. The same year, hurricanes and cyclones devastated Hawaii and Tahiti, while the midwestern United States experienced a cycle of severe flooding followed by drought, causing $3 billion in damage. Flooding on the west coast in California caused $200 million in agricultural damage.

Scientists are beginning to understand the association between these events and El Niño episodes. Data shows that 93% of the droughts in Indonesia have coincided with El Niño intensifications. The rare failure of the monsoons in India also occurred at approximately the same time as the El Niño phenomenon off the South American coast. In 1982—the year before the severe El Niño episode of 1982 to 1983—a severe drought in West Africa began, leading to the death of 100,000 people and millions of animals. Scientists are now recognizing the interconnectedness of global climate patterns. Associated events, such as the onset of tropical cyclones in Australia and rain in southeast Africa, are thought to precede the start of a pronounced El Niño episode.

ELTON, CHARLES S.

(1900–1991), British zoologist and biologist. Charles Sutherland Elton was a pioneer in animal ecology and a major figure in the establishment of the Nature Conservancy in England in 1948. His first book, *Animal Ecology*, published in 1927 when he was only 26 years old, gave rise to a whole new discipline. His later studies of animal habitats in Wytham Woods near Oxford, England, following World War II, were published in 1966 in a second landmark book, *Pattern of Animal Communities*. This laid the foundations for ecological research that had impact throughout the world, for it led to definitions of ecosystems and, particularly, understanding of distinctions between renewable and non-renewable resources. When Elton died May 1, 1991, The Times of London said he had been "the prime stimulus to young biologists all over the world to put animal ecology on the map."

Indeed, Elton was better known outside Britain than in his home country. He had so many students and disciples in the United States that many in Britain thought he was an American. He thus shared relative anonymity at home and renown abroad with his long-time friend and colleague, Frank Fraser Darling, with whom his work is often paired as well as compared.

Sometimes described as a "population biologist," Elton is credited with illuminating the interplay of forces in animal life, the role of food chains, the growth and waning of animal populations, and the selection of habitats. He first proposed the notion of an ecological pyramid, which scientists refer to as the Eltonian pyramid, in which a large number of plants support a smaller number of herbivores, which support a still smaller number of carnivores. He identified three kinds of ecological pyramids: a pyramid of numbers, i.e., population; a pyramid of biomass, usually expressed in size; and a pyramid of energy, usually expressed in activity, including aggressiveness.

Elton's findings were credited with giving new force to game conservation practices. For example, Aldo Leopold's influential textbook on game management in 1933 drew heavily on Elton's ecological studies. Elton's work also bridged research in botany and zoology in a way that produced ecological information of use in interpreting the human environment. His trip to Lapland in 1930 led to a book, *Moles, Mice and Lemmings*, in 1942, on the population dynamics of rodents, lagomorphs, and their predators. This prepared him to direct a successful rodent control program during World War II in London and other bombed cities.

Born and educated in Liverpool, where his father was a professor of English, Elton graduated from New College, Oxford, in 1922 with first class honors in zoology. By then he had already had experience as a student assistant to the naturalist Julian Huxley on a 1922 Oxford University expedition to Spitzbergen in the Arctic Circle. Encouraged to carry out an ecological survey on that trip, Elton followed it up on two additional Arctic trips in 1923 and 1924 before publishing *Animal Ecology* in 1927. A third trip to the Arctic, in Lapland, in 1930, was the basis of his second great work, *Animal Ecology and Evolution*, and his work on rodents.

More of a naturalist than most of his contemporaries, Elton rejected traditional academic research in the laboratory. He set out in his early Arctic surveys to devise a new field of animal ecological studies, but he did not claim discovery. "Ecology is a new name for a very old subject," he said. "It simply means scientific natural history." Elton's paramount contribution was not so much his description and tabulations of plants (what grew, where they grew, when) and of animals (what they ate, where they found food, their excretions, even the contents of stomach and intestine) although these were crucial to the science. He saw these related to the human fate but, he also made clear, so was almost everything else so related.

"Ecology is a branch of zoology which is perhaps more able to offer immediate practical help to mankind than any of the others," the young man wrote in his pathfinding classic. But he also wrote, "it might be worthwhile getting to know a little about geology, the movements of the moon or of a dog's tail, or the psychology of starlings, or any of those apparently specialized or remote subjects which are always turning out to be at the basis of ecological problems encountered in the field."

Nearly forty years later, in a new introduction to this durable book, he remarked that "population analysis has become more sophisticated with the use of models and computers." He did not oppose that, but he was more comfortable with hands-on field studies, often based chiefly on good observation and simple arithmetic. In his work in Wytham Woods Elton had catalogued the relationships of some 4,000 species of animals.

Elton shunned public gatherings, including professional meetings. He once refused to read his presidential address to the British Ecological Society, preferring to sit in the audience while a colleague read it from the podium.

JACK RAYMOND

For Further Reading: Anne Chisholm, *Philosophers of the Earth* (1971); Charles S. Elton, *Animal Ecology*, with new Introduction (1967, orig. published 1927); Charles S. Elton, *Pattern of Animal Communities* (1966).

EMERSON, RALPH WALDO

(1803–1882), American preacher and lecturer, poet and essayist, philosopher of religion and nature. Emerson was so eminent in each of his endeavors that even his admirers had trouble classifying him. His friend Henry Wadsworth Longfellow called him the Buddha of the West; others, the American Plato. Robert Frost said Emerson was one of the three greatest Americans, along with Jefferson and Lincoln. Emerson called himself a poet and naturalist, not to identify his profession but because, he said, poets were naturally selected to interpret reason.

Emerson began his career as a Unitarian minister but broke with the church ritual regarding "the Lord's supper" and resigned his ministry for a career as lecturer and writer. Emerson offered an optimistic philosophy of life based on an exaltation of the harmony of man in nature. His most ardent contemporary disciple was Henry David Thoreau, who lived on Emerson's property in the woods near Concord, Massachusetts, when he wrote *Walden*, a work that hails man's place in nature. Emerson was a propounder of transcendentalism, a movement that taught the revelation of truth through emotion and reason, rather than observation and experience. He inspired and supported the Brook Farm community, a socialist-like undertaking by prominent writers and editors. It failed in five years.

Emerson was descended from several generations of New England clergymen. His father was pastor of the First Unitarian Church in Boston. His mother, the daughter of a distiller, was also very devout. Illness wracked the family; of eight children, only two lived to maturity. Ralph Waldo himself struggled with tuberculosis and suffered bouts of near blindness all his life.

At Harvard College Emerson began his life-long journal, which, when ultimately published, was declared "the most remarkable record of the 'march of the mind' to appear in the U.S." Beginning to doubt church teachings, he raised money for a trip to Europe and sought counsel from prominent figures including Goethe. When he returned, he received a license to preach, largely on the reputation of his clerical forebears, and ultimately graduated from Harvard Divinity School and was ordained. He was 26 when he married Ellen Tucker. In little over a year she was dead. Riven by doubts, he concluded that "in order to be a good minister, it is necessary to leave the ministry."

Once again he sailed to Europe and sought out leading thinkers, especially in England where he met Coleridge, Carlyle, and Wordsworth. In 1833—his wife's estate having given him financial security—he acquired a house where he presided as "the sage of Concord" for 50 years until he died. He was married a second time in 1835 to Lydia Jackson, with whom he had three children. Their first-born, a son, died at age 6.

Yet Emerson displayed wit and bravura in a veritable Niagara of lectures and publications, among them *Nature* in 1836 which was called the bible of American transcendentalism; two volumes of essays (1841 and 1844); two addresses at Harvard Divinity School at Harvard, the first of which was published as *The American Scholar* and called for America's cultural independence from Europe (both lectures in 1837); two volumes of poetry in 1847 and 1867 in which such poems as "The Rhodora," "The Humble-Bee," and "Woodnotes" celebrated nature; and *Representative Men* (1849) in which he cited "great men" through the ages who exemplified the greatness in everyone.

Emerson enjoyed lecturing. He even served as his own agent, contracting for a hall and keeping the proceeds. Subjects as well as venues ranged far and wide but were unified by themes of optimism, individualism, and a mystic sense of God's design of man in nature. God is everywhere, now, he taught. Historically based Christianity is too obsessed with the biblical era. Man is not redeemed by church ritual but by his comprehension of his affinity with nature. The church made too much of Christ's miracles; man's life itself was a miracle.

Emerson paired abstract ideas with clever rhymes that might have served as song lyrics. Young people quoted him: "Whoso would be a man must be a nonconformist." In one of his lectures at Harvard his assertion that redemption could be found only in one's own soul was taken to be atheistic and he was not invited again for 28 years—then honored with an LL.D. Long before modern psychiatrists identified the creative psyche, Emerson identified the "over-soul," within which every person's particular being is contained.

In his later years, as the Civil War loomed and played out its tragedy, he was found by some to be inadequately supportive of the abolitionist cause; yet he excoriated Daniel Webster for supporting compromises, ostentatiously attended John Brown's lectures, and urged his fellow townsmen to disobey the Fugitive Slave Act. After the Civil War, Emerson reduced his travel but not his output. He contributed regularly to *Atlantic Monthly.*

He was almost Shakespearean in his vast production of enduring epigrams: *Consistency is the hobgoblin of little minds. Every man is a quotation from all his ancestors. No great men are original. Beauty is its own excuse for being. Cut these words and they would bleed. This is the drop which balances the sea. To be great is to be misunderstood.* And, of course, the climactic line in "Concord Hymn," commemorating the first battle of the American Revolution:

> By the rude bridge that arched the flood,
>   Their flag to April's breeze unfurled,
> Here once the embattled farmers stood
>   And *fired the shot heard round the world.*

Emerson's shots, too, were heard round the world and echo to this day.

JACK RAYMOND

For Further Reading: Ralph Waldo Emerson, *The American Scholar* and *English Traits* (1909); Oliver Wendell Holmes, *Ralph Waldo Emerson* (1885, reprinted 1980); Ralph L. Rusk, *The Life of Ralph Waldo Emerson* (1949, reissued 1964).

EMF

See Energy, Electric.

EMISSIONS STANDARDS

See Clean Air Act; Clean Water Act; Fuel Economy Standards; Quality Standards; South Coast Air Quality Management District.

ENDANGERED SPECIES ACT

The Endangered Species Act was enacted in 1973 to protect animal and plant species threatened by extinction. Amended in 1978, 1982, and 1988, this act provided that any endangered or threatened (likely to become endangered) species or subspecies be listed with the Secretary of the Interior, or if a marine species, with the Secretary of Commerce. The Fish and Wildlife Service and the National Marine Fisheries Service were then to develop regulations to protect those species listed as endangered or threatened.

The act required that all federal agencies assure that their actions do not jeopardize the continued existence of an endangered or threatened species or result in the destruction or modification of the critical habitat of such species. The Secretary of the Interior, in consultation with the affected state, was to determine what habitat was critical to a species' existence. The act also prohibited the killing, capturing, importing, exporting, or selling of any species listed as endangered or threatened.

A major stipulation which contributed to the strength of the act was its provision for enforcement on the public level. Citizen law suits could be filed to enjoin a federal project if that project did not provide for the protection of endangered or threatened species. This provision for citizens' suits was a powerful weapon for the defense of endangered species and their habitats since so many environmentally altering activities, such as recreational development, dam building, logging, and mining, either take place on federal lands or are subject to federal financing or federal regulations. The case of the Snail Darter vs. Tellico Dam was one of the first widely known cases to invoke the Endangered Species Act to protect a species and its habitat. The 1978 amendments to the act somewhat limited this powerful piece of legislation. It created the Endangered Species Committee that was given the authority to grant exemptions to the act when no reasonable alternative could be found, when the proposed project was of national and regional significance, or when the benefits of the project clearly outweighed those of any alternative action. This committee became known as the "God Squad" because of its authority to make decisions which could give life to, or eliminate, a species. The amendments also required that before a species could be listed, the boundaries of that species' critical habitat must be delineated, an economic impact statement be made, and public hearings on the issue be held.

The 1982 amendments strengthened the act by allowing a species to be listed without any consideration of the economic ramifications. In 1988 amendments were passed which allowed for emergency listing procedures for those species in urgent need of protection. It was also mandated that a species be monitored for three years after it had been de-listed so that its recovery was assured. The 1988 amendments allocated funding for the act for five more years. In 1992 Congress raised concerns regarding the social and economic impacts of preservation actions during discussions to reauthorize the act. In addition recommendations included the need to enhance the act's capacity to foresee the need to conserve and protect species by fostering partnerships with state and local governments, industry, individuals, and landowners which would include "active management programs, private species enhancement programs, and efforts by the federal government to protect habitats containing more than one species," according to Representative W. J. Tauzin of Louisiana.

ENERGY

One of the earliest and most distinguishing traits of human beings was their ability to harness energy for their own use through the control of fire and the labor of themselves and other animals. Some early machines were powered by flowing water, after people learned that *potential energy* stored in an elevated lake or reservoir could be converted into *kinetic energy* that could be extracted from the downhill flow.

Before the European industrial revolution in the 1700s, the world's energy came mostly from hydropower and wood. Windmills pumped water and waterwheels were used as mechanical drives. In the 1700s, knowledge about steam power increased, as inventors found that boiling water in a confined system was a way to raise pressure—which could then be used to do work, thus converting *thermal energy* to *mechanical energy*. In 1784, a steam engine invented by James Watt was first operated in a mine in the United Kingdom, with subsequent developments leading to much greater industrial growth and use of steam power. Even though coal was then used as a fuel, it was not until the mid-19th century that the coal-based Bessemer process for making steel and the growth in the railroad network motivated much larger-scale mining and use of coal. In the latter half of the century, Edison's invention of practical electric lighting created a demand for electricity. By the start of the 20th century, electric power distribution networks were expanding beyond local areas eventually to form national grids, fed primarily by coal-fired power plants with some hydropower.

Although petroleum was discovered in the mid-19th century, its use grew slowly until the introduction of the automobile in the early 20th century. Crude petroleum is a mixture of hydrocarbon compounds, ranging from heavy tars and asphalts to associated lighter gases such as butane/propane (LPG—liquefied petroleum gases), ethane (an important chemical feedstock), and methane (natural gas). Refining technology allows the separation of the crude oils (which vary widely in composition depending on the source) into a variety of products, including industrial fuels, gasoline, diesel fuel, heating oils, and raw materials for manufacturing chemicals. Many oil fields contain associated gas, which initially was burned in a flare as a waste product. However, discovery of gas fields and applications that needed a "cleaner" gaseous fuel encouraged major expansion of natural gas energy and development of a national natural gas pipeline network in the 1950s.

The technical feasibility of generating energy from nuclear fission power was demonstrated with the operation of the first atomic pile in 1942. Consumers, conscious of finite global resources of coal, oil, and gas, hailed nuclear energy as a major energy source for the future. Uranium-fueled fission reactors were designed and built for electric power generation in the 1950s and 1960s. Other non-fossil energy sources were also studied with particular interest in harnessing solar, geothermal, wind, and wave energy. Solar energy was captured and stored for heating; also discovered were photovoltaic materials that transform solar energy to electricity. In the past decades, improvements in solar conversion efficiencies, geothermal exploration and reservoir stimulation, and the development of more sophisticated wind machines encouraged wider use of these non-fossil resources.

Worldwide annual consumption of energy during the 20th century increased 17-fold, from about 20 quads in 1900 to over 340 quads in 1992. (There are many equivalent units for energy. For global-scale energy quantities, *quads* or *exajoules* are frequently used. Conversion factors include: 1 quad = 10^{15} Btu = 0.95 × 10^{18} joules = 0.95 exajoule.) Part of this essentially exponential growth was driven by a 3.5-fold increase in world population from 1.5 to 5.3 billion over this period. On average, the annual per capita energy consumption worldwide has increased more than 4-fold, from about 13 to 60 quads per billion people (or the equivalent of from about 480 to 2200 kg coal consumed per person per year). The U.S. per capita con-

sumption is now about six times the world average; it is approximately double the consumption rate in other developed countries in Europe and Japan. The higher rate in the U.S. does not indicate poor energy efficiency, *per se*, but is fundamentally linked to the U.S.'s larger land area and longer transportation distances, higher heating and cooling requirements due to a more severe climate, a higher standard of living, and larger average size of dwellings.

In 1990 about 40% of U.S. energy was supplied from oil, 24% from natural gas, 24% from coal, 7% from nuclear, and 4% from hydroelectric and other renewable power. The worldwide usage was 39% from oil, 27% from coal, 22% from natural gas, 7% from renewables (mostly hydroelectric), and 6% from nuclear. In the U.S., the energy is used in roughly equal proportions in the transportation, buildings, and industrial sectors.

For the first two thirds of this century, the environmental effects associated with energy supply and consumption, employing primarily hydrocarbon and biomass combustion processes, have been of minor concern. However, as we enter the 21st century, the impact of energy generation and use on the environment has become a major factor both in influencing the development of new technologies and in formulating strategies to meet future energy needs.

In developing countries, new industries that produce primary materials and consumer goods require significant increases in energy production and use. Often coal is the least expensive and most available fuel for development. Simplest and cheapest utilization technologies often are inefficient and highly polluting. More efficient technologies with emissions controls add costs; alternative fuels are usually more expensive. With the world population projected to reach an estimated equilibrium level of 10 to 15 billion people in the 21st century, there is a mismatch between the energy technologies now in use and those technologies that will be necessary to meet the world's needs for economic development and sound management of environmental resources. These new technologies will be optimized to minimize polluting wastes and to use energy efficiently. Energy conservation is becoming increasingly important and will be motivated by increasing energy prices.

Predicting our energy future and developing strategies for managing it involve estimating demand as well as how energy will be supplied, stored, converted, and used. The majority of the environmental problems that we face today on a local, regional, or global scale are fundamentally linked to worldwide energy supply and its use. These problems are aggravated by our strong reliance on fossil fuels and combustion process-

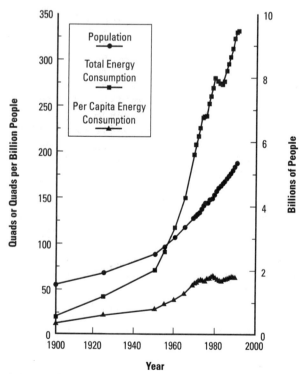

Figure 1. Historical data and projections of annual worldwide primary energy consumption (total and per capita) and worldwide population. Source: Tester et al., 1991.

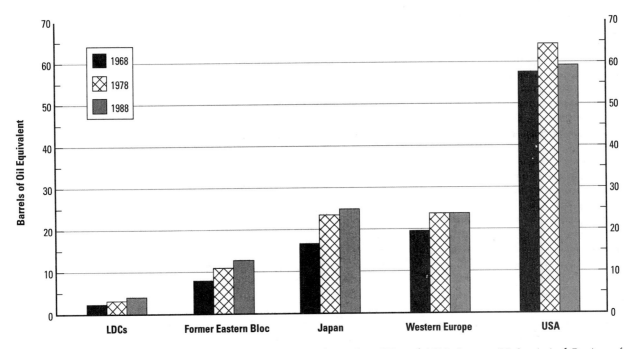

Figure 2. Annual per capita energy consumption by country for 1968, 1978, and 1988. Source: *BP Statistical Review of World Energy*, June 1989.

es as our primary energy sources. As we move into the 21st century, we will face the need to develop new energy sources to supplement and eventually replace fossil fuel use. At present use rates, we have proven reserves of gas and oil to last about 40 years, with unproven resources that might extend that time to 100 years or more. Proven coal reserves are projected to last about 400 years, with unproven resources likely to last several thousand years. However, each year we are using an amount of fossil fuel that required about a million years to produce, so these resources will eventually be depleted. In addition, if the increases in CO_2 in the atmosphere due to fossil fuel use are found to be an imminent threat to climate, or if other emissions (e.g., SO_x, NO_x, particulates, or volatile hydrocarbons) are shown to adversely affect human health or the world's ecological balance to a higher degree than currently suspected, then even more emphasis will be needed on energy conservation and efficiency, as well as on fuel switching.

Non-fossil fuel energy options available for the next century all have challenges along with potential benefits. Major options include the following:

Nuclear fission energy: Longer-term, large-scale fuel supply without greenhouse gas emissions, but currently more expensive than fossil fuel, with major public concerns about operational safety and disposal of radioactive wastes. Advanced reactor designs suggest the potential for designing safer, "passively stable" power units. Nuclear waste reprocessing technology exists, but transport of wastes is required. Technology for long-term waste stabilization exists, but the siting and acceptance of secure repositories for storage over tens of thousands of years is difficult. Misuse of fuels for weapons manufacture remains a concern. Strong public safety concerns in many countries about nuclear energy will have to be addressed before nuclear power will be allowed to become a larger portion of world energy supply.

Solar energy: Viewed by many as the most environmentally sustainable renewable energy option, but currently more expensive than nuclear. Subject to diurnal and seasonal cycles, as well as variation with latitude. The capture of low-incident solar flux requires considerable collector area. On a large scale, energy storage to even out energy or power production cycles is too costly to be practical. Research is likely to increase efficiency and reduce capture area requirements and cost. Solar energy will be an important supplemental source of power in some regions, but it is not widely enough available to become a major continuous portion of our energy supply.

Hydropower: A clean source of energy, with minimal emissions, available in many parts of the world. Limited availability; development impacts environment through significant land use requirements associated with damming of rivers and back-flooding to create reservoirs. Much of the accessible, large-scale hydropower has already been developed.

Wind power: Intermittent with limited availability, but useful for distributed utility supply systems. New wind machines are much more efficient but do require considerable land area and generate noise. This source is likely to be used where available, but not as a major future source of new energy.

Geothermal energy: Hydrothermal energy is available in a few areas such as northern California, Italy, Ireland, the Philippines, New Zealand, and other places where natural steam or hot water is trapped in shallow reservoirs. Potential for recovery of methane and thermal energy exists in geopressured basins such as those in the Gulf Coast region of Texas and Louisiana. Environmental impacts include the need to control emissions of hydrogen sulfide, ammonia, and other toxic trace gases. **Hot dry rock (HDR) geothermal energy** is a potentially large future energy source. Thermal energy deep below the earth's surface could be tapped through deep wells using water as a heat transfer fluid. Although this energy resource is large and environmental impacts appear low with minimal emissions, barriers include the high costs of deep drilling and challenges in managing the deep geological heat reservoirs. Current projections suggest that only those HDR resources with hot rock at reasonable depths (such as those in the western U.S. where rock at 250°C or higher is reached at 4- to 6- kilometer depths) would be economically viable. New, lower cost drilling methods and reservoir stimulation techniques could lead to long-term economic development of lower-grade HDR areas, providing a major worldwide energy source.

Fusion power: Power from nuclear fusion offers promise for large scale energy generation in the future, but is a long way from practical demonstration. Development of much higher temperature materials and containment technology will be necessary. Major investments will be required and commercialization will probably take many decades. Major environmental issues have yet to be defined, but will probably include decontamination of neutron-irradiated materials and some need for waste disposal.

Biomass: Wood and crop wastes were among our earliest energy sources. Today, interest in biomass fuels is again increasing. Biomass is considered a renewable fuel, since CO_2 generated by its combustion has already been taken from the atmosphere during its growth. Growing biomass for fuel can be considered as a way of converting solar energy to storable energy. Problems include cost, large land use and transportation requirements, low energy conversion efficiencies, emissions from combustion, and requirements for water, fertilizer, and sun. In the far future, genetic engineering may create organisms that directly produce specific fuels. Research on artificial photosynthesis may also lead to technologies that generate fuels more efficiently than natural processes involving chlorophyll.

Conservation and **increased energy efficiency** will be major factors in finding future solutions to our energy supply and consumption needs. More efficient use of energy makes the saved energy available for other uses. Some promising areas for improving energy use efficiency are:

In the transportation sector: advanced high-energy automobile and aircraft engines, continuously variable transmissions, levitated high-speed trains, improved scheduling and control, smaller lighter vehicles, fuel cell/electric powered cars, and aerodynamic vehicle design.

In the buildings sector: improved building insulation, window insulation, heat pumps, advanced space-conditioning control systems, high-efficiency lighting, improved refrigeration systems, air-to-air heat exchangers, innovative furnace and water heater designs, and improved cooking appliances.

In materials production: improved durability, materials substitution (non-Portland-based cements, ceramics, composites, superconductors), advanced recycling technology (for paper, plastics, aluminum and steel, and other primary metals).

In industrial processing: heat and work integration, cogeneration, re-engineering processes for energy efficiency, advanced separation methods, variable speed drives and motors, and advanced automated process control.

To fully exploit these opportunities, scientists and engineers must work proactively with social scientists, environmentalists, and policymakers to help understand and implement a mix of environmentally sound, economic energy technologies to fuel our future.

JEFFERSON W. TESTER
ELISABETH M. DRAKE

For Further Reading: Michael Brower, *Cool Energy* (1992); John H. Gibbons and William U. Chandler, *Energy: The Conservation Revolution* (1981); Jefferson W. Tester, David O. Wood, and Nancy A. Ferrari (eds.), *Energy and the Environment in the 21st Century* (1991).

See also Coal; Cogeneration of Electricity and Heat; Fuel, Fossil; Natural Gas; Nuclear Power, History and Technology; Petroleum; Population; Solar Cell; Transportation.

ENERGY, CHEMICAL

Chemical energy is the driving force behind everyday facets of life, from combustion of wood and fossil fuel to deriving electricity from batteries to biological metabolism. Any time a chemical reaction occurs, energy is either absorbed or released.

Chemical energy processes are governed by the laws of thermodynamics. The first, most basic, law of thermodynamics states that the total energy of a system and its environment must remain constant. Energy changes in a chemical reaction are distributed among the newly formed products or in their temperature and pressure. The solar-carbon cycle is a good example of the storage and release of energy via chemicals. In this cycle, green plants store solar energy in carbohydrate molecules by photosynthesis of carbon dioxide and water. In this way energy from the sun is stored and passed through the food chain, powering life on Earth.

In more detail, the process of photosynthesis takes carbon dioxide and water molecules, both simple, low-energy structures, and converts them into a more complex, high-energy molecule of glucose. The biochemical processes involve an intricate system that energizes chlorophyll electrons with sunlight and incorporates their energy into the glucose end product. That energy allows the characteristic ringlike structure of glucose to be assembled.

Metabolism of glucose then takes it apart step by step, de-energizing the electrons and using their energy for the functions of life. Photosynthesis and metabolism, although they differ in the specifics of the reactions, are essentially mirror images, with one undoing what the other has done. Together they make a process for the controlled, efficient storage and retrieval of chemical energy.

Biological metabolism is not the only way to retrieve chemical energy stored through photosynthesis. Consider the example of the carbon cycle. Any process that converted glucose to water and carbon dioxide would liberate the same amount of energy, if the process were equally efficient. A wood fire, for example, converts the complex carbohydrates of a log into essentially those components.

Similar to burning wood, the combustion of fossil fuels is the release of chemical energy in storage. The photosynthetic products of ages past are being used to power modern society. The energy they provide is solar energy, stored for millenniums and concentrated through geological activity. Over millions of years, geothermal temperatures and pressures altered the organic remains into what we now burn as oil and coal.

Released chemical energy can be stored for future use. Batteries are structures that store energy in chemical form, taking electrical energy (originally from fossil fuels) and using it to convert other chemicals into higher-energy states for later retrieval. Every time the energy is converted from one form to another some is lost in the reaction, usually showing as heat.

The energy that powers life and the energy that powers society are essentially from the same source—the sun—but are retrieved in different ways. People can readily retrieve organic chemical energy via combustion and can store electrical energy. The finer and incredibly efficient methods of controlled energy release—conducted every moment by every living thing—remain beyond human mastery.

ENERGY, ELECTRIC

Humans produce electric energy by transforming other forms of energy. The radiant heat of the sun, for example, captured by ancient plant life and preserved in the chemical bonds in coal, can be transformed into heat energy through burning. That heat can be used to produce steam, which in turn moves turbine blades, which then rotate an electric generator. The generator changes the mechanical energy in the turbine to electric energy. Another source is the potential energy of water stored behind a dam, which can also be changed to kinetic energy—the energy of an object in motion—when the water spills over the top of the dam. Kinetic energy can in turn be converted to electric energy when it spins turbines that generate electric power.

Electricity can also be generated from the energy contained in the atom. Atoms are composed of protons and neutrons surrounded by orbiting electrons. Some atoms can be split, a process called fission, which releases energy that can be converted to heat. The heat is used to produce steam, which then turns turbines to produce electricity.

Like water contained by a dam, charged particles possess energy, which is called **voltage.** When positive and negative ions are brought together, the voltage produces **electric current.** In an automobile battery, for instance, positive charges are gathered at one electrode, negative charges at the other. When a conductor connects these two electrodes, negatively charged electrons migrate toward the positive ones, forming a circuit. The flow of electrons represents an electric current.

Both energy and electricity can be measured in a variety of ways. The amount of energy consumed, which is also referred to as energy demand, is typically measured in British thermal units, or **Btu.** One Btu is the amount of heat energy required to raise the temperature of one pound of water by one degree Fahrenheit. Btu is a useful measure for describing how much energy is used on a relatively small scale. But the energy consumed by the nation and the world is so large that engineers use an informal measure called a **quad,** which equals one quadrillion, or 10^{15} Btu. A quad is the equivalent of 172 million barrels of oil or 40 million tons of coal. Each year, the U.S. consumes approximately 80 quads of energy.

Electricity demand is measured in **watts.** A 60-watt bulb uses 60 watts of electric power. To describe the amount of electricity consumed over time, engineers use the watt-hour. One watt-hour is the use of one watt over one hour. Because the watt is a very small unit, most measurements take the form of kilowatts and kilowatt-hours. The average home in the U.S. uses about 8,900 kilowatt-hours of electric power each year. The U.S. as a whole consumes an enormous amount of electricity—more than 2.8 billion kilowatt-hours annually.

From the beginning of the electric age more than a century ago, the amount of electricity consumed around the world has grown dramatically and continues to grow. Over the past two decades alone, electricity utility generation has increased 86%. From 1970 to 1990, average annual kilowatt-hour use per residential customer increased 28.5%. Electricity's share of total energy used has also steadily grown—from 28% in 1974 to more than 36% in 1990. By 2010, electricity is expected to account for about 45% of all energy use in the U.S.

Unique Attributes of Electricity

Electricity is a refined energy source, generated for the most part from primary energy sources such as coal, nuclear fuel, natural gas, and oil. Water power, wind, geothermal, biomass, and solar power also serve as sources of energy to generate small amounts of electricity. Because considerable energy is required to convert a primary fuel such as oil into electricity, one unit of primary energy from coal, oil, or gas yields only about one-third unit of electric energy from today's typical power plant. In contrast, oil may lose only about 16% of its primary energy content through refining and delivery of liquid products to customers. As a result, primary energy sources deliver more of their original energy at the point of use.

At first glance, that would seem to make them superior to electricity. But electric power provides unique advantages. Electricity is an orderly energy form, in contrast to thermal energy, which is random. Because it has no inertia, electric energy can be instantly varied to meet changing energy requirements. Electric power can be concentrated and controlled far more precisely than other forms of energy. Lasers and electron beams, for instance, can concentrate energy a million times more intense than an oxyacetylene

torch onto a surface area no bigger than 0.001 square inch. The precise controllability of electricity makes it possible to open and close the circuit gates of a computer microsecond-by-microsecond. In industrial processes, electricity can achieve much higher temperatures than combustion processes fueled by oil and gas. The practical limit for fossil fuels is about 3,000° F. In contrast, arc-produced plasmas powered by electricity routinely achieve temperatures of 10,000° F. Unlike fossil fuels, which heat by radiation or convection, electricity also can be used to heat within the material itself. Using microwave or radio-frequency radiation instead of surface heating methods to dry moist materials such as paper or textiles, for example, can save significant amounts of energy by cutting drying time and improving product yield.

In general, electrical processes and applications involve three distinct electrical phenomena. **Electromotive effects** occur when a current flows in a conductor located next to a magnetic field, producing a force in the conductor. The most common example of electromotive effects is the use of electric current to turn a motor. Electric motors convert electric energy to mechanical energy at extremely high efficiencies, often exceeding 90%. **Electrothermal effects** occur when electricity is used to produce heat, which in turn brings about a desired physical or chemical change. The simplest method involves attaching electrodes to a material and sending a current through it—the way the heating element in an electric stove produces heat. Materials that do not conduct electricity can be heated dielectrically, a process that occurs when a material containing polar molecules (water, for instance) is placed in a rapidly alternating electrical field. The most common example is a microwave oven. Finally, electricity applied to materials can produce specific chemical changes on a molecular level, called **electrolytic effects.** One of the most important uses of electrolysis is in the production of basic materials such as aluminum and chlorine from natural resources such as ores or brines.

In some applications, electrical phenomena can combine to produce synergistic effects that add to its value. In aluminum processing plants, for instance, electrothermal effects keep the cryolite bath in a molten state while electrolysis separates the aluminum for collection at the cathode.

Together, the unique attributes of electricity often combine to make electric appliances or industrial processes far more energy-efficient than those powered by other energy sources, even when the energy lost in converting primary fuels such as oil or gas into electricity is taken into account. Recent studies have shown that electric buses are about 85% more efficient than conventional diesel buses. Electric rail systems are 50% more energy-efficient per passenger mile than gasoline-powered automobiles. Electric-powered trains use only 70% of the primary energy consumed by semi-trailer trucks to carry the same amount of freight the same distance.

Electricity and Economic Progress

The beneficial effects of electrification are seen most clearly in the way overall electricity use, energy use, and industrial productivity have changed over the past century. During the 1920s, widespread electrification of the nation's industries, principally in the form of electric motors, rapidly transformed production processes. During that decade, productivity climbed an average 5.2% a year, a fivefold increase over the average gains recorded during the two preceding decades. Yet during the same period, the ratio of total energy input to manufacturing output, a measurement called intensity of total energy use, declined by an average of 4.3% a year. In short, the nation was using less energy to do more work—in large part because of the exceptional energy efficiency of electric-powered processes and technologies.

The same trend continues today. In 1992, the U.S. used only 7% more primary energy than it did in 1979, yet the gross national product (GNP) had increased 46%. Some estimates suggest that the wider use of electricity in the future could actually reduce the consumption of primary energy by 28% by the year 2010.

Another measure of the range of electricity's effects is the growing number of electric end-use processes and technologies that have been developed over the past century. Electric home appliances have taken over labor-intensive tasks and offered convenience and comfort. Electricity has become crucial to many industries, from steel-making and pulp and paper to food processing. The development of microprocessors has sparked another revolution in manufacturing. Computers provide far more precise and flexible control of many kinds of industrial and commerical operations. In some industries, robots powered by electricity have taken over repetitive or dangerous tasks. Inexpensive microprocessors are transforming the communications, information, and entertainment industries.

Uses of Electricity

Electricity customers are typically divided into three main sectors—residential, commercial, and industrial. Residential customers account for 88% of all electricity users, although they consume only about 33% of total electricity generated. Commercial customers, which include stores, hospitals, office buildings, restaurants, supermarkets, and hotels, represent 11% of all utility customers and consume about 30% of the

electricity generated. Industries account for only 0.5% of all utility customers but consume more than 34% of the nation's electricity. Among these sectors, some of the most important applications of electricity include:

Electric Motors

In 1900, only 5% of industry used electric motors. Today, more than 90% of the nation's industrial plants depend on them. Used to turn fans and blowers, run pumps and compressors, and propel conveyors and process lines, electric motor drives consume 67% of all the electricity used by the nation's industrial sector. Recently, microprocessor technologies applied to motors have resulted in adjustable-speed drives that can precisely match their speed to any task, saving significant energy. Studies by the Electric Power Research Institute (EPRI), which conducts research and development on behalf of the nation's electric power utilities, suggest that the widespread use of electronic adjustable-speed drives could reduce industrial energy use by 20%.

Lighting

Edison's invention of the electric light inaugurated the beginning of the electric century, and 25% of all electricity sold in the U.S. is still used for lighting. A new generation of highly efficient electric lighting technologies has been developed that includes compact fluorescent lamps, low-wattage metal halide lamps, and occupancy sensors and photocell controls that turn lights on and off as they are needed. EPRI estimates that new lighting technologies already available could improve energy efficiency by 39% to 55%.

Cooling and Refrigeration

The nation's supermarkets consume 4% of all electricity generated, one-quarter of which is used to operate refrigeration units. In 1987, the Montreal Protocol established a timetable for the worldwide phaseout of chlorofluorocarbons (CFCs)—substances used in refrigerators, freezers, some forms of foam insulation, halone fire extinguishers, and industrial solvents. New regulations have also established stricter minimum energy efficiency standards for refrigeration units and other appliances that use CFCs as the working fluid in a vapor compression cycle. Research is currently underway to find energy-efficient non-ozone-depleting alternatives to CFCs.

Space Conditioning and Water Heating

Space heating and cooling and water heating account for nearly 40% of all residential electricity use and 33% of commercial electricity use. Electric heat pump technology has significantly improved energy efficiency. Heat pumps use electricity to collect and concentrate heat from ambient air or the ground, both of which are warmed by solar radiation. As a result, heat pumps are far more energy efficient than conventional heating and cooling systems. By the year 2010, the increased use of electric heat pumps in the residential sector alone could reduce net primary energy consumption by more than one quad.

Information Technologies

The rise in information technologies has spawned whole new industries and dramatically transformed others. Information technologies currently account for about 4% of total electricity use. The use of information technologies is expected to grow significantly in coming years, but because new technologies are increasingly energy efficient, most analysts believe the percentage of electricity used by information technologies is likely to remain stable.

Electric Transportation

One of the most important uses of electricity at the turn of the century was to power electric trolley systems that operated in many American cities. Electric passenger cars were also popular. In 1899, only 936 gasoline-powered cars were produced in the U.S., compared to 1,575 electric cars. But within a few years, combustion engines gained dominance, and by the 1940s gasoline-powered vehicles had taken over as the nation's principal means of transportation. In many cities, electric buses and rail systems still operate, and in recent years there has been renewed interest in electric vehicles as a solution to the problems of environmental pollution connected with the use of internal combustion engines. Several electric vehicle models are nearing commercial production. Accelerated research into advanced batteries is currently underway that will extend the driving range and convenience of electric vehicles.

Other Uses

Electricity has made possible a variety of potentially life-saving medical technologies, from x-rays to computer-assisted tomography, or CAT scans. Electric power has also revolutionized entertainment, from home video players to compact disk players. A variety of electric technologies are also proving useful to remove environmental pollutants from air and water.

Energy Efficiency

As the efficiency with which electricity is generated and delivered improves, more power is tapped from primary energy sources such as coal and gas. Improv-

ing energy efficiency also reduces the impact of energy use on the environment. The efficiency of coal-based electric generating plants has steadily improved—from 5% during the Edison era to 35% or more today. Advanced generating technologies under development today are expected to achieve efficiencies up to 50% or more from coal-based generation and to dramatically reduce emissions. Improved delivery systems will help cut losses that occur as electricity is transmitted and distributed, further improving system efficiency.

Perhaps the most dramatic progress in energy efficiency, however, is being made in end-use applications of electric power. Many household appliances, from refrigerators to water heaters, are several times more energy efficient than they were several decades ago. Because electric power can be controlled and focused far more precisely than many other forms of energy, applications of electricity typically use only as much energy as they need. Instead of heating an entire piece of steel in a coal-fired furnace, for instance, electric current can be applied just where it is needed to shape the metal.

The widespread use of electric vehicles would dramatically reduce urban air pollution. In Los Angeles 60% of the air pollution is from petroleum-burning transportation. Electric vehicles produce virtually no emissions at the point of use. Emissions restricted to power plants are far easier to control than emissions from hundreds of thousands of tailpipes. The California Air Resources Board has mandated that by 1998, 2% of all cars and light trucks sold in the state must be powered by electricity. Five years later, 10% of all new cars sold—100,000 a year—will be required to run on electricity.

New electric-based technologies may also offer revolutionary new ways to perform work that save enormous amounts of energy. Not long ago, letters and documents had to be transported by truck or bus. Today, fax technology transmits documents electronically at the speed of light—using from one-half to one-seventh the primary energy required to send a letter by overnight courier. The energy from a barrel of oil can send roughly 25,000 pages by courier. That energy, converted to electricity to run a fax machine, can send 175,000 pages.

These and other electric technologies could help us significantly cut our consumption of primary energy, thereby reducing the environmental impact of fossil fuel use. EPRI estimates that the primary energy savings offered by a variety of highly efficient electric technologies available today—from electric vehicles and electric-based industrial processes to information technologies—could reduce carbon dioxide emissions by 220 million tons a year by 2010.

The Future of Electric Power

Current research and development suggests that several technological advances will have a significant impact on electric use. Advanced nuclear systems and coal-based technologies will generate electricity at high efficiencies and with very little impact on the environment. The development of low temperature superconductors, materials that conduct electricity over long distances with virtually no losses, could significantly improve the efficiency with which electricity is transmitted, stored, and used. Superconducting power transmission lines may be able to transmit electricity more efficiently, at lower voltages, and at lower cost than conventional lines. Superconducting magnetic energy storage systems use superconductors to store large quantities of electricity, providing utilities with greater flexibility in how and when they generate and supply electric power.

The growing use of microprocessors will also increase energy efficiency. Semiconductors and power electronics switches are currently replacing conventional mechanical switches on utility transmission and distribution systems, providing far more precise control of current flow across the nation's vast power grid. Computer-controlled devices in homes will provide a way to optimize electricity use by integrating home appliances and taking advantage of off-peak power.

Renewable sources already contribute a share of the world's electricity, and rapid advances in several renewable energy technologies, including wind and solar power, suggest that the contribution of renewable sources to the overall generating mix will continue to increase — ensuring a reliable and sustainable supply of electric power in the future.

Electricity will be very important in improving the quality of life in the developing nations of the world for at least three reasons. First, it will improve the efficiency by which natural resources can be used, therefore easing population pressure on the environment. For example, cooking a dinner over an open fire is only about 2% efficient. If the same wood were burned in a boiler to make electricity, it would cook 10 to 15 dinners. Second, electricity will improve the efficiency of human labor by substituting machines, permitting faster economic development. Today, one agricultural worker in the developing countries supports about 1.2 laborers on average, whereas in the U.S. one agricultural worker can support about 37 laborers. Third, electricity will provide the means by which technical innovation can be introduced and sustained. Electricity is the lifeblood of a modern society, a unique form of energy required for efficient manufacturing, as well as for the computers and telecommu-nications needed to establish links to the global economy.

BRENT BARKER

For Further Reading: Brent Barker, "Energy Efficiency: Probing the Limits, Expanding the Options," *EPRI Journal* (March 1992); Peter Jaret, "Electricity for Increasing Energy Efficiency," *EPRI Journal* (April/May 1992); Sam H. Shurr, *Electricity in the American Economy* (1990).
See also Air Conditioning; CFCs; Electric Utility Industry; Energy; Nuclear Power; Superconductivity; Technology; Vehicles, Alternative Energy.

ENERGY, GEOTHERMAL

Geothermal energy is acquired from the naturally occurring internal heat remaining from the earth's creation. Once used to heat Roman baths and homes, it currently heats public buildings, residential dwellings, and greenhouses, and it also generates electricity. First used in 1904 in Italy to produce electricity, it is now used in New Zealand, Japan, Iceland, Mexico, Italy, and the United States. Geothermal energy sources are widespread and renewable, and they require relatively simple technology. These sources can also operate uninterrupted regardless of changes in weather, unlike other renewable energy sources.

Geothermal power is most often exploited by harnessing steam (or water) that has been superheated by underground geothermally active rock. These "hydrothermal" systems use naturally occurring geysers (liquid-dominated) and fumaroles (vapor-dominated) to drive electrical turbines. Often these systems "reinject" used water to sustain underground reservoirs for continuous energy production.

Though benign compared to other, waste-producing energy sources, currently operating geothermal systems do have an environmental impact, usually as a result of the naturally occurring substances associated with geological activity. Geothermal plants may release considerable amounts of carbon dioxide, in addition to small amounts of poisonous toxins, such as dihydrogen sulfide, arsenic, boron, mercury, and radon gas. Release of these toxins into the environment has already been curbed considerably and will require some further technological improvements for system containment. Landslides and increased seismic activity have also been attributed to geothermal exploitation. These effects usually result when an adequate analysis of the geological conditions beneath and near plant locations has not been completed prior to plant construction. Very thorough analysis of geological activity around plants will deter future occurrence of geologically related accidents. However, the newer Hot

Dry Rock (HDR) system will bypass this problem by dealing exclusively with solid rock.

HDR and magma geothermal systems, researched since the late 1970s, employ a closed-loop system similar to the hydrothermal system. Holes are drilled into impermeable rock to depths of three to seven km (10,000 to 25,000 ft), where temperatures range from 200° C to 300° C. The rock is then hydraulically fractured to create artificial permeability, and water is injected into the rock. The water, now able to pass through freely, is converted to steam, enabling the system to recover energy. These synthetic systems are inherently simple and safe to operate, posing little danger of catastrophe or leakage. The HDR resource is also widely distributed and accessible virtually everywhere on the earth's surface with a tremendous long-term potential of producing huge amounts of energy. The experimental difficulty is maintaining flow through the fractured rock. Problems still facing HDR installation beyond the experimental level are primarily economic and depend on technological advancements for methods of controlling reservoir size and reducing costs of drilling deep wells.

In 1991, the United States produced 58% of the world's geothermal output. Presently (1992), the largest geothermal field site, at The Geysers in California, produces 69% of United States output and provides 7% of California's energy. Costs are generally competitive with other inexpensive fuel sources, though technological improvement will be necessary to make use of low- to mid-grade resources before they can become competitively accessible. There is much newfound interest in utilizing this barely recognized, highly promising energy source.

ENERGY, HEAT

That heat is a form of energy was not known until late in the 18th century. Up to then heat was thought to be a substance, called the caloric. Around 1800 Benjamin Thompson (Count Rumford), a native of Woburn, Mass., observed that during the operation of boring cannons, a dull tool produced about the same amount of heat as the sharpened drill. He became convinced that heat could not be a substance released by the iron chips.

Following this reasoning James Prescott Joule (1818–1889), a British scientist, realized that heat and work are interconvertible at a fixed ratio. He churned up water with a paddle driven by a falling weight. The temperature rise in the water and the known amount of work of the weight enabled him to calculate the mechanical equivalent of heat.

The experience of converting work to heat led to two fundamental scientific laws, the first and second laws of thermodynamics. According to the first law, work or heat added to a system cannot be destroyed within that system. In other words, the total amount of energy is conserved. Work can usually be converted to heat, but the reverse, a complete conversion of heat into work, is not possible. This was expressed as the second law of thermodynamics with its various corollaries. One of them is the fact that heat does not flow spontaneously to a system of higher temperature. Only when air conditioners, refrigerators, and heat pumps are added is such "upward" flow possible.

Heat can move in space from one point to another by radiation, conduction, and convection. Radiation, for instance, transfers heat energy from the sun to the Earth. The sun is a continuous fusion reactor in which hydrogen atoms combine to helium, and mass is converted to energy at temperatures of many millions of degrees.

The motion of radiant heat in space is similar to the propagation of light. Paradoxically and similar to visible light, radiant heat is emitted by a body in the form of finite batches, or quanta. When these strike a material surface, the molecules at that location acquire a higher kinetic energy and hence a higher temperature. Subsequently, molecules with that added energy will transmit part of their energy to molecules in a lower-temperature region—a process called conduction.

Solar collectors make use of all three phenomena of heat transfer. In collectors radiant solar energy is absorbed and then conducted along the collector surface to a conduit where a moving fluid (air or water) carries the energy—via convection—to its final location of usage. The efficiency of solar collectors is increased by taking advantage of the greenhouse effect. The glass panes covering collectors are transparent to the short-wave radiation coming from the sun but opaque (impenetrable) to the infrared radiation from the collector surface. Unfortunately, the same principle comes into effect in global warming, only the collector is now the surface of the Earth and the glass is replaced by gases such as carbon dioxide.

Radiant energy from the sun also plays an important part in the global water cycle and the process of photosynthesis. Heat energy from the sun evaporates sea water, which forms clouds and then rain. Through photosynthesis green plants join together water and carbon dioxide, converting them to oxygen and new plant material in the form of carbohydrates. Fossil fuels were formed from decomposed plant matter over millions of years under influence of heat and pressure.

WALTER LOSS

ENERGY, NONRENEWABLE

See Coal; Fuel, Fossil; Petroleum; Solar and Other Renewable Energy Sources.

ENERGY, SOLAR AND OTHER RENEWABLE SOURCES

See Solar and Other Renewable Energy Sources.

ENERGY, TIDAL

Tides have been used as a source of energy since the 12th century, when the first known tide-powered mills were built along the coastal areas of Brittany in France and in the British Isles. One built in Woodbridge, England, in 1170 operated for 800 years. Similar mills were built by the Dutch in Holland and in New Amsterdam (New York). In the 20th century, many ideas for harnessing tidal energy emerged, though none became technically or economically feasible until 1966, when French engineers constructed a plant that used tides to create electricity at the Rance Estuary on the coast of Brittany. Currently the world's only large tidal plant, the Rance plant has the capacity to produce 240 megawatts (MW) of electricity, though it usually operates at 62 MW with a yearly output of about 500 million kilowatt-hours.

Tidal plants employ one or more natural tidal basins, which fill and empty with the tide. Dams built between these basins and the sea contain hydraulic turbines that release water into or out of the basins. Since normal tidal periods (about every 13½ hours) seldom coincide with peak use of electricity, plants meet peak load demands by pumping water to higher reservoirs, from which turbines can operate continuously.

Tides, predominantly caused by gravitational pull from the sun and moon and by the earth's rotation, are greatest at the poles and in narrow bays and estuaries, where flow is concentrated. Concerns over coastal topography and the volume of tidal basins, the distance electricity generated must be transmitted, and the impact of plants on coastal ecologies have created controversy over tidal plant design and economic feasibility.

Although potential exists for thousands of megawatts of tidally produced electricity worldwide, both economics and the possibility of ecological damage remain as barriers to tidal plant development. Engineers and scientists in Canada and the United States have considered exploiting locations on the northeastern Atlantic coast since the 1930s, though plans were not considered economical until the 1970s. The first tidal generator in North America was completed in 1984 at Annapolis Royal, Nova Scotia, on the Bay of Fundy (where tides rise as much as 17 meters), and it now produces 20 MW of electricity. Environmental concerns have so far prevented the building of a large tidal plant damming up the entire bay, which could theoretically produce energy equivalent to that of 250 large nuclear power plants, although three small plants in Canada and another in Passamaquoddy, Maine, have been proposed. Concerns about possible disruption of the normal estuary flow and trapping of pollution have delayed plans by the British government to build a massive 4,000-MW tidal station in the Severn Estuary in southwest England. At present and for the foreseeable future, tidal energy is used on a very limited scale.

ENERGY, WIND

Wind, a source of energy since the advent of the sail, was not harnessed for land use until about the seventh century A.D., when it began to be used in China and Persia for milling grain and irrigation, uses that persist today. Wind was first employed as a source of electricity in Denmark in 1890, but until the late 1970s it was considered inefficient on a large commercial scale.

Today wind energy has the potential to be an inexpensive source of electrical energy for both industrial and developing regions. Endlessly available, it has little detrimental impact on the environment. With an estimated future generation potential of 20 million megawatts of electricity worldwide (not including offshore wind power), wind energy is considered one of the most promising nonpolluting energy alternatives. It is, however, an intermittent source, dependent on wind duration. At present levels of technology, production is considered sustainable where wind speeds average above 7.5 meters per second.

Wind turbines have two standard designs. The earliest and most common is the two- or three-bladed model, similar to an airplane propeller, which is most efficient in strong, steady winds. For efficiency in low winds, the inexpensive, "egg-beater" shaped Darrieus rotor, developed in Canada in the 1960s, is used.

Wind energy is not yet fully viable, technologically or economically. Improved battery storage or ready accessibility to other energy sources remains necessary to cover energy needs during windless periods. Although improving, the cost of the manufacture, sale, and maintenance of wind turbines remains prohibitive for many markets. Other problems concern the impact of turbines on wildlife (particularly birds), electromagnetic interference with radio and television, and safety (for example, the possibility of broken blades in strong winds and storms).

Seeking solution to these problems, current research focuses on ways to increase production efficiency, including generating energy at variable wind speeds; designing airfoils that adjust to wind speed while decreasing erosion and increasing energy output; and devising control systems that start and stop according to wind speed, adjust to high-speed winds with minimal mechanical fatigue, and adjust turbine direction with slight wind shifts.

Since the 1970s, wind energy in Europe and the United States has become significantly more economical. Compared with a 1981 cost-efficiency of 25 cents per kilowatt-hour (kWh), wind-generated electricity currently costs from 7 to 9 cents per kWh. This improved cost-efficiency makes wind-power competitive with petroleum energy, which presently costs 2 to 6 cents per kWh. California promoted turbine sales through federal and state tax incentives and now uses more than 16,000 wind turbines to generate 90% of U.S. and 75% of world wind power. Wind-generated electricity markets have been emerging in Denmark, Sweden, Holland, Great Britain, Greece, and India. With hundreds of thousands of potential generating sites worldwide and continual decrease in costs, wind promises to become our most valuable alternative to the nonrenewable energy sources now in use.

ENERGY EFFICIENCY

Energy efficiency is the ratio of energy delivered to energy supplied. In most cases energy efficiency is significantly less than 100% because so much energy is "lost" in the form of waste heat. Electric motors, generators, and batteries are highly efficient energy converters; internal combustion engines, jet engines, and coal-fired steam turbines are not. The system efficiency for lighting an incandescent light bulb from coal-fired generation is 1.3%, compared to 5.2% for a fluorescent light. The illuminating energy (the amount converted to visible light) of the fluorescent bulb itself is 20%, in contrast to 5% for the incandescent bulb. The energy system efficiency of heating water by natural gas is 60%, compared to 24% by coal-fired electric heating.

Energy efficiency is just as important in natural ecosystems as it is in systems created by humans, and the same operating principles apply to both—namely, the first and second laws of thermodynamics. Energy inflow must be balanced by energy outflow, and energy is "lost" as unusable heat during each energy transfer. In photosynthesis usually 1% or less of the solar energy that reaches green plants is converted to energy available to herbivores. At each transfer of energy in the food chain, approximately 10% of the energy is available. The efficiency of energy transfer in a food chain is known as the Lindeman's efficiency. Lindeman's efficiencies vary greatly; they usually range from 10% to 25%, but 70% efficiencies have been reported for some marine food chains.

More power companies are becoming involved in the process of improving energy efficiency in the cultural ecosystem. In 1989 rules were proposed that would change the way utilities make profits. The regulations would remove a utility's profit incentive to sell more energy by compensating utilities for revenue lost through efficiency. The Pacific Gas & Electric Company in California has already turned to investing in efficiency and is keeping as an incentive 15% of the money saved through new efficiency programs. The New York Niagara Mohawk Power Corporation has planned twelve efficiency programs for 1990, at a cost of $30 million. If the programs achieve the state's goal of saving 133 million kilowatt-hours, however, the power company will be allowed to recover those costs as well as a $1 million profit. Participating consumers also benefit. Although the price per kilowatt-hour is higher, the efficiency programs will enable consumers to use less energy, resulting in lower bills. Cogeneration systems are another way of increasing the efficiency of power generation. At least 60% of the energy generated by traditional power plants is lost as heat. Cogeneration systems capture this waste heat and use it for steam, hot water, or space heating.

Due to improved technology, the potential to save energy through efficiency has at least doubled since 1985. In addition, it has been estimated that the amount of energy saved through efficiency since 1973 has been far more than the amount of energy the world gained from new sources.

ENERGY-EFFICIENT BUILDINGS

An energy-efficient building is one that provides the various amenities offered by a building—shelter, comfort, light—while using a minimum amount of energy. Energy in buildings is used in various ways: for heating, cooling, and ventilating; for lighting; and for powering equipment such as elevators. Energy efficiency is achieved by addressing both the building shell and the building systems.

The building shell includes the physical features of a building's exterior, including the roof, walls, windows, doors, and floors. The shell contributes to energy efficiency by keeping cold air out (infiltration) and keeping warm air in (heat loss)—or vice versa, depending on the season and climate region.

The energy efficiency of a building shell can be improved by using materials and methods that reduce infiltration and heat loss—for example, insulation in the walls, floor, and roof; storm windows or double- or triple-glazed windows; and weatherization materials such as caulking or weather stripping. These can be designed into a new building or added to or installed in existing buildings in place of less efficient materials.

Energy is consumed by equipment in buildings known collectively as *systems*. The most common examples of building systems are the lighting system and the heating, ventilation, and air conditioning (HVAC) system. Building systems use electricity (for lighting and to run fans and motors as part of the air-handling system), and may also use gas or oil for space conditioning. Energy-efficient options are available for lighting, heating, and cooling tasks, and controls—such as timers, thermostats, or energy management systems—are available for regulating the use of building systems to maximize efficiency.

The concept of improving a building's energy efficiency first became popular with consumers and businesses in the late 1970s as a result of the oil crisis. At that time the focus was on reducing the consumption of heating oil. The federal government adopted a policy of encouraging efficient energy use through legislation passed in the late 1970s that mandated the dissemination of information and helped building owners identify energy-saving opportunities. Tax credits were also made available for investments in building improvements that resulted in energy savings. More recently, as public attention has shifted from oil to electricity (as a result of concerns over the risks of nuclear power and potential consequences of burning fossil fuels), the focus has shifted to ways of reducing electricity consumption in buildings. Since the mid-1980s electric utility companies have begun offering services and financial incentives to customers willing to install efficient technologies, such as high-efficiency light bulbs. The objective of these programs has been to postpone the need for building additional power plants, which can be both costly and environmentally risky. Also, building industry associations such as the American Society of Heating, Refrigeration, and Air Conditioning Engineers have established efficiency standards for new buildings and equipment.

LUISA FREEMAN

ENERGY LAW

See Law, Energy.

ENERGY RESOURCES, ALTERNATIVE

See Solar and Other Renewable Energy Sources.

ENERGY TAX

Like other consumption taxes, an energy tax—which may be assessed against a fuel's Btu content, polluting properties, monetary value (ad valorem), or simply its weight or volume—both curbs the activity being taxed and generates revenue for the public treasury. Whether such a tax is more effective at restricting consumption than augmenting revenues depends on its relative magnitude and on the consumption pattern of the energy users being taxed. In designing and imposing such a tax, the government must have one or the other of these outcomes in mind as its principal policy objective. As with other fiscal policy initiatives, it must also weigh the consequences of any such action for economic activity: a very large tax on oil imports into the United States may promote "energy independence" while throttling an economy unable to replace such imports except at very high cost.

Energy taxes are scarcely a new idea. Motor fuel taxes have been assessed in many countries for decades. In the United States, federal and state gasoline taxes have averaged around 35 to 40 cents per gallon in recent years; in many European countries gasoline taxes are six times as high. For the most part, these taxes have yielded substantial government receipts—their primary historic purpose. In Europe and elsewhere, however, they have also been steep enough to encourage use of fuel-efficient cars and public transport, thereby helping promote a better environment. In the United States they have helped finance highway construction.

Public concern in the last few years has not overlooked the revenue-raising side of energy taxes; witness the United States debate over a possible gasoline tax increase as a deficit reduction measure. But there has been a growing recognition—especially in the more prosperous industrial countries—that new and stiffer energy taxes of some sort may be justified to achieve desirable environmental goals. There are many environmentally harmful energy-using activities whose social cost is simply not captured in ordinary market transactions—whether it involves a coal-burning facility whose carbon dioxide emissions contribute to global warming or a large hydroelectric plant that displaces populations, diminishes fish runs, or encroaches on parkland. Energy taxes are a strong candidate among the policy instruments that can make the private costs incurred in ordinary market transactions more nearly reflect social costs. In addition, they do so in an economically efficient way.

Virtually all economists agree that, to achieve a given environmental end, it is more efficient and politically less intrusive to tax the offending energy source at an appropriate level than to impose an absolute ban on the activity in question. Alternatively, the *quantity* of environmental discharges can be fixed with emissions trading among polluters. Taxes that drive up costs and prices induce substitutions and adjustments by those firms and consumers readily able to make them. Others may find it cheaper to pay some taxes rather than alter production or consumption patterns. Society as a whole will have attained its overall goal in the most cost-effective way.

One prospective energy tax much discussed in recent years, for example, would assess the carbon content of fuels. This tax, which has a reasonable prospect of adoption by a growing number of countries in the future, would inhibit carbon dioxide emissions that contribute to greenhouse warming. Electric power plants burning coal, which is rich in carbon, would have the incentive to switch to nonfossil fuels; electricity customers facing a tax-induced price hike would be encouraged to conserve. Although its widespread acceptance would entail complicated international negotiations, a carbon tax could advance worldwide environmental aims. At the same time, the economic pain inflicted by a carbon tax (or other energy taxes) might be mitigated by "recycling" the proceeds throughout the economy or by reducing other burdensome taxes on business or household earnings.

JOEL DARMSTADTER

ENVIRONMENT

The word *environment* became part of everyday language in the 1960s and its meaning is far from clear. In fact, it has evolved and continues to evolve, in part due to an increase in scientific knowledge but even more as a result of changes in the mood of the general public. During the 1960s, for example, the word *environment* evoked chiefly pollution, the depletion of natural resources, overpopulation, and crowding: the thousand devils of the ecological crisis. In contrast, there is greater emphasis in the 1990s on the positive qualities of environments—on those physical and social characteristics that contribute to the quality of life.

The scientific language of the environment presents more fundamental difficulties. At first glance there seems to be no problem in differentiating between *organism* and *environment*. The word *environment* is used to denote the setting in which the organism develops and functions, and refers to all the factors of the external world that affect biological and social activities. In other words, it is the sum of the abiotic (physical), biotic (living), and cultural (social) factors and conditions directly or indirectly affecting the development, life, and activities of organisms and populations, in the short and long term. From a narrow scientific point of view, therefore, the human skin or the cellular wall sharply separates the living organism—human or microbial—from the external world.

The distinction between organism and environment becomes blurred, however, when one considers living things and external world, not as separate static entities, but as interacting components of complex dynamic systems. The very process of living transforms environments profoundly and lastingly, and this is particularly true of human activities. All organisms impose on their environment characteristics that reflect their own biological and social nature. This is particularly true of the human environment, viewed by the biologist, the physician, the farmer, the forester, the engineer, the urbanist, the sociologist, the painter, the poet—from their own specialized points of view, and referring to it in their own specialized languages.

From the humanistic point of view, environments are places where the natural and human order undergo some kind of fusion. The interplay between human beings and environments can be considered from two complementary points of view: (1) Human beings always transform the environments in which they live and function; in fact, practically all inhabited environments are artificial in the sense that they have been profoundly altered by human cultures. (2) Human beings are shaped by the environments in which they develop, each culture reflecting the influence of the environment in which it has been created and has evolved.

We expect more of the environments in which we live, however, than conditions suitable for our health, resources to run the economic machine, and whatever is meant by good ecological conditions. We want to experience the sensory, emotional, and spiritual satisfactions that can be obtained only from an intimate interplay, indeed from an identification with the places in which we live. This interplay and identification are the spirit of place. Environments acquire the attributes of a *place* through the fusion of the natural and the human order. All human beings have approximately the same fundamental needs for biologic and economic welfare, but the many different expressions of humanness can be satisfied only in particular places. The English hedgerow or moor countryside, the European bocage, the Mediterranean hill towns, the Pennsylvania Dutch country, the Chinese mountain and water landscapes call to mind ecosystems intimately associated with certain ways of life. The fitness they exhibit between local people and nature entitles them to be called places. The catalyst that converts an environment into a place is the process of experiencing it deeply—not as a thing but as a living organism. Fitness is achieved only after slow progressive reciprocal adaptions and therefore requires a certain stability of relationships between persons, societies, and places.

Environmental factors exert such a powerful governing influence on the development of all human characteristics that they literally shape the body and the mind. The adaptive responses that humans make to the physiochemical, behavioral, cultural, and even historical stimuli that they experience during the formative stages of their development constitute the mechanisms through which they achieve biological and mental fitness to their surroundings.

The word *environment* does not convey the quality of the relationships that humankind can ideally establish with the Earth. Its widespread use points in fact to the present poverty of these relationships. In common parlance, as well as etymologically, the environment consists of things around us, out there, that act on us and on which we act.

The words *surroundings* and *events* are both used in an attempt to convey the multiplicity and complexity of the factors covered by the word "environment." The total environment refers of course to the complete physiochemical and social setting in which the organism develops and functions; it includes elements that may have no biological effect whatever, and also elements that are biologically active but have not yet been defined. In describing the physiochemical environment, one usually ignores the whole electromagnetic spectrum except for the wavelengths of light. Yet taking a radio or a Geiger counter into a natural situation shows that there are in the environment all sorts of forces that are not detected by the senses; some of them presumably have important biologic effects. It is only since the 1950s that reliable observations have been made, for example, on the responses of plants, animals, and humans to the cosmic forces that are responsible for the diurnal, lunar, and seasonal cycles.

Biologists have so far studied chiefly the environmental factors that are intercepted by the sense organs, and that constitute therefore the perceptual environment. But our perceptual environment should not be regarded as representing the total environment. Each living thing inhabits a perceptual world of its own. A dog sniffing the breeze or the traces of a rabbit on the ground lives in a world that a human or a frog hardly perceives. An insect moving at night toward a potential mate, a salmon crossing oceans toward its mating ground, or a bird exploring the soil or a dead tree for an insect: each uses clues that are nonexistent for another species. Much of animal and human behavior is thus influenced by stimuli that make the perceptual environment differ from species to species and indeed from one individual organism to another.

It is being realized more and more that the responses of organisms to their total environment embrace much that seemed unusual only a few decades ago. Animals have been shown to receive information through many unfamiliar ways such as pheromones (substances excreted from the body and perceived by animals of the same species), ultrasound waves (in bats), and infrared waves (in moths and pit vipers). Weather changes have been reported to affect the human autonomic nervous system and various physiological processes, such as blood clotting and blood pressure. There is no doubt, in any case, that various unfamiliar channels of communication, once dismissed as non-existent and indeed impossible, enable us to acquire information from our environments and from each other without awakening in us a conscious awareness of the process.

In addition to the factors that are inherent in nature and can—or eventually could—be identified by physiochemical methods, the total environment includes elements that exist only in human minds. For most archaic people, on a Micronesian atoll for example, the environment includes not only sea, land, and sky, but also a host of "spirits" that lurk everywhere. These factors of the Micronesian conceptual environment do not have less influence on the inhabitants of the atoll for having no concrete existence. The "spirits" are generally harmless, but they become injurious if not properly treated, and can then elicit behavioral responses that may be even more dangerous than wounds inflicted by sharks or moray eels.

Nor is the conceptual environment of less importance in industrialized societies. Whether learned and sophisticated or archaic and ignorant, every human being lives in a conceptual environment of his own which conditions all his responses to physiochemical, biological, and social stimuli. These responses eventually contribute in turn to the manner in which he shapes his surroundings and ways of life. In our societies, the conceptual environment is becoming increasingly powerful as a mediator between man and external nature.

The phrase "conceptual environment" is almost synonymous with what psychoanalysts call "superego" and even more with what anthropologists call "culture." Its only merit, perhaps, is to help make clear that the total environment involves much more than the effects of natural forces on the human body: it is a determinant of human behavior and evolution as well as a product of human intervention.

Darwinian evolution through natural selection implies that an organism cannot be biologically successful unless it is well adapted to its external environment. Genetic science has defined furthermore how this adaptation is achieved through selective processes that are under the control of environmental forces. As is now well understood, mutation and selection provide the mechanisms that allow adaption to the environment and that progressively become incorporated in the genetic apparatus of the species.

While mutation and natural selection account for the evolution of species, these processes contribute little to the understanding of the precise mechanisms through which each individual organism responds adaptively to its environment. This complementary knowledge has evolved in large part from Claude Bernard's visionary concepts regarding the interplay between the external environment and what he called the internal environment. Bernard formulated the hypotheses that organisms could not maintain their individuality and could not survive if they did not have mechanisms enabling them to resist the impact of the outside world. Whether human, animal, plant, or microbe, the organism can function only if its internal environment remains stable, at least within narrow limits. The constancy of the internal environment determines in fact the organism's individuality. In the case of man, it involves not only biological attributes but also mental characteristics.

The recognition that the internal environment must remain essentially stable, even when the external environment fluctuates widely, constitutes such an important landmark in biological thought that Bernard's phrase *milieu interieur* has gained acceptance in the English language. The most commonly quoted expression of his law is: "The constancy of the *milieu interieur* is the essential condition of independent life."

William James was one of the first to recognize the philosophical importance of the *milieu interieur* concept and he referred to it in an editorial in the *North American Review* as early as 1868. But it was probably Lawrence J. Henderson who did most to make scientists aware of the concept. Bernard had guessed that the maintenance of stable conditions in the body fluids and cells was in some way dependent upon control by the nervous system. This view was developed much further by the physiologist Walter B. Cannon in his classical work on the role played by the sympathetic nervous system in maintaining the internal equilibrium of the body. Cannon introduced the word *homeostasis* to describe this phenomenon in his book, *The Wisdom of the Body*.

Responses to biological and mental environmental changes must naturally help the organism to function adequately under the changed conditions; but the ad-

justments must remain within limits precisely defined for each organism. These two demands apply to populations as well as to individual organisms. Whatever its complexity, a biological system can continue to exist only if it possesses mechanisms that enable it, on the one hand, to maintain its identity despite the endless pressure of external forces, and, on the other hand, to respond adaptively to these forces. The complementary concepts of homeostasis and adaptation are valid at all levels of biological organization; they apply to large social groups as well as to unicellular organisms.

A further elaboration of the *milieu interieur* concept was formulated by Norbert Wiener, providing a link with cybernetics—the theory of control and communication in machines and organisms. He wrote:

> Cannon, going back to Bernard, emphasized that the health and even the very existence of the body depends on what are called homeostatic processes . . . that is, the apparent equilibrium of life is an active equilibrium, in which each deviation from the norm brings on a reaction in the opposite direction, which is the nature of what we call negative feedback.

Both the theory of evolution (Darwin, Mendel, and the Neo-Darwinians) and the cybernetic theory of physiological responses (Bernard, Cannon, Wiener) provide a dynamic approach to some of the problems posed by the interplay between humans and their environments. But neither theory deals with the mechanisms through which each individual person becomes what he or she is and behaves as he or she does in response to the environmental forces that have impinged in the course of development. The shape of our biological and mental individuality is influenced by forces that do not affect genetic constitution, but act on our organism at the critical periods of development.

Cannon's and Wiener's self-correcting cybernetic feedback represents sophisticated expressions of Bernard's constancy of the milieu interieur concept. Unfortunately, an uncritical belief in the effectiveness of homeostatic processes tends to create the impression that all is for the best in the best of all worlds. The very word *homeostasis* seems indeed to imply that nature in its wisdom elicits responses that always bring the organism back to the same ideal condition. This is, of course, far from the truth.

Furthermore, if it were true that the cybernetic feedback always returned the organism to its original state, individual development would be impossible. But in fact, the situation is very different. Most responses to environmental stimuli leave a permanent imprint on the organism, changing it irreversibly. If an organism were truly in a state of complete equilibrium, it could not develop.

The effects of environmental factors on the development of individuality are complicated by the fact that humans tend to symbolize everything that happens to them and then to react to the symbols themselves as much as to external reality. All perceptions and apprehensions of the mind can thus generate organic processes of which the environmental cause is often extremely indirect or remote. In many cases, furthermore, individuals do not create the symbols to which they respond; they receive them from their group. Their views of the physical and social universe are impressed on them very early in life by ritual and myth, taboos and parental training, traditions and education. These acquired attitudes constitute the basic premises according to which they organize their inner and outer worlds, in other words — their conceptual environment.

RENÉ DUBOS
(EDITED BY WILLIAM R. EBLEN)

For Further Reading: René Dubos, *A God Within* (1972); René Dubos, *The Wooing of Earth* (1980); Gerard Piel and Osborn Segerberg, Jr., eds., *The World of René Dubos* (1990).

ENVIRONMENTAL DEFENSE FUND

See Environmental Organizations.

ENVIRONMENTAL DISASTERS

See Bhopal; Chernobyl; Love Canal; Natural Disasters; Three Mile Island.

ENVIRONMENTAL IMPACT STATEMENT

The Environmental Impact Statement (EIS) is a document whereby policy makers and the general public evaluate the potential impact of federally initiated or permitted actions and weigh all reasonable alternatives. The EIS is designed to assure Congress and the public that all federal agencies are abiding by the goals and principles of the National Environmental Policy Act (NEPA). NEPA, enacted by the United States Congress in 1970, established a body of law and guidelines for environmental protection and directed all federal agencies to incorporate these standards into their decision-making process. Section 102 of the act requires that all federal agencies proposing to engage in or permitting private parties to engage in activities that significantly affect the environment — such as mining, oil drilling and exploration, highway construction, and hazardous substance disposal — prepare an EIS.

In every EIS document a federal agency must demonstrate the degree to which its proposed actions constitute a "major" undertaking with a "significant" impact on the human environment and provide a detailed exploration of alternatives and actions to mitigate potential environmental effects. EISs are by definition interdisciplinary, requiring agencies to examine a variety of scientific and socioeconomic factors in order to fully analyze the potential impacts of a project. Since the passage of NEPA federal agencies ranging from the Department of Agriculture to the Department of Defense have prepared thousands of impact statements; annually, nearly 700 EISs are submitted.

In cases where uncertainty exists as to the significance of a proposed action, federal agencies perform an environmental assessment (EA) to determine the legal requirement for an environmental impact statement. The EA is a concise document that serves as an early decision tool for identifying potential environmental impacts associated with proposed actions. On the basis of its EA an agency declares whether or not its proposed action merits the preparation of an EIS by issuing either a finding of no significant impact (FONSI) or an affirmative decision to prepare the EIS. By law all agencies that conclude "no significant impact" must present their case to the public for a 30-day review period before commencing action. The final decision is incorporated into the administrative record and may be used in future legal challenges. Approximately 30,000 environmental impact assessments are prepared by federal agencies each year.

Despite the impressive gains made through the administration of environmental impact statements, over 2,000 NEPA lawsuits have been brought against various federal agencies by private citizens and environmental organizations. The bases for private party lawsuits range from failure of federal agencies to present adequate alternatives (in an EIS) to the general public to a dispute over whether federal agency actions warrant the preparation of an EIS. Private party involvement in the EIS process represents an important avenue for challenging public institutions whose actions violate the spirit of NEPA. Such legal actions help to clarify and interpret NEPA guidelines and establish precedents for future cases.

ENVIRONMENTAL MOVEMENT

The organizational and ideological roots of contemporary environmentalism are commonly traced to the progressive conservation movement that emerged in the late 19th century in reaction to reckless exploitation of natural resources. While conservation faded from attention with World War I, it left a legacy of organizations (e.g., the Sierra Club and National Audubon Society) and government agencies (e.g., the National Park Service and Forest Service). The next great wave of concern developed momentum in the 1960s, as the Sierra Club and other organizations called attention to growing threats to natural beauty, and quickly evolved into modern environmentalsim. The emergence of the environmental movement is typically marked by the Earth Day celebration on April 22, 1970, which drew millions of participants.

The modern environmental movement arose during an era of widespread political activism and reform, and quickly achieved high levels of support from broad sectors of society. The growth of scientific evidence on environmental degradation, coupled with well-publicized disasters such as the Santa Barbara oil spill of 1969, stimulated widespread concern. Post World War II affluence enabled larger numbers of people to spend leisure time in the outdoors, heightening their commitment to preserving areas of natural beauty. Affluence, combined with increased urbanization and education, also stimulated changes in social values, lessening concern with materialism and generating interest in the quality of life, including environmental quality.

At the same time, traditional conservation organizations, such as the Sierra Club and the Wilderness Society, were aggressively battling threats to natural areas, and broadening their agendas to incorporate a variety of newer issues. Legal changes enabled organi-

zations to fight battles on several fronts, from congressional offices to courtrooms, where legal standing was becoming easier to achieve. Furthermore, new organizations, such as the Environmental Defense Fund and the Natural Resources Defense Council, began to develop, often aided by funding from corporate foundations.

Despite predictions of demise and evidence of a slight dip in public support after Earth Day, the movement remained strong throughout the 1970s, maintaining relatively high public support and continuing to gain in organizational membership. Ironically, the anti-environmental agenda of the Reagan administration generated renewed support for environmentalism in the 1980s, and organizations such as the Sierra Club grew dramatically.

The twentieth anniversary of Earth Day in 1990 proved to be even more successful than its predecessor. In addition to being a media event and stimulating a "greening" of corporate and political rhetoric, it encouraged consumers to seriously consider the environmental impacts of their lifestyle choices. Recycling programs and "ecologically correct" products became increasingly popular, and environmental organizations experienced another major membership growth spurt. While this dramatic growth has subsided, the organizations seem unlikely to suffer significant declines. Environmental issues remain at the center of public attention and are increasingly becoming significant factors in political campaigns.

Table 1
Three stages of environmental activism

	Conservationism	Environmentalism	Ecologism
Approximate beginnings	Late 19th c.	Middle 20th c.	Late 20th c.
Primary goals	Conservation of natural resources	Protection of environmental quality	Maintenance of ecological sustainability
Dominant ideology	Natural resources should be used efficiently for the good of all society	Environmental quality should be protected for a high quality of human life	Ecosystems should be protected for the benefit of all species
Worldview	Anthropocentric	Anthropocentric	Ecocentric
Nature of issues	Geographically bounded (typically rural), specific, unambiguous	Geographically dispersed (often urban); delayed, subtle and indirect effects; potentially harmful to human health	Extremely diverse; systemic and synergistic effects; potentially irreversible and harmful to all life on Earth
Example	Over-logging in a specific forest	Urban air and water pollution	Global environmental change
Cost of solution	Relatively small, localized	Often substantial	Potentially infinite
Tactics	Lobbying	Lobbying, litigation, citizen participation	Lobbying, litigation, electoral action, direct action, lifestyle change
Opposition/culprits	Natural resource industries; local economic interests (loggers, hunters)	Corporations, modern lifestyles, economic growth	Status quo; excess human production, consumption, and population

The Evolution of Environmental Activism

The progressive conservation movement represents some of the earliest attempts in American history to collectively address issues of natural resource management. Although modern environmentalism clearly encompasses the concern and tactics of conservationism, it goes beyond them in substantial ways. A third phase of activism, "ecologism," has recently emerged, partly in response to perceived weaknesses within mainstream environmentalism. This evolution of the movement has been marked by a gradual broadening of issues and growth in the diversity of groups dealing with such issues.

A key aspect of each phase of the environmental movement is the distinction they exhibit between an anthropocentric worldview, in which humans are the center of concern, and an ecocentric (or biocentric) worldview, in which ecosystems, other species, and all life on Earth are important. Despite the efforts of preservationists such as John Muir, conservationism was predominantly anthropocentric because it emphasized the wise use of resources for human benefit. Similarly, environmentalism has also been largely anthropocentric in its emphasis on environmental quality as crucial for a high quality of *human* life. Concerns for human health, outdoor recreational opportunities, and an aesthetically pleasing natural world were all motivated by an underlying concern with the welfare of humans.

The recent emergence of ecologism, however, broadens earlier concerns by incorporating an ecocentric worldview largely stemming from two noteworthy branches of the modern movement: deep ecology and radical environmentalism. Adherents of ecologism believe that nature has a right to exist in and of itself, apart from human desires. Although this position existed in the conservation movement via John Muir and his followers, and was given new impetus by Aldo Leopold's "land ethic" in the mid-1900s, only recently has it become a potent voice in the evolving movement. Advocates of ecologism hold as their primary goal the maintenance of ecological sustainability, for the entire Earth and all of its inhabitants.

The issues addressed by environmental activists have expanded over time, as environmental problems and society's awareness of them have grown. Not only has each stage seen new issues incorporated into a growing environmental agenda, but the issues themselves have been perceived quite differently over time.

Conservationism typically dealt with specific and unambiguous issues, such as the protection of particular forestlands from logging, whereas air and water pollution are typical concerns of environmentalism. Pollution can diffuse over a broad area, its sources are often ambiguous, and its effects may be delayed and difficult to identify. In short, pollution is more subtle than logging, and controlling or stopping it is typically more difficult and costly.

Ecologism incorporates an even wider range of issues (often brought to our attention by scientific research), as well as a significantly different perception of these issues. Especially notable is a greater emphasis on both micro and macro concerns. At the micro level, grassroots groups around the country are mobilizing NIMBY ("Not-In-My-Backyard") protests against garbage incinerators and hazardous waste sites. Minorities in particular are mobilizing against "environmental racism," the practice of locating environmentally noxious facilities in their communities. At the macro level, ecologism is concerned with issues of international import, such as global warming, ozone destruction, and the loss of rain forests. In addition, the growing radical wing of the movement promotes an ecocentric worldview and an emphasis on global ecological sustainability. In fact, radical groups like Earth First!, as well as many of the grassroots groups, have developed explicitly in reaction to what they refer to as the "reform" environmentalism and "shallow" ecology embodied by the mainstream national environmental organizations.

Ecologism recognizes the inherent interdependence of all life systems. Even NIMBYites are becoming concerned with the sustainability of current production systems, as witnessed by the growing NIABY ("Not-In-Anyone's-Backyard") attitude. Ecologism thus broadens environmental concern in various ways, fighting on a greater number of fronts—from local to global—and viewing other species and ecosystems as having rights to exist independent of human interests. Concern for the equity between species parallels a growth in concern for equity within the human race and across generations, as advocates of sustainable development attempt to rectify poverty as well as environmental devastation throughout the world.

Ecologism entails an expanded critique of the status quo, based on a systemic and large-scale view of human impact on the natural environment. The pivotal concerns of ecologism are typically those that involve long-term, irreversible, synergistic, and often unpredictable consequences of human actions. Global warming and loss of biodiversity, for instance, are not only long-term and irreversible, but they stem from a complex interplay of factors. While environmentalism has often addressed these issues, it has looked at them in a piecemeal fashion. Ecologism, on the other hand, views them in the context of the larger ecological-evolutionary global system. Advocates of ecologism talk about the end of nature, mass extinction, and the halt of evolution—unless human practices are altered, and soon. The purported causes and consequences, as well as the costs needed to remedy them, are therefore truly colossal.

Just as the issues have broadened, so have the tactics employed by environmental activists. Tactically, conservationists relied heavily and quite successfully on lobbying government officials. The solution to resource problems, they felt, came with governmental and scientific management of resources. In addition to traditional lobbying, environmentalists added litigation, research, and citizen participation to its repertoire, through the development of research- and legal-oriented groups as well as via letter writing campaigns and mass protests.

Ecologism builds upon this tactical legacy—grassroots activists march and petition government officials, organizations engage in litigation and lobbying—but it increasingly employs more aggressive tactics such as consumer boycotts and various forms of "direct action." Members of the growing radical wing of the movement have engaged in sit-ins, tree-spiking, "monkey-wrenching" of equipment, and other forms of "ecotage" (from pouring quick rice into the radiator of a bulldozer to ramming a drift-net ship on the high seas) that are disavowed by mainstream environmentalists intent on "working within the system" in order to reform it.

In recent years environmental organizations have become increasingly active in political elections, publicizing candidates' records and publicly supporting selected candidates. In addition, advocates of ecologism, like their predecessors, promote lifestyle change, ranging from recycling and purchasing "green" products, to reduced consumption and dietary changes.

The opposition to environmental activists and the source of their grievances have traditionally been the same, for those who benefit from environmentally harmful practices are most likely to oppose attempts to end those practices. However, as our understanding of ecological problems has progressed, so has recognition of their embeddedness in the status quo. No longer are just a few "robber barons" to blame for our problems; rather, as Pogo said: "We have met the enemy and he is us."

Conservationism laid the blame for resource depletion at the feet of a relatively small group of people, and thereby stimulated limited opposition. Environmentalism blamed entire industries, modern lifestyles, and "growthmania" in general, and in the process engendered broader opposition. Ecologism issues a critique that leaves few unscathed, for our entire species and the status quo, at least within industrialized societies, are to blame, albeit some aspects more so than others. Simply reforming current practices via legislation is therefore unlikely to suffice. For example, rather than favoring installation of scrubbers on factory smokestacks to reduce pollutants, advocates of ecologism would argue for alternative production techniques or even giving up the product completely. Because the economic and social costs of halting human-induced environmental change could prove enormous, and leave little of modern life untouched, ecologism has the potential of stimulating enormous opposition.

Summary and Conclusion

The environmental movement has been an exceptionally enduring and influential social movement. In little more than two decades, it has grown substantially in its organizational base and number of supporters. It is widely considered to be one of the most successful social movements of this century, especially in terms of gaining widespread societal acceptance of its goals.

Indicators of the movement's success are numerous: the establishment of environmental agencies and passage of environmental laws and regulations at all levels of government; the proliferation of environmental science programs at all levels of schooling; enormous media attention to environmental problems; and gradual changes toward more ecologically responsible behavior (e.g., recycling) and the emergence of "green consumerism." In the U.S., and most other industrialized nations, environmentalism has clearly had a pervasive impact throughout society.

Yet the fact that ecologism has arisen out of frustration over environmentalism's failure to halt ecological degradation, suggests an uncertain future for this latest phase of human efforts to protect the environment. Ultimately, the true test of a social movement is its success in achieving its goals, not simply in maintaining its own survival.

RILEY E. DUNLAP
ANGELA G. MERTIG

For Further Reading: Riley E. Dunlap and Angela G. Mertig (eds.), *American Environmentalism: The U.S. Environmental Movement, 1970–1990* (1992); Robert Cameron Mitchell, "From Conservation to Environmental Movement: The Development of the Modern Environmental Lobbies," in *Government and Environmental Politics: Essays on Historical Developments Since World War Two,* ed. Michael J. Lacey (1989); Roderick Frazier Nash, *American Environmentalism: Readings in Conservation History,* third edition (1990).
See also Conservation Movement in the U.S.; Ecofeminism; Ecology, Deep; Environmental Organizations; Green Movement; Radical Environmentalism; U.S. Government.

ENVIRONMENTAL MOVEMENT, GRASSROOTS

In 1978 residents of the Love Canal neighborhood in Niagara Falls, New York, convinced the state government to evacuate a neighborhood contaminated by a hazardous waste dump, propelling the issue of toxic chemical pollution into the national media, the agendas of Congress and state governments, and the minds of the U.S. people. Since that time, thousands of other communities have organized to confront environmental hazards.

Several factors help to explain the dramatic expansion of local environmental activity since the 1970s. These include the growth of the petrochemical industry in the post-World War II period with factories, waste sites, and transportation routes throughout the U.S.; greater public awareness of environmental hazards as a result of environmental activism and its media coverage; and awareness of new strategies and tactics for social change brought about by the social movements of the 1960s and 1970s.

By the early 1990s grassroots environmentalism had evolved into a loosely structured movement with three overlapping but distinct levels of organization: community-based groups, regional coalitions, and national organizations. Local community organizations are the foundation of the movement. While precise numbers are not available, the Citizen's Clearinghouse for Hazardous Waste, a national support organization, works with 7,000 grassroots environmental groups to protect their communities against perceived hazards.

These groups are often formed by a few people who are directly affected by a perceived health hazard in their community. In some cases victims of pollution and their families play an important role. Typically groups begin by attempting to document a hazard and link it to a health problem, often leading to extensive interactions with public health officials, scientists, and sometimes lawyers.

Members of these groups include a broad cross section of class and occupational categories. Unlike the mainstream environmental organizations, women, African-Americans, Native Americans, and Latinos are represented among members and leaders. Many founding members are experienced community activists, but new leaders also emerge. Most of these groups receive no funding, relying exclusively on volunteers.

Local community groups join together to form regional or statewide networks and coalitions in order to find new allies in their struggles and to present their case more effectively in both scientific and political forums.

Since 1981 a few national organizations have emerged to provide scientific, legal, and political support to the smaller groups and to lobby and advocate at the national level. These groups organize national conferences for activists, offer leadership training, publish newsletters and manuals, provide technical assistance to local groups, and develop policy initiatives. They also channel grassroots concerns to legislators and lobby for laws to protect local environments and strengthen the role of community organizations.

Although the grassroots environmental movement is not a homogeneous grouping, its members and leaders generally share certain values. First, they espouse the right of citizens to participate in making environmental decisions. This emphasis on citizen participation reflects a mistrust of government officials, often based on negative experiences in a local struggle.

Second, human health rather than environmental esthetics, wilderness preservation, or other such issues, is the primary concern of most grassroots environmentalists. Their main motivation is to protect themselves, their families, their community, and future generations against some perceived hazard.

Third, grassroots environmental groups are ambivalent toward scientific expertise. On the one hand, most groups interact with scientists and many report positively on such relationships. On the other hand, surveys and anecdotal reports suggest widespread mistrust of scientists and health officials. This skeptical attitude may reflect activists' experience with scientists hired to defend industry or government or a preference to contest environmental issues in the political arena—where most community groups are more skilled—rather than in the scientific arena where they perceive they are at a disadvantage.

Fourth, implicit in the movement's action is a challenge to the belief that economic growth is itself good and benefits all sectors of society equally. Grassroots groups challenge the assumptions of traditional cost-benefit analysis by asking who pays the cost and who gets the benefit. These groups question the right of government or corporations to make decisions that have health or social consequences without community involvement.

These characteristics distinguish the grassroots environmental movement from the more established environmental organizations that tend to focus on the federal government, are less concerned about human health, and seldom reach people of color. In the last decade, however, national environmental organizations have begun to change, devoting greater attention to health issues (such as lead poisoning) and seeking to establish a dialogue with community organizations and low-income communities.

Because the national environmental groups and the grassroots movement have overlapping agendas and strategies, it is not possible to make conclusive distinctions between their separate accomplishments. The following achievements, however, seem more closely associated with the grassroots movement: (1) forced the cleanup of contaminated waste sites, blocked the construction of hazardous facilities, won bans on spraying of pesticides and forced corporations to upgrade pollution control equipment; (2) forced corporations to consider more closely the environmental consequences of their actions; (3) created new political and economic pressures for a preventive approach to environmental contamination; (4) won legislative and legal victories that expand the rights of citizens to participate in making environmental decisions; (5) had an important impact on communities affected by toxic disasters, helping people to cope with the stress and loss of community often associated with these events; (6) brought environmental concerns and action to working-class and minority Americans; and (7) played a role in increasing public support for environmental protection by helping to educate Americans about the links between the environment and public health.

NICHOLAS FREUDENBERG

For Further Reading: Murray Edelstein, *Contaminated Communities: The Social and Psychological Impacts of Residential Toxic Exposures* (1988); Nicholas Freudenberg, *Not in Our Backyards! Community Action for Health and the Environment* (1984); Adeline G. Levine, *Love Canal: Science, Politics and People* (1982).
See also Environmental Organizations; NIMBY; Public Health.

ENVIRONMENTAL MOVEMENT, RADICAL

Radical environmentalism is the name given to the militant beliefs of several small but influential environmental groups that formed during the 1970s and 1980s. Earth Day 1970 ushered in a decade of disquieting change for the environmental movement. In the United States national groups such as the Sierra Club, Wilderness Society, and National Wildlife Federation saw their membership rolls and budgets swell to historic levels. This growth bolstered their influence in Washington, D.C., where many environmental organizations began to cluster, but it also altered the character of the movement's leadership.

By the late 1970s many grassroots activists perceived the mainstream environmental groups to be remote, bureaucratic, and too willing to make political compromises. For these activists, the political strategy of traditional environmentalism seemed increasingly ineffectual in protecting natural ecosystems from the threats of overpopulation and industrial growth. Disenchanted with the status quo, they began to form more militant groups whose ideas and confrontational tactics greatly influenced the course of environmentalism in the 1980s.

Among these groups was Earth First!, founded in 1980 by five environmental organizers, including Dave Foreman, a former Wilderness Society lobbyist. With its war cry of "No Compromise in Defense of Mother Earth," Earth First! used guerrilla theater, civil disobedience, and "ecotage" (the sabotaging of machines harming the ecology) to disrupt logging and mining operations. In Canada, a co-founder of Greenpeace, Paul Watson, organized the militant Sea Shepherd Society, which utilized its 200-foot-long flagship to interfere with, even ram, whaling vessels. The Australian John Seed established the Rainforest Information Centre, which not only alerted the public to the ecological consequences of deforestation, but mobilized protesters to occupy trees about to be cut in Australia's remaining rain forests. Splinter Green groups in Europe initiated a clandestine campaign against nuclear power by toppling hundreds of electrical towers near reactor sites.

Although these and other insurgent ecology groups differed from one another in many respects, their militancy set them apart from mainstream organizations and earned them the label of "radical environmentalism."

Philosophically, radical environmentalists tend to reject the notion that humans have a right to exploit or manage the natural world for human ends. Influenced by the writings of Thoreau, Muir, and Leopold, they argue that nature has a right to exist *for its own sake*. This position, often termed Deep Ecology or biocentrism, diverges from the ethics of mainstream environmentalism. Mainstream groups accept human control of nature, but advocate a more enlightened management that includes benefits often ignored by policymakers, such as recreation and wildlife protection. In contrast, radical environmentalists envision the restoration of large areas to a natural state in which nonhuman biological communities can evolve without the intrusions of industrial society. They rely especially on the findings of conservation biology, which suggest that small, isolated wilderness areas are inadequate to preserve species diversity. In short, mainstream environmentalism seeks to reform society's ecological policies; radical environmentalism supports a fundamental change in how society relates to nature. It urges that we remake our institutions to conform to the natural environment, rather than the other way round as has historically been the norm in Western cultures.

The tactics of radical environmentalists have been as controversial as their beliefs. Disillusioned with the political process embraced by traditional environmentalists, radical activists have engaged in ecological civil disobedience by chaining themselves to logging equipment or climbing into giant redwoods slated for cutting. Perhaps the most publicized of these protests occurred in 1981 when members of Earth First! unfurled a 300-foot-long plastic "crack" on the face of Glen

Canyon Dam to decry the damming of the Colorado River.

Ecotage, or monkeywrenching (so-called after Edward Abbey's novel, *The Monkey Wrench Gang*, which narrates the adventures of a band of ecologically minded saboteurs), is most often directed at saving wild lands and includes pulling up survey stakes on proposed logging roads, introducing abrasives into bulldozer engines, and placing metal spikes in trees to damage saw blades in mills. By the late 1980s resource industries operating in national forests were suffering approximately $20 million in damage annually due to ecotage.

Radical environmentalists argue that militant tactics have a twofold purpose: to publicize and to discourage ecological destruction by inflicting costly damage on those who perpetrate it. While radical environmentalists unanimously endorse civil disobedience, many reject ecotage, especially tree-spiking, seeing it as potentially dangerous to people.

During the 1990s radical environmentalism lost some of its ardor, and incidents of ecotage declined sharply. This was due in part to law enforcement efforts, including an FBI operation that led to the arrest of Foreman and others. Ideological differences between Leftist-oriented and biocentric activists contributed to the break-up of Earth First!, while disagreements over tactics provoked resignations by Sea Shepherd Society leaders.

Nevertheless, radical environmentalism has been immensely influential. Its media-oriented agitations changed the face of public lands policy in the U.S. Newspaper and television coverage of protesters confronting bulldozers or dangling 100 feet up in trees attracted national attention to controversies such as overcutting in the nation's forests. Agencies such as the U.S. Forest Service, which traditionally made major environmental decisions in obscurity, now must take into consideration the public scrutiny radical activism may bring to their policy choices.

Perhaps the greatest impact of radical environmentalism has been on the character of the environmental movement itself. Its militant demands made mainstream environmental groups look moderate by comparison, enhancing their credibility and blunting the charges of extremism industry often leveled at them. The flamboyant exploits of Earth First! and the Sea Shepherd Society also prodded mainstream organizations to act more aggressively, for fear of being overshadowed. Finally, the ecological ideals radical environmentalists brought to the debate have caused many traditional environmentalists to reexamine the direction of the environmental movement and to call for more ambitious goals.

CHRISTOPHER MANES

For Further Reading: Dave Foreman, *Confessions of an Eco-Warrior* (1991); Christopher Manes, *Green Rage: Radical Environmentalism and the Unmaking of Civilization* (1990); John Davis, ed., *The Earth First! Reader: Ten Years of Radical Environmentalism* (1991).

ENVIRONMENTAL ORGANIZATIONS

Modern social movements are strongly identified with the organizations that represent their causes, and this is especially true of the environmental movement. U.S. national and international environmental organizations have combined memberships numbering in the millions; they solicit and control multimillion-dollar budgets; and they employ large professional staffs of lobbyists, lawyers, and scientists. These organizations have played prominent roles in the development of the environmental movement, and are the most visible aspect of contemporary environmentalism worldwide. In addition, there are thousands of organizations dedicated to environmental protection at the local, state, and regional levels.

The modern U.S. environmental movement emerged in the 1960s, although it had strong roots in the conservation movement that began at the turn of the century. Despite significant cleavages among early conservationists, they shared a common concern for the protection of natural resources, especially land and wildlife. These conservation-era issues typically focused on protection of a particular geographical area. Modern-day environmentalists emphasize issues such as pollution that span large areas; directly affect humans as well as nature; and have complex causes and indirect, delayed, and elusive consequences. Rachel Carson's book *Silent Spring* (1962) exemplified this newer environmental perspective by describing the impacts of DDT on ecosystems. As the movement moved from a "conservation" focus to an "environmental" one, the older organizations incorporated the newer issues into their agendas, while new organizations developed specifically in response to environmental issues.

The National Environmental Lobby

Environmental organizations engage in a wide range of activities, including information campaigns, environmental research, litigation, and lobbying for the development and implementation of environmental legislation. Political lobbying, however, can threaten an organization's nonprofit status, and is thus avoided by many organizations. Twelve influential national organizations openly lobby for environmental policies, and can be considered the core of the national "environmental lobby" (see Table 1).

Table 1
Major Environmental Organizations in the U.S.

Lobbying Organizations				Non-Lobbying Organizations			
Era/Organization	Year Founded	1990 Membership[1] (thousands)	1990 Budget ($ Million)	Type/Organization	Year Founded	1990 Membership[1] (thousands)	1990 Budget ($ Million)
Progressive Era				**Direct Action**			
Sierra Club	1892	560	35.2	Greenpeace USA[2]	1971	2300	50.2
National Audubon Society	1905	600	35.0	Sea Shepherd Conservation Society	1977	15	0.5
National Parks and Conservation Association	1919	100	3.4	Earth First!	1980	(15)	0.2
Between the Wars				**Land & Wildlife Preservation**			
Izaak Walton League	1922	50	1.4	Nature Conservancy	1951	600	156.1
The Wilderness Society	1935	370	17.3	World Wildlife Fund	1961	940	35.5
National Wildlife Federation	1936	975	87.2	Rainforest Action Network	1985	30	0.9
				Rainforest Alliance	1986	18	0.8
Post-World War II				Conservation International	1987	55	4.6
Defenders of Wildlife	1947	80	4.6				
				Toxic Waste			
				Citizens' Clearinghouse for Hazardous Waste	1981	7	0.7
Environmental Era				National Toxics Campaign	1984	100	1.5
Environmental Defense Fund	1967	150	12.9				
Friends of the Earth	1969	30	3.1	**Other Major Organizations**			
Natural Resources Defense Council	1970	168	16.0	League of Conservation Voters	1970	55	1.4
Environmental Action	1970	20	1.2	Sierra Club Legal Defense Fund	1971	120	6.7
Environmental Policy Institute	1972	NA[3]	NA	Cousteau Society	1973	264	16.3
				Earth Island Institute	1982	32	1.1

[1]Membership data are for individual members. Data in parentheses are estimates.
[2]Greenpeace created a sister lobbying organization, Greenpeace Action, in 1988. Membership overlaps considerably between the two organizations.
[3]Not a membership group.

The three oldest organizations, the Sierra Club, the National Audubon Society, and the National Parks and Conservation Association, are early products of the conservation movement. Both the Sierra Club and the National Audubon Society currently rank among the top three environmental lobbying organizations in size, visibility, and influence. The Izaak Walton League and the National Wildlife Federation, the largest of the lobbying organizations, were both founded by sportsmen during a subsequent wave of conservationism between World Wars I and II. The Wilderness Society, also founded at this time, fought for the preservation of wildlands outside the newly developed national park system. Defenders of Wildlife, the last conservation-era organization, initially focused on the welfare of individual animals, but gradually broadened its attention to include wildlife habitat and endangered species.

The transition from the conservation era to the environmental era is often demarked by the first Earth Day in April 1970. However, from an organizational perspective, the transition is represented by the 1967 founding of the Environmental Defense Fund (EDF), the first of the new breed of national environmental organizations. Both EDF, whose origins stem from the battle to ban DDT, and the Natural Resources Defense Council (NRDC) were assisted greatly by Ford Foundation grants. Both are membership organizations staffed by scientists and lawyers devoted to environmental research and litigation.

Friends of the Earth (FOE), founded by David Brower after he was fired as the Sierra Club's executive director for acting too independently of its elected board of directors, and Environmental Action, created by the student organizers of Earth Day, have been regarded as much more radical than the other major organizations. This is because of FOE's uncompromising opposition to most forms of development, big business, and big energy, and because of Environmental Action's campaigns against businesses and members of Congress with poor environmental records. The Environmental Policy Institute (EPI), a non-membership lobbying organization backed by wealthy patrons, was organized by a group of FOE's Washington lobbyists who felt that FOE leadership and organization were inefficient. EPI has recently merged with FOE and the Oceanic Society.

During the "environmental decade" of the 1970s, these twelve organizations dominated the environmental movement. All except Environmental Action and the Defenders of Wildlife belong to an informal coalition called the "Group of 10" whose leaders meet periodically to discuss common problems and strategies. However, Environmental Action, along with the League of Conservation Voters and the Conservation Foundation, tends to work closely with the coalition. A few years ago the coalition produced a volume entitled *An Environmental Agenda for the Future* (1985) in an effort to set an agenda for the entire environmental movement.

From the first Earth Day in 1970 to its twentieth anniversary in 1990, most of the national environmental lobbying organizations experienced tremendous growth. It is significant that the older organizations also grew substantially in the years just before the first Earth Day, demonstrating that this event was as much the culmination of a process as it was the launching of a new movement. From 1960 to 1969, the total membership of the older organizations increased almost sevenfold, from 123,000 to 819,000. Their growth in popularity paralleled their increased aggressiveness in support of environmental protection.

The publicity surrounding the first Earth Day initiated further spurts in membership growth. In just three years, 1969 to 1972, nearly 300,000 new members were added to these conservation-turned-environmental organizations. These older organizations played a prominent role in mobilizing Earth Day. Having strong name recognition, in addition to substantial financial resources and expertise, they were able to capitalize on mass concern for environmental quality, gaining five out of every six new members of environmental organizations at this time.

Many observers predicted the decline of environmentalism following the fervor of the first Earth Day. However, despite some slowing of growth rates after 1972, a majority of the twelve organizations, especially the newer "environmental era" organizations, continued to experience significant growth during the 1970s. Further, the Reagan administration's attempts to limit environmental protection in the 1980s stimulated a surge of membership in the national environmental organizations. Several organizations utilized the publicity generated by Secretary of Interior James Watt's anti-environmentalist rhetoric and the Superfund scandal at the Environmental Protection Agency to launch successful recruitment campaigns.

Subsequent growth has been strong, so that by 1990 these organizations counted a total membership of more than 3,100,000. Highly publicized recent events, such as ocean beach contamination and the 1989 *Exxon Valdez* oil spill, combined with the substantial mobilization efforts made by these organizations in conjunction with the twentieth Earth Day celebration, enabled most of them to continue growing into the early 1990s. While some of the organizations suffered a decline after 1990, judging from past experience it would be surprising if their memberships did not grow over the rest of this decade.

Reasons for Success

The success of the national environmental lobby is due to several factors, a major one being broad-based public support for its goals. Public opinion polls have recorded strong and increasing support, albeit with some ups and downs, for environmental protection from 1970 to the present. The national organizations exploited this support by successfully recruiting mass memberships through direct mail campaigns. Although an efficient way to solicit members, recruitment by direct mail reduces the degree of commitment required for membership, the only requirement being to pay dues. This results in limited involvement in organizational affairs, and a high rate of annual turnover in memberships. The Izaak Walton League chose to forego developing such a "mass membership," preferring a relatively stable membership base. Environmental Action and Friends of the Earth tried but failed to achieve the level of membership growth that other organizations achieved during the 1980s, primarily due to financial and organizational difficulties.

A proliferation of environmental laws, changes in legal standing allowing non-economic interests into the courtroom and the revision of tax laws enabling organizations to maintain nonprofit status while engaging in non-educational advocacy, also strengthened environmental organizations' influence in national policy-making. Their success has turned the "environmental lobby" into an influential interest group, commanding considerable respect in national policy-making.

Consequences of Success

As is typical among enduring and expanding organizations, the success of these national environmental organizations both stimulated and stemmed from organizational attempts to become more professional, stable, and efficient. In contrast to earlier eras, when charismatic leaders attracted and led small groups of volunteers, today national environmental organizations have established complex organizational structures. They have lines of authority and responsibility formalized by constitutions and bylaws; they employ full-time, salaried staff who develop managerial and professional careers; they command the resources of legal and scientific experts; they have become masters at marketing their causes via a wide array of goods, from glossy magazines to recreational clothing; and they can mobilize considerable financial and political support, similar to other power brokers on the Washington scene.

The large increase in the number of full-time environmental lobbyists in Washington is one indicator of how professionalized the environmental movement has become. In 1969, before Earth Day, only two full-time lobbyists served the environmental movement. By 1975, the twelve major organizations employed 40 lobbyists; a decade later the number of environmental lobbyists had swelled to 88, and it continues to increase.

The tendency for organizations to become more bureaucratic and professionalized results from the need to adapt successfully to changing circumstances. As the environmental movement grew in terms of members and resources at its disposal, management was stretched beyond the capabilities of a few volunteers. The complexity of environmental issues, the expansion of environmental regulations, and the growth of government bureaucracy to implement the regulations further necessitated the employment of full-time scientists, lawyers, and administrators to enable organizations to operate in an effective and sustained manner. Another characteristic of enduring organizations, centralization, has also occurred within the environmental lobby. The migration of most organizational headquarters to Washington, DC, has given the organizations greater access to federal policymakers and also fosters cooperation among them.

Not surprisingly, these developments stimulated severe criticisms. Increased centralization and professionalization result in small cadres of bureaucrats who share little in common with mass members, and who tend to become more interested in organizational and job stability than in movement goals. Also, to maintain credibility within the political system, organizations and their staffs pursue "reasonable" goals, necessitating continual compromise within the political arena. An inevitable result has been that the national organizations are increasingly seen as being unresponsive to their members, too eager to engage in political compromise, and more concerned about their budgets and survival than about the state of the environment. One result of this perceived ossification and political co-optation of the "mainstream" organizations has been the development of new, more aggressive organizations pursuing environmental causes. Thus, the very resources that have made the environmental lobby a significant force in national politics have also generated increased strains within the movement.

Non-Lobbying National Organizations

Several other national environmental organizations are notable for the fact that they do not explicitly lobby. These non-lobbying organizations, while frequently overlapping in interests, focus on a broad range of concerns and employ a variety of tactics (see Table 1). Their activities include conducting research, litigation, education programs, grassroots organizing, land purchase and maintenance programs, and—in a few cases—direct action on behalf of environmental goals. Many developed explicitly in response to perceived weaknesses in the national lobbying organizations and represent some of the newer strains in the environmental movement: the rise of direct action or radical groups, the growth of international concerns, and the increasing prominence of grassroots groups.

The use of direct action, or tactics that go beyond conventional political action, has become increasingly popular. The first organization to pursue direct action as its primary strategy was Greenpeace. Its tactics, which have won Greenpeace an enormous following, have included plugging industrial effluent pipes and maneuvering inflatable boats between whaling vessels and their prey. This style of action was taken a step further by Earth First! and the Sea Shepherd Society, both of which are not averse to damaging property (e.g., tree spiking, "monkey-wrenching" bulldozers, and sinking whaling ships) and using other forms of "ecotage" (see Environmental Movement, Radical). Both groups were founded by disgruntled members of older, established environmental organizations who felt that those organizations lacked aggressiveness and were too eager to compromise.

Land and wildlife preservation organizations pursue a concern that has been important to environmentalists since the early conservation movement. Their focus, however, now reaches far beyond the U.S. Most of the organizations in this category focus primarily on other countries, especially those with tropical rain forests. Conservation International, which pioneered debt-for-nature swaps, was founded by members of the Nature Conservancy who felt that the Conservancy's program of purchasing land for preservation was not sufficiently international. The increasingly global nature of environmental problems has led to the development of organizations devoted specifically to the global environment, while organizations that previously had a national focus have increasingly developed international agendas, affiliates, and memberships.

The growth of local, grassroots groups reflects growing concern in communities around the nation over the location of environmentally hazardous facilities, such as toxic waste dumps or garbage incinerators.

Such groups often develop because community activists perceive a lack of interest in local problems by the national environmental organizations. Further, grassroots activists often distrust national organizations because the latter often seem willing to compromise the interests of local communities.

The widespread growth of grassroots environmental groups around the nation has prompted the development of national umbrella organizations that provide support and training and, more generally, facilitate the concerns of local groups. Two such national organizations promote grassroots activism on toxic waste issues. The Citizen's Clearinghouse for Hazardous Waste was founded in 1981 by Lois Gibbs, former Love Canal activist, on behalf of victims of toxic pollution by Love Canal in Niagara Falls, NY. This organization attempts to enhance the organizing skills of local groups across the nation. The National Toxics Campaign is somewhat more removed from the grassroots than the Clearinghouse, but provides technical, legal, and political assistance to local groups.

The proliferation of national environmental organizations in recent years prohibits an exhaustive listing of them, but other organizations illustrate some of the additional varieties. The Sierra Club Legal Defense Fund, an offshoot of the Sierra Club, engages exclusively in litigation. The League of Conservation Voters specializes in electoral politics, publishing environmental score cards for political candidates. The Earth Island Institute, founded by David Brower after his split with Friends of the Earth over its move to Washington, DC, supports various projects. The Cousteau Society is well known for its nature documentaries and environmental research projects, dealing especially with the world's oceans.

These non-lobbying organizations have expanded in number since the first Earth Day, and half were established in the 1980s. Most have also grown in size. Greenpeace's intensive direct mail campaigns enabled it to capitalize on its reputation for uncompromising and effective action but, some would argue, in the process it followed other national environmental organizations in becoming more professionalized (with all the attendant problems).

Summary

Overall, the national environmental organizations in the U.S. have clearly prospered during the past two decades. Not only have many earlier conservation-era organizations grown in size since Earth Day 1970, but the number and diversity of national organizations have dramatically increased (even aside from the large number of regional, state, and local organizations that have emerged since 1970). Although this inevitably results in some redundancy and competition for resources among organizations, overall this organizational network (enhanced by sub-national organizations) provides the environmental movement's greatest political strength. Undoubtedly, the national organizations will continue to provide a strong organizational base for the evolving environmental movement, and a potent political force for environmentalism, well into the 21st century.

ANGELA G. MERTIG, RILEY E. DUNLAP,
ROBERT CAMERON MITCHELL

For Further Reading: Susan D. Lanier-Graham, *The Nature Directory: A Guide to Environmental Organizations* (1991); Robert Cameron Mitchell, Angela G. Mertig and Riley E. Dunlap, "Twenty Years of Environmental Mobilization: Trends Among National Environmental Organizations," in *American Environmentalism: The U.S. Environmental Movement, 1970–1990,* eds. Riley E. Dunlap and Angela G. Mertig (1992); John Seredich, ed., *Your Resource Guide to Environmental Organizations* (1991).

ENVIRONMENTAL POLICY INSTITUTE

See Environmental Organizations.

ENVIRONMENTAL PROTECTION AGENCY, U.S.

The Environmental Protection Agency (EPA) was created in 1970 by combining many existing government organizations. The environment had become an important item on the American political agenda. Media coverage of the subject increased dramatically; public opinion polls showed the average citizen to be deeply concerned about environmental quality; and Edmund Muskie, who had made his reputation in large part chairing the Senate subcommittee on air and water pollution, was the leading contender for the 1972 Democratic presidential nomination.

Under these political circumstances, in late 1969 President Richard Nixon charged his Advisory Council on Executive Organization, headed by Roy Ash, with recommending a reorganization plan for the federal environmental agencies. The Ash council's first preference was creation of a Department of Environment and Natural Resources encompassing all the federal land management functions as well as pollution control. When the department proposal proved politically impractical, the council recommended creation of an Environmental Protection Agency. The EPA reorganization plan was approved by Nixon and ratified by

Congress, and on December 2, 1970, the new agency came into existence.

The main components of the agency were the Federal Water Pollution Control Administration from the Department of the Interior and the Air Pollution Control Office from the Public Health Service. The pesticides program in the new EPA was pieced together from four organizations, some of which had a long history of conflict with each other. The task of making disparate organizations work together was made more difficult by the absence of any overriding mission for the new agency. Unlike most major federal agencies, EPA has never had a charter, organic law, or any statutory statement about its overall goals and purposes.

The impelling force behind agency actions has been a complex and detailed set of laws aimed at specific environmental problems. This was part of EPA's inheritance, since the separate organizations from which it was formed came to the new agency with their own separate and unrelated legislative mandates. At the time the reorganization plan was submitted to Congress, the intent had been that the new agency would be organized on the basis of functions. William Ruckelshaus, the first administrator of the new agency, created functional offices for research, enforcement, and planning. A functional organization would have cut across the problem-based organizations (air pollution, water pollution, pesticides, etc.) of which the original agency was composed.

However, the Clean Air Act of 1970 began a trend of fragmented, detailed, command-and-control legislation that would continue to be the approach, accepted by both Congress and the executive branch, to dealing with environmental problems. The act, typical of those to come, contained numerous deadlines, specific standards, and detailed prescriptions for how the agency should handle various air pollution problems. It and the beginning of debate on what became the Water Pollution Control Act Amendments of 1972 created such an avalanche of work for the new agency that Ruckelshaus decided that, at least for the moment, the major part of the air, water, and other problem-based organizations had to be retained intact. Thus, in addition to the functional offices, he created an office for air and water programs and another office that encompassed pesticides, radiation, solid waste, and noise control. Over the years there have been many reorganizations and many permutations of offices, but the half-functional, half-programmatic organization remains a basic feature. The visibility, size, and importance of the programmatic offices have always overshadowed the functional offices.

Another continuing legacy of the agency's genesis has been a high degree of regional decentralization.

More than 75% of EPA personnel are located in the ten regional offices and the dozen EPA laboratories. Regional administrators, who are political appointees, have a large amount of freedom in dealing with problems in their states, although they are constrained by the budgetary and other ties between the regional personnel in a particular program and the Washington headquarters staff of the same program. For example, water people in the EPA regions work as much for the Washington leaders of the water program as for the regional administrators.

Ruckelshaus was the first of a generally distinguished line of EPA administrators. A state legislator and senatorial candidate from Indiana, he was sensitive to the political requirements for getting the new agency on its feet. As a lawyer and former assistant attorney general he considered regulations and enforcement as the logical methods for dealing with pollution. Several enforcement actions and the banning of the pesticide DDT established a "Mr. Clean" image for Ruckelshaus and a tough anti-pollution image for the agency.

In the wake of the Watergate scandal, Nixon asked Ruckelshaus to temporarily assume directorship of the FBI. The president named Russell Train to become EPA administrator. Also a lawyer, Train had served as undersecretary of the Interior Department in 1969, and as the first chairman of the new Council on Environmental Quality in 1970. Train became EPA administrator in 1973 and served until the Carter administration took office in 1977. Train successfully steered the agency through a difficult period, a time when the enthusiasm of the 1970 Earth Day had waned and the administration was concerned with the environment primarily as a drag on the economy or an obstacle to dealing with the energy crisis.

Douglas Costle, Train's successor, continued the tradition of lawyer-administrators. Costle had headed the Ash council team that wrote the reorganization plan creating EPA and had gone on to serve as the innovative head of Connecticut's environmental agency. Although President Carter personally was more supportive of environmental concerns than either Nixon or Ford had been, the rapid rise in energy prices in 1976 and 1977 colored environmental policy throughout the Carter term. EPA was still viewed by the White House primarily as an obstacle to achieving energy efficiency and to increasing domestic energy production, and also as an agent of cost increases in an economy bedeviled by high rates of inflation. Environmental concerns were viewed as peripheral nuisances and EPA as a maverick agency that interfered with smooth policymaking in other areas. These views have characterized the executive branch context in which the agency has operated for most of its history.

For the administration of President Ronald Reagan, elected in 1980, environmental regulation was not just a nuisance, it epitomized governmental interference with the efficient functioning of the free market. Consistent with this view, Reagan appointed Anne Gorsuch (later Anne Burford) as EPA administrator. Gorsuch was a lawyer, a former member of the Colorado legislature, and a protégée of James Watt, Reagan's outspokenly anti-environmental Secretary of the Interior. Under the Reagan-Gorsuch regime, EPA's budget and personnel levels were sharply cut. Morale within the agency plummeted. The number of enforcement actions declined sharply. By every measure, the Reagan administration's goal of reducing environmental regulation to the maximum extent was being successfully implemented.

Implementation of the Reagan goals, however, produced a backlash. Congressional Democrats and some Republicans were outraged. Membership in environmental organizations skyrocketed. The press began featuring stories about EPA political appointees doing favors for industry and maintaining blacklists of environmental scientists. Major segments of industry that were dependent on the agency taking action (to register new pesticides, for example) began protesting about EPA paralysis. Other members of the business community worried that Gorsuch's political ineptness and the pro-environment backlash would result in stronger environmental regulation than had existed before Reagan took office.

In December 1982 the House of Representatives voted to cite Gorsuch for contempt of Congress, making her one of the highest ranking government officials ever subjected to a contempt citation. The following February, Gorsuch asked Rita Lavelle, EPA assistant administrator in charge of hazardous waste, to resign. When Lavelle refused, she was fired by the president. Three weeks later Gorsuch fired the agency's inspector general and the assistant administrator for administration. On March 1, 1983, John D. Dingell, one of the most powerful Democratic congressmen, wrote the president stating that he had evidence of wrongdoing and criminal conduct at EPA. On March 9, Gorsuch submitted her resignation to the president, saying she hoped it would "terminate the controversy and confusion that has crippled my agency."

The agency had indeed been crippled by the decline and fall of Gorsuch, and its reputation for integrity and capability, nourished during the first 12 years of its existence, had been badly tarnished. To try to repair the damage, President Reagan persuaded EPA's first administrator, Bill Ruckelshaus, to return to the position. Ruckelshaus brought with him a half dozen of the top officials who had served on the first Ruckelshaus team. While they made good progress in restoring mo-

rale within EPA and public confidence in the agency's integrity, the wounds of the Gorsuch era would take many years to heal.

In 1985 Ruckelshaus was succeeded by Lee Thomas, a career government manager who had been the assistant administrator for hazardous waste programs. During the second term of the Reagan administration there was little that an EPA administrator could do except defend against periodic assaults against the agency's authority. Thomas proved politically adept at conducting such defenses. Also, he continued the process of repair of the agency's reputation that had been started by Ruckelshaus, involved EPA in one of its first major successful international negotiations (the Montreal Protocol dealing with stratospheric ozone depletion), and started the agency on the path of risk-based decision making.

George Bush came into office in 1989 and appointed William K. Reilly, head of the Conservation Foundation and World Wildlife Fund, to be EPA administrator. Reilly, a lawyer and city planner by training, was the first career environmentalist to head the agency. Several of the most visible and important problems facing the agency were global in scope—notably global warming and depletion of the stratospheric ozone layer. This was congenial to Reilly, who had extensive international experience and a keen sense of the importance of cooperation with other nations. One of his first decisions was to elevate the status of the EPA International Office.

For most of its history the agency had perceived itself purely as an agent of statutory mandates without any independent judgment or responsibility. Reilly instituted a strategic planning process, encouraged risk-based priority setting, and tried to strengthen the scientific capability of EPA, all in an effort to better equip the agency to make difficult choices and to take some responsibility for setting the environmental agenda. In part, the effort was made necessary by the continuing great disparity between the tasks given to EPA by legislation and the resources given to the agency to perform the tasks, although Reilly did succeed in obtaining significant increases in the EPA budget.

The effort also was spurred by proposed legislation to make EPA a regular cabinet-level department. The legislation enjoyed widespread support but was bedeviled by political problems, fights over specific language, and disputes over what provisions would accompany elevation to cabinet status.

Given its difficult history, especially the disparity between demands and resources, EPA has accomplished a great deal. The contribution of the agency's programs to environmental improvement cannot be measured definitively, but environmental conditions, as measured by conventional indices, have improved since 1970. It is reasonable to assume that EPA's efforts have been an important factor in bringing about the improvement. In some cases, such as the 95% reduction in lead in the air, the connection between EPA's programs and environmental progress is clearcut. The combination of EPA's regulations reducing (eventually eliminating) the lead content of gasoline and its automobile emission control requirements clearly account for the sharply reduced levels of lead in ambient air.

The levels of all air pollutants emitted in large volumes have been reduced since 1970, with the exception of nitrogen oxides, which have increased somewhat. Emissions of particulates and carbon monoxide have been cut in half. Emissions of sulfur dioxide and volatile organic compounds have been reduced about 20%.

Data on water quality are not good enough to answer whether the nation's water is getting cleaner. Gross pollution from industrial sources probably has been reduced (although to what extent is not known) and the proportion of the population served by adequate sewage treatment has increased. However, contaminated water entering lakes and streams from agriculture, city streets, mines, and other diffuse sources has not abated and has probably grown worse.

The federal government's role in dealing with solid waste has grown from almost nothing to a multi-billion dollar program since the late 1970s. Most visible of the EPA efforts is the Superfund program to clean up abandoned hazardous waste sites. More than 30,000 potentially hazardous waste sites have been identified, about half of which have been determined not to require any federal action. By the fall of 1992, EPA had completed all clean-up work on less than 150 sites, leading to criticism that the pace of the program is too slow and the costs of the program are excessive.

EPA's success in dealing with some of the conventional air and water pollutants has been offset by its inability to deal with some newer problems. Toxic substances in all parts of the environment have been a problem for the agency, in part because the structural separation of the air, water, and waste offices of the agency does not lend itself well to dealing with substances that transfer easily from one part of the environment to another. Indoor air pollution is a significant source of adverse health effects, and levels of indoor pollution generally are unrelated to outdoor levels. However, neither Congress nor EPA has shown much inclination to deal with the issue, although the agency has conducted and made available studies about the adverse effects of tobacco smoke on nonsmokers.

EPA's current resources of about 17,000 people and a budget of almost $7 billion make it by far the largest

of the federal regulatory agencies. However, almost every year the agency's responsibilities have grown at a faster pace than its resources. In the coming years it will likely have to face a new and broadened agenda including a vastly expanded set of international responsibilities and the need to develop the capability to influence policy in agriculture, transportation, energy, and other sectors. Major emphasis will be placed on pollution prevention, other integrated cross-media approaches to dealing with environmental problems, and market-based mechanisms such as effluent charges and trading of emission permits. These new responsibilities will make the agency's activities more diverse, transcending the promulgation of regulations. They also will strain EPA's skills and resources to the limit.

TERRY DAVIES

For Further Reading: Environmental Protection Agency, *Securing our Legacy* (1992); Marc J. Roberts, *The Environmental Protection Agency: Asking the Wrong Questions* (1990); Walter A. Rosenbaum, *Environmental Politics and Policy,* 2nd ed. (1991).
See also Pollution, Non-Point Source.

ENVIRONMENTAL REGULATION

Economic regulation for environmental goals is not a new idea. The U.S. has long been concerned about protecting its air and water supplies. It was the "environmental revolution" of the late 1960s and early 1970s, however, that sparked the creation of the Environmental Protection Agency (EPA) and passage of the National Environmental Policy Act. These early milestones were soon followed by a host of major environmental laws concerning air, surface water, drinking water, pesticide use, hazardous waste management, and control of toxic chemicals.

The need for environmental regulation arises from the notion that the natural environment is a "public good." No one owns the air, the water supply, or the wildlife population; they are common property resources. There is an incentive for individuals, industry, and government to use these "free" resources as much as they want. Unfortunately, this often leads to overuse and degradation of environmental resources. Unguided market forces are not good at pollution control and protecting the environment. Government policy is needed to restore an efficient allocation of these resources and to protect the environment from the effects of overuse and pollution.

U.S. environmental laws are written by Congress, with specific regulations developed and implemented by the EPA and other relevant agencies. The process of developing regulations is a public one with participation by environmental and health advocates, manufacturing and pollution control industries, and state and local environmental officials. Scientists from a variety of disciplines submit and validate data on the health and safety risks of the issue under review; engineers impart their expertise on the technical feasibility of proposed regulatory control options. The voice of the public is heard as well, particularly over community issues such as the siting of an incinerator or sludge facility or the remediation of an abandoned toxic waste site.

State and local governments also play an increasingly important role in the regulatory process. Not only are they often responsible for monitoring and enforcing federal regulations, but state and local governments are taking on more responsibility for drafting their own environmental regulations. A key example of this is California's Safe Drinking Water and Toxic Enforcement Act of 1986. Better known as Proposition 65, this law goes beyond current federal regulations in prohibiting certain discharges of toxic chemicals and requiring that those exposed to them be warned of the risks.

Historically, many environmental regulations have used "command and control" strategies to set emissions or disposal limits. Regulations under the Clean Air Act, for example, include such "end-of-pipe" controls as scrubbers on factory smokestacks and catalytic converters on automobile exhaust systems. These types of regulations often call for use of the best available technology to reach the lowest achievable emissions rates.

Command and control strategies are seldom cost-effective, however. All firms are forced to achieve the same level of control regardless of their individual control costs. To ameliorate this problem, there has been growing interest in more decentralized incentive-based regulations, in which firms are given greater flexibility to choose the least expensive methodology to reach emissions targets. Types of such market-based regulatory approaches include marketable permits, emissions charges or taxes, and deposit-refund systems. By emphasizing economic incentives, firms and individuals are also encouraged to explore pollution prevention options such as source reduction, substitution of nonhazardous chemicals, and recycling.

Environmental regulation is not cheap. The costs to the government of implementing and enforcing the regulations are substantial. The need for resources at EPA and state agencies has grown significantly over the past several decades as public demand has led to a burgeoning number of environmental responsibilities.

The costs to those that must comply with the regulations are steep as well. Industry spends billions of dollars annually on technology and other efforts to heed the emissions, disposal, and other regulations. Some industries are forced out of states with tight environmental standards, while others leave the U.S. altogether. Numerous firms establish industrial plants in nations—often developing nations—where the regulatory controls are less strict.

Environmental regulation raises distributional and ethical issues as well. As industry passes regulatory costs onto consumers in the form of higher prices, lower-income populations may suffer disproportionately. Income that might otherwise be used for health care, education, or other benefits goes instead to pay for environmental protection. Similarly, as regulatory costs force some industries to go overseas, there is not only the loss of American jobs and income, but also the exportation of our pollution to poorer nations. Compensating to some extent for these income losses is a growing environmental cleanup industry that offers many high-skilled jobs and business opportunities in the U.S. and abroad.

Environmental protection is an international issue. Concern for the global environment has led to the development of several major international treaties calling for protection of the Earth's resources and reductions in environmental toxins. The 1987 Montreal Protocol treaty calls for a 50% reduction by 1998 in worldwide production of chlorofluorocarbons (CFCs) to protect the stratospheric ozone layer that shields the Earth from excessive levels of ultraviolet radiation. The treaties signed at the 1992 United Nations Conference on Environment and Development, or Earth Summit, in Rio de Janeiro, Brazil, include such key efforts as controlling emissions of carbon dioxide and other greenhouse gases believed to be instrumental in global warming and developing voluntary principles aimed at preserving the world's forests. A treaty on biodiversity and protecting endangered plant and animal species was also signed by many of the 170 nations present. While many important issues remain in the international environmental arena, these types of agreements symbolize the recognition of environmental protection and regulation not only as an issue to be grappled with by individual countries but as a global concern as well.

SUSAN W. PUTNAM
JOHN D. GRAHAM

For Further Reading: Jessica Tuchman Mathews (ed.), *Preserving the Global Environment* (1991); Paul R. Portney, (ed.), *Public Policies for Environmental Protection* (1990); Tom Tietenberg, *Environmental and Natural Resource Economics* (1992).
See also Clean Air Act; Economic Incentives; Economics; Environmental Protection Agency, U.S.; Law, Environmental; Law, International Environmental; National Environmental

Policy Act of 1969; United Nations Conference on Environment and Development; United States Government.

EPA

See Environmental Protection Agency, U.S.

EPIDEMIOLOGY

Epidemiology is generally defined as the study of the distribution of disease in human populations and the factors that determine the observed disease distribution. Characteristics of people and their environment are examined to uncover associations that are causally linked with the occurrence of human disease. When these studies focus on specific problems related to environmental exposures they fall in the domain of environmental epidemiology. Examples of current issues are studies of the human health effects of exposure to hazardous waste sites, pesticides, lead, asbestos, and nonionizing radiation, such as electric and magnetic fields. These exposures may occur via a number of routes such as air pollution, water pollution, contact with industrial chemical wastes, or ingestion of contaminated foodstuffs.

There exists a tendency to consider environmental epidemiology as a relatively young discipline. This view, however, is not supported by the history of medicine and public health. Epidemiology is the core science of public health, arising from the earliest concerns regarding the effects of the environment on human health. In fact, one of the earliest medical scholars, Hippocrates, wrote a treatise around 400 B.C. entitled "Airs, Waters, and Places," wherein he cautioned people about the health-related risks associated with polluted air and water and other environmental influences on health. The ancient Greeks were the first to look at disease scientifically. They rejected notions of disease as punishment for sin and instead studied the association of certain diseases with aspects of the environment and lifestyle. For example, they observed that it was unhealthy to live near swamps, engage in certain lifestyles, and eat certain foods.

One of the best known early community studies of an environmental exposure is the investigation by a London physician named John Snow of the cholera epidemics in England in the 19th century. We know today that cholera is a bacterial disease, but before the identification of microbial agents at the end of the 19th century, the belief that disease was a consequence of evil behavior co-existed with the recognition of contagion.

Cholera is characterized by severe diarrhea, vomiting, and muscular cramps. The diarrhea can be so severe as to cause extreme dehydration, collapse, and even death within a few hours after the onset of the illness. What was not known at the time of Snow's observations on cholera is how cholera is transmitted. There existed evidence that it could be transmitted by close personal contact, but there was also evidence that some who had close personal contact with the sick, such as physicians and undertakers, rarely were afflicted. Furthermore, outbreaks could occur at places located at considerable distances from already existing cases of the disease.

Snow's observations of cholera in London led him to conclude that cholera can be transmitted through the water supply if the water is contaminated with sewage carrying the excretions of cholera victims. The distribution of the cholera cases in the Soho district and surrounding districts of London suggested to Snow that sewage was leaking into the water supply of the pump located at the corner of Broad and Cambridge Streets. He recommended that the handle of the Broad Street pump be removed so that the people living in the vicinity of the pump, where the epidemic was raging, would be forced to draw their water from other pumps located several streets away. The handle of the pump was removed, but, as Snow noted, by this time the epidemic had subsided, perhaps because many neighborhood residents had either succumbed to the disease or fled. Thus, this historic public health intervention did not produce a dramatic effect on the number of new cases of cholera in the district, and did not provide confirmatory evidence for the hypothesis regarding the transmission of cholera.

The confirmatory evidence was provided by Snow's carefully controlled observation of the cholera experience in a single neighborhood which was supplied by two different water companies. The two water companies, the Lambeth Company and the Southwark and Vauxhall Company, both drew their water from the River Thames, originally from locations that could be expected to be contaminated with sewage. After the cholera epidemic of 1849, the Lambeth Company moved its waterworks upstream, whereas the Southwark and Vauxhall Company remained in the original location. Both companies continued to provide drinking water to this single district in London, which consisted of approximately 66,000 households. Snow determined the water supply for each individual house in the district and the houses where cholera deaths occurred. Of the 1,361 deaths from cholera in this district during the 1853 epidemic, 1,263 occurred in the 40,000 households receiving water from the Southwark and Vauxhall Company, and 98 deaths occurred in 26,000 houses serviced by the Lambeth Company.

Snow expressed the company-specific cholera mortality rates as deaths per 10,000 houses. In this manner Snow noted that the mortality rate for houses supplied by the Southwark and Vauxhall Company (315 deaths per 10,000 houses) was 8 to 9 times greater than the mortality rate for houses supplied by the Lambert Company (37 deaths per 10,000 houses). This comparison provided strong evidence that the household source of water was causally associated with the risk of cholera death, and more importantly provided the evidence needed to intervene effectively to prevent the further spread of this disease and future epidemics.

Snow's observations utilized what is termed an "experiment of nature" in that one group in the community was exposed to what was believed to be the cause of cholera, and another group in the same community, which was similar in every relevant characteristic to the first group, was not exposed to the contaminated water supply. The use of "natural experiments" in epidemiology allows an investigator to utilize a research approach in an observational setting analogous to a controlled experiment. Epidemiologic research methods incorporate this basic contrast into the design of most studies. Since obvious ethical and practical prohibitions on experimentation with humans exist, epidemiologic data must be collected in human populations on the natural occurrence of the disease and the suspected environmental exposures under study. Numerous historical examples exist where an epidemic was traced to an environmental cause. For example, the gastrointestinal effects of lead were studied by Baker in the mid-18th century. He associated the colic experienced by cider drinkers in Devonshire, England, with lead intoxication. At that time, it was the practice to transport cider in leaden containers or to process the cider in lead-containing presses. The acidity of the cider caused sufficient lead to be leached from the containers and presses into the cider to cause acute lead toxicity.

Because epidemiology draws its conclusions from observations of the natural distribution of disease, epidemiologic studies possess both unique strengths and limitations when compared to controlled laboratory studies. When humans are the subjects of study, the problem of extrapolating from animal experiments is avoided. On the other hand, epidemiologic research generally provides less conclusive findings than laboratory research. In a laboratory investigation of a suspected environmental toxin, it is assumed that the animals under study differ only on the basis of their exposure regimen. Any ensuing differences that are found between exposed and non-exposed animals can then reasonably be attributed to the exposure itself.

A broad array of individual health indicators may be studied to determine the impact of environmental agents on the health of a community. These indicators range from excess deaths, increases in illness, aggravation of the health status of those already ill, production of symptoms, impairment of functions, and annoyance. All of these indicators may demonstrate increases or decreases with or without any association with environmental exposures. The challenge in environmental epidemiology is deciphering which of these fluctuations in health outcomes are due to the adverse effects of specific environmental exposures, and therefore, which are preventable.

Age-adjusted cause-specific rates of certain cancers (e.g., liver, lung, or stomach) vary substantially by geographic area. Thus, it is often assumed that the excess rate above the rate of the lowest area is due to "environmental factors." Until such environmental factors can be specified and other modifying or causal factors such as genetic susceptibility, exposure to infectious disease agents, and access to medical care can be considered, such assumptions should be regarded cautiously.

Today, it is generally accepted that epidemiologic studies can be appropriately used to study all diseases, conditions, and health-related events in human populations. In studying disease as a phenomenon of human populations, individuals are classified as "diseased" or "non-diseased." The development of disease is often an evolving process and the stage at which an individual is labelled "diseased" rather than "non-diseased" may be arbitrary. All diseases, especially chronic diseases which persist for many years, have a natural life history. The natural history of disease describes the course of disease over time unaffected by treatment, and includes the identification of factors which favor the development of disease. These factors (known as "risk factors") may antedate the appearance of clinical disease by many years, even decades.

Risk factors may or may not be susceptible to change. Some factors such as age, sex, race, and family history of disease are major determinants of risk and not subject to change. Other risk factors (e.g., cigarette smoking, exposure to environmental agents at work or in the home, and elevated blood pressure) can be altered. Understanding disease risk factors provides a basis for determining the most effective approaches for disease prevention and control. Understanding the natural history of disease allows one to consider the stages of disease along a spectrum from pre-disease — when susceptibility and risk factors are important considerations — to later stages of presymptomatic and clinical disease. At the stage of presymptomatic disease there is no manifest disease but pathogenic changes have started to occur, such as premalignant, and sometimes malignant, tissue changes or alterations in lung function. At the stage of clinical disease sufficient tissue changes have occurred so there are observable signs and symptoms of disease.

As disease evolves over time the pathologic changes may become fixed and irreversible. Since the implicit goal of epidemiology is the prevention of disease, the aim of epidemiologic studies and public health interventions is to push back the level of detection and intervention, and to reduce exposure to internal and external factors which increase the probability of disease.

Typically persons with the same disease are grouped together according to the cause of the illness. Prior to the advances in bacteriology in the late 19th century, human diseases were classified on the basis of symptomatology. Thus, there were the poxes, fluxes, and fevers, which today might be described as rashes, gastrointestinal disturbances, and febrile illnesses. With the increased knowledge of pathology and physiology, disease classification progressed from a basic description of symptoms to a classification based on the manifestations of disease such as abnormal physiology or morbid anatomy. The most significant advance in disease classification, however, occurred when specific organisms were linked to major disease entities such as cholera, plague, and tuberculosis. As a result an etiologic classification (based on the cause or origin) was added to that based on the manifestations of disease. For example, in the 19th century various clinical entities of tuberculosis were described such as "scrofula" of the cervical lymph glands and the "consumption" of respiratory tuberculosis. When the role of the tubercle bacillus was understood, these seemingly discrete clinical entities were then recognized as manifestations of the same disease and related to each other by the presence of a single agent, the tubercle bacillus.

The etiologic basis of many infectious, nutritional, and metabolic diseases are known, and this knowledge has resulted in rational approaches to disease prevention.

Unfortunately, with some exceptions, such as environmental asbestos, lead, and ionizing radiation, environmental exposures do not usually cause a specific type of illness or impairment, but rather are nonspecific, and increase the risk of diseases caused by other factors. For example, asthma, cancer, heart disease, and impaired lung capacity have many causes, and when studying the association of these conditions with a specific environmental exposure it will be necessary to consider the effects of all other relevant exposures. In fact, seldom can disease in human populations be attributed to a single agent or to the operation of any one factor. Rather, disease occurs as a consequence of the interaction of two sets of factors; some intrinsic, which operate to increase the susceptibility of the human host to disease, and some extrinsic, which influence the opportunity for exposure in the external environment. The susceptibility of an individual to disease at any given time results from the interactions of genetic factors with environmental factors over the entire lifespan. For some conditions the relative contributions of genetic and environmental factors are quite readily distinguished, whereas for others it is not as clear.

The relationship between genetics and environmental agents embraces two distinct but complementary phenomena. The first, referred to as ecogenetics, concerns inherited variation in susceptibility to exogenous (environmental) agents. The second approach, genetic toxicology, studies damage to the genetic apparatus of an individual resulting from environmental exposures. In addition, previous genetic damage may increase susceptibility to subsequent environmental exposures, so these two processes are clearly interrelated.

A rapidly increasing number of genetically determined factors have been identified which relate to the susceptibility to certain diseases. For example, individuals with a genetically acquired skin condition called *xeroderma pigmentosum* also have a genetically determined inability to repair damage induced by ultraviolet light and are thus at unusually high risk of developing multiple skin cancers, including melanomas, in areas of skin exposed to sunlight. Additionally, recent studies have shown that individuals with certain genetically determined enzymes metabolize chemicals to which they are exposed differently than individuals lacking such enzymes, and that these metabolic differences may increase or decrease susceptibility to cancer.

Although the focus of environmental epidemiology in recent years has been on physical aspects of the environment, the environmental or extrinsic factors considered by epidemiologists in studies of disease causation include biological and social factors, as well as physical factors. The physical aspects of the environment include heat, light, air, water, radiation, gravity, atmospheric pressure, and chemical agents of all kinds. Environmental factors may also indirectly affect susceptibility. For example, the increased risk of lung cancer associated with asbestos exposure is approximately 10 times greater for smokers compared to nonsmokers.

In developed areas of the world, a great deal of control has been established over the physical environment through the provision of adequate shelter, clean drinking water, sewage treatment, and indoor heating and cooling. New environmental problems, however, continue to arise as old ones are solved. Air pollution, for example, continues to represent an urgent threat to health in heavily polluted areas. When weather conditions are unfavorable, masses of polluted air can be trapped and hang over a city for several days, thereby exposing the residents to high concentrations of a number of toxic substances. A more recent concern, however, is the air pollution in indoor home and office environments. The increasing costs of fuel have, in many instances, led to complete insulation of homes and the reduction in fresh air circulation. These decreases in ventilation often result in concomitant increases in indoor pollutants.

The most challenging aspect of environmental epidemiology is that concerned with estimating individual exposure to the environmental agent(s) under study, especially when the exposures occurred in the past and direct measures of these exposures are not available. The emerging field of "exposure assessment," which is a crucial component of environmental epidemiologic studies, encompasses numerous techniques to measure or estimate the contaminant, its source, the environmental media of exposure, avenues of transport through each medium, chemical and physical transformations, routes of entry to the body, intensity and frequency of contact, and spatial and temporal concentration patterns.

Much greater emphasis used to be placed on the measure of disease than on the measure of exposure in epidemiologic studies of environmental (physical) agents. This may be due in part to the clinical orientation of most epidemiologists and the greater familiarity with biological and social causes of disease than with physical causes. Now environmental epidemiologists are collaborating effectively with toxicologists and environmental scientists to improve the quality of exposure data and to develop methods that provide better estimates of cumulative exposure by accounting for uptake, metabolism, and excretion of toxic materials. This approach has improved the validity of exposure-response estimation over a much broader range of exposure levels, allowing for more effective prevention strategies.

PATRICIA A. BUFFLER

For Further Reading: Charles Hennekens and Julie Buring, *Clinical Epidemiology* (1990); J. M. Last and R. B. Wallace, eds., *Maxcy-Rosenau-Last Preventive Medicine and Public Health,* 13th ed. (1992); K. Rothman, *Modern Epidemiology* (1986).
See also Exposure Assessment; Health and Disease; Indoor Air Pollution; Lead; Public Health; Risk; Toxicology; Water Pollution.

EQUITY

It is generally agreed that equity means treating equals equally, but this simple formula yields a number of approaches to equity and the environment. The issue can be broken down into two major questions: Equity toward whom? Equity in what time period?

Equity Toward Whom?

Citizens

Until recent times it has been conventional to assume that issues of equity arise only with respect to persons, and that a government's duties of justice apply only to its own citizens. The focus of environmental policy in the wealthy industrialized countries such as the United States has been on the health of their own citizens: air and water quality and the disposal of hazardous substances have had top priority. Two conceptual frameworks have been applied to the distribution of risk in the national population: price internalization, or efficiency, and vulnerability. Most environmental regulations attempt to protect the highly vulnerable in society: air quality must be safe for those with impaired lungs, and water must be safe for all to drink. Recently, increased attention has been paid to the distribution of environmental risk within industrialized nations, the argument being that the poor and those from racial minorities often have much higher exposure to dangerous substances, and that this both derives from and compounds other instances of injustice. In general, the benefits of environmental regulations are seen to accrue more to the middle and upper classes than to the poor. When the key variable is considered ethnicity rather than poverty, the policy is labeled "environmental racism."

Neo-classical economists have argued that these laws aimed at protecting the vulnerable produce inefficiencies—that the costs exceed the benefits. They argue that the dollar value of the benefits is outweighed by the losses in consumption caused by expenditures on pollution control and similar devices. They assert that there is an optimal or efficient level of pollution: the amount that occurs when the economic costs of marginally reducing the pollution equal the benefits. They advocate a sharp line between equity and efficiency: environmental policy should aim at efficiency, and issues of equity should be treated as problems concerning the distribution of income. Critics of this view argue first, that this point of view is incapable of distinguishing between needs and preferences since it looks only at market or market-like values which do not discriminate along this dimension; and second, that individuals in the neo-classical tradition typically have little systematic to say about what the distribution of income should be—hence *de facto* ratifying the current distribution of wealth.

Duties Beyond Persons

In the Judeo-Christian cultures of western Europe the dominant view until around 1800 was that moral concerns applied only to persons. Theological doctrines from the Bible concerning the uniqueness of human beings were reinforced by the Cartesian cosmology that held that human cognition was different from all of the rest of the world, and that animals were incapable of experiencing suffering.

This perspective began to weaken in the early part of the 19th century. Common sense and advances in comparative anatomy showed that the Cartesian view had no basis in fact, and humane societies in both Britain and the U.S. successfully outlawed certain treatment of animals. Further (and slightly later), the account of natural selection Charles Darwin applied to human beings in *The Descent of Man* depended on the idea that persons had evolved from other animals and were not different in kind from them.

The doctrines of utilitarianism held by Jeremy Bentham and later by John Stuart Mill argued that morality required maximizing pleasure and minimizing pain. Bentham was especially explicit in noting that morality included all animals capable of such emotions. These concerns are carried forward today in the animal rights and animal liberation movements, and form the basis of certain forms of vegetarianism. In general these groups are concerned with individual animals such as those used in the testing of cosmetics and drugs; as well as the conditions of animals on farms such as crowding, and the means—and in some cases even the practice—of animal slaughter.

Early in the twentieth century another set of considerations began to weaken the human-centered focus of environmental equity. In the U.S., concern arose that resources, particularly forests, were being overused or misused, and that reliable and inexpensive forest products would run out, either in the near or the long term. Looking at it from a utilitarian perspective and anticipating the neo-classical school described above, Gifford Pinchot—the founder of the U.S. Forest Service, found this over exploitation undesirable because it undercut human happiness.

Parallel to, and standing in criticism of, the Pinchot utilitarian perspective was a school most identified with John Muir, the founder of the Sierra Club. Beginning a reversal of earlier interpretations of biblical texts, Muir argued that the earth and the creatures on it were sacred because they were creations of God. Nature thus was to be respected as such without regard to utilitarian considerations. It was possible to treat nature herself unfairly. Modern ecologists, most notably Aldo Leopold, have made common cause with Muir—though without his explicit theological premises. Drawing on a Darwinian perspective members of this school argue that humans are part of a broad community of life, and that this community is to be respected—with humans being simply members of it among others. Though they would appear to have much in

common with the animal rights movement, the view that nature is to be respected in her own right leads in quite different directions: the former arguing that the suffering of animals is to be prevented or mitigated, the latter that the processes of nature that involve predation—and hence suffering—are not to be disrupted.

Duties Beyond Borders

Another development that has eroded the idea that environmental equity is a concern only for fellow citizens is the vast expansion in world trade since World War II, which has made formerly exotic plant and animal products commonplace, at least for the rich. Large human migrations, too, have become more pronounced during the same period, often with individuals moving back and forth between new countries and those of their origin. Many immigrants send money and consumer goods to their families left behind, rendering national borders more porous to moral concerns. It has simply become less clear who is "us" and who is "them."

This de facto blurring of national borders is reinforced by traditional moral theories, which customarily refer to obligations among and to all persons. From the teachings of Jesus to the doctrines of classical utilitarianism, the language we use to describe our obligations does not limit them to members of our own nation. The convergence of porous borders with the universalist features of our moral theories has meant that issues of environmental equity have become global in the minds of many. Hence environmental groups, implicitly invoking the vulnerability model of obligation, often criticize trade in pesticides and the disposal of hazardous wastes in developing countries as inequitable, voicing the same concerns over differential risks as are experienced by the domestic disadvantaged. On the other hand, economists invoking the neo-classical efficiency standard argue that in many cases it is more efficient to put polluting industries and dispose of hazardous wastes in developing countries, since the market typically places a lower value on goods and services there.

The expansion of our moral concern to species other than our own has an international dimension. With the great increase in the loss of biodiversity—accelerated since World War II—and the gradual expansion of moral concern beyond the human species, many groups now consider the actions of persons in other countries or by other governments to be inequitable if they threaten plants, animals, or ecosystems within their country. This school of thought essentially regards the biosphere as the common property of all mankind, limiting the rights of individuals and nations by strong obligations of preservation and, where appropriate, restoration.

There has been a counterreaction to the gradual expansion of moral concern to other species, as well as other nations. In the U.S. one response has been the property rights movement, which stresses that individual property owners have strong rights to use their property as they wish—rights that can be curtailed only by showing that there is a clear and tangible harm to another human individual or his/her property. In a parallel fashion, seeing the state as the individual writ large in the tradition of foreign policy realism, there is a counter movement which asserts that attempts to argue that certain biological resources are the common property of mankind usurps the rightful prerogatives of national governments to use their resources for the benefit of their own people.

Obligations to Future Generations

There are two schools of thought with regard to obligations to persons not yet conceived concerning environmental productivity and diversity. The view of neo-classical economics and the property rights movement recognize no explicit obligations to future generations. Neo-classical economics often insists on discounting the future—assigning a lower value to events that will occur in the future based on their dollar value to persons living today. Discounting expresses a common sense notion that we would rather have a dollar today than one tomorrow for a variety of reasons including the alternative investment opportunities foregone by the delay. Because of the nature of compound interest this has the effect of assigning no practical value to the distant future, and is therefore rejected by some economists.

Neo-classicists often assume that there will be no overall resource scarcity because technological innovation will allow substitutes for products as their prices rise, and thus there is no need for an explicit policy to protect natural resources for future generations. In related arguments believers in strong property rights argue that scarce resources will be conserved in order to obtain a higher price later, and that enforceable patents will stimulate the technological advances to make better use of existing resources or to develop substitutes.

Both the neo-classical and property rights views are rejected by modern communitarian thinkers for four reasons. First, they point out that levels of human population and economic activity are having unprecedented impacts on the biosphere, and that the assumptions that climate and other natural systems will continue to behave as they now do may be wrong with possible catastrophic results for food production, water supplies, and the like. Second, the neo-classical and property rights movements simply ignore all claims that

might be made on behalf of other species. There is an obligation to size the human population to the biosphere in such a way as to allow other species to flourish, an obligation which is being violated in light of species loss now occurring. Third, property rights and neo-classical views concentrate too exclusively on the role of the environment in meeting consumption needs and ignore the public, symbolic, and cultural meaning of the environment. Fourth, those holding religious beliefs that God created nature make common cause with this argument, holding that the natural world belongs to God and that humans may simply use it, providing they do not degrade it.

Members of this school of thought stress that in order to discharge our obligations to future generations we should: (1) reduce the use of raw materials in the economy, thus reducing the impact on the biosphere particularly when it comes to the disposal of wastes such as carbon dioxide in the earth's atmosphere; (2) restrict, and in some cases even reverse, the rise in the human population so as to allow other species to flourish; (3) restore damaged ecosystems wherever feasible; and (4) modify or abandon altogether current measures of human well-being that rely heavily on consumption indexes such as the gross national product.

PETER G. BROWN

For Further Reading: Edith Brown-Weiss, *In Fairness to Future Generations* (1988); James Rachels, *Descended From Animals* (1990); World Commission on Environment and Development, *Our Common Future* (1987).

EROSION, SOIL

Soil erosion and soil creation are natural processes. As rains plummet to the earth or as winds race across the surface of the land, soil is removed from the land. In natural environments undisturbed by humans, about an inch of soil will be removed every 100 to 250 years. Soil also forms naturally. As vegetation dies and decays, it gradually turns into soil, building slowly back that which wind and rain have earlier removed.

However, when people disturb soil, as we must in order to have agriculture, the process of erosion can be quite rapid. Erosion may take place as sheet erosion, in which thin layers of soil are stripped from the land. If the rate of sheet erosion is only four or five tons per acre per year, the natural process of soil creation will usually offset the loss. But if the erosion accelerates to 40 tons of soil per acre per year, then four inches of soil are lost during every 15 years of cultivation and natural soil production is unable to provide replacement soil. Unfortunately, an erosion rate of 40 tons per acre or more is not unusual for many areas throughout the world.

Soil erosion also appears as more noticeable rills or gullies. Rills look like small channels or incisions on the face of the land, and they usually result from rain runoff or from melting snow. Gullies are severe forms of rills that appear as deep scars across the land; they can inhibit cultivation.

The severity of erosion varies throughout the world. This variation is due to many factors, of which one of the most important is climate. The soft, gentle rains and mists that envelop Britain, for example, create the green lush vegetation so often pictured on travel posters. These soft rains cause few erosion problems. Contrast those rains with the hard, driving rains of the southeast U.S., where violent thunderstorms can dump inches of rain in a few hours. If these rains fall on lands that have recently been plowed, soil can be severely eroded.

Another set of factors that can influence soil erosion is soil type and topography. Land best suited for agriculture is level and well drained, and has deep fertile soil. If the lands are steep or if the soil is thin or sandy, the soil will erode at a far greater rate. Thus the steep hills of Nepal, for example, experience some of the world's worst cases of human-induced erosion.

Another factor influencing the severity of erosion has to do with the number of people using available cropland. It is estimated, for example, that the People's Republic of China (PRC) is feeding over a billion people on approximately half the cropland available in the U.S. At one time much of the highlands of the PRC were covered with forests and grasses. In order to feed its people, the PRC cleared and cultivated much of this land. Because the area has steep hills, it has exceptionally high erosion rates. This erosion creates gullies and results in serious siltation of PRC's rivers such as the Yellow River, which is so named because of the vast amounts of yellow soil carried in its waters.

The final major factor influencing the severity of soil erosion is the farmer's choice of farming practices. Farming practices that can erode the land include leaving plowed land fallow, overgrazing, planting crops up and down a hill rather than using contour planting, and removing all vegetation after harvest. Other farming practices, such as planting cover crops, using rotational grazing, plowing and planting on the contour, and leaving harvest residue on the field can reduce soil erosion. Also, certain crops, such as soybeans or tobacco, are more erosion-prone than others, such as hay or alfalfa.

There are two reasons why soil erosion is of concern: (1) its effect on future harvests, and (2) its effect on water and air quality. Plants rely on soil. The relationship of soil erosion to future harvest abundance, or productivity, is a much debated issue. There is no doubt that soil erosion eventually will cause a decline in crop yields. Some decreases in crop yields can be offset by using more fertilizers or pesticides, but these are expensive and not always available throughout the world.

Furthermore, there comes a point when erosion's negative effects on crop yields cannot be totally offset. Erosion ultimately causes a loss in organic matter and moisture-holding capacity of the soil. Ultimately such losses result in a decline in yields. Thus soil erosion that exceeds the rate of soil formation will slowly reduce productivity. For example, researchers at the U.S. Department of Agriculture have estimated that in the U.S., 100 years of erosion at 1982 erosion levels would lower the productivity of the crop and fiber sector by 3.6%. These estimates assume that farmers will be using fertilizer and lime to offset the effects of erosion. The researchers note that the figure of 3.6% is just an average. Many soils will be little affected, but a few will lose much of their productivity.

The situation appears to be much worse for other areas of the world. United Nations researchers have estimated that an area approximately the size of the PRC and India—almost 11% of the world's vegetated surface—has suffered moderate to extreme soil degradation in the last 45 years. For example, the Indonesian island of Java is considered one of the most eroded places in Asia. The Indonesian government estimates that over 2.5 million acres, about 8% of their cropland, is critically eroded. Some of it is so badly eroded that it cannot even sustain subsistence agriculture. Researchers from the World Resources Institute estimate that every year over 4% of the total agricultural output for all of Java is lost due to soil erosion and there are several areas where permanent soil productivity loss has exceeded 20%.

Similar stories can be found elsewhere. It is estimated that about 22 million acres of the 32 million acres of cropland in the Philippines are eroded. One province, Zamboanga del Sur, has reportedly lost as much as 3.5 feet of soil, in some places exposing boulders. Local corn farmers have complained of yield declines of as much as 80% in 15 years. Other researchers estimate that erosion has reduced permanent soil productivity on the hills of Lebanon, Jordan, and Israel by at least 20%.

The second major concern about soil erosion is the impact on environments. As rain or snow-melt washes across cropland, the water carries soil and pollutants with it. These pollutants include fertilizer residues, insecticides, herbicides, fungicides, and dissolved minerals such as salt, as well as bacteria from animal waste. These pollutants can cause severe damage to streams, lakes, or even underground stored water. Not only are such pollutants health hazards, they also can destroy the value of water for recreation and cause serious siltation problems in waterways.

A dramatic case of the substantial problems that can result from agricultural runoff can be found in Reelfoot Lake in Tennessee. This lake was formed by a large earthquake in 1811. It is a shallow lake with a maximum depth of only 25 feet. Soybeans are grown near the lake and sedimentation of the lake from the nearby fields has become a major problem. Some recreation areas have closed because boat docks that were useful only a few years ago have been left dry. Some excursion boats continue to provide scenic nature tours, but the surface of the lake is increasingly difficult to navigate because of shallows and weeds. Reelfoot is literally disappearing under the increasing sedimentation. It is estimated that at the current rate of sedimentation, the lake will be entirely filled by the year 2032.

Similar tales can be found in other regions. The silt in the PRC's Yellow River destroys turbines in dams and fills water reservoirs as well as the river's channel. As the channel of the river is choked by silt, it creates the potential for massive floods.

Soil erosion by wind also can cause serious damage to air quality. In the U.S., the severe dust storms of the 1930s are legendary. These "black blizzards" blew soil and dust hundreds of miles across the land. Dust from the Midwest could be seen in the nation's capital. In western India, damage is caused by sandblasting of crops and human settlements, as well as by wandering sand dunes.

In the U.S., several measures of the effects of soil erosion have been developed. For example, researchers of the Department of Agriculture estimate that over half of the cropland will maintain soil productivity indefinitely, but over half is exceeding recommended soil erosion rates. Worldwide estimates are far more difficult to obtain, but are most likely much larger than those of the U.S. Such erosion rates present serious ecological and economic problems.

Fortunately there are many solutions that exist for soil erosion problems. Steep lands can be terraced and especially eroding lands can be planted to grass or trees or left idle. Farmers can follow crop rotation patterns that are less erosive, or use techniques such as stripcropping, retaining harvest residues on the land, or leaving forested buffers between fields and water bodies. Appropriate selection of crops to plant as well as the timing of plowing and harvesting activities can greatly affect rates of erosion, as can the use of certain plowing practices such as conservation-tillage or ridge-tilling.

With all these solutions why is there so much soil erosion? There are many reasons: technical, financial, institutional, and personal. The technical reasons may have to do with the farmer being unaware of the existence of soil erosion on his or her farm or being unaware of the techniques available to alleviate it. There may not be good technical assistance available to help a farmer implement soil erosion control. Also, a farmer may be financially unable to adopt soil erosion control. Or there may be institutional factors such as insecure land tenure, limited access to credit, or taxes and agricultural policies that inhibit the adoption of soil erosion control.

Because of these barriers to the adoption of soil erosion prevention, there is a need for public policies to encourage soil erosion control. Generally, these policies fall into three types: (1) education and technical assistance, (2) economic incentives, and (3) regulation. Education and technical assistance involve informing the farmer of soil erosion problems and providing technical assistance in the adoption of soil conservation practices. The U.S. program of soil erosion control has been built around education and technical assistance for over a half century. Education and technical assistance are frequently coupled with economic incentives in order to make it profitable for a farmer to adopt soil conservation practices, often taking the form of cost-sharing, in which tax dollars are used for part of the cost of adoption of soil conservation practices.

Increasingly, the U.S. and Europe are turning to regulatory policies to enforce soil conservation. With regulation, farmers must follow certain standards of protection of the soil and neighboring water bodies or find themselves subject to civil penalties. For example, in Barbados, burning sugar cane after harvest is banned. The reason is that when the harvest residue is left on the land, sugar cane yields increase by four tons per hectare and soil erosion declines.

U.S. President Franklin Roosevelt said that "a nation that destroys its soil destroys itself." While fertilizers and lime can compensate for some lost soil, fertilizers are expensive or unavailable in much of the world. Fertilizer is also a potential pollutant of surface and groundwater. As world populations grow, the protection of the soil resource for both future harvests and the protection of the environment is increasingly important.

SANDRA S. BATIE

For Further Reading: Sandra S. Batie, *Soil Erosion: Crisis in America's Croplands?* (1982); John M. Harlin and Gigi M. Berardi, *Agricultural Soil Loss: Processes, Policies, and Prospects* (1987); The World Resources Institute, *World Resources 1992-93* (1992).

See also Agriculture; Herbicides; Pesticides; Pollution, Non-Point Source; Soil Conservation.

ESTHETICS

The word *esthetic* comes from the Greek *aesthanesthai* which means to perceive. The definition of the word has evolved from Plato's dialogues and theories of beauty to contemporary art criticism of both the "fine" and "applied" or "useful" arts, including the design of buildings, parks, gardens, towns, and cities. The contemporary concept of environmental esthetics encompasses all environments: natural, designed, and vernacular—the everyday rural and urban environments in which most people live, work, and play. While esthetics was originally part of the intellectual domain of philosophers, contemporary environmental esthetics is also of interest to art historians, legislators, resource managers, architects, landscape architects, and researchers in the field of environmental perception.

Concepts of esthetically satisfying environments have changed throughout history. For example, architectural historian Vincent Scully, in describing relationships between temples and sacred landscape features in ancient Greece, suggests that because of concepts that developed at the beginning of the modern age when landscape was romanticized in art, it is difficult for us to see and understand the Greek esthetic qualities of Greek temple-landscape relationships. These relationships were profoundly different from those developed in Roman times and in the Middle Ages when structures and natural landscapes were frequently separated by walls and concepts of beauty were related to utility and orderliness in the environment.

British art historian Kenneth Clark has suggested that the average person in Europe during the Middle Ages would not have considered it wrong to enjoy the natural environment, but would have simply found it unenjoyable. Whether working on farms, at sea, or in forests, people associated the natural environment with hard work, long hours, and frequently uncomfortable working conditions. The medieval perception of nature as not being enjoyable and of wilderness as something to be conquered was carried to North America by the early Pilgrims. William Bradford, governor of Plymouth Plantation, upon setting foot in the New World, questioned whether this landscape was anything more than "a hideous and desolate wilderness, full of wild beasts and wild men?" In 1653 Edward Johnson described the settlement of Concord, Massachusetts, and commented on how the settlers, "placed downe their dwellings in this Desart Wildernesse. . . ." When these early settlers built homes and gardens, they created orderly, utilitarian environments similar to those they had left behind in England.

Historical antecedents of contemporary environmental esthetics can be found in painting, garden design, and perceptions of nature and landscapes. Secular paintings that depict natural landscapes, humanized with the inclusion of a simple structure or a person, have been found in China from the 2nd century B.C. and in Japan from the 9th century A.D. Secular landscape paintings of similar context did not appear in the West until the 14th century. The art of painting in Europe during the Middle Ages was in service to the church. Paintings depicted religious scenes in which nature, if included, was symbolic rather than realistic. This may have reflected a Middle Ages' mistrust of nature. At that time, gardens in Europe were geometrically organized and walled, while those in Japan, although also walled, were abstractions of nature, designed to represent elements of nature, simplified and at a much reduced scale.

A useful benchmark in the evolution of Western concepts of environmental esthetics is found in the works of Sienese painters of the 14th century, particularly the frescoes of Ambrogio Lorenzetti that depict the effects of "good and bad government in town and country." Lorenzetti broke from tradition and depicted secular landscapes as the central topic of his paintings. His work suggests that neat, orderly, and productive landscapes, both urban and rural, are sources of pleasure and satisfaction, while wild and disorderly landscapes are associated with danger, crime, and lack of productivity.

Throughout Europe the association of beauty with geometrically organized landscapes, both gardens and productive rural areas, prevailed into the 18th century. During the 18th and early 19th centuries, an awareness of the loss of natural landscapes emerged and people began to see beauty in and derive pleasure from the remaining natural landscapes and mountains. This shift has been associated by some historians with growing appreciation of the 17th century paintings of Salvator Rosa, Claude Lorrain, Nicolas Poussin, and others. These painters romanticized the landscape, combining the rugged with the pastoral, creating scenes that were categorized as "picturesque": that looked good in a picture and contained variety and intricacy. By contrast, "beautiful" landscapes were gentle and pastoral; "sublime" landscapes created a sense of awe, mystery, and power.

The romantic vision of landscape beauty that developed in England appeared in the United States in works of writers and painters in New England, upstate New York, and the Hudson River Valley. These included James Fenimore Cooper's Leatherstocking stories, Henry David Thoreau's nature writings, and the paintings of Thomas Cole and Frederick Edwin Church.

Painters and writers alike extolled the beauties of nature as one of the great resources of the United States. In the built environment the romantic vision was manifest most dramatically in the design of urban parks, starting with the design of New York City's Central Park in 1858 by Frederick Law Olmsted and Calvert Vaux.

Landscape and nature as a source of esthetic pleasure was discussed in 1896 by Harvard philosophy professor George Santayana in *The Sense of Beauty*. He suggested a cultural basis for the perception of environmental beauty when he stated that only when individuals were conscious of the act of perceiving would the "forest, the fields, all wild or rural scenes" be considered beautiful and full of "companionship and entertainment."

At the beginning of the 19th century there was little mention of beauty in legislation or judicial decisions. Nevertheless, the ceding of the Yo-semite Valley (later called Yosemite) by the United States Congress to the state of California for a state park in 1864 and the establishment of Yellowstone National Park in 1872 implicitly recognized their scenic qualities. Explicit recognition of the importance of visual quality in national parks appeared in 1916 when Congress, in establishing the National Park Service, recognized the "conservation of scenery" as one of the primary purposes of the new service. Since the enactment of that momentous legislation, the number of national parks and related areas has increased through executive orders and legislative acts to approximately 360, all under the 1916 mandate to "conserve the scenery." Furthermore, park scenery includes natural areas, prehistoric and historic sites, cultural landscapes, urban sites, and recreation landscapes.

During the 1930s laws were enacted to protect historic areas in New Orleans and Charleston. These actions set precedents for protecting developed urban landscapes. Visual quality is not specified as one of the resource values to be protected in these laws. It is implicit, however, in the expressed concern with the appearance of new development and of additions to and remodeling or renovation of existing structures within the historic district. A benchmark acknowledging esthetic values in built environments is the 1954 U.S. Supreme Court case of Berman versus Parker for which Justice Douglas wrote, "it is within the power of the legislature to decide that the community should be beautiful as well as healthy . . ."

Official protection of specific landscapes continued with the Wilderness Act of 1964, and was expanded significantly following the 1965 White House Conference on Natural Beauty, which focused on the appearance of highways, junkyards, mined areas, rural areas,

suburbs, towns, and urban waterfronts. The 1965 Highway Beautification Act, one of the earliest federal attempts to address issues of visual quality on private lands, was intended to eliminate and regulate visual eyesores such as billboards and auto junkyards along the federal highway system. Although its impact was reduced by the lobbying power of the billboard industry, it did reflect a growing public concern with esthetic issues in developed as well as natural areas. The 1977 Surface Mining Control and Reclamation Act required surface mined areas to be restored to their approximate original contour.

Public concern about the shortage of outdoor recreation areas and facilities led to the 1968 Wild and Scenic Rivers Act and a National Trails System Act, which included provision for recreational, scenic, and historic trails. The Coastal Zone Management Act of 1972, the Federal Land Policy and Management Act of 1976, and the National Forest Management Act of 1976 extended federal responsibility for visual quality to more lands.

The National Environmental Policy Act of 1969 (NEPA) probably had the greatest impact on visual values of any federal legislation. NEPA required that all major projects undertaken or funded by federal agencies consider the effects of those proposed projects on the environment, including visual impacts in natural and developed landscapes. NEPA called for the use of the environmental design arts, such as landscape architecture and architecture, in addressing these impacts. Similar legislation has been enacted by a number of states, counties, and municipalities. NEPA was also the stimulus for a substantial body of research about public perceptions of beauty in the environment and about how to inventory and evaluate the elements of the environment that are related to perceived beauty. Some of this research has focused on perception of air and sonic qualities. At Grand Canyon National Park, for example, pollutants have been found to decrease the perceived quality of the spectacular views across the Canyon. The objective of the sonic environment perception research has been to identify the level of noise people are willing to accept rather than to identify and protect pleasurable environmental sounds such as the singing of birds.

Results of landscape perception studies suggest that the time-worn adage "beauty is in the eye of the beholder" is misleading and overly simplistic. Several studies have demonstrated agreement among persons from similar ethnic, economic, and age groups, notably white, middle-class, college students and middle-aged adults. Their preferred environments tend to be reminiscent of parklike 19th-century romantic, picturesque, and sublime landscapes. Missing from this research has been an equivalent emphasis on all age groups, and on the broader cross-cultural and cross-ethnic studies required for a fuller understanding of perceived environmental esthetics. The limited cross-cultural perception research suggests that when cultures have notably different lifestyles, economic status, and native landscapes, perceptions of environmental beauty are different. Such differences have been found, for example, between Eskimos living in an Arctic landscape and Caucasians in a temperate landscape, and between Caucasians and persons of African origin living in a tropical landscape.

Several theoretical frameworks have been proposed to explain human esthetic responses to environments. Drawn from the history of painting, literature, scholarly writings, and perception research, they represent evolutionary and cultural orientations.

The evolutionary orientation suggests that human esthetic preferences are for landscapes that provide information and settings important for human survival. American psychologists Rachel and Stephen Kaplan suggest that humans prefer landscapes that are coherent and legible, and offer promise of more information. British geographer Jay Appelton's "prospect-refuge" theory suggests that people prefer places where one can see without being seen, places that afford views and protection. In contrast, philosopher Marcia M. Eaton suggests that the foundation of environmental esthetics lies in traditional beliefs about things that are considered delightful. She also suggests that esthetic values differ across time and culture. Eaton's view parallels that of George Santayana and supports conclusions drawn by Erwin H. Zube from cross-cultural research.

ERVIN H. ZUBE

For Further Reading: M. M. Eaton, *Aesthetics and the Good Life* (1989); J. L. Nasar, ed., *Environmental Aesthetics Theory, Research, & Applications* (1988); E. H. Zube, *Environmental Evaluation: Perception and Public Policy* (1984).

ETHANOL

See Fuel, Current, Near-Term, and Futuristic.

EUTROPHICATION

Natural eutrophication is the process that transforms lakes from oligotrophic (lacking in plant nutrients) to eutrophic (enriched with plant nutrients). The nutrients include minerals, such as nitrogen and phosphorus, that support plant and animal life. In most freshwater lakes, phosphorus is the primary limiting

nutrient and usually serves as the regulator of the lake's trophic state. The time period for natural eutrophication to complete this transition varies: in small ponds, it takes only a few years; in large lakes, thousands of years. The meaning of the term eutrophication has evolved to mean the overnourishment of any body of water, with the associated consequences. For example, phosphates, nitrates, and other excess nutrients from agricultural runoff, industrial discharges, and sewage greatly accelerate the natural process. When the addition of excess nutrients is the result of human activity, it is call cultural eutrophication.

Because they are fed by melting snow and ice and receive few nutrients, mountain lakes are clear, unproductive, and oligotrophic—harboring few life forms. Most lakes naturally gain an increased amount of nutrients and organic deposits each year as sediments wash in from surrounding land and organisms die and decompose. However, if the inputs from terrestrial sources remain low and the sediment/water interface at the lake bottom remains oxygenated, a lake may remain oligotrophic, or nearly oligotrophic, until it completely fills. As long as the sediment/water interface remains oxygenated, the primary limiting nutrient, phosphorus, is chemically bound to sedimentary material and cannot migrate to surface waters. If the amount of nutrients added from terrestrial sources should cause the bottom waters of the lake to become anoxic, then the exchange of nutrients between the bottom and surface water accelerates, causing the surface water to become enriched with nutrients, particularly phosphorus. This process can be accelerated in late summer when excess plant growth on the water's surface blocks sunlight from reaching the deeper levels and prevents photosynthesis from replenishing the oxygen supply. As the animals die off due to the lack of oxygen, their decomposing remains, along with plant remains and sediment, fill in the lake bed. Eventually the lake, whether in a eutrophic or oligotrophic state, fills in and becomes a wetland, and possibly a forest.

Cultural eutrophication, which is associated with pollution, occurs at an incredibly rapid pace compared to the normal geological time scale that measures the life of a lake. This process has diminished the quality of many of the American Great Lakes just within the last 100 years. Those lakes with large human populations nearby have suffered the most. It has been found, however, that many lakes can be returned to their natural trophic state by stopping the pollution and allowing the lake to "cleanse" itself. This cleansing occurs naturally when the bottom waters of the lake become oxygenated again and the nutrients remain bound to the bottom sediments and are prevented from circulating in the surface waters.

EVAPOTRANSPIRATION

Evapotranspiration is the combination of evaporation—the physical conversion of water to water vapor—and transpiration—the physical loss of water from leaves to the atmosphere. Both processes represent principal flows in the water, or hydrologic, cycle on earth, adding water vapor to the atmosphere by drawing moisture from other parts of the cycle. Evaporation draws moisture from bodies of water and from the soils, and transpiration draws moisture from plants—both terrestrial and floating aquatic. The two processes are often combined because it is not easy or especially useful to set apart the contributions of each on land masses. They both represent vertical outflows in the water cycle, in contrast to inflows in the form of precipitation, which includes rain, sleet, hail, or snow.

On an annual average the outflows and inflows for land, sea, and air are balanced. The average water balance between precipitation and evapotranspiration differs greatly between land masses. The difference between precipitation and evapotranspiration is measured by the amount of runoff, or precipitation that finds its way back to the sea after falling on land. This difference indicates the amount of water that could be used. The runoff of South America, for example, is approximately three times that of North America because of the Amazon River, which represents almost one seventh of the earth's total runoff. Using average annual figures of the U.S. Geological Survey, approximately two thirds of the total precipitation reaching the continental U.S. is returned by evapotranspiration from moist surfaces and plants. The other third—runoff—eventually returns to the atmosphere by evaporation from the oceans. The reason the water cycle uses a large amount of the solar energy that reaches the earth is that a great deal of energy is needed to evaporate water—50 times as much as it would take to lift an equal amount of water 5 kilometers, or approximately 3 miles. Evapotranspiration, in combination with the other principal flows in the water cycle, plays both unique and essential roles in the complex interactions between abiotic, biotic, and cultural factors in the total environment.

EVOLUTION

Nothing on Earth is constant. The continents move, the climate changes from warm periods to ice ages, mountains rise and are again eroded. All is in a flux, as a Greek philosopher said. The weather changes from day to day, and so do the seasons and the climate. Beginning in the 18th century, a growing number of naturalists began to realize that floras and faunas were changing over geological time, and that species likewise were not constant, but had the potential to evolve and to give rise to new species. The publication in 1859 of Charles Darwin's *On the Origin of Species* heralded the beginning of the general acceptance of the concept of organic evolution, and this concept has been ever more generally applied in the ensuing 130 years.

There has been considerable diversity of opinion and actual confusion about the meaning of the word *evolution*. The term was introduced into science in the 18th century by Charles Bonnet for the preformational theory of embryonic development. This theory held that the adult form of an individual is "preformed" in the fertilized egg and "evolves" (becomes unfolded) during embryonic development. The term *evolution* is no longer used in embryology, but was transferred by some authors around 1850 to the history of life, even though it was not yet used by Darwin in 1859.

There have been three major theories concerning the mode of organic evolution.

(1) Life on Earth changes, according to the typologists (essentialists), owing to the sudden origin of an individual representing a new species and owing to the extinction of earlier species. The origin of such a new individual occurs through a major mutation or **saltation,** and this individual through its descendants becomes the progenitor of the new species. Such saltational evolution is also referred to as transmutationism. Such ideas, although not called evolution, were proposed from the Greeks to the 18th century. After the publication of Darwin's *Origin* they were adopted by T. H. Huxley and by various anti-Darwinians. The rediscoverers of Mendel's work (William Bateson, Hugo De Vries, and Wilhelm Johannsen) adopted saltational evolution and later so did the geneticist R. Goldschmidt, the paleontologist O. Schindewolf, and the botanist J. C. Willis. Such theories of saltational evolution are now no longer supported in biology because no evidence for their occurrence has been found.

(2) Jean-Baptiste Lamarck, the first author to propose a consistent and well-elaborated theory of evolution in his 1809 *Philosophie Zoologique*, promoted the concept of **transformational evolution**. According to this theory, a given object gradually changes in the course of time, as in the development of a fertilized egg into an adult. Transformational evolution is also found in the inanimate universe as when a white star evolves into a red star; indeed, nearly all changes in the inanimate world are of this nature, like the rise of a mountain range from the action of tectonic forces or its subsequent destruction by erosion. According to Lamarck, organic evolution was due to the origin by

spontaneous generation of new simple organisms, called "infusorians," and their gradual change into higher, "more perfect" species.

(3) Darwin in 1859 introduced a drastically new concept of evolution, **variational evolution**. According to this theory an almost inexhaustible amount of genetic variation is produced in every generation (usually by sexual recombination), but only a few of the vast number of offspring survive and give rise to the next generation. The individuals that, owing to their combination of genes, are best adapted to their environment have the highest probability to survive and to reproduce. Which genotypes in a population or species are "best" changes from generation to generation. Since all changes take place in populations of virtually all genetically unique individuals, evolution is by necessity gradual and continuous. After 1859 the majority of biologists resisted Darwin's concept of natural selection and continued to support Lamarckian transformational evolution. It was not until the 1930s and 1940s, during the so-called evolutionary synthesis, that Darwinian variational evolution was more or less universally accepted.

Two major aspects of organic evolution are of importance for the study of the environment: (1) the achievement of adaptedness and (2) the rise of diversity.

Adaptedness

The environment changes from season to season and from year to year, not only in its physical components (temperature, humidity, etc.), but also in its biotic factors, such as the invasion of new competitors, predators, parasites, and pathogens. Every population of every species is exposed in every generation to a slightly different set of environmental factors. As each individual (except identical twins) in a sexually reproducing species is genetically unique, some individuals are better adapted to answer the new challenges of the changed environment than others, and these individuals have a greater probability of surviving and reproducing. It must be emphasized that it is the potentially reproducing individual that is normally the target of selection. It can never be predicted in advance what would be the best constellation of genes for next year, and the whole process of natural selection therefore is not teleological but, so to speak, an *a posteriori* reward for the chance appearance of the best gene combination.

When thinking of natural selection it is helpful to distinguish between selection *of* and selection *for*. Natural selection is the "selection of" a reproducing individual, and the "selection for" a particular gene or gene combination that gives this individual its selective superiority.

The observable appearance of an organism, known as its phenotype, as a whole is the actual target of selection and hence all the developmental factors involved in translating the genetic program into the adult phenotype make a contribution to the ultimate fitness of the individual. Any change in a developmental process that leads to a greater adaptedness of the resulting individual will be favored by selection.

Diversity

The second major aspect of evolution is the origin of organic diversity. How did millions of species of animals and plants evolve, and why is there such great diversity?

One of the major criticisms of Darwin's *Origin* was that it did not explain the origin of life. All endeavors by microbiologists to demonstrate the spontaneous generation of new life turned out to be failures. Then, said Darwin's opponents, how could life have ever originated? It is now clear that the Earth's atmosphere at the time when life originated, 3.5 to 4 billion years ago, was totally different than it is now. It was a reducing atmosphere, devoid of oxygen. In such an atmosphere life indeed could originate spontaneously. The oxygen in the present atmosphere would at once destroy newly arising molecules needed for the origin of life. However, they could survive in a reducing atmosphere. There are now three or four different scenarios for the steps between inanimate molecules and the first life, but it may never be possible to decide which is the right one. The most primitive organisms now found on Earth are already so highly evolved compared with their presumed ancestors that we cannot determine either from their structure or from any fossils of that period which of the scenarios is the most likely.

It is, however, highly probable that all life now existing on Earth, from the lowest bacteria to the highest plants and animals, is the result of a single origin of life. This is indicated by the fact that all living things have the same genetic code by which the nucleic acid base pairs are translated into amino acids.

Let us now start at a later stage and ask, How do new species originate? A paleontologist, looking at the fossil record, might observe the gradual transformation of a species over the course of thousands and millions of years into another species, a process usually referred to as phyletic evolution. However, when speaking of speciation, an evolutionist seeks to explain the multiplication of species: how one species can give rise to several derived species. The following modes may be recognized:

(1) **Sympatric speciation** is the origin of a new daughter species within the range of a parental species through the acquisition of isolating mechanisms by

certain individuals. Sympatric speciation occurs when certain individuals of a population acquire characteristics or mechanisms that prevent them from interbreeding with the parental population. Such speciation might occur when a host-specific insect or parasite switches to a new host.

(2) **Geographic** or **allopatric speciation** is the evolution of a population into a new species while isolated geographically from other populations of the parental species. This occurs in two forms: **dichopatric speciation**, which occurs when a parental species is separated into two separate portions by the development of a geographic, vegetational, or other extrinsic barrier; and **peripatric speciation**, which occurs when the descendants of a founder population establish themselves beyond the periphery of the parental species range and thus take a divergent evolutionary path to the status of a new species.

In rare cases, chromosomal changes (particularly polyploidy) may cause instantaneous speciation when the new chromosomal type is reproductively isolated from the mother population. This process is largely restricted to plants.

The environment is crucially involved in all aspects of speciation. No new species population can become established and survive unless it can find a suitable niche in which it is not threatened by competitors. The restructuring of the genotype that often accompanies a speciation event is believed to be a particularly auspicious situation for major evolutionary shifts. A new species that has successfully entered a new adaptive zone may become the ancestor of an important higher category.

The adaptive radiation and diversification of most higher groups of organisms was invariably correlated with a particular environmental factor. For instance, the extinction of the dinosaurs opened up tremendous opportunities for the ancestral mammals, enabling them to diversify in the Paleocene and Eocene eras into all the existing major orders of mammals. After the songbird type had evolved in the first third of the Tertiary it found such a favorable environment that this one group of birds now comprises 5,300 species, far more than all other kinds of birds put together. Why some seemingly highly successful groups like the trilobites, dominant in the Paleozoic, eventually became extinct, is still a puzzle. Competition by the mollusks might have been a contributing factor, but one can ask why natural selection did not help the trilobites to withstand such competition successfully.

Even though at first sight there seems to be a complete break between speciation and the origin of new higher groups of organisms, called taxa, there is actually a complete continuity. It is always a species that gives rise to a new higher taxon, and the genetic char-

acteristics of this species must have originated in some of the individuals of which the species is composed. Thus, ultimately, genetic changes in an individual are the starting point not only of micro-, but also of macro-evolution.

Darwin's theory of common descent gives the ultimate explanation of the total diversity of life. All cat-like species were derived from an ancestral cat species. Cats, dogs, weasels, bears, and other carnivores were derived from one ancestral carnivore species; and, going further back, one can say the same about mammals, vertebrates, and animals — all the way back to one-celled protists and the most primitive bacteria (prokaryotes). All types of organisms, no matter how richly diversified they are now, go back to a single ancestral species.

With new types originating as a single species and its descendants, it is highly improbable that all of these ancestral connecting types can ever be found in the fossil record. These absent forms have been referred to as "missing links." However, it is amazing how many of such formerly missing links have been found as research advances. Examples include *Australopithecus*, a species embodying a developmental stage between humans and the higher apes, *Archaeopteryx* between birds and the reptiles, and *Ichthyostega* between amphibians and fishes.

Not only species and higher taxa evolved, but also entire biota (floras and faunas). Their evolution is sometimes quite gradual but also can be characterized by drastic periods of extinction, such as at the end of the Permian era, and by a reorganization of the biota in the ensuing geological period. The cause of such sudden extinctions is unknown in most cases, although it is reasonable to speculate that it must have been the result of a drastic change of the environment. The great mass extinction at the end of the Cretaceous is now plausibly explained as the result of an asteroid impact on the Earth. Other great extinctions, like that at the end of the Permian era, may have been due to an extreme drop in temperature on Earth, or, on the contrary, to an increase in temperature due to some kind of greenhouse effect. Much research in this area is currently underway. The great changes the biota has undergone demonstrate how fragile life on Earth has always been and how subject to devastation and change when adverse conditions developed.

The changes brought about by the human population explosion, if not soon brought to a halt, may reach the dimensions of the mass extinction at the end of the Cretaceous. The biota on islands seem to be particularly vulnerable, and at least 80% of recent extinctions of birds caused by humans have occurred on islands. Species with insular distributions on continents owing to physiographic or vegetational conditions are likewise highly vulnerable. At one time, the major threat to faunas was actual killing by humans, such as the destruction of the megafauna of Asia and North America, or on Hawaii and other islands in the Pacific. The most spectacular recent instance was the extermination of the passenger pigeon in North America, a species once represented by many millions of individuals. The near extermination of the bison is another illustration. In the 1990s, changes in the environment such as the deforestation of vast areas in the tropics, the draining of wetlands, and pollution, seem to be the major causes of extinction.

Evolutionary thought, from the 18th century on, has played an increasingly larger role in human philosophy. At first, reinforced by Lamarck's theories, it consisted of a belief in universal gradual advance, a trend toward betterment if not perfection, a viewpoint particularly reflected in social theories. It was often referred to as the idea of progress.

In more recent times, the realization that Darwinian evolution is not a gradual transformation toward perfection, but rather an alternation between the production of variation and the survival of the best adapted, began to permeate social theory as well as literature. A distorted version of this type of thinking was the so-called social Darwinism of the late 19th century, a theory actually owing more to Herbert Spencer than to Charles Darwin. It overemphasized ruthless struggle for existence and overlooked that in a structured society cooperation and "mutual help" (as it was called by the opponents of social Darwinism) are more important social factors than unbridled competition. Darwinian evolution simply calls attention to unequal survival of competing systems, a realistic description of the real world.

ERNST MAYR

For Further Reading: Douglas J. Futuyma, *Evolutionary Biology* (1986); Ernst Mayr, *The Growth of Biological Thought* (1982); Ernst Mayr, *One Long Argument* (1991).

EXOTIC SPECIES

Ever since life appeared on the Earth, species have invaded new habitats, exploited resources, altered species associations, and then disappeared. The dynamics of this continuing process are complex and varied, and operate on very different time scales. These include competition, predation, and coevolution, all occurring against a background of environmental variation and change. These are inexorably linked to the development of ecosystem organization and its vulnerability to invasion.

Species invasions, the colonization of an environment by non-native or exotic species, are directly linked to extinction and speciation events. For example, the variety of finch species found on the Galapagos Islands derived from changes that occurred after the arrival of one non-native finch species. Invasions such as this are responsible for some of the species distribution patterns we now see across the Earth. Recently, however, the rate of species invasion has increased enormously—due largely to human commerce and travel. Invasions no longer reflect the natural course of species spread, but rather a growing ecological catastrophe with many untold consequences.

The earliest human-aided invasions undoubtedly occurred during the migrations of hunter-gatherer cultures. However, it was not until regional and global exploration became a reality that invasions became noticeable. For example, ocean-going ships often carried soil as ballast. Upon arrival the ballast was removed and replaced with cargo. This ballast ultimately found its way to shoreline depositories where diverse plants and animals became established. While human-aided invasions are indeed a cause for alarm, many species invade without human assistance. For example, during the last 100 years, nearly 80% of the known bird invaders of Great Britain colonized without help from humans.

Currently, there are about 350 exotic species inhabiting the coastal waters of the U.S., and over 2,000 invading insect species on the mainland. In many areas invaders have replaced entire ecosystems, while in others the number of invading species rivals that of native species. For example, New Zealand has almost as many exotic plant species as natives. In many areas throughout the world the number of invading plant species has rapidly approached 50% of the total native plant species.

While the number of invading species throughout the Earth is indeed staggering, cataloging the number of invaders is not a good measure of the prospects for ecosystem damage. In some cases, invaders live alongside native species with few dramatic effects. Conversely, a single invading species can cause serious ecological damage to the ecosystem it colonizes. The zebra mussel (*Orisena polymorpha*), which invaded the Great Lakes in the late 1980s, has exacted a significant economic toll on the region. The extent and direction of any ecological impact caused by the zebra mussel remains as yet unclear. Other invasive species which have caused extensive damage are the European gypsy moth, starling, house sparrow, rat (*Rattus* sp.), and house mouse (*Mus musculus*).

Often invading species like the zebra mussel are conspicuous because they possess a novel growth form compared to native species. For this reason such species have been labeled "exotic" species. In reality all invaders are exotics. Nevertheless, species which possess a novel growth form often have the potential to severely disrupt the structure and function of an ecosystem. For example, around 1990 a large freshwater zooplankter (*Daphnia lumholtzi*) invaded the southeastern U.S. This species is novel in that it possesses very long spine-like structures. These spines aid in defense against predators. At the same time, the large size of this zooplankter confers some degree of competitive advantage against smaller species of zooplankton. The potential for large shifts in community organization exists wherever this species invades.

A direct consequence of the increased rate of species invasions is an ever increasing homogenization of the Earth's biota. At one extreme, invasion by exotic species has resulted in the wholesale replacement of entire ecosystem types. For example, species which compose the Mediterranean-climate grasslands of northern California have been replaced by counterpart species of true Mediterranean origin. At the other end of the spectrum, some invasions have occurred with little or no observable effect. Estimates of the proportion of invaders which have affected the native communities they colonized range from 6% to 21%. Of these invaders, 8% led to the extinction of one or more community members.

Why do some invaders cause a serious impact while others appear benign? First, not all species make good invaders. When species find themselves in a new environment some are incapable of substantial spread and population growth. Second, some ecosystems strongly resist invasion, and many ecosystems appear to be vulnerable to invasion only under certain conditions. The recent invasion of the Great Lakes by the zebra mussel provides an instructive case study. The mussel apparently hitched a ride in freshwater ship ballast from Europe to North America about 1988. Since then the mussel has severely impacted industry in the region and has spread into the southeastern U.S. While it didn't take long for the sea lamprey (*Petromyzon marinus*) to exploit the newly completed canal system in the 1960s and enter the Great Lakes, for some reason the zebra mussel was unable to invade until much later. Clearly, the transport vector was in place for many years before the invasion actually took place.

The key element in understanding variation in invasion success and differences in ecosystem resistance to invasion lies in the interplay between species biology, ecosystem properties and processes, and pure chance. The many coexisting species within an ecosystem share and compete for numerous resources, forming numerous tightly woven relationships. Many ecologists argue that particular patterns of intercon-

nectedness enhance resistance to invasion. However, an ecological system may or may not be susceptible to invasion, depending on the fluctuating availability of resources within the system, and the ability of the invader to exploit these resources more efficiently than the native species. A disturbance to the system may modify resource availability by adding or shifting resource distribution within the system. For example, a forest fire or tree fall might create open space, allowing more sunlight to reach the soil surface. These new resources may be utilized by surviving species, or may permit invasion by other native or exotic species into the community.

While government attempts to legislate invasions by imposing quarantines (e.g., California has extensive bans on the transport of many plants into the state), the rate of new invasions continues to accelerate with serious ecological and economic tolls. The zebra mussel is expected to cost industry from 10 to 15 billion dollars during the decade of the 1990s. Another exotic species that has exacted a serious socioeconomic and ecological toll is a submerged, aquatic plant, *Hydrilla verticillata*. *Hydrilla*, found in Africa, Asia, and Australia, was recently introduced into Florida where it quickly replaced native species by forming dense stands, rendering the waters less suitable for recreation and lowering their esthetic appeal. More commonly seen is the terrestrial plant species affectionately known as honeysuckle. This species was introduced to the U.S. as an ornamental—much like carp, which was introduced as a game fish. Honeysuckle, like the exotic kudzu of the southern U.S., now infests many ecosystems and has become the bane of gardeners and land managers.

The global consequences of increasing invasion pressure by exotic species are nothing short of the possible elimination of many species and entire ecosystem types. In the 1950s Charles Elton called attention to this problem when he surmised, "We must make no mistake; we are seeing one of the great historical convolutions of the world's fauna and flora." Fortunately, Elton's call for action has not gone unheeded. Scientists, politicians, and the general public have realized the magnitude of the threat posed by exotic species and, armed with recently gained knowledge about the invasion process, they are slowing the rate of new invasions.

JAMES A. DRAKE
K. JILL MCAFEE

For Further Reading: A. W. Crosby, *Ecological Imperialism: The Biological Expansion of Europe* (1986); C. S. Elton, *The Ecology of Invasions by Animals and Plants* (1958); H. A. Mooney and J. A. Drake, eds., *The Ecology of Biological Invasions of North America and Hawaii* (1986).

EXPOSURE ASSESSMENT

An exposure assessment involves the application of numerous measurement and modeling techniques to identify contaminant concentrations and sources; media of exposure; transport for each medium (air, water, soil, food, or injection); chemical and physical transformations; routes of entry into the body; intensity, frequency, and duration of contact; and relevant human activity patterns. It is necessary to accurately address the issue of human contact with a contaminant for the purposes of risk assessment, risk management, epidemiology, or clinical intervention and prevention. One important component of an assessment is the identification of the distribution of exposures in order to define the mean and high-end values. All these techniques need to be placed within a scientific framework that is now referred to as the discipline of exposure analysis. It includes the acquisition of information on a contaminant, from its emission by a source or formation in the environment to its health effects. The fundamental principles are derived from the definition of *exposure:* "an event consisting of contact at a boundary between a human and the environment at a specific contaminant concentration for a specified interval of time." The mathematical description of exposure is an integral:

$$E = \int_{t_1}^{t_2} C(t)dt$$

where E is the exposure (units of concentration-time); C(t), the concentration; t, time of contact; and t_1 to t_2 is an interval consistent with the expected biological response to the compound (requiring knowledge of the contaminant toxicity).

An operational form of the above that delineates the exposure of an individual in different microenvironments (locations or situations) where an individual spends time is:

$$E_{i,j,k} = \sum C_{i,j} t_{j,k}$$

where the kth person is exposed to a concentration $C_{i,j}$ of the ith contaminant in the jth microenvironment. The concentration $C_{i,j}$ is considered to be constant during the $t_{j,k}$ time interval. The values can be summed over all possible microenvironments (single or multiple media), or contaminants or people.

The approaches used to collect the data needed for an exposure assessment include both direct and indirect measurements. Direct measurements actually employ sampling the person, using monitors attached to the individual, or sampling biological media such as blood or urine. Indirect measurements include collection of information about where, when, and how people spend their time and the concentrations of a contaminant associated with a medium that contributes to important routes of exposure. The data from indirect measurements are then used to estimate a person's exposure using simple or complex mathematical analyses or simulations. Most exposure assessments will use a combination of direct and indirect measurements. Unfortunately many assessments have had to use disparate data that usually have not been collected in the same study or at the same time. Since the mid-1980s the situation has begun to change, and based upon the theoretical advances of the early 1990s, integrated multimedia and multi-route studies should become the norm within the general population and high-end exposure subgroups.

A major concern of exposure assessments is the identification of the uncertainties in the data base. The limitations frequently encountered are the lack of information on the distribution of exposure, the most important media and routes of exposure, and the activity patterns used to predict a mean and/or the high-end exposures within a population. Such deficiencies will propagate through an assessment and can be reduced only by the acquisition of appropriate population exposure data and the validation of models.

The construction of models of individual or population exposure to contaminants is essential since it is nearly impossible to measure all exposures experienced by an individual or by the general population. Therefore we have to select statistically representative groups of the population to measure and use these data to estimate the exposure with deterministic or fundamental models. The data bases used in such models, however, must continually be updated for use in exposure assessments to ensure that exposure estimates for the 21st century are not being made using data that were appropriate for the 1970s. It must be emphasized that population exposures will change over time based upon changes in lifestyle, implementation of source control strategies, and the introduction of new industrial and consumer products.

After the information on actual or estimated exposure is obtained, it must be used to estimate a dose in the body for risk characterization analyses. At this point the absorption and adsorption, deposition, bioavailability and transformation data are required to estimate the potential internal or biologically effective dose to the body. Depending upon the data available the dose estimate will have its own uncertainty. Usually the minimal estimate sets most parameters to a default value and applies a range of contact rates to the estimated or measured exposures. This value is called the potential dose.

Once the calculations of exposure and dose are completed, results of the exposure assessment can be used for risk assessments and risk management decisions.

The exposure values alone can be used to define exposure-response relationships in epidemiological studies, clinical interventions, and exposure characterization studies.

PAUL J. LIOY

For Further Reading: Committee on Biological Markers of the National Research Council, "Biological Markers in Health Research," in *Environmental Health Perspectives* (1987); Paul J. Lioy, "Assessing Total Human Exposure to Contaminants," in *Environmental Science and Technology* (1990); National Research Council, *Human Exposure to Air Pollutants: Advances and Opportunities* (1991).
See also Dose Response Relationship; Risk Assessment.

EXTERNALITIES

Social and private costs are equal for most market decisions, e.g., the price of potatoes reflects the social cost of producing and marketing them. The market works well in these cases because both producers and consumers face the social costs. In contrast, some decisions involve externalities, incidental conditions, which drive a wedge between private and social costs. As a consequence an individual making decisions on the basis of private costs will not make decisions that are best for society. For example, the private cost of disposing of sulfur dioxide from burning coal is small; just put it into a smokestack and release it into the environment. The resulting air pollution imposes little or no cost on the polluter, even though it can impose a large cost on society. Many aspects of environmental degradation are viewed as externalities of economic transactions.

A classic example of an externality is the foul air in Pittsburgh in 1945, London in 1952, and Mexico City in 1992. Each mill or automobile emits pollutants that are seen by the owner as no more than a small annoyance. In aggregate, however, severe air pollution results which, in turn, injures the health of residents, destroys property, and raises production costs.

To "internalize" externalities and make markets efficient, a tax can be imposed that is the estimated social harm imposed by the externality. This idea was suggested by the English economist A. C. Pigou more than eighty years ago. When deciding how much pollution to abate, this tax aligns the individual polluter's incentives with the social costs. For example, emissions of greenhouse gases are expected to increase the average global surface temperature, which in turn could change precipitation, ocean currents, and ecology for centuries in the future. If this externality is not considered, coal will continue to be the most attractive fuel because it costs less than renewable energy resources in most cases. However, once a carbon tax is imposed to reflect the cost of global climate change, in many cases renewable energy resources become cheaper than coal. Firms would then use renewable energy resources, diminishing the chance of climate change.

F

FAMILY PLANNING PROGRAMS

Family planning programs are "organized efforts to assure that couples who want to limit their family size or space their children have access to contraceptive information and services and are encouraged to use them as needed." In view of the illegal status of artificial birth control in most countries only a century ago, the worldwide increase in the numbers of family planning programs and users of contraception is an impressive accomplishment in social engineering. Today publicly funded services for birth control, often through maternal and child health care programs, are available to an overwhelming majority of the world's couples. Close to 500 million or about half the world's couples are estimated to be practicing family planning.

Since 1965 there has been rapid expansion of large-scale public programs in the developing world, notably in China, India, Indonesia, Brazil, Thailand, Bangladesh, Mexico, and Kenya. In the first half of the 1900s, however, important initiatives were taken to secure the legal status of birth control, advance biomedical research on contraceptive methods, and begin public funding to deliver birth control services. Intrinsic to establishing publicly funded family planning programs has been debate over its goals, an issue that persists into the present. The debate centers on the appropriateness of the public's management of childbearing, as a private behavior. Government intervention into couples' decisions about their childbearing has been justified for three types of reasons: demographic, health, and human rights. The demographic rationale argues the importance of reducing fertility rates to slow population growth and improve opportunities for economic development. The health rationale promotes birth spacing and limiting to safeguard the health of mothers and children. The human rights rationale asserts that couples are entitled to access to birth control information and services to pursue their childbearing intentions. Consensus on the relative primacy of these goals has not evolved easily in national, regional, or international dialogue.

Public services for birth control in industrialized countries are offered as part of the overall health care programs. In the U.S, current annual federal and state funding for contraceptive services totals about $500 million. The share of federal funding, which began in 1970 with Title X of the Public Health Service Act, has been steadily declining. In both the U.S. and Europe, about 65% to 75% of sexually active couples practice birth control, and the majority of contraceptive practice is privately serviced. Due to legal and regulatory constraints, the range of contraceptive choices has been more limited for U.S. than European couples.

The evolution of policies supportive of family planning programs in the developing world largely originates from governmental perceptions of excessive fertility levels. India was the first country to adopt such an official policy, in 1951. Between 1960 and 1965 nine other countries followed suit, among them Pakistan, the Republic of Korea, China, Egypt, Singapore, and Sri Lanka. In past years, the health rationale has been used to support expanded public initiatives for family planning in sub-Saharan Africa. A recent study of nearly 100 developing countries shows that during the 1980s the level of family planning effort across a range of areas—policy support, the design of service delivery, contraceptive method availability, and monitoring and evaluation—has increased. With family planning well established in Asia, the greatest change is being registered by African, Latin American, and Middle Eastern countries.

The delivery of family planning services in the developing world has diversified over time. The total "program" now engages the full efforts of governments as well as private groups to provide contraceptive information and services and encourage their use, through public, private non-profit, and commercial sector outlets. Whereas in earlier years family planning services were offered through a few clinics and limited to one or two methods, a family planning program can now offer five or more methods in clinics and two or three outside clinics. Programs extend the reach of their services through visiting by community-based field workers, subsidizing contraceptive sales by local pharmacists and retailers, and fostering services offered through private clinics at workplaces. These efforts augment fixed-site services to better serve the family planning needs of urban poor, rural, and male populations.

The structure of family planning programs in developing countries has also become more complex over time. A modern family planning program requires co-ordination of a number of organizational subsystems to manage and mobilize medical and nonmedical personnel, contraceptive supplies, funds, and equipment in performance of the standard program functions, such as service delivery, training, promotion, counselling, and evaluation. Management information and service data systems are increasingly necessary to monitor and evaluate the performance, impact, and costs of program functions.

Partly a result of the expanded availability of family planning services, contraceptive prevalence (use of contraception among couples where the wife is of childbearing age) in the developing world has steadily risen from under 10% estimated for the mid-1960s to over 50% in 1991. The practice of modern forms of contraception (sterilization, intrauterine devices, subdermal implants, oral pill, hormonal injectables, and condoms) predominates. Among users, an estimated 30% practice sterilization, an especially prominent method in China, India, the U.S., and many Latin American countries. The next most popular method of contraception is the IUD, used by almost one quarter of couples in the developing world and less than 10% in the developed world. Contraceptive use in Africa is principally of the pill.

The greater the practice of effective contraception, the lower the fertility rate. Parallel to the increase in contraceptive use between 1965 and 1990, the total fertility rate (the average number of births a woman would have if she experienced fertility at present rates throughout her childbearing years) has declined 36%, from 6.1 to 3.9. In spite of the fertility decline, the annual rate of population growth has shifted downward only moderately from 2.4% to 2.1%. In this same period, global population size rose from 3.0 to 5.3 billion, with population in the developing world nearly doubling from 2.1 to 4.1 billion, an outcome of past high fertility contributing to the population base.

Standard indicators, such as literacy (especially of females), life expectancy, per capita income, and infant mortality, show social and economic improvement in the human condition. The expanding reach of mass media has contributed to a global sharing of lifestyle images and ideas that accelerate societal transformations and alter the perceived value of children. Adjustments in the motivations of couples to limit family size and space children are key to ensuring the effective use of family planning services. The joint effect of these social changes and the availability of family planning services is significant in explaining the rise in contraceptive use and decline in fertility since the late 1960s.

While the prospects of further family planning program accomplishments are favorable, the same may not be true of the outlook for resources to sustain them. Government funding for family planning in the developing world continues to increase but is constrained in growing sufficiently to service the projected levels of contraceptive demand. Funding from international donor agencies, led by the U.S., annually approximates $600 million but has not grown appreciably in recent years. Increasing attention is being given to cost recovery efforts through privatization of family planning, a trend that may be inevitable but risks the contraceptive choices available to the rural poor. Strengthened commitment to and funding of family planning programs can ensure that services continue to be available to meet the continued decline in demand for children and avoid increases in unwanted childbearing.

AMY ONG TSUI

For Further Reading: Peter J. Donaldson and Amy O. Tsui, "The International Family Planning Movement," *Population Bulletin* (1990); James F. Phillips and John A. Ross (eds.), *Family Planning Programmes and Fertility* (1992); George Simmons, "Family Planning Programs," in Jane Menken (ed.), *World Population and U.S. Policy: The Choices Ahead* (1986). *See also* Contraception; Human Reproduction; Population.

FAO

See United Nations System.

FDA

See United States Government.

FEDERAL ENERGY REGULATORY COMMISSION

Established in 1977, the Federal Energy Regulatory Commission (FERC) is a branch of the Department of Energy. It is responsible for issuing licenses for hydroelectric power and providing for recreational opportunities, flood control, and the efficient and safe operation of project dams. In addition, a large part of the commission's work is the control of the pricing and sales of 60% of the natural gas and 30% of the interstate electricity used in the United States. Comprising only five members, FERC affects national energy policy through its decisions on how natural gas and interstate electricity should be regulated.

FERC is the successor to the Federal Power Commission (FPC), the agency that regulated the prices of interstate sales of natural gas and electricity from its establishment in 1920 until 1977. Under the direction of the Natural Gas Act of 1938, the FPC was to ensure an adequate supply of gas while keeping consumer prices low. Yet the FPC kept prices so low for such a long period of time that the energy market became distorted. Gas producers claimed it was unprofitable to search for new sources, and gas supplies dwindled to the point that public utilities refused new customers. Industry, homebuilders, and homeowners switched to electricity, which was more expensive, and to oil, which was vulnerable to embargoes and price hikes.

In the late 1970s gas producers wanted the ceiling price on gas lifted, and an environmentally conscious American public wanted greater supplies of this clean-burning fuel. Reacting to this pressure, the FPC voted to raise gas prices in 1976 in order to encourage gas exploration, which had come to a virtual standstill.

In 1977, Congress turned the FPC into FERC. FERC accelerated the increase of interstate gas prices in the 1980s due to the administration's policy of industry deregulation. Consumers saw their home gas bills rise as much as 60% in 1982, and pipeline companies were forced to pay higher prices to the gas producers. In 1985 FERC, intent on changing energy prices from being regulated by the government to competing freely on the market, proposed that pipelines be allowed to buy and sell gas free from government regulation.

FEDERAL INSECTICIDE, FUNGICIDE, AND RODENTICIDE ACT

The Federal Insecticide, Fungicide, and Rodenticide Act (FIFRA), as passed in 1947, required that pesticides be registered with the U.S. Department of Agriculture, and that they be labeled, but nothing more. After the publication of Rachel Carson's *Silent Spring* in 1962, there was an onslaught of public opinion against pesticides such as DDT, aldrin, dieldrin, and the herbicide 2,4,5-T. Authority to administer FIFRA was transferred to EPA in 1970, and the act was rewritten in 1972. It has been amended several times since then. As amended, FIFRA sets standards for registration of pesticides, regulates their use, and provides authority for removing them from the market.

Before a pesticide can be used in the U.S., it must first be registered with the EPA. Pesticide registration involves submittal of the pesticide formula, a proposed label, and extensive test data. The submittal must show that the pesticide will not cause "unreasonable adverse effects on the environment," including health, and take into account economic, social, and environmental costs and benefits. A registration applies only to specific crops and insects, and is effective for five years. For other uses, and for renewals after five years, additional safety data must be submitted.

A controversial aspect of FIFRA registration is its treatment of trade secrets. When a registrant submits test data under a claim of trade secrecy, EPA is required to preserve the confidentiality of the data. Pesticide testing is extremely expensive. When an applicant wishes to rely on data submitted previously by another applicant, FIFRA recognizes that it would be unfair to allow free-loading, and provides a mechanism for the later applicant to reimburse the earlier applicant for use of the data. The Supreme Court decided in a 1984 case, *Ruckelshaus v. Monsanto Co.*, that pesticide data could be protected as a trade secret, but that the earlier applicant could not preclude disclosure to the later applicant. The anti-free-loading provisions put the applicant on notice that its data would be disclosed, and provided for compensation.

FIFRA classifies pesticide use as general or restricted. Certain pesticides that are improper for general use may be used by "certified" users who are presumably experienced. Experimental uses may be permitted.

When there are "substantial questions of safety" regarding a pesticide, EPA may issue a cancellation order initiating a review of whether the approval of the pesticide should be revoked. If there is an "imminent hazard" to health or the environment, then EPA may issue a suspension order, an immediate ban on production and distribution of the pesticide. The registrant has an opportunity to contest the order. In the case of an emergency suspension, however, the use, sale, and distribution of a pesticide may be stopped before there is a hearing. For example, EPA used its emergency authority to suspend sale and use of 2,4,5-T and silvex for certain uses. In general, however, use of existing stocks is permitted after a suspension because a recall could seriously disrupt agriculture.

MICHAEL A. GOLLIN

FEDERAL WATER POLLUTION CONTROL ACT

See United States Government.

FERAL ANIMALS

Feral animals are wild animals descended from previously domestic ones. Several species are commonly found in the wild including dogs, cats, horses, burros, goats, and pigs. The wild horses and boars that roam the Americas today were originally introduced by Spanish settlers and explorers in the 16th century.

Many wild burros, which are numerous in the southwestern U.S., Central America, and parts of South America, are descendants of animals that were abandoned by miners in the 19th century.

As introduced species feral animals often have a profound impact on their new host environments. They compete with native animals for food and water, can be destructive to local plants that have not developed adequate protection against them, and often do not have any natural predators that would keep their populations under control. In the Great Smoky Mountains of the U.S. wild pigs foul springs and destructively rout forests. The large decline in songbird populations in North America, known as the "silent spring" phenomenon, has been linked to feral cats, one of the major predators of birds.

Feral animals pose a particularly grave threat to island ecosystems whose isolation makes them especially fragile. Many of Australia's marsupial species such as kangaroos, koalas, and opossums are now threatened by large populations of feral rabbits, camels, and donkeys, which have destroyed habitats and spread some types of diseases to the marsupials. In the Galapagos Islands off the coast of Ecuador, wild pigs have been blamed for the decline of the famous Galapagos tortoise and the disappearance of the land iguana. The islands of New Zealand have had a long history of sport and domestic animal introduction, which has nearly eliminated hundreds of native plants, reptiles, birds, and mammals. Yet another dramatic example of habitat destruction by feral animals can be seen at Hawaii Volcanoes National Park where wild pigs and as many as 20,000 feral goats led to a pronounced destruction of forests. All but a handful of the goats were killed by park rangers and the forests have made a remarkable comeback. After the goats were eradicated, researchers in the park found a species of native plant sprouting that had not been seen for decades. Similar problems exist in Grand Canyon National Park but few control measures have been undertaken due to social and political pressures.

How to manage ecosystems that are impacted by feral animals is a subject of heated debate among conservationists, ecologists, animals rights activists, hunters, and farmers. Many of the programs designed to rid areas of feral animals require that the animals be killed. Park managers and biologists fear that leaving them alone would result in irreversible damage to the habitats in which they live. Several animal rights groups have objected to killing the animals and have successfully halted or delayed some feral animal eradication programs. Pressure from these groups, as well as hunting groups, has also led to alternatives. A program launched in England to control feral cat populations involved catching the cats, neutering them, and

releasing them. This method has gained popularity in the U.S. and has also spawned "adoption" programs in which homes are found for the wild animals. Some conservationists and ecologists argue, however, that this method is too costly and too time-consuming. They say that some ecosystems may not survive the decades it may take to implement population control programs of this kind.

FERC

See Federal Energy Regulatory Commission.

FERTILIZERS

Fertilizers are soil amendments, either organic or inorganic, that are applied at strategic points in the production cycle to ensure adequate nutrition and to raise yields. They provide nitrogen, phosphorus, potassium, and sulfur, as well as trace minerals (including magnesium, boron, copper, zinc, and calcium) that are otherwise in short supply in the soil. Examples of organic fertilizers include human wastes, animal manures, green manures (plant material), compost, and blood meal. Inorganic fertilizers are largely produced from petroleum. Organic fertilizers tend to release nutrients slowly into soils, allowing plants to gradually absorb minerals and proteins throughout their life cycles. Inorganic nitrogen fertilizers tend to be quick-releasing and are intended to feed the plant, rather than the soil.

Natural soils are generally deficient in phosphorus and nitrogen, although nitrogen is relatively more available through the nitrogen-fixing processes of legumes, organic matter in the soil, and nitrogen from the atmosphere. In most agricultural systems, some level of fertilizers must be added to replenish the soil and to stimulate plant growth. Fertilizers are a component of fertility management practices, which include rotation of nitrogen-fixing legumes or cover crops, fallow periods (to allow soil organic matter to stabilize), and minimal mechanical tillage (which eventually destroys soil structure). Both organic and inorganic fertilizers can be overused, and the effect on soils and ground water can be identical: nitrogen toxicity. Fertilizer application rate and timing—which affect the absorption of minerals into soils and plant roots—are critical factors.

Despite the many impressive gains realized through the application of chemical fertilizers, questions remain as to the long-term environmental and economic sustainability of such practices. It is estimated that chemical fertilizer consumption represents 30% to 50% of the total energy expenditure in U.S. agriculture, and is roughly equivalent to 1% to 2% of society's total use of fossil fuels. A frequent criticism of the total agricultural energy "budget" (which includes the necessary energy to extract petroleum and to produce, transport, and process food products) is the relatively poor net energy/calorie performance ratio. In addition, critics point to several negative environmental impacts that result. These include depletion of nonrenewable resources; the eventual destruction of farmland through continuous cultivation; and degradation of other ecosystems, through eutrophication of surface and ground water and soil salinization.

FIBER OPTICS

Fiber optics is the technology of the transmission of light through a bundle of very fine flexible glass or plastic fibers. Medical uses of fiber optics include endoscopes and fiberscopes that enable physicians to examine body cavities, photograph inaccessible places, perform surgery, and sample tissues for analysis. The application of fiber optics in communications for transferring information is transforming the world into a global village.

Fiber optic systems were developed in the late 1960s and early 1970s. The work grew out of a paper published in 1966 by Dr. C. K. Kao of International Telephone and Telegraph Corporation in England. He proposed that an optical fiber might carry a light signal with minimal loss. By 1970 Corning Glass Works produced an optical fiber made from silica glass capable of relaying telecommunications signals.

However, it wasn't until 1983, when the next generation of fiber optic systems appeared, that they became cost-effective. A year later American Telephone and Telegraph (AT&T) agreed with the government to divest itself of 22 local telephone companies in exchange for the opportunity to enter new business arenas. The local companies were arranged into seven regional phone companies: Ameritech, Bell Atlantic, Bell South, Nynex, Pacific Telesis, Southwestern Bell Corporation, and US West. They too began investing in fiber optics. Other companies such as MCI and Sprint also became involved, challenging AT&T, which has spent billions of dollars replacing its vast collection of existing equipment including copper wire pairs—the traditional phone lines—with this new high technology.

Fiber optics, by being able to carry a billion or more bits of information, has precipitated a merging of computers, television, video and audio systems, by virtue of its ability to transmit pulsating light waves. Fiber optic systems are capable of carrying sound, data, and video when it has been digitized.

Fiber optic cables can transmit the digital signals worldwide into the home allowing a user to select from a large database of programs, video telephone calls, or sophisticated computer-generated entertainment, all on television. Telephone and cable television companies are expected to offer similar services allowing users to select from hundreds of channels of digital information. New products are being developed and many materials already have been digitized on computer databases, CDs, and laserdiscs.

The fiber optic network may bring about radical changes in the workplace. Some workers may do their jobs at home—a practice called telecommuting. Businesses will be able to link together headquarters and facilities worldwide, allowing workers access to vast amounts of information. They will be able to tie individual computers into a company's network, accessing documents and files, and conducting videoconferences, all from the comfort of homes. With many employees working out of their homes, the environmental implications will involve tradeoffs and benefits in transportation, electrical consumption, and changed social interactions.

However, the transformation to a fiber optic society is expected to be gradual. Cost prohibits rapid replacement of the seemingly endless miles of telephone lines, particularly in remote areas.

FIFRA

See Federal Insecticide, Fungicide, and Rodenticide Act.

FIRE

Using fire to alter natural environments represents an extremely important dimension of human evolution. Beyond the acknowledged significance of domestic fire, the setting of habitat fires represents the considerable impact that prehistoric humans had in being able to both change and maintain many of the important habitats and related successional stages exploited for hunting and gathering.

Arguments and counterarguments have been made by anthropologists and others about the first clear evidence of fire use and its eventual production by protohumans. The earliest suggestions for domestic fires are reported to be hearths from east Africa, dating more than 1.5 million years ago, while fires within cave sites occupied by *Homo erectus* a million years ago have been claimed for France and Spain. The more widely accepted presence of hearths comes from Choukoutien near Beijing in China, dating between 500,000 and 400,000 years ago. Hearths demonstrate that protohumans were able to transfer and maintain fire or embers across space and over periods of time. By extension, it is argued that means of maintaining fires were necessary for the regular occupation of caves. Clearly, however, the techniques for husbanding fire preceded the invention of fire-making tools by hundreds of thousands of years.

The earliest devices for generating fire are assumed to have been based on friction (e.g., fire drills and fire saws) and would have been made of wood. These are not preserved or not recognizably artifacts. Nonetheless, it is reasonable to assume that by the time Neanderthals occupied Europe during the Mesolithic 200,000 years ago, one or more means of generating fire had been discovered. As with more recent arctic and subarctic adaptations, fire-making techniques would have been necessary for surviving Europe's Ice Age.

In addition to the origins and uses of domestic fires, prehistorians and paleo-environmentalists have acknowledged that foragers would have set habitat fires to stands of grass, brush, and trees in areas adjacent to campsites, occupied caves, and in selected areas further afield. Though no unequivocally clear evidence exists to show that prehistoric peoples set habitat fires, the wide scattering of ash within and around human settlements favors the very high probability that they did. Also, as historical and contemporary hunter-gatherer uses of fire have shown, it is difficult to imagine how prehistoric peoples would have adapted to a range of environments and local habitat types without using fire as an integral component of hunting and gathering technologies. However, unlike stone or fossilized bone tools, it is unlikely that the ad hoc, disposable artifacts used for igniting habitat fires in surrounding areas—fire sticks, firebrands, or sheaves of grass—would be found within the remains of human settlements or even recognized as tools.

Ecologically, habitat fires are infinitely more significant than those which provide heat and light within the contained perimeters of campsites, shelters, and caves. In contrast to the irregular occurrence and random effects of natural fires, prescribed habitat fires would have been particularly important for prehistoric adaptations in maintaining a more reliable and predictable resource base. Both the technological means and the ecological consequences of cultural fires have been demonstrated from a number of ethnographic studies showing how fire has constituted an integral part of hunting-gathering adaptations across a wide range of different environmental zones.

With the exceptions of the northernmost arctic regions and the wettest tropical rain forests, hunter-gatherers universally employed habitat fires for a variety of interrelated reasons and in functionally similar environments. Parallel considerations undoubtedly motivated prehistoric peoples to undertake similar practices: to drive or simply move animals from one site to another; to induce new, more nutritious plant growth; to encourage greater productivity of a range of plants exploited by humans (e.g., legumes, seed grasses, forbs, and berries); to attract game; to increase or maintain the carrying capacities of preferred animal species, especially medium to large ungulates; to clear older, unproductive stands of brush and trees; to intensify and maintain a greater mix of habitat types than what occurs naturally, representing a wider diversity of ecological stages of succession (a cultural fire mosaic); to increase the extent of ecotones, those areas favored by edge species, many of which are important game animals; to clear campsites and reduce the numbers of undesirable insect pests and snakes, especially important in tropical and sub-tropical regions; and to maintain corridors of grass and grassland clearings ("yards," or patches) within temperate and in some cases tropical rain forests.

An important consequence of a greater number of small fires, set on a more frequent basis than those which occur naturally, is the reduction of accumulated fuels and the lessening of disruptions that derive from highly unpredictable, randomly distributed, and dangerous perturbations that follow from less frequent, much hotter natural fires. At the same time, such conflagrations can temporarily degrade hunting-gathering areas and disrupt the movements, distributions, and adaptations of local populations. As the ethnographic evidence shows, while hunter-gatherers know how, when, why, and where to set fires—and, alternately, not set fires—they also understand and, for long-term adaptations, must understand the ecological relationships among plants, animals, and fire, including the consequences of both natural and cultural fire regimes. Such understandings involve indigenous "theories" about the systems of cause and effect relating to fires in specific environments and local habitats, all of which is an extremely important dimension of traditional ecological knowledge.

As is reflected in the evolutionary record on material cultural change, contemporary and historically recent populations of foragers clearly represent much more sophisticated technologies of fire than those that would have been found during the Paleolithic or Mesolithic. Pleistocene hunter-gatherers, even Australopithicenes, must have recognized, and correspondingly benefited from, some of the more important and obvious environmental changes that derive from burning. During the earliest stages of human evolution, cultural patterns need not have differed significantly from the ways that non-human species, particularly ecotone species, recognize and exploit fire-influenced areas in the weeks and months after sites have been renewed by fire—for some species actually during and immediately after burning. However, unlike other animals, protohumans could have transferred fire from place to place, at selected times of the year, employing fire frequencies and intensities that increasingly facilitated human adaptations and favored many of the plant and animal species upon which they depended. In this respect it is probable that protohumans would have used fires to enhance local adaptations, however crude such practices may have been by comparison with modern *Homo sapiens* subsequent to the Pleistocene epoch.

Initial practices would have set the stage for millennia of cultural trials and environmental errors leading to the technological and ecological understandings of fire as employed by modern hunter-gatherers during the past 10,000 to 20,000 years. Unfortunately, science—and within it anthropology—has been slow to recognize and acknowledge either the meaning or significance of indigenous uses of habitat fires. In most industrialized societies, knowledge about environmental fires has become highly specific, primarily the concern of foresters and other environmental agencies. Overwhelmingly dominated by a preoccupation with the dangers rather than the advantages of fire, it has involved assumptions and attitudes that have carried over and adversely influenced the amounts of scientific study and research support.

Paradoxically, most developed nations have been unwilling or unable to act upon the knowledge that the major dangers and environmental disruptions of fire arise from the accumulation of fuels in overaged, overprotected forests and brush fields—areas which periodically erupt in holocausts. Only in the past few decades has the importance of prescribed uses of fire begun to be emphasized by government agencies, though based more on attempts to replicate natural fire regimes than the burning practices used by populations of hunter-gatherers.

While agriculture is acknowledged as a cultural means for altering natural processes—the earliest forms of which would have necessarily included the use of fire for clearing vegetation, recycling nutrients, and controlling weeds—the widespread deployment of fire by hunter-gatherers for changing and maintaining local environments has been considered technologically and ecologically insignificant. Despite a growing documentation on the burning technologies and ecological understandings of hunter-gatherers, neither the importance of what these practices have meant to human evolution nor the significance of how such practices would have influenced the related evolution of "natural" environments is sufficiently understood. The view that foragers and other so-called primitive

societies lacked the knowledge, means, and numbers of people to significantly alter natural processes—in what Europeans erroneously conceived as "untouched wilderness"—has undoubtedly contributed to the neglect of seriously considering cultural fire regimes as the most important ways in which indigenous people directly influenced the growth and reproduction of natural systems. "Modern" societies are generally far too ethnocentric to accept the idea that "primitive" societies could have been—and in some cases still are—more knowledgeable about something as universally important as managing environments.

A small number of anthropologists and others have contributed to understanding indigenous burning practices and local knowledge about the effects of prescribed fires in two ways: by presenting specific examples, most derived from the study of Native North Americans and Australian Aborigines, plus the use of comparison, both in contrasting ethnographic examples and in comparing folk practices with what science has shown from studies in fire ecology. In this respect, specific ethnographic examples can be compared against, and examined within, the contexts of scientific generalizations about the multiple role of fire within a range of environments.

In addition to the significance that traditional fire regimes have had for the adaptations of societies in different times and geographic settings, the growing number of ethnographic studies have begun to reveal parallel solutions in the ways that humans have developed similar technological responses in the multiple uses of prescribed burning. The practices involved are found in the differences that broadly characterize the patterns of cultural and natural fire regimes, patterns which are at the core of how burning has accommodated human adaptations, particularly the adaptations of hunting and gathering peoples.

Currently, the most destructive uses of fire are found in tropical regions where forests are cleared for logging, farming, and cattle ranching. These are effectively pioneering practices that stand in marked contrast to indigenous systems of shifting agriculture, which have traditionally involved cutting and burning small separated stands of forest, followed by long periods of fallow and natural reforestation.

The most common and most ancient uses of agricultural burning are found with cereal farming and the removal of crop stubbles, involving a variety of reasons that include: the reduction of toxins and plant diseases; a temporary check on competition from weeds; the reduction of insect larvae and host plants; the regeneration of hard-seeded legumes; the recycling of organic matter; and the incorporation of potash and chemical trace elements. Stubble fires and government attempts to enforce burning schedules are widely reported for both ancient Rome and Greece. The same basic concerns and reasons for firing crop residues can be found from equatorial rice farming areas to the subarctic regions in which wheat, barley, and other hard-grained cereals are grown.

Considering the extent of research carried out by agronomists on other aspects of farming technologies, it is surprising how relatively little study has been made of agricultural uses of fire. Popular views about the dangers of fire and concerns about smoke pollution have influenced a number of governments (e.g., in California) to introduce anti-burning legislation. With the consequences of prohibiting agricultural fires not fully known, and with pressures on farmers to substitute mechanical and chemical retardants, the consequences of excluding post-harvest burning practices could be environmentally more damaging than allowing the continuation of practices that are more in keeping with the natural recycling of nutrients.

The earliest studies on indigenous uses of fire in North America and Australia were published shortly after World War II by a small number of writers—most prominent among them Gordon M. Day, a biologist; Carl O. Sauer, a geographer; and Omer C. Stewart, an anthropologist—and they and more recent writers are referred to in Stephen J. Pyne's works. As yet, however, no single, comprehensive overview of indigenous uses of fire exists, with most published works scattered in a range of inter-disciplinary journals, topical monographs, and published symposia.

Despite the work done on traditional uses of fire in North America and Australia, indigenous systems of ecological knowledge of fire have been little examined elsewhere, with only a few studies on South America, fewer still on East and Southeast Asia, and an even greater dearth in reporting from Africa and the Near East. Given the importance of understanding and effectively using fire as an integral component of environmental systems, the long-term uses and indigenous knowledge systems assume an increasing importance for comparative and scientific study.

HENRY T. LEWIS

For Further Reading: P. J. Crutzen and J. G. Goldammer (eds.), *Fire in the Environment: Its Ecological, Climatic, and Atmospheric Chemical Importance* (1993); Stephen J. Pyne, *Fire in America: A Cultural History of Wildland and Rural Fire* (1982); Stephen J. Pyne, *Burning Bush: A Fire History of Australia* (1991).
See also Anthropology; Solar and Other Renewable Energy Sources; Forests, Deciduous; Fuelwood.

FISH AND WILDLIFE SERVICE

See United States Government.

FISH FARMING

See Aquaculture.

FISHING

Fishing for food as well as pleasure is probably as old as humanity. The use of rudimentary fishing gear including hooks probably extends to ten thousand years in the dim past. Pictographs, five thousand years old, depict aristocrats fishing. Clearly fishing was engaged in for pleasure or recreation in that era. In the 15th century Dame Juliana Berner published *Treatise of Fishing with an Angle*, the forerunner of today's wealth of written material on fishing as a sport. Some 150 years later, Izaak Walton published *The Compleat Angler*, which is considered by many to be the first authoritative work on sport fishing.

A rare combination of ready availability and positive social values blend in recreational fishing in ways that attract the participation of one in four Americans each year. Several notable angling enthusiasts have occupied a large home on Pennsylvania Avenue in the nation's capital. The most notable among them in this century were Herbert Hoover, Franklin Roosevelt, Dwight Eisenhower, Jimmy Carter, and George Bush.

Sport fishing challenges the mind, body, and spirit to scale the highest peaks to entice native cutthroat trout to take the smallest dry fly offered with space-age composite rod and line. The enormous range of intellectual, financial, and physical challenges makes sport fishing among the world's most popular recreational activities.

Social science investigators have conducted many studies of what motivates people to fish for pleasure. They have identified stress relief, spending time with friends and family, enjoying the outdoors, and many other positive attributes. While most sport fishermen do not need to catch fish on every fishing trip, the anticipation of catching fish, either for the table or to release, is the defining attribute that distinguishes fishing from hiking, picnicking, and other outdoor activities.

It is this essential reliance on renewable fish communities and the watershed characteristics that support the fishery that have crystallized anglers into leading and powerful conservationists over the last half century. Many vitally important environmental laws were championed by anglers unwilling to accept the continued abuse of waterways and landscapes. A key example is the U.S. Clean Water Act, which includes language that establishes a national goal of "fishable" for the nation's surface water. This term reflects the leadership and involvement of anglers in the legislative struggles that resulted in passage of this pivotal federal legislation.

Recreational anglers commonly experience a maturing of personal angling philosophy as they gain experience and skills. Many anglers seek to catch fish for the family table while capturing associated social benefits in the company of fishing friends and family in picturesque locales.

With experience, other anglers seek progressively more challenging fishing experiences. This evolution commonly takes one of two forms. Some anglers gradually progress to competition, to pit their fishing skills against their peers. This may eventually take the form of competing for prizes. An entire fishing culture, including television shows, has emerged around the competitive angler and the celebrity status of the more consistently successful of these competitors.

The second manifestation of adding greater challenge to angling adventures is to adopt more difficult gear such as fly fishing, and to practice these fishing techniques for wild (nonstocked) fish in far-away or difficult-to-reach locales. Catching fish for the table is generally of much lesser importance, and catch-and-release is commonplace for both of these evolving angling philosophies.

Fishermen willingly support vitally important fishery resource enhancement and habitat protection programs through the annual contribution in the U.S. of $600–$700 million in fishing fees, including fishing licenses and taxes on fishing equipment and boat fuels.

Beyond the enormous financial investments in conservation work made possible by fishermen, each angler has an increasingly important personal responsibility to fisheries resources and other resource users. A stringent code of personal behavior or conduct has become absolutely critical as more outdoor recreationists seek to extract benefits and enjoyment from an essentially stable water base. An excellent code has been developed by the U.S. recreational fisheries leadership. The code calls upon each and every angler to:

- Keep only the fish needed
- Do not pollute—properly dispose of trash
- Sharpen angling and boating skills
- Observe angling and boating safety regulations
- Respect other anglers' rights
- Respect property owners' rights
- Pass on knowledge and angling skills
- Support local conservation efforts

- Never stock fish and plants into public waters
- Promote the sport of angling

To protect future generations' privileges to enjoy the satisfactions and fulfillments of sport fishing, this code must become more than a written credo. The principles must become uniformly practiced standards of behavior.

Some anglers not only contribute financially to resource conservation and practice high standards of personal behavior that are sensitive to other users and the resource, they also contribute time and energy to hands-on habitat restoration projects in cooperation with their state and federal fisheries management agencies. This takes the form of "adopting" a particular reach of lake, river, or stream, or participating in habitat restoration such as stream bank stabilization.

It has been widely accepted that fishing for recreation recharges the human spirit. Recreational fishing provides an excellent source of high-quality protein for the table, families are bonded, and friendships are forged. These attributes, when combined with personal participation in the protection, restoration, and conservation of aquatic resources, make sport fishing important not only to the individual participant but to the well-being of the world and the sustainability of its aquatic resources.

NORVILLE S. PROSSER

For Further Reading: Roderick Haig-Brown, *Bright Waters, Bright Fish* (1980); Albert J. McClane, *Fishing with McClane, 30 Years of Angling with America's Foremost Fisherman* (1975); Albert J. McClane, *McClane's New Standard Fishing Encyclopedia* (1974).

FISHING INDUSTRY, U.S. COMMERCIAL

Throughout the world commercial fishing is a diverse industry employing thousands of workers and generating valuable products for domestic use and foreign trade. The United States, blessed with vast coastlines on two oceans, the Gulf of Mexico, the Arctic seas, the Great Lakes, and extensive rivers, lakes, and reservoirs, is among the leading fishing nations, annually landing more than 10 billion pounds of fish and shellfish with a value of $3.9 billion. In 1990 the U.S. ranked sixth in total harvest behind the Soviet Union, China, Japan, Peru, and Chile. Harvestable fishery resources within its 200-nautical-mile exclusive economic zone make up about 15% of the world's total.

The U.S. commercial fishing industry is composed of harvesting, processing, and marketing segments, each with a support infrastructure. About 31,000 fishing vessels (five net tons and over) are documented with the federal government and 80,000 smaller craft are registered with coastal states. Most of these operate from ports carrying out fishing operations in nearby waters. Larger vessels, up to 400 feet in length, operate in waters distant from home port, such as off Alaska in the Bering Sea and Gulf of Alaska, or in the tuna fishing grounds of the central and western Pacific Ocean. Fish processing and fish tender vessels operate almost exclusively in North Pacific waters, receiving fish from the catcher vessels. These larger vessels and some smaller vessels employ electronic devices for finding fish, determining the physical characteristics of the bottom, precise navigation, and rapid communication between vessel and port. Aboard the largest craft high-powered deck machines deploy and retrieve the fishing nets and handle the catches.

The number of individuals who fish commercially is not known exactly. An estimated 230,000 jobs are available aboard fishing vessels. Turnover is high at the entry level so the total number of individuals employed seasonally or part-time in the fisheries is significantly higher. Working conditions throughout the fishing fleet are very difficult as fishermen work long hours, often under severe environmental conditions. Fishermen are not paid wages. They share a percentage of the proceeds from the sale of the catch after the vessel owner deducts expenses: fuel costs, groceries, electronic leasing fees, owner's share, and bonuses for the skippers and other more highly skilled personnel.

Although U.S. fishermen use diverse gear such as hooks, spears, weirs, traps, pots, and dredges, most of the U.S. catch is taken by bottom and midwater trawls and purse seines capable of taking a large catch in a brief period of time. The development of high-strength plastic fibers has made possible the production of extremely strong fishing nets that can hold catches weighing thousands of pounds.

There are about 4,500 establishments within the processing and wholesaling sector that employ about 80,000 people. Some processing and packaging is conducted aboard ships operating in the North Pacific waters and the Bering Sea, both off Alaska, and the Confederation of Independent States. U.S. processors also import edible seafood products for further processing before marketing. Overall the U.S. commercial fishing industry contributed about $16.6 billion in value to the U.S. gross national product in 1990.

International trade in seafood products is a dominant factor that shapes this economic performance of the U.S. commercial fisheries. Imported seafood products comprise 41% of total U.S. consumption. Domestic landings have increased about 2% annually. Per capita consumption has grown from 11.8 pounds in 1970 to 15.5 pounds in 1990. As consumption increases, the U.S. industry may not be able to meet the need

so suppliers of seafood may increasingly call upon imports and aquaculture production to meet U.S. demand. Exports of seafood products also have grown significantly, valued at $5.6 billion in 1990. Despite this growth, the balance of trade remains high.

Prior to 1976, states managed fisheries and federal management was limited to international treaty obligations covering fish and shellfish of the North Atlantic, Pacific halibut, salmon, and the migratory tunas, whales, and Pacific fur seals. Overfishing was attributed to foreign fishing activities and efforts at control focused on foreign fleets. In 1976, Congress enacted the Fishery Conservation and Management Act, beginning modern fishery management by creating eight fishery management councils and authorizing the federal government to conserve all fishery resources except the tunas within the U.S. fishery conservation zone of 200 nautical miles from the coast. These eight councils have principal responsibility for developing fishery management plans. Developed to meet the requirements of the law, the plans are approved and implemented by the Department of Commerce's National Oceanic and Atmospheric Administration (NOAA). Enforcement of the management regime is the joint responsibility of NOAA and the U.S. Coast Guard.

The fishery management councils have continued the traditional management practices of the past, including fishing gear limitations, quotas, closed seasons, closed areas, fishery time limitations, and prohibited species. Because fishing efficiency has grown significantly, declines in abundance, largely from over-exploitation, have taken place. Examples of overexploitation include cod, haddock, and flounder off New England, red drum and groupers in the Gulf of Mexico, and salmon stock along the Pacific coast. As a result, management institutions set up more restrictive conservation regimes designed to arrest declines, begin stock rebuilding, or preclude overfishing. The restrictive measures and keen competition for the fish and shellfish have resulted in economic pressures upon the fisherman. In an effort to reverse this situation, the fishery management councils have begun to explore controversial conservation measures designed to limit fishery effort to a level sufficient to rebuild depleted stocks and restore a greater economic return on the capital and labor invested. Such measures presently include Individual Transferrable Quotas, shares in the resource that may be sold, traded, or bartered, enabling vessel operators much greater freedom of how, when, and where they may fish. Such measures are expected to reduce the cost to the government of carrying out its management responsibilities.

Other conservation laws such as the Marine Mammals Act and Endangered Species Act also regulate fishing activities. Regulations are designed to protect marine mammals such as seals and whales and other threatened or endangered marine life such as sea turtles, and constrain the fishing industry from fishing activities that might harm this marine life. The principal fisheries affected by these laws are tuna in the eastern Pacific—where U.S. fishermen must use modified purse seines to save porpoises—and shrimp fisheries along the Gulf of Mexico and southern Atlantic coast—where excluder devices are used in the shrimp trawls to reduce the incidental catches of endangered sea turtles. Amendments to the Fishery Conservation and Management Act require that fishery managers address the incidental take of small fish that are discarded, resulting in losses of future generations of fish. These amendments also require the federal government to improve management of species such as bluefin tuna and swordfish. These and other conservation issues will continue to confront U.S. fisheries.

WILLIAM G. GORDON

For Further Reading: National Oceanic and Atmospheric Administration (NOAA), National Marine Fisheries Service (NMFS), *Fisheries of the United States 1990* (1991); National Research Council (NRC), *Fishing Vessel Safety Blueprint for a National Program* (1991); J. G. Sutenen and L. C. Hansen, eds., *Rethinking Fisheries Management* (1986).

THE FIVE E'S OF ENVIRONMENTAL MANAGEMENT

The human presence has had such profound influences on practically all ecosystems that the continuation of human life implies in fact humanization of the Earth. We have far to go before understanding all the subtleties of the relationships between humankind and Earth, but we can at least begin to analyze them in the terms of several categories. We define these categories by five words arranged alphabetically but all beginning with the letter *e*—ecology, economics, energetics, esthetics, and ethics. Following are illustrations for each in simple life situations.

Ecology

Almost all human activities result in alterations of natural ecosystems, but these alterations need not be destructive and many of them have indeed been highly creative. Some of the world's most productive and beautiful ecosystems are now very different, biologically as well as visually, from the areas of wilderness out of which they were created. This is illustrated, for example, by the famous agricultural landscapes and parks of Europe, Asia, and North America, most of which have been developed from different types of primeval forests.

It is an obvious advantage, on the other hand, that artificial ecosystems should be as compatible as possible with the prevailing ecological characteristics of the regions in which they are created. Some day botanists will cooperate with landscape architects to develop not only grasses but other kinds of ground cover suited to different types of rainfall, soil composition, and human use.

Since agriculture also implies a constant struggle against nature, it is most likely to be successful in the long run when it takes account of ecological constraints. All over the world many types of agricultural land that were initially created from the wilderness have remained productive for centuries and even for millennia because farmers have utilized them with ecological wisdom, in other words, they have used agricultural methods ecologically suited to natural local conditions. Certain contemporary practices, however, stand in sharp contrast with this ancient ecological wisdom. In several parts of Texas and of the American West, for example, very high yields have been achieved by irrigating semi-desert lands with underground water (in many places fossil water from the Ogallala aquifer) which must be pumped at great cost. Such reserves of underground water are being depleted, and since irrigation water contaminates the soil with various salts as it evaporates, this type of farming will eventually have to be abandoned, leaving behind a legacy of degraded land.

Economics

Ecological thinking is needed to deal intelligently with nature, but economic considerations usually intervene in ecological choices—often with disastrous results. The only justification for the costly and destructive type of irrigation farming and aquifer-mining is the possibility of achieving great financial rewards over the short period of time—regardless of future consequences.

Many questions of value, furthermore, must be introduced into the economic aspects of pollution control and of environmental restoration. Limiting the control of pollution to the removal of 90% of the pollutants is usually an inexpensive operation resulting in marked environmental improvement. As environmental criteria become more exacting, however, pollution control becomes more and more difficult and costly while the increase in beneficial effects becomes less and less evident. Granted that the law of diminishing returns applies to all aspects of pollution control, there are types of benefits for which it is impossible to give a monetary value. Would it be an unreasonable criterion, for example, to require that Manhattan air quality make it once more possible for lichens to grow on the boulders and tree trunks of Central Park and for the Milky Way to become visible on cloudless nights?

Energetics

All relationships of humankind to the Earth are affected by the levels of energy consumption. One example will suffice to illustrate the wide range of influences that the level of energy consumption can have on human life and the environment.

An isolated, free-standing house, surrounded by as much open ground as possible, has long been one of the ideals of American life. This ideal used to be compatible with the social and economic conditions of the past when there was much unoccupied, inexpensive land and when the family house was essentially self-sufficient: with its own water supply from a well or a stream; its own fuel supply from the woodlot; its own food supply from the garden, domestic animals, and wildlife; and few problems about waste disposal. But conditions have changed. The free-standing house is now increasingly dependent on public services for electricity, telephone, water, and sewage systems; and dependent also on fuel for heating, transportation, road maintenance, snow removal, and practically all conveniences of modern life.

In the modern world, the way of life implied by the free-standing house involves such high social costs, especially with regard to labor and energy, that it will inevitably become an economic burden, too heavy for the average person and perhaps socially unacceptable. Social constraints are therefore likely to favor a trend toward some form of cluster settlements.

The increase in energy cost may act as the catalyst for a restructuring of human settlements and particularly for greater clustering of habitations. This would make for economies in fuel consumption, in the maintenance of roads, in access to utility lines, shops, and schools; it would furthermore facilitate group activities and thereby foster community life.

Many architects and social planners are trying to design human settlements that provide both the technological advantages of cluster housing and the sense of privacy and space associated with the free-standing house. If they are successful, the new types of planning will provide opportunities for larger woodlands, meadowlands, playgrounds, and other areas for community life. The village green and public square will not be recreated in their traditional forms but new concepts of land use and of architectural design may favor the revival of community spirit.

Esthetics

Human beings have created artificial environments out of the wilderness over most of the Earth, wherever they have made their homes. These humanized environments have become so familiar to us that we tend to forget their origin; we contemplate them in a mood of casual acceptance and reverie without giving thought to those parts of primeval nature that had to be profoundly transformed, or destroyed, to make them fit not only our biological needs, but also our esthetic longings. Fortunately, a combination of woodland, open space, waterscape, and horizon is compatible with many different types of cultural expression. It can be satisfied by the classical French landscape style, by the more romantic English treatment of the land, by the complex and symbolic design of Oriental parks—and also by the tremendous creations of nature in the American national parks.

We are only now beginning to acquire accurate information concerning what people really find attractive in scenery. Tests carried out with different social groups have revealed an almost universal preference for orderly sceneries, in which "nature" has been tamed, and even disciplined. It is probable that most of us long not for real wilderness, but for a taste of the small-scale wildness that gives additional interest to the sceneries they admire.

Except for several famous examples of real wilderness, some of the most esthetic manifestations of nature are found today in prosperous farming areas and on large private estates. Under present economic and social conditions, however, farming can hardly survive in the vicinity of urban areas and large estates are also destined to disappear—the likely outcome of both of these changes being a visual degradation of the environment.

Long-term planning for the maintenance of esthetic quality of the countryside around large urban centers must consider not only the present condition and use of the land but also the potentialities inherent in the soil, topography, climate, and water. Such knowledge might suggest new ways of managing the environment that would be compatible with both its welfare and that of humankind.

There are many possibilities of management for the areas of open land that still exist near urban centers, for example: allowing them to return progressively to a state approaching that of wilderness; establishing greenbelts; creating parklands designed for public use; developing human settlements combining high population density with large open public spaces; or reintroducing agricultural production, especially of perishable crops such as vegetables and fruits. A mix of these different types of management cannot be achieved without generating controversial problems of zoning.

Zoning policies have generally aimed at achieving some form of socioeconomic and occupational segregation. It is probable that in the future they will increasingly incorporate considerations of ecology and of environmental perception. Instead of being based on segregation, the new philosophy of environmental zoning should aim at creating areas where appropriate groups of uses can coexist in a suitable setting. Ideally, zoning should be considered not as a restrictive but as a constructive process; its goal should be to integrate different types of land uses that would interplay interestingly in planned environments.

Ethics

In the last chapter of *The Sand County Almanac*, Aldo Leopold formulated his view of "land ethic" in statements which are probably the most famous and the most frequently quoted expression of ecological wisdom in all environmental literature. "We must quit thinking about decent land use as solely an economic problem. . . . A thing is right when it tends to preserve the integrity, stability, and beauty of the biotic community. It is wrong when it tends otherwise." These statements are often interpreted to mean that Leopold was opposed in principle to any transformation of the land by human activities. However, he did not state anywhere in his writings that natural biotic communities are necessarily the most desirable ones. In fact, he described several ecosystems of which he obviously approved even though they are the products of accidental or intentional human intervention: for example, the man-made blue-grass country of Kentucky and the European farmlands.

Leopold properly stated that the land is healthy when it has "the capacity . . . for self-recovery. Conservation is our effort to understand and preserve this capacity."

It is as a contribution to this understanding that the five "e" categories are introduced to evaluate the various aspects of human intervention into nature. The emphasis on "human" interventions acknowledges that human activities are now among the most powerful and most ubiquitous forces acting on natural as well as on man-made ecosystems.

There is now widespread acceptance, in theory if not in practice, that the behavior of our societies toward the Earth must be based on a new kind of ethics, and there are reasons to believe that this new kind of ethics has begun to result in environmental improvement. Unfortunately, there is much less agreement on aspects of environmental ethics that concern the rights of human beings with regard to the Earth. In Sweden, for example, all people have free access at any time of the year to meadowlands and woodlands, and indeed to most non-built areas. Similarly, in several European nations the general public can have access to most of the seashore. The experience of the

American national parks reveals, however, that public use may be devastating. The most difficult environmental problems will not be those related to ecology, economics, energy, or even esthetics, but to those involving ethics: the rights and duties of people with regard to the Earth. In principle, we all have the same rights, but in practice we can enjoy these rights only to the extent that the Earth is maintained in a healthy state. This is, of course, an ecological problem, but one in which the effects of natural forces must be evaluated in the light of human needs, activities, tastes, and aspirations.

RENÉ DUBOS
(EDITED BY RUTH A. EBLEN)

For Further Reading: Benjamin Dysart and Marion Clawson, *Managing Public Lands in the Public Interest* (1988); Benjamin Dysart and Marion Clawson, *Public Interest in the Use of Private Lands* (1989); Aldo Leopold, *The Sand County Almanac* (1970).

See also Ecology as a Science; Economics; Energy; Equity; Esthetics; Human Ecology; Humanized Environments; Land Use; Quality of Life.

FLOODPLAINS

All river valleys in which stream flow fluctuates have areas that are inundated only at times of abnormally high flow. Those areas are known as floodplains. They are defined basically in terms of the magnitude and frequency of flood flow. The occurrence of such floods is described by the probability of recurrence of flows of specified size. Floods are often divided into four classes according to the cause of high discharge of water. The land areas in each class differ in their landform, their natural habitat, and their current land use. For each land use the occupants have open to them a wide range of adjustments to the extreme flows, with accompanying consequences for the welfare of the people and societies affected.

Floodplains may range in size and use from narrow strips in wilderness areas to large, densely settled cities, with many sorts of combinations of environments and human use between those extremes. Some of the more dramatic landscapes in remote areas — such as in the Grand Canyon of the Colorado or along small streams of the Adirondacks — center on floodplains. Likewise, some of the great industrial areas — such as the Pittsburgh metropolitan area — and some of the more productive agricultural areas — such as the lower valley of the Mississippi River — are floodplains. Many of the great civilizations of world history have had their locus in floodplains: the Yellow, Yangtze, Me-

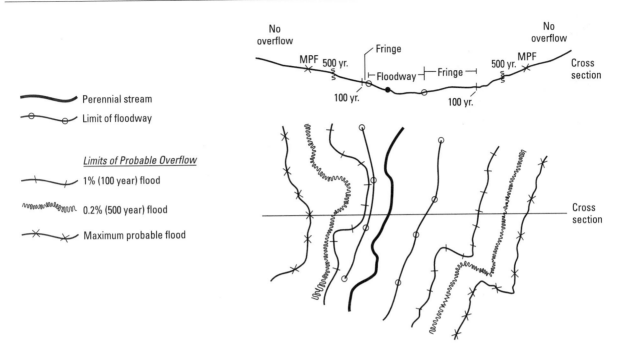

Figure 1. Cross section and sectors of a typical river reach.

kong, Ganges, Indus, Tigris-Euphrates, Nile, Rhine, and Mississippi are a few. They all have in common their susceptibility to high water that may bring rich harvests as well as catastrophic losses.

Sectors of a Floodplain

A cross section across the direction of flow in any valley can divide the land into four major sectors. (See Figure 1.) Central is the stream channel, the land that is covered by water during times of minimum flow or in most of the year. At the other extreme is the land never known to be inundated by the stream. In between are the "floodway" portion of the valley that carries flood water outside the stream channel, and the "fringe" portion that stores water from the high flow in ponds but does not carry water rapidly downstream.

While these four sectors or zones can be delimited for any given discharge of water in the stream, their boundaries shift with the volume of water flowing down the valley. Thus, if a given stream has a minimum flow of, say, 200 cubic feet per second (cfs), it may not flow over its banks until the flow reaches 500 cfs. At 1,000 cfs the floodway may be wide or narrow according to the slope of the valley away from the channel. At 20,000 cfs the area serving as floodway may encompass much of the land that was previously in the fringe sector. What was free from inundation at the lower flows becomes flooded with the higher flows. The actual limit of a floodplain for a given valley is a function of the flow.

Occurrence of Flooding

Because the actual or prospective flow determines the extent of a floodplain and its constituent sectors for a given valley, the estimates of the flow are basic to describing its nature and use. The estimates are of three major types: historical, probability, and maximum possible.

The most common statement is from the historical data. For example, Mud Creek records over 50 years show how many times it was out of bank. The maximum flow of record then delimits the clearly proven floodplain. But records are often incorrect and they rarely extend as far back as might show the full range of extremes. In addition to historical hydrologic data, evidence from soils or sedimentary deposits or clay varves (layers) is analyzed.

For purposes of estimating the past and future size and probability of floods, hydrologists use a variety of methods. These typically yield a calculation that a flood of a specified discharge has a probability of occurring once in 100 years. Any probability can be calculated, but it is common to estimate 4%, or 1%, or .02% floods. These computations are drawn in several ways from records of past floods in the basin or in similar basins, from meteorological records in compara-

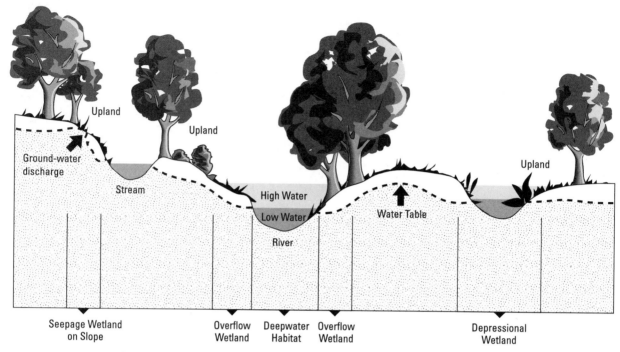

Figure 2. Cross section of typical floodplain habitat.

ble areas, and from assigning frequencies to other events, such as severe storms.

In some instances, calculations are made of the "maximum probable flood," to which a frequency may or may not be assigned. The aim is to delimit the maximum areas that might possibly at some time in future be inundated, but hydrologists are cautious about asserting that there never will be a larger flood.

Causes of Flooding

Most floods are from one of four major causes. The more common inundations are from precipitation or snow melt, and are divided into flash flows that rise sharply in a matter of minutes and more gradual peaks that gather force over periods of many hours or days. A third cause is the jamming of ice in channels, leading to ponding of water upstream. A fourth set of causes is human action in abrupt release of water from reservoirs or through the failure of dams upstream. The latter events may generate high water flows solely as a result of planning or negligence, and may extend the limits of a floodplain far beyond expected natural boundaries.

The coasts of lakes and the ocean have zones that, like inland streams, are subject to inundation from high winds exceeding normal tidal variations. These lands in the United States are often treated for regulatory and insurance purposes like riverine floodplains. So, too, are areas subject to mud flows or to liquefaction generated by seismic waves.

Landforms

Because of wide differences in terrain of valleys in which they occur, floodplains may be distinguished by the degree to which the stream system is eroding sediment from the valley floor or is depositing sediment. The first factor is erosional and the second is alluvial. Whether the stream is cutting down or is building up, it may develop delta landforms in its lowermost reaches, where the velocity decreases as the stream flows into a receiving lake or the ocean.

The depth and seasonal availability of groundwater from the surface may be different from one sector of a floodplain to another. It is common for alluvial valleys to have wetland areas such as old stream channels or areas behind natural levees where the water table is above or near the surface.

The proportion of the total land area in a river basin that is subject to occasional inundation is influenced by the type and frequency of flooding, the configuration of landforms, and the degree and type of human intervention. About 7% of the total area of the United States is at risk from a 1%-probability flood. The proportions differ greatly from one region to another. (See Figure 2.) In many parts of the alluvial valley of the Mississippi River, all of the land is natural floodplain and—before the construction of the present levee, channel, and reservoir protection works—would be rated entirely vulnerable to a 1%-probability flood. In sections of arid and semi-arid mountain terrain, less than 1% would be considered vulnerable. In north central Iowa, 9% to 17% would be at risk.

Floodplains as Habitat

Floodplains typically display at least two vegetation zones within what are designated as palustrine (marsh) systems: the corridor associations of plants immediately adjacent to the channel, where the water supply is perennial and flora are highly diverse, and sectors farther removed, where soil moisture and groundwater table may be more ephemeral and highly variable. (See Figure 3.) In both erosional and alluvial valleys there may be linear bands of vegetation along and back from the channel, and in many alluvial valleys the pattern of varieties of soil and natural vegetation may be complex. Some floodplain soils may be extremely fertile for plant growth; other recent deposits may be relatively unproductive. Riparian corridors typically provide habitat for a large variety of birds and mammals, and this is especially true of arid and semi-arid ecosystems. Floodplain wetlands serve as feeding and resting areas for migratory waterfowl throughout temperate and subtropical zones.

Patterns of Land Use

Related to the hydrology, configuration, and biota of the floodplain are human factors affecting the productivity of the land for possible use in each climatic region. Least intense are uses for recreational open space purposes. Most intense are central city uses, such as dense commercial and industrial activities. Agricultural uses in many drainage areas account for a large proportion of all land within reach of overflows.

The particular use is shaped by the prevailing climate, culture, economic system, political organization, and technical interventions by society. Thus, in the lower Mississippi alluvial valley, some land is protected and managed as wilderness used by migratory waterfowl, some is managed for production of timber from riparian hardwoods, some is used for cropping cotton, and some is occupied by cities clustered on the river's bank. The lower floodplain of the Yellow River is among the more densely settled and cultivated agricultural areas in the world. The floodplain of the Amazon River until recently encompassed some of the less disturbed of the world's ecosystems in which indigenous peoples made sophisticated use of the diverse resources. In all of these areas, human interventions play an influential role in setting the use pattern.

Adjustments to Flood Hazard

For any set of land uses there is a particular and distinctive combination of measures that have been taken by local people and organizations to cope with flood occurrence. These measures include:

- Bearing the losses from flooding when it occurs and then drawing upon the community and individuals affected to help compensate for the costs.
- Public assistance to flood sufferers from outside.
- Insurance for vulnerable property owners or occupants to indemnify them against loss.
- Forecasting and associated warning and emergency response systems to enable people subject to flooding to reduce the hazard by preparing to evacuate vulnerable areas or to take other action when a flood is predicted.
- Readjustment in land uses in the floodway or fringe sectors so as to minimize vulnerability to extreme flows.
- Treating present and prospective buildings in the floodplain so as to reduce their susceptibility to damage when flooding occurs, a package of possible measures known as flood proofing.
- Protecting the area subject to flood by either reducing flood flows or confining the flows so as to avoid their reaching vulnerable areas. Such engineering measures include dams, levees, and channel modifications.
- Reducing the volume or duration of flood flows by changing the vegetation and land use of upstream lands in the drainage area, as with forest management and soil conservation practices.

The adjustments vary in efficacy according to the degree to which the unique combination of policies and projects is suited to local conditions. For example, the distinctive type of land use and adjustments practiced for a given floodplain, such as along the Ohio River, may have utterly destroyed the virgin habitat and replaced it with intensive agriculture or city structures. Public policy preserves some areas, such as waterfowl refuges, from further human use by establishing reserves and by prohibiting encroachment upon both the floodway and the fringe. The flow of a stream in extreme cases may have been regulated to eliminate flooding, although the floodplain soils and landforms remain. More often, the efforts to reduce or confine flood flows do not surely prevent all floods or the very large floods. Where there is less than complete protection, the engineering works may encourage people to settle in areas in which they feel safe but may be exposed to a rare catastrophe.

When societies ask what has been the effect of their interventions in their floodplains, they can inquire into the alterations in the natural systems of water, soil, and biota and how those have been enhanced or degraded. They also can assess what the consequences have been for human use. This involves appraising the net increases in amenities, goods, and services, along with the costs of protection and the remaining losses from flooding. Thus, the lower Nile floodplain retained its old landform after great floods were controlled by building and operating the High Dam at Aswan, but the groundwater conditions, bank erosion, deposition of sediments and plant nutrients, and the intensity of agriculture were altered radically.

In the United States, the heavy investment in flood protection works prior to 1990 eliminated the hazard of losses in numerous areas while the overall toll of losses for the country as a whole continued to climb. Additional vulnerable fringe areas were occupied. Some fringes were settled more densely and, although protected from lesser floods, became subject to disaster from the very rare floods. Meanwhile, urban development increased the frequency and peak flow of local streams in many stream reaches. In these and other ways the stability of floodplain habitat and use continued to change.

GILBERT F. WHITE

For Further Reading: Thomas Dunne and Luna B. Leopold, *Water in Environmental Planning* (1978); W. G. Hoyt and Walter Langbein, *Floods* (1955); U.S. Federal Interagency Floodplain Management Task Force, *Floodplain Management in the United States: An Assessment Report*, Vol. 1 (1992).

FLOODS

A flood is a flow that overtops the natural or artificial banks in a reach of river channel. Where a floodplain exists, a flood is any flow that spreads out over the floodplain. Long before the Neanderthals began flaking flint, the overflow of floodplains was occurring much as it does today. Flood damage is a consequence of human utilization of the floodplain.

When rain falls upon the earth, an astounding amount of water is involved, even in a light shower. A two-day rain is commonplace in the central U.S. and western Europe, and in that period an inch of fall is very common. An inch of rain over the whole state of Ohio would be 2,190,000 acre-feet, or 712,000,000 cubic meters. A comparable rain of 25 mm falling over Belgium would be 760,000,000 cubic meters.

Such a rainstorm is not unusual and would not necessarily cause a flood. Ordinarily, flood runoff constitutes less than half the water falling in a given storm. Most either is retained in the soil, from which it is drawn by plants and thus returned to the atmosphere, or drains gradually to ground water storage and then in part to river channels.

The volume of water which may fall as rain is large but so also is the volume of channels of the river sys-

tem. W. B. Langbein computed that in a 300-square-mile basin above St. Paul, Indiana, if at every point in the channel system the flow were 10 cfs (cubic feet per second) per square mile, which would be within the bank-full stage (full to the top of the banks) at most points, the volume in channels would be about 6,000 acre-feet, almost 2 billion gallons, or about 0.4 inch over the drainage basin. Under the same flow conditions, a drainage area ten times the size, or 3,000 square miles near Spencer, Indiana, would have in its channel system 120,000 acre-feet, 39 billion gallons, or twenty times as much as in the basin above St. Paul. Thus the volume of the channel system increases rapidly downstream.

Within the banks of stream channels there is a great amount of space capable of holding water. But the space is not utilized efficiently because storm rainfall is seldom distributed geographically in a uniform manner. The channel system may be overtaxed in one place while in another the streams flow at less than full capacity. So, despite the capacity to handle large total volumes of water without damaging overflows, floods still occur.

The river channel is constructed by the river. On most days only the bottom part of the channel carries water. On several days each year the channel is three-quarters full, and once or twice a year, on the average, the river flows bank-full. A discharge expected once in 50 years, the 50-year flood, will cover the floodplain to a depth about equal to the mean channel depth at bank-full.

Flows so large that they cannot be contained in the channel must spread out over the adjacent floodplain. The floodplain is where nearly all flood damage occurs because there people have chosen to grow crops or build buildings. They have encroached on an area that the river must at times cover with water.

The river channel does not contain an unusually high discharge. The bank-full condition is reached with a recurrence interval of 1.5 years. That is, the highest flow in any year will equal or exceed bank-full two years out of three. On the average, the water that is discharged at rates in excess of channel capacity is only about 5% of the total annual discharge of the basin.

The recurrence interval is the average interval of time within which a flood of a given magnitude will be equalled or exceeded only once. It is also a statement of probability. A flood with a recurrence interval of ten years has a 10% chance of recurring in any one year. Such a flood might occur two years in a row. But on the average it will occur once in a ten-year period.

Extreme floods have been noted in historical records or in the mythology of nearly all countries. There are some very long and extremely well documented records of flood heights, not only on the Nile, where the record is a famous one, but on several rivers in Europe and Scandinavia. The record of measured flood discharges, in contrast to flood heights, is much more limited. In the U.S. river gaging in a modern sense did not even begin until 1895. There were in 1992 7,590 gaging stations operating. The records of floods would, at first glance, make it appear that extreme floods are increasing in magnitude, but this is merely apparent. The records are getting more numerous and longer so the recorded floods of large magnitude increase with time.

The U.S. Geological Survey has studied the flood records within each broad geographic area and derived regional flood-frequency curves. The results of this effort are now published for all of the U.S. except the arid regions.

Great floods occur when a drainage basin is incapable of infiltrating additional water and all precipitation or snow melt must run off the surface. Such a basin condition may occur after prolonged rainfall, or when freezing makes the surface impermeable. Forests and heavy vegetation cover tend to spread runoff over time, thus reducing peak discharge. Another important influence is the alteration of basin surface that speeds water runoff. Urbanization is the major contributor. Roads, street gutters, roofs, paved parking areas, and storm drains all move water downhill faster than would occur on natural slopes. The speeding of runoff means that a given volume of water must be discharged in a shorter period of time. This increases the peak rate.

Though there has been a modest climatic change in the U.S. beginning about 1950, the trend toward more raining and cooler conditions has been too small to account for the important change of runoff, whereas the expansion of urban areas is obvious. To the urbanization one may logically assign the change in peak flows observed.

The probability of floods determines financial as well as organizational preparedness. If culverts were too small and many of them washed out, the expense would be great. If overdesign were the rule, money for construction would be needlessly spent. Culverts constructed with U.S. federal money are designed to carry a 50-year flood discharge. Such a decision places an enormous burden on hydrology to develop methods to estimate flood discharge.

Despite knowledge of the importance of flood estimates, people continue to build on floodplain areas needed to carry discharges about once a year. So despite billions of dollars spent on controlling floods, flood damage continues to rise year after year.

Gaging stations have been operated for short or long periods of time at more than 20,000 sites in the U.S. In 1992 the number of operating gages was 7,590. At these sites the computation of discharge for any chosen recurrence interval can be determined. But it has been estimated that there are approximately 3,250,000 miles of river channels in the U.S. so the number of gaged locations is a minuscule part of the locations where flood discharges might potentially be needed. Hydrologists have spent much effort developing methods of estimating discharge values for ungaged locations. These have concentrated on using combinations of precipitation, topography, vegetal cover, soils, geology, and sediment materials as indicators. Some of these procedures are remarkably good predictors but all have restrictions on applicability and require information not readily available.

The recently developed procedures based on simple measurements of the river channel are now available for nearly all parts of the country. Nearly all depend on field measurement of bank-full width or on width of the action channel. The equations permit the computation of discharge for recurrence intervals of two to fifty years. The method is based on the logical notion that the channel is carved and maintained by the discharges it experiences and thus reflects those events. Such studies have been published for the western U.S. differentiated on the basis of alpine, northern plains, southern plains, and Rocky Mountain areas. Relations of channel width to discharges of various recurrence intervals have been published for California, Missouri basin, Kansas, Idaho, Colorado, Utah, Piedmont region, Wyoming, Nevada, and New Mexico.

LUNA B. LEOPOLD

For Further Reading: Thomas Dunne and Luna B. Leopold, *Water in Environmental Planning* (1978); William G. Hoyt and Walter B. Langbein, *Floods* (1955); Luna B. Leopold, *Water—A Primer* (1974).
See also Coastlines and Artificial Structures; Dams and Reservoirs; Rivers and Streams; Water Cycle.

FLUE GAS CLEANUP

Flue gas is waste gas that results from combustion processes such as incineration and is released into the atmosphere through a chimney or stack (flue). It contains oxides of sulfur, nitrogen, and carbon as well as particulates and other unwanted materials. Flue gas cleanup involves systems—desulfurization and denitrification—and devices—flue gas scrubbers and electrostatic precipitators.

Since sulfur is a significant factor in the production of acid rain, many processes or controls exist to reduce sulfur emissions before, during, or after combustion. Flue gas desulfurization is a post-combustion control method that is widely used all around the world. Sulfur oxides are removed from the flue gases while still in the chimney by scrubbing the flue gas with a chemical reagent or absorbent such as lime, limestone, or magnesium oxide. In nonregenerative flue gas desulfurization systems, which dominate the market in terms of both number and size, the reagent is used and discarded. Large amounts of waste in the form of sludge, which is difficult to get rid of, are produced. Regenerative systems, on the other hand, recover and then reuse the operating reagent.

Several experimental flue gas denitrification systems work by turning nitrogen oxides into nitrogen using selective catalytic reduction. This system can reduce up to 90% of the nitrogen oxides. Noncatalytic reduction methods are less expensive but not as effective.

Flue gas scrubbers are specialized equipment that remove fly ash (fine solid particles) and other unwanted materials. The processes are termed either wet or dry, depending on the type of end product, and serve to lower the temperature of the outgoing stream. Wet scrubbers operate by wetting the gas particles that are trapped as the gas travels across the face of the reagent. Normally the liquid reagent is sprayed so that the droplets collide with the particles in the gas stream, which then adhere or attach to the droplets and are removed with the liquid. Dry scrubbing processes produce a waste product that is more manageable than sludge material that is difficult to dispose of.

An electrostatic precipitator is a device based on the electrostatic attraction and repulsion of charged bodies and is used for removing solid particles from gas streams. Precipitators are either one-stage or two-stage in design. One-stage precipitators combine the ionization of the gas stream and particle collection into a single step. In the two-stage devices, the steps operate in sequence. Both designs are very efficient at removing submicron-size particles.

FLUIDIZED BED COMBUSTION

Fluidized bed combustion was invented in the 1920s as a method of burning coal and low-grade fuels. A mixture of crushed coal and limestone particles about a meter deep (three feet) is distributed on a perforated grid in the combustion chamber. Small jets force air through the bed, suspending the particles above the grate. Oil is then sprayed into the suspended mass and set on fire. The incombustible material stores heat (at about 900° C), instantly igniting coal particles as they enter the bed. The coal is volatilized and produces fuel gases that promptly burn. Fresh coal and limestone are fed continuously into the top of the bed while ash and residue are drawn off from below. Steam for electricity or other applications can be produced by pumping water through coils immersed in the bed.

Typical uses for fluidized bed combustion include industrial boilers used to fire coal and mining wastes, and the burning of hazardous wastes such as chemical plant waste, heavy metal waste, and oily water sludge. The incineration of food processing wastes, lumber waste, industrial tailings, agricultural waste, and municipal refuse is facilitated by fluidized bed combustion.

Fluidized bed combustion is less expensive and cleaner than conventional combustion methods. Both high- and low-grade coal, regardless of ash or moisture content, can be burned more completely and at a higher rate of thermal efficiency than with conventional systems. The possibility of using low-grade coal such as lignite or unwashed subituminous coal reduces the costs involved. Also, the thermal efficiency allows for more energy to be extracted from each pound of coal. In this process the coal does not have to be finely crushed, and only a few coal particles (coal makes up one to three percent of the bed) are required in the bed at a time.

A fluidized bed can be operated at high pressures, expanding hot flue gas through a gas turbine and further increasing power-generating capacity. This efficiency allows coal to be used as a substitute for oil and gas. Oil shale can also be burned using this technology.

Air pollution is reduced in the process of fluidized bed combustion. More than 90% of sulfur dioxide is captured by the limestone particles. Sulfur-calcium compounds that form during combustion can be removed with the ash. Nitrogen oxide formation is reduced by keeping temperatures in the chamber around 900° C (1,600° F), as opposed to temperatures in other furnaces that are twice as high. Thermal pollution to rivers and other water sources is also avoided. The relatively low temperatures, constant motion, and air supply of the system eliminate the likelihood of residue buildup, further enhancing the burning efficiency. In addition, the volatilization of alkali salts is reduced. This decreases the incidence of corrosion and consequently reduces maintenance costs. The fluidization process eliminates ash buildup, although some minerals are released into the air as fly ash.

FLUORIDATION

Fluoridation is the practice of adding fluoride to public supplies of drinking water to reduce tooth decay. In small concentrations in drinking water, fluoride can be beneficial for the protection of teeth. It combines chemically with tooth enamel during the formation of permanent teeth to form harder, stronger teeth that are more resistant to decay. Although fluoride is a minor natural constituent in many waters, it may be added at water treatment plants when it is not present in sufficient amounts.

Fluoride is effective for improving dental health at concentrations of approximately 1 mg/L. When concentrations exceed 2 mg/L, teeth may become discolored, an effect called mottling. Although both strengthening and mottling of teeth occur during tooth development, their effects carry over into adulthood. Fluoride has no effect on adult teeth. When fluoride levels exceed 5 mg/L, skeletal abnormalities may develop, resulting in a crippling condition of the joints.

Acceptable drinking water levels of fluoride are related to temperature, with higher levels allowed at lower temperatures under the assumption that people drink more water in warm weather. The Environmental Protection Agency limits drinking water levels of fluoride to between 1.4 and 2.4 mg/L, while the World Health Organization recommends 0.6 to 1.7 mg/L. Because relatively small increases in the concentration of fluoride can cause harmful effects, fluoridation is not without controversy. Periodically, groups will attempt to stop utilities from practicing fluoridation because of health concerns. Indeed, water fluoridation coupled with the use of fluoridated toothpaste and fluoride dental treatments is known to create mottling of teeth, even when fluoride levels in drinking water are deemed acceptable.

It appears, however, that based on 30 years of widespread successful use in the U.S., fluoridation will continue to be used in the treatment of drinking water.

JOHN T. NOVAK

FOOD ADDITIVES

See Labeling.

FOOD AND AGRICULTURE ORGANIZATION OF THE UNITED NATIONS

See United Nations System.

FOOD AND DRUG ADMINISTRATION, U.S.

See United States Government.

FOOD CHAINS

The energy which plants capture from the sun during photosynthesis may end up in the tissues of a hawk, via a bird the hawk has eaten, the insects eaten by the bird, and the plants on which the insects fed. The plant-insect-bird-hawk system is the food chain. Each stage is called a *trophic level.* More generally the trophic levels are called *producers* (plants), *herbivores* or *primary consumers* (the insects), *carnivores* or *secondary consumers* (the bird), and *top carnivores* or *tertiary consumers* (the hawk).

Food chains may involve parasites as well as predators. The lice feeding in the feathers of the hawk are yet another trophic level. While living plants form the basal level of the example just given, decaying vegetation, dead animals, or both provide the energy sources of other chains. These are called *detrital food chains.*

Food chains are never as simple as the above example suggests. Species are rarely arranged into neat, clear trophic levels because some species feed on more than one trophic level. Ecologists who study food chains call these species *omnivores.* (This is a special use of the term. It is often used to mean species that feed on many different species, even though the species may be at the same trophic level.)

We can represent the complex patterns of the actual feeding interactions within a community with a *food web.* A food web is a map that describes which kinds of organisms in a community eat which other kinds. Figure 1 shows a food web describing the feeding relationships among the species in a pitcher plant. Pitcher plants contain water into which insects fall and drown. The plant gains nutrients from the decaying insects. Many species thrive in these tiny ponds, however. Their size and simplicity make them popular with those studying food chains.

Notice that at the base of the various food chains are drowning insects, dead insects, and older organic debris. This is a detrital-based system. Feeding on the decaying material are several species. Some of these have predators, others do not—at least within the pitcher plant. Notice, too, that some species receive energy from more than one level. The predator *Misumenops* feeds on another predator, *Lestodiplosis,* and both of these species feed on *Endonepenthia.*

In this example, the shortest food chains have two levels: the live, but drowning insects that have fallen into the water, and *Misumenops.* The longest chains have five levels: old organic debris, the bacteria and protozoa that feed on it, two species of mosquito (*Culex*), *Dasyhelea,* and *Misumenops.* One way to describe and simplify all of the various food chains is to count the most common number of levels from the top to the bottom of the web. Thus, although *Misumenops*

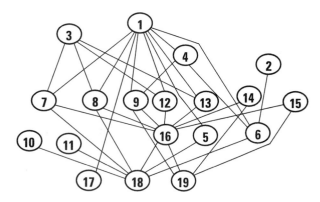

Figure 1. A food web of the insects in the pitcher plant *Nepenthes albomarginata* in West Malaysia. Each line represents a trophic linkage; predators are higher in the figure than their prey. Key: (1) *Misumenops nepenthicola*, (2) Encyrtid (near *Trachinaephagus*), (3) *Toxorhynchites klossi*, (4) *Lestodiplosis syringopais*, (5) *Megaselia* sp. (?*nepenthina*), (6) *Endonepenthia schuitemakeri*, (7) *Triperoides tenax*, (8) *T. bambusa*, (9) *Dasyhelea nepenthicola*, (10) *Nepenthosyrphus* sp., (11) *Pierretia urceola*, (12) *Culex curtipalpis*, (13) *C. lucaris*, (14) Anotidae sp. 1, (15) Anotidae sp. 2, (16) bacteria and protozoa, (17) live insects, (18) recently drowned insects, (19) older organic debris.

sits atop chains of two and five levels, the most common chains in the web have three levels.

The example is quite typical in having only three levels. Most food chains are three to four trophic levels long (if we exclude parasites), though there are longer ones.

Why are food chains so short? There are several possible explanations. The first involves energy. In many ecosystems there are more plants than insects, more insects than insectivorous birds, and more insectivorous birds than hawks. These ecological pyramids reflect an underlying energetic constraint. The first law of thermodynamics tells us that when we convert energy from one form to another, the amount of energy remains constant. An approximate way of stating the second law is that the amount of useful energy decreases at each conversion. When the insects eat the plant, they convert the energy locked up in the plant tissues into insect tissue. Yet they also pay an energetic cost in doing so. Mammals and birds pay an even greater cost. Only about 1% of the energy of the food we consume goes to produce new tissues. The rest is "lost" as heat, both to keep us warm and also as a by-product of the conversion process.

Between each trophic level much of the energy is lost as heat. As the energy passes up the food chain, there is less and less to go around. There may not be

enough energy to support a viable population of a species at trophic level five or higher.

This energy-flow hypothesis has several supporters. It also has critics who notice that it predicts that food chains should be shorter in energetically poor ecosystems such as bleak arctic tundra or extreme deserts. These systems often have food chains similar in length to energetically more productive systems. So what else might limit food chain length?

Another hypothesis has to do with how quickly the species in food chains recover from environmental disasters. Consider a lake with phytoplankton, zooplankton, and fish. What happens when the phytoplankton decline for some reason? The zooplankton will also decline without their food, and then the fish. The phytoplankton may recover, but will remain at low levels, kept there by the zooplankton. At least transiently, the zooplankton may reach higher than normal levels because the fish, their predators, are still scarce. Simply, the phytoplankton will not completely recover until all the species in the food chain have recovered. Mathematical models can expand such arguments. These models show that the longer a food chain, the longer it will take its constituent species to recover. (The phytoplankton could recover quickly in the example if they were the only trophic level.) Species atop very long food chains may not recover before the next disaster. Such arguments predict that food chains will be longer when environmental disasters are rare, short when they are common, and not necessarily be related to the amount of energy entering the system.

STUART L. PIMM

For Further Reading: J. E. Cohen, F. Briand, and C. M. Newman, *Community Food Webs* (1990); S. L. Pimm, *Food Webs* (1982).

FOREST PRODUCTS INDUSTRY

Old records of the Virginia Company describe a sawmill established at the Jamestown colony in 1607—America's first commercial enterprise. The forest and paper industry has since grown to become one of the largest manufacturing sectors of the United States economy. Hundreds of pulp and paper mills and thousands of sawmills produce products that are ubiquitous in American life. The industry accounts for about 7% of all U.S. manufacturing output, employs 1.6 million workers directly, and generates more than $200 billion in sales, including $16 billion in exports. Newsprint, pencils, paper, toilet paper, packaging, and lumber to build houses and make furniture are among the industry's many products. The U.S. is the world's largest producer of forest products, manufacturing about 25% of the world's lumber and solid wood products, 30% of the world's paper and paperboard, and 35% of the world's pulp.

Few industries are affected more by environmental concerns than the forest industry. The interaction between this industry and the environment begins at the point where its products are derived—the forests—and continues through the manufacturing process.

American Forests

In the 1600s forests blanketed about half of the continental U.S. and Alaska. The popular myth is that American forests in pre-settlement days existed in vast, undisturbed tracts, but in fact, the forest landscape was a mosaic of forest types and age classes, frequently altered by natural events and the intervention of indigenous populations. Large natural fires were commonplace before and after the early settlement of America, and fire was the principal means by which forests changed and evolved. Natural fires killed older, less vigorous trees and from the ashes emerged new trees, continuing the cycle of the natural ecosystem. Native Americans also used fire to clear forestland for grazing, for crops, and—knowing that game species required forage to propagate—to promote wildlife.

To most Americans in the 17th, 18th, and 19th centuries, progress was defined as establishing new settlements and expanding agriculture, to which the forested wilderness represented a barrier. Clearing land proceeded at what today would be regarded as an alarming pace, and by 1900 most of the eastern and north central U.S. had been cleared.

Fear that wood supply would run out led to concerns about forest depletion. As early as 1691, fearing that the colonists would use the best trees, the British crown required that New England's tall, straight pine trees be reserved for exclusive use as masts for the royal fleet. The reserved trees were marked by three cuts of an axe, called the "King's Broad Arrow." That regulation was probably the first environmental law applied in America. Other regulations to deter excessive tree-cutting were attempted, but they were largely unenforced. Timber was a plentiful commodity. As the cities expanded, however, shortages of fuelwood were not uncommon, and periodically during the nation's building booms, lumber was in short supply. Benjamin Franklin is known to have worried about shortages of wood, which used to be available "at any man's door."

In the late 1800s the chorus of concerns about "timber famine" was joined by prominent writers and thinkers such as Henry Thoreau, John Muir, Robert Underwood Johnson, and George Perkins Marsh, and an organized conservationist movement began to emerge. Early foresters like Bernhard Fernow and Gifford Pinchot believed in wise use and argued for the creation of forest preserves.

Today the U.S. possesses 737 million acres of forestland, representing about one third of the continental land area. Of this amount, 493 million acres are classified as commercial timberland, that is, available for timber production. The fear that America is running out of trees is unfounded. The inventory of standing timber—the number of trees—has been increasing since World War II. According to the U.S. Forest Service, the total forest inventory increased 38% between 1952 and 1987. The increase results from idle agricultural land reverting to productive forest and intensive forest management leading to higher growth rates, particularly on industry-owned land. Approximately 70 million acres of commercial forestland, or 15% of the total, are owned directly by companies involved in processing forest or paper products.

Environmental Balance

The most contentious environmental conflicts involving the forest products industry occur over the management of publicly owned forestland. The federal estate comprises approximately 97 million acres, of which 85.2 million acres are contained in the National Forest System and managed by the U.S. Forest Service, an agency of the Department of Agriculture. They contain 47% of all of the "large" (sawtimber) softwood trees in the U.S. A substantial portion of the national forests has been officially designated as wilderness or has been classified as unsuitable or unavailable for timber harvesting. But the national forests in the Pacific Northwest contain the last virgin stands of old-growth forest in the continental U.S. Protection of old-growth forest values is an important, some would say critical, environmental objective. Preventing harvesting in these forests has become a central political objective of many environmental groups. The industry, in turn, questions how much old-growth acreage should be permanently set aside compared to how much should be available for active management, including management for old-growth ecosystems. The debate is intertwined with controversies surrounding the protection of endangered or threatened species, such as the northern spotted owl.

Perhaps no environmental law has had such a pervasive impact on forests and hence the forest products industry as the Endangered Species Act (ESA). Originally enacted in 1973, the ESA was intended as a means to preserve species threatened by specific federal actions, such as the building of a dam or construction of a federal highway. Courts have interpreted the act to apply in situations where the existing or potential habitat for a given species might span an entire geographic region. The listing under the act of the northern spotted owl as a threatened species is one highly publicized example. Other forest-dwelling species which have been listed, or are being proposed for listing, include the red-cockaded woodpecker, the Mexican and California spotted owls, four subspecies of salmon, and the gopher tortoise.

Protection of endangered species is linked to a broader environmental debate over biological diversity. Forest management can either enhance or diminish biological diversity. Some species prosper in early successional forests, such as a forest that has been recently harvested, others prefer late successional stages. The widest diversity of species would seem to require a broad array of land, forest, and vegetative types, including forest stands of varying age classes. On the other hand, intensively managed forest plantations consisting of single-species trees are potentially less biologically diverse than are uneven-aged stands. Increased mechanization and the use of chemical fertilizers and pesticides may also have detrimental impacts on terrestrial and aquatic communities. Modern silviculture attempts to minimize adverse impacts. Foresters have been able to identify the needs of specific species and allow for the maintenance of suitable habitats even under intensively managed conditions. Retention of some old stems dispersed throughout a stand, for example, provides habitat for species that require older trees for nesting. Many wild animals, including deer, elk, wild turkey, and certain migratory birds, benefit from harvesting activity, which promotes the growth of the vegetation these animals prefer. In many cases, the forest industry has played a pivotal role in studying and managing wildlife. Industry-sponsored management programs have been instrumental in the recovery of some species, including the bald eagle, peregrine falcon, and grizzly bear.

Forestry Operations

Forests are harvested using one of several techniques. Selective or shelterwood harvesting is employed where uneven-aged management is practiced. Under intensive, even-aged management, clear-cutting is used. Put simply, this is a harvesting system where a stand of trees is completely removed and the site is then prepared for planting. Increasingly, where clear-cutting is employed, seed trees are left to naturally regenerate the site.

Clear-cuts are unattractive and, more importantly, can contribute to sedimentation and soil erosion, par-

ticularly on steep slopes. Extremely large clear-cuts are now unusual, but even smaller areas that are clear-cut leave a patchwork of cleared land that is often objectionable for visual or environmental reasons. On the plus side, clear-cutting generates a more healthy, more productive regenerated forest in its aftermath. It enables the regeneration of shade-intolerant species such as most varieties of pine and Douglas fir. For the forest industry, it is also a question of economics. Clear-cutting allows the removal of the highest volume of fiber at the lowest per unit cost.

"New Forestry" or "New Perspectives" is a movement to incorporate more diverse objectives into forest management decisions. Its proponents emphasize selective or shelterwood harvests over clear-cutting and advocate an ecosystem approach to forest management.

Land that is clear-cut or otherwise harvested is usually reforested by planting or is reseeded naturally as other surrounding trees seed the newly cleared site. In 1992 more than 2.5 million acres of cut-over land were planted. The forest products industry is the largest planter of new trees. Every acre of industry-owned land is regenerated after harvesting, and many companies operate landowner assistance programs to encourage farmers and other private owners to reforest their lands. In 1992 timberland owners, including the forest industry and government agencies, planted 1.6 billion seedlings—over seven trees for each American. In addition, tens of thousands of acres of forest (usually hardwoods) are managed to ensure they regenerate naturally. Partly because of this aggressive planting and forest management effort, one-third more wood is grown each year in the nation's forests than is harvested or lost to fire, insects, and diseases.

Since forestry operations can affect water quality, stricter regulations and voluntary guidelines have been developed for private lands. In the West, several states have Forest Practice Acts which require permits and compliance with strict environmental guidelines for any forestry activity. The 1977 Clean Water Act requires states to develop voluntary Best Management Practices (BMPs) to reduce water quality impacts from forest operations. BMPs include environmentally sound design of logging roads, buffer strips along streams and waterways, and other practices designed to minimize the impact of forest activities on environments. The act also recognized the compatibility of forest management with wetlands conservation and granted an exemption from permit requirements for certain types of activities defined as "normal silviculture." In recent years, however, the federal government has defined wetlands more strictly and has moved to broaden its regulatory purview of jurisdictional wetlands under federal and state law.

Manufacturing

Engineered wood products increase utilization of forest resources by using lower-value wood or wood residue. Particleboard and hardboard, traditional engineered wood products, have been in use in making furniture and in other industrial applications for many years. Newer products include oriented strand board, laminated veneer lumber, i-joists, roof trusses, and parallel strand lumber. These products are frequently stronger and more uniform than traditional dimension products.

Some lumber (and plywood) is chemically treated with preservatives to prevent decay and rot in outdoor or marine applications. The most commonly used chemicals are formulations of copper arsenate in water solutions. Oil-based preservatives such as creosote and pentachlorophenol are also used, although the use of the latter has decreased due to regulation stemming from its persistence in the general environment. The wood-preserving industry, under the Federal Insecticide, Fungicide, and Rodenticide Act (FIFRA), is required to adhere to specific labeling and licensing requirements. The treatment process and the wastes it generates are also regulated under the Resource Conservation Recovery Act (RCRA).

Paper is typically manufactured by pulping wood fiber, separating the cellulose from the lignin (the substance that binds it together), and then drying and matting the resulting material which is then ready for more refined processing. Lignin, recovered as spent pulping liquor, is used as an energy source. Both the solid wood and paper segments of the industry are highly energy efficient. More than 75% of the energy consumed in the manufacture of lumber and other solid wood products is generated from burning residues, sawdust, and scraps. In paper manufacturing, more than 55% of the energy consumed is derived from spent pulping liquors, wood residues, or other self-generated sources.

During the 1980s the pulp, paper, and paperboard industry averaged $400 million per year in capital expenditures for pollution abatement. During the 1990s those expenditures have exceeded $1 billion annually. Since 1975 the discharge of the most common water pollutants—biochemical oxygen demand (BOD), total suspended solids (TSS), and acute toxicity—have been greatly reduced in conformance with federal and state regulations.

Modern paper-making uses bleaching to enhance brightness, strength, absorbency, and softness and to prevent discoloration. Chlorine also disinfects and removes substances that can cause disagreeable odors or bad taste in some packaged foods, such as milk. The fate of chlorine and related compounds in the general environment is monitored closely. According to a 1991 study by the Pulp and Paper Research Institutes of Canada and Sweden, 90% of the chlorine used in pulp and paper bleaching ends up as common salt and 10% forms chlorinated organics, another pollutant associated with the use of chlorine. Scientific studies have not shown any adverse environmental or health effects associated with chlorinated organics from pulp and paper processing or any negative effects on growth or reproduction in fish or other aquatic organisms.

In 1985 the carcinogen dioxin was unexpectedly found to be an unintended by-product of chlorine bleaching. Since then, the industry has successfully reduced dioxin discharges by 75% as a result of concerted research efforts and adjustments to manufacturing processes. The total dioxin discharges from all bleaching mills in the pulp and paper industry are less than 1% of the total released into the environment annually from all natural and man-made sources, but it remains a high-priority concern.

Environmental Concerns

Forest products are among the most environmentally friendly materials available for construction, packaging, or communications. Forest products are derived from a renewable resource, and the energy consumed in manufacturing wood products is a fraction of that consumed for making steel, masonry, or other substitutes. The wood used to construct a typical framed wall consumes 2.541 million Btu in its manufacture, while making concrete building blocks for the same wall uses 17.087 million Btu. In addition, forest products store carbon, a greenhouse gas; to the extent that forest products are used in place of other materials which emit carbon in their manufacture, forest products contribute positively to the earth's carbon budget.

The forest industry recycles its own by-products and utilizes recyclable materials from the nation's waste stream. In manufacturing lumber, for example, chips and sawdust are used to make other wood products such as particleboard or are shipped to a pulp mill to be pulped for paper-making. The paper industry recovers and reuses the chemicals used in the pulping process and uses the spent pulping liquors for energy production.

To manufacture paper, the industry uses as raw material both virgin fiber (harvested trees) and waste paper. While the industry has for a long time utilized waste paper for certain products, recent concerns about waste disposal have resulted in expanded curbside recycling programs. The paper industry has set a goal to recover 40% of all paper used in 1995. While this goal is now expected to be exceeded, there are some practical limits to how much paper can be recovered through recycling and how many times individual fibers can be re-pulped before losing the bonding strength necessary for paper-making. For these reasons it will always be necessary to inject virgin fibers into the system.

The release of carbon dioxide into the atmosphere through the burning of fossil fuels has raised concern about climate change. The problem of climate change has spurred proposals to prevent the harvesting of forests since trees absorb and store carbon dioxide, a "greenhouse gas." However, by reforesting harvested areas and continuing to grow trees (younger trees absorb more carbon than older, slower growing trees), and by producing products that store carbon, the carbon balance of the forest industry is highly favorable— more carbon is stored than is released through manufacture. When products are made from trees, such as lumber used to build a house, carbon is stored in the product. Perhaps the greater environmental peril is conversion of forestland to other non-forest uses such as grazing or development. Land-use changes in the tropics are certainly worrisome. To maximize carbon sequestration, policies should encourage increased forest productivity.

Demand for forest products will continue to grow domestically and internationally. Thus, strains on the world's forests are likely to continue unabated. Sustainable forests—forests that continue to grow and produce—are vital to the future of the forest products industry. Increased recycling of paper products, energy conservation, innovative building techniques, and other new technologies will greatly stretch the use of resources. But more intensive management of forestland, incorporating environmentally sensitive practices, will be needed to meet the world's growing wood demands. There are no easy answers to the conflicts that pit society's need for products against its desire to protect the total environment, only a constant challenge to find an acceptable balance.

ALBERTO GOETZL

For Further Reading: Kenneth D. Frederick and Roger A. Sedjo, eds., *America's Renewable Resources—Historical Trends and Challenge* (1991); Douglas W. MacCleery, *American Forests: A History of Resiliency and Recovery* (1992); Michael Williams, *Americans and Their Forests, A Historical Geography* (1989).

FOREST SERVICE, U.S.

See United States Government.

FORESTS, CONIFEROUS

Forests dominated by conifer trees constitute some of the most widespread and spectacular vegetation on Earth. Broad expanses of coniferous forest are the epitome of northern and rugged wilderness, encompassing some of our most valuable wildlife habitat, scenic vistas, and recreational potential. But these same conifers are also the basis for a worldwide market in softwood lumber, pulp fiber, and paper products. Coniferous forest climates may be subject to some of the most extreme changes if global warming occurs.

The Conifers

Conifers are typically needle-leaved, evergreen, cone-bearing trees and shrubs of the taxonomic category known as the Order Coniferales. Forming the climax (equilibrial) vegetation typical of cold subarctic latitudes, snowy subalpine elevations, and wet temperate coastal zones, conifer forests are also found in semi-arid savannahs, scattered tropical islands and mountains, and temporarily within the deciduous forest biome. Consisting of over 550 species in six plant families, conifers are especially important in the Northern Hemisphere. Podocarp and araucaria (e.g., Norfolk Island pine and monkey-puzzle) forests of the Southern Hemisphere are much less widespread and less understood. Pine or spruce forests also dominate much of the southeastern United States, central Europe, and New Zealand, although these forests are early successional (temporary, non-climax) or artificial plant formations.

Conifers and the flora and fauna associated with them often have remarkable adaptations to severe environments; they typically occupy climates and soils less hospitable than those dominated by deciduous forests. Northern conifers have foliage which is needle-like and evergreen, excepting the deciduous habits of the larch and a few other species. This is not to say that conifers rarely shed their leaves, but rather that each leaf has a lifetime of several years. This represents an adaptation to the low availability of soil nutrients, reflecting the low rates of decomposition and nutrient mobilization characteristic of both cold and droughty sites.

Coniferous Forests

The coniferous forest biome can be divided into three major sections: (1) the subarctic, circumpolar boreal zone (12 million square kilometers); (2) the subalpine elevations of temperate and subtropical mountain ranges (2.7 million square kilometers); and (3) the winter-wet coastal forests restricted to western North America (0.8 million square kilometers). Together, these forests (which occupy 36% of the world's forested area, or 12% of the global land area) are estimated to contain 130 billion cubic meters of softwood timber, 42% of the world's total timber resource.

The boreal zone represents the archetypical coniferous forest: broad expanses of closed canopy forest with relatively few tree species (usually representatives of the pine, larch, spruce, or fir genera) in a cold, continental climate with severe winter temperatures and a short growing season. Where absolutely no broadleaf trees are found, this forest is termed *taiga*. The boreal zone typically has gently rolling terrain in which shallow-soiled or infertile uplands (having soils dominated by *podzolic* [leaching and impoverishing] processes) alternate with poorly drained (*gleysolic*) and organic soils. Soils may be permanently frozen (*permafrost*), especially where water accumulates in the soil profile and organic matter insulates the soil from warming. Extensive organic terrain with scattered conifers is often called *muskeg*. Such conditions characterize much of Canada, Alaska, Siberia, and Scandinavia.

Subalpine and montane forests (found in the European Alps, the Caucasus Mountains, the Himalayan Range, and the Western Cordillera, for example) are often extensions of the boreal zone or drier variants of coastal forests, with many shared plant and animal species. Snow accumulation is greater than in most of the boreal zone and snow can fall any month of the year. Subalpine terrain is usually much steeper than in the boreal zone; soils are typically shallow and recently derived from rocky parent materials (and hence are known as *regosols*).

The wet coastal conifer forests stretch from California to Alaska and on some windward slopes and valleys inland to the Continental Divide. These forests might also be considered a coniferous subset of the temperate rain forests. This zone has heavy precipitaion (especially in the winter), as rain at lower elevations on the coast, or as snow at higher elevations and in the interior. Soils are variable and include all of the categories mentioned above. In addition to spruces, firs, and pines, these forests also have hemlocks, giant arborvitae (also called western redcedar), Douglas-fir, false cypress, incense cedar, and redwoods. Where the growing season is long and sites rich, several of these tree species can attain tremendous stature (70 to 100 meters tall).

Coniferous Ecosystems

Forests are not only stands of trees, but complex ecosystems of plants, animals, and microorganisms interacting with local soils and climates. These ecosystems vary considerably with the local climate, soils (depth, texture, moisture regime, nutrient regime), and age of the forest stand. Many of the forests described above also have significant representation by deciduous tree species such as poplars (which include aspens and cottonwoods), birches, alders, and maples. Common shrubs include willows, alders, junipers, viburnums, roses, spiraeas, and numerous members of the raspberry genus and the heather family (especially heathers, salal, blueberries and cranberries, labrador teas, and laurels). Many understory species have creeping woody stems and are so small that they form a *dwarf-shrub* layer. Characteristic herbaceous species in coniferous forest regions include wide-ranging, early successional species such as fireweed, horsetails, and bracken fern, as well as more restricted shade-loving species of ferns, orchids, wintergreens, and assorted saprophytic (nonphotosynthetic) species such as pinedrops and Indian pipe. Grasses (especially fescues and *Calamagrostis* species) or sedges can be found on the periphery of wetlands, dry south-facing ridges, and where the forest grades into semi-arid grasslands. Mosses often cover rotting logs and the forest floor; the feather mosses (*Hylocomium*, *Pleurozium*, and *Ptilium* species) are especially abundant. The *Sphagnum* (peat moss) species characterize the bogs that are so widespread in the boreal zone. Lichens, especially the *Cladina* or "reindeer moss" species, can be found on the ground where soils are depauperate and the conifer canopy opens up as it grades into the Arctic tundra. Other lichen genera (such as *Usnea*, *Lobaria*, *Parmelia*, and *Alectoria*) can be found draping from the branches of older trees.

Much less is known about the distribution, taxonomy, and ecology of microorganisms and invertebrate animals in coniferous forests. It is conceivable that the ratio of below-ground to above-ground biodiversity is much greater than in temperate and tropical forests. Conditions for decomposition are typically much more exacting than in temperate forests, because of the more resistant structure and chemistry of conifer foliage and the prevalent limitations of either heat or moisture. Decomposing bacteria tend to be less dominant than fungal decomposers. Fungi also include several species which initiate root rots and mycorrhizal (symbiotic) relationships with trees and other vascular plants. The invertebrate fauna is rich and largely unknown. Particularly prevalent are nematodes, springtails (collembola), bark beetles, and predatory ground beetles (carabids). Some herbivorous insects (especially the spruce budworms and tent caterpillars) are often responsible for widespread defoliation of susceptible conifer or associated hardwood species, respectively.

Numerous songbirds forage on insects found on the foliage or under the bark of coniferous and broadleaf trees. Most bird species migrate to regions with more moderate climates during winter, but several are year-round residents. Woodpeckers are notably dependent on the presence of standing dead trees (snags) in old-growth forest stages. The Corvidae (ravens, crows, and jays) are especially characteristic of coniferous forests but are more flexible in their needs. Magnificent birds of prey (hawks, falcons, eagles, and especially owls) are top carnivores in many coniferous food webs. Wetlands within the boreal forest constitute the breeding grounds for innumerable ducks, geese, cranes, and swans. Beavers are especially important because of their role in altering drainage. Other widespread rodents include seed-eating mice, cone-caching tree squirrels, flying squirrels, and widely fluctuating populations of voles and lemmings. These small mammals serve as food for several carnivores, especially Mustelids (weasels, marten, fisher) and birds of prey. Hares undergo regular (10-year) population cycles, and serve as food for lynx. Characteristic ungulates include the moose and caribou or reindeer species in the boreal zone, and the mountain sheep of the subalpine. Large carnivores include the wolf, cougar, and bears.

Prehistory

Conifers have been around for a long time, much longer than most of the herbs, shrubs, and trees (largely angiosperms or flowering plants) that dominate the other biomes of the world. The fossil record indicates that conifers or their immediate progenitors had already evolved during the Permian period, over 230 million years ago. Gymnosperm forests (consisting of conifers and their relatives) came to dominate most terrestrial areas of the globe during the age of dinosaurs, and persisted for 100 million years. Unlike the dinosaurs, the conifers did not die out at the end of the Cretaceous period, but continued evolving into modern forms. There appears to have been less speciation (evolution of new species) in the conifers than in many orders of flowering plants. This is probably due to broad ecological adaptations within species, and extensive gene flow through the widespread dispersal of pollen and seeds. Ecological tolerance and effective dispersal have resulted in coniferous forests being one of the more persistent plant formations on the face of the Earth. Their composition and distribution have waxed and waned with evolution, extinction, and cycles of changing climate, but this vegetation type has generally been quite extensive.

Most recently, the ice ages of the Pleistocene epoch compressed the ranges of many northern species into the more southerly parts of Europe, Asia, and North America. During this period (2,500,000 years ago to as recently as 10,000 years ago), coniferous forests covered many areas which now support deciduous or grassland vegetation. These paleohistoric dynamics have been deduced from the composition of pollen which has settled into the mud at the bottom of lakes, ponds, and bogs. Many conifer species are still expand-

ing their range in response to the relatively recent retreat of the continental ice sheets. Elsewhere (notably at the tree line with the Arctic and alpine tundra), conifers are failing to regenerate over portions of their range because particular marginal habitats are no longer as favorable as they had been. The boundaries of the coniferous forests are thus one of the most sensitive and well-studied indicators of climate change.

Vegetation Dynamics

Like all forests, coniferous forests also undergo successional dynamics over a time scale of decades to centuries. This succession of dominance by different species over time often characterizes the biological response to disturbance: some species can survive disturbance, others are colonizers but are short-lived or unable to regenerate in their own shade, and still others may slowly establish and maintain viable populations indefinitely. Being able to disperse long distances by means of winged seeds, and having a preference for germination on bare mineral soil, many species of conifers are adept at colonizing substrates newly exposed by retreating glaciers, volcanic eruption, landslide, or floodplain deposition. Establishment is often further enhanced, however, if the site is first ameliorated (through microclimatic protection and the recycling of nutrient-rich plant litter) by the growth of other colonizers such as dryad, alders, or cottonwoods.

If the catastrophic disturbance is somewhat more modest, several structural and functional legacies of the original forest stand remain on a site and promote its recovery. Wildfire is pervasive in most coniferous forests, whether started by lightning or by man. It is likely that most locales within the boreal or subalpine forests would naturally be subjected to wildfire at least once every 100 to 200 years, perhaps every 400 years in the wet coastal forests. Much of the forest floor (recent plant litter and partially decomposed humus) usually remains after fire, along with the seeds, rhizomes, invertebrates, microbes, and nutrients therein. Several tree species (especially jack pine, lodgepole pine, and black spruce in North America) have high proportions of serotinous (late opening) cones which retain viable seeds for years until the heat of a forest fire triggers their opening. The "secondary succession" which follows is thereby much more rapid than the "primary succession" described above, because the colonization and soil-building phases of ecosystem recovery can proceed more rapidly. These rich, open sites are often colonized by dense stands of fireweed and raspberries, as well as by willows and aspens which resprout from underground stems or root crowns. Many species of wildlife, such as deer and moose, thrive in these biologically productive habitats. The rapidly colonizing pines (in the boreal and subalpine forests) or Douglas-fir (in much of the North American West Coast forest) may quickly dominate the site, or it may go through several decades of dominance by deciduous tree and shrub species. Eventually the conifer trees tend to prevail, if only by virtue of their stature and longevity. But unless the site is dry and the tree canopy sparse, seedlings of these sun-loving conifers will not be able to regenerate in the shade of their parents. If fires do not erupt in the interim, the forest understory and the older successional stages of coniferous forests become dominated by the more shade-tolerant species, such as the firs, hemlocks, and sometimes spruces. If the fire return interval is short (less than 50 to 100 years), components of the coniferous forest biome may remain perpetually in dense pine or heathland vegetation.

Even the shade-tolerant tree species can rarely dominate without disturbance, however. In contrast to the stand-level disturbance of wildfire, there are also small scale disturbances such as the windthrow of individual trees and the mortality of small groups due to root rot or insect attack. Such events can be characterized as "releasing disturbances," in that they open the canopy without killing the younger trees, accelerating their successional replacement of the original dominants. Tree mortality and replacement at the scale of individual trees becomes more pronounced during the old-growth stages of stand development. The complex composition, canopy structure, and forest floor (usually characterized by much coarse woody debris) of this stage often supports a diverse flora and fauna.

Environmental Roles and Issues

Human utilization of the coniferous forests of the world has historically been fairly sparse. Indigenous cultures have traditionally emphasized subsistence hunting (of ungulates, fish, and waterfowl) and animal husbandry (of sheep and reindeer), supplemented by the collection of other forest products such as berries and mushrooms, and the harvesting of furs for cash. These traditional cultures persist in northern Canada, Alaska, Siberia, and Lapland, but are threatened by industrial development, a history of government efforts at assimilation, and the difficulties of making the transition to a cash economy. Educational and employment opportunities are typically poor, and fur trapping remains a controversial issue in terms of its sustainability and its morality.

Due to their harsh climate, these forest regions have rarely supported commercial agriculture. Even where the climate is more hospitable, soils are often too shallow, peaty or too poorly drained for cultivation. As a result, major land-use conversions from natural forest

to agriculture have never threatened coniferous forests to the same extent as in most temperate and tropical biomes.

The industrial world has exploited coniferous forests primarily for timber. Conifer wood is light, strong, and rather uniform; when grown at high densities under natural conditions, the wood typically has a tight grain and is clear of knots. Many conifers have long, straight, columnar trunks, well suited for poles and for milling into dimensional lumber. This lumber currently serves as the mainstay for the residential construction industry in North America. More recent product developments have centered on reconstituted and laminated wood products such as particleboards, plywood, and laminated strandboard. During the twentieth century, conifers have also served as the feedstock for the pulp and paper industry. Conifer fibers (termed "tracheids") are long, narrow, and strong, allowing the manufacture of high quality paper using chemical and/or mechanical pulping processes. Various conifers can also serve as the source of industrial chemicals such as methanol and turpentine, and specialty products such as cedar oil and Canada balsam.

The flexibility of conifer wood as a raw material has led to the development of an extensive timber industry, with annual commercial consumption of more than 1.6 billion cubic meters worldwide (more than 90% of it softwood), and revenues of approximately 100 billion dollars. The demand for wood fiber continues to grow with increasing world population and affluence. But most of the world's wood supply still comes from wild forests rather than managed plantations, and herein lie the greatest environmental controversies regarding coniferous forests. Many forests are being harvested at rates greater than their ability to regrow that fiber. Conditions of cold or shallow soils, thick organic layers, and intense competition from brush often mean that regeneration and growth are difficult without massively subsidized interventions from humans. It can be argued that such lands should not be considered commercial forests because continued conifer production is not sustainable, at least not within the 80- to 120-year rotations preferred by foresters. Moreover, the conversion of old-growth forests to managed plantations can result in unacceptable compromises to biodiversity and wilderness values.

Another prevalent industrial use of the northern and mountain forest lands has been for the generation of hydroelectric power. Massive dams and reservoirs have been constructed throughout this zone, flooding some of the most productive valley bottoms. Since the demand for electricity resides in the populations and industries of more temperate regions, extensive networks of high-voltage power lines link many major cities to forested wilderness. The rivers which drain the coniferous forests of the world are also important spawning and rearing habitats for commercial, subsistence, and recreational fisheries, especially for salmon and trout species. Water, whether in rivers, lakes, or as snow, is also the focus of much recreational use in these forests. The need for intact forest cover to regulate the melting of snowpacks, the infiltration of water, and the minimization of soil erosion (thereby ensuring the quantity and quality of water available for a multitude of uses) is thus often at odds with the demands for timber harvesting.

Coniferous forests also play a major role in the global carbon cycle and may have significant feedback effects on the global climate. Through photosynthesis, these ecosystems fix atmospheric carbon dioxide into structural biomass at rates typically ranging from 500 to 1,000 grams per square meter per year. Some of this carbon is released through death and decomposition, but much of it accumulates in wood and in the forest floor. In addition to dictating much of the biology and dynamics of these forests, the conditions which result in slow decomposition of this material further result in coniferous forests and their associated peatlands being a net carbon sink. Huge reservoirs of carbon are thus embedded in the conifer trees, forest floors, and peatlands of the globe. Peatlands are also being exploited to supply peat as a plant growth medium for the horticulture industry.

The net flux of carbon dioxide and other products of decomposition (such as methane) is highly sensitive to changes in temperature and moisture. Slight warming may promote the release of more of these greenhouse gases, potentially exacerbating global warming. This could be a real danger, as boreal forests are located at latitudes predicted to see mean annual temperatures rise by several degrees Celsius. On the other hand, plant growth (i.e., carbon assimilation, which is often temperature-limited in coniferous forests) may be enhanced enough to buffer some of the increased concentration of carbon dioxide in the atmosphere. The interplay and consequences of these and related phenomena (such as the albedo or reflectance effect of tree and snow cover) are currently being explored through the use of satellite technology and simulation modelling of biophysical processes over large areas.

PHILIP J. BURTON

For Further Reading: N. Nakagoshi and F. B. Golley, eds., *Coniferous Forest Ecology, From an International Perspective* (1991); E. C. Pielou, *The World of Northern Evergreens* (1988); H. H. Shugart, R. Leemans, and G. B. Bonan, eds., *A Systems Analysis of the Global Boreal Forest* (1992).

FORESTS, DECIDUOUS

Deciduous forests are forests in which most of the trees lose their leaves for several months each year in response to seasonal climatic stress. In temperate regions broadleaved trees such as oaks, maples, and birches drop their leaves to avoid frost injury during the cold months. Deciduous forest also occurs in seasonally dry tropical regions such as India and northern Australia where trees lose their leaves in the dry season to avoid drought injury.

Although deciduous forests are often dominated by broadleaved trees, many stands are intermixed with evergreen conifers such as pines. Broadleaved trees tend naturally to predominate on sites with a fertile soil and mild climate, while conifers often have a competitive advantage on dry or infertile sites or areas with a harsh climate. In colder regions a mixed deciduous-coniferous forest prevails in which conifers such as spruce, hemlock, and pine may make up half or more of the trees.

Temperate deciduous forest is found only in three parts of the world—eastern North America, western Europe, and eastern Asia (including the northern parts of China, Korea, and Japan). There are striking floristic similarities among these regions. For example, all three regions contain species of maples, birches, ashes, elms, and oaks. Thus the first Europeans to reach eastern North America found and described forests quite similar to ones they had known at home. The forests of eastern North America and eastern Asia are even more similar and share several genera not found in Europe. Deciduous forests harbor a high diversity of animal species, although many of them can be found in other vegetation types as well. Climatically all regions of temperate deciduous forest are characterized by warm summers, cool or cold winters, and relatively high rainfall that is more or less evenly distributed throughout the year.

Natural and Human History

Broadleaved deciduous forests have been in existence for at least 50 million years, but their geographic distribution has frequently shifted in response to climatic change. In the Eocene period, just after the age of dinosaurs, deciduous forest was present at the polar regions, while the eastern U.S. and western Europe were covered by tropical rain forest. Subsequent cooling caused deciduous forests to shift southward to the midlatitude positions. The uplift of great mountain ranges also caused local climatic drying and shrinkage of suitable habitat; much of its former range became occupied by grassland and desert. During the Pleistocene Ice Age, 16 separate periods of glaciation pushed the deciduous forest even further south. In North

America boreal pine-spruce forest was located as far south as the Carolinas; temperate deciduous forest found a refuge on the coastal plain of the Gulf of Mexico. In Europe southward migration was blocked in part by mountain ranges. Large numbers of species, such as sweetgum, silverbell, and tulip-poplar, became extinct. Only hardy generalists such as birches, oaks, maples, and pines survived. To this day the deciduous forests of Europe are relatively impoverished compared to those of North America and Asia.

As in all forests, natural disturbance may also occur at a more local scale even during periods of relatively stable climate. While broadleaved deciduous forest is not nearly as prone to fire or insect epidemics as many coniferous forests, large-scale destruction, especially by wind, does occur periodically. For example, land surveyors working in Wisconsin in the 1850s recorded over 400 "windfalls," averaging 230 acres in size and ranging up to 9,000 acres, where the forest had been flattened by tornadoes and thunderstorm winds. In New England, a hurricane in 1938 destroyed more than a million acres of forest.

Humans have generally had a much greater impact on temperate deciduous forests than on most other kinds of temperate forest because deciduous forests are concentrated in areas where the climate and soil are conducive to agriculture. Farming was already widespread across central Europe 6,500 years ago. Although some parts of Europe were still heavily forested in the early Middle Ages, further population increases led to a rapid decline in forest area. By 1000 A.D. only 15% of England was still covered by forest and these remaining tracts were heavily modified by humans as a result of cutting, coppicing, and grazing. Animals dependent on large forest tracts, such as bear, lynx, and even certain bird species, became extinct in England well before 1000 A.D.

Although the forests of eastern North America seemed extensive and comparatively wild to the first explorers, Native Americans had a substantial impact on the forests in most regions. Colonist Thomas Morton wrote in 1632 that the Indians in coastal Massachusetts "set fire of the Country in all places where they come," and as a result, "the trees growe here and there as in our parks." This custom of annual burning, mentioned by many early explorers, helped improve deer habitat and nut production on which the Indians were partly dependent for food. Burning also made travelling easier by getting rid of underbrush. Indians also practiced a shifting "slash-and-burn" agriculture. The explorer Samuel de Champlain wrote in 1604 that they were "constantly making clearings."

Widespread use of fire almost certainly shaped the species composition of the eastern deciduous forest as well as its appearance. Fire tends to favor oak-hickory forest over the fire-sensitive maple species; in fact, many recent studies suggest that oaks are often being replaced by maples in the absence of fire. Had it not been for frequent burning by Indians, it is likely that maple rather than oak would have been the predominant forest type in much of the mid-Atlantic and midwestern regions.

Disturbance of the eastern deciduous forest greatly increased under the influence of European settlers. Most areas suitable for agriculture had been cleared of forest by the mid-19th century. In response to increased demand for furniture, flooring, railroad ties, charcoal, and other products made from deciduous "hardwood" trees, nearly all of the remaining forest was cut heavily between 1900 and 1940. Only a few sizable tracts in remote areas escaped this concentrated wave of logging. Most existing stands in eastern North America are therefore made up of trees 60 to 90 years old that originated after this episode of logging. They are what foresters call "even-aged" stands because all the trees are approximately the same age.

Despite encroachments upon the forest by suburban and rural development, the amount of forested land has greatly increased since 1850 as a result of the abandonment of agricultural fields on steep or infertile sites. Large areas of former crop and pastureland in the Northeast and former tobacco and cotton fields in the Southeast are now covered with dense forest. Forest now occupies about 65% of the land in the Northeast and 60% in the Southeast.

Contemporary Management of Deciduous Forests

Deciduous forest currently occupies about 250 million acres in the eastern U.S. About 70% of this acreage is privately owned as small woodlots by nonindustrial landowners, including farmers. The other principal owners are the forest industry (15%), the federal government (7%), and state and county governments (6%).

As in other temperate forests there are two basic types of harvesting that are considered sustainable in deciduous forests. Selective cutting (known as *uneven-aged management*) removes scattered individual trees or groups of trees at regular intervals of about 10 to 15 years in each stand. Gaps in the forest created by harvest are filled by new seedlings and saplings, usually by natural regeneration rather than planting. Uneven-aged management is most appropriate for tree species that are very tolerant of shade, such as maples, hemlock, and beech. It is not generally suitable for species that require high light intensities for seedling survival and growth.

With *even-aged management* all trees are removed over a fairly short period of time. Clear-cutting is probably the most widely known form of even-aged management, but other forms of even-aged management have a more extended period of harvest. The general public often confuses clear-cutting with deforestation, but the two are quite different. Clear-cuts in deciduous forest are quickly reoccupied by young trees, generally by natural means. Clear-cutting mimics heavy disturbance that occurs periodically in natural forests and upon which many species of plants and animals are dependent. Even-aged management is generally necessary for species such as white birch, aspens, oaks, tulip-poplar, ash, and pines because seedlings of these species do not survive or grow well in small openings.

Throughout the eastern deciduous forest, only about half of the annual growth is being harvested, so the forests are not being depleted. However, many small private woodlots are subjected to exploitive logging in which only the high-quality trees and valuable species are removed. Repeated "creaming" of the best trees leads to "junk hardwood stands" of little commercial value. This practice has several causes: lack of knowledge about forest management (both logger and landowner often believe they are simply removing mature trees to make way for younger trees of similar potential), limited markets for small and low-quality trees, and limited economic incentives for owners to invest in long-term management. Government assistance and cost-sharing programs have stimulated better management among small landowners, although it appears that a large majority of timber harvests are still conducted without the assistance of a forester.

Clear-cutting on public lands has been a controversial practice for several decades. In 1964, after years of research showing that selective cutting in hardwood forests was not successful in regenerating most of the dominant species, foresters in several eastern natural forests began to use clear-cutting. The controversy became especially pronounced on the Monongahela National Forest in West Virginia, which was heavily used for recreation and where some of the clear-cuts were quite large. The public debate eventually led to the passage of the National Forest Management Act of 1976. Recognizing that clear-cutting was often the best method for regenerating certain tree species and beneficial for many wildlife species, Congress did not prohibit clear-cutting, but restricted its use to cases where it would be considered the optimal method. Individual clear-cuts were also limited in size to 40 acres or less. Passage of the act, however, did not eliminate the controversy. In 1992, in response to public pressure, the U.S. Forest Service announced that clear-cutting would be reduced by 70%, and other even-aged methods used instead where the more light-demanding species were involved. Some environmental groups continue to lobby for a total ban on all forms of even-aged management.

Environmental Issues and Outlook

Two centuries of western civilization in North America have had variable effects on the integrity and sustainability of deciduous forest ecosystems. In most areas that were never cleared for agriculture, site productivity is probably comparable to that in presettlement times. However, many areas on the eastern piedmont once cultivated for cotton, tobacco, and other crops lost large amounts of topsoil. Many originally good sites were probably degraded to average sites as a result.

Forestry operations generally have only minor effects on soil and nutrient loss because, unlike agriculture, little soil is directly exposed to the impact of rain, the period of exposure is brief, and intervals between harvests are quite long. Most cases of significant erosion after logging can be traced to poorly designed logging roads. Some states have specific laws restricting the placement of logging roads and other states have established voluntary guidelines.

In spite of widespread human disturbance in the 19th and 20th centuries, few plants and animals of eastern deciduous forests became extinct during recorded times (the passenger pigeon is a notable exception). However, some animals such as wolves and mountain lions were extirpated over most of their original range as a result of bounties and illegal hunting. These animals are protected by the Endangered Species Act of 1973, but reintroduction of predators into areas of former habitat has been controversial. Some animals, such as the white-tailed deer, have become too numerous. Overbrowsing by deer is even causing conversions of entire forests from one species type to another.

While many animal populations have made remarkable recoveries in recent decades, the outlook is not as promising for a large group of migrant American songbirds. Many species that winter in the American tropics but breed in temperate deciduous forests have shown sharp declines in recent decades. Some may be on the brink of extinction. Evidence points to nest parasitism by cowbirds as the major cause, with tropical deforestation as a secondary cause. Cowbirds were originally rare in most of the deciduous forest but have expanded their range greatly as forests became fragmented into isolated woodlots and winter food sources became available on agricultural fields.

Acid rain has had unfavorable impacts on aquatic life, but appears to cause relatively little damage to forest vegetation and soils. Most experts believe that air pollution has reduced deciduous forest growth by about 5%, with ozone rather than acid rain being the major cause. A much greater threat to forest health is the accidental introduction of foreign insect and disease pests for which native trees have little resistance. Diseases have already nearly eliminated American chestnut, elm species, and butternut from the canopy of eastern forests. Numerous other introduced pests, such as beech bark disease, balsam woolly adelgid, dogwood anthracnose, gypsy moth, white pine blister rust, hemlock woolly adelgid, and maple thrips have decimated forests over large areas. Damage will increase as these pests spread throughout the range of their new hosts.

While deciduous forest ecosystems are currently in better condition than they have been at any time in the past 100 years, they face an uncertain future. Decimation by foreign insects and diseases, exploitive logging by small private landowners, and continued fragmentation of extensive forests into isolated woodlots surrounded by human development are probably the greatest threats to the continued existence of a biologically diverse and economically valuable resource.

CRAIG G. LORIMER

For Further Reading: Malcolm L. Hunter, *Wildlife, Forests, and Forestry: Principles of Managing Forests for Biological Diversity* (1990); John Terborgh, "Why American Songbirds Are Vanishing," *Scientific American* (May 1992); Raymond A. Young and Ronald L. Giese (eds.), *Introduction to Forest Science*, 2nd ed. (1990).
See also Clear-Cutting; Deforestation; Exotic Species; Pest Control; Pollution, Air; Tropical Dry Forests.

FORESTS, OLD-GROWTH

Forests undergo a progression of successional stages following their initial establishment, culminating in the development of a relatively stable and complex ecosystem often referred to as "old growth." Old-growth forests tend to differ significantly from young and mature forests in composition (the organisms that are present), function (such as in paths and rates of carbon and nutrient cycles), and structure (such as presence of multiple canopy layers and a broad range of tree sizes). Not surprisingly a distinguishing feature of old-growth forests is the presence of old and — where environmental conditions are favorable — large trees.

Old-growth forests typically have an associated set of animals and plants that strongly prefer or require the distinctive structural and microclimatic features found in these ecosystems. Recognition of these organisms — particularly vertebrate animals, such as the northern spotted owl (*Strix occidentalis caurina*) — has

greatly increased scientific and public interest in old-growth forests, especially in temperate regions.

Old-growth forests are extremely varied in age and ecological characteristics depending upon their location within the world. Indeed, the character of old-growth forests may vary dramatically even within a relatively small area, depending upon local variations in soil and other site conditions. It may take a forest 80 to over 200 years to begin developing old-growth characteristics. In North America shorter periods of time are characteristic of the deciduous hardwood forests of the Northeast and the pine forests of the Southeast; longer periods of time are characteristic of the evergreen coniferous forests in the coastal Pacific Northwest.

The coniferous forests of the Pacific Northwest provide the old-growth archetype; in fact, they are often used to define the concept. The immense size and age of the old-growth trees, extensive scientific study of the forest ecosystems, and the presence of dependent animal and plant species have all contributed to the widespread recognition of these particular forests as well as the old-growth concept in general. These forests are dominated by evergreen coniferous tree species, such as Douglas fir (*Pseudotsuga menziesii*), western hemlock (*Tsuga heterophylla*), and western redcedar (*Thuja plicata*). Old-growth conditions typically begin to develop after about 200 years but many old-growth forests in this region are 500 to 700 years of age and some have existed for over 1,000 years.

Old-growth forests in the Pacific Northwest have high biological diversity in groups as varied as plants, vertebrates, invertebrates, and fungi. Although species diversity is also high early in forest succession (due to the occurrence of many weedy generalist species), the species diversity associated with old-growth forests includes many highly specialized species that prefer or require the specialized conditions found there. Well-known examples include the northern spotted owl, which has been officially listed as an endangered species, and the Pacific yew (*Taxus brevifolia*), which is the source of a drug (known as taxol) being utilized for the treatment of cancer.

Old-growth forests in the Pacific Northwest also have distinctive functional attributes, function referring to the various activities or processes carried out by an ecosystem and the rates at which they are accomplished. Old-growth forests are typically very productive forest ecosystems, photosynthetically fixing and processing large amounts of solar energy. This productivity is assured by the huge leaf masses found in a typical old-growth conifer forest; a single old-growth Douglas fir tree may have 4,000 square meters of leaf (needle) surface. Much of the energy fixed annually is used in respiration, however, to maintain the large mass of living tissue. These old-growth forests have the largest accumulations of organic matter (biomass) in the world; hence, they sequester large amounts of carbon that might otherwise be adding to the load of carbon dioxide in the atmosphere. Typically, the northwestern old-growth forests are also very efficient at conserving nutrients, reducing erosion, and providing stable streamflows.

The structural characteristics of old-growth forests are extremely important because many of the old-growth-related species and processes depend upon specific structural elements found there. Large and old trees provide critical habitat for many animal and plant species because of their diverse forms and conditions; for example, trees with areas of wood decay or broken tops are mixed with completely sound trees. Large standing dead trees, called snags, are particularly important for wildlife species that create and utilize cavities, such as woodpeckers. Large logs on the forest floor may persist as structures for many centuries in the forest of the Pacific Northwest and perform many critical functions as habitat for other organisms, as long-term sources of energy and nutrients, and as contributors to the physical stability and diversity of associated streams and rivers.

Old-growth forests in other regions display the characteristics of the old growth of northwestern coniferous forests to differing degrees. Tree species composition, which determines such characteristics as maximum attainable tree sizes and ages and levels of log decay resistance, is an important variable. Many temperate deciduous hardwood species decay rapidly following death so that snags and logs are not nearly as prominent as structural elements in the old-growth hardwood forests of northeastern North America, except where they contain a substantial component of eastern hemlock (*Tsuga canadensis*) or eastern white pine (*Pinus strobus*).

Human Interactions with Old-Growth Forests

Prior to interactions with humans, all forests were of natural origin and some of these were probably in a late successional or old-growth state. Of course, the proportion of a forested region in an old-growth condition at any given time depended upon many variables including the history of natural disturbances: for example, the frequency, extent, and intensity of wildfire. In the Pacific Northwest it appears that at least 25% of the forests have been in old-growth condition throughout the last several millennia; about 65% of the forests would have been classified as old-growth when western settlement began about 1825.

Humans have dramatically reduced the distribution and extent of old-growth forest ecosystems. They have been cleared to provide land for agricultural and urban activities. They have been cut to provide wood products, from firewood to fleets of ships; because of the size, density, and quality of many old-growth trees they have often had extraordinary utility and value. More recently old-growth forests have been cut and converted into tree farms for the intensive production of wood fiber.

As a consequence of these human activities, essentially all old-growth forests have been eliminated from long-settled temperate regions, such as Europe, eastern North America, and most of moist, low-elevation Asia. In addition to western North America (including Canada), remnants of old-growth temperate forests remain in Australia, New Zealand, Chile, and small areas in northern China. Extensive old-growth forests remain on less productive sites, particularly in sparsely settled regions. The boreal or high-latitude forested zones (taiga forests) over much of Asia and North America are an example. Many tropical rain forests also qualify as old-growth forests although the concept has generally not been applied there.

The future of these forests—whether they will be cut down as sources of wood products or preserved as repositories of biological diversity—is typically at the center of a major political controversy in every locale where they are present.

JERRY F. FRANKLIN

For Further Reading: M. E. Harmon, W. K. Ferrell, and J. F. Franklin, "Effects on Carbon Storage of Conversion of Old-Growth Forests to Young Forests," *Science* (1990); C. Maser, et al., *From the Forest to the Sea: A Story of Fallen Trees,* (1988); L. F. Ruggiero, et al., eds., *Wildlife and Vegetation of Unmanaged Douglas-Fir Forests,* (1991).

FORESTS, URBAN

An urban forest is a collection of trees (and related plants and natural resources) in and around a city, including street and yard trees, park trees, and those in newly expanding suburbs. It is a natural resource that provides considerable value to people in cities and towns, in terms of its ecological and social benefits, though not in wood products as with rural forests. Some of these benefits are only partially understood, but the scientific measures are developing rapidly. The benefits can be categorized as ecological, social, and human health related. Ecological benefits include water quality, soil conservation, air quality, and wildlife habitat. The benefits produced are proportional to the tree cover: the larger the biomass produced by the trees, the larger the benefits produced. Social benefits include community building and aesthetic improvement. Trees give a sense of pride to a community and bring people together for planting, care, and recreation. Public health benefits are also being recorded. In recent years, researchers have documented the healing effects of trees on hospital patients. Patients with a view of trees come out of the hospital faster and use less pain medication. The public health value can be expanded to local parks and greenways that provide soothing scenery, recreational space, and a more pleasant environment.

Urban forests differ from rural forests in that the growth and development of the tree cover is limited by city structures rather than by forest succession and competition with other trees. The places where trees can thrive in the city are limited, and design changes are needed if communities are to reap the many benefits urban forests provide. The quality of tomorrow's urban forests will be measured by the physical space, public attention, and quality of the care we provide for trees.

Urban forests in the United States cover 70 million acres of municipal land and an uncountable number of acres in suburbia and small towns. Geographically, urban forests exist in four zones. The suburban fringe, where new subdivisions are found, begins at the rural edge. Here tree cover is removed and topography reshaped to make room for roads and house lots. In the suburbs, where subdivisions are numerous and the natural forest has been removed except along streams and wetlands, most of the trees were planted after construction. City residential space is tightly packed, established housing with yards. There are trees in almost all the yards, and in public spaces such as parks, schoolyards, and along the streets. City center trees grow in pots, in holes in the sidewalk, in vacant lots, and even on buildings. Almost every tree has to be "engineered" into this zone, or a space cut out of the concrete if the planners neglect to allot one. In the city center, people outnumber the trees by a large margin, and life for trees is difficult and usually short.

The average American city has a tree canopy covering about one third of its area. However, the health of the trees forming this green umbrella is failing. Canopy cover can be increased by planting more trees and by increasing the size of existing trees. The average life span of a tree declines as one moves closer to the center of the city; a downtown tree lives an average of 7–10 years, while street trees in more open areas of the city live to be an average of 32 years. Trees in parks live much longer, about 50 years, because they have more space to grow. The skill of the urban foresters can make a big difference in the life span of trees. In a few cities, with the best programs, street trees have an average life span of 60 years.

Street trees are the most widely understood because in most places they are planted and maintained by the municipal government. A 1989 street tree survey of U.S. cities by J. Kielbaso estimated 60 million existing street trees, with open spaces for planting 65–75 million more. Street trees are estimated to make up 10% or less of the total urban tree count.

Parks and greenways accommodate about 40% of the nation's urban forest trees. They provide more space and better growing conditions than the small strips of soil between the sidewalk and curb called street lawns. Trees live longer in these open spaces, but intensive recreation, soil compaction, and physical damage also creates stress and shortens tree life. Yard trees, about 50% of the urban forest, provide variety by reflecting the interests of the homeowners.

To extend the life span of newly planted trees, old planting techniques have been challenged. Researchers have learned that 90% of a tree's roots are in the top 30 centimeters of soil, and the more compact the soil, the closer to the surface the roots tend to grow. Old planting techniques, designed a century ago for planting in well developed soils, are unfortunately still widely used in hardpacked urban soils. New planting techniques have been designed that loosen soil in a wide but shallow planting area for better rooting and longer tree survival.

Urban forestry involves the growth, management, and care of trees in and around urban areas. Urban foresters are challenged to grow healthy trees in ecologically difficult urban conditions. Urban forestry started as street tree management with urban foresters working for parks departments. Their focus was on planting and care that fit the existing streetside conditions. Today, urban forestry is a movement that includes more issues and a mixture of people ranging from citizens to tree managers and scientists. The view of the urban forest, its potential, and its value have expanded. Cities need to make space for trees and develop management techniques that maintain a healthy and abundant tree canopy.

GARY MOLL

For Further Reading: Robert W. Miller, *Urban Forestry: Planning and Managing Urban Greenspaces* (1988); Gary Moll and Stancey Young, *Growing Greener Cities* (1992); Gary Moll and Sara Ebenreck, *Shading Our Cities: A Resource Guide for Urban and Community Forests* (1989).

FORESTS, U.S. NATIONAL

In the second half of the 19th century, Americans in increasing numbers settled on arid portions of the West. To protect crucial water supplies, on March 30,

1891, Congress authorized the president to create forest reserves from the public domain. Within a decade and a half, several presidents set aside nearly ten percent of the nation's lands by designating it what after 1907 were called national forests. Today these lands include 191 million acres.

During the six years following 1891, Congress debated the purposes of these reserves and on June 6, 1897, determined that the resources on these lands were to be used, that the needs of those who lived on or around them were to be treated with sympathy, and that the national interest in these federal lands was to prevail. This law governed national forest management until superseded by the 1960 Multiple Use–Sustained Yield Act and the 1976 National Forest Management Act.

From 1891 until February 1, 1905, the national forests were administered by agencies of the Department of the Interior. For several reasons centered on corruption in that department, jurisdiction was transferred to the Bureau of Forestry in the Department of Agriculture, and the name of the agency was changed to Forest Service. The conservation movement gained momentum at this time, and the national forests and conservation became synonymous.

During the second decade of the 20th century, again to protect forested watersheds, Congress authorized the purchase of land that would be added to the national forest system. The forests on these lands, which generally lay east of the Mississippi River, had been mostly logged off and the land had been effectively abandoned by its owners. A long rehabilitation effort was so successful that some of these lands have recently been declared wilderness. These purchases now exceed 24 million acres; essentially all of the national forests in the eastern United States have been purchased while those in the West were created by reserving parts of the public domain.

Initially the national forests were to provide water for users—farmers, towns, industries—that were located downstream and outside the forest. But there were resources in the forests themselves, such as minerals, forage, and wood, that were essential to economic growth. Controversy raged over the proper accessibility of these resources and even the constitutionality of federal regulations for the use and sale of materials owned by the public. In the end, the federal view prevailed; all of the national forest resources are available for economic use according to federal rules and regulations, and in line with national priorities.

Recreational use and esthetic values were also germane to national forests, and gradually they too were given legitimacy. In 1917 the Forest Service completed a detailed study of recreational opportunities and values in the national forests. During the early 1920s chief and staff debated the merits of establishing wilderness areas, and on June 3, 1924, the world's first wilderness area was set aside in the Gila National Forest of southern New Mexico.

National parks in the Department of the Interior are examples of extraordinary beauty or uniqueness, and commercial activities are limited to providing for the needs of visitors. National forests also contain examples of extraordinary beauty, but commercial extraction of their resources is integral to their mission, and at times there are conflicts and the need to set priorities. Parks contribute recreation, wildlife habitat, and watershed protection; national forests contribute the same, but also forage, wood, and minerals. At times members of the public, not realizing the distinction between the two types of public reserves, have criticized the government for allowing commercial operations.

During the Great Depression of the 1930s millions of Civilian Conservation Corps workers built trails, campgrounds, roads, and firelines, and developed an essential national forest infrastructure. Specialization spawned sister agencies—the Soil Conservation Service, the Grazing Service, and Tennessee Valley Authority—all of which dealt in some fashion with forested lands and thus overlapped the mission of the national forests. In 1946 the Grazing Service merged with the General Land Office in the Department of the Interior to form the Bureau of Land Management, with responsibility for all federal land not in national forests, national parks, wildlife refuges, or military reservations.

Postwar prosperity and significant decline in privately owned timber supply prompted a great demand for all resources found on national forests, setting the stage for an unanticipated conflict. This day had been long awaited, but some segments of the public were concerned about misplaced priorities and perhaps permanent degradation of the forested environment. This concern became visible during the 1950s, and exploded in the 1960s.

The Multiple Use–Sustained Yield Act became law on June 12, 1960. National forests were to be managed for "outdoor recreation, range, timber, watershed, and wildlife and fish purposes." Big in concept, the act is only a page and a half in length.

Since 1924 the Forest Service had been designating wilderness areas, and by 1964 there were 13 million acres in the wilderness system. Land managers felt that wilderness had been adequately treated and that environmental conditions had been handled responsibly. Some people felt, however, that there was insufficient wilderness and that, unless protected by act of Congress, the areas could also be "undesignated" by executive edict. Thus, the wilderness bill became law on September 3, 1964. By 1990, more than 33 million acres had been classified as wilderness.

Laws to conserve endangered species and wild and scenic rivers followed before the end of the decade. On the first day of 1970, the National Environmental Policy Act became law, authorizing establishment of the Environmental Protection Agency, the Council of Environmental Quality to advise the president, and making impact statements a requirement on federal activities that could affect the environment. The national forests were directly influenced by all of these statutes.

The courts, too, were to influence national forest management. In 1973 the U.S. District Court for the Northern District of West Virginia reexamined the 1897 Forest Management Act. The West Virginia case reviewed logging practices on the national forests and determined that clear-cutting was in violation of federal law. In response Congress passed the 1976 National Forest Management Act, of a much more prescriptive nature than the earlier Multiple Use Act, to provide corrective language, and clear-cutting was allowed under certain conditions. The act also stipulated the need to sustain biological diversity. This act made the forests permanent by law; until then they had existed only by presidential proclamation. Thus, the national forests with their multiple uses and multiple controversies are here to stay.

HAROLD K. STEEN

For Further Reading: Thomas R. Cox, et al., *This Well-Wooded Land: Americans and their Forests from Colonial Times to the Present* (1985); Dennis C. LeMaster, *Decade of Change: The Remaking of Forest Service Statutory Authority during the 1970s* (1984); Harold K. Steen, *The U.S. Forest Service: A History* (1976, reprinted 1991).

FORMALDEHYDE

See Pollution, Air.

FOSSIL FUEL

See Fuel, Fossil.

FRANCIS OF ASSISI

(1181–1226), Saint Francis, founder of the Franciscans. In *A God Within*, René Dubos wrote: "The Judeo-Christian peoples were probably the first to develop on a large scale a pervasive concern for land management

and an ethic of nature. Among the great Christian teachers, none is more identified with nature than Francis of Assisi, who treated all living things as if they were his brothers and sisters."

Francis was the son of a rich merchant. Gallant, high spirited, and generous, he assisted his father in business until he reached the age of twenty. During his twenties, however, a number of events caused him to become more and more dissatisfied with his comfortable life. Eventually, he discarded all wordly goods and set about a life of prayer, simplicity, and service to the poor. Before long, Francis's example and the power of his spirit drew a band of like-minded followers, who became the first Franciscan monks. He drew up a simple rule of life to guide them, the "Regula Primitiva." This rule became the foundation for Franciscan attitudes towards social justice, the poor, and nature. In the years following the formation of the Franciscans, their ideals were embraced by one of the leading noblewomen of Assisi, Clare, who started a sister order for women.

Francis's theology, and by extension the founding theology of the Franciscan order, was based on a universally inclusive principle of love and an unyielding devotion to the image of a father/creator. As a young man, Francis had been a poet and an enthusiastic admirer of many ladies in Assisi. With his conversion to spiritual life, the erotic instincts of his youth were transformed into a consuming affection for all creation. Poverty became Lady Poverty, virtue became Queen Wisdom, and simplicity became for Francis a pure and holy sister. It was not far from these visions to considerations of Brother Sun and Sister Moon, Brother Fire and Sister Death. The love instinct for Francis became a universal principle that sought to include all creation and experience in a horizontal family of anthropomorphised brothers and sisters to be loved and celebrated.

In the midst of this affectionate family, Francis maintained a vertical devotion to a father/creator whose power preserved and maintained the horizontal family he had created. Under the father, all lived in the same great house. All acquired a deep intimacy with the rest of creation. There were no enemies and no threats, only a profound respect for the father to whom all owed their being.

The axes of horizontal love and vertical devotion that dominated Francis's theology served as an apt model for the power of the cross and as a basis for the Franciscan ethic of nature. In this ethic, an extensive and varied creation is ruled over and cherished by a loving father and by human beings, who have been made the crown of that creation and to whom the father has given a special responsibility as stewards to conserve and maintain creation. One of the primary spiritual and physical responsibilities of any Franciscan is to celebrate and work to maintain the beauty and diversity of nature.

Thomas of Celano, his first biographer, wrote in 1229, "Who can explain the joy that arose in Francis's spirit from the beauty of the flowers. Finding himself in the presence of many flowers, Francis would preach to them, inviting them to praise the Lord, as if they enjoyed the gift of reason."

Stories of Francis's attitude towards nature border on the absurd. He is said to have avoided stepping on rocks so as to not wear them down, gathered worms from muddy roads to preserve them from harm, and left out stores of honey during the winter so that the insects would not perish from hunger. Looking past their absurdity, however, these stories describe a distinct way of being in the world, a way dedicated not to dominating or disturbing nature, but to living together with nature as brother or sister.

As Francis gathered followers around him, his way of being in nature became the foundation for an ethic that is central to Franciscan communal life to this day. Franciscan brothers are prohibited from cutting any living tree at the roots. They are commanded to always leave a plot of uncultivated land in the middle of their gardens so that all types of wild plants, including weeds, can grow. In Franciscan orchards, fruit trees are intermingled with herbs, flowers, and wild plants so that the creator's wise and varied hand may be clearly seen.

Francis's ethic of nature is a radical expression of gratitude for the love the creator has shown to each person by putting them in such a beautiful and varied world. Individuals show gratitude by doing all they can to maintain and preserve the beauty and variety that surrounds them. Francis's ethic teaches people to embrace all living things as brother and sister and say, from the deepest level of their being, "yes" and "thank you."

The ethic of nature that Francis created became one of the cornerstones of the movement for conservation in the Western world. When it is most effective, conservation is based on a clear understanding of human values and human needs. The way of Francis of Assisi helps individuals to see that only by prayerfully cooperating with and respecting all forms of nature can they reach their full potential as human beings.

FRANCIS H. GEER

For Further Reading: Leonardo Boff, *Saint Francis, A Model for Human Liberation* (1989); René Dubos, "Franciscan Conservation vs. Benedictine Stewardship," in *The World of René Dubos* (1990); *St. Francis of Assisi: Omnibus of Sources* (1973).
See also Benedict of Nursia; Christianity; Religion.

FREE MARKET

A market is free if each potential buyer or seller of a good (or service) gets to decide: (1) whether to buy or sell this good at all, (2) the quantity of the good each individual would like to buy or sell, (3) the price at which to offer to buy or sell the good, (4) to whom to offer to sell the good or from whom to buy it, and (5) whether to vary the attributes of the good. The first three attributes are much more important than the other two.

A free market gives each individual (producer or purchaser) the maximum freedom without interfering with the same freedom for other producers or purchasers. Producers will choose to make a good and offer it for sale if they are more attracted by the potential revenue for this good than for other activities they could perform. Buyers will purchase this good if the price and other attributes make it more attractive than other goods being offered. No one is forced to make or buy a good; rather each individual is guided by Adam Smith's "invisible hand" to make a contribution to the economy.

Even for people who have grown up in free market economies, it seems astonishing that individuals will produce food, clothing, shelter, and hair cutting in just the amount that buyers want. How can a free market figure out the number of size 6 AAA women's brown pumps to produce? The number of airline seats from San Jose to Sacramento at 8 A.M. on Friday, June 12?

Free markets are in stark contradiction to the Marx-Engels notion of "from each according to his ability, to each according to his need." Marx and Engels envisioned an economy where workers would put in about the same level of effort, working at tasks that made use of individual abilities. This production would result in food, clothing, shelter, and other goods and services that would be made available to people according to their needs. The Soviet economy in 1990 required an elaborate planning apparatus to tell firms what to produce and where to ship it, and individuals where to work. Furthermore, "from each according to his ability" didn't provide sufficient motivation to get individuals and firms to provide the goods that consumers "needed."

In a free market, individuals earn income by supplying their labor or property. This income is used to purchase products, giving firms the ability to hire workers and buy raw material. Need or desire is not the primary determinant of who gets the goods that have been produced. Rather, income plus desire determines the demand for each good. A combination of income and desire determines whether individuals choose to purchase a good, and if so for how much.

Picture a vegetable or fish market on Saturday afternoon, when everyone knows the market will be closed Sunday. If it becomes apparent to sellers that they will not be able to sell their vegetables or fish at current prices, they will be motivated to lower prices in order to clear the market. Similarly, if they find they are running out of the goods, they will be motivated to raise prices. If sellers find they are rarely able to sell romaine lettuce for more than $0.50 per pound, they will offer no more than about $0.25 per pound to growers. If a grower expects to earn more from raising iceberg lettuce than from romaine lettuce, she will switch to the more profitable product.

Free markets are marvelously efficient for producing goods cheaply and satisfying consumer demands. However, market failures mean that the free markets will not accomplish everything that society desires. Thus, all advanced economies have a "mixed" economic system that reins in the free market in areas such as the environment.

LESTER B. LAVE

FREE-RIDER

A free-rider is anyone who is able to get the benefit of someone else's economic action without paying any of the cost. For example, Smith may enjoy the benefits of clean air, available because everyone else has controlled their emissions, even though he refuses to bear the expense. Clean air and unpolluted water are "public goods" that all may enjoy; no individual can be excluded from their benefit even if she refuses to do her share in the environmental cleanup. Unfortunately, many individuals may shirk the costs of voluntary cleanup, hoping to "free ride" on the efforts of others. What this means in practice is that voluntary cleanup is unlikely to work, even though the social benefits of cleanup may exceed the social costs.

Can we overcome the free-rider problem and act in the social interest rather than in our individual self interests? If only two individuals were involved, presumably they could negotiate a joint maximizing solution. If property rights are well-defined, exclusion is feasible, and bargaining costs are low, the two should be able to negotiate a socially beneficial solution. In the free-rider problem, exclusion is not feasible; in almost all real cases, many more than two individuals are involved.

There are many instances where some individuals choose not to free ride—as in donating to a charity. If all people are to participate, the more likely solution is for individuals to agree that all must contribute. For example, voters decide that a city's sewage must be treated and that all who use the sewage system must pay for treatment costs. In other words, government can overcome the market failure. There remain problems of deciding what programs are in society's interests and how to enforce the agreed-upon charges on those who do not want to pay.

DON FULLERTON

FRIENDS OF THE EARTH

See Environmental Organizations.

FUEL, FOSSIL

Fossil fuels, including coal, petroleum, and natural gas, are hydrocarbon remains of organisms buried during an earlier geologic period and changed through geologic processes. The combustion of fossil fuels supplies 88% of the planet's energy needs, with nuclear energy accounting for most of the remainder.

Coal, a solid material formed primarily of plant remains, is the most abundant of the fossil fuels. It came into wide use and sparked the Industrial Revolution in the eighteenth century with James Watt's development of the coal-burning steam engine. Today many nations continue to use coal-fired steam plants as their main source of electrical energy. Coal is also used in steel smelting processes and can be distilled into petrochemicals or liquified into oil or gas. Although it has been supplemented by oil as an energy source, especially for transportation, coal still provides as much energy as oil worldwide. Coal provides three fourths of the energy needs of China and is a major energy source in other developing nations.

Petroleum, the remains of decomposed single-cell organisms, accumulates in folded layers of permeable, porous sedimentary rock. Crude oil, which is a mixture of many different hydrocarbons, may be separated into a variety of compounds such as gasoline, kerosene, and fuel oil. In its various forms, petroleum provides most of the energy needs of the developed world. In addition, it is used in synthetics and petrochemical agricultural products.

Natural gas is a mixture of hydrocarbon gases, primarily methane and ethane. It may be produced through the decay of anaerobic organisms, during petroleum "cracking" (the separation of petroleum into smaller, lighter hydrocarbons), or as part of the coalification process. Until around 1960 natural gas was considered a waste product of oil recovery. Recently, however, it has become an important source of energy. The accessibility of natural gas depends on the characteristics of the reservoir in which it is found. Gas contained in underground coal seams is known as *unconventional gas.*

Fossil fuels are at the center of a host of political and environmental concerns. Oil and natural gas are distributed unevenly throughout the world, with three fourths of the planet's reserves located in the Middle East. The dependence of the United States and Europe on this oil has been a recurring source of diplomatic and military conflict, including the energy crisis of the 1970s and the Persian Gulf war.

Coal use is plagued with environmental problems. Underground mining entails serious worker health hazards, such as pneumoconiosis (black lung disease); surface mining can irrevocably destroy the land. Coal is full of impurities, and its combustion results in high sulfur dioxide emissions that contribute to acid rain problems. Gasification and liquification of coal avoid some of these problems but may be economically unfeasible.

Natural gas consumption, clean by comparison to coal or even oil, is approaching the limit of reserves that are readily accessible. Turning to unconventional gas may be uneconomical as well. Because of the dwindling supply of natural gas, the United States continues to regulate its use to preserve it for the future.

Fossil fuels are essentially a finite resource, by definition unsustainable and unreplenishable. Moreover, the combustion of fossil fuels releases carbon dioxide and thus contributes to global warming through the greenhouse effect. Conservation and efficiency programs, as well as the development of alternative energy sources, are all necessary.

FUEL, LIQUID

Liquid fuels are liquids that produce usable energy in the form of heat or light when burned. They are used by machines, such as automobiles, jet airplanes, and agricultural and construction equipment, that require a fuel lighter in weight than a solid and yet more condensed than a gas.

In the 1960s technology was developed that allowed natural gas to be liquefied. This expensive process consists of cooling natural gas to a temperature of $-259°$ F and transporting and storing it in specially made cryogenic tanks. Unlike its bulky parent, liquefied natural gas (LNG) can be transported all over the world. It later is regasified and piped to customers. The extreme flammability of LNG, however, prompts residents to resist the location of storage tanks in their communities.

Liquid petroleum gas (LPG) is another liquid fuel. Usually in the form of propane or butane, it is the by-product of natural gas, separated at the wellhead or from other parts of petroleum. It is generally used as fuel for tractors, buses, trucks and as domestic fuel for remote places. Since propane and butane are liquid at normal temperatures, they are shipped by pipelines, railway, and tank trucks and do not have to be regasified to be used by customers. Yet, like LNG and other fossil fuels, LPG is extremely flammable, is a nonrenewable resource, and produces pollutants when it is burned.

Liquid hydrogen, formed by cooling hydrogen to $-421.6°$ F, is being researched as an alternative to gasoline for powering automobiles. Its most attractive attribute is that it is the cleanest burning combustion fuel. When burned, its only byproduct is water vapor. If all modes of transportation could be switched from fossil fuels to hydrogen, one of the major sources of air pollution would be eliminated. Some disadvantages currently exist, however. Liquid hydrogen must be stored in expensive cryogenic tanks three to four times as large as an ordinary gas tank. In addition, the tanks currently made do not prevent fuel evaporation, which could result in an empty tank. Despite these problems, however, liquid hydrogen proponents are optimistic that solutions can be found to enable this fuel to be one of the answers to the energy problem.

Eythl alcohol, or ethanol, is a liquid fuel that derives its energy from the solar energy latent in organic matter. It can be distilled from fruits and grains that have been fermented with yeast. Ethanol can also be produced from biomass such as sewage, manure, garbage, and agricultural waste, thereby helping to reduce waste. Farmers in the Corn Belt are using the abundant grain to produce large amounts of ethanol, which is then combined with gasoline to form a liquid automobile fuel called gasohol. Gasohol is generally 10% ethanol and 90% gasoline.

FUEL, SYNTHETIC

Synthetic fuels are fuels that are not of natural origin. The U.S. synthetic fuels program in the 1970s sought to increase domestic production of petroleumlike products to reduce the nation's dependence on foreign sources. In 1980 the U.S. Congress established the Synthetic Fuels Corporation, an independent federal corporation created to finance the research and development of synthetic fuels. The goal was to create fuel from domestically available substances such as oil shale, tar sands, and coal.

Coal gas is produced by adding crushed coal to steam and oxygen at extremely high temperatures and pressure, a technique called the Lurgi process. The resultant gas has the same amount of energy as natural gas.

Coal can also be used to produce another type of synthetic fuel. It is first gasified and then sujected to zinc or copper catalysts, resulting in methanol. Methanol has successfully been mixed with and converted into gasoline to be used by automobiles and jet airplanes. Liquefication processes have been developed by combining different substances with the crushed coal and changing the intensity of heat and pressure to produce a variety of liquid coal solvents.

Oil shale—a dark brown rock that in rich concentrations actually burns—is another source of synthetic fuel. It is found in Colorado, Wyoming, and other parts of the western U.S. When the oil shale is crushed and heated to temperatures of about 500° C (930° F) in the absence of oxygen, liquid oil, gas, and residual carbon are formed. Differing conversion technologies allow oil to be recovered above ground or in the deposit itself. A synthetic fuel can also be produced from tar sands through the extraction of bitumen, a dense, sticky semisolid that is 83% carbon. Bitumen is mined or pumped from the ground, since it is sometimes found in deep, warm reservoirs. When heated and liquefied, bitumen floats on water. Most of the recovery processes, therefore, use steam or even direct burning. The liquid bitumen is then skimmed from the top of a water-filled separation tank.

Many environmental problems are associated with synthetic fuels. Those sources that are recovered in situ, such as oil shale and tar, risk the possibility of contaminating groundwater with toxic pollutants. The mining of oil shale causes massive surface destruction of mountainsides where the shale is located, making land reclamation difficult. Large amounts of crushed rock and highly toxic sludge residues from the extraction procedures require disposal. Dust from mining and crushing also contributes to air quality degradation. Properly disposing of such byproducts raises the cost of these synthetic fuels.

FUEL CELL

A fuel cell is a device in which the energy of reaction between a fuel and an oxidant is converted directly into electrical energy. A fuel cell consists of a fuel anode and oxidant cathode separated by an ion-conducting electrolyte. The fuel is oxidized to produce carbon dioxide and water. When the electrodes are connected through a load by an external metallic circuit, such as a simple motor, an electric current is transported through the external circuit. Fuel cells are valued because of their high level of energy efficiency, their low environmental impact, and their wide range of possible applications. They can power an entire city or an elec-

tric car, and can operate efficiently on both fossil fuels and renewable energy sources.

Fuel cells generally derive oxygen from the air and hydrogen from a fossil fuel. Although many different kinds of fuel can be used with fuel cell technology, a fuel cell plant usually uses clean-burning hydrogen gases such as natural gas or methane, while for cars and buses a liquid fuel such as methanol may be more practical. The hydrocarbon fuel is heated and mixed with steam to produce hydrogen-rich gas. Another environmental advantage of fuel cells is the relative ease of recovering waste water. Because the combination of hydrogen and oxygen produces water, no external sources of water are necessary. Fuel cell plants not only burn cleaner than conventional plants, but also tend to operate relatively quietly.

Fuel cell construction is similar to that of a dry-cell battery. The major difference between the two is that the reactants, air and gas, are stored outside of the fuel cell structure and are fed to reaction areas only when electric power generation is needed. Fuel cells do not run down or need to be replaced as normal batteries do.

The six or seven types of fuel cell fit into two basic categories, low- and high-temperature models. The low-temperature models already in limited use operate at temperatures below 400° F; models operating at 900° F or higher are still in the research and development stage. Higher-temperature cells are more efficient and may be cheaper when they are available commercially some time in the mid-1990s.

When used to produce both electricity and heat, a fuel cell can operate at 80% efficiency. Little waste heat is produced because the conversion of energy to heat is direct. Because of their small size, fuel cells can be placed close to the final destination of the electric power, decreasing energy loss during transfer. The output can be easily regulated, making them useful for peak-load facilities and practical for use in conjunction with renewable but variable energy sources such as solar or wind power. The absence of moving parts and the modular construction of a fuel cell plant in stacks of individual cells make them easy to maintain, expand, repair, and replace.

FUEL ECONOMY STANDARDS

In many countries, governments establish standards for fuel efficiency in motor vehicles based on the consumption of fuel per mile or kilometer travelled, and on the amount of waste produced by the vehicle. Fuel economy standards in the United States became law in 1975 under the Motor Vehicle Information and Cost Savings Act. The standards have evolved into a complex set of statutes and regulations that works to mandate fuel efficiency for domestic and imported cars based on average fuel economy for the fleet produced by the manufacturer in a given year. Therefore, they are known as Corporate Average Fuel Economy (CAFE) standards. As problems such as global warming and smog come to the forefront, fuel efficiency standards are alternately lauded and vilified as either a promising solution or an ineffective and unnecessary restriction.

Congress enacted the current fuel economy standards in 1975 and provided for average fuel efficiency for passenger cars to increase to 27.5 miles per gallon (mpg) in 1985. Since then, Congress has approved no change in the statutory level, allowing for consumer choice among vehicles, while still achieving the goal of overall efficiency. The statute calculates fleet efficiency for that car model, adding together the results for all of the manufacturer's models, and then dividing the total number of vehicles produced by that sum.

The law is somewhat flexible, allowing the Department of Transportation to adjust the standards to the "maximum feasible" level. After petitions by manufacturers, the Secretary of Transportation reduced the requirements in 1986 to 26.0, the statutory floor, and raised them to 26.5 in 1989. The Secretary of Transportation determines maximum feasibility by considering technological and economic feasibility, effects on fuel economy of other government automotive standards, and energy conservation requirements. Manufacturers may earn credits by exceeding the standards for one year and carrying over the excess to the next or by producing alternative fuel or energy vehicles.

If CAFE standards are raised, manufacturers may meet the statutory goal by building more fuel-efficient engines or lighter vehicles, or balance these and other considerations. Critics of the mechanism, most notably the Motor Vehicle Manufacturers Association, argue that a more market-based approach, such as a gasoline or carbon tax, would produce more efficient results. Manufacturers also contend that reductions in vehicle size and weight to improve fuel economy would reduce vehicle safety. Advocates of increases in CAFE standards counter that market-based methods lose effectiveness as fuel efficiency increases and that they provide manufacturers with only indirect incentives to improve efficiency. Safety, they say, actually improved during the first 15 years of CAFE standards, due more to sound design than to heavy vehicles.

FUELWOOD

Fuelwood is wood that ranges in size from kindling size to larger pieces of mature trees and is used to sup-

ply heat or power. It is one example of a biofuel, a fuel source made of biological matter. About 40% of the world's population, or approximately 1.5 billion people, the majority of them in developing countries, use wood as their primary fuel for heating and cooking. In some places wood is also increasingly being used by industry for power generation.

Although nearly 29% of the global land area is covered with forests, this natural resource is rapidly being depleted due to increasing demands for lumber, pulpwood, fuelwood, and food. Both the commercial and subsistence benefits derived from the world's forest ecosystems place forested areas, especially those in the tropics, at risk.

One of the primary causes of deforestation is the gathering of fuelwood. As forests bordering communities are consumed through subsistence activities, the people in these communities need to travel farther to obtain fuelwood. The first wood used is fallen wood. As fallen wood becomes scarce, however, branches are cut off, trees are felled, and even seedlings, bushes, and herbaceous plants are uprooted. In other communities, the scarcity of fuelwood may not directly result from fuelwood consumption but may be due to factors such as seasonal availability. One example of this is in Nepal, where the existing stock of fuelwood may be depleted during the monsoon season. Areas where demand for fuelwood has already exceeded the supply of trees include the arid and semiarid regions of Africa, heavily populated areas of Southeast Asia, and the mountainous regions of Latin America. Supplies of fuelwood in these areas are not likely to increase greatly. A positive effect of these fuelwood shortages, however, may be that a more careful management of fuelwood resources will emerge.

Attempts to ease the scarcity of fuelwood include reforestation efforts, such as replanting the original or a faster-growing species in a forest or introducing a commercial timber industry, with the harvested timber designated to be sold as fuelwood. Local reforestation efforts yield some positive results, and cultivating fast-growing tree plantations can alleviate the pressure to cut natural forests while increasing the supply of fuel. The commercial markets, however, often lead to the overharvesting and exploitation of forest resources. Also, the timber crops intended for sale as fuelwood often carry a higher cash value when they are used for a purpose other than fuel. For this reason, animal dung and crop residues such as rice straw or maize stalks and leaves are used as secondary fuel. In some places these may replace the use of wood as fuel.

FULLER, R. BUCKMINSTER

(1895–1983), American architect and inventor. Richard Buckminster Fuller was an early champion of the principle of *synergy* and a pioneer in the exploration of whole systems and their relevance to understanding and addressing the problems of Earth. Decades before others, he drew pictures of a "One Town World," and developed a distortion-free world map to illustrate our "one-world island in a one-world ocean." He coined the phrase "Spaceship Earth" early in the 1950s before NASA photographed the planet. Between 1927 and the 1960s, as part of his ongoing research, he made inventories of the world's resources, trends, and needs, concluding that humanity has the resources and technological capacity to support 100% of the human population, long before others even asked the question. He extolled the importance of concepts like recycling, years before they achieved fashion. He designed and prototyped "artifacts," i.e. inventions, such as mass-produced shelter, developed to address specific global problems. He was the author of 28 books, and recipient of dozens of awards, including the Presidential Medal of Freedom and 47 honorary doctorate degrees in arts, sciences, engineering, and humanities.

Such a man defies categories. Fuller was called a "genius," a "crank," the "first poet of technology," a "Renaissance man," the "Leonardo Da Vinci of our times" (by Marshall McLuhan), and a "scientific idealist whose innovations proceed not just from technical dexterity but from an organic vision of life." *The New York Times* obituary called him "a preacher as much as an architect, a town crier as much as a scientist. . . . He was among the century's first minds to see that every aspect of the physical environment was connected to every other." He was one of the first environmentalists to feel that humanity's success was dependent upon using technological tools in harmony with nature—as he said, "nature is nothing but the most extraordinary technology."

For "Bucky," as he was often called, *ecology* meant the whole system, Universe—large and small, visible and invisible, physical and metaphysical. As such he felt that in order to be effective as a problem-solver one must always "start with the whole." He defined his own work in various ways, disciplining himself to be "an effective explorer" in the principles of what he described as *comprehensive anticipatory design science.*

It makes sense that Fuller came from an old New England family of high-minded visionaries, including Margaret Fuller—friend of Emerson and Thoreau—a major force in the transcendentalist movement. Born in 1895, he lived through a time of great technological advances and social optimism. In his 88 years he saw humans first learn to fly, then go to the moon. He often said that his lifelong experience as a sailor shaped his work significantly, suggesting that to survive at sea, one must understand and cooperate with nature's principles.

With this as a grand strategy Fuller developed in the 1930s and 1940s a series of *livingry* artifacts that were his attempts to turn the world away from "weaponry" (i.e. war-based economy) by re-tooling war industries to life-supporting products. All of these inventions used his trademark *Dymaxion*, coined from *dynamic* and *maximum*, and embodied his engineering philosophy of "doing more with less." In 1927 he designed and modeled his first Dymaxion House. Called a "Machine-for-Living," the house was based on tension and triangulation. Designed to weigh 6,000 pounds, the house could be mass-produced, and would sell for 25 cents per pound. In 1933 he followed with his Dymaxion Car, a three-wheeled, rear-steered, fuel-efficient auto that carried eight people comfortably. In 1938, while working with the Phelps Dodge Company, he developed the Dymaxion Bathroom, an elegant aluminum prototype developed as a mass-producible, high-efficiency unit using almost no water. In 1946 he developed at Beech Aircraft in Wichita, Kansas, the first actual prototype of his Dymaxion House concept. He wanted to find ways to re-tool post–World War II aircraft companies to build housing, solving the housing shortage while providing permanent employment, "because there is no basic difference between the fabricating of aluminum parts for the Dymaxion House and for B-29's."

Fuller also developed a number of large-scale solutions to global problems that were never implemented. They included a floating city for Baltimore Harbor, developed for the Johnson administration; a domed city for citizens of East St. Louis; a global energy grid that would allow international sharing of electrical power; a "world game" computerized data base for electronically displaying an inventory of resource trends and needs worldwide; and plans such as offering tax credits to businesses for recapturing sulfur pollutants from smokestacks and recycling the collected materials.

In 1954 Fuller patented his most famous invention, the geodesic dome, which the U.S. Marines called the first "major improvement in mobile military shelter in the last 2,600 years." *Time* magazine called the dome "a massive mid-century breakthrough." His domes are in use around the world; some estimates suggest there are over three hundred thousand. They can be built easily and effectively in aluminum, plastic, wood, cardboard, and bamboo. As *Time* also said, "it uses less structural material to cover more space than any other building ever devised," combining the structural virtues of the triangle and the sphere. Whatever their size, geodesic domes are incredibly strong and energy efficient. Early prototype domes withstood 180-mile-per-hour winds on Mount Washington, New Hampshire, and protected workers on or near both of the Earth's poles. Others provide shelter for football fields, train stations, and people in places like Africa. American exhibits in various international Expos have been housed in domes, to the delight of the citizens in Istanbul, Bangkok, Lima, New Delhi, Tokyo, and Moscow. The magnificent dome that housed the U.S. Pavilion in the 1967 Montreal Expo established Fuller's reputation.

Fuller's development of the geodesic dome derived from his larger exploration of the principles of nature's design and geometry, comprehensively described in his *Synergetics*, two volumes of over fourteen hundred pages. Since his death in 1983, scientists have discovered that the geodesic dome defines, among other things, the structure of viruses and the human eye. Also it is now known that a carbon molecule called C-60, which looks like a soccer ball, is the same dome shape that Fuller patented in June 1954. It may be the most abundant element in the universe. How fitting that Professors Smalley and Kroto, the chemists who discovered the molecule, chose to call it *buckminsterfullerene*—now known in scientific circles as the *buckyball*.

JANET BROWN

For Further Reading: Buckminster Fuller, *Cosmography* (1992); Buckminster Fuller, *Critical Path* (1981); Buckminster Fuller, *Operating Manual for Spaceship Earth* (1963).

FUNGI

Fungi are simple plantlike organisms that lack chlorophyll. They obtain their nutrition from living on or in other organisms (parasitically), with other organisms (symbiotically), or by breaking down dead organic materials (saprophytically). Most fungi grow as well in darkness as in light. They include mushrooms, molds, and yeasts. They can exist as single cells or filaments (hyphae) that make up a mass (mycelium), and they reproduce asexually by spores.

Fungi have an especially important role in the decomposition of wood. Mushrooms feed on both living and dead trees but predominantly on old, dying, or dead ones. Their conspicuous fruiting bodies, such as puffballs, toadstools, or shelves, are attached to the hidden mycelium that has penetrated throughout the wood. Along with many other organisms, fungi are both waste reducers and nutrient recyclers. In the soil, for example, they make nitrogen more readily available through the process of decomposition. In addition, they are the main humus producers: there may be as many as 400,000 fungi in one gram of rich agricultural soil.

Both plants and animals depend on fungi for specific interactions. Soil fungi are the principal food of earthworms, soil mites, and insects. Symbiotic relationships exist both between fungi and other plants and between fungi and specific animals. For example, there are so-called mycorrhizal associations between soil fungi that get essential nutrients from the roots of forest trees and orchids, and in turn enable the roots to take minerals from the soil. More than 80% of all plant species with roots have mycorrhizae, and mycorrhizal fungi facilitate the acquisition of scarce nutrients, such as phosphate. Similarly, lichens represent a unique partnership between fungi and algae, with the alga providing food by photosynthesis and the mycelium of the mold holding water and protecting the alga.

Other mutually protective relationships involve termites with fungi that digest wood; leaf-cutting ants that provide food for fungi in their underground gardens and in turn feed on the growing mycelium; and wood-boring beetles with fungi that soften the tough wood fibers in the cavity where both live. In order to make sure that the unique partnership endures, the female wood-boring beetle even smears each egg she lays with the spores from that fungus.

Fungi are both harmful and helpful to human activities. Molds, mildews, rusts, blights, and smuts spoil food and cause plant diseases, resulting in the devastation of hundreds of millions of dollars worth of crops. Molds grow on bread, fruit, and vegetables; rusts on white pines and wheat; smuts on corn, oats, and onions. In 1970, 17% of the anticipated U.S. corn crop was lost to the southern corn-leaf blight. The high occurrence of liver cancer in tropical regions is related to a fungus that affects both grain and root crops stored in humid conditions. Chestnut blight and Dutch elm disease have reduced both American tree species to near extinction. Both athlete's foot and ringworm are fungal infections of the skin. On the beneficial side, Camembert and Roquefort cheeses are ripened by molds. The antibiotic penicillin is derived from penicillium molds. Through fermentation, yeasts are responsible for making bread, beer, and wine. Selected mushrooms, puffballs, and truffles provide foods favored by many people.

FUNGICIDE

A fungicide is any substance that kills fungi or inhibits the growth of spores. Fungicides are included in the more general category of pesticides, which also include insecticides and herbicides. The more common fungicides are captan, chlorothalonil, dicloran (DCNA), folpet, and quintozene (PCNB). These chemicals are useful in combating fungi and their spores on a variety of fruits and vegetables, including apples, peaches, plums, grapes, cherries, almonds, melons, onions, potatoes, and an assortment of small grains. It is estimated that approximately 35% of all crops worldwide are lost due to pests such as fungi and other microorganisms.

The use of fungicides has become widespread throughout the world as a method of inhibiting pests and fungi. Particularly in the U.S., fungicides have been used in response to consumer demand for unbruised, fresh-looking fruit products. The increased use of fungicides and pesticides also has occurred throughout Asia, particularly in China during the past twenty years. However, the use of fungicides has gradually diminished as medical institutions and other public health organizations have heightened public concern over the health effects of using such chemicals in food production.

Public concern does not stop with public health issues relating to direct consumption of foods treated with fungicides. There is also concern over the run-off of fungicides into underground and aboveground water sources, affecting drinking water in many regions as well as stimulating other negative environmental effects. The concern is warranted, since fungicides are known or suspected to cause cancer, mutagenesis, and various reproductive and prenatal problems.

Internationally the use of pesticides, including fungicides, in agricultural production has become a problem of which people are increasingly aware. Although agricultural yield is admittedly greater during the first few years of fungicide use, this increase in yield decreases again as the land and crops adjust to the use of chemicals. Eventually more chemicals are needed to produce the same yield.

Yield differences between countries can be attributed to the intensity of agricultural activity and discrepancies in the quality of agricultural inputs. For example, it is useful to compare Nigeria and Mexico, two countries with similar total hectares of cropland per capita. Mexico's relatively intensive use of pesticides (27,630 metric tons in 1982–1984 vs. 4,000 in Nigeria) and agricultural equipment (165,333 average number of tractors in 1987–89 vs. 11,033 in Nigeria) yields a lower level of agricultural production per capita than Nigeria (95 in 1988–90 vs. 113).

FUSION

Since the 1950s, a large and productive research effort has been made to explore the potentially tremendous amount of energy produced by an atomic process called fusion. Fusion occurs when two atomic nuclei are forced together to form a single nucleus of greater mass, but with a small reduction in the total mass. This reduction appears as energy. This type of powerful reaction occurs in the sun, and it is the energy source for the thermonuclear hydrogen bomb. Unlike fission, in which the nucleus of large atoms is split in half to release energy, the fusion reaction combines two nuclei of deuterium (a form of hydrogen with about twice its ordinary mass) to make an atom of helium. The energy released from this process can be used to drive an ordinary steam turbine. Deuterium can be derived from ordinary water with relative ease, making the fusion energy source endlessly available. One gram of deuterium can yield as much energy as ten tons of coal.

The energy yield from fusion is considerably greater than that from fission, but requires significantly more control. Fusion reactions reaching temperatures in millions of degrees Celsius have been achieved with large, complex, and costly experimental devices. The two major approaches to controlling the fusion process are magnetic confinement and inertial confinement. The latter uses lasers to compress a frozen fusion pellet (made of tritium—an even heavier form of hydrogen—and deuterium) to tremendous density in order to create a fusion reaction. Magnetic confinement, by far the most common and most heavily funded method, incorporates deuterium atoms (often with tritium), which can be compressed with the use of powerful magnets. The most common of these reactors is the Russian "tokamak," which has dominated confinement research as the most successful design by reaching a "steady-state" production level (in which as much energy is produced as is used to create and contain a reaction). Future tokamak reactors should be capable of self-heating and of producing much more energy than is required for start-up procedures.

The production of energy by fusion is relatively benign in its impact on the environment. Like fission, fusion has the advantage of emitting no hydrocarbons. However, the relative seriousness of a nuclear accident is greatly reduced with fusion. Fusion reactors operate without any possibility of runaway reactions. The radioactive materials produced with fusion are less volatile than fission products and generally much less likely to escape, except for gaseous tritium, which is a radioactive component requiring containment. However, eventual disposal of radioactive material will be necessary after extensive use of certain reactor parts, such as the inner wall and core elements.

Fusion systems as presently conceived are likely to be technologically and economically prohibitive. When these systems eventually become competitive, their use will also depend on public acceptance of risk, as with any method of energy production involving radioactive materials.

GAIA HYPOTHESIS

The expression "Spaceship Earth" to designate our planet has led to some faulty thinking about environmental problems. The word *spaceship* calls to mind a mechanical structure carrying a limited supply of fuel for a defined trip and with no possibility of significant change in design. The Earth, in contrast, has many attributes of a living, evolving organism. It constantly converts solar energy into innumerable organic products and increases in biological complexity as it travels through space. James Lovelock in *Gaia: A New Look at Life on Earth* put forward the idea that the surface of the Earth behaves as a highly integrated organism capable of controlling its own composition and its environment. He used the name of the Greek Earth goddess Gaia to symbolize this complex biological behavior.

It is not the first time that the Earth has been regarded as a living organism. In a book entitled *The Land Problem*, Otis T. Mason, one of the early American environmentalists, wrote in 1892, "Whatever our theory of its origin, the earth may be discussed as a living, thinking being. . . . We are in the presence of something . . . that has come to be what it is, has grown, that can be sick and recover." The Gaia hypothesis can best be formulated in Lovelock's own words: "The physical and chemical condition of the surface of the Earth, of the atmosphere and of the oceans has been and is actively made fit and comfortable by the presence of life itself. This is in contrast to the conventional wisdom which held that life adapted to the planetary conditions as it and they evolved their separate ways."

The Gaia concept has its origin in the fact that the chemical composition of our atmosphere is profoundly different from what it would be if it were determined only by lifeless physico-chemical phenomena. For example, it would produce an atmosphere consisting of approximately 98% carbon dioxide with very little if any nitrogen and oxygen, whereas the corresponding figures for our atmosphere are approximately 0.03%, 79%, and 21%. Lovelock presents numerous other examples of such profound departures from chemical equilibrium and postulates the existence of a global force that is able to bring about and keep fairly constant a highly improbable distribution of molecules.

He believes that this hypothetical global force consists in the countless forms of life which create and maintain disequilibrium situations through cybernetic systems.

The fitness of an organism to its environment is, of course, an essential condition of its biological success and even survival. All living things seem to be endowed with a multiplicity of mechanisms that enable them to achieve fitness by undergoing adaptive changes in response to those of the environment: genetic changes that occur during the evolution of the species, and physiologic and anatomic changes that occur during the existence of each particular organism. In higher species and especially in humankind, these biological mechanisms are supplemented by adaptive social forces. Early in the present century, however, the Harvard physiologist L. J. Henderson pointed out that there was more to fitness than the adaptive potentialities of living things. As he claimed in his famous essay, "The Fitness of the Environment," the achievement of fitness depends not only on adaptive mechanisms but also on the fortunate fact that the terrestrial environment has certain physico-chemical attributes which happen to be just right for life. But this theory now seems essentially meaningless for the following reasons.

Many present forms of life would certainly be annihilated if the surface of the Earth were very different from what it is now. They could not adapt rapidly enough if the salinity, the acidity, the relative proportions, gases, minerals, and organic substances — or any of the other physico-chemical characteristics of the Earth — were to deviate far from their present values for any length of time. In other words, the present environment does indeed exhibit fitness for the present forms of life but — and this is what Henderson had overlooked — it would have been unsuited to those of the distant past. An atmosphere with 21% oxygen, for example, would have been extremely toxic for the earliest forms of life. During the past 3.5 billion years, the global environment has progressively changed, probably as a consequence of the activities of living things, and living things also have undergone corresponding changes through a feedback process. The Gaia system postulated by Lovelock thus appears to be a result of co-evolution. In Lovelock's words: "The air we breathe can be thought of as like the fur of a cat and the shell of a snail, not living but made by living cells so as to protect them against an unfavorable environment." During the 1970s, NASA considered the possibility of making the planet Mars suitable for human life which included: providing water, oxygen, moderate temperatures, and protection from ultraviolet radiation. The conclusion was that Mars could be made habitable only through the progressive introduction of living species capable of creating, over an immense period of time, more and more complex ecosystems similar to the ones that have evolved on Earth during more than 3 billion years. This analysis has helped us recognize the profound and innumerable changes that life had to bring about on the surface of primitive Earth to create, for present living things, the fitness of the terrestrial environment that L. J. Henderson had taken as the normal state of affairs.

According to the Gaia hypothesis, the Earth's biosphere, atmosphere, oceans, and soil constitute a feedback or cybernetic system that seeks an optimal physical and chemical environment for life. At any given time, this system results in a relative constancy both in the composition of the environment and in the characteristics of living things. Lovelock refers to this equilibrium situation as "homeostasis," a word invented by the Harvard physiologist Walter B. Cannon to denote the remarkable state of constancy in which living things can maintain themselves despite changes in their environment.

The word *homeostasis*, however, does not do full justice to the Gaia concept, which implies in addition that living things have profoundly transformed the surface of the Earth while themselves undergoing continuous changes, in a co-evolutionary process. Practically all the examples that Lovelock discusses refer, in fact, to creative evolution rather than to homeostatic reaction. For example, the accumulation of oxygen in the air which became significant two billion years ago (a result of biological photosynthetic activities) probably destroyed many forms of life for which this gas was poisonous, but species emerged that were capable of living in the presence of oxygen and of using it for the production of energy. In Lovelock's words, "Ingenuity triumphed and the danger was overcome, not in the human way by restoring the old order, but in the flexible Gaian way by adapting to change and converting a murderous intruder into a powerful friend." In this case, as in most other environmental changes, the Gaian way was not an automatic homeostatic reaction but a creative co-evolutionary response.

It seems worth considering that the Gaian control results in global homeostasis only over a period of time that is short on the evolutionary scale. One figure will suffice to illustrate the magnitude of the terrestrial changes that are continuously caused by life. In their aggregate, all the green plants now fix approximately 840 trillion kilowatt hours of solar energy per year in the form of biomass. This is more than 10 times the amount of energy that all of humankind uses annually, even with its most extravagant technologies. Who can doubt that this continuous turn-over of organic matter and energy must have modified and goes on changing the surface of the Earth. Lovelock predicts that the process of change may pick up speed and complexity as a result of human interventions, and he quotes René Dubos in stating that, on a local level, profound co-evolutionary changes have occurred in certain terrestrial environments and in their biological systems during historical times.

In the last chapter of his book, Lovelock explores the relevance of the Gaian hypothesis to the effects of human interventions into nature. He agrees with René Dubos that environmentalists often aim at wrong targets because the resiliency of the Earth, considered as an organism, probably makes ecosystems more resistant to pollution than commonly believed.

The Gaia hypothesis has not only homeostatic aspects but also creative aspects. This is emphasized by Lovelock's statement that the Gaia concept is an alternative to the "depressing picture of our planet as a demented spaceship, forever traveling driverless and purposeless, around an inner circle of the sun."

RENÉ DUBOS
(EDITED BY RUTH A. EBLEN)

For Further Reading: J. E. Lovelock, *GAIA: A New Look at Life on Earth* (1987); J. E. Lovelock, *Healing GAIA: Practical Medicine for the Planet* (1991); Norman Myers, ed., *GAIA: An Atlas of Planet Management* (1993).
See also Atmosphere; Ecological Stability; Remote Sensing; Space Environment.

GARBAGE

See Waste, Municipal Solid.

GENETIC ENGINEERING

See Biotechnology, Agricultural; Biotechnology, Medical.

GENETICS

Genetics is the science of inheritance. Geneticists seek to understand the mechanisms that underlie two of the most curious phenomena of the living world: the preservation of species over many generations, and the slow but steady appearance of new species. Together these have covered the Earth with an astonishing diversity of living organisms. Aspects of quantitative genetics are related to molecular biology, biotechnology, evolution, and biodiversity.

Quantitative genetics began with the controlled mating of pairs of inbred strains of garden peas that were identical except for single, discrete differences in appearance. From studies in the mid-1800s of the hybrid offspring of inbred strains of pea plants differing by single traits (yellow peas in the pods vs. green ones, for instance), the Moravian monk Gregor Mendel made two discoveries, and from these he was able to construct the first accurate model of inheritance. First, he found that the appearance, or phenotype, of the hybrid offspring was never an "average" or "mixture" of two parental phenotypes, but was always identical to the phenotype of one parent: this was named the dominant phenotype. The parental phenotype that disappeared from view in the hybrid plants was named recessive. Second, he found that although the recessive phenotype was invisible in the hybrid, hybrid plants mated to themselves gave rise to offspring of recessive as well as dominant phenotypes, and did so at a constant ratio of one recessive out of four offspring.

Building on these and other experimental results, Mendel's model for inheritance envisioned the presence in hybrid plants—and by extension in all living creatures—of particles of information encoding the various phenotypes that together make up the form and function of an organism, entities subsequently given the name *genes*. Pollen, sperm, and egg cells—all called gametes—are formed in the bodies of sexually mature individuals by a series of cell divisions called meiosis. In the Mendelian scheme, a complete set of genes is donated to each individual organism by each of its parents as the female parent's egg cell fuses with the male parent's pollen or sperm cell at the moment of fertilization, assuring that each adult organism contains two copies of each gene from the moment of conception. Recessive phenotypes result from the inheritance of two copies of the recessive version, or allele, of a gene. A dominant phenotype may result from two possible genotypes, or pairs of alleles: two copies of the dominant allele or one dominant and one recessive allele would each generate the dominant phenotype.

For this model to work, gametes must contain only one copy—not two—of each gene, and equal numbers of gametes formed in the body of a hybrid organism must contain each parental allele. Then, after the random union of pollen and egg at fertilization, the recessive genotype would occur, at random, one in four times, which was the observed ratio.

Mendel's model, which depended on the existence of stable genes, was first reported in the 1860s but not recognized by European science until the beginning of the 20th century. Recessive alleles, no less than dominant ones, are stable through time. They need never be "diluted" away, even when they are not visible in the phenotype; a hybrid genotype can be inherited in any number of generations, each generation looking dominant but carrying, undiluted, the capacity to give rise to a recessive offspring. Mendel's insights remain valid today, and they provide clear evidence that earlier—but still prevalent—notions of inheritance operating through the "dilution" of "blood" or some other fluid, cannot be right.

Insightful as it was, the Mendelian paradigm of inheritance was limited in two ways by the laboratory systems in which it was first studied. First, students of population genetics soon showed that most phenotypes seen in nature are not the results of differences in single genes. The ones that are most interesting in humans, height or skin coloration for instance, result from the activities of dozens or hundreds of different genes, each of which might have two or more different possible functional alleles. Even for traits that are the result of a small number of genes, in nature it is usually the case that their gene(s) encode a range of responses to their environments, not an absolute phenotype: inadequate food over a lifetime will keep a person short regardless of whether he or she carries alleles for shorter or taller stature.

Second, Morgan and other geneticists working with simpler organisms like the fruit fly showed that genes are not marbles in a bag, autonomously passed from generation to generation by random assortment in gametes. Rather, they occupy specific positions in the chromosomes that fill the nucleus of each living cell. Thousands of genes on the same chromosome are linked to each other, but this linkage may be interrupted by crossing over during meiosis, so patterns of inheritance are not always as predictable in advance as the one-in-four ratio observed by Mendel.

Molecular Genetics

Genes and genotypes remained mathematical abstractions linked in an obscure way to chromosomes, and the way in which genes and chromosomes were copied at each generation remained a complete mystery, until the mid-1940s, when Oswald Avery at the Rockefeller Institute for Medical Research showed that a bacterial gene was made of the long polymer DNA. Within a decade, in 1953, James Watson and Francis Crick had shown that DNA was a double-stranded helix of indefinite but enormous length, a twisted ladder made of four different kinds of rungs called base-pairs, held in a fixed sequence by two stretches of repeating sugar-phosphate groups. They also showed that while the outer contour of a DNA molecule was an information-free repeat of these simple sugar-phosphate groups, the four different sorts of inside "rungs" or base-pairs could appear in any order, and that the genetic information of a DNA molecule was contained within its sequence of base-pairs. Finally, and all within a single, 900-word paper in the scientific journal *Nature*, they presented a model for DNA's ability to replicate itself, showing that DNA met this requirement of the genetic material. When one molecule of DNA unzips by the separation of base-pairs, each single-strand of DNA can serve as a template for the synthesis of a new, double-stranded DNA whose base-pair sequence will be identical to the parent model. As they put it:

> The sequence of bases on a single chain does not appear to be restricted in any way. However, if only specific pairs of bases can be formed, it follows that if the sequence of bases on one chain is given, then the sequence on the other chain is automatically determined. . . . It has not escaped our notice that the specific pairing we have postulated immediately suggests a copying mechanism for the genetic material.

Within another decade scientists had worked out the RNA-based mechanisms and the genetic code used by cells to read the genetic information in a stretch of DNA and translate it into sets of proteins, the molecular machines that create the phenotypes we see. From this work the four great generalities of molecular biology arose: first, that the mechanisms of DNA replication and translation into protein were common to the cells of all living things; second, that a gene could be redefined as a stretch of DNA that encoded a protein; third, that each gene was surrounded by regulatory DNA sequences that did not encode protein but rather determined whether or not the protein encoded by that gene would be synthesized in a given cell; and fourth, that changes in DNA base sequence—even as small a change as one base-pair—were the cause of the sudden appearance of stable differences in genotype and phenotype called mutations.

Evolutionary Genetics

The genetic explanation for the universality of basic molecular mechanisms is the simplest: given a common ancestry of all living things, these mechanisms would have been active in the last common ancestor of us all. Then, providing that mutation did not cause their loss, they would be retained in all living things to this day. The notion of evolution, that all living species share a common ancestor, was not first generated by molecular biology. The living world today is separated into tens of millions of populations that are reproductively isolated and genetically distinct, populations we call species. In 1859, the same decade as

Mendel reported on his work with peas, Charles Darwin published *The Origin of Species*, his observations on the universality of variation from individual to individual within a species, and his model of natural selection. In this model, new species may arise from isolated groups of variant individuals within an old species, but only if they have a greater than average capacity to generate offspring and carry them to an age where these can, in turn, inherit variant skills or strengths—called adaptations—and pass them on to yet another generation.

Natural selection requires an environment of limited resources, and a great deal of time so that many generations can be subject to its force. Darwin, working in the absence of information about genes and mutations, and with a very much foreshortened time scale for natural selection (millions of years, versus the 3 billion years life actually has been present on Earth), could not explain how adaptive variations arose, and posited that stress and deprivation may have generated them. There is no conclusive evidence for the notion, also championed by Jean Lamarck, that adaptive mutations can be induced to order by manipulation of an organism's environment. Rather, mutations—even those induced by chemical mutagens—occur at random in any sequence of DNA, and many more variant phenotypes are generated at random than can possibly be adaptive for survival in any set of circumstances.

Darwin also presumed that natural selection would be uniform and gradual in its effects. Today it appears that in more than one instance many species arose rather rapidly (in only a few million years) soon after mass extinction of a fair percentage of all large living creatures. Evidence is increasing that these rare events resulted from impacts of large asteroids. The most recent of these mass extinctions occurred at the Tertiary-Cretaceous boundary about 65 million years ago, killing dinosaurs and other large animals and plants, and allowing the rapid radiation of new species of birds and mammals, including the shrew-like animals that were our earliest mammalian ancestors.

Darwin's original model lacked any mechanism for either the generation of variation or its persistence through inheritance. Mendel's genes provided the latter, and the discovery of mutations—sudden changes in genotype that expressed themselves as new, stably inherited phenotypes—provided the former, at the turn of the century. Mutations occurring at random sometimes generate diversity of phenotype, but always create new alleles. New alleles that are adaptive in a given environment, and those that are merely neutral and not detrimental, will survive over time. As first shown by Ernst Mayr, natural selection can be understood to work by changing the frequency of survival of alleles in small, isolated populations within a species. Members of a species may become the genetically isolated precursors of a new species either through physical isolation, or by the inheritance of physiological or behavior barriers to successful mating with other members of their species, or by a mixture of both isolating mechanisms.

While the detailed mechanism of species formation remains incompletely understood, the current vast diversity of living forms is clearly the result of millions of years of differential reproduction of genetic change in isolated groups of interbreeding individuals. Darwin's major insight, that all living creatures are related to each other by descent from common ancestors, remains as valid today as it was 140 years—or 3 billion years—ago.

Genetics and Biotechnology

Just as the discovery of a common genetic code and a common genetic molecule—DNA—shows that all living species are linked to each other through long-lost common ancestors, so the tools developed to manipulate DNA allow genes of all living species to be shuffled into combinations never seen in any ancestral organism. Novel assortments of DNA sequences from different species are called recombinant DNA molecules, and the assembly of them is called gene splicing. The tools and techniques scientists use to carry out gene splicing collectively amount to a molecular form of "word processing." Here the "text" is DNA, a long stretch of information in the form of a unique sequence of base-pairs. The molecular versions of "cut" and "paste" keys are restriction enzymes and hybridization. Restriction enzymes find specific sequences and cut the DNA at precise points, generating short DNA fragments with identical, defined ends. Hybridization of single-stranded DNA re-zips two strands into one double-stranded DNA when the two strands can form an uninterrupted run of proper base-pairs.

The molecular word processor's "software" comes from bacteria and the viruses that live within them. These microbes were discovered relatively recently, not long before Mendel's discovery of stable recessive alleles. Only a century ago the notion that a disease might be the result of a microscopic organism growing inside of people had the same novelty that the notion of a disease caused by a latent allele has today. But current biotechnology uses microbes to carry and reproduce novel genes and the proteins they encode. Molecular engineers "search," "cut," and "paste" DNA sequences by using hybridization to find a gene, then splicing it into a bacterial chromosome for propagation as a recombinant DNA. To learn how genes work, scientists can get them into a fertilized egg cell so that they will be replicated into every cell of an embryo, and their effects over the organism's lifetime can be

observed. Recombinant genes that reach this state are called transgenes.

The polymerase chain reaction, or PCR, "copies" long pieces of DNA that lie between two short, known sequences, generating billions of identical copies of the entire sequence between the two known ends. The root of a hair or the tiniest drop of blood has many copies of an animal's or a person's entire genome and much more to tell—with hybridization and PCR—than a fingerprint. PCR has also democratized the genome. Once a pair of short sequences bracketing a gene are published, anyone can synthesize them at relatively low cost and then use PCR to pull as many copies of the entire gene out of a chromosome, as might be needed for any purpose.

Genetics and Biodiversity

Biodiversity is greatest in the tropical forests, and these are disappearing at an alarming rate. Only a small fraction of all species are known to humankind; humans are killing off species faster than we can even identify them. In a rearguard action, DNA technology can be used to save the genomes of animals, plants, and microorganisms even as the living creatures are killed off. Gene banks, as they are called, are but a pale shadow of the real species, but they are better than nothing, and they will give scientists of the future a chance to reconstruct at least a partial record of the disappearing species. Museums of science and natural history are now being rediscovered as libraries of ancient species, as DNA from preserved specimens and even from fossils is amplified and recovered by PCR. The oldest DNA recovered from a fossil so far is from a plant that lived about 250 million years ago; the oldest animal DNA is from a termite that lived about 30 million years ago. It is sadly ironic that the technology of DNA research has used the universality of this molecule to generate powerful new ways of examining the genes of all species, just as other human technologies are destroying species at an ever-increasing rate. This is a problem the next generation of geneticists must confront.

ROBERT E. POLLACK

For Further Reading: Horace F. Judson, *The Eighth Day of Creation: The Makers of the Revolution in Biology* (1979); Robert E. Pollack, *Signs of Life: The Language and Meanings of DNA* (1994); James D. Watson, *The Double Helix: A Personal Account of the Discovery of the Structure of DNA* (1968).

See also Biodiversity; Biotechnology, Agricultural; Biotechnology, Environmental; Biotechnology, Medical; Darwin, Charles; DNA; Evolution; Mendel, Gregor.

GEOGRAPHY

Geography is the study of the spatial distribution of natural processes and human activities on the surface of the globe and the resulting character of places and their interrelationships. Too simply put, as history is the study of human activities in time, geography is the study of human activities in space. Geography has traditionally been concerned with four themes: (1) the relationship of human activities to the land or natural environment; (2) the character of regions or places; (3) the concept of location, or the structure of spatial relations involving location, networks and their interactions including movements of people, goods, and services; and (4) mapping or cartography, the display of spatial relationships on maps.

Over time, people and societies modify the natural environment and the landscape created by their predecessors. For thousands of years, much of the globe has been altered primarily by agriculture for crops and grazing. The vast grain belts of the United States, Russia, and Canada, as well as fields and pastures of Europe, were created from natural grasslands and forests. Since the European Industrial Revolution of the 19th century, urban areas have transformed the landscape, and the networks of transportation and communication that connect modern urban centers have further transformed world geography. Landscape, then, reflects both nature and humankind.

Geographers study the way nature and humans influence changes in the landscape and the way these changes take place. Geography involves a synthesis of information from the natural and social sciences and from the humanities, a synthesis reflecting the many influences which alter the landscape over time. The key concepts of geography are process, transformation, landscape, and spatial distribution.

The traditional theme of geography, the relationship of humans and land, has become the modern theme of the relationship between humans and the environment. The distribution of human activities and technology on the globe today still reflects the possibilities or potential provided by the natural environment. Changes in perception of resources by human beings, and changes in the technology they possess, result in the creation of new values in the environment. Human beings are modifying the land, hydrosphere, biosphere, and atmosphere over the entire globe. The study of the way in which this happens and the resulting degradation or enhancement of the available resources, are now part of modern geography.

New mapping technology has made great contributions to the social and natural sciences and to fields such as environment, planning, and economic development. Geographic Information Systems (GIS) is a computer system that codes mapping information in digital form, that is, as points on a grid covering a given area. This permits the delineation of many kinds of spatial data at any scale and allows comparisons of the spatial distributions of any number of variables. For example, in cities one can look at the spatial distribution and interactions among such factors as the quality of housing, transportation patterns, water supply, wastewater, ethnicity of the population, incidence of crime, and other variables important in explaining the dynamics of social and physical processes in cities. Similarly, in the natural scene, factors that control the distribution and habitats of plants and animals, such as geology, soil, climate, vegetation, population numbers and species, topography, and human activity, can all be located in space. Until the development of this technology, displaying these phenomena, characteristics, or processes was difficult, and, more importantly, the ability to compare and relate spatial distributions was exceedingly limited. Geographic information systems have become major bases for planning and analysis of the infinite variety of natural and human activities distributed across the globe.

Geography is an old discipline that described where places were and what they looked like, and geographers sought to explain the origin and character of these places and the reasons for their distribution. Geography today continues this tradition but it is also concerned with the processes that transform the landscape. These in turn relate to pressing problems of the environment and the habitability of the globe by humans and other organisms.

M. GORDON WOLMAN

For Further Reading: Ronald F. Abler, Melvin G. Marcus, and Judy M. Olson, eds., *Geography's Inner Worlds: Pervasive Themes in Contemporary American Geography* (1992); Peter Haggett, *Geography: A Modern Synthesis*, 2nd ed. (1975); R. J. Johnston, *A Question of Place: Exploring the Practice of Human Geography* (1991).

GEOLOGY, MARINE

Marine geology is the broadly inclusive study of the materials composing the ocean basins, the processes that have formed and modified those basins, the waters of the oceans, and the forces that constantly move the waters covering 70% of the earth's surface. Although the ocean basins were long believed to be static and relatively featureless depressions, marine geologic studies have now shown that they are constantly changing and contain a vast variety of prominent features.

Marine geology can be conveniently subdivided into the study of the physiography of the basins, the marine rocks and sediments, the chemistry of sea water, and the processes that form waves, tides, and currents.

Physiography: The ocean basins contain several distinct provinces. Continental shelves are the very gently sloping extensions of the coastal plains that lie at the margins of continents and serve as the sites of deposition for most sediments eroded from the continents. Continental slopes at the outer edge of the shelves are the true edges of the continents; here the slopes steepen and the water depth increases from about 200 meters to approximately 3,000 meters. Deep ocean floor lies at depths greater than 3,000 meters and may reach 6,000 meters. Mid-ocean ridges are vast mountain chains that extend for thousands of kilometers across several of the major ocean basins. They remain relatively unknown and poorly understood, because all but the highest peaks, visible as islands like the Canaries and Iceland, are covered by water. These ridges have formed at the margins of diverging crustal plates, where volcanism constantly injects new lava and thrusts up the edges of the plates. The oceanic trenches are deep canyonlike structures at crustal plate margins, where the ocean floor is being drawn down under continents as if it were on a conveyer belt. They vary in depth but reach as much as 13,000 meters near the Marianas Islands in the southwestern Pacific.

Marine rocks and sediments: The formation of new rocks at the mid-ocean ridges and the consumption of older rocks, and any sediments on them, at the deep sea trenches is relatively rapid, so that the oldest rocks in the ocean basins are only about 200 million years old, even though oceans have probably existed since very early (4 billion years ago) in earth history. Many kinds of rocks and sediments that form in the oceans are similar to those on the continents, but the deposition of large masses of evaporite minerals, such as salt, and organic-rich sediments that form petroleum only occur in the marine environment. Furthermore, the volcanic rocks of the mid-ocean ridges are distinct from those on continents, and some types of sediments composed primarily of marine organisms can only form in the oceans.

Chemistry of Seawater: Seawater contains an average of 3.5% dissolved salts but can exhibit significant variability.

Waves, tides, and currents: The constant motions of the oceans are described in detail under Oceanography.

JAMES CRAIG

See also Seawater.

GLACIERS

Glaciers are masses of flowing ice. They form from compacted snow in areas where snow accumulation exceeds melting and sublimation.

Glaciers have probably existed somewhere on Earth for hundreds of million of years. Climatic cooling initiated the current episode of fairly extensive glaciation only about three million years ago. At the same time, human evolution began its final thrust toward becoming *Homo sapiens*. The fluctuations in climate, and to a lesser extent even the existence of the glaciers themselves, profoundly influenced the course of human evolution over this critical time span.

Continental ice sheets have advanced and retreated many times during the past three million years. Major advances have occurred every 100,000 years or so, and have lasted about 90,000 years. The latest advance, called the Wisconsinan, ended about 10,000 years ago, and it was during its latter phases, when ice pushed as far south as central Illinois and Ohio (Figure 1), that humans began to occupy the New World. How long it will be before the next such invasion of ice to mid latitudes is presently impossible to say, but when that time comes the advance will be just as inexorable as the previous ones and life on Earth will have to adapt.

To appreciate the role that glaciers play in human affairs, it is necessary to understand the principles of glacier mass balance.

Mass

Winter precipitation in high latitudes and at high altitudes is commonly in the form of snow. If summer temperatures in these areas are above freezing, some or all of the snow melts. Where temperatures are not warm enough to melt all of the winter snow, the snow pack becomes thicker, and deeper layers are gradually compacted into ice. Such areas are the source areas, or **accumulation areas**, for glaciers.

Although ice is a crystalline solid, migration of imperfections in the crystals enables it to flow, rather like a viscous liquid. The flow rate is proportional to the fourth power of the ice thickness, so a 5% increase in thickness results, approximately, in a 20% increase in flow rate. The flow transports ice to lower elevations or, in the case of continental ice sheets, to more equatorial latitudes. Here temperatures are warmer and summer melt exceeds winter snowfall. Thus all of the snow and some of the underlying ice melt during the summer. The area in which this occurs is called the **ablation area**.

When the volume of ice that flows from the accumulation area to the ablation area during a year exceeds the volume melted during the summer, the glacier advances to still lower elevations or more

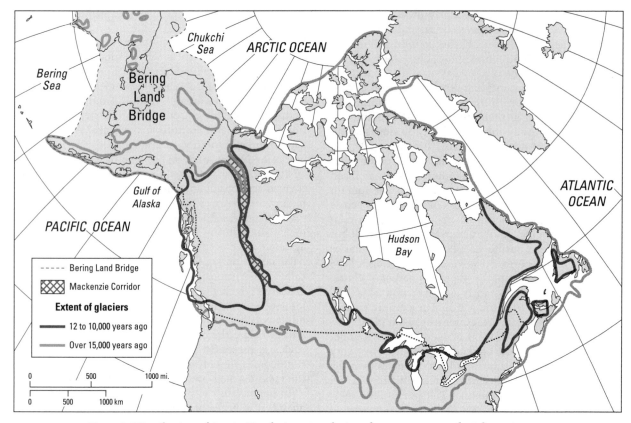

Figure 1. Distribution of ice in North America during the most recent glacial maximum.

temperate latitudes, and is said to have a positive mass balance.

Central to our understanding of the interactions between glaciers and humans are three characteristics of this system: (1) a change in temperature of only a few degrees can lead to significant change in mass balance; (2) over time, such changes in mass balance may result in appreciable changes in thickness, which have a multiplied effect on flow rates; and (3) the time required for a glacier to reach a new steady-state geometry after a change in mass balance ranges from decades to centuries, depending on its size and flow rate.

Glaciers and Humans

When the climate cools and glaciers begin to form, water is removed from the oceans and stored on land. Sea level thus falls. In many areas around the world, submerged shorelines, detected by sonic profiling, indicate that sea level was as much as 100 meters lower during the last phase of the Wisconsinan, say from 28,000 to 10,000 years ago.

Large areas of the Bering and Chukchi seas that separate Siberia from Alaska are less than 50 meters deep, so these areas were frequently dry land during the gla-

cial periods of the Pleistocene. This land, and in fact much of Alaska, was not covered by ice. This was probably because winter precipitation was low and northeastwardly-flowing oceanic currents made the summers comparatively warm. Thus tundra plants occupied the newly exposed land, and fauna adapted to arctic conditions followed. During the most recent glaciation, paleolithic humans followed game animals onto this tundra plain and eventually reached what we now know as Alaska.

The Bering Land Bridge, as this plain is called, was not so much a bridge as a vast lowland, comparable in extent to the central plains of the U.S. Humans did not intentionally cross it as one would a bridge. Rather they diffused into it as sea level fell, some perhaps along the coast following the rich sea life on which they depended, and others inland moving with the larger game animals—mammoth, bison, elk, and caribou. Some researchers believe the former gave rise to the present Eskimo and Aleut populations, while the latter were the progenitors of the American Indian.

Between 22,000 and 13,000 years ago these people would have found their way blocked by ice in central and eastern Alaska, as the continental ice sheet expanding out of Hudson Bay merged with alpine glaciers in the Rocky Mountain cordillera, closing what is known as the Mackenzie Corridor. The primitive character of some stone implements found in the southwestern U.S. has led certain scholars to suggest that some people managed to migrate southward before the Mackenzie Corridor was closed, perhaps as early as between 70,000 and 40,000 years ago when the "bridge" probably became open for the first time in the Wisconsinan. However, the period from about 30,000 to 22,000 years ago is perhaps more likely. The lack of archeological materials in the Southwest yielding radiocarbon dates earlier than about 12,000 years ago, however, has led others to suggest a rapid migration southward through this corridor as soon as it became open upon retreat of the ice.

This brief discussion provides only a glimpse of the impact of Pleistocene climates and their attendant continental ice sheets on humans. Surely, human evolution was strongly influenced in many other ways by the thousands of years that people lived within a few hundred kilometers of an active continental ice margin.

Following the retreat of the last continental ice sheets, there was a period of warmer wetter climate until about 6,000 years ago. During this time agriculture rapidly emerged, supplanting the primitive hunter-gatherer societies. Climatic swings impacted humans through their effect on crops.

There were several significant climatic oscillations over the next millennia, including a warm phase from about 850 to 1500 A.D., during much of which the Vikings maintained settlements in Greenland. The climatic cooling that increased the concentration of sea ice off these settlements, hindering resupply and thus contributing to their demise, culminated in what is now known as the Little Ice Age. This was a period, lasting from about 1500 to 1900 A.D. during which glaciers advanced in many parts of the world.

It is from the Little Ice Age that we have some of the earliest historical records of glacier fluctuations. Between 1680 and 1750, farmers around Jostedalsbreen, an ice cap with numerous outlet glaciers in western Norway, frequently sought relief from taxes as advancing ice buried their farms and destroyed farm buildings. Records tell of buildings being crushed into small pieces. When the ice retreated, the ground exposed was either bouldery rubble or bare bedrock, not suitable for farming.

Advancing Little Ice Age glaciers also threatened human settlements further south in Europe. The Mer de Glace in France presented a particular problem, and several times during the 17th century exorcists were sent out to deal with the situation. They appear to have been successful, as glaciers here were then near their Little Ice Age maximum and beginning to retreat.

The advances of glaciers during the Little Ice Age were slow and measured. Humans had time to contemplate the situation and adjust to it, though not necessarily willingly. Such is not always the case.

In glaciated alpine areas, small valleys are typically well above the bottoms of larger ones to which they are tributary. This results in what is known as a hanging valley. In some situations, the tributary valleys are short and steep, and their accumulation areas may, even today, support sizable glaciers. Ice advancing out of such tributaries over the steep sides of the main valley may become unstable and release large blocks that cascade down into the main valley. Thus were farms and farm buildings destroyed on numerous occasions in Norway and France during the onset of the Little Ice Age. Likewise, during retreat of such a glacier from the main valley, a section of the end of the glacier may break off. Allalin Glacier in Switzerland failed in this manner in 1965, killing 88 construction workers. Having thus become shorter than its stable equilibrium length, such a glacier may gradually extend itself again until the process is repeated.

What appear to be only snow fields high on mountainsides may actually be underlain by slowly flowing ice. These glaciers too are subject to catastrophic failure. Such an ice avalanche descended the steep slopes of the Weisshorn on December 27, 1819. After becoming combined with snow on Bis Glacier, the total volume of the resulting avalanche was 13 million cubic meters. The village of Randa was not as large then, and the avalanche missed most of it, but two people were killed. An ice avalanche from Mt. Huascarán in Peru on January 10, 1962, was similar in origin; it killed about 4,000 people.

When the potential for such catastrophes is recognized, the speed of the ice can be monitored, using laser distance-measuring systems, and warnings can be given if appropriate. A gradual acceleration of the ice near the glacier terminus suggests that failure may be imminent. Evacuation of any people in its path is then in order.

It is not unusual for a glacier in one valley to extend across the mouth of another that is occupied by a river. Lakes form behind such ice dams. However, because ice floats, the rising water does not overflow but rather forces its way out beneath the dam. When water begins to escape in this manner, the energy that it dissipates melts ice and enlarges the conduit. As the lake level falls, however, the pressure in the water decreases, and ice begins to squeeze into the developing conduit, tending to close it. If the lake is small and the dam is thick, the water pressure may drop so fast that the conduit closes faster than it is enlarged by melting. However, if the lake level falls slowly, the conduit can grow rapidly in size, resulting in a flood of catastrophic proportions.

Such floods are probably the single greatest hazard to humans resulting from processes related to glaciers. They occur without warning, often far from human habitations, and the flood waves can be high enough to cause significant damage tens or hundreds of kilometers from the source. The largest such floods known occurred in the Pleistocene when a 2,000-cubic-kilometer lake in Montana, called Lake Missoula, broke its glacier dam repeatedly. The resulting floods inundated 40,000 square kilometers in western Washington, carving huge valleys and forming immense gravel bars.

Glaciers and Industry

Glaciers and ice sheets act as reservoirs of fresh water. These reservoirs, however, store and release water in response to climate rather than to human requirements. This becomes significant in planning projects utilizing water from glacierized watersheds. Obviously the water yield from such watersheds is seasonal and planners recognize this. Less obvious is the fact that if the glaciers have a negative mass balance, the yield will exceed the precipitation input, and conversely. Eventually the lengths of the glaciers adjust to the prevailing climate and the mass budget comes into balance. Thus, if planners use stream records from a period when glaciers are retreating, the water yield will decrease as the glaciers approach their equilibrium length. Initiation of an advance will decrease the water yield still further. Planners of long-term hydrological projects in glacierized basins thus must consider the mass balance of the glaciers and potential future climate trends.

Glaciers and Mining Activities

Mineral deposits sometimes occur near glaciers, and exploitation of such deposits may require some understanding of the glacier. In the late 1960s, a copper mine was planned in British Columbia, Canada, for example, and the entrance to the mine was to be near the snout of Berendon Glacier. An advance of the glacier could cover the entrance to the mine. Evaluating this possibility involved assumptions about possible climate change and an analysis of the time scale upon which the glacier might respond to these changes.

Another study involved an iron deposit that lies partly beneath the margin of the Greenland Ice Sheet. The plan was for an open pit mine, part of which would be in ice. By drilling to study temperatures and deformation rates in the ice, it was found that where the ice was more than 100 meters thick, liquid water was present at the base of the glacier. Rapid removal of the marginal ice could lead to release of this water, and

would lead to high ice flow rates that would complicate the mining program.

There are also valuable mineral deposits in Antarctica, and humans may eventually feel the need to exploit these. New technology will then be necessary to deal with the glaciological consequences. However, an international protocol currently (1993) in the ratification process bans mineral exploration and exploitation in Antarctica for 50 years, so at least in that part of the world the time is not imminent.

Glaciers and Hydroelectric Power

Hydroelectric power is a major source of energy in small countries with ample precipitation and high relief. The amount of energy that can be recovered from water depends on the volume of water and on the elevation difference between the generators and the inlet to the conduits leading to the generators. In order to collect water at the highest possible elevations in mountainous regions of Europe, tunnels have been excavated in the rock beneath glaciers and then up to the glacier sole to tap subglacial streams.

A number of engineering problems are associated with such installations. The most important is how to find the water. Glaciers are commonly several hundred meters wide while subglacial streams are only a few meters across and, owing to the pressure distribution in the ice, the streams are not necessarily in the deepest part of the valley. Furthermore, because flowing water melts ice, the course of the stream may change during the melt season, and long-term changes in glacier geometry can also result in changes in the water flow paths. Beneath Glacier D'Argentière in France, the water was "lost" for three years, and eventually found 340 meters from its original location and 25 meters above the valley bottom.

Future Prospects

Emissions of carbon dioxide, methane, and chlorofluorocarbon compounds may have initiated a global warming that may be measurable by early in the next century. Computer models suggest that such warming could be roughly three times as large in polar regions as in temperate latitudes. As has been the case for millennia, the response of the glaciers will be slow but inexorable; they will retreat and sea level will rise. If humans have to deal with this relentless invasion of the sea, the challenges glaciers have posed heretofore will pale by comparison.

ROGER LEB. HOOKE

For Further Reading: John R. Gribbin and Mary Gribbin, *Children of the Ice* (1990); Jean M. Grove, *The Little Ice Age* (1988); David M. Hopkins, *The Bering Land Bridge* (1967). *See also* Antarctica; Anthropology; Climate Change; Greenhouse Effect; Hydropower; Lakes and Ponds; Mining Industry; Natural Disasters; Rivers and Streams; Water; Water Cycle.

GLOBAL CLIMATE CHANGE

See Climate Change.

GLOBAL WARMING

See Climate Change; Greenhouse Effect.

GRAINS

Cereal grains have been the essential basis of civilization ever since the first Neolithic farmers planted special selections of the wild grasses growing in the hills and valleys of western Asia. Wheat has been the preeminent food grain of the Western world, rice of the East, and maize (corn) of the New World. Barley is fourth in importance and quantity, globally. Other grains are important in specialized growing regions: rye and oats in poorer soils of northern regions, and sorghum and millets in hot, dry climates. In the industrialized countries, maize and sorghum (and oats and rye) now are used primarily as feed grains for cattle, pigs, and chickens, and barley is important for brewing as well as for animal feed. Maize, on the other hand, has become an important food grain in many parts of the developing world.

The grain crops require plentiful supplies of nutrients, especially nitrogen, for bountiful yields. Through the millennia, nitrogen has been supplied in limited amounts from animal manures or through crop rotation with legumes such as beans or clovers. But since about 1950, low-priced synthetic nitrogen fertilizers have become the chief source of nitrogen fertilizer for grain production. Starting in developed countries, the practice of applying synthetic nitrogen fertilizer is now widely used in many developing countries, as well. It has given greatly increased yields of grains, especially when farmers use specially bred new varieties that prosper with abundant fertilizer and water.

But the practice of providing plentiful supplies of nitrogen for grain production has introduced problems. When crops are overfertilized with nitrogen fertilizers, either synthetic or natural, excess nitrates can find their way into underground water supplies, and may reach levels that are dangerous to human health or disruptive to the environment. For this reason, efforts are underway to establish levels and methods of fertilizer application that will be profitable for farmers and safe for the environment.

Since the 1920s plant breeders have transformed grain production by development of high-yield varieties, able to yield several times as much as varieties in use at the turn of the century. Development of hybrid varieties of maize, sorghum, and millet has increased yield potentials of these crops even more. Improved varieties make better use of abundant fertilizer and water supplies, are more tolerant of weather problems such as heat and drought, can better tolerate low soil fertility, and are more resistant to insect and disease problems. For all of these reasons, they yield more and have broader adaptation than the old varieties they replaced.

As with fertilizers, use of new, professionally bred varieties started in the industrialized countries, but the practice now has spread to many parts of the developing world, as well. Farmers have adopted the new varieties and plant them exclusively in all of the developed world and in most major production areas of the developing countries. The term "Green Revolution" was coined to describe the widespread and rapid adoption of new, high-yield varieties of wheat and rice in developing countries, starting in the late 1960s.

This change has brought problems, also. Concentration of farm production on only the most productive new varieties has greatly reduced the genetic diversity of crops in the field. This brings danger of uniform susceptibility to a new kind of disease or insect, or to an unexpected pervasive weather problem. Although plant breeders can breed new varieties to counter new problems, it takes time to develop the new varieties and to multiply their seed for farm use. And of utmost importance, breeders must be able to test a broad range of old varieties for new kinds of pest or weather tolerance, to be inserted into the next generation of new high-performance varieties.

Emphasis now is placed on two measures to counter the dangers from lack of genetic diversity in grains: (1) Farmers are encouraged to plant several varieties of each crop rather than just one or two. Breeders are encouraged to broaden the range of genetic differences among new varieties they produce. (2) Collection and long-term storage of farm varieties from all parts of the world is seen as an essential service performed by national governments and intergovernmental organizations. Farmers who still grow the diverse varieties handed down from their ancestors will be encouraged to continue growing them, as essential parts of the world's genetic heritage.

DONALD N. DUVICK

For Further Reading: Congress of the United States, Office of Technology Assessment, *Beneath the Bottom Line: Agricultural Approaches to Reduce Agrichemical Contamination of Groundwater* (1990); Jack R. Harlan, *Crops & Man* (1992); H. A. Wallace and W. L. Brown, *Corn and Its Early Fathers* (1988).

GRASSLANDS

Grasslands are terrestrial biomes where grasses are the dominant vegetation. Earth's natural grasslands covered about 30% of the terrestrial land surface. In addition to grasses, the vegetation includes scattered trees, if only along water courses and escarpments, and a variety of herbaceous dicotyledonous plants that, although conspicuous, are not common. Particularly evident dicots include composites, such as the asters and sunflowers, and legumes, with symbiotic nitrogen-fixing bacteria in their roots that enrich the grassland soil.

The major areas of Earth's grasslands included the central North American prairies (a French word derived from the Latin for meadow), the central to southern South American pampas (a Spanish word derived from the Incan word for plain), the central to eastern Eurasian steppe (from an Old Russian word for lowland), African and north to central South American tree-studded grasslands called savannas (from a Native American word transliterated into Spanish), and the varied grasslands of Australia. As all the local names for grasslands suggest, the topography is flat, ranging from the almost table-like pampas to the rolling hills of the prairies. Ecological niches of grasslands are occupied by similarly adapted species: for example, the ostrich in Africa, the rhea in South America, and the emu in Australia.

Grasslands occupy the interiors of the continents in the North Temperate Zone but extend across the continents in the Southern Hemisphere. Their occurrence is principally determined by the spatial and temporal distribution of precipitation. Precipitation ranges from 10 to 40 inches (250 to 1000 centimeters), with grasslands giving way to desert in drier areas and forest in wetter locations. The principal distinguishing characteristic of grassland climates, however, is the sporadic and unpredictable nature of precipitation. It varies from month to month during the year, usually with a pronounced dry season, and from year to year. Since grasses are shallow-rooted plants usually incapable of drawing upon groundwater deep below the surface, they depend on rainfall during the growing season. The dust bowl of the 1930s on the North American Great Plains, for example, was economically devastating because it followed an extended period of higher than normal precipitation that allowed agriculture to expand into areas that could not normally support it. Such periods of extended drought are characteristic.

Gradients of grass stature accompany the gradients of rainfall. Tall grasses, often over 6 feet (2 meters) in height, occur in the wettest regions adjacent to forests. Early settlers in Illinois and Iowa were amazed at grasses so tall that a horse and rider might disappear from view when passing through them. Short grasses,

no taller than 20 inches (50 centimeters), occur in the driest regions adjacent to deserts. And medium height grasses occur over intermediate ranges of precipitation.

Fire is also a characteristic feature of grasslands. Unlike most plants, grasses lack an abscission layer—a band of weakened tissue at the leaf base that results in leaf fall during dormant periods. Instead, grass leaves just dry in place at the end of summer or during drought. This produces huge quantities of tinder that ignite readily. The grasses are resistant to fires but tree saplings and shrubs often succumb, eliminating them from the vegetation. Among the most frightening features of the American prairies to early settlers were the extensive and violent grass fires each year.

Geologically, grasslands are a comparatively recent phenomenon. According to fossil records, grasses originated about 65 million years ago (Paleocene epoch) in South America and Africa when those continents were still joined. They appeared first in glades in a forested landscape. By 55 million years ago (Eocene epoch), separation of the continents and orogeny (mountain-building) were leading to increasing aridity in continental interiors. Developing mountain chains in the Americas created rain shadows (regions of reduced rainfall) by draining the moisture of Pacific storms as they increased in altitude over the mountains, cooled to the dewpoint, and then produced rain or snow. This led to an expansion of grasslands that has continued, with fluctuations associated with glaciation, into the present.

Hoofed herbivores—the ungulates—coevolved with the developing grasslands and were found everywhere except Australia. The first ones appeared in the Eocene. There are two groups: odd-toed ungulates, such as horses and rhinoceroses, with the middle nails greatly enlarged; and even-toed ungulates, such as cattle and antelopes, with the two central nails expanded into hooves. Early in the development of grasslands, the odd-toed ungulates were the most common. Beginning about 35 million years ago (Oligocene epoch), the even-toed ungulates underwent spectacular adaptive radiation and came to dominate the fauna of grasslands into historical times. Accompanying them in all grasslands were numerous large predators—in North America, wolves and cougars. In Australia, however, placental herbivores never arrived and the large grassland mammals were kangaroos.

The first humans originated in the savanna grasslands of eastern and southern Africa where they combined gathering of roots and vegetables with the hunting of evolving large ungulates. This primordial small-group, hunting-gathering society extended into historical times and still exists as tiny isolated remnants, principally in Africa and Australia. The invention of weapons with shafts, such as arrows and spears, and the utilization of fire led to very effective hunting.

Fires were used to drive animals into topographic traps or attract them to the fresh green regrowth, where they were killed. Some scientists believe this combination of fire-driving and weapons with handles contributed to major extinctions of ungulates beginning about 1.8 million years ago (Pleistocene epoch).

Agricultural societies first developed in the grasslands of the Anatolian and Iranian plateaus in the Middle East. Hunter-gatherers there learned that returning to previous campsites would provide them with a ready source of grains during the rainy season from the plants derived from germinated waste seeds of wild grasses harvested previously. Those "garbage grains" became our modern wheat, rye, and other cereals. The cereals were important as energy sources and cultivation soon followed. Cultivation of legumes, such as lentils, provided much-needed vegetable protein in the diet. But the agricultural diet was still unbalanced due to lack of certain amino acids abundant only in animal products. Therefore, the first towns typically developed along the routes of migratory ungulates where hunting could supplement the agricultural diet. Domestication of goats for milking closed the dietary loop and full-fledged agriculture was underway by about 12,000 years ago.

Agriculture based on irrigation was soon devised in the desert basins of the major river systems of the Middle East in the Fertile Crescent of the Tigris-Euphrates Valley, and in the very fertile Nile Valley. Principal crops, however, were derived from grassland plants, and livestock also were grassland species. Cattle were domesticated for milk and meat, horses for transport.

As the grassland-derived agriculture rapidly radiated into Asia and Europe, hunter-gatherer societies began to disappear. On the Eurasian steppes a new type of society developed, a pastoral system based on herding cows, sheep, and goats. The mobility conferred by riding horses allowed this social system to exploit the sporadic rainfall of the steppes. By 900 B.C., the Scythians had learned that this mobility had substantial advantages in another activity—warfare. Over the succeeding centuries, conquering nomads from the steppes ruled many of the agricultural societies.

Mercantilism, trading, and a money-based economy arose out of the development of agriculture. Wherever agricultural societies matured, domesticated grasses provided the energy base of the community. Particularly important were rice in Asia, maize (corn) in the Americas, and the small grains in Europe and the Middle East. Those grasses retain their importance in modern human society.

American colonists transplanted European agricultural practices into the New World while adapting some of the crops and technologies of Native Americans, particularly the cultivation of maize, squash, and other vegetable crops. From the beginning of permanent settlements in North America in the early 17th century, the occupied habitat was originally forested and then cleared gradually but extensively as the population expanded. In South America livestock raising became a major enterprise as cattle were introduced onto the pampas of Argentina and Uruguay.

The North American Great Plains were first seen by Europeans about 1530 when Cabeza de Vaca and three companions skirted its southern edges in Texas and learned of, but did not see, bison. The Coronado expedition of 1540–1542 was the first true reconnaissance of the Great Plains by Caucasians. He was searching for the fabled Seven Cities of Cibola, imaginary locations of immense wealth. He and thirty men from his army penetrated well into the grasslands, probably as far as Kansas, but found it markedly inhospitable, and highly unlikely to support the legendary cities of gold. His report to the king explicitly commented on the abundance of "cows," i.e. bison, and the shortage of wood and water.

So it remained for over two centuries. Regions below 20 inches (50 centimeters) of annual precipitation are commonly incapable of supporting non-irrigated agriculture. That rainfall level corresponds with the 98th meridian, passing from the eastern Dakotas through central Texas. Major Stephen H. Long of the U.S. Army was dispatched to survey the lands between the Mississippi River and the Rocky Mountains in 1820. He returned to describe a featureless, treeless expanse that he named the Great American Desert. Mountain men traversed the plains to the Rocky Mountains in search of the riches of beaver pelts, wagon trains crept westward toward the Pacific Coast—particularly after the discovery of gold in 1848 in California—but settlement remained unthinkable in the North American grasslands.

Grasslands east of the 98th meridian also were resistant to settlement. Although rainfall is higher than on the Great Plains, this prairie peninsula stuck like a thumb of tall grass into forests, extending from western Ohio through Illinois to Iowa. Both prairie fires and the heavy nature of rich grassland soils acted as deterrents to agriculture. Increasing settlement along the shores of Lake Michigan and the valleys of the major rivers gradually led to a diminution of fire, but the rickety plows transported from the Old World to the New were incapable of turning over the prairie sod.

Perfection of the all-steel plow in the 1830s was the first of four technological innovations in rapid succession that changed the relationship between humans and grasslands forever. That plow was instrumental in expanding agriculture from the lighter soils previously exploited into the high clay, high organic matter, and therefore sticky, heavy soil formed under grassland. The fertility of grassland soils was unmatched due to millennia of accumulation of nutrients in a climatic regime where production by plants outstripped decomposition by soil microbes. The steel plow opened that fertility to human utilization.

Still, lack of water and a means of fencing where trees would not grow acted as barriers to grassland settlement. Three additional technological innovations sounded the death knell for natural grasslands, not only in North America, but around the world. Well-drilling, rather than digging, was first used in 1832 to provide a well for Paris. The technology was rapidly transported to the U.S. In 1854 Daniel Halliday invented the self-regulating wind pump (commonly referred to by the general name of windmill) in Connecticut. He soon realized there was little market for it there, and moved to Illinois to establish a factory in the grassland region. The transcontinental railroads first combined drilled wells with wind pumps to tap the deep water tables beneath the plains. Cattle growers almost simultaneously realized the potential of the technology and implemented it. Thus, ranching was invented in a dry grassland climate, for the first time uncoupling the availability of water from rainfall.

There still was no effective means of confining livestock and the period of open range ranching lasted about two decades after the Civil War. Finally, an Illinois inventor, J. F. Glidden, devised barbed wire in 1873. This was the first practical means of confining livestock on vast expanses of open grassland. These American technologies were disseminated throughout the world with the practices of grassland ranching and farming, and remain today virtually as they were conceived in the mid-19th century.

Farmers soon followed the ranchers. Farming spread first through lowland areas that were well-watered, with small plots irrigated by wind pumps. Then, several decades of above average rainfall led to widespread cultivation of upland soils, principally for wheat. Thus was born the first of the Earth's modern grain belts, harkening back to the original grain belt of the Middle Eastern Fertile Crescent. Just as ranching was invented through innovative technology, so farming on the broad expanses of grassland led to increasing mechanization, fueling industrialization in the U.S. First the reaper, then the binder, then horse- or mule-drawn riding plows, gang plows, multiple-row cultivators, and other machinery followed.

The lack of stones, stumps, and tree sprouts in the fertile grassland soils made such technology useful. The acreage farmed, much larger than in the East, demanded it. The drought of the 1930s disrupted, but did not destroy, these developments.

Now, petroleum-powered pumps draw water from deep beneath the grasslands to irrigate cropland. Nebraska has the largest proportion of irrigated cropland of any state, using water pumped from deep aquifers to transform a grassland climate into an oasis of highly productive cropland.

The highly productive soils of grasslands, once water was brought to them, created a revolution in agriculture, just as the invention of ranching created a revolution in livestock husbandry. Agricultural acreage in the eastern U.S. reached a peak in the late 19th century and declined steadily thereafter, displaced in the marketplace by the bounty from grassland soils.

As was inevitable, both water and wind erosion depleted the nutritional stock of the soils, so new methods of conservation such as stubble preservation and terracing were developed. Nevertheless, much grain belt agriculture is now a procedure of adding nutrients as fertilizer, mining the nutrients with a crop, and repeating the process year after year. Depletion of the aquifers that underlie the Great Plains is a distinct hazard to the long-term viability of this practice.

Overgrazing also has been a chronic consequence of combining a reliable water source, independent of rainfall, with fences. As American ranching methods spread around the world, the detrimental consequences followed them. Degradation of the grasslands, erosion, and encroachment of shrubs and unpalatable plants became a worldwide phenomenon.

Today, natural grasslands are very rare, to the point of non-existence. There is nowhere that human influence has not modified them — except perhaps in some of Africa's largest and most remote game reserves. Alien plants, often weeds from the Eurasian steppes, are commonplace. So, also, the immense herds of native herbivores preceded the vegetation that supported them into near extinction throughout the world.

Few sizable remnants of the once overwhelming North American grasslands survived. The Flint Hills in Kansas is perhaps the largest expanse and Yellowstone National Park preserves areas of short, montane grassland. But many attempts are underway to restore the prairie. The arboretum at the University of Wisconsin in Madison was one of the first attempts at such reconstruction, and many are now underway on small parcels of land throughout the grassland region.

Rural areas of the North American Great Plains are being depopulated at an increasing rate in the latter half of the 20th century. Increasingly energy- and capital-intensive farming and ranching methods displaced the labor-intensive grassland economies of the late 19th and early 20th centuries. Through much of the region windmills, if present, are rusting hulks, replaced by gasoline- or electrically-powered pumps.

Fences remain in the ranching regions but are now gone from most farming areas where livestock are scarce and fences have been removed so every arable square foot can be farmed in continuous expanses of maize and soybeans. The Earth's vast grasslands of less than a century and a half ago have disappeared, to be seen only as tiny relics in the lifetime of present humans.

SAMUEL J. MCNAUGHTON

For Further Reading: Mary L. Grossman, Shelly Grossman, and John N. Hamlet, *Our Vanishing Wilderness* (1969); Time-Life Books, *Grasslands and Tundra* (1986); Walter Prescott Webb, *The Great Plains* (1931).
See also Agriculture; Aquifer; Biomes; Desertification; Energy, Wind; Erosion, Soil; Fire; Grains; Land Use; Ranching; Rangeland; Soil Conservation; Tundra.

GRASSROOTS ORGANIZATIONS AND COMMUNITY GROUPS

See Environmental Movement, Grassroots.

GREAT BARRIER REEF

The Great Barrier Reef is the world's largest biological structure, both now and throughout the geologic past. The reef extends for more than 1,250 miles (2,000 km) off the northeastern coast of Queensland, Australia, at 10 to 100 miles offshore. It actually comprises thousands of individual reefs that form a more or less continuous chain extending parallel to the Queensland coast and acting as a breakwater to the waves that come in from the Coral Sea. The major zones and positions of the various reefs are thought to have been established approximately one million years ago.

The reef complex is home to many species, including coral and their associated algae; lobster, crab, crayfish, shrimp, and other crustaceans; octopus and other mollusks; marine worms; sponges; anemones; fish; starfish; sea cucumbers; and other marine animals. The reef complex also includes coral sand islands (cays) and continental islands, some of which hold patches of mangrove forest and rain forest. The islands house a diverse array of bird, insect, and reptile species.

The waters surrounding the Great Barrier Reef show little seasonal temperature variation; mean temperatures range from 81.5° F at the northern end to about 75.2° F at the southern end. The oxygen content of the water is high and nutrients are generally well mixed. This combination allows a steady growth of phytoplankton (floating vegetative matter), which provide a constant supply of food to the zooplankton (floating animal forms), which in turn act as a major food source for the carnivorous corals.

For thousands of years Australian aboriginal groups lived on or near continental islands and obtained food from nearby reefs. Although Captain James Cook is credited with discovering the reef during his 1770 voyages around Australia, it is considered likely that other navigators (Portugese, Indonesian, Malaysian, Chinese, and possibly Egyptian) preceded him.

In 1922 the Great Barrier Reef Committee was formed to carry out scientific investigations of the reef. The committee conducted the Great Barrier Reef Expedition of 1928–1929, which contributed a great deal of information about the ecology and taxonomy of the many species present on the reef. During the 1970s attempts to initiate oil drilling and limestone mining on the reef sparked controversy and concern over the reef's preservation. Research stations on Heron, One Tree, Lizard, and Orpheus islands continue to conduct scientific investigations of the reef.

Major problems affecting some areas of the Great Barrier Reef include the overharvesting of many species of flora and fauna for food, tropical fish collections, and tourist curios; increased siltation as a result of poor agricultural practices and deforestation on the mainland; and massive coral die-offs due to population outbreaks of the crown-of-thorns starfish, a predator of coral species. Many scientists attribute the population outbreaks of the crown-of-thorns starfish to the removal by humans of key species that prey on the starfish. Other scientists contend that the outbreaks are part of natural fluctuations in the population cycles of the starfish. Regardless of their origin, the outbreaks, which began in 1961, have resulted in massive coral loss in areas where coral populations have been unable to renew themselves.

GREEN BELTS

The term *green belt* originally signified a relatively wide band of rural land or open space surrounding a town or city. In the United States, the term has come to mean, generally, any swath of open space separating or interrupting urban development. Land so designated is controlled through regulation or public or quasi-public ownership (such as the Nature Conservancy) to retain its natural character and provide a semblance of rural ambiance in urban or urbanizing areas.

British social reformer Ebenezer Howard advanced the green belt concept in 1898, in connection with the planning of "new towns" located outside the orbit of London, which even then was sprawling far into the countryside. Howard proposed "garden cities" as an antidote, each of which would be surrounded by an agricultural "country belt." It was British architect and planner Raymond Unwin, a new town designer and contemporary of Howard's, who actually coined the term *green belt*.

In Britain, Howard's concept took two forms: the green belts surrounding the new towns in rural Britain (the first of these was Letchworth, built in 1903); and, beginning in the 1930s, the application of the idea to London itself. Here, the green belt had to be on a heroic scale compared to the new towns, whose populations were rarely greater than 50,000. The first step was the London Green Belt Act, passed by Parliament in 1938, which gave recognition to publicly owned open spaces surrounding the city. A more elaborate plan was created in 1944 by Patrick Abercrombie, who proposed a belt five or more miles wide, consisting of both public open spaces and private holdings that would be regulated to preclude suburban development. At length, in 1955, this expanded plan was adopted and the green belt has been added to since. Upon its establishment, private landowners were offered compensation on a one-time basis for any loss of development value which they could effectively demonstrate. Today the London Green Belt appears to be a permanent fixture, though often attacked as socialistic by political conservatives and developers.

In the United States, the administration of Franklin Roosevelt tried to adapt Howard's new town concept as part of its resettlement program. Three such towns were built, with Greenbelt, Maryland, being the best known. Here a partial green belt does in fact separate the community, now a part of the Washington, D.C., metropolitan area, from its neighboring subdivisions. The green belt concept was also proposed in the plan for Radburn, New Jersey, a well-known privately financed new town built in the 1920s, but was not fully implemented. Possibly the only fully realized garden city–style green belt is in Boulder, Colorado, where a band of largely publicly purchased land now encircles this city of 83,000.

In the United States, where strict land-use controls are politically difficult to promulgate and the purchase of developable land too expensive for most jurisdictions, green belts on the British models of Letchworth and London are virtually unknown. Some postwar American new towns, such as Columbia, Maryland, and Reston, Virginia, have surrounding open space areas, but the American new-town movement appears to have foundered for lack of long-term capital. Moreover, modern American urban forms tend not to be circular, and thus are less containable by "belts" of open space. Without the influence of original medieval walled cities as in Europe, American urban areas have become linear and interconnected, growing outward along the routes of commerce: rivers, coasts, and, nowadays, interstate highways.

Accordingly, true green belts encircling major American cities are non-existent, although both Boston and San Francisco have citizen-led programs to encourage the formation of London-style green belts on a de facto basis. In Boston, a "Bay Circuit," proposed in 1929 by Benton MacKaye, a regional planner and father of the Appalachian Trail, planner Charles Eliot II, and others, has been struggling to come into being for well over half a century. The circuit is made up of strung-together parks, nature sanctuaries, and historic sites in a 100-mile-long open space corridor around the city. In San Francisco, open space proponents have advocated the establishment of a green belt since 1959. The Bay Area Greenbelt, a hoped-for 3.8-million-acre girdle of parks, watershed lands, farms, rural estates, and ranches was given a boost in public awareness in 1988 by the establishment of a multi-governmental project to build a "Ridge Trail" through the proposed green belt area in the highlands surrounding San Francisco Bay.

Given these exceptions, the concept of Howard, Unwin, and Abercrombie has not been successful in America.

CHARLES E. LITTLE

For Further Reading: Ebenezer Howard, *Garden Cities of Tomorrow,* (1898); reprinted 1965; Charles E. Little, *Greenways for America* (1990); Lewis Mumford, *The City in History* (1961).
See also Greenways.

GREEN CONSUMERISM AND MARKETING

The late 1980s and early 1990s have seen increasing public interest in environmental issues. Global issues such as acid rain, global climate change, and stratospheric ozone depletion; national news stories, such as the Exxon Valdez oil spill and the odyssey of the New York City garbage barge that could not unload; and countless local stories have increased public awareness and concern about environmental issues and the consequences of our individual actions on the environment. As a result, many consumers, and not just the most environmentally conscious, are seeking ways to lessen the environmental impacts of their personal buying decisions through the purchase and use of products and services perceived to be environmentally preferable — **green consumerism**.

Marketers, in turn, have responded to consumer demand in several ways: by promoting the environmental attributes of their products; by introducing new products; and by redesigning existing products — all components of **environmental marketing**. According to a recent report from the U.S. Environmental Protection Agency, the percentage of new grocery store products marketed in the U.S. with environmental claims on their packages or in their advertising increased from 5.9% in 1989 to 11.4% in the first half of 1992.

There have been a number of calls for federal-level action in the U.S. to ensure that environmental marketing can and does result in real environmental benefits. Manufacturers, environmental groups, and consumer protection groups have called for establishing national consensus definitions for environmental marketing terms as well as establishing guidelines for their use. In addition, several private groups in the U.S. and government agencies abroad are pursuing a different approach using seals of approval, **ecolabeling**, as a means of augmenting environmental marketing efforts and promoting an environmental agenda.

Environmental marketing differs from other forms of advertising in three important ways. First, unlike price, quality, and other features, the environmental impacts of a product are not always apparent and may not affect the purchaser directly. Thus environmental marketing claims are often more abstract and offer consumers the opportunity to act on their environmental concerns. Second, unlike most advertised product attributes, environmental claims may apply to the full product life cycle, from raw material extraction to ultimate product disposal, reuse, or recycling. Third, and most important, environmental marketing provides an incentive for manufacturers to achieve significant environmental improvements, such as toxics use reduction and recycling, by competing on the basis of minimizing environmental impacts of their products. Proponents of environmental marketing as a policy tool feel that environmental marketing guidance that incorporates sound scientific analysis and reflects the national environmental agenda would reward manufacturers for reducing the environmental burden of the manufacture, use, and disposal of their products. To be effective, however, marketing claims must be truthful, must be readily understood by consumers, must be linked with an education program, and must reflect the environmental agenda of policy makers, scientists, and other knowledgeable members of society.

The environmental marketing claims used to describe products and packaging range from vague, general terms, such as "Earth-friendly," to more specific claims, such as "made with 20% postconsumer recycled materials." The proliferation of poorly defined or ambiguous environmental terms has resulted in legal actions against some marketers for deceptive advertising and calls for federal definitions of environmental marketing terms.

On July 28, 1992, the Federal Trade Commission issued guidelines for the use of environmental marketing claims. These guides spell out what claims the FTC might consider to be false or misleading and give examples of appropriate claims. They issued the guides for several reasons. First, a large percentage of consumers appeared to be ill-informed and were skeptical about the meaning and implications of environmental marketing claims. Second, in the absence of federal definitions and guidelines, a number of states, cities, industry groups, and standards-setting organizations had developed their own definitions, some of which are legally binding within certain jurisdictions. As a consequence, marketers face a patchwork and sometimes costly marketplace where relabeling, legal actions, and negative publicity can create additional costs, can cause market share losses, and may deter some from making credible claims altogether.

Another vehicle to pass environmental information to consumers is the use of certification or seal-of-approval programs known as *ecolabels*. Typically, ecolabeling efforts are voluntary, third-party expert assessments of the relative environmental impacts of a product. By performing a thorough evaluation of a product, but awarding only a simple logo on packages, ecolabels offer consumers clear guidance based on expert information. Government-sponsored ecolabeling programs have been used successfully in Europe, Canada, and Japan to identify products or services that have the least environmental impact of similar products.

There are several important distinctions between ecolabeling and the use of environmental marketing claims. Ecolabels are, in general, third-party certifications of the environmental attributes of products and services with the goals of raising consumer and manufacturer awareness of the impacts of manufacturing, use, and disposal, and creating marketplace incentives for producers to develop more environmentally benign products and processes. Environmental marketing, on the other hand, simply involves unverified environmental claims made by marketers, so long as it does not constitute false advertising. In addition, environmental marketing claims usually address one or two attributes whereas ecolabels are awarded to products within a specific category based on a set of criteria (e.g., energy use, pollution, natural resource use) and encompass most of the product's life cycle, from manufacture to disposal. A further difference is that while environmental marketing at best follows national environmental policies, an officially sanctioned ecolabeling program can be used as a tool to identify and actively lead the marketplace toward those policy goals.

Regardless of how consumers receive information on the environmental burden of their purchases, without improved and sustained consumer confidence in environmental marketing, the opportunity to realize the benefits associated with environmentally oriented products may be lost.

ANDREW STOECKLE, JULIA WORMSER,
BENTHAM PAULOS, HERBERT HAN-PU WANG,
RICHARD P. WELLS

For Further Reading: U.S. Environmental Protection Agency, *Evaluation of Environmental Marketing Terms in the United States* (1993); John Elkington, Julia Hailes, and Joel Makower, *The Green Consumer* (1993); Carl Frankel (ed.), *Green Market Alert* (monthly newsletter).
See also Ecolabeling; Labeling.

GREENHOUSE EFFECT

The greenhouse effect is the result of certain atmospheric gases letting sunlight through to the Earth's surface while trapping energy radiated outward from the surface. This alters the balance of energy received from the sun and radiated from the Earth, resulting in a warming of the planet. The principal gases in the atmosphere with this property are water vapor, carbon dioxide, methane, chlorofluorocarbons (CFCs), hydrogenated chlorofluorocarbons (HCFCs), tropospheric ozone, and nitrous oxide. Without the naturally occurring greenhouse gases (mostly water vapor and carbon dioxide), the Earth's average temperature would be almost 33°C (59°F) colder. These greenhouse gases thus help make Earth a suitable place for life.

Greenhouse Warming

Atmospheric concentrations of carbon dioxide, methane, and CFCs are increasing. This suggests the possibility that the average temperature of the Earth will rise even further. The added increase in global average temperature is usually called greenhouse warming to distinguish it from the natural greenhouse effect. It is possible that further increases in atmospheric concentrations of these gases could raise the surface temperature enough to cause ecological damage and make human life more difficult.

The temperature rise accompanying a given increase in these gases depends on many interacting processes in the atmosphere, in the ocean, and on land. These processes are only partly understood and scientists do not agree about the warming that will occur due to continued emissions of greenhouse gases.

The Earth's Radiative Balance

Although a small amount of heat reaches the surface from the Earth's center, most of the warmth we experience comes from solar radiation. Over time, the Earth radiates the same amount of energy into space that it absorbs from the sun. If it radiated less than it received, it would slowly warm up; if it radiated more, it would cool down.

The Earth cooled during past ice ages, and warmed during interglacial periods. We are currently in such an interglacial period. The major cause of the ice ages is variations in the Earth's orbit around the sun that reduce the amount of sunlight reaching the Earth.

Climatic Change Due to Greenhouse Warming

Continued increases in emissions of greenhouse gases will eventually warm the planet. However, it is not known how much the Earth will warm as a result of a particular increase in atmospheric concentrations of greenhouse gases. Nor are the consequences of this warming for humans and for systems of plants and animals known with precision.

The basic priniciples are understood, however. About 25% of incoming solar radiation is reflected back into space by the atmosphere, 25% absorbed in the atmosphere, and 50% transmitted to the Earth's surface. The surface also reflects some of this light, so that 45% of the solar radiation striking the top of the atmosphere is absorbed by the ocean and land. Nearly 70% of the Earth's surface is ocean, so most of this energy is absorbed by water in the ocean. The heat absorbed by the atmosphere and the ocean provides the force that drives weather and storms. But some of the heat absorbed in the ocean is transmitted to water hundreds of meters beneath the surface rather than affecting weather immediately. Part of the energy absorbed in the ocean therefore does not change weather and climate until the entire ocean is warmed, many decades later. Thus the increased atmospheric concentration of greenhouse gases since the beginning of the industrial revolution has committed the Earth to some future rise in temperature. But it is very difficult to tell just how big this temperature change will be.

Determinants of Global Warming

The amount of climate warming depends on several things. First, and most important, is the amount of sunlight reaching the Earth. As mentioned above, this is the main driving force for ice ages. Second is the atmospheric concentration of greenhouse gases. Their concentrations depend, in turn, on the rates at which greenhouse gases are emitted, on chemical transformations in the atmosphere, and on the rates at which they decompose or are removed from the atmosphere. Third is the effectiveness of positive or negative feedback mechanisms that enhance or reduce warming, such as the extent of cloud cover or ice and snow. Finally is the effect of human actions on all of these factors except the amount of light from the sun.

Feedback Mechanisms

It is hard to know how much temperature will change, because the climate system is made up of many interacting processes. Some of these processes, called feedback mechanisms, either increase or decrease the temperature change compared to what it would otherwise be.

One major feedback mechanism involves clouds. As air warms, it can contain greater concentrations of water vapor. Increased water vapor usually results in formation of more clouds. So rising average temperature should be accompanied by more clouds. If the clouds are high convective clouds like thunderheads, their greenhouse properties will dominate and they will enhance warming. But if they are low thin clouds like those typically found over the ocean, their reflective properties will be more important and they will cool the planet. Satellite measurements suggest that the overall effect of clouds today is to cool the planet slightly. Unfortunately scientists do not yet know whether a rise in global average temperature will produce more high convective clouds or more low reflective clouds. So they cannot say with certainty whether clouds will enhance or diminish the warming caused by greenhouse gases.

Another feedback mechanism involves snow and ice. Global warming might be expected to increase the melting of snow and ice in the high latitudes. Since ice and snow are highly reflective, this would reduce the amount of light reflected from the surface and thus enhance warming. However, warming also increases the amount of moisture that can be absorbed in air, and is expected to increase precipitation. Increased precipitation in the form of snow might expand snowfields, at least during part of the year. Scientists are unsure whether warming will increase or decrease the extent and thickness of snow cover and ice fields.

Projecting Climatic Changes

Scientists use huge computer simulation models to project the temperature changes associated with greenhouse warming. These models are called general circulation models (GCMs) and include equations that describe relevant processes in the atmosphere, in the ocean, and on land. The models are so complex that they can be run only on the biggest supercomputers. A typical simulation imposes atmospheric concentrations of greenhouse gases equivalent to twice the preindustrial level of carbon dioxide. The computer then calculates the climatic effects that would result from the associated radiative and atmospheric conditions.

Various GCMs project global average temperature increases ranging from about 1.5°C (2.7°F) to 4.5°C (8.1°F) associated with conditions equivalent to doubling the preindustrial levels of carbon dioxide.

Because the factors are not completely understood, projections reflect different levels of confidence. For example, most scientists believe that atmospheric concentrations of greenhouse gases such as those modeled by GCM experiments would be accompanied by global average warming, global average precipitation increase, reduction in sea ice, and winter warming in high latitudes. It is possible that there would be sea level rise, drier summers in temperate mid-continental regions, and increased precipitation throughout the year in high latitudes. The local details of climate change, regional distribution of precipitation, regional changes in vegetation, and changes in tropical storm intensity or frequency are very uncertain.

Assessing the Impacts of Climatic Changes

The uncertainty in these projections, especially the lack of regional and local detail, makes it quite difficult to estimate precisely the likely consequences of changes in climatic conditions. In addition, the effects of climate change will not emerge until decades into the future. A careful evaluation of these consequences must therefore take into account changes that occur in the intervening years. Nevertheless, it is possible to develop rough estimates of the likely consequences in many areas.

Direct Effects of Carbon Dioxide

Carbon is an essential element to life on Earth, and plays a major role in photosynthesis in green plants. It is primarily available to plants in the form of atmospheric carbon dioxide. Increased concentrations of carbon dioxide can modify the physiological behavior of plants. Increased atmospheric carbon dioxide increases plant growth in a number of ways: stimulation of photosynthesis; depression of respiration; relief of water and low light stresses; relief of nutrient stress; and prolongation of the growing season. About 95% of the plant matter on Earth is in species that are highly responsive to changes in concentrations of carbon dioxide. However, it is not known what effects it would have for prolonged periods in natural ecosystems. Elevated atmospheric carbon dioxide could lead to long-term absorption of carbon in terrestrial ecosystems, but there is not yet evidence demonstrating such a carbon dioxide fertilization effect in forests.

Water Supply

Hydrologists know a great deal about how water resource systems operate under climatic stress. Increasing annual air temperature by 1° to 2°C (1.8° to 3.6°F) and decreasing precipitation by 10% decreases river flows by 40% to 70% in regions with already low precipitation. Exactly what will happen as climate changes depends on topography, plant cover, geologic formations, and other factors such as use of water by people. The Pacific Northwest may experience increased annual runoff and flooding. California may not see much change in total annual runoff, but have increased river flows in winter and decreased flows in summer. The Great Lakes Basin may have decreasing runoff and increasing evaporation.

Natural Ecosystems

The projected climatic changes would dramatically alter natural ecosystems. The projected warming would produce higher temperatures than any time within the last 100,000 years. More important, the rate of temperature increase would be 15 to 40 times faster than past natural changes. These rates of change may exceed the ability of many species to adapt or disperse into more favorable locations. Some plant and animal species may become extinct, and the makeup of virtually all natural ecosystems would be altered. Species with the following characteristics are at greatest risk: those at the edge of or beyond their optimal range; those geographically constrained on islands, mountain peaks, or remnant patches of vegetation in rural areas; specialized organisms that are highly adapted to specific conditions; species that disperse or migrate slowly; species that reproduce slowly; and local concentrations of annual plants.

The projected rate of climatic changes will almost certainly cause the breakup and reconfiguration of ecosystems. Only in rare instances, however, would land be denuded. Rather, a new pattern of plants and animals will emerge. Some locations could benefit ecologically, ending up with biologically more productive systems. The overall effect of climatic changes, however, may be substantial destruction of habitats and loss of plant and animal species.

Agriculture

Three types of effect from climatic changes are considered most important for agriculture: the physiological effect of elevated concentrations of atmospheric carbon dioxide, the effect of changes in various climatic conditions, and the effect of changes in sea level.

In most cases, the effect of carbon dioxide on crop productivity is beneficial. Most crops grown in cool, temperate, or moist climates (e.g., wheat, rice, barley, root crops, legumes) are highly responsive to elevated levels of carbon dioxide. Only about one fifth of the world's food production involves plants that are less responsive to carbon dioxide concentrations (e.g., maize, sorghum, millet, sugar cane). These latter crops, however, are important sources of food in tropical regions.

Changes in climatic conditions can be quite important for agriculture. Temperature increases can be expected to lengthen the growing season, especially in the Northern Hemisphere where more crops are grown at higher latitudes. Precipitation changes are also likely to have marked consequences for production, especially in combination with heat stress and other extremes.

Areas of high vulnerability with regard to sea-level change include deltas, polders, areas subject to marine flooding, small islands and atolls, and wetlands including mangrove swamps and marshes.

Since climatic effects will take several decades, agricultural practices will adjust as those changes emerge. Changes in land use, management, and infrastructure are considered most important. In high latitude and high altitude areas, warming may reduce climatic constraints on agriculture, so the area of cultivation could be expanded. In other regions reduced moisture may lead to decreased productive potential and possibly to substantial reduction in crop acreage. Changing the strains grown in a particular location can also help adjust to different conditions. Calculations suggest, for example, that rice yields utilizing the current varieties grown in northern Japan would increase about 4% with the increase in the number of growing days projected by GCMs. Adoption of late-maturing varieties now grown in central Japan, however, might increase yields in northern Japan by 26%. Other important adjustments include improving irrigation efficiencies and matching fertilizer use to changed conditions.

In general, agricultural regions most at risk are those where climatic changes are expected to produce decreases in soil moisture and where current resources are inadequately supporting existing populations. In these terms, the agricultural regions thought to be most vulnerable to greenhouse warming are Maghreb, West Africa, Horn of Africa, southern Africa, western Arabia, southeast Asia, Mexico and Central America, and parts of eastern Brazil. North America should be able to make the adjustments necessary to assure continued food supply.

Controlling Greenhouse Gas Emissions

Greenhouse gases are affected by virtually all human activities, from subsistence agriculture to the most modern industrial production processes. By far the largest contribution, however, comes from burning fossil fuels. Carbon dioxide accounted for about two thirds of the enhanced radiative forcing that occurred from 1880 to 1980. Current global carbon dioxide emissions due to human activity are estimated to be between 5.8 and 8.7 billion metric tons of carbon per year. Of this, between 5.2 and 6.2 billion metric tons is due to burning fossil fuels, and between 0.6 to 2.5 billion metric tons is due to deforestation. Reducing fuel consumption would reduce emissions of carbon dioxide. Reducing deforestation would increase the number of trees and woody plants capable of absorbing carbon dioxide from the atmosphere.

Chlorofluorocarbons (CFCs) and halons are long-lived synthetic chemicals containing chlorine, fluorine (and for halons, bromine), and carbon. They are extremely effective greenhouse gases. Some are nearly 20,000 times more effective on a molecule-by-molecule basis than carbon dioxide at trapping infrared radiation from the Earth's surface. CFCs are released into the atmosphere from many sources, including venting from refrigerators and air conditioners, production of "open-cell" foams, and during use as aerosol propellants, solvents, and fire extinguishers. CFCs also deplete ozone in the stratosphere, which counters some of the greenhouse warming caused by CFCs because ozone is also a greenhouse gas. The so-called ozone layer prevents ultraviolet radiation from reaching the Earth that can cause premature aging and skin cancer, enhance cataracts, and suppress immune system responses. In 1987, 47 nations agreed to control CFC emissions. Continued effective elimination of CFC emissions would also reduce greenhouse warming.

ROB COPPOCK

For Further Reading: J. T. Houghton, G. J. Jenkins, and J. J. Ephraums, eds., *Climate Change: The IPCC Scientific Assessment* (1990); National Academy of Sciences, *Policy Implications of Greenhouse Warming (1991)*; U.S. Congress, Office of Technology Assessment, *Changing by Degrees: Steps to Reduce Greenhouse Gases* (1991).
See also Atmosphere; Carbon Cycle; CFCs; Climate Change; Deforestation; Desertification; Energy; Fuel, Fossil; Meteorology; Methane; Ozone Layer; Pollution, Air.

GREENHOUSE EFFECT, ECONOMICS OF

Scientists believe that as carbon dioxide (CO_2) and other greenhouse gases (GHGs) accumulate in the atmosphere, they act like a blanket to insulate the planet, warm its surface, and alter the climate. Scientific monitoring has firmly established the buildup of the main greenhouse gases — CO_2, methane, nitrous oxides, and chlorofluorocarbons (CFCs). Not all greenhouse gases

are created equal, however, for CO_2 has been and will be mankind's largest contributor to global warming.

Relying on climate models and historical temperature records, scientists predict that the global average surface temperature will rise by from 1½°C to 4½°C over the next century. The warmer climate will increase rainfall, and some models foresee hotter and drier climates in mid-continental regions, such as the American grain belt. However, much is conjectural in such forecasts.

In terms of economics, climate has a modest impact upon advanced industrial societies. The most vulnerable sectors are those that are dependent on unmanaged ecosystems, particularly agriculture, forestry, and coastal activities. Even so, farmers have historically shown great ability to adapt to different climatic zones and to changing environmental conditions. Humans thrive in a wide variety of climatic zones as life and livelihood are increasingly climate-proofed by technological changes such as air conditioning and shopping malls. Most economic activity in industrialized countries, however, operates independently of climate.

Less is known about the impacts of climate change in the non-farm sectors or in poor countries. Non-farm sectors will probably experience mixed impacts. Developing countries are more vulnerable to greenhouse warming than are advanced ones. In the world's low-income countries, with a total population of about 3 billion, one third of GNP originates in agriculture. Possible mitigating factors are the enhancement of plant growth due to higher levels of CO_2 and the relative decline in agriculture through economic development.

Of the policies to cope with the threat of global warming, prevention has received the greatest public attention. Reducing or phasing out production of CFCs would be particularly beneficial: besides their other harmful effects, they are extremely powerful greenhouse gases. Some people have proposed planting trees as a method of removing carbon from the atmosphere. Slowing or stopping tropical deforestation is highly cost-effective in slowing greenhouse warming; afforestation in temperate climates, particularly when land has high alternative values, is of questionable value.

Any large program to cut emissions of greenhouse gases will require a significant reduction in the burning of fossil fuels, especially coal. CO_2 emissions can be reduced through a wide variety of measures, from energy conservation to using new technologies. Energy studies indicate that between 10% and 40% of CO_2 emissions can be reduced at modest costs, but beyond this reduction further cuts are likely to become sharply more expensive. Recent estimates suggest that an efficient program to reduce GHGs by 50% below the level that would otherwise occur would cost between

1% and 2% of world output (or about $220 billion annually at current output levels).

The option of climatic engineering has been neglected. Possibilities include shooting particulate matter into the stratosphere to cool the Earth, altering land-use patterns to change reflectivity, and cultivating carbon-eating organisms in the oceans. Although such measures would raise profound legal, ethical, and environmental issues, they would also probably be far most cost-effective than deep cuts in energy use.

A third option is to adapt to the warmer climate. This would take place gradually on a decentralized basis through the automatic response of people and institutions, or through markets, as the climate warms and the oceans rise. In addition, governments can prevent harmful climatic impacts by land-use regulations or investments in research on living in a warmer climate.

Several concrete steps have been suggested for slowing greenhouse warming in an efficient manner: (1) As climate change is a global issue, efficient policies will involve steps by all countries. In order to induce international cooperation, affluent nations will need to expand the concept of foreign aid to include subsidizing environmental efforts by poor nations. Unilateral action may be better than nothing, but concerted action is better still. (2) Policy makers need better information. Physical scientists must improve their understanding of greenhouse warming, and social scientists must enhance their understanding of the economic and social impacts of past and possible future climate change. (3) If energy consumption must be curbed, new technologies could reduce economic costs substantially. Governments must strengthen research and development on new technologies that will slow climate change, particularly in the energy sector. Some people have suggested an international Manhattan Project to develop safe nuclear power. (4) There are countless "no-regret" policies, ones that would be beneficial on other grounds and would also tend to slow global warming. These steps include efforts to strengthen international agreements that severely restrict CFCs, to slow deforestation, and to slow the use of fossil fuels. (5) Some people have proposed "carbon taxes," which would levy environmental taxes or fees on the emission of greenhouse gases, particularly on CO_2 emissions from the burning of fossil fuels. A carbon tax would be preferable to regulatory intervention because taxes can harness markets to minimize the costs of slowing climate change and strengthen the incentive to develop new technologies.

WILLIAM D. NORDHAUS

For Further Reading: National Academy of Sciences, *Policy Implications of Greenhouse Warming: Mitigation, Adapta-*tion, and the Science Base (1992); William D. Nordhaus, "Greenhouse Economics: Count Before You Lead," *The Economist* (July 7, 1990); William D. Nordhaus, *Managing the Global Commons: The Economics of Greenhouse Warming* (1992).
See also Economics; Greenhouse Effect.

GREEN MOVEMENT

The term *green* has increasingly become associated with movements and political parties that campaign on environmental issues. Originally, the "green" label was linked to a radical ecological outlook. Going beyond the single-issue reformism typical of environmental pressure groups, the "green movement" combined protest against environmental deterioration with a broader critique of society and state institutions. Greens—defined in those terms—share a perception that the world is afflicted by a major ecological crisis that requires major political and social changes. However, *green* has been used increasingly for every type of group or behavior associated with environmental positions. The new ubiquity of the green label may be seen as a sign of the rising influence of environmental ideas, but for most "original" Greens, it is purely an exercise of symbolic politics in which the radical edge is blunted by the semantic occupation of the new "green" ground.

Many green activists had been socialized in the peace movement of the late 1950s and early 1960s, the anti-Vietnam war protest, and the student movement. The environmental movement of the late 1960s and early 1970s, and particularly the movement against nuclear energy, provided the background for the emergence of radical ecology. The new peace movements of the 1980s provided a further key political experience for greens.

The first major organization to use the term *green* was Greenpeace, formed in 1971 to organize direct actions against nuclear weapons tests. In its combination of environmental and peace activities and its direct challenge of state authority with its actions, Greenpeace anticipated many of the facets of green politics of later years. Another early use of the label *green* occurred in Australia in the mid-1970s when trade unions protesting against the destruction of nature conservation areas for new building projects formed so-called green bans. The term *green* attained its current worldwide importance, however, when a small new political party called "The Greens" entered the West German parliament in 1983.

In Germany environmental parties emerged first at the local and regional level in 1977. The first national party, the *Grüne Aktion Zukunft* (Green Action Future), was formed in 1978 by the conservative deputy Herbert Gruhl who, one year later, joined various regional parties and independent ecological campaigners, such as Petra Kelly, in a provisional national formation called *Die Grünen* (The Greens), campaigning in the 1979 European elections. In 1980 the provisional formation was turned into the political party. With the debate about the stationing of U.S. nuclear missiles in Germany and other European countries in the early 1980s, the Greens reached international prominence, and after they entered the German parliament in 1983, the term *green* acquired a new political meaning also outside Germany.

However, the German Greens were by no means the first "green" party to be formed or to enter a national parliament. The first local environmental parties were set up in Switzerland and Sweden in 1971, the first regional ecologically oriented party was formed in 1972 in Tasmania, and the New Zealand Values Party formed in the same year is commonly regarded as the first national green party. In Europe the first ecological party was formed in Britain in 1973, and strong ecological organizations participating in elections emerged in France and Belgium in the 1970s. The first ecological candidate to enter a national parliament did so in Switzerland in 1979.

In some countries green parties have become permanent features of political life; in Germany, Belgium, the Netherlands, Switzerland, and Austria, Greens usually command between 5% and 10% of the vote. Despite a set-back in national elections in 1990, the German Greens continue to be an important factor in state government. In France the Greens in recent years have kept an opinion poll rating of between 10% and 15%. Green parties in other countries have been less successful; British Greens polled 15% in 1989, but the majority voting system in the U.K. leaves them with no opportunity to make a real political impact. Attempts by the Citizens Party—formed in 1979 with Barry Commoner as their 1980 Presidential candidate—to portray itself as the U.S. Green Party in the early 1980s did little to stop its demise. The present U.S. green movement is quite diverse and its strengths lie in state green parties, which have had some minor successes, particularly in California, Alaska, and Hawaii.

Green parties provide an outlet for radical ecological politics in countries that offer many opportunities for small parties to make an impact on government policy, as is the case in most continental European countries. But there are many more organizations that may describe themselves as "green." The wave of environmental concerns has boosted the membership of both long-established environmental groups (Sierra Club in the U.S. and National Trust in Britain) and newer groups (Friends of the Earth and Greenpeace). In the early 1990s, internationally, grassroots mobilization with street demonstrations and spontaneous direct actions has been replaced by a more controlled, pressure group–oriented approach. This relies on a passive membership and wealthy sponsors to provide resources; direct action is well planned and controlled for maximum benefit from media reporting. The "movement" character of these organizations has largely been lost; they have become institutionalized and are run virtually like businesses. In a sense this institutionalization repeats cycles where challenging groups first emerge through grassroots mobilization and then become integrated in the political process. The establishment of green political parties can also be seen as part of this institutionalization process. Many of these parties are going through a difficult transition in which the original aims are shed in favor of achieving results in terms of electoral success and policy impact. It remains to be seen whether this can be achieved without a fatal loss of core support.

WOLFGANG RÜDIG

For Further Reading: Fred Pearce, *Green Warriors: The People and Politics Behind the Environmental Revolution* (1991); Wolfgang Rüdig (ed.), *Green Politics One* (1990); Wolfgang Rüdig (ed.), *Green Politics Two* (1992).
See also Environmental Movement; Environmental Movement, Radical; Environmental Organizations.

GREENPEACE

See Environmental Organizations.

GREEN REVOLUTION

The green revolution was a period of time during which great increases in agricultural productivity were achieved. New high-yielding wheat and rice varieties were developed to help feed millions of people. The green revolution was a concerted effort on the part of Western industrialized nations to stimulate staple food production in India and other countries through improved hybrid seeds, chemical fertilizers, and pesticides. Low-input, labor-intensive agricultural practices were transformed into modern, mechanized farming systems that required increased energy and capital.

At the request of the U. S. government, in 1944 the Rockefeller Foundation funded a study to find ways of increasing Mexico's production of basic food crops. Subsequently Norman Borlaug, head of the Mexican research program, developed high-yielding, semi-dwarf wheat varieties and received the 1970 Nobel Peace

Prize for his efforts in overcoming world hunger. In 1962 new rice varieties were developed at the International Rice Research Institute, adding to the success of the green revolution. By the early 1970s wheat and/or rice varieties were growing in Mexico, India, the Philippines, Iran, Algeria, Morocco, Tunisia, Iraq, Saudi Arabia, Turkey, Kenya, Egypt, Pakistan, Brazil, Indonesia, and other countries. Crop harvests increased more than two and one-half times between 1950 and 1980.

Fertilizer, especially nitrogen fertilizer, was a crucial ingredient in the success of the high-yielding varieties. India increased its import of fertilizers by 600% between 1960 and 1980, spending more for fertilizers than for food imports in the worst years of famine. Other costs incurred to foster the green revolution included tractors, combines, irrigation pumps, and pesticides.

Green revolution proponents maintain that when the use of modern inputs offers producers the potential of harvesting more food on less land, several economic benefits accrue to both wealthy and poor farmers alike. By raising yields per unit of land, this intensified land use theoretically frees up acreage otherwise devoted to subsistence crops for the alternative production of more lucrative cash crops. In addition, early-maturing hybrids make it possible to harvest more than one crop a year. Ideally, this increased production results in lower prices for basic foods, thus benefiting the society at large.

There are several fundamental contradictions to the purported goal of eliminating social and economic inequalities. In order to effectively grow high-yielding plants, a farmer must possess irrigated land and have access to fertilizers and other energy inputs. Ninety percent of the world's subsistence farmers are faced with additional obstacles such as small parcels of land on poor soil and lack of access to capital, credit, information, and markets. Many small farmers went into debt or went out of business due to the costs incurred with the energy-intensive green revolution. Large farmers and wealthy landowners were more likely to benefit from green revolution technologies. Income inequalities are aggravated as machinery replaces human labor in poor countries, and small farmers who produce traditional or even high-yielding varieties fail to compete adequately with the more "efficient" modern system. In addition, the green revolution caused land degradation and ground water contamination. Consequently, a worldwide shift is occurring toward sustainable agriculture.

GREENWAYS

Greenways are corridors of undeveloped land that link developed areas to outdoor recreational areas. The greenway concept has developed most completely since the 1970s as a synthesis of two earlier landscape designs known as the greenbelt and the parkway. Thus, greenways are intended to combine the greenbelt idea of separating developed areas with undeveloped tracts of land and the parkway plan of providing for wooded or landscaped thoroughfares. Designed as linear open spaces, greenways are often established along natural corridors such as river valleys and ridgelines, or along scenic roads, historic trails, and railroad rights-of-way that have been modified for recreational use.

Two of the primary purposes of greenways are to add to the quality of life in developed areas and to provide for the maintenance of important ecological functions. They typically connect otherwise isolated parks to areas of development and to one another, allowing urban dwellers to take advantage of a greater area for recreation. Some greenway projects have succeeded in attracting business ventures in outdoor recreation. The linear configuration of greenways has aesthetic as well as practical value. They create the illusion of depth and expansiveness without covering large areas of land. In keeping natural corridors free from extensive development, greenways help to preserve natural ecosystems. Although intended primarily for human use, many greenways also serve as routes for wildlife migration.

Landscape projects that fit to the flexible definition of greenways can be found around the U.S. Notable areas include the Willamette River Greenway in Oregon, the Bay and Ridge Trails in the San Francisco Bay area, Boston's Bay Circuit (a part of that city's "Emerald Necklace"), and the greenway project of Boulder, Colorado. The Canopy Roads Linear Parkway in Tallahassee, Florida, is a modern version of a corridor of high-arching trees originally built by Creek Indians. The Brooklyn-Queens Greenway connects over forty miles of parks and parkways in the New York City area.

GROUNDWATER

All water found under the earth's surface is known as groundwater. Water falling on the earth that does not run off or evaporate continues downward, filling pore spaces in the soil until it reaches an impermeable layer, usually composed of materials such as clay or unfractured bedrock. Depending on where these layers are located, groundwater can be found at depths ranging from a few inches to more than a thousand feet below the surface of the earth. Groundwater often replenishes surface water through springs and wells, and during droughts it plays a critical role in maintaining the health of ecosystems, including fish and wildlife populations.

Compared to surface water, groundwater moves very slowly. When contained in aquifers, the rate of movement is influenced by the relative density of the aquifer, the quantity of water it can store, and the depth at which it occurs. Groundwater flow occurs when the pressure of the aquifer is greater than the atmospheric pressure at the earth's surface. In a shallow porous aquifer, groundwater may circulate completely in less than a year, whereas water stored at a great depth in a low-porosity aquifer may remain there for thousands of years.

Because groundwater lacks the flushing action which helps cleanse surface water, concentrations of contaminants are often several hundred times greater in groundwater than in surface water and may remain that way indefinitely. This contamination may lead to the abandonment or restricted use of the polluted groundwater. Remedial techniques to reduce contamination are extremely costly, may require many years to complete, and, in most situations, can only achieve limited success.

Groundwater is used in a variety of human activities. Although virtually unseen, 98% of the world's freshwater supply is groundwater. In the United States groundwater is the primary source of drinking water for 50% of the general population and 97% of rural residents. Vast quantities of groundwater are used in municipal water supply, in mining processes, and in the manufacture of paper, chemicals, and petroleum. In arid regions of the world, housing developments, agriculture, and industry are largely supported by groundwater.

Groundwater is mineral-rich because, as it percolates through the soil and bedrock, it tends to dissolve and absorb minerals. It also contains a wide range of contaminants. Inorganic ions such as chloride, nitrate, and heavy metals, synthetic organic chemicals such as pesticides, and pathogens such as bacteria, viruses, and parasites are but a few of the contaminants in groundwater that cause water to be unsuitable without treatment for drinking or for other applications requiring pure water. Common sources of these pollutants include industrial waste disposal sites, mining and petroleum production, old or improperly constructed municipal landfills, animal feedlots, and septic tanks. Agricultural pesticides and fertilizers also contain toxic chemicals and nutrients that can pollute groundwater.

Although there is no federal regulatory framework for groundwater protection in the United States, several laws do address related issues. Some federal regulations focus on pollution prevention, standards for maximum contamination levels for specific toxins in public drinking water supplies, and emergency cleanup of hazardous contaminants posing a threat to public health. Additionally, some states have acted to protect isolated aquifers that serve solely as sources of drinking water. However, because the total acreage of many aquifers spans vast regions, and their uses can be so diverse, the current trend in groundwater protection is toward a comprehensive, coordinated, basinwide approach.

H

HAZARDOUS MATERIALS

See Toxic Chemicals.

HAZARDOUS WASTE

A waste is considered hazardous if it causes or significantly contributes to an increase in the human death rate or in serious irreversible or incapacitating illness; or if it poses a substantial present or potential hazard to human health or the environment when improperly treated, stored, transported, or disposed of, or otherwise managed. Of the 6 billion tons of industrial, agricultural, commercial, and domestic wastes generated annually in the U.S., about 250 million tons are considered by the U.S. Environmental Protection Agency to be hazardous. Hazardous waste was defined in 1976 and the definition was expanded in 1984, by the U.S. Congress under the Resource Conservation and Recovery Act (RCRA). As the definition implies, the harmful effects of hazardous waste can often be prevented through proper planning and management.

Hazardous wastes may be liquid, solid, semi-solid (sludge), or containerized gas. They are created as byproducts of manufacturing and agricultural processes, and in the form of trash, when consumers discard commercial products such as household cleaning fluids or battery acid.

There are tens of thousands of wastes that may be hazardous. They may be: ignitable (readily catch fire), such as solvents and friction-sensitive substances; corrosive (acidic or capable of corroding metal tanks, containers, drums, and barrels); reactive (unstable under normal conditions)—when mixed with water, they can create explosions, toxic fumes, gases, and vapors; or toxic (harmful or fatal when ingested or absorbed). This definition includes many things that the average American stores at home. The Water Environment Federation estimates that the average U.S. household contains between three and ten gallons of materials that are hazardous to human health or to the natural environment. Homeowners should check with local or state public health officials about how to dispose of these wastes. Pouring them down the drain or putting them in the trash may be dangerous to the community.

Another common source of hazardous waste is industry. Many of the material goods our society relies upon are made through production processes that create hazardous wastes as byproducts. For example, petroleum refining and wood preserving use chemically reactive solvents and specific commercial chemical products such as creosote that may be dangerous when discarded. Generators of hazardous wastes include large and small industries and businesses, universities, and hospitals. (See Table 1.) Agriculture also creates

Table 1

Examples of hazardous waste generated by businesses and industries

Waste generators	Waste type
Chemical manufacturers	Strong acids and bases Spent solvents Reactive wastes
Vehicle maintenance shops	Heavy metal paint wastes Ignitable wastes Used lead-acid batteries Spent solvents
Printing industry	Heavy metal solutions Waste inks Spent solvents Spent electroplating wastes Ink sludges with heavy metals
Leather products manufacturing	Waste toluene and benzene
Paper industry	Paint wastes containing heavy metals Ignitable solvents Strong acids and bases
Construction industry	Ignitable paint wastes Spent solvents Strong acids and bases
Cleaning agents and cosmetics manufacturing	Heavy metal dusts Ignitable wastes Flammable solvents Strong acids and bases
Furniture and wood manufacturing and refinishing	Ignitable wastes Spent solvents
Metal manufacturing	Paint wastes containing heavy metals Strong acids and bases Cyanide wastes Sludges containing heavy metals

Source: U.S. EPA

hazardous wastes through the use and demand for production of pesticides.

Hazardous waste is also found at facilities owned by various federal agencies, such as the Department of Defense and Department of Energy (DOE). Waste at these sites is a result of the nation's post-World War II weapons research, military operations such as munitions testing, and use of jet fuel. Sites owned by DOE present especially difficult challenges to clean up because of their large size and contamination by complex mixtures of hazardous chemicals and radionuclides.

Hazardous materials may become wastes inadvertently, through spills or leaks. In thousands of communities across the U.S., spills and leaks of toxic solvents or gasoline from underground storage tanks (USTs) and associated piping systems have severely contaminated the groundwater. An estimated one to two million USTs in the U.S. are used by a variety of industries, including motor fuel marketers and chemical manufacturers.

Once the generator of a waste determines that it is hazardous, he or she must decide whether to treat or dispose of the waste on-site or transport it off-site where the waste is managed by a commercial firm or a publicly owned and operated facility. Records that verify the path of hazardous waste from its point of generation to disposal must be presented to the U.S. Environmental Protection Agency (EPA) or the authorized state environmental agency. Proper packaging ensures that no hazardous waste leaks from containers during transport. Detailed labeling is required to enable transporters and public officials to rapidly identify the waste and its hazards in case of emergencies.

Hazardous waste may be disposed of in engineered facilities such as landfills, surface impoundments, or underground injection wells. Landfills are permanent disposal facilities where hazardous waste is placed in or on land that is lined with clay or impermeable liners and covered with soil. Surface impoundments — natural or man-made depressions of diked areas, properly lined — can be used to treat, store, or dispose of hazardous waste; these are often referred to as pits, ponds, lagoons, and basins. Underground injection wells are specially designed steel or plastic pipes placed deep in the earth into which liquid hazardous wastes are injected under pressure.

As people become more aware of the dangers of hazardous waste and regulations become tighter, it is increasingly difficult to open or operate a land disposal facility. As an alternative to or as pre-treatment before land disposal, hazardous waste may be treated, or remediated, utilizing a growing number of technologies that involve physical, chemical, and/or biological processes. Incineration, solidification, and vitrification are three such processes. Incineration destroys most or-

ganic substances and results in an ash or residue which must then be disposed of on land; however, through treatment by solidification, vitrification, and other processes that change the form of the waste, the resulting product will be less hazardous or even non-hazardous.

Dangers from hazardous wastes persist long after their disposal. For example, at Love Canal in New York State, disturbance of a hazardous waste landfill years after the site had been closed caused leakage of hazardous chemicals into the basements of nearby homes. Closed and abandoned hazardous waste sites in the U.S. are regulated by the EPA under the federal Superfund program, established by the Comprehensive Environmental Response, Compensation, and Liability Act of 1980. Superfund sites include former landfills, mining areas, manufacturing facilities, and illegal hazardous waste dumps.

To deal with concerns about public health and the environment resulting from failed or improper land disposal practices in the past, U.S. law now requires companies to minimize hazardous wastes by reducing the volume, recycling, and treating them. Waste volume can often be reduced through manufacturing process changes, source separation, recycling, and raw material substitution. The latter may offer the greatest opportunity for waste reduction; by replacing a raw material that generates a large amount of hazardous waste with one that generates little or no hazardous waste, manufacturers can substantially reduce the waste volume. Examples of waste minimization possible through substitution of key raw materials include: aerospace industry—replace cyanide cadmium plating bath with a non-cyanide bath; Air Force installations—reduce use of solvent-containing paint strippers by switching to mechanical means of stripping; pharmaceutical industry—replace solvent-based tablet-coating process with a water-based process; and printing industry—substitute water-based ink for solvent-based ink.

THOMAS P. GRUMBLY

For Further Reading: Richard C. Fortuna and David J. Lennett, *Hazardous Waste Regulation: The New Era—An Analysis and Guide to RCRA and the 1984 Amendments* (1987); U.S. Environmental Protection Agency, *The RCRA Orientation Manual* (1990); U.S. Environmental Protection Agency, *Does Your Business Produce Hazardous Waste? Many Small Businesses Do* (1990); U.S. Environmental Protection Agency, *Catalog of Hazardous and Solid Waste Publications* (1992). *See also* Deep Well Injection; Hazardous Waste, Household; Incineration; Landfills; Source Reduction; Superfund; Toxic Chemicals; Waste Disposal; Waste Treatment.

HAZARDOUS WASTE, HOUSEHOLD

Household Hazardous Waste (HHW) consists of products with constituents that are defined in the Resource Conservation and Recovery Act (RCRA) as hazardous or have the characteristics of either flammability, reactivity, corrosivity or toxicity. In the U.S. EPA's report on HHW, these products fall into five broad categories—paint and other household improvement products, pesticides, automotive products, household cleaners, and "other" products (pharmaceuticals, pool cleaners, explosives, household batteries, etc.). Used oil and paint are the two most common materials brought to HHW collections. Pesticides are usually the third largest category, and a significant number of car batteries are also received.

Annually over 90% (176 million gallons) of used oil discarded by do-it-yourself oil changers and at least 260 million gallons overall are improperly disposed of—equivalent to 27 *Exxon Valdez* spills. The environmental damage resulting from the improper disposal of used oil, the largest single source of oil pollution (62% of the total) in the nation (with most ending up in the oceans), cannot be overestimated. Improperly discarded oil from one DIY (do-it-yourself) oil change can foul a million gallons of fresh water, the amount needed to supply 50 families for a year. One part per million (ppm) of oil in water can be detected by odor and taste; 35 ppm causes a visible slick; 50 ppm can foul a water treatment plant by disrupting the bacteria that break down organic wastes. For all these reasons, used oil has become the focus of EPA and congressional initiatives to regulate its management and disposal. In addition, used oil is a valuable non-renewable natural resource that is rich in energy or, when properly cleaned, can be used indefinitely as a lubricant. The yield from refining one gallon of used oil is 2½ quarts of lubricating oil—the same as the yield from a 42-gallon barrel of crude oil. Recovery of used oil could reduce our petroleum imports by 25.5 million barrels a year and save altogether 1.3 million barrels of oil per day. Some communities offer curbside used oil pickup; in several states garages and service stations act as voluntary collection centers for DIY used oil; the states of Maryland and Rhode Island are installing collection tanks that will be serviced by state-approved contractors for exclusive use by DIYs. Vermont will use specially installed tanks at service stations and other sites paid for with state subsidies and grants from the oil overcharge fund. All of these states use extensive public education and awareness efforts aimed at the DIY, stressing the environmental damage that results from improper disposal of used motor oil.

Household batteries are a significant source of the heavy metals in municipal solid waste (MSW)—espe-cially mercury, nickel, zinc, and cadmium in incinerator ash and emissions. The use of mercury in household batteries is declining, but in 1987 it accounted for 587 tons, or 37% of the metal used in the U.S. The two largest sources of lead in MSW are lead-acid batteries (65%—138,043 tons of lead discards in 1986) and consumer electronics (27%—58,536 tons) from soldered circuit boards, leaded glass in television picture tubes (the largest contributor), and plated steel chassis. Most of the remainder comes from glass and ceramics, plastics, and pigments. Rechargeable nickel-cadmium batteries account for more than half of cadmium discards in the U.S. In 1986 this was 930 tons or 52%. Many of these batteries are sealed inside small cordless appliances. Plastics contribute most of the rest—28% (502 tons) of cadmium discards in 1986. Cadmium is used as a stabilizer in polyvinyl chloride and as a pigment in a wide variety of resins.

When HHW is thrown into the trash, there is the potential for explosions or fires in the trash truck; exposure of sanitation workers, landfill operators, or resource recovery facility operators to corrosive, reactive, flammable or toxic materials; landfill leachate formation resulting in contamination of ground water; pollution releases from landfills or incinerators; and contamination of mixed MSW compost and MSW incinerator ash. When such products are thrown down the drain, they can cause damage to the pipes or the wastewater treatment plant, air pollution during the wastewater treatment process through the release of volatile organics, or sludge contamination thus limiting its further use.

The products that become HHW are also of concern even before they become wastes. The greatest number of household poisonings comes from these products. Many contribute to indoor or ambient air pollution problems and also are a hazard when a dwelling catches on fire and firemen unsuspectingly enter the burning home. HHW has been measured in various ways. Trash has been sorted and the amount of HHW has been calculated. Various sorting studies show that an average of 15.5 pounds of HHW per household per year has been thrown into the trash, about 1% of MSW.

HHW, however, is also disposed of in other ways. Homeowners have been interviewed as to what they do with unwanted products with hazardous constituents. Certain products, such as used oil, have been most frequently poured down storm drains. When the types and percents of these products that are disposed of in some other way are added to the amounts and types found in MSW sorts, a yearly generation rate of 21.5 pounds per household per year is established.

When residents bring HHW into a collection program, an average of 100 pounds is brought in, five times what one would anticipate from the yearly generation rate. This shows that there is a considerable amount that may be left, poured down the drain, or put into the trash. This figure is also important to potential HHW collection program sponsors who are trying to anticipate how much HHW they will receive on the collection day.

Another way in which HHW generation is measured is the contribution of households to the toxic loadings of the wastewater stream entering the wastewater treatment system. In Orange County, California, 20% of the heavy metals in the wastewater were calculated to come from the home.

Contaminants such as arsenic have been traced to the home. As noted earlier, heavy metals not only can interfere with the operation of the publicly owned treatment works, but they can influence the quality of the sludge and limit the options one has for composting it, applying it to soil, or disposing of it.

DANA DUXBURY

For Further Reading: Joyce Kathan, "The Role of Private Organizations and Citizen Groups in Controlling Health Hazards" in D. LeVine and A. Upton, *Management of Hazardous Agents,* Vol. 2: *Social, Political, and Policy Aspects* (1992); Peg Kocher and Anita Sieganthaler, *The World of Waste* (1988); The Waste Watch Center, *Proceedings of the U.S. E.P.A. Conference on Household Hazardous Waste Management* (1992).

HAZARDS FOR FARM WORKERS

Occupational hazards associated with farming have been recognized at least since 1555. In recent years, agriculture has consistently ranked among the industries with the greatest number of injuries and illnesses. In an estimated population of 5 million U.S. farmers, 6.5 million farm family members, and 2.7 million hired hands, nearly 200,000 disabling conditions and 1,600 deaths related to farming occur every year. Of these approximately 24,000 non-fatal injuries and 300 deaths affect children. Costs for health care and rehabilitation are estimated to exceed $3 billion.

Agricultural workers are exposed to a wide variety of hazardous agents and work under adverse climatic and economic conditions, most often without the benefit of sophisticated health and safety services available to other industries. In addition, the population at risk in agriculture is unique among occupational groups. Because the farm represents both a workplace and a home, a large proportion of farm workers are younger than 16 years or older than 65.

Traumatic death and injury are related primarily to farm machinery. Sprains, strains, cuts, fractures, amputations, and deaths can occur when humans interact with machines. The tractor, in particular, is in-

volved in more than half of all farm fatalities. Most tractor-related deaths are due to rollovers. Other farm machines and livestock are also common causes of injury.

In addition to mechanical hazards, farmers are exposed to a variety of organic dusts, irritant gases, agricultural chemicals, biological organisms, and other environmental conditions that can result in illness or injury. Compared to workers in non-agricultural industries, farm workers suffer from increased rates of noise-induced hearing loss, respiratory disease, dermatitis, and certain cancers.

Noise-induced hearing loss, resulting from exposure to farm machinery, has been found to afflict a quarter of all young farmers and more than half of all older farmers. Significant numbers of those affected develop a communication impairment by age 30, often resulting in social isolation.

Farmers are exposed to respiratory hazards during activities including soil preparation, grain handling and transport, pesticide mixing and application, feed mixing, and handling animals. Increased use of livestock confinement buildings has resulted in exposure to elevated concentrations of dusts and gases over extended periods of time. Hazardous airborne agents are typically a complex mixture of plant material, including pollen, grains, and fragments of leaves and stems; pathogenic and non-pathogenic bacteria, fungi, viruses; microbial toxins; soil particles; insects; animal-derived particles; irritant gases (ammonia, hydrogen sulfide, nitrogen oxides); agricultural chemicals including medicated feed additives and pesticides; and diesel exhaust. Farmers exposed to these hazards have been found to exhibit increased rates of chronic bronchitis, asthma, organic dust toxic syndrome (an acute febrile condition), and hypersensitivity pneumonitis or farmer's lung (a respiratory immune syndrome).

Dermatitis is one of the most common medical conditions affecting farmers, accounting for two-thirds of all agricultural illnesses reported. Farmers are four times more likely to have skin disorders than all non-agricultural employees, with a rate three times that in manufacturing. Physical, chemical, and infectious agents that can cause or exacerbate dermatitis include: plant and animal products; agricultural chemicals (pesticides, fertilizers); food products; sunlight; temperature extremes; and zoonotic infections (ringworm, anthrax).

Although agricultural workers' mortality rate for all cancers combined is lower than normal, attributable largely to low smoking rates, particular types of cancers tend to occur more frequently among farmers than non-farmers. Excess cancers of the lip and skin are linked to increased exposure to sunlight. Less is known, however, about the potential causes of increased rates of hematologic cancers (leukemia, Hodgkin's disease, non-Hodgkin's lymphoma, and multiple myeloma), and cancers of the stomach, prostate, and brain. Suspected causative agents include nitrates in drinking water, certain herbicides, solvents, and viruses.

Degenerative musculoskeletal syndromes result from chronic exposure to vibration from farm machinery, and performance of repetitive traumatic tasks. Commonly reported conditions include low-back pain and arthritis. Operation of tractors and other heavy equipment, lifting of heavy weights with bending, twisting, pushing, and pulling all contribute to low-back pain. Dairy farmers, in particular, appear to have an increased prevalence of degenerative knee disease, probably related to squatting and kneeling during milking.

Exposure to agricultural chemicals has been associated with acute and chronic chemical toxicity. In addition, the persistence of many herbicides and pesticides in the environment contributes to chronic exposure of farmers and their families. The term *pesticides* is broadly applied to a diverse group of chemical compounds targeted at a variety of insect, animal, and plant pests. Although pesticides may be absorbed through inhalation and ingestion, the primary route of exposure is via skin contact. Acute and some chronic neurotoxicity are major hazards from exposure to the widely used organophosphate and carbamate pesticides. Other adverse health effects associated with pesticides include liver and kidney damage, reproductive disorders, birth defects, dermatitis, and possibly certain cancers. Skin contact or inhalation of ammonia fertilizers can result in chemical burns and pulmonary edema. Nitrates in drinking water and food are responsible for methemoglobinemia (blue baby syndrome) in infants, and may be related to chronic effects in adults.

With increased economic pressures, cultural changes, the loss of family farms, long work hours, and isolation, stress-related mental disorders play a significantly increased role among farmers and their families. Depression, drug abuse, and domestic violence are common in the rural environment. Recent studies in the rural Midwest report significantly increased rates of suicide among farmers. All of these conditions are further exacerbated by the lack of readily available health services.

STEVEN J. REYNOLDS
JAMES MERCHANT

For Further Reading: R. A. Aherin, D. J. Murphy, and J. D. Westaby, *Reducing Farm Injuries: Issues and Methods* (1992); J. A. Dosman and D. W. Cockcroft, *Principles of Health and Safety in Agriculture* (1989); D. J. Murphy, *Safety and Health for Production Agriculture* (1992).

HEALTH AND DISEASE

The term *health* is generally interpreted to denote physical, mental, and social well-being. It also denotes the capacity to function effectively in a given environment and to adapt successfully to the challenges presented by the dynamically interacting components of the environment, which include a wide variety of physical, chemical, biological, and social influences.

Human health and survival in prehistoric times were limited primarily by the availability of adequate food, shelter, and protection against predators. Later, as people gathered in villages, towns, and cities, infections and parasitic diseases became the leading causes of illness and death. During the past century, these diseases have decreased dramatically in frequency in the industrialized countries of the world, enabling life expectancy to double, so that it now surpasses the biblical "three score and ten" years. Consequently, the major causes of death in the industrialized world today are diseases of the elderly, notably heart disease, cancer, and other degenerative disorders (Figure 1).

The latter diseases, formerly regarded largely as hereditary or inevitable accompaniments of aging, are now linked increasingly to unhealthful lifestyles, stress, and various disease-causing chemical and physical agents in air, water, food, consumer products, the workplace, the home, and other environments. Because these diseases take years or decades to become manifest, their environmental causes can be demonstrated only by appropriate epidemiological or toxicological methods.

Evaluation of Effects on Health

The human body is made up of trillions of microscopic cells, each of which is governed by interactions between its genes and the environment in which it resides. Most of the cells are specialized for reacting to specific environmental stimuli. For example, cells of the retina of the eye are capable of reacting to visible light, while those of the skin can react to touch and temperature. Some environmental stimuli act by switching specific genes on or off in certain cells, without necessarily harming the cells in the process. Other types of environmental stimuli, however—for example, ionizing radiation—are capable of damaging genes at virtually any level of exposure and may, therefore, be potentially hazardous even in the smallest doses.

If enough cells in a particular organ are killed, the result may be immediate incapacitation followed by death, as can happen when the skin over a large part of the body is badly burned. Although the body can normally compensate for the loss of a few cells, damage

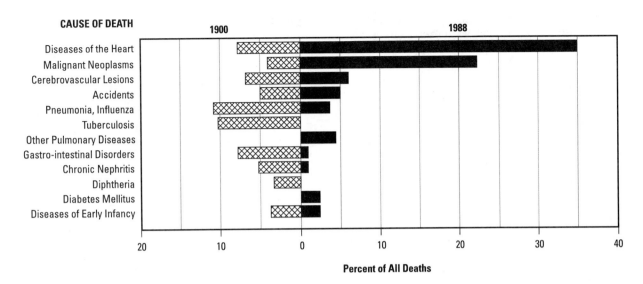

Figure 1. Leading causes of death, 1988 versus 1900. Source: National Center for Health Statistics.

to a single cell may suffice to cause disease under certain conditions. For example, the production of a mutation in a single reproductive cell can, if transmitted to a child via a fertilized egg, cause disease in that child. The production of cancer, likewise, is thought to result from the occurrence later in life of such a change in the genetic apparatus of some other susceptible cell.

Epidemiological studies of human populations provide the only direct means of detecting and measuring such environmental influences on health, but important supporting data are provided by complementary studies with laboratory animals and other nonhuman surrogates. The latter, which afford the only means of identifying potentially toxic agents in advance of human exposure, are now required by law for the premarket testing of newly synthesized chemicals. Such studies include: (1) predictive assessments of the toxicity of chemicals from analysis of their physical and chemical properties; (2) short-term tests for the toxicity of chemicals in laboratory animals and cultured cells; and (3) long-term tests for the toxicity of chemicals in laboratory animals exposed to the agents throughout life. Such studies are not necessarily predictive of risks to humans, but they have proven to be sufficiently reliable to become indispensable in public health. They also provide the only means for investigating in detail the mechanisms of disease causation, experimentation on human subjects for that purpose being unethical.

An active area of inquiry in the 1990s includes the identification of chemical and physical agents in the environment that may contribute to the causation of cancer, elucidation of their modes of action, and de-

velopment of methods for preventing or counteracting their effects. This constitutes one of the most rapidly growing branches of oncology, the field of science devoted to the study of tumors, including benign and malignant growths of all types. Oncology thus encompasses studies of the nature and causes of tumors, their rates of occurrence in different populations, biological and clinical properties, detection, diagnosis, management, and prevention. As a result of the rapid growth of knowledge in these subject areas, oncologic subspecialties have been established in various branches of medicine.

Environmental Factors Influencing Health

Air Pollutants

Illness and mortality rates have long been known to be increased in association with episodes of severe air pollution. Long exposure to certain pollutants at high concentrations in the air of the workplace (e.g., carbon monoxide, vinyl chloride, coke oven emissions, radon, lead, mercury, arsenic, nickel, asbestos, silica, cotton fibers, and coal dust) is also known to cause disease. The health effects on the general population of chronic exposure to the levels of combustion products, oxidants, and other pollutants that are prevalent in the atmosphere are less well known but are estimated to account for thousands of fatalities each year. Certain indoor air pollutants also constitute health hazards, as is exemplified most dramatically by cigarette smoke, the leading avoidable cause of death in the industrial world today. Other common indoor air pollutants (e.g., radon, asbestos, lead, and allergens of various kinds) also account for many cases of illness and death. The effects of indoor air pollutants are not limited to specific diseases, such as asthma and lung cancer, but also include relatively nonspecific reactions, such as nausea, headache, and irritation of the eyes, nose, and throat.

Water Pollutants

Microbial contamination of drinking water, although still a major health problem in the third world, no longer constitutes a leading cause of disease in industrialized countries. Nevertheless, water supplies in a growing number of areas in such countries have been found to be contaminated with heavy metals, toxic wastes, pesticides, and other chemicals, necessitating the closure of thousands of wells each year. The health impacts of small quantities of such chemicals in drinking water cannot be assessed fully without further research, but costly remediation measures, along with steps to prevent the further pollution of groundwater, are clearly called for.

Toxic Waste

All living things produce wastes; however, the kinds and amounts of hazardous waste generated today far exceed those of the past, severely taxing available disposal methods and resources. The number of toxic waste sites posing potential threats to public health in the U.S. is estimated to exceed 400,000. While substantial quantities of toxic waste have been released into the surrounding environment at many such sites, the resulting exposure of persons in the area has generally not been sufficient to produce toxic reactions. Nevertheless, the risks of carcinogenic or other long-term effects in such persons have been a subject of growing concern. In some instances, clusters of cancers, birth defects, or other diseases in the vicinity of a site have been interpreted as site-related effects, but further research will be necessary to exclude other possible causes of such clusters.

Food

It is well known that proper nutrition is essential to normal growth, development, and reproduction, and that certain constituents or contaminants of food can cause sickness. Much less is known, however, about the optimal diet for preventing disease and prolonging healthy life in old age. Although chemical contamination of the food supply may, under certain conditions, pose threats to particular subgroups—for example, those subsisting on fish from bodies of water contaminated by heavy metals, PCBs, or other toxic wastes—such contamination is not thought to constitute a major health hazard for the population at large. Similarly, pesticide residues in food are thought not to pose significant risks to the general population. The greatest threat from food is caused by its contamination

with microorganisms or viruses, which account for millions of cases of "food poisoning" each year.

Socioeconomic Status and Lifestyle

Of the various environmental factors influencing human health, socioeconomic status is one of the most important, as exemplified by the plight of the poor, who suffer from malnutrition, congested and stressful living conditions, and relegation to hazardous working conditions. There is evidence, moreover, that members of racial and ethnic minorities are more likely than other persons to reside in polluted areas. It is not astonishing, therefore, that the rates of many common causes of death vary inversely with socioeconomic levels. The importance of wholesome living conditions and health habits in all persons is well established. Regular meals, physical exercise, adequate sleep, weight control, abstinence from smoking and habit-forming drugs, and moderation in the use of alcohol all are associated with below-average rates of illness and death. Cigarette consumption, which accounts for nearly 400,000 deaths per year in the United States from cancer, respiratory ailments, and cardiovascular disease, varies inversely with educational level, exemplifying the importance of educational and behavioral factors in shaping the environment for better or for worse.

Occupation

Studies of occupational groups have identified a wide variety of chemical, physical, and microbial disease-causing agents, most of which are not confined to the workplace but affect populations in other environments as well. In spite of the fact that a large percentage of work-related diseases go unrecognized and unreported, as many as 400,000 new cases of work-related disease and 100,000 work-related deaths are estimated to occur in the United States each year.

Ecology

Since human health depends on the integrity of the global ecosystem, anything disturbing the latter may ultimately jeopardize human health, although the time required to do so or to reverse a trend may span centuries. A number of ecological changes now occurring have prompted growing concerns. These include: (1) the depletion of stratospheric ozone, (2) the build-up of greenhouse gases and the resulting threat of global warming, (3) the progressive destruction of natural habitats, and (4) the accelerating loss of species diversity. Because of their complexity and global scope, these problems have captured the attention of scientists and decision-makers throughout the world.

ARTHUR C. UPTON, M.D.

For Further Reading: J. F. Fries, "Aging, natural death, and the compression of mortality," *New England Journal of Medicine* (1980); National Academy of Sciences/National Research Council, *Diet and Health. Implications for Reducing Chronic Disease* (1989); United States Environmental Protection Agency, *Reducing Risk: Setting Priorities and Strategies for Environmental Protection* (1990).

HEALTH AND NUTRITION

No part of the environment to which human beings are continually exposed is as chemically complex as the foods and beverages we consume. In addition to the substances that provide nutrition—proteins, carbohydrates, fats, vitamins, and minerals—the human diet contains huge numbers of other chemicals that are natural constituents of the various plant and animal products that comprise it. Some of these substances, which are mostly organic chemicals of diverse molecular structures, impart flavor and color and aroma, but most are present because of their roles in the lives of the plants and animals that are the sources of our foods. These natural constituents comprise by far the largest numbers of chemicals we take into our bodies from any environmental source. A cup of coffee, for example, contains nearly 200 distinct organic chemicals that are natural components extractable from the bean. Relatively few of the hundreds of thousands of natural food constituents have been investigated for their effects on health.

Human beings have never been content to leave their diets unmodified. Since ancient times humans have added various substances both natural and synthetic to preserve, to emulsify, to improve taste or texture, to sweeten, and to color food. In recent times we have introduced most food processing and packaging technologies, most of which result in the addition, albeit in tiny amounts, of new chemical substances. Treatment of crops with pesticides often creates residues of them that persist to the dinner plate. Drugs used in animal husbandry sometimes end up in meat, milk, and eggs. These various types of "additives" increase substantially the numbers of chemicals present in the human diet.

Finally, there are some substances that contaminate the diet. Various microbial agents and their chemical metabolites can sometimes contaminate foods that are improperly treated or inadequately protected, and some of these can cause illness. Certain unwanted industrial chemicals, such as PCBs, lead, and mercury, may reach segments of the food chain, although many foods also contain a natural background level of the metals.

Attempts to understand the role of diet in human health focus on six main questions: (1) What levels of calorie and nutrient intake provide the maximum health benefits and lowest health risks? (2) What naturally occurring, non-nutritive constituents of the diet pose risks to health, and which enhance health, and at what levels of intake? (3) What contaminants of the diet, either of natural or industrial origin, pose risks to health and at what levels of intake? (4) Do substances directly or indirectly added to the diet pose health risks, and at what levels of intake? (5) What factors (e.g., age, sex, disease states, genetics) affect an individual's susceptibility to risk factors of environmental or dietary origin? (6) What overall dietary pattern provides maximum promotion of health and maximum protection against risk of disease?

Three investigative tools are used to study these questions: (1) Epidemiological studies help uncover relationships between increased risks of some chronic diseases and diet, and trace the roles of microbial agents in outbreaks of food poisonings. (2) Clinical studies in human subjects are useful in clarifying the benefits of nutrients. (3) Experimental animal models identify the hazardous properties of individual dietary constituents, including nutrients, additives, pesticides, and contaminants, and set safe limits of human intake. Each of these tools has limitations and no one of them can provide adequate answers. Closer integration of the efforts of epidemiologists and clinicians attempting to understand the role of the diet as a whole and of certain of its macroconstituents, and toxicologists concerned with individual constituents, would seem necessary.

Nutrients and Other Macroconstituents

Of the ten leading causes of death in the U.S., eight are in part related to diet or excessive alcohol ingestion (Table 1). In 1987 these accounted for roughly 1.5 million deaths, about 75% of the total deaths. It is not possible to estimate accurately the fraction of these deaths in which dietary factors and alcohol abuse play a direct role, but it seems clear that dietary modifications will significantly reduce the public health toll associated with these causes. Significant reductions in heart disease rates have occurred since the peak year of 1967, probably in large part due to improved eating habits. The British epidemiologist Sir Richard Doll, together with Richard Peto, estimated in the 1980s that dietary modifications could reduce the risk of death from human cancers by about 35% (their estimate included a possible range of 10% to 70%). Sir Richard recently noted that, "Nothing that has happened in the last 10 years leads us to think that diet is less important than we thought 10 years ago. On the contrary, the evidence of its importance as a cause of cancer, or as a defense against it, has been strengthened."

The benefits of a diet higher in fibers (non-digestible

Table 1

Cause of death	Estimated number of deaths (% of total)	Estimated total cases of individuals suffering diseases	Implicated dietary factors
Heart disease	760,000 (35.7%)	6.7 million	Excessive calorie intake (obesity) Excessive fat intake High fiber diet appears beneficial
Cancers	480,000 (22.45%)	1.0 million new cases/ year	Excessive fat intake (colon, breast cancers) High fiber diet appears beneficial Excessive calorie intake (obesity)
Strokes	150,000 (7%)	2.7 million	Excessive calorie intake (obesity)
Motor vehicle and other accidents	94,000 (4.4%)	—	Alcohol abuse
Diabetes mellitus	38,000 (1.8%)	10 million	Excessive calorie intake (obesity) High carbohydrate/low fat diet is beneficial
Suicide	29,500 (1.4%)		Alcohol abuse
Chronic liver disease and cirrhosis	25,000 (1.2%)		Alcohol abuse
Athero-sclerosis	23,000 (1.1%)		Excessive calorie intake (obesity) Excessive fat intake

carbohydrates) seem apparent, but it is not entirely clear that such benefits are due directly to the presence of fibers in the diet or rather to the fact that diets high in fiber tend to be low in fat. The American Cancer Society and the National Cancer Institute urge increased consumption of whole grains, fruits, and vegetables rather than the use of fiber supplements, which might do little to reduce fat intake.

Other health problems, some afflicting very large portions of the population, are also significantly due to dietary factors. Obesity, osteoporosis, gallstones, and dental caries are conditions that can be substantially controlled through the use of an appropriate diet. There is even evidence to suggest a beneficial effect on longevity when caloric intake is restricted.

Vitamins and Minerals

Evidence is accumulating that some vitamins and minerals may help prevent some chronic diseases. Carotenoids occur in dark-green leafy vegetables and in yellow and orange vegetables and fruits, and are enzymatically converted in the small intestine to Vitamin A. The most abundant carotenoid is beta-carotene. Epidemiology studies have provided substantial evidence that diets high in carotenoids decrease the risks of several types of cancers, including those of the lung, cervix, bladder, and upper digestive tract. These early indications have prompted several large-scale studies on beta-carotene's possible anticarcinogenic properties.

Vitamin C and other antioxidants may provide some protection against stomach cancer and atherosclerosis. The risk of developing osteoporosis may be reduced by Vitamin D. Calcium supplementation of the diet would seem a reasonable approach to reducing the risk of osteoporosis, but the available data on this matter are not consistently supportive of this conclusion. In fact, most of the evidence concerning a significant role for vitamins and minerals in chronic disease prevention (beyond that needed to prevent the well-known deficiency diseases) is still not well developed, and many experts believe it is too early to translate the evidence to specific dietary recommendations.

Additives

The Food, Drug, and Cosmetic Act and its 1958 amendments make several important legal distinctions among substances that are directly or indirectly added to food. Some have been designated GRAS substances ("generally recognized as safe"); these ingredients, such as salt, had been used for many years prior to 1958, and were exempted from the extensive premarket testing requirements that were imposed upon newly introduced additives. These and other legal distinctions — those regarding color additives, pesticides, and animal drugs — are particularly important in regulations issued by the Food and Drug Administration (FDA) and EPA. Except for a few major ingredients such as sugar, most additives are present in food in

Table 2
Some food "additives" and their functions

Function	Example	Legal classification
Flavoring agent	Ginger, licorice	GRAS
Flavoring agent	Elder tree leaves	Food additive
Preservative	BHA, BHT	GRAS for some uses, food additive for others
Anticaking agent	Silicon dioxide	Food additive
Non-nutritive sweetener	Aspartame	Food additive
Processing aid	Acrylate/acrylamide resin (to clarify juices)	Food additive
Adhesives	Acrylate esters	Food additive[1]
Multipurpose ingredients	Caramel, lecithin	GRAS
Coloring agent	FD&C Green No. 6	Color additive
Veterinary medicine	Ronidazole (residues in pork)	Animal drug

[1]Food contact may result in migration of small amounts of certain substances to food. This is an *indirect* food additive.

very small amounts. Table 2 lists some additives and their functions.

Of all the substances present in foods, additives have been the most thoroughly investigated for their possible adverse effects on health. Federal laws require that additives (except for GRAS substances) be subjected to testing to establish their safety *prior* to their use in foods. Testing is the responsibility of those who wish to sell the additive for food uses. To varying degrees, additives are put through a series of toxicity tests in laboratory animals and the intakes needed to cause adverse biological effects are identified. A large safety margin is then introduced to set an upper limit on human intake (called the Acceptable Daily Intake, or ADI) that is well below the intake that is without adverse biological effects in the most sensitive laboratory animal species. An additive can not be legally used unless its maximum dietary intake for humans is below its ADI, as described in additive-specific regulations issued by FDA. There is little evidence that additives to the diet pose risks. The FDA and the World Health Organization consider additives to be the least important health concern among the categories of substances in foods.

Pesticides

Because many are used on or near crops, residues of pesticides can be found in foods of both plant and animal origin. Such residues are regulated by the EPA un-

der provisions of the Food, Drug, and Cosmetic Act (residues in processed foods) and the Federal Insecticide, Fungicide and Rodenticide Act (residues on raw commodities). Like additives, pesticides are required to meet certain safety criteria prior to their EPA registration and introduction into commerce. The testing requirements and safety criteria are similar to those used for other additives, although EPA also requires extensive studies of environmental effects.

Pesticides have been of concern to the public for many decades. One of the first major public outcries about cancer-causing chemicals concerned a herbicide found in small amounts in cranberry sauce and declared by the federal government in 1959 to be carcinogenic in animals. In 1989 the animal carcinogen called Alar was reported to be present in some apples and apple products. Such cases have engendered significant public concern and confusion, usually because of new data not available when the pesticide was first registered for use.

Contaminants of Natural Origin

Most of the directly measurable mortality and morbidity associated with food consumption arises from microbiological contamination. The most important pathogenic bacteria are various species of *Salmonella*, *Staphylococcus*, and *Clostridium* (this last organism, now rarely encountered, can in improperly canned vegetables having low acidity produce the most deadly of known poisons, botulinum toxin). Foods of animal origin are most commonly affected by these bacteria. Most outbreaks of illness are associated with failures during food processing or handling. Education of food handlers and constant vigilance during processing and handling can prevent significant problems, but foodborne illnesses are still relatively common in industrialized societies and a major public health problem in developing countries.

Another class of food contaminants of natural origin are metabolic products of fungi that are called mycotoxins (from the Greek word for fungus, *mykes*). The historically most important of these substances are the ergot alkaloids (of which LSD is a derivative). Ergotism is the disease produced by ingestion of flour made from grains contaminated by metabolites of the fungus *Claviceps purpurea*. It was first recognized during the Middle Ages and periodic outbreaks have occurred all over the world ever since. The disease is characterized by hallucinations, convulsions, and ultimately gangrene of the extremities. Ergot contamination can be avoided by the removal of grains upon which the easily recognized fungus has grown. This is not the case with what is nowadays the most important of the mycotoxins, a collection of related chemicals called aflatoxins that are produced by certain species of *Aspergillus.* Affected commodities include corn, peanuts, other oilseeds, and certain tree nuts. If they are present because of mold growth on raw commodities, the aflatoxins can survive food processing and end up in marketed products, where they can be detected only by chemical analysis. Aflatoxins are potent animal carcinogens and are also carcinogenic in humans. The effects in human populations are exacerbated by the presence of hepatitis B virus. Regulatory controls have been instituted on a worldwide basis to limit human intake of aflatoxins, but everyone is exposed to some degree and, in certain populations that rely heavily upon affected commodities, exposure is probably excessively high.

Certain toxins may be present in fish and shellfish. Paralytic shellfish poison is a highly neurotoxic chemical that is produced by certain protista, the ocean blooms that cause some of the so-called Red Tides. The poison may accumulate in shellfish and make them poisonous.

Contaminants of Industrial Origin

Certain industrial products and byproducts have accumulated in various segments of environments, including foods of both animal and plant origin. Among these chemicals are a variety of chlorinated pesticides (e.g., DDT), commercial products such as polychlorinated biphenyls (PCBs), and dioxins, which are byproducts of certain manufacturing and combustion processes. These organic chemicals are not readily degraded in environments and are highly fat soluble, so they tend to accumulate in animal fats. Most people, even those not exposed occupationally, carry some body burden of chemicals in these classes, and the diet appears to be the major source.

Heavy metals such as lead, mercury, and cadmium, as well as other inorganic substances such as arsenic, may migrate from soils to plants to people. These substances are always present in foods because of their natural occurrence in soils and water, but may accumulate to excessive levels in areas that have been subjected to industrial contamination.

Non-Nutritive Natural Components of Foods

Foods naturally contain immense numbers of chemicals of no known nutritional value. Although most of these have not been characterized, some are known to exhibit serious forms of toxicity. Among these are neurotoxic alkaloids in white potatoes, toxic terpenes in nutmeg and other spices, and some animal carcinogens in certain herbs and vegetables. Plants generate such substances in response to stresses such as insect and

fungal invasion; some are natural pesticides, and some are toxic to mammals. It is not yet known whether natural toxicants present a threat to humans at their natural level of occurrence.

It is evident that dietary factors play a highly significant role in the development of those diseases that are the principal causes of mortality and morbidity. Presently available evidence would attribute most importance to certain of the nutrients and major constituents, and would find less risk associated with additives, pesticides, industrial and natural contaminants, and non-nutritive natural components. While most experts would support such a statement, no comprehensive evaluation of the relative risks posed by dietary constituents and contaminants has been undertaken.

In many of the developing countries, poor sanitation creates opportunities for microbial growth and high rates of foodborne illness. Inadequate nutrition, including but not limited to an insufficient supply of calories, still plagues more than 500 million people around the world. Both over- and under-supplies of calories and other dietary factors are substantial risk factors in the most important human diseases.

JOSEPH V. RODRICKS

For Further Reading: Catherine E. Woteki and Paul R. Thomas, eds., *Eat for Life: The Food and Nutrition Board's Guide to Reducing Your Risk of Chronic Disease* (1992); Joseph V. Rodricks, *Calculated Risks: The Toxicity and Human Health Risks of Chemicals in Our Environment* (1992); *FDA Consumer* (periodical).
See also Bacteria; Cancer; Carcinogens, Natural; Dioxins; Epidemiology; Food and Nutrition; Fungi; Health and Disease; Neurotoxins; Pesticides; Public Health; Risk; Salmonella; Toxic Chemicals; Toxic Metals; Toxicology.

HERBICIDES

Herbicides are chemicals used to destroy or inhibit plants, especially those considered pests. Other pesticides include insecticides and fungicides.

The history of herbicide development dates to the late 19th century with the use of inorganic compounds. For example, copper sulfate was used for the destruction of charlock (an annual weed in the mustard family) in cereal crops in the late 1890s. The first use of organic chemicals for weed control dates to 1935 when nitrophenols were used as selective herbicides. Perhaps the most important finding occurred in 1945 when it was noted that 2,4-dichlorophenoxyacetic acid (2-4-D) was effective against broad-leafed plants such as charlock and sugar beet, but it did not harm cereal plants. In 1949 about 10,000 tons of 2-4-D was manufactured in the U.S. In addition, several other classes of chemicals had expanded manufacturing in the late 1940s. The number of herbicides in general use in the U.S. and Canada by 1949 was about 25, used on 23 million acres of agricultural land. This increased to approximately 100 herbicides applied to 53 million acres by 1959. By 1976 the totals had increased to 165 herbicides applied to 200 million acres. However, recent tendencies have been toward slightly reduced use because of environmental concerns and health risks to applicators.

In spite of such concerns, there are many benefits from herbicide use. They reduce costs of hand and mechanical tillage, fertilizer, irrigation, harvest, grain drying, transportation, and storage, as well as the number of laborers required, crop yield losses, and acres needed for crop production. But the most important benefit is generally regarded as increased yields.

The increase in yields per acre of corn produced in the United States over the past fifty years has been very dramatic. The average production was 28.4 bushels per acre in 1940, 86.4 bushels in 1975, and 116.2 bushels in 1989. Smaller increases have been accomplished in soybean production (16.2 bushels per acre in 1940, 28.9 bushels in 1975, and 32.4 bushels in 1989) and in wheat production (15.3 bushels per acre in 1940, 30.6 bushels in 1975, and 32.8 bushels in 1989).

It should be emphasized that herbicide usage is only one of several reasons for the increases in production. Other factors include use of more powerful machinery, availability of improved crop hybrids, and use of more fertilizers. On the other hand, because of year-to-year variation in yields, the 1989 productions per acre are not maximum yields. In fact, the 1985 yields per acre are higher than the 1989 yields by approximately two bushels per acre for corn and soybeans and by nearly five bushels per acre for wheat.

The percentage of acres treated with herbicides is much higher for corn and soybeans than for wheat. In 1985, 97% of corn acres and 96% of soybean acres were treated with herbicides, compared to 79% and 68% respectively in 1971. In contrast, 47% of acres planted with wheat were treated with herbicides in 1989, compared to 41% in 1971. Of course, the comparable percentages in 1940 were virtually zero.

The use of herbicides does have a possible negative effect, however. Although mortality from all types of cancer is less in farmers than in nonfarmers (primarily the result of sharply lower deaths from smoking-caused cancers in farmers), mortality from stomach, lip, and prostate cancer, as well as from leukemia, multiple myeloma, and non-Hodgkin's lymphoma (NHL), has been shown to be higher in farmers than in nonfarmers. The latter three cancers of the blood and blood-forming organs have been associated with possible increased exposure of farmers to herbicides.

In particular, a study completed in Kansas showed that non-Hodgkin's lymphoma increased with increased exposure to the herbicide 2-4-D. The increased exposure was expressed primarily as days of use per year. NHL was also negatively associated with increased use of protective clothing, such as gloves and long-sleeved shirts. This result is important because it indicates that rather simple and inexpensive measures can be taken to reduce risk. Consistent with this conclusion is the fact that NHL was associated with the type of spraying utilized by farmers. Of particular concern was the use of hand-held sprayers and back-pack sprayers. These modes of spraying greatly increase the risk of dermal exposure. Subsequent studies of exposure modes indicate that exposure to unprotected hands when using hand-held devices is much greater than exposure to farmers on a tractor, particularly if the tractor has a cab and the farmer is wearing protective clothing.

Similar studies in states producing more corn than Kansas have shown similar results. The findings from Kansas were verified by a study completed in Nebraska. Of importance was the finding that NHL was associated with increased days of exposure to 2-4-D. Additional studies in Iowa and Minnesota also conclude that care should be exercised in the handling and application of herbicides.

Concern has recently been expressed about the findings of detectable levels of herbicides in groundwater and in wells, especially those whose depths are less than 100 feet. The concern is partly based on the possibility of interaction of some types of herbicide with nitrates, found in a greater percentage of rural wells than herbicides. This concern is based on the fact that possible cancer-causing agents can be formed in the pH conditions found in the human stomach. It is very likely that this route of exposure is far less serious than the dermal exposure to herbicides resulting from unprotected handling and application. Nevertheless, efforts are being made to reduce usage of nitrogen and herbicides without greatly affecting yields through more reliance on soil testing and continued careful study of application rates and yields.

In summary, increased use of herbicides is one factor that has led to increased yields of grains in the U.S. Care must be taken in the application of the herbicides because of possible health effects. However, with careful attention given to reduce exposure, it is likely that herbicides will continue to greatly aid farmers in their efforts to feed an ever-increasing world population.

LEON F. BURMEISTER

For Further Reading: Charles Benbrook and Phyllis Moses, "Engineering Crops to Resist Herbicides," *Technology Review* (November/December 1986); Leonard P. Gianessi and Cynthia Puffer, *Herbicide Use in the United States* (1990); James A. Dosman and Donald W. Cockcroft, *Principles of Health and Safety in Agriculture* (1989).

See also Agriculture; Fungicides; Insecticides; Pest Control; Pesticides.

HERODOTUS

(ca. 484–ca. 425 B.C.), ancient Greek historian. Acclaimed through the ages as the "father of history" because of his account of the Greco-Persian wars, Herodotus is of special interest to environmentalists. Although his accounts liberally embraced legends and tales, he described cultural traits and physical environments. He cited weather patterns and seasonal changes in assessing times past. He noted the shells and saline efflorescence of the desert and deduced the prior presence of the sea. He was thus the first to emphasize cause-effect links between humans and the environment.

Herodotus devoted almost all of one of the nine books in his *History of the Greco-Persian Wars* to Egypt and it best exemplifies his environmental interests. His musings on the rhythmic flooding of the Nile and a grand speculation on the effect of an alteration of its course, millennia before his own time, were later praised for being scientific conceptually, even though often wrong.

Herodotus is believed to have been born in Halicarnassus, in Asia Minor, the son of a prominent Asiatic Greek family. By the time he was 35 or 40, in his then comparatively long life span of about 50 to 60 years, he had traveled over much of the known world. He also engaged heavily in politics. He was exiled from his birthplace but returned to help overthrow its tyrant. He later himself became unpopular in his native city and left it to settle in Athens, the capital of his people and his culture. He subsequently helped found a new Greek city in Thurii, in southern Italy. There he completed his history and died.

Herodotus traveled by ship along the coast of Asia Minor to the northern islands. He scouted the shores of the Black Sea. He covered parts of Mesopotamia, Babylon, and Egypt, where he entered the western delta at Canopus and left Pelusiam in the eastern delta after about three months. He traveled up the Nile River to the first cataract.

Herodotus criticized earlier explanations of the river's recurrent flooding and offered his own: that the flooding was due simply to the rain-swollen lakes and streams in the Ethiopian highlands. This was proven true in the 19th century with the mapping and geological examination of the Upper Nile's tributaries, the White and Blue Niles.

Herodotus challenged accepted historical estimates of when the pyramid kings held sway. He calculated that the pyramids could not have been built when supposed, as the sites would still have been under water. His calculations were wrong but his method was scientific.

Herodotus also evaluated other environmental impacts. He compared estimates of the height that the Nile River waters were said to have reached in earlier floodings with the height to which the river would have to rise to have the same impact in his own time. He estimated how long it took for silt to build up along the river as well as in the delta. He speculated that at one time the Mediterranean might have had a huge gulf near the present Egyptian coast. He suggested that a change in the Nile's course might have caused it to empty into a previously larger Red Sea. He thus surmised that much of Egypt was a product of millennia of river silt. If that were so, he reasoned, it would have taken 5,000 years to create a land the size of Egypt itself.

Herodotus said the Egyptians were lucky to have their river. He compared the ease of the Egyptian farmer's life on the banks of the Nile with the hardscrabble life of the Greek farmer. All the Egyptians had to do, Herodotus wrote, was to wait for the flooding to subside and then move in easily to farm the rich, silted residue. Known for his mix of probable fact and fancy, Herodotus was criticized by some scholars until recently for his accounts of how Egyptian farmers used herds of swine as dray animals to cultivate the river banks. Not possible, it was said, in view of the Arabs' abhorrence of the animal as food. Subsequent research uncovered writings and drawings that confirmed that the Egyptians of Herodotus' time had indeed raised swine on the river banks.

Herodotus' observations on the Nile River had a fascinating resonance thousands of years later when the Great Aswan Dam was built in the late 1960s. For once again the course of the river, the flooding of its banks, the pace and quantity of its silt deposits were the stuff of politics, economics and history—this time at the height of the Cold War between the United States and the former Soviet Union. Just as Greek and Persian ambitions met in Egypt, so did American and Soviet diplomacy converge there. Given his predilection for observation beyond the battlefield, a latter-day Herodotus might well have speculated upon the environmental as well as political impact of the great new dam over which the Communist Soviet Union successfully contended with the United States for the privilege of providing engineering and economic assistance.

Since ancient times, Egyptians have sought to enhance the Nile's blessing with dams to control the flooding and expand the arable land. Senusret II built a wall 27 miles long to gather the waters of the Fayoum basin and reclaim 25,000 acres of marshland. The new Aswan dam was constructed in the early Sixties four miles south of an old one. The new artificial lake allowed the cultivation of two million additional acres of land—one-third more than in 1965. The new dam's hydroelectric plant tripled Egypt's electricity output.

But the new dam also has had its environmental downside. Critics note it has limited the annual deposits of the silt Herodotus measured so many centuries ago and it has forced increased use of artificial fertilizer. This is not only less efficient than the natural silt deposits but more expensive. The new silt formations in the delta also have led to a proliferation of a water snail that carries the bacteria of schistosomiasis, an eye disease that is a growing scourge in the area.

Herodotus' environmental interests were not as specific as the concerns that make up the environmental agenda today. Yet he found it pertinent to examine the merits of the ancient Egyptians' use of tightly woven fishnets to protect against gnats and mosquitos. He suggested the Greeks emulate them. Thus Herodotus made one of the earliest recorded proposals for environmental technology transfer.

JACK RAYMOND

For Further Reading: Herodotus, *History of the Graeco-Persian Wars,* in Greek and with an English translation by A. D. Godley, Books I, II (1981); Wilhelm Spiegelberg, *The Credibility of Herodotus' Account of Egypt,* translated by A. M. Blackman (1927); Kimball Armayor, *Herodotus' Autopsy of the Fayoum: Lake Moeris and the Labyrinth of Egypt* (1985). *See also* Aswan High Dam.

HIGHWAY TRUST FUND, FEDERAL

The Federal Highway Trust Fund (HTF) was created by the Federal-Aid Highway Act of 1956 to provide a stable funding source for constructing the 40,000-mile interstate freeway network. Since there was no "Interstate" funding—only regular federal highway funds and the regular 50% federal funding ratio—less than one percent of the system (authorized in 1944) was built by 1952. Interstate construction moved forward rapidly after passage of the 1956 law, which set the federal share of interstate costs at 90%; authorized $25 billion specifically for Interstates; and captured for road purposes many road-use taxes that previously had gone to general government purposes.

The per-gallon federal motor fuel tax began as 1 cent in 1932, rising to 2 cents in 1951, 3 cents in 1956, 4 cents in 1959, and 9 cents in 1983. Since 1984 the diesel fuel tax has been 6 cents a gallon higher than the gasoline tax.

In December 1990 the gasoline tax rose from 9 cents to 14 cents, but half of the increase went to "deficit reduction," so HTF revenues rose only to 11.5 cents. (The same law imposed the first-ever federal fuel tax on railroads—2.5 cents for deficit reduction.)

Even adding in state and local taxes, the U.S. average 34-cent-a-gallon total is far smaller than the per-gallon gasoline taxes of the following countries: Italy—$3.56; France—$2.80; Germany—$2.59; Great Britain—$2.21; and Japan—$1.63 (averages for the first half of 1992).

Moreover, the U.S. is unusual in earmarking most road fuel tax revenues for road construction. According to the U.S. Department of Transportation's *National Transportation Strategic Planning Study* (1990): "In Europe, road tax policy is influenced by the desire to raise general revenues, suppress automobile and truck use, and reduce the level of subsidies to local public transit. . . . In both the Netherlands and Great Britain, road users get back about 25% of their road fees in road spending. Most other European countries spend about $1 on highways for every $3 received from highway users."

To fully appreciate the relative advantage U.S. road-building interests enjoy, however, the growing volume of nonuser tax dollars spent on U.S. roads should be noted: $20.5 billion in 1991 (up from $4.0 billion in 1971 and $12.5 billion in 1981). The 1991 level was more than the highway trust fund's $15.7 billion and more than 27% of the $48.8 billion total U.S. highway expenditures that year. The biggest single nonuser tax category was municipal general fund appropriations ($7.7 billion in 1991). In 1983 a mass transit account was created within HTF, funded from 1 of the HTF's 9 cents and rising in late 1990 to 1.5 out of every 11.5 cents. However, the $2.4 billion in total highway user taxes spent on mass transit in 1991 was far less than the $20.5 billion in nonuser taxes spent on roads.

The highway trust fund does have some disadvantages. First, although recent changes in the law may lead to more use of nontransit-account trust fund monies for mass transit, the continuing popularity of spending most road funds on roads reflects a faulty definition of the user. Under this system the gasoline tax paid for local driving can go to distant freeways but not to mass transit improvements in the local community. Many people have called for a unified transportation trust fund and policy to replace existing highway and aviation trust funds and policies.

Second, the trust fund distorts overall priority-setting efforts by giving the purposes it supports an artificial edge over public functions lacking a trust fund. Also, generous federal spending ratios similarly tilt localities and states toward road spending, as reflected in the growth of nonuser road expenditures already noted.

ROSS CAPON

HINDUISM

The teachings of Hinduism were first recorded in the Vedic hymns. The exact period of their composition is uncertain, but the present texts are over three thousand years old. They address the immanent deities of the natural elements, such as Indra, the god of rains, Vayu, the wind-god, and Agni, the god of fire. These hymns were followed by the philosophical poems of the Upanishads. Pervading these were the great deities of Shiva and Vishnu, around whom developed the two principal branches of Hinduism: Vaishnavism, which focuses on Vishnu and his avatars; and Shaivism, which follows Shiva. In essence these traditions are two sides of one coin.

The two great epics, Ramayana and Mahabharata, which tell the stories of Rama and Krishna and include the essential teachings of the Hindu tradition, are today the most popular of the Hindu scriptures. Contained in the Mahabharata is the Bhagavad Gita, the "Song of God," the essence of Hindu wisdom. In it Krishna describes the eternal self and the all-pervasive presence of God in this world. He also teaches the path of bhakti, devotion to God, as the easiest path of spiritual enlightenment.

The Hindu universe is conceived as the form of the Cosmic Person. Vedic cosmology divides the space inside the universe into fourteen layers of planetary systems, from the Patala planets, which are the soles of His feet, to the heavenly planets called Satyaloka, which are His one thousand heads. An ancient Vedic hymn called Purusha Sukta describes this universal form of the Cosmic Person and relates how all within this world is part of it. This hymn is recited every day by priests and devout Hindus as part of their worship of Vishnu.

The daily appearance of the sun is greeted as an auspicious moment for meditation. At sunrise, noon, and sunset brahmanas recite the *gayatri* mantra, calling upon the sun, which illuminates the earthly and heavenly realms, to enlighten the mind of the meditator with divine inspiration. The sun is the eye of Vishnu, which sees all; by its energy all living things flourish.

The Vedic literatures personify Earth as the goddess Bhumi, or Prithvi. She is the abundant mother who showers mercy on her children. Her beauty and pro-fusion are vividly portrayed in the beautiful Hymn to the Earth in the Atharva Veda:

> May you, our motherland, on whom grow wheat, rice, and barley, on whom are born five races of mankind, be nourished by the cloud, and loved by the rain. O mother, with your oceans, rivers and other bodies of water, you give us land to grow grains, on which our survival depends. Please give us as much milk, fruits, water, and cereals as we need to eat and drink.

Although Hindus believe in many lesser gods called demigods, they worship one Supreme Being, the source of all others. Vaishnavas know Him as Vishnu, the Lord of all creation. Before this world existed, before there were any demigods or sun, moon or stars, Vishnu existed in His own eternal realm.

It is said that the oceans are Vishnu's waist, the hills and mountains are His bones, the clouds are the hairs on His head and the air is His breathing. The rivers are His veins, the trees are the hairs on His body, the sun and moon are His two eyes and the passage of day and night is the moving of His eyelids. In the words of the Bhagavad Gita:

> Everything rests on Me as pearls are strung on a thread. I am the original fragrance of the Earth. I am the taste in water. I am the heat in fire and the sound in space. I am the light of the sun and moon and the life of all that lives.

For Hindus, this world is not made of inanimate matter, to be wasted and exploited for selfish ends. When they see the sunrise and feel its scorching heat, when they taste water or smell the Earth in the monsoon rains, they are reminded of Vishnu. Vishnu is both inside and outside this world, and it cannot be separated from him. All is sacred, God-given and mystically created.

Hindus did not know that they were Hindus until Europeans told them so, nor that their land was called India. To them it was, and still is, Bharat, and has been so named since the time of the great emperor Bharat, whose life is recounted in their ancient histories. As for their religion, it did not have a name, any more than existence itself, because to live in Bharat meant to share a way of life common to all, a spiritual and material culture so all-pervading as to be invisible to those within it, like the air they breathed.

The word *Hindu* was used by the British rulers of India in the nineteenth century. It came from the Persians, whose Muslim descendants ruled India for close to a thousand years. They derived it from the river Indus, which flowed through the northwestern plains of the sub-continent and gave its name to the land and its people. How apt that in naming the religion of India, we should call it after its bio-region. Hindus, with their reverence for sacred rivers, mountains, forests, and animals, have always been close to nature.

If Hinduism can be given a legitimate name it is "Sanatan Dharma," which is used by many Hindus today. Roughly translated from the Sanskrit, this means "the eternal essence of life." This essence is not limited to humans. It is the essential quality that unites all beings — human, animal, or plant — with the universe that surrounds them and ultimately with the original source of their existence, the Godhead. This perception of underlying unity is what causes Hindus to steadfastly refuse to separate their religion from their daily life, or to separate their own faith from the other great faith traditions of the world.

Today there are over 800 million Hindus, and their number and influence are increasing. As India gains in stature in the world as an independent democracy and as Hindu communities multiply in other countries it is only natural that Hindus will want to revive their cultural and religious identity, which has been partially obscured for so long, first by Islamic and then by British rule. The world has little to fear from a revitalized Hinduism. Hindus uphold the ideal of non-violence even to the point of vegetarianism, and have always believed in religious tolerance, and in a multiplicity of forms of God and paths of worship. To Hindus, all religions are part of the process of discovering the unity of God, humanity, and nature.

RANCHOR PRIME

For Further Reading: Kerry Brown, ed., *Essential Teachings of Hinduism* (1988); Ranchor Prime, *Hinduism and Ecology* (1992); Alan W. Watts, *Nature, Man and Woman* (1970).

HISTORICAL ROOTS OF OUR ECOLOGICAL CRISIS

The role of Judeo-Christian teachings with regard to environmental problems came to be seen in a new light following a statement made in 1953 by the Japanese Zen Buddhist Daisetz T. Suzuki and later reformulated by Lynn White, Jr., in an article entitled, "The Historical Roots of our Ecological Crisis." It is a measure of its popular success that this article has been reproduced extensively. Whether valid or not, White's thesis demands attention because it has become an article of faith for many conservationists, ecologists, economists, and even theologians.

The thesis runs approximately as follows: The ancient Oriental and Greco-Roman religions took it for granted that animals, trees, rivers, mountains, and other natural objects can have spiritual significance just like humans and therefore deserve respect. In contrast,

according to the Judeo-Christian religions humans are apart from nature. The Jews adopted monotheism with a distinctly anthropomorphic concept of God. The Christians developed this trend still further by shifting religion toward an exclusive concern with human beings. It is explicitly stated in Genesis 1:28 that man was shaped in the image of God and given dominion over creation; this has provided the excuse for a policy of exploitation of nature, regardless of consequences. Christianity developed, of course, along different lines in different parts of the world. In its Eastern forms its ideal was the saint dedicated to prayer and contemplation, whereas in its Western forms it was the saint dedicated to action. Because of this geographical difference in Christian attitudes, the most profound human impacts on nature have occurred in the countries that make up what is known as Western civilization. To a large extent modern technology is an expression of the Judeo-Christian belief that humans have a rightful dominion over nature. Biblical teachings thus account for the fact that Western people have had no scruples in using the earth's resources for their own selfish interests or in exploring the moon to satisfy their curiosity, even if this means raping nature and contaminating the lunar surface.

Since the roots of the environmental crisis are so largely religious, the remedy must also be essentially religious: "I personally doubt," White wrote, "that ecologic backlashes can be avoided simply by applying to our problems more science and more technology." For this reason, he suggested that the only solution may be a return to the humble attitude of the early Franciscans. Francis of Assisi worshiped all aspects of nature and believed in the virtue of humility, not only for the individual person but for humans as a species; we should try to follow in his footsteps, so as to "depose man from his monarchy over creation, and abandon our aggressive attitude toward Nature." White concluded, "I propose Francis as a patron saint of ecologists."

This hypothesis seemed so completely at odds with historical facts to René Dubos that he read more than 100 articles and books on the subject by lay scholars and theologians to better understand its meaning and implications. In Appendix III of *The Wooing of Earth* Dubos listed these publications "in the hope that they will stimulate a more rigorous analysis of the influence that religious beliefs or doctrines have exerted on human attitudes toward the Earth." From his survey, Dubos drew four conclusions, which will be treated in turn below.

1. Extensive and lasting environmental degradation occurred long ago in many places where people had not had any contact with biblical teachings — in many cases long before the biblical writings.

2. Ruination of the land was usually the consequence of deforestation, exploitative agriculture, and ignorance of the long-range consequences of farming practices.

The process of environmental degradation began ten thousand years ago. A dramatic extinction of several species of large mammals and terrestrial birds occurred at the very beginning of the Neolithic period, coincident with the expansion of agricultural culture. Humanity's eagerness to protect cultivated fields and flocks may account for the attitude, "if it moves, kill it," which is rooted deep in folk traditions over much of the world. Nor was the destruction of large animals motivated only by utilitarian reasons. In Egypt the pharaohs and the nobility arranged for large numbers of beasts to be driven into compounds where they were trapped and then shot with arrows. The Assyrians, too, were as vicious destroyers of animals—lions and elephants, for example—as they were of people. Ancient hunting practices greatly reduced the populations of some large animal species and in some cases led to their eradication. This destructive process has continued throughout historical times, not only in the regions peripheral to the eastern Mediterranean, but also in other parts of the world. In Australia the nomadic aborigines with their fire sticks had far-reaching effects on the environment. Early explorers commented upon the aborigines' widespread practice of setting fires which, under the semiarid conditions of Australia, drastically altered the vegetation cover, caused erosion, and destroyed much of the native fauna. Huge tracts of forest land were thus converted into open grasslands and the populations of large marsupials were greatly reduced.

Plato declared his belief that Greece was eroded before his time as a result of deforestation and overgrazing. Erosion resulting from human activities probably caused the end of the Teotihuacan civilization in ancient Mexico. Early people, aided especially by that most useful and most noxious of all animals, the Mediterranean goat, were probably responsible for more deforestation and erosion than all the bulldozers of the Judeo-Christian world.

Nor is there reason to believe that Oriental civilizations have been more respectful of nature than Judeo-Christian civilizations. As shown by the British scientist and historian Joseph Needham, China was far ahead of Europe in scientific and technological development until the 17th century A.D., and used technology on a massive and often destructive scale. Many passages in T'ang and Sung poetry indicate that the barren hills of central and northern China were once heavily forested, and there is good reason to believe that, there as elsewhere, the loss of trees and soil erosion are results of fires and overgrazing. Even the Bud-

dhists contributed largely to the deforestation of Asia in order to build their temples; it has been estimated that in some areas temple-building was responsible for much more than half of the timber consumption.

One of the best-documented examples of ecological mismanagement in the ancient world is the progressive destruction of the groves of cedars and cypresses that in the past were the glory of Lebanon. The many references to these noble evergreen groves in ancient inscriptions and in the Old Testament reveal that the Egyptian pharaohs and the kings of Assyria and Babylon carried off enormous amounts of the precious timber for the temples and palaces of their capital cities. In a taunt against Nebuchadnezzar, king of Babylon, the prophet Isaiah refers to the destructive effects of these logging expeditions. The Roman emperors, especially Hadrian, extended the process of deforestation still further. Today the few surviving majestic cedars are living testimony to what the coniferous forests of Lebanon were like before the ruthless exploitation, which long preceded the Judeo-Christian and technological age.

3. The early teachers of the Judeo-Christian-Moslem religions expressed great concern for the quality of Nature and advocated practices for its maintenance.

In ancient Judaism, for example, a religious commandment ordered farmers to take fields out of cultivation every seventh year, a practice of great ecological value since it helped to maintain the fertility of the soil. Similar recommendations can be found in the writings of many Christian and Moslem teachers.

René Dubos hastened to acknowledge that the professed ideals of a culture, like those of its politicians, are rarely translated into actual practice, but this is at least as true of Oriental as of Western peoples. China was one of the countries that experienced the worst deforestation and erosion, despite the professed reverence of Chinese scholars, artists, and poets for wilderness sceneries and for nature in general. Similarly, the teachings of St. Francis of Assisi probably had little if any influence on the destruction of wildlife by Italians and other Europeans.

4. People destroyed more of Nature as the centuries went by, not because they had lost respect for it, but because the world population increased and also because technological means of intervention became more and more powerful.

The plains Indians lived in "harmony with Nature" as long as their impact in the hunt was limited to what they could do with bows and arrows, but they decimated the herds of bison once they could hunt them from horseback with firearms. As long as the Caucasians had only axes of stone and of metal, it took them centuries and even millennia to destroy a large percentage of their forests, but power equipment now makes it possible to clearcut immense areas in a very few years.

René Dubos concluded that the historical origin of the present ecological crisis is therefore not in Genesis 1:28 but in the failure of people to anticipate the long-range consequences of their actions and, in recent times, this shortsightedness has been aggravated by the power and misuse of modern technology.

RENÉ DUBOS
(EDITED BY WILLIAM R. EBLEN)

For Further Reading: René Dubos, *The Wooing of Earth* (1980); Gerard Piel and Osborn Segerberg, *The World of René Dubos* (1990); David Spring and Eileen Spring, eds., *Ecology and Religion in History* (1974).

HOMELESSNESS: CAUSES

Homelessness is often defined as the absence of a place to sleep and receive mail. However, the condition is typified by a range of housing and social arrangements. Homeless people sleep on streets or sidewalks but also in shelters, cars, abandoned buildings, or welfare hotels, or doubled-up with others. Thus a more inclusive definition suggests a continuum between "homed" (or domiciled) and "homeless" street dwellers, with those living in temporary, insecure, or physically substandard dwelling units falling somewhere between the two extremes.

Today's homeless population is diverse and varies considerably from place to place. In comparison with past periods when most homeless people were older white males, the homeless population is dominated by younger individuals and people of color. Many of the men are Vietnam-era veterans. Most have never married or are currently single, but homeless families, especially female-headed families with children, are the fastest growing segment of the homeless population. Rates of mental disability and substance abuse are high. Most homeless people are as well educated as the general population, and are long-term residents of their communities.

Most observers recognize that homelessness is the result of a complex process of social and personal disadvantage, and/or progressive isolation from networks of family and friends. A high proportion of homeless people are extremely poor, distanced from social ties, or have personal vulnerabilities or adjustment problems (e.g., mental disability, substance abuse, criminality). But homelessness also has its roots in deep-seated social, economic, and political trends. Since the 1970s, the demand for social services, welfare, and assisted housing has greatly increased, largely as a consequence of economic recession and restructuring, and demographic change.

National unemployment rates rose from 5.9% to 10.7% between 1979 and 1982, leaving a total of 12 million workers jobless; the number of discouraged workers doubled, to 1.2 million. Between 1981 and 1989, the minimum wage remained at $3.35 per hour, while the cost of living rose 39%. Over one million jobs *net* were lost in the manufacturing sector during 1980 to 1988; between 1979 and 1984, 44% of net new jobs created paid poverty-level wages. Personal income of the poorest one-fifth of American families fell, especially among female-headed households and minorities.

With respect to demographics, a bulge in the size of the working-age population depressed their earnings and increased their likelihood of unemployment. Between 1968 and 1984, the real earnings of workers under 35 declined, those of older workers rose; unemployment rates among the younger cohort grew from 5% to 15% while rates for older workers remained stable. The number of single-person households grew 112% between 1970 and 1990, increasing demand for small, inexpensive housing units. Female-headed households grew faster than any other household type: 91% between 1969 and 1979, and another 15% between 1979 and 1988; rates of unwed parenthood rose from 5.1% to 14.5% between 1969 and 1986, and marriage rates fell from 83% to 62% between 1960 and the mid-1980s. These factors were linked to loss of spousal support and thus higher poverty rates among women, especially women with children (46% compared to 10% among other household types). Growth in women's employment meant greater economic independence, but their relatively low wages reinforced the feminization of poverty. Lastly, continued growth in the elderly population meant higher demand for specialized housing and social service supports.

At the same time that demand for public assistance grew, the volume of services and affordable housing available to people in need declined, as government retreated from earlier commitments to the welfare state. Between 1982 and 1985, federal programs targeted to the poor fell by $57 billion. Large numbers of public assistance recipients lost their benefits as a result of welfare policy shifts, and the real value of AFDC payments dropped by one-third between 1970 and 1985. Some states eliminated or reduced their General Assistance programs for single poor people; for example, the real value of an Illinois General Assistance payment in 1985 was less than half its 1968 level. In addition, deinstitutionalization policies discharged hundreds of thousands of people from psychiatric hospitals and other institutions into the community, where without adequate support many became destitute.

Other human service programs were privatized and their public funding reduced.

The supply of affordable housing dramatically declined throughout the 1970s and 1980s, largely as a consequence of gentrification, urban renewal, and the loss of Single Room Occupancy housing and other kinds of affordable units. Between 1973 and 1983, 4.5 million units were removed from the nation's housing stock, half of which had been occupied by low-income households. Loss of low-cost units continued unabated throughout the decade. In addition, the federal government drastically reduced its role in subsidized housing. Budget authorizations for housing were cut from $28 billion in 1977 to $19.8 billion in 1981, and fell to $7.5 billion in 1991. The public housing program shrunk, essentially replaced by a voucher system that forced recipients to compete for a shrinking number of affordable housing units on the open market. Exclusionary zoning practices and community opposition (the "Not-in-My-Backyard" or NIMBY syndrome) limited the expansion of social service and affordable housing opportunities in many (especially suburban) jurisdictions.

Taken together, these trends have created a crisis for many economically marginalized and precariously housed people. The experience of an adverse life event is frequently sufficient to cast such individuals into homelessness. The five most commonly cited precursors of homelessness are: eviction, job loss, release from an institution with nowhere to go, loss of welfare support, and personal crises (e.g., domestic violence). Such events are best understood as the climax of an extended period of cumulative personal difficulties.

Numerical estimates of homelessness vary widely in scope and quality. These range from 350,000 to 3 million homeless. Recent enumeration efforts have produced middle-range estimates of between 446,000 and 600,000 during a one-week period in 1987, and 700,000 on a specific day in 1991. Using estimates of the median duration of homeless episodes and the 1991 single-point-in-time estimate of 700,000, between 840,000 and 1.1 million Americans experienced an episode of homelessness during 1991.

Despite debate over baseline figures, service providers report dramatic increases in demand for emergency food and shelter. For example, the largest soup kitchen in Washington D.C. reported serving fewer than 100 meals per day in 1980, but was serving more than 1000 a decade later. The number of adults and children housed in emergency shelters rose from 98,000 in 1983 to 275,000 in 1988—a 180% increase in just five years.

JENNIFER R. WOLCH

For Further Reading: Michael Dear and Jennifer Wolch, *Landscapes of Despair: From Deinstitutionalization to* *Homelessness* (1992); Marjorie Robertson and Milton Greenblatt (eds.), *Homelessness: A National Perspective* (1992); Peter Rossi, *Down and Out in America* (1989).
See also Demography; Homelessness: Urban Environment; Housing; Communities, Low and Moderate Income.

HOMELESSNESS: URBAN ENVIRONMENT

The nature of the urban environment shapes the ways homeless people cope with their circumstances, and come to see and value themselves as individuals. Historically, homeless people have clustered in older "Skid Row" zones of American cities, and they continue to do so today. Typically on the fringe of the central business district, Skid Row zones are places of environmental degradation, and economic and social marginalization. Changing land-use pressures have led to the demolition of Single Room Occupancy (SRO) housing and inexpensive retail and personal services that once characterized these neighborhoods, but human service agencies—missions, soup kitchens, drug/ alcohol treatment centers, and shelters—remain rooted there. They serve an increasingly diverse homeless population that includes young people, people of color, deinstitutionalized mentally disabled, women, families with children, substance abusers, ex-offenders, and economically dislocated persons.

As physical environments, Skid Row zones are usually landscapes of despair, hard-edged and inhospitable. They are often devoid of trees or other landscaping; sidewalks are stained and dirty, and in some areas lined with old trash cans used as firepits. Parking lots are barren expanses surrounded by cyclone fencing or barbed wire. Despite widespread demolitions, Skid Row districts retain SRO hotels, many in disrepair. In addition to hotels there are typically low-rent apartments, missions and emergency shelters, and social service agencies. Sidewalks and open spaces, especially parks, are heavily used by homeless residents. Surveillance activity is typically omnipresent. Police vehicles cruise the streets, and other social control mechanisms may also operate. For instance, overhead sprinklers outside some service agencies wet the sidewalk periodically to prevent loitering and curbside encampments. The signs and symbols of despair and deprivation are everywhere: used needles and syringes, cocaine pipes, and liquor bottles along with cast-off clothes, broken-down cookstoves, and cardboard-box shelters.

Such urban environments support Skid Row populations having several components. In any Skid Row area, there are two residential groups. One consists of homed (or domiciled) residents, living in SRO hotels and cheap apartments. Homed residents include day laborers and low-wage downtown workers. A second resident group is the homeless. The demographic profile of Skid Row residents varies by metropolitan region, but in any one Skid Row community profiles of homed and homeless residents tend to be similar. Indeed, a significant share of Skid Row residents cycle back and forth between homed and homeless status. Lastly, in addition to residents, there is a significant daytime population of human service providers and workers in the Skid Row businesses.

Despite the common perception of Skid Row districts as transient no-man's lands, residents and workers together clearly constitute an urban community. As such, these areas are characterized by social networks, composed of people who offer one another material, emotional and/or logistical support. These networks link homeless and homed residents; connect residents to service providers and the business sector; and tie service providers to the business community.

Skid Row social networks are especially crucial to homeless people, whose social supports are predominantly community-based. The networks of the homeless consist of *homed* and *street* elements. The homed portion includes remnants of the social network from prior (domiciled) periods; panhandling "clients" or donors; employers and workmates in casual labor; social workers and other service providers. The street component, in contrast, includes other homeless people: friends and family; homeless lovers/spouses; informal communities based in street encampments; and members of homeless political organizations. Both homed- and street-based ties provide finances, food, transportation, psychological support, and a variety of social services. However, networks can also be damaging, for example abusive spousal relationships, conflict between the homeless and hostile business owners, and victimization of homeless residents through crime.

Whether positive or negative, community networks of the Skid Row homeless serve to bring continuity to their daily activity paths as they seek out the same people at the same general places on an everyday basis. Moreover, social relationships can substitute for what would be fixed stations in the daily path of a domiciled person, such as home or work. For example, one homeless woman leaves her husband with all their belongings at their sleeping area, allowing her to depart and secure resources for survival. At day's end, she returns to the security and protection of her husband. Over time, their sleeping place varies but her day begins and ends wherever her husband is staying. Similarly, homeless people who panhandle often see this activity as analogous to a job, with set hours and habitual donors or "clients," although the panhandling spot may change periodically due to police interventions.

In general, community networks and daily paths in Skid Row environments affect the identities and self-esteem of the homeless populations living there. Social ties and the environment may have both positive and negative effects on self-definition and morale. A "self-as-homeless" identity may be readily adopted, if the experience of homelessness creates a clearly defined role, recognition (as a leader or nurturing figure, for instance), notoriety, or other forms of attention previously unavailable to the individual. Devastating impacts on identity and self-esteem are more common, however. The immediacy of survival needs, precariousness of social resources, and the devalued and degraded physical environment of Skid Row can lead homeless people to postpone strategies to improve their life chances and to resign themselves to a negative "self-as-homeless" identity, deteriorating self-esteem, and hopelessness. In the words of one homeless Skid Row man, "A slum area is a slum attitude, they can keep you in a slum attitude by keeping you in slum places."

JENNIFER R. WOLCH

For Further Reading: Michael Dear and Jennifer Wolch, *Landscapes of Despair: From Deinstitutionalization to Homelessness* (1992); Jonathan Kozol, *Rachel and Her Children* (1988); Stacy Rowe and Jennifer Wolch, "Social Networks in Time and Space: Homeless Women in Skid Row, Los Angeles," in *Annals, Association of American Geographers* (1990).
See also Homelessness: Causes.

HORTICULTURE

See Agriculture; Organic Gardening; Urban Parks.

HOST-COMMUNITY BENEFITS

Compensating communities for hosting various types of land use, such as electricity generation, hazardous and other material landfilling, or nuclear waste handling, is referred to as host-community benefits. Since the host community must bear the costs associated with some land uses, current payments are required to prevent the host community from becoming worse off as a result of the land use. The costs vary by land use and by location.

One type of cost is amenable to insurance or bonding schemes, because its occurrence is uncertain. A second type of cost is not amenable to insurance or bonding schemes, because it is borne by the host community when the siting occurs, rather than becoming the subject of uncertainty at some future date.

An example of an insurable cost is dealing with water contamination. If monitoring reveals that a landfill (or other land use) has contaminated the host community's water supply, the landfill operator may promise to pay the cost of refreshing the water supply. Bonds can be required as a demonstration of the depth of the promise to pay, or the landfill operator may pay for insurance against such contingencies. Another example of an insurable cost is a decline in property value as a result of negative public perception toward the land use. An example of uninsurable costs is when perceptions of diminished quality of life by host community residents must be compensated by direct payment, so that local residents do not suffer a loss in well-being.

When the siting process fails to account for these costs, host communities face living with the land use without any compensation. As a result, local residents may balk with a NIMBY (not-in-my-backyard) response, and the siting process often devolves into an argument over the rationality of local concerns. But, regardless of whether expert and local opinion on the risks associated with the land use converge, NIMBY sentiment causes these land uses to be more expensive. If the land use is socially valuable, policy makers are challenged with reducing the cost and facilitating the siting.

One method of reducing these costs is a negotiated package of host community benefits (HCBs). By acknowledging that compensation can overcome NIMBY sentiment, HCBs enhance the probability of siting socially beneficial land uses. Examples of the various types of HCBs that have been used in siting solid waste landfills include free water tests, water quality guarantees, paying present owners property value losses on sale, and landscaping to hide the landfill.

However, HCBs are not without opponents. The scientific issue is whether concerns among host-community members about land use risks are justified. The distributional issue is whether cost generators or host communities should bear the cost of the land use. Although the scientific issue is important, land use demands often cannot wait for the scientific issue to be resolved. The distributional issue is actually a question of fairness. HCBs are viewed as enhancing efficiency and fairness. Efficiency is enhanced because the actual costs borne by host communities are included in the benefit-cost comparison accompanying land-use siting. Fairness is enhanced because host-community members participate in the decision-making process. Increased use of HCBs can be expected to reduce the cost of socially beneficial land use, given current knowledge about the risks associated with such land use.

RODNEY FORT

HOUSING

Functions of Housing

The provision of housing includes the assemblage of physical components to make a domicile. However, housing transcends narrow technological considerations and fulfills a much broader range of functions. Here, a distinction can be made between functions for users and producers of housing. Among users several needs prevail. A good house provides shelter from the cold, heat, rain, and other adverse climatic conditions. It offers protection against undesirable intruders as well. Through its location, housing provides access to neighbors, friends, relatives, jobs, shops, and community facilities and services that support the household's functioning. In addition, housing is many times seen as a symbol and reflection of the occupant's status and identity. Over time, sentiments of attachment to housing and its environment frequently develop, further enhancing the psychological meaning of housing as a home. Often housing also represents a significant financial investment.

Among housing producers, economic considerations tend to be of overriding importance. This is particularly so in market systems as found, in various forms, in the U.S., Canada, Australia, and Japan. In these systems, housing provision is driven by the profit motive of developers, builders, real estate agencies, lending institutions and other actors with a stake in the production or trading of housing. Resulting risk-aversive investments may find expression in discrimination against economically and politically marginalized population groups who cannot translate their housing needs into an effective market demand. For architects and planners, housing is also a potential source of professional prestige. This may produce environments that are ill suited to the needs of the users.

Housing provision has rarely been a high priority for governments. More typically, housing is treated as a tool to attain other public policy objectives, considered more important. For example, residential construction has numerous forward and backward linkages with multiplier effects in national economies. Hence, it can be a powerful lever to influence levels of unemployment and inflation. Following World War II, industrialized countries incorporated house building into schemes for economic development. The aim was to create, strengthen, and direct consumer demand in order to absorb rejuvenated productive capacities, previously diverted to the war effort. An outcome has been the rapid expansion of home-and-garden and do-it-yourself industries.

Housing has also been an instrument in various population-related policies. Dominant among them have been plans for (re)distribution of the population. Examples include Britain, the Netherlands, Sweden, Israel, Egypt, Tanzania, Hong Kong, India, Cuba, and China. Such plans have often been elaborated as part of broader strategies for urban and regional development. Objectives in this connection have been to curtail adverse effects of overurbanization, to protect scarce agricultural land, to reduce regional disparities, and to enhance military capabilities. In a related vein, there have been attempts (for example, in the U.S., Britain, and Israel) to achieve the integration of diverse population groups through the judicious distribution of housing. In the U.S. during the 1950s there were limited attempts at racial integration in public housing and later the new town of Reston (VA) included a deliberate private sector effort along similar lines. In 1988 the defiance by the city council of Yonkers (NY) of a federally ordered desegregation plan calling for the construction of housing units for low- and moderate-income households in mostly white neighborhoods became a symptomatic illustration of the limited national success of efforts at residential integration in the U.S. In Singapore housing has served as an instrument of social control to help quell popular dissent. In South Africa it has been effectively used to help implement apartheid policy.

Evolution of Housing Provision

Prior to the Industrial Revolution, and still in many parts of the less industrialized world and rural areas in Eastern Europe, house builders were close to their materials and techniques of construction. They usually inhabited the shelters they produced and subsequently altered them as necessary.

A fundamentally different situation emerged following industrialization. The integrated roles of traditional inhabitant builders were increasingly assumed by a growing cadre of professional specialists employed by newly formed production entities, for example, architectural firms, development corporations, and construction companies. At the same time, innovative technologies relaxed physical constraints, making possible the manufacture of prefabricated components, high-rise building, and large-scale residential development. Another aspect of modern housing provision is its encapsulation in complex government regulations which selectively guides development through mandatory material standards, building codes, zoning ordinances, taxation, and subsidization. All of these processes have combined to produce nationally distinct systems of housing provision.

Over time, the functional differentiation led to a growing gap between the environments needed or desired by users and those created by the producers. In recent decades there has been increasing research on the needs of users and ongoing work aims to translate this information into guidelines for the design, planning, and management of housing. However, observed mismatches between residents and their housing have not only been the result of a lack or deficiency of information about user needs. As noted, a number of rationales can underlie housing provision; resident needs are just one of them. Nevertheless, not all systems of housing provision are alike, and several basic models can be distinguished.

Models of Housing Provision

In most countries around the world today government acknowledges certain responsibilities in assuring citizens agreed-upon minimum living standards, including the provision of decent and affordable housing. The precise nature of these responsibilities, the scope of the standards, and the modes as well as effectiveness of government intervention vary significantly from one country to another. Factors such as political ideology, economic resources, level of technological development, and extent of institution building underlie such variation. A global distinction places capitalist market systems, characterized by a minimal role of national government, at one end of the spectrum and socialist redistributive systems, where national government plays a key role, at the other end.

Housing provision in capitalist nations has been patterned according to a "free" market model. It treats housing as a commodity. Provision of housing is expected to result from the interplay of supply and demand, where transactions are driven by the profit motive of private investors. A fundamental assumption underlying the capitalist model of housing provision is that households that are unable to compete for the more desirable housing stock will occupy units that are vacated and "filter down" as more affluent households move into new housing of higher standards. Government's role is minimal, restricted to actions intended to ensure smooth market functioning.

By contrast, in centrally planned socialist nations—such as those found in the East Bloc before its recent transformation and, for example, China, Vietnam, Cuba, and, until recently, Nicaragua in the Third World—a major theoretical premise of societal organization is that the state distributes costs and benefits, resulting from national functioning and development, equally among all segments of the population. According to this egalitarian ideology, the state must maintain full administrative control over rationally conducted planning, production, management, and consumption processes. In line with these normative principles, housing is viewed, and often legislated, as an entitlement, and construction, distribution, and management of housing are essential state responsibilities.

In reality, capitalist systems do not operate as genuinely free markets. Instead, markets are segmented and distorted owing to, for example, regulation and discriminatory practices. As a result, filtering does not occur as hypothesized. In practice, socialist systems also do not conform to the theoretical model. Households with privileged access to bureaucratic mechanisms of housing allocation—for example, the party cadre, the intelligentsia, and the military elite—tend to receive preferred housing. As housing is seen as a right, rents are very low. Hence, revenues are low, cost recovery is low, and investment has often been in inexpensive, low-quality materials.

In welfare states in Northern and Western Europe, national governments emerged from World War II with a legitimate and active responsibility in spheres of collective welfare, including housing. This responsibility derived in part from traditional legacies. Additionally important was the collapse of private markets. In the context of more encompassing recovery programs, government stepped in to infuse much-needed finance. Accordingly, public and subsidized housing came to represent significant proportions of the housing stock in these countries: for example, Britain (31%), Germany (20%), France (16%), the Netherlands (44%), and Denmark (20%). In more recent years, privatization policies have reduced the social housing sector in a number of welfare states. However, the relative size of this sector still remains in sharp contrast to that in more market-oriented systems as found in the United States (2%), Canada (4%), Japan (7%), and Australia (7%).

Most Third World countries contain many of the elements of capitalist "laissez-faire" market systems; others, like Cuba, Vietnam, and China, display features prominent in centrally planned socialist systems. However, among the common characteristics that tend to set Third World nations apart are a low level of industrial development, an inadequate organizational and physical infrastructure, a position of high economic dependency, a large low-income population, formidable population growth, and massive urbanization. Some countries fit this profile better than others, and the differences among Third World countries are as important as the elements they share in common.

Rapid household formation during the postwar period and its concentration in urban centers, particularly as a result of rural-urban migration, presented a challenge to housing provision that few Third World countries have been able to meet effectively. Typically, squatters have undertaken "self-help" building in response to the inability of the public and private sectors to deal effectively with the overwhelming demand for housing in cities. Spontaneous settlements of different types have sprung up in the Near East, Far East, Africa, and South and Central America. They frequently house close to half or more of the urban population. These settlements are commonly located on land that is inexpensive because it is unsuitable for profitable development, often in remote locations on the urban fringe from where access to employment and community facilities is difficult.

Information on housing in Third World nations is typically fragmented and incomplete. This situation is in part a reflection of the relatively low priority that is generally accorded to housing in overall schemes of national development. If data on housing are collected, it is usually on a local basis rather than nationwide. By the time they become available, they have often been overtaken by subsequent developments. Residential construction frequently takes place in an irregular and piecemeal manner and without systematic monitoring. Much building also occurs illegally, without legal title or permit and in contravention of prevailing building codes, zoning ordinances, and land-use plans. As a result, the volume of officially authorized construction may seriously underrepresent the number of units actually constructed.

Few Third World countries have been able to develop effective national strategies to cope with the housing emergencies with which they have been faced. In most of these countries, there is little or no realistic prospect that formal housing provision, either through the state or through the private sector, will eliminate the severe shortages that exist. The quantitative deficits are exacerbated by the pervasive lack of basic amenities and poor quality of dwellings. There are, however, important variations within countries. Especially discrepant is the inferior quality of rural housing compared with that of urban housing, save only for density levels, which tend to be lower in rural areas. There are also differences reflecting social hierarchies where positions of privilege may derive from tribal lineage, ethnic origin, caste membership, or economic class.

Policy Issues

In most of the industrialized world the number of housing units now exceeds the number of households and absolute shortages have been eliminated. However, relative housing shortages remain. Such shortages have several dimensions. Spatial shortages emerge when there exists a local disequilibrium between labor and housing markets. Other shortages occur when available units are too expensive for households in need of housing, and when available units are not accessible owing to discriminatory practices. In the U.S., housing audits continue to show that discrimination on the grounds of race and ethnicity remains widespread, among owners as well as renters. However, discrimination is not limited to racial and ethnic minorities. Restrictive rental practices exclude families with children from one out of every four rental units in the U.S. Exclusionary zoning and so-called risk-aversive lending policies also produce discriminatory outcomes.

As absolute housing shortages have declined in significance, questions of quality and affordability have gained in prominence. The most recent American Housing Survey found a little more than 8% of the housing stock having either severe or moderate physical problems. Many more housing units that satisfy minimum standards of habitability are nevertheless inappropriate for their occupants because of their layout or other design characteristics. For example, it has been widely accepted, and several countries have adopted pertinent legislation, that high-rise buildings are not ideal environments for families with children. Similarly, walk-up apartment complexes are ill suited to persons with impaired mobility. There is an increasing interest in adaptable housing whose features can be modified according to the changing needs of the household. This development parallels a growing concern about the housing needs of the fast increasing elderly populations.

In the recent period, most industrialized countries have engaged in privatization. This trend is apparent in all stages of housing provision: planning, financing, construction, and management. Although the full impacts remain to be determined, a likely consequence is heightened affordability problems for many households that are marginal to contemporary economic and political processes.

A general rule-of-thumb is that households should not spend more than 25% of their income on housing costs. Many low-income households pay in excess of this proportion, often leaving insufficient resources for other necessities of life such as food, medical care, education, and clothing. In many developed countries, there has been an increased targeting of housing assistance programs to particular population groups. This reorientation has occurred within the broader framework of a shift from object (unit) subsidies to subject (household) subsidies. Accompanying this development has been the tightening of eligibility standards. Many benefits have also been curtailed or eliminated, although not all countries have adopted an identical set of policies. With the transition of most former centrally planned housing systems to market systems and the increased global economic integration, questions of policy convergence and divergence in housing finance are likely to become more salient.

WILLEM VAN VLIET

For Further Reading: Peter Boelhouwer and Harry van der Heijden, *Housing Systems in Europe* (1992); Valerie Karn and Harold Wolman, *Comparing Housing Systems: Housing Performance and Housing Policy in the United States and Britain* (1992); Willem van Vliet, ed., *International Handbook of Housing Policies and Practices* (1990).

See also Architecture; Communities, Low and Moderate Income; Homelessness: Causes; Homelessness: Urban Environment.

HOUSING AND URBAN DEVELOPMENT, U.S. DEPARTMENT OF

See United States Government.

HUMAN ADAPTATION

The ability to adapt to a wide range of environments is not peculiar to humans. Adaptability is found throughout life and is perhaps the one attribute that distinguishes most clearly the world of life from the world of inanimate matter. Living organisms never submit passively to the impact of environmental forces; however primitive they may be, all of them attempt to respond adaptively to these forces, each in its own manner. The characteristics of this response express the individuality of the organism and determine whether it will experience health or disease in a given situation.

For the general biologist, the best and perhaps the only measure of adaptive fitness to a particular environment is the extent to which the organisms of the species under consideration can occupy this environment, make effective use of its resources, and therefore multiply abundantly in it. The biologically successful species can invade new territories by broadening its adaptive range. These criteria satisfy the tenets of Darwinian evolution, especially in the light of population genetics, but they are obviously inadequate when applied to humans because neither survival of the body, nor continuance of the species, nor fitness to the conditions of the present suffices to encompass the richness of the nature of humans. To be really relevant to the human condition, the concept of adaptability must incorporate not only the needs of the present, but also the limitations imposed by the past, and the anticipations of the future.

Throughout prehistory and history, human societies have utilized many different hereditary, physical, mental, and social mechanisms in order to adapt to new environmental situations. This biological and social versatility accounts for the spectacular and continued success of the human species. Now that people can alter their physical environment so profoundly and modify it so rapidly to their ends, there is a tendency to believe that the biological mechanisms on which they have depended for adaptation in the past have become of negligible importance. It is commonly stated that the human species can without danger afford to lose the physical and mental qualities that used to be essential for its survival, because it can create an environment in which these attributes are no longer needed.

Because they can manipulate so many aspects of the environment, and also govern to some extent the operations of their bodies and minds, modern humans have entered a phase of evolution in which many of their ancient biological attributes are no longer called into play and may therefore atrophy through disuse. The caveman, for example, for all his strength and primitive resourcefulness, would not get along well in a modern city. The present environment demands a different kind of human. Natural selection cannot possibly maintain the state of adaptiveness to an environment that no longer exists, any more than it can adapt human populations to environments that have not yet been created. Yet because of technological advances, such new environments will continue to appear at an accelerated rate. In order to keep pace with them, human evolution will therefore have to depend even more than in the past on cultural and social evolution.

Darwinian fitness achieved through genetic mechanisms accounts for only a small part of the adaptive responses made by humans and their societies. This does not mean, however, that purely genetic phenomena are no longer of importance. Now as always, the human body undergoes natural mutations, spontaneous genetic changes which, although not adaptive themselves, become so through the operation of selective pressures. Depending upon the local circumstances, these pressures determine which genetic type tends to predominate, the result being usually a nearer approach to Darwinian fitness.

It is tempting to believe that the problem of adaptive fitness in the case of humans can be restated simply by supplementing the classical biological criteria with new social and cultural criteria. This would be consonant with the fact that, even though people are still slowly evolving anatomically and physiologically, so-

cial and cultural forces have become of increasing importance in their evolution because they determine in a large measure the goals toward which they are moving. In other words, sociocultural forces are now more powerful than biological ones in orienting the evolution of their ways of life.

Biologists, physicians, sociologists, and laypeople use the word *adaptation* in their own ways, to denote a multiplicity of genetic, physiologic, psychic, and social phenomena, completely unrelated in their fundamental mechanisms. These phenomena set in motion a great variety of totally different processes, the effects of which may be initially favorable to the individual organism or social group involved, and yet have ultimate consequences that are dangerous in the long run. Furthermore, an adaptive process may be successful biologically while undesirable socially. The concepts and the consequences encompassed by the word *adaptation* as it is used in practice are therefore extremely diversified.

Laypeople are naturally unconcerned with the diverse concepts and the complex processes associated with the word *adaptation*. For them, to be well adapted simply means to be able to function effectively, happily, and as long as possible, in a particular movement. In many respects a word is a living organism; its meaning changes with the conditions of its use. Like other manifestations of human behavior, the word *adaptation* has adapted itself to the changing needs of the human condition.

The aspect of the problem of adaptation that is probably the most disturbing is paradoxically the very fact that human beings are so adaptable. This very adaptability enables them to become adjusted to conditions and habits that could eventually destroy the values most characteristic of human life.

Human beings can survive, function, and multiply despite malnutrition, environmental pollution, excessive sensory stimuli, ugliness, boredom, high population density, and its attendant regimentation. But while biological adaptability is an asset for the survival of *Homo sapiens* considered as a biological species, it can undermine the attributes that make human life different from animal life. From the human point of view, the success of adaptation must be judged in terms of values peculiar to humanity.

RENÉ DUBOS
(EDITED BY WILLIAM R. EBLEN)

For Further Reading: René Dubos, *Man Adapting* (1980); René Dubos, *So Human an Animal* (1968); Gerard Piel and Osborn Segerberg, *The World of René Dubos* (1990).

HUMAN DIMENSIONS OF GLOBAL ENVIRONMENTAL CHANGE PROGRAMME

Because of the importance of anthropogenic factors in bringing about global environmental change and the potential magnitude of the consequences of global change for humankind, the International Social Science Council (ISSC) decided that there should be an international social science research program that would parallel and complement the natural science global change research efforts, the World Climate Research Programme (WCRP) and the International Geosphere-Biosphere Programme (IGBP). The Human Dimensions of Global Environmental Change Programme (HDP) was launched by the ISSC at its Eighteenth General Assembly in November 1990. It identified seven broad areas for research: (1) the social dimensions of resource use; (2) perception and assessment of global environmental conditions and change; (3) impacts of local, national, and international social, economic, and political structures and institutions; (4) land use; (5) energy production and consumption; (6) industrial growth; and (7) environmental security and sustainable development. The first three topics deal with the fundamental social science research issues in the human dimensions of global change. The fourth through the sixth topics cover the proximate anthropogenic causes of global change. The final topic deals with the impacts of global change and strategies to mitigate change or adapt to it.

The purposes of the Human Dimensions of Global Environmental Change Programme are: to stimulate and encourage research conducted by individuals and groups; to sponsor a limited number of focused research programs that involve topics that can be studied more effectively under international auspices and within an interdisciplinary approach; and to facilitate the training of scholars throughout the world.

To accomplish its first purpose, HDP facilitates communication among individuals and groups doing research on similar topics, especially through scientific symposia and a publishing program.

To fulfill the second purpose HDP will sponsor a limited number of focused programs. By late 1992 two had been established. One is a program on land use and cover change that will be undertaken jointly with the International Geosphere-Biosphere Programme. The second is the Global Omnibus Environmental Survey (GOES), which will become an instrument for regularly monitoring knowledge, attitudes, and behavior about global environmental change. In addition, HDP is deeply committed to ensuring that data are available throughout the world for research on the human dimension of global environmental change. The Consortium for International Earth Science Information Network (CIESIN) will serve as HDP's Data and Information System.

Some training will occur as a consequence of participation in HDP's networking activities and its focused programs. The primary instrument for training, however, will be the Global Change System for Analysis, Research and Training (START). IGBP, WCRP, and HDP jointly sponsor START. START will have Regional Research Centers covering all parts of the world. The Centers will first be established in developing countries.

HAROLD K. JACOBSON

HUMAN ECOLOGY

Human ecology is an attempt to understand the interrelationships between the human species and its environment, by whatever definition. The real origins of human ecology are obscured in antiquity, but it is certain that all humans, of whatever species and of whatever time, have of necessity tried to understand their surroundings and how they, as humans, related to those surroundings.

For many generations, humans were closely fitted into their natural surroundings, trying to get along as best they could, dealing with an often overwhelming natural environment. Impacts on that environment were minimal, at least on a relative scale, when compared to the impacts of contemporary *Homo sapiens*: over-exploitation of resources, accelerating urbanization, and massive industrialization.

Human Ecology as a Social Philosophy

Much of philosophical inquiry—whether West or East, ancient or modern—has been oriented toward trying to understand the role of human beings in nature. Roots of human ecological thought can be traced through thousands of years in both eastern and western philosophy.

Human ecological wisdom is difficult to trace chronologically in Eastern philosophies, but the similarities to ecological concepts in Western philosophy are striking. One of the central teachings of Buddhism is the *anatman* doctrine, essentially that no self exists independently and that egocentric doctrines are false. The tangle of beings in the cycle of existence—including the nature of individual relationships to environment—is explained in Buddhism by the chain of interdependent or "conditioned" arising, the *Pratitya-samutpeda*. The *mandala* in Buddhism symbolizes a synthesis of many distinctive elements into a unified scheme that links together center and periphery.

Tao explains the source of all being in the root of Heaven and Earth; Taoism uses the concept of *Fu* or "return" and, with Confucianism, the I-Ching or book of changes, to describe the process of being changed into its opposite: at the peak of the Yin, the Yang returns. Yin and Yang form a whole only in combination and their cyclic mingling creates the five elements and the "ten thousand things" of Earth and Heaven.

The *Bhagavad-Gita* in Hinduism focuses on the conflicts within each individual and instructs how to achieve "union to the highest reality." The most comprehensive and venerable symbol in Hinduism is the OM, a concrete manifestation of the truth that no concepts or objects of this universe are independent, all are connected with one another. Hinduism teaches that truth is one, but has many names.

In Western thought, a human ecological thread can be traced from the pre-Socratian dialogue between Parmenides and Heraclitus, both Greek philosophers of the 5th century B.C. Each propounded a conception of reality, i.e., the reality of human relationships to the world around: heracliteanism resulted in a philosophy based on the theory that everything is in flux and that the world is made of the four elements fire, water, earth, and air, which are continually transmuted into one another; parmenidean philosophy emphasized a conception of reality as an absolute, unchanging eternal. Heraclitus was the true ecologist, recognizing a balance in all things, but a balance in flux, symbolized by the comings and goings of fire, by the material changes in the process of combustion. To Heraclitus the world *was* a process. He defined that as the *Logos*, the orderly process whereby all change takes place. The *Logos*, in the heraclitean doctrine, is the true *One*. The ecological theme emerges in traditional philosophical concerns with the antinomy of "being" and "becoming."

Plato emphasized Being, but admitted that perfect reality cannot be "devoid of change and life" and, in his dialogues, recognized the interconnection between the self, the community and ultimately the world. Aristotle was closer to Heraclitus, recognizing the reality of change and diversity. European belief-systems seesawed over the centuries: Leibniz and Kant for eternalism and against change; Berkeley and Hume for change; Hegel ambivalent. Contemporary European thought, especially the process philosophy of Whitehead and others, has reaffirmed the idea of change, advancing the notion of succession through history (Whitehead's "creative advance of nature") and the continuity of the irreversible process of becoming.

The basis of human ecology as a social philosophy in many cultures is a question especially of how individual humans or human groups (as a part in the context around them) relate to their surroundings (the whole). Much of Chinese thought, for example, builds on the need for harmonious relationships between contrasting part-whole entities: the relationship between Yin and Yang "produces all things and phenomena in the world . . . creating interdependencies and transformations that result in the dynamics of nature, body, and human society." One result is the relationship between individual Chinese peasants and the complexities of an ancient land, sustaining an agriculture through a 7,000-year history that still supports a quarter of the world's population on 7% of the Earth's arable land.

Follow the same part-whole dialectic through the recent history of Western philosophy, beginning with Pascal in the 1600s: an individual human "is but a reed, the weakest thing in nature, but . . . a thinking reed." We can, then, think about Hume's dictum—that individuals cannot change their nature, but can change, to some extent, their situation . . . and their behavior changes accordingly—and relate it to the current human ecological problems reported in every day's newspaper.

In the 19th century, Schopenhauer believed that each human must be the "dust of the other," a belief rooted in ecological wisdom, whether called that or not. John Stuart Mill explained the phenomena around him in terms of the "association of atomistic images." Individual atoms—by extension, individual humans—in association provoke analysis of connective relationships, the crux of social philosophy. Kierkegaard continued in a similar vein by emphasizing the individual and the crowd in his dialectic stages and in his idea of existence. Finally, it is a short step from Nietzsche's "biological pragmatism" to the ecological pragmatism espoused today. Humans, like all organisms, must be explained, at least in part, in materialistic terms.

Those concerned with human-ecological problems can benefit from John Dewey's 20th-century statement that the resolution of a problem demands the modification of environmental obstructions to human projects as well as modifications of human behavior. Many contemporary environmental problems can be approached from that perspective. Dewey, in a human-ecological mode, asked the question for all philosophers: How shall we deal with a changed planet—and therefore, he thought, with changing humans—actively and intelligently?

Human ecology emphasizes relationships that lead to associations as communities, a central concept in social philosophy. A long-held theme in Chinese philosophy, for example, is the attempt to understand the three corners of relational community: *Tian* (heaven or nature), *Di* (Earth or resource), and *Ren* (human or society). Community is also a repeated theme in Western philosophy, especially in the 20th century.

Connecting humans, through community interactions, is clear in Dewey's declaration that the relationship between organism and environment is not unilateral but bilateral. Individual-community/environment interactions are involved in the realization of an individual's *Lebenswelt* ("life-world"), analyzed early by Husserl and more recently by Merleau-Ponty. Note Max Scheler's concern with community as the primordial given: the human self and "the other" emerge as a result of differentiation within the community. Jacques Maritain focused on the person in the context of "a larger community of society." To Karl Jaspers also, the basic question of human life was how people, through interaction, formed into communities. Gabriel Marcel also emphasized that "existence is co-existence," the very nature of "being in the world." Marcel suggested that by "an act of the mind," individuals can grasp the union between "person—engagement—community—reality."

Even Heidegger's "philosophical anthropology" recognizes that human knowledge is concerned with relationships to nature, "with that which is actually given in the broadest sense of the term." Today, in human ecological terms, that which is actually given is interpreted as the environment surrounding individuals and communities of human beings. Heidegger also interpreted "worldhood as that referential totality which constitutes significance," not a bad definition of environment in the ecological sense as interpreted today. Merleau-Ponty extended his concept of lifeworld by pointing out that "to be born is both to be born of the world and to be born into the world."

What can be learned from the many truths of human ecology as a unified social philosophy? Begin with the radical idea that "independence" is ecologically impossible! As Stephen Strasser noted, "something of the being comes from the surrounding world and something of it returns to the surrounding world." All people are dependent on a community—natural and human—of some kind, interdependent with community. People can learn that total freedom is a myth, that total freedom means a lack of belonging; that our communities themselves are interdependent; that we are all caught in a web of relationships—and would not be human without them!

Emergence as a Scientific Discipline

The modern study of human ecology emerged from and still depends for its basic constructs on biological ecology. Humans are, after all, subject to the same principles that govern the rest of nature: They are dependent on producer organisms to translate sunlight into biologically useable forms of energy, and, like all animals, they have to eliminate wastes back into the environment. Human ecology then, like the ecology of other organisms, begins with trophic relationships, energy flows, and the movement or cycling of materials through the system. Statistics on birth and death rates, on migrations and movements, etc., are of central interest to all ecologists.

Most of the concepts used in biological ecology derive their utility from differences in species. The entire scope of human ecology is focused on a single species, *Homo sapiens*, and other organisms are brought into that scope of study only superficially. Individual human ecologists may not necessarily be trained in biology but may come from any of the social sciences: sociology, anthropology, geography, psychology, economics, political scinece, etc. An extensive literature can be found in several of these, but a really successful human ecology is yet to emerge in any of them. Exchange among the disciplines has been minimal, with development parallel rather than interdisciplinary.

Sociology can claim the strongest ecological tradition among the social sciences in the sense that it is perhaps the oldest and most substantially represented in the literature. Human ecology in sociology usually dates from the Park and Burgess work at Chicago in the early 1920s, but some claim can be made that Spencer, Ward, Bagehot, and Durkheim in the 19th century were the real pioneers of the ecological approach. The ecological perspective has been weakened over the years in sociology by the increasingly "environmental-less" approach. A splinter group under the rubric "environmental sociology" is seeking to regain a strong connection with biological ecology and to bring "environment" back into sociological human ecology.

Geography as a social science employs the concepts of ecology in the study of spatial patterns in human communities in a manner similar to some of the urban sociologists. Other geographers focus on culture groups in a manner similar to anthropology.

An ecological perspective in anthropology is more recent but implicitly dates to the discipline's origins. Anthropology's traditional concern is with preindustrial societies that are more intimately involved with their natural environment than groups studied by sociology and the other social sciences. The physical or natural environment is of direct day-to-day concern and anthropologists consider it as a contributing factor in their explanations of group culture and behavior.

The ecological approach in psychology is even more recent, dating from Lewin's work in the 1930s and 1940s. Ecological psychology builds on mainstream psychology by focusing on individual behavior. It differs by admitting as tenets the complexity of daily life and the context for individual behavior—individual-context linkages identified as "behavioral settings" or "physical environment," including the influence of context on behavior.

Major Themes in Contemporary Human Ecology

Topics and themes in human ecology are almost as diverse as the disciplines and practitioners involved in creating the field. Often terms and concepts from bioecology are given different definitions and interpretations in other fields, tending to prevent ready understanding across disciplinary boundaries. Nonetheless, four basic thematic topics and three basic units can be identified: the thematic concepts of interaction; hierarchy theory; functionalism; and holism; and the unit concepts of niche, community, and ecosystem.

Human ecology studies the *interactions* between human organisms and others (other humans, other organisms, and extant components of an environment or environments of some kind). Constant, variable, and ongoing interactions create a complex system of levels of relationship for humans everywhere. Interactions along a series of levels can always be analyzed in terms of part-whole or functional relationships: between individual and family, between individual or family and community, between community and other communities or region or state, between state and region or state and world. And, finally, for a complete look at the complex of relationships possible, a holistic approach is mandated.

Interaction takes place within a field that, in human terms, can be actual, "merely" perceived, or conceptualized, even purely symbolic. From this interactional field can be derived the basic units of ecology. The field space carved out by an individual human or an individual species is commonly referred to as a niche; the field space occupied by a group of organisms of a variety of sizes and make-ups can be labeled a community; and the community of organisms together with its environmental base (including energy flows and movements of materials) is the ecosystem.

From this framework emerges such considerations as concern for the globe as a commons; studies of carrying capacity, including load factors and tolerances for use or abuse; estimates of sustainability and how to achieve it; the role of personal or aggregate actions in environmental impacts; questions of diversity; and comparisons of the relative role and responsibilities of developed nations and developing nations.

Recent Trends in Human Ecology

Since about 1970, scholars in human ecology around the world have become much more aware of each oth-er and have started meeting regularly in international conferences. A number of successful scholarly societies have been founded and established in various countries in Europe, in India, Japan, the U.S., and Canada, with several others in the planning stage, including China and Australia. New journals have emerged in the late 1980s and early 1990s in Sweden, Spain, India, and the U.S. Established journals in Scandinavia and Poland have been refurbished in form, content, and audience and are taking on new life.

New vigor in applied human ecology is exemplified by a vast range of studies and projects, e.g., carrying capacity studies in the western U.S. by the Institute of Human Ecology; in a range of applied work by Wang's group in China and by Wolanski and associates in Poland; and by increased awareness among practitioners of all disciplines. Human ecology is gaining new strength and identity in interdisciplinary and global domains through dramatically increasing communication across disciplinary boundaries—with biology, the social sciences, home economics, and various design and applied fields all meeting and contributing and receiving new insights.

GERALD L. YOUNG

For Further Reading: Rusong Wang (ed.), *Human Ecology in China* (1989); Gerald L. Young, "A Conceptual Framework for an Interdisciplinary Human Ecology," in *Acta Oecologiae Hominis* (1989); Gerald L. Young (ed.), *Origins of Human Ecology* (1983).

See also Anthropology; Religion; Sociology, Environmental.

HUMANIZED ENVIRONMENTS

The farming country of the Ile de France north of Paris, where René Dubos was born and raised, has been occupied and profoundly transformed by human beings since the late Stone Age. Before it was inhabited, the region was covered with forests and marshes, and if it were not for the human presence it would return to this state of wilderness. Now that it has been humanized, however, it consists of a complex network of prosperous farmlands, tamed forests and rivers, parks, gardens, villages, towns, and cities. While it has repeatedly experienced destructive wars and social disturbances, it has remained ecologically diversified and economically productive. From the human point of view, it is more satisfying visually and more rewarding emotionally—for most people—than it would be in the state of wilderness. It provides a typical example of symbiosis between humankind and the Earth. The historical development of this region convinced Dubos that human beings can profoundly alter the surface of the Earth without desecrating it and they can indeed create new and lasting ecological values by working in collaboration with nature.

Among people of Western civilization, the English are commonly regarded as having a highly developed respect and appreciation of nature; but the English landscape, admirable as it is, is far different from what it was in the state of wilderness. The prodigious and continuous efforts of settlers and farmers have created an astonishing diversity of ecosystems that appear natural because they are familiar, but are really of human origin. Roadsides and riverbanks are trimmed and grass-verged, trees no longer obscure the views but appear to be within the horizon, foregrounds contrast properly with middle distances and backgrounds. Much of the English landscape is indeed so humanized that it might be regarded as a park or a vast ornamental farm.

For example, the patchwork of semirectangular fields so characteristic of East Anglia was created in the 18th century to facilitate agricultural improvements. The farming country was divided by law into plots of 5 to 10 acres, often without much regard to the natural contour of the land. The fields were divided by drainage ditches and hedges, and trees were planted in rows. When this landscape came into being, it shocked farmers, nature lovers, and landscape architects. Within a very few generations, however, it evolved into a pleasing and highly diversified ecosystem; its ditches and hedges harbor an immense variety of plants, insects, birds, rodents, and larger mammals. It has come to be regarded as a "natural" environment.

This landscape is poorly suited to modern practices. Ditches, hedges, and trees must now be sacrificed in order to create larger tracts of land more compatible with the use of high-powered agricultural equipment. This change is destroying the habitats for the many kinds of wild animals and plants that lived in the hedged-in fields, but the open fields will develop their own fauna and flora and, furthermore, have the advantage of permitting large sweeps of vision.

Thus, the ecological characteristics of an environment are determined not only by geographic and climatic factors, but also by sociocultural imperatives. In addition, the genius of the place is profoundly affected by purely cultural values, as is illustrated by the great English parks created in the 18th century. The landscape architects took their inspiration from bucolic but imaginary landscapes painted by Claude Lorrain, Nicholas Poussin, and Salvatore Rosa. They did not believe that "nature knows best," but instead tried to improve on it by rearranging its elements. They eliminated vegetation from certain areas and planted trees in others; they drained marshes and channeled the water into artificial streams and lakes; they organized the scenery to create both intimate atmospheres and distant perspectives. In other words, they invented a new kind of English landscape based on local ecological conditions but derived from the images provided by painters.

The English parks are now the envy of the world. However, as can be seen from 18th-century illustrations, they were then far less attractive than they are now. The planted trees were puny, the banks of the artificial streams were bare and raw, the masses of vegetation were often trivial and poorly balanced. The marvelous harmony of scenic and ecologic values that is now so greatly admired did not exist in the 18th century except in the minds of the landscape architects who created the parks. The sceneries composed from the raw materials of the Earth acquired their visual majesty and came to fruition only after having matured with time. Their present magnificence symbolizes that human interventions into nature can be creative and indeed can improve on nature, provided that they are based on ecological understanding of natural systems and of their potentialities for evolution as they are transformed into humanized landscapes.

Every part of the world can boast of humanized lands that have remained fertile and attractive for immense periods of time. From China to Holland, from Japan to Italy, from Java to Sweden, civilizations have been built on a variety of ecosystems that have been profoundly altered by human intervention. Many of these artificial ecosystems have proved successful even in regions not highly favored by nature. In Greece, for example, a large olive grove in a valley near Delphi has been under continuous cultivation for several thousand years; many rice paddies of tropical Asia also have been successful for millennia. Israel, which was once the land of milk and honey, then became largely desert after Roman times, has once more achieved agricultural prosperity as a result of skillful ecological management, including irrigation and reforestation.

There are different kinds of landscapes that are satisfactory to humans: on the one hand, the various types of wilderness still undisturbed by human intervention; on the other, the various humanized environments created to fit the physiological, esthetic, and emotional needs of modern human life.

There will be less and less wilderness as the human population increases, but a strenuous effort must be made to preserve as much of it as possible, for at least two reasons. The wilderness is the greatest producer of renewable sources of energy and of materials—as well as of biological species—and is therefore essential to the maintenance of the ecosystems of the Earth. Furthermore, human beings need primeval nature to reestablish contact now and then with their biological origins; a sense of continuity with the past and with the rest of creation is probably essential to the long-range sanity of the human species.

In practice, however, most people spend most of their time in humanized nature. They feel most at ease in landscapes that have been transformed in such a way that there exists a harmonious interplay between human nature and environmental forces, resulting in adaptive fitness. The quality of this interplay requires a constant expenditure of effort because any environment, left to itself, is no longer adapted to the physiological and mental needs of modern man. Even though a landscape has been economically productive and esthetically attractive for many generations, it will be invaded by brush and weeds as soon as it is neglected. The rapid degradation of abandoned gardens, farmlands, or pastures is evidence that humanized nature cannot long retain its quality without constant human care. Conservation practices are as essential for the maintenance of humanized nature as they are for the protection of the wilderness.

The stewardship of the Earth, however, goes beyond good conservation practices. It involves the creation of new ecosystems in which human interventions have caused some changes in the characteristics of the land and in the distribution of living things, to take advantage of potentialities of nature that would remain unexpressed in the state of wilderness. Throughout history and even prehistory, humans have tampered with blind ecological determinism. Forests have been cut down or managed, swamps have been drained, and agricultural productivity has been increased by practices designed to modify the physical structure, chemical components, and microbial life of soils. The fauna and flora have also been managed by introduction of new plant and animal species, selection and improvements of strains, crop rotation, and control of weeds. Ever since Neolithic times, human life has taken place in managed environments.

Experience shows that most natural situations can be converted into several different ecosystems involving different kinds of relationships to humankind. In the American Southwest the Navajos, the Zunis, and the Mormons have established viable relationships with nature based on very different ways of livelihood and different social relationships. These three ethnic groups relate to the same kind of soil under the same sky but march to different social drums in their own artificial ecosystems.

Nature is like a great river of materials and forces that can be directed in this or that channel by human intervention. Such intervention is justified because the natural channels are not necessarily the most desirable, either for the human species or for other species.

It is not always true that "nature knows best." Nature often creates ecosystems that are inefficient, wasteful, and destructive. By using reason and knowledge, people can shape ecosystems that have qualities not found in wilderness. Many potentialities of the Earth become manifest only when they have been brought out by human imagination and toil.

Just as the surface of the Earth has been transformed into artificial environments, so have these in turn influenced the evolution of human societies. The reciprocal interplay between humankind and the Earth can result in a true symbiosis, a relationship of mutualism so intimate that the two components of the system undergo modifications beneficial to both. The reciprocal transformations resulting from the interplay between a given human group and a given geographical area determine the characteristics of the people and of the region, thus creating new social and environmental values.

Symbiotic relationships mean creative partnerships. The Earth is to be seen neither as an ecosystem to be preserved unchanged nor as a quarry to be exploited for selfish and short-range reasons, but as a garden to be cultivated for the development of its own potentialities of the human adventure. The goal of this relationship is not the maintenance of the status quo, but the emergence of new phenomena and new values. Millennia of experience show that by entering into a symbiotic relationship with nature, humankind can invent and generate futures not predictable from the deterministic order of things, and thus can engage in a continuous process of creation.

RENÉ DUBOS
(EDITED BY WILLIAM R. EBLEN)

For Further Reading: René Dubos, *A God Within* (1972); René Dubos, *The Wooing of Earth* (1980); Gerard Piel and Osborn Segerberg, *The World of René Dubos* (1990).
See also Agriculture; Anthropology; Esthetics; Land Use; Urban Parks.

HUMAN REPRODUCTION

The rapid population growth of the past half century is caused not by an increased birth rate but by a decline in the death rate, resulting from many aspects of modernization, including improved food production and distribution and the mass application of public health measures such as water sanitation, antibiotics, and vaccines. Death rates have come down because significant advances in modern science have been exported successfully from technologically advanced countries to less privileged countries. But factors leading to a reduction in the birth rate have not been transferred so easily from country to country.

The human reproduction process is an intricate series of events which must proceed in perfect succession. The role of the male ends with fertilization. The female, however, continues with the complex process of harboring the newly fertilized egg in a remarkable nutritional and protective matrix.

The human female, from the time of sexual maturation, undergoes monthly cyclical events to prepare for the possible occurrence of a pregnancy. Each lunar month two different sequences of events are repeated. From among the tens of millions of immature eggs in the ovaries, several will start to develop, but after about ten days only one will continue to flourish and become fully mature, ready to be released at about mid-cycle on approximately the fourteenth day. Ovulation occurs and the released egg is swept from the surface of the ovary by the undulating end of the Fallopian tube.

Fertilization may occur if, around the time of ovulation, sperm ascend to the Fallopian tube. The egg's genetic material has been reduced by half, so that fusion of sperm nucleus and egg nucleus produces the normal complement of hereditary factors. Next, a slow series of cell divisions begin as the fertilized egg passes through the Fallopian tube into the uterus. Four days after fertilization, the egg is a cluster of 32 or 64 cells, beginning to divide more rapidly. Very little further development occurs during the two days that the egg remains free within the uterus. It assumes the form of a microscopic signet ring, with an inner cell mass that is encircled by a single row of cells in alignment. This pre-embryo state is called the blastocyst. Under proper conditions, the blastocyst will nestle into the uterine wall on the sixth day and begin to form the placenta. The inner cell mass, after several more days of cell divisions and internal arrangements, will become a human embryo.

Meanwhile, a second sequence is taking place to assure a safe and supportive nesting place in the uterus. Early in the menstrual cycle, before ovulation, the ovary secretes in ever-increasing amounts the female sex hormone estrogen, which enters the bloodstream and reaches the uterus. It stimulates the uterine lining, the endometrium, to proliferate and become much more vascular.

At about the time that ovulation is occurring, the ovary begins to produce a second hormone, progesterone, and the production of estrogen falls off considerably. By day 20 of the cycle, the entire uterine wall is ready to accept and nurture a fertilized and dividing egg.

The synchrony of these two sets of events, egg development and the preparation of the uterus for the attachment of a fertilized egg, is the result of the highly integrated functions of the hormones that are involved in each process. For example, the same pituitary hormones that stimulate the ovary to begin the maturation of an egg signal other cells of the ovary to produce estrogen. Progesterone, the hormone that maintains the uterus in a receptive state throughout pregnancy, prevents further cycles of egg development and release, thus assuring against superimposed pregnancies.

The male reproductive system is not cyclical. From the time of puberty, the human testis, stimulated by the same pituitary hormones that control ovarian events in the female, produces every day tens of millions of spermatozoa. Unlike the female whose ovary ceases to function at the time of the menopause, the human male continues to produce sperm in large numbers throughout life.

Sperm formation in the testis involves reduction by one-half the number of its chromosomes, a massive reduction in cell-size and the development of a whiplike tail. Once released from the testis, hundreds of millions of sperm assemble in the adjacent epididymis, where a further maturation process occurs, imparting to the sperm the ability to move its tail to create forward motility, without which it is incapable of fertilizing an egg. At the time of ejaculation, hundreds of millions of sperm are propelled, in a chemically intricate seminal fluid, through the vas deferens and urethra into the vagina. Even after entering the female reproductive tract, the sperm cells require a final maturation process, conditioned by substances produced by the female.

With a world population of 5.4 billion and approximately one billion couples of reproductive age, human fertilization, the successful interaction of egg and sperm, occurs about 300 million times each year. About 100 million of these fertilized eggs do not implant in the uterine wall to establish a pregnancy. They simply pass out unnoticed with the menstrual flow. Fifty million that do implant and begin to develop are either spontaneously or voluntarily aborted. The remaining 150 million proceed through the entire nine months of gestation and the birth process.

SHELDON J. SEGAL

For Further Reading: Sheldon J. Segal, "The Physiology of Human Reproduction," *Scientific American* (September 1974); United Nations Department of International Economic and Social Affairs, *World Population Prospects* (published annually).

See also Contraception; Family Planning Programs; Genetics; Population.

HUMAN RIGHTS

Belief in innate human rights was allegedly derived from natural law, known intuitively or instinctively by humans. This concept of human rights was basic to the so-called Enlightenment of the 18th century, and found political expression in the American and French Revolutions. Life, liberty, and the pursuit of happiness are inalienable rights in the Declaration of Independence. Rights inherent in nature were inalienable and not joined to obligations. Rights regarded as socially determined and conditional, however, would not be inalienable and would be vulnerable to the exercise of coercive power by political authorities. Natural law and derivative human rights became a moral philosophy derived by intuition and affirmed by positive or civil law embodied in declarations, codes, statutes and constitutions. On this basis civil law was valid only if consistent with natural law.

The concepts of human rights of the Enlightenment in 18th-century Europe were reinterpreted at the national level as civil rights. But entitlements defined by law through politics reflect the ethos of political winners, and invariably select for some rights as against others. Thus when asserted rights conflict, law will protect some rights and deny others. Legislation—not intuition—determines what rights are recognized. The notion that rights, whatever their origin, may be socially defined and hence denied, opens the way for advocates of extension by law of the rights concept to nature. During the last half of the 20th century the animal rights movement gained popular recognition and was often associated with the movement for environmental protection.

In modern societies in which human life has been regarded as unique and separate from all other life forms, no restraint has been placed on human behavior toward nature. Humankind has been regarded as an exception to the rest of life, separated from nature by natural law. Thus there were no rights inherent in nature that humans were obliged to respect. In paradoxical practice, however, human rights declared to be universal and inalienable were not applied equally to all humans in modern society. Women, children, servants, slaves, and alien peoples did not enjoy equal rights. Such rights as were allowed them were socially derived—albeit alleged to be consistent with the designs of God or nature.

Environmental science (notably ecology) and physiology broke down the conceptual barrier between mankind and the rest of nature. Humans, while increasingly seen as unique, were not seen as exceptions to the rest of nature. A rights of nature movement paralleled (or included) the concept of animal rights and demanded recognition of the survival rights of all living things, and respect for inanimate nature. Christopher Stone's book *Should Trees Have Standing? Toward Legal Rights for Natural Objects* (1974) epitomized the rights of nature movement. But defense of the rights of nature required denial of a human right to exploit or to unnecessarily destroy nature. Quite apart from an ethical injunction against a human right to degrade nature have been civil restrictions necessitated by relative scarcity on traditional rights to hunt, fish, or gather. Thus licenses to hunt, fish, timber, or harvest have been required because the resources of the natural world are more limited than are human demands.

Possibly the most significant redefinition of human rights has been the reintegration of humans and nature through the environmental movement and through holistic concepts in physiology and medicine. An expression of this emergent view of life has been the concept of environmental rights, given legal status through positive law with penalties against their violation. Where rights to a sanative sustainable environment are enforced, some alleged rights over nature—formerly taken for granted—must be restricted. A consequence of this necessity has been the projection of the rights issue into the arena of political controversy.

In relation to the environment, three aspects of human rights may be identified. They are (1) right to alter or exploit nature, (2) right to maximize economic return from property, and (3) right to reproduce.

The first right may be denied by positive laws and policies to protect living species and the quality of the environment. Rescission of the once tacit right to take from nature that which was not the property of others has been undertaken by legislation such as the Endangered Species Act and International Treaty. By vesting the custody of natural objects in public agencies as common property rights, the question of whether human rights are "natural" or socially derived becomes pertinent, especially when perceptions regarding public health and welfare are seen to change.

The second aspect of rights—to own and maximize property—encounters opposition when that right is exercised to diminish or destroy environmental quality or to obstruct or prevent the "taking" of private property for a legitimate public purpose. The right to own or to dispose of property is now encumbered by a number of legal, ethical, and ecological qualifications. Environmental concerns call into question the validity of some alleged property rights once regarded as sacrosanct.

The human right to reproduce the species has become the most contentious and least understood aspect of the human rights debate. In earlier times, the right to conceive and bear children was seldom absolute, and was governed by custom and consensus selective for survival of the tribe or community—a practice still followed by some societies. In modern society a legacy of the Enlightenment concept of universal human rights has been assertion of the absolute right of human reproduction. The morality of this position has now been challenged by arguments against overpopulation, against perpetuating severe genetic defects, against giving birth to and imposing costly care for hopelessly handicapped infants, and against the tragedy of unwanted children. The position on the right to reproduce associated with the environmental movement is one that requires limits and restraint to population growth, thus subjecting this right to societal mediation.

Even though its implementation has been difficult and often ineffectual, the human rights concept has been adopted by the United Nations and promoted by its Human Rights Commission. Growing numbers of non-governmental organizations now monitor the human rights performance of governments. Meanwhile the emerging concept of humankind as a responsible tenant on the planet with obligations to future generations and other life forms is necessitating new ways of understanding human rights. Efforts have been made to declare by written law human rights to a sanative environment. In 1969 the right to a healthful environment was written into the U.S. National Environmental Policy Act but failed of inclusion in final passage. In 1973 the Working Group on Environmental Law of the Council of Europe proposed a protocol to the European Human Rights Convention declaring a fundamental human right to an environment conducive to health. These among other efforts have failed to achieve their purpose because of difficulty of enforcement. More protection for nature may be gained by denial of some alleged human rights than by asserting rights for which no satisfactory remedy is available. The concept of crimes against the environment has been proposed as a restraint on human behavior. The U.N. Conference on Environment and Development adopted the Rio Declaration on Environment and Development in 1992. This statement, consisting of 27 principles designed to guide national and international policies, acknowledges that poor countries have a "right to development" and that rich countries bear a special responsibility "in view of the pressures their societies place on the global environment."

LYNTON KEITH CALDWELL

For Further Reading: Mary Anglemyer and Eleanor R. Seagraves, *The Natural Environment: An Annotated Bibliography and Values* (1984); Diana Vincent-Davis, "Human Rights Law: A Research Guide to the Literature—Part I International Law and the United Nations," in *New York University Jour-*

nal of International Law and Politics (Fall 1981); *Human Rights: An International and Comparative Law Bibliography* (1985).

See also Animal Rights; Contraception; Family Planning Programs; Law, Environmental; Property Rights.

HUNGER AND FAMINE

Hunger, in the biological sense, is the insufficient intake of food required for an individual's growth, activity, and maintenance of good health. Scholars from different professional and disciplinary backgrounds study hunger from one of three perspectives: the supply and availability of food; the social, economic, and cultural conditions determining who has access to food; and the effects of undernutrition on individuals. Integrating these perspectives at different scales of human organization are three distinctive hunger situations: *regional food shortage,* in which there is not enough food available within a bounded area; *household food poverty,* in which there may be sufficient food available within an area but some households do not have means to obtain it; and *individual food deprivation,* in which there may be adequate food, but food may be withheld from individuals, their special nutritional needs may not be met, or illness may prevent proper absorption of the food that they take in. These situations are linked; in times of regional food shortage, a cascade of troubles plunges food-sufficient households into food poverty and adequately fed individuals into food deprivation.

Global Food Shortage

Considering the world as a single region, global food shortage depends on the standard of food sufficiency adopted and the assumed purchasing power of the population. For example, it is widely believed that there is actually plenty of food in the world and that hunger results mainly from its maldistribution. This is certainly true for a world universally content with a basic vegetarian diet. Distributed equally according to need, the vegetarian food supply plus the naturally grazed animals could feed about 120% of the world's current population. But for a world whose diet contains a modest amount of animals fed with cereal grains, there is enough food produced at present for only three-quarters of the world's population. And to feed people with a healthy but animal-rich industrialized nation diet, there would be enough food for only 50% of the world's population.

These figures do not necessarily imply that people cause hunger by eating cereal-fed animals or that there is or will be a global food shortage. Economists rightly point out that if poor countries and peoples had greater purchasing power then there could easily be greater production of food as the world has much unused capacity for raising food. And without such purchasing power, cereal products not used in animal feed would still not be available to poor people unless given away as food aid. Studies of poor people's expenditures also show that with increased income, most poor people want to spend some of that income for a diverse diet that includes animal products except where restricted by religious preference.

While there is underused capacity today, these considerations also provide caution for the future. To feed a future population that will be at least double that of today will require a food supply three to four times greater than that of today in order to provide a healthy and desired diet for everybody. It is appropriate, therefore, to question not only whether such increased production can be achieved, but also whether the global environment can sustain the increased irrigation, fertilizers, pesticides, land conversion, and energy use that would be required under current agricultural technologies.

Famine

Regional food shortage and subsequent starvation is what is popularly conceived of as *famine*, an absolute shortage of food within a bounded area, primarily caused by crop failure or destruction, or wartime sieges or blockades. But studies of great modern famines—Soviet Union, 1932–34; Bengal, India, 1943; China, 1958–61; Ethiopia, 1972–73, 1984–86—indicate that widespread hunger and starvation can occur even when food is available if large numbers of people lose their capacity to produce, purchase, exchange, or receive food. Thus a sudden rise in food prices, a drop in laborers' incomes, or a change in government policy can create hunger for millions even in the absence of droughts, floods, pests, or armed conflict.

A Profile of Hunger

No one really knows how many hungry people there are in the world. Hunger differs in each of the situations of food shortage, poverty, and deprivation and in the multiple ways it affects human growth, activity, and health. Efforts to improve data collection and analysis have been hampered by the minimal resources devoted to the measurement and reporting of hunger. Estimates of the numbers of hungry are especially sensitive to assumptions about population age, size, and physical activity levels, allowances for household and institutional food losses, and estimates of national dietary energy supply.

To extend the available data and to capture the varied situations, a broad-based profile of hunger using ten indicators of food shortage, poverty, and deprivation has been constructed. In the early 1990s, the Hunger Profile showed that 1.5 billion people lived in countries that cannot provide them with sufficient food, even if the available food were distributed solely by need. A billion people lived in households too poor to obtain the food they need for work; half a billion lived in households too poor to obtain the food they need to maintain minimal activity. One child in six was born underweight and almost two in five children were underweight by age five. Hundreds of millions of people suffered anemia, goiter, or impaired sight, or died from diets with too little iron, iodine, or vitamin A.

The Hunger Profile does not provide a means to identify the total number of people affected by hunger, because numbers affected by shortage, poverty, and deprivation overlap. It is likely, however, that more than a billion of the world's 5.5 billion people experience some form of hunger during the year. Most of the world's hungry are found in Africa and South Asia (Bangladesh, India, Pakistan, Sri Lanka). Proportionately, food shortage and food poverty are most prevalent in sub-Saharan Africa, food deprivation in South Asia.

Viewed over time, the estimates of hunger since the 1950s offer both limited comfort and continuing concern. The proportion of extremely food-poor individuals in the world has decreased by more than half, from 23% to 9% of the world's population, yet the number of food poor has stayed relatively constant because of rapid population growth. Although threat of famine dominates the news of world hunger, famine has decreased since the end of World War II, reflecting both a lessening in the incidence of famine and a major shift from populous Asia to less-populated Africa. Exceptions to this are the major famines of 1958–1961 in China.

Causes of Hunger

Hunger has always haunted humankind and it is to history that one can turn to examine the persistence of immediate causes and underlying processes of hunger. The history of hunger is marked by occasional plenty and frequent food shortage and poverty. Hunger appears when population numbers grow more quickly than food production, when agricultural productivity declines or slows, when groups compete for limited resources, when those in power appropriate too great a share of agricultural production or maintain large populations at the margin of existence, and when environmental change or deterioration limits what can be produced. Although underlying causes endure, the mix of proximate causes appears to change in important ways.

Over time, natural environmental fluctuations as a cause of hunger diminish, and social causes dominate as the nature of food entitlement changes from access to natural resources and the help of kinfolk to a complex set of productive resources, exchanges, and gifts. Hunger created in the course of warfare persists, even as the scale and technology of warfare changes. Famines and extreme food shortages diminish as the availability of food extends to the entire globe from the earlier limits of a day's walk, a hunting trip, or a seasonal migration. This enlargement of scale however, so important to the reduction of scarcity, renders some areas marginal and courts catastrophe when errors in food-system management occur. For many isolated areas, integration into regional, national, and global markets has reduced hunger; but many also seem to have traded seasonal hunger for chronic malnutrition.

Historical studies also allow us to trace the sometimes paradoxical history of food crops, their manipulation by humans, and their impact on hunger. The movement of staple food between the Western and Eastern Hemispheres has expanded the absolute numbers of people who can be supported globally but has created the context for occasional disasters. For example, the potato, native to South America, originally served to extend the range of human habitation into the high-altitude Andes. In Europe, however, its cultivation promoted steady growth in population but ended in the Irish famine of 1846–48 when its monoculture collapsed because of disease. Transfer of crops without the foods that have served as their dietary complements has been problematic. Maize and cassava from the New World were distributed in Africa and Asia without protein-rich complements, such as beans or nutrition-enhancing processing, and resulted in niacin-deficiency pellagra, protein malnutrition, and occasionally poisoning and death.

Ending Hunger

Many opportunities exist to prevent famines, reduce food poverty, lessen undernutrition for mothers and children, and virtually eliminate some nutritional deficiencies.

Preventing Famine

A framework for famine prevention already exists in national and international early-warning and response systems and in international efforts to provide safe passage of food in zones of armed conflict. A major early-warning system coordinated by the U.N. Food and Agriculture Organization (FAO) was established in 1975, and national and regional systems are in operation as well. Two to four million metric tons of emergency food aid have been distributed annually in re-

cent years and more could be made available if food aid were better correlated with need.

It is the continued use of hunger as a weapon of war through the destruction or interdiction of civilian food supplies that is the major remaining obstacle to eliminating famine deaths. Humanitarian interventions such as efforts in the early 1990s to deliver food to civilian populations in Northern Iraq, Southern Sudan, Somalia, and former Yugoslavia, have had some success in alleviating mass starvation. Efforts by the U.N. to protect civilian food supplies and provide for the safe passage of emergency food and disaster relief, with military force if needed, will become more common.

Reducing Food Poverty

Programs providing food subsidies to broad populations have proven either ineffective or extremely costly. But more limited programs, that distribute or subsidize foods that are consumed primarily by those who are poor and hungry; distribute food, coupons, or ration shops in poor neighborhoods; or target mothers and children with special needs, have proven quite effective. Providing opportunities to earn income can lead to substantial improvements in nutrition because among the very poor, a large proportion of additional income goes into purchasing foodstuffs. Thus, programs like the Grameen Bank in Bangladesh that provide, especially to rural women, revolving loan funds, with which to start small enterprises, manufacture local products, or provide needed services, have proven very effective. Rural income-producing opportunities have been combined with efforts to increase agricultural productivity in places as diverse as Maharastra, India and Botswana, southern Africa by providing wage or food income in return for labor to construct needed agricultural infrastructure and to restore environmental resources.

Lessening Undernutrition
Among Mothers and Children

The critical nutritional times for children are the prenatal months and the early years. Such hunger is readily treated if pregnant women have access to a protein-containing diet and are educated about its importance, breast-feeding is sustained, underweight children are identified and provided with additional sustenance, and the effects of childhood illness on undernutrition are reduced. Recent surveys demonstrate, despite impressions to the contrary, that breast-feeding continues or even increases in many developing countries, possibly as a result of international efforts to encourage breast-feeding and to limit the use of manufactured infant foods. Regular weighing of young children can promote detection of early signs of wasting and stunting. Mothers can use information about their children's growth to supplement or increase the number and duration of feedings, especially following bouts of diarrhea and high fever, and to wean children from breast-feeding to the family diet. Immunization and low-cost home treatment of diarrhea promise to limit the impact of disease on nutrition for most young children. From the early 1980s to the early 1990s, immunization of children has risen from 10% worldwide to 80%, and the use of oral rehydration therapy (ORT) for treating diarrheal disease has increased from 1% to 36%, progress that will continue. The best of these new programs also address mothers' needs for help with child care, for community-based support, for access to health services, and for encouraging greater spacing between births.

Eliminating Nutritional Deficiencies

A worldwide effort is underway to eliminate the hidden hunger of micronutrient deficiencies: iodine-deficient goiter and cretinism, vitamin A–deficient blindness and illness, and iron-deficient anemia. The campaign joins efforts to supplement dietary vitamins and minerals, to fortify existing foods such as adding iodine to salt, and to increase the consumption of iron- or vitamin A–rich sources of locally available foods. With these efforts, it is possible to virtually eliminate the 190 million cases of goiter or to protect the 280 million children at risk of vitamin A deficiency. Anemia caused by iron deficiency is more difficult to address because it requires frequent iron supplements or long-term dietary change, but an effort is underway to reduce anemia by a third among the half of the world's pregnant women who suffer it.

Hunger and Environment

A recent set of studies examined the relationship between hungry peoples and threatened environments, comparing thirty case studies of hunger and poverty in threatened tropical forests, hill lands, and dry grasslands in Africa, Asia, and Latin America. A common theme in these studies is how the poor and hungry are displaced, how their resources are divided, and how their environment is degraded.

Hungry people are displaced by activities which in the name of development or commercialization deprive them of their traditional access to the common property resources of land, water, and vegetation so essential to their survival. Their resources are divided and reduced by their need to share them with their children or to sell off bits and pieces of their resources to cope with extreme losses (crop failure, illness, death), social requirements (marriages, celebrations), or simple subsistence. The resources of the hungry and poor are degraded by excessive or inappropriate use, by

failure to restore or to adequately maintain protective works, and by the loss of productive capacity from natural hazards. Driven by forces of development-commercialization, population growth, poverty, and natural hazards, hungry and poor people are trapped in spirals of household impoverishment and environmental degradation.

Driven by development activities, commercial interests and population growth, richer claimants or poor competitors displace poor people from their resources. For the displaced, either division of the remaining resources follows or forced migration to other, usually more marginal, areas. Driven by population growth and existing poverty, meager resources are further divided. Resources are then degraded by excessive use of divided lands, or inappropriate use of environments unable to sustain the requisite resource use. Even in the absence of further displacement and division, poverty creates poor households unable to maintain needed protective works or to restore resources, while natural hazards of disease, drought, flood, soil erosion, landslides, and pests further degrade these natural resources.

Thus, for the poorest fifth of humankind, maintaining access to the natural resource base and the inputs needed for agriculture, herding, or fishing is becoming increasingly difficult in the face of growing population, increased competition for land, and "development" itself. Increasingly, food-poor households have had to cope with the deterioration of their resources, the loss of crucial access to common resources, and restriction to all but the most ecologically marginal land. Faced with growing numbers in land-short countries, the children of the hungry become landless laborers or move onto marginal lands, degrading them in the process. Even in land-rich countries of Africa, this process is well under way in the regions of highest land productivity. And in small ways everywhere, the desperate search for fuelwood for energy, for grazing in times of drought, and for additional land taken from poorly protected reserves, makes life more difficult, degrades the commonly shared resources, and culminates in extensive deforestation and desertification.

There are a variety of remedies for the environmental plight of the hungry. At the local level, agroforestry and other agricultural techniques have demonstrated their ability to sustain productivity, provide fuelwood energy, limit soil erosion, and increase food and income. Everywhere, there is a heartening increase in the organization of the hungry and dispossessed in their own behalf and the emergence of local advocacy groups to strengthen their voice. But, while promising methods exist for limiting the environmental damage of human activities, these cannot replace a basic commitment to alleviate the root causes of unsustainable development: poverty, population growth, and the activities undertaken in pursuit of development that neither benefits the hungry nor sustains the resource base.

ROBERT W. KATES

For Further Reading: Jean Drèze and Amartya Sen, *Hunger and Public Action* (1989); Arline T. Golkin, *Famine: A Heritage of Hunger. A Guide to Issues and References* (1987); L. F. Newman, W. Crossgrove, R. W. Kates, R. Matthews, and S. R. Millman (eds.), *Hunger in History: Food Shortage, Poverty, and Deprivation* (1990).
See also Agriculture; Health and Nutrition; Population.

HUNTING, U.S. SPORT

Most hunting in the 20th century, especially in developed countries, is sport hunting, in which sport hunters pursue animals for recreation. Sport hunting began among the leisured upper classes in preindustrial society; with urbanization and industrialization, and the consequent increases in leisure, the sport came to be more widely practiced. In the U.S., sport hunting developed in the latter half of the 19th century, and resulted in the development of laws, institutions, and customs to support the activity. Sport hunting differs widely across societies. Sport hunters place restrictions on themselves to make the pursuit more difficult and to preserve the base level of the prey species for future pursuit. Hunting ethics refers to these cultural and legal limits and to how well the individual hunter adheres to these rules and conventions in the pursuit of prey.

Sport hunters pursue wildlife for recreation and diversion, and on most hunting trips they do not kill an animal. Hunters' motives vary: they say they hunt to be close to nature, to be with friends and family, to participate in traditions, to provide wild meat for the table, and to obtain display trophies. Although incidental to some of these motives, killing animals remains an important part of the activity in that it defines the action. The Spanish philosopher Ortega y Gassett says "the death of the game is not what interests [the hunter]; that is not his purpose. What interests him is everything he had to do to achieve that death—that is the hunt . . . one does not hunt in order to kill; on the contrary, one kills in order to have hunted." Surveys of hunters show that they report higher levels of satisfaction if they see, shoot at, and actually kill animals.

Since most people do not hunt, we can not say that sport hunting is instinctive behavior among humans. Hunters are taught to hunt and to enjoy hunting. Studies point to a social-learning explanation for hunting.

Males raised in rural areas and whose fathers hunt are much more likely to become hunters, because they are taught by their fathers and live in a culture that values animals for their utility to humans. Females living in urban areas and whose fathers do not hunt only very rarely become hunters. Interaction with an animal in a predator role is a complex and powerful experience. One has to be taught how to do it and how to interpret the experience positively.

In 1990 there were 14 million hunters 16 years old and older in the U.S., 7% of that age group and a much higher proportion of hunters than is found in most developed countries. Hunting participation varies across the U.S., from a high of 13% in the west north central region to a low of 4% in the Pacific and New England regions. The top five states for numbers of hunters are Texas (1,060,000), Pennsylvania (1,027,000), Michigan (826,000), Wisconsin (747,000), and New York (742,000). Together these states account for nearly one third of the hunters in the nation. In 1990, hunters spent an average of 17 days hunting. Over three quarters hunted big game, such as deer and elk; over half hunted small game, such as squirrels and pheasants; and over one fifth hunted waterfowl.

In 1985 nine out of ten hunters were male. About 2% of the U.S. female population hunted and 18% of the male population. Two thirds of the hunters were less than 45 years old. People living in non-metropolitan areas had a participation rate of 16% and contributed just over half of the hunters in 1985. Although people living in metropolitan areas are much less likely to hunt than those in rural areas, metropolitan areas produced 47% of all hunters because the vast majority of Americans live in cities. Hunters have higher levels of income and education than the nation as a whole and are more likely to be white.

Sport hunting has many impacts. In 1990 it provided 235 million days of outdoor recreation for more than 14 million Americans. Many of these individuals developed a close association with and learned much about wildlife and nature through hunting. Hunters spent more than $12 billion on this activity or nearly $900 each in a single year. Since hunting takes place in rural areas, these dollars tend to help rural economies.

Hunting has both positive and negative environmental impacts. On the negative side, millions of individual animals die at the hands of human predators. Many are consumed by humans, although some are lost through crippling and wounding. Many of these animals would, of course, die of other causes, but some would have lived much longer lives in the natural world if it were not for sport hunting. Sport hunters and the institutions they support manage ecosystems to produce large numbers of huntable species. These very large populations have strong influences on other plant and animal species which tend to reduce the diversity of the ecosystem. On the positive side, funds provided by hunters and scientific management systems, including federal agencies and state departments of wildlife management, enhance the environment and provide habitat protection through wildlife refuges and other preserved areas. These systems benefit both hunted and non-hunted individual species. In addition, these management systems have successfully restored or dramatically increased the numbers of hunted and non-hunted individual animals.

The future of sport hunting in the U.S. is uncertain. During the 20th century the number of sport hunters more than tripled, increasing even in percentage of population, despite urbanization and declines in agriculture. In the last quarter of the century, there was a decline both in the proportion of the population that hunted and in absolute numbers of hunters, and it is expected to continue and possibly accelerate as the U.S. population ages. This is attributable, in part, to decreasing rural populations and to changes in wildlife populations. In addition, land use and land ownership patterns make it more difficult for hunters to gain access to hunting opportunities. Finally, a small but significant proportion of the non-hunting population strongly opposes hunting. These factors create concerns about the future of hunting and the viability of the institutions created by hunters to provide for wildlife. Given the key role of hunting in human history; the political and economic influence of the more educated, well-off, white male hunters; and the need, even in the most urban societies, to control certain wildlife populations, hunting may decline but will not disappear in the foreseeable future. Sport hunting will remain one of the many acceptable ways that humans interact with non-human animals, especially in rural regions of the U.S.

THOMAS A. HEBERLEIN

For Further Reading: Thomas A. Heberlein, "Stalking the Predator," *Environment* (September 1987); José Ortega y Gasset, *Meditations on Hunting* (1985); U.S. Department of Interior, *1985 National Survey of Fishing, Hunting, and Wildlife Associated Recreation* (1988).

HYDROGEN

Hydrogen (H), the lightest of all elements, is a gas (H_2) at ordinary temperatures. It is colorless, odorless, and tasteless. It is abundant in the universe, on Earth, and in living things. Its size and reactivity determine its crucial role in the living world. It is insoluble in water. It does not support burning of other substances but in the presence of oxygen it burns quietly with a pale blue flame, producing water. When a spark is brought to a mixture of hydrogen and oxygen, they react explosively. A sample of hydrogen gas can be distinguished from other colorless gases by testing with a lighted wooden splint; only hydrogen yields a tiny explosion, an audible "pop" sound. It is a reactive element that combines with many other elements to form compounds. Hydrogen condenses to a liquid at the extremely cold temperature of $-252.87°C$, and liquid hydrogen freezes to make metallic solid hydrogen at $-259.14°C$.

In 1776 Henry Cavendish prepared, collected, and investigated a new and distinct substance, and called it "flammable air." Given the phlogiston theory of the time, he was also convinced that he had isolated pure phlogiston. In the 1780s, the Father of Chemistry, as Antoine Lavoisier is often called, identified that substance as an element and named it hydrogen, meaning "water producer."

The hydrogen atom is the smallest possible atom, having one proton and one electron; this structure accounts for its chemical activity. Naturally occurring hydrogen exists as diatomic molecules of the isotope H-1, sometimes called protium; 1 out of 6700 or so hydrogen atoms being a deuterium atom, which has an extra neutron. The atomic weight of naturally occurring hydrogen is 1.0079; the atomic number is 1. Tritium, the H-3 isotope, is radioactive. Water made mostly of molecules with deuterium atoms is called "heavy water" because it is denser than ordinary water.

A sample of hydrogen gas can be readily obtained in the laboratory by liberating the hydrogen of acids, water, and hydroxides. Typically the reaction of zinc and dilute sulfuric acid is used to produce hydrogen. The gas can be collected over water. More reactive metals, such as sodium or calcium, yield hydrogen in a violent reaction with water, and iron will release hydrogen from steam. Sodium and potassium hydroxide solutions also give up their hydrogen in a reaction with aluminum. An electric current passed through water causes the molecules to decompose into hydrogen and oxygen—a process called the electrolysis of water. To obtain the large amounts of hydrogen needed in industry, however, there are three other methods: the reaction of heated steam and carbon to yield hydrogen and carbon dioxide, the decomposition of hydrocarbons, and as a byproduct of the electrolysis of brine to make chlorine and sodium hydroxide.

Hydrogen is the most abundant element in the universe; more than 90% of all atoms are believed to be hydrogen, contributing about three fourths of the total mass of the universe. Hydrogen forms 99% of the mass of our sun. Because of its low density it does not stay in Earth's atmosphere and less than 1 part per million, by volume, of our atmosphere is hydrogen. Nevertheless, in considering the air, the oceans and rivers, the living world, and the surface crust of the Earth, 15% of all atoms are hydrogen. Two thirds of the atoms of all water molecules and 11% of the weight of oceans, lakes and rivers are hydrogen. There are more atoms of hydrogen in the human body than of all other elements combined, contributing 10% of human body weight. Living things are made of a lot of water and virtually all organic molecules, such as carbohydrates, fats, proteins, nucleic acids, hydrocarbons, and alcohols, have hydrogen atoms as their most numerous constituent. A second key role of hydrogen in organic molecules is to hold the coils and folds of these huge molecules together. Because hydrogen is such a small atom, the bonds that it forms with other atoms are highly polar and create a residual electric charge. The sphere of influence of this charge promotes the temporary and loose hydrogen bond to form among different parts of a very large molecule, for example in the folds of proteins.

Hydrogen provides the nuclear energy of stars by undergoing nuclear fusion. In the extraordinary temperatures of stars, 10 million to 100 million degrees Celsius, groups of four H-1 atoms fuse together to make one helium atom. This process is enabled by the carbon-nitrogen cycle. The four hydrogen nuclei, which are simply protons, are captured by carbon nuclei and in a complex series of steps become the two protons and two neutrons of the helium nuclei. Two positrons are produced, which are annihilated by two electrons, yielding tremendous energy; energy in the form of gamma rays is produced at each step. This fusion reaction converts 10 times as much mass to energy as the fission of uranium, U-235.

Industrial uses of hydrogen include the making of ammonia with nitrogen in the Haber process and the production of hydrochloric acid. Hydrogenation is a process of bubbling hydrogen through oils and fats to turn them into solid cooking fats and oleomargarines. Liquid hydrogen is so cold, so close to absolute zero ($-273°C$), that it is used both in cryogenics and in enabling superconductivity. An oxyhydrogen torch yields temperatures of 2400°C and is used for high-temperature welding. Hydrogen still provides the lift for most meteorological balloons. It is extremely attractive as a possible alternative fuel in place of gasoline in hydrogen-powered cars, for example, because the product of combustion is simply water, and in nuclear fusion power plants since there is no radioactive waste from the fusion of hydrogen.

JEAN LYTHCOTT

For Further Reading: P. A. Cox, *The Elements: Their Origin, Abundance and Distribution* (1989); Ralph E. Lapp and Editors of *LIFE, Matter* (1965); Bernard Jaffe, *Crucibles: The Sto-*

ry of Chemistry (1960); Glenn T. Seaborg and Evans G. Valens, *Elements of the Universe* (1958).
See also Atmosphere; Automobile Industry; Petroleum; Water.

HYDROLOGIC CYCLE

See Water Cycle.

HYDROPONICS

Hydroponics is the cultivation of plants in nutrient solution rather than soil. The plant's roots are immersed in water containing essential elements. In a variation—aeroponics—a fine mist of nutrients is sprayed onto the root system suspended in the air. Humidity, heat and light still need to be provided.

The Aztecs used a variation of hydroponics on rafts known as *chinampas* on which they grew trees, vegetables and flowers. The rafts were covered in sediments from the lake bottom and the plants took root through the raft into the water. Plants growing in water can also be seen in Egyptian hieroglyphics. Research conducted since the seventeenth century has provided a greater understanding of the role of soil and water in providing nutrition to the plant.

The term *hydroponics* was coined by W. F. Gerick at the University of California who in 1936 developed its principles in a commercially viable way and grew 25 foot high tomato plants. After Gerick's efforts, hydroponics was relegated to the laboratory as promoters sold poorly designed units and both the Great Depression and World War II diverted resources, time and expertise. However, the U.S. Army established large scale hydroponic farms in the Pacific to help feed soldiers stationed on non-arable islands.

Hydroponics continued to blossom over the decades as small-scale farming and greenhouses refined the growing techniques and showed consistent commercial success. A large portion of the plants grown in greenhouses is grown hydroponically and hydroponic farming exists around the world.

This method has numerous benefits over traditional farming. By placing the plant in a controlled environment, growers can eliminate diseases, soil-borne pests and weeds. In addition, it requires less space in which to grow more plants. Plants grow more rapidly and produce greater yields with more uniform results. Water and nutrients can also be recycled and conserved.

Growing mediums and methods vary greatly. Commercial hydroponic growers often use plastic rain gutters or polyvinyl chloride piping to grow the plants in. Wooden or plastic boxes can also be used with a sheet of styrofoam with holes cut out to support the plant, allowing the roots to reach the water and nutrients. Home units can be constructed of pots, milk jugs, or a modified version of a commercial unit. Materials used to anchor roots are gravel, sand, vermiculite, perlite, crushed rocks or cinder blocks, styrofoam, sawdust, pumice and rice hulls. Nutrient-filled water is fed through the system and collected on the other end where, if the nutrients are adequate, it can be reused. Commercial hydroponic systems automate the entire process using pumps and monitors to control water and modify the nutrient balance with predetermined amounts and concentrations. Since nutrients are the key, essential elements to be provided to the plant are nitrogen, phosphorus, potassium, calcium, sulfur, iron, manganese, zinc, boron and copper—found in a number of commercially available formulas. Organic mixtures include manure, compost, or seaweed.

Hydroponic cultivation provides a proven, commercially viable, and cost-effective alternative to traditional farming and growing techniques—adapted to work on large-scale farms or in small urban apartments.

HYDROPOWER

Hydropower, or hydroelectric power, is the generation of electricity from water, usually using a dam to store water and divert it to turn a turbine.

Fluctuating fossil fuel prices and associated energy crises, as well as concerns about global warming, have brought hydropower once again to the forefront of energy policy development. Yet today hydropower is mired in controversy as a result of negative environmental impacts associated with well publicized megaprojects. The best known is the Aswan Dam in Egypt, which has altered the downstream ecology of the entire region. There are other projects that have not had the attention of the public but that have had similar impacts on the local environment such as the Tarbela and Mangla projects in Pakistan. The development of large dams has decreased because most favorable sites have been developed and because of the major environmental impacts, including opposition by environmental groups and people living in the immediate area. The current exception is Three Gorges in China, a 13,000 megawatt (MW) scheme that will require the relocation of 1.3 million people. This project has been under study for several years and the Chinese government decided to move forward because of the need for large amounts of electric power, which would otherwise have to be generated from coal-fired plants. This dilemma is common to many developing countries that are rapidly growing and in need of increasing amounts of electrical energy.

A dichotomy lies in the fact that developing countries need electrical energy for continued development and often have favorable hydropower sites. But the development of those sites may require the flooding of agricultural areas and small communities. On the other hand, generation of electrical energy from fossil fuel plants entails negative environmental impacts as well.

Historical Perspective

Waterpower has contributed to the development of humankind since biblical times, with references to the use of water wheels for milling and pumping dating back to early Greece. Between these early uses of the water wheel and the advent of the industrial revolution, running water and wind were the only sources of mechanical power. Improvements in power recovery from flowing water were steadily introduced as exemplified by the sophisticated water works designed in the 1600s for the palace of Versailles outside Paris. As they evolved, waterpower systems, and ultimately hydroelectric generating plants, were developed from attempts to improve the efficiency of the water wheel. Early work occurred in France in the years between 1750 and 1850 when various turbine designs were investigated. At the beginning of the industrial revolution, France did not have access to large coal deposits and had to rely on its water resources to generate the energy needed for industrial expansion. Much theoretical work was done during this period by mathematicians and engineers, resulting in significant improvements in turbine efficiency. Early turbines attained an efficiency of 10% to 20% and by the mid 1850s, 60% to 65%. Modern turbines, the Francis and Kaplan types, achieve efficiencies of 90% to 95%.

In the early 19th century, water provided mechanical power for industrial applications. With the invention of the dynamo or generator in the 1880s, and the popularization of electricity as a source of energy, many water turbines were converted to electricity production. The first hydroelectric unit in the U.S. is reported to have been a 12.5 kilowatt (kW) plant installed in 1882 on the Fox River at Appleton, Wisconsin. The ability of hydroelectric plants to provide electricity for large population centers was recognized as early as 1890: A hydro project at Niagara Falls was being studied to supply electricity to the city of Buffalo. Inadequacies in electrical transmission technology, however, delayed the implementation of this and other projects. With the development of high-voltage transmission lines in the early part of the 20th century, a shift occurred from small-scale plants serving local electricity markets to large-scale plants feeding into extensive distribution grids. The same trend was observed in thermal plants fired by coal and later by petroleum. In most industrialized countries, therefore, the 20th century has been characterized by a proliferation of large-scale (hundreds of megawatts) hydroelectric projects serving wide areas.

Hydropower technology utilizes the difference of potential energy between different parts of a water body at a rate that is roughly proportional to the product of the water level difference (head) and the volume of water flow per unit time (discharge). Hence, hydropower planning and design is directed toward increasing these two quantities both by proper site selection and by construction measures. With regard to the development of head and control of discharge, different plant types can be distinguished: (1) river power plants where the head is created by weirs or dams; (2) diversion schemes that basically utilize naturally available heads; (3) run-of-river plants with little or no control of discharge; and (4) storage power plants with high dams and large reservoirs for flow regulation. Since the amount and reliability of discharge, as well as the possibility of creating additional head by structural methods, depends entirely on local hydrological and geomorphological conditions respectively, hydropower planning and design does not result in standard solutions but rather in highly site-specific solutions. Because of this capital costs tend to vary greatly.

Nevertheless, independent of their ability to provide reserves of generating capacity and respond to peak demand, large-scale plants—both thermal and hydroelectric—benefit from economies of scale. These economies can be quite significant in terms of capital investment and financing costs. On the basis of a scaling factor of 0.7, generally accepted in many engineering projects, the capital investment in a 200 MW generating plant would only be about 1.6 times that of a 100 MW plant. Prior to the drastic increase in fossil fuel costs in 1973–74, economies of scale were particularly important for hydroelectric projects in competition with thermal plants fired by low-cost fossil fuels.

The case for hydro is often damaged by the high cost of power transmission, as well as environmental impact. The location of a hydroelectric plant is determined by water resource availability, site potential, and environmental acceptability. Many such locations are relatively remote from population and industrial centers. Bringing the electricity generated to the end users and integrating that capacity into a country's distribution grid requires extensive and expensive transmission lines. By contrast, thermal plants are more flexible in terms of siting: fuel can be transported to a convenient location close to markets.

On the other hand, there are often opportunities for the development of small-scale hydro plants for remote areas having adequate hydrologic resources. These may provide energy to a local grid which may be backed up with diesel power or interconnected with the national grid. In many developing countries the use of small hydro (mini-hydro) systems is common. Depending on site conditions, these plants tend to be costly compared to large-scale hydropower facilities but are cheaper on a life-cycle basis than diesel. Diesel fuel in remote areas of the world costs up to $4 per gallon.

World Hydropower Resources

Various estimates of hydropower resources have been made over the years: worldwide, regional, and for specific river basins. Since 1929 the World Energy Conference (founded in 1924 as the World Power Conference) has periodically published a survey of world energy resources, both potential and developed. Energy information supplied by the United Nations and by the 80 member countries of the Conference is published in a comprehensive volume at 6-year intervals, the first in 1962. In 1974 the World Energy Conference established a Commission on Energy Conservation whose responsibility it was to evaluate the world's energy resources in light of periodic price escalations.

The results of the Commission studies in the late 1980s show that water provides 23% of the world's electric power. This figure varies from 100% in remote areas to negligible amounts in some Middle East oil-rich countries. This was not the case in 1960 when 29% of world electricity production was attributed to hydro and geothermal power. The percentages of fossil fuel inputs in electricity production were 71% and 75% in 1960 and 1970 respectively. By 1990 hydro and geothermal power accounted for 20% to 25% of world electricity production compared to 51% for fossil fuels.

Based on the Commission studies, the total potential considered feasible for development is 2.2 million MW of generating capacity with a probable annual production of 9,700 million megawatt-hours (MWh). A 50% capacity factor, about average for today's hydropower production, was assumed. These figures were determined from the capacities at various current and future installations. To produce the same amount of energy, thermal plants would burn the equivalent of about 40 million barrels of crude oil per day. The present annual production of hydropower is about 2,100 million MWh, which is approximately 22% of the Commission's estimated expected potential. To produce the same amount of electricity in oil-fired plants would require burning approximately 8.7 million barrels of crude oil per day.

In the U.S., the most desirable sites for large-scale hydropower have been developed, producing approximately 13% of all electricity used domestically. Concern over environmental impacts, long delays in licensing, and increasing construction costs reduce the attractiveness of additional large-scale hydro plants. By contrast, fluctuating oil prices, enabling federal legislation, and the emphasis on independent power generation during the 1980s have brought small hydro systems into a competitive position with other sources of electricity. Improvements in manufacturing, standardization, and automatic control in small hydro systems have also increased their economic attractiveness. However, forecasts for ever-increasing amounts of electrical energy have not been borne out in the U.S. Many power plants currently under construction may not be fully utilized; most of these are coal- and gas-fired.

Environmental Impact

Much has been written about the environmental impacts of large dams and megaprojects such as Aswan. The popular literature is replete with environmental disasters created by the damming of great rivers and the flooding of thousands of acres. How environmental impacts are perceived depends to a large extent on the point of view of the analyst. In the past those responsible for implementing a project also carried out the environmental assessment. Now a more diverse range of opinion is sought, in particular the opinions and needs of directly affected parties. The operating premise is to provide a level of compensation to affected parties that will restore them to the standard of living they had prior to the project.

Mitigation measures for negative physical and biological impacts are incorporated into project planning to the extent possible. Irreversible impacts are addressed by compensating for the loss. For example, fishways are created to allow fish to swim upstream for spawning. If certain types of habitat are lost, new ones must be created elsewhere. If nutrient-rich silts are no longer available to nearby farms, then organic fertilizer must be provided free of charge. Major steps have been taken by governments and financial institutions to ameliorate the negative impacts of hydropower projects. Not all of these measures have been successful.

The following issues have been identified as the major impacts planners need to address when building hydropower facilities. Some of these occur during construction and some during subsequent plant operations.

Flow Disruption

Flow disruption occurs during construction and during operation depending on how the reservoir behind the dam is operated. Impacts include channel degradation, downstream sedimentation, and, if water is diverted from one basin to another, a net loss of water in one drainage basin and a gain in another. Other possible impacts include decreased downstream flooding and subsequent loss of fertilization of downstream agricultural land, variation in groundwater levels, and disturbance of saltwater/freshwater balances in estuaries. If pond elevation fluctuates greatly because of peaking power production, shoreline erosion and habitat destruction may occur. The least disruptive operation is "run-of-the-river"—no water storage takes place and the water is allowed to flow into the downstream channel at a lower elevation.

Sedimentation

Sedimentation results from both project construction and operation. The impacts of sedimentation are many. It can cause turbine damage, bury fish-spawning areas and bottom-dwelling organisms, and shift the composition, abundance, and distribution of aquatic biota. Additionally, the retention of sediment by reservoirs can result in premature siltation. It can also trap nutrients in the reservoir and upset the nutrient balance, because these nutrients otherwise would have traveled to downstream wetlands and estuaries. The increase in the reservoir's nutrients may lead to the growth of aquatic weeds and algae—both of which lead to decreased dissolved oxygen and other water-quality problems. Many large dams are plagued with sediment problems cutting short their expected lifespan.

Water Quality

Generally, a run-of-the-river operation with no impoundment will have little or no effect on water quality in streams. Water quality may be affected, however, if there is diversion from one watershed to another (particularly if it is of poor quality or changes the chemical composition of the receiving water) or if there is an impoundment. The impoundment can affect water quality if the water undergoes thermal stratification, causing a temperature differential, a decrease in bottom dissolved oxygen, an increase of hydrogen sulfide, and a reduction of ionic forms of iron and manganese. Impoundments of water can also increase the exposure time of organisms to contaminants. In impoundments some pollutants may be absorbed by suspended settling particles and eventually be buried on the bottom. Oils, greases, and chemicals entering the stream from the construction site can also affect water quality.

Aquatic Organisms

There are many possible effects on aquatic organisms at hydropower sites. Examples include turbine damage to fish, blockage of upstream migrating fish; impingement (trapping of organisms against intake screens); destruction of downstream habitats due to decreased flow; damage downstream due to reservoir releases containing suspended material, low dissolved oxygen, toxic contaminants, and low-temperature water; changes in the composition and abundance of species due to change from flow to reservoir habitat (particularly important to economically important fisheries); and gas bubble disease to fish caused by supersaturation of water by atmospheric gas as water is released over a spillway.

Water-Borne Diseases and Parasites

Water impoundments provide breeding grounds for mosquitoes (which carry such diseases as malaria, yellow fever, and dengue) and snails (part of the chain of schistosomiasis and other diseases). Impoundments also decrease the natural purifying elements of running water—turbulence, sunlight penetration, and aeration. On the other hand, impoundments also decrease flow in certain sections of a stream and lead to a possible localized reduction in onchocerciasis (river blindness).

Land Use and Development

The direct impact of a hydro project on land use is the loss of land for access roads, canals, turbine-generator building, transmission lines, and reservoir pools. The land required for the reservoir can be significant, requiring the flooding of large areas. Another significant impact may be new development created by the project. For example, colonization in the headwaters region of a project may lead to increased erosion and subsequent siltation of the reservoir.

Terrestrial Impacts

Terrestrial impacts may include alteration of wildlife habitat, due to land use changes and creation of an impoundment, and interference with nesting activities of birds due to construction noise. Construction and maintenance of transmission lines can disrupt natural landforms and vegetation and hinder wildlife migrations. There may also be impacts on local communities through which the lines pass.

Socio-Cultural Impacts

Potentially the most damaging impacts of large-scale hydropower development involve the resettlement of large numbers of people from areas to be inundated for the reservoir. Historically, the largest resettlement program will be at Three Gorges in China with 1.3 million affected. Planned projects in Brazil to build nine dams will require the relocation of 50,000, mostly Indians. The Mahaweli scheme in Sri Lanka when complete will require the resettlement of close to one million. A project that has the attention of the World Bank and its critics is the Sardar Sarovar project in India, which will require the relocation of at least 100,000 and the submergence of 245 villages. Forced resettlements have created controversy in that compensation for lost homes and farmland has in the past not been adequate. Many displaced people have not been provided with replacement living standards equivalent to what they gave up. Efforts have recently been increased among financial institutions and governments to assure that displaced people are adequately compensated.

Conclusion

Hydropower as an energy source provides a number of environmental benefits over conventional fossil-fuel-fired plants, primarily the absence of air pollution. However, the development of large hydropower plants with attendant dams, reservoirs, and long-distance power lines often requires the sacrifice of valuable natural resource areas such as river valleys and may require the relocation of large numbers of people. Planners need to consider all the possibilities, including energy conservation, in determining the power system appropriate for national circumstances. Most important is that a range of options be considered that accurately weigh environmental costs and benefits along with more traditional indications of financial and economic performance.

JACK J. FRITZ

For Further Reading: Jack J. Fritz, *Small and Mini-Hydropower Systems* (1984); Edward Goldsmith and Nicholas Hildyard, *The Social and Environmental Effects of Large Dams* (1984); Calvin C. Warnick, *Hydropower Engineering* (1984).
See also Aswan High Dam; Dams and Reservoirs; Dissolved Oxygen; Electric Utility Industry; Energy, Electric; Floods; Rivers and Streams; Sea and Lake Zones; Water Cycle; Watersheds.

HYDROSPHERE

See Water.

I

ICE AGES

See Climate Change.

IMMUNIZATION

See Health and Disease.

INCINERATION

Incineration is the conversion of the organic component of waste to energy, through high temperature, controlled combustion. High temperature incineration provides a thorough and permanent method of destroying the organic portion of the waste and getting rid of the harmful, pathogenic and toxic materials in the waste. Incineration can reduce the volume of the waste significantly: by close to 100% for pure hazardous organic wastes; 90% for municipal solid waste. This lessens the need for new landfills and extends the life of existing ones. Incineration plants also generate electricity or steam, which is then sold to partially offset the cost of building, maintaining and operating such facilities, while decreasing the need to supply this energy from conventional, e.g., oil, coal or nuclear, sources.

An incinerator is a device in which wastes are burned at a temperature typically greater than 1500°F, with a proper amount of air and with adequate time to ensure destruction of the wastes. Many types of incinerators are in operation worldwide. Although each is designed somewhat differently, each accomplishes essentially the same results. A state-of-the-art incinerator is equipped with operating controls and monitoring systems which assure good combustion to destroy the waste, and an effective method of cleaning the air and the water which are byproducts of the process. The primary purpose of the air pollution control system is to remove inorganic acid gases, metals and particulate matter. The costs of incineration are high, ranging from $75-$100/ton for municipal solid waste to over $1000/ton for some hazardous wastes.

Opposition to incineration as a waste-management option and to siting of incineration plants is strong.

Concerns regarding incineration focus on potentially harmful air emissions that may cause health and environmental problems and the disposal of the incinerator ash and air pollution control system residues. Incineration does not destroy all of the organics in the waste. An extremely small fraction of the organics is emitted from the stack and trace quantities remain in the incinerator ash. In particular, dioxin and furan emissions from incinerators are a major concern.

Dioxins and furans make up a group of 210 compounds with similar chemical structure. Although most of these compounds have been found to have little or no effect on humans, one of the compounds and a mixture of two others are extremely toxic to certain animals and are probable human carcinogens. Small amounts of these chemicals cause different responses in many animals; current information about the impact on human health is inconclusive.

Emissions of toxic metals such as cadmium, chromium, lead, and mercury that can cause adverse health effects if significant amounts enter the body, are also a concern. Incineration does not destroy toxic metals, and a portion of them enters the exhaust gas stream as fly-ash particulates. Most particles are removed by the air pollution control equipment; however, some escape to the atmosphere where they have the potential for negative environmental and health impacts.

By using theoretical worst-case assumptions, the highest possible risk from an incinerator can be estimated. That risk is calculated for an imaginary person, termed the Maximum Exposed Individual or MEI, who stays in the same spot continuously for 70 years at the point where the concentration of pollutants at ground level is the highest. The worst-case increased cancer risk to the MEI as a result of emissions from properly designed and operated incinerators ranges between one chance in 100,000 and one in 100,000,000. This risk is comparable to or less than the risk of smoking one cigarette in a lifetime or crossing the street one time, or the possibility of being struck by lightning.

The ash and scrubber sludges from the incineration process contain the toxic metals from the original waste. These metals have the potential to leach from the residue and contaminate groundwater. Hence ash and scrubber residues must be handled and disposed of in a secure landfill that is designed to prevent groundwater pollution.

When properly implemented and controlled, incineration is an effective and efficient component in a total program of waste management and disposal and poses minimal risk to public health and the environment.

RICHARD S. MAGEE

For Further Reading: Richard Denison, "Health Risks of Municipal Solid Waste Incineration: The Need for a Comprehensive View," in *The City as a Human Environment* (1990); John R. Ehrenfeld, "Social Impacts of Incinerating Hazardous Materials," in *Management of Hazardous Agents—Vol. 2: Social, Political and Policy Aspects* by Duane G. LeVine and Arthur C. Upton (eds.) (1992); Norton Nelson et al., "Cleanup of Contaminated Sites," in *Toxic Chemicals, Health, and the Environment* by Lester B. Lave and Arthur C. Upton (eds.) (1987).

See also Cogeneration of Electricity and Heat; Dioxins; Pollution, Air; Risk; Toxic Metals; Waste Disposal; Waste Treatment; Waste, Municipal Solid.

INDICATOR SPECIES

An indicator is a characteristic of the environment that, when measured, quantifies the magnitude of stress, habitat characteristics, degree of exposure to the stressor, or degree of ecological response to the exposure. An indicator species is an organism that can perform one of these tasks.

In the original sense of the term, indicator species are those used to characterize ecological response of the community to pollution exposure. The indicator species is one whose environmental requirements have been well-defined. If the indicator species is present, environmental conditions are within defined tolerance limits for that species.

More recently, the term *indicator species* has been used to refer not to an organism whose tolerance limits have been defined, but instead to an organism that accumulates pollutants in its tissues. Measurements of tissue concentrations are used to estimate cumulative environmental exposure to a pollutant rather than community response. Because the organisms are assigned no indicator value, but are instead used to quantify pollution levels, this technique is alternatively referred to as bioaccumulation monitoring.

Characterizing the effects of sewage on aquatic organisms was among the first uses of indicator species. Early in the 20th century Kolkwitz and Marrson developed the Saprobian System. Based on censuses of degraded and clean waters, organisms able to tolerate sewage discharges and the consequent low oxygen and related water quality changes were labeled saprobic.

There are a number of problems with the use of indicator species as a tool in environmental management. First, organisms' responses to pollution are not uniform across stress type. The saprobian system was developed before the modern industrial age, before there were a multitude of substances quite different from natural compounds degrading natural systems. Some species are very tolerant to one pollutant but extraordinarily sensitive to another. So saprobic organisms that tolerate low dissolved oxygen may be quite sensitive to heavy metals or pesticides or habitat degradation. Separate indicator status is required for each pollutant or combination of pollutants. G. W. Suter III points out that the indicator-species concept has been effective in assessing oxygen-demanding pollution, but not for other types. Second, the indicator species approach is based on having a solid knowledge of the organisms' response to pollution. With many of the species on Earth not yet named, the occurrence of species for which there is no indicator status could well be a major problem, particularly in developing countries where flora and fauna are not well described or characterized. Third, the presence of a species does not assure that the community of organisms is robust and functioning well rather than merely surviving. Fourth and finally, the approach is reactive rather than predictive—that is, it records existing damage but does not prevent damage. Effective environmental management requires prediction and prevention of damage to ecosystems.

JOHN CAIRNS, JR.
B. R. NIEDERLEHNER

INDIGENOUS PEOPLES

Most Americans think indigenous peoples are Indians. In fact the term includes groups inhabiting about every environment on the planet—the San of the Kalahari desert, the Sami of Scandinavia, the Yanomami of the Amazon, the Maasai herders of East Africa, and the Kanaks of New Caledonia in the Pacific. All these groups have been around for hundreds, even thousands of years. They are generally characterized by distinct language, culture, history, territorial base, and self-government that predates the creation of modern states. These groups have lived in and maintained some of the earth's most fragile ecosystems. Today, the survival of the groups and their resources is threatened.

During this century, more indigenous peoples have disappeared than during any previous one. Brazil, for example, has "lost" one Indian culture per year since the turn of the century (a third of all remaining cultures) while some 12% of the Amazon has been degraded. To put it another way, indigenous peoples are disappearing faster than the often fragile resources they have used, yet maintained, for centuries. Human rights violations precede environmental degradation.

Indigenous Peoples and Development

There are about 600 million indigenous people in the world who retain a strong identity as well as an attachment to their homeland. Indigenous peoples are distinguished from ethnic groups who, though much larger in absolute numbers, have traded away political autonomy to states for the right to retain and practice other cultural beliefs.

Indigenous peoples account for 10 to 15% of the world's population but claim 25 to 30% of the earth's surface area and the resources on it. This is increasingly a cause of conflict between indigenous peoples and states.

After being decimated from 1500 to 1900 through contact and colonization, the population of indigenous people has increased tremendously since 1900, precisely when most states were created. Thus, states were established exactly when many indigenous peoples felt that it was possible to regain the political autonomy that was denied them under colonialism.

Indigenous Peoples and Natural Resources

Since World War II a growing awareness of the finite nature of the earth's resources has led to the invasion of remote areas and the appropriation of indigenous peoples' resources. In some cases the resources are taken by force, in other instances, they are appropriated after denying the rights of local inhabitants. A whole body of law has been developed to deny the rights of indigenous people to their natural resources. For example, in some countries indigenous people can own the land but not the subsoil rights, the land but not the trees, the land but not the water, etc. These foreign concepts are strange for some indigenous groups who do not understand ownership or for others who see ecosystems as indivisible.

Yet Indians throughout the Americas have lost their lands to state-sponsored colonists. Pastoralists in Africa have lost land to agriculturalists from groups large enough to control "democracies." Oil has been taken from the Kurds in Iraq and from other indigenous peoples without compensation because such groups do not hold subsoil rights. While groups such as the Penan of Malaysia have land rights, they do not own the timber that grows on it. (But for a hunter-gatherer group, the land without the timber is worthless.) Even constitutionally guaranteed resource rights of indigenous people are often subsequently denied. Through-

out history, indigenous peoples have found "paper" rights to be worthless.

Profits from the sale of natural resources go to those who control states. Those who rule gain considerably. Developed countries are not unwitting observers in these developments, however. The West created most of the highly centralized, top-down systems of government in the Third World. These were trading partners. While such governments are said to promote security and stability, in reality they function to facilitate the international trade of natural resources without the high costs of governing far-flung empires. Investments from North America, Japan and Europe—along with their political support and manipulation—help ruling groups maintain their power bases. Foreign and military assistance support the dictatorships and single-party states that dominate the world today and provide the "free" flow of resources needed by multinational corporations to feed the world's voracious appetite for consumer goods.

Still, the Third World political reality is that, with or without outside assistance, indigenous groups that resist the authority of the state are destroyed. Guatemalan soldiers armed with fence posts and rifles destroyed Mayan Indian villages when the U.S. cut military assistance to that country. Likewise, the Tutsi in Burundi used machetes and simple firearms to put down a rebellion by the country's Hutu (who constitute 85% of the population) when foreign assistance was not forthcoming. Measles-infected blankets have been effective weapons in Brazil when military decrees did not wrest legal land rights from Indians. It does not take nuclear weapons, poison gas (although Iraq used it against the Kurds), or other sophisticated weapons to eliminate indigenous peoples.

Those who rule see the control of indigenous people as essential to their survival; yet their policies fuel hostility, making control of indigenous people difficult. State programs such as relocation (e.g., in the former USSR, China, the U.S., Israel, Ethiopia, Nicaragua, and South Africa), colonization (e.g., in Australia, the Americas, and New Zealand), and resettlement and villagization (e.g., in Ethiopia, Guatemala, and Indonesia) ensure the control of indigenous peoples and their lands and resources by states. Food and famine become weapons in conflicts between indigenous peoples and states (e.g., Angola, Ethiopia, Guatemala, Mozambique, Nicaragua, Somalia, and Sudan). Displacement, malnutrition, environmental degradation, refugees, and genocide become commonplace.

Indigenous Knowledge and Resource Rights

Recent attacks on indigenous peoples have expanded to include the more subtle theft of their knowledge.

The recent rush to discover, inventory and save the world's biodiversity is often done with the assistance of indigenous peoples, but not for their benefit. In the name of salvaging information before it disappears, the knowledge of curers, shamans, and indigenous peoples is stolen without a second thought. Basic agreements (contracts, licensing agreements, or royalties) that Western researchers would routinely negotiate are denied to groups that provide culturally specific knowledge or intellectual property.

Perhaps indigenous peoples should not have *all* rights to genetic materials or even necessarily to medicines and cures that they have discovered and developed. But, they, too, should have *some* rights just as do scientists, states, and corporations. Without such cooperation, few raw materials would end up as new products. However, indigenous peoples are unique; they are excluded from profiting from their information.

Indigenous Peoples and Resource Management

Through the centuries and throughout the world, indigenous peoples developed sustainable systems of resource management long before sustainability was in vogue. Such groups have long recognized that without a resource base future generations will not survive. Thus they have developed systems that use the resources of their area while protecting the base for future generations. Their systems are complex; they often combine root crops, vegetable crops and select tree crops which, in turn, improve hunting, fishing and gathering. More recently, domesticated animals, cash or marketable crops and even imported varieties have been added to the mix.

The various world views and beliefs about environments that distinguish indigenous peoples from us as well as each other lead to culturally-specific systems of resource management. These systems are rarely random or even mostly opportunistic. Indigenous peoples are not preservationists; they constantly manipulate and change their environments. Rather, they are conservationists because they know that they must use their resources and protect them for future generations.

Some indigenous resource management systems are sustainable over time, others are not. Some are sustainable under certain conditions but destructive if those conditions change. In a single group, some individuals are more cautious and conserving of resources than others. And, more recently, in many indigenous societies undergoing rapid change, many young people no longer want to learn the methods by which their ancestors maintained fragile ecosystems.

There are a number of reasons why indigenous peoples' resource management systems are not always sustainable. The two most important, however, are population pressure and the loss of land or resources. As local populations grow, resources are used more intensively. For slash-and-burn agriculturalists this means that areas may be farmed every seven years rather than letting the land recover for 20 years between plantings. For pastoralists, population pressure means that herds may graze areas for longer periods and may rotate through the same pastures more times each year. For hunters and gatherers, increased numbers mean a more rapid depletion of the main plants and animals that provide most of a group's food. This is also true of fishing communities.

Population pressure is compounded by the loss of or reduction of resource/land rights. More people on a smaller area is a recipe for ecological degradation without significant changes in resource management.

But, the traditional resource management systems of indigenous peoples work when there is not undue pressure either from within or from outside. Indigenous peoples have maintained fragile ecosystems throughout the world. Their sophisticated and extensive knowledge of the local resources is based on the view that environments are the source of life and should therefore not be pillaged for short-term gain. Unlike farmers in mid-latitude areas who depend on machinery, specialized seeds, fertilizers and pesticides and who increasingly view the land as their adversary, indigenous people see the land and other resources as the life blood of their culture. Without a resource base, they are lost.

Yet indigenous peoples, because of the romantic views concerning their pristine lifestyles, are often forced to adhere to a different standard than everyone else. Many scientists now argue that indigenous peoples themselves degrade their resources and that the world cannot stand by and allow this to happen. Indians in Latin America overhunt species until they are nearly extinct. Indigenous peoples' agricultural practices are the main cause of deforestation in parts of Africa. The next cure for a disease, it is argued, could be going up in smoke.

Such discussions are often couched in the language of the greatest good for the greatest number. This argument has frequently been used to strip indigenous peoples of their assets in the name of the majority, but only a minority benefit. This line of reasoning also plays into the hands of states that would like nothing better than to further their control of such areas by removing indigenous peoples altogether.

In the final analysis, then, what is the record of indigenous peoples as conservationists? Do they manage resources effectively? Anthropological research, which may at times err on the side of romanticism, indicates that their use of resources is consistently more balanced than our own. Many practices are indeed conservationist even though their "scientific" basis may not yet be understood. It should be noted, too, that indigenous peoples have domesticated most of our basic foods (60% coming from the New World alone). In fact, their field trials of new crops and management systems continue.

The basic question, then, is what conditions encourage indigenous people to conserve resources? The most important factors appear to be resource rights (e.g., to land, timber, and water), basic political organization to protect themselves and their land and resource base, and resource management systems that can be transformed to meet modern needs. The success of indigenous peoples in managing their resources shows the importance of involving local people, whose futures depend on a specific resource base, in the wise management of the planet's resources. Not every goal is best accomplished by international laws and treaties signed by people thousands of miles from the areas in question. This is why the current, centralized state system is such a threat to the maintenance of the resources of this planet.

Even if states granted indigenous peoples resource rights, however, few would be able to survive in the modern world managing their resources as their ancestors did. To survive groups will have to adapt traditional resource management systems, rather than attempt either to keep them intact or abandon them totally in favor of foreign agricultural technologies. The adaptation of traditional resource management systems will allow indigenous peoples to maintain rational land-use patterns. By discovering the extent to which traditional management practices can be altered, more cash crops can be generated to meet the increasing material demands of indigenous peoples.

Much of the pressure on indigenous peoples and their resource bases comes from the need of Third World states to generate cash and from the insatiable consumption patterns and nonsustainable resource utilization practices in industrialized countries. Whether indigenous peoples survive will depend in large part on halting or reducing consumption practices in the industrialized regions of the world that threaten both the world's cultural and biological diversity.

JASON W. CLAY

For Further Reading: Jason Clay, *Indigenous Peoples and Tropical Forests* (1988); Suzanne Head and Robert Heinzman, eds., *Lessons from the Rainforest* (1990); Mark Miller, ed., *The State of the Peoples* (1993).

INDUSTRIALIZATION

Industrialization is a process of transforming economies, societies, regions, and nations through the development and application of technology and the systematic use of energy sources. It is often interpreted as being at the heart of modernization, a process that describes how today's nations and cultures developed into urbanized, energy-intensive, consuming, well-educated, and centrally-governed peoples. Industrialization, with its driving forces of technological and organizational change, has brought many societies an extraordinarily high standard of living. But it has also brought great costs, many of which have been environmental. Even waterpower, thought of as environmentally benign, caused extensive environmental damage when the flow of rivers was altered. These effects are best understood when viewed locally, regionally, and globally, even though they have not been distributed evenly, geographically or temporally.

Industrialization is marked by fundamental shifts in the energy and material bases of economies and societies. Traditional societies have relied on renewable sources of energy and materials and in order to survive have strived to maintain a balance between them. When that balance has been disturbed and resources become scarce, especially in the context of population increases, some societies have undergone tremendous instability. Others, however, have sought substitutes for diminished resources, thereby launching the processes leading to industrialization.

The first industrialized nation was England, and the process it followed—although a source of debate by scholars—clearly illustrates one path to industrialization. For centuries the British had depended on the land for most of their materials, but in the 16th century growing energy demand led to timber shortages, especially in large cities such as London. Timber was a vital resource, depended upon for energy (i.e., firewood), industrial chemicals (e.g., potash), and construction and shipbuilding materials. Under the stress of resource depletion, the English opted for resource substitution, turning to a mineral fuel, coal. The process of finding ways to substitute coal for timber unleashed a technological and organizational wave of innovation and invention that eventually transformed society and greatly increased environmental stress as the nation moved towards full industrialization.

Initially the environmental effects of substituting coal for timber were confined locally, and manifested, for instance, as heavy smoke palls and the depletion of nearby coal outcroppings. Increasing population, decreasing timber supplies, and the growth of heat-using industries, however, demanded more coal and produced more extensive environmental effects. Coal mining shifted to deeper mines, causing water drainage problems. The development of the steam engine—a direct result of the move to deep mines—helped solve the problem of mine flooding but also resulted in the pollution of local streams by acidic mine wastewater. Coal-burning locales grew into coal-burning industrial regions and air pollution became a regional problem. Overall, the English economy shifted from a low-energy to a high-energy base and from a society based on renewable energy and renewable resources to one relying on nonrenewable fossil fuels and energy-intensive materials. These patterns were fully visible by the end of the 18th century.

With increasing knowledge about physical and chemical processes and with continued growth of textile and metallurgical industries, the British economy became distinguished for what Lewis Mumford called its "carboniferous capitalism." Through the development of techniques such as puddling and the hot blast, the British adapted their iron-making processes to the use of mineral fuels rather than charcoal. Coal also served as a source of illumination with the development of the coal gas industry and became critical as both a fuel and a feedstock for the development of the inorganic and the organic chemical industries.

The chemical industry played an important role in the substitution process, supplying inorganic chemicals to the textile and glass industries as substitutes for previous wood-based inputs such as potash. Chlorine, nitrogen, sulfur, and phosphorus all entered the factory either directly or in various acid formulations. In 1856 with the discovery of mauvine (a purple dye and the first synthetic organic dyestuff) the English launched an organic chemicals industry based on coal tar, a byproduct of the process of distilling coal into coke. The emissions of these new chemical factories had unprecedented noxious effects on the countryside. Most serious were the emissions of the alkali industry, whose operations discharged large amounts of hydrochloric acid into the atmosphere. This pollution was so extensive that in 1863 Parliament enacted the first legislation attempting to regulate an industry on a national basis—the Alkali Act.

At different times, with varying labor and material inputs, and with varying degrees of success, all the nations of Western Europe and North America underwent processes similar in their broadest outlines to the industrialization of England. That is, nations engaged in industrialization shifted energy and resource inputs and material outputs. Over time in these nations, as in England, scale of production grew as did the society's ability to manipulate nature to its own ends. The growth in the worldwide production and consumption of coal (from 26 million metric tons in 1825 to 762 million metric tons in 1900) suggests the scale of potential environmental damage to air, land, and water. In addition, vast amounts of local resources such as timber and minerals were consumed in production, driving industrial nations to seek new sources of resources by engaging in imperialistic ventures and seeking colonial empires. In these colonies Western nations frequently repeated the destructive exploitive processes that had already marked their home locales. The vast scope of Western industrial production thus produced local, regional, and even global environmental effects.

The 20th century, like the 19th, has been marked by a dependence on fossil fuels. Some industries, such as iron and steel, continue to be coal-dependent. During the latter part of the 19th century, a revolution in steelmaking occurred, beginning with the Bessemer process first developed in the 1850s and the open hearth process developed several years later. Starting in the mid-19th century, these new steelmaking processes transformed the industrialization of transportation (the railroads, streetcars, and subways) and the extension of the sinews of urbanization (bridges, skyscrapers, water supply systems, and sewers). Steel was increasingly produced in large integrated iron and steel mills that included coke works, blast furnaces, and open hearth furnaces. In addition, chemical firms located nearby in order to utilize the various coal tars as feedstocks. By the beginning of World War I, gigantic iron and steel mill complexes dominated the river banks in industrial regions such as Pittsburgh (U.S.), Manchester and Sheffield (Britain), and the Ruhr (Germany). These iron and steel complexes polluted the air, water, and land with their gaseous and solid effluents and produced some of the world's most environmentally scarred and polluted industrial landscapes.

The development and subsequent growth of electric light and power beginning in the 1870s added to coal demand. The electric power industry moved from local or isolated power plants to central stations in urban areas and to massive interconnected power plants that provided electricity to whole regions through extensive power grids. Although some of these plants are powered by water, most burn coal or other fossil fuels. Over time, coal-burning power plants addressed local air pollution problems by building higher and higher smokestacks. Utilities thus shifted the problem from the local to the regional level and, ultimately, to the global dimension.

Increasingly, however, in the 20th century, the energy source of choice became petroleum, first commercially produced in 1859 in western Pennsylvania. World production moved from a few hundred thousand metric tons in 1900 to 2829 million metric tons in 1973. Unquestionably, the development of the internal combustion engine (Otto cycle, 1876; Diesel cycle, 1896) and its application to transportation—the automobile and the motor truck—have had a profound influence on the course of industrialization and on the total environment. Until the development of the internal combustion engine, the market for petroleum was largely limited to the production of kerosene for illumination, but the rise of the automobile propelled the production of petroleum in the form of gasoline (for fuel) and oil and grease (for lubrication). The automobile industry became the largest single consumer of iron and steel and a major consumer of products such as plastics and paints manufactured by an increasingly diversified chemicals industry, thereby increasing the demand for coal (in the form of coke for smelting and coal tar for chemical feedstocks). The process of industrialization was dynamic and interactive; one thing led to another.

The growth of the chemicals industry in the 20th century, also increasingly linked to petroleum as feedstock, had profound implications for environments: locally, regionally, and globally. The industry developed a rapidly growing array of new products, from synthetic dyestuffs to plastics and synthetic fibers, from quick-drying lacquers and paints to synthetic rubber, and from synthetic ammonia to organic pesticides. The story of the creation, development, and widespread use of several chemicals demonstrates the complex character of industrialization and its often unanticipated environmental consequences. The cases used to demonstrate these developments involve the automobile, Freon (or CFCs), and agriculture.

The automobile was first developed in the late 19th century, but further improvement in automobile engine performance was stymied by the problem of "knock"—uncontrolled explosions of fuel in engine cylinders. Researchers at General Motors discovered that the addition of a small amount of a compound first synthesized in the 19th century, tetraethyl lead (TEL), to gasoline, would suppress engine knock. Although numerous workers at General Motors, Standard Oil, and DuPont died from exposure to the highly toxic compound, a formal investigation by the U.S. Surgeon General of TEL's safety found that it posed no undue public health hazard when added to gasoline in minute quantities. Consequently, TEL was adopted as a gasoline additive, and its use grew as automobile manufacturers introduced higher and higher performance engines. With increased production of automobiles and increased consumption of gasoline, however, the effects of TEL's adoption eventually became manifest. Scientists such as Clair Patterson from California Institute of Technology pointed to a world-wide increase in the environmental lead burden, while intensified air pollution (smog) in cities such as Los An-

geles increased the concern of public health officials about resultant negative health effects. The identification of TEL as a major cause of the problem led to the enactment by California of the first state air pollution control statutes and eventually to the first federal Clean Air Act in 1963. Eventually, in 1973, the EPA banned TEL completely for use in new automobiles in the U.S. and leaded gasoline has gradually disappeared from the market although lead residues have often persisted in the soil and sediments of metropolitan areas.

Another chemical that solved a major problem in the 1920s turned out to pose global problems in the 1970s, and the world has only begun to deal with those problems. When mechanical refrigeration was developed in the late 19th century, such chemicals as ammonia, sulfur dioxide, and methyl chloride were used as refrigerants. But these chemicals are toxic in their gaseous phase, and therefore a leak in a refrigeration system could be deadly. A number of serious accidents in the 1920s caused a search for alternatives, and in 1929 General Motors researchers discovered that chlorine- and fluorine-substituted hydrocarbons made safer refrigerants because they were both nontoxic and nonflammable. Consequently, in conjunction with DuPont, they developed Freon—dichlorodifluoromethane (CCl_2Fl_2)—which effectively eliminated the dangers of mechanical refrigeration and thus allowed the development of household refrigeration and air conditioning that utilized Freon as the coolant. Freon appeared to be so safe and useful that it became the propellant employed in many personal aerosol products and was also widely adopted as a solvent in the manufacture of electronic equipment. Although it appeared to be the perfect chemical and was safe locally and regionally, Freon turned out to be dangerous globally. In 1974 two University of California chemists published a paper in *Nature* maintaining that the whole family of Freons—chlorofluorocarbons—were depleting the ozone layer in the earth's upper atmosphere, thereby threatening the earth's flora and fauna with increased solar radiation. Their work was initially viewed with skepticism, especially by the CFC producers, but during the next decade and a half the international scientific community built a massive case for ozone depletion by CFCs, leading to a program of phasing out their production.

The third case involves the chemical industry's participation in the industrialization of agriculture. Mechanization of agriculture first took place in the 19th century, but it was the development of the internal combustion engine and the gasoline powered tractor in the 20th century that brought agriculture into the world energy economy. The increasing adoption by farmers of commercially produced synthetic fertilizers drew agriculture even more deeply into the framework of industrialization. Beginning in the 1930s, the chemical industry initiated large-scale research and development programs aimed at controlling pests, weeds, and fungi in agriculture. By mid-century, new "miracle" chemicals such as DDT (credited with helping to keep malaria and other diseases in check during World War II), and 2-4D, a highly active herbicide with limited selectivity, had been developed and were in wide use. The effects of these chlorinated hydrocarbons, however, proved far more environmentally devastating than predicted because of both their toxicity to many biological species and their persistence. By the late 1950s and 1960s, environmental scientists proved that DDT was the cause of declines in wildlife populations and then demonstrated the mechanism by which the pesticide entered the food chain, moving from the targeted insects all the way up to the breast milk in humans. Even though the EPA banned the most toxic of the pesticides in the 1970s, runoff from agricultural chemicals has polluted streams, lakes, estuaries, and groundwater supplies in whole regions and systems of regions. Although it has brought impressive gains in productivity (the so-called Green Revolution), the linking of agriculture to the chemical industry has had extreme environmental costs.

Near the beginning of the 21st century, the larger patterns of the effects of industrialization on the environment have become clearer. Some of these effects, such as land and groundwater contamination at present or former industrial sites, are a result of past industrial waste disposal practices by firms looking for low-cost solutions, while others result from ongoing current disposal methods. Some of the worst examples of industrial environmental pollution exist in the nations of the former Soviet bloc in Eastern and Central Europe such as East Germany and in the former Soviet Union itself, revealing that the environmental abuses of industrialization were not confined by economic system. While the older and more advanced industrial nations have been attempting to control the environmental effects of industrialization, the newer and developing countries, in their haste to increase production, often repeat past errors.

Industrialized nations and societies depend upon hydrocarbons for their survival and are in fact built around the production, manipulation, and use of hydrocarbons. Coal, natural gas and petroleum—all depletable fossil fuels—provide the basis for this civilization. These fuels are critical in materials, transportation, manufacturing, agriculture, and domestic consumption. The environmental effects of the hydrocarbon civilization are apparent locally, regionally, and globally. They include local pollution, such as land fouled by the overuse of herbicides and beaches ruined and wildlife killed by Alaskan and North Sea

oil spills; severe regional pollution in smog-ridden urban basins such as Los Angeles or Mexico City; the death of large bodies of water such as the Aral Sea in the former Soviet Union; and global pollution, such as ozone depletion by decades of CFC production. Large-scale consumption of hydrocarbons has introduced increasing amounts of carbon monoxide and carbon dioxide into the atmosphere that contribute to increased concerns about global warming whose environmental effects are potentially catastrophic.

Some efforts to lessen our dependence upon nonrenewable fossil fuels have generated environmental hazards. Nuclear power, for example, promised what H. G. Wells called "a world set free" from carbon-based energy. But the accident at Three Mile Island (1979) and the far more serious accident at Chernobyl (1986) demonstrated the grave local, regional, and global dangers inherent in nuclear power plants. Moreover, the processing of nuclear fuels and their disposal when spent pose major problems to a clean, healthy, and safe environment.

Industrialization has thus been a paradox. It has greatly increased knowledge about the world and used technology to provide greater material abundance and improved health. At the same time, the environmental costs of industrialization have often been high in terms of pollution of the air, land, and water; disfigurement of the landscape; and unanticipated effects for the long term health of the planet.

DAVID A. HOUNSHELL
JOEL A. TARR

For Further Reading: Jesse H. Ausubel and Hedy E. Sladovich, eds., *Technology and Environment* (1989); Merril Eisenbud, *Environment, Technology, and Health: Human Ecology in Historical Perspective* (1978); B. L. Turner, et al., *The Earth As Transformed by Human Action: Global and Regional Changes in the Biosphere over the Past 300 Years* (1990).

INDUSTRIAL METABOLISM

At the most abstract level of description, industrial metabolism is the whole integrated collection of physical processes that convert raw materials, energy, and labor into finished products and wastes (Figure 1). The stabilizing controls of the system are provided by its human component, which has two aspects: (1) direct, as labor input, and (2) indirect, as consumer of output. The metabolic system is stabilized, at least in its decentralized competitive market form, by balancing the supply of and demand for products and labor through the price mechanism. Thus the market system is, in essence, the metabolic regulatory mechanism.

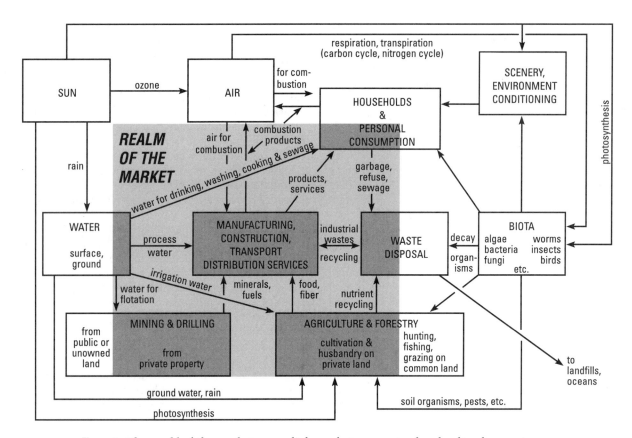

Figure 1. The world of the market: some linkages between natural and cultural ecosystems.

The word *metabolism*, as used in its original biological context, refers to the internal processes of a living organism that are necessary for the maintenance of life. There is a compelling analogy between biological organisms and industrial activities, not only because both biological and industrial systems process materials into "useful" forms, using an external source of free energy, but also because both are examples of self-organizing "dissipative systems" in a stable state.

Industrial metabolism can be identified and described at a number of levels. The concept is applicable to nations or regions, especially isolated ones such as watersheds or islands. The key to regional analysis is the existence of a well-defined geographic border or boundary across which physical flows of materials and energy can be monitored.

The concept of industrial metabolism is equally applicable to another kind of self-organizing entity, a manufacturing enterprise or firm. A firm is the economic analog of a living organism. This analogy between firms and organisms can be carried further, resulting in the notion of "industrial ecology." Just as an ecosystem is a balanced, interdependent, quasi-stable community of organisms living together, so its industrial analog may be described as a balanced, quasi-stable collection of interdependent firms belonging to the same economy. The interactions between organisms in an ecosystem range from predation and/or parasitism to various forms of cooperation and synergy. Much the same can be said of firms in an economy. Some of the differences are interesting, however. In the first place, organisms reproduce themselves; firms produce products or services, not other firms (except by accident). In the second place, firms need not be specialized and can change from one product or business to another. By contrast, organisms are highly specialized and cannot change their behavior except over a long (evolutionary) time period. In fact, the firm (rather than the individual) is generally regarded as the standard unit of analysis for economics.

It is useful to distinguish industrial metabolism from industrial ecology. Although the two concepts have considerable overlap, there is an important difference. An ecosystem recycles most substances internally. Producers (plants) convert solar energy into carbohydrates and sugars which, in turn, become food for animals. The complex hierarchical relationships among grazing animals, predators, omnivores, parasites and decay organisms are called a food web. A true industrial ecosystem would have a similar structure. Most materials would be recycled within the system, with minimal addition of extractive resources to replace inevitable dissipative losses. Companies, like organisms, would "eat" each other's wastes. However, it must be recognized that, at present, the industrial system is almost exclusively dependent on extractive resources, with very little reuse or recycling.

R. U. AYRES

INFESTATION

Infestation is the presence of animal parasites such as insects or worms in or on another organism. The invaded organism is known as the host. A parasite lives its life, either in part or in whole, at the expense of the host, feeding on the host's blood or tissue. Some parasites are internal; others attach themselves on the outside. They flourish in areas where the climate is hot and standards of hygiene are low.

Worms are the largest type of parasite that infests humans. Parasitic worms include roundworms, such as hookworms and pinworms; and flatworms, such as flukes and tapeworms. Parasitic worms are commonly found in swine and may also be found in beef and some types of fish. Travelers unaccustomed to conditions in developing countries may inadvertently expose themselves to infestation by worms. Proper handling of food is necessary to prevent infestation by worms. Worm infestations and their diseases are treated by emetics (substances that cause vomiting), purgatives, and other drugs.

Parasites such as aphids and caterpillars can cause injury to plants by feeding on them. Aphid populations increase very rapidly in early summer. Infestation by aphids, as well as by insects in general, can cause damage to plants and crops. Insects may be controlled by insecticides, by the introduction of natural enemies of pests, or by integrated pest management techniques. For example, ladybugs are a natural enemy of aphids and can be introduced to control aphid infestation. Similarly, caterpillars can be controlled by cultivating and releasing a parasitic wasp in infested areas. The wasps lay their eggs in the caterpillar. The wasp eggs grow into larvae by feeding on the caterpillar and then burrow their way out, killing the caterpillar host.

Most insects that attack stored foods are types of beetles. Granary weevils and rice weevils eat grains both in their larval stage and as adults. They breed quickly and can produce several generations in one year. Infested grain is unfit for use. It is contaminated with insect remains. The milled grain is discolored, tastes bad, and may cause illness.

Infestation in grain warehouses is controlled by cleaning out storage areas with vacuums, using insecticides, and sterilizing grain sacks. Sealing grain in airtight storerooms or containers helps to reduce the likelihood of infestation. Inside a sealed container, the amount of oxygen is decreased by the natural chemical processes of the grain. This in turn increases the amount of carbon dioxide, killing insects in the grain. For these same reasons, grain stored in the home should be kept in airtight containers.

Another form of infestation occurs when insects live on or feed upon inanimate objects such as wood or furniture. If the insects are not too numerous, the wood can be treated with a penetrating insecticide. Severely infested wood should be replaced with insecticide-treated wood to prevent insect attack.

INSECTICIDES

Translated literally, insecticides are chemicals or other agents that kill insects. In recent years, the term has been broadened to include materials of all types, of natural or synthetic origin, that provide economic control of insect pests (also sometimes extended to other arthropods such as spiders and mites) by a variety of mechanisms including direct toxic action, repellency, disruption of mating and other behavior, antifeeding activity, and interference with reproduction or development.

Early chemical insecticides, reportedly used as early as biblical times, included such inorganic materials as sulfur, arsenic, and mercury. These were followed by a number of naturally-occurring, plant-derived materials such as pyrethrum (from chrysanthemum flowers), nicotine (from tobacco), and rotenone. These types of materials, along with several tars and oils (creosote, kerosene, etc.) and a few synthetic organic chemicals (dinitrophenols, thiocyanates) were almost all that were available for insect control up until the late 1930s. Discovery of the insecticidal properties of DDT by Paul Muller in 1938 (for which he received the Nobel Prize) demonstrated the remarkable potential of synthetic organic chemicals for the control of insect pests. During the 1940s and 1950s, the use of DDT to control mosquitos and other vectors of human disease as well as insects of agricultural importance markedly reduced human suffering and resulted in enormous economic benefits worldwide. During this same period the search for other synthetic insecticides led directly to the discovery and commercial development of additional groups of chlorinated hydrocarbons (e.g., the cyclodienes such as aldrin, dieldrin, chlordane, heptachlor, and hexachlorocyclohexane), the organophosphorus compounds (e.g., parathion, tetraethylpyrophosphate) and, a little later, the carbamates (e.g., carbaryl).

The introduction of DDT and the other chlorinated hydrocarbons had a profound impact on pest control practices and there was speculation at the time that some important pests might even be eradicated. The chlorinated hydrocarbons were inexpensive, highly effective, broad-spectrum pesticides and, in retrospect, there is no doubt that they were both overused and misused. Several problems soon became apparent. First, their general lack of selectivity had a serious adverse impact on populations of many beneficial insects (such as predatory ladybeetles and parasitic wasps) that provided natural control of many pest species; this was directly responsible for sudden resurgences of pest populations following applications of DDT. Secondly, the ability of insects to develop resistance to synthetic organic insecticides first became apparent following the introduction of DDT. This phenomenon results simply from the genetic selection of those individuals in a population that possess specific biochemical (metabolic), physiological (thicker cuticle), or other characteristics that allow them to survive exposure to the insecticide. The buildup of insecticide resistance is by no means restricted to the chlorinated hydrocarbons and, in many situations, continues to be a source of concern with respect to effective and economic insect control.

Other non-arthropod, non-target species also suffered deleterious effects. The stability and persistence of chlorinated hydrocarbons in the environment and their ability to accumulate in the tissues (especially fat) of living organisms caused them to undergo biomagnification as they moved through food chains (they often concentrated hundreds or thousands of times). This caused serious problems for many raptoral birds and carnivorous fish. The ability of DDT/DDE to cause eggshell thinning and adverse reproductive effects in several species of birds (especially eagles, hawks, and owls) placed populations of several species at risk. The use of DDT in the U.S. was discontinued in the early 1970s although, even now, it remains widely used in other parts of the world.

The discovery of residues of DDT and other chlorinated hydrocarbons in human fat tissues also initiated concern over the potential long-term effects of pesticides on human health, an issue that has continued to be highly controversial. Indeed, even today, there remains a high level of fear in a large segment of the U.S. population that, despite the absence of any supportive evidence, trace residues of insecticides in food, water, or the environment are responsible for a variety of human ills including cancer. It is most unfortunate that the use of pesticides has become such a controversial topic and that there remains so much public misunderstanding of the real benefits and risks associated with their use.

There is little doubt that the widespread use and misuse of the chlorinated hydrocarbon insecticides raised awareness of the potential adverse effects of pesticides, stimulated the promulgation of new federal and state legislation to regulate pesticide use, led to more rigorous testing of new products, and challenged the agrochemical industry to develop new and better pesticide products.

Despite the technical difficulties and expense (up to $40 million per product) of developing new chemicals that meet the more stringent modern requirements for environmental and health safety, the agrochemical industry has been able to generate a constant flow of new products. These more sophisticated products are the result of years of innovative basic research to enhance efficacy against insect pests and at the same time improve other characteristics. In contrast to DDT, which was applied at up to five pounds per acre, some of the new synthetic pyrethroid insecticides require just a few grams per acre. Such low application rates reduce the likelihood of residues in food, water, and the crop commodity, and decrease the chances of accidental poisonings. Many of these new products have been developed through chemical modification of existing compounds. Thus chemists have discovered the keys to increasing biodegradation and mammalian safety without destroying insecticidal efficacy and formulators and engineers have improved the efficiency of spray application techniques and equipment.

A major problem with the early insecticides was their general lack of selectivity between target and nontarget species including mammals. This resulted from fundamental similarities in the biochemical and physiological systems with which the materials interfere to exert their toxic effects. Thus, at the molecular level, the insect and mammalian systems for energy production (e.g., oxidative phosphorylation) and nerve conduction and synaptic transmission are quite similar, so that an insecticide affecting such systems adversely frequently exhibits broad-spectrum, nonselective toxicity. Examples of such materials are the chlorinated hydrocarbons that act on nerve conduction, the organophosphorus compounds and carbamates that inhibit cholinesterase, and the dinitrophenols that block oxidative phosphorylation. In recent years basic research on potentially vulnerable biochemical and physiological systems unique to insects has led to the development of several new and more selective materials. Examples are dimilin, which inhibits chitin synthesis, and a number of compounds that interrupt insect development (larval molting, ecdysis) through interference with critical hormonal regulation. Other compounds have been developed that interrupt insect feeding behavior (antifeeding agents), and identification of the chemicals used by insects to communicate with one another has led to the use of sex pheromones to confuse natural populations and interrupt mating behavior.

Insecticides will continue to be important adjuncts to modern agriculture. It is likely, however, that the insecticides of the future will be used increasingly in a manner analogous to prescription drugs. They will be more selective in their action, rapidly biodegradable, and relatively safe for humans, and they will be viewed as just one of several alternative pest control agents to be incorporated into integrated pest management (IPM) programs.

CHRIS F. WILKINSON

For Further Reading: GRC Economics, *The Value of Crop Protection: Chemicals and Fertilizers to American Agriculture and the Consumer* (1990); Duane G. LeVine and Arthur C. Upton, *Management of Hazardous Agents*, Vol. I and II (1992); National Academy of Sciences, *Regulating Pesticides in Food: The Delaney Paradox* (1987).
See also Biotechnology, Agricultural; DDT; Evolution; Integrated Pest Management; Pesticides.

INSECTS, SOCIAL

See Social Insects.

INSECTS AND RELATED ARTHROPODS

Number of Species and Characteristics

There are more species of insects, spiders, mites, lobsters, shrimp, and other related arthropods than all other plants and animals combined. Insects and related arthropods make up about 90% of the approximately 10 million species that are estimated to be on Earth. In contrast, mammals and birds comprise only about 0.1% of the total species found in nature. The phylum Arthropoda includes several classes: insects, crustaceans (e.g., lobsters and shrimp), spiders, and centipedes and millipedes. Table 1 shows the various orders of insects.

Although insects and other arthropods have no one characteristic that separates them from all other animals, they do possess three features that are unique to this group: an outside shell or exoskeleton, a segmented body, and jointed legs. The segmented body of insects is divided into three distinct sections: head, tho-

Table 1
The orders of insects

Order	Name or examples	Comments
Thysanura	silverfish, firebrat	Some occur in warm moist areas of houses.
Collembola	springtails	"Spring" on tail allows them to jump long distances. Frequently occur in soil; a few feed on plants.
Ephemerida	mayflies	Larvae aquatic, important fish food. Adults delicate with four veined wings.
Odonata	dragonflies	Predacious. Larvae live in water. Adults large, winged; prey on flying insects such as mosquitoes.
Isoptera	termites	Live in colonies, important pests feeding on wood.
Orthoptera	grasshoppers, crickets, cockroaches	Large group of insects.
Dermaptera	earwigs	Usually feed on decaying organic matter. Rarely pests.
Plecoptera	stoneflies	Larvae aquatic, adults have four large veined wings.
Corrodentia	booklice	Very small, feed on paper or other organic matter.
Mallophaga	birdlice	Often parasitic on birds.
Thysanoptera	thrips	Many are serious pests of crops.
Anoplura	headlice, bodylice	Sucking lice of mammals including humans.
Hemiptera	true bugs: squash bug, harlequin bug	Large group including the true bugs. Some feed on plants and are pests, others are predacious and beneficial.
Homoptera	aphids, leafhoppers	Small sucking insects.
Coleoptera	ladybird beetle, potato beetle	This group contains the most species of insects. Some are beneficial, some are pests.
Neuroptera	lacewings	Some have very delicate wings with many veins. Most are predacious and beneficial.
Lepidoptera	butterflies, moths	Some are serious pests of crops and trees.
Diptera	flies, mosquitoes	Many are pests of humans and animals. Some are predacious and beneficial.
Hymenoptera	bees, wasps, hornets	Most are beneficial. A few are pests.

rax, and abdomen. All insects also have three pairs of legs (the Hexapoda).

External and Internal Structures and Adaptations

The surface of most insects and other arthropods is covered by a chemical substance that hardens into a shell when exposed to air. This external skeleton (exoskeleton) is made of a nitrogenous substance called chitin, which is hard but flexible and is resistant to most chemicals and degradation. Thus, it provides an effective armor for the protection of the arthropods.

In contrast to the human body, insects have no internal structures that support the soft organs within the exoskeleton except as ingrowths from the outside. For example, arthropods breathe through air tubes that pass through the exoskeleton and extend to every organ inside. However, a few arthropods, such as crustaceans, use gills to provide oxygen for respiration.

The exoskeleton limits the size of arthropods and their ability to regulate their body temperature. Arthropods are cold blooded and do not regulate the temperature of their bodies as humans and other mammals (homeotherms) do. Instead their body temperature is governed by their surroundings.

Insects are extremely adaptable to varied environments. Their small size and their short generation time, sometimes as short as six or seven days, as well as their genetic diversity facilitate their adaptability. In addition, most insects have tremendous reproductive capacities, which further enhances their ability to survive in varied environments. A female cricket may produce 6,000 offspring during her lifetime. A biologist has calculated that, if all the offspring of a single pair of houseflies survived to reproduce all summer long, the Earth would be covered with houseflies several feet deep. Fortunately, natural enemies and environmental factors limit populations of flies and other related arthropod populations from increasing at these extreme rates.

Habitats and Occurrence

An examination of natural habitats suggests that insects and related arthropods not only outnumber all other species, but also occur in the most habitats. They are present in the soil, on the soil surface, on and inside plants, on and inside other animals, in and on organic matter and various animal and plant wastes, in freshwater and saltwater ecosystems, and even in oil. Their presence extends from the tropics into the arctic regions. They survive in moist habitats and in desert habitats.

Because of the abundance of arthropods, their diversity, and their extensive distribution in most habitats,

they serve as the prime source of food for many species of fish, including trout and salmon; for toads, frogs, and other amphibians; for many birds, such as the bluebird and house wren; and for moles, shrews, skunks, and many other small mammals.

Foods of Insects and Related Arthropods

Insects and other arthropods feed on a greater variety of substances than any other group of organisms, which is undoubtedly a major reason that they are the most widespread and successful group of organisms in nature. For example, gypsy moths feed on plant leaves, and apple maggots on the insides of plants. Lice feed on the surface of animals, and bot flies on the inside of cattle. Houseflies feed on dead plant material, and blow flies on dead animals. Fruit flies feed on microorganisms, and ladybird beetles on other insects. Carpet beetles and clothes moths feed on wool and dead insects. Dung beetles feed on dung. A recently discovered fly species even feeds on oil.

Many species, especially those that feed on dung or dead plant and animal matter, play a vital role in recycling wastes in nature. For example, when cattle were introduced into Australia, cattle dung accumulated on the surface of the land and was becoming a major problem. To solve their problem, Australians imported several species of dung beetles from Africa to bury and recycle cattle dung.

Economic Benefits

Pollination

Insects and related arthropods not only keep the natural life system functioning by eating and being eaten, but they also serve a most important role in plant reproduction, that of pollination. In agriculture, honey bees and wild bees are responsible for pollinating $30 billion worth of fruits and vegetables annually.

The role of just honey bees alone in pollination is impressive. For example, a single honey bee may visit and pollinate 1000 blossoms in a single day (10 trips with 100 blossoms visited per trip). In New York State, with about 3,000,000 beehives each with 10,000 worker bees, honey bees could visit and pollinate 30 trillion blossoms in a day. Wild bees pollinate a number of blossoms equal to or more than those pollinated by honeybees. Thus, on a bright, sunny day in June, more than 60 trillion plant blossoms may be pollinated—a Herculean task carried out by the "busy bees."

Despite the many technological advances made by humans, no substitutes for insect and related arthropod pollination of both cultivated and wild plants are available to insure the production of fruits and vegetables.

Honey and Wax Production

Bees make several products of value to humans, two of which are honey and wax. About 200 million pounds of honey and 4 million pounds of wax are produced in the United States annually, with a value of about $130 million and $12 million, respectively. To produce 1 pound of honey, 1500 bees work 10 hours, and to make 1 pound of beeswax (an ingredient of most lipsticks), honeybees consume about 8 pounds of honey.

Dyes, Silk, and Shellac

Cochineal is a beautiful carmine-red dye that is produced by a mealybug. This natural red pigment is made from the ground-up bodies of mealybugs that feed on prickly pear cacti and is used as biological dye.

Shellac is made from the secretions of the lac insect. The paint made from lac secretions is often diluted with alcohol and typically dries rapidly to a hard shiny surface that is resistant to water and other materials.

Perhaps one of the most valued and most widely used products of insects is silk produced by the silkworm moth. Silk is lightweight, fine, but extremely strong. It also has the unique characteristic of being able to stretch without breaking. These characteristics have made silk a prized cloth.

Beauty and Religious Objects

Various insects and other arthropods are used by different societies as individual items of beauty and for religious purposes. For example, scarab beetles are worshipped in Egypt and are considered to be good luck charms. They or their likeness are made into all kinds of jewelry. Drawings of scarabs have been found in the ancient pyramids.

Food for Humans

A wide variety of insects, usually large insects, are used as food by various human cultures. For example, locusts, grasshoppers, grubs, termites, ants, and other insects are valued as nutritious food for a great many people in the world.

As for the arthropods, no food is more highly prized than lobster and shrimp as human food. These two arthropods are considered a delicacy by most people and sell for extremely high prices in the United States and world market.

Biological Control of Pests

In addition to providing food for humans, beneficial insects and other arthropods help protect crops, forests, and livestock from the attack of a wide array of pest insects, weeds, and plant pathogens. In both natural and agroecosystems, many species, especially predators and parasites, control or help control plant-feeding pest populations. Indeed these natural species make it possible for the ecosystem to remain "green." It is estimated that about half of the pest control in crops and natural vegetation is carried out by beneficial insect and other arthropod species, which are enemies of the pest species.

The impact of insect feeding pressure for weed control is illustrated throughout nature. For example, the plant pest known as Klamath weed was introduced into California about 1900, and by 1944 it had become a dominant weed on about 2 million acres of valuable pasture and rangeland, crowding out the preferred forage. To control the weed's spread, a small beetle that feeds on it in its native habitat was introduced into California in 1945. The beetle population increased, dispersed, and eliminated 99% of the weed in California. With the reduction of the Klamath weed, previous forage vegetation has regrown, making pasture and range once again suitable for cattle production.

Crop Losses to Insects and Related Arthropods

Despite the heavy use of insecticides and miticides, insects and other arthropods destroy approximately 13% of potential crop production in the United States and world. An estimated $1 billion is spent annually on pesticides for insect/arthropod control in the United States. It is estimated that this use of pesticides saves about $4 billion in crops. However, this positive economic assessment does not include the environmental and public health costs of using pesticides.

DAVID PIMENTEL

For Further Reading: H. E. Evans, *The Pleasures of Entomology: Portraits of Insects and the People Who Study Them* (1985); B. Klausnitzer, *Insects: Their Biology and Cultural History* (1987); T. M. Peters, *Insects and Human Society* (1988).

INTEGRATED PEST MANAGEMENT

Integrated pest management (IPM) is a comprehensive, multidisciplinary approach to pest control that seeks to minimize — if not eliminate — the use of chemical pesticides in agricultural systems. The underlying premise of IPM is that pesticide usage in commercial agriculture is generally excessive, injurious to the environment, and unsuccessful in controlling crop damage caused by pests. IPM is a farm management system that strives to balance food production and pest control with species diversity and ecological sustainability.

As an alternative or complement to pesticide usage, IPM requires intimate knowledge of the local ecosystem and careful planning and monitoring of the cultivated system. Farmers employing IPM use a broad range of crop husbandry practices including soil fertility, minimum tillage, crop rotation, timing of planting, intercropping, and using beneficial organisms to control a pest population. This combination of techniques, combined at times with minimum pesticide usage, helps to disrupt pest life cycles and maintain their populations at nonthreatening levels, while enhancing the overall biotic and abiotic stability of the system.

Biological diversity is a key component of IPM. If allowed to flourish, biological diversity mitigates pest outbreak through cultivating a beneficial habitat for predators of crop pests. Biological diversity reduces the risk of total crop failure due to insect attack by raising more than one species in the field. This diversity also provides greater economic security in the event one crop suffers total loss. However, the cost and effectiveness of IPM vary from system to system and therefore cannot be readily calculated at the outset of a production year.

Although the original intent of IPM is to eliminate pesticides through intensive, organic methods, large-scale production systems have successfully modified the practice to include chemical pesticides. In this case, pesticides applied at strategic points in the production process destroy specific insects at peak reproductive cycles. With this approach, pesticides may be managed and applied at more judicious rates than ordinarily observed in conventional large-scale systems.

INTERMODAL SURFACE TRANSPORTATION EFFICIENCY ACT

The Intermodal Surface Transportation Efficiency Act of 1991 (ISTEA) governs surface transportation spending, the largest U.S. infrastructure investment program, providing the nation with a potential $155 billion federal investment through 1997. It provides for three key reforms in the areas of planning, public participation, and flexible funding. ISTEA reinforces metropolitan planning requirements for transportation and requires state planning for the first time to "address the overall social, economic and environmental effects of transportation decisions" and conform to state air quality plans, among other important factors. It mandates broader earlier involvement of the public in decision making. And it abolishes old, mode-oriented categories of funding in favor of new programs available for all modes of transportation.

Flexible funding replaces the old approach to transportation funding, which was based on mode (e.g. highway, transit). Title I of ISTEA contains two new programs open to all modes of transportation: the Surface Transportation Program (STP) and the Congestion Mitigation and Air Quality Improvement Program. Both programs stress the need to select the best projects for overall mobility, taking into account other important community and environmental factors. States may also transfer some or all funds to STP from traditional categories such as the Interstate and Bridge programs and from the new National Highway System program, making most Title I funds fully flexible.

In addition to the flexibility provided for by ISTEA, the law institutes a uniform federal match for most road and transit projects of 80% federal funds to 20% state or local funds. Under previous law, federal-aid road projects were eligible for up to a 95% federal match, compared to a 70% match for transit.

Before ISTEA the majority of transportation decisions were made, project by project, with little overall coordination and virtually no coordination with other important community goals, such as regional growth and land use, energy use, access to jobs and housing, environmental protection and recreation. ISTEA creates a framework for comprehensive planning by requiring metropolitan and state planners to consider a full set of factors in order to better coordinate these decisions. It also requires that the public be involved in all aspects of transportation decisions, from long-range planning to project selection.

ISTEA reserves 10% of the funds authorized for the STP for a category of activities called transportation enhancements. These include facilities for bicycles and pedestrians; scenic improvements, easements, and byway programs; landscaping and beautification; historic preservation; preservation of abandoned railway corridors; archaeological planning and research; and some water quality protection activities. To be eligible for enhancement funds, a project must go beyond mitigation of negative impacts on cultural or natural resources, or routine access for bicycles and pedestrians. And enhancement activities must be incorporated into the transportation planning process.

LISA WORMSER

INTERNAL-COMBUSTION ENGINE

The internal-combustion engine is a device that converts thermodynamic energy (from burning fuel) into mechanical power. There are several different types of automotive spark-ignition (SI) internal-combustion engines. Examples include the four-stroke engine, which is the most common; the Wankel engine, which is sold in small numbers; and the two-stroke engine, which may become popular in the 1990s. Although the mechanical construction of each of these engines is quite different, all SI engines operate under the same thermodynamic cycle, known as the Atkinson cycle.

The Atkinson cycle includes four thermodynamic processes: (1) compression of the fuel-air mixture; (2) combustion of the compressed fuel-air mixture; (3) expansion of the combustion products to atmospheric pressure; (4) purging of the exhaust gas and refilling of the combustion chamber with a fresh charge.

For a compression ratio—the volume of the cylinder of a combustion chamber when the piston is at bottom dead center divided by the volume when the piston is at top dead center—of 10:1, the idealized Atkinson cycle has a thermodynamic efficiency of approximately 65%. The average efficiency of most automobile engines sold today, however, is much lower, approximately 17%. The causes of engine efficiency loss include mechanical friction, pumping losses, heat loss, and mechanical limitations of the engine. These losses are highest when the engine is operated at light load. Current automobile engines are relatively inefficient because they are operated at a light load most of the time.

The maximum power of most new car engines is approximately 130 horsepower. The average power actually used when driving, however, is only approximately 8 horsepower. At this horsepower much of the power generated is used to run the engine itself.

Technologies that can improve light-load engine efficiency include turbocharging, variable valve control, lean burn combustion, and advanced transmissions. In 1992 one manufacturer (Honda) introduced a lean-burn variable valve control engine with an average efficiency of 26.5%. This represents a 56% improvement in fuel economy above today's typical 17% efficient automotive engines. Significant gains in automotive fuel economy can be expected during the next decade as several of these relatively new technologies mature.

INTERNATIONAL DECADE FOR NATURAL DISASTER REDUCTION, THE

Each year the world suffers hundreds of incidents of tropical cyclones, drought, locust infestations, and volcanic eruptions, and thousands of earthquakes, tornadoes, wildfires, and landslides, while devastating storms and floods number in the tens of thousands annually.

The cost in human lives is tragically great—too great—over a million lives are lost to natural forces each decade. More than three times that number are left homeless, sick, or injured. Disasters cause tens of billions of dollars in property damage each year and rob nations of hard won investment and productivity. For developing countries the burden is particularly heavy. They suffer over 90% of the lives lost, and incur relatively greater damage to their infrastructure and economies. Long-term sustainable development is threatened as gains are wiped out and government and private expenditures must be diverted from development to relief and reconstruction.

The United Nations has declared the 1990s to be the International Decade for Natural Disaster Reduction (IDNDR). The Decade was launched by Dr. Frank Press, President of the U.S. National Academy of Sciences, who first proposed the idea at the Eighth World Conference on Earthquake Engineering in 1984. Dr. Press pointed out that there had been great advances in the scientific understanding of natural hazard causality and of technological means for limiting its impact, but that application of this knowledge—particularly in developing nations—was lagging and that disasters were viewed with a misplaced sense of fatalism. Further, he noted that natural hazards did not recognize national boundaries, and there was already a strong history of cooperation in the scientific community internationally. He felt that the time was right for a "Decade"—an opportunity to confront and overcome fatalism, by putting into practice new advances in science and engineering, global communications, and other tools. "What better way to start the new millennium than a world better organized to reduce suffering?" he asked the participants of the conference.

Originating in the science and technology communities around the world, the concept spread over the next few years to gain support from representatives of local and national governments, professional associations, nongovernmental organizations, members of the business community, and many others. The U.N. General Assembly came to view the Decade as an opportunity to bring the benefits of science and technology to the world's people. In response to a General Assembly request, Secretary-General Javier Perez de Cuellar established the Ad Hoc Group of Experts which clarified the Decade's activities and goals, as well as the structure a U.N.-based Decade would have. With minor changes, the General Assembly approved the Group of Experts' plan, launching the IDNDR in 1990 as an extra-budgetary U.N. activity in cooperation with the world's scientific and technological community.

The U.N. created a Decade Secretariat to oversee the international program, and appointed a Scientific and Technical Committee (STC), as well as a Special High-Level Council. The STC, consisting of 25 of the world's leading technical experts, advises the Secretariat and promotes international efforts, particularly in the professional community. The Council, composed of highly visible international figures (and chaired by Miguel de la Madrid, the former President of Mexico), brings the message of preparedness and mitigation to the attention of decision makers and world leaders.

On the national level, where primary responsibility for disaster reduction rests, committees to implement Decade activities have been formed (as of Spring 1993, there are more than 100 such groups). The U.S. National Committee for the Decade for Natural Disaster Reduction (USNC/DNDR), which brings together academic, industry, and government experience, was formed in the National Academy of Sciences' National Research Council at the request of federal agencies and the Congress. The U.S. National Committee reflects a broad national commitment to disaster mitigation and serves as spokesman to the world community regarding the activities of the U.S. in the IDNDR. In the federal government, program coordination is accomplished via the Subcommittee for Natural Disaster Reduction (SNDR) which was established in response to the Decade. The membership consists of representatives of the various agencies that address the diverse aspects of hazards and disaster mitigation. The SNDR works closely with the USNC/DNDR.

National committees for the IDNDR are working toward the overarching IDNDR goal: to reduce the loss of life, property, and social and economic disruption caused by disasters.

The IDNDR has challenged all countries to reach three main targets by the year 2000: (1) To have in place national assessments of the natural hazards they face and of the vulnerability of their most important concentrations of population and resources; (2) To adopt prevention and preparedness plans, including mitigation measures, emergency-response plans, and public education programs regarding awareness of risks and actions to take; and (3) To insure access to warning systems to allow countries the time to avoid or reduce disaster impacts.

By raising the visibility of disaster reduction, the International Decade aims to change the professional and policy focus of disaster planning from a reactive to a proactive stance. Prevention is less costly and more effective than engaging in a disaster recovery cycle. Where it is possible to avoid or reduce risk, the goal is to marshal knowledge and technology to do so; and where prevention is not possible, the goal is to prepare for emergency response and recovery. Success in prevention and preparedness means that risk management must be integrated into all aspects of life. In the process of development, governments, private firms, communities, and individuals all make decisions that can either add to or reduce vulnerability to disasters. By knowing the hazards and adopting appropriate measures, in the normal course of lives and work, individuals can protect their communities and investments and insure a more sustainable future.

Reducing the world's vulnerability to the forces of nature and making disaster-conscious future investment decisions are the continuing challenges through the 21st century. The IDNDR asks each nation to take stock of its present situation, reorder priorities, and look for tangible actions both at home and abroad that will reduce disaster vulnerability in the short term and help build a disaster-resistant society for the decades ahead.

CAROLINE CLARKE GUARNIZO
STEPHEN RATTIEN

For Further Reading: IDNDR, *International Decade for Natural Disaster Reduction: Overall Programmes for Disaster Reduction in the 1990s* (March 1992); National Research Council, *A Safer Future: Reducing the Impacts of Natural Disasters* (1991); UNESCO, *Standing Up to Natural Disasters: UNESCO Contributions to the IDNDR 1990–2000* (1991). *See also* Natural Disasters; United Nations.

INTERNATIONAL ENVIRONMENTAL REGIMES

One theory that attempts to explain why things work out as they do in modern international relations is the theory of international regimes. There are several working definitions of what constitutes a regime. The first could be called the "This is the way we do things around here" theory, in that it considers an international regime to be a set of norms, rules, or decision-making procedures that serve to regulate how states will act in a certain issue-area based on established patterns of behavior.

The establishment and enforcement of this first type of regime relies heavily on a history of relations in an issue-area against which a state can judge its actions. An example in the environmental arena would be international activity in drift-net fishing. Despite the absence of an enforceable international treaty banning drift-net fishing, a large number of states (including economically powerful ones such as the U.S. and some European countries) have imposed restrictions upon themselves and have urged the same on others. As a consequence, drift-net fishing has been virtually eliminated. Actions like these put heavy pressure on states to comply with a regime's way of doing things. Punitive measures may even be taken against "rogue" states. With the passage of time, the prescribed (or proscribed) patterns of behavior may become second nature to most states, thus firmly establishing the regime's effectiveness. In other instances, however, a majority of states may be able to circumvent or otherwise ignore the behavior the regime is trying to enforce, thus undermining it.

The second type of international regime is one that is encompassed by the body of a negotiated and signed international agreement or treaty. In environmental politics, these kinds of rules and norms are generally set down through the agreements known as conventions. The convention itself may contain the specific rules and details that regulate a state's actions in the issue-area, or it may simply be a codified "agreement to agree" (called a "framework convention"), the specifics of which can be established after further study and through the negotiation of protocols. This latter example was most evident in the negotiation of the Vienna Convention for the Protection of the Ozone Layer (1985), which was made effective only by the 1987 and 1990 Montreal Protocols specifying percentage reductions in ozone-depleting emissions and time frames within which the reductions were to be made.

A broad range of environmental issue-areas, including whaling, trade in endangered wildlife and ivory, certain types of transboundary air pollution, marine pollution, and ozone protection, have been and are being considered by some sort of multistate agreement. Most recently, framework conventions concerning global climate change, species diversity, and the protection of tropical rain forests were signed. These last examples demonstrate well how the effectiveness (or forcefulness) of international regimes may vary. The growing study of international regime theory, especially the mechanics of regime formation and strengthening, focuses on methods and approaches to further enhance that effectiveness and thus to allow scholars and policymakers to consistently apply lessons learned in one issue-area to another.

INTERNATIONAL GEOPHYSICAL YEAR

The International Geophysical Year (IGY) spanned the months from July 1, 1957 to December 31, 1958. This period of unprecedented cooperation among the members of the international scientific community was dedicated to gathering data on the physical processes that shape the earth. During the IGY more than 60,000 scientists from 66 countries systematically explored the earth's atmosphere, oceans, surface, and interior.

IGY scientists were not the first to recognize that areas of scientific study such as weather, earthquakes, terrestrial magnetism, and solar and atmospheric influences on the earth require a global perspective. The International Polar Years, which preceded the IGY (1882–83 and 1932–33), were established to provide a system of coordinated observations on these and other phenomena. No other undertaking, however, has

matched the scale and reach of the IGY investigations. Newly developed technologies, such as radar and other observational tools, created opportunities for data collection, and expanded communications systems enabled the rapid dissemination of findings. For the first time, outer space could be included in a geophysics program as both a subject of investigation and a vantage point for observation. Satellites and the rockets that propel them were first used during the IGY, and data collection from far outside the confines of the earth became possible.

Indeed, the existence of this capability constituted a major reason for creating the IGY, as did an anticipated period of strong solar activity between 1957–58. To carry out the program, scientists established research stations in their own countries, paying particular attention to Antarctica, the Arctic, the equator, and three pole-to-pole meridians along which a maximum number of observations were made. The legacy of the IGY begins with these research stations themselves. Many still function and in Antarctica, for example, the effort involved in establishing the stations provided the impetus for setting it aside as an international laboratory for scientific research.

World Data Centers were established to disseminate the results of the IGY and the geophysical research that has followed in the years since. Yet, the most significant result of the IGY may be the other international efforts it has spawned. For example, in the 1960s the International Upper Mantle Program and the International Year of the Quiet Sun generated data for comparison with the comprehensive record from the IGY. International studies of the earth's climates, waters, and biological systems have also followed, culminating in the International Geosphere-Biosphere Program (IGBP) established in 1986. The program seeks, in the IGY's global spirit, to unite physical and biological knowledge of the earth by "directly addressing the dynamic and transdisciplinary nature of the Earth system, its past changes and uncertain future." The World Climate Research Program, established in 1980, seeks "to improve our knowledge of the physical controls of climate and its variability." The Human Dimensions of Global Environmental Change Program was set up in 1990 "to advance understanding of the socio-economic causes and impacts of global environmental change."

INTERNATIONAL GEOSPHERE-BIOSPHERE PROGRAMME

The International Geosphere-Biosphere Programme (IGBP) was launched by the International Council of

Scientific Unions (ICSU) in 1986 to address the dynamic and transdisciplinary nature of the biogeochemical aspects of the Earth system, its past changes, and its uncertain future. In terms of its interdisciplinary scope, IGBP is the most ambitious research program ever undertaken by ICSU. The objective is to describe and understand the interactive physical, chemical, and biological processes that regulate the total Earth system, the unique environment that it provides for life, the changes that are occurring in this system, and the manner in which they are influenced by human activities.

Human activities have become a significant force affecting the functioning of the Earth system. Our use of land, water, minerals and other natural resources has increased more than tenfold during the past 200 years, and the future increases in population and economic growth will intensify such pressures. Since climate, the global cycles of carbon and water, and the structure and function of natural ecosystems are all closely linked, major perturbations in any one of the components of these systems will affect the others, with potentially serious consequences for humanity and other life on Earth.

A major problem is the uncertainty of present forecasts of future climate changes. It is not enough to predict global warming and mean changes in precipitation. It is necessary to make predictions at the regional and perhaps even local level. But climate is a global system and must be understood and analyzed at that spatial level. Further research is urgently needed that will bridge the global and local scales and assess the effects of change on the biosphere at various spatial scales.

The IGBP is an evolving program that selects those questions that are deemed to be of greatest importance in contributing to our understanding of the changing nature of the global environment on time scales of decades to centuries, that most affect the biosphere, that are most susceptible to human perturbations, and that will most likely lead to practical predictive capability. The specific objectives of IGBP, and the way they could be achieved, were developed by the international scientific community. The research strategy for the next decade was published in 1990 and most components are underway.

While the research questions and the projects that make up the program are expected to evolve with new insights and understanding, the initial operational phase of the program involves core projects focusing on the following key questions:

(1) How is the chemistry of the global atmosphere regulated and what is the role of biological processes in producing and consuming trace gases? The International Global Atmospheric Chemistry (IGAC) Project

was initiated in 1990. The objectives of this project are: to develop a fundamental understanding of the processes that determine the chemical composition of the atmosphere; to understand the interactions between atmospheric chemical composition and biospheric and climatic processes; and to predict the impact of natural and anthropogenic forcing on the chemical composition of the atmosphere.

(2) How do ocean biogeochemical processes influence and respond to climate change? The Joint Global Ocean Flux Study (JGOFS) was initiated in 1990. The objectives of this project are to determine and understand on a global scale the processes controlling the time-varying fluxes of carbon and associated biogenic elements in the ocean and to evaluate the related exchanges with the atmosphere, sea floor, and continental boundaries; and to develop a capacity to provide on a global scale the response of ocean biogeochemical processes to anthropogenic perturbations, in particular those related to climate change. Global Ocean Euphotic Zone Study (GOEZS), a project to study the upper layers of the oceans, is currently under development.

(3) How do changes in land-use, sea level, and climate alter coastal ecosystems and resources and what are the wider consequences? Land-Ocean Interactions in the Coastal Zone (LOICZ) was initiated in 1993. Its overall aim is to develop a predictive understanding of the dynamic behavior of the land-ocean interface, and hence the response of the coastal ecosystem to global change. Research under this project is likely to focus on land-ocean exchanges of matter and energy, carbon fluxes and trace gas emissions, responses to changes in relative sea level, and human dimensions of coastal system change.

(4) How does vegetation interact with physical processes of the hydrological cycle? Biospheric Aspects of the Hydrological Cycle (BAHC) was initiated in 1991 with an overall aim to improve our understanding of how ecosystems and their components affect the water cycle, fresh water resources and partitioning of energy on Earth. Research under this project focuses on patch scale studies of fluxes between soil, vegetation and the atmosphere, regional scale studies of land surface properties and fluxes, and large scale studies of the effects of the biosphere on the hydrological cycle.

(5) How will global changes affect terrestrial ecosystems? Global Change and Terrestrial Ecosystems (GCTE) was initiated in 1990 with an overall aim to develop a predictive understanding of the effects of changes in climate, atmospheric composition, and land-use on terrestrial ecosystems, and to determine the feedback effects to the physical climate system. Research under this project focuses on: ecosystem physiology, changes in ecosystem structure, global

change impact on agriculture and forestry, and ecological complexity.

(6) What significant climate and environmental changes have occurred in the past and what were their consequences? Past Global Changes (PAGES) was initiated in 1991. This project is to be conducted along two specific lines: Stream I focuses on reconstructing the detailed history of climatic and environmental change for the entire globe for the past 2,000 years, with temporal resolution that is at least decadal and ideally, annual or seasonal. Stream II focuses on a reconstruction of the history of environmental and climatic change through a full glacial cycle, in order to improve our understanding of the natural processes that invoke global climatic changes.

In addition, three ongoing IGBP activities of an overarching and integrative nature are operational. The Task Force on Global Analysis, Interpretation and Modelling (GAIM) focuses on how our knowledge of components of the Earth system can be integrated and synthesized in a numerical framework that provides predictive capacity. The aim of this task force is to advance the development of comprehensive prognostic models of the global biogeochemical system, and to couple such models with those of the physical climate system.

The global Data and Information System (IGBP-DIS) will develop effective data management within the IGBP as a whole, and facilitate access to global data sets collected by other research groups and agencies, particularly those obtained through remote sensing.

IGBP's System for Analysis, Research and Training (START) promotes research on regional origins and impacts of global environmental changes. START focuses on the establishment of interconnected Regional Research Networks, Sites and Centres, with special emphasis on the needs of developing countries, where projects of relevance to overall IGBP objectives and regional priorities will be undertaken. Training and exchange programs will be one of the mechanisms to involve the scientists from the region in IGBP project activities. START promoted regional efforts are underway in South America, southeast Asia, and Africa.

Core Projects as well as the overarching activities of the IGBP will require the support of many participating nations if they are to be successfully implemented. Each participating nation, while subscribing to the overall priorities of the IGBP, will emphasize the various program elements in a manner consistent with its own needs and capabilities. For example, the U.S. effort in this regard is focused under the interagency U.S. Global Change Research Program, which is funded at a level of around $1.3 billion. Similar but relatively less well supported efforts are being pursued by other nations, especially in the western hemisphere.

National IGBP Committees constitute the crucial link between the national and international efforts. Many such committees have been established worldwide, but further efforts need to be made to involve more developing countries in the IGBP planning and implementation, and to ensure that financial resources are made available to scientists in the developing world.

IGBP research will normally be financed through regular national channels. In some instances, countries have developed global change programs for research with designated funding; in other instances, projects are evaluated solely on individual merit. In order to promote dialogue between national funding agencies to create an informal mechanism of consultation, the International Group of Funding Agencies for Global Change Research (IGFA) has been established. Funding for international planning and coordination as well as funding for the START Regional Research Networks (RRNs) will have to be secured from the nations participating in the program. It is estimated that the core IGBP planning and coordination needs an annual budget of about $1–2 million; the total cost for the Core Project Offices is on the order of $3–5 million; funding for the RRNs once fully operational is needed on the order of $50–100 million; the cost for actual research needed may be $1 billion. In addition, satellite observations, deployment of automatic submarine platforms to collect ocean data, and a full Global Climate Observing System, as proposed during the Second World Climate Conference during 1991, might be on the order of $10 billion annually.

To summarize, the IGBP has launched a world-wide research effort, unprecedented in its comprehensive interdisciplinary scope, to address the functioning of the Earth system and to understand how this system is changing. The IGBP will design and implement research projects to produce global data sets on properties and processes central to global change. These will include observations and studies at the Earth's surface as well as from an array of Earth-sensing satellites. This research will make use of a network of Regional Research Networks in forging a new understanding of the interactions among biogeochemical cycles and physical processes of the Earth system. The body of information generated by the IGBP effort will form the scientific underpinning for predictions relating to future causes and effects of global changes. Through its observational network and process studies, and the effective communication of the resulting data to scientists in all nations committed to this endeavor, the IGBP will help provide the world's decision makers with input necessary to wisely manage the global environment.

In the course of this endeavor, the IGBP will promote an interdisciplinary approach to studies of the Earth system. It is essential to educate the next generation of scientists in such a manner that they will more fully understand the complexities of this system. This knowledge will be the key to success in the wise use of the Earth's resources for generations to come.

HASSAN VIRJI

For Further Reading: IGBP, *Global Change News Letter* No. 13 (1993); IGBP, *The International Geosphere-Biosphere Programme: A Study of Global Change* (1990); J. A. Eddy (ed.), *The PAGES Project: Proposed Implementation Plans for Research Activities* (1992).
See also Atmosphere; Climate Change; Coastlines and Artificial Structures; Ecology, Marine; Glaciers; Meteorology; Water Cycle.

INTERNATIONAL MONETARY FUND

The International Monetary Fund (IMF) is a specialized agency of the United Nations, founded at the Bretton Woods Conference in 1944. Originally forty-four countries took part in the conference, although membership in the IMF grew as former colonies of Britain, France, and other nations gained independence. Headquartered in Washington, D.C., the IMF is the official organization for maintaining international monetary cooperation, stabilizing exchange rates for nations, and expanding international liquidity (the ability to convert property and assets into cash). The IMF Executive Board, its primary voting body, has twenty-two elected and appointed representatives of various nations and regions. Loans to countries must be approved by the Executive Board.

Because of the extent of its global involvement, the IMF has done work in various fields, acting as an outstanding source of research and publications on world monetary issues. The Fund was also created to improve economic equity between nations and to strengthen world trade by offering financial support and advice to nations having economic difficulty. The IMF is important as a depository of knowledge on international economics and trade, acting with technical expertise and without political interests. Through careful, systematic analysis of nations' economies, the IMF can offer sound economic advice and support to the world's nations.

Many observers, however, consider aspects of IMF decision making to be politically motivated. There has been criticism of the IMF's influence over developing nations—occasionally the IMF seems to act with more political than economic interest. However, misunderstandings about the best advice for a developing country are likely because many different issues are involved in international economics.

The IMF initially focused on stabilizing exchange rates between nations. Starting in the 1970s, however, its primary interests shifted to offering financial resources to nations with weak or stressed economies. IMF funds come from member countries. Depending on its economic strength and voting power, each member country has a certain quota to pay. Countries are given borrowing or drawing rights that they can use under certain conditions to help relieve their economic difficulties. Members are required to repay their drawings over a three to five year period, although loan periods are sometimes extended to ten years. Repayment has not always been simple, however, and many developing nations have had difficulty in repaying their debts even after two or three decades.

In the 1970s borrowing became quite heavy among certain developing countries, and their external debt expanded at a rapid, unsustainable rate. The result was an international financial crisis. These poorer nations were able to reschedule their debts to lessen the payment burden, but still had staggering debts to repay—more than $500 billion by the mid-1980s. The debts created substantial stress on the entire world economy and also deepened some problems within poorer nations. Economic problems in some countries encouraged many reckless practices, notably in agriculture and energy generation. These practices increased environmental problems and exacerbated civil wars.

INVARIANTS OF HUMANKIND, THE

The invariants of humankind are fundamental characteristics and needs that all human beings have in common. These play essential roles and are found throughout the species *Homo sapiens*. People become only one of the many persons they are capable of becoming. All are born with a wide range of potentialities that enable them in theory to develop an immense diversity of attributes, but in practice they develop only those aspects of their nature that are compatible not only with the conditions to which they are exposed, but even more with the choices they make in the course of their lives. The marvel is that nature and nurture can become so completely integrated that they generate a unique socialized identity, the human person, out of the biological organism, *Homo sapiens*. Each person is unprecedented, unrepeatable and unique. Not even homozygotic—so-called identical—twins are identical in real life. They have the same genetic constitution, but nevertheless become different persons because they are exposed to different environmental conditions, first *in utero* and even more so after their birth. On the other hand, while human beings come to differ profoundly from each other as they live in different ecological niches, they can all interbreed and thus remain members of the same species (*Homo sapiens*). Different from each other as they are, they all have in common many fundamental characteristics and needs that can be called the biological and behavioral invariants of humankind and that play essential roles in all the sociocultural expressions of human life. These invariants are found throughout the species regardless of economic, social, ethnic or national status.

One of the greatest scientific achievements has been the demonstration that the hereditary biological characteristics of all living organisms—which are their invariants—are transmitted by DNA molecules which constitute genes. All DNA molecules have the same fundamental chemical structure, but minor differences among them and in their arrangement along the chromosomes are sufficient to account for the phenomenal diversity of living species and for the peculiarities of each individual organism in a given species. On the one hand, the general pattern of the DNA molecules and of their arrangement determines whether the creature will be a hare, a horse or a human being. On the other hand, subtle differences in the DNA patterns of each species determine the hereditary characteristics of each individual hare, horse or human being. The fundamental potentialities and needs referred to as the invariants of human nature can be satisfied in so many different ways that they may be difficult to recognize in the usual manifestations of human life. A few examples illustrate how the fundamental uniformity of *Homo sapiens*—the invariants of human nature—can express itself in the prodigious diversity of human life under the influence of nurture—the environmental and sociocultural conditions.

The nutritional habits of vegetarians appear at first sight diametrically opposed to those of carnivorous people. Many African people and a very large percentage of Asians feed almost exclusively on vegetables, tubers and fruits. In contrast, the East Africa Masai people are nourished almost exclusively from what they derive from their herds of cattle, including the blood they draw from these animals every day. Despite these profound differences in nutritional habits, however, all human beings, whether vegetarian or carnivorous, have essentially the same requirements with regard to intake of calories, carbohydrates, fats, amino acids, minerals, vitamins and other essential chemical nutrients. Diverse as they may appear, all human diets provide similar mixtures of these chemical nutrients, granted that the total intakes differ according to age and the ways of life. From the nutritional point of view, it makes little difference whether the essential nutrients are derived from plants or animals, or even

from products obtained by chemical synthesis; they can provide adequate nourishment if consumed in adequate amounts and proportions, provided of course that they are not contaminated with dangerous microbes or chemicals. The required mixture of nutrients is thus an invariant of human nature whereas the kinds of foodstuffs we eat are its sociocultural expressions. That these can take many different forms is illustrated by the fact that cooking recipes are among the most distinctive characteristics of national, regional, social and cultural groups.

Another invariant of human nature is that all normal human beings need enclosed shelters or at least protected areas into which they can retreat either for safety and comfort or simply to withdraw from public contact. Stone Age people commonly had access to caves and they built simple huts; countless kinds of shelters have been used or built throughout the ages. On the other hand, human beings enjoy open vistas and probably even have a psychological need for them. In the not too distant past, a common way to punish a child for misconduct was to have him or her face a wall for a certain length of time as this was known to be an unpleasant experience. The essential visual needs of humankind can be met in different ways—by clearing areas for settlements in the temperate or tropical rain forest, by lawns in front of homes, by extensive views from the top of a hill, a mountain or a skyscraper. Whether nutritional requirements are satisfied with a vegetable dish or a steak, the need for shelter by retreating into a natural cave or behind the closed shutters of a cozy room, the longing for open space by contemplating a seascape, an undulating field of grain or a classical parterre, humans deal with fundamental and universal invariants of human nature in as many different cultural ways. And so it is for other invariants.

The brains of all humans are genetically endowed at birth with special structures located in the so-called Broca area that make it possible to learn any of the existing thousands of human languages. Some persons have mastered more than twenty languages, but the immense majority learn only the language of the particular social group in which they are born and raised. As has always been the case, young people crave adventure and sexual satisfactions, adults are eager for achievement, most old persons long for quiet and stability; but these universal urges are governed by countless codes of behavior peculiar to each ethnic group, each society, and each period. In the course of time, the games of love have been played in the countless ways which have been represented in as many forms of literature and art; the eagerness for achievement may find its expression in political power, the accumulation of wealth, or the discovery of a natural law; quiet and stability may be found in the management of a garden, in a daily visit to a park or within a community. While the fundamental basis of behavior is still what it was millennia ago, its social expressions are culturally determined and have changed in the course of history.

From time immemorial, human life has derived its color from dancing, music, poetry, fiction, painting, sculpture, tattooing, body painting and festivities that transcend obvious biological needs; but these expressions and celebrations have greatly differed from time to time and from one social group to another. Pageantry has involved situations as different as those symbolized by the Lascaux paintings, the Stonehenge circle, the Buddhist temples, the Greek agoras, Mayan plazas, the Gothic cathedrals, the Renaissance palaces, the Victorian ceremonies and the Arches of Triumph everywhere—all manifestations that would have different meaning or no meaning whatever for members of the species *Homo sapiens* who had been raised in a different culture. It is by incorporating an immense diversity of sociocultural patterns into biological life that *Homo sapiens* becomes truly human. Biological uniformity is readily explained by assuming that all members of the species have the same origin, and that they have continued to interbreed despite the differences generated by life in their various ecological and sociocultural niches. Social diversity, in contrast, has multiple causes, most of which are poorly defined. It derives from minor genetic differences between groups and persons, from the influence exerted on development by environmental forces, from the uniqueness of individual experiences, from the artifacts and institutions created by each society, from traditions, imaginings and aspirations—all of which are the consequences of innumerable choices.

The social contrasts between Athens and Sparta, between the Vikings and the troubadours, between the Zuni and the Apache obviously depend on factors more numerous and complex than racial characteristics, or than the climate, topography and geology of the regions where these people developed and lived. Neither do economic patterns account for the cultural differences among the world's peoples. City-states and nation-states emerged, not from natural forces but from historical and social occurrences that created various conceptual environments in the different populations. Views of the physical and social universe are impressed upon individuals by rituals and myths, taboos and parental influences, traditions and education—all mechanisms which provide the basic premises according to which the inner and outer worlds are conceptualized. The process of socialization through which *Homo sapiens* becomes really human consists precisely in the acquisition of the collective symbols characteristic of one's social group, with all their associated values. Admittedly most symbolic systems change with time. In general, however, symbolic systems persist for many generations within a given culture, even though they may change form. Concepts about the universe and behavioral patterns are thus transmitted as a social heritage which minimizes individual differences within a given group, or at least masks them, and thereby gives greater homogeneity to the group. Practically all factors responsible for human diversity are interrelated. Despite its diversity, humankind consists in the manifestations, endlessly ramified, of the various aspects taken by *Homo sapiens* under the combined influences of the cosmic, biologic, and cultural forces that have generated the seemingly endless spectrum of human societies.

RENÉ DUBOS
(EDITED BY WILLIAM R. EBLEN)

For Further Reading: René Dubos, *So Human an Animal* (1968); Paul Shepard and Daniel McKinley, *The Subversive Science* (1969); Evon Z. Vogt and Ethel M. Alber, *People of Rimrock: A Study of Values in Five Cultures* (1966).
See also DNA; Genetics; Health and Nutrition; Language; Nature vs. Nurture Controversy; Shelters: Environmental Conditioning; Shelters: Social Relevance.

IN VIVO AND IN VITRO TESTS

In vivo and in vitro tests are used to evaluate health and environmental problems of virtually every kind. *In vivo*, which means "in life," refers to conditions that exist within living organisms. *In vitro*, which means "in glass," refers to artificially maintained conditions, as in a test tube or laboratory.

In pregnancy, for example, in vivo and in vitro tests are commonly used to detect fetal abnormalities and to gather basic information on the general development and growth of the fetus. One in vivo test is ultrasound, the only noninvasive technique that allows one to see the fetus in vivo—inside the uterus. In testing for HIV, the virus that causes AIDS, an indirect assay is done in vitro to detect the presence of HIV antibodies in a blood sample.

In vivo tests using animals are often employed to gather estimates of health risks, with the results then extrapolated to human populations. Such studies have varying degrees of uncertainty associated with them. Experimental animals are usually exposed to extremely high concentrations of specific substances in order to ensure that any acutely toxic effects will be seen at statistically significant levels. The concentrations of hazardous substances to which people are generally exposed are usually much lower and over different time scales than the ones employed in animal studies. In addition, toxicologists have uncovered many differences between humans and other animals in the manner in which substances are metabolized and ultimately in the mechanisms by which cancer or other harmful effects occur.

Assessments of the risk associated with exposure to hazardous substances can vary in subject and in time scale. Some studies attempt to project the number of cancers that will develop in a population as a result of exposure to a certain agent over decades, while other studies attempt to quantify the health risk posed to an individual after consuming a certain material. Risk assessment studies are often very complex and rife with uncertainty. Researchers must determine how a hazardous substance moves in and out of the environment; how much of a substance people are exposed to by breathing, eating, or absorbing through the skin; and the mechanisms through which a substance affects the body. There are often large differences between individuals in the amount of exposure to a substance. The amount of time spent outdoors, breathing rates, dietary habits, and other factors vary widely in a population and affect overall exposure levels.

IRRIGATION

Irrigation for agriculture is the application of water to land to enhance crop production. It may vary from total dependence on imported water in rainless areas, to supplementing rainfall with ground or surface water, to managing seasonal flood waters.

Irrigation has been practiced for millennia and has been closely associated with the rise and fall of several civilizations. Egypt has depended on the Nile for 8,000 years. The Sumerian empire in Mesopotamia reached its peak 4,000 years ago; its decline went together with the siltation of canals and salination of the soil, and consequent failure of the irrigation system. The Chinese, who may have learned the art from the Sumerians, developed irrigation in the same timeframe. Evidence of irrigation in the Americas dates back at least 2,000 years.

Recent estimates place the total area irrigated in the world between 230 and 280 million hectares. Most of that is in India, China, Commonwealth of Independent States (the former USSR), United States, and Pakistan; over 60% is in Asia and over 70% in developing countries. Especially since the Second World War, there has been a fast rate of expansion reflected in a growth rate of 2.7%. Also, water storage capacity increased ninefold between 1950 and 1985. However, in recent years, the rate of growth has declined substantially to around 1%.

In the U.S., significant irrigation goes back at least to the Hohokam Indians in Arizona in the 5th century A.D.; by the 14th century, the network of irrigation canals extended over more than 240 kilometers. In more recent times, irrigation became an important component of the development that settled the West. Early private attempts to develop irrigation projects, in the late 19th century, were not always successful. The substantial capital requirements tended to be overwhelming. This led to the Reclamation Act of 1902, originally billed as an attempt to provide federal loans to facilitate irrigation development that would encourage homesteading. However, the program soon changed character from just providing loans to a program of federal financing, design, construction, and operation; over time, the stipulation of family farms less than 160 acres (64 hectares) came to be circumvented, if not totally ignored.

The U.S. Bureau of Reclamation, organized to implement the Reclamation Act, today provides water to less than 25% of irrigated land; just the same, it has had a tremendous impact on water development in the West. Irrigation expanded rapidly from 1.4 million hectares in the 1890s to 8 million in 1940 and 24 million by 1990. Expansion has recently come to a halt and, in some areas, even been reversed.

Irrigation has contributed greatly to the production of food and fiber and to a richer and more diversified diet. Year-round availability of fruits and vegetables depends on irrigation and, in 1982, 32% of the value of crops produced in the U.S. came from the 13% of the cropped land that was irrigated; essentially the same ratio holds worldwide. As the world population continues to grow and the demand for food expands, the need for irrigation is also expected to increase. However, irrigation has not been a universal blessing; its costs to the environment and to society are not always fully recognized.

Because irrigated plants consume pure water and leave behind in the soil those salts contained in the water, irrigation always degrades water quality. All irrigation water contains salts, which must be removed from the soil through leaching; without some drainage, irrigated soils would become too salty in a very few years. This drainage water requires disposal. Quite often, disposal is by means of rivers into the ocean; in that sense, irrigation may be seen simply as an acceleration of a natural process, transporting salts from geologic formations into the ocean. Sometimes, however, salinized river water is harmful to downstream users; in the case of the heavily used Colorado River, where about 37% of the salt load has been attributed to irrigation, substantial damage is done to municipal and industrial water users downstream, as in Los Angeles. In other cases, the drainage water ends up in inland seas or lakes, accelerating their salinization; the Dead Sea in the Near East and the Salton Sea in California are examples. Without adequate drainage, soil salinization soon reduces production; in the Punjab plain in Pakistan, lack of drainage caused 30–40,000 hectares per year to be taken out of production during the 1950s and 1960s.

The discovery in the 1980s of severe adverse impacts on fish and birds from selenium carried in irrigation drainage water to a wildlife refuge in California has added a new dimension to environmental concerns. Studies have demonstrated similar problems with trace elements at several other sites across the western U.S. and shown that these potential toxins, of geologic origin, are dissolved from the soil after irrigation is introduced.

The heavy demand for water has reduced available wetlands—in California by more than 90%—and put pressure on remaining wetlands to support local and migratory birds; water development has reduced, if not eliminated, much of the wildlife habitat. Irrigation, by far the largest user of water, is often blamed for these adverse effects on natural ecosystems and the reduction in recreational opportunities. Other adverse effects have been identified in the form of disruption of community structure and inequitable distribution of wealth derived from irrigation. All of these factors must be weighed against the benefits as society modifies the institutions that govern water resource use and development.

Irrigation no doubt started by attempts to retain spring flood waters in river deltas to recharge soil water. Where rainfall was sufficient, irrigated areas complemented dryland areas for a mixed farming system. Already in ancient times, however, extensive canal systems were often developed to divert water from streams. Elsewhere, qanats, or water tunnels, that transported water many kilometers underground from high-lying ground water bodies to lower-lying lands were dug with amazing skill many thousands of years ago; many are still in use today.

Major engineering works collect water in reservoirs and transport it over often huge distances for use. The most common method of water application to the land is surface irrigation. Methods of surface irrigation include contour ditch, basin, border, and furrow irrigation. For all but the contour ditch system, some land shaping must be done to enable more or less uniform water distribution.

A smaller, but growing fraction of the area irrigated uses pressurized systems; they include various kinds of sprinkler systems and micro-irrigation or drip systems. The older portable sprinkler systems were labor intensive and have been replaced with designs requiring less labor. The most popular of the new systems is the center pivot sprinkler, in which a raised distribution line pivots around a water source—generally a well. Discharge devices are calibrated to provide the same amount of water per unit land area at different distances from the pivot. Micro-irrigation delivers water at very slow rates at many points in a field frequently. One form, drip irrigation, consists of small plastic tubing placed in or on the ground; constant discharge devices that operate satisfactorily over a range of pressures are spaced every meter along the tubing. The capital investment in drip irrigation is relatively high and a clean water supply must be carefully filtered to avoid plugging of the system. However, the control afforded by these systems permits very high water use efficiencies; when combined with fertigation (injection of nutrients with the water), high crop yields can make these systems both profitable and environmentally friendly.

The demand for water from outside the agricultural sector and the relative ability to pay, together with increasing environmental concerns, have put pressure on irrigationists to increase water use efficiency. It is easier to obtain high efficiencies with pressurized systems. Indeed, in the last decade, the area irrigated by gravity has decreased from 65% to 57% of the total, and that by pressurized systems has increased from 35% to 43%; of this, 3% is micro-irrigation. In some areas, irrigation is in steep decline; in the Texas High Plains, the area irrigated dropped 30% and the volume of water pumped 44% in 15 years. In others, there continues to be an increase. Nationwide, the area irrigated is starting to drop.

There is a clear change towards more water- and labor-efficient systems and management. New systems, computer programs to automate management, and greater concern with supply limitations, together with a better understanding of crop response, have combined to impact irrigation practice substantially. One can expect to see further improvements in irrigation systems and management, and in the U.S., a further small decline in area irrigated. A relatively small decrease in water use by irrigated agriculture can continue to provide adequate water for municipalities for many years. Drastic changes are not anticipated. In contrast, in the developing world one can expect a continued expansion of irrigated areas.

JAN VAN SCHILFGAARDE

For Further Reading: G. J. Hoffman, T. A. Howell, and K. H. Solomon, *Management of Farm Irrigation Systems* (1990); Donald Worster, *Rivers of Empire* (1985); National Research

Council, *Water Transfers in the West: Efficiency, Equity, and the Environment* (1992).
See also Agriculture; Dams and Reservoirs; Desertification; Water; Wetlands.

ISLAM

Islam means submission to the will of God (Allah in Arabic) and it is derived from the root *salm* which means peace. A person who professes the Islamic faith, that is one who submits to the will of the creator, is known as a Muslim. A process of true submission thus creates peace within the hearts of the believers leading to peace in the community (Ummah) of believers.

To be a Muslim a person makes a two part declaration (Shahadah), the first part of which is to affirm that there is No God But God. By making this declaration a Muslim accepts that there is only one true God. This is the foundation of Islamic monotheism known as *Tawheed*. By accepting that there is only one creator and sustainer of the universe a Muslim accepts that the order, regularity, and balance in the natural state (Fitra) come only from one source. From this stems the belief that all things in God's creation are interrelated and connected and to modify or tamper with one aspect of creation will have effects on others. It therefore becomes humankind's responsibility to maintain the eco-balance of our environment; for this purpose God has given our species the role of guardianship and stewardship (Khalifa).

The second part of the declaration is that Muhammad is The Prophet of God. A Muslim by accepting this also accepts that Muhammad is the final Prophet. This also means that Muslims accept that other prophets appeared in different ages, in different places for different people. Thus Islam is the final expression of the primeval, universal, natural faith of humankind.

Historical Islam bridged the ancient and modern worlds and emerged in the seventh century C.E. Its founder, the Prophet Muhammad, was born in Mecca in 570 C.E. and the Qur'an, the book of guidance for Muslims, was revealed to him by God from the time he was 40 years old until his death when he was 63. In the twelfth year of his mission, 622 C.E., Muhammad was forced by persecution to flee to Medina from Mecca. This flight, known as the Hijrah, marks the first year of the Islamic Calendar. Encapsulated within the Qur'an is knowledge for the individuals and societies of all epochs, which is still in the process of being unlocked. For example, the profound explanation for the prohibitions on usury is only now beginning to become clear as the effects of banking practices on debt, development, and pollution are being recognized.

When Muhammad died in 632 C.E., he left behind the revealed Qur'an and his practice (Sunnah) on which a voluminous literature (Hadith) has developed since his death. The Qur'an and the Hadith literature together form the basis of the Islamic code (Shariah) upon which Islamic law (Fiqh) is based. The Shariah contains comprehensive guidelines for the preservation of the environment ranging from desert reclamation to anti-pollution measures, the protection of animals, and economic guidelines that confine human greed to a level that is environmentally sustainable.

Muslims created seats of learning, developed the sciences, translated Greek philosophy, and created a whole new artistic milieu. It is now coming to be recognized that the European renaissance was to a large extent built on what Muslims had created.

The major Muslim schism developed as a result of disputes over the successor to Muhammad. The mainstream (eventually to be known as Sunnis) chose the path of election by consensus. The Party (Shi'a) of Ali, the cousin of Muhammad and the fourth Caliph of Islam, led the faction that unsuccessfully campaigned for hereditary succession. This was originally a political split and in time developed a theology of its own, but nevertheless still retains its basic Islamic roots. Shi'as form ten percent of the world Muslim population today; the rest are almost entirely Sunnis.

Within 100 years of the death of Muhammad, Muslims had defeated the Byzantine and Persian armies and were ruling from the Atlantic coast of Africa in the west to the borders of China in the east. Now there are over fifty independent Muslim states, most of whom are members of the Organisation of Islamic Countries, founded in 1969. World Muslim population is now variously estimated as being between 1,100 and 1,300 million. Some estimates put it as high as 1,500 million. The majority, about four-fifths, live outside Islam's original homeland, the Middle East, and are non-Arab. Its growth areas are sub-Saharan Africa, Europe, and North America.

As we have seen, belief (Iman) begins with the Shahadah. From this is derived the belief in angels (described in the Qur'an as functionaries or agencies); the revealed books of God, i.e. the Qur'an and sacred scriptures of the Judeo-Christian tradition (as described in the Qur'an and not revisions); the prophets; and the day of judgment, which will be the last day.

The Shahadah is the first of five pillars or essentials. The other four are Prayer (Salat), which is performed five times a day; Fasting (Sawm) in the month of Ramadan the ninth month in the Muslim lunar calendar, which is a process of spiritual cleansing; Charity (Zakat), which is the only compulsory tax in Islam and is paid on one year's accumulation of wealth of any kind; and Pilgrimage (Hajj) to Mecca at least once in a lifetime by those who can afford it. The five pillars provide a value framework within which the individual Muslim and the Ummah function. It begins with the Shahadah, which is a treaty an individual establishes with God, and culminates in the ultimate act of the community, the Hajj.

Islam provides humankind with a unitary explanation of life. It provides a value framework that fuses the individual and the community. While it recognizes and affirms the individual it protects the community, for without community the individual is left unprotected. Furthermore, each individual bears the responsibility for his or her own actions and is encouraged to strive (Jihad) to overcome the distractions of the ego (Nafs) and also to establish justice and fairness in the community. There is no priesthood or intercession in Islam. As Khalifa each individual person also has the responsibility to protect the Earth and save it from abuse and degradation.

Although popularly known as a religion, Islam does not neatly fit this category and is described in the Qur'an as Din, which may be interpreted as a divinely inspired code of life transactions based on justice. This provides the Ummah with its political base and recognizes the fact that all decisions made by a group however small, or a community, or a state, are in the final analysis political by nature. It also recognizes the fact that all political decisions sooner or later tax the finite resources of creation: the environment.

FAZLUN M. KHALID

For Further Reading: A. J. Arberry, *The Koran Interpreted* (1983); Fazlun M. Khalid and Joanne O'Brien, eds., *Islam and Ecology* (1992); Seyyed H. Nasr, *Man and Nature* (1991).

J

JUDAISM

The Jewish understanding of the environment is rooted in its belief in the goodness of all that God has created and of the special role humanity has to play within this world.

The drama of creation and its consequences for human behavior and responsibility are clearly spelled out in the Torah and further elaborated in the Talmud, the teachings and commentaries upon the Torah. Fundamental is the belief that creation is good and that it reflects the glory of its creator. Genesis 1:31 says, "God saw everything he had made, and indeed it was very good." Judaism is not a world-denying faith. It does not see the physical world as bad or the provenance of evil forces or spirits. Everything comes from God and is therefore inherently good. Judaism enjoys life because of this. There is a saying that on the Day of Judgment, you will be judged and condemned for all the pleasures of this world that you could legitimately have enjoyed but did not!

It is no accident that one of the most common images evoked by conservationists is Noah's Ark. Judaism believes that God cares for all His creation, regardless of their 'usefulness' to humanity. In the Books of Leviticus and Deuteronomy, laws are laid down that insist upon a respect for the diversity of creation and forbid careless or casual mixing of species. The Ark took on board all animals, regardless of whether they were considered clean or unclean. The Bible and Jewish teaching witness to the divine purpose and pattern in creation, each species being in its rightful place, and warn humanity against disturbing this purposeful structure of creation.

This belief in the hierarchy of creation means that Judaism puts humanity at the top of the hierarchy. We have been created to play a special role and to do so in obedience to God and with responsibility for the rest of creation. Some environmentalists have assumed that the commands in Genesis to rule the Earth meant that Judaism believed in humanity's right to abuse nature. This is not so. The 12th-century scholar Abraham ibn Ezra said: "The ignorant have compared man's rule over the Earth with God's rule over the heavens. This is not right, for God rules over everything. The meaning of 'he gave it to the people' is that man is God's steward over the Earth and must do everything according to God's word."

This emphasis on humanity's steward role is even more strongly expressed in the teachings and commentaries of the Kabbalistic writers who emerged in the 12th and 13th centuries. For instance, in this tradition, as Adam named all of God's creatures, he helped define their essence. Adam swore to live in harmony with those he had named. Thus at the very beginning of time, humanity accepted responsibility before God for all of creation.

This acceptance of responsibility carries with it the demands of justice and compassion. When humanity acts unjustly, then creation suffers as the prophet Hosea commented in Hosea 4:1–3. Injustice in human actions causes distress and death to the rest of creation. Conversely, justice means that nature thrives. It is both the problem and the blessing of humanity that we have been given such power and it is God's mercy that He has given us the conscience to know the difference between good and bad.

This teaching is not just for Jews, of course. For while the Jewish people live according to the 613 commandments revealed by God, the whole of humanity has the seven Noahide laws believed to have been given by God at creation, for all humanity. These contain clear teachings on humane treatment of animals and illustrate the concern for the proper treatment of animals that is spelled out in the laws in Leviticus and Deuteronomy concerning domestic animals.

The emphasis on justice and moral integrity is a message that Judaism has preserved throughout its history, even when it had no land of its own. Now, this message is urgently needed by the peoples of the world. There can be no justice for the environment unless there is also justice for the poor. There can be no hope of justice unless the people in power act morally and know what it means to be moral. Judaism does not believe that simply following laws or commandments is enough. It believes that what we are called to do is to live good lives in justice and morality rather than to seek to proscribe and dictate good behavior through penal codes and legislation. These may be necessary for a while, or to curb the wider excesses of human behavior, but ultimately the future security of the natural environment and of humanity cannot be determined by prohibition of the bad, but by invocation of the good within humanity.

Judaism is particularly bound up with land: in particular, of course, with the chosen land. But Judaism has a deep understanding of the needs of the land. It is from Judaism that the idea of resting the land, letting it lie fallow once every seven years, comes. This was not just so the land could rest. It was also so that the wild creatures would always have somewhere to go for shelter and food. The land is there to be used, but not only for human needs. The rest of creation must also be considered in any land use.

Finally, Judaism has always taught that waste was abhorrent to God as is wanton destruction. In Deuteronomy 20 there is a strict injunction against cutting down fruit trees of the enemy when besieging a city. The trees are not your enemy, thus there is no grounds for destroying them. Judaism has a deep love of trees. One of its oldest festivals, now increasingly celebrated as a festival of the environment, is Tu B'Shevat, the festival of the trees. Today, tens of thousands of trees are planted every year on this festival.

To end, let us recall a story told over 2,000 years ago by a wise rabbi. Two men were in a rowing boat. Suddenly one of them produced a saw and started to cut a hole in the bottom of the boat. He claimed he had a right to do as he wished. His companion pointed out that his actions would cause the sinking of both of them. That is our world today.

MARTIN PALMER

For Further Reading: Arthur Hertzberg, "The Jewish Declaration on Nature" and "The Jewish Celebration" in Lewis Regenstein, *Replenish the Earth* (1991); Aubrey Rose, ed., *Judaism and Ecology* (1992); Ismar Schorsch, "Learning to Live with Less: A Jewish Perspective" in *Spirit and Nature: Why the Environment Is a Religious Issue: An Interfaith Dialogue,* Steven Rockefeller and John Elder, eds. (1992).

K

KEYSTONE SPECIES

When Robert T. Paine of the University of Washington coined the term *keystone species*, he meant species that are exceptionally important to the structure of communities, and help maintain their organization and diversity. When Paine removed certain predatory starfish (*Pisaster*) from intertidal rocks on the northwest coast of the U.S., one or two voracious mussel species soon devoured all other creatures and dominated the entire zone. Only when the mussel population was held in check by starfish could other types of bottom-dwelling creatures flourish there. In other words, some species are not merely adapted to environments—they create optimum living conditions for themselves or others.

Charles Darwin was fascinated by the tiny coral polyps that build reefs, lagoons, and even islands. Coral colonies provide anchorage for sea anemones, sponges, and thousands of other invertebrates, which in turn attract and support many species of fish. This varied community could exist without some of its members, but not without the coral. By creating a favorable environment, corals are the keystone of the ecosystem—just as an architectural keystone keeps a stone arch from falling apart.

Sea otters have been called keystone predators because they help to create undersea kelp forests. Otters feed on sea urchins, which destroy the kelp. In two Aleutian Islands of Alaska, one with sea otters and one without, ecologists could clearly observe the alternatives. Where otters were present, the kelp forests thrived, providing shelter and food for many other marine organisms. Without otters, the urchins ate the kelp, destroying the entire marine forest community.

Some ecologists divide keystone species into categories. Keystone prey species may be so numerous that they can survive while supporting an increase in predators. Their sheer numbers may provide some relief to other prey, especially when the more common species is the preferred prey. Under different circumstances, however, an increase in predators may be detrimental to the rarer prey species. In either case, removal of the keystone prey dramatically changes community dynamics.

Some keystone species are mutualists, whose interdependence supports otherwise separate food webs.

Examples include many pollinating insects, hummingbirds, and even giraffes, who feed on acacia tree flowers and inadvertently distribute pollen from tree to tree. Keystone hosts include such plants as palm nut and fig trees, which are critical in supporting many primates, rodents, and birds. As much as three quarters of the tropical forest's mammal and bird biomass may be supported by wild fruit trees.

Perhaps the most spectacular kinds of keystone species are the modifiers: those such as corals that can produce major changes in an environment. The North American beaver, for instance, dams streams to create ponds and lakes. These new habitats enrich biodiversity by providing niches for thousands of plants and animals. Entire communities grow up around the beaver's radical transformation of the landscape.

In East Africa, recent decimation of elephant herds by ivory poachers has had profound and unexpected effects on many organisms. Important modifiers of environments, elephants routinely transform their habitats. Herds can clear thousands of acres of trees and underbrush, creating new grasslands. During droughts, elephants dig water holes, supporting dozens of other species. Ecologists have only an inkling of the biodiversity that elephants foster, from tiny flies to hippopotamuses. Groves of a certain palm tree species spring up from patches of elephant dung. Their large, tough seeds have to pass through elephants' stomachs to be softened before they can germinate.

A precise definition of keystone species is elusive, because effects on other organisms are relative. Ecologists now use the term loosely for species that dramatically affect their communities, particularly when one species must be present to ensure the survival of several others. Often keystone species are not obvious. The American chestnut tree was thought by some ecologists to be a possible keystone, since it was a major source of food and shelter for many birds and mammals. When blight disease practically wiped out the species, however, their disappearance did not cause major disruptions among American forest creatures. Many of their functions were taken over by increased populations of oaks and red maples.

Where feasible, a species' importance to its ecosystem is best determined by removing it from habitat samples. One such field experiment, conducted in the 1980s, focused on the ability of a small saltwater snail to alter coastal habitats. Mark D. Bertness, a marine ecologist at Brown University, was interested in the advance of the periwinkle snail along the coast of the northeastern U.S. Introduced from Europe a century ago, the periwinkle's spread seemed to be correlated with the conversion of coastal marshes to rocky beaches, accompanied by changes in other animals and plants. Bertness wondered whether these changes were caused by climate, human factors, sea-water salinity— or by the presence of the periwinkle.

A coastal marsh community contains many distinctive grasses, algae, and the creatures that use them for food and shelter. Marsh grasses need to be rooted in mud and silt that adhere to algae-covered rocks. When periwinkles feed on the algae, they keep the beach rocks clean and smooth. Mud and algae layers cannot adhere or build up and the marsh habitat soon disappears. When the beach consists only of cleaned rocks and pebbles, a different set of organisms arrives: hermit crabs, for instance, which often move into the empty shells of periwinkles.

On a rock shore at Bristol, Rhode Island, Bertness constructed eight bottomless cages, each enclosing several square feet. All periwinkles were picked out of the caged areas, which were then left alone for several months. Within the test cages, algae formed on the rocks, silt clung to it, and marsh plants took root. A small marsh habitat developed in each of the cages, distinct from the surrounding area. Eventually, mud worms, fiddler crabs, and mussels thrived in the small, silty marsh plots.

Knowledge of keystone species is crucial for understanding how to preserve habitat and perhaps even to restore some that have been destroyed. The idea of a natural balance is popular and reassuring, but many ecologists now view nature as unstable and in flux. Extinction of some species may cause little harm to the survivors, while others are absolutely crucial to the web of life. Ecologists are trying to discover which particular microorganisms, for instance, may be essential for renewing the air and water upon which all life depends.

RICHARD B. MILNER

For Further Reading: Daniel B. Botkin, *Discordant Harmonies: A New Ecology for the Twenty-first Century* (1990); Daniel H. Janzen, *Guanacoste National Park: Tropical, Ecological, and Cultural Restoration* (1986); L. Scott Mills et al., "Questioning the Utility of the Keystone-Species Concept," *BioScience* (April 1993).
See also Biodiversity; Coral Reefs; Ecological Stability; Wetlands.

L

LABELING

Labeling is the written, printed, or graphic matter attached to or on a consumer product or its container or wrapper. Labeling has three major purposes: to ensure that manufacturers do not compete unfairly by misleading consumers about their products, to allow consumers to make informed purchase decisions, and to inform users of the risks products might pose.

In the United States, most product labels display a combination of information that is required by federal law and information that the manufacturer voluntarily includes. Information offered voluntarily must be accurate. Although states had earlier begun to require labeling of some products, the first federal labeling law was passed in 1906. Federal labeling requirements gradually expanded both in the kinds of products covered and the kinds of information required. At present, pharmaceuticals, cosmetics, pesticides, most food products, and many consumer products must be labeled according to regulations promulgated by regulatory agencies, while alcoholic beverages, cigarettes, and products containing ozone-depleting chemicals must be labeled with messages prescribed by Congress.

Pesticides constitute one of the most environmentally important categories of products that must be labeled. Because their purpose is to kill certain organisms, pesticides may pose hazards to human health or to non-target species. In order to market a pesticide, manufacturers must register the product with the federal Environmental Protection Agency (EPA). EPA requires manufacturers to submit test data showing the efficacy of the product for its intended use as well as effects on humans and animals. Based on this information, the agency and manufacturer develop an appropriate label, which describes the crops and pests on which the pesticide may be used, the manner in which it may be applied, and any warnings about human health or environmental effects. It is against federal law to use a pesticide in a way that differs from the labeled instructions, which are intended to protect human health and the environment. Common environmental warnings prohibit washing containers with water (because the water would become contaminated with the pesticide) and disposal of the pesticide onto the ground. This is intended to avoid runoff of unabsorbed pesticide into streams or groundwater. Although EPA formerly emphasized human health in reviewing pesticides for registration, ecological and environmental concerns have received increasing attention more recently.

For humans, food constitutes the largest and most important category of labeled products. The federal Food and Drug Administration (FDA) is responsible for overseeing labels on most food products, but meat and dairy products are generally regulated by the United States Department of Agriculture (USDA). The safety of prepared food products is largely achieved through regulations concerning cleanliness and procedures in the factory and by FDA review of the health effects of proposed additives, as well as by EPA-mandated limits on the amounts of pesticide residues in food. Food labels provide information about quantities and ingredients. In 1972, the FDA also instituted a voluntary program of nutrition labeling, prescribing certain information to be displayed if manufacturers made a nutrition claim for the product or added nutrients. About 40% of prepared food products and 10% of processed meat and poultry carried the labels by 1992, when the Nutrition Labeling and Education Act of 1990 made nutrition labeling mandatory. Manufacturers are now required to group additives according to function (e.g. sweeteners), explain that ingredients are listed in descending order of weight, list ingredients of a large category of formerly exempt products, and limit or justify nutrition and health claims such as "fresh" or "low fat."

Labeling provides information for consumers to use, first in deciding whether to purchase a product and later in how to use it. To be effective, therefore, label information must be read, understood, and acted upon. There are barriers to its effectiveness at all three stages. Consumers may be overwhelmed by quantity of product information as they shop and simply filter out or ignore much of it. Congress required four different messages on cigarettes in an effort to overcome the tendency of the eye to ignore messages previously read. Although federal regulators have some rules about minimum type size and format, many product labels use small type, impeding reading by the growing number of elderly and others with limited vision.

Pictograms, or small pictures, are often eye-catching and also help convey the message quickly.

Even consumers who choose to read labels may not understand them. For example, many food ingredients have complex chemical names that mean little to most readers. For this reason, the FDA requires that manufacturers identify the purpose of the ingredient, such as "preservative." Even this language may be difficult to comprehend: for example, readers must know what "emulsifiers" do in a food product. Moreover, the nutrition label in use before 1992 required consumers to perform arithmetic calculations to acquire the information they needed: Information was provided by a manufacturer-defined serving size rather than in some standard format, such as per ounce. Even pictograms may be difficult to understand: While it is easy to depict fire or indicate a hazard to the eyes, more abstract concepts such as irradiation cannot be illustrated directly and users must learn what the symbol means.

Finally, consumers must act upon the information provided. When the risk to be conveyed is a short-term risk such as flammability and the label provides explicit instructions about avoiding the risk (e.g., do not store or use near open flame), people are likely to act as the label suggests. For longer-term and more abstract risks, including environmental risks posed by improper use of pesticides or health risks posed by improper nutrition, labels require more of the consumer. For example, the nutrition labels in use until 1992 required a person trying to consume the Recommended Daily Allowance (RDA) of iron to keep track of the numbers of servings and percents of iron in each food eaten until the total reached 100 percent. The new food labels reduce the difficulty of this task by allowing manufacturers to include evaluative words (high, low) and by requiring them to provide a chart for converting grams to calories as well as to state the amounts per serving and recommended total daily consumption of several desirable and undesirable dietary components for two different levels of calorie consumption.

Despite their limitations, labels are an important component of any product. Consumers have a right to know the ingredients of their products, so that they can avoid those to which they are allergic or that pose unacceptable risks. They have a right to be given enough information to be able to evaluate nutrition claims and claims of efficacy, and to know whether a food or consumer product poses any risks to their health or to the environment. Labeled information is most useful when supplemented by appropriate consumer and nutrition education and when information is provided in a consistent and easy-to-use format. Al-though the number of different federal and state agencies and the many different laws that mandate labeling are a barrier to complete consistency, they still offer consumers a powerful means of ensuring that consumer products and food bear appropriate and useful labeling.

SUSAN G. HADDEN

For Further Reading: Susan G. Hadden, *Read the Label: Reducing Risk by Providing Information* (1986); Wesley A. Magat and W. Kip Viscusi, *Informational Approaches to Regulation* (1992); U.S. Food and Drug Administration. Notice of Proposed Rulemaking (Nutrition Labeling). 57 Federal Register 32058 (July 20, 1992) and 58 Federal Register 2079 (January 6, 1993).

LABOR

Labor ranks as the nation's most important renewable resource. In 1991 61.6% of the population aged sixteen and older was employed, and labor services accounted for about three-fourths of the gross national product.

The quality of the nation's labor resource can be increased through education and training, both "human capital" investments that make labor more productive. The rate of return on human capital investments ranges between 10% and 15% and compares favorably with other types of capital investment. It is estimated that over half of all national wealth is in the form of such human capital.

Labor productivity can also be increased by managerial practices that motivate effort and ingenuity. Financial incentives such as bonuses and commissions, penalties such as suspension or termination from work for poor performance, and elaborate systems of social and psychological work-force motivators are examples of management practices that increase labor productivity. Education and training, improvements in work processes, and management incentive practices have raised labor productivity by over 3% a year since the late 1930s, although there has been a lessening of productivity growth in recent years.

Unsafe and unhealthy working environments endanger the labor resource. There are about 4.5 million deaths and disabling injuries in the workplace each year. The highest injury rates are in manufacturing, where approximately one-third of all workers experience a disabling injury each year. The service sector is a close second.

In principle, competition for labor provides incentives for employers to correct unsafe and unhealthy work environments. Where workplaces are risky, employers should have to pay higher wages to attract workers and to compensate them for health and safety problems. However, workers often have poor information about health and safety risks in the workplace. Even where good information about such risks is available, workers routinely underestimate the economic consequences of unhealthy and unsafe working conditions. As a result, labor markets charge too low a wage tax on workplace risks, and workers remain exposed to levels of health and safety risks that are too high in terms of their economic cost to society.

Two types of public policy are used to address this labor market failure. Employer liability for workplace injury is governed by state-based workers' compensation laws. About $30 billion was paid out in worker compensation claims in 1988, although one in six workers is still not covered by such laws. Because workers' compensation laws concentrate primarily on safety issues (to the neglect of health problems such as toxins, carcinogens, and electronic emissions from video displays) and rely on voluntary action for preventive programs, a federal Occupational Health and Safety Act (OSHA) was passed in 1970. Under OSHA, the federal government sets regulatory standards for workplace health and safety and employers can also be required to adopt preventive programs to promote workplace health and safety.

OSHA regulation has been controversial. Unlike safety risks, unhealthy environmental factors often affect the quality of life or shorten the life span of workers without causing immediate earnings loss. Such quality-of-life issues make it difficult to assess the costs and benefits of OSHA programs.

PETER B. DOERINGER

LABOR MOVEMENT

A Common Community Base

The key to understanding the relationship between labor and environmental movements worldwide is to understand that for each the basic cultural unit is a geographically-bound community peer group. Both grew from socially-oriented, leisure-time local organizations: workers' social and mutual aid clubs and sporting-conservation groups. Both memberships did and do overlap each other and another peer group: local voluntary health agencies.

At the turn of the century in the U.S., the organization of a national park and forest system included in its leadership national and local labor leaders. The organizer of the National Parks Association was the Assistant General Counsel of the CIO.

At the same time the public health movement in the U.S. grew from the efforts of citizens' committees such as the New York State Committee of State Charities Aid on Public Health and Tuberculosis (which assisted the organization of the American Public Health, Cancer, Heart, and Lung Associations). The committee had a history of collaboration with unions originating from a common interest in stemming the tide of tuberculosis generated in poor work environments. Together they strongly supported the creation of the Adirondack and Catskill State Parks, which required wresting control of the land from railroad and timber interests.

State Charities Aid was also instrumental in the success of New York's campaign to finance the building of sewage and water treatment plants through state bonds. Unions financially and politically supported another initiative of the committee: Action for Clean Air Committees in every county of the state organized for the purpose of generating public participation in regulatory decisions.

In 1961, stimulated by the success of the "action" committees in New York State, the clean air movement in the U.S. achieved national coordination through the National Air Quality Commission of the American Lung Association. Jack Sheehan of the United Steelworkers of America was an early member. He later also joined the Board of the Natural Resources Defense Council.

Sheehan symbolizes widespread involvement by his union in environmental affairs, often expressed by local action in support of community efforts. Canadian steelworker Paul Falkowski, for example, operated an air sampling station in Sudbury, Ontario, which provided data forcing the province of Ontario to take action against gross pollution from a smelter of the International Nickel Company. I. W. Abel, then President of the Steelworkers, was the first national leader to call a meeting of a national organization in support of clean air legislation (in 1969). From that meeting came the initiative by Abel, who was also president of the Industrial Union Department of the AFL-CIO, to press for passage of the Occupational Safety and Health Act of 1970. The OSHAct was modeled on environmental law mandating maximum feasible participation by an impacted public.

The Role of Labor Organizers

Industrial union leaders, because of their organizing abilities and strong ties to community structure, often provide leadership to local environmental movements. Tony Mazzocchi in 1964, then a local leader of the Oil, Chemical, and Atomic Workers union, converted the Nassau County (NY) Planning Commission (on which he served) into the prime mover for the control of pollution in the Great South Bay of Long Island.

As a national leader in 1975 Mazzocchi organized a network of labor-student coalitions and national labor and environmental organizations (including the machinists, steelworkers, chemical workers, the AFL-CIO through its Industrial Union Department, Friends of the Earth, Environmental Action, Sierra Club, Environmental Defense Fund, and the Health Research Group) that led to passage of the Toxic Substances Control Act. He also organized the "right-to-know" movement that not only affected the future shape of environmental law and regulation in the U.S., but through the ICEF (International Federation of Chemical and Energy Workers), stimulated similar action in Europe, especially in Scandinavian countries, leading to legislation in the European Community consistent with American laws.

The international labor movement provided the medium for much of the stimulus for the organization of environmental political action in Europe, as in the case of chemical and metalworkers assisting in the creation of an NGO meeting paralleling the 1970 Stockholm Environmental Conference and in supporting environmental leaders such as Petra Kelly, a founder of the Green Party in Germany. This kind of relationship has its historic precedent in the founding of the Sierra Club by a Scottish machinist, John Muir. The Sierra Club maintains that tradition through a board-appointed Labor Committee headed by legislative activist Les Reid, a retired UAW machinist. Basque steelworkers, and later their brethren in Central and Eastern Europe, were at the heart of the nascent clean air movements in their countries.

In totalitarian regimes, e.g., Guatemala and Nicaragua, the informal tie between labor and environmental leaders is best characterized as "The Green Resistance." When the Sandinistas took power in Nicaragua, the first person they assassinated was a labor leader closely tied with the underground environmental movement. Student environmentalists in Guatemala were endangered when informing Indian peasants of the dangers of pesticides aerially sprayed on the cotton fields while whole families were cultivating the crop.

Mutual Support and Conflict

Early support from labor for the fledgling international environmental movement forged ties that persist today. Environmental activists provide critical support to labor on protecting the work environment of mines, factories, hazardous waste sites, and even offices. Examples of this are to be found in Sierra Club support for the OCAW-initiated Shell boycott and in the struggle between Phelps-Dodge and the steelworkers in the

Southwest. For years, a primary source of information about toxic substances in the Malaysian workplace was Friends of the Earth in that country.

Perhaps the most dramatic example of mutual labor/environmental support is in the founding of the OSHA/Environmental Network in 1980 by Michael McCloskey, Chair of the Sierra Club, and Howard D. Samuel, President of the Industrial Union Department. At its height, the network involved 18 unions and 7 national environmental organizations in 28 states. It fought successfully to prevent the repeal of the OSHAct and a weakening of a broad range of local, state and federal environmental regulation.

Differences of opinion between leaders in American labor and environmental organizations over key provisions of the Clean Air Act of 1990 perceived to be troublesome to the construction, auto, and coal industries, coupled with local disputes ranging from recycling to forestry practices, created strained relations between the two movements. Similar differences exist in Britain and in Europe over nuclear power. Coalitions between environmentally-active groups (which unlike their American counterparts tend to be overt political parties) and labor unions in the liberated countries of Eastern and Central Europe are tenuous due to the worldwide recession and underlying cultural factors.

Class, Caste, and the Environment

Class consciousness separates the environmental and labor movements to a greater degree outside the U.S., where class mobility historically has been greater. Conflict between environmentalists and 'workers' in the U.S. often is between self-perceived peers over economic and land use issues. Perceived social distance between 'workers' and bourgeoisie intelligentsia was more important in learning circles on the work environment in the Folkshüset of the Norwegian Labor Organization in Stavanger, the operations center of the North Sea oil fields, than with oil workers in the U.S.

These social differences may be growing. The appellation "worker" drew more negative comment from union members twenty years ago than now in the U.S. But increased class consciousness does not mean a weakened commitment to collaboration with "middle class" environmentalism as it once did. Even socialists, who once decried environmentalism as a 'middle class distraction from the real issues,' now embrace environmental issues. Much of this change came about through the influence of socialist labor leaders such as William Winpisinger, former President of the International Association of Machinists and a forceful leader in the OSHA/Environmental Network.

Still another overlay is the color caste factor. Black activists once typically characterized environmental issues as the same sort of distraction once perceived by socialists. Through the intercession of black American labor leaders, such as William Lucy, Secretary-Treasurer of the massive American Federation of State, County and Municipal Employees, there is now an active "urban environmental movement" in the U.S. For these environmentalists, lead-based paint in slums and asbestos insulation in schools may be more important than endangered species, but both kinds of activism reinforce each other as evidenced in joint demonstrations and picketing organized against deregulatory policies espoused in the Nixon and Reagan administrations.

The same changes may be less likely in European settings and in American agriculture, where "guest" workers do the most hazardous jobs and are returned to their own countries prior to the end of the latency periods for most toxic agent disease. This is the case of North Africans in France, Turkish workers in Germany and Mexican migratory workers in the American southwest, truncating visibility for issues that would otherwise draw environmentalists and workers together.

Time — tradition — is a factor in differences in risk acceptance between the labor and environmental movements. This is best evidenced by common acceptance of massive river pollution, mostly silting, in Asia and North Africa over millennia. It is also seen in worker attitudes toward toxic agents in Europe. Swiss, French, German, and Dutch unions since the turn of the last century have come to terms with the owners of the refineries, plants, and mills on the Rhine. They are much more tolerant of toxic river pollution than their American counterparts on the Ohio who are still largely at odds with their employers.

Near the uranium mines of Saxony and Bohemia, first opened in the 13th century, miner's disease (cancer, silicosis) was ritualized and thus made culturally acceptable by the miners' guilds dominating community life. Their communities still reveal a lower level of overt anxiety about the effects of radioactive mine tailings than is found in similar communities on the Colorado Plateau. However, in the aftermath of the withdrawal of Soviet armed forces from Saxony, church, university, labor and community leaders formed a loose environmental group to deal with the tailings, raising visible concern for at least a year to a level observed some 20 years ago in Grand Junction, Colorado. The lack of a free trade union movement in Eastern Germany at the time hindered organized labor participation.

Common Enemy or Common Cause?

It would be incorrect to see a common enemy as the unifying principle in the history of labor-environmen-

talist relations. It is true that union-environmental coalitions often have been built on the basis of mutual desire to "fight" a company or trade association. After each of these episodes, however, both "blue collars" and the "executives" (of relatively unimpacted industries) and "professionals" who dominate environmental groups learn mutual respect. The environmentalist is learning that massive world-wide training programs conducted by unions on the work environment create "spillover" knowledge and skills not available to them, and labor leaders gain confidence in their ability to work with environmental organizations in response to this new respect.

However, the common enemy, "industry," is no longer a monolithic opponent of environmental protection. Increasingly, the corporation is entering the environmental movement. A new structure is emerging in which a common cause is the unifying factor.

SHELDON W. SAMUELS

For Further Reading: Harriet L. Hardy, *Challenging Man-Made Disease* (1983); David Rosner and Gerald Markowitz, *Dying for Work* (1987); Sheldon W. Samuels, *Worker Participation* (1972).
See also Clean Air Act; Collective Bargaining; Environmental Movement; Environmental Organizations; Forest, U.S. National; Green Movement; Hazards for Farm Workers; Occupational Safety and Health; Parks, U.S. National; Public Health; Toxic Substances Control Act.

LAISSEZ FAIRE

Laissez faire usually denotes distrust and hostility toward state intervention in the natural workings of a market economy. However, it is also frequently used to describe policy beliefs in other fields — such as the social and administrative — that seek to reduce or eliminate intervention and regulation of the private by the public sector.

The term *laissez faire* comes from the French expression "laissez faire, laissez passer," which means "Let events go ahead and happen as they might." The concept is thought to have originated in the protest of a late 17th-century French businessman who was strongly opposed to state intervention in the then nascent capitalist market economy. The concept developed and flourished into a principle among French, British, and American classical liberals in the mid-19th century and has also enjoyed a degree of formal resurgence in the last few decades.

Several of the more recent advocates of "laissez faire economics," such as Milton Friedman, could be identified as "purists," or unequivocably opposed to state intervention in the market. However, those who developed the concept in the mid-19th century are recognized to have been more qualified in their positions. Specifically, they often provided lists of particular circumstances where the principle of laissez faire should be avoided in practice. These included instances when public services such as sanitation, transportation, education, and infrastructure need to be provided; the formation of monopolies should be prevented; and patent laws enforced.

The nature and extent of intervention and regulation the government should be allowed in order to address problems of environmental degradation, pollution, and resource depletion are also controversial issues. The controversy centers on the extent that state environmental policy should be regulatory in nature, and the degree that it should allow market forces to correct particular problems—the traditional laissez faire position. A third viewpoint is that government policies should play a significant role in regulating the broad framework of market interaction, while allowing market forces to correct the problems without direct regulatory intervention. One example of this approach is a system of tradeable pollution permits that allows companies to reduce emissions through the most cost-effective market-based mechanism.

LAKES AND PONDS

Standing waters are usually referred to as lakes, although some of the largest, such as the Caspian, are called seas, while the smallest may be called ponds. The water quality and biota of these waters are largely determined by the nature of the watershed. Most of the water entering a lake or pond is from surface runoff and rainfall on the water surface, although some have significant inflow of ground water. Consequently, human activities that affect the soil and vegetation of a drainage basin as well as air quality will largely determine the kinds and amounts of substances entering lakes. Lakes in regions where evaporation exceeds precipitation can have large concentrations of salt, e.g., Great Salt Lake. Lakes in rocky basins with little weathering have the very low mineral content of soft water. Lakes in forested areas may receive much organic material and therefore have darkly stained waters typical of bog lakes. Hence there is a broad spectrum of lakes from those that have soft water with few nutrients (oligotrophic), to hard-water, nutrient-rich, eutrophic lakes.

Water has unique characteristics. In addition to being a nearly universal solvent it makes life possible and governs the distribution of chemicals, temperature, movement of water, and the presence and abundance of biota. Fresh water has its greatest density at 4°C and becomes less dense above and below 4°C. Consequently in warm months cold, dense (hypolimnetic) waters occur in deeper areas of lakes and less dense, warm (epilimnetic) waters make up an upper layer. Furthermore, it is important that density decreases below 4°C, otherwise ice would form on the lake bottom rather than on the surface. Density differences between cold and warm water result in thermal stratification, which greatly reduces mixing and results in a seasonal cycle in temperate lakes. These lakes usually have two mixing periods, spring and fall. Mixing is limited by ice cover in winter since most mixing is wind-induced. As the ice cover disappears in spring, water temperature and density are relatively uniform from top to bottom and the lake mixes (spring turnover). The lake warms and vertical density differences occur. Wind energy becomes insufficient to mix the warm surface waters with the deeper colder waters and the lake becomes stratified during summer. Stratification persists until cooling in the fall reduces density differences enough to permit wind-generated mixing (fall turnover). Some large lakes that are ice free mix most of the year, as do shallow lakes and ponds unless they become ice covered. Even tropical lakes can stratify despite small vertical temperature gradients. Density differences for a few degrees' temperature change in warm water are much greater than density changes for several degrees in cold water. The relationship between temperature and density change is not linear.

Stratification can provide a refuge for cold-water species, such as trout, during the summer in the hypolimnion (the lower layer that is noncirculating and perpetually cold). Consequently a deep lake can have an assemblage of cold-water organisms as well as warm-water species. Stratification also results in gradients in concentrations of chemicals. For example, dissolved oxygen, essential for most life, enters the water at the lake surface and from photosynthesis of plants in the epilimnion (the warmer upper circulating layer). Respiration of organisms in the hypolimnion may deplete the dissolved oxygen. Thus dissolved oxygen will be plentiful in the epilimnion but may be exhausted in the hypolimnion.

The periods of mixing are critical for redistribution of nutrients essential for algal growth as well as for mixing oxygenated water into the depths. A few lakes are permanently stratified due to density differences associated with salts or humic material. These lakes never mix and are called meromictic.

Each kind of lake has its particular assemblage of plants and animals. Oligotrophic lakes usually have a diverse biota, but relatively few organisms such as trout. Eutrophic lakes are usually rich in algae, large plants, and warm-water fishes. Saline lakes may have only a few species, e.g., brine shrimp. Bog lakes usually are very acid or alkaline and have few organisms other than insects. A diversity of habitats occur in a lake. Large plants, e.g., rushes and lilies, are found in the shallow areas. Submerged plants are farther offshore. The open surface waters provide a favorable habitat for some organisms, while others occur only in the deep waters. Organisms inhabiting the sedimentary environments of lakes and ponds make up the benthic (bottom) community of attached algae, bacteria, worms, clams, snails, insect larvae, and some crustaceans, e.g., crayfish. This community might include abundant populations of sponges, hydras, and other invertebrates. Organisms with limited locomotion that float in the water are the plankton, which includes many algae, rotifers, and crustaceans. Nekton consists of animals that are swimmers, primarily fish.

Human interactions with lakes have been both positive and negative. Lakes have played a prominent role in the dispersion of humans because they have provided transportation, water supply, and food. It is not happenstance that many of the world's large cities are on lake shores. Human activities often have led to major, even catastrophic, changes in lakes. The damming or diversion of tributaries can result in loss of certain fishes or even drying up of a lake. The Aral Sea, which was the world's fourth largest lake, has lost about 40% of its surface area due to diversion of tributaries.

Overfertilization of lakes from sewage, runoff from farms, and atmospheric inputs of nitrogen and phosphorus has resulted in eutrophication (nutrient enrichment). The rapid eutrophication of Lake Erie was especially alarming because it was assumed that such a large lake (9,930 square miles) could not easily be affected by human activity. "Clean" water benthos was replaced by pollution-tolerant forms, large algal blooms occurred, some fish disappeared, nutrients and other chemicals increased, and dissolved oxygen depletion occurred. Billions of dollars have been spent to limit nutrient inputs to lakes in North America and Europe. As a result, some lakes have shown reversal in eutrophication, e.g., Lake Erie and Lake Washington at Seattle.

The reversal of eutrophication in some lakes, restoration of viable fish populations in others, and lessened destruction of wetlands and shorelines give encouragement that lakes will continue to exist for the enjoyment and use of future generations of humans.

ALFRED M. BEETON

For Further Reading: William Ashworth, *The Late Great Lakes* (1987); David G. Frey, *Limnology in North America* (1963); Robert G. Wetzel, *Limnology* (1983).
See also Algae; Dissolved Oxygen; Eutrophication; Glaciers; Oxygen; Plankton; Pollution, Water: Case Studies; Pollution, Water: Processes; Water.

LAND

Land is the solid part of the Earth's surface, also known as the lithosphere. It is created by a combination of abiotic and biotic processes. Physical phenomena including climate, geology, physiography, and hydrology sculpt the land through time. Human activities originate on, and are sustained by, the physical and biological properties of land. The human use of land has come to have economic, legal, and ethical consequences.

Geology affects the lay of the land strongly. Rock is a mineral material of consolidated or unconsolidated composition. It may originate from the cooling of molten liquid, as when volcanic lava cools to form basalt; from the deposition of layers of sediment; and from the metamorphosis of existing rocks that have been subjected to heat or pressure changes in the crust of the Earth.

Climate and geology together fix the physiography of land, that is, the physical conditions of the land surface. The soil of a desert differs from that of a tundra because of climate. Temperature, precipitation, and wind affect the fundamental structure of regions and localities. Physiography varies widely; the world is a myriad of peaks and depressions, ridges and valleys, rolling hills and flat plains, mesas and canyons, buttes and ranges.

This variation alters the flow of water at the surface and underground. Fluvial processes, in turn, reshape the land. Flooding, for example, from streams and rivers or from abnormally high tidal water or rising coastal water resulting from severe storms, hurricanes, or tsunamis can quickly change the character of land.

Lying atop the lithosphere, soils provide the interface between abiotic and biotic elements. Their properties result from the integrated effect of climate and living matter acting upon parent material over periods of time. Each soil can be described in terms of a profile, defined as a sequence of layers or horizons from the surface downward to rock or other underlying material. These layers include: organic horizons, which form above the mineral soil from litter derived from dead plants and animals; eluvial horizons, which are characterized by leaching; illuvial horizons, which are the zone of maximum accumulation of materials including iron, aluminum oxides, and silicate clays; unconsolidated horizon, which is outside the areas of major biological activities and is the zone of the least weathering and accumulation; and bedrock.

Soils change through deposition and erosion. Deposition is the addition of soil forming material by wind and water. Erosion is the searing away of the land surface by running water, wind, ice, or other geologic agents and by processes such as gravity. Tolerable soil loss can be defined as the maximum rate of annual soil erosion that will permit a high level of crop productivity, or alternative use, to be sustained economically and indefinitely.

Flora and fauna are the major biotic processes fashioning the land. For example, eutrophication occurs where nutrients are added to a water body, causing excess plant growth, which kills animal life by depriving it of oxygen. As organisms die, they sink to the bottom of the water body and decompose, which further decreases the oxygen. The process results in the eventual silting up of the water body, changing wet land to dry.

The lithosphere plays a primary role in the production of human food, energy, and materials. As a result, the use of land has received considerable attention from human societies. Each culture defines its relationship with the land in its own unique manner, giving rise to characteristic settlement patterns, economic structures, laws or rules, and land ethics. For example, the Navajo, who call themselves *Dineh* meaning "the People," contend land was created by God for everyone's use. Therefore, land should not be bought or sold. They believe that if land is owned by one person, then it will be wanted by someone else and conflict will result.

All of the world's major religions, as well as less extensive belief systems, address human relationships with nature and the land. For example, a traditional Chinese approach to guiding land use is *feng-shui* (wind and water), which is based on the concept that the Earth has energy fields. *Feng-shui* is used to analyze these energy fields in the Earth and plan human use accordingly.

In the Judeo-Christian Bible, the book of Genesis directs people to "Be fruitful, and multiply, and replenish the Earth and subdue it: and have dominion over the fish of the sea, over the fowl of the air, and over every living thing that moveth upon the Earth." The book of Leviticus gives clear requirements for people to be stewards of the Earth, "The land shall not be sold in perpetuity; for the land is mine," God says in Leviticus. This book also contains guidelines for giving the land a year of rest every seven years and for public ownership of pasture land.

Today, land economics is concerned with the various uses: agriculture, recreation, business, industry, housing, transportation, mining, forestry, and wilderness.

The U.S. Geological Survey classifies land use and land cover in the most general way as urban or built-up land, agricultural land, rangeland, forest land, water, wetland, barren land, tundra, and perennial snow and/or ice.

How land is used determines the wealth, health, and power of individuals, communities, and societies. Differing views of land use and control can lead to conflict, even war. As a result, most societies have adopted rules regarding how land can be used. In the U.S., the legal foundation is based on principles set forth by the English political theorist John Locke, who viewed a major purpose of establishing a government as the preservation of property. He defined property as "lives, liberties, and estates." However, it has been the view of property as possession, rather than Locke's predominant version—life, liberty, and estate—that has prevailed. Although property is not exactly synonymous with land, land is a type of property. The Fifth Amendment of the U.S. Constitution contains the clause: "No person shall . . . be deprived of life, liberty, or property, without due process of law; nor shall private property be taken for public use without just compensation." Protecting property rights was seen as a fundamental necessity by those in the new republic who had fought against the landed elite of the mother country.

In the U.S., certain rights inhere in land ownership: the right to use it for farming, forestry, and grazing; the right to mine; the right to fence to keep out members of the public; the right to sell, give away, or bequeath; and the right to develop for other uses. With these rights come responsibilities, including the requirement to pay taxes to local governments. The Tenth Amendment of the Constitution enables state and local governments to use their police powers to regulate the use of private property to protect the public's health, safety, welfare, and morals.

Thomas Jefferson especially felt democracy was based on a close tie between people and land. As he stated in his well-known letter to John Jay: "Cultivators of the Earth are the most valuable citizens. They are the most vigorous, the most independent, the most virtuous, and they are tied to their country, and wedded to its liberty and interests by the most lasting bonds."

Beginning with the Northwest Ordinance of 1785, partly written by Jefferson, the federal government initiated a series of actions to encourage private land ownership. The ordinance established the system of six-square-mile (15.5-square-kilometer) townships in what were then the northwest states (from Ohio to Minnesota). The system of square townships facilitated the subdivision of land into farms of 160 acres (64.8 hectares). The 160-acre family farm was further established as a national goal in the Homestead Act of 1862 and its amendments. Under this law, a settler could receive title to 160 acres of land by meeting certain requirements. Later amendments enlarged the amount of land which could be claimed to reflect the larger tracts of land needed for a family farm in the more arid American West. Also beginning in 1862, a series of transcontinental railway acts were passed. Through these acts, land grants were provided to private developers to construct the nation's system of railroads.

Although most of the land base in the U.S. was converted to private ownership after it was taken from the Native Americans, much remained in public control. In the U.S., the largest single owner of land is the federal government, which administers 32% of the 2.3 billion acres (931,500,000 hectares) of land in the nation. Roughly 40% of the land in the U.S. is owned by federal, state, and local governments. Most of this land is in the western states. As a result, the land planning programs of governments, especially at the federal level, have a significant impact on western rural landscapes. All federal agencies are subject to environmental planning laws, most notably the National Environmental Policy Act. Agencies that are especially large and important land holders include the U.S. Forest Service, the U.S. Bureau of Land Management, and the U.S. National Park Service. Each of these federal agencies has its own land planning and management procedures.

As the U.S. became more densely settled through the 20th century, land-use conflicts increased and the need for more regulation of private property arose. The American system for land-use planning and management became more like those in European democracies, which have been settled more intensively and longer. For example, in the Netherlands land planning rules are established at the national level. Individuals and local government must conform to these rules. In addition, in its historic struggle with the sea and flooding, the Dutch have had to create new lands. They have also developed an elaborate system of reallocating parcels of land when through inheritance they become too small to farm economically. Lands are pooled in a region, then reallocated for farms, urban expansion, recreation, and nature areas.

In 1933 the American ecologist Aldo Leopold articulated a way of regarding landscapes that transcended the visual to encompass all the senses. His aesthetic was closely linked to ecology—that is, the reciprocal relationship of all living things to each other and to their physical and biological environments. Leopold also saw land as a community to which we all belong. Such a land ethic is vital for the future. As the lithosphere is of primary importance for the production of human food, energy, and materials, that production has profound impacts on the biosphere.

Soil erosion and eutrophication, for example, are natural processes, but human activities can alter the normal rate. Farming, grazing, home-building, and road construction activities disrupt the soil. If the best possible management practices are not used to conserve the soil, its fertility can be reduced and it can be transported off site, often into waterways as sedimentation. Such eroded sediments are a major source of water pollution. Humans can also speed up the eutrophication process by adding excess nutrients to the land. For example, fertilizers used in backyard gardens can be transported during rainstorms into nearby streams and aquifers.

Changes in land use can cause major changes in both the regional and local climate. For example, when forests are cleared for urban growth, the albedo (reflected solar radiation) and evapotranspiration (the sum of evaporation and transpiration during a specified period of time) of a region are greatly altered, thereby affecting its temperature cycles. Because human use of land can have such profound environmental effects, choices and decisions concerning land use involve ethics.

How land is used determines the fate of societies. The flourishing of many early civilizations was made possible by the fertile soils of river valleys such as the Nile in Egypt, the Tigris and Euphrates in Mesopotamia, the Indus in India, and the Yangtze and Hwang-Ho in China. In the Tigris and Euphrates valleys, the lands were mismanaged and once-productive soils became useless. As a result, this cradle of civilization was abandoned. Land ethics hold that each generation has the responsibility to use the land wisely, and to pass it to the next generation in a healthy state.

FREDERICK STEINER

For Further Reading: Susan L. Flader and J. Baird Callicott (eds.), *The River of the Mother of God and Other Essays by Aldo Leopold* (1991); Frederick Steiner, *The Living Landscape* (1991); Peter Wolf, *Land in America* (1981).
See also Biosphere; Economics; Erosion; Eutrophication; Floods; Land Use; Law, Environmental; Leopold, Aldo; National Environmental Policy Act of 1969; Property Rights; Religion; Soil Conservation; United States Government.

LANDMARKS

A landmark can be thought of in physical terms that, at the narrowest, can delimit a special arrangement, i.e. landmarks as posts which mark the perimeter of a piece of land. More broadly, a landmark can be an element of the environment—a butte, an ancient tree, a water hole—that gives us a feeling of orientation. But a landmark can also illustrate a transition or chart a direction or development so that it can be a literary work that marks an evolution.

The German word for landmark is *Denkmal,* which can be translated as "monument," or more poetically broken into syllables that read literally, "reflect for a moment." This interpretation suggests an expansive definition of landmark as having the capability to calibrate our relation to time and space. A landmark, then, can constitute an act of arousal, evoking a passion or conviction that empowers or enables successive generations to reimagine the events that occurred in or around it, or merely to grasp, through the cognitive physicality of what remains, a palpable sense of place.

Such an expansive definition helps to explain the commitment of the Polish people to rebuilding and replicating at great cost the core of their capital city, Warsaw, when almost nothing was left standing after the Nazi demolition of it during World War II. The dedication of Japanese Shinto worshipers to rebuild exactly the traditional wooden temples every 20 years—as they have been doing at the Iso Shrine since the 10th century—may not evoke the same richness of historical associations of medieval Warsaw, but it does make this powerful connection to physical condition.

As a term of reference, *landmark* might appear to be a dry, antiquarian word relating to a special category of historic sites that are seen as more important than some others. For example, there are, under the National Register of Historic Sites as defined by the Historic Preservation Act of 1966, 60,000 listings of buildings, sites, objects, and districts in the U.S. Only 2,000 of them in 1993 were deemed by the Department of the Interior, the final arbiter following nomination through state preservation offices, to be nationally significant.

Indeed, the National Association of State Preservation Offices has resisted the concept of special protection for National Landmarks for fear that it would lead to increased threats for other less hallowed sites under the National Register. This approach can encourage a type of trophy hunting where the best example remains in splendid isolation, perhaps situated in a sea of parking lots or high rise buildings, or even stuck in a heritage park with other landmarks. It contrasts with the British approach under the Civil Amenities Act of 1967, which empowered the creation of environmentally attractive districts that encompassed a broader definition of excellence, and where the powers of the Act included efforts to further enhance that condition of livability.

Yet the evocation of the term *landmark* can conjure up a sense of mythical power, which challenges the capacity of those who experience it to reinvent the meanings of place in their own mental landscapes. The recently created highway-scale monuments that Radnor Township commissioned near Philadelphia demonstrate how new elements can evoke a mental landscape of associations. The dramatic excavation of a new arterial highway, the Blue Route, across Philadelphia's "Main Line" prompted Radnor Township to engage in a process of enhancement. The design strategy for this five-mile corridor reimagines the neolithic stone landscape of Wales, home of Radnor's original Quaker settlers, while recalling the 18th century stone walls and milestones of Lancaster Pike, America's oldest turnpike. Consequently, rocks excavated from the Blue Route are grouped in megalithic sculptures including a 22-foot-high cairn and a 90 by 100 foot griffin that mark key entry points along the turnpike. A rhythm of plinths, 8 feet high, supplement the old (18 inch) milestones and are sandblasted with representations from the Township seal, including a tree, dragon, lion, and wheat sheaf. These designs were also stenciled onto 14-foot-high sound barrier walls encasing the Blue Route bridges. This comprehensive approach suggests how an enhancing strategy can self-consciously create a sense of resonance with place that sets about to realize the broadest definition of landmark evoked by the translation of that German word *Denkmal*: 'reflect for a moment.'

RONALD LEE FLEMING

LAND OWNERSHIP

To own land is to possess a set of rights, including the core right to exclude others. An owner also has some—usually not complete—rights to erect a building or otherwise to use the land. A further important element of land ownership is the freedom to transfer the rights—usually to an heir or by selling them. There are also typically obligations associated with ownership, such as the requirement to maintain the land in a safe condition.

In most primitive societies, land was not owned in this sense; use was instead controlled through informal mechanisms based on custom or the fiat of a leader. In Europe, as recently as the feudal era, possession of land was similarly granted by a king or other ruler who thereby entered into a set of reciprocal military and other obligations with his subjects. The lands of a feudal manor were no more saleable than is the command of a general in the U.S. Army today.

In Africa, Asia, and the Americas before European settlement, the range of land tenures found was very wide. Where there were large amounts of land relative to population, as in many parts of the Americas and

Africa, there was less pressure to develop formal systems of ownership rights to land. In Asian countries such as India and China, the scarcity of arable land and the demands of feeding large populations in many areas promoted the creation of systems of individual ownership rights to peasant farms.

In Europe formal property rights to land emerged as part of the long process by which the institutions of a market economy evolved. The use of land would be determined by the preferences of that buyer willing to pay the highest price. Land could be transferred among users in the same way that other commodities in the market were bought and sold. For all this to happen it was essential to develop a legal system with a well defined concept of land ownership.

In the late 17th century, John Locke provided a philosophical justification for land ownership. According to Locke, it was the law of nature that the individual should possess the product of his labor on the land (and other resources). Through a social contract, governments were formed for the purpose of facilitating the enforcement of property rights. If land were instead left in common and were freely available to all, the incentive for individual efforts to improve the land would be greatly diminished.

Much of the history of the Americas revolved around the conquest and the subsequent establishment of ownership rights to land. In the United States the Pre-emption Act of 1841, the Homestead Act of 1862, and the railroad land grants—all landmark events of 19th century history—provided for the transfer of land from government possession into a system of private ownership. The era of disposal largely came to a halt in the early 20th century. American progressives argued that government could manage the land more efficiently and responsibly than private owners. Today, the federal government still retains 30% of the land (and more than 60% in Alaska, Idaho, Nevada, and Utah). State governments hold another 8%. Government agencies such as the U.S. Forest Service (in the Department of Agriculture) and the National Park Service and Bureau of Land Management (in the Department of the Interior) are responsible for managing the huge federal domain. In many areas, the task is complicated by the fact that private and government lands are thoroughly intermingled.

At about the same time that the U.S. created its system of public land management, local governments moved to assert greater control over the use of private land. Introduced in New York City in 1916, zoning was approved by the U.S. Supreme Court in 1926. The court found in part that the implementation of municipal plans for land use required greater public control over the actions of private landowners. Zoning requirements—along with building codes, en-vironmental controls, and various other forms of local regulation—limit the types of uses and the building designs that are permissible. In historic districts, governments exercise regulatory authority over even house color, shrubbery, and other minor details of aesthetic concern.

Individual rights of ownership are also subordinate in many places to collective private rights. In a condominium, the individual unit-owner can make some changes in use as a matter of right, but many other features are controlled by fellow residents. Easements similarly transfer to another party the control over particular aspects of use. There are also many places in the United States where ownership of land is divided into surface rights and mineral rights held separately. Land can thus be owned individually, in common privately, or by government—and each of these possibilities may coexist with respect to different rights involving the same parcel of land.

The rise of environmental concerns since the 1960s has brought new forms of control, further restricting the rights of land ownership. Rather than the old nuisance law remedies of private suits against particular polluters, environmental regulations have been applied across the board to whole classes of industries and land uses. For many environmentalists the very idea of land as a marketable item—as a "commodity"—has also come to be seen as objectionable. Inspired partly by the writings of Aldo Leopold, it is said that we need a new "land ethic" that recognizes more of the sacred and less of the commercial in land.

Others, however, argue that individual landowners are being asked unfairly to bear costs for the benefit of society at large. A better balancing of benefits and costs and more efficient land use will result, they contend, if governments are required to find the revenue to pay for the "opportunity costs" they impose through environmental regulations. This issue is being joined today in the courts in controversies concerning when and where government must pay compensation in order to avoid an unconstitutional "taking" of land-owner rights.

In the 20th century, the rights of land ownership have often fared poorly around the world. Governments acting in the name of scientific planning and management of society have curtailed individual freedom in favor of collective purposes in the use of land. In the extreme, communist social systems eliminated private ownership of land altogether. Yet, as the end of the century nears, there is evidence of a contrary trend. Since 1978, China has curtailed the role of collective agriculture, allowing households to sell significant portions of the outputs from farm lands for which they are now given individual responsibility. Governments in Eastern Europe and the former Soviet Union are moving in a similar direction. Throughout the world government management of land too often did not suppress the role of self interest, but directed it in new and sometimes much less socially constructive ways.

The role of land ownership in a society is a litmus test for the basic values and self-concept of that society. Current debates around the world concerning the meaning of land ownership involve social issues as fundamental as the proper role of the competitive market, the desirable extent of individual freedom, and the degree of obligation to society which an individual landowner must assume.

ROBERT H. NELSON

For Further Reading: Richard F. Babcock, *The Zoning Game* (1966); William A. Fischel, *The Economics of Zoning Laws* (1985); Frederick Pollock, *The Land Laws* (1979; 1st ed. 1883).

LANDSAT

See Remote Sensing.

LANDSCAPE

The word *landscape* is now often used to indicate agreeable rural or natural scenery. But for many centuries *landscape* had a precise meaning: the territory of a small homogeneous working society of farmers and stockmen. It has been part of our language from prehistoric times, and its many variations, its many metaphorical usages tell us that we have always needed a term to indicate a certain workaday relationship with the environment.

In our remote Germanic past *land* meant open land: meadow or field or heath as opposed to forest or wilderness. The second syllable, *scape*, is a collective suffix indicating a collection of similar objects. Thus at one time landscape meant a composition of fields cut out from its forest surroundings.

It was man-made, and this in turn implies a community of farmers or stockmen using the fields or grazing their cattle in the surrounding wasteland. Such a setup presupposes northern Europe. In the Mediterranean region the Latin equivalent of landscape was *pagus*, meaning a small rural settlement, a social unit based on an agreement, a pact. That Latin term reminds us of an important fact: that a landscape was once something more than an economic or agricultural space, that it was social, even political in origin, and that its reason for existence was to allow the small society to survive and develop.

One consequence of that early emphasis on the social function of the landscape was that within a given region all landscapes resembled one another: not in topography but in way of life. It was as if they were all variations on some historical or mythic prototype, no matter where they were located. The Pueblo landscapes of the Southwest, though widely scattered and in very different climatic zones, show a remarkable uniformity in architecture, in the crops they raise, and in their religious practices. It is likely that the primitive landscapes of northern Europe of two thousand years ago showed the same indifference to environmental factors and the same loyalty to family or tribal associations.

The impulse to intensify landscape evolution came slowly and from several sources: from the draft horse as a new kind of energy, from deforestation, from collaborative farming, and not least of all from the scholarly and autocratic establishment. In the Middle Ages landowners introduced for their own benefit new crafts, new tools, new markets, and new plants from Asia and later from America. The village became a center of new imported skills. The new plants and farming techniques focused attention on soils and climate and work routines—factors hitherto ignored. The consequence was that many landscapes discovered their unique capacities, social as well as environmental, and acquired a new and special identity. It was in the late Middle Ages that what we now recognize and cherish as distinct landscapes, each with its own local culture, its own field layout, its own housetype, its own calendar, and even its own dialect, began to multiply throughout Europe. It was as if we had discovered the importance of place, the link between environment and the social order, for this flowering of regional landscapes came at a time when the European establishment became aware of how to organize space—in architecture, in gardens, in national policy—and it was also when land ownership acquired importance. The landscape, once legally defined as "a small rural administrative subdivision," was discovered by artists and poets. To the thoughtful it was seen as a counterbalance to the artificiality of the city: a bucolic world of natural beauty and simple, unchanging ways.

But what makes landscapes natural (and human) is that they continue to change and even to invite change. The 19th century witnessed the gradual but complete destruction of the preindustrial landscape, both agriculturally and socially speaking; for what was destroyed was that belief in the sanctity and permanence of place as the true basis of any well-balanced landscape. The development of chemical fertilizers, the increasing of the yield of many plants, and the introduction in Europe of corn, all had the effect of transforming the agricultural aspect of the landscape. Time-honored rotation practices were abandoned; poor quality land was upgraded; and new mechanical implements, new work routines, and new markets changed field layouts and reduced the role of the homestead. The abandoning of local traditional customs and environmental restraints produced a more mobile population and new forms of community.

It is scarcely necessary to mention the reaction of many elements in the urban population to these changes. The squalor and ugliness of early industrialization were what first caused public indignation, but the Romantic movement with its worship of unspoiled nature concentrated much criticism on what was happening in the rural environment. When it was said that the landscape was being desecrated and destroyed, the emphasis was on the visual, on the disappearance of the picturesque and the traditional. The human consequences of the agricultural revolution were less spectacular than the deforestation, the surface mining, the rows of uniform workers' houses, the smoke and pollution and slag heaps. For middle class urban men and women this was an understandable reaction: the countryside was not their place of work, it was where they sought esthetic pleasure and often solitary enjoyment. But as a result, the word *landscape* came to mean *scenery*, not a viable social unit.

In any case, the impact of the Romantic Landscape Movement on the broad workaday landscape of America has been beneficial but very limited. We have acquired magnificent national parks, urban parks, landscaped highways, attractive suburbs, college campuses, and cemeteries—all inspired by a largely imaginary pretechnological pastoral tradition—and none of them meant for work. As the technological transformation of our countryside and our cities—and even of our wilderness areas—proceeds apace, the question we might ask is: what new concept of landscapes is being formulated to suit the modern world?

What we lack, particularly in the design professions, is any awareness of landscape history: Our collective memory in this respect extends no further back than two centuries and is confined to northwestern Europe and the northeastern United States. When confronted with industrialization, urbanization, the vast increase in minorities and the influence of the automobile and other forms of communication, all we can think about are the esthetic and ecological consequences: the political and social functions of the generic landscape are ignored.

The fact is, there have been far more radical landscape changes in the past, and geographers are contributing valuable insights into landscape history in their studies of the conquest of pre-Columbian landscapes by Spanish and English invaders. Ninety percent of the native population died. Radically new forms of energy,

new kinds of land ownership, new work processes, new settlement patterns were ruthlessly introduced. Many native landscapes were totally wiped out, but many survived and managed to assimilate the changes, and are now not only productive, but beautiful.

The instinct to produce landscapes with a cultural as well as an economic function is as strong as ever. Students of landscape history can learn much by serious examination of the past, and by looking around the modern world. If they look with discernment and without romantic hangups they will see the beginnings of a post-industrial landscape in many parts of the United States.

JOHN B. JACKSON

For Further Reading: John B. Jackson, *Discovering the Vernacular Landscape* (1984); John B. Jackson, *A Sense of Place, A Sense of Time* (1994); Donald W. Meinig, *The Interpretation of Ordinary Landscapes* (1979).
See also Ecology as a Perspective; Esthetics; Geography; Humanized Environments; Land; Rural Communities; Shelters: Social Relevance.

LANDSCAPE ARCHITECTURE

Landscape architecture is a design discipline centered on the creation, management, and preservation of a broad spectrum of places in the environment. Embedded within the creative and artistic foundations of the practice of landscape architecture is the primary mission of the stewardship of the Earth in all its human and natural contexts. As a profession landscape architecture serves as the essential connecting tissue among related disciplines: architecture, planning, civil engineering, horticulture, and ecology. It draws on philosophy, religion, literature, aesthetics, sociology, anthropology, geography, art, art history, agriculture, forestry, and recreation to inform design and management decisions. It is dedicated to establishing symbiotic relationships among artistic, scientific, and cultural ideas that have the potential to influence the quality and character of the land.

The domain of landscape architecture extends from garden design to town, city, and even regional design. Within this domain design is understood to be an expression of society's or an individual's values in built form. Horace W. S. Cleveland, an early pioneer in landscape architecture, described the then-emerging profession in 1873 as "the art of arranging land as to adapt it most conveniently, economically, and gracefully to any of the varied wants of civilization."

Landscape architectural design also incorporates preservation and management plans or strategies for historic, cultural and natural resources. Considering its ethic of stewardship, it is ultimately about expressing a set of values that encompasses a humanized world without compromising the natural world that we are a part of.

Landscape architecture is a discipline challenged by the paradox presented by environmental ethics and human settlement. It attempts to reconcile our desire to preserve and even celebrate wilderness and wildness with an equally strong penchant for enhancing our own immediate environs. Human appreciation for wildness and cultivation lies at the heart of the practice of landscape architecture.

While a significant part of landscape architecture emerged from the discipline of architecture, equal attribution may be given to developments in the arts (landscape painting in particular), literature, and the science of horticulture. The still emerging fields of ecology and environmental history continue to further expand the definition of landscape architecture.

Landscape architecture is a spatial art, explored initially through two-dimensional drawings (diagrams, sections, axonometrics, and perspective), further explained through three-dimensional modeling, and finally acted out in the third dimension of space and the fourth dimension of time. It is the impact of time and change that most distinguishes the practice of landscape architecture from other forms of design and art.

Considering that the scope of landscape architecture ranges from the smallest of gardens to the scale of regions, the constituent elements remain remarkably consistent: soil, stone, water, and plants. More abstractly, light and air, sun and shade, climate and geology can be influenced by a work of landscape architecture. The greatest works tend to succeed because of the designers' understanding, acknowledgment, and incorporation of natural systems. Cycles of weather, temperature, plant growth, water flow; the specific and general qualities of every site—its orientation to the cardinal points, to views, the lay of its topography, the make-up of its geologic underpinnings, and its mantle of soil and vegetation, the presence and distribution of wildlife, previous settlement and agricultural patterns, the utilities of modern conveyance and convenience: paths, roads, water, sewer, electricity, telephone, cable, and countless more human ones (the needs and desires of the owner/client, the future inhabitants of the site), will influence the design evolution of a place.

At the scale of the master plan for the design of a campus, neighborhood, place of employment, park, public institution, town, city, or transportation corridor, the siting, composition, orientation and proportion of all elements of the landscape are subject to the discretion of the landscape architect. At times it may

be difficult to distinguish the work of architecture from the work of landscape architecture, as the design of landscape often entails the mediation between building and its natural setting.

Water may be molded into lakes, ponds, pools, canals, fountains, rills and runnels. The plasticity of soil allows it to be shaped into mounds, terraces, and berms. Most often plants represent the quintessential ingredient of landscape architectural practice, though examples can be cited of important landscape designs where living plants play a limited role. The landscape architectural palette recognizes the form and change-giving properties provided by plants, from groundcovers to shrubs to trees. It is in the use of plants that landscape architecture most frequently crosses boundaries with horticulture and garden design.

Landscape architecture requires a fundamental grasp of design and compositional principles that are in great measure shared with other creative fields. Traditional and contemporary landscape architectural texts stress primary design considerations such as form, texture, and color as these are subject to the 'artistic' principles of proportion, repetition, harmony, balance, rhythm, and scale. While these principles and ingredients do persist, design is not often successfully achieved through a recipe approach. Critical to design is an understanding of values, concepts and essential qualities that might infuse a place with enduring meaning.

Landscape architectural practice may include any or all of the following:

Site planning, for many, constitutes the core of traditional practice. It is directed toward planning sites to accommodate a variety of human activities and interventions including the location and design of buildings, roads, parking, service areas, paths, courtyards, and drainage systems. Such design may be at the scale of house and garden or of larger institutions: schools, museums, recreation centers, commercial centers, housing, resorts, and public parks. Garden design and park design are specific project types at the heart of traditional landscape architectural practice. The nature of these projects varies depending on the scale, purpose and scope of each. Garden design may be primarily about plants and planting design, or the extension of architecture into the landscape through paving, walls, steps, and terraces. Gardens are often designed to reflect the character and style of their associated buildings, or they may be designed to serve as a natural counterpoint to the built environment. Park design varies similarly in context and purpose. The spectrum ranges from parks designed with minimal human intervention such as Yellowstone and Yosemite National Parks to completely designed and manipulated urban parks such as Central Park in New York or the Mall in Washington, D.C.

Town and community planning, as practiced by landscape architects, is essentially larger-scale site planning that takes into account all systems and elements in the siting and composition of neighborhoods around one or a series of civic and community centers. Town planning encompasses circulation systems and the design of streets and blocks into some hierarchical order: avenues, boulevards, parkways, streets, ways, alleys, promenades, walkways, and paths. It incorporates virtually all facets of human life into a composed setting and must account for the location and design of places for work, dwelling, play and recreation, worship, socializing, and education. Such projects range from heterogeneous communities to quite homogeneous ones such as company towns, resort communities, and retirement villages.

Urban design focuses more directly on the making of cities and the design of various parts of cities. Projects within this venue include the design of streets, sidewalks, parks, courts, gardens, roof terraces, and promenades. The revitalization of former industrial areas, railroad yards, waste dumps, derelict waterfronts, and transportation corridors provides an increasing range of opportunities for landscape architects in contemporary practice.

Regional planning addresses ever greater land areas, encompassing counties or even states. Often politically or geographically defined, these areas share common concerns and goals regarding future development. Transportation planning, town planning, location of major industries, airports, landfills, regional park systems, linear parks and greenways, agricultural development, tourism, preservation planning—preserving and protecting natural, historic, and cultural resources—are all dependent on coordinated strategic planning that often includes the expertise of landscape architects.

Additional areas of concentration in landscape architectural practice include land reclamation, visual resource assessment, environmental impact assessment, wetlands delineation, preservation and mitigation, landscape and plant community restoration, real estate development, geographic information systems, international development and ecotourism, urban forestry, golf-course design, wildlife habitat design and management, and water and natural resource planning.

Practitioners work within the private or public realms, for small or large firms, for public agencies at the local, county, regional, state, or federal level. Agencies such as the National Park Service, U.S. Forestry Service, highway departments, and city and county planning departments typically employ landscape architects at all levels of responsibility.

The origins of landscape architecture as a profession can be traced through many sources: horticultural, agricultural, artistic, and spiritual. These roots reach to the cultivated lands of Mesopotamia and Egypt, to the ritual and ceremonial landscape of native and aboriginal cultures in virtually all the inhabited continents of the globe, and most obviously to the evolution of garden design, ceremonial temple site planning, park design, and town planning in Southeast Asia, the Americas, and the Mediterranean basin.

The city of Babylon gave us one of the more enduring of landscape architectural images: the Hanging Gardens, conjectured to have been terraced gardens 75 feet high. The Assyrians, from 1350 B.C., contributed the hunting park and royal gardens to landscape architectural history. These proved, in essence, to be the progenitors of today's great city public parks.

One of the earliest visual records of a designed garden is depicted in a tapestry of Thebes in Egypt. It shows a walled rectangular enclosure containing pavilions, pools and vine-covered trellises. Flowers, palm trees, and lotus plants are depicted throughout. Other Egyptian contributions to the evolution of landscape design include numerous monumental sacred sites along the Nile River; primary among these is the temple of Queen Hatshepsut (1503–1482 B.C.).

As civilization expanded throughout the Mediterranean, Greek site planning marked the next remarkable era of landscape-related design achievement. The powerfully interlocking unity of architecture with landscape manifested in Olympia, Delphi, the Acropolis, Epidauros, and other sites remains an accomplishment unequaled since.

The Roman Empire contributed rational town planning and the development of the garden as an immediate and essential extension of the house. From the courtyard gardens of the urban house to the lavish country villas of the wealthy landowner, Romans built a tradition of garden design out of both agricultural and architectural roots. Of particular influence from this era have been the Laurentian Villa of Pliny the Younger (A.D. 23–79) and Hadrian's Villa near Tivoli (A.D. 118–38).

After the decline of the Roman Empire, the primary developments related to landscape architecture occurred during the Middle Ages in the cloister gardens of monasteries in Europe. As human settlement and religion turned inward, gardens served primarily as sources for the collection of medicinal and herbal plants and places for reflection and contemplation. These cloistered gardens were derivative of the earliest paradise gardens, and, in their attention to collections of plants for specialized purposes, presaged the arrival of the first botanical garden in renaissance Italy. Other contributions to site planning came from the dramatic siting of towns, castles, and particularly monasteries. Often deliberately located in remote rural areas, for purposes of both protection and isolation, these architectural compositions would often appear to grow organically out of their steep and rocky settings.

The 1400s through the 1600s saw remarkable landscape and garden developments in places as diverse as Italy, Spain, China, Japan, and India. The nations of Islam produced town plans and gardens reflecting the structure and qualities of earlier civilizations, but with increased attention to new discoveries in mathematics, the sciences, and agriculture. Walled gardens, traversed by canals, laid out in orthogonal configurations, were the standard. The highest design achievements of this era would be found in Isfahan, a city of gardens, mosques, terraces, canals and palaces (1598). In Spain, the Court of the Oranges at Cordova, the summer palace of the Generalife (prior to 1319), and the palace of the Alhambra (1250–1500) in Granada are enduring testaments to Moorish garden design. The Mughul gardens of Kashmir (after 1620) are a vast system of verdant terraces celebrating water in the form of fountains, rills, pools and cascades. The Taj Mahal (1632–54), as the ultimate serene expression of memorial architecture and landscape architecture in India, restates the Paradise garden and garden of Eden theme of four squares divided by four rivers or canals originating from the center.

The landscape and religious cultures of Eastern Asian civilizations provide an entirely different set of precedents. But like Western and Middle-eastern civilizations, the expression of architecture and landscape architecture arose directly from philosophical and religious foundations. At Katsura Imperial Palace in Kyoto (1620), water, stone, architecture, and plants were woven into compositions both symbolic and sacred; designed to evoke images of distant treasured landscapes, they served simultaneously as stroll gardens and contemplative retreats. Zen Buddhism infused the principles by which this and other gardens were designed and experienced.

Other remarkable and influential gardens from the environs of Kyoto include the stone and sand garden of Ryoan-ji, where raked sand represents the sea, and seven stones judiciously placed within represent some greater landscape beyond the walls. At the Saiho-ji temple the garden is composed primarily of more than a hundred species of moss.

The Renaissance that swept Europe from the 1400s onward had its most expressive origins in Italy. Villa gardens celebrated mathematical proportions, ideal geometrics, extensions from architecture, the discovery of new plants, and engineering and hydraulic advancements, all within a harmoniously designed framework that heralded talk of a return to some mythical Golden Age. The gardens of the Medicis in Florence (ca. 1460s), the Villa d'Esté near Rome (1550), and Villa Lante at Bagnaia (1566), remain to this day some of the greatest examples of gardens mediating between the built environment of the town or house and the larger agricultural and natural environment beyond the enclosures.

Two consecutive eras of significant achievement in landscape architecture followed the Italian Renaissance. In the 17th century attention shifted to France and the monumental landscape compositions of André Le Notré. His designs at Vaux Le Vicomte, Chantilly, and Versailles reshaped the countryside into a planar and three-dimensional geometry that extended through the great houses, on such a grand scale that they became minor parts of the composition.

In deliberate reaction to a landscape expression that appeared to control and dominate a vast nation, the 18th century saw the development, primarily in England, of a picturesque and romantic landscape, as designers and writers alike eschewed geometry, symmetry, and architectural extensions of buildings into landscape. Instead, the designs sought an aesthetic, no less manipulated than Le Notré's, that mimicked the natural lines, curving hills, water courses, random clumps and spacing of trees to, in effect, replicate the idealized and romanticized paintings of artists such as Claude Lorrain. Designers such as Humphrey Repton and Capability Brown worked on a scale equal to or greater than Le Notré, and set in motion the landscape transformation of an entire nation. This English romantic landscape school served as the primary influence for the development of public parks throughout the world.

Thomas Jefferson, in his design for his estate at Monticello in Virginia (late 1700s, early 1800s) and in his design for the University of Virginia (1819), proved to be one of the important forerunners of American landscape architecture. His interpretation and adaptations of European design ideas to his own circumstances always transformed them into a uniquely American expression.

Andrew Jackson Downing combined a horticulturalist and designer's perspective with appropriations from England's picturesque landscape in targeting the rise of an American middle class. His *Treatise on the Theory and Practice of Landscape Gardening Adapted to North America* (1841) proved to be one of the most influential and enduring books on American landscape design.

The profession of landscape architecture evolved from a design collaboration between Frederick Law Olmsted and Calvert Vaux in their creation of New York City's Central Park in 1848. Landscape architecture as an accepted term and practice found its official origins in America, for as the founding nation for democratic ideals and, ultimately, as the progenitor of cultural diversity in its population, the foundation of this field as a profession lay in the creation and setting aside of public parks for social as well as aesthetic purposes. The profession was seen by Olmsted and his contemporaries as a moral imperative toward the betterment of society. The partnership of Olmsted, a man of many interests, and Vaux, an architect, foretold much about the evolution and character of the profession. It suggested the importance of collaboration within this complex discipline. More revealing, the two contributors reflected different positions on the inherent qualities of landscape architecture, Vaux on the artistic side of practice, Olmsted on the social side. The relative merits of these two "sides" of the profession continue to be hotly debated.

Following Olmsted's pioneering achievements in civic, park, and estate design were other practitioners in the late 19th and early 20th centuries such as H. W. S. Cleveland, George Kessler, O. C. Simonds, William Lebaron Jenney, Beatrix Farrand, and Jens Jensen. Jensen, concentrating in the Midwest, established a regional ethic of design in both public and private projects, emphasizing the use of native plants and drawing on indigenous qualities of local landscapes for primary inspiration.

The turn of the century saw significant developments in the evolution of American landscape architecture including the formation of the American Society of Landscape Architects (ASLA) in 1899, the Chicago World's Fair in 1893, and, in 1904, the MacMillan Plan for Washington, D.C. The last two events signaled a revival of classical landscape monumentality in city planning. And in 1916 the early efforts of Olmsted, John Muir, and others to establish a nationwide system of national parks were officially assured with the congressional authorization of a National Park Service (NPS). This system of national parks set a precedent that continues to be the model worldwide. State and regional park systems grew from these origins. Design work in parks ranged from master plans to development plans for roads, parking, campgrounds, interpretive areas, picnic areas, and trails. Franklin Roosevelt's New Deal initiatives in the 1930s and 1940s provided a considerable boost in establishing, expanding, and developing state and local park facilities and scenic parkways.

After World War II, the field of landscape architecture expanded exponentially. Prominent figures from 1940 to the present include Roberto Burle Marx from Brazil; Dan Kiley, Garrett Eckbo, Thomas Church, Lawrence Halprin, Hideo Sasaki, and Ian McHarg from the United States; Sir Geoffrey Jellicoe and Dame Sylvia Crowe from England; and Luis Barragan from Mexico. The work of these designers varies from garden-scale projects to town and regional planning.

The diversity of their work, their 'styles,' and approaches to projects reflect the richness, complexity and breadth of the profession. Their sources of inspiration come from art and nature, modern and traditional architecture, agriculture, philosophy, and an increased awareness of ecology.

As we seek understanding of what is an appropriate modern or contemporary designed landscape, and what an ecologically-driven landscape architecture is, the profession evolves. The call for sustainable living has at its core the very tenets of sound landscape architectural practice, a practice that cherishes, interprets, and productively transforms the land for the good of all its creatures great and small.

WARREN T. BYRD, JR.

For Further Reading: Geoffrey Jellicoe, Susan Jellicoe, *The Landscape of Man* (1987); William Tishler (ed.), *American Landscape Architecture* (1989); Stuart Wrede and William Howard Adams (eds.), *Denatured Visions: Landscape and Culture in the Twentieth Century* (1991).
See also Architecture; Esthetics; Green Belts; Humanized Environments; Land Use; Landscape; Olmsted, Frederick Law; Parks, U.S. National; Suburbs; Urban Parks; Waterfront Development.

LAND USE

Land use indicates the spatial relationship between people and their environment by describing the occupation or reservation of a land or water area, including the overhead air space, for any human activity or purpose. Land use can be broadly described in four categories: crop land, grazing land, forest land, and other land.

Each of these broad categories covers a vast number of subdivisions. For example, the "other" category in general land use descriptions includes glaciers and deserts as well as urban lands, which may be separated into residential, industrial, commercial, institutional, and recreational areas. Within the residential area, we often find delineations between neighborhoods of single-family homes, multi-family dwellings, and mobile home parks. Thus, like any system of classification, land use categories consist of a many-tiered set, ranging from the general to the very specific.

Land use choices are made by the people and institutions who own or control the lands, but those choices may be affected or limited by a number of factors. These include the physical and biological qualities of the lands themselves, the economic situation of the people, and the institutional framework that society has constructed for controlling land use.

The physical and biological capacity of the land is termed *land use capability*. The most widely utilized capability classification system has been developed by the USDA Soil Conservation Service to indicate the land's capacity to support cultivated agriculture. This system consists of eight classes, expressed as roman numerals (I–VIII). Class I indicates the best land, with deep, fertile, level soils that can withstand intensive cultivation, render high yields, and suffer little or no damage in the process. Classes II and III are progressively less able to support intensive cultivation, and Class IV can stand only occasional cultivation such as would be associated with haylands or pasture. Class V is too wet for cultivation. Class VI and VII lands are too steep or their soils are too shallow for cultivation, and they are limited to grazing and woodland management. Class VIII is too rough, rocky, and steep even to be used for grazing or forestry without special precautions against soil damage. This system recognizes the current condition of the land, taking into account any prior alterations that may have affected the land's capacity (such as excessive topsoil loss from a previous use).

In considering land for a particular use, another concept—land use suitability—comes into play. Here, the decisionmaker considers the potential capability of the land if changes are made. For example, trees can be cleared from forest land so that crops can be grown or houses built; or the introduction of irrigation, fertilization, or drainage can dramatically change the capacity of land to support intensive cropping or development.

The economic situation that affects land use decisions can change dramatically, often altered by events or decisions that seem far removed. A far-away drought that raises the world price of wheat will induce people to plow up grasslands. A bank collapse that dries up loan funds may slow home building for a few years, causing a forest or farm field to remain in its current use.

Institutional factors affecting land use can be indirect or direct. Indirect factors such as tax laws, export policies, or a decision to locate a new road are examples of public policies that, while enacted for a variety of other reasons, may be critical in influencing land use decisions. Direct policy influences include legislative designation, such as the creation of a park, a wild and scenic river, or a wilderness area; or zoning and other controls that regulate what private owners can do with their land.

As these factors change, people tend to adjust land uses accordingly. Over the long term, however, it is critical that land uses remain reasonably consistent with the physical and biological capacity of the land itself. A land use that degrades the physical or biolog-ical quality of the land (such as growing crops on steep lands where soil erosion removes the topsoil) is unsustainable over the long term. Any land use that is incompatible with the natural forces that affect the land (such as building a house on a floodplain) will eventually be overwhelmed. Thus, an understanding of the current "fit" between land use and land capability can be helpful in assessing future environmental opportunities and challenges.

Some land uses, such as a cotton field or a homesite, dominate the land physically to the point that they preclude all other uses. Many lands, however, can be managed for multiple uses. Some forest lands, for example, can support a combination of timber production, grazing, minerals extraction, public recreation, watershed protection, and wildlife habitat. On other forest lands, perhaps equally suited to multiple use, a decision to dedicate them for wilderness, wildlife habitat, or a biological reserve may mean that many uses have been prohibited by law.

Multiple-use management involves a constantly-shifting balance, in which some uses may be suppressed in some places or at some times. In the multiple-use forest, for example, wildlife, recreation, and watershed values may be adversely affected during the several-year interval while a mature forest is harvested and a new forest established. In that same forest, the existence of a nesting tree for bald eagles may cause foresters to leave a grove of mature trees standing during a harvest cut, solely for the protection of the eagle's nesting habitat. Multiple-use requires balance and compromise.

In the United States, about one-third of the land remains under federal ownership, and use is guided by federal law. Such laws can range from general multiple-use statutes, such as those that guide the majority of the lands managed by the Forest Service and the Bureau of Land Management, to very restrictive and specific guidance, such as might be contained in the designation of a wildlife refuge, historic park, or military reserve.

On the two-thirds of the land that is in non-federal ownership, economic and institutional factors are important determinants of land use. Land use controls on private land are designed and administered mainly by local governments, such as cities, townships or counties. In many rural areas, land use controls have historically been weak or absent, and economic considerations have prevailed as the main determinant of land use. This market-driven, individually-determined system of land use allocation seldom formed a coherent pattern. Neither planned nor designed, the outcome was the result of a myriad of individual, unconnected, often competitive decisions.

Leaving land use entirely to the vagaries of the economic market does not always result in outcomes that people find acceptable. If there are only limited lands that can grow a special crop, converting all that land into housing developments may not be the best solution. Continued rebuilding of houses on a flood plain results in repeated human and economic costs, both to individuals and governments. At such points, people begin to intervene in land use, usually through land use planning and control.

Land use planning is a political process. It begins with an inventory of the status, potentials, and limitations of the land and its resources, then involves the public to determine their needs, wants, and aspirations for the area. Planners then hypothesize patterns of land use compatible with both the area's physical capabilities and the people's needs and wants, and propose a set of incentives and controls to achieve desired outcomes. Political authorities select policy and program options, which are implemented in the form of legislation, regulations, or other programs. The planning process documents the ongoing land use situation, comparing expected and real outcomes, then repeats the process over and over to account for changing environmental conditions, new technologies, and evolving human needs and wants. Other institutional pressures, such as tax laws, water laws, or pollution control regulations, add to the institutional framework within which land use decisions are made.

History of Land Use in the United States

The "lower 48" states consist of about 1.9 billion acres. When the first European settlers arrived, it has been estimated that about 900 million acres were covered by forest, 900 million by grass or shrubs, and the remaining 100 million were ice fields, deserts, and other barren terrain. The Europeans began immediately to change that pattern, as they built towns and cleared farm fields near rivers or bays that provided access for the ships that brought essential supplies from Europe. A great deal of land clearing and road building was required along the coast before people began to move inland, often to higher, better-drained, and more healthful sites. But, even as they moved inland, people congregated along the navigable rivers and their tributaries.

As the major transportation routes, waterways were the arteries that shaped settlement patterns, but land quality was also held in high esteem. Land suitable for water development sites, towns, or of obvious high quality for farming was in high demand. As railroads provided access to lands distant from water routes, the development process began to fill the lands between the water arteries. By the end of the 19th century, people began talking about the end of the American fron-

tier. At that time, as Figure 1 shows, there were over 300 million acres in cropland, much of which had been carved out of the immense forests that blanketed the eastern part of the nation.

But the end of the frontier days did not mean the end of the process of land use change. Changing demography, economic conditions, and technology continued to drive new land use patterns. By 1950, cropland had risen to around 400 million acres, while both forests and grasslands had been significantly reduced. In the last half of the 20th century, "other" land uses, such as urban and wilderness areas, continued to expand at the expense of forest and grassland, while the area of cropland remained fairly constant.

Figure 1 suggests that after 1950, land use in the United States has remained fairly static. But the lack of change, particularly in the amount of cropland, tends to camouflage a process that is still very active. For example, the Soil Conservation Service determined that total cropland acreage increased only 1.4 million acres between 1982 and 1987. But this change was the net result of converting 16.1 million acres of 1982 cropland to pasture, range, forest, and urban uses, while 17.5 million acres of pasture, range, and forest were converted to crop use. Thus, while the overall total indicated a net change of only 1.4 million acres, we see that over 33 million acres—an area the size of the entire state of Arkansas—was affected by land use change in only a 5 year period! Clearly, land use remains a dynamic process, reflecting an ongoing change in the relationship between the people and their environment.

From the first European settlements in the United States, development has been the major theme driving land use. Government policies have treated land as an economic resource, best used by private exploitation. Thus, millions of acres of public lands were sold at prices of $1–$2 per acre, and homesteaders claimed more than 381 million acres between 1891 and 1935

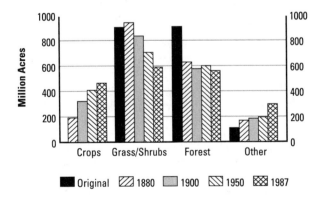

Figure 1. Trends in major land uses in the 48 contiguous United States, 1600–1987.

by promising to establish a home on the land and farm it for five years.

Railroads that tied the huge land mass together in the 19th century were subsidized by large government grants of land, often consisting of alternate sections (a square unit of land, one mile on each side, containing 640 acres) extending several miles on each side of the railroad right-of-way. These lands, given to the railroads to provide timber for the construction of railroad ties and bridges, today constitute some of the larger private forest landholdings in the western United States.

Just as development remained the major theme driving land use in 20th-century America, transportation was the major determinant of how land use patterns changed. Started in 1954, a massive national defense highway system turned into the familiar freeways that now criss-cross the United States. The new freeways bypassed some rural towns, virtually assuring their demise, and created crossroad interchanges where new collections of gas stations, fast-food stores, and motels sprang up to become the "new towns" in rural areas.

Agriculture, no longer tied to river barges or rail cars for shipment of its goods to market, became more specialized and centralized in large-scale operations that could take economic advantage of the new transportation network. Refrigerated trucks carried produce from Florida, Texas, California, or Mexico to northern cities, making year-round fresh fruits and vegetables available to most American tables, and making northern vegetable farms obsolete because they could only produce on a seasonal basis. Prime northern cropland, no longer economically competitive in agriculture, reverted to pasture or forest if it was not immediately taken up for housing or other urban land uses.

As the 20th century draws to a close, transportation systems and economics continue to form the skeletal patterns of changing land use. Around major cities, the "beltways" created to hasten automobile travel around congested downtowns have turned into "ring cities," where new metropolitan centers now compete with the central core, both in terms of commerce and congestion. Mass transit systems, such as the Bay Area Rapid Transit (BART) system in San Francisco and the Washington, D.C., Metro, have defined new commuting, living, and development patterns.

People the world over have been drawn to cities from the country, and this has dramatically affected land use patterns. In the United States, for example, over 90% of the people lived on farms in 1790. By 1920 that had dropped to less than 30%, and today it is slightly over 2%. Rural states, particularly in the center of the continent, have seen populations drop in recent decades, while urban areas, particularly those in the South and West, have grown.

In a reversal of earlier trends, vegetable and specialty-crop agriculture is again emerging in some northern urban regions, drawn back by the increased highway transportation costs associated with the fourfold rise in petroleum costs since 1970, and an added willingness on the part of consumers to pay higher prices for fresh produce. Lands that were abandoned or little-used pasture and woodland a decade ago are now being reclaimed as higher-value croplands.

Another important factor in determining land use in the waning years of the 20th century is the increasing public concern for the environmental integrity of land and water resources. Farmers, grazers, developers, and foresters alike are being challenged to demonstrate that their use of the land is sustainable, not just in economic terms, but in ecological terms as well. Wetlands were once viewed as worthless, and it was considered good business to drain them for crop production or fill them in to support development. Today, wetlands are recognized as important systems upon which other ecosystems depend, and their destruction has been slowed by increasingly strict regulation. Prime croplands, even when they are also prized as homesites, are recognized as a rare and precious resource that can feed generations of people yet unborn if they are kept in agriculture.

Environmental integrity represents a new ideology that challenges development as the dominant theme controlling land use in the United States. Such a challenge is not easily resolved, and intense battles between environmentalists and developers over the proper use of disputed land are common. While it sometimes seems that there are only two points of view involved (growth versus no-growth) the real outcomes fall somewhere between, as the new and old ideas blend, the land and its limits are considered, and new technologies and economic pressures sway the balance. The outcome is not predictable, except to say that land use will remain one of the most important and contentious aspects of the relationship of humans to their environment.

R. NEIL SAMPSON

For Further Reading: Stuart Chase, *Rich Land, Poor Land* (1936); Walter C. Lowdermilk, *Conquest of the Land Through 7,000 Years* (1953); and R. Neil Sampson, *Farmland or Wasteland: A Time to Choose* (1981).

LANGUAGE

Human language is a system of abstract signs that enables its users to represent, define, create, and manipulate the social world by means of strings of sounds, which in turn can be represented visually. The ability to use linguistic signs is unique to the human species and is one of the most important attributes that separate humans from other animals and from machines. All systems of communication—from animal cries to road signs to gestures—consist of signs, i.e., concrete or abstract entities (objects, actions, ideas) that can stand for other entities. As a system of signs, language is extremely adaptable, complex, and flexible, which distinguishes it from other sign systems. Gestures can convey only a limited range of meaning and must frequently co-occur with linguistic signs to acquire precise meaning. However, in contrast to gestures, the sign-language systems with which deaf people communicate are on a par with verbal language in terms of their complexity and flexibility. Animal communication is much more dependent on context for its meaning than human language, and does not allow animals to be as creative and analytic as humans are with language. Computer "languages" are comparatively simple systems of representation confined to a small sub-range of the communicative capacity of human language.

The ability to speak evolved in humans between 30,000 and 100,000 years ago, although even this wide range of dates is open to controversy. Difficulties arise in dating precisely the emergence of language because language is a predominantly abstract system, and the ability to use it has few physiological correlates. While human language as we know it today requires that the jaw, mouth cavity, and larynx be structured in a certain manner and that certain areas of the brain can develop enough, the capacity to develop language is distinct from the actual ability to use it. Thus physical anthropologists and paleontologists can only work with inconclusive criteria for deciding when language might have emerged in human evolution.

The ability to acquire and use a spoken or sign language, which all humans share, must be distinguished from literacy, or the ability to represent language visually. While verbal communication emerged in humans as part of the general evolution of human intellectual capacity, literacy is an invention, which only dates to about 3000 B.C. Furthermore, writing is a secondary sign system: Written signs stand for spoken signs, which in turn represent the non-linguistic world. Barring physiological or cognitive impairment, all children acquire the ability to speak relatively effortlessly with minimal coaching from adults; in contrast, learning to write is a conscious process that requires a more or less formal pedagogical infrastructure, which is not available in all social groups.

There are about 5,000 different languages in the world today, of which only a handful are spoken by large numbers of people. Most of the world's languages are spoken in very small speech communities and in highly confined areas. But the structural complexity of a language is completely unrelated to the size of the community that speaks that language or to the technological and socio-political complexity of that community. Some of the most complex syntax in the world is found in languages spoken by tiny communities in Papua New Guinea whose traditional technology is basic. Linguistic complexity itself is a difficult notion to define precisely in that there are many different components to language (e.g., vocabulary structure, syntax, social uses), each of which can exhibit different degrees of elaboration, and can do so in different ways. Consider vocabulary size, for example; while fishing people are likely to develop a more elaborate vocabulary to refer to fish species than a group of horticulturalists, vocabulary complexity does not invariably correlate with the cognitive or cultural salience of the objects or ideas that words denote. Thus, while humans everywhere are exposed to a comparable range of colors in the social and natural environments in which they live, some languages have very small color vocabularies, while fine color distinctions are encoded in others. However, color vocabularies are not structured haphazardly, but are governed by specific structuring rules that apply to all languages. Universal rules can be shown to affect many other aspects of linguistic structure, such as syntax and sound structure.

All languages change over time, a process that eventually leads to the universal situation whereby earlier forms of a language are virtually unrecognizable and unintelligible to speakers of latter forms. For example, speakers of 20th-century English must learn Old English as a foreign language in order to read a text in that language, even though contemporary English is a direct descendant. Language change is a natural phenomenon related to many different factors—some of which are of a linguistic nature—while others are related in a very complex manner to social and cultural change. Change affects all areas of language structure and use, not only the most obvious like the meaning of words and the sound structure of language, but also syntax, rules of stylistics, and the social evaluation of certain ways of speaking as more or less acceptable. While some manifestations of language change are random and unpredictable, most linguistic change follows systematic principles that govern the historical evolution of all of the world's languages.

In short, linguists, anthropologists, and psychologists have discovered both universality and diversity in language. Where commonality and heterogeneity are attested, which of the two is more important is the subject of considerable disagreement between disciplines and theories. However, most scientific approaches recognize that language is both a cognitive phenomenon and a social tool used in the conduct of social interactions. Indeed, the development of language requires certain cognitive prerequisites. At the same time, language arose in prehistory to answer social needs for efficient communication, and it emerges in every child in response to an interactional environment; thus children who are never exposed to language fail to develop the ability to speak. These two facets of language constantly interact and are often difficult to isolate.

NIKO BESNIER

For Further Reading: David Crystal, *The Cambridge Encyclopedia of Language* (1987); Edward Finegan and Niko Besnier, *Language: Its Structure and Use* (1989); Philip Lieberman, *Uniquely Human: The Evolution of Speech, Thought, and Selfless Behavior* (1991).
See also Anthropology; Communication; Risk Communication.

LAW, ENERGY

As first conceived, the role of federal and state utility commissions was to protect the public interest with respect to energy pricing, to ensure reliable service to utility customers, and to prevent unreasonable profit by what were essentially monopolistic entities. While this mandate is undiminished, the increased emphasis on environmental protection and the preservation of natural resources—and the obvious relationship between energy and the environment—have widened the regulatory scope.

State and federal regulators have exhibited increasing concern and control regarding the environmental impact of energy production and transmission facilities. This shift both reflects today's social agenda and drives it. Events such as the Arab oil embargoes of the 1970s and more recently the Persian Gulf war have demonstrated the necessity of encouraging alternatives to oil-fired electric generation. By encouraging the use of alternative fuels—whether natural gas, coal, nuclear, hydro, or any of the emerging technologies—regulators have established an environmental agenda for the nation's utilities.

Fuel for electric generation is by no means the only item on that agenda. A variety of regulations have been promulgated to require extensive environmental review prior to the construction of either generating or transmission facilities. In the early 1960s, the permitting requirements for generating facilities were fairly simple. A Federal Power Commission (FPC) construction license for hydroelectric projects and a municipal building permit would generally suffice, with little en-

vironmental review. The regulatory review associated with the construction of an electric generating facility today, however, focuses as much on environmental impact as it does on necessity and cost.

This focus arose, in part, from litigation in the mid-1960s, when Consolidated Edison Company of New York was granted a Federal Power Commission license to build a pumped storage hydroelectric project at Storm King Mountain on the Hudson River. In remanding the matter back to the FPC in response to a suit filed by the Scenic Hudson Preservation Conference, the New York Court of Appeals ruled, among other things, that the FPC's renewed proceedings "must include as a basic concern the preservation of natural beauty and of historical national shrines, keeping in mind that, in our affluent society, the cost of a project is only one of the several factors to be considered." After years of further litigation, the Storm King Project was ultimately canceled, but its impact was far-reaching as regulators and utilities nationwide recognized the emerging role of environmental protection in the siting process.

Today, a host of regulations control the siting process. The National Environmental Policy Act (NEPA), enacted in 1969, requires federal agencies to prepare an Environmental Impact Statement prior to the commencement of any major federal action that will significantly affect the environment. If construction of a generating facility requires federal permits, a NEPA Statement also may be required.

The Federal Clean Air Act Amendments of 1990 established new standards for allowable emissions of pollutants into the atmosphere, and delegated to the states the responsibility for created implementation plans to achieve the Act's goals. Utilities, with their fossil fuel plants, are significantly affected by this legislation, and as the matter enters into the states' regulatory arena the definition and implementation of standards will have a major impact on utilities and their customers.

Of course, air quality is not the only environmental standard for which utilities are under regulatory control. The Federal Clean Water Act requires that any applicant for a federal license to construct or operate a facility that may discharge into the navigable waters of the U.S. must provide the licensing agency a certification from the state in which the discharge originates that the facility would comply with the state's applicable water quality standards. Since most electric generating facilities discharge into navigable waters, they require such certificates.

In addition to federally-driven regulatory requirements, most states now have legislation requiring the preparation of an Environmental Impact Statement for any action that may have a significant effect on the environment. Clearly, the construction of generating facilities must meet these state requirements.

It is axiomatic, of course, that reducing energy production has great environmental benefits in the form of resource preservation and pollution reduction. The trend to focus on demand reduction began in the early 1980s, when conservation, now generally called demand-side management (DSM), emerged in some circumstances as an attractive alternative to the construction of generating facilities. Public utility commissions and utilities throughout the nation have now recognized the value of using demand-side management measures as an adjunct to—or instead of—continued construction of power plants.

In New York, for example, since 1984 electric utilities have filed, and the Public Service Commission has approved, a series of DSM plans. Generally speaking, the utilities' plans have grown increasingly ambitious. For example, in the New York plans filed in 1990, utilities projected that they would reduce 1992 energy consumption from what it would otherwise have been by over 1.6 million megawatt-hours—the equivalent of over 260,000 homes!

Completing the spectrum of supply and demand alternatives is the increased use of integrated resource planning, by which utilities and regulators consider all available options to determine the most cost-effective and environmentally sound means of fulfilling their mandates. Thus, utility-owned facilities, purchased power from other utilities, purchased power from non-utility generators, and demand-side management are all considered. In considering the choices, greater weight is now being given by regulators to environmental externalities: how much of a cost premium for energy is acceptable for the enhancement of environmental quality. Regulators are increasingly recognizing that the reduction of environmental impact has economic value, and that a utility should be allowed to recover the costs of providing energy in a way that reduces or avoids pollution—even if that way may cost more than alternatives. Whether it is burning cleaner fuel or relying more heavily on demand-side management, many utilities are now being allowed to pay higher prices for generating or securing energy if there is a demonstrable environmental benefit and if the cost is considered prudent.

The trend is unmistakable and irreversible: The environmental impact of energy production will continue to be a driving force as state and federal regulators and utilities fulfill their mandates to provide safe, reliable energy at the lowest possible cost.

VICTOR A. ROQUE

For Further Reading: Charles F. Phillips, Jr., *The Regulation of Public Utilities* (1988); A. J. G. Priest, ed., *Principles of*

Public Utility Regulation (1969); Nicholas A. Robinson, ed., *New York Environmental Law Handbook* (1988).

See also Clean Air Act; Clean Water Act; Demand; Electric Utility Industry; Energy; Energy Conservation; Environmental Impact Statement; Fuel, Future; Law, Environmental; National Environmental Policy Act of 1969; Supply.

LAW, ENVIRONMENTAL: ENFORCEMENT

Environmental laws and regulations are based on the setting of standards and the implementation and enforcement of those standards. Standards set by the congressional and executive branches reflect the goals sought in abating pollution. The Clean Air Act, for example, includes many specific air quality standards that limit the emissions of hazardous air pollutants. Standards alone, however, are merely ideals. A stated goal is not a realized goal until activities designed to reach that goal are implemented and enforced.

Enforcement often depends on whether the industry being regulated is able and willing to comply with the standards. Many factors impede voluntary compliance. Congress may choose to send a message to industry by setting stringent standards that require the purchase of advanced technology. If the technology is not available within the given time frame, the industry cannot meet the standards. In addition, standards are usually set according to the scientific information available at the time. For example, technical data are used to determine what levels of various pollutants are safe. This information is often imprecise and unreliable, since the risks of many pollutants depend on local environmental factors. Standards are sometimes set based on technical information without any consideration of the costs to industry or whether the benefits are worth the costs. Industries often legally oppose the standards and willfully neglect to obey them, charging that the standards are scientifically faulty and too costly.

Since voluntary compliance is the most efficient and quickest means of assuring that pollution standards are met, enforcement officials are often flexible and generous with industries attempting to comply. Officials would rather see some pollution control than none. However, when regulated interests fail to voluntarily comply with standards, officials rely on a variety of enforcement methods that have increasing levels of severity. Penalties range from the assessment of a small fine to litigation and criminal penalties. Yet litigation is slow and expensive, and the desire of the enforcement agency is to curb pollution as quickly and efficiently as possible, not to tie up all its resources in court proceedings. Therefore, enforcement officials commonly bargain with the regulated interest on how best to achieve the standards, with each side compromising on the use of required technologies or on meeting certain deadlines.

As with any government procedure, enforcement is filled with opportunities for political influence. When laws are made, the wording and intent of certain clauses may be unclear; as scientific findings change, technical requirements for pollution control may change also; and when standards are too rigorous, compliance deadlines are often moved back—as has occurred many times with the Clean Air Act and the Clean Water Act. These elements all provide the opportunity for political groups to influence the effectiveness of the enforcement of environmental laws.

LAW, ENVIRONMENTAL: EVOLUTION

The Environmental Movement out of which modern Environmental Policy and Environmental Law have evolved began in 1962 with the publication of Rachel Carson's *Silent Spring*. The Movement has sought to implement as current public policy and law important aspects of the morality and political philosophy of the English Lake Poets, Thoreau, Walt Whitman, John Muir, John Burroughs and Aldo Leopold. Its precursor, the Conservation Movement of the early 20th century, led by Theodore Roosevelt and Gifford Pinchot, had as its narrower objective the "wise use" of natural resources.

Shortly after the publication of *Silent Spring* a controversy arose over plans by the Consolidated Edison Company of New York for a pumped storage hydroelectric power plant at Storm King Mountain, approximately 50 miles north of New York City, on the western shore of the Hudson River. When the Company began the proceedings before the Federal Power Commission to secure the necessary permit, residents on the mountain, together with members of clubs of hikers in the nearby park areas of the Hudson Highlands, affiliated in the New York-New Jersey Trails Conference, organized the Scenic Hudson Preservation Conference to oppose the permit in the hearings before the Commission. When it granted the permit they appealed to the United States Court of Appeals for the Second Circuit.

The historic 1965 decision of the Court on that appeal marked the beginning of modern environmental law, although that term did not come into general use until approximately five years later. The Court reversed the grant of the permit and remanded the matter to the Commission. In doing so it first granted "standing in court" to Scenic Hudson as an "aggrieved party," able to complain in the Court of errors by the Commission in the grant of the permit. The Court ordered new hearings by the Commission, stating: "The renewed proceedings must include as a basic concern the preservation of natural beauty and of national historic shrines, keeping in mind that, in our affluent society, the cost of a project is only one of several factors to be considered."

Launched by the morality and social science of *Silent Spring* and the law of the Scenic Hudson case, particularly that of the standing in court of an environmental group to represent the public interest, environmental movement advocates were engaged between 1965 and 1970 in a number of other historic controversies, including those involving an oil spill on the California Coast at Santa Barbara; a proposal for dams in the Grand Canyon of the Colorado River as part of a massive water plan for the Southwest; an attempt by a German chemical company to build a dye plant on the coast of South Carolina, opposite Hilton Head Island; a proposal by Walt Disney Enterprises to develop a ski center in Mineral King Valley in the Sierra Nevada Mountains; and highway projects that would have paved over parkland in the cities of San Antonio and Memphis and been built on landfill along the Hudson River.

During the same period a number of widely read books and articles appeared that focused on much broader and deeper problems than those involved in the earlier conservation movement. Environmental associations began to bring other lawsuits based on existing statutes, particularly the Rivers and Harbors Act of 1899, under which the Army Corps of Engineers was empowered to grant permits for structures in navigable waterways. In most of the early legal actions environmental advocates sought to expand and deepen the scope of the judicial review of administrative action, in order to compel agencies with developmental missions to consider environmental factors.

The courts were looked upon as the instrument of reform and the agencies, looking with skepticism and fear upon persons asserting environmental rights, resisted expansion of the power of the courts. The agencies included the Federal Power Commission, the Army Corps of Engineers, and the Federal Department of Transportation, the developers of or granters of permits for hydroelectric plants, dams, and highways (particularly those of the Interstate Highway System).

A 1969 case concerning a proposed expressway along the Hudson River was the first major environmental suit resulting in the permanent enjoining of a major project. In it and other early suits Federal District Courts and Circuit Courts of Appeals confirmed the standing of citizen environmental groups. It was confirmed by the Supreme Court in 1972 and 1973.

By late 1969 a considerable number of the nation's

law schools were teaching environmental law as a separate subject. In the fall of 1969 an important conference, sponsored by the Conservation Foundation, was held at Airlee House in Virginia. It was the first important assembly of practicing attorneys, law school faculty, environmental organization leaders, and other citizen activists involved in this emerging field of law. Papers were presented on the several aspects of the still very limited body of environmental law. An important result of the conference was the creation of the Environmental Law Institute, for the purpose of advancing the growth and study of environmental law by, among other activities, the publication of an Environmental Law Reporter.

Another important development in the evolution of environmental law in the period was the creation of three major environmental public interest law firms: the Environmental Defense Fund (EDF), the Natural Resources Defense Council (NRDC) and the Sierra Club Legal Defense Fund (SCLDF). EDF was started with a grant of funds by the National Audubon Society to bring legal actions to restrict the use of DDT. NRDC was started by leaders of Scenic Hudson. SCLDF was at first a division of the Sierra Club; it later became a separate corporation, representing the Club and other environmental groups in legal proceedings.

During the period 1965–1970 the principal environmental organizations greatly expanded their political activities. The Sierra Club began to appear in legal proceedings, particularly the second round of Federal Power Commission hearings in the Storm King Mountain controversy.

The first of the major environmental statutes, the National Environmental Policy Act (NEPA), became effective on January 1, 1970. It created a Presidential Council on Environmental Quality (CEQ), declared broad environmental findings and policies and mandated their implementation by all Federal agencies "to the fullest extent possible." It required the preparation of environmental impact statements (EISs) for all major Federal projects significantly affecting the "human environment."

In April 1970 President Nixon created by executive order the Environmental Protection Agency. During that year an important movement took place in several states to revamp traditional, narrowly-oriented Conservation or Fish and Game Departments or Commissions to become environmental protection agencies with substantially broadened powers and responsibilities.

Throughout much of 1969 the Senate Environment and Public Works Committee, led by Senator Edmund Muskie of Maine, worked on the drafting of the second major environmental Federal statute—the Clean Air Act of 1970. The Act provided for air quality standards and their attainment by "State Implementation Plans," approved by EPA. Its enforcement provisions included a "Citizen's Suit" section, granting standing in Court under certain conditions to enforce such standards and enjoin other unlawful actions by EPA.

By the end of 1970 environmental policy was acknowledged to be one of the most important areas of public policy. "Environmental law" became a household phrase and those courses became part of the standard curricula at most of the nation's law schools.

Spurred by the acceptance of environmental protection as a national policy, Congress in the 1970s and 1980s enacted a number of additional major environmental statutes, including the Clean Water Act of 1972, the Toxic Substances Control Act, the Noise Pollution Control Act, the Resources Conservation and Recovery Act, the Endangered Species Act, the Comprehensive Environmental Response Compensation and Liability Act (Superfund), the Coastal Zone Management Act and the Safe Drinking Water Act. In the great majority of Federal environmental statutes Congress included a Citizen's Suit provision fashioned after that of the 1970 Clean Air Act.

The Federal statutes were complemented and followed by state laws, particularly those dealing with air and water pollution. In each of those fields, under the Clean Air Act and Clean Water Act, respectively, states were required to adopt programs of regulation consistent with EPA.

Beginning also about 1970, by legislation or court decision, most states granted standing in state courts to citizens and environmental groups similar to that accorded by the Scenic Hudson case and the later rulings in Federal courts. A number of states, including most of the more populous ones, also began to pass State Environmental Policy Acts ("little NEPAs"), following closely the Federal Act. The little NEPAs declared broad environmental policies and objectives and created environmental quality councils to adopt more detailed regulations and govern the implementation of the acts, or vested the governance in existing state environmental departments. They required EISs for state agencies, and in most cases also municipal agencies, for projects and permits that might result in significant environmental impacts.

The grants of standing in Federal and State courts and the application of NEPA and little NEPAs to Federal, State and municipal projects have profoundly enhanced the power of opponents of governmental actions impacting substantially upon the environment to enjoin or delay such actions. The power extends in many states to traditional zoning and other land use actions in the urban, suburban and rural communities that have land use controls.

Because the "environment" is the "human environment," including that of urban and suburban areas, NEPA, little NEPAs, and other environmental laws have sometimes been used in efforts to insulate existing communities from changes. The changes may be construction of housing for minorities or the relatively poor, public facilities to serve disadvantaged groups, or waste treatment or other undesirable facilities. The power to oppose them by court action and delay or enjoin the projects may involve what is now often referred to as the "NIMBY" (not in my back yard) problem.

The additional processing of governmental actions under NEPA, little NEPAs, and other environmental laws often adds appreciably to the time required for most such actions. In many cases courts have enjoined projects temporarily—to require EISs or the correction of defective EISs or other action under environmental laws—and thereafter permitted such projects to proceed. Very few projects have been permanently enjoined. A large number have been modified to lessen their environmental impacts and some, after such temporary injunctions, have been abandoned.

There are differing views as to whether the effects of environmental laws are a net benefit or detriment. There is, however, no real force in any political effort to repeal or significantly weaken the controls of such laws. There are important movements to expedite and simplify the processing under such laws and to reduce the number of controversies which reach the courts.

The NIMBY problem is closely related to what has often been referred to as the "elitism" problem. Many of the early major controversies were over the protection of the scenic beauty or wildness of areas far from centers of population. Environmental advocates have constantly been charged with caring more for animals and trees than for people. The membership of the mainline environmental organizations is primarily executive or professional, college educated, and white, much out of proportion to the population as a whole.

The elitism aspect of the Environmental Movement has occasionally resulted in differences with civil rights and minority groups. Beginning, however, in the late 1970s and increasingly through the 1980s, the focus of environmental activism and the development of the law has been upon the total environment and the protection of health and life, particularly from toxic materials in the workplace and the inner cities. Environmental organizations have made major efforts to broaden their membership to include more minorities, working-class people, and labor union groups. The mainline groups largely support the move for "environmental equity," which seeks to reduce the location of health-threatening and other undesirable facilities in minority and poor neighborhoods. Newer and more radical environmental organizations often depart from the traditional methods of lobbying and court action; alternative tactics include barricades, other civil disobedience, and destruction of property.

As the number of environmental statutes and regulations have increased, more and more questions have been raised over whether, and to what extent, "command and control" methods are superior to economic or market mechanisms. For example, the Clean Air Act amendments of 1990 allow utility companies that reduce emissions of pollutants below the levels permitted to sell the allowances not needed by them to others that have difficulty meeting their limits.

Another question that is receiving increasing attention is, "How clean is clean?" As laws and regulations mandate a more pollution-free environment, the relative costs of the stricter and stricter standards increase sharply. The issues are similar to those present in the evolution of a national health program, which cannot provide for every treatment available for every person who can be helped by it.

The diversification of the Environmental Movement and the absence of purely ideologic solutions are products of its maturity. The "messianic" phase of the Movement passed when it became clear that the principal problems addressed by it could not be resolved by ideology alone. The development of Environmental Policy and Law requires balancing many factors and interests—including costs, economic development, the rights of minorities, and a complex of constitutional and other legal issues.

DAVID SIVE

For Further Reading: Philip Shabecoff, *A Fierce Green Fire* (1993); Council on Environmental Quality, *Environmental Quality* (annual reports); Joseph Sax, *Defending the Environment* (1970).

See also Clean Air Act; Clean Water Act; Economic Incentives; Environmental Impact Statement; Environmental Movement; Environmental Organizations; Environmental Protection Agency, U.S.; Labor Movement; Law, Environmental: Present Scope; Law, Environmental: Relationship to Other Bodies of Law; Law Firms, Public Interest; National Environmental Policy Act of 1969; NIMBY; Tradable Permits; United States Government.

LAW, ENVIRONMENTAL: PRESENT SCOPE

As other bodies of public law administered largely by governmental agencies, Environmental Law is in three parts: statutes (Federal, state and municipal); administrative rules and regulations; and court and administrative decisions. The legal authorities upon which administrative and court decisions are made may also include critical materials such as treatises, law review articles and notes and other published works, particularly in a relatively new field of law such as Environmental Law. The law is both civil and criminal. Each of the major federal statutes and most of the state and municipal laws and ordinances contain criminal provisions as well as providing for civil penalties which may be levied against violators of the statutes.

Most of the federal statutes are enacted under Congress' constitutional power to regulate interstate and foreign commerce. The limits of that constitutional power determine the reach of the governance of waters and wetlands. This is because the Water Pollution Act of 1972 (known as the Clean Water Act) speaks of "Waters of the United States including the territorial seas." The "Waters" referred to are those included under the broadest permissible interpretation of the interstate commerce clause. The Act thus governs much of the nation's wetlands.

The Clean Air Act governs the ambient air everywhere, for which the Environmental Protection Administrator sets standards for air pollutants. The standards are enforced primarily by the states under state implementation plans approved by the Federal Environmental Protection Administrator (EPA). Neither the Clean Water Act nor the Clean Air Act zones or otherwise directly governs the uses of land. Zoning and other direct land use laws are mainly by municipalities. The principal exceptions are the states of Vermont and Hawaii. The state of Florida governs critical areas and New York State's Adirondack Park Agency governs the large Adirondack Mountain area.

Under the Clean Air Act, however, the approval by EPA of state implementation plans requires that the plans include transportation controls. Such controls profoundly, albeit indirectly, govern the uses of lands. Under the Clean Water Act states must promulgate area-wide waste treatment plans for approval by EPA. The plans must prescribe a regulatory program for both *point sources*, discrete points such as pipes emptying into waterways, and *non-point sources* (such as agricultural lands). The governance of area-wide waste treatment plans is thus an important land use planning aspect of the Federal Clean Water Act.

Under the property clause of Article IV of the federal Constitution, Congress is empowered to govern "the Territory or other Property belonging to the United States." Federal lands aggregate approximately 725 million acres (one half in Alaska). They include national parks, national monuments, wildlife preserves and the large areas of the West administered by the Bureau of Public Lands. Environmental law includes the law of environmental protection under pre-existing federal land laws.

Of the federal statutes and their state analogues, the most basic and broadest in scope are the National Environmental Policy Act (NEPA) and the state analogues, so-called little NEPAs. The scope and importance of NEPA were not realized by its principal sponsor, Senator Henry Jackson, or by Congress in enacting or President Nixon in signing it. Almost immediately upon its becoming effective, however, on January 1, 1970, its principal action forcing provision, the requirement of an environmental impact statement (EIS), to be included "in every recommendation or report or proposals for legislation and other major Federal actions significantly affecting the quality of the human environment," became the subject of many court actions. It was quickly established that the term *actions* included not only those directly undertaken by federal agencies but permit actions and grants-in-aid.

Most of the little NEPAs, enacted now in sixteen states, including those with large populations and the principal commercial states, also apply to permit actions and grants by the states and municipal agencies, as well as projects directly undertaken by them. The actions include the great bulk of significant zoning and other land use actions governed by older land use laws. The environmental assessment process of NEPA and little NEPAs law thus extends to the vast majority of all direct governmental actions and non-governmental actions and programs of potential environmental significance.

Environmental law has created the business and institution of environmental auditing, which may be defined as the review by a business organization of its operations and facilities to determine whether it meets all of the requirements of all applicable environmental laws and regulations. The frequency and importance of such auditing have expanded sharply because of the liabilities of land owners and operators of facilities on lands which may harbor hazardous wastes, particularly under the Comprehensive Environmental Response, Compensation and Liability Act (Superfund). The liabilities, which may be independent of fault and extend to the maker of loans secured by property containing such wastes, may be for the whole of the costs of cleanup, though with the possibility of securing contributions from other responsible persons. Environmental audits, and the assignment of liabilities in case they disclose any adverse conditions, are provided for in vast numbers of sale and lending transactions.

DAVID SIVE

For Further Reading: Environmental Quality (23rd Annual Report of The Council on Environmental Quality) (1993);

Frank P. Grad, *Treatise on Environmental Law* (1992); William H. Rodgers, Jr., *Environmental Law* (1986).

LAW, ENVIRONMENTAL: RELATIONSHIP TO OTHER BODIES OF LAW

The relationship of environmental law to other bodies of law requires first a definition. A purely literal definition—"the law governing the 'environment'"—is clearly much too broad if one starts with a comprehensive definition of *environment*—the sum of the physical, living, and social factors and conditions directly or indirectly affecting the development, life, and activities of organisms—and includes all of the law governing it. A far more useful definition, derived from the objectives of the EPA, defines environmental law as the modern body of law which protects natural ecosystems and human health and welfare from pollution or its effects.

The adjective *modern* is critical. Many other and older bodies of law have governed natural and cultural resources. They include administrative law, constitutional law, the nuisance aspects of the law of torts, zoning and land-use law, natural resources law, the law of water rights, and public utilities law. The term "Environmental Law" first came into use in the late 1960s. It was the product of the environmental movement which began with Rachel Carson's *Silent Spring* in 1962. The beginning of environmental law was in early court decisions in the mid and late 1960s which were really decided as aspects of administrative law. They expanded the scope of judicial review of administrative action by requiring the consideration of the environmental effects of agencies' actions.

Much of environmental law to this day is administrative law. The basic Federal Administrative Procedure Act, enacted in 1946 to govern the practices of administrative agencies and the judicial review of their actions, applies to many actions of the Environmental Protection Agency, which administers and enforces most of the environmental statutes, beginning with the Clean Air Act of 1970. More than half of all environmental law cases involve judicial review of actions of EPA or other agencies rendering environmental decisions.

Much of environmental law is constitutional law. The first issue in the development of environmental law, that of environmentalists' standing in court, is in large part an issue of the limits of the judicial power under the "Cases and Controversy" clause of Article III of the federal Constitution. The regulation of the uses of property under modern Environmental Laws, if it goes too far, may constitute an "inverse condemna-

tion" and "taking" under the due process clauses of the Fifth and Fourteenth Amendments. The dual regulation by the federal government and the states of air, water, and land pollution poses issues of federalism and the division of powers. The administrative processes of the many different actions of environmental regulatory agencies may pose issues of procedural due process. Problems of the separation of powers between the legislative and the judiciary have arisen frequently because many important environmental controversies are dealt with simultaneously in the two *fora*. Environmental law may also involve the limits of Congress' interstate commerce power, sovereign immunity under the Eleventh Amendment, the equal protection clause under the Fourteenth Amendment and the reserved rights of states under the Tenth Amendment.

Nuisance is a tort. The centuries' old law of nuisance is part of the law of torts. Nuisance law has often been referred to as the common-law zoning of real property. Cases and materials on the law of nuisance comprise a section of any book covering environmental law. The development of environmental law has shaped the law of nuisance to deal with modern environmental issues. A classic 1963 case of nuisance law, concerning air pollution from a cement plant, is studied as an important environmental law case even though it predated by several years the general use of the term.

Zoning and other land-use law also govern the use of natural and cultural resources. In all but a small number of states, zoning is primarily by municipal ordinances. Legislative bodies enact and amend zoning regulations; planning boards and zoning boards of appeals rule on site plan and subdivision approvals, variances, and special use permits; and courts review these actions. The thrust was mainly toward development until environmental advocates began to intervene in and institute proceedings to prevent or limit development. This could not be done until they achieved standing in state courts, beginning around 1970, after the grant of standing in federal courts, beginning in 1965.

Since 1970 much of the province of older zoning and land-use law and of its practitioners has become that of environmental law and environmental lawyers. This has occurred in large part through the superimposition of the environmental impact assessment process on many zoning and other land-use actions, under state and municipal acts and ordinances fashioned after the environmental impact statement process of the National Environmental Policy Act (NEPA).

Environmental law may also be deemed to be natural resources law. The distinctions are historical and ideological. American natural resources law is concerned primarily with the commercial uses of natural resources, in large part the public lands of the West.

The extraction of oil and gas, the mining of minerals, grazing, and lumber cutting are governed by rules of natural resources law. As environmental advocates, however, beginning in the 1960s, intervened in administrative and court decisions seeking to limit the use of resources, in large part to preserve wilderness and other critical areas, the law thus created became part of the body of ideology-driven environmental law.

The law of water rights, particularly of riparian rights, is likewise a very old body of law. Until the advent of the environmental movement the rights dealt with were primarily those of private users. The assertion, by government agencies and by private parties asserting a public interest, of claims to the protection and conservation of water bodies, has created an important part of environmental law.

Public utilities law has for many years been a separate subject of study in law schools and its practice a recognized specialty. The permit process for hydroelectric plants falls under that law. Since modern environmental law began with the upholding in 1965 of a challenge by an environmental association of a pumped-storage hydroelectric plant, much of the field of public utilities law has become environmental law. That aspect of environmental law stems from the revision by both statutes and court decisions of the law governing the permitting and siting of energy plants, legal requirements for energy conservation, and the regulation of air pollution.

DAVID SIVE

For Further Reading: Environmental Quality (23rd Annual Report of The Council on Environmental Quality) (1993); Frank P. Grad, *Treatise on Environmental Law* (1992); William H. Rodgers, Jr., *Environmental Law* (1986).

LAW, INTERNATIONAL ENVIRONMENTAL

The body of international environmental law has grown rapidly since the early 1970s and increasingly expresses human attempts to regulate their impacts on the Earth's biosphere—the complex matrix of water, air, land, and living things that supports all life. The reason lies in the explosive growth of human population and human economic activities. The world's population is about 5.2 billion people and is increasing by about one billion people per decade. Economic activity has grown even faster. It is estimated that world economic activity (the equivalent of a nation's Gross National Product) tripled between 1950 and 1990. It is predicted to triple again between 1990 and 2040, a ninefold increase in less than a century.

The first human impacts on the environment that were considered serious enough to deserve international attention were threats to migratory wildlife. Between the late 19th and the mid-20th century, countries entered into a number of agreements to protect migratory species such as fur seals, whales, fish, and birds. In 1954, growing coastal pollution caused nations to negotiate a treaty aimed at controlling marine oil pollution. In 1959, many countries signed a far-reaching agreement designating Antarctica as an international scientific reserve. Because of the need for international actions to deal with environmental problems that could not be solved by any one nation, the U.N. Conference on the Human Environment was held in 1972 in Stockholm, Sweden. The Conference and the best-selling "conference book," *Only One Earth*, greatly increased popular and governmental awareness of international environmental problems and the need for cooperative actions to address them. More concretely, the Conference established the framework for much of the international environmental law that exists today.

In the international sphere, a nation is legally bound to do something only if it formally agrees to do so by becoming a party to a treaty or other international agreement. There is no higher power that can compel nations to act. However, international environmental law has dimensions that extend beyond legally binding agreements. As shown by the Stockholm Conference itself, the "hard law" of treaties and other agreements is surrounded by a body of "soft law" that is hortatory rather than legally binding but nevertheless exerts great influence. Because of the consensus-building process that must precede even the beginning of negotiations on an international treaty, some form of soft law commonly foreshadows and makes possible the hard law that follows.

For example, Recommendation 86 of the Stockholm Conference called for the convening of an intergovernmental conference to agree on the text of a treaty to control the dumping of wastes in the oceans. The British Government convened such a conference almost immediately, and the Ocean Dumping Convention was ready for signature on December 29, 1972. On the other hand, Stockholm Conference Recommendation 70 simply called on governments to "be mindful of activities in which there is an appreciable risk of effects on climate." Twenty years later, the issue of climate change had moved high on the international agenda, and a treaty on the subject was signed in June 1992.

Sometimes soft law provisions set in motion a process of study and analysis that is necessary before it is possible to devise a treaty. For example, Stockholm Conference Recommendation 79 called for the strengthening of international studies of the atmosphere.

In addition to the distinction between hard and soft law, international environmental law falls into geographic categories. There is a large body of hard and soft law concerning bilateral relations between neighboring countries, dealing with such problems as transboundary air and water pollution and shipments of hazardous substances. There is a body of law concerning the management of international lakes and rivers. Some of these water management agreements are the basis for important institutions, such as the International Joint Commission which was created by the United States and Canada to help manage the Great Lakes.

International agreements also have created major regional institutions, many of which have growing environmental responsibilities. Examples include the European Community (EC) and the Association of South-East Asian Nations.

The Body of International Environmental Law

The Atmosphere

Since the 1970s, a great deal of attention has been devoted to human impacts on the Earth's atmosphere. The problem of stratospheric ozone depletion due to emissions of manmade chemicals was first identified in 1974. The U.S. and a few other nations took unilateral steps to reduce their emissions of ozone-depleting chemicals before 1980, but most of the other industrialized nations refused to do so at that time. In 1985, the Vienna Convention for the Protection of the Ozone Layer established a framework for dealing with ozone depletion. Spurred in part by the frightening discovery of the "ozone hole" over Antarctica, nations in 1987 added to the Convention the Montreal Protocol. In the spring of 1992, 56 nations endorsed amendments to the Montreal Protocol that would require industrialized nations to phase out their harmful emissions by 1996. Other amendments have created a fund to assist developing nations in making the transition to non-ozone-depleting chemicals.

The other major global atmospheric issue is climate change due to the greenhouse effect. Negotiations on a climate change treaty began in 1990 under the auspices of the U.N. The issue attracted wide interest, and negotiators from 143 nations agreed in May 1992 on the text of a treaty, the U.N. Convention on Climate Change, aimed at stabilizing the concentration of greenhouse gases in the atmosphere in order to minimize global warming. The Convention, which was signed by 155 nations at the Earth Summit in Rio de Janeiro in 1992, requires industrialized nations to limit their emissions of greenhouse gases. However, it leaves up to individual nations the decision as to what those limits will be; as a result, it does not impose firm targets and timetables for reducing greenhouse gas emis-

sions. The Convention states that the obligations of developing nations to control emissions will depend on the transfer of technological and financial resources from industrialized nations. However, the question of how those transfers will occur remains to be resolved by further negotiations.

At the regional level, the most important atmospheric problem is acid rain, which affects much of North America and Europe and is emerging as a concern in developing regions. The International Treaty on Long-Range Transboundary Air Pollution, negotiated under the auspices of the U.N. Economic Commission for Europe, went into effect in 1983. It promotes the international sharing of information and encourages research, but it does not require nations actually to reduce transboundary pollution. In part, this is due to the fact that the causes of acid rain are not yet fully understood.

Oceans and Regional Seas

The oceans have been a subject of concern for international environmental law at least since nations agreed in 1911 on a treaty to protect migratory fur seals. In addition to treaties protecting wildlife, there are many global and regional agreements aimed at preventing marine pollution. Among the most important are the 1972 Ocean Dumping Convention, the 1975 International Convention on the Prevention of Pollution from Ships, and agreements concerning liability and compensation for damage due to oil pollution.

In 1982, after many years of negotiations, 117 nations signed the U.N. Convention on the Law of the Sea. The Convention deals with almost every human use of the oceans, including environmental aspects such as pollution, fishing, and conservation of marine resources. The U.S. and a number of other industrialized nations have refused to become parties to the Convention because they disagree with its provisions concerning deep-sea mining. Nevertheless, the Convention is very influential because even these nations agree with and adhere to most of its provisions.

At the regional level, the Regional Seas Program of the U.N. Environment Program has fostered separate agreements aimed at cleaning up ten marine regions: the Mediterranean Sea, the Persian Gulf, the Red Sea and Gulf of Aden, the Kuwait region, the Wider Caribbean, the East African region, the West and Central African region, the East Asian seas, the South Pacific, and the Southeast Pacific. Outside of this framework, important agreements have been negotiated for cleaning up the North Sea and the Baltic Sea.

Toxic Substances and Hazardous Wastes

During the 1970s, many nations adopted domestic laws governing the transportation and disposal of hazardous materials. Concern grew about the effects of international shipments of hazardous wastes. In 1989, nations completed negotiation of the Basel Convention on the Control of Transboundary Movements of Hazardous Wastes and Their Disposal. The Basel Convention, which went into effect in May 1992, prohibits the export of hazardous wastes unless there is a bilateral agreement between the exporting and importing countries that ensures safe disposal of the wastes.

Wildlife and Natural Areas

International environmental law began in the late 19th and early 20th centuries with agreements to protect migratory species of birds, seals, fish, and whales. A treaty protecting polar bears was agreed to in 1976. The wildlife treaty that has produced the greatest controversy is the International Whaling Convention, first agreed to in 1931. After years of argument, the parties to the Convention agreed in 1982 to a moratorium on all commercial whaling. However, in 1992 some nations, dissatisfied with the moratorium, threatened to resume commercial whaling.

A major advance in wildlife protection occurred with the negotiation in 1973 of the Convention on International Trade in Endangered Species (CITES), in which nations undertook to cooperate by banning or restricting trade in animal and plant species that are designated by CITES as endangered or threatened. In 1979, CITES was supplemented by the Convention on Conservation of Migratory Species of Wild Animals, intended to protect animals that regularly migrate across national boundaries.

There also have been international actions to protect especially valuable natural areas. In 1971, the General Conference of the U.N. Educational, Scientific, and Cultural Organization (UNESCO) endorsed the Convention Concerning the Protection of the World Cultural and Natural Heritage, under which nations nominate natural areas "of outstanding universal value" for inclusion on the World Heritage List. The nations where these areas are located are obligated to protect them, and are eligible for financial assistance from the World Heritage Fund in doing so. The Ramsar Convention, under which nations agreed to make efforts to conserve wetlands, went into effect in 1975. Under UNESCO's Man and the Biosphere Program, initiated in 1971, hundreds of important natural areas in many nations have been designated as "biosphere reserves."

The Antarctic Treaty of 1959, which protected Antarctica as an international scientific reserve, generally has worked well to safeguard the fragile Antarctic environment. In 1991, 31 nations signed an amendment to the Treaty that would essentially prohibit mining in Antarctica for 50 years.

Recognition of the urgent need to protect species and their habitats, especially in the developing nations of the tropics, caused nations in 1989 to begin negotiating a broad, worldwide treaty to conserve biological diversity. The resulting Convention on Biological Diversity was signed by more than 150 nations at the 1992 Earth Summit in Rio de Janeiro. Those nations agreed to make efforts to protect their biological resources by a variety of means. By signing the convention, countries promise to share "in a fair and equitable way the . . . benefits arising from commercial and other utilization of genetic resources" with the nations that provide those resources. Developed countries agree to provide additional financial resources to help developing countries meet the costs of conservation, and the treaty states that a mechanism shall be established for providing those resources.

Desertification and Deforestation

Desertification due to human misuse of rangelands and croplands affects large and growing areas of the Earth's surface. In the Sahel region of Africa, desertification has reduced agricultural productivity and has brought additional suffering to millions of the world's poorest people. In 1977, the U.N. Conference on Desertification adopted a Plan of Action to Combat Desertification, with the aim of supporting a wide range of specific measures. However, funding for the Plan depends on voluntary contributions from nations and has fallen far short of official targets. The 1992 Earth Summit called on the U.N. General Assembly to set up negotiations for a more effective treaty to prevent desertification.

Because of concern about the rapid loss of tropical forests, 64 nations agreed in 1985 to establish the International Tropical Timber Organization (ITTO). Although the ITTO is concerned mainly with timber production and marketing, it is supposed to take account of the need for conservation. At the 1992 Earth Summit, the nations represented agreed on an authoritative Statement of Forest Principles.

The Rio Conference and Beyond

The 1992 U.N. Conference on Environment and Development, or Earth Summit, in Rio de Janeiro was the occasion for the signing of major treaties concerning climate change and conservation of biological diversity. The Rio Conference, attended by representatives of more than 150 nations, produced other documents that are sure to lead to future advances in international environmental law.

One result of the Conference was the Rio Declaration, which outlines the relationship between environmental protection and economic development and states broad principles of environmental law, both national and international. For example, Principle 14 declares that nations "should effectively cooperate to discourage or prevent the relocation and transfer to other States of any activities and substances that cause severe environmental degradation or are found to be harmful to human health."

Another major product was Agenda 21, a set of more than 30 detailed statements containing recommendations for action on virtually every subject of concern in the fields of environmental protection and economic development. The subjects of these statements range from combatting poverty to technology transfer among nations to preventing deforestation. The Agenda 21 recommendations are soft rather than hard law, but they appear certain to exert great influence on the future development of international environmental law, much as the Stockholm Declaration did 20 years earlier.

THOMAS B. STOEL, JR.

For Further Reading: Lynton Keith Caldwell, *International Environmental Policy: Emergence and Dimensions* (1990); David A. Kay and Harold K. Jacobson (eds.), *Environmental Protection: The International Dimension* (1983); United Nations Environment Program, *Selected Multilateral Treaties in the Field of the Environment.* Vol. 1 (1983) and Vol. 2 (1991). *See also* Antarctica; Desertification; Geology, Marine; Greenhouse Effect; Oceanography; *Only One Earth;* Pollution, Air; United Nations Conference on Environment and Development; United Nations Conference on the Human Environment.

LAW, WETLANDS

Wetlands are low-lying areas that are frequently flooded, such as marshes, swamps, and river banks. Wetlands provide nesting and breeding grounds for hundreds of bird and other wildlife species; they control floods by absorbing extra water flow, moderate droughts by releasing water during low-flow periods, and recharge ground water; and they protect water quality by filtering sediment and pollutants before they reach other waters. In addition, coastal wetlands protect human settlements by absorbing storm swells. Wetlands are very important in themselves and as a part of the larger water system. However, more than 50 percent of the nation's wetlands have been destroyed in the last 200 years. Drainage and development still destroy hundreds of thousands of acres of this critical ecosystem each year.

The Water Bank Act of 1970 was the first legislation designed to protect wetlands from drainage and development by authorizing the federal government to pay landowners to leave their wetlands in a natural state. The act helped to set aside over half a million acres of wooded prairie pond areas in the upper Midwest, where many ducks and waterfowl breed. Yet, because of rising land prices, thousands of acres of wetlands continued to be lost to agriculture and development. To combat this loss, Congress passed legislation in 1979 increasing the types of lowlands eligible for Water Bank funding, including 20 million acres of wetlands in the South.

In addition to the Water Bank Act, the protection of wetlands was addressed in the Coastal Zone Management Act of 1972, which was enacted to protect coastal waters from increasing population and overdevelopment. One of its major provisions provides for the preservation, protection, management, and restoration of the nation's coastal zones, including wetlands, and funds were appropriated to serve this goal. The 1990 revision of the Coastal Zone Management Act includes a provision for the improvement of state coastal management programs in wetland areas.

The Clean Water Act is another major law affecting wetlands. This act explicitly mandated the restoration and maintenance of the chemical and biological integrity of the nation's waters. Section 404 of the act requires permits for discharging dredged or fill materials into the nation's navigable waters and adjacent wetlands and restricts such discharges in fish spawning and breeding areas and wildlife and recreational areas that might be adversely affected by use as disposal sites.

As of 1992, the Clean Water Act is being reviewed by Congress for reauthorization. Developers and agribusiness interests are seeking legislative changes that would remove millions of acres of wetlands from federal jurisdiction. Environmentalists, on the other hand, are endorsing the preservation of wetlands through the strengthening of Section 404.

LAW FIRMS, PUBLIC INTEREST

A lawyer's highest calling is advocacy on behalf of underrepresented but compelling societal interests. By the early 1970s concerns about pollution had produced a handful of law firms dedicated solely to the public interest in environmental quality. Also emerging were regulatory agencies with statutory mandates to protect the environment, along with judges prepared to enforce those mandates at the request of the new public interest lawyers.

By 1972, for example, an aroused public had spurred the U.S. Congress to enact extensive restraints on air and water pollution, while making environmental analysis a prerequisite to many of the federal agencies' most significant decisions. The generosity of charitable foundations and private donors, in turn, allowed small teams of lawyers to help enforce the new laws and expand their coverage; often pressing local and regional concerns also animated these efforts.

Key elements of the judiciary, particularly the intermediate federal appellate courts, were willing to hold federal officials to account for inadequate environmental stewardship. Cases began to proliferate that named as victorious parties groups such as the Natural Resources Defense Council (NRDC), Environmental Defense Fund (EDF), and the Sierra Club Legal Defense Fund. Staff scientists often were a necessity for the new firms, given the prominence of technical disputes in environmental forums.

This advocacy was not confined to the federal court system. Many states passed their own versions of the National Environmental Policy Act, which required detailed environmental assessments prior to actions that might significantly affect the environment. National organizations like NRDC and EDF increasingly made common cause with local and regional counterparts, such as 1000 Friends of Oregon, the Conservation Law Foundation of New England, and the Southern Environmental Law Center.

Initially some commentators worried that public interest lawyers would be denied access to courts for want of "clients" in the conventional mode. Legal theorists argued that attorneys should be permitted to sue in the name of injured lands or species. But this proved unnecessary, as even the relatively inaccessible federal court system opened its doors to those who could show that they had aesthetic, recreational or professional interests in areas threatened with degradation.

Courts looked well beyond traditional economic interests when deciding whether—and at whose request—to protect the environment. If a lawyer represented someone who was clearly among the injured, judges at all levels would accept suits seeking relief for widely shared injuries rooted in environmental degradation. Some public interest firms assembled rosters of dues-paying members throughout the nation, in part to help ensure access to suitable parties regardless of the source and location of environmental insults; other firms made themselves available to independent organizations with the requisite membership base.

The early successes of the public interest litigators, coupled with their growing expertise, created important opportunities outside the courtroom. Both staff attorneys and scientists were called upon to help craft public- and private-sector solutions to environmental problems. Once the threat of litigation had become credible, litigation itself often proved unnecessary as administrators and corporations sought settlements in order to avoid the vagaries of the courtroom. As a means for lending greater weight to environmental interests throughout society, negotiation and lobbying became at least as effective as litigation.

For the environmental groups an increasingly dominant theme has been finding ways to avoid pollution at the source rather than simply trying to control it at the point of emission. For example, as an alternative to installing costly scrubbing devices on the smokestacks of fossil-fueled power plants, public interest firms convinced many utilities to invest in technologies and systems for saving energy more cheaply than it could be produced. And they convinced state regulators and legislators to link the utilities' overall profits to their success in delivering those cost-effective energy savings to consumers.

By the 1990s public interest law firms could look back over two remarkably productive decades. Their achievements spanned much of the framing and administration of the national Clean Air and Clean Water Acts; the creation and preservation of substantial wilderness areas in the continental U.S. and Alaska; a phaseout of many hazardous substances including lead and a series of pesticides, starting with DDT; reorientation of national and regional energy policies toward more efficient uses of energy as an alternative to increased energy production; and significant progress in preserving the protective global ozone layer from destruction.

As this partial list suggests, the agendas of the public interest law firms span at least three major missions today. The first and most traditional is to guard local and regional ecosystems by preserving habitat and watersheds. At least equally prominent has been a campaign to protect human health by regulating the use and disposal of toxic substances, including pesticides and hazardous industrial wastes.

The third grouping, which draws in some measure on each of the other two, involves efforts to harmonize economic development with environmental quality. Like the 1992 Earth Summit in Rio de Janeiro, these initiatives seek to demonstrate that societies need not choose between healthy economies and environments. The greatest opportunity here may lie in attacking the threat of global warming, by removing barriers to economically productive activities that also reduce greenhouse gas emissions (for example, energy efficiency improvements and renewable energy resources). At the same time, public interest law firms are exploring a role in promoting population control.

The public interest law firm model for the 1990s includes an agenda that is both domestic and international in scope and a staff whose expertise far transcends anything taught in law schools. The scale of these enterprises remains modest by corporate standards; the largest, NRDC, employed 35 attorneys and 40 scientists in 1992. Prospects for the public interest firms depend on their continuing capacity to integrate litigation, science, media relations, negotiation and lobbying into effective campaigns for social change.

Some see a threat to those contributions in a few recent U.S. Supreme Court defeats, overlooking the fact that the Court's orientation has been almost unremittingly hostile since the onset of the "new" environmental litigation a generation ago. The environmental community's long list of achievements since 1970 includes no Supreme Court landmarks granting requests by NRDC or EDF or the Sierra Club, although sometimes they could take comfort in the rationale for a rejection. In the environmental arena, the public interest legal movement has been built on victories in the state courts and administrative agencies, the Congress, the lower federal courts, the private sector, and the realm of public opinion. That remains a promising combination for the future, no matter what the Supreme Court's membership or disposition.

RALPH CAVANAGH

For Further Reading: "Comment: The New Public Interest Lawyers," 79 *Yale Law Journal* 1069 (1970); Oliver Houck, "With Charity For All," 93 *Yale Law Journal* 1415 (1984); D. Rhode and D. Luban, *Legal Ethics* (1992).
See also Environmental Organizations; Law, Environmental.

LEAD

Humans have used the heavy metal lead in a wide range of applications, from making bullets and automobile batteries to use as a gasoline additive, in paints, and in ceramic glazes. Lead is toxic, however, as well as useful. Depending on the amount of exposure, lead's effects can range from subtle damage to the red blood cells, kidneys, and the nervous system to coma and death. Lead poisoning also has been associated with high blood pressure, infertility, and spontaneous abortion.

Children are particularly vulnerable to lead poisoning. According to the Centers for Disease Control and Prevention (CDC), lead poisoning is the most common—and most preventable—pediatric health problem in the U.S. today. In other industrialized nations, particularly those with lax pollution controls or those that use leaded gasoline, it is an even more important problem. Lead poisoning spares no socioeconomic class, geographic area, or ethnic group. Because millions of children in the U.S. may be exposed to enough lead to cause learning disabilities, prevention is a priority.

Recent research has shown that if a child's blood contains as little as 10 to 15 millionths of a gram of lead in 100 milliliters of blood, before or after birth, the child may suffer delayed mental development, lower IQ, hearing loss, speech and language handicaps, or poor attention span. Quick action to stop exposure may enable a child to overcome the effects of low-level lead poisoning, but high-level or long-term exposures can cause irreversible damage.

Lead levels of 10–20 millionths of a gram per 100 milliliters (μg/dl) of blood can subtly affect IQ, hearing, and growth in fetuses and young children. Anemia begins above 20 μg/dl and becomes pronounced above 40 μg/dl. Acute poisoning of the nervous system, with coma, convulsions, irreversible damage to the intellect, and even death, can take place at 80 μg/dl for children and at 120 μg/dl for adults.

The sources of lead exposure are myriad. In the U.S., an estimated 57 million homes contain lead paint; when the paint deteriorates, lead particles remain. During the 1960s and 1970s, automobiles burning leaded gasoline deposited 4–5 million metric tons of lead in the soil near roads. Dust tracked in from lead-contaminated soil from auto exhaust, paint, or emissions from smelters or battery plants, or dust carried home on the clothing of parents exposed to lead at work can be major sources of exposure.

Lead also appears in drinking water, leached from lead pipes or solder. Lead solder in imported canned goods and lead glazes in ceramics can leach into food. The use of folk medicines or cosmetics and such hobbies as making pottery or stained glass or refinishing or burning lead-painted wood can also cause lead exposure.

Babies and small children are the most likely to develop lead poisoning because they explore the world by putting objects in their mouths and thus might swallow lead dust or flaking paint. Their bodies may absorb as much as 50% of the lead they swallow. If a pregnant woman has high levels of blood lead, her fetus will be exposed as well. Because pregnancy can mobilize lead stored in bone, it is important for girls and young women to avoid lead long before they consider becoming pregnant. A 1993 study has suggested that the decline in intelligence caused by an accumulation of lead in children's blood can be at least partly reduced when steps are taken to reduce the blood levels of the toxic metal, and examined the benefits of reducing lead levels in children who have no obvious symptoms of lead poisoning.

BARBARA SCOTT MURDOCK

LEOPOLD, ALDO

(1887–1948), American wildlife biologist. Over a forty-year career, Leopold made fundamental contributions to conservation policy and philosophy, the environmental sciences, and the natural resource management professions. As an early proponent of the need to adopt an ecological perspective, he played a pivotal role in expanding the scope of the conservation movement. As a forester, wildlife manager, scientist, writer, educator, and advocate, he worked to instill that perspective among both his professional colleagues and the general public.

Leopold was born and raised in Burlington, Iowa. Inspired by the early American conservation movement and its preeminent leaders, President Theodore Roosevelt and forester Gifford Pinchot, he chose a career in forestry. After attending Lawrenceville Preparatory School in New Jersey, he entered Yale University. In 1909 he received his Master of Science degree from Yale, the principal training ground for the first generation of American foresters.

Upon graduation Leopold joined the United States Forest Service. From 1909 to 1924 he served as a field officer and administrator in the national forests of the American Southwest. His wide-ranging interests made him an innovator in several areas of forest management. His concern over the effects of accelerated soil erosion led him to undertake early studies of the environmental impact of forest management practices in the semi-arid Southwest. He directed a region-wide game protection movement that made wildlife conservation a higher priority not only in the Southwest, but throughout the U. S. Forest Service. Similarly, he helped to establish recreational needs as an important consideration in national forest policy and planning.

In the early 1920s Leopold began his lifelong efforts to protect America's remaining wilderness lands. As a leading spokesman for what became known as "the wilderness idea," he articulated the recreational, cultural, historical, and scientific value of wildlands. His work led to the creation in 1924 of the Gila Wilderness Area within New Mexico's Gila National Forest, the first such area in the world to be so designated. His advocacy also stimulated the Forest Service to adopt, in the late 1920s, its first service-wide wilderness protection policy.

In 1924 Leopold became assistant director of the Forest Products Laboratory in Madison, Wisconsin. He left the Forest Service in 1928 to devote himself to the emerging field of game management. As head of a privately sponsored research institute, he undertook a three-year survey of game populations, habitat conditions, and management needs in the American midwest.

Through his work on the game survey, he began to lay the foundations of wildlife management as a distinct profession. At the time, wildlife conservation consisted largely of legal restrictions on hunting, the ad hoc establishment of reserves, and the captive rearing and release of game animals. By contrast, Leopold's approach emphasized the protection, restoration, and management of habitat so that game and other forms of wildlife could perpetuate themselves. The need to provide favorable habitat conditions was, in his words, "the fundamental truth [that] the conservation movement must learn if it is to attain its objective."

Increasingly, ecological concepts came to undergird Leopold's work. The culmination of this phase of his development came in 1933 with the publication of *Game Management.* The first text in the field, it provided a synthesis of ecological theory and management practices applicable to all forms of wildlife. Leopold gained new opportunities to contribute to the underlying science when in 1933 he was appointed Professor of Game (later Wildlife) Management at the University of Wisconsin. He held this position for the remainder of his life.

As ecology matured in the 1930s, Leopold and other American scientists and conservationists faced a variety of environmental dilemmas, including the devastating conditions in the Dust Bowl on the western plains. In response, Leopold's early management focus on forests and game animals broadened to include all members of what he called "the biotic community," as well as the ecological processes that determined the stability of land. He began to redefine conservation as "a state of health in the land." Ecology, as "a new fusion point for all the natural sciences," became key to defining and understanding that state of environmental health.

Leopold's new approach to conservation had broad implications for human land use, conservation policy, and the natural sciences. It suggested that greater attention needed to be given to the status of biological diversity within ecosystems. It highlighted the need to manage resources in a more integrated manner (as opposed to the traditional division of responsibilities among soil scientists, foresters, wildlife managers, and other specialists). It placed greater emphasis on ecosystem restoration; Leopold himself became a pioneer in restoration ecology through his work at the University of Wisconsin Arboretum and on his family's land in Wisconsin's "sand counties." Wilderness areas also took on new importance as places in which scientists could study the processes by which healthy land maintained itself.

Leopold clarified and promulgated these ideas through his leadership in more than a hundred scientific, professional, and conservation organizations. He was a charter member of The Wilderness Society, which came together in 1935 under the direction of forester Robert Marshall. He served as president of the Ecological Society of America and The Wildlife Society (which he helped establish in 1937). After World War II, he became more active in international conservation as an advisor to the Conservation Foundation and to the International Scientific Conference on the Conservation and Utilization of Resources, sponsored by the United Nations.

In addition to two books that appeared in his lifetime, Leopold published more than 500 articles, essays, scientific papers, reviews, and editorials on a wide range of conservation-related topics. In the early 1940s he began to compose a series of literary essays describing his field experiences as a naturalist and his concerns as a conservationist. With the posthumous publication of these essays as *A Sand County Almanac and Sketches Here and There*, his ideas began to reach a broader audience. In *A Sand County Almanac*, Leopold combined the voices of the scientist, historian, poet, and philosopher to produce what has often been described as "the bible of the environmental movement."

Leopold's conservation philosophy attained its most fully developed expression in "The Land Ethic," the capstone essay of *A Sand County Almanac* and a seminal statement in environmental ethics. The ecological perspective, he argued, required that the concept of community be expanded "to include soils, waters, plants, and animals, or collectively: the land." Under this expanded concept, simple utilitarian standards no longer sufficed to gauge decent land use. A "land ethic" was needed to guide people in their land relations, and to affirm the right of other components of the community "to continued existence, and, at least in spots, their continued existence in a natural state." He concluded that individuals should "examine each question in terms of what is ethically and esthetically right, as well as what is economically expedient. A thing is right when it tends to preserve the integrity, stability, and beauty of the biotic community. It is wrong when it tends otherwise."

Leopold died on April 21, 1948, suffering a heart attack while fighting a grass fire on a neighbor's farm in Wisconsin.

CURT MEINE

For Further Reading: J. Baird Callicott, ed., *Companion to "A Sand County Almanac": Interpretive and Critical Essays* (1988); Susan L. Flader and J. Baird Callicott, eds., *The River of the Mother of God and Other Essays by Aldo Leopold* (1991); Curt D. Meine, *Aldo Leopold: His Life and Work* (1988).

LIFE CYCLE ANALYSIS

A life cycle analysis (LCA) of a product considers both the direct and indirect effects of its manufacture, use, and disposal. These effects occur at every stage of the "life cycle" of the product, from the extraction of raw materials from the ground, through the various processing, manufacturing, fabrication, and transportation steps, and to consumption, disposal, or recovery for recycling. For example, the direct effect of recycling newspapers is avoiding the problems of creating more solid waste. However, the indirect effects include, among many other things, the burning of coal to generate electricity for the recycling operation. The burning of coal results in many consequences such as the generation of substantial amounts of ash, which must be handled as a solid waste. A life cycle analysis quantifies these indirect effects as well as the direct consequences and provides valuable information for decision-making by consumers, manufacturers, and governments.

The first life cycle analysis of products was initiated in 1969 to compare several beverage containers to determine which produced the least effects on natural resources and the environment. The quantities of various effluents and the amounts of natural resources consumed were measured. The result was an accounting of things that create a burden on the environment without actually assessing those effects directly. The direct effects are degradation of human, animal, and plant health, and the depletion of finite resources. These are quite difficult to quantify in an objective fashion, and not all scientists can agree on how to do so. The concept became known as Resource and Environmental Profile Analysis (REPA). Between 1970–75 several REPA studies were conducted and the research protocol was developed.

REPA/LCA studies quantify energy requirements and emissions—not human health and ecosystem impacts. These impacts must be dealt with separately through risk assessment, ideally before the REPA/LCA is conducted. However, the quantification alone can be used to identify the most significant opportunities for the reduction of resource use and emissions. REPA/LCA studies are used primarily by private companies examining their own products in comparison to a modified version of the same product, or new or proposed products. This type of study is ideal for use in examining proposed changes in raw materials selection or process changes. Companies can readily examine options to minimize the environmental impact of a product or its packaging.

The Concept of Equivalence: REPA/LCA studies focus on specific products or services and are usually based on the number of uses delivered to the consum-

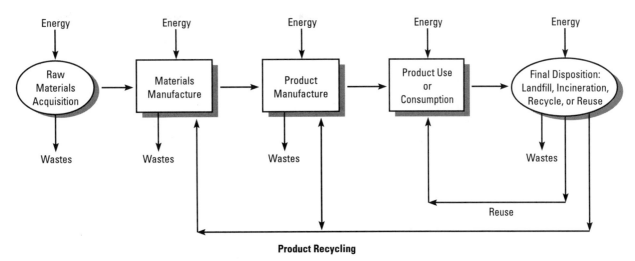

Figure 1. General materials flow for "cradle-to-grave" analysis of a product distribution system.

er. For more than one product, the performance of the products must be equal. For example, comparing frozen concentrated orange juice with ready-made orange juice, an appropriate comparison is volume of the product as drunk by the consumer. Often consumer research must be conducted to determine the amount of different products used to complete the same task.

Material Flows: Once the materials and specific products are identified, the complete "cradle-to-grave" system is developed to provide a blueprint for data gathering. The figure below is a generalized flow diagram of a complete system, showing the key elements. All important processes, from extraction of raw materials through final disposition of the product, must be considered, including all transportation of material from each block including any recovering of materials for recycling or reuse.

Secondary impacts may be important. For example, there are energy and environmental consequences resulting from the manufacture of the steel used in the assembly of trucks that transport products. These operations need to be included for completeness.

Each block in the flow diagram is defined as a module system. The data are collected for each module based on the given weight of product output from that module. Modules are developed to include material balances after pollution controls are in place. In other words, only those effluents actually discharged from the plant boundaries after internal pollution control, reuse, and recycling measures have been applied are included. The basic categories for inputs and outputs for the modules are natural resources, such as energy and water; and environmental emissions, such as air pollution, water pollution and solid waste. Obtaining data is the most time consuming and difficult part of a REPA/LCA study. Detail is the key to the quality of

the study. Many key data points needed for a REPA/LCA study are not accurately reported in public sources. In many cases an analysis for a specific product may need the accuracy of private data to be valid. It is not always possible to find good data sources for all of the thousands of values required.

Scope and Boundary Conditions: Considerations in setting the boundaries of a REPA/LCA study include: geographic scope; manufacturing issues, such as personnel activities and capital equipment; fuel/energy issues, such as secondary effects of combustion and generating electricity; fuel sources and efficiencies (nuclear power, hydropower); the energy content of material resources; and litter.

Subjective Issues: While the focus in REPA/LCA studies is on quantitative analysis, there are many subjective issues concerning levels of resource and energy usage, emissions, etc., that are acceptable and what levels represent a significant opportunity for improvement. These issues include resource issues, including renewable versus nonrenewable natural resources and the generation of energy from residues; weighting or determining the impact of the results, including weighting within a category, which includes air and water pollutants, energy and solid waste, and hazardous waste.

Recycling Issues: In a broad view, all of the natural resource and environmental emissions of a given industry must be assignable in some way to the products of that industry. While this is quite clear and easy for some products, it becomes quite complex for others when materials recovery and recycling occurs.

REPA/LCA studies have been suggested as a basis for "green" labeling of products or as a basis for legislatively banning or restricting products or packaging. However, although REPA/LCA studies are a useful source of information on resource usage and environmental emissions, they do not quantify all environmental consequences of a product. They are based on hundreds of assumptions that need to be carefully evaluated. These studies should not be used to decide that one particular product is "good" or "bad," but rather the information presented should be used as part of a much more comprehensive decision process or used to understand the broad or general trade-offs.

MICHAEL H. LEVY

For Further Reading: Richard A. Denison, *Toward a Code of Ethical Conduct for Life Cycle Assessment* (1992); Society of Environmental Toxicology and Chemistry, *A Technical Framework for Life-Cycle Assessment* (1991); U.S. Environmental Protection Agency, *Life-Cycle Assessment: Inventory Guidelines and Principles* (1992).
See also Ecolabeling; Economics; Recycling; Resource Management; Transportation; Waste, Municipal Solid.

LIMITS TO GROWTH

The phrase *limits to growth* refers to the basic pre-analytic vision of the economy as an open subsystem of an ecosystem that is finite, nongrowing, and materially closed. A closed system is one in which materials neither enter nor exit the total system, although energy does flow through it. As the economic subsystem grows it incorporates an ever larger proportion of the total ecosystem, and must therefore gradually conform to the basic characteristics of the total ecosystem, namely that it is nongrowing, finite, and closed.

The total ecosystem is further characterized by a complex network of positive and negative feedback loops. Thanks to these solar-powered biogeochemical cycles, waste (high-entropy) materials are reconstituted into useful (low-entropy) resources that are again capable of supporting life. The total ecosystem serves the economic subsystem as an absorber of inevitable wastes and regenerator of newly usable resources. Its capacity to do this on a sustainable basis is clearly limited, and in turn limits the physical growth of the economic subsystem.

Because an expanding Gross National Product (GNP) requires assimilation or accretion of materials, economic growth, as measured by GNP, is limited. However, efficiency in the use of materials and energy to satisfy human wants can increase independently of any increase in the actual quantity of matter and energy throughput, so economic growth is not limited in

this sense. If the latter qualitative improvement is called "development," to distinguish it from the quantitative increase termed "growth," then the idea of limits to growth means "development without growth," which during the late 1980s came to be called "sustainable development."

In addition to the biophysical limits to growth, there are also ethical and social limits. If, in wealthy societies, increase in satisfaction comes from relative wealth rather than absolute wealth, then aggregate growth in wealth is powerless to increase aggregate satisfaction since someone's relative position must diminish whenever someone else's relative position increases. Growth may still be physically possible, but it becomes a self-canceling activity, as when all spectators at a football game stand and crane their necks to get a better view—and in the end no one is able to see more than when they were all seated.

There is also a strictly economic limit to growth stemming from the law of diminishing marginal utility coupled with the law of increasing marginal costs. As the economic subsystem grows we must sacrifice ever more important natural services (such as photosynthesis and purification of water) in exchange for extra production devoted to the satisfaction of ever less pressing wants. At some point, growth increases environmental costs faster than it increases production benefits, and consequently makes us poorer rather than richer. The ideal goal is for the economic subsystem to grow to an optimal physical scale relative to the total ecosystem, and then stop growing, although it may continue to develop.

The "limits to growth" view of economics distinguishes three aspects of the economic problem: allocation, distribution, and scale. However, standard neoclassical economics deals mainly with allocation (the division of a given resource flow among alternative product uses: cars, shoes, etc.) and distribution (the division of a given resource flow as embodied in products among alternative people). It does not address the question of scale. (*Scale* refers to the physical scale of total resource throughput, and may be broken down into population times resource use per capita.) The likely reason for this is that standard neoclassical economics does not view the economy as an open subsystem of a finite, nongrowing ecosystem. Rather it sees the economy as an isolated circular flow of exchange value between firms and households, with no essential dependence on the environment. For an isolated system, unconstrained by relations to any larger system, there are no limits to growth. Ideologically this apparent lack of any limit to growth has allowed neoclassicals to conceive of growth as the solution to all problems. Without growth the solution to poverty must be sought in sharing and in population control. Both of these solutions raise many volatile political and social issues. Hence the ideological fervor of the debate.

HERMAN E. DALY

For Further Reading: Herman E. Daly, *Steady-State Economics*, 2nd ed. (1991); Donella H. Meadows, D. L. Meadows, J. Randers, and W. W. Behrens, *The Limits to Growth* (1972); Donella H. Meadows, Dennis L. Meadows, and Jorgen Randers, *Beyond the Limits* (1992).

LIMNOLOGY

Limnology is the study of physical, chemical, and biological processes and their interactions in fresh waters. Limnologists typically study lakes or streams, but the science also extends to groundwaters, temporary pools, and other freshwater habitats.

Dissolved chemicals and suspended particles that enter from the catchment, or watershed, strongly influence the characteristics of lakes and streams, affecting the types of organisms that occur, their interactions, and their biological productivity. For this reason, limnologists are attentive to processes in terrestrial ecosystems and wetlands that determine the composition of fresh waters.

During the past century, limnologists have demonstrated many interactions of physical, chemical, and biological processes that determine the functioning of freshwater ecosystems. An extensive literature describes the biology and interactions of microbes, phytoplankton, protozoa, periphyton, higher plants, benthic invertebrates, zooplankton, and fishes. The cycling of carbon and essential nutrients has been quantified. Currently there is great interest in biological feedbacks to nutrient cycles and productivity, interactions within microbial communities, coupling of groundwaters and surface waters, and changes that occur from the headwaters to the sea.

On a global scale, fresh water is a limited resource of great importance to humans. Applied limnology is concerned with problems of water quality, contaminants, species invasions, and fisheries. A variety of contaminants affect fresh water, including acid precipitates. Mercury and organochlorine contaminants are especially troublesome because they can enter fresh water through the air and are therefore hard to control. These substances accumulate through the food chain, sometimes reaching excessive concentrations in fish. Eutrophication results from enrichment of lakes by human sewage or runoff from urban and agricultural areas. This process leads to nuisance growths of algae and higher plants and, in extreme cases, oxygen depletion and fish kills. The discovery that excess phosphorus was the primary cause of eutrophication was a crucial step in mitigating its effects. Species invasions are a worldwide problem in lakes and streams. Some invading species, like Eurasian watermilfoil and the zebra mussel, build enormous populations and have significant ecological and economic impacts. Freshwater fisheries are a significant food source in some parts of the world. In the United States, recreational fishing is a major enterprise, with annual values of billions of dollars.

S. R. CARPENTER

LIVESTOCK

See Animals, Domesticated.

LONGEVITY

The Longevity Revolution

One of the most extraordinary changes in the human environment during the 20th century has been the great expansion of longevity. That life is growing longer in itself is not unusual: life expectancy has been gradually increasing throughout recorded history. From an estimated average life expectancy of 22 in Roman times, to Renaissance averages somewhere in the 30s for aristocratic males, Western nations progressed by 1900 to an average life expectancy in the late 40s—47 in the U.S.—a gain of about 1.2 years every century.

And then something unexpected happened. From 1900 to 1992 Americans have added 28 years to average life expectancy, rising to approximately age 75 in less than one century—adding more than the total increase of the preceding 2,000 years. Moreover, our longevity has continued to increase, so that by the year 2000 the total may well be higher.

Just as surprising is the fact that scientists do not really know what has caused this leap forward in the length of our lifetimes. Although there are various factors that may have contributed—for example, life-saving medical techniques that help prevent premature adult death, such as bypass surgery and new treatments for cancer—the best biological scientists in the field of aging say that there are no data to explain this trend conclusively. But even without an explanation scientists acknowledge that it is indeed taking place.

A Second Middle Age

Initially, it appeared that the 20 or 30 years added to the average life expectancy would be an extension of old age—an extra measure of illness, disability, and dysfunction. But in a development as mysterious as the explosion of longevity itself, as lifetimes have lengthened, the beginning of physical old age has been increasingly postponed. People are simply getting physically old later and later in life. The result is something completely unprecedented: the time gained through longevity has been added to middle age. A new period of adult life has been created, one which has never existed before as a widespread phenomenon: a second middle age. If the old concept of "middle age" was the period from 35 to 50, then the "second middle age" occupies roughly the third quarter of a 100-year lifetime: ages 50 to 75. Like longevity itself, the boundaries of the second middle age are still expanding: the existence of healthy and youthful people in their late 70s and early 80s suggests that at some future point the second middle age for some of us may last until 80 or 85.

The most comprehensive scientific study that has been mounted to date describing and validating this phenomenon—which is occurring in all the industrialized countries—is a massive study designed by Dr. Alvar Svanborg and his associates at the University of Gothenburg, Sweden, of three successive 1,200-person groups of 70-year-olds. Contrary to his expectations, Svanborg, now professor of Geriatrics at the University of Illinois, Chicago, found that younger groups were healthier in late life than the oldest group—despite the fact that they had been the beneficiaries of more life-saving technologies and so might have been expected to be sicker and frailer.

The Long Careers Study

The result is unprecedented in history as we know it: generations in which millions of people are living into their 70s, 80s, 90s, and 100s in good health, still engaged in the same kinds of activities that preoccupy adults at earlier ages. Despite their chronological ages, in terms of their interests and activities, they are not "elderly"; they are normal adults. It is a picture very different from our previous image of aging.

From 1987 to 1992, 150 active people from age 65 to 102 were interviewed for the Long Careers Study to see what this new longer lifetime was like. The group was not a scientifically random sample. There was no easy or economical way to obtain one. Moreover, since the objective was to study success and health, not illness, as many previous studies have done, about half of the

group were well known people: e.g., Isaac Asimov, Julia Child, Norman Cousins, Hume Cronyn and Jessica Tandy, Barbara McClintock, Linus Pauling, Jonas Salk, Margaret Chase Smith, and Cyrus Vance. Another 25% were accomplished but not well-known, and the remaining one-quarter were ordinary people, still actively employed as secretaries, clerks, and small businesspeople.

The results were provocative. The single largest most common feature was that all of them were involved in work that they enjoyed. Almost all of them saw their work in terms of helping others, a reinforcement of the Eriksonian concept of generativity. Many participants reported that they felt that what they did was very important, whether it was helping people find a home (real estate), manage their money (finance), or preserve their health (medicine).

The average age of the group was 79; there were three centenarians, fourteen people in their 90s, fourteen people in their 60s, and the remainder in their 70s and 80s. Of those interviewed 70% said they exercised regularly; 60% exercised every day. As in the Svanborg study, many of them have had serious illnesses or life-saving medical procedures, from which they recovered.

The study provided two new models of aging, in addition to the traditional model of a long period of decline and illness at the end of life. As of this writing, 20 out of the 150 participants have died. Of those, 60% had only a very brief period of illness or no illness at the end of their lives—what Dr. James Birren of UCLA's Borun Center on Aging calls the "Plateau Model" of aging: the person achieves a plateau of stable good health at some time during the 60s or 70s and remains at that level of functioning for the rest of his/her life, even into the 90s or 100s. Of those who died, 30% had a period of long decline, in accordance with the traditional "Cumulative Decline" model of aging. And 10% had what Bronte calls the "Episodic Pattern," alternating episodes of good health with illnesses from which they completely recovered.

It is provocative to think that perhaps, as successive generations who have paid more attention to diet and exercise continuously throughout their lives move into their older years, the percentage of people who follow the Plateau model might increase. New scientific discoveries may also improve our prospects; new research at the University of Kentucky Medical School, for example, has correlated death from Alzheimer's disease with the presence of brain tissue accumulations of mercury, a highly toxic metal, and has found that the amount of mercury correlates to the number and size of "silver" mercury amalgam dental fillings in the individual's mouth. If causes could be found for Alzheimer's, the most-feared disease of old age, and if the three other major killers of adults (heart disease, arteriosclerosis, and cancer) could be reduced in frequency, the prospects for an even more active long lifetime would undoubtedly improve.

The Rebirth of Wisdom

What is even more provocative, however, is the potential for both the human and planetary environment of so many people with 50 years or more of active life experience. Among many Study participants there was a kind of progression in their careers, from youthful work at the grass roots early in life to working for the well-being of the nation or the planet in later life; or from early concern with a single issue, to integrative work with a number of different disciplines or issues in the post-60s decades. It is possible that if such progressions occur in even 20 or 30% of long-lived people, then the presence of so many older people who hold uppermost in their minds the interests of the community may ultimately improve the living conditions of us all. Scientific research shows that many older people score higher than younger persons in "crystallized intelligence," a contemporary scientific term for wisdom and good judgement. Ours is a society which needs as much wisdom as we can get, both now and in the foreseeable future.

The suggestion is that the capacity for growth is innate in human beings and that it does not diminish as a result of chronological age. Indeed, it is possible that it may remain constant or increase, as experience is gathered, analyzed and put to good use. In a society that until recently has looked upon old age as a tunnel with no light at the end, the participants of the Long Careers Study suggest that the time added to our lives in this century may be instead a great gift—both for the individual and for the good of this Earth.

LYDIA BRONTE

For Further Reading: Lydia Bronte, *The Longevity Factor: How the New Reality of Long Careers Is Leading to Richer Lives* (1993); William Evans and Irwin H. Rosenberg, with Jacqueline Thompson, *Biomarkers: The Ten Determinants of Aging You Can Control* (1991); Linus Pauling, *How to Live Longer and Feel Better* (1986).

LOVE CANAL

Love Canal is a neighborhood in Niagara Falls, New York, where chemical contamination of the environment posed serious health threats to community residents during the 1970s. The canal, originally excavated in the 1880s as part of a model city scheme envisioned by entrepreneur William T. Love, was eventually abandoned and later purchased by the Hooker Chemical Company in 1946. From 1947 to 1952,

Hooker Chemical used the canal to bury legally steel drums containing approximately 22,000 tons of industrial chemical wastes. In 1953 Hooker sold the land to the Niagara Falls Board of Education for one dollar, upon condition that the company not be held liable for any damages incurred because of the chemical wastes. The city board built an elementary school on the land and sold the remainder to real estate developers.

After several years of unusually heavy rains, the canal's groundwater level rose and chemicals began leaching out onto the surface. Community residents had complained to city and state officials for several years prior to 1978 that children had suffered chemical burns from playing in their backyards. These complaints prompted state officials to investigate and dioxin and other poisonous chemicals, many of them suspected carcinogens, were found in soil, water, and basement air samples. In some instances the levels of contamination were from 250 to 5,000 times those which are deemed safe by the Environmental Protection Agency (EPA). Lois Gibbs, a Love Canal resident, along with other concerned citizens, formed the Love Canal Homeowners' Association (LCHA) and conducted an informal survey of over 1,100 residents with the assistance of Dr. Beverly Paigen, a cancer researcher. Although the survey revealed higher than average rates of miscarriage, birth defects, kidney disorders, asthma, and nervous disorders among area respondents, it was criticized by state officials as unscientific. State officials soon conducted independent surveys that found abnormally high rates of miscarriage and birth defects among residents closest to the site.

In 1978, New York State Health Department officials declared a health emergency in the area and recommended that pregnant women and children under two years of age be evacuated from the area immediately surrounding the former dump site. President Jimmy Carter declared the site a federal disaster area and allocated federal aid to help with cleanup and relocation efforts. New York state and federal governments would eventually spend 12 years and $275 million studying the site, cleaning the contaminated area, and relocating seven hundred families. The Love Canal incident also prompted Congress to create the Superfund cleanup program in 1980 to deal with over 1,000 hazardous waste sites across America.

In 1990, the EPA ruled that most of the Love Canal neighborhood was safe enough from chemical contamination to be populated once again. A containment system was built to seal off the 16-acre dump and the adjoining areas, and periodic testing of air, water, and soil samples was proposed to ensure the safety of the new Love Canal. Many citizens' and environmental groups argued that the EPA studies that assessed the area's safety were inadequate, and that further scientific study, especially concerning the possible long-term genetic damage suffered by residents, should precede resettlement of the area. Love Canal, as much as any other environmental disaster known in the United States, has come to symbolize the dangers associated with hazardous waste disposal.

LUMINOUS ENVIRONMENT

Solar radiation is the source of energy, establishing (providing) the luminous and thermal environments. It is one of three interacting subenvironments (the others are the atmospheric and the thermal). Vision, of course, depends on light and artificial lighting and daylighting can influence the way occupants see a space and give that space a unique character.

Artificial light is produced by a variety of sources: incandescent, tungsten halogen, fluorescent, high intensity discharge (high and low pressure sodium, mercury vapor, metal halide), neon, and cold cathode lamps. Artificial light sources offer varying degrees of color, brightness, and quality (control). The use of these sources varies from fixture to fixture and the placement of these fixtures defines the visual relationship with the interior architectural elements.

The Illuminating Engineering Society (IES) researches and publishes recommended light levels for specific tasks and spaces. These recommendations are categorized according to a range of light levels. A target light level is selected depending on the occupant's age, speed and/or accuracy required in performance of a task, and the reflectance of the task background.

Daylight is a significant source of illumination. Even though it costs nothing, it varies with location, season, weather, and time of day. Daylight is one of the most pleasant forms of illumination when it is controlled, otherwise direct daylight can cause uncomfortable glare. Maximizing the presence of daylight in interior spaces with windows, skylights and/or clerestories has an appreciable effect on reducing the need for artificial lighting.

Unlike daylight, artificial lighting has the advantage of being able to be controlled by switches, local dimmers, occupancy sensors (infrared or ultrasonic), or building management systems (BMS). Switches control light on or off at a convenient location at the entry/exit of a space. A local dimmer, a further modification of switching, allows a range of light levels and produces them in a smooth transition. Occupancy sensors turn on light fixtures when occupants are present, and turn them off when they leave by sensing heat (infrared–line of sight) or sensing sound (ultrasonic–volumetric). Daylighting sensors register the amount of daylight present and adjust light fixtures when com-

bined to a predetermined light level. Light fixtures could be connected into a BMS with an astronomical time clock to turn light fixtures on and off at designated times of the day.

Intelligent lighting design, through the use of artificial and natural light, influences the way in which occupants feel and behave. Office lighting is designed for maximum productivity, arenas and airports for safety and circulation, and residences for hospitality and relaxation. An understanding of the activity in a space and its requirements will yield an appropriate lighting design.

CHARLES A. STARNER

LYELL, CHARLES

(1797–1875), British geologist. Lyell's persuasive arguments for a steady-state evolution of the Earth, keen sense of the importance of theory erected only on "true causes" consistent with observations, and enormous ability to continue to learn and revise even his most cherished views throughout his life, made him one of the most influential geologists in Britain and a world leader of this science. He also contributed significantly to the emancipation of geology from the confines of religion. He raised Hutton's legacy from near oblivion to the main cornerstone of modern Earth science and expanded its scope. Often in scientific conflict with his contemporaries, some of whom, such as Sir Roderick Murchison, were then undisputed leaders of geology, Lyell's vision of geology as a testable theoretical enterprise ultimately triumphed over his opponents' view of geology as a static collector's pursuit or a set of unfalsifiable existential statements.

As a child Lyell developed an interest in natural history and geology. At Oxford he attended a lecture by William Buckland, one of the great pioneers of geology in Great Britain, which fired his enthusiasm. His first observations of the Yarmouth Delta, which had also influenced Hutton's thinking, indicate his rich and dynamic imagination, capable of picturing to himself the gradual evolution of geological processes in time and space. In 1819 he was elected a fellow of the Geological Society of London.

Between 1823 and 1830, when the first volume of his magnum opus, *Principles of Geology*, was published, Lyell made geological tours in Europe. His review in 1827 of G. P. Scrope's book on the volcanoes of Central France—in which Scrope described the gradual evolution of river valleys—gave Lyell the idea of writing a book to marshal evidence for the adequacy of present geological processes to explain the geological record, a startling proposal at the time. Despite his belief that the world was generally intelligible by reason, he retained a religious core in his geological thinking and believed in the special creation of man. Only late in life did the ideas of his friend Charles Darwin force him to change his views, albeit still reluctantly.

In the *Principles* Lyell sought to create a new science of geology freed "from the clutches of Moses" by considering Earth's past in terms of its present. He adopted Hutton's teachings and argued that given enough time present geological agencies could explain the entire geological record. To account for warmer climates in the past, Lyell suggested ingeniously that if all the present continental surface were gathered around the equator, the Earth would have a much warmer climate. Unlike Hutton, he underestimated the potency of fluvial erosion and too commonly opted for marine erosion. Lyell was much troubled by the weighty evidence for the progression of life through geologic time and devoted the entire second volume of his *Principles* to a careful scrutiny of the present-day changes in the organic world. His interpretation was that his steady-state world (called "uniformitarian" by W. Whewell in a review) housed immutable species that were generated and became extinct at random. He used this view in the third volume to create a statistical palaeontology to subdivide the Tertiary strata in Europe into Eocene, Miocene, and the older and newer Pliocene on the basis of the proportion of living mollusc species to extinct species. Lyell thought that all classes of rocks—volcanic, plutonic, sedimentary, and what he for the first time called metamorphic—were being generated all the time and that the older term "primitive rocks" reflected an error in interpretation.

In the *Principles* Lyell laid down the rules for reasoning in geology, in other words, a uniformitarian theory of geological evolution that was testable. When his theory failed to explain certain phenomena such as past glaciations, Lyell reassessed the evidence and revised his opinions at the sacrifice of some of his theories, but not of his overall view of the intelligibility of geological history in terms of causes now in operation, or of those that may be reasonably deduced from the present laws of nature. The greatness of Lyell lies not so much in his competence in commonplace geological techniques as in his rational and critical uniformitarian approach to geological processes, assuming no presupposed order and regularity in nature (or being always ready to abandon any such presupposition) into which the inferences had to be fitted. This made him the true successor of Hutton, although he was not quite as original, as clear, or as versatile a thinker as Hutton.

Lyell's great accomplishments made him the best-known geologist in the world of his time and earned him numerous honors. He was knighted in 1848. A great promoter of young merit, he was instrumental in Darwin's scientific development. His own work became the main foundation of modern geology and, despite himself, he laid the groundwork for Darwin's theory of organic evolution. Lyell left a rich legacy both as a scientist and as a human being and was buried in Westminster Abbey.

A. M. C. ŞENGÖR

For Further Reading: E. B. Bailey, *Charles Lyell* (1962); D. Dean, "Lyell (ch. 10)" in *James Hutton and the History of Geology* (1992); Proc. Lyell Centenary Symp., *British Journal for the History of Science* (1976); L. G. Wilson, *Charles Lyell—The Years to 1841: The Revolution in Geology* (1972).

M

MALARIA

Malaria is a serious infectious disease of humans that causes fever, chills, anemia, and sometimes death. It occurs in over one hundred countries and causes between one and two million deaths each year. Malaria in humans is caused by four protozoan species of the genus *Plasmodium*. The infection is transmitted through the bite of an infected female anopheline (of the genus *Anopheles*) mosquito. The protozoan parasites establish a cycle of infection first in the liver and then in the bloodstream, where red blood cells are invaded by the parasites, which then reproduce asexually. The characteristic recurrent fevers of malaria are caused by the synchronized destruction of infected red blood cells and the liberation of new parasites. When a female anopheline mosquito bites an infected human, it ingests red blood cells containing the parasites in the sexual stage of development. Sexual reproduction occurs within the gut of the mosquito, and the infectious forms of the parasites then migrate to the mosquito's salivary glands, ready to be transmitted to a new host.

The distribution of mosquitoes and the transmission of malaria are affected by changes in the mosquito's environment, e.g., humidity, rainfall, and altitude. Temperature appears to be the most important, affecting the time required for a mosquito to develop to adulthood, its life span, and the time necessary for the malaria parasite to become infectious within the mosquito's gut.

Campaigns to eradicate malaria began in the 1950s, after the discovery of DDT and other insecticides appeared to promise a quick solution to this ancient scourge. However, global eradication efforts were not successful, and the disease has since made a substantial comeback in many parts of the world. DDT is still used to control the anopheline mosquito in affected countries such as Brazil.

Human-induced changes in the environment can also significantly alter the transmission of malaria in an area. For example, agricultural development, construction projects, deforestation, and rice cultivation have all been reported to increase the prevalence of malaria in surrounding areas, generally due to the creation of new breeding sites for mosquitoes. In addition, human migration has led to a much higher prevalence of malaria in many parts of the world. When a large proportion of immigrants with little or no immunity to malaria enter an area with a previously low level of the disease, transmission can increase tremendously and result in severe outbreaks.

MALNUTRITION

See Health and Nutrition.

MALTHUS, THOMAS ROBERT

(1766–1834), English political economist. T. Robert Malthus was born in Wooton, Surrey. He attended Jesus College, Cambridge, where his principal subject was mathematics. In 1793 he became curate at Okewood in Surrey. The same year, he was elected fellow at Jesus College, a position which he resigned when he married in 1803. In 1805 he was appointed professor of history and political economy at the East India College.

In addition to writings on population, which made him famous internationally, Malthus published a series of pamphlets on economic subjects, including his 1815 *Inquiry into the Nature and Progress of Rent*, which started a long and friendly controversy with David Ricardo, a contemporary English economist. The *Principles of Political Economy* appeared in 1820.

In 1798 he published anonymously *An Essay on the Principle of Population, as it affects the future improvement of society, with remarks on the speculations of Mr. Godwin, M. Condorcet, and other writers.* The book expressed pessimism on the perfectibility of mankind, the possibility of social equality, and the spread of liberty and happiness. British intellectuals familiar with the ideas of the Enlightenment were then confronting a particular emergency, "that tremendous phenomenon on the political horizon, the French revolution," which threatened social order and private property by its excesses. Malthus based his defense of the *status quo* on the necessary balance between population and the availability of subsistence. Nature was setting limits to production that could not be exceeded through the reform of society. He stated two postulates: that food was necessary to human existence, and that the passion between the sexes was necessary and would not diminish. Room and nourishment were regulated by scarcity, while the sex drive would inevitably lead to marriage and population increase. The tendency of population to grow in a geometric progression while subsistence could only grow in arithmetic progression would be checked by vice and misery. Upper classes would be guided in their behavior by the desire to maintain status, but any increase in the mass of laborers would be cut back by misery inherent in a level of real wages depressed by the supply of labor. Malthus followed the ideas of Adam Smith and earlier economists who had assumed that the reproduction of the laboring classes was limited by the ability of the worker to support a family, and that the excess number of children was checked by mortality. Individuals could improve their lot by avoiding marriage and by hard work, and necessity would provide them with incentives to better their conditions; but social inequality would prevail.

The work was an immediate success, and Malthus spent the following years reading and traveling in continental Europe to buttress his arguments with more facts. The much bulkier second *Essay on the Principle of Population; or, a view of its past and present effects on human happiness* . . . (1803) went through seven editions, the last one posthumous. The discussions in the second *Essay* ranged widely over the world to illustrate how the operation of the principle was prevented by social organization. Common property would lead to shared poverty; the institution of private property while preserving inequality was insuring a surplus in favor of social classes able to further the arts and sciences that made technical progress possible. Upper classes served necessary functions: Landlords benefiting from the rent of the land purchased the luxuries and services that sustained industry and city life, while the profit made by entrepreneurs stimulated productivity. The imperfections of society made for less than optimal utilization of resources. This was more so in the Ottoman Empire or the Spanish possessions of America where much of the land had been transformed into desert by reckless exploitation, than in Europe or the United States. China, on the other hand, was an example of what would result from the maximum use of resources made possible by the egalitarian policies of a benevolent ruler: life at a bare minimum of subsistence, overpopulation tempered by the resort to infanticide. In England the bad government of the past had allowed a reserve of land; this allowed the future progression of subsistence and population. Malthus granted that the fate of the British laboring classes had been improving. The length of life had increased, thanks to the practice of late marriage and permanent celibacy—what he called the *preventive check* of moral restraint. The growth of population be-

yond what could be supported would cause the alternative of the *positive check* of higher mortality through vice and misery. Malthus was opposed to the Poor Laws: the system of parish-based assistance that encouraged dependence, improvident marriages, and the generation of more paupers.

In Malthus's time the basis of the economy was agricultural, and he did not foresee the release of productive powers by the industrial revolution and the use of fossil fuel. The picture painted by the *Essay* has been confirmed by demographic historians as a description of the pre-industrial economy of England, with the joint operation of nuptiality and mortality to adjust population size to available resources. Malthus's model operated at the family level within a stratified social system utilizing the limited resources of a finite world. It fitted the times but would prove inadequate as a guideline for future policy. He reasoned on a closed economy and underestimated the role of trade and migration in detaching population growth from a close relation to agricultural production. Moreover, he did not foresee that fertility limitation within marriage would check the growth of European populations, and was opposed to contraception which he considered vicious. Malthus's views on property alienated later socialist thinkers such as Marx and Engels, who were more aware of the possibilities of economic expansion and wanted to reform society.

Malthusianism has maintained its intellectual appeal because it posits the existence of limits. These limits once involved the availability of cultivable land, but the idea can be generalized to the scarcity of raw materials or even living space. Limits can be pushed back in the short run by social arrangements, but Malthus's message is that they will inevitably reappear.

ETIENNE VAN DE WALLE

For Further Reading: Patricia James, *Population Malthus: His Life and Times* (1979); William Petersen, *Malthus* (1979); E. A. Wrigley, "Elegance and Experience: Malthus at the Bar of History," in David Coleman and Roger Schofield (eds.), *The State of Population Theory: Forward from Malthus* (1986). *See also* Economics; Human Ecology; Land Use; Limits to Growth; Population; Smith, Adam.

MARINE BIOLOGY

See Ecology, Marine; Oceanography.

MARINE PROTECTION, RESEARCH, AND SANCTUARIES ACT

See United States Government.

MARSH, GEORGE PERKINS

(1801–1882), American statesman, diplomat, scholar, and father of the conservation movement. George Perkins Marsh was born in Woodstock, Vermont in 1801 to Charles and Susan (*nee* Perkins) Marsh. Charles Marsh was a lawyer, politician, and leading businessman in the town, who also had a strong interest in science and the natural world. The Marsh home was beautifully situated at the bend of the river on the lower slopes of Mount Tom, and the family's comfortable economic status gave the young Marsh the leisure to study and appreciate his natural surroundings from the beginning. The seeds of his magnum opus, *Man and Nature; or, Physical Geography as Modified by Human Nature*, "the fountainhead of the conservation movement," were planted here as he watched lumbermen stripping frontier Vermont of its forest cover for fuel while sheep grazed the eroding hillsides and silt clogged the rivers and streams. As he matured he began to understand the effects of these human actions on the droughts and floods which were becoming ever more common.

This studious youngster was forced to abandon his books because of an eye affliction which recurred intermittently all of his life. Instead of reading he accompanied his father on excursions through the countryside, during which he developed a capacious memory and an enduring sympathy with nature. It was on one of these excursions that his father explained the concept of the watershed to him, and this concept would become the germ of the great work to come. The greatness of *Man and Nature* is due to the evolution of Marsh's thought and the synthesis of study and observation gained over a lifetime. The book was published in 1864 when he was 63 years old.

Marsh was educated at Dartmouth College where he developed an inclination toward the humanities, particularly linguistics. After graduation in 1820, he taught at Norwich Military Academy for a few years before turning to law. The self-taught attorney had moderate success in a Burlington, Vermont, practice from 1825, but as was common at the time, he pursued various other business opportunities to help make his fortune. He bred sheep, sold lumber, edited a newspaper, developed a marble quarry, and speculated in real estate. All of these ventures were failures because of lack of interest and attention: Marsh's love was scholarship, and he found business boring. After only five years of marriage, his first wife, Harriet Buell, died in 1833, and he turned his attention to the study of Scandinavian languages to escape his grief. He published the first Icelandic grammar in English and pursued his studies on the "Gothic origins of the New England race." In 1839, he met and married Caroline Crane, a kindred spirit in scholarship and love of nature. Shortly thereafter he was encouraged to run for Congress, and he was elected as a Whig Republican in 1843, serving three terms. His greatest achievement was his contribution to the founding of the Smithsonian Institution as a center for original research.

Marsh petitioned President Zachary Taylor for a diplomatic post in Europe where he could study and travel, and he was pleasantly surprised by an appointment as Minister in Residence to the Ottoman Empire in 1849. He used these years to travel extensively in the Middle East where he was able to witness the civilization's longterm effects on nature: he saw "The very earth, the naked rocks and sands of the desert . . . the meadows levelled and the hills rounded . . . by the assiduous husbandry of hundreds and hundreds of generations." Here was the advanced result of the incipient process he had watched as a child in Vermont.

The election of Democrat Franklin Pierce as president in 1852 brought Marsh reluctantly back to Vermont, where he lectured, completed major studies on the English language, and published *The Camel* (1856), a fascinating study which resulted in the importation of over 100 camels to the deserts in the U.S. southwest and prefigured the lessons Marsh would later learn from his survey of lands laid waste through the hand of man.

In 1857, Marsh was appointed simultaneously as head of three Vermont commissions: the Statehouse, Railroads, and Fish. As Statehouse Commissioner, he presided over the design of the finest specimen of Greek Revival architecture among state capitols. As Commissioner of Railroads, he advocated public ownership of transportation and communication services—a reversal of his former *laissez-faire*, utilitarian views, and a view that would figure in his later writings. But it was as Fish Commissioner that he gave intimations of his philosophy of natural history and his indictment of man's treatment of the land and its creatures. His *Report on the Artificial Propagation of Fish* was published in 1857, and by the spring of 1860 he had substantially conceived the book that became *Man and Nature*. He would write this book in Italy, following his appointment in 1861 by President Abraham Lincoln as minister to the new Kingdom of Italy. Marsh served in this post until his death in 1882.

Marsh divided *Man and Nature* into three principal sections—"The Woods," "The Waters," and "The

Sands"—with an introduction and a concluding chapter entitled "Projected or Possible Geographical Changes By Man." The introduction ascribes the fall of the Roman Empire to bad laws and human-induced disturbances to the physical environment. The conclusion attempts to analyze the potential impacts of major land projects, such as the Suez Canal, underway in 1864. Marsh's thesis was that man disrupts the fundamental balance or harmony of nature, which if left undisturbed is basically stable. Natural forces such as storms inflict superficial damage that heals quickly. Man, on the other hand, abetted by modern technology, can permanently transform the earth. Destroying the forests, plowing the soil, draining the bogs, and channeling the streams had already devastated much of the Old World. "The earth," he said, "is fast becoming an unfit home for its noblest inhabitant, and another era of equal . . . human improvidence" would perhaps result in the "extinction of the species." Unlike Darwin, whose *Origin* is often compared to *Man and Nature*, and who felt that man was a part of nature, a major tenet of Marsh's book was to enforce the opposite opinion, that man "so far from being . . . a soul-less, will-less automaton, is a free moral agent working independently of nature." The problem as he saw it in the mid-nineteenth century was the abuse of this stature. Marsh proved man to be a "disturbing agent" rather than a neutral presence as was believed before the publication of his work.

Marsh was a man of his time. He was a humanist like the Transcendentalists before him, seeking to preserve the wilderness for its own sake, but he was also a utilitarian. Where the romantics failed to convince a generation that was bound for Manifest Destiny, he provided the economic reason to preserve the wilderness: "[M]an has too long forgotten that the earth was given to him for usufruct alone, not for consumption, still less for profligate waste." This was not an academic question to Marsh, but involved the earth's ability to support mankind. Primarily because it made protecting wilderness compatible with progress and economic welfare, his argument became a staple for preservationists.

Marsh was particularly concerned about deforestation because it led to the drying up of watersheds and the resultant erosion. The most famous lines from *Man and Nature* warn: "Even now we are breaking up the floor and wainscoting and doors and window frames of our dwelling for fuel to warm our bodies and seeth our pottage." The book balances this warning with the pragmatic optimism that reform would follow understanding, and the purpose of the book was to provide understanding. Man could reverse this process of destruction and deterioration through conscious effort and through the powers of science. This optimis-

tic view of the future is the basis for present day restoration ecology.

CONNELL B. GALLAGHER

For Further Reading: Larry Anderson, "Nothing Small in Nature" in *Wilderness* (Summer 1990); David Lowenthal, *George Perkins Marsh* (1958); George Perkins Marsh, *Man and Nature*, ed. David Lowenthal (1965).

MASS-ENERGY EQUIVALENT (E = MC²)

The famous formula $E = mc^2$ was derived by Albert Einstein (1879–1955) in his Special Theory of Relativity (1905), which explored the relation between the energy (E) and mass (m) of a particle moving at the velocity of light (c). Thus, if the mass of a body changes by m, the energy change is $E = mc^2$. Nuclear weapons, nuclear power plants, and the sun's energy are all consequences of mass being converted into energy.

The concept was originally proposed in 1900 by H. Poincaré, but proven by Einstein. Einstein was interested in describing how a particle of mass at rest undergoes change when it is set in motion: Does it acquire energy, or does it contain energy that is released when set in motion? He discovered that mass and energy are the same, but in different forms, and that all particles contain "resting energy," which is released under certain circumstances and becomes moving energy. In a nuclear reaction, mass is converted into energy by using neutrons to split the nucleus of an atom into its components. Concurrently, there is a release of neutrons from the atom and a large amount of energy. A chain reaction results from the release of neutrons, which collide with and split other atomic particles, thus releasing significant quantities of energy. Einstein's theory provided the basis for understanding both nuclear fission and fusion and led to the development of the atomic bomb and the hydrogen bomb.

Up to the time of Einstein's publication of his theory in 1905, scientists thought that light and other forms of electromagnetic radiation did not follow the principles of classical physics. Previously, the principles of Newtonian physics (based on the work of Sir Isaac Newton, 1642–1727) dictated that absolute values existed for space and time, which could be used as fixed points against which acceleration and force could be measured. Later, however, other scientists refuted this notion in its application to light, and proved (in 1881 and 1887) that the speed of light is constant and does not vary with the motion of either the source of light or the observer. Einstein examined this idea and concluded that only the speed of light was absolute regardless of the movement of the observer. He further postulated that nothing can equal or exceed the speed of light; if it did, its length would be zero, its mass would be infinite, and time would stop. If one accepts that the velocity of light is invariable, then the speed of light (c) can be used as a means of comparing time by observers who are in uniform relative motion. In other words, Einstein discovered that the measurement of the speed or velocity of all objects is related to the location and movement of the observer. He further deduced that, as the speed of a particle approaches the speed of light, its energy increases. Because c^2 in the mass–energy equation is so large, only a small amount of mass is equivalent to a vast amount of energy. The conversion of mass into energy occurs in nuclear reactors, nuclear weapons, and stars.

LUISA M. FREEMAN

MATERIALS BALANCE

Mass and energy are independently conserved in most processes on Earth's surface. Mass conservation implies that the weight of all material inputs to every process must exactly equal the weight of all material outputs, including wastes. The material inputs to, and outputs from, every process must balance. This concept, known as materials balance, is a powerful analytical tool for chemical engineering and is equally powerful, though little used, for process analysis at the plant level, the industry level, the regional level, and even the national or global levels.

The consolidated inputs and outputs for the U.S. steel industry illustrate the materials balance principle at the industry level of analysis. Among other things, such analysis can be used to determine how much of each kind of raw material must be extracted and processed to produce a single ton of "average" steel. Data of this kind is increasingly important in so-called life cycle analyses. Only by carefully comparing consumer products in terms of the entire chain of materials is it possible to resolve such questions as whether styrofoam or paper plates are more environmentally friendly.

The materials balance principle also has important global economic and environmental implications. The word *consumption* is much used in economics. However, it is a slippery and potentially misleading term. Materials and material products (unlike pure services) are not really consumed. When so-called consumption goods (except food and drugs) are used up, they are actually dispersed or chemically transformed. For instance, fuels are burned in air and converted into carbon monoxide, carbon dioxide, and other gaseous pollutants. So-called durable goods are worn out or become obsolete. All of the physical materials extracted

and processed each year by the economic system are either incorporated into some long-lived structure or eventually discarded as waste. Only a small percentage of the total mass of materials processed is embodied in products or structures that last for more than a decade, such as books, machines, or buildings.

Because mass is conserved, however, the discarded materials do not physically disappear after their utility is exhausted in the economic sense. Instead, they become waste residuals. In fact, it is not difficult to show that the tonnages of waste residuals are actually greater than the tonnages of crops, timber, fuels, and minerals recorded by economic statistics. Waste residuals tend to disappear from the market domain but they reappear in the external environment, where they can have many adverse effects.

R. U. AYRES

MATERIALS RECOVERY FACILITY

A materials recovery facility (MRF) is a processing plant where commingled used bottles and cans are separated using mechanical and manual techniques. The materials are then prepared for shipment to recycling markets by crushing, shredding, or baling.

Since first introduced in the late 1980s, the term has taken on other, less frequently used meanings. Some operators of source separation facilities label their plants MRFs. Facilities where recyclables are separated from mixed municipal solid waste are also labelled MRFs. (Some recycling industry members term these "dirty MRFs.")

The purpose of a commingled MRF is two-fold. First, recycling preparation and collection requirements are improved over a source separation program. Residents merely set out commingled bottles and cans at the curb, thus minimizing the slight inconvenience of storing these recyclables separately in the home. Additionally, fewer recycling trucks and crews are needed, because the recyclables can be placed in one bin on the recycling trucks, rather than in separate metal, glass, and plastic bins at each stop.

The second benefit of a MRF is the facility's economy of scale. Rather than requiring a single community to establish a processing center, MRFs allow numerous communities to use one large, efficient, cost-effective plant. For instance, most New Jersey counties have established a MRF for use by local communities. As a result, hundreds of Garden State cities and towns now have curbside collection of commingled recyclables.

The typical MRF uses a variety of techniques to accomplish the processing needs. After the commingled bottles and cans are dumped on the floor, a loader pushes the materials onto an incline conveyor. A magnet is used to extract the iron-containing food and beverage cans, which are then shredded or baled. The remaining aluminum, glass, and plastic containers fall onto a vibrating belt that has a moving chain hanging over it. The heavier glass bottles pass through the chain, while the lighter aluminum and plastic containers are pulled off the belt by the chain. Glass is manually sorted by color and crushed. Aluminum cans are separated and then briquetted or baled. Plastics are sorted by resin type and baled.

MRF operations are changing rapidly. Whereas the first MRFs in the United States and Canada were fairly large (above 250 tons per day in processing capacity), numerous mini-MRFs (with 25 to 50 tons per day of processing capacity) are now in operation. Also, MRF operators are developing new systems to reduce the number of workers performing tedious sorting jobs. For instance, several MRF operators are experimenting with optical-glass-sorting units that detect bottle color. In addition, several firms are employing infrared and ultraviolet fluorescence technology to distinguish plastics by resin type.

Presently more than 100 commingled sorting plants operate in the United States, with the largest plants approaching a capacity of 400 tons per day. The majority of the facilities are in the Northeast and Mid-Atlantic regions, where the population densities support the program's economics. (Large MRFs can cost from $3 million to $8 million and require $50 per ton to operate.) Nearly all MRFs are privately operated, with many plants owned by local governments—often a county—and built and operated under contract by a private firm.

JERRY POWELL

MCCALL, THOMAS

(1913–1983), Oregon governor and conservationist. Thomas McCall, a Republican, used his position as Oregon's governor (1967–75) to introduce a doctrine of conservation in balance with economic concerns. Rejecting strict preservationism, he showed how environmental controls actually encouraged economic growth by preserving his state's livability and its forest and agricultural base. Under his leadership, Oregon became a pioneer in statewide land-use planning, coastal protection, strict industrial pollution standards, recycling and energy conservation.

McCall's doctrine relied on a slow-growth policy that rejected unbridled economic boosterism. To dramatize his views, in 1971 he told a national television audience, "Come visit us again and again. Oregon is a land of excitement. But for heaven's sake, don't come here to live."

McCall's declaration was characteristic of his brash style. His political strength lay in his eloquent use of language to describe environmental threats to his state, and in his ability to marshal public support for his reforms. His independence made him a modern folk hero in Oregon and a national environmental spokesman for land-use reform, bottle and can deposit laws and energy conservation.

Thomas William Lawson McCall was born March 22, 1913, in Scituate, Massachusetts, into a family of wealth and influence. His paternal grandfather, Samuel Walker McCall, was a Massachusetts congressman and governor. His mother's father was Thomas W. Lawson, a flamboyant Wall Street raider who became one of the world's richest men. This heritage invested in McCall a commitment to public service and an eccentric manner. Raised in part at Lawson's Dreamwold estate in Scituate, he spent most of his upbringing in the central Oregon desert after Lawson's fortune collapsed. His parents had established a dairy ranch along the Crooked River, and the vast, arid landscape of the Oregon desert forged his concern for nature.

McCall began his career in 1936 as a sports reporter in Idaho and later joined the Portland *Oregonian*. Later, he entered radio and television as a political commentator. During the infancy of television news, he used documentaries to highlight the conflicts between industry and the environment. In 1958 he warned of the imminent clearcutting of the massive Klamath Indian forest in southern Oregon, which supported the region's economy and wildlife. His documentary sparked support for bringing the forest under federal control and sustained-yield practices. His most important work was "Pollution in Paradise," a 1962 documentary on the polluted Willamette River, Oregon's main waterway. State officials had led Oregonians to believe a four-decade effort to clean up the river had succeeded. McCall vividly documented unchecked pollution by pulp mills and local sewage systems. The award-winning documentary ignited public outrage over the Willamette and prompted a 10-year clean-up effort that culminated under his governorship.

His family's heritage of public service drove McCall toward politics, and McCall's fame as a broadcaster propelled him into the governorship in 1966. He forged a new doctrine that put economic and environmental concerns in balance. While hoping to protect his state's aesthetic beauty, he believed foremost that uncontrolled growth and pollution threatened Oregon's natural resources base. He angered many business leaders with his stiff opposition to polluting industries coming into the state. "Oregon has not been a lap dog to the economic master," he told business leaders during a 1969 industrial recruitment tour. "Oregon has been wary of smokestacks and suspicious of rattle and bang." At home his statements drew praise for their environmental sensibility while generating criticism for encouraging a provincial outlook toward Oregon's economy. Nevertheless, Oregon's population and economy grew steadily during his administration.

McCall helped Oregon create a national model for environmental reforms. Stringent pollution standards on pulp mills and municipal sewer systems turned the turbid Willamette into a thriving river by 1972. In 1967 McCall successfully championed the protection of Oregon's beaches from private ownership and development, guaranteeing that the 429-mile Oregon coastline remained public and accessible. He was one of the first political leaders to encourage recycling when he led efforts to win approval of the nation's first mandatory beverage container deposit law, known as the Bottle Bill. He preached energy conservation and introduced energy-saving standards long before the Arab oil crisis of 1973 shocked America. Oregon's energy conservation plans, which included the U.S.'s first voluntary gasoline rationing, served as a national standard during that crisis.

The hallmark of McCall's career was Oregon's land-use planning laws, the first comprehensive planning reform of its kind in America. These laws gave the state control over development standards that emphasized protection of farm and forest lands. The reforms passed in two stages in 1969 and 1973—over opposition from developers who argued the laws usurped private property rights. He again rallied public support, denouncing the opponents as "grasping wastrels of the land" who hoped to overrun Oregon with "sagebrush subdivisions" and "coastal condomania." The resulting Land Conservation and Development Commission has since regulated land use based on statewide goals.

McCall left office in 1975, but remained prominent as a spokesman for land-use planning and mandatory deposit legislation. Between 1975 and 1982 he served on the boards of the Nature Conservancy, the Conservation Foundation and the Center for Growth Alternatives. In 1981 he was named executive chairman of The René Dubos Center for Human Environments.

McCall tried to make a political comeback, running again for governor in 1978. However, his progressive views by then had made him a pariah within his own Republican party, and he lost his bid in the GOP primary.

The economic recession of 1980–83 battered Oregon's timber industry, and many business leaders blamed McCall's environmental legacy for the state's economic problems. He steadfastly stood by his doctrine, however. Dying of cancer, he spent 1982 fighting a ballot initiative that would have repealed Oregon's land-use laws. Opponents of land-use planning used the state's economic problems to whip up support for the measure. Yet McCall's impassioned pleas to protect Oregon's environmental standards led to the measure's defeat. He died two months later, on January 8, 1983, in Portland.

BRENT WALTH

For Further Reading: H. Jeffrey Leonard, *Managing Oregon's Growth: The Politics of Development Planning* (1983); Charles Little, *The New Oregon Trail* (1974); Tom McCall, with Steve Neal, *Tom McCall: Maverick* (1977).

MEAD, MARGARET

(1901–1978), American anthropologist. Margaret Mead became world famous in 1928 with her book *Coming of Age in Samoa*, based on her first field trip. The book became an instant classic, bringing new scope to the field of anthropology by incorporating psychological themes, utilizing interviews more than statistics and, not least, by being provocative and well-written. Its descriptions of the sexual mores of Samoan teenagers titillated, outraged, and sold books to a huge public that would never otherwise have read a scientific report.

Mead stayed world famous—and controversial—in more than a half century's association with the American Museum of Natural History in New York City. She involved herself in most subjects of public interest, explaining they were "all anthropology." She courted attention. As a young scientist, 5 feet 2 inches tall, weighing barely 85 pounds, she dressed in native style to win the cooperation of local leaders and the confidence of the adolescents she interviewed; in later years, as a veteran campaigner on social issues, her tiny frame grown heavyset, she took to wearing dramatic capes and using a unique shoulder-high walking stick that she wielded like a scepter.

When she died, of cancer, November 15, 1978, it was front-page news. She was publicly mourned by the President of the United States, the Secretary-General of the United Nations, many professional organizations, and probably millions of people who rallied to her causes around the world.

Mead was born in Philadelphia, December 16, 1901. Her father was an economics professor, her mother a sociologist, and her paternal grandmother, who babysat and tutored her, a school teacher and principal. Margaret was the oldest of five children in a family that was education-oriented. As a child she wrote poetry and started a novel. At age 8, she was taught by her grandmother to draw and identify plants; at age 11, she helped her mother in a study of Italian immigrants; also at 11, although her parents were indifferent to religion, she had herself baptized an Episcopalian and was an ardent church-goer all her life. At age 16, she became engaged to her first husband whom she married six years later. She married and divorced three times. Her husbands were Luther Cressman, minister turned sociologist; Reo Fortune, an Australian anthropologist; and George Bateson, a British anthropologist, with whom she had a daughter who became a fellow anthropologist and one of her parents' biographers.

Mead first attended her father's college, DePauw, in Indiana, but switched to Barnard, a part of Columbia University in New York City, where she majored in psychology. She took an anthropology course with the renowned Franz Boas, whose assistant, Ruth Benedict, became her intimate friend and mentor. Mead received her B.A. in 1923, an M.A. in anthropology in 1924 and a grant from the National Research Council for a field study in 1925. The rest became history.

Coming of Age in Samoa reinforced Boas's long-held contention that basic human capacities were not, as many scholars asserted, racially determined. Mead added her own findings that the emotional vexations of adolescents were cultural rather than biological. She contrasted the relative calm of Samoan girls, exposed uninhibitedly to sex, childbirth, and death, with the distress of adolescents in Western societies.

Mead published over the next several years *Growing Up in New Guinea, Sex and Temperament in Three Primitive Societies, Balinese Character*, and *New Lives for Old*. These presented her conclusions that social mores could not be altered simply by imposing behavior patterns on children; sought-after changes required adults to change their own values and conduct. There were no significant personality differences between men and women due to biology; when traditional male-female roles were reversed, temperamental differences did not occur; there was no "right" or "wrong" way to be male or female. Mead said afterward the "sex and temperament" studies were her most important.

Mead wrote 24 books based on her research and co-authored or edited an additional 18 books. She wrote innumerable scientific papers, but also articles for the popular press. She was called "the busiest, most influential American female writer and thinker in the social sciences of this century." But her tendency to use flippant "one liners" reinforced a somewhat widespread criticism that she sacrificed scientific accuracy for sweeping assertions. Despite such attacks, she was

elected president of the American Association for the Advancement of Science in 1974, the first anthropologist since Boas, in 1931, and the first female to achieve that honor. She was president of several anthropological organizations.

Not surprisingly she became an ardent leader in the environmental movement, characterizing it as a "revolution in thought comparable to the Copernican revolution" that taught that the Earth revolved around the sun and not the other way around. She was a founding member and president of the Scientists Institute for Public Information (SIPI). She was a founder of International Earth Day. She went to Stockholm in 1972 to speak and demonstrate at the first U.N. Conference on the Human Environment. She went to Bucharest in 1974 for the U.N. Conference on Population. She and René Dubos headed an 18-member committee that successfully called upon the World Council of Churches to oppose large scale use of plutonium in nuclear power plants. She served as President of Constantinos Doxiades' World Society for Ekistics, a term the Greek city planner devised to represent the science of human settlements. She was a regular on Doxiades' annual yacht cruises in the Aegean Sea on which he invited prominent professionals from throughout the world to philosophize on current issues.

JACK RAYMOND

For Further Reading: Jane Howard, *Margaret Mead, A Life* (1984); Mary Catherine Bateson, *With a Daughter's Eye* (1984); Margaret Mead, *Coming of Age in Samoa* (1992 17th printing; originally published 1928).

MEDICAL WASTE

Medical waste, often called "biohazard waste" or "infectious waste," is a broad term used to describe a variety of solid and liquid wastes produced by health care facilities, funeral homes, veterinarians, research facilities, and private citizens.

Interest in medical waste peaked during the summers of 1987 and 1988, when needles, syringes, tubings, blood vials, and other waste associated with medical care washed ashore on the northeastern coast of the United States. These washups resulted in aesthetic degradation of the beaches, increased public concern regarding perceived health hazards, and beach closures, with associated economic loss to communities. However, investigations in New York and New Jersey concluded that the majority of the wastes (99%) were not mismanaged hospital wastes but improperly discarded boating trash, household waste that had been intended for the Fresh Kills landfill in New York City's harbor, and needle and syringe disposal due to illegal drug use. In addition, changes in the normal patterns of prevailing winds and tide currents contributed to the washups.

In response to public concern, the United States Congress quickly enacted the Medical Waste Tracking Act of 1988 (MWTA) to regulate the disposal of medical wastes and thus avoid similar washups. This law, which was an amendment to the Resource Conservation and Recovery Act (RCRA), Subtitle J, became effective on June 22, 1989. It required the Environmental Protection Agency (EPA) to promulgate regulations, to set standards for proper medical waste management, and to establish a "cradle-to-grave" tracking system for medical waste in certain areas of the country. The resulting EPA regulations defined seven categories of regulated medical waste (RMW) and regulated generators, transporters, and other waste handlers; disposal of home-generated medical waste was excluded from regulation. The MWTA, which was intended to be a two-year demonstration program, involved Connecticut, New Jersey, New York, Rhode Island, and Puerto Rico. The program was not reauthorized by Congress and expired on June 22, 1991.

The MWTA, along with public perception of potential health risks associated with improper medical waste disposal, spurred interest in medical waste issues in many states not covered by the Act, resulting in passage of various state regulations regarding medical waste management. In addition, other federal agencies such as the Centers for Disease Control and Prevention, the Department of Transportation, and the Occupational Safety and Health Administration, became involved with medical waste regulation.

Disposal of most regulated medical wastes is currently accomplished by incineration. Other disposal methods include landfilling (with or without prior treatment to render the waste noninfectious) and sewer discharge of liquid wastes. There are several methods available for the treatment of medical wastes, including dry heat, autoclaving (steam sterilization), chemical disinfection (e.g., chlorine), and microwave and radiofrequency technologies. Other treatment technologies currently in the experimental stage include electron beam irradiation and electropyrolysis.

Due to the trend in using disposable items to avoid disease transmission, approximately 20% of medical waste consists of plastic. This has led to the development of methods for recycling the plastic portion of the waste; however, such recycling is not a widely instituted practice. Reduction and reuse strategies are being investigated by facilities as options to decrease their medical waste stream.

DIANN J. MIELE

MEDICINE

See Health and Disease.

MEDICINE, ALTERNATIVE

Alternative Medicine refers to a number of therapies, techniques, and systems of healing not widely accepted within the conventional Western medical model. However, many alternative therapies — acupuncture, yoga, meditation, homeopathy, guided imagery, hypnotherapy, and biofeedback, for example — have become more widely used as a result of increasing interest on the part of health consumers and growing scientific evidence of their effectiveness.

Much of this interest can be traced to the holistic health movement that began during the 1960s and 70s. Holistic medicine espouses a more comprehensive approach to the treatment and prevention of physical ailments, taking into account psychological, emotional, social, economic, and environmental causes of illness (and health) that can act alone or in concert. This view has always been a theme — though often neglected — within Western medicine itself since the time of Hippocrates and a central tenet within diverse traditional cultures around the world. The holistic approach to health care also assigns responsibility for health to the individual as much as to the physician.

Alternative Medical Systems and Techniques

There are numerous alternative practices. Some represent complete systems of medicine; others are specific techniques. Those most widely applied in the U.S. have their sources in traditional beliefs (for example, in the close relationship between mind and body, and in the body's inherent healing powers), traditional techniques (such as massage, acupuncture, herbal remedies, and meditation), and the clinical proof and application of these principles through modern scientific methods.

Acupuncture was developed by the Chinese. All Chinese medicine is based on maintaining the proper balance and harmony of the body's vital energy, known as Qi (pronounced CHEE), and the body's five basic elements (wood, fire, earth, metal, and water). Qi flows through channels or meridians of the body, which can be therapeutically altered through an acupuncturist's application of needles or by massage (acupressure) at precise points on these meridians. Chinese herbal medicine is also aimed at restoring the proper balance of energy (as between the complementary principles of Yin and Yang) and the elements within the body, and is an extraordinarily complex art, involving a pharmacopeia of thousands of natural remedies. Acupuncture is becoming more widely used in the U.S., especially for pain relief, arthritis, and drug addiction, although its applications are much more numerous and diverse in China.

Ayurveda is based on ancient healing techniques of India, many of which developed in tandem with Hindu religion and philosophy. As in Chinese medicine, Ayurvedic diagnosis and therapy sees the body as composed of basic elements (earth, fire, water, and so on) and vitalized by energy (prana) that travels through various channels of the body. Diagnosis involves examination of pulses as well as body fluids. Yoga and diet are important elements of treatment. Ayurvedic medicine in general views the individual's body and mind as integrated aspects of the whole organism.

Herbal medicine, a part of many traditions, from Chinese to Native American to shamanic cultures around the globe, effectively uses balms and medications prepared from the roots, flowers, leaves, and other parts of plants. The Chinese medical system includes one of the most complex and elaborate systems of herbal medicine.

Homeopathy was devised by the German physician Samuel Hahnemann in the 19th century, and was used widely by American physicians until the beginning of this century. Its basic tenet, "like cures like," postulates that a substance that would ordinarily produce symptoms of illness in a healthy person can cure those same symptoms in someone who is ill. The other scientifically controversial tenet of homeopathy is its use of highly diluted substances as medication—the more diluted, it's believed, the more powerful. Homeopathy is still widely practiced in England, France, India, and elsewhere, and is once again growing in popularity in the United States.

Naturopathy is an approach to the prevention and treatment of illness with diet, exercise and other "natural" remedies, including herbs and homeopathic medicines, rather than conventional drugs or surgery. As such, naturopathy is a hybrid form of medicine rather than a complete system (like homeopathy).

Biofeedback, first used in the early 1970s, is a particularly good example of cross-fertilization between traditional and modern principles. Based on the understanding that individuals can control "involuntary" body functions through various forms of mental concentration, a biofeedback apparatus uses modern technology to measure these responses and signal (with sounds or visual displays) the patient's progress, thus aiding in the conscious control of these physiological functions.

Guided imagery (or visualization) is the use of mental imagery deliberately conjured up to facilitate healing or pain relief.

Hypnotherapy is a method of inducing a mental state characterized by extreme suggestibility that allows patients to relax, control pain, and alleviate symptoms.

Meditation takes hundreds of forms around the world, but all involve focusing the mind's attention on either the breath, parts of the body, an image, words or a sound, or an external object in order to calm and relax the mind. Many forms of meditation stem from religious traditions and are considered spiritual practices. However, the relaxation, calm, and mental well-being cultivated by meditation were also found to have a beneficial effect on human physiology, as demonstrated in the research of Dr. Benson and many others in the West. Dr. Benson's "relaxation response" is one Western application of basic meditative techniques for alleviating illness and symptoms.

Yoga derived from physical exercises used originally in the Indian Hindu tradition in the context of spiritual practice; the breathing and movement involved in yoga have long been known in the East to be beneficial to both the body and the mind. Along with physical movement, yoga incorporates aspects of meditation, like breathing and mental concentration.

Alternative and Western Medicine

During the past several decades, and particularly since the 1970s, a number of persuasive exponents of the holistic approach have helped bring various alternative medicines into the mainstream of Western medicine. The terms *behavioral medicine* and *mind-body medicine* are often used to describe the study of the connections between mind and body in relationship to physical health; there are now departments of behavioral medicine in many medical centers and teaching institutions.

In 1979 Norman Cousins published his landmark book, *Anatomy of an Illness as Perceived by the Patient: Reflections on Healing and Regeneration,* in which he described his recovery from an "incurable" degenerative disorder through a regimen of high doses of ascorbic acid (vitamin C) and, equally important, drawing upon his own recuperative powers by marshaling positive emotions, including laughter. *Anatomy of an Illness* argued cogently for modern medicine to take into greater account the role of the mind in health and healing and took the medical establishment to task for its over-reliance on technology and drugs. Cousins's writings built upon the work of medical researchers such as Walter Cannon, who, in *The Wisdom of the Body* (1963), wrote about the body's innate ability to heal itself without medical intervention, and Hens Selye, author of *The Stress of Life* (1956), whose ground-breaking work described the effects of psychological stress on physical health.

In the 1970s several other physicians contributed important research that gave greater credence to alternative approaches to medicine. Cardiologists Meyer Friedman and Ray H. Rosenman in *Type A Behavior and Your Heart* (1974) suggested that personality traits such as impatience and irritability could increase a person's risk of heart disease. Cardiologist Herbert Benson demonstrated in studies and in his book *The Relaxation Response* (1975) the power of meditation to lower blood pressure, heart rate, and other supposedly involuntary physiological functions.

Clinical studies have examined the effectiveness of techniques such as meditation, acupuncture, yoga, hypnotherapy, guided imagery, massage, and homeopathy in ailments such as arthritis, chronic pain, heart disease, some cancers, diabetes, asthma, infertility, and skin conditions. Basic research in neurobiology, cardiology, and immunology has also opened new doors.

In 1981, University of Rochester professor Robert Ader published his textbook, *Psychoneuroimmunology*, helping to launch an entirely new and important field of biomedical research. Ader conducted a series of careful experiments with rats demonstrating that the animals' immune response could be conditioned. In one set of experiments, rats were given a saccharin-sweetened solution to taste and simultaneously administered an injection of an immunosuppressive drug. The rats became conditioned to exhibit the same immune-suppressant response by drinking the solution without the injection of the drug. The saccharin solution, in other words, began to have a physiological effect resembling that of the actual drug.

Mental conditioning of the immune system has significant implications for the treatment of human illness. It casts an interesting light on the "placebo effect," wherein the placebo, an inactive substance, can sometimes equal the power of an active medication or other medical intervention. Placebos in the latter part of this century have been used primarily in clinical experiments; a drug or therapy that is no more effective than a placebo is deemed ineffective. But, increasingly, researchers are rediscovering that the placebo effect itself can have a legitimate and powerful curative effect and are working to discover its physiological mechanism. The placebo effect may have a central role in nearly all successful medical treatment, whether administered in a modern hospital or by a tribal shaman.

Many researchers are trying to identify the biochemical mechanisms responsible for communication between the mind and the body. Neuroscientist Candace Pert discovered neuropeptides, natural chemicals that relay signals between the brain, the immune system, and the endocrine system. These molecules and their receptors are found in the limbic system, a part of the brain believed to be strongly connected with emotions; the thymus gland; on the surface of immune system cells, and in other areas of the body, providing evidence of a biochemical pathway between our emotions and physiological functions.

The practical applications of harnessing the mind-body connection in a clinical setting have been the focus of much research. In 1989, psychiatrist David Spiegel demonstrated that women with metastatic breast cancer who underwent conventional medical care and "psychosocial treatment" — including support group meetings and training in self-hypnosis, guided imagery, and relaxation techniques — lived twice as long after the time they entered the study than a control group who received only the standard medical care. Cardiologist Dean Ornish, in a 1990 study, demonstrated that behavioral interventions such as yoga, meditation, and group support, as well as diet and exercise, can halt or even reverse heart disease without drugs or surgery. One experiment has demonstrated that stress increases a person's chance of catching the common cold virus.

There are currently thousands of health practitioners — many of them medical doctors — who apply some of these alternative or "complementary" methods to their patients. ("Complementary" medicine, the term now preferred by many practitioners, implies the integration of conventional and alternative medicine, rather than one precluding the other.) More than 2,000 physicians reportedly use acupuncture in conjunction with conventional medicine, approximately 5,000 use hypnotherapy, and another 1,000 physicians practice homeopathy — and the numbers appear to be growing. Of the more than 500,000 doctors in the U.S. involved in patient care, at least 10% use some form of alternative therapy. Many more non-physicians practice these therapies, as well as chiropractic (therapeutic adjustment and manipulation of the spine), naturopathy, herbal medicine, shiatsu (a Japanese massage technique using pressure points), reflexology (therapeutic massage of pressure points on the soles of the feet), acupressure, and many others, some scientifically valid, others as yet unproven, and still others unlikely ever to be proven scientifically legitimate. Scores of hospital and clinic-based programs now use alternative methods that were once ignored or dismissed. Medical colleges at many major universities have introduced holistic principles and alternative medicine into their curricula.

The influence of alternative medicine may ultimately revitalize principles of Western medicine long ignored: preventive medicine; the healing effect of a good patient-doctor relationship; the powerful role of the placebo effect; the body's natural recuperative powers; and the importance of treating illness by addressing the whole person—a patient's body, mind, and life circumstances. Other techniques not previously used in Western medicine—acupuncture, yoga, visualization, and certain forms of meditation—may enhance the armamentarium of the Western physician as well as enhancing the patient's ability to cope with illness. The overall effect of alternative medicines may be to re-humanize Western medicine at a time when many patients and physicians feel overwhelmed and alienated by the abundance of technology and the over-specialization and depersonalization of health care. The holistic themes central to most alternative approaches to medicine remind practitioners and health care consumers alike that the body is not merely a machine; that the mind and the emotions, one's environment and social relations, are relevant to physical health; and that physical health has diminished value without emotional and psychological well-being.

DOUGLAS S. BARASCH

For Further Reading: Norman Cousins, *Anatomy of an Illness* (1988); Daniel Goleman and Joel Gurin, *Mind/Body Medicine: How to Use Your Mind for Better Health* (1993); Andrew Weil, *Health and Healing* (1988).
See also AIDS; Cancer; Drugs; Health and Disease.

MELTDOWN

See Nuclear Power: History and Technology.

MENDEL, GREGOR

(1822–1884), Austrian biologist. Gregor Mendel's reports on experiments with garden peas in the mid-19th century led to major advances in agriculture, botany, horticulture, and animal husbandry. They helped provide an understanding of the laws of heredity and evolution, and they laid the basis for the modern science of genetics.

The story of Gregor Mendel and how it took 35 years for the world to realize what he had discovered are part of the lore of modern science. Mendel was a monk. He was born Johann Mendel, July 22, 1822, in Heinzendorf, Austria, on a family farm. His family was poor and young Mendel sickly, but he was encouraged to attend secondary school, the *gymnasium*, and the Phil-

osophical Institute at Ulmuetz (now Olomouc, Czech Republic).

At age 21, troubled by recurring bouts of stress and unable to afford further education anywhere else, Mendel entered an Augustinian monastery at Brno and took the name Gregor. As a young monastery priest he taught Greek and mathematics in a nearby high school but failed biology in his examination for a regular teaching certificate. He was sent to study the natural sciences at the University of Vienna. Mendel returned to teach again, but still never obtained his teaching certificate.

Beginning in 1856, eight years after he was ordained a priest, Mendel began experimenting in the monastery's garden. Seeking hybrids, he cross-pollinated varieties of garden peas, by size, color, shape and other observable characteristics. Long before anyone else, Mendel recognized that each plant carried two trait units, later to be called genes, and that these could be transmitted in reproduction from one generation to another. Mendel theorized that these trait units (genes) existed in pairs, one from each parent, and that they copied themselves exactly. Occasionally there were surprises, mutations, but these also copied themselves exactly until they mutated again.

Mendel described his experiments at meetings of the Natural Science Society in Brunn in 1862. His lectures were published in the society's proceedings the following year. Although Mendel's findings were circulated in the usual places and Mendel himself corresponded with other scientists, there was no meaningful reaction. Mendel continued to experiment but became occupied with church affairs. He rose to the elective office of Abbot and struggled with government authorities over taxation.

Then in 1900, three other European botanists separately proclaimed what they had independently learned regarding inherited plant characteristics. Each cited Mendel's papers from more than three decades before, and the monk became world famous. Unfortunately he had died 16 years earlier.

The foregoing is widely accepted, but in recent years, scholars have suggested that Mendel's experiments actually were fairly well known among scientists. But it took decades before other findings in cell biology and, especially, the acceptance of Darwin's findings on evolution made Mendel's reports on plant hybridization pertinent.

In any case, Mendel was hailed as the first to depict a mathematical formula governing heredity. And he was credited with establishing a new science, Mendelism, which soon became known as genetics. So constantly could his experiments be duplicated in almost all organisms, Mendel's conclusions took hold as biological "laws."

Mendel's first law, the *law of segregation*, stated that as reproductive cells were formed, the pairs of parental character traits separated, one going to each reproductive cell, with each trait uninfluenced by the other. When a pure strain of one trait, or gene, was inbred for many generations and crossed with a pure strain of the alternate gene, one of them prevailed over the other. Mendel termed one dominant and the other recessive. This was the *law of dominance.* Mendel also found that traits were inherited independently of each other. Tallness, color, and other genetic characteristics appeared in accordance with their being dominant or recessive on their own and didn't necessarily turn up in the same groupings. This was Mendel's *law of independent assortment.*

Early scholars had long assumed that hereditary traits were carried in the blood. Some people still emphasize "bloodlines" in human heredity (thus the term "blueblood," denoting some higher quality). In Mendel's time there was a wide assumption that inherited characteristics, wherever they might be carried, were related to transfers in the sex act. But no one before Mendel credibly depicted the process. Mendel demonstrated a mathematically reliable and repeatable procedure for breeding and cross breeding with a particular result. Within a few years, follow-up experiments with mice and poultry showed that the phenomena were not limited to plants. Tests of metabolism verified the hereditary phenomena in humans.

Mendel's laws of inherited traits have shaped comprehension and applications in evolution, physiology, biology, biochemistry, medicine, animal culture, agriculture and even social science. Indeed, the modern Human Genome Project, in which scientists hope to "map" all the genes that make up a human being—and thus learn "what makes a human human"—surely had its paradigm in Mendel's "mapping" of inheritable trait units in garden peas.

JACK RAYMOND

For Further Reading: William Bateson, *Mendel's Principles of Heredity* (1908); Hugo Iltis, *Life of Mendel* (1932); Peter J. Bowler, *The Mendelian Revolution* (1989). *See also* Genetics.

MERCURY

Mercury, a liquid metallic element, has had a wide variety of uses in medications, paper manufacture, thermometers, electrical switches, thermostats, latex paints, dry-cell batteries, and fluorescent lights. It is toxic, as are other heavy metals. Because mercury is volatile at low or moderate temperatures, the manufacture and disposal of mercury-containing products can release mercury vapor into the air. Once in the atmosphere, mercury can be transported far from its origins, returning to Earth through rain, snow, and dry deposition.

Recently, mercury levels have been rising in both sediments and fish in remote lakes far from any city or industrial source. In bodies of water that have certain chemical or bacterial conditions, such as softwater lakes that are vulnerable to acid rain, mercury can be converted to the nerve toxin methylmercury. Research has found that flooded areas behind hydroelectric dams also can have methylmercury problems, possibly because the large amounts of rotting vegetation provide the chemical and bacterial conditions that mobilize naturally-occurring mercury from the underlying soils. Methylmercury readily enters the food chain and can become concentrated in fish. Typically, large game fish carry the highest levels of methylmercury.

Because of its volatility, collecting and controlling all mercury emitted to the air can be difficult. In general, however, mercury levels in the atmosphere are no direct threat to human health. Indoors, however, mercury vapor can be a problem. While most high-level exposures are occupational, in homes mercury from broken thermometers or from phenyl mercuric additives in some latex paints made before 1990 may become concentrated enough to cause health problems, since the lungs absorb mercury with almost 100% efficiency.

Mercury vapor levels that can cause subtle effects on human health begin at 50 millionths of a gram in a cubic meter of air (concentration averaged over the course of an eight-hour day). Most people are exposed to approximately 1 millionth of a gram each day from air, less than 2 millionths of a gram from water, and between 20 and 75 millionths of a gram from food, depending on how much fish they eat. Swallowing metallic mercury usually does not pose a health problem since the body absorbs little mercury from the intestinal tract and excretes it readily. But mercury vapor and methylmercury are absorbed easily and excreted slowly over several months.

Most methylmercury problems have been associated with industrial pollution, such as an outbreak of nervous system disorders among people who ate fish from Japan's Minamata Bay during the 1950s and 1960s. Mercury exposure can damage both the nervous system and the kidneys. Behavioral effects include loss of short-term memory, inability to concentrate, tremors, numbness, emotional instability, and insomnia. High exposure can cause personality disturbances, blindness, deafness, disequilibrium, and even death. All forms of mercury cross the placenta, and babies also may take in mercury when they nurse.

Treatment for mercury poisoning involves stopping the exposure and using drugs to hasten mercury's excretion from the body. Because it is not always possible to achieve complete recovery, some of the neurological effects may persist.

BARBARA SCOTT MURDOCK

METALS, TOXIC

See Lead; Mercury; Toxic Chemicals.

METEOROLOGY

Meteorology is the study of weather: atmospheric conditions at a given time and changes in them over a period of a few days. The fundamental parameters of weather are clouds, precipitation, pressure, temperature, humidity, and wind velocity. The last three are easily sensed and closely related to body comfort. Air pressure, although not easily sensed by humans, has also been measured from early times because it has long been known that large pressure drops indicate approaching storms.

Troposphere

The atmosphere has layers that are distinguished by temperature and other physical characteristics. The lowest layer, the troposphere (the surface contact layer), extends up to an average altitude of 10 kilometers at mid latitudes, higher at the equator and lower at the poles. Most weather phenomena occur in the troposphere. Like land and water at the earth's surface, the troposphere is unevenly heated; denser cool air tends to sink, but sunlight and the earth's rotation prevent the air motion from stopping when it reaches the earth's surface. Instead, the atmosphere is in constant motion, seeking the equilibrium motion of heated air (convection); vertical convection currents form as cool air sinks, is heated by the warm ground, and rises again. Also, there are horizontal currents of fast-moving air in the upper troposphere—the polar jet stream and the tropical jet stream, both moving roughly from west to east.

Air Masses

Meteorology involves the study of air masses, which develop their characteristic physical properties—moisture and temperature—from their place of origin. Formation occurs when the air masses are under the in-fluence of high pressure with little horizontal motion. A dry air mass is called continental; a moist one, maritime. Warm, cool, and cold air masses are tropical, polar, and arctic, respectively. Different air masses may be designated as continental polar (dry and cool), maritime tropical (moist and warm), and so on. When wind patterns change, portions of the high or low pressure systems move and interact with bordering air masses. This movement produces the weather changes studied by meteorologists. Our present knowledge of air mass origins is based on the more or less continuous record of documented measurements taken over the last 150 years.

Clouds

Clouds, made of water in the form of droplets and ice crystals, distinguish the earth from the other planets and help make conditions suitable for life. Clouds form when air rises and its temperature falls to the dew point. Just as dew forms on the ground, a cloud droplet requires a nucleus on which to form. There are usually many suitable particles in the air. The droplets or ice crystals in a cloud are supported by vertical air currents as they either grow in size or evaporate and scatter (refract) light, making clouds visible.

There are 25 classifications describing the appearance of clouds, how they are formed, and the convective state of the atmosphere. Stratus clouds are hazy or layer-like (fog is a stratus cloud based at the ground) and represent stable or stratified atmospheric conditions. Cumulus clouds have a distinctive billowy shape with vertical development and represent unstable or convective atmospheric conditions. Cirrus clouds are filamentous in appearance and are composed of ice crystals. Stratus and cumulus clouds can exist at any altitude, but cirrus clouds exist only at high altitudes where temperatures are below freezing. Middle-level clouds have the prefix *alto-* and are about 2 to 5.5 kilometers high. High-level clouds have the prefix *cirro-*. Low clouds may be stratus or cumulus, stratocumulus (layers of cumulus), or fractocumulus (cumulus torn apart by wind). The cumulonimbus (thunderstorm cloud) is considered the most majestic of clouds, and although classified as a low cloud, its height extends to the top of the troposphere.

As clouds form, condensing vapor gives off considerable heat. (Water droplets freezing into ice give off less heat.) This heat warms the surrounding air, causing it to rise. As more and more moisture is condensed and lifted, large convective clouds are formed. All clouds constantly evaporate at their edges, so they will dissipate unless they continually regenerate by upward motion of moist air. When the droplets or ice crystals in a cloud become too large to be supported, they fall as precipitation. Surface water from the raindrops evaporates, cooling the surrounding air so that it moves downward. The cold air gusts near thunderstorms are the result of this process. Downbursts and the more intense microbursts result from sudden evaporation and changes in wind velocity.

Cloud cover exists over some part of the Earth at all times, and clouds generally move in the direction of, but at slightly slower speeds than, the winds at their altitude. Because temperature decreases with height, cloud temperatures, determined remotely from infrared sensors on satellites, permit the estimation of these heights. Satellites, such as the Geostationary Observing Environmental Satellites (GOES) located 22,700 miles above the equator, exist over various longitudes and can scan the entire earth about once every half hour. The measured displacement of clouds between the half-hour scans provides very useful wind information at many different altitudes.

Storms

The greatest frequency of thunderstorms is in the southeast corner of the U.S. and in the mountainous regions of the East Indies—curiously at almost opposing poles on the earth. The reason for the large number in the Indies is the proximity of the warm moisture in the surrounding Pacific waters to the volcanic mountains with prevailing winds that transport the moisture from the water to the mountains. The same combination of prevailing winds and surrounding warm water helps generate thunderstorms year round in the southeast U.S. However, most of the very severe thunderstorms with large, damaging hailstones occur in the Western Hemisphere. While thunderstorms occur most frequently in Florida, the location of the severest local storms is in the Great Plains area of the U.S., where the prevailing warm, moisture-laden southeast winds encounter the Rocky Mountains.

Most of the world's tornadoes (typically 750 out of 1,000 annually) occur in the U.S., with the greatest number of them (often accompanied by giant hailstorms) occurring in the Midwest. This phenomenon is caused by the combination of the jet stream, guiding the weather in its U-shaped path over the part of the U.S. east of the Rocky Mountains, and the surface flow of warm, moist air from the Gulf of Mexico. The area of greatest tornado intensity moves in a regular pattern with the seasons and is related to hurricane occurrences.

An abrupt increase in the observation and reporting of tornadoes coincided with the arrival in the U.S. of principal tornado scientist Dr. T. Fujita. Largely due to his efforts and lifelong interest in the elusive, yet intensely destructive meteorological phenomenon, public, political, and scientific awareness has increased to

the extent that new instrumentation has evolved to improve tornado warnings. In the 1990s, a special doppler radar weather network was designed and is currently being installed in the U.S. to extend early warning periods.

Annual reported tornado deaths per capita have decreased from about 2.5 per million population in the 1880s through the 1920s to about one per million in the 1950s to less than 0.2 per million in the 1980s. This decrease in death reports occurred while the reported number of tornadoes gradually increased from about 100 per year to about 300 per year in 1952 and then rapidly to about 800 in 1970; it has remained at about 750 since. Categorization of tornadoes by intensity from 1916 to 1990 indicates that with the most destructive tornadoes (F5, with wind speeds 261 to 319 mph), the number of annual deaths decreased from 36 to 10; with the next most destructive (F4, with wind speeds 207 to 260 mph), annual deaths decreased from 12 to 3.

Air Pollution Meteorology

Atmospheric scientists are developing and using mathematical models to study the behavior of air pollutants. Potential problems for study range from local factory fumes to greenhouse gases that affect global warming. Researchers use equations that characterize mixing and diffusion. The stirring agent of the atmosphere is the wind. When the air is nearly calm, stagnation occurs. When air speeds increase near the ground, interaction with the ground and ground-based objects causes turbulence. Studies of mixing show that the dominant mixing process near the source of the contaminant is molecular-scale diffusion; beyond 100 meters, mechanical turbulence dominates. Meteorologists continue to work on predicting with acceptable precision the spread of a pollutant whose source is known and on determining the sources of pollutants detected at specific locations.

Meteorological Measurements

Parameters are measured at surface stations all over the world in order to locate and follow weather systems. In the U.S., the National Weather Service maintains monitoring stations about 200 kilometers apart—generally at airports—which make frequent measurements. More closely spaced stations (daily summary stations) are tended by citizen volunteers. They record the daily total precipitation and the maximum and minimum temperatures and transmit the data to government stations. These networks make daily weather information available for a variety of purposes.

Supplementally, balloon-borne instrument packages measure pressure, temperature, and humidity at particular altitudes and radio the information to surface stations for decoding and interpretation. Radar detects and monitors storms. Since the early 1960s, satellites have observed weather and clouds in both the visible and infrared part of the solar spectrum. The capability of global monitoring has inspired scientists to organize several international research projects, beginning with the World Weather Watch in 1963 and continuing through the 1990s in the form of Global Change programs. The study of higher layers of the atmosphere, a meteorological specialty called aeronomy, continues as orbiting bodies, containing both live creatures and robotic mechanisms, increase.

Weather Prediction

Weather forecasts are so familiar that the public sometimes takes them for granted. They are important not only for decisions affecting daily life, but also for the planning of major projects such as space missions or humanitarian relief airlifts. Forecasts for such undertakings require monumental coordination to gather data from all over the earth, perform computations, make forecasting decisions, and disseminate results in time for deadlines.

Numerical weather prediction is performed using computers programmed with mathematical models of the atmosphere. With frequently updated data, the computers solve sets of equations representing the known physical principles governing atmospheric behavior. The complexity of the equations—and thus the precision of the model—is limited by the amount of time between the transmission of the data and the occurrence of the predicted event.

Forecasts 12 to 24 hours in advance verify with the greatest precision. Forecasts of weather sooner than 12 hours are less precise. Beyond 24 hours, forecast verification decreases steadily to about 72 hours. Beyond three days, intermediate-range or long-range forecasts or outlooks use increasing amounts of historical statistical information from climatological studies. Long-range forecasts include probabilities of events occurring over broad areas and departures from normal conditions, rather than the specific values, points, and times found in short-range forecasts.

Numerical weather prediction is a purely computational aspect of meteorology, permitting comparison of quantifiably measured input with objectively verifiable product. Improvements in the verification of predictions have paralleled the increase in physical knowledge of the atmosphere's nature and the increase of measurement coverage density—geostationary satellites, which remain fixed above the earth, provide near-continuous remote surveillance of the atmosphere at a resolution of a few kilometers (as compared to the 200 kilometer resolution of ground stations).

This noninvasive means of measurement has vastly improved weather monitoring, especially over the oceans and deserts which previously were data-blank areas, and allows for more immediate weather prediction by extrapolating from current events. With geostationary satellites, predictions taking into account factors contributed by human activity are feasible; although the extent of human effect on Nature is not yet fully understood, "Nowcasting" includes human modification of the landscape and atmosphere, again by extrapolation of current events. The physical computations used in traditional forecasting are based only on classical fluid dynamics.

JOSEPH L. GOLDMAN

For Further Reading: Horace Robert Byers, *General Meteorology* (1974); Vincent J. Schaeffer and John A. Day, *A Field Guide to the Atmosphere* (1981); Glenn T. Trewartha, *An Introduction to Climate* (1968).
See also Atmosphere; Climate Change; Greenhouse Effect; International Geosphere-Biosphere Programme; Natural Disasters; Ozone Layer; Pollution, Air; Seasons; Troposphere.

METHANE

Methane is a colorless, odorless, flammable gas that occurs as the natural product of the decomposition of organic matter by anaerobic bacteria. With a formula of CH_4, it is the most basic compound in the alkane or paraffin series of hydrocarbons. Methane is lighter than air and burns readily, producing carbon dioxide and water vapor with a very hot flame. It is relatively stable; however, mixtures in air of 5% to 14% are explosive, and such explosions have occurred in coal mines. Large increases in methane concentrations have raised concern about its role in global warming.

The bacterial production of methane occurs in many locations, including the digestive systems of cattle and termites, rice paddies, and marshes and wetlands. It is naturally produced by bacteria as a major product from the oxidation of simple organic compounds, often cellulose, coupled with the reduction of carbon dioxide. Such decomposition of vegetation similarly produced methane in large fossil-fuel deposits created over hundreds of millions of years. This accounts for the high degree of methane present in natural gas, which is associated with coal and petroleum. Hence, geothermal disturbances produced by mining and drilling for fossil fuels, as well as volcanic activity, often release large amounts of methane.

Although methane is not a major component of the present atmosphere, it made up a significant portion of Earth's original atmosphere, which resembled those that still exist on the four Jovian planets. Methane played a role in complex reactions with ammonia, water, and hydrogen, activated by electrical charges (lightning) or solar radiation, that spontaneously created the basic components of all life on Earth: amino acids. In 1953 Stanley Miller and Harold Urey repeated this process, involving methane in an experimental environment that simulated the primeval conditions on Earth and the "origin of life."

Although methane existed in conditions favorable for creating life, its increase in the present global atmosphere may be detrimental to the climate. Its excess production by human activities, such as agriculture (cattle, rice), mining, and industrial processes has led to an annual increase of about one percent, with more than half of the increase coming from energy uses. Methane contributes roughly one-tenth of the global warming caused by atmospheric gases and is much more heat-absorptive than carbon dioxide.

Efforts to trap escaping methane, mostly from garbage landfills, have managed to accumulate an energy supply equivalent to over six thousand barrels of oil per day. A similar biogasification process can be carried out in so-called "methane digesters." Methane digesters can be small and inexpensive for processing the organic waste from homes and villages or larger and more elaborate for larger communities and farms. They produce methane fuel for cooking, heating, generating electricity, and use as a gasoline substitute. This technology continues to be developed in different parts of the world, especially in developing countries.

MICROWAVES

Microwaves are a form of electromagnetic radiation whose wavelength ranges from one millimeter through one meter. On the electromagnetic spectrum they lie between 1,000 and 300,000 megahertz (MHz). That position places microwaves between infrared and shortwave radio emissions.

Microwave radiation was first produced artificially and studied by German physicist Heinrich R. Hertz in 1886, but its numerous applications were not realized until the invention of adequate microwave generators. Among the various types of microwave generators, the vacuum-tube devices, the klystron, and the magnetron continue to be most widely used, especially where large power output is required. Klystrons are primarily used in radio relay stations and for dielectric heating, whereas magnetrons have been adopted for radar systems and, most familiarly, microwave ovens.

The microwave oven has been a household appliance since the 1970s. It cooks food quickly by bathing it in a microwave field of about 2,450 MHz. The microwaves typically agitate water and fat molecules,

thus heating the food evenly. This reduces cooking time by about a hundredfold. Although metals reflect and scatter microwaves, the ovens do not heat dry materials such as glass or ceramic, used to contain food. Because of the health risks involved with microwaves, ovens are subject to certain safety standards to ensure minimal exposure.

Microwaves can harm humans when intense exposure (at least 10 to 20 milliwatts of power per square centimeter of body surface) heats body tissues to temperatures exceeding 43°C (109°F). Furthermore the lens of the human eye is particularly sensitive to waves with a frequency of 3,000 MHz, and repeated and extended exposure can result in cataracts.

Transmitted and received in parabolic dish antennas, microwaves are the primary carriers of high-speed telecommunication in an extensive network of earth-based stations, satellites, and space probes. A system of satellites orbiting about 36,000 kilometers above the Earth is used for international broadband telecommunication that includes such media as television, telephone, and telefacsimile (FAX). Microwave beams are also useful as radar, in determining distance by measuring the time it takes a pulse to travel to and from an object, such as an airplane or ship. Similarly, the speed of objects can be determined in order to guide airplanes and ships or detect speeding motorists. Since microwaves are scattered by water droplets in the atmosphere, they are also used for charting and forecasting meteorological activity. The cosmic background radiation, whose discovery confirmed the big bang theory of the universe's origin, consists of microwaves.

MIGRANT LABOR

Migrant workers are persons who move for the purpose of finding employment. Migration for employment across national borders is increasing as workers migrate from poorer to richer countries around the world, but migration for employment is decreasing within aging industrial countries.

Migration for employment can involve a temporary or a permanent move away from home. Agriculture is the industry that has traditionally relied most on temporary migrant workers to fill seasonal jobs. More labor is usually needed during the summer and fall when crops are ready to be harvested than during the winter months.

The United States has more migrant farm workers than any other industrial country. These 800,000 migrant workers are about 40% of the two million hired workers employed on U.S. crop farms. They are mostly immigrants from Mexico, and they earned in the early 1990s an average $5 hourly, $200 weekly, and $5,000 annually for the half of the year in which they find farm work.

There is no standard definition of migrant farm worker, and data on them are not published regularly by the federal government. However, all definitions consider migrant farm workers to be persons who move in order to do farm work. A Presidential Commission defined a migrant as a "worker whose principal income is earned from temporary farm employment and who in the course of his year's work moves one or more times, often through several States." This definition captures the familiar image of a person moving from farm to farm, and on each farm, doing work for two weeks to two months.

Other descriptions are often used to complete the picture: hard work, low wages, few benefits, poor housing, child labor, and abusive employers. Migrants have been near the bottom of the U.S. job ladder for over a century, and their place there has been a concern for governments and organizations concerned about the well-being of all workers. Their plight has been chronicled in some of the nation's most enduring literature, such as John Steinbeck's *The Grapes of Wrath*, some of the best-remembered television documentaries, such as Edward R. Murrow's *Harvest of Shame*, and in numerous popular, scholarly, and government publications. Many of these publications have self-explanatory titles, such as *Migrant and Seasonal Worker Powerlessness, A Caste of Despair*, and *The Slaves We Rent*.

Migrant workers provide much of the labor needed to cope with the peak labor needs of a relatively small slice of U.S. agriculture. The need for migrant workers is greatest on large fruit, vegetable, and horticultural specialty farms; these farms account for only one-sixth of U.S. farm sales, but they employ two-thirds of the nation's migrant farm workers. Although 75,000 U.S. fruit and vegetable farms hire labor, migrant workers are concentrated on the largest 5,000 factories in the fields.

Migrant workers "come with the wind and go with the dust" in order to meet the seasonal need for labor on larger than family-sized farms. There have been three responses to the seasonal need for labor in U.S. agriculture: family farms, slavery, and migrant farm workers. Family farms are those in which the farmer and his family own most of the land that they farm and supply most of the labor needed on the farm. Family farms today tend to be diversified grain and livestock enterprises that, with mechanization, can do their farm work without hired hands.

Thomas Jefferson promoted family farming, believing that such farms were healthy for the economy and essential for American democracy. However, Jefferson and many of America's first political leaders owned plantations that relied on slaves. Plantations could keep workers busy for most of the year in labor-intensive crops such as cotton or tobacco, so that plantation owners could justify an investment in slaves and their upkeep throughout the year.

Migrant farm workers emerged as the major answer to the seasonal need for labor as agriculture spread westward. In western states such as California, farmers needed large tracts of land in order to graze cattle or produce crops in the hope that there would be enough rain. When irrigation systems developed in the 1870s, and the transcontinental railroad became available to transport fruit and other specialty crops to eastern U.S. markets, these large farms were expected to be sub-divided into family-sized farms. But the Chinese workers who had been imported to build the transcontinental railroad were excluded from urban jobs, and they became migrant farm workers.

A succession of immigrants followed the Chinese as mainstays of the migrant work force: the Japanese, Mexicans, and Filipinos, small farmers displaced by drought in the 1930s, and Mexicans since the 1940s. A Presidential Commission in 1951 summarized this history as follows: "Migrants are the children of misfortune . . . we depend on misfortune to build up our force of migratory workers and, when the supply is low because there is not enough misfortune at home, we rely on misfortune abroad to replenish the supply."

Most migrant farm workers today are immigrants. However, instead of following the ripening crops from state to state, most migrants shuttle between Mexican homes and U.S. fields. The trend toward a shuttle immigrant farm work force should intensify in the 1990s. Over 90% of the new entrants to the farm work force are immigrants, so that the immigrant share of the farm work force should rise above the current 60% share as retiring U.S. citizen workers are replaced by immigrants. U.S. agriculture should continue to employ about 2.5 million workers sometime during the year in 2000, including 500,000 migrant workers who shuttle into and out of the U.S., but the number who follow the crops within the U.S. may shrink below its current level of about 300,000.

The federal government began programs in the mid-1960s to help migrant workers and their families. At that time there were almost 500,000 U.S. citizen migrant workers, many of whom traveled across state lines to harvest crops. Federal programs for migrant workers and their families multiplied during the 1970s and 1980s, and 12 programs spend over $600 million annually to assist migrant and seasonal farm workers (MSFWs) and their families. Federal expenditures under these programs are equal to 10% of what these workers earn, the equivalent of $600 per MSFW per year, but federal assistance efforts have been unable to nudge migrant workers up the U.S. job ladder, largely because they help migrants to escape from agriculture.

The most prominent myth is that poor migrant workers are the price that the society must pay for cheap food. But hired workers do only one-third of the nation's farm work. Migrant farm workers, the poorest hired workers, do about half of the work done by hired workers. The migrant labor system that impoverishes hundreds of thousands of workers holds down the average family's food bill very little. Even in the case of the fruits and vegetables that migrant workers often harvest, farm wages account for less than 10% of the retail price of a head of lettuce or a pound of apples. Doubling farm wages, and thus practically eliminating farm worker poverty, would raise retail food prices by less than 10%.

Instead of reforming the migrant labor system, the U.S. government has been persuaded to open the border gates to immigrant farm workers. These immigrants, who usually had no other U.S. job options, are willing to accommodate themselves to the low wages and migratory lifestyles of seasonal agriculture. However, so long as additional workers without options are available to be migrant workers, there is little effective pressure from them to persuade farmers to improve wages and to eliminate migrancy. Thus a vicious circle is created: vulnerable migrant workers are available and so migrancy continues; migrant workers aspire to nonfarm jobs, and some of them escape; and the resulting farm labor shortages are used to justify the admission of more immigrant workers.

PHILIP L. MARTIN

For Further Reading: Philip L. Martin, *Harvest of Confusion: Migrant Workers in U.S. Agriculture* (1988); Philip Martin and David Martin, *The Endless Quest: Helping America's Farm Workers* (1993); Presidential Commission on Migratory Labor, *Migratory Labor in American Agriculture* (1951). *See also* Agriculture; Migration, International.

MIGRATION, INTERNATIONAL

Migration is an adaptive mechanism for the survival of human beings. Hunters and gatherers and herders of domesticated animals have led nomadic lives since the beginning of human history. More settled life developed with the adoption of agriculture, perhaps as early as 8000 B.C. Even then, population pressure and expansion of political and military dominance by powerful warlords, city states or imperial centers led to migrations. These population transfers sometimes led to opening frontier areas that were unsettled or sparsely settled. In other instances, migration was experienced as invasions by "hordes" as a ripple effect spread its way across large land masses.

Contemporary international migration has some of the same underlying causes but takes place in a geo-political world organized much differently from that predating the 19th century. Like hunters and gatherers, people still move for the opportunity to make economic gains. So too, political persecution and the devastation of war push millions of refugees to seek safety.

The global system of nation states, developed in Europe and spread from there due to European domination and influence, is the basic organizing political structure. The nation state is built on the presumption that every nation (a homogeneous, self-reproducing group with a shared history, culture and identity, with historical attachment to a territory that they claim as uniquely their own) has the right to a state. The particular form of government in states varies widely, but states usually seek to preserve and defend the nation, the community and its culture. In this idealized system, international migration is seen as deviant behavior. Ordinarily people are not expected to abandon their cultural group and states feel that they have to guard against being overwhelmed by foreign influences, that may destroy them or their culture. States discourage migration and certainly try to control it at levels considered safe. Cultures differ on the degree to which they think they can absorb newcomers. If migrants are accepted, there is also a tendency by modern nation states to make such entry temporary, to import migrant workers while withholding settlement rights.

Nevertheless, because of unequal distribution of endowments, different lifestyles and different histories of economic and political development, there is both a supply and a demand for international migrants. The inability of some economies in the developing world to create jobs in proportion to the growth of their labor forces due to past (and in many cases continuing) rapid population growth leads to a supply of people mobilized for international migration.

The idea of mobilization is important because the mere lack of a job or being desperately poor does not necessarily translate into international migration. To leave one's culture and homeland requires incentive, saleable skills, and a certain sophistication or integration into a network that will smooth migration to a different society. Those mobilized for migration often have education and work experience and live in a society with economic, political, historical or other links to the destination countries. A supply of migrants does not move randomly to any country with a labor demand or higher wage rate. Migrants are also usually not the poorest of the poor, but are more likely to be from middle or upwardly mobile strata of their societies, who have education or who worked in the modern sectors of the economy and gained skills usable in the international economy. Or they are people who are in a kin or territorial network that recruits workers and helps find them jobs, housing, and generally "shows them the ropes" in a new society. Recruitment by networks is accompanied by recruitment by employers in host countries. This took place in 19th century Europe, when American railroads sent recruiters throughout Europe, to entice immigrants to the United States. German recruiters brought in guest workers from Turkey and Yugoslavia in the 1960s, and recruiters were used by contractors seeking workers for Middle East development following the oil price rises of 1973.

Contemporary international migration, therefore, is a complex web of movements with different causes, different legal policies, and streams of people with varied characteristics. Its complexity does not mean it is random. International migration is highly structured but the operative structures vary.

A highly visible component of international migration is the 18 million refugees around the world. Refugee production since the 1960s has been concentrated in the developing world. Much of it was the result of decolonization that led to the expulsion or pressured flight of former colonials and third country nationals brought in by colonial powers, such as Asian Indians in Uganda. Former colonies often had boundaries fixed by imperial powers that included many nationality groups. For example, the newly independent African countries agreed in the charter of the Organization of African Unity to respect the inherited borders in order to avoid prolonged territorial wars and incessant attempts at secession. The price of this reasonable agreement has been a series of civil wars among nationalities vying for control of the levers of power of weak states, inherited from the colonial legacy. This weakness made rebellion not extremely difficult to start and quite hard to put down. Somalia, Ethiopia, Sudan, Mozambique, Uganda and Angola are part of a litany of countries that have produced, and received, refugees.

Besides decolonization, ideological differences about the proper form of government and economic structure have led to revolutions of the left, right and religious variety. Often these disputes were used as proxies for the Cold War competition of the great powers who interfered politically, militarily and economically with their clients in government or in the rebel groups. Vietnam, Afghanistan, Angola, and Nicaragua are among the states where protracted civil wars have produced refugee flows.

The reemergence of nationalism in East and Central Europe after the demise of Communist governments is a continuation of a process of nation building. Begun after World War I and built on the principle of self-determination enunciated in Wilson's fourteen points, ethnic unsorting has resulted once again in violence and refugee production. Yugoslavia is most notable but in the early 1990s Europe feared a wholesale unleashing of ethnic violence. Czechoslovakia seemed to be splitting up in a peaceful settlement, but it was the exception. The effects of Balkanization were once again felt in Europe, and it is feared in the not distant future throughout Africa.

In the 19th century, international migration was away from the developed countries of Europe to the relatively undeveloped peripheral states in the Western Hemisphere and Southern Africa. In the 20th century, temporary worker movements have generally been toward the developed countries and the oil exporters. Most movements were due to demand for workers by labor-importing countries. The size, direction, and composition of the migrant streams were primarily dictated by the importers, not the labor suppliers. In Europe, former colonies often supplied workers, especially in the cases of France and Britain. Germany called on Yugoslavia and Turkey, where prior economic, political and military ties existed. The United States has used workers from Mexico, a continuation of a long history dating back to the opening of mines and building of railroads in the 19th and early 20th centuries and continued in the temporary worker or *bracero* program that operated from 1942 to 1964. Arab oil exporters followed a policy of mixing labor sources to balance Arab and Islamic unity goals, labor costs and increased assurance that workers would leave. Thus Palestinians, Pakistanis and Filipinos, among others, were recruited in an attempt to orchestrate achievement of competing political and cultural goals as an adjunct of labor recruitment.

The few countries that accept immigrants for permanent settlement are primarily the traditional immigrant countries of Canada, Australia, and the U.S. All three continue to accept significant numbers of immigrants and refugees for resettlement. The policies are not without controversy internally due to concerns about population size, consumption and cultural absorption capacity. Politically there does not seem to be enough support to curtail current flows severely, but opposition could mount and a precipitating event may change the course in any of the countries.

In the meantime, there is also significant South-South international migration within broad regions of the world. Labor migration to coastal countries of West Africa continues. This migration is primarily dictated by economic opportunity due to differential growth rates and sizes of economies. In Latin America there are established migration streams in the Southern Cone, in the Andean countries, and along the Amazonian frontier. Much of this movement, like internal migration in the participating countries, is due to population pressure expressing itself in unemployment and underemployment and the search for opportunities in frontier regions. The potential for refugee production is high given the statistical dominance of multinational states in a world organized by adherence to the principles of the nation state and self-determination.

Finally, globalization of goods and capital markets is accompanied by a global labor market. States try to control the operation of that market by entry and exit laws but do so imperfectly. There is little serious support for migration control akin to free trade because of the importance of nationality and cultural integrity, to say nothing of political hegemony of dominant groups in the various countries. Nevertheless, migration will continue and the contentions will be over the degree of regulation. A globalized labor market may tend toward more "temporary workers," especially skilled workers in the management, professional, and technical areas. They will spend part of their working lives in a foreign country with no expectation of settling or changing political or cultural allegiance. The development of these streams, much like the temporary workers of the mid-century decades, will be more a function of labor demand than of labor supply. How many move, to where, and with what skills and experience will be a response to demand rather than to labor supply.

CHARLES B. KEELY

For Further Reading: Daniel Kubat, *The Politics of Migration Policies* (1993); Silvano M. Tomasi (ed.), *International Migration: An Assessment for the '90s* (1989); Aristide Zolberg, Astri Suhrke, and Sergio Aguayo, *Escape from Violence: Conflict and the Refugee Crisis in the Developing World* (1989).
See also Migrant Labor.

MINERAL CYCLES

The cycles of the mineral elements of biological importance are referred to as elemental cycles, mineral cycles, or nutrient cycles; the field of science involved with the study of these cycles is called biogeochemistry or nutrient cycling. Energy flows through the biosphere, but nutrients (chemical elements of biological importance) cycle within ecological systems. In the broadest sense, the term *mineral* can refer to any element in the periodic table. Ecologists frequently deal with cycles of the common nutrient elements—carbon, hydrogen, oxygen, nitrogen, phosphorus, and sulfur—but the cycles of calcium, magnesium, sodium, potassium, silicon, iron, aluminum, and molybdenum,

among others, are also important. A mineral cycle can be described for each element in the periodic table.

Elements can exist in a variety of physical states (solid, liquid, or gas), and in either elemental form or associated with other elements in compounds, both inorganic and organic. Within an ecosystem, elements are segregated in compartments (e.g., plant or microbial biomass, detritus, soils or sediments, water, and air). Each different form of an element represents a different "pool." The growth, stability, or decline of each pool is determined by the rates of various inputs (processes that form additional pool mass) and outputs (processes that deplete the current pool mass). At so-called "steady state," formation and degradation rates are balanced and pool sizes are stable.

Mineral cycles are often illustrated by box and arrow diagrams; the boxes represent nutrient pools, and are connected by arrows indicating input and output pathways (Figure 1). When numerical values for pool masses and transfer rates are added, these diagrams become mathematical models of mineral cycles. Biogeochemists attempt to quantify the pool sizes and transfer rates of mineral cycles, and then draw inference using such models.

Mineral cycles can be described at various scales. At the ecosystem level (e.g., a temperate forest), plants take up inorganic nutrients from the soil or the atmosphere and transform them to organic form (assimilation). Nutrients in plant biomass can be: (1) returned to the soil in litterfall or by leaching, (2) transferred to animal biomass through herbivory, or (3) lost to the

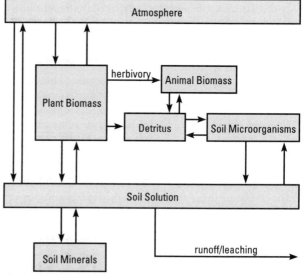

Figure 1. Box and arrow diagram of a generic nutrient cycle for a forested ecosystem. Boxes represent pools within the ecosystem; arrows represent potential fluxes between compartments.

atmosphere. Most organically-bound minerals eventually enter the soil detrital pool, where they are mineralized (transformed to inorganic form) by heterotrophic microorganisms called decomposers. Following mineralization, inorganic nutrients can either be immobilized (assimilated into microbial biomass) by these microorganisms or released to the soil environment, where they may be taken up by plants, stored in inorganic form, exported from the ecosystem in water, or transformed to gaseous form and released to the atmosphere. In most undisturbed ecosystems, the mass of a nutrient that cycles internally usually exceeds external ecosystem inputs and outputs by several orders of magnitude.

Ecosystem-level studies of mineral cycles were popularized in the late 1960s and early 1970s in experiments using small mountain watersheds as model ecosystems. Precipitation inputs and nutrient exports in stream flow were monitored, and comparisons were made between undisturbed (control) watersheds and watersheds subjected to tree harvesting and the suppression of vegetation regrowth using herbicides. Internal cycles of associated forests were studied by measuring forest biomass, nutrient content, productivity, and litter production.

The existence of atmospheric inputs to, and atmospheric and hydrologic exports from, these small watershed ecosystems illustrates the fact that ecosystem-level nutrient cycles are open (rather than closed) systems. At steady state (e.g., a mature forest), external inputs and outputs are balanced, and no net increase or decrease of internal nutrient pools occurs. Many ecosystems are not at steady state, however, but are in the process of recovering from some past natural or anthropogenic disturbance.

Ecosystems receive inputs of minerals from other ecosystems upwind, upstream, or upslope, and export nutrients to other ecosystems downwind, downstream, or downslope. "Landscape-level" nutrient cycles focus on the transfers between ecosystems. Landscape-level nutrient cycles can vary greatly in scale, from the small-scale processes that occur along hillsides as minerals move with the flow of water, to large-scale processes such as the production of sediment and nutrient loads in the Susquehanna River valley in interior Pennsylvania that affect mineral cycles in the Chesapeake Bay. Like ecosystem-level nutrient cycles, landscape-level nutrient cycles are open systems, and can usually be extended in at least one direction on the landscape.

Global cycles encompass all the transformations associated with an element that occurs on Earth and exemplify closed cycles, yet even some global cycles are not completely closed. Earth receives small amounts of some minerals via meteorites, while small amounts

of the lightest elements can escape Earth's gravity. For the most part, however, global cycles are characterized by fixed masses of nutrients that cycle continuously. In the absence of disturbance, global pools should reach steady state over geologic time.

While it is convenient to think of mineral cycles as pools and pathways associated solely with natural processes, humans significantly influence rates of mineral cycling. For example, the mining of phosphate rock for fertilizers is an important anthropogenic pathway associated with the global phosphorus cycle. Phosphate fertilizer represents an additional input to fertilized agricultural or forest ecosystems. Excess inputs often leave target ecosystems in runoff, and are deposited as additional inputs in other ecosystems.

Mineral cycles can be divided into "sedimentary" and "atmospheric" cycles, based on whether an element possesses a significant gas phase. Sedimentary cycles include the cycles of phosphorus, calcium, magnesium, sodium, and potassium. Carbon, hydrogen, oxygen, nitrogen, and sulfur have atmospheric cycles.

The Phosphorus Cycle

The phosphorus (P) cycle is a sedimentary cycle. At the ecosystem level it has distinct biological and geochemical subcycles. The biological subcycle includes processes that involve the activities of living organisms and the transformations associated with organic P (P associated with carbon). The geochemical subcycle is driven by chemical processes that operate independently of living organisms, although chemical equilibria can be influenced by the activities of living organisms as they affect pH and concentrations of dissolved phosphate in solution. Phosphate in the soil solution is the sole contact point between these two subcycles.

Primary mineral P (associated with calcium in apatite minerals) is the ultimate source of phosphorus to the biosphere. Apatites are tricalcium phosphates and are relatively unstable under conditions found at the Earth's surface, so they tend to weather rapidly, particularly in acid soils. Apatite weathering releases phosphate ions (PO_4^{3-}) to solution. Phosphate is the only stable form of dissolved inorganic P in the environment.

Phosphate in solution can be taken up by plants, immobilized by soil microorganisms, or removed from solution by adsorption or precipitation reactions with soil minerals. In addition to primary mineral P, geochemical phosphorus pools include labile, secondary mineral, and occluded P. Labile P is loosely bound to colloidal surfaces and equilibrates rapidly with dissolved phosphate in solution. In acid soils phosphate ions can become "fixed" in secondary minerals, binding with aluminum and iron exposed on clay surfaces, or precipitating as insoluble aluminum and iron phosphates. In calcareous soils P tends to precipitate as calcium phosphate. These secondary mineral forms are relatively stable in their respective environments and replenish depleted solution phosphate pools slowly. Maximum P availability occurs between pH 6 and 7, coincident with maximum solubility of aluminum, iron, and calcium minerals.

Over geologic time, geochemical phosphorus transformations cause the depletion of primary minerals and the formation of resistant secondary minerals. Secondary mineral P is eventually transformed to occluded form, which is completely unavailable to living organisms. As these processes occur, some of the P weathered from apatite is retained within the biological subcycle, and some is exported from the ecosystem in drainage waters and eventually deposited in ocean sediments. In the absence of additional inputs, amounts of available and total phosphorus in the ecosystem decline over geologic time.

Most phosphate taken up by plants or microorganisms is transformed to organic form. In biological systems, phosphorus is an important component of: (1) ATP (adenosine triphosphate), the energy currency of the cell; (2) coenzymes (e.g., NAD, NADP) and phosphorylated sugars, involved in biological oxidation and metabolism; (3) phospholipids, the primary matrix component of membranes; (4) nucleic acids, which carry genetic information; and in plants (5) phytin, the calcium-magnesium salt of inositol hexaphosphate. Plant organic P can be transferred to animal biomass through herbivory. When organisms die or when plants shed above- or belowground plant parts in litterfall, organic phosphorus associated with plant and animal biomass enters the detrital pool.

Soil microorganisms transform organic phosphorus compounds in detritus to inorganic form by a process called mineralization. This can occur as microorganisms digest associated organic carbon (biological mineralization), but can also occur independently, through the action of enzymes that release inorganic phosphate to the soil solution, but leave the associated carbon skeleton intact (biochemical mineralization). Microorganisms take up at least a portion of the phosphate they mineralize (immobilization). When microbial cells lyse (disintegrate), about 40% of the microbial P released is either inorganic phosphate, or is rapidly mineralized to inorganic phosphate by soil enzymes. Mineralized phosphate replenishes the pool of solution phosphate, where it may again be taken up by plants, immobilized by other soil microorganisms, or form labile or stable secondary minerals.

Not all organic P is mineralized, however. Stable forms of organic P resist mineralization and tend to accumulate in soils. Like inorganic phosphate, some

organic phosphates can also be adsorbed by aluminum and iron compounds in soil. Over geologic time, both stable organic P and occluded inorganic P accumulate in soils, while the total amount of P in the ecosystem declines. Very old soils, such as those found in some areas of the tropics and Australia, tend to be P deficient.

There are no gaseous transformations associated with the phosphorus cycle. Atmospheric P inputs occur either as dust, or as phosphate dissolved in precipitation. Ecosystem losses occur by wind and water.

The Sulfur Cycle

Sulfur (S) has several significant gaseous forms. Hydrogen sulfide (H_2S) and sulfur dioxide (SO_2) are reduced and oxidized forms of gaseous inorganic sulfur, respectively; methyl disulfide and dimethyl disulfide are organic sulfur gases. Thus the sulfur cycle is atmospheric. Like phosphorus, sulfur transformations are affected by both biological and geochemical processes. Distinct biological and geochemical subcycles do not exist within the sulfur cycle, however, and most of the important transformations are mediated by microorganisms. Sulfur is the sixth most abundant element on Earth. Because of its global abundance, incidences of sulfur limiting plant or microbial growth in ecological systems are more rare than for P, but such deficiencies do occur in some soils.

Pyrite (FeS_2) in igneous rocks represents the original primary mineral sulfur pool. Other naturally occurring mineral sulfur forms include gypsum ($CaSO_4$), common in arid environments, and elemental sulfur. Weathering of sulfur minerals releases sulfate (SO_4^{2-}) to the soil solution. Weathering of reduced sulfur minerals is coupled with oxidation by chemoautotrophs of the genus *Thiobacillus*, and by purple photosynthetic bacteria.

Like phosphate, sulfate in solution can be taken up by plants, immobilized by soil microorganisms, or adsorbed by soil colloids. Sulfate can also be reduced to sulfide (S^{2-}) by soil microorganisms. Sulfate adsorption and phosphate adsorption are competitive processes. When concentrations are equivalent, phosphate binds more strongly, but sulfate concentrations generally exceed phosphate concentrations in soil. Sulfate that is neither taken up by plants or microorganisms, adsorbed by soil minerals, nor reduced to sulfide will leave the ecosystem as dissolved sulfate in runoff.

Most sulfate taken up by plants and microorganisms is reduced internally to sulfide and incorporated into organic sulfur compounds. Sulfur is an important component of proteins, vitamins, hormones, and the plant growth regulators thiamine and biotin. The carbon-bonded organic sulfur compounds represent the majority of both plant and microbial sulfur, and their mineralization requires respiration of the associated carbon structure (biological mineralization). Like phosphate, sulfate can also bond to carbon via ester linkages, and can be mineralized independently of carbon (biochemical mineralization).

In reduced environments such as anoxic wetland sediments, sulfate is reduced to sulfide by bacteria during the oxidation of organic carbon. The sulfide ions produced can combine with reduced iron to form iron sulfide, with hydrogen to form H_2S, and with organic compounds to form organic sulfur gases.

The oceans are a major source of dimethyl sulfide (DMS), which is produced from decomposing phytoplankton. It has been proposed that DMS production may provide a negative feedback mechanism ameliorating global warming. By this mechanism, warmer temperatures and higher atmospheric carbon dioxide concentrations would stimulate phytoplankton productivity, causing increased DMS production. Atmospheric oxidation of DMS to sulfate would increase cloudiness, by increasing atmospheric concentrations of condensation nuclei, increasing the reflection of incoming solar radiation and lowering global surface temperatures.

As H_2S rises towards the soil surface and oxygen availability increases, it can be reoxidized to elemental sulfur, and eventually to sulfate, by processes similar to the oxidative weathering of reduced sulfur minerals. In some wetland environments, sulfate can cycle repeatedly between oxidized and reduced zones in the soil. Each cycle is associated with the oxidation of organic carbon stored in the wetland soil (peat), and it has been proposed that increased sulfate inputs from acidic deposition could negatively affect the carbon balance of some peatlands.

In addition to the biogenic sulfur gases, atmospheric sources of sulfur include direct inputs of sulfate, from dust in arid regions or by evaporation from the oceans, and inputs of SO_2 from volcanic eruptions and the burning of fossil fuels. In the atmosphere, SO_2 dissolves in water to form H_2SO_3, which is rapidly oxidized by microorganisms to sulfuric acid (H_2SO_4), a major component of acid rain. Anthropogenic sources represent an additional sulfur input to the atmosphere of approximately 50% of natural sources.

Sulfur dioxide can injure plants directly when it is taken up through the stomates because it is converted to sulfuric acid in plant tissues. H_2SO_4 in precipitation can leach nutrients from plant foliage. Hydrogen ions also replace nutrient cations on soil exchange sites, causing their release into the soil solution. In weakly buffered systems, increased acidity can solubilize soil aluminum, which then enters nearby lakes or streams in runoff, killing fish. Sulfate in runoff can enhance nutrient leaching from terrestrial ecosystems and ac-

celerate mineral weathering, illustrating the important interactions that frequently occur among mineral cycles.

MARK R. WALBRIDGE

For Further Reading: Gene E. Likens, F. Herbert Bormann, Robert S. Pierce, John S. Eaton, and Noye M. Johnson, *Biogeochemistry of a Forested Ecosystem* (1977); William H. Schlesinger, *Biogeochemistry: An Analysis of Global Change* (1991); Frank J. Stevenson, *Cycles of Soil* (1986).
See also Acid Rain; Bacteria; pH; Phosphorus; Sulfur.

MINING INDUSTRY

Viewed in global terms, mining is an elemental force. Estimates are that the minerals industry accounts for 5% to 10% of world energy use and strips 28 billion tons of material from the Earth every year—more than is moved by the natural erosion of all the planet's rivers. Regulation and the concerns of the public are forcing mining companies to reappraise the way in which they operate. But such is the scale of the industry, that efforts to ameliorate its environmental performance raise fundamental questions about who should pay, and about the economics of exploiting the world's natural resources. Given the magnitude of its operations, it is hardly surprising that the excesses of large mining companies have become synonymous with environmental degradation. The industry acknowledges that times have changed and so it devotes huge resources to environmental measures.

The mining process starts with exploration. Some environmentalists believe that there should be 'no-go' areas for mining, while the industry maintains that it is essential to have wide-ranging access in order to be able to determine what resources are available. The availability of the world's mineral wealth is further complicated by the changing geopolitical map. Deposits in the former U.S.S.R. and Eastern Europe, for example, have become more accessible, although the impact of these deposits on the market for minerals is uncertain. There is a general drift towards privatization that is also opening up new horizons for mining, for example, in Mexico and Chile.

Having identified a viable ore body, by exploration, the development phase begins, with the Environmental Impact Study (EIS) being a central element. The EISs carried out by some mining companies are often more rigorous than is required by the relevant authority, and these companies are investing considerable sums in such studies. Before extraction at a new site, plans are compiled in conjunction with government agencies that take into account all environmental considerations. A rigorous examination of drainage patterns and existing water quality leads to assessment of the likely effects of the proposed site's hydrology and the most effective ways to countering them. Flora and fauna baseline surveys establish existing ecosystem characteristics, and form a basis for determining site-specific rehabilitation objectives.

After a mining project has negotiated the development phase and becomes operational, its environmental problems are only just beginning. During its operational life a mine generates huge quantities of materials. Gold mining alone produces an estimated 620 million tons of waste annually. In order to create the Bingham Canyon copper mine in Utah, 3.3 billion tons of material have been evacuated—seven times the amount removed to make way for the Panama Canal.

In open cast (pit) mining, the rock that covers the mineral deposit is stripped away, then the ore is mined and transported by trucks—with capacities that can exceed 240 tons—or conveyed to crushers. Here the material is finely ground and separated, and the host rock, or the tailings, often diverted to settling ponds. The concentrated ore is sent for smelting. This process generates a variety of pollutants. Tailings can contain contaminants such as arsenic and cadmium, as well as traces of organic chemicals such as toluene. This fine-ground material is easily penetrated by water, with the potential for leaking toxins into the general environment. If there is a high sulfur content—associated with minerals such as copper, lead, nickel, and zinc—then acid drainage can result, a major problem that is not fully understood. A technology known as heap leaching involves extracting gold by spraying piles of low-grade ore with cyanide solution.

There is also the physical impact of mining. The mining area attracts population, and the impact on the ecosystem can not only be huge but also difficult to control. While different ores pose varying degrees of risk, there is no linear relationship between the type of mining and the environmental threat involved. Each mine has its own characteristics, and a range of factors, such as the composition of the ore, local geography, and the waste disposal methods used, determine the facility's risk profile.

The industry is investing in research programs with the aim of developing more effective ways of removing pollutants from the mining process. A promising development is the use of biotechnology to clean up waste materials. Certain microorganisms absorb toxic substances and with genetic engineering can become very efficient natural scrubbers. Moreover, these techniques can be self-financing. One application involves the use of microorganisms to remove traces of copper from waste materials, thus increasing the mine's mineral yield.

The phosphate mines in Florida operate in wetland ecosystems, utilizing a hydraulic process to extract the phosphate, and the conditions combine to present complex reclamation problems. Particularly troublesome are the slimes, a unique residual resulting when the phosphate is separated from clay and sand by washing. Approximately two-thirds of the processed matrix is returned to the field as a waste product. About 60% to 70% of the area mined is covered by clay settling ponds up to one square mile in size. Primary concerns over these settling areas are their size, potential of surface water pollution from dam failure, and the amount of water used or entrained in the clay impoundment process.

The use of environmental audits is increasing with most of the large companies adopting "eco-audits" as part of their standard operational procedures. Such practices are not only prudent in an operational sense, but also commercially, because more stringent auditing requirements are likely to become mandatory in more countries. Companies maintain that they set environmental standards in line with the toughest codes, and that these are applied regardless of which country they are operating in. Moreover, banks are becoming increasingly particular as to which mining projects they will invest in, and the potential environmental impact of the facility is a deciding factor.

The level of environmental control is also increasing. Within the European Community, rules in 1993 concerning the monitoring of toxic wastes and the trans-boundary movement of sludges have impacted the mining industry. More specific regulation concerning acid rock drainage is also expected. There is a general perception that, in a global sense, mining-related regulations in the South are much more lax than those in the North. Metals Minerals and Research Services (MMRS) in a survey conducted in 1992 examined over 400 pieces of environmental law. At the top of the rankings was the U.S., whereas Zaire—which failed to report any regulations—was at the bottom. However, the global picture is changing with a definite movement towards more international homogeneity. In Indonesia, for example, a new regulatory framework is based on Canadian practice.

In Chile, decree 182 came into force in 1992; this legislation pertains to the release of sulfurous anhydride, particulate material, and arsenic. A tacit environmental policy is being applied by so-called Regional Commissions of the Environment (COREMAS). These bodies evaluate mining permits and such aspects as decontamination plans. Chile has 718 regulations of varying kinds covering environmental activities.

But assessing the impact of mining regulations can be a difficult task. It is difficult to calculate regulatory costs because mining companies do not separate environmental from ongoing operational costs. Basically, the industry does not have an answer to what is the cost of compliance—it could be between 1% and 15% of operational expenses. Smaller mining concerns are finding it increasingly difficult to absorb these costs.

The final phase of a mining project, rehabilitating the site after the deposits have been exhausted, is one of the most expensive. In some rehabilitation activities, mine sites are revegetated, for example, as timber plantations. Sophisticated methods of soil and seed preparation, direct seeding, and research into cloning technology have made it possible to reestablish whole forest communities.

For example, within the Darling Plateau, Western Australia, are low-grade bauxite reserves which supply 18% of the western world's alumina. The resource is found in less than 5% of the nearly two million hectares covered by a mixture of eucalyptus trees. The soil in this area is laterized, weathered in such a way that silica and alkalies have been removed, leaving high concentrations of iron and aluminum oxides. These forests persist on the deep, infertile soils but grow very slowly—taking two hundred years from germination to maturity. They have adapted through an extensive rooting system to the mediterranean-like climate of rainfall during the cooler season and drought during the warm season. Mining of the bauxite may take five years from the start of mining to the end of rehabilitation. Forests are cleared, topsoil is retained for nutrients and seed store of undergrowth, and tree species are developed. Following mining, topographic grading takes place to minimize water effects during the rainy season and maximize water availability during the dry season. Overburden and topsoil are replaced and four-month seedling trees are hand planted. Use of modern techniques of reclamation began in 1976 and provide protection for 85% of the animal species of the original forest.

A spectacular example of reclaiming of open mines is that of the lignite (brown coal) mine near Cologne, West Germany, known under the name Fortuna. The mine is 300 meters deep, occupies a surface of approximately fifteen square kilometers, and has been operated since the late 1950s. As the mine advances, agricultural fields, villages, and roads are completely destroyed but rebuilt almost immediately. To this end, the topsoil is set aside before the mining operations are begun. For each ton of lignite mined after the removal of the topsoil, there are two tons of useless soil, which is also set aside and then returned immediately to the hole. The topsoil is replaced on the surface and enriched with fertilizers, then planted with trees and crops. The fields achieve a normal state within ap-

proximately five years. Since 1964 3,000 hectares have thus been reconstructed with lakes, beaches, forests, amusement grounds, roads, villages, and even a canal. Agricultural fields have been created and more than 60 million trees have been planted. According to the overall plan, 12,000 hectares will eventually be divided into 4,000 hectares for agriculture, 700 hectares for forty-five lakes (of which seven will be suitable for sailing), 5,000 hectares for trees, and the balance of 2,300 hectares for villages and roads.

Similarly, mineral extraction and reclamation of mined land proceed simultaneously in the modern Belle Ayr and Eagle Butte open pit coal mines on the plains of the Powder River basin near Gillette, Wyoming. Modern surface mining on this scale involves a 20–25 year program of computer-simulated advance planning of the volumes of overburden and coal to be moved each year: at the beginning of the mining operation there is already a schedule for the overburden that will be replaced on the last day of mining 20 years or more ahead.

Actually, reclamation begins with the first steps of mining. Valuable topsoil is stripped from the land by scrapers and either spread immediately on restored land or stockpiled for later use in the reclamation process. In areas where the land has been restored to its natural contours following mining, topsoil is placed on the rebuilt land in depths of 18 to 24 inches. It is carefully prepared as a seedbed, seeded, fertilized, and mulched. Seed mixtures are chosen from over 40 species according to land use, moisture availability, and other considerations. Full revegetation takes approximately three years from the time of seeding.

The second activity is removal of the dirt and rock which lie under the topsoil but above the coal (the "overburden") and any materials between seams of coal (the "parting" layers). This "spoil" is hauled immediately to the areas from which coal has been removed and is used to fill the pit; the term "backfill" is applied to this part of the process. Shovels dig the material in bites of about 22 cubic yards, loading it in 120 ton trucks for the trip to the dump sites. The trucks haul the material wherever it is needed: sand is hauled to the reconstructed creek channel, brown subsoil is hauled to reconstructed hilltops; gray shale is placed in the bottom of the backfill. Once the material has been placed in its basic position, dozers and graders mold its final shapes, creating hills, valleys, creeks, and other features akin to the natural topography. The restored land is then ready for the final activity—restoring topsoil and revegetation.

Removal of topsoil, overburden, and parting has only the final purpose of exposing coal seams for mining. Both overburden and coal are removed and transported by the same shovels and trucks, providing flexibility in handling and placing material. Coal is hauled out of the pit to preparation plants, where it is reduced to chunks no larger than two inches maximum diameter, and stored in silos holding up to 15,000 tons of coal. Finally, it is loaded by computer control into 100-car unit trains for shipment to electric utility plants, often 1,000 or more miles away. At Belle Ayr the land can progress from agricultural use (such as grazing) to active coal mining and back to agriculture in as little as four years.

Abandoned mines not only represent a stark visual reminder of the impact of mining, but are a source of pollutants that leach into the general environment for years to come. Estimates available from the Environmental Protection Agency suggest that in the U.S. there are 125 sites, where mining is listed as one of the uses, in need of remedial action.

The main regulatory problem is the retrospective nature of legislation driven by society's attempt to pass on to mining companies the full costs of cleaning up mining wastes of past generations, whose attitudes and practices were very different. It is a difficult problem, because the costs involved are colossal, and even if the industry is penalized, it seems inevitable that society will pay through more expensive raw materials. The cleanup costs are particularly onerous in developing countries. For example, in order to attract foreign investment in mining operations, some governments in these countries offer to take on the environmental liabilities associated with mining installations. However, these liabilities can be complex, and the countries concerned may not possess the level of expertise required to discharge their responsibilities.

A fundamental review of how the world's mineral resources are exploited is needed, looking at such aspects as recycling and tax mechanisms since policies tend to favor the use of virgin materials. The industry maintains that it already supports recycling, pointing out, for example, that over 50% of lead is now recycled, and that this figure should reach 60% by the year 2000. The movement towards more regulation is unlikely to abate and the problems will become more urgent as the demand for mineral products continues to increase, driven by industrializing economies such as those of South Korea, Taiwan, India, and certain South American countries.

KEN COTTRILL

For Further Reading: John Cairns, *Rehabilitating Damaged Ecosystems*, Volume 1 (1988); Robert Cahn, "Environmental Adventures at Amax," in *Footprints on the Planet* (1978); John E. Young, Worldwatch Paper No. 109, *Mining the Earth* (1992).

See also Bacteria; Bioremediation; Biotechnology, Environmental; Coal; Environmental Impact Statement; Erosion, Soil; Land Use; Law, Environmental; Mineral Cycles; Resource Management; Restoration Ecology; Surface Mining Control and Reclamation Act; Toxic Metals; Wetlands.

MONOFILL

A monofill is a place of disposal for municipal waste combustion ash (MWC ash), the residue from municipal incinerators or combustors of solid waste. Since many municipalities have chosen incineration as one step in handling solid waste, proper management of the MWC ash must take place.

The presence of heavy metals such as lead, cadmium, and mercury in MWC ash is of particular environmental concern because they may lead to the contamination of groundwater. To provide groundwater protection and meet regulatory requirements, monofills are lined and equipped with a collection system for the leachate, the soluble materials contained in the ash that dissolves out by water. Ash monofill leachates typically contain equal or lesser concentrations of metals and greater concentrations of salts, chloride, sodium, and sulfate than leachates from municipal solid waste landfill. Monitoring of the amount and properties of the leachates in the leachate collection system and the groundwater surrounding the monofill is generally conducted. Monofill field leachate chemical analyses show that the concentrations of heavy metals in the leachates are very low and the tested water frequently meets the primary drinking water standards.

Regulatory Extraction Tests have shown ashes from MWC facilities on occasion to exhibit a hazardous waste characteristic. For example, in such tests, periodically, lead and cadmium have exceeded the limit beyond which a waste is regulatorily defined to be hazardous. The debate regarding the applicability and representativeness of these tests in regard to MWC ash leachability is ongoing in the scientific and regulatory community.

Unlike Municipal Solid Waste landfills, which are usually designed for an active use period of 20 to 30 years, monofills are designed to accept ash for only 3 to 5 years. As the monofill reaches capacity, it is capped and revegetated to prevent precipitation from entering. The design of the monofill cap depends on regulatory permit requirements and site, topographical, geological, and climatological conditions. Monitoring of groundwater and collected leachates usually occurs before and after the monofill closes.

Continued diligence aimed at proper siting and design of the monofill, proper management of the disposed ash in the monofill, and continuous environmental monitoring is necessary in order to ensure that potential environmental impacts associated with MWC ash disposal in monofills remain below environmental concern.

HAIA K. ROFFMAN

MORTALITY AND LIFE EXPECTANCY

Mortality is the demographic process for accounting for deaths that take place in a population. It is often measured mathematically by the *crude death rate,* or the number of deaths in a given year per 1,000 population. When data are available, an age-specific death rate (for example, aged 55 to 59 per 1,000) is a much better method of accounting.

In Sweden, a country with excellent health care, about 11 people die per year per 1,000 population, a crude death rate that is higher than that of a country such as Guatemala, whose crude death rate is 7. Sweden has a higher crude death rate than Guatemala because it has a much higher proportion of older persons in its population. The risk of dying at every age, however, is higher in Guatemala than in Sweden. This difference in risk can be seen by comparing age-specific death rates. The death rate for males aged 70 to 74 is 58 per 1,000 in Guatemala, while in Sweden it is 41.

One of the better-known summaries of a population's mortality experience is *life expectancy at birth,* which is defined as the number of years a newborn infant can be expected to live based upon the mortality rates in effect at the time of birth. Life expectancy at birth varies widely throughout the world. In the United States, life expectancy for a baby born in 1990 is 75.4 years. In Nigeria, however, life expectancy is much lower, at only 49 years. Japan has reported the world's highest life expectancy, at 79 years.

Worldwide, life expectancy has been rising steadily. Since the end of World War II, some of the more dramatic increases have occurred in countries in Africa, Asia, and Latin America. These sharp drops in mortality—without a compensating drop in birth rates—are the cause of the postwar population "explosion." The improvement in mortality rates in Asia, Latin America, and Africa has been attributed to the rapid postwar spread of public health and sanitation programs, including antimalarial campaigns and inoculations.

The study of mortality often focuses on causes of death: what they are and how they vary from place to place and from time to time. Generally, causes related to contagious disease and hunger still predominate in Africa, Asia, and Latin America, while degenerative diseases, such as heart disease and cancer, claim more lives in Europe and North America. In the United States in 1990, 33.5% of all deaths were from heart disease and 23.4% from cancer. In New York City in 1866, the leading cause of death was tuberculosis, and, at the turn of the century, pneumonia was the leading cause of death in New York and nationwide.

The following representation of one family's experience vividly illustrates the mortality revolution:

> Abraham Lincoln's mother died when she was thirty-five and he was nine. Prior to her death she had three children: Abraham's brother died in infancy and his sister died in her early twenties. Abraham's first love, Anne Rutledge, died at age nineteen. Of the four sons born to Abraham and Mary Todd Lincoln, only one survived to maturity. Clearly, a life with so many bereavements was very different from most of our lives today.[1]

CARL HAUB

MOTOR VEHICLE EMISSIONS

See Pollution, Air.

MUIR, JOHN

(1838–1914), Scottish-born American naturalist, conservationist, explorer, and writer. John Muir lived in Scotland until age 10, when his family moved to a farm in the North Woods of Wisconsin. John, with a flair for invention and a budding love of and passionate curiosity about the natural world, enrolled in the University of Wisconsin, where he spent four years (studying Latin, Greek, botany, geology, and chemistry) but earned no degree. He went to Canada to avoid the Civil War draft, then made his way via the Gulf of Mexico

Table 1
Life expectancy at birth, 1950–1990

	1950–1955	1985–1990
World	46.4	63.3
Africa	37.7	51.7
Asia	41.0	62.9
Latin America	51.4	66.5
North America	69.0	75.2
Europe	65.7	74.4
Oceania	61.1	71.5
U.S.S.R. (former)	64.1	69.1

Source: United Nations Population Division, *World Population Prospects,* 1993

1. David M. Heer, *Society and Population,* 2nd ed. (1975), p. 56.

and the Isthmus of Panama to San Francisco and the Sierra Nevada, arriving in 1868.

Muir spent the next decade mainly in the mountains, climbing, hiking, and conducting informal scientific studies of flora, fauna, and geology. He offered his writing to Eastern journals and quickly became a prominent and admired correspondent. His first published work was a heretical attack on the theory of how Yosemite Valley had come to be. Muir insisted it was carved by glaciers although the accepted theory was that Yosemite had been created by cataclysm. Muir was right and proved it.

After writing voluminously for a decade, Muir spent the next ten years with his wife and two daughters. He became established as an orchardist, with such success that he could spend the last third of his life as a full-time writer and conservationist.

With an East-Coast magazine editor, Robert Underwood Johnson, Muir conceived of creating Yosemite National Park, essentially a vast expansion of the Yosemite Reserve, set aside during the Lincoln administration to preserve Yosemite Valley and the Mariposa Grove of giant sequoias in the central Sierra Nevada of California. At the urging of Muir and Johnson, Congress established Yosemite National Park in 1890. A counterattack ensued immediately, led by loggers, miners, and sheepmen, who urged drastic shrinkage of the park. In response, Muir teamed up with fellow lovers of the Sierra—hikers, climbers, and scientists—to found the Sierra Club, in 1892.

Thus, Muir's greatest tangible legacy may be an organization established "To explore, enjoy, and render accessible the mountains of the Pacific Coast." ("Render accessible" was changed to "protect" in the 1940s, when the club argued about whether to support paved roads into the back country of the Sierra). The Sierra Club is arguably the most powerful environmental organization in the world, and it still hews to Muir's ideals. It is certainly one of the most democratic of the large national organizations, with direct election of directors, and hundreds of chapters, groups, issue committees, and the like. Total membership in its hundredth year was in excess of 650,000.

Muir's impact on American history—particularly on the history of conservation—continues to be enormous. He has been idealized by Stephen Fox as the epitome of the "passionate amateur" as opposed to the professionally trained resource manager, or policy-making bureaucrat, or land use lawyer. Muir, like thousands who followed him, fell into conservation through love of the land, and pursued his passion with his heart. His attachment was to wild nature. He argued with Gifford Pinchot, first Chief of the Forest Service, over preservation versus conservation, an argument that rages to this day. Muir's thesis was that certain supreme examples of wilderness should be left wild: visited, studied, and enjoyed by people but left unexploited, undamaged, undeveloped. As he wrote in 1867, "The world we are told was made for man. A presumption that is totally unsupported by facts. . . . Nature's object in making animals and plants might possibly be first of all the happiness of each one of them, not the creation of all for the happiness of one. Why ought man to value himself as more than an infinitely small composing unit of the one great unit of creation, and what creature of all that the Lord has taken the pains to make is less essential to the grand completeness of that unit?"

Muir was thus an early leader of what would become a large wilderness preservation movement that would be carried on by Aldo Leopold and Bob Marshall and lead to the creation of The Wilderness Society in the 1930s, the passage of the Wilderness Act by Congress in 1964, and the founding of the radical Earth First! in the 1970s.

Though without formal degrees, Muir also had a considerable impact on the development of natural science, through his deduction of how glaciers had carved the Sierra Nevada and his intuition of what would become known as ecology. "When we try to pick out anything by itself," Muir wrote, "we find it hitched to everything else in the universe."

Along with the Sierra Club, Muir's legacy includes his writing, virtually all of which is in print 80 years after his death. He is most often compared with Thoreau and Emerson, though his writing is more purely natural history. It has very little to do specifically with human beings, and focuses far more on the West, with his seminal studies of the mountains of California and the landscape of Alaska.

The best measure of Muir's enduring influence may be the frequency with which people still invoke his name. Candidates for the Sierra Club's board of directors routinely don the mantle of their illustrious founder. Others insist that Muir would be appalled at what the club has become; these are of the school that holds that the club has drifted too far from the "passionate amateur" ideal to a professionals-dominated staff.

As one further measure of his endurance, Muir's name appears all over California on schools, roads, mountains, meadows, trails, and passes. Even a union of Sierra Club employees proposed in early 1992 called itself the John Muir Local.

TOM TURNER

For Further Reading: Stephen Fox, *John Muir and His Legacy: The American Conservation Movement* (1981); Frederick Turner, *Rediscovering America—John Muir in His Time and Ours* (1987); Tom Turner, *Sierra Club: 100 Years of Protecting Nature* (1991).

MUMFORD, LEWIS

(1895–1990), American philosopher, literary critic, historian, city planner, cultural and political commentator, and writer. Mumford is known for his books on architecture and urban planning, notably *The Culture of Cities* (1938) and *The City in History* (1961). He was a guiding force in the development of urban planning in America. His early admiration of American philosophy, literature, and architecture—coupled with a gift for creative synthesis—led him to a unique overview of civilization and the problems of the modern American city.

Mumford grew up in New York City sensitive to the clash of classical and modern elements that were so characteristic of the turn of the century when the comforts and trappings of the Victorian Age were giving way to the faster, more functional, and progressive climate of the Modern Age. He began to question the effect of the change—to determine what parts of the physical structure of the city were worth saving and what parts should be bulldozed to make way for the new. These questions led to larger ones about how "we" use the city and how the city cultivates the cultural life within it. The more he studied and articulated his feelings, the more he saw the tie between mass production and the exploitation of human and natural resources; and the decline, in his eyes, of the cultured, humanistic world he admired.

When Mumford discovered the work of Patrick Geddes, he found a model for his future endeavors. Geddes was an early environmentalist, trained as a botanist who used his scientific training as an approach to human environments. He was a social innovator and an activist who led in revitalizing the grim, sooty industrial slums of Edinburgh, Scotland. Geddes taught Mumford to experience cities both for what they are and for what they had been. To find what they are, Geddes walked for days or weeks through the city just observing and taking notes the way a botanist would walk through a forest, noting where there is life, what sorts of activity support that life, and where the ecology has been disrupted.

Mumford also adopted Geddes's ideas about using the past to suggest possible approaches to configuring the modern city. The notion was that the city is constantly evolving in an organic way and that the past might offer more successful models for planning urban areas. One of Geddes's methods was to organize masques and festivals so that people could experience the vitality of their city at a different point in history. Mumford learned from Geddes to approach the city as a humanistic ecosystem whose well-being depends on a whole world of social and physical relationships.

In 1923 Mumford joined with three of America's great planners—Clarence Stein, Henry Wright, and Benton MacKaye—to form the Regional Planning Association of America. They organized in an effort to stop unplanned growth and restore a sense of human scale. They advocated limits to existing large cities by creating smaller, satellite cities separated from the parent city by open space. Mumford lived for a time in housing built by the R.P.A.A. in Sunnyside, Queens, that featured clustered buildings and large common gardens.

In looking for models upon which to base his cities of the future, Mumford looked to other eras. He found such a model in Pompeii: "When one considers the amount of space and fine building given to Pompeii's temples, its markets, its law courts, its public baths, its stadium, its handsome theatre, all conceived and built on the human scale, with great nobility of form, one realizes that American towns far more wealthy and populous than Pompeii, do not except in very rare cases, have anything like this kind of civic equipment, even in makeshift form."

For similar reasons, Mumford admired the life of medieval cities where the physical layout was much more conducive to human contact, expression, and ultimately the growth of culture. Medieval processions, pageants, and other sensual and spiritual ceremonies indicated to him that there was a great public show of human spirit, not found in today's cities. In *Technics and Civilization*, one of four volumes in his "Renewal of Life" series, he wrote that the many technological advances during the medieval ages that set the stage for the scientific and industrial revolutions—such as the clock, telescope, printing press, and blast furnace—were put to more humanistic ends, such as building cathedrals, than the technology that followed.

In the medieval city Mumford also found support for his theory of "organic planning," which takes into account the various forces in the historical process by which a city evolves, generation after generation building on the work of the last. He believed "organic cities" could be created today if biological, social, and personal criteria were established and used as guidelines.

Mumford believed that modern people had become so obsessed with technology that they had forgotten about human values, the very biological, social, and personal criteria that marked the great civilizations of the past. The ramifications of this ran through much of his work in over thirty books and more than a thousand essays and reviews. He challenged planners, architects, and the general public to look at the cost on the human environment of projects pushed by shortsighted politicians and industrialists. He lumped their promises of "progress" under the "myth of the machine."

Mumford's greatest challenge was New York City. He wrote a regular column for *The New Yorker* called the "Sky Line" and used it to attack all advocates of "bigger is better" growth policies, especially Robert Moses, the New York Commissioner of Public Works. His fights with Moses were legendary and bitter. Mumford thought building a highway system into the city was good for cars and bad for the people who lived there. He also did battle in his "Sky Line" column with the city and developers who were changing the look of midtown forever with huge glass and steel structures. He called skyscrapers the "architecture of Organization Man" and "elegant monuments to nothingness."

Mumford lost most of his fights, but succeeded in raising consciousness over urban issues, preservation issues, and instituting community dialogue in the planning process. Through his outrage and activism, planning was no longer solely the work of planners working behind closed doors.

Throughout his life Mumford became more and more despairing about the dehumanization caused by huge social machines, and the desocialization that comes from unemployment, especially among the young. These and other social tragedies were much greater environmental problems in his mind than air and water pollution, which were only symptoms of the social decay. He feared that civilization might undergo a total breakdown unless we rediscover the art of putting people first, ahead of "progress." The riots in the 1960s reinforced his belief that the cities could not hold up when the values of individuals, families, and communities were breaking down.

At the end of his life Mumford thought that no less than a miracle would save civilization. But he wrote, "Impossible? No: for however far modern science and technics have fallen short of their inherent possibilities, they have taught mankind at least one lesson: Nothing is impossible."

JAKE EHLERS

For Further Reading: Lewis Mumford, *The Pentagon of Power* (1970); Lewis Mumford, *Interpretations and Forecasts* (1973); Lewis Mumford, *My Works and Days* (1979).
See also Architecture; Technology; Transportation and Urbanization; Urban Challenges and Opportunities; Urban Planning and Design; Urban Renewal.

MUNICIPAL SOLID WASTE MANAGEMENT

See Waste Management, Municipal Solid.

NATIONAL AUDUBON SOCIETY

See Environmental Organizations.

NATIONAL ENVIRONMENTAL POLICY ACT OF 1969

The National Environmental Policy Act of 1969 (NEPA) was the first comprehensive federal environmental law in the U.S. It marked "an effort for the first time to impress and implant on the Federal agencies an awareness and concern for the total environmental impact of their actions and proposed programs." In enacting NEPA, Congress established a national policy of using "all practicable means and measures . . . to create and maintain conditions under which man and nature can exist in productive harmony and fulfill the social, economic, and other requirements of present and future generations of Americans."

To carry out this new environmental policy, Congress directed the agencies of the federal government to, among other things, insure that environmental amenities and values are given appropriate consideration in decision-making by preparing detailed statements on the environmental impacts of major federal actions significantly affecting the quality of the human environment. These detailed statements are called environmental impact statements (EIS), and the preparation of those statements by federal agencies has come to be called the environmental impact assessment (EIA) process.

NEPA also created the Council on Environmental Quality (CEQ) to oversee the implementation of the Act, to serve as the in-house environmental advisor to the President, and to coordinate executive branch environmental policy. CEQ has issued regulations that implement the procedural provisions of NEPA.

The EIA Process Under NEPA

The EIA process has two principal purposes: integrating environmental considerations into the federal decision-making process, and encouraging participation by other federal agencies with specialized expertise and by the public in the preparation of environmental analysis and in the eventual decision.

Integrating environmental concepts into decision-making demands that an agency begin the preparation of a NEPA analysis as close as possible to the first development or presentation of a proposal. NEPA analysis must not be a justification for a decision already made.

Each federal agency promulgates NEPA procedures, consistent with the CEQ regulations, which address how NEPA is to be applied to that agency's activities. Among other things, the agency procedures identify the appropriate level of environmental analysis required for the agency's normal activities. Actions that typically have a significant impact on the quality of the human environment require preparation of an EIS. Actions that may or may not have a significant environmental impact, depending upon the situation, are the subject of briefer documents known as environmental assessments (EAs). Agencies usually follow EAs by either a Finding of No Significant Impact (FONSI) or a decision to prepare an EIS. Actions which normally do not have, either individually or cumulatively, a significant environmental impact are categorically excluded from NEPA documentation.

In preparing an EIS, the agency must fully consider the proposed action and its environmental consequences and all reasonable alternatives and their environmental consequences.

Finally, when an agency makes a decision on a proposal requiring an EIS, the agency must demonstrate that it has adequately considered environmental values by preparing a record of decision (ROD). The ROD must state what the decision was, identify all alternatives considered by the agency in reaching its decision, specify the alternative that was considered to be environmentally preferable, and state whether all practicable means to avoid or minimize environmental harm from the alternative have been selected.

The second important purpose of the EIA process is to promote public participation. Following publication of a notice of intent to prepare an EIS, the agency must develop "an early and open process for determining the scope of the issues to be addressed and for identifying the significant issues related to a proposed action." As part of this "scoping" process, the federal agency must invite the participation of affected agencies, Indian tribes, and other interested persons (including those who may be opposed to the action on environmental grounds); determine the scope of the issues and alternatives to be considered in the EIS; and indicate the relationship between the timing of the preparation of environmental analyses and the agency's tentative planning and decision-making schedule. Further, the agency may set page limits, time limits, and hold a public scoping meeting.

Environmental impact statements are prepared in two stages, draft and final. Draft EISs must be made available for public comment for a minimum of 45 days. A federal agency with responsibility for preparing an EIS must make an effort to circulate the draft EIS to interested parties and relevant federal agencies with jurisdiction by law or special expertise over any portion of the proposal, the alternatives, or their environmental impacts. Comments on a draft EIS should be as specific as possible, and may address either the adequacy of the statement or the merits of the alternatives discussed or both.

The agency must review the comments it receives on the draft EIS and include responses to those comments in the final EIS. Such responses might include modifying the alternatives, evaluating alternatives not previously considered, supplementing or improving its analyses, making factual corrections, and explaining why the comments do not warrant further agency response. Once the agency considers and responds to the comments, it then issues the final EIS and circulates it to those who provided comments. After a minimum 30-day waiting period following the issuance of the final EIS, the agency may then issue its record of decision and implement that decision.

When an EA is being prepared, agencies have more flexibility for coordinating public participation, but they are required to involve environmental agencies, applicants and the public, to the extent practicable. In some circumstances, Findings of No Significant Impact must be circulated for public review and comment for a minimum of thirty days.

Agency Responsibilities

NEPA requires agencies of the federal government to "[u]tilize a systematic, interdisciplinary approach which will insure the integrated use of the natural and social sciences and the environmental design arts in planning and in decisionmaking. . . ." A lead agency (the agency proposing to take the action) must carefully select the staff who will prepare a particular EIS, and must make sure that their areas of expertise are appropriate to the issues identified in the scoping process. At the request of a lead agency, other federal agencies with jurisdiction by law or special expertise may become "cooperating agencies." Cooperating agencies are to participate in the scoping process, as-

sume responsibility for developing information and preparing environmental analyses for the EIS at the request of the lead agency, and make available staff support to enhance the lead agency's interdisciplinary capability.

CEQ oversees federal agency implementation of the NEPA process. In addition to promulgating and interpreting regulations to implement that process, CEQ works with the federal agencies to insure full compliance with NEPA. CEQ is responsible for resolving interagency environmental disputes. It may also publish findings and recommendations and submit them to the President for action.

EPA files, reviews, and comments on all draft and final EISs prepared by the federal agencies. Using a rating system, EPA's comments relate to both the adequacy of the analysis and the potential environmental impacts. These comments are available to the public. EPA also has the authority to refer to CEQ federal legislation, federal projects, or proposed federal regulations which EPA believes to be "unsatisfactory from the standpoint of public health or welfare or environmental quality. . . ."

The federal courts have played a major role in the interpretation and implementation of NEPA. Many concepts that are considered integral to NEPA today are not specifically referred to in the statute, but were rather developed through a synergism of case law and CEQ guidance and regulations. Much of the most significant NEPA case law was developed in the first half of the 1970s, as agencies struggled with compliance questions and the courts faced the challenge of interpreting a broad but vague statute. The courts have consistently viewed strong enforcement of NEPA's procedural provisions as appropriate, but they have been quite reluctant to find in NEPA an obligatory requirement to make decisions based on NEPA's substantive goals. Instead, the U.S. Supreme Court has repeatedly stated that: "NEPA does set forth significant substantive goals for the Nation, but its mandate to the agencies is essentially procedural. . . . It is to insure a fully informed and well-considered decision, not necessarily a decision the judges of the Court of Appeals or of this Court would have reached had they been members of the decision-making unit of the agency." Or, as the Supreme Court stated more recently, "Other statutes may impose substantive environmental obligations on federal agencies, but NEPA merely prohibits uninformed—rather than unwise—agency action."

As expected by the sponsors of NEPA, a number of states have implemented a system of EIA in the early and mid-1970s. Frequently referred to as "little NEPAs," these laws vary a great deal in respect to their legal basis, scope and requirements. Nineteen states, the District of Columbia and Puerto Rico have enacted "little NEPAs" by statute or executive order. Some states, such as California, New York, and Washington, have established vigorous EIA systems, supported by comprehensive regulations and active judicial enforcement. Other states have systems that apply to a narrower range of actions and that appear to be less dynamic in their relationship to state decision-making.

In 1979 Executive Order 12114 was issued to further the purposes of NEPA by addressing the assessment of environmental effects abroad of major federal actions. Agencies must prepare environmental analyses for actions affecting the environment of the global commons outside the jurisdiction of any nation, such as the oceans and Antarctica, and for actions affecting the environment of a foreign nation that is not participating with the U.S. in the action.

Recently the U.S. signed the Convention on Environmental Impact Assessment in a Transboundary Context, drafted and sponsored by the Economic Commission for Europe (ECE). The Convention establishes a system of notification, shared information and consultation among ECE member signatories.

The U.S. has also been a strong advocate of establishing EIA in multilateral organizations. The U.S. supported the formulation and adoption of Principles and Guidelines for Environmental Impact Assessment developed by the United Nations Environment Programme. In 1991 the U.S. signed a Convention on Environmental Impact Assessment in a Transboundary Context, developed under the auspices of the Economic Commission for Europe. The U.S. also strongly supports the application of EIA to projects funded by multilateral development banks, such as the World Bank.

DINAH BEAR

For Further Reading: Lynton K. Caldwell, *Science and the National Environmental Policy Act: Redirecting Policy Through Procedural Reform* (1982); Environmental Law Reporter, *NEPA Deskbook* (1989); Daniel Mandeltier, *NEPA Law and Litigation* (1992).

See also Environmental Impact Statement; Environmental Protection Agency, U.S.; Law, Environmental.

NATIONAL INSTITUTE OF HEALTH

See United States Government.

NATIONAL WILDLIFE FEDERATION

See Environmental Organizations.

NATIONAL WILDLIFE REFUGE SYSTEM

Intertwined within the complex patterns of North American biomes like reinforcing threads in a gigantic multitextured tapestry, National Wildlife Refuges (NWRs) shelter multitudes of wildlife, including caribou, muskoxen, and snowy owls on arctic tundra; moose and woodcock in north woods; fringed orchids, waterfowl, and buffalo on remnants of the great plains; Sonoran pronghorns, gila monsters, and pupfish in the desert; spoonbills and ocelots in subtropical forests; and whales, sea turtles, and seabirds in marine areas. Refuges range from vast wetland areas such as the immense joined deltas of the Yukon and Kuskokwim Rivers in western Alaska, hosting over 200 million migratory birds during the brief Arctic summer, to very small areas protecting individual bird colonies or endangered endemic species. America's National Wildlife Refuge System (the System) was among the world's first, and is today its largest, nationwide land system managed for wildlife conservation needs over other land uses.

With at least one refuge located in each state and six U.S. commonwealths and territories, the System offers opportunities to view America's flora and fauna in their native habitats. Spanning entire watersheds and nearly entire ecosystems, the sixteen refuges in Alaska contain over 76 million acres or 84% of the System's land and water area. There opportunities exist to gain vital knowledge of future benefit to mankind from intact arctic ecosystems. The Alaska Maritime NWR, for example, spans portions of three arctic marine ecosystems. Along with polar bears, sea otters, and many other mammals, it hosts a large variety of fish and endangered species such as the Aleutian shield-fern. It also has the Northern Hemisphere's largest concentrations of nesting seabirds, portions of which contribute to the renowned biological productivity of Bristol Bay. Other refuges in the western states conserve important segments of desert and arid grassland ecosystems, and numerous refuge islands in the Atlantic and Pacific Oceans, Bering and Chukchi Seas, Gulf of Mexico, and Caribbean Sea serve as biological havens for their respective marine zones.

Many intensively managed refuges have been located across the nation to meet the special needs of waterfowl and other migratory species. At some time in their annual movements 90–95% of all species of migratory birds found in North America utilize NWRs. More than 162 federally listed endangered or threatened species depend upon refuges in some way, with an increasing number of refuges established primarily for their conservation. Still other refuges have special capabilities to offer environmental education through hands-on outdoor experiences and wildlife observation near major cities such as San Francisco, New Orleans, Minneapolis, Boston, New York City, Philadelphia, and Washington.

The System was initiated as a result of deepening concerns in the late 1800s about the disappearance of colony-nesting birds such as the terns, egrets, and herons that were being killed for hat feathers, and the pelicans, whose feathers made good ink pens. Following two decades of work to stop market hunting and enact state bird protection laws, the Committee for the Protection of North American Birds of the American Ornithologists' Union (AOU) brought this plight to the attention of President Theodore Roosevelt who, on March 14, 1903, established Pelican Island Reservation in Florida to protect the last brown pelican nesting colony on America's east coast. With this precedent, and a commitment by the AOU and National Association of Audubon Societies to help pay and equip wardens to protect refuge wildlife, refuge advocates investigated and reported on potential reservations along America's coastlines and in many other locations. Before leaving office, Roosevelt reserved 51 bird refuges, the Wichita Mountains (bison) Reserve in Oklahoma and a moose reserve in Alaska. Under a Congressional Act in 1908, Roosevelt's Administration also had begun purchase of the National Bison Range in Montana. Succeeding presidents also supported the refuge concept, enlarging the System to 86 refuges covering over 5 million acres by 1930.

The 1916 Migratory Bird Treaty with Canada recognized the many benefits of protecting birds moving internationally and established a foundation for federal management of migratory wildlife. In 1929, the Migratory Bird Conservation Act provided fresh authority for the Bureau of Biological Survey (now U.S. Fish and Wildlife Service or FWS) to establish refuges for all migratory birds covered by the treaty and other wildlife.

Severe declines in waterfowl populations during the 1930s, caused by unrelenting drought and exacerbated by extensive wetland drainage, resulted in concentrating refuge efforts on saving wetlands and associated water-dependant species. This emphasis was reinforced by the 1934 Migratory Bird Hunting Stamp Act, which created a special "Duck Stamp Fund" to buy refuges and conduct related programs. Subsequent land acquisitions expanded the concept begun earlier of establishing refuges at key stopover points along the major waterfowl migration routes, called flyways, that cross the nation.

By the 1950s it was apparent that continuing wetlands loss, primarily due to agriculture, was outstripping efforts to provide adequate protection for many wildlife species. In 1961, Congress enacted a special "seven year" wetland program authorizing funds to accelerate refuge purchases. Unfortunately, the program failed to gain momentum and the total area protected, even after 25 years, was about one-half of the original goal.

Following the environmentally enlightened years of the 1970s, the System grew primarily from such Congressional events as passage of the Alaska National Interests Land Conservation Act in 1980, which established an additional 54 million acres of refuges in Alaska. Appropriations from the Land and Water Conservation Fund, used for acquiring endangered species habitats and refuges with important biotic representation, enlarged the System by a million acres between 1970 and 1992. Increasingly, the public has strongly supported efforts to establish refuges for a broad array of native species from crocodiles and red wolves to bats, condors, small fish, and plants. As of September 30, 1992, the System contained 90,946,765 acres within 485 designated National Wildlife Refuges and other management units.

As this century began, the Refuge System was born to fulfill distinctive but unmet wildlife needs through man's beneficence and specific habitat reservations. Recent evidence of drastic fragmentation, degradation, and loss of wildlife habitats coupled with steeply declining wildlife populations provide enduring *raisons d'etre* to maintain and thoughtfully enlarge America's National Wildlife Refuge System.

WILLIAM C. REFFALT

For Further Reading: Noel Grove, *Wild Lands for Wildlife: America's National Refuges* (1984); George Laycock, *The Sign of the Flying Goose: The Story of the National Wildlife Refuges* (1973); Laura Riley and W. Riley, *Guide to the National Wildlife Refuges* (1993).

NATURAL DISASTERS

For most of human history, sudden and unexpected meteorological or geophysical happenings, which resulted in drastic misfortunes for human beings and communities impacted by them, were attributed to supernatural forces. In the Western world they came to be known as "acts of God," a designation which attained legal status for insurance and other purposes. The occurrence of disasters was seen as beyond human control or intervention, and was thought of as essentially unpredictable and unknowable.

However, as secularism spread in European cultures, such happenings increasingly were attributed to natural, physical forces whose operative principles were discoverable by the physical sciences. The advent of the 20th century brought the view that the conditions that created natural disasters were fully knowable and explainable by natural forces. Thus, although the older notions still exist in popular thought and some religions, an earthquake is seen not as the expression of God's wrath, but as the result of the slippage of tectonic plates beneath the Earth's crust. This scientific understanding allowed limited forecasting, prediction, and the issuance of warnings, especially for those natural disasters involving a meteorological agent. But this still limited the use of human intervention to minimize negative effects. Being able to predict a blizzard or a hurricane could not stop it from happening.

However, in the middle of this century, natural disasters began to be seen as the result of interplay between physical agents and the social setting. For example, on flood plains where there was no human occupancy or activity, rising waters would be considered no more than a physical phenomenon explainable in physical terms, not a natural disaster. Natural disasters are now increasingly viewed as relatively sudden (but often anticipated) crisis occasions, involving a physical agent, which result not only in human casualties and property damage, but also in disruptions of community life and negative ecological effects. The term *crisis* refers to the fact that immediate preparations for and responses to such occasions require nontraditional and extraordinary measures. This formulation from a social science perspective about what constitutes a disaster also suggests that people need not passively wait to be hit by a physical agent but can do something ahead of time to alleviate the impact.

Among the major physical agents involved in natural disasters are earthquakes, hurricanes, floods, tornadoes, volcanic eruptions, blizzards, and tsunamis (what are popularly but technically incorrectly labeled as "tidal waves"). Not included in natural disasters are happenings such as forest fires, avalanches and landslides, dam failures, and land subsidence, which primarily result from human activities; as well as droughts and famines, which also are not sudden in onset of impact.

The shift to a view of natural disasters as social crises has extended the range of possibilities for dealing with them. Thus, current disaster planning emphasizes four phases or steps: (1) prevention or mitigation, (2) preparedness, (3) response, and (4) recovery. In addition, it has become clear that planning in prior phases affects the later ones. For example, if preparedness planning for tornadoes does not include effective warning systems, there will be more casualties to handle in the response phase. If during the period of disaster recovery following an eruption, evacuees are allowed to return to living on the slopes of the volcano, this is inconsistent with the land use zoning or restriction on human habitation that should be part of disaster mitigation planning.

Statistics about the effects of disasters can vary considerably. The exactness of any figures, official or otherwise, is subject to the narrowness or broadness of the criteria used for inclusion, the inherent difficulties of obtaining valid data about emergency time period activities, the bias of political consideration in making assessment of losses, and how and whether secondary and indirect effects should be estimated. Variations in any given set of statistics thus can represent different uses for different purposes by different sources.

That said, natural disasters, whether looked at in aggregation or even just one catastrophic occasion, can have massive consequences. Between 1964 and 1983, natural disasters throughout the world killed nearly 2.5 million people. Between 1900 and 1976, disasters left 3 million injured or homeless on the average each year. Such massive effects are more likely to occur in developing countries. For example, the United States has had only three natural disasters in its history that have killed more than a thousand persons, the worst being the hurricane that hit Galveston, Texas in 1900 killing 6,000 and injuring another 5,000. These numbers pale in significance compared with disasters in Latin America or southwest Asia where the dead or the displaced typically may number in the tens of thousands. The Tangshan, China earthquake of 1976 killed over 300,000 persons. A flood in Bangladesh in 1987 left 25 million homeless.

The negative disaster results in the United States are much more likely to take the form of massive economic losses and social disruptions. Hurricane Hugo and the Loma Prieta earthquake in California (which killed at most 90) in 1989 resulted in direct losses of approximately $15 billion and indirect losses of $30–45 billion. Property losses from floods alone amounted to $1.1 billion in 1989; during 1979–1988, economic losses averaged $2.4 billion each year.

There is relatively little correlation between losses and frequency of the appearance of the physical agent involved in natural disasters. For example, in the United States there were 23,488 tornadoes between 1959 and 1988, but casualties, economic losses, and social disruptions are on the average much higher for earthquakes and hurricanes, or even an isolated occasion such as the Mt. St. Helens volcanic eruption.

Good planning can make a significant difference in reducing the effects of disasters, particularly in saving lives and preventing injuries. For example, the Loma Prieta earthquake was slightly higher on the Richter magnitude scale than the one that occurred in Armenia in the same year. However, the California earthquake left fewer than 90 dead, while more than 30,000 died in Armenia. Even taking into account differences in the two localities in population density and other factors, the difference in the fatality figures can almost entirely be attributed to the stronger building codes, housing regulations, and land use regulations in California.

However, it might be possible even to reduce or at least stabilize property losses and social disruptions if there is implementation of some of the planning that is being considered in connection with the International Decade for Natural Disaster Reduction. This United Nations declared decade for the 1990s is intended to encourage all countries to improve their planning for disasters. In particular, there is an emphasis on implementing measures—ranging from stronger land use regulations to better preparedness stance of emergency response organizations—that will mitigate or lessen the negative effects of natural disasters when they occur.

E. L. QUARANTELLI

For Further Reading: Reducing Disasters' Toll: The United States Decade for Natural Disaster Reduction (1989); Principal Threats Facing Communities and Local Emergency Management Coordinators (1991); A Safer Future: Reducing the Impacts of Natural Disasters (1991).

NATURAL GAS

Natural gas is a mixture of hydrocarbons. Methane is the principal constituent, but ethane, propane, and heavier hydrocarbons are present. Inert compounds, like nitrogen and carbon dioxide, and impurities, such as hydrogen sulfide, are also often present. It is distributed by pipelines and widely used in homes, businesses, electric utilities, and industry for cooking, space heating, water heating, electricity generation, and as an industrial fuel. Natural gas is found underground under pressure, often associated with crude oil, and is produced by drilling a well and collecting the gas at the wellhead. It was called "natural" gas to distinguish it from "town" gas, which it has now completely replaced. Town gas was a synthetic gas usually made by the partial combustion of coal and distributed by small pipelines to the central part of a city or town. In the early 20th century, many cities in the U.S., Canada, and Europe had a town gas works. Town gas had a heating value from one-third to one-half that of natural gas.

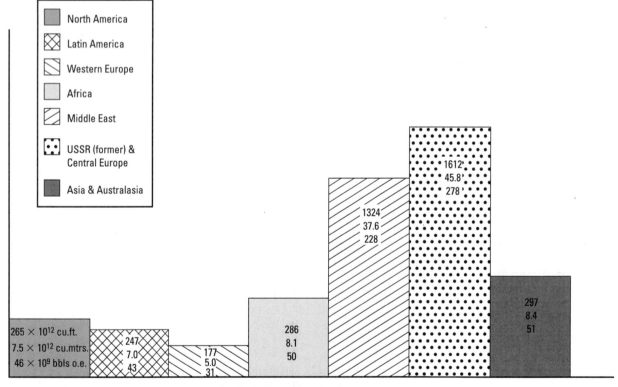

World = 4208 × 10^{12} cubic feet
119.4 × 10^{12} cubic meters
727 × 10^9 barrels oil equivalent

Figure 1. Proved natural gas reserves at the end of 1990. Source: *BP Statistical Review of World Energy*, June 1991.

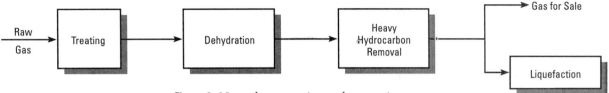

Figure 2. Natural gas treating and processing steps.

The occurrence of natural gas is widespread but not uniform. Of the world's proven natural gas reserves, 38% is in the Confederation of Independent States (of which Russia is dominant) and 31% is in the Middle East. Only 6% is in North America, roughly 4% in the U.S. (largely in Texas, Louisiana, and Oklahoma) and 2% in Canada (northern Alberta). Natural gas is a significant hydrocarbon resource. Proved reserves at the end of 1990 were 4,208 trillion cubic feet (TCF). This is equivalent to 72% of the world's proved crude oil reserves of 1,009 billion barrels.

Natural gas in the U.S. is measured in standard cubic feet (SCF). The heat content of 1 SCF of natural gas is 1000 Btu. Elsewhere, the volume measure is cubic meters; the heat content of one cubic meter is 35,300 Btu or 9,000 kilocalories. A heat content of 5.8 million Btu per barrel of oil has been assumed to convert natural gas quantities to crude oil equivalents. Liquefied natural gas (LNG) is bought, moved, and sold worldwide, including the U.S., by metric tonne with 1 metric tonne = 52 million Btus.

Production, Treating, and Processing

Natural gas is found underground at various depths and pressures. A gas well drilled in the U.S. may be more than 20,000 feet deep. Most gas is found not mixed with crude oil, especially that at great depths, but 35% of all natural gas is "associated" with crude oil.

Table 1
Natural gas compositions, % mole

	Field product	Ready for sale
Methane	79.5	93.8
Ethane	6.5	1.4
Propane	4.0	0.2
C4 & Hvy.	3.0	-
Nitrogen	4.0	4.6
Carbon Dioxide	2.2	-
Hydrogen Sulfide	0.8	-

A series of treating and processing steps are performed in a natural gas separation plant (usually located at a gathering point close to several gas fields) to prepare pipeline quality gas. The steps in a typical gas separation plant are:

Treating

A scrubbing operation on the raw gas stream uses a liquid to remove carbon dioxide and contaminants such as hydrogen sulfide.

Dehydration

The method of water removal depends on what will follow in the processing series. Usually cooling, which will condense much of the water, is sufficient.

Heavy Hydrocarbon Removal

A gas expansion turbine is used most often in a gas separation plant. The high-pressure treated and dehydrated natural gas is expanded rapidly when the gas is passed through the turbine. The gas temperature falls and liquid hydrocarbons condense at the expander. As much as 90% of the incoming ethane and all the propane and heavier hydrocarbons can be removed. The liquids collected are called Natural Gas Liquids (NGLs). NGLs are important feeds to refineries and chemical plants.

Liquefaction

After water, hydrogen sulfide, carbon dioxide, and heavy hydrocarbons are removed, natural gas can be liquefied by cooling below its boiling point (about −256°F at atmospheric pressure). The cooling is accomplished in two refrigeration stages. The first stage might use propane as refrigerant and bring the natural gas to about −30°F. The second stage uses a lower-boiling refrigerant (such as a mix of nitrogen, methane, and ethane) to cool the natural gas to its liquefaction temperature. Liquefied natural gas (LNG) is handled, pumped, and stored as a liquid at a temperature of about −256°F. LNG storage tanks – shipping tanks, those on board LNG ships, and receiving tanks – are designed to hold their contents at slightly above atmospheric pressure. All tanks and transfer piping are made of special low-temperature steel alloys, but despite thick insulation, some heat reaches the contents; a small quantity of LNG boils off and is used as fuel in the liquefaction process or as ship's fuel.

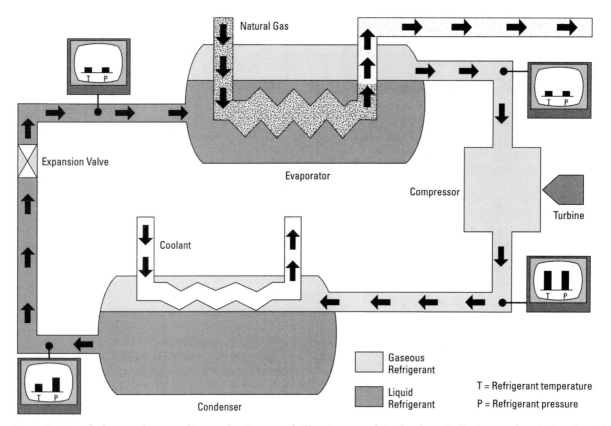

Figure 3. Liquefied natural gas cooling cycle. Source: *Shell Science and Technology Bulletin*, number 6, October 1986.

Distribution and Pricing

Large-scale consumption of natural gas had to wait for long-distance transportation technology to bring it from where it was found to places where people can use it. That technology is the high-pressure, large-diameter, natural gas pipeline with compressors spread along the line that intermittently boost the gas pressure to push it along the way. Natural gas in such pipeline systems can be economically moved thousands of miles.

The long-distance pipeline systems for the distribution of natural gas in the U.S. were put in place during the late 1940s and early 1950s. Natural gas consumption rose from less than 8 TCF in 1950 to over 20 TCF by 1970. Consumption in the U.S. has since plateaued at a little less than that level. At the start of this period, consumption was almost exclusively an American phenomenon and as late as 1970, the U.S. still consumed 65% of the world's natural gas. Today, the U.S. consumes no more than 27%. The former Soviet Union, Europe, and Japan have become major consumers.

Natural gas pricing is different from oil pricing. At a given time the price of crude oil around the world var-

ies by only about 10%, depending on tanker transport distances and the quality of the oil. The consistency of price is because oil is relatively easy to transport and store.

Table 2
U.S. natural gas prices (averaged over 1990)

Market	$ per million BTU
Wellhead price	1.72
Trunkline purchases	2.17
Electric utilities	2.38
Industrial users	2.92
Commercial purchases	4.83
Residential purchases	5.77

With natural gas pricing, the type of market and pipeline transportation linkages play large roles. Table 2 shows U.S. natural gas prices averaged over 1990. Wellhead prices were about $0.30 higher in winter and $0.30 lower at off-peak times. Trunkline price, the price at the end of a major pipeline, was about $0.50 higher on the east and west coasts. The lower prices paid by electric utilities and industrial users reflect

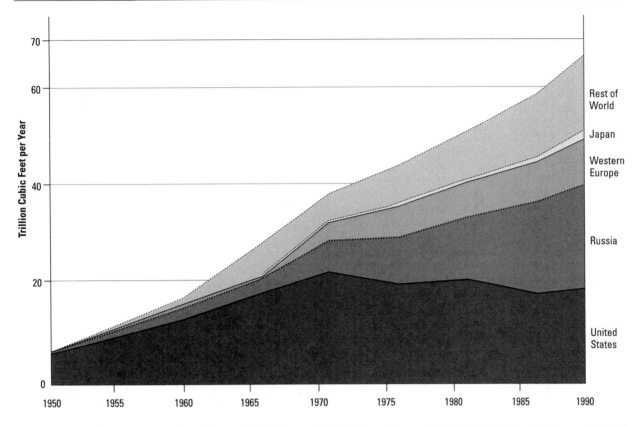

Figure 4. Natural gas consumption. Source: *BP Review of World Gas*, August 1991; *BP Statistical Review of World Energy*, June 1986 and 1976; and *Basic Petroleum Data Book*, VIII, number 3, September 1988, API.

their command of over half of the natural gas market and their ability to buy from either a trunkline marketer or a producer. Many industrial customers also have the flexibility to switch to fuel oil, so the cost of natural gas to industrial users tends to stay close to the price of 1% sulfur residual fuel oil. Commercial and residential purchasers are at the end of the distribution chain provided by trunklines and local distribution companies.

Trunkline prices are higher in Europe. LNG costs are higher still because of the cost of the liquefaction facility, the LNG tanker fleet, the receiving and vaporization terminal, and whatever distribution pipeline systems are required on the customer end.

Natural gas has increased its share of world primary energy consumption from 13% in 1961 to 22% today while total energy consumption itself was increasing by more than 2.5-fold. The share of natural gas in future world energy consumption is not likely to increase to more than about 24% during the next ten to twenty years, because most growth will have to be supplied as LNG, far more expensive than traditional pipeline expansion. This might seem surprising since natural gas is the most environmentally friendly fuel.

Environmental Considerations and Future Use

When natural gas is burned, provided air is in slight excess, its combustion products are only carbon dioxide and water. (Some nitrogen oxides are formed by the reaction of the nitrogen in the air.) There are no sulfur oxide emissions, as with most oil and coal, that potentially can contribute to acid-rain formation. There are no solid particulate emissions that require control by collection as when coal is the fuel. The transport, storage, and use of natural gas do not contribute to the formation of urban smog. Even when methane is inadvertently leaked, unlike most heavier hydrocarbons, it does not enter into the photochemical reaction with nitrogen oxides to make ozone and smog.

One environmental problem is that its combustion product, carbon dioxide, is a greenhouse gas. However, natural gas fuel contributes less carbon dioxide per million Btu released than does oil or coal. Another

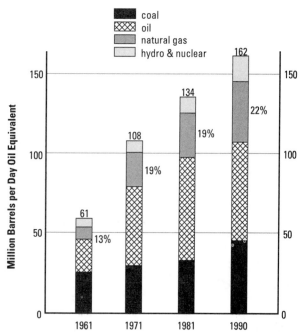

Figure 5. World energy consumption by fuel type. Source: *BP Statistical Review of World Energy,* June 1991 and 1976; and *Basic Petroleum Data Book,* VIII, number 3, September 1988, API.

debit is that methane itself is a greenhouse gas, 20 times as efficient in trapping radiated heat as carbon dioxide.

There are a number of uses where natural gas might be the fuel of choice. In the U.S., because of the 1990 Clean Air Act Amendments (CAAA), two niche markets are electric utilities and motor vehicles.

Electric Utilities: Under the CAAA, the 110 dirtiest power plants in 21 states are required to reduce their sulfur emissions by 50% by the year 2000. One choice a utility has (rather than stack-gas scrubbing or switching to low-sulfur coal) is to co-fuel with natural gas. Only a few of these plants have co-fueling capability now, but additional utilities are likely to elect this route to lowering sulfur emissions.

Motor Vehicles: The CAAA requires that in Los Angeles—an Extreme Smog area—certain classes of vehicles, such as transit vehicles and delivery fleets, must be fueled with alternative fuels that have lower emissions than gasoline. Other locations, Texas for example, are enacting similar regulations. One alternative fuel is Compressed Natural Gas (CNG). Hydrocarbon and carbon monoxide emissions are much lower than with gasoline and nitrogen oxides seem no higher. However, at present, a CNG vehicle has a range of only 150 miles even with three times the volume of tank space found in gasoline cars.

The most significant increase in natural gas consumption in the U.S. may be in generating electricity. Based on cost, the method of choice will be cogeneration using a natural-gas-fired turbine. Non-utility companies (such as chemical plants and refineries) are already providing over 40% of new generating capacity and are selling their excess power to utilities, which are required by law to take it.

There is major concern about greenhouse gas emissions and their contribution to possible climate change. The thrust of most of the proposed actions to limit greenhouse emissions is to stabilize carbon dioxide emissions. One proposal for achieving stabilization is to impose a large tax on the carbon content of fuels. A carbon tax has two effects. First, it raises the cost of all energy forms, which reduces energy consumption because consumers use energy more efficiently. Second, and here natural gas would be favored, the carbon tax discriminates between fuels and provides incentives to switch from high-carbon fuels (coal = 1.6 carbon/Btu; oil = 1.4 carbon/Btu) to low-carbon fuels (natural gas = 1.0 carbon/Btu) or even non-carbon fuels such as hydro- and nuclear-power.

Any estimate of future natural gas consumption must be placed in the context of future energy consumption, which is already largely determined by population increase. Developing countries needing economic growth and energy consumption growth will tend to opt for the least expensive and most flexible energy increase option, oil. An energy consumption increase of 50-60 million barrels per day (oil equivalent) over today's consumption of 162 million barrels per day is a conservative estimate of what might occur in the years until 2012. The share for natural gas will be 11-13 million barrels per day if natural gas retains its present 22% share. If gas increases its share of the market to 27%, natural gas consumption would increase from today's 72 TCF per year to between 95 and 106 TCF per year—a very major and costly undertaking since much of the expansion will need to be as LNG.

— NORMAN E. DUNCAN

For Further Reading: American Gas Association, *Natural Gas Production Capability* (1990); Scientific American Magazine, *Energy for Planet Earth* (1991); Jefferson W. Tester, David O. Wood, and Nancy A. Ferrari, *Energy and the Environment in the 21st Century* (1991).
See also Automobile Industry; Carbon Cycle; Carbon Tax; Clean Air Act; Electric Utility Industry; Energy; Fuel, Fossil; Greenhouse Effect; Methane; Petroleum; Pollution, Air; Smog-Tropospheric Ozone.

NATURAL HABITATS

What Is a Habitat?

The term *habitat* has come to be used in a variety of ways, but there are two primary scientific uses. One is generic, referring to an environment of a particular kind, such as a coniferous forest, a prairie pond, or a coral reef. It can also refer to the environment in a particular place, such as the alpine meadows of the Rocky Mountains or East African tall-grass savanna.

The other use of habitat is specific to a particular animal or plant and refers to the specific environment where that organism is found and to the environmental conditions which that organism requires for its survival. This formulation of habitat as home was the original and is still probably the most widespread use of the term, particularly in wildlife management and allied fields of natural resource management.

Concern with environmental quality and quality of life has also led to the increasing use of the term habitat in relation to humans. The 1976 UN Conference on Human Settlements in Vancouver was called the Habitat Conference, and the UN Center for Human Settlements in Nairobi has become the Habitat Center. Habitat for Humanity is a current international program to provide housing for the poor.

How Natural Is Natural?

Natural is a relative term because it is very difficult, if not impossible, to find any area of Earth which has not been affected in some way by human activities. We are familiar with the often-catastrophic modification of natural habitats caused by man's current activities involving land use changes, pollution, and exploitation of natural resources. Much less well known is the profound impact which our ancestors exerted on the face of the Earth. Intentionally or otherwise, humans have used fire to modify vegetation for many thousands and possibly millions of years.

The effect of fires is to remove woody vegetation and encourage grasses, and fires have created, spread, or maintained grasslands, prairies, and savannas throughout much of the world. Even today most grasslands or savannas are maintained by fire, mostly caused by humans.

When the first Europeans visited Yosemite Valley in California, the natural habitat of the valley floor was largely open grassland, maintained that way by Indians who burned regularly to facilitate hunting and food-gathering and to maintain open visibility to see possible enemies. Regular fires were stopped when the Indians were removed and the area was declared a national park, and consequently the trees grew tall and thick. In recent years the National Park Service has had to cut trees and reinstate controlled burning to regain the so-called natural conditions which prevailed when the valley was first discovered by Europeans.

In addition to fires, shifting cultivation and other human activities have been altering the forested parts of the Earth for millennia. In the absence of human activities, most of southeast Asia would be dense tropical forest. However, for thousands of years humans have practiced shifting cultivation throughout the area. This process involves burning or otherwise clearing a patch of forest, cultivating it briefly, then abandoning it and moving to a new area. This process has continually created ideal habitat for a great variety of wild animals including deer, wild cattle, elephants, rhinos, and pigs. None of these animals could survive in a closed tropical forest habitat, yet until recent years this great diversity of larger wildlife was found throughout the region, largely because of human actions.

Even some of the most pristine rain forests in places such as Sri Lanka and Indonesia have been found to be growing on the sites of ancient cities or fields. Consequently, when we speak of natural habitats, we usually are referring to habitats which have not been obviously or recently altered by human activities, or, as in North America, to habitats which existed when the first Europeans arrived.

The Habitat Web

No plant or animal lives alone in isolation. Instead each organism may be considered to live at the center of a web. Deer provide an example. The web represents all the components of the deer's habitat, including such things as other deer, other herbivorous animals, predators, water, climate, and the plants which provide the deer's food and cover. Each are interconnected, so that a change in any one may affect some or all of the others. The welfare and survival of the deer depend on maintaining its habitat web intact. If any part of the web is damaged it may not be possible for the deer to survive in that place. But the situation becomes much more complex when we realize that every other living thing in the deer's habitat web has its own habitat web. The webs of the plants, for example, may involve the soils and the rock from which they are derived, algae, fungi, a variety of invertebrates ranging from the soil microfauna to pollinating insects, and the right conditions of light, temperature, and moisture.

However, some species are more rigid in their habitat requirements than others. Species such as coyotes and crows have shown remarkable flexibility in their habitat requirements. Stated another way, they have shown themselves able to adapt to new conditions, to new parts of their changing webs. Because of this adaptive ability, such species have much more chance of surviving today's pervasive habitat changes than do species which are much more rigidly linked to their habitat web. But even for flexible species, their fundamental requirements for food, water, and cover must be met.

The future for less adaptable species is directly linked to the future of their natural habitats, except in the few cases where manmade habitats substitute adequately for the natural ones. For example, North American deer require an *edge habitat,* as discussed below. When the European settlers and their successors changed the face of much of North America, they destroyed many natural habitats, but they also created edge habitats in many areas where none had existed previously. Consequently, there is far more suitable habitat for deer today than in pre-settlement times, even though relatively little of it is natural habitat.

The Diversity of Habitats

There is an immense variety of habitats. The major habitat types, called *biomes* by ecologists, include such general categories as tropical rain forests, tropical deciduous forests, tropical savannas, deserts, coniferous forests, deciduous forests, temperate grasslands, and tundra. In addition, fresh water habitat types include various kinds of rivers and streams, lakes, ponds, swamps, and other wetlands. And there are major salt water habitats ranging from mangrove swamps, estuaries, rocky shores, continental shelves, and coral reefs to deep trenches. There are also special habitats, significant from the standpoint of biodiversity, but more limited in area than the major types; these include caves, springs, mountain tops, and even the hot vents found on the ocean bottom.

Each major habitat type contains a great variety of more specific habitats, from the soil to the tree tops. For example, as one climbs through the largely coniferous forest on the west side of California's Sierra Nevada mountains, one encounters a series of distinct forest habitat types, each dominated by a different species or group of species of trees adapted to the climatic and other conditions at just that altitude and exposure. The lower slopes are dominated by species such as oaks and digger pines which are adapted to the hot, dry conditions found there. Ponderosa and Jeffrey pines form tall, even stands at the mid levels which are the yellow pine or transition zone habitat types, with yet other distinct forest habitats at their upper and lower boundaries. And the higher slopes toward timberline are the home of the juniper, mountain hemlock, and mountain pine, all hardy trees able to survive in the near arctic conditions of low temperatures, high winds, limited moisture, and short growing seasons.

Each of these habitats has its own distinct assemblage of birds, mammals, and other organisms which find there the conditions they need to survive and prosper. But there are many smaller or micro-habitats within these larger habitats. At an elevation of around 5,000 feet, fallen yellow pine logs provide a specialized habitat for a mountain salamander, which finds its needed cover and feeding grounds beneath the logs and under their loose bark. The talus slopes near and above tree line provide the habitat for two characteristic alpine rodents, the marmot and the pica or rock rabbit. Both use the rocks as observation platforms or places to sun, rest, or take cover from predators. The loose soil under and between the rocks accommodates their burrows, and the adjacent grassy slopes provide food.

Habitats Provide the Necessities of Life

Every organism is adapted to its particular habitat, and it follows that the habitat must provide the necessities for life for that organism. This is true whether the habitat is terrestrial, marine, fresh water, subterranean, or aerial. Normally the necessities include, at a minimum, food, water, and cover, all provided in a suitable climate. Further, the necessities must be available at the right time and in the right place.

The spatial relationship between the food, water, and cover is particularly important. In the case of the marmot and pica, if the feeding area was not close enough to the cover provided by talus rocks, the animals would need to travel considerable distances between the rocks and the food during their daily foraging and they would be much more vulnerable to being eaten by the golden eagles, hawks, and coyotes who share their habitat.

This interspersion also explains what wildlife managers call the edge effect. A closed forest has little if any food or cover available at ground level to large herbivores such as deer or wild cattle. Prairies or other open areas do not provide cover, and they can only provide adequate food for animals such as deer when the grass is green in the wet season, not in the dry season. But the edges between forest and grassland provide both suitable food and cover in close proximity. The food includes both grassland plants, which are nutritious in the wet season, and brush and small trees at the edge of the forest, the leaves and twigs of which provide adequate nutrition the rest of the year. Consequently an edge habitat is required by virtually all large terrestrial herbivores which eat both grass and browse. This includes, for example, all the North American deer, red and roe deer in Europe, most other deer worldwide, antelope such as the impala in Africa, and most of the large array of ungulates in southeast Asia.

Food

Food requirements must be met all year. The food must be in adequate supply and of suitable nutritional value at the right season and stage of the animal's life cycle. Wildebeests in the Serengeti-Mara area of Tanzania and Kenya illustrate this principle. The long rains from March through late May provide the only statistically reliable time when fresh green grass is continually available. Wildebeests give birth to their young during a very brief period between February and March, at the start of the rainy season, thus ensuring them the highest quality nutrition during the critical months following the birth of their young and leading up to the rut, which takes place three months later. In a normal year it is unlikely that any wildebeests born at different times will survive.

Food is as much a major habitat factor for marine species as for all other species, but it is not evenly distributed in salt water habitats. The nutrient-rich, upwelling waters off the coast of Peru feed the stock of anchovetas and form the basis for a fishery which once represented 20% of the world's total marine fishery catch. Krill in polar waters provide the basic food for many of the world's whale stocks, along with other krill predators such as seals, penguins, and other polar birds. Coral reefs are a habitat particularly rich in food sources for fishes and other reef organisms, from the herbivores which graze on the tiny plants growing over the corals, the organisms which fasten to the reefs and capture phytoplankton or zooplankton which drifts past to the abundant predators which feed on each other and on the herbivores. The deep benthic regions provide food for those species, including fishes such as orange roughy off New Zealand or the coelacanths in the depths off of Madagascar, which are specially adapted to survival at those depths.

Water

Water may occur in terrestrial habitats as open, standing, or running water, as dew, or as deposits in succulent plant parts. Desert-dwelling rodents throughout the world receive much of their water from dew drops; they lick it up in the morning, running from plant to plant. Rodents and much larger dryland animals such as the Arabian oryx, the addax of north Africa, and east Africa's gerenuk get much of their dry season water from succulent plants. And in turn, predators in the driest areas get much of their water from the flesh of the animals they eat. Water in the habitat may even come from fog, as in the Peruvian coastal desert where night fog condenses on scattered trees, drops to the ground, and creates small circular areas of green in otherwise parched landscapes; or on Namibia's Skeleton Coast where beetles plow furrows in the sand across the prevailing fog-laden winds, thus catching moisture which they then drink each morning.

Cover

All habitats must provide cover for protection from predators, for shelter from the elements, and for nests, dens, or beds. Habitat cover may mean vegetation for hiding, used by deer and many birds, or suitable soil for digging burrows, as is done by woodchucks, prairie dogs, and ground squirrels in North America; badgers in Europe; and aardvarks and squirrels in Africa. Habitat cover may even require a particular texture of sand, as in the case of desert lizards in southeastern California which, when pursued, dive into the sand and, with a rapid swimming movement, actually disappear to just below the surface. Or, as discussed below, it may require suitable trees for nest cavities for a variety of birds and small mammals or holes or other shelter for fresh water and marine species.

Special Habitat Needs

In addition to food, water, and cover, habitats must fulfill particular species' special needs. There is a bewildering variety of such needs. Many birds need grit for their digestive system, so their habitat must include a suitable source. Prairie grouse require an open area with relatively short grass for so-called booming sites where the males display to attract mates. For the same purpose, the southern Africa widow bird requires a flat area with grass about two feet high so it can trample a circle and put on its spectacular mating dance, bouncing some meters above the grass to display its remarkably long tail.

Pacific salmon and sea-run trout are anadromous fishes which spend part of their lives in the open ocean but return to spawn in the section of fresh water stream where they were born. Each of the five species of salmon and two of sea-run trout which spawn in the streams and rivers of Oregon requires a somewhat different set of spawning conditions. All require streams with some form of gravel bottom, but they must have the right size stones, be the correct depth, and have the right velocity of current. Further, different species spawn in different parts of the river basins, some in the small tributary streams high in the mountains and others in the main river. Consequently, even though all seven species are anadromous and many share the same rivers, the specific habitat requirements differ for each one.

Bears need a suitable place for hibernation, such as a cave or hollow log. Many animals require a hole for a nest, denning site, refuge from predators, or, as in the case of the moray eel, a base for hunting. The habitat for moray eels, small octopuses, and a number of reef and other fishes must include a suitable hole. Holes in trees provide nests for many birds, ranging from tropical African hornbills to northern woodpeckers. There is a whole suite of birds in North America which require dead snags for this purpose, and as a result the U.S. Forest Service may require that loggers leave standing a certain number of old dead trees. Many animals which are not accomplished diggers appropriate the holes made by animals which are. In east Africa, holes originally made by aardvarks may be used by spotted hyenas, bat-eared foxes, and jackals. In the American west, burrows originally made by marmots and badgers may be appropriated and enlarged by coyotes and foxes, and holes dug by ground squirrels may become homes for the diminutive burrowing owl or even for rattlesnakes.

These examples show that an appropriate habitat for one species may require the presence of other species. This principle is obvious where species interact as predator and prey, but it has been less often recognized where one species facilitates the needs of another, as in the case of hole diggers. Herds of wildebeest, zebra, and buffalo in east Africa graze down taller grasses and thus facilitate the feeding of others such as the Thomson's gazelle, which feed on short grass and the small herbaceous plants beneath the tall grasses. In the drier parts of Africa, elephants dig holes to reach water below the surface of seasonally dry watercourses. These holes then provide the water that allows a variety of other animals, from antelope to sand grouse to monitor lizards, to survive in these areas. Similarly, American alligators dig holes in shallow water courses; in dry periods when the surface water is gone these "gater holes" provide water for many of the other animals in that habitat.

Residents and Migrants

Some animals are resident within a relatively small habitat which provides all their needs on a year-round basis. The Devil's Hole pupfish is an extreme example, with the total population of the species swimming in one desert spring in Nevada. Others migrate through a single habitat type, as with the annual wildebeest migrations through the savanna ecosystem of the Serengeti-Mara plains, or they migrate between habitat types, as with the wapiti in Wyoming which annually travel between summer ranges in high forest and timberline habitats in Yellowstone National Park to the lower grasslands in Jackson's Hole. Probably the champion long distance migrant is the Arctic tern, which migrates annually between the poles. Often less well understood are the migrations of marine mammals, pelagic fishes, eels, and other marine fauna. Humpback whales migrate from feeding grounds off Alaska to the Hawaiian Islands, where they deliver their calves, and many other whales migrate far north of the krill-rich waters of Antarctica for the same reasons. Strictly speaking, the habitat of such migratory species includes all the habitat types through which they pass during their annual peregrinations.

Habitats Are Not Static

Species continue to evolve; living things continue to adapt. Nature is not in a static balance but is constantly changing. The habitat of many species is in a successional stage. Left alone, that habitat will move into other successional stages; unless the species can find and move to another location at a similar successional stage, it may be lost. This is true for virtually all grassland species, since most grasslands are successional stages, usually maintained as grassland by fire or a combination of fire and grazing. Other habitats are at the most mature stage of a succession. The future of the spotted owl in the U.S. Pacific northwest, for example, is directly linked to the future of the old growth Pacific fir forests.

While it is true that most if not all natural habitats have been modified to some degree by human activities, virtually all natural habitats are now under threat of much more catastrophic change and consequent loss. And with the loss of habitats goes the loss of most of the species which depend on them. This threat affects both resident species and migrants. Even in the U.S., where there have been outstanding conservation efforts to protect natural habitats, there are increasing threats to species which migrate to or from elsewhere. Populations of most migratory song and other birds which summer in the U.S. have been declining for some years, not so much because of habitat threats or lack of protection in the U.S., but because of habitat destruction in their winter ranges in the Caribbean and the southern hemisphere.

Ultimately the success of efforts to maintain the remaining natural habitats will determine the fate of the species which depend on them, and that may include the human species.

LEE M. TALBOT

For Further Reading: D. Halpern, ed., *On Nature: Nature, Landscape and Natural History* (1987); M. E. Soule and B. A. Wilcox, eds., *Conservation Biology: An Evolutionary-Ecological Perspective* (1980); E. O. Wilson, *The Diversity of Life* (1992).

NATURAL RESOURCES

See Resource Management.

NATURAL RESOURCES DEFENSE COUNCIL

See Environmental Organizations.

NATURE VS. NURTURE CONTROVERSY

The nature (heredity) vs. nurture (environment) controversy involves two entirely different theories of human development. One is that the characteristics of the adult person are the expressions of heredity and are already apparent during childhood; they merely continue to unfold. The other is that the experiences of very early life shape the physical and mental attributes of the child in an almost irreversible manner and thereby determine what the adult will become.

There is no real conflict between these two interpretations. Both are correct, because each corresponds to one of the two complementary aspects of development in all living things. Whether the organism be microbe, plant, animal, or man, all its characteristics have a genetic basis, and all are influenced by the environment. Genes do not inexorably determine traits; they constitute potentialities that become reality only under the shaping influence of stimuli from the environment.

The governing influence of environmental factors is particularly effective during the formative stages of life, both in the course of gestation and for a few years after birth. Effects of such early influences commonly last so long that they determine to a large extent the characteristics of the adult. The most important effects of the early environment may be the ones that convert the child's inherited potentialities into the traits that constitute his or her personality. In this regard, it must be emphasized that mere exposure to a stimulus is not sufficient to affect physical and mental development. The forces of the environment act as formative influences only when they evoke creative responses from the organism.

Prenatal, neonatal, and other early influences thus constitute a continuous spectrum through which the environment conditions the whole future of the developing organism. The rates of physical or sexual maturation and the final adult size are not the only, or the most important, effects of these early influences. Physiological characteristics, tastes, interests, and social attitudes are also shaped early in life by environmental facts. Physically and mentally, individually and socially, the responses of human beings to the conditions of the present are always conditioned by the biological remembrance of things past. William Wordsworth's (1770–1850) statement in "The Rainbow" that

"the Child is Father of the Man" is a poetical expression of a broadly conceived biological Freudianism.

The view that man is the product of his environment, so forcefully stated by Hippocrates (460?–377? B.C.) in *Airs, Waters, and Places*, has long remained influential not only among physicians but even more among philosophers. John Locke (1632–1704), Jean Jacques Rousseau (1712–1778), and other partisans of the "nurture" theory of human development believed that the newborn child is like a blank page on which everything is consecutively written in the course of life by experience and learning. Thomas Huxley (1825–1895) gave a more biological expression to this thesis when he asserted that the newborn infant "does not come into the world labelled scavenger or shopkeeper or bishop or duke, but is born as a mass of rather undifferentiated red pulp the potentialities of which can be revealed only by education."

The "nurture" theory has taken many forms in our times. Sigmund Freud (1856–1939) and his followers believed that the peculiarities of each person's mind, and thus most of the difficulties that plague human existence, can be accounted for by early influences, including and especially those around the time of birth. This was Margaret Mead's (1901–1978) general view, and that of Columbia University's School of Social Anthropology of which she was a part when she gained fame with her book *Coming of Age in Samoa*. B. F. Skinner (1904–1990) still holds the most extreme position with regard to the effect of the environment on behavioral determinism, as illustrated by his laboratory techniques for the conditioning of experimental animals and by his books—such as *Walden II* or *Beyond Freedom and Dignity*—where he states that any type of social behavior could be created by shaping the proper kind of social environment for humankind.

In contrast to the partisans of "nurture," Thomas Hobbes (1588–1679), Herbert Spencer (1820–1903), and the social Darwinists upheld the view that nature (heredity) determines to a very large extent the characteristics of the person, young or adult. On the basis of very inadequate statistical evidence, Francis Galton (1822–1911) concluded that this genetic view accounted satisfactorily for the stratification of English society. As he saw it, judges begot judges, whereas workmen, artisans, and even businessmen were not likely to be born with the innate mental ability required for a successful performance in the intellectual world. From Joseph-Arthur Gobineau (1816–1882) to Adolf Hitler (1889–1945), a narrow interpretation of genetic determinism has given rise to attitudes concerning the existence of inferior and master races.

On the other hand Carl Jung's (1875–1961) writings early in the twentieth-century implied that humankind can be understood only by exploring the many factors which played a part in the genesis of the human mind during the remote past; according to him, much of individual behavior is influenced by archetypes as old as the human race itself. Contemporary advocates of genetic determinism are most specific. In their views, we are nothing but naked apes; our relationships with other human beings and with the rest of the cosmos are governed by territorial imperatives and other aggressive and even destructive attributes inherited from our Stone Age ancestors who had to be "killers" because they derived their living from the hunt. According to Professor Edward O. Wilson (1929–), the leader of this school of sociobiology, even altruism and religious feelings are the consequence of genetic mechanisms that once had and usually still have a selective survival value. According to Wilson, religious practices confer biological advantages because, in the midst of the potentially disorienting experiences of each person in daily life, religion leads to membership in a group and thereby provides a driving purpose in life compatible with self-interest.

Although behaviorism and sociobiology are scientifically poles apart in the biological mechanisms they invoke, they derive from a similar attitude with regard to human life. With either explanation, the human being loses identity as *subject* since it is shaped and governed by forces over which it has no control. The person becomes a mere *object*, whose behavior and fate do not involve conscious choices. Human beings become products of "chance and necessity," beings for whom freedom and dignity are just meaningless concepts.

Humans are unquestionably shaped to a very large extent not only by their genetic constitution but also by the environments in which they function and by their ways of life. Further, human development is determined less by the forces to which people are passively exposed than by the choices they make concerning their personal lives and the organization of their societies. For this reason it is of extreme importance that individuals have as much freedom and as many options as practically possible in selecting or creating the environmental conditions that shape them. New solutions cannot come from a transformation of human nature because it is not possible to change the genetic endowment of the human species. But they can come from the manipulation of social structures because these affect the quality of behavior and of the environment and therefore the quality of life.

In summary, both the genetic constitution and the total environment play roles in all aspects of human development and behavior, but these deterministic mechanisms do not fully account for human life. Persons and societies do not submit passively to surroundings and events. They make choices about where to live and what activities to pursue—choices based on what they want to be, to do, and to become. Furthermore, persons and societies often change their goals and their ways; they can even retrace their steps and start in a new direction if they believe they are on the wrong course. Thus, whereas animal life is prisoner of *biological* evolution which is essentially irreversible, human life has the wonderful freedom of *social* evolution which is rapidly reversible and creative. Wherever human beings are concerned, trend is not destiny.

RENÉ DUBOS
(EDITED BY WILLIAM R. EBLEN)

For Further Reading: René Dubos, *Man Adapting* (1980); Lee R. Dice, *Man's Nature and Nature's Man* (1955); George C. Williams, *Adaptation and Natural Selection* (1966).
See also Human Adaptation; Invariants of Humankind, The.

NELSON, NORTON

(1910–1990), American toxicologist. Norton Nelson, who began his professional life as a biochemist, is regarded by many as the father of modern environmental health. Throughout his long and distinguished career, he pioneered the recognition of the importance of chemical and physical agents as potential causes of human disease, and led in mobilizing the talents and resources needed for identifying and controlling such agents. Dr. Nelson sparked developments that have reshaped attitudes in public health and have benefited people in all parts of the world.

In 1947 Nelson joined the New York University School of Medicine, where he served as Director of the Institute of Environmental Medicine from 1954 to 1979. Under his leadership, the Institute developed into the largest and most prestigious academic unit of its kind in the world, being noted particularly for its research on cancer, pulmonary disease, and environmental radiation. As Director, Nelson contributed actively to research on the metabolism of carcinogens, the deposition of inhaled particulates in the respiratory tract, and the experimental induction and epidemiology of lung cancer.

Under Nelson's chairmanship, in 1965 the National Advisory Environmental Health Committee submitted a Special Report to the Surgeon General of the United States Public Health Service, which recommended establishing a federal system for developing standards to limit occupational exposure to hazardous agents, and which ultimately led to the creation of the National Institute for Occupational Safety and Health. Many similarly important contributions resulted from his Chairmanship of the Executive Committee of the World Health Organization's Scientific Group on

Methodology for the Safety Evaluation of Chemicals, and from his participation in over 100 other advisory committees and professional organizations. The importance of his contributions to legislation that "benefited and promoted the field of environmental health" was recognized by the *National Journal* in its June 14, 1986 issue, which listed him as "one of the 150 people best able to influence the Federal Government" in ways that serve the national interest.

David Rall, former Director of the National Institute of Environmental Health Sciences, called Nelson the father of the second generation of environmental public health. The first generation was concerned mainly with illnesses caused by organisms such as mosquitos or fleas, poor sanitation, or impure water or food. The second generation has been concerned with the health hazards associated with the industrialization of society. Thanks to Nelson, a whole generation of scientists are able to identify such new environmental threats to human health, determine how they act, and exploit scientific advances to prevent or mitigate their impacts.

ARTHUR C. UPTON, M.D.

For Further Reading: S. Belman and T. Kneip, eds., "Proceedings of the 40th Anniversary of the Institute of Environmental Medicine, N.Y.U. Medical Center," in *Environmental Health Perspectives* (1989); The René Dubos Center for Human Environments, *Environment and Human Health: Toxic Chemicals* (1985); Lester Lave and Arthur Upton, *Toxic Chemicals, Health and the Environment* (1987).

NEUROTOXINS

The human nervous system is responsible for interactions involving both external and internal environments. It serves functions as fundamentally primitive as regulating body temperature and as profoundly complex as creating cosmological theories. Interference with its functional integrity by neurotoxins (poisons that affect the nervous system) can range in severity from disabling neurological disease to subtle reductions in intellectual capacity. The severity of the problem depends upon the nature of the neurotoxins and the dose.

Many chemicals are neurotoxins, including heavy metals, pesticides, organic solvents, food additives, and polychlorinated biphenyls (PCBs). Even natural, uncontaminated foods contain neurotoxins. The workplace is also a major source of neurotoxic hazards. About 25% of workplace exposure standards are based on neurotoxic reactions, ranging from the inability to respond to emergencies to decreased performance efficiency.

The effect of neurotoxins can take the form of reductions in sensory function, such as the damage to vision from the industrial chemical acrylamide, impaired coordination from mercury compounds, or cognitive deficiencies, such as lowered scores on psychological tests, from chronic exposure to organic solvents.

Even more subtle deficiencies may occur. Exposing a fetus or young child to relatively low levels of lead shifts the IQ distribution to lower scores. The developing brain appears to be extremely sensitive to other contaminants as well. Because of pollution, many freshwater and marine species of fish contain enough methylmercury (an organic compound of mercury), so that even moderate consumption of these species during pregnancy poses a threat to human fetal brain development.

The other end of the life cycle has also prompted concern and speculation about neurotoxins. For example, some researchers speculate that Parkinson's disease (a chronic progressive nervous disease) may be due to the combined effects of exposure to an environmental neurotoxin and of aging, and that aging reduces the brain's ability to compensate for the toxin-induced damage to the nerves controlling certain muscles. Others are searching for environmental factors that might contribute to Alzheimer's disease.

The Environmental Protection Agency views adverse effects on the nervous system as a primary target of regulation and has proposed guidelines for assessing neurotoxic risks. These guidelines rely primarily on behavioral tests in animals, which have two aims: first, to identify neurotoxins, and, second, to predict levels of environmental exposure that may pose risks to humans.

BERNARD WEISS

NGO, ENVIRONMENTAL

See Environmental Organizations.

NICHE

An ecological niche is defined as: (1) the function, or occupation, of an organism within an ecological community, and (2) the specific area within the habitat occupied by an organism. The ecologist Eugene Odum wrote in 1959 that an ecological niche is "the position or status of an organism within its community and ecosystem resulting from the organism's structural adaptations, physiological responses, and specific behavior (inherited and/or learned)." An organism's niche is thus defined by its habitat, nutritional sources, behaviors, relationships with other organisms, and many other dynamic characteristics. In the famous example of the Galapagos finches described by Charles Darwin, their niche can be compared in terms of a single attribute—beak size—and the way in which that attribute differs among the many species.

Niches are formed by competition and environmental pressures. It may be instructive to separate out two aspects of the broad concept of *niche* that focus on such attributes. The term spatial niche is referred to when analyzing differences in the physical location (microhabitats). When considering the energy and food chain interactions of an organism, scientists refer to the trophic niche. For Darwin's finches, competition among the ancestral finches either led to specialization in an environment, eliminating existing niches, or to the introduction of new species, bringing about niche overlap. In niche overlap, species are forced to share resources or other niche characteristics. If resources are unlimited, the species may coexist. If not, competition will ensue, which will either drive one species out or force it to adapt to another niche where it may be more able to compete or coexist.

One prevailing model for niche analysis from a theoretical perspective is called the hypervolume model. This model, originally proposed by G. E. Hutchinson in 1957, provides the foundation of modern niche ecology. In the hypervolume model, niches are depicted graphically by plotting coordinates that represent the range of positions possibly occupied by individuals of a species along an environmental gradient—for example, the tree height at which a bird resides, or the time of day an animal searches for food. Overlap among the possible ranges available to two different species restricts the range that the species may actually use. Thus, there is a distinction between a species' fundamental niche—the size of the niche that a species could occupy without competition—and the realized niche that the species actually inhabits when constrained by its neighboring species. This analysis allows ecologists to consider the influence of competitors and predators on the niches they affect.

NIH

See United States Government.

NIMBY

NIMBY is an acronym for Not In My Back Yard. It gained currency in the 1980s as a battle cry against environmentally or locally objectionable facilities—for instance, hazardous waste dumps, nuclear sites, polluting factories, or low-income housing. In actual usage the term can refer to the objectionable facility itself ("The highway is a NIMBY"); the objectors to the facility ("She is a NIMBY"); or the political impulses animating the objectors ("It is a NIMBY group").

NIMBY objectors to NIMBY facilities use several basic arguments against them. Opponents may contend, for example, that the NIMBY is not needed anywhere, as in the case of nuclear power plants or strip mines. Alternatively, they may argue that the NIMBY is needed, but not where it is proposed—say, in a wetland. Then again, they may contend that its siting or operating procedures—for instance, the relevant citizen participation mechanisms—are inadequate. Or perhaps its effects, such as its air pollution, will be or are harmful. In many NIMBY disputes these arguments often appear simultaneously.

NIMBY facilities pose the seemingly impossible problem of having in our midst large numbers of big projects no one wants nearby. The U.S. routinely resolves this dilemma through familiar environmentalist devices such as land-use regulations, pollution controls, citizen participation, impact statements, preservation areas, and lawsuits. These are some of our rituals to purify tainted, taboo NIMBY objects. Resolving the NIMBY dilemma has always been a central concern of environmental planning and environmental law, perhaps *the* central concern. However, the task is getting harder.

The difficulty is that in recent years NIMBY objectors have succeeded in blocking many NIMBY facilities. No large, new, free-standing hazardous waste facility was sited anywhere in the United States during the 1980s. Only two large metropolitan airports—in Denver and Austin—have appeared since the early 1960s. The lack of locations for new prisons has caused such overcrowding in many existing jails that some systems—for instance, in New York City, Chicago, Pittsburgh, and the entire Texas, Florida and Connecticut state systems—have had to release convicted criminals. Most big cities, even relatively wealthy ones, have not begun a major low-income housing, mass transit, or highway project in fifteen years. NIMBY blockage is much of the reason.

Some environmentalists are now retreating from their previous hard-line positions against NIMBY facilities. It is beginning to seem obstructionist, selfish, elitist or racist, and to make NIMBY a pejorative term. Instead of resisting every instance of NIMBY facilities, environmentalists are starting to find (and to encourage their opponents to find) middle-ground positions: new, practical, working relationships between valid local (that is, back-yard) goals and valid regional and national ones, and innovative ways to mesh environmental and economic imperatives.

Thus environmentalists are increasingly willing to simplify the bureaucracy of regulatory requirements or to allow existing NIMBY facilities to expand in exchange for agreements that will prohibit new ones. Environmentalists are also paying more attention to the racial, ethnic, and class behaviors that site NIMBY facilities; they are reevaluating the equity of whose back yards actually end up with NIMBYs.

Other acronyms have become associated with NIMBY facilities or NIMBY impulses: for instance, LULUs (Locally Unwanted Land Uses), TOADS (Temporarily Obsolete Abandoned Derelict Sites), NIABY (Not In Anybody's Back Yard) and NIMTOO (Not In My Term Of Office).

FRANK J. POPPER

NITRATE/NITRITE

Nitrate, a compound of nitrogen and oxygen (NO_3), occurs naturally in plants and forms from ammonia in animal wastes and chemical fertilizers. Nitrate can enter ground water or surface waters from faulty septic tanks or from runoff from animal feedlots or overfertilized fields. While people in the United States normally ingest between 75 and 100 milligrams of nitrate each day in vegetables without harm, nitrate in drinking water can raise that intake significantly and be a health problem.

Bacteria in a very young infant's intestinal tract can convert nitrate to nitrite (NO_2). When nitrite is absorbed in the intestine, it enters into a complex chemical reaction with hemoglobin (the red pigment in blood that carries oxygen to the body's cells and tissues). This interaction produces methemoglobin—a form of hemoglobin that cannot carry oxygen. Because an infant's stomach has relatively low acidity, it offers an excellent environment for the bacteria that convert nitrate to nitrite. Young infants also lack the highly developed enzyme systems that adults have for converting methemoglobin back to hemoglobin. An exposed infant can develop symptoms of oxygen starvation or even suffocate. The pale lavender or bluish color of an infant with methemoglobinemia gives the illness the name "blue-baby syndrome."

Boiling nitrate-contaminated water only concentrates the nitrate; it does not remove the contamination. The EPA and the U.S. Public Health Service have defined the maximum nitrate concentration in safe drinking water as 45 milligrams of nitrate per liter. Drinking water that exceeds this standard is often from shallow or poorly constructed rural wells.

Several studies indicate that nitrate levels in drinking water seem to be on the rise. In the early 1980s an EPA survey reported that 2.7% of the nation's rural water supplies—603,000 households—had nitrate levels above the standard; in 1985 a U.S. Geological Survey study found 6% above the standard. Other health concerns arose from the fact that nitrate, nitrite compounds used to preserve meat, and nitrites could combine with some pesticides or with amino acids in the body to form cancer-causing compounds, such as nitrosamine.

The risks from increased exposure to nitrates are hard to determine, however, in part because there are two steps necessary for toxicity: first, the conversion of nitrate to nitrite, and second, the reaction of nitrite with other nitrogen-containing compounds. Although some studies have suggested that high levels of nitrate and nitrite may be associated with higher risks of stomach and esophageal cancer, they have not shown that nitrate and nitrite cause cancer. The long-term health effects of exposure to these compounds are also not known.

BARBARA SCOTT MURDOCK

NITROGEN

Nitrogen (N), an element abundant in the atmosphere and the life forms of planet Earth, is a colorless, odorless and tasteless gas (N_2) at ordinary temperatures. It is a necessary constituent of protein and the nucleic acids, DNA and RNA. It is slightly less dense than oxygen and is less soluble in water. Nitrogen condenses to a colorless liquid, that looks like water, at $-195.8°C$; at $-209.9°C$ the liquid freezes to solid nitrogen. It is chemically stable and unreactive. It does not burn, it does not support burning, and it reacts readily only with the most reactive of metals, such as sodium and calcium, forming nitrides. When a powerful electric spark is passed through a mixture of nitrogen and oxygen, however, nitric oxide does form, and this reacts further with oxygen, producing nitrogen dioxide. Nitrogen exists as diatomic molecules each held together by a triple covalent bond and it is this structure that explains its stability. Naturally occurring nitrogen is 99.6% the N-14 isotope with some N-15 isotope. Its atomic weight is 14.008; its atomic number is 7.

In 1772 Daniel Rutherford at the University of Edinburgh recognized a gas that was a portion of air that was incapable of sustaining life. Others, notably Karl Scheele and Joseph Priestley, also isolated and recognized this gaseous substance. It was Antoine Lavoisier, however, who identified the gas as an element and named it "azote," meaning "without life." The English name "nitrogen" came from the Greek term meaning "niter producing" because of the presence of the element in niter (potassium nitrate).

A sample of nitrogen gas can be obtained in the laboratory by heating an aqueous solution of ammonium nitrite and collecting the ensuing gas over water. The large amounts of nitrogen needed by industry are generated by cooling air until it is a liquid and then allowing the gases to evaporate slowly. In this process, called fractional distillation, the nitrogen evaporates to a gas at a lower temperature than the oxygen and so is readily collected.

The air of planet Earth is 78.09% nitrogen by volume, which amounts to about 4,000 billion tons of nitrogen. Mars, by contrast, has an atmosphere that is 2.6% nitrogen. Some can be found dissolved in oceans, lakes and rivers; 66% of the air dissolved in the waters of Earth is nitrogen, the other 34% is oxygen. There are large deposits of compounds of nitrogen, called nitrates, especially where large lakes dried and left salty crusts. In the Sun, carbon, oxygen, and nitrogen isotopes cycle hydrogen nuclei resulting in their fusion to helium nuclei. Eventually the carbon, oxygen, and nitrogen isotopes are processed into stable nitrogen, the N-14 isotope.

Nitrogen, along with carbon, oxygen and hydrogen, is an essential building block of the molecules of which living things are made. Proteins are distinguished from carbohydrates and fats because of the nitrogen. The base constituents of nucleic acids also necessarily contain nitrogen. When there is a deficiency of nitrogen in the soil, plants are stunted in their growth and have yellow leaves. The nitrogen in the air is not available for plants to use since it is chemically inert. During episodes of lightning, nitrogen in the air reacts with oxygen, and oxides of nitrogen are washed into the soil with the rain. Now in compound form, the nitrogen is available to plants. Through the action of bacteria, nitrogen from the air and from decaying organic matter is also made available to plants. Nitrogen from dead material is returned to the air by the action of bacteria.

Plant manufacture of nitrogen-containing molecules is the ultimate source of animal protein and nucleic acids. Proteins provide the structural basis of animals in skin, nails, and muscle; the enzymes that drive the reactions of life for energy, building up, and breaking down materials are proteins. The nitrogen in proteins is present in the amine group, $-NH_2$, and this group along with the organic acid group $-COOH$ confers the capacity of amino acids to polymerize into proteins. The nucleic acids DNA and RNA have a structure that involves five organic bases that are the elements of the genetic code. These bases have nitrogen atoms as a key structural component of their molecules.

Industrial activity with nitrogen is directly related to the need for nitrogen by crops; through the Haber Process of combining nitrogen and hydrogen to make ammonia, and the Ostwald Process of making nitric acid, nitrogen compounds are made into fertilizers. Other industrial applications use nitrogen's inertness to prevent fires, to protect light bulb filaments from oxidation, to prevent spoilage of foods (such as bacon and peas) by packaging them in nitrogen, and to prevent oxidation of coinage metals by heating them in nitrogen. A third industrial use is refrigeration by liquid nitrogen. Compounds of nitrogen are very chemically active. All explosives, other than nuclear devices, involve compounds of nitrogen, e.g., gunpowder, dynamite, and TNT (trinitrotoluene). The cyanides, which are poisonous, also contain nitrogen.

JEAN LYTHCOTT

For Further Reading: Isaac Asimov, *The World of Nitrogen* (1958); Bernard Jaffe, *Crucibles: The Story of Chemistry* (1960); Ralph I. Freundenthal and Susan Loy, *What You Need to Know to Live with Chemicals* (1989); C. H. Snyder, *The Extraordinary Chemistry of Ordinary Things* (1992).
See also Atmosphere; Bacteria; Fertilizers; Nitrate/Nitrite; Nitrogen Cycle.

NITROGEN CYCLE

Nitrogen is cycled through the biosphere and participates in long-term geologic and geochemical cycles. It differs from many other elements of biological interest in that its cycle is more complex than most for a number of reasons. It not only has a structural role in all life forms but it also has a catalytic function in virtually all enzymic reactions and is itself a reactant in many energy-yielding processes. Most of this vital element is in the atmosphere as comparatively inert nitrogen gas (N_2) or in sediments and so the processes of its cycling have been of considerable scientific interest. The distribution of nitrogen in the various compartments of the Earth is shown in Table 1.

The nitrogen cycle is not a single methodical sequence but rather a combination of processes and possibilities, the most fundamental of which entail the cyclic movement of nitrogen from the soil to plants, possibly to animals, and thence through death and decay back to the soil. This cycle is modulated by processes of nitrification, denitrification, and nitrogen fixation (conversion into stable, biologically assimilable compounds). There is also the leakage of nitrogen from the soil system by leaching to ground waters, transport to the oceans and sediments, and a variety of other processes, even including the slow movements

Table 1
Estimated distribution of nitrogen between various compartments of biogeochemical interest

Compartment	Species	Mass (teragrams, 10^{12} grams)
Atmosphere	N_2 gas	3,800,000,000
	N_2O	1,300
	Other combined N	1.3
Land	Organic (including coal)	800,000
	Inorganic	140,000
Oceans	N_2 dissolved	22,000,000
	Organic	340,000
	Inorganic	580,000
Sedimentary rocks		750,000,000
Crust		650,000,000
Mantle		56,000,000,000

of plate tectonics. Here we consider the nitrogen cycle in its more narrow sense as it is usually understood, involving the biological cycle and its modification by human activities with only brief mention of these other long-term processes in order to place them in perspective.

Nitrogen Uptake and Assimilation

The most likely fate of nitrogen compounds in the soil is to become part of the pool of nitrate ion (NO_3^-) most immediately available to plants. This is the most oxidized form of nitrogen, its lowest energy level in an environment in equilibrium with atmospheric concentrations of oxygen. As a negatively charged ion, nitrate is readily mobile in the soil solution, and thus available for assimilation by the plant. In the plant it can be reduced to the level of amino or amide nitrogen, the level most commonly appearing in biological compounds, including amino acids, proteins and other nitrogenous structural elements of plants and animals. Reduction of nitrate ion in an aerobic environment is an energy-requiring process and so is governed by nitrogen demands of the system. If nitrate ion in excess of that needed for immediate use is present, it need not be reduced and can accumulate within the cell for future transport or use. Reduction can take place in either roots or leaves. In either case energy is required, but the direct availability of photosynthetic energy in the leaf may permit a more direct and more efficient reduction.

Once synthesized into organic structures or otherwise incorporated into plant tissues, the nitrogen is part of the nutrient budget available to foraging animals or microorganisms for their own structural or energy uses, so entering the trophic (food) sequence.

Mineralization

On the death and decay of plants, animals, or microorganisms, nitrogenous organic residues are returned to the soil for further microbial decomposition. Nitrogen in excess of that required for the consuming microorganisms is released to the soil, usually as ammonium ion (NH_3^+), and in an aerobic environment becomes available for yet another energy-yielding process called nitrification in which it is oxidized, first to nitrite ion (NO_2^-) and eventually to nitrate ion, thus completing the most rudimentary portion of the nitrogen cycle. The term *mineralization* has come to mean the totality of reactions whereby the organic materials are oxidized releasing their mineral constituents to the soil: nitrogen as ammonium ion, phosphorus as phosphate ion, and other elements as their inorganic, usually ionic form.

Nitrification

The ammonium ion released in the mineralization process is itself an energy source for a group of autotrophic (meaning that they can subsist on an inorganic medium) microorganisms capable of oxidizing ammonium ion to nitrate ion as their sole energy source. These nitrifiers are commonly viewed as two groups because the reaction takes place in two steps. First, nitrite ion is formed from ammonium: $NH_3^+ + 2\,O_2 + 2\,e^- \longrightarrow 2\,H_2O + NO_2^-$. Then nitrite is oxidized to nitrate ion: $NO_2^- + 0.5\,O_2 \longrightarrow NO_3^-$.

Although these equations show the overall process of nitrification, there are other possible side reactions of interest. Nitrous oxide (N_2O) can be produced in the process under some conditions, particularly if the concentration of ammonium ion is high. Nitrous oxide, popularly known as "laughing gas," is one of the greenhouse gases—atmospheric constituents that absorb infrared radiation and thereby alter the Earth's heat balance.

Denitrification

In the absence of oxygen and with other conditions suitable for microbial growth, nitrate ion can serve as oxidizing agent (electron acceptor) for the oxidation of organic or other materials, also with the release of energy, by a group of microorganisms known as denitrifiers. The overall reaction can be expressed by the following equation although it takes place in several steps with various intermediate nitrogenous compounds, including nitrite ion: $[HCHO] + NO_3^- + H^+ \longrightarrow H_2O + N_2 + CO_2$. The $[HCHO]$ in this reaction is a generic representation of organic matter but a wide range of organic compounds are subject to oxidation in this manner. The gaseous N_2 produced is released to the atmosphere; this reaction is the major source of

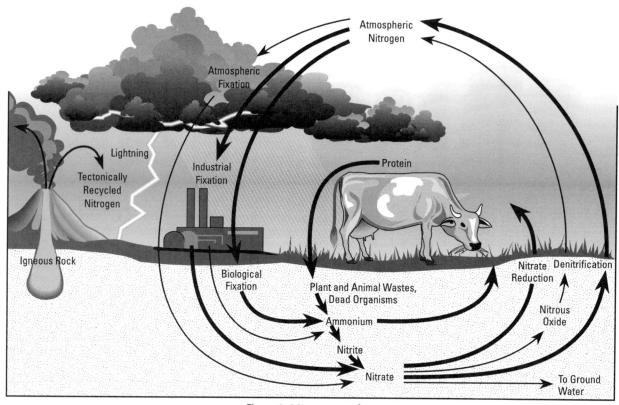

Figure 1. Nitrogen cycle.

atmospheric nitrogen. In the absence of the biological cycle, nitrogen would tend to become oxidized and accumulate in the ocean as nitrate ion.

Alternatively, the denitrification reaction can stop at the oxidation level of nitrous oxide: $[HCHO] + NO_3^- + H^+ \longrightarrow 1.5\ H_2O + N_2O + CO_2$. The nitrous oxide can be further reduced to nitrogen gas or can be released directly to the atmosphere, adding to the pool of greenhouse gases.

The denitrification reaction results in the loss of chemically combined or "fixed" nitrogen from the system. When the reaction was first demonstrated in the mid-19th century, there was some concern that the Earth would eventually lose its supply of fixed nitrogen and life forms would become nitrogen starved, in spite of the fact that the processes of biological nitrogen fixation were demonstrated at about the same time.

Nitrogen Fixation

Although the denitrification reaction annually involves only a small fraction of the pool of nitrogen available to plants (and therefore animals), it has resulted in making the gaseous N_2 of the atmosphere the

major reservoir of the Earth's nitrogen aside from that locked in sediments, a kinetically quasi-stable condition. There are a number of processes that return atmospheric nitrogen to the soil-plant system as outlined in Figure 1. Atmospheric nitrogen can be fixed by ionizing processes in the atmosphere such as lightning and other electric discharge, meteor trails or cosmic rays. Most of the fixation takes place in biological reactions. There is also some return of fixed nitrogen from combustion processes such as forest fires or by volcanic activity. This latter source has been looked upon historically as "juvenile" nitrogen from the outgassing of the planet. We now realize that most of this probably is recycled nitrogen from sedimentary materials returned by plate tectonic activity, representing probably one of the longest time constants in the combination of events that constitute the nitrogen cycle.

Biological nitrogen fixation can be carried on by free-living microorganisms or microorganisms in symbiotic association with higher plants or animals and involves some complex enzymic reactions, but under most biological conditions involves an expenditure of energy. Hence in the competitive biological world, nitrogen-fixing organisms or associations tend to occupy nitrogen-deficient environments; where the nitrogen supply is not limiting, as in older, stable ecosystems, they are frequently crowded out by more energy-efficient species or associations.

Although in a thermodynamic sense the fixation of nitrogen should not be energy-expensive, it appears that with biological systems, the energy required for the splitting of the N-N bond of N_2 (activation energy) is never recovered, making fixation energetically expensive. Thus even in anoxic systems where reducing capabilities are no problem, there still is required this energy for splitting the bond. Consequently although nitrogen-fixing systems can compete well in a nitrogen-deficient environment, if adequate fixed nitrogen is available they lose their competitive advantage.

The best-known examples of nitrogen fixation are the symbiotic associations of leguminous plants with *Rhizobium*. Here the specific microorganism is harbored in a nodule in the plant root. There are other associations such as that of actinomycetes in the root nodules of some plants or other plant associations with microorganisms or algae, or in the digestive tracts of animals including insects such as termites. All of these associations have in common the provision of an energy resource to the microorganism by the host plant or animal in exchange for which fixed nitrogen is provided by the microorganism. In free-living microorganisms the fixation of nitrogen may be to meet the immediate needs of the organism, as with some algae, or it may be a more loose synergistic plant-microbial association in which fixed nitrogen in excess of that required by the organism itself is released to the soil and supports the local plant population which, in turn, releases organic materials that support the microorganism.

Industrial Fixation

Agricultural practices cause extra demands for additional fixed nitrogen. Many agricultural crops leave land fallow during part of the year and nitrate ion in the soil is subject to loss to ground waters by leaching. Additionally, agricultural crops are commonly removed from the growth site and transported to urban locations where they are utilized and their nitrogenous residues or wastes disposed of locally, eventually to ground waters or to the sea or lost by volatilization or denitrification. This removal of nitrogen from the soil is countered by the provision of new nitrogen fixed industrially or by the use of leguminous crops. Some nitrogen from domestic use is returned to soil as organic matter, but this is a relatively small fraction of the total lost. Additionally, it requires the expenditure of energy for its transportation and distribution, so there are economic factors discouraging this route of recycling.

Industrial nitrogen fixation requires an energy input, either from hydroelectric sources or by the use of fossil fuels as in the Haber-Bosch process for producing ammonia (NH_4). Both of these processes require the input of energy for the breaking of the N-N bond at the rate of about 600 kilojoules per mole of nitrogen gas converted. Alternatively it is possible to effect the direct catalytic reaction of nitrogen and oxygen in the atmosphere at high temperatures with the production of oxides of nitrogen and eventually nitrate ion, overcoming this expenditure of energy, but the resultant nitrate ion and its accompanying cation (such as potassium) have a much greater mass than the ammonia coming from the Haber-Bosch process. This in turn requires the expenditure of energy for its transportation and emplacement, negating the energy advantage of the fixation. Consequently the anhydrous ammonia or ammonium compounds are favored industrial sources, both from an energetic and convenience standpoint.

Other combustion reactions including forest and prairie fires, internal combustion engine exhaust, and industrial processes other than those designed specifically for nitrogen fixation also make a significant additional contribution to the annual input of new fixed nitrogen.

Human Intervention

In attempting to determine the contribution of human activities to the overall nitrogen cycle, it is difficult to arrive at accurate comparative figures, but it appears that the combined annual input of new fixed nitrogen from human sources, including the agricultural use of legume crops, approximately equals that occurring historically before human intervention. It is even more difficult to evaluate the significance of this human contribution. There is no concern regarding the availability of nitrogen for future uses. The most pressing problems have to do with the quality of ground and surface waters because of the export of fixed nitrogen to them by leaching or waste disposal or by volatilization, and release of nitrogen compounds to the atmosphere from which they are precipitated. Additionally nitrous oxide (N_2O), resulting from both denitrification and nitrification reactions, is radiatively active in the atmosphere and hence a potential contributor to climate change. Although atmospheric concentrations are low compared with other greenhouse gases, it could also influence the stability of the protective ozone layer.

The release of nitrogen oxides and other oxides to the atmosphere from engine exhaust and industrial sources results in an increased atmospheric burden of acidic and other toxic substances, resulting in acid precipitation, which can injure vegetation and animal life, particularly in poorly buffered soils and waters. The extent of these effects is only poorly cataloged. Some plant and animal populations with limited ability to deal with these environmental changes are particularly vulnerable.

The accumulation of nitrate ion in ground water by leaching from fallow fields or from sewage disposal systems is a particularly troublesome problem. Some ground water reservoirs represent the storage from tens of thousands of years. Aside from their depletion for irrigation applications, they are accumulating undesirably high levels of nitrate ion. The nitrate content of drinking water that can be tolerated by animals, including humans, is limited, particularly for some individuals with metabolic or other health disorders. Ground water in some areas already is well over levels considered "healthful" or even safe by public health standards.

In order to lessen the potential harm that a continually expanding human population can do in perturbing the nitrogen cycle it would be necessary to alter agricultural practices, waste management procedures, industrial procedures and others of the many contributors of nitrogen compounds to the atmosphere, to soils and to waters. All of these revisions will require policy action at the highest level. Many of the available procedures or practices will be expensive of energy and effort or otherwise difficult to attain. The alternative degradation of conditions that can result if corrective action is not taken also has a high human price.

C. C. DELWICHE

For Further Reading: C. Delwiche, *The Nitrogen Cycle* (1970); M. Schidlowski, J. M. Hayes, and Isaac R. Kaplan, "Isotopic Inferences of Ancient Biochemistries: Carbon, Sulfur, Hydrogen and Nitrogen," in *Earth's Earliest Biosphere, Its Origin and Evolution,* J. William Schopf (ed.) (1983); R. Soederlund and B. H. Svensson, *The Global Nitrogen Cycle* in B. H. Svensson and R. Soederlund (eds.), *Nitrogen, Phosphorus and Sulfur Global Cycles* (1978).
See also Acid Rain; Atmosphere; Fertilizers; Nitrogen; Pollution, Air.

NITROGEN OXIDES

See Pollution, Air.

NONGOVERNMENTAL ORGANIZATIONS

The importance of nongovernmental organizations (NGOs) has increased significantly since the Stockholm Conference in 1972. There are nearly five thou-

Table 1
Nongovernmental organizations

Founding date	Name	Based in
1875	American Forestry Association	USA
1913	The Garden Club of America	USA
1919	International Council of Scientific Unions (ICSU)	France
1922	International Council for Bird Preservation	UK
1929	National Council of State Garden Clubs	USA
1942	Oxfam (Oxford Committee for Famine Relief)	UK
1952	International Planned Parenthood Federation (IPPF)	USA
1965	Australian Conservation Foundation (ACF)	Australia
1967	Promocion del Desarrollo Popular (PDP)	Mexico
1968	Zero Population Growth (ZPG)	USA
1972	Institute for Research on Public Policy	Canada
1973	International Institute for Environment and Development	UK
1975	Worldwatch Institute	USA
	Environment Liaison Centre International (ELCI)	Kenya
	The René Dubos Center for Human Environments	USA
	Muslim World League	Saudi Arabia
1976	Society for International Development (SID)	Italy
1978	Rural Development Foundation of Pakistan	Pakistan
1981	Centro Feminista de Informacion y Accion (CEFEMINA)	Costa Rica
	The Centre for Science and Environment (CSE)	India
	Women's Environment and Development Organization	USA
1983	Earth Search	Nigeria
1984	National Rainbow Coalition	USA
	Thailand Development Research Institute (TDRI)	Thailand
1985	Native Americans for a Clean Environment	USA
1987	Japanese NGO Center for International Cooperation (JANIC)	Japan
1988	World Conservation Union (formerly IUCN)	Switzerland
1992	American Forests (formerly American Forestry Association)	USA
	World Eco Reform	Russia
1993	MWENGO (Swahili for "NGO vision")	Zimbabwe

sand NGOs representing millions of supporters who play an expanding role in decision-making from the local to the global level. The growing partnership between governments and NGOs—demonstrated at the Earth Summit (1992)—continues with the International Conference on Population and Development (1994). The increasing diversity of these organizations reflects their interest in environment and development issues throughout the world. This representative sampling of these NGOs arranged in chronological order according to their date of origin illustrates the expanding emphasis on the human environment.

NUCLEAR POWER: HISTORY AND TECHNOLOGY

Nuclear power comes from fission, a nuclear reaction in which the nucleus of a heavy atom splits into two parts, with the emission of neutrons and enormous energy.

In 1939 four scientists working in Berlin (Meitner, Frisch, Hahn, and Strassman) discovered the basis for nuclear fission. Lise Meitner fled to Denmark and told Neils Bohr about it. When Bohr reached the U.S., he told Enrico Fermi, who had arrived from Italy. Fermi and other nuclear scientists from Europe sought to convince the U.S. government about the implications of these discoveries. Perhaps if a major war had not been on the horizon, nuclear power would have developed more slowly and quite differently. But 1939 was a time in which world war was about to break out and the early interest in fission in the U.S. was driven not by the potential for generating electricity but rather by the potential for a destructive weapon—the atomic bomb. A famous letter from Albert Einstein to President Franklin D. Roosevelt triggered a massive effort by the U.S. and its allies to develop nuclear weapons. This effort, code-named the Manhattan Project, mobilized thousands of U.S. scientists and brought others from Canada and Europe to harness the power of the atom in a way that could be used for mass destruction.

During the early 1940s plants were built in isolated sites in Washington, Tennessee, and Georgia to extract the materials necessary to make a nuclear weapon: plutonium and uranium. Uranium comes in two forms (isotopes): U-235 and U-238; as found it is more than 99% U-238 and less than 0.7% of U-235. Only U-235 is useful for fissioning, so massive plants were built to separate the U-235 from the U-238. The development of the weapon itself was done under the guidance of Robert Oppenheimer at Los Alamos, an isolated site in New Mexico. This concentrated effort was successful and two nuclear weapons were built and exploded. Controversy still surrounds the question of whether those weapons should have been dropped on populated areas. They were, and a new era in world history unfolded. Following World War II the weapons program was continued, with many tests of larger weapons. The Soviet Union developed its own nuclear bombs and also began a major series of tests.

Development of nuclear power for controlled use began with the early work on fission done by scientists such as Enrico Fermi and Eugene Wigner, who built the first reactor, called a pile, under the football stadium (for security reasons) at the University of Chicago in 1942. It consisted of 56 layers of uranium-charged graphite bricks that produced the first man-made controlled chain reaction. After World War II ended, atomic energy was seen as having great potential for good and harm. The Atomic Energy Act of 1954 established the Atomic Energy Commission (AEC), a federal agency responsible for controlling the use of nuclear power. The AEC was to develop nuclear power technology and also own the nuclear material. The Division of Naval Reactors in the AEC, led by Hyman Rickover, was responsible for the early design work. The U.S. Navy recognized that a major problem with submarines was that running on diesel engines was possible only on the surface. Batteries required for below-surface operation had a limited life. Nuclear power offered the possibility of long-time underwater operation, so the Navy initiated a major effort to design small, simple reactors to power submarines. The rapidity of success owed a great deal to Rickover's leadership. However, this focus may have restricted the types of reactors that were developed. Early interest in reactors that might have greater commercial application for land-based use was not pursued. This may have led to missing major opportunities. The AEC did encourage the development of commercial reactors and several vendors accepted this as an opportunity.

Nuclear Reactors

A principal reactor design developed in the U.S. uses regular water, called "light water." In a nuclear reactor, neutrons collide with the nuclei of fissionable atoms. These atoms split, breaking into two nuclei of much lower atomic mass, producing gamma rays and other radiation, and more than one neutron. The particles have considerable kinetic energy. When the atom splits, the resulting nuclei have lower total mass than the original nucleus and the mass difference is converted to energy. There are three important factors in this fissioning.

The first factor is that the particles have considerable energy of motion of the nuclei, which is converted into heat as they collide with other particles. The high temperature is then transferred to a cooling liquid, usually water, which is heated far past the boiling point at normal pressure. In one type of reactor (boiling water), the water boils and becomes steam. In another (pressurized water), high pressure keeps the water from boiling. This pressurized hot water is sent to a heat exchanger in which a second body of water becomes steam. In both reactors, the steam is sent to a turbine in which the high temperature steam turns the turbine and generates electricity.

The second factor is that radiation is produced in fissioning. The two new nuclei themselves are unstable, giving off more radiation as they decay into eventually stable materials. This radiation release makes the fissioned material highly radioactive and dangerous.

The third factor is that two or three neutrons come out for one neutron going in. This leads to a chain reaction: one nucleus fissions, leading to more nuclei fissioning, leading to many more fissioning. This chain reaction proceeds extremely rapidly: in one hundredth of a microsecond. This extremely rapid release of tremendous energy (that makes a nuclear bomb so devastating) enables nuclear power plants to run.

The uranium fuel, in pellets, is encased in metal rods. Fission energy heats the fuel and the fuel rod to a high temperature. In U.S. plants water as a coolant flows across the rods and carries the heat away. Deuterium is an isotope of hydrogen with double the mass; water with deuterium is called "heavy water." U.S. reactors are called light water reactors, or LWRs. Canadian reactors use heavy water as the coolant; other reactors use a liquid metal. The few metal designs operating use sodium. Each reactor type has advantages and disadvantages.

Seventy-eight percent of the operating reactors in the world are LWRs, with the basic U.S. design having been replicated in both Japan and France. In the U.S., nuclear power can be dated by the 2 December 1942 operation of Enrico Fermi's pile at the University of Chicago, but a more significant date is 20 December 1951, when a small research unit of the Argonne National Laboratory in the desert of Idaho generated enough electricity to light four light bulbs. The first commercial reactor in the U.S. (Dresden 1 in Illinois) went into operation in June, 1960. The United Kingdom led in applying nuclear energy for electric generation: Calder Hall Unit A went into operation in 1956, followed by six more units by the end of 1959. All were small (50 Megawatts electric or MWe). Electric generating stations are usually described by the maximum electric power they can generate. This contrasts with the total power generated, Megawatts thermal, which usually is about three times the MWe. The first Japanese power reactor was Tokai 1, going into operation in 1966; the first French reactor was Chinon A1 in 1964; and the first Soviet reactor was Novovoronezh-3 in 1972. At the beginning of 1961 there were two operating reactors in the U.S.; by the end of 1969, there were twenty. Nuclear power continued to grow in this country until the early 1970s, coming to a complete halt by the mid-1970s.

Problems of Nuclear Power

The problems nuclear power has had in the U.S. are not unique, and are being seen or have been seen in other countries. These problems are based upon four major factors: diminished need, high cost, public anxiety, and poor operations.

Demand dropped following the oil embargoes and rapidly rising cost of oil in 1973 and 1978. Prior to that time, the AEC and the utility industry were confident that electricity use would continue to grow at the rate following World War II, about 7% per year. With this growth rate, the AEC confidently forecast that, by the year 2000, there would be at least 1,000 large nuclear power plants operating in the U.S. — there will be about 110. Utilities, convinced both by themselves and by the government that nuclear power would be the preferred option for generating electricity, ordered nuclear power plants at a very high rate. When the oil shock hit, energy conservation became fashionable, energy use dropped, and forecast growth plummeted. From 1970–75, utilities filed requests to build 12 plants per year, and from 1975–78, 9 plants per year. There have been no orders since 1978. Other plants that had received approval to be built were being canceled. In 1975–1983, nine plants per year were canceled; the rate of cancellation exceeded the rate of beginning construction. The principal reason no new nuclear power plant has been ordered in the U.S. since 1978 is that the demand was not there: Utilities had overbuilt.

A second major problem, which will keep many utility planners and executives from choosing nuclear power even if demand forecasts reach the point where new generation capacity is necessary, is the cost of nuclear power plants. In 1954, the head of the AEC, Lewis Strauss, said that "It is not too much to expect that our children will enjoy in their home electrical energy too cheap to meter . . ." However, by the 1980s nuclear power became unacceptably expensive for most utilities. Nuclear power plants going into operation in the later part of that decade were costing four and five billion dollars. When new plants started generating power and their costs were included in electricity charges, the rates sometimes went up by more than 25%.

The third major problem faced by the reactor advocates was public anxiety about nuclear waste, nuclear weapons, and nuclear accidents. Nuclear waste was identified as a problem before 1950, but most advocates believed that it was a minor issue. The technical problems in handling nuclear waste did not look very great. In some sense, the technical problems still do not look large. What has perhaps become insurmountable is the problem of where to put this waste.

Nuclear waste from power plants is labeled either high-level or low-level. Low-level waste consists of material contaminated with radioactive particles that remain dangerous for less than a few decades and whose total hazard is small. The material includes tools, some parts, and worker coveralls. More than 50% of low-level waste generated in the U.S. comes from nuclear power plants. Other sources include industrial testing devices and medical applications. The used fuel from nuclear power plants, called spent fuel, constitutes high-level waste. It is hot in both temperature and radiation hazard. After being removed from a reactor, the waste is stored at the power plant. The hypothesis on which these plants were built was that, after about 10 years, this fuel would be sent to a reprocessing facility that would chemically separate the unused uranium and the plutonium that had been produced in the fission process. The dangerous radioactive fission products would be solidified and disposed of underground. However, reprocessing is not being done in the United States, although it is being done in France, the United Kingdom, and Russia, and will be, in Japan. Furthermore, there is no repository for the high-level waste, reprocessed or not.

In 1977, when President Carter came into office and identified energy as the most important issue for his administration (directly related to the energy shocks), a review was undertaken of the disposal of high-level waste. That review concluded reprocessing was not advisable and that the used fuel should be disposed of directly underground. U.S. policy since then has been to keep the fuel at the reactors until a permanent location can be found.

The Energy Department, the successor to the AEC, has the responsibility of siting and constructing a final location for the high-level waste, currently planned to be a repository. This program has been unsuccessful. Congress finally selected Yucca Mountain in Nevada as the repository site. Criticism of both the Department of Energy and the Congress has focused on the process by which the site was selected, arguing that it was a political decision to locate the site at an existing federal operation, the Nevada Nuclear Test Site, and on the unfairness of locating it in a state that does not have a single nuclear power plant.

Attempts to locate both high- and low-level waste sites have led to strong public opposition. The opponents frequently argue that the government has not done adequate surveys of underground conditions, such as the potential for earthquakes or the hazard to underground water, as well as arguing that a location should be chosen that benefits directly from nuclear power.

Opposition is epitomized by a California law passed in 1976 that identifies the condition for a nuclear power plant being built in the state: a successful solution be demonstrated for permanent disposal of nuclear waste.

Proliferation refers to the spread of nuclear weapons. The devastation caused by the weapons dropped in Japan in 1945 and the enormous power demonstrably released in weapons tests, particularly those in the Pacific, built fear of nuclear weapons. The International Atomic Energy Agency (IAEA), established in 1953, has preventing the spread of nuclear weapons and related technology as a fundamental goal. The efforts of those countries having nuclear weapons to prevent others from getting them have been reasonably successful. Although in the 1950s political analysts forecast dozens of nuclear powers by the 1980s, in 1992 there were only five: the Soviet Union, the United States, France, the United Kingdom, and China. India set off a nuclear detonation in 1974, describing it as a peaceful nuclear explosive. Israel is suspected of having nuclear weapons, Iraq and Pakistan were charged with trying to develop nuclear weapons, and several other countries at one time or another apparently attempted to develop a nuclear weapons program. The largest proliferation challenge came recently when the Soviet Union broke up, leaving several of the independent states with nuclear weapons. Many people associate nuclear power with nuclear weapons. Consequently, part of the public anxiety about nuclear power is related to fear of nuclear weapons.

Accidents are another major cause of the public's concern. Some small accidents caused damage to U.S. reactors in the 1960s, but the nuclear industry in general believed that accidents could not happen. They based this faith on engineering design and on confidence in the operators, who were thought to be well trained.

That all changed on March 28, 1979, when a reactor in Harrisburg, Pennsylvania at the Three Mile Island (TMI) plant suffered a series of events which led to destruction of its core, essentially a meltdown. The triggering event was caused by a mistake. By a series of mistakes, the event became the worst U.S. nuclear accident. No significant radiation was released in this accident, but a reactor was destroyed, nearly bankrupting the utility. Because of lack of understanding of what was happening, the Nuclear Regulatory Commission and the State of Pennsylvania issued bulletins which, with the leverage of media, panicked the public. More than 200,000 people evacuated the region. "TMI" became an acronym for "nuclear power is dangerous." As a result of the TMI accident, the nuclear industry significantly improved its operations and the Nuclear Regulatory Commission aggressively implemented regulatory changes. Changes since the Three Mile Island accident include installing systems to measure how much water is in the reactor; upgrading the emergency systems at all nuclear power plants so that they could successfully shut down in the presence of the steam, water, and radiation of an accident; adding a technically competent person in the control room and significantly upgrading the qualifications of operators; requiring utilities to have simulators on which operators could train; and improving the maintenance of the plants. Although these improvements are estimated to have decreased the risk of an accident by perhaps a factor of 10, the TMI accident did demonstrate that accidents can happen, which may lead to radiation release.

The U.S. nuclear industry struggled to overcome the public perception that nuclear power was unsafe. With time passing since the accident at TMI, the public may have begun to forget that nuclear reactors could suffer accidents. On April 26, 1986, a new name came into the public's consciousness: Chernobyl. A reactor near the small town of Pribyat, near Kiev in the Ukraine, blew up. It was not a nuclear explosion in the sense of a nuclear weapon, but the explosive release of energy threw out 50 million curies of radiation in a plume that spread throughout Europe, massively destroyed a huge reactor, and led to the deaths of at least 31 people. It is uncertain how many more have died or will die from this accident. The nuclear industry correctly notes that the Russian reactor bore no resemblance to those in the U.S. It was of a completely different design, had no containment, and was primarily a weapons material production reactor design being used to generate electricity. The industry also points out that the Russian operators were poorly trained, violated many safety regulations, and that such a reactor could never have been licensed to operate in the U.S. or in many European countries. While all of this is true, it did not reassure the European public and left a sense of uneasiness in the U.S. public.

The final problem with nuclear power has been that many plants in the U.S. have not run well. One measure of efficiency of a generating plant is how much time it generates electricity compared with how much time it could generate electricity if it were running all the time. This is called the capacity factor, or load factor. Lifetime nuclear power load factors range from 60.5% in the U.S. to 78.2% for Canada, with Sweden and West Germany above 70% and Japan and France at 68%. These percentage differences represent an enor-

mous amount of money. With 100 nuclear power plants running, the difference between 60% and 68% is equivalent to 8 large nuclear power plants. Since these power plants cost many billions of dollars, that is the equivalent of at least $20 billion of investment. Many studies have focused on how to improve the design of nuclear power plants, but few on how to improve the operation. The conclusion of those few of the second type is usually to improve the management. U.S. plant load factors have been improving; the 1991 average was 70.2%.

Nuclear power has been embroiled in environmental controversies in the U.S. since the passing of NEPA (National Environmental Protection Act of 1969) in 1970. This required all government actions, such as licensing a nuclear power plant, to be preceded by an Environmental Impact Statement (EIS). EISs have provided a powerful lever for opponents of nuclear power plants. In the U.S. a utility desiring to build a nuclear power plant must get a construction permit from the federal authorizing agency. From 1954 until 1975, this was the Atomic Energy Commission (AEC); since 1975 the Nuclear Regulatory Commission (NRC). Following completion of the construction, the utility must obtain an operating license from the same federal agency. (The Energy Act of 1992 eliminated the second step, providing for a combined construction permit and operating license.) NEPA allowed opponents to raise environmental issues, such as thermal pollution of rivers and lakes and salt spray on crops near cooling towers.

The federal agency operates under the Administrative Procedures Act (APA). The AEC/NRC uses adjudicatory hearings, in which lawyers present arguments before a panel of judges for and against granting the permit or license. The formalities of this process have led to frustration of both license applicants and intervenors (the name applied to those who oppose the utility and intervene to block granting a permit or license). The regulatory system is often blamed by the industry for the problems that nuclear power has had in the U.S.

At the local level strong opposition has come from environmental activists. Following the TMI accident the NRC promulgated a regulation that has been used to slow and, in the case of Shoreham on Long Island, to halt a nuclear plant. This regulation requires emergency plans to be developed by the utility and the local government to handle accidents at the plant, including, if necessary, evacuation of the nearby areas. This requirement for local government cooperation has been used to oppose plant operation.

In the 1990s nuclear power is a vital element of U.S. and international electricity supply, but its growth has slowed and, in the U.S., halted. At the end of 1991, 420

Table 1
Major nuclear power countries (end of 1991)

Country	No. of operating reactors	Total net MWE (megawatts electric)	No. of reactors under construction
U.S.	111	99,757	3
France	56	56,873	5
(USSR)	45	34,673	25*
Japan	42	32,044	10

*Status uncertain since some of the independent states have halted construction of reactors in their territory.

nuclear power plants were in operation in 25 countries (counting the former Soviet Union as one) around the world, and 76 more were under construction. In 12 countries, more than one-quarter of the electricity was generated by nuclear power. Five countries generated more than 47% of their electricity from nuclear power, with France the highest, at 72.7%. The U.S. generated 21.7%. Table 1 shows the major nuclear power countries.

Further development will depend primarily on a growing need for electricity, constraints on fossil fuel burning, safe operation, and favorable economics.

JOHN F. AHEARNE

For Further Reading: Bernard L. Cohen, *The Nuclear Energy Option* (1990); Committee on Future Nuclear Power Development, *Nuclear Power: Technical and Institutional Option for the Future* (1992); Stanley M. Nealez, *Nuclear Power Development in the 1990's* (1990).
See also Chernobyl; Decommissioning; Electric Utility Industry; Environmental Impact Statement; Fusion; Mass-Energy Equivalent ($E = mc^2$); National Environmental Policy Act of 1969; NIMBY; Plutonium; Pollution, Thermal; Radiation, Ionizing: Biological Effects of; Strontium-90; Three Mile Island; Uranium; War and Military Activities: Environmental Effects; Waste, Radioactive.

OCCUPATIONAL SAFETY AND HEALTH

Occupational health has rarely received much attention from the American public. Historically the primary commitment has been to economic advance through technology, resulting in little attention to the related toll on workers' health. Through much of U.S. history, workers themselves have been engaged in the more pressing task of making a living for their families. Consequently they too have paid little attention to widespread occupational safety and health problems. Furthermore, the labor movement in the U.S. has not been strong enough to force public attention to these issues on a continual basis.

In Europe the tradition of occupational medicine is much stronger. In the 16th century the occupational health problems of miners and foundry and smelter workers were studied by Paracelsus. The classic text on the occupational diseases of workers was authored by the Italian physician Bernardino Ramazzini (1633–1714). While Europe underwent two centuries of economic development leading to the creation of the modern welfare state, occupational health traditions evolved into approaches now considered exemplary, particularly those in the Nordic countries. By contrast, in the 19th century, the host of safety problems that accompanied the industrial revolution in the U.S. led to a more erratic public response, addressed primarily at the state rather than the federal level. Massachusetts created the first factory inspection department in 1867 and in subsequent years enacted the first job safety laws in the textile industry. The Knights of Labor agitated for safety laws in the 1870s and 1880s and by 1900 some minimal legislation had been passed in the most heavily industrialized states.

After 1900 the rising tide of industrial accidents resulted in passage of state workers' compensation laws, so that by 1920 virtually all states had adopted this no-fault insurance program. A leading figure of the time was Dr. Alice Hamilton, whose path-breaking work on lead poisoning received national attention and won her the first faculty appointment for a woman at Harvard University.

Throughout the 1920s the rise of company paternalism was accompanied by the development of occupational medicine programs. Much attention was paid to pre-employment physical exams, however, rather than to industrial hygiene and accident prevention. Occasional scandals reached the public eye, like cancer in young radium watch dial painters in New Jersey, but not until the resurgence of the labor movement in the 1930s was important national legislation enacted: the Walsh-Healey Public Contracts Acts of 1936 required federal contractors to comply with health and safety standards and the Social Security Act of 1935 provided funds for state industrial hygiene programs. The Bureau of Mines was authorized to inspect mines.

Although there were further national programs and funding to protect worker health and safety during the Second World War, the end of the war saw a rapid decline in attention to workplace health and safety. An exception to the general neglect of the field was passage of the Atomic Energy Act in 1954, which included provision for radiation safety standards. Not until the 1960s when organized labor regained some political power under the Democratic federal administration did the issue reemerge as significant. Occupational injury rates rose 29% during the 1960s, but it was finally a dramatic mine disaster in 1968 in Farmington, West Virginia, when 78 miners were killed, that captured public sympathy. In 1969 the Coal Mine Health and Safety Act was passed which established the federal Mine Safety and Health Administration (MSHA). This was rapidly followed in 1970 by passage of the broader Occupational Safety and Health Act (OSHAct).

With this new federal emphasis on workplace health and safety, the last twenty years has led to work environments that are generally much safer places than at the close of World War II. However, the rate of improvement is still uneven and there is evidence that the oldest of occupational hazards (lead, silica and noise) are still inadequately controlled.

The Extent of the Problem in the U.S.

The absence of sufficient attention to workplace hazards and their controls in the U.S. is reflected in the fact that work-related injuries and disease are widespread problems whose full dimensions have yet to be charted. The best estimates of the number of occupational injuries per year are based on the U.S. Bureau of Labor Statistics' (BLS) Annual Surveys, although these estimates are limited to private industrial employment. The most recent data are based on the 1991 survey, which estimated a total of 6 million injuries with almost half (2.8 million) resulting in work limitation or lost work-days averaging 22 days per disabling injury. The BLS reports that their estimate of 2,800 occupational injury fatalities is significantly low. Other estimates range from 3,500 to 11,000.

Occupational diseases are even more poorly documented and go largely unreported, especially those diseases with a long latency. The BLS report for 1991 identified 368,000 new occupational illnesses (predominantly shorter duration illnesses such as dermatitis and repetitive trauma disorders). Epidemiologic estimates of work-related cancer in the U.S. have ranged between 4% and 10%. Perhaps the best known occupational cancer hazard in the last two decades results from the variety of occupational exposures to asbestos which causes asbestosis in addition to mesothelioma, lung and other cancers.

Indirect approaches to estimating the magnitude of the occupational disease problem add dimension to the picture. During the 1980s approximately one quarter of the chemical hazard samples collected by federal and state Occupational Safety and Health Administrations (OSHAs) exceeded exposure limits. There are also a number of materials and processes in use that have not yet been fully evaluated for public health impact. Examples include evidence that electrical and magnetic fields increase cancer risk and the fact that the newest semiconductor, gallium arsenide, has been reported a suspect human carcinogen. In addition, there is inadequate understanding of disease risk associated with exposures to such common agents as machining fluids, hydrocarbon solvents, fiberglass, and plastic polymers or polymer systems.

There is also a large, less well identified problem of illness and injury due to physical hazards. Seven of the ten most common hazards identified by the National Institute for Occupational Safety and Health (NIOSH) in the National Occupational Exposure Survey are physical, rather than chemical hazards. These are problems which have historically received inadequate attention—probably because they are both prevalent in a wide variety of settings and not life-threatening.

Among these problems which have gained increasing importance are cumulative trauma disorders (CTD) of the musculoskeletal system. These disorders are common and intermittent, and symptomatic workers may work with pain for extended periods after their onset. Some of the physical factors responsible for these disorders of the upper extremities and the back are generic: awkward postures, repetitive movements, high force requirements, and static postures. Better methods are needed to identify additional ergonomic risk factors susceptible to control.

An example of the importance of these disorders is provided by the Bureau of Labor Statistics' annual survey of injury and illness. Between 1982 and 1987 the BLS documented that the number of CTD disorders in their survey approximately tripled from 22,600 to 72,930. The problem is not limited to manufacturing industries, but includes many non-manufacturing industries such as newspapers, insurance companies, service operations and other white-collar occupations. The risk, whether new or merely recognition of an old one, requires extensive revisions in work processes and work organization.

On the horizon is an occupational health problem that does not fit neatly into concern with either chemical or physical hazards. Increasingly investigators are becoming aware of the impact of work organization on worker health. Since the days of the scientific management movement the literature of labor relations has discussed the importance of work organization on productivity and efficiency. There is now awareness of the impact of work organization on physical and mental health as well. Studies have shown that work pace is associated with increased musculoskeletal disease, that work that is paid on a piece-rate basis is associated with more health complaints, and that there are added health costs of machine-paced work noted in industries with jobs as diverse as poultry inspection and postal letter sorting.

Finally, attention is only recently being focused on work organization and occupational cardiovascular disease. For example, there is convincing epidemiologic evidence that cardiovascular disease is associated with jobs where there is high psychological demand matched with low worker control of the demand. The effect has been noted in studies in the U.S., Europe, and Japan. Other characteristics of work organization that may be equally important are job instability and shift work.

Regulatory Efforts

In 1970 the Federal OSHAct created a new regulatory agency in the U.S. Department of Labor, OSHA, with the authority to require employers to provide safe and healthy workplaces and to promulgate and enforce safety standards. In addition, the OSHAct established NIOSH (as part of the U.S. Public Health Service) to do research and health hazard evaluation of the work environment. The OSHAct also created a federal commission to make recommendations for the reform of workers' compensation laws—the state no-fault occupational industry insurance systems. Unfortunately action on this latter problem remains stalled.

When OSHA was first established, the Agency adopted a host of so-called "consensus" standards. In addition to extending the Walsh-Healey regulations for government contractors to the rest of industry, OSHA adopted large numbers of voluntary guidelines that had been developed by the American National Standards Institute and the American Conference of Governmental Industrial Hygienists. Essentially this enabled OSHA to enter the field "running" with standards to enforce. Unfortunately, many of the guidelines were inappropriate as legal standards. Some were contradictory, some were overly detailed, some were anachronisms. (For example, OSHA adopted a requirement that toilet seats be split in the front, an idea that came from when it was thought that syphilis was caught in bathrooms.) When Dr. Eula Bingham became head of OSHA in the Carter Administration in 1977, one of her earliest and most important tasks was standards simplification, that is, throwing out inappropriate, ineffectual, or silly standards.

The process of developing new standards, however, has been slow and cumbersome, uniformly involving substantial litigation before any new worker protection is extended. Until 1989 only 24 standards controlling toxic substance exposures had been adopted. Perhaps the most tortuous path was that of the field sanitation standard for farmworkers, which took 14 years to develop and ultimately was promulgated only because the courts required OSHA to do so. On the other hand, in 1989 OSHA adopted hundreds of new permissible exposure limits for air contaminants in one heroic effort to update its standards, only to have this procedure set aside by the courts in 1992.

Enforcement of standards, moreover, has left much to be desired because OSHA has been seriously understaffed. Perhaps the most active OSHA office under Thorne Auchter, Reagan's first OSHA Director, was the State Plans office, which delegated authority to enforce OSHA to state governments. In 1992 a tragedy in North Carolina in a chicken processing plant, in which workers were killed, highlighted the inadequacies of that state's program and the general failure of OSHA to oversee the implementation of state plans to protect worker health and safety.

While the OSHAct covered most workers in the private sector, the Mine Safety and Health Act established a special regulatory body to deal with the high-risk mining industry. Perhaps the most important accomplishment of this law was the virtual elimination of new cases of black lung—coal workers' pneumoconiosis. MSHA through the 1970s was well-staffed and effective. Regulation of pesticide exposures of agricultural workers was placed in the hands of the EPA, although other aspects of farm employment, such as migrant labor camp conditions and field sanitation, were in the hands of OSHA. ￼

In summary, the tension between economic policy and regulatory policy that was present throughout the

1980s illustrates how public health regulatory activity can falter if not adequately protected.

The most important extensions of worker protection in recent years have been linked to growing public concern with general environmental issues. For instance, 1987 amendments to the federal acts concerning hazardous waste required OSHA and EPA to adopt safety and training requirements for a broad range of hazardous waste workers and emergency responders dealing with hazardous materials. Labor and community environmental coalitions forced passage of "right to know" laws in many states and localities and ultimately resulted in OSHA promulgating a "hazard communication" standard.

Federal government policy since 1980, however, has been characterized by a neo-conservative anti-regulatory stance. Public health advocates complain of the slow pace of OSHA standards promulgation, the federal ceding of enforcement authority to states, the failure to protect worker-complainants from employer discrimination, and the decimation of NIOSH's budget. Further, the decline of the American trade union movement has removed the political impetus for OSHA enforcement activity. Recent efforts at legislative reform stress streamlining OSHA standards development procedures and enhancing workers' "right to act."

Innovative Approaches

Currently there are a number of activities directed at reforming OSHA, including such actions as mandating the formation of joint labor-management health and safety committees. Such committees already exist in countries other than the U.S. (e.g., Sweden and certain Canadian provinces) but would mark a major departure in regulatory approach in the U.S. OSHA has experimented with such committees in the construction industry and has been pleased with this less adversarial approach to improving the work environment. Researchers, however, are skeptical of the efficacy of such committees unless there is high-level management support for improving health and safety.

Another important development in the U.S., beyond government agency activity, has been the creation of joint labor-management health and safety endeavors in research and in training. In the automobile industry collectively bargained funds have supported major research initiatives supervised by an independent scientific advisory panel along with jointly developed and administered worker health and safety training programs. In the construction industry an extensive network of hazardous waste worker health and safety training programs is maintained by the joint efforts of the Laborers' International Union of North America and the Associated General Contractors with assistance from the federal government.

A different type of organizational innovation has evolved to direct attention to worker health and safety. This is the development of grassroots coalitions or committees for occupational safety and health at the local or state levels. These groups are coalitions of worker health and safety activists and public health professionals that engage in occupational health advocacy and training.

Finally, the political coalition of environmental and occupational health advocates has resulted in state pollution prevention legislation focusing on the reduction in use of toxic substances in industry. Following pioneer legislation in Massachusetts, a number of states have passed similar laws that would protect both workers and the general public by mandating that companies inventory their production and use of toxic materials and develop plans for their reduction.

Perhaps one of the most pressing international problems in occupational health concerns the increasing integration of the world economy. In North America the development of a continental free trade agreement has to reconcile the more advanced work environment standards of Canada and the U.S. with a host of new hazards to Mexico. The export of hazardous technologies, products, and waste represents increasing challenges for public health worldwide. On the one hand, our understanding of the nature of health hazards to workers has been improving; on the other hand, the restructuring of the world economy may tax the efforts to control these hazards.

DAVID H. WEGMAN
CHARLES LEVENSTEIN

For Further Reading: Barry S. Levy and David H. Wegman, eds., *Occupational Health: Recognizing and Preventing Work-Related Disease*, 3rd ed. (1994); Robert Karasek and Tores Theorell, *Healthy Work: Stress, Productivity, and the Reconstruction of Working Life* (1990); David Rosner and Gerald Markowitz, eds., *Dying for Work: Workers' Safety and Health in Twentieth Century America* (1987).
See also Hazards for Farm Workers; Labor Movement; United States Government.

OCEAN DUMPING

Ocean dumping is the process by which pollutants, including sewage, industrial waste, consumer waste, and agricultural and urban runoff, are discharged into the world's oceans. These pollutants arise from a myriad of sources. Concern regarding ocean dumping in-

creased in the U.S. when miles of sewage, medical waste, and garbage washed up on shorelines across Long Island, Rhode Island, and Massachusetts during the summer of 1988. Although the impact of pollution is widespread in coastal waters, contamination has also been found in the open ocean and in deep water fish.

Nutrients such as nitrogen and phosphorus in treated sewage, effluent, and solids—commonly known as sludge—can directly or indirectly encourage the growth of algae when emitted into coastal waters. The algae prevent sunlight from reaching living organisms below the surface, causing the supply of oxygen in the surrounding water to be depleted. Consequently, these areas are unable to support life. In addition, sewage wastes can contain toxics passed on from industrial dischargers to the sewers. These toxics can have a detrimental effect on fish, mollusks, crustaceans, marine mammals, birds, and humans.

Chemical pollutants such as polychlorinated biphenyls, heavy metals, petroleum, and industrial hydrocarbons have been introduced to the ocean from a variety of sources: electrical power plants, industries, shipping accidents, commercial and recreational ships, agricultural lands, waste incineration, municipal storm sewers, and sewage treatment plants. The toxic and pathogenic contamination of fish and shellfish has led to closure of commercial and recreational fisheries and shellfish beds to protect human health.

Ocean dumping accounts for 10% to 20% of global ocean pollution. International agreements to protect the marine environment and manage its resources date back to 1958 with the Convention on the High Seas. The London Convention of 1972, to which the U.S. is a signatory, is a global treaty governing disposal of wastes and other matter in the ocean. It has resolved to phase out the dumping of all industrial wastes by 1995; most parties have already ended the practice. In international forums, such as the London Convention of 1972, the principles of precaution, prevention, and clean production are now taking precedence over managed ocean dumping. As a result of continued degradation of coastal waters and the fisheries they support, Congress passed the Ocean Dumping Ban Act of 1988, which prohibits the dumping of industrial waste, sewage sludge, and other contaminated materials in the U.S.

OCEANOGRAPHY

Oceanography—the exploration and scientific study of the ocean and its phenomena—cannot be discussed as a science as is the case with physics, chemistry, or biology. It is the study of the major fraction of the Earth's surface and near surface along with its interactions with the atmosphere that yield weather and climate. It includes the study of the orbital oscillations of the solar system for the elucidation of the tides and phenomena such as the ice ages and paleontological oceanography.

We know the solid surface of the Earth in great detail—animal, vegetable and mineral—because that is where the overwhelming fraction of humankind lives and has lived. Knowledge has accumulated over the millennia by intimate contact and exploration for vital materials, including living space.

By contrast, at any given time, only a minuscule fraction of humankind is on the ocean and even only a small fraction of those live there. Even today, the accumulated knowledge of the ocean bottom is far less than the surface of the moon, even its back side. The ocean edges, which are available, presented problems to the early scientists. A twenty-three-century myth still persists that Aristotle failed to understand the source of tides and reputedly committed suicide because of his failure. Even Galileo perpetrated absurdities in his explanation of ocean tides.

The first systematic, though fragmented, knowledge of the oceans derived from fishing and trading activities. Knowledge of the major oceanic current systems and winds developed from these pursuits. In fact, the boundaries of major currents, such as the Gulf Stream, were first established by noting the changes in the biota. Benjamin Franklin made the first physical (temperature) measurements to delineate the Gulf Stream. Much, and correctly, is made of HMS *Challenger*'s voyages over a century ago toward developing an outline of a systematic approach to the study of the oceans, but in the 18th and early 19th centuries, Russian and Chinese admirals were so involved; Matthew Fontaine Maury's monumental work on winds and currents marked an epoch in the study of the world's oceans.

Oceanography is thus the study of the world's oceans and their interactions with their environments using all the available scientific and engineering disciplines. It has become an organized campaign to understand all aspects of the oceans. It is an operation that is still developing with unexpected major developments occurring almost yearly, such as the discovery of the benthic thermal vents. It is best outlined by its current major intellectual divisions: physical oceanography, marine biology, biological oceanography, marine geology, chemical oceanography, and coastal oceanography.

Marine biology is perhaps the oldest subdiscipline and has become well fractionated; marine microbiology, for example, has become a well-recognized specialization. Speaking very generally, marine biology is the study of the individual plants and animals that exist in the oceans or on the ocean floor and shores. Marine biology is also a basic subdiscipline of biology and it can be viewed as either the tracer element or the goal of marine ecology.

As on land, it has two divisions—plant and animal—but, unlike terrestrial biology, the ocean biological kingdom is divided into two distinct groups, the nekton and the plankton, in recognition of the fluid environment. The plankton is that grouping of plants and animals whose movements are largely determined by the motion of the fluid, and the nekton is composed of those animals that propel themselves largely independently of their environment. The phytoplankton and the zooplankton are established in the upper several hundred meters of the oceans where photosynthesis occurs. It has been estimated that the phytoplankton is generated and consumed in a two-week cycle.

Biological oceanography is the professional title of what can be described in more modern terms as marine ecology. Ocean fisheries are the primary practical connection to biological oceanography and the fundamental phenomenon is the food chain. The chain starts with the phytoplankton and goes through five or six trophic levels before meeting the climax predator. It is estimated that the energy stored in each level is reduced by a factor of six in progressing upward. No success has been achieved in recouping even a part of this loss in commercial fisheries by going to lower levels.

The major exception to this picture is the existence of benthic vent colonies around thermal vents along the zones of tectonic plate separation. Fountains of chemically charged water spew forth that are hot enough to melt lead. They establish a local environment that is highly active biologically and where the original energy source is not the sun's radiation. The knowledge of the existence of these vents dates only from the middle 1960s—a fact that underlines our still relatively poor knowledge of the oceans and the rapidity with which we are expanding it.

The practical social side of biological oceanography is the maintenance of the productivity of the world's fisheries. A significant fraction of human protein consumption derives from the oceans. In the 20th century major fisheries such as those off Western Europe have slowly eroded because of overfishing. There have been two spectacular collapses of fisheries, the sardine off of California and the much greater Peruvian anchovetta fishery, both probably due to a combination of overfishing and ocean climate change. Nevertheless the oceans function as a valuable food resource that largely operate under individual national controls called EEZ (Exclusive Economic Zones).

On a scientific basis it is interesting to note that the mathematical science of ecology began with the work of Lotka and Volterra immediately after World War I using the Mediterranean fisheries as a model. The non-linear equations they used to model the interaction of species led to the understanding of catastrophe theory and chaos.

Marine geology is a relatively new discipline originating just after World War II. This was partly due to technical advances in seismic and magnetic antisubmarine operations but the search for minerals, particularly oil, also played a substantial part. Marine geology is the study of the ocean's solid basins including the islands and shore formations. It is clearly the most difficult and arduous of all the ocean's disciplines and the world's ocean bottoms are still largely unmapped. Nevertheless, in just the few years from 1945 to 1975 enough was found to revolutionize our entire understanding of the Earth's crust. The resulting complex of plate tectonics is one of the great scientific breakthroughs of the 20th century.

The current commercially valuable resources of the oceans are largely in the shallow regions of the continental shelves where the large petroleum deposits are found. Tin, sand, and gravel are other mineral resources of consequence. For a long time there were those who proposed certain benthic minerals such as manganese nodules as ores of great promise, and much development was undertaken to exploit them. The possibility of great financial returns was a major stimulus to the Law of the Sea negotiations. To date no economic benefit has derived from these efforts, and this and other factors have effectively stalled action on the international front.

The initial developments on plate tectonics were based on the bathymetric observations of the world-girdling mid-oceanic ridge, the variation of the thickness of the bottom sediments (from near zero at the ridges to kilometers at the continental edges), and the observation of magnetic stripe anomalies across ocean basins, among other indicators. Some of the principal contributors to this early development were Harry Hess of Princeton, Victor Vacquier and Russell Raitt of the Scripps Institution of Oceanography, Tuzo Wilson of Toronto, and Ronald Mason of Oxford. It was all brought together by the research conducted by the drilling ship *Glomar Challenger* operated by the Scripps Institution of Oceanography for the National Science Foundation. The mid-ocean drilling cores obtained by this vessel verified the hypothesis of sea floor spreading in exquisite detail.

Among the many discoveries that were made in parallel to their development were the thermal vents. This phenomenon is consistent with the picture of volcanic activity being concentrated near the edges of the plates in the tectonic model. One phenomenon that would appear at first sight to be in contradiction is the set of so-called "hot spots." However, the most famous set—the Hawaiian Islands—actually add another triumph to the theory. The Hawaiian Islands, including their extended chain of sunken seamounts, form a continuous line whose ages perfectly match the motion of the Pacific Plate on which they are located. Their diminishing sizes due to the erosional forces of time are clearly in evidence and a new island is being built up below the surface just southeast of the large island of Hawaii.

One very important phenomenon connecting marine geology and physical oceanography is the tsunami (often, incorrectly, called a tidal wave). Tsunamis are giant waves that can tower 100 feet or more when impinging on the shore line and can cause great damage. They are most often the result of powerful earthquakes thousands of miles away that generate surface ocean waves of great energy and velocity. Warning alerts are regularly made but damage is unpredictable since the resulting effects depend heavily on the direction of arrival, the state of the natural tide, and the local shape of the shoreline.

Among the many bizarre but consistent discoveries of the modern *Challenger* is the closing of the Mediterranean Sea between five and twelve million years ago. This happened at least twice at the Straits of Gibraltar. The evidence is clear in two layers of salt found in the sediment cores of the Gibraltar basin.

Chemical oceanography is a much less well defined field. One major objective is to trace the history and development of the world's oceans which were not always a feature of the planet. The aim is also to explain the chemical composition of the oceans and the special features that are found there such as the origin of the manganese nodules. That the research is incomplete is evident from the fact that the thermal vents play a major role in this process yet they only appeared in the discussions recently.

Chemical oceanography plays a vital role in mapping the movements of the deeper ocean masses. The chemists follow "tracers"—often radioactive isotopes—to determine the birth and death of masses of ocean waters and their ages. This is a vital element in the study of the Earth's climate, which, in the end, is determined by the oceans, although in the longer term of tens of thousands of years solar radiation and orbital variations must be taken into account as well. The measurement of carbon in all its forms, biological and inorganic, sets a very major role in the geophysical, geochemical, and biological cycles but also in the climate problem because of the phenomenon of greenhouse warming. This is a matter of great concern because of anthropogenic emissions from power generating sources.

Physical oceanography is a more fundamental discipline because it concerns the movements of the waters at all depths, including surface wave motion and the less obvious but important internal wave oscillations. These motions are generated through a complex interaction with the atmosphere that produces currents by direct friction and, in conjunction with solar heating, produces changes in density through evaporation which also generate currents. As in the case of the atmosphere, the Earth's rotation plays a major role in the configuration of the currents. Two of the outstanding examples are the Gulf Stream off the coast of the North American continent and the Kuroshio, off Japan, that parallels the Asian continental edge.

Perhaps the largest in volume of flow is the Cromwell current, which flows west to east across the Pacific for 6,000 miles from Asia to South America and is just subsurface at the equator. One of its characteristics is a line of biological productivity that yields a valuable fishery in an otherwise infertile ocean. In general, the productive areas of the oceans are at the continental edges where the effect of the Earth's rotation generates an upwelling that continually replenishes the nutrients in the upper ocean layers. The observation that the oceans are blue away from the continental edges and green near the coastlines stems from this physical phenomenon. Most of the ocean away from the continental shelves can be described as a "desert."

Another large-scale phenomenon is the aperiodic El Niño. This happening, which is observed every three or four years and is very variable in intensity, is a global climate change that normally lasts for four months but has lasted for eighteen months in an extreme case. It is a joint atmospheric-oceanic event but it is initially observed as a major change in the dominant current system off the coast of Peru. It has had a profound effect on the anchovetta fishery, which was once the world's greatest but is now much reduced. As an indirect effect, the flocks of guano birds that also depend on this fishery resource are greatly reduced. A large El Niño has a global effect on climate; much effort is expended in developing beneficial predictive models, such as winter rainfall estimates for California.

The Arctic and Antarctic ice formations and associated waters play a vital role in climate and the formation of the "deep water." These benthic waters average about 1,000 years in age, that is, since they have been at the surface of the oceans. This represents the average "turnover" of the oceans.

Since about 1960 the nations of the world have tried, so far unsuccessfully, to establish a "Law of the Sea regime" that would encompass international regulation of the ocean resources, including unhindered passage through "classical" straits. Much of the interest has concentrated on the recovering and processing of the manganese nodules that carpet large areas of the ocean bottoms. These nodules contain, in addition to the manganese, copper and nickel in appreciable amounts. The negotiations have stalled for a variety of reasons including questions of "transfer of technology" and related intellectual rights.

Coastal oceanography is the study of the coastal region, the most valuable area of the oceans in commercial terms. It is rarely appreciated that the overwhelming fraction of the ocean's economic value lies in this zone. Localized there is tremendous competition between fishing, mineral exploitation, shipping, and the inevitable concentration of human beings and their activities, particularly recreation. This leads to the almost insurmountable problem of pollution of the sea near shore due to this concentration of occupations and interests.

Coastal engineering is a vital part of the adjacent country's existence. Huge expenditures are made to preserve or rebuild beaches that are eroded because of interior flood control and irrigation dams. These structures block the normal flow of sand that usually replenishes sand lost to the oceans due to wave action and subsurface surges. The problem is compounded by ill-founded engineering structures such as groins that block the normal longshore flow of sand.

In regions where the surface temperature of the waters can exceed 28.4°C (80°F) for some seasons of the year, hurricanes are possible that can cause enormous property damage and loss of lives. Where there is a concentration of people and activities, the results can be and have been frightening. The Caribbean area (including Florida), the Western Pacific and Southeast Asia are all struck with regularity. In some areas such as Bangladesh the situation is one of increasing gravity because the population explosion is driving millions of people to living in the low lying coastal regions that are especially vulnerable.

Among other research in oceanography and meteorology the most active research area is climate prediction in the early term—on the order of months to a year. This is not yet in sight. In fact, despite the great contribution of Earth satellite monitoring in tracking hurricanes, the exact point where the hurricane will make landfall is not yet predictable, requiring the evacuation of large areas to save lives.

No survey discussion of the ocean can be complete without some mention of its role in modern warfare. Submarine warfare was critical in both World War I and II. It was the technical efforts in antisubmarine warfare that supplied civilian oceanography with the research tools that mark the rapid advances in the field after World War II and, conversely, it was just that antisubmarine warfare defense need that encouraged governments to invest heavily in oceanography. The introduction of nuclear weapons to the oceans via the long-range missiles of the Polaris class submarines formed the basis for reliable deterrence against nuclear warfare but only if the invulnerability of the submarine could be assured. This goal also establishes a strong symbiosis between military needs and civilian oceanographic research.

WILLIAM A. NIERENBERG

For Further Reading: Elizabeth M. Borgesi, *Ocean Frontiers: Explorations by Oceanographers on Five Continents* (1992); Margaret Deacon, *Scientists and the Sea* (1971); George Pickard, *An Introduction to Oceanography* (1982).

See also Antarctica; Atmosphere; Carbon Cycle; Climate Change; Coastlines and Artificial Structures; Coral Reefs; Ecology, Marine; El Niño; Fishing Industry, U.S. Commercial; Food Chains; Geology, Marine; Greenhouse Effect; Law, International Environmental; Meteorology; Natural Disasters; Ocean Dumping; Oil Spill; Petroleum; Plankton; Pollution, Water: Processes; Sea and Lake Zones; Seawater; War and Military Activities: Environmental Effects; Water Cycle; Wetlands; Whaling Industry.

OIL

See Petroleum.

OIL SPILL

An oil spill occurs when crude or refined oil is discharged or leaked on land, inland water, or the open seas. Most of the world's oil resources are found in locations distant from markets. This makes the transport of huge quantities a necessity and greatly increases the chances for accidents involving spills. On land, spills occur when pipelines or tanks leak or break or when wells blow out. The largest and most publicized spills, however, are caused by tanker accidents during maritime transport. Not all spills are accidental; up to forty percent of the oil found in waterways is from open sea bilge pumping and the rinsing of tanks by maritime transporters in order to take on new cargo. These processes are illegal but commonplace.

The extent of damage from marine oil spills depends on many factors: the type and amount of oil, the location, the weather, the water temperature, and the nature of the affected ecosystem. Crude oil is heavy and difficult to remove but is less toxic than some other types of oil. Spills in open seas are easier to clean than those near coastal areas involving inlets, coves, and beaches. Heavy storms and windy conditions make

treatment difficult but help disperse the oil. Warmer water temperature increases the rate of evaporation. Spills near coral reefs, marsh grass, and other delicate habitats are very damaging to life in these ecosystems and are almost impossible to treat effectively.

The disposal or treatment of marine oil spills relies on a combination of natural and controlled methods. Under average conditions only ten to fifteen percent of the oil from a spill is recovered; however, as much as half of the oil evaporates before it can be skimmed. Burning the surface oil has been tried but is not a particularly effective removal method. Among the natural processes, evaporation is the most effective, especially immediately following the spill. A small fraction of the oil dissolves directly into the water. The remaining oil disperses, incorporating into a dilute oil-in-water suspension. Chemicals can be added to the water to enhance this process, increasing the surface area of the spill to further dilute the oil. This is a controversial method because the addition of another foreign substance to the spill may cause more damage than the oil itself. But these chemicals also help prevent the emulsification of oil into a mousselike substance that is difficult to remove from the surface and shore. Bioremediation is becoming an accepted treatment. It involves adding natural microbes (such as bacteria, yeast, and fungi) to spills to metabolize the oil through their normal life processes. This helps eliminate and not simply disperse the oil.

In addition to these methods, workers are sent to spill sites with oil booms, surface skimmers, and vacuum tanks in attempts to collect the oil before it reaches the shore. Straw and sawdust are added to spills as absorbants. Oil on beaches is scrubbed off or sprayed with high-pressure, hot-water hoses. These treatments are also controversial, for they may contribute further damage to habitats.

The spilled oil presents a great hazard to wildlife and local economies dependent on fishing and/or tourism. Animals that rely on insulating properties of fur or feathers to survive, such as sea otters and waterfowl, become coated with oil and are no longer able to keep warm. They eventually die of hypothermia or intestinal problems due to ingestion of oil. Attempts have been made to clean these animals, but success rates have not been high. The economies of entire regions suffer as fishing industries struggle to survive, tourism declines because of ugly polluted seashores, and property values fall. Fortunately, the combination of natural processes and treatments seems to cause the environment to recover eventually. The timing and the extent of recovery to be expected are also controversial topics.

The petroleum industry has attempted to mitigate the impact of spills by rerouting tankers away from sensitive areas and by switching to double-hulled tankers. Legislation such as the Oil Spill Prevention Act of 1990 has been instituted to help prevent further problems, and strike forces have been set up to act immediately when spills do occur. Further preventive measures will undoubtedly be required. As long as the world economy is dependent on oil, oil spills will continue to be a problem.

OLMSTED, FREDERICK LAW

(1822–1903), American landscape designer. Frederick Law Olmsted, more than anyone else, established the profession of landscape design in America and the practical discipline of environmental planning. Though he came to this in a roundabout way, everything he did in his first 35 years led him, in 1857, into the job of building a park in the middle of Manhattan: a project that was a first for him and, in concept and execution, a first for the country.

His father's interest in nature had made young Frederick a countryside rambler, getting to know, he wrote, "interesting rivers, brooks, meadows, rocks, woods . . . "—the raw material of his later work. A long voyage to China on a clipper ship gave him the discipline and stamina that would carry him through his arduous struggles to implement his art and ideas. A few years at what was called "scientific farming" gave him the rudiments of an environmental approach— how to adapt nature for human needs while still respecting and preserving it. A trip to Europe opened his eyes to the use of public parks as social instruments. Friendship with the socially-conscious intellectuals of New York and a stint as editor of a liberal magazine opened his mind to what Albert Fein has called Olmsted's "constant search for means by which to translate humane ideas into environmental forms." His reporting on social changes, notably his classic work *The Cotton Kingdom*, deepened his understanding of how democratic processes were being changed and warped by the country's shift to an urban society.

All this gave Olmsted the tools to design or work on such environmental landmarks as Central Park in Manhattan and Prospect Park in Brooklyn, the park and parkway systems of Boston and Buffalo, the woodlands of Lynn, Massachusetts, the Arnold Arboretum, the community of Riverside, Illinois, the campus of Stanford University, and many others. In 1882, the U.S. Commissioner of Education wrote Olmsted that in his travels from Maine to Oregon he never saw "any mark of woodland or landscape taste and training that I do not refer to your work." Today's attitudes toward the environment still refer to Olmsted's seminal thoughts and sensible solutions.

In designing Central Park with his gifted partner, Calvert Vaux, Olmsted took on what Oliver Wendell Holmes called the "hips and elbows and other bones of nature" to create a breathing space in a city that was on the way to suffocating itself. A quagmire of swamps, squatters' colonies, and abattoirs was transformed into a place of pleasant ponds and promenades while an area of rugged ledges and outcroppings was left largely in its natural state. The romantic, pragmatic design set criteria for American public spaces. It set the park apart from the city while at the same time making it a part of it. Acutely aware of the physical and psychological stresses in a hardening urban environment, Olmsted and Vaux aimed to turn "democratic ideas into trees and dirt" by attending to the needs of a mixed population while fighting off privileged demands and inappropriate intrusions.

During the Civil War Olmsted left landscaping to organize the Sanitary Commission, which reformed the Army's lethally inefficient medical corps. He went on to manage a mining enterprise in California, which led to his helping preserve Yosemite as a wilderness and to draft what would become a charter for the national park system.

Going into private practice in 1877, he became an original regional planner, devising for Boston and Buffalo a series of parks along a network of parkways. For a private developer, he created Riverside, Illinois, one of the country's early planned communities. Taking on the new problems of urban sprawl and suburban isolation, he laid out a self-sustaining community linked to the city by good highways and woven together by its own roads.

In a time when *ecology* was a brand new word in the language and its principles were unformed, Olmsted was applying its precepts. He maneuvered against the fashionable practice of introducing exotic plants by artfully using native species. In designing a campus for Stanford University he managed with some success to have it fit California's microclimates rather than have unsuitable traditions imposed on the site. As a setting for Biltmore, a grandiose estate near Asheville, North Carolina, Olmsted provided not a baronial park but a managed forest. His choice of young Gifford Pinchot to run it was to have an environmental impact all its own.

Always flooded with work, Olmsted laid out the grounds for the capitals at Washington and Albany, for the Chicago World's Fair (Columbian Exposition) of 1893, for scores of estates and institutions. Though sometimes dismissed as "a romantic enthusiast . . . wholly impractical" he was a shrewd self-advocate and manipulator of public opinion. To save Niagara Falls from destructive exploitation he used public committees, press agentry, political lobbies, influential friends, and a petition signed by 900 eminent men of many nations. Of all the legacies Olmsted left the environmental movement, this technique may be the one it finds most handy.

Though dedicated to the democratic process, Olmsted himself could be imperious in his methods. He took an elitist stance in carrying out what the 19th century called the "civilizing" effect of parks on the working classes, emphasizing the pure pleasures of nature and only grudgingly providing recreation areas. His success lies as much in the man as in his art and philosophies. He knew that detail was as important as concept, that compromise could be as important as confrontation. He had, one admirer wrote, "absolute purity and disinterestedness . . . a monomania for system." He himself said his work "occupied my whole heart." Twice under the stresses of his work and personality he broke down; he spent the last years of his life in an asylum.

After his death in 1903, he was all but forgotten for decades—made, as has been said, into a footnote. But he knew his own worth. "The result of what I have done," he wrote prophetically, "is of much more consequence than anyone but myself supposes."

Today cities trying desperately to retrieve some open space for themselves find a text in his insights into urban ecology: that people have not only a physical but a psychological need to be in touch with nature. His formulas for regional planning have been used well and also misused: his parkways, like all those that followed, have become traffic-fomenting speedways. His innovative ideas for forest management became, in the first years of this century, a focus for the country's new environmental consciousness.

Radical in his time, he recognized that people and the environment inexorably shape each other. But his approach does not align with radical sectors of present-day environmental movements where preservation is all. Olmsted was concerned not just with the preservation of nature but also with the prudent use of it. His aim, he said, was to "make improvements by design which nature might make by chance."

JOSEPH KASTNER

For Further Reading: Albert Fein and George Brazziler, *Frederick Law Olmsted and the American Environmental Tradition* (1972); Elizabeth Stevenson, *Park Maker: A Life of Frederick Law Olmsted* (1977); Laura Wood-Roper, *F.L.O.: A Biography of Frederick Law Olmsted* (1983).

ONLY ONE EARTH

The book *Only One Earth: The Care and Maintenance of a Small Planet* by Barbara Ward and René Dubos is the unofficial report of the 1972 United Nations Conference on the Human Environment. The book was the first widely read, authoritative discussion of the notion that just as human beings must tend their own patch of real estate, humanity collectively must take care of the planet as a whole.

The authors warned of global warming as carbon emitted by cars and factories accumulated in the atmosphere, intensifying the natural greenhouse effect that keeps most of the planet in the human comfort range. They looked to car-clogged Los Angeles as the urban future unless automobile travel ceased to be an end in itself. They even gave notice that theories predicting an ozone hole bore watching.

Ward and Dubos also warned that environmental issues were becoming too scientifically complex for ordinary people—and less than extraordinary leaders—to judge. They observed that indifference to the squalor and hopelessness afflicting growing numbers of people would breed humiliation, strife, and—finally—riot since how the "other half" lives is no longer a secret in the age of TV. They recommended, in the first instance, the political courage to act on good science and to pause periodically to invest in the future even if it means slower economic growth for a while; and, in the second, the power of family, community, and church or belief to help at least some children born into poverty to escape its life sentence.

The thesis of *Only One Earth* is that "the two worlds of man—the biosphere of his inheritance, the technosphere of his creation—are out of balance" and that the human race is "in the middle . . . on the hinge of history." Science, at once dedicated to discovering the laws of nature and to toying with them, has given us the power to manipulate and destroy what we do not fully understand and cannot replace. We have removed the natural constraints on energy and matter and now—with no hope of returning the genie of technology to its containment vessel—we must devise our own checks and balances.

Ward and Dubos made a poignant call for applying greater insight and wisdom to economic and political affairs as well as to science. Two of their particular concerns were how relying solely on markets to keep resources available invites environmental degradation and how the inward pull of nationalism fueled by narrow self-interest works against the communications revolution, the globalization of markets, and the other centrifugal forces creating "one world."

To these and other issues the authors brought to bear not only a wealth of experience and fact, but also faith in the human enterprise. They were more optimistic about science and the search for the truth than about politics and people's ability to stop making war. (Or perhaps, more accurately, Dubos the scientist saw more "progress" in his lifetime than Ward the expert on politics and economics saw in hers.) In any case, they seem convinced that knowledge of human nature is the beginning of political wisdom and the best defense against the Faustian excesses of science.

In an era of heightened divisiveness between the rich in the industrialized "North" and the poor in the agrarian or industrializing "South" few since Ward and Dubos have ventured to make it an article of faith that our cultural differences may define us but need not keep us from acting collectively for the common good. But if an unabashedly humanistic approach to economic development and environmental conservation now seems to belong to a more innocent age—certainly to one before deep ecologists questioned the centrality of human life in the biosphere—the debate over how to live is the poorer for abandoning it. For Ward and Dubos have been proven right that people whose children are hungry will not allow themselves to worry about the environment, that in any but the most affluent nations the need for work will always win out over the need to conserve nature's capital, temporarily, and that people who are rich will not give up—even by inches—hard-bought leisure, mobility, or comfort unless they are smitten with a vision of something more deeply satisfying than material wealth.

Possessed of an enviable certainty about the sacredness of life on Earth and the importance of the human experiment, Ward and Dubos were not afraid to plumb the heart along with history and the scientific record looking for the keys to survival. Their work endures because it exemplifies what they came to consider "the astonishing thing about our deepened understanding of reality over the last four or five decades"—"the degree to which it confirms and reinforces so many of the older moral insights of man."

KATHLEEN COURRIER

For Further Reading: M. W. Holdgate, M. Kassas, G. F. White, *The World Environment 1972–1982* (1982); Shridath Ramphal, *Our Country, the Planet* (1992); Barbara Ward and René Dubos, *Only One Earth* (1972).

OPEC

OPEC (Organization of Petroleum Exporting Countries) is a loose entity of 12 oil-exporting countries pledged to concerted action to promote their oil interests and, particularly, oil price and oil-export revenues. The organization was started in 1960 but it was not until the early 1970s that its activities had a major impact on the oil world. Since the first oil shock in 1973, OPEC may be judged to have had considerable success

in achieving and maintaining oil prices than might otherwise have obtained (Figure 1).

OPEC's capacity to bolster oil price derives from two basic facts: it possesses 77% of the world's proved oil reserves (Table 1, column 1); and it is willing, however unevenly, to restrain oil production through production allocation quotas among its members. The oil market has heeded these cartel actions because of OPEC's overwhelming dominance of supply. Core OPEC members can, and have, significantly reduced and increased the supply of oil with immediate impact on oil prices. No other oil producer or group of producers has such power and capacity.

An oil barrel contains 42 U.S. gallons. The first oil production in 1859 in Pennsylvania was moved in barrels by teamsters. These barrels were soon standardized at 42 gallons. The U.S. consumes about 17 million barrels of oil each day. Each American's share is thus just under 3 gallons per day. Some 10 million of the 17 million barrels the U.S. uses each day is for transportation: automobiles, trucks, diesels, aircraft, etc. Two 11-gallon auto fill-ups require refining one barrel of crude, which yields gasoline at a rate of 53%.

OPEC: A Most Disparate Group

OPEC is not an organization of equals. Saudi Arabia, for example, has oil reserves of 260 billion barrels; Gabon has less than 1 billion (Table 1). The five core OPEC members of the Gulf—the first five of Table 1—have 84% of OPEC reserves and 64% of world reserves. These same core members have production time horizons—at their 1989 production rates—of 90+ years. (Should production levels later be raised, it is as likely that added reserves will be proved.) Accordingly, these nations tend to take a long-term perspective on matters pertaining to oil production. On the other hand, the five OPEC countries with the lowest reserves must take a short-term view. Their oil reserves and their oil revenues are under threat as fast-diminishing assets. Population differences among OPEC members are large as well, with oil revenues per capita

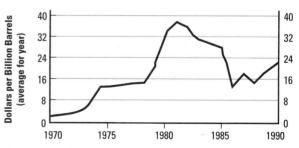

Figure 1. Landed cost of OPEC crude oil to U.S. Source: Energy Information Agency, *Monthly Energy Review*, April 1993.

Table 1
OPEC countries oil reserves, production rates, and revenues

	Proved reserves (billion barrels) end 1990	Production (thousand barrels) per day, 1989	Oil revenues (million $ US) 1989	Population (millions) 1990	Time horizon (years) 1990
Saudi Arabia*	260	5,465	24,093	14.1	130
Iraq	100	2,825	14,500	18.9	97
United Arab Emirates	98	2,035	11,500	1.6	132
Kuwait*	97	1,825	9,306	2.1	146
Iran	93	2,845	12,500	56.6	90
Venezuela	60	2,010	10,020	19.7	82
Libya	23	1,160	7,500	4.5	54
Nigeria	17	1,635	7,500	113.0	28
Indonesia	11.1	1,415	6,059	180.5	21
Algeria	9.2	1,205	7,000	25.4	21
Qatar	4.5	410	1,955	0.4	30
Gaber	0.7	215	1,200	1.2	9
Total OPEC	773	23,045	113,140	438.0	91
Total World	1,009	63,895		5,292.0	43
United States	34	9,160		249.2	10

*Includes neutral zone.
Note: 1989 Production and revenue numbers used since 1990 and later impacted by Iraq and Kuwait suspension of exports.
Sources: Columns 1 and 2: *BP Statistical Revue*
        Column 3: Arthur Andersen and Cambridge Energy Research Associates
        Column 4: UNESCO Data

varying enormously. Each Indonesian's share in 1989, for example, was about $33; each Nigerian's, $68; an Iraqi's, about $7,500; a Saudi's, about $17,000. Seven OPEC countries—Saudi Arabia, Iraq, Kuwait, United Arab Emirates, Libya, Algeria, and Qatar—are Arab nations. These countries, plus Iran and Indonesia, are overwhelmingly Islamic in religion.

Neither OPEC membership, nor Arab brotherhood, nor a shared Islamic faith has prevented war between some members. Iraq and Iran fought the long and bloody contest of 1980 to 1988. Iraq overran Kuwait in 1990 and threatened Saudi Arabia until Iraq was repulsed by the combined United Nations force led by the U.S. Yet even during these war periods, the oil ministers from the warring countries often attended OPEC meetings.

OPEC: Its Start and Why

Oil was first produced commercially in 1859, after the Pennsylvania discovery of Edwin L. Drake. However, 80 years later, at the start of World War II, consumption in the entire world had barely reached 6 million barrels per day (60% of which was in the U.S.). The boom in consumption, to 65 million barrels per day in the early 1990s, took place after World War II when oil replaced coal as the world's industrial fuel and when the use of the auto became a world phenomenon and not just an American one (Figure 2).

A fascinating aspect about oil has been the remarkable steadiness of its price. Figure 3 shows this history in deflated dollars since the 1870s. The price of oil has been about $10/barrel with much smaller variations than for other commodities. The reason for this steadiness in price is not without dispute but one explanation is that with few exceptions (1973 and 1979, for example) oil has always been in excess supply. The problem with a product in perennial excess is avoiding ruinously low prices. Over the history of the oil industry, various organizations have been identified with trying to maintain or bolster oil price:

Rockefeller's Standard Oil Trust (1875–1910): From the 1870s to 1910, the Trust dominated the U.S. refining industry, imposing stability on crude oil price.

Texas Railroad Commission (1933–1960): The early 1930s were times of reduced oil demand because of the Depression and times of prolific oil finds. As a response to the oil glut, Texas and other states started conservation commissions that limited their states' oil production.

Seven Sisters (1930–1975): After World War II, the seven largest international oil companies held most of the world's oil concessions (in countries that later formed OPEC) and could thus control and allocate production and price.

OPEC (since 1960): OPEC countries, mainly in the early 1970s, nationalized concessions given the inter-

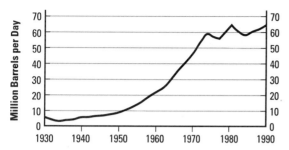

Figure 2. World oil consumption. Source: *BP Statistical Review of World Energy*, June 1991, 1981, and 1976; U.S. Bureau of Mines; and *Shell Petroleum Handbook*, 4th edition.

national oil companies and took over the role collectively of setting oil production and, hence, price. As can be seen in Figure 3, no form of oil production control has worked for long.

The impetus for the founding of OPEC in 1960 was a unilateral cut by Seven Sister companies in the posted price of crude oil from $1.92 to $1.82 per barrel. (The posted price was the price on which the concession 12.5% royalty was assessed and, after deducting royalty and production expense, the basis for the 50% "tax take" of the producing governments.) The market price for comparable crude oil was 20¢ to 30¢ per barrel lower at the time, in part because of the availability of distress Soviet oil. Nevertheless, the loss in revenue was ill-received by producer countries, who already felt exploited by the oil companies. Five countries, Venezuela, Saudi Arabia, Kuwait, Iraq, and Iran, reacted by founding OPEC in September 1960. The principal architects were Juan Pablo Perez Alfonso, the oil minister of Venezuela, and Abdullah Tariki, the Saudi oil minister. OPEC resolutions and strong protests on the part of these governments, however, went largely unheeded. During the 1960s, OPEC membership was expanded but its principal accomplishment was to forestall further erosion in the posted price. OPEC's lack of bargaining power was due primarily to surplus oil production in the market. Despite its failure to increase revenue per barrel, OPEC did increase its production from just under 8 million barrels per day in 1960 to 23 million in 1970, tripling its collective income.

OPEC: A Defining Act

In the early 1970s, there was a continuing increase in the world's consumption of oil. Between 1960 and 1970, for example, consumption rose from 22 to 47 million barrels per day (Figure 2). And, as just noted, OPEC production accounted for 15 million barrels per day or 60% of this increase.

The sense of exploitation in many of the OPEC

countries was intensified as the production amounts were raised. First Libya and Algeria in the early 1970s, and then Iraq, Iran, and Venezuela, threatened to nationalize their concessions. After Algeria did nationalize 51% of the assets of the French companies operating there and as the call on OPEC production rose from 23 million barrels per day in 1970, to 25 in 1971, to 27 in 1972 and to 31 in 1973, the price began to move upward. OPEC's production capability was widely believed to be about that 31 million barrels per day. With the world producing at what was believed to be full capacity, the situation was ripe for change.

The occasion for change was the October War, the fourth Israeli-Arab war. On October 6, 1973, Israel was attacked by Egypt and Syria attempting to regain territory lost to Israel in the Six-Day War of 1967. Initially, the war went badly for the Israelis until large amounts of armaments began to be received from the U.S. Meanwhile, the market price for oil had risen above $5 per barrel bid up by an anxious world. On October 16, the core members of OPEC—that is, the Gulf members—unilaterally raised the posted price to $5.12 per barrel (from $3.01) in the wake of the market price increases. Then, on October 19, U.S. President Nixon proposed a $2.2 billion aid package to Israel. The Arab response was to embargo the production of their oil, cutting output by 5% in October and announcing further 5% cuts for the next months. The U.S. was specifically embargoed but because oil is easily transported and traded, the U.S. did not suffer from shortage more than other consumers. Of course, the price of oil was bid up further. Figures 1 and 2 have earlier provided broad oil price and oil consumption histories. Figure 4 focuses on world oil consumption (production) and OPEC's share of it from 1970 to 1990. This chart can be the roadmap to understanding OPEC workings and how its dynamics have evolved during the last twenty years.

OPEC Workings

The $11–$13 Period: In the first months of 1974, Arab and Israeli forces disengaged and the embargo was ended. Few were more surprised than OPEC countries that spot oil markets in that winter bid the price of available oil to an "unbelievable" $20 per barrel. OPEC, however, did acknowledge this by raising the price of the OPEC marker crude, Saudi Light, to $11.65 per barrel. During the next years, OPEC countries were mostly busy with the nationalization of their concession-properties and the establishment of their national oil companies. OPEC ministerial meetings were held every six months or so. The marker price was raised to $12.09 in January, 1977, to $12.70 in July of that year and to $13.43 in January of 1979, ostensibly to keep

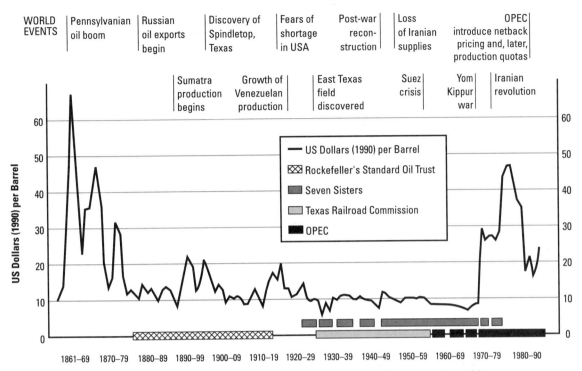

| WORLD EVENTS | Pennsylvanian oil boom | Russian oil exports begin | Discovery of Spindletop, Texas | Fears of shortage in USA | Post-war reconstruction | Loss of Iranian supplies | OPEC introduce netback pricing and, later, production quotas |
| | | Sumatra production begins | Growth of Venezuelan production | East Texas field discovered | Suez crisis | Yom Kippur war | Iranian revolution |

Figure 3. U.S. wellhead price of crude oil. Source: *BP Statistical Review of World Energy*.

pace with inflation. Certain OPEC countries, such as Algeria, Libya, and Iran, urged even higher prices but the "price hawks" were kept in check by producers that had long time horizons, particularly Saudi Arabia. Accordingly, OPEC's most obvious and official act in these days was to set the price of the marker crude.

Production levels and country quotas were not an issue since the demand for OPEC oil was 31 million barrels per day and any country that wished to produce full-out could do so. Saudi Arabia was producing between 8 and 9.5 million barrels per day and it is not inaccurate to say for this time that OPEC = Saudi Arabia = Price.

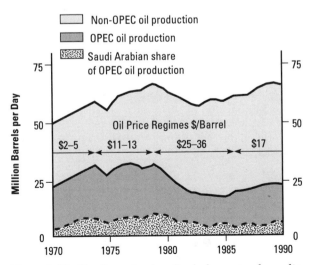

Figure 4. World oil production (includes natural gas liquids) per million barrels a day. Source: *BP Statistical Review of World Energy*, June 1991, 1986, and 1981.

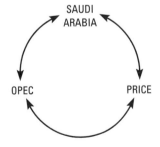

The $25–$36 Period: The 1973 pattern of world oil price escalation was repeated in 1979 when the Shah of Iran was deposed by revolution to be succeeded, eventually, by a religious government headed by the Ayatollah Khomeini. The threatened loss of 4.5 million barrels per day of Iranian oil exports again propelled prices upward, this time to the $40 per barrel level. Again, OPEC acted in the wake and as a kind of price

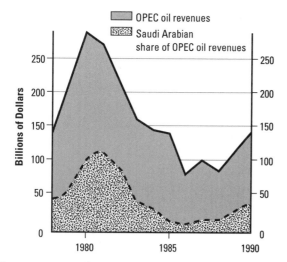

Figure 5. OPEC oil revenues (1978–1990). Source: Arthur Andersen and Cambridge Energy Research Associates, *World Oil Trends*, 1992.

ratchet: raising price to $14.54 on April 1; to $18 on July 1; and to $24 on December 1. In successive actions in 1980 and 1981, as Iran's exports stayed off the market, OPEC raised the official price to $26, $28, $30, $32, and $34. These actions can be seen in the increasing cost of imported oil to the U.S. (Figure 1).

But the higher oil prices of the 1970s, and now the very high oil prices of the early 1980s, had their reaction: a decline in world oil consumption and production (Figure 4) of 8 million barrels per day between 1979 and 1985, and an increase in non-OPEC oil production, stimulated by the higher prices, of some 6 million barrels per day in the same period. Both developments resulted in a decrease of OPEC's production—down a stunning 14 million barrels per day by 1985 from the 31 million barrel level of 1979. Now, how to divide 17 million barrels per day among the then 13 OPEC members? As it turned out, 17 was not easily divisible by 13. Production norms and quotas were started in the early 1980s, but exceeding quotas was typical of this period as the control approach was shifted from price to a quota discipline.

To sustain OPEC price in the early 1980s, Saudi Arabia acted in a "swing" producing role, reducing its production and absorbing a disproportionate share of the revenue loss. The Saudi production went down from over 10 million barrels per day in 1980 to about 3.5 million barrels per day in 1985, while its revenue declined from $105 billion to $24 billion per year (Figure 5). In late 1985, however, Saudi Arabia renounced the swing role and said it would produce at what it claimed was its fair share in OPEC, 25% or, at that time, 4.5 million barrels per day. The result was the

predictable response to excess oil in the market—a plummet in the oil price. (See Figure 1.)

The New $17 Period: The rest of OPEC reluctantly decided to adhere to its quotas, thus maintaining the price at $17 per barrel. OPEC volumes and revenues crept upward from the lows of 1986 as a consequence (Figures 4 and 5).

The invasion of Kuwait by Iraq in August 1990 removed 4.5 million barrels from the market. Again, there was an upward price spurt, which boosted the 1990 average price (Figure 1) and OPEC revenue (Figure 5). However, other OPEC producers were able to quickly replace the lost Iraqi and Kuwaiti exports. For example, Saudi Arabia reached and maintained a production level of 8 million barrels per day from which, even at the $17 level to which prices returned in January 1991, it achieved revenues of $50 billion per year.

OPEC's Future

The resumption of Kuwaiti and Iraqi oil production and exports will require adjustment of volumes by the other 10 OPEC members. These 10 will certainly wish to maintain their recent collective revenue of $150 billion (including Saudi Arabia at $50 billion) as their volumes are reduced. The only way for that to happen is for oil prices to increase.

Energy consumption and oil consumption will increase in the next years because of world population growth and economic development. This assures a growing market for OPEC oil even if concern for global climate change and carbon dioxide emissions increases. (In the spring of 1992, OPEC convened its first seminar in twelve years to consider the environment and carbon dioxide emissions.)

Certain OPEC producers, principally Kuwait and Venezuela, and now Saudi Arabia, have sought more assured markets for their crude oil by the purchase of refining, distribution, and marketing assets in major industrial countries. Several additional arrangements seem possible but no strong trend is likely. OPEC's core producers, taking a long-term perspective, will no doubt continue to set a price on oil that will produce a growth in oil consumption equal to the growth in world energy consumption, that is, one that keeps oil's share at about 40% of world energy consumption.

NORMAN E. DUNCAN

For Further Reading: Anthony Sampson, *The Seven Sisters: The Great Oil Companies and the World They Shaped* (1975); Ian Skeet, *OPEC: Twenty-five Years of Prices and Politics* (1988); Daniel Yergin, *The Prize: The Epic Quest for Oil, Money, and Power* (1991).
See also Automobile Industry; Carbon Cycle; Fuel, Fossil; Greenhouse Effect; Petroleum Industry.

OPPORTUNITY COST

In a market economy, cost is usually measured in dollars, but that tends to obscure the real cost of an activity—the cost of foregone opportunities. For example, the cost to a consumer of purchasing a television set for $500 is that he or she cannot purchase the next best alternative, such as a camera. The cost to a firm of spending $500 on a new machine is that it cannot use the money to purchase a new computer program. Different consumers or firms experience different foregone opportunities—one consumer may forego a camera, another may forego a short vacation; one firm may forego a machine, another a different piece of equipment. But since the foregone opportunities in our example would both cost $500, it is convenient to speak of the cost of the chosen purchases as $500, rather than specify for each consumer what he or she could have purchased instead of the television.

The main purpose of calculating opportunity cost is to aid in the selection from among various activities. Other concepts of costs (such as accounting costs) have different objectives, such as evaluating performance, rewarding employees, or determining tax liabilities. Therefore, opportunity costs may differ greatly from costs as calculated by accountants. Thus, some expenses often counted as costs are not necessarily opportunity costs. For example, a project that improves access to a nearby park may reduce the profits earned by hotel owners near a more distant park, but the reduction in profits is not considered a cost. Indeed, it can be a measure of the benefits to consumers near the new facility who do not have to travel as far. Most importantly, historical or acquisition costs need not reflect opportunity costs. For example, the opportunity cost of using land as a park is the benefit that could be obtained from its best alternative use. The price at which the land was purchased, which may be significantly lower or higher, is irrelevant because that historical price does not reflect the possible alternative uses of the land today. In contrast, the price the land will fetch in the market today often does represent its opportunity cost. The main qualification has to do with unpriced externalities (as when residents benefit from an unspoiled view, or suffer from smoke emitted from a factory on that land).

Opportunity cost can also apply to non-market situations. Consider regulations designed to lower health effects from toxic chemicals in the environment. Should the Environmental Protection Agency (EPA) direct its efforts to revising the standards for benzene, a known human carcinogen that is already stringently regulated, or to reducing indoor tobacco smoke or radon in buildings? The opportunity cost for revising the benzene standard is the inability to deal with indoor smoke or radon. The EPA administrator is often faced with the conflict of diverting attention from a situation in which lives could be saved to one where many more lives are at risk. In such cases, environmental groups must understand that resources should be directed where the opportunity costs are the greatest, even if their particular situation is neglected.

AMIHAI GLAZER

ORGANIC GARDENING

Organic gardening is the cultivation of food crops for human consumption without the use of processed chemicals. The general goal of organic cultivation is to maintain a balance between the alteration of an ecosystem to achieve a reliable food source, and the confines and laws of the natural ecosystem itself. The technical approaches to working within a natural ecosystem include inter- and intra-species diversity; intensive or close planting of crops; use of organic fertilizers; recycling of garden and household wastes as compost; using raised beds; and encouragement of beneficial microbial organisms, insects, and weeds within the system. In this way, organic gardening, which is generally practiced on a relatively small scale, is an intensively managed system.

Humans have practiced sedentary agriculture for at least 10,000 years, and have done so without the benefit of chemical fertilizers and biocides. Thus by definition, traditional human societies have always practiced some form of organic cultivation.

Today's concept of organic gardening derives from observations first made by Rudolf Steiner, an Austrian scientist, who noted the negative impacts associated with chemical fertilizers in the late 19th century. He began experimenting with close plant spacings and animal manures, and he found a beneficial symbiotic relationship from that association. As these early experiments took place outside of Paris, organic cultivation, as it is known today, is often referred to as the "French" method. In the mid-20th century, further studies carried out in the United States by Alan Chadwick, Robert Rodale, and others used raised-bed cultivation, which provides greater aeration of the soils and mobility and interaction of nutrients, microbial organisms, water, and plant roots. The combination of raised-bed and "French" techniques is referred to as the French biointensive method, which became popular during the counterculture movement of the 1960s in the United States and Europe, and is widely practiced and scientifically respected by professional gardeners throughout the world today. Research and experiments in the biointensive method have shown crop yields per unit of land as high as four to six times the average yield of conventional cultivation. Furthermore, by utilizing local resources and recycling nutrients back into the soils, the biointensive system should maintain its higher yields due to its relative biological sustainability.

Public interest in chemical-free food and concern for environmental degradation are steadily increasing, as is the demand for organic fruits and vegetables. At present, because organic produce is relatively scarce in the U.S. market, higher prices make organic production a profitable venture.

With the passage of the 1990 Farm Bill (Title XXI, *Organic Certification*, P.L. 101–624), the U.S. Congress established federal and state criteria for organic certification programs; standards for general organic production (soil fertility, chemical inputs, crop and livestock production); domestic and imported food marketing and distribution; as well as enforcement powers. The act requires each state to establish organic certification programs to regulate production and handling. Several states, including California and Minnesota, have adopted organic certification programs establishing legal parameters for marketing organic foods, and technical norms for organic production.

ORGANIZATION OF PETROLEUM EXPORTING COUNTRIES

See OPEC.

OSHA

See Occupational Safety and Health.

OUR COMMON FUTURE

Our Common Future, published in 1987, is the final report of a 22-member commission set up by the United Nations to sound out citizens, grassroots organizations, and experts around the world on two questions that have confounded leaders since the strong links between economic development and environmental protection were first recognized decades ago: What is sustainable development, and how do we achieve it? The book has sold more than 500,000 copies in some 20 languages. Its proposals have sparked widespread debate and its impacts have been so numerous and diverse that tracking the responses of governments, business, and private organizations has become a cottage industry.

The Commission described sustainable development simply as "development that meets the needs of the present without compromising the ability of future generations to meet their own needs." To make thinking about global environmental and development problems manageable, the Commissioners identify the core issues, sketch the trends behind them, and point out the dangers of letting predictions play out unaltered. The clear message of *Our Common Future* is that while their dimensions, severity, and interconnections may come as news to some, the problems before us are by now well-enough understood to attack.

The Commission's first concern is population stabilization. Instead of focusing narrowly on family planning, the report's authors urge governments to husband natural resources and boost their productivity, mainly by offering users wide-ranging economic incentives to take care of private and common property, and "to provide people with forms of social security other than large numbers of children" by, for instance, improving workplace safety, broadening access to clean water and basic sanitation, educating more girls, and giving minorities and tribal groups greater say over how local lands and waters are used.

The Commission's responses to its second broad challenge—making sure that all the world's more than five billion people get enough to eat—also cut a wide swath across economic, health, education, industrial, and agricultural policy. The remainder of the report mirrors questions that U.S. environmentalists are raising here. How can we save species and ecosystems without destroying livelihoods and traditional ways of life? How can economic growth continue uncoupled from the prodigious use of energy required in the past? How can industry use fewer resources to produce more goods? And how can our burgeoning cities again be made livable?

Proposing answers to these questions, *Our Common Future* makes three signal contributions to the debate on sustainable development. One is addressing both the immediate causes of environmental decay and economic stagnation with short-term "techno-fix" proposals and the root causes—poverty, institutional inertia, and the legacies of colonialism—with longer-term proposals for deeper social change. The second is showing how economic and social problems feed upon each other, and thus how economic and social policy are ultimately at each other's service. The third is making it painfully clear that if environmental problems were once merely the unfortunate side-effects of economic growth, now they are impediments to further growth as well.

As an offspring of the United Nations, the Commission envisioned a strong U.N. role in the institutional and legal changes ahead. But it also pressed all nations to develop "a foreign policy for the environment" and to keep close tabs on their own environmental quality. It invited governments to help scientists and non-governmental activists participate more fully in debates over environmental and development decisions. It urged regional organizations (such as the Arab League or the Organization of American States) to plan for environmental emergencies. And it proposed new international laws that would spell out the rights and responsibilities of both nations and individuals and help resolve environmental disputes.

Two proposals that the Commission pinned particular hope upon—a new Global Risk Assessment Program and a Universal Declaration on sustainable development—reveal both the staying power of the Commission's ideas and the high drag co-efficient of international decision-making.

At the 1992 U.N. Conference on Environment and Development (or "Earth Summit") in Rio de Janeiro, the Global Risk Assessment Program to spot, report, and assess threats to survival manifested itself variously as calls for improved data collection and assessment, for better environmental research and monitoring facilities, for a greater role for the U.N.'s trend-tracking Earthwatch, and, to stay on top of socio-economic trends, a companion Development Watch. The Universal Declaration on sustainable development called for in the report resulted in the "Rio Declaration"—a dilute and less courageous expression of government resolve than either the conference delegates or the authors of *Our Common Future* originally had in mind, but still an outgrowth of the debate given shape and steam by the Commission.

Our Common Future has come in for sharp criticism. The most damning complaint has been that its proposals are predicated on rapid economic growth, which has been a *cause* of many of the problems the Commission was looking to solve. Others have claimed that the plan paid only lip service to the roles and needs of women in socio-economic development and reflected the viewpoint of the privileged few already in power. But if it did not go far enough in asking whether vigorous growth is preferable to slower growth paired with attempts to share wealth more equitably, or if it did not give women or the dispossessed their full due, this book has still managed to integrate an enormous number of practical ideas for surviving the next century peacefully and to dispel the misguided notion that environmental ills can wait until economic growth creates wealth enough to subdue them.

KATHLEEN COURRIER

For Further Reading: The Stockholm Initiative on Global Security and Governance, *Common Responsibility in the 1990's* (1991); World Commission on Environment and Development, *Mandate for Change* (1985); World Commission on Environment and Development, *Our Common Future* (1987).

OUR COUNTRY, THE PLANET

Shridath Ramphal's book *Our Country, The Planet* begins with the point that even though it is increasingly clear that all nations and peoples must make common cause to save the global environment, if those with the most to gain during a transition period don't soon come to terms with those hardest hit, common sense and even long-term mutual interest will fall victim to political deadlock.

Calling for some sort of planetary bargain between the North (shorthand for the industrialized countries) and South (the developing countries), he excites hope that sterile speculation over how bad things will have to get before governments get serious about global pollution and biological impoverishment may finally be giving way to more constructive attempts to spell out the political compromises needed to curb both.

Ramphal's premise is that what we call human progress has passed most people by. The material wealth widely equated with quality of life has been enjoyed by one fourth of humanity, partly at the expense of the other three fourths, and from the perspective of the have-nots, "this colossal gap in the quality of human life is the most serious flaw in the record of human achievement."

Buried in this chasm, says Ramphal, are the seeds of the global environmental crisis now unfolding. On one side of the great divide, industrialization and rapacious consumption of resources by the rich are giving rise to acid rain, the ozone hole, the buildup of greenhouse gases, and other border-blind environmental menaces. On the other, absolute poverty is driving its victims to strip forests, overcrop fragile soils, and flock to out-of-control cities to compete for work and social services.

It's scarcely news that the rich and poor are on different economic trajectories or even that both lead to environmental dead ends. Ramphal's underappreciated point is, rather, that developed and developing countries contribute to environmental threats "in such unequal measure, and have such markedly different economic experiences and economic capacities, that the crisis itself is perceived quite differently—threatening relations between countries and blocking a convergence of response to the crisis."

As evidence of the perceptual differences that separate North from South, Ramphal cites a heady communique issued at the Group of Seven summit in 1988. In it, the decade's surprise boom in the OECD countries was hailed as "the longest period of economic growth in postwar history"—never mind that low commodity prices, high interest payments on debt, and trade barriers all but crushed most of the South during the 1980s. Even more pointedly, Ramphal sums up what poor countries hear when rich ones talk about saving tropical rainforests: that maintaining Northern economic growth justifies continued wholesale use of fossil fuels, even though the greenhouse gases given off when these fuels are burned contribute to the global warming that the South is being asked to avert by preserving its forests. In short, is the North to suffer no losses while the South postpones gratification indefinitely? As Ramphal notes, the same question pervades current debates on "intellectual property rights" to the biological wealth found in those same forests and throughout the tropics.

Reconfiguring such lopsided bargains will be hard. Yet the need to end what Ramphal calls "the pollution of poverty" and to make money-making less ecologically dangerous leaves no other choice. According to the author, simply muddling through on equal parts of inertia and optimism serves nobody's long-term interests.

For the North, Ramphal believes, a major psychological hurdle is parting with the notion that it's much harder for rich countries to accept any drop in living standards than for poor ones to abstain from ecologically harmful practices. A related challenge is following the "middle path" staked out by the environmental movement on energy use and other lifestyle issues, steering a course "between ascetic self-denial and complacent self-indulgence." As for basic purse-string concessions, greater trade liberalization and more development assistance for poor countries are essential.

The South also has some reckoning to do, counsels Ramphal. Developing countries can't take the high ground against northern authoritarianism in international relations unless they respect democratic values and human rights at home. They must also drop their strident confrontational tone in global forums and refrain from both putting all their eggs in one basket (as in some trade negotiations) and pinning every hope and grievance onto one agenda (as, some would say, happened at the Earth Summit). And with help from the North, they must make family-planning programs part of all economic development plans.

The "one world" Ramphal has in mind resembles that envisioned in 1972 by Barbara Ward and René Dubos in *Only One Earth*, source of the phrase "our country, the planet." In such a world, solar and other forms of renewable energy would replace fossil fuels, the bloated global defense budget would be converted into environmental plowshares, and "the strong tide of democracy and market economics" would be backed by local decision-making and self-help initiatives. Women everywhere would find schools open to them, and the poor would find work within mainstream society's margins.

But while he reaffirms the wisdom of the earlier book and tracks many of the same demographic, economic, and social trends that led Ward and Dubos to conclude that "in every alphabet of our being, we do indeed belong to a single system . . . depending for its survival on the balance and health of the [whole]," Ramphal doesn't believe that the "one world" they had in mind was a highly ordered world. "International community" can easily become a euphemism for a small group of powerful industrialized countries, he contends, and somehow it's always the other guy's rights that get overridden in the name of the common good. But Ramphal also believes that interdependence is irreversible and that the nation-state may have outlived its utility, locking us into a dysfunctional "mind-set of separateness."

If neither an ordered world with its implicit threat of imperialism nor a neo-feudal state with its back to the future can serve the human interest or preserve the total environment, what can? Ramphal's answer is that some measure of national sovereignty must be traded away for enlightened global governance based on enforceable international law. Such ideals as neighborliness and environmental justice can be converted to legal obligations without destroying diverse cultures and trampling on human rights, claims Ramphal, and can offer a timely alternative to a prevailing ethic predicated on the principles of economics and engineering.

The new "common law" Ramphal calls for can come about only through compromise and consensus. As the sole veteran of all five U.N. commissions created to size up postwar social, economic, and environmental prospects, Ramphal thus urges all countries—led by the richest and strongest—to build on *The Compact for a New World*, *Agenda 21*, *Caring for the Earth*, *Our Common Future*, and other recent multinational reports that lay out new priorities and initiatives for sustainable development. He also urges member countries to make the United Nations live up to the high hopes and ideals of its founders. But Ramphal is most persuasive as a moralist who reminds us of what he says we all know in our bones anyway—that human fate is collective and tied unalterably to that of our country, the planet.

KATHLEEN COURRIER

For Further Reading: Shridath Ramphal, *Our Country, The Planet* (1992); Barbara Ward and René Dubos, *Only One Earth* (1972); World Commission on Environment and Development, *Our Common Future* (1987).

OXYGEN

The element oxygen (O) at ordinary temperatures is a gas (O_2) that has no color, smell, or taste. It is the most abundant element on Earth and has the capacity to react with most elements and many compounds. It is slightly heavier than air. When the gas is bubbled into water, about 1 milliliter of oxygen will dissolve in 25 milliliters of water under standard conditions. The identifying characteristic of this gas, however, is its ability to support combustion. When something that is smoldering, such as a piece of wood, is put into oxygen it immediately bursts into flame, but the gas itself does not burn. At the low temperature of −182.962°C oxygen gas condenses into a pale blue liquid that is magnetic; cooled even more to −218.4°C it freezes into a pale blue solid.

There is evidence that the gas was known by its properties in China in the 8th century. Its generation, collection and identification were reported in 1774 by Joseph Priestley. Karl Scheele isolated the gas independently at the same time. In the next decade, Antoine Lavoisier identified it as one of his simple substances (i.e. elements) and from Greek roots named it, according to his theory about acids, as *oxygen*, meaning "acid former."

There are essentially two ways to obtain a sample of oxygen. It can be separated from the elements with which it is combined or from the mixture of gases that is air. Strongly heating potassium chlorate causes the oxygen part of the chlorate to be released as oxygen gas. Manganese dioxide catalytically speeds this oxygen release. To separate oxygen from the hydrogen in water molecules, electrical energy is used—this is called the electrolysis of water. To collect the large amounts of oxygen needed in industry, however, first air is cooled until its gases condense into liquid air, and then the oxygen is left behind as the nitrogen evaporates out of the liquid—a process called fractional distillation.

The rocky surface, the water, and the air covering of planet Earth each contain oxygen, either as the element or in chemical combination with other elements. About 60% of the atoms in the Earth's lithosphere are oxygen, making up just over 49% of the weight of the crust. About 21% of the volume of dry air is oxygen. In the hydrosphere not only is one of the three atoms that make up each water molecule an oxygen atom providing nearly 89% of water's weight, but also the water has oxygen gas dissolved in it. Oxygen is also one of the four basic building blocks of organic matter and so is a crucial constituent of the materials of which the myriad life forms of Earth are made. About two thirds of the human body is oxygen, and it is a necessary part of every living thing. It is found elsewhere in our solar system as well. On Mars, for example, it forms about 0.15% of the atmosphere, and in the Sun it is the third most abundant element. Naturally occurring oxygen is 99.7% the O-16 isotope; the relative atomic weight of naturally occurring oxygen is 15.999. Its atomic number is 8.

Living things use oxygen in a complex series of reaction steps to release the energy needed for the myriad functions and processes required to maintain life. This process is called respiration. A constant, massive, and regular replacement of oxygen, then, is necessary. Since a byproduct of photosynthesis is oxygen, it is green plants that generate the oxygen that sustains life.

Oxygen reacts with virtually every other element and also with itself, making ozone (O_3), an unstable triatomic form of oxygen that is found particularly in the upper atmosphere. Elements become constituents of compounds often through their first reaction with oxygen. Almost all combustion is an oxygen reaction. Metals react with oxygen forming basic oxides; other common oxygen-containing compounds are hydroxides, silicates, nitrates, carbonates, sulfates, phosphates, and all other compounds whose name ends in either "ate" or "ite." Non-metals react with oxygen to make acidic oxides. Oxygen is a constituent of hundreds of thousands of organic compounds, for example, alcohols, organic acids including amino, nucleic, and fatty acids, and carbohydrates. Since living things are not able to make use of most elements in their elemental form, the capacity of oxygen to move other elements into chemical combination is crucial to life. The nitrogen of the air becomes available to plants by reacting with oxygen during lightning to form nitrogen oxides. Carbon and hydrogen from the abiotic environment are appropriated by living things in the compounds carbon dioxide (CO_2) and water (H_2O).

Industrial production of oxygen mostly enhances steel production, but it is also used in hospitals for aiding the respiration of patients, for oxyacetylene welding, to speed up the biodegradation of sewage and in brewing to speed up the early stages of fermentation. It was one component of the fuel cells of the Apollo space program and is consistently used in continuing experimental efforts to produce a reliable fuel cell. Industrial importance also centers on the effects of pollutants on the formation of ozone in the atmosphere, and on the production of non-metal oxides, such as sulfur and nitrogen dioxides, as air pollutants responsible for acid rain.

JEAN LYTHCOTT

For Further Reading: Ralph I. Freudenthal and Susan Loy, *What You Need to Know to Live with Chemicals* (1989);

Ralph E. Lapp and Editors of *Life, Matter* (1965); H. M. Leicester, *The Historical Background of Chemistry* (1971); Jerome S. Meyer, *The Elements* (1957).
See also Atmosphere; Carbon Cycle; Dissolved Oxygen; Fire; Fuel Cell; Nitrogen Cycle; Oxygen Cycle; Ozone Layer; Water.

OXYGEN CYCLE

The oxygen cycle is one of the complex and interwoven biogeochemical cycles that continuously circulate elements essential for life back and forth between living and nonliving components of the biosphere. Oxygen occupies a prominent place in the composition and processes of the biosphere. It accounts for about one fourth of the atoms that make up the biosphere—that is, the entire crust of the Earth that participates in biological processes, including the lithosphere, the hydrosphere, the atmosphere, and the biomass. About a fifth of all the molecules of the atmosphere are the molecular form of oxygen, O_2. In the Earth's crust oxygen atoms account for more than 50% of all the atoms.

The chemical reactivity of oxygen results in the formation of a large variety of compounds, usually referred to as oxides. Most of the oxygen atoms of the biosphere are present as oxides of silicon, iron, calcium, magnesium, sodium, and hydrogen. The most familiar and biologically most important of these compounds are water (H_2O), a compound of oxygen and hydrogen, and carbon dioxide (CO_2). Of the many compounds of oxygen, only water and carbonates, and perhaps sulfates and phosphates, participate actively in the cycling of oxygen.

Oxygen of the Atmosphere

In the atmosphere the most abundant form of oxygen is its diatomic molecule O_2, making up about 20% of today's atmosphere. The molecules of oxygen can absorb ultraviolet (UV) light from the sun so that at sea level sunlight contains only traces of this potentially harmful radiation. Other oxygen-containing molecules, including water and oxides of carbon, nitrogen, and sulfur, also absorb ultraviolet radiation.

The absorption of radiant energy causes many chemical transformations of molecules in the upper atmosphere. An important product of these photochemical reactions is ozone (O_3). Ozone molecules in turn absorb ultraviolet radiation of the wavelength range that causes the most severe biological effects (around 260 nanometers). Ozone also absorbs in the infrared region of the spectrum and thus contributes to the green-

house function of the atmosphere. Ozone is continuously generated due to the absorption of UV by oxygen molecules. It is continuously destroyed by absorbing radiation and also by chemical interactions with other photodynamic molecules such as halogens. The amount of ozone in the upper atmosphere is therefore maintained in a steady state. Substances such as halogens that can accelerate the rate of breakdown of ozone, or those such as oxides of nitrogen that accelerate the rate of ozone synthesis, can cause an imbalance in the steady state. This permits either the deeper penetration of ultraviolet rays through the atmosphere or the warming of the biosphere due to excessive absorption of infrared radiation by the atmosphere.

Carbon dioxide strongly absorbs the infrared radiation from the warm Earth and is perhaps the most important contributor to the greenhouse effect. It also plays a major role in regulation of the photosynthetic activities of plants. Water molecules in the atmosphere absorb infrared radiation and thus function as a greenhouse compound. Water also absorbs short UV radiation, which causes the molecular disintegration and liberation of free oxygen and hydrogen in the upper atmosphere. This nonbiological reaction is one minor source of oxygen of the atmosphere. D. L. Gilbert estimates that of the total oxygen produced per million years, 15×10^{21} moles, only 7×10^{15} moles are produced by photodecomposition of water in the upper atmosphere. Most of the molecular oxygen originates in the biological process of photosynthesis.

Oxygen of the Lithosphere

More than half of all the atoms that make up the lithosphere are those of oxygen in the form of various oxides. The most abundant of these is silicon (SiO_2), which makes up about 64% of the total. Oxides of aluminum, magnesium, and water are the other major oxygen compounds of the lithosphere. Of special importance to the dynamic activities of the biosphere are the salts of carbonic acid, which serve also as reservoirs for carbon dioxide.

An important geochemical factor is the remaining potential of components of the lithosphere to bind additional oxygen, i.e. become oxidized. Thus the erosion of sedimentary rocks exposes new surfaces, which can combine with free oxygen. Especially prone to oxidation are sulfur, iron, and manganese. It is estimated that by this process all of the present atmospheric oxygen could be used in four million years. Since atmospheric oxygen is maintained constant, there must be an active process in the biosphere that replenishes the lost free oxygen in the atmosphere.

Oxygen of the Hydrosphere

More than 99.9% of the oxygen of the hydrosphere is in the molecules of water. Other minor constituents are the soluble oxides such as sulfate, borate, and carbonate. The soluble carbonate serves as a reservoir and buffer for the carbon dioxide of the atmosphere.

Molecular oxygen also dissolves in water to an extent determined by pressure and temperature. The total soluble molecular oxygen amounts to one one thousandth of a percent of the total oxygen of the hydrosphere. Dissolved oxygen is essential for the respiration of the living organisms in the oceans.

Oxygen enters the hydrosphere by diffusion from the atmosphere and is also generated in the illuminated zone of the ocean by photosynthesis. The consumption of oxygen by reactions of the biomass, especially in locations where there is an accumulation of organic substances, can lead to the creation of local anoxic conditions. For example, the deep waters of the Black Sea are anoxic because of the high input of organic substances by the river waters.

Oxygen of the Biomass

The biomass, the portion of the biosphere that includes living organisms and their immediate products, is about 25% oxygen. The bulk of the biomass is at an oxidation level similar to that of the carbohydrates, in which the atomic ratios are CH_2O. Carbohydrates can be further reduced to the level of hydrocarbons or oxidized to CO_2. The oxidation of carbon compounds is usually associated with liberation of energy, as in combustion. The uptake and release of energy associated with the processes of the biosphere are reactions in which reduced carbon compounds are being synthesized or oxidized.

Biological systems derive energy from the decomposition and oxidation of organic molecules. In the presence of oxygen (aerobically) reactions such as the burning of methane ($CH_4 + 2O_2 = CO_2 + 2H_2O$ + heat energy) result in complete oxidation of the molecule. In the absence of oxygen (anaerobically) energy can be released by the oxidation of some molecules or parts of a molecule and the simultaneous reduction of other molecules or other parts of the same molecule. For example, the sugar glucose can be completely oxidized aerobically in a process called respiration ($C_6H_{12}O_6 + 6O_2 = 6CO_2 + 6H_2O$ + energy). Under anaerobic conditions, glucose undergoes a fermentation reaction in which parts of the molecule are reduced to alcohol and parts are oxidized to CO_2 ($C_6H_{12}O_6 = 2C_2H_5OH + 2CO_2$ + energy). The complete oxidation of glucose yields twelve times as much energy as the partial oxidation and partial reduction reaction.

About 2,000 million years ago, early microbial organisms evolved the process of photosynthesis, in which energy from sunlight is captured by biological catalysts and used to split water molecules, combining the hydrogen with CO_2 to form carbohydrates $(CH_2O)_N$ and releasing oxygen as waste. The generalized reaction of photosynthesis can be expressed as $CO_2 + H_2O$ + light energy = $CH_2O + O_2$. This process is thought to be the source of most of the free oxygen in the biosphere.

The generalized equations for respiration and for photosynthesis are identical but they proceed in opposite directions. Thus for the biosphere to remain in a steady state the respiratory combustion of organic substances and the photosynthetic generation of organic substances should balance each other. An increase in the rate of photosynthesis will increase the amount of organic substances and free oxygen and diminish the amount of free CO_2. On the other hand, increased combustion of organic substances increases the free CO_2 in the atmosphere.

Molecular oxygen is a reactive substance that is harmful to many living organisms. These organisms had to adapt to the presence of oxygen in their environment by producing special antioxidant substances. The carotenoid pigments which are found in the photosynthetic cells of green plants are examples of such antioxidants.

D. L. Gilbert estimates the following rates in moles per million years (Myr) for the active reactions in the cycling of oxygen: Photosynthetic O_2 production: 15×10^{21}; O_2 consumption in respiration and decay: 14.94×10^{21}; O_2 used in methane oxidation: 0.05×10^{21}; O_2 used in weathering of the lithosphere: 0.01×10^{21}.

This balanced estimate of O_2 production and consumption assumes that no accumulation of organic substances is occurring. However, in the past, excess organic material accumulated to form coal and oil deposits. We are now utilizing these organic substances and are burning these fossil fuels at increasing rates. Gilbert estimates that 0.35×10^{21} moles of oxygen per Myr are being used in the combustion of fossil fuels and these appear as added CO_2 in the atmosphere. The excess CO_2 in the atmosphere increases the greenhouse effect due to the increased absorption of infrared radiation from the Earth's surface. Procedures for bioremediation of this phenomenon require the removal of the excess CO_2 from the atmosphere to form a deposit of carbon, which will be removed from direct exchange with the biomass.

It is apparent that the major dynamic processes involved in the cycling of oxygen occur in the biomass. The generation of O_2 in the process of photosynthesis and the use of oxygen in the oxidation of organic substances are the major reactions of the oxygen cycle. These exchange reactions occur mostly via the atmosphere and hydrosphere.

AHARON GIBOR

For Further Reading: Preston Cloud and Aharon Gibor, "The Oxygen Cycle," *Scientific American* (September 1970); D. L. Gilbert, ed., *Oxygen and Living Processes* (1981); J. E. Lovelock, *Gaia* (1979).

OZONE DEPLETION

See Pollution, Air.

OZONE LAYER

Ozone is a chemically reactive and relatively rare form of oxygen in which three oxygen atoms are bound into a single gas molecule. The common and much less reactive form of oxygen has two atoms in its molecule. Single oxygen atoms are also present in the atmosphere but they are very reactive and very rare.

Atmospheric ozone is important mainly because it absorbs harmful ultraviolet radiation emitted by the sun. The biological effects of this radiation include skin cancer, eye cataracts, and disruption of the immune system in humans and reduction of growth rates in plants. Quantitative estimates of biospheric disruption caused by ozone depletion are hard to make because of the many variables that augment or diminish the effects of ultraviolet radiation.

Ozone is produced in the atmosphere by a chemical reaction in which an oxygen atom combines with an oxygen molecule to yield an ozone molecule. The oxygen atom entering this reaction is most often a product of the destruction of an ozone molecule by ultraviolet light. This sequence of reactions, in which destruction of an ozone molecule is followed by formation of an ozone molecule, causes no change in the total amount of ozone in the atmosphere, but it serves to maintain a low concentration of very reactive oxygen atoms. Less frequently, however, new oxygen atoms are supplied by the ultraviolet disruption of an oxygen molecule. These new oxygen atoms usually combine with oxygen molecules to add to the total amount of ozone in the atmosphere. Occasionally, instead, an oxygen atom reacts with an ozone molecule to form molecular oxygen. The average amount of ozone in the unpolluted atmosphere depends on the balance between production by the reaction of new oxygen atoms with oxygen molecules and destruction occurring mainly in the reaction between oxygen atoms

and ozone molecules. Because solar ultraviolet radiation provides the new oxygen atoms that form ozone and this radiation is absorbed in the atmosphere above the ground, ozone is most abundant in a layer centered at altitudes around 20 kilometers.

The natural balance between production and destruction of ozone is upset by very small amounts of several pollutants, the most important of which is chlorine. A chlorine atom can destroy an ozone molecule by taking away one of its oxygen atoms to form a molecule of chlorine monoxide. The chlorine monoxide, in turn, reacts with an oxygen atom to release the original chlorine atom back to the atmosphere, where it can attack another ozone molecule. In this way, chlorine destroys atmospheric ozone without itself being consumed, so a little pollution can cause a lot of destruction.

The chlorine that destroys ozone is carried to the upper levels of the atmosphere in industrial compounds, principally the chlorofluorocarbon (CFC) gases invented in 1928 for use in refrigerators. These gases are wonderfully unreactive and therefore harmless to humans, animals, and plants. Unfortunately they are so unreactive that they accumulate in the atmosphere until air motions carry them to the level of the ozone layer, where ultraviolet light can break them apart, releasing the chlorine atoms that destroy ozone. This upward transport of pollutant is so slow that the average chlorofluorocarbon molecule will remain in the atmosphere for about a century. Nearly all the chlorofluorocarbons ever released are still in the atmosphere, leaking slowly into the ozone layer, where their destructive impact will continue many years after they were manufactured.

The destruction of ozone would be much worse were it not that most of the chlorine in the ozone layer reacts with other atmospheric gases and is converted into molecules that do not react with ozone. This process is reversed by reactions on the surface of cloud particles that drive destructive chlorine atoms back out of the unreactive molecules. In the ozone layer, cloud particles form mainly at the very low temperatures encountered during the winter above Antarctica. For this reason, the destruction of the ozone layer by pollutants is most apparent in what has become known as the Antarctic Ozone Hole.

The way in which chlorofluorocarbons would damage the ozone layer was deduced theoretically by Mario Molina and Sherwood Rowland and described in a landmark paper published in 1974. Their analysis was repeatedly endorsed by high level scientific panels, including several convened by the United States National Academy of Sciences, during several years of acrimonious debate, which culminated in the 1978 ban by the United States, Canada, and several

Scandinavian nations on the use of chlorofluorocarbons in aerosol sprays.

Plans for further restrictions on chlorofluorocarbons in the United States were abandoned in 1981. In 1985, however, scientists of the British Antarctic Survey reported their discovery of the Antarctic Ozone Hole, the first unambiguous detection of ozone destruction by pollutant chlorine. Political opposition to restrictions on chlorofluorocarbons did not long survive these results. By 1990, most nations had agreed to cease production of these gases before the end of the century. By 1992, the global rate of production had fallen by 50% from its maximum value, achieved in 1988.

We have known since 1974 that chlorofluorocarbons damage the ozone layer. Respected scientific voices have called repeatedly since then for restrictions on their use. What are the environmental costs of the delay of more than 15 years in the decision to regulate these pollutants?

Three factors affect the answer to this question. First, both the rate of production of chlorofluorocarbons and their abundance in the atmosphere have been increasing steadily. In 1990, both were more than twice what they had been in 1974. Damage to the ozone layer is more than twice as severe. Second, the situation will continue to deteriorate for several decades as chlorofluorocarbon production is gradually phased out and as the pollutants already in the lower atmosphere are carried up to the ozone layer. Third, recovery will take centuries because chlorofluorocarbons are removed so slowly from the atmosphere.

Because of these factors, the levels of ozone destruction encountered in the 1990s are comparable to the worst that would have occurred if regulations had been imposed in the 1970s. Instead, damage will become more than twice as severe before concentrations of pollutant chlorine begin to decline, several decades into the 21st century. It will take more than 200 years for the ozone layer to recover to what would have been the worst level of destruction, had regulations been imposed in the 1970s.

JAMES C. G. WALKER

For Further Reading: P. Hamill and O. B. Toon, "Polar Stratospheric Clouds and the Ozone Hole," *Physics Today* (December 1991); S. L. Roan, *Ozone Crisis* (1989); F. S. Rowland, "Chlorofluorocarbons and the Depletion of Stratospheric Ozone," *American Scientist* (1989).

P

PACKAGING

See Source Reduction; Waste, Municipal Solid.

PAINT TOXICITY

Paint toxicity can stem from three of its four basic components: pigments, solvents, or additives. Many pigments, if disposed of improperly, are toxic to the environment or to human health. For example, some pigments contain lead, a toxic substance that, if ingested, can cause a wide range of effects—from irritability and limited muscle control to irreversible brain damage, seizures, and death. In addition, the use of lead in paint was significantly restricted in the United States in 1978, and only 0.06% of paint currently produced contains lead.

Solvents, used during manufacturing to thin paint, can be either volatile organic compounds (VOC) or water. Alkyd, or oil-based paints, use volatile organic compounds, particularly hydrocarbons. When these paints dry, the solvents evaporate, combine with other compounds present in air, and react with sunlight to produce ozone and other pollutants of smog. In the 1990s, several states in the United States passed laws to limit volatile organic compounds. In November 1991, the first international legislation to reduce VOC emissions was agreed upon under the auspices of the United Nations Economic Commission for Europe. The Geneva Protocol mandates a 30% cut in VOC emissions in Europe and North America by 1999. Many companies within the coatings industry are now using water-based paints instead of organic-solvent-based types.

Although water-based latex paints do not pose danger to the environment due to their solvents, they may do so because of various additives. For example, some latexes made before 1991 contained mercury compounds to fight mildew and bacterial growth, which, when volatilized, may pose a hazard to humans and the environment. The catalysts or additives used in oil-based paints to promote drying consist of one or more compounds of certain organic acids with metals such as lead, manganese, or cobalt that may cause detrimental effects in the environment. Additives in antifouling paints—tributyl tin (TBT) and copper oxide— repel barnacles and other organisms from boat bottoms, but are toxic to other marine life. In response to these findings, France restricted the use of TBT in 1982, and the United Kingdom followed with its own limits in 1986. The United States has proposed restrictions, while many individual states already have set their own. A new paint additive of copper mixed with epoxy that eliminates barnacles without harming other organisms is being developed to replace TBT and copper oxide.

PALEONTOLOGY

Paleontology is the study of past life forms through fossils. A fossil is the preserved remains of a living thing or traces of its activity. Most fossils are found in rocks that formed from sediments and contain the hard parts of an organism, rather than its organic soft tissue. Soft parts have been found preserved in bogs, tar, permafrost, and other environments that can prevent tissue decay. The traces of an organism's activity include burrows constructed for shelter, feeding trails, footprints, and, in the case of early humans, tools. The oldest known fossils are microscopic organisms from three and a half billion years ago that are related to modern bacteria. Each fossil reveals information about the physical, biological and, with the advent of humans, cultural factors that made up the environments of its time. Nicholas Steno (1638–1687), famous for investigations in the field of geology, was the first to demonstrate the nature of fossil animals. Gideon Mantell (1790–1852) discovered the first dinosaur fossils (teeth) in England. Othniel Marsh (1831–1899) developed methods in the western U.S. that are still in use for taking large fossil bones of dinosaurs out of the ground.

Paleontologists analyze fossils for evidence of plants (paleobotany) and animals (paleozoology) long extinct and trace the ancestry and evolution of present day life forms. For example, paleoanthropology is the comparative study of fossils of monkeys, apes, and other humanlike creatures more primitive than *Homo sapiens.* The work of classifying and recreating how ancient organisms looked and lived produces an ever-growing catalogue of life on Earth. Baron Georges Cuvier (1769–1832)—called "the father of comparative anatomy"—

was the first person to apply methods of study already developed for recent animals to skeletons of fossil vertebrates. His work was important because it showed for the first time that many fossil animals were of types that no longer exist. William Smith (1769–1839), an English surveyor, established the principle that each sedimentary layer contains its own characteristic fossils. This study (stratigraphy) is used by geologists and archaeologists as well. Along with Steno's concept of superposition it is used to establish time relations in working out historical sequences. Paleontology shows that the history of life as a whole is marked by both gradual and radical change, by continual interactions and adaptation, and by success, failure, and accident.

Paleontological information contributes to knowledge of past environments through the assumption that fossils of similar biological structure lived under similar conditions (paleoecology). The rise, fall, and diversity of different populations of organisms are used to indicate prevailing environmental conditions at the time those plants or animals lived. Certain molluscs, for instance, thrive in slightly warmer waters than others. Variation in the mollusc population is then an indicator of fluctuations in water temperature over time (paleotemperature), which in turn is an indication of climate change.

As paleontology contributes to our understanding of the past, it provides information for environmental decision-making. The repeated extinctions, both mass and gradual, that characterize the fossil record, and the bursts of new species that follow, give insight into the ability of ecosystems, communities, and individual species to respond to environmental disturbances.

The testing of theoretical models of environmental change is still another area to which paleontology contributes. Computer models are now used to anticipate the effects of potential global warming on the ecosystems of the world. However, periods of global warming and global cooling have occurred throughout millions of years of geologic time and the fossil record provides data with which computer models can be tested (paleoclimatology).

PAMPAS

See Grasslands.

PARASITISM

Parasitism is an intimate association between two different kinds of organisms in which one, the host, provides food and shelter for the other, the parasite. The parasite cannot exist in nature without its host; the relationship is obligate for the parasite but not for the host. All organisms—from bacteria to whales, from one-celled algae to giant trees—have their parasites. Furthermore, the parasitic way of life can be found among virtually all groups of organisms. Viruses, which are acellular assemblages of DNA or RNA with a limited number of proteins, are all intracellular parasites dependent for their replication on the metabolic and protein-synthesizing machinery of a living host cell. Among bacteria, the prokaryotes, many kinds are parasitic, including the agents of such major human diseases as tuberculosis, leprosy, typhoid fever, cholera, and syphilis. Bacteria are also responsible for economically important diseases of animals and plants. Among eukaryotic organisms, parasitic species are particularly numerous in fungi, the protozoa, nematodes, flatworms, insects, and crustacea. Fungi are of special significance as parasites of crop plants; wheat rust and corn smut for example wreak great havoc. Protozoan parasites include the agents of malaria, still the world's principal infectious disease. Among the parasitic nematodes, we find the hookworm and the worm causing "river blindness" in West Africa, as well as major parasites of domestic animals. Most important of the parasitic flatworms are the schistosomes, cause of bilharziasis, a disease affecting millions of people. Parasitic insects usually parasitize plants and other insects, but we find here ectoparasites of humans, as fleas and lice. Insects furthermore play major roles as vectors in the transmission of parasites of vertebrates. For parasitic crustacea the hosts are typically aquatic animals such as other crustacea or fish.

Parasites play major roles in the balance of nature; they affect host populations, the expression of genetic traits, and even host behavior. Malaria and sickle cell disease in West Africa provide an outstanding example of the effect of a parasite on the genetic composition of a human population. Sickle cell disease is a genetic disorder resulting from a small mutation in the gene for hemoglobin. Individuals who receive this mutant gene from both parents (homozygotes, designated SS) suffer from a serious abnormality of their red blood cells that produces severe anemia and painful crises; they rarely survive to reproductive age. Yet in many regions of West Africa the frequency of this gene is over 10%, sometimes as high as 20%. The basis for this high frequency of a deleterious gene was a puzzle. Clearly, there must be some strong selective pressure favoring the survival of those who have received the mutant gene from one parent (heterozygotes, designated SA) over those homozygous for the normal hemoglobin (designated AA). The geneticist J. B. S. Haldane suggested in 1949 that the disease malaria was the selective force. There is a striking geographic correlation between the prevalence of the gene for sickle hemoglobin and the prevalence of malaria caused by *Plasmodium falciparum*, most pathogenic of the four species of human malaria parasites. In holoendemic regions such as West Africa, falciparum malaria is a principal cause of child mortality. Recent research has shown beyond question that SA children have 90% less chance of either severe anemia or cerebral malaria caused by *P. falciparum* than AA children. Accordingly, though not protected from infection, they are much less likely to die from it. Both in the intact host and in culture *in vitro* falciparum malaria parasites do not multiply as much in the red blood cells of SA individuals as they do in those of AA persons. There are a number of other genetic red blood cell abnormalities that occur in malarious regions and most likely have been maintained by the selective pressure of malaria.

Malaria also serves well to illustrate the close relationship between physical environment and the occurrence of particular parasitic infections. Malaria in Europe was traditionally associated with marshy regions, but this is not necessarily so elsewhere. It all depends on the breeding requirements of the local species of anopheline mosquitoes serving as vectors of the parasite. The very efficient African vectors of the *Anopheles gambiae* group develop as larvae in temporary water puddles; these are abundant during the rainy season in the African savannah, and that is when most of the malaria occurs. There are regions, as in some areas of the Philippines, where the vector breeds in streams in hilly country and malaria is more a disease of the uplands than the lowlands. Each ecological situation must be investigated if one is to attempt to control malaria by environmental manipulation, as by control of mosquito breeding.

The ecological situation becomes even more complex if the parasites, unlike malaria, are not restricted to human hosts but also infect other species of animals. Chagas' disease in South America, responsible for much serious illness, provides a good example. As for malaria, the etiologic agent is a single-celled eukaryotic microorganism that depends for its transmission on an insect, any of a number of species of blood-feeding bugs in the family Triatomidae. Unlike malaria, however, this parasite infects many species of wild mammals, which serve as a reservoir of the infection. When infection bugs take up residence in a house, as in its thatched roof or in the cracks of its mud walls, the infection becomes established among the humans of the village. Person-to-person transmission via the infected bugs then occurs. Such an infection, transmitted to humans from a reservoir among other species of animals, is called a zoonosis. Chagas' disease could be controlled just by providing better housing that would ensure an environment free from

the triatomid bugs that are the essential vectors. But the costs of doing so for the large populations involved would be staggering. This disease illustrates the impact of socioeconomic conditions, often as important as the physical environment.

Appropriate manipulation of the environment can break the chain of transmission of many parasitic infections. When transmission is effected by an insect or via an intermediate host such as a snail, any measure that reduces the numbers of the vector or that protects people from contact with it will reduce transmission. Such measures have been highly effective in some areas, but in others they may be difficult or prohibitively expensive. Good plumbing, sanitation, and a good water supply are the most successful of preventive environmental measures. They protect large populations from the many kinds of parasites transmitted by means of resistant cysts or eggs that are excreted by one host and infect another when ingested with food or drink. When prevention fails, reliance must be placed on chemotherapy or on attempts to develop a vaccine.

Parasitism and disease of the host are not synonymous. Many parasites exploit a host without doing any apparent damage. Destruction of the host is not necessarily to the advantage of the parasite. Even with parasites that produce disease, an equilibrium is often established in which small numbers of parasites persist in a host whose immunological reactions prevent further infection. Thus, a few pairs of schistosome worms can live for many years in a person, all the while producing eggs that reach the environment and serve to infect snails and in this way disseminate the species.

The special adaptations that parasites must make to be able to survive and develop in their living host, to get from one host to another, to locate particular tissues or cells within the host, and to switch quickly from one developmental pathway to another, are attracting increasing attention of biochemists, geneticists, and cell and molecular biologists to the study of parasitism. Such work is already revealing fundamental phenomena in cell biology and immunology. We may expect that it will also reveal new ways to deal with some of these parasitic diseases that have plagued humanity so long.

WILLIAM TRAGER

For Further Reading: Paul C. Beaver, Rodney C. Jung, and Eddie W. Cupp, *Clinical Parasitology* (1984); Paul T. Englund and Alan Sher, eds., *The Biology of Parasitism* (1988); William Trager, *Living Together: The Biology of Animal Parasitism* (1986).

See also Bacteria; Epidemiology; Health and Disease; Malaria; Pest Control; Symbiosis; Vector-Borne Diseases; Viruses.

PARKS, U.S. NATIONAL

America's national parks—over 360 units from the Arctic Circle to the Tropic of Cancer and a third of the way around the Earth—represent the diversity of the ecosystems and the breadth of the cultural history of the nation. The idea of national parks has been called our greatest export; over 100 other nations have duplicated the concept of Yellowstone, the world's first national park, established in 1872.

On August 25, 1916, in the act that established the National Park Service, Congress verbalized a unique vision that has guided the creation and preservation of our system of national parks to this day: " . . . to conserve the scenery and the natural and historic objects and the wild life therein and to provide for the enjoyment of the same in such manner and by such means as will leave them unimpaired for the enjoyment of future generations." Those words captured two principles of the National Park System. The first principle is the 19th century heritage of Emerson and Thoreau, who saw that nature was essential to an individual gaining self-respect and inner freedom. The second principle is a precept of democracy: that these special places be preserved for and accessible to all, not just a privileged few.

Well before Yellowstone National Park was established, there were forces at work that served as the foundation of national parks. The industrial revolution was taking its toll on the well-being of people, as well as the rest of nature. Henry Thoreau's *Walden* (1854) documents the connection between a specific natural place and the discovery of self. During the same era, painters were driven to capture the majesty and mystery of the American landscape, creating America's first school of painting, the Hudson River School. After the Civil War, many were drawn westward to find economic and social opportunity. With the movement west, the nation discovered the awesome beauty and bountiful resources of the region. John Muir took Thoreau's philosophy on the relationship of the human spirit to nature one step further. Muir saw the value of nature in its own right, irrespective of its relationship to mankind. Twenty-two years old at the outbreak of the Civil War, he fled eastern industrialization and in 1890 he led the movement to create Yosemite National Park out of state park lands.

This century of change also created a new profession, landscape architecture, with its special blend of artistry and science personified in Frederick Law Olmsted. He and his associates designed and built Central Park out of a New York slum. His work also survives in a dozen cities as well as Yosemite National Park. By 1900 there were a handful of national parks including Yellowstone, Yosemite, Sequoia, and Mt. Rainier.

President Theodore Roosevelt made conservation a public purpose. In addition to establishing forest reserves and wildlife sanctuaries, he expanded on the concept of natural wonders by designating historic and scientific areas such as Devils Tower, Montezuma Castle, and El Morro national monuments. When the National Park Service was created in 1916, it was made part of the Department of Interior and headed by Stephen T. Mather. In its first decade, it struggled to survive. Short of staff, it operated on money borrowed from Mather's own pocket for basic office expenses, and faced congressional pressure to satisfy special interests, such as timber companies, concessionaires, and dam-builders. Assembling the presidents of 100 American universities and associations, Mather created the National Park Association to "defend the National Parks and monuments fearlessly against assaults of private interests and aggressive commercialism." He further sought to build an alliance between the newly evolving state parks and national parks.

In the 1930s, then director of the National Park Service, Horace Albright, accepted responsibility for battlefields and historic places that were under the financially strapped military, expanding the park system to practically every state, and therefore every senator.

In the 1940s, another Interior agency, the Bureau of Reclamation, proposed to build reservoirs in Dinosaur National Monument (Colorado). The ensuing battle, finally ending in the 1950s, reflected the inherent tension between preservation and use. Citizens joined forces with others inside and outside the Park Service, and using the front pages of America's newspapers, they prevailed in unraveling the political agreement of a President, the Congress, and the dam-building industry. It was the beginning of the confrontational approach taken by environmentalists to this day.

For the first 50 years of the National Park Service, parklands were set aside because of scenic value. In the 1970s, the Nixon administration thrust the Park Service into urban America to provide park facilities where a majority of people live. Golden Gate in San Francisco, Gateway in New York City, national lakeshores, long distance trails and seashores, all gave the Service new access to the contemporary agenda of the people. By the 1970s, the National Park System comprised 21 categories of units. The world park movement, under the leadership of the International Union for the Conservation of Nature and Natural Resources, established a five-category system: scientific reserve/strict nature reserve, national park, natural monument/natural landmark, nature conservation reserve, and protected landscape. Moving through the list, each category allows for increasingly more resource use, such as grazing and energy development. They reflect the public and political acceptance of national parks around the world.

In 1980 the Alaska National Interest Lands Conservation Act doubled the holdings of the National Park Service from approximately 40 million acres to 80 million acres. The 1980s brought a conservative political swing in the nation that halted the expansion of the park system. Some raised the idea of limiting the number of parks and suggested that private property rights overruled public need. Some, including President Reagan's Secretary of Interior, James Watt, even called for turning National Park Service responsibilities over to the private sector. By the election of President Bill Clinton in 1992, park usage had doubled to more than 268 million visitors a year and a severe decline had occurred in park infrastructure. The National Park Service's ability to manage and conserve the resources it was charged to protect had come under intense criticism from a wide variety of experts including the National Academy of Sciences. President Clinton's Secretary of Interior, Bruce Babbitt, told the nation that preserving and expanding the parks were his highest priority.

America's national parks—the personnel and the natural and historical resources—are experiencing overwhelming changes. Each ranger deals with more than 90,000 daily visitors a year, nearly twice the 1980 level. Representative of problems facing the national parks nationwide, Everglades National Park has witnessed a 90% decline in its bird populations and the near-extinction of the Florida panther. Cultural parks are having similar problems; the Statue of Liberty was severely deteriorating until private citizens raised funds to rebuild it. Equally disturbing is the lack of a plan for the parks other than the one offered by the private citizen organization the National Parks and Conservation Association (formerly the National Parks Association). Nearly 2 million of the 80 million acres of the park system remain in private hands without a conscious commitment to purchase these inholdings. Development on these inholdings threatens the well-being of park resources. America's national parks are the model for the world. Without resolution of these and other crucial issues, the parks may become mere postcard images devoid of nature and of the historic structure that has shaped the nation.

Many are calling for the National Park System to focus on a new mission of preservation of biological diversity of North America. Others express concern that it commemorate the heritage of all cultures that contribute to the fabric of America's national culture.

PAUL C. PRITCHARD

For Further Reading: Conservation Foundation, *National Parks for a New Generation* (1985); Michael Frome, *Regreening the National Parks* (1992); Ronald A. Foresta, *America's National Parks and Their Keepers* (1984).
See also Conservation Movement in the U.S.; Emerson, Ralph Waldo; Environmental Movement; Forests, U.S. National; Muir, John; National Wildlife Refuge System; Roosevelt, Theodore; Thoreau, Henry David.

PASTEUR, LOUIS

(1822–1895), French chemist and microbiologist. To view Louis Pasteur's professional achievements gives one the impression that he led an enchanted life. His contributions to science, technology, and medicine were prodigious and continued without interruption from his early twenties to his mid-sixties. His skill in public debates and his flair for dramatic demonstrations enabled him to triumph over his opponents. His discoveries had practical applications that immediately contributed to the health and wealth of humankind. His worldwide fame made him a legendary character during his lifetime; he was, and remains, the white knight of science.

His extraordinary successes, however, had been achieved at the cost of immense labor and against tremendous odds, including the stroke that paralyzed him on the left side at the age of 46. Writing as if he had not been complete master of his own life, Pasteur stated time and time again that he had been "enchained" by the inescapable logic of his discoveries; he had thus been compelled to move from the study of crystals to fermentation, then to the then still debated hypothesis of the spontaneous generation of life, on to infection and vaccination.

One can indeed recognize a majestic ordering in Pasteur's scientific career. Yet, the logic that governed the succession of his achievements was not as inescapable as he stated. At almost any point in the evolution of his scientific career, he could have followed, just as logically, other lines of work that would have led him to discoveries in fields other than fermentation and vaccination. Some of his actual remarks indicate that he was aware of the potentialities he had left undeveloped.

Early in his scientific life he predicted, for example, that a day would come "when microbes will be utilized in certain industrial operations on account of their ability to attack organic matter." Microbial processes are now used on an enormous scale to produce organic acids, solvents, vitamins, enzymes, and drugs. In 1877 he observed that the anthrax bacillus loses its virulence when placed in contact with certain soil mi-

Pastoralism **521**

crobes and he suggested that saprophytic organisms might be used to combat infectious agents. This was of course a vision of antibiotic therapy, more than sixty years before its actual beginning. Such lines of investigation, and others that he suggested, were within Pasteur's technical possibilities and he could have followed them, if he had time. He had good reasons indeed to ask himself whether "the road not taken" might not have been the better road.

Even though Pasteur's name is identified with the "germ theory" of fermentation and disease—namely, the view that many types of chemical alterations and of pathological processes are caused by specific types of microbes—he was intensely interested in what he called the "terrain," a word he used to include the environmental factors that affect the course of fermentation and of disease. The magnitude of his theoretical and practical achievements derives in large part from the fact that his conceptual view of life was fundamentally ecological.

From the very beginning of his biological investigations, Pasteur became aware of the fact that the chemical activities of microbes are profoundly influenced by environmental factors. Furthermore, he developed very early a sweeping ecological concept of the role played by microbial life in the cycles of matter. During the 1860s he wrote letters to important French officials to advocate support of microbiological sciences on the grounds that the whole economy of nature, and therefore human welfare, depended upon the beneficial activities of microorganisms. He boldly postulated that microbial life is responsible for the constant recycling of chemical substances under natural conditions: from complex organic matter to simple molecules and back into living substance. In language that was more visionary than scientific, he asserted that each of the various microbial types plays a specialized part in the orderly succession of changes essential for the continuation of life on earth. Long before the word *ecology* had been introduced into the scientific literature, he thus achieved an intuitive understanding of the interplay between biological and chemical processes that brings about the finely orchestrated manifestations of life and of transformations of matter in natural phenomena.

Pasteur's ecological attitude can also be recognized in his repeated emphasis—to the point of obsession—on the fact that the morphology and chemical activities of any particular microbial species are conditioned by the physicochemical characteristics of the environment. He pointed out, for example, that molds can be either filamentous or yeastlike in shape, depending upon the oxygen tension of the medium in which they grow. He demonstrated also that the gaseous environment determines the relative proportions of alcohol,

organic acids, carbon dioxide, and protoplasmic material produced by microbes from a particular substrate. Observations of this type give to the book in which he assembled his studies on beer (published in 1876) an importance that far transcends the practice of beer-making. In that book he approached the problem of fermentation from an ecological point of view. By demonstrating that "fermentation is life without oxygen," he introduced the first sophisticated evidence of biochemical mechanisms in an ecological relationship.

Furthermore, the ecological attitude in Pasteur's laboratory certainly helped his associate Emile Duclaux, who eventually became director of the Pasteur Institute, to recognize that microbes can be modified at will by altering the composition of the culture medium. This was the first demonstration of a phenomenon that opened the way for other key discoveries and thus constitutes another fundamental link in the understanding of the ecological relation between environmental factors and biological characteristics. In addition, Pasteur's recognition of the effects that environmental factors exert on metabolic activities is now incorporated into theoretical microbiology and technological applications.

In summary, his immense practical skill in converting theoretical knowledge into technological processes made him one of the most effective men of his century; he synthesized the known facts of biology and chemistry into original concepts of fermentation and disease and thus created a new science that dealt with the urgent needs of his social environment. The other side of his genius, although less obvious, is more original and perhaps more important in the long run. His emphasis was on the essential role played by microorganisms in the economy of nature, and on the interplay between living things and environment. He contributed to scientific philosophy by perceiving that all forms of life are integrated components of a global ecological system.

RENÉ DUBOS
(EDITED BY RUTH A. EBLEN)

For Further Reading: René Dubos, *Louis Pasteur: Free Lance of Science* (1976); René Dubos, *Pasteur and Modern Science* (1960); René Dubos, "Louis Pasteur: An Inadvertent Ecologist," in *The World of René Dubos*, Gerard Piel and Osborn Segerberg, Jr., eds. (1990).
See also Antibiotics; Ecology as a Science; Health and Disease.

PASTORALISM

Pastoralists are members of kin-based or tribal societies, involved in raising migratory livestock (typically cattle, goats or camels) and traveling within a defined geographic area over the course of a year. This seasonal pattern of movement, or transhumance, is largely defined by the prevailing climate and the availability of high-quality pasture, water points, and markets. Pastoral cultures are especially prevalent in western and eastern (sub-Saharan) Africa, a transition zone between hot desert and savanna ecosystems. Pastoralists also thrive in other world regions, including the Middle East and Australia.

The expansiveness of pastoral territory and cultural world transcends not only ecological zones but international borders. As a result, pastoralists rarely consider themselves members of any single nationality, which presents a growing source of conflict between governments and transhumant societies. Pastoral autonomy and resistance to settlement has unfortunately increased the level of political and ethnic strife between these traditional people and the nations through which they travel and to which they historically belong.

Since pastoralists typically inhabit fragile, semiarid regions, the sustainability of their herds and culture rests on a profound knowledge of local and regional environments and the production capacities of varying ecosystems throughout the year. For hundreds of years pastoral families have adapted to their precarious environment (which is largely dictated by variable annual rainfall) with relatively low impact on the surrounding environment. In Africa, for example, rainfall can vary between 100 and 600 mm/year in the Sahelian region (western sub-Saharan) and between 100 and 250 mm/year in eastern sub-Saharan countries. Rainfall recharges wells and streams, which provide watering points for people and cattle, and gives life to ephemeral and annual grasses and shrubs for grazing.

The carrying capacity of rangeland varies; in Somalia, for example, seven to twenty hectares are required to support one animal. However, transhumant patterns are increasingly constrained by modern land-use changes, forcing historically noncompetitive herders to compete with one another for a diminishing resource base. In this situation the quality and abundance of rangeland deteriorates, leading to weight loss, susceptibility to disease, and eventually death of herds. Low herd weight as a function of falling range capacity is a blow to pastoral income, as these cattle cannot compete with those raised in controlled, sedentary conditions. In order to counter such risks, pastoralists in central Africa, for example, increase the size of their herds in order to maximize their prospects.

Overgrazing is not an inevitable feature of traditional pastoralism, but rather a twentieth-century phenomenon precipitated by forced settlement or the curtailing of transhumant migratory patterns. The growing scarcity of high-quality rangeland in Africa is a function of widespread settlement, land-use changes over the past century, and recent pronounced climatic anomalies. The most profound changes in rural Africa, as in other regions with pastoral populations, have come about through the privatization of previously communal lands. Increasing portions of historic grazing areas are preempted by these widespread transformations or are rapidly being depleted. It is therefore unfair to equate pastoralist production with "overgrazing" and desertification.

While pastoral livestock production is predominately subsistence-oriented, a high percentage of the population in countries such as Somalia is engaged in this practice. Therefore, a significant percentage of the gross domestic product of a given country may be generated by pastoralism. This mode of production is essentially pre- or noncapitalist, driven more by tradition, cultural identity, and ecological constraints than by a competitive market. As African development has become increasingly capitalist, however, pastoral production is colliding ever more with sedentary agriculture and with a changing social and economic system that considers this mode of production both marginal and backward.

PCBs

PCBs, or polychlorinated biphenyls, are a class of synthetic organic compounds widely recognized as toxic contaminants that are highly persistent in the environment. They are prepared by reacting a biphenyl molecule (a double-ring structure of six carbon atoms) with chlorine in the presence of a catalyst. As the number of chlorine atoms increases, PCBs become more toxic and persistent in the environment. Commercial PCB products range from light, oily fluids to heavy, honeylike oils, to greases and waxes. These products were primarily used in capacitors and transformers because of their flame-resistant and dielectric properties. They have also been mixed into paints, plastics, flame retardants, hydraulic fluids, dyes, lubricants, protective coatings for wood, metal, and concrete surfaces; and sprays for unpaved roads to reduce dust and stop plant growth. In 1979 the Environmental Protection Agency (EPA) prohibited the manufacture of PCBs in the U.S. and provided severe restrictions and an orderly phaseout over a five-year period.

Early concerns, which were later shown to be invalid, initiated regulatory action. These concerns are based on the Yusho human poisoning incident in Japan; the finding by the Center for Disease Control (CDC) in 1975 that PCBs caused liver cancer in rats; and the spread and biodegradation of spilled or disposed PCBs in the environment.

The Yusho incident involved about 1,500 people in Japan who became ill after eating rice oil that was accidentally contaminated with PCBs. Later it was discovered that high levels of polychlorinated dibenzofurans (PCDFs)—which are much more toxic than PCBs—were in the mixture. Both the National Institute of Environmental Health Sciences and the EPA concluded that most of the effects were due to the PCDFs.

Dr. Renate Kimbrough, who conducted the research in 1975 at the CDC on whether PCBs caused liver cancer in rats, concluded in 1987 that no evidence thus far reported shows that occupational exposure to PCBs causes an increased incidence of cancer in humans. EPA's Cancer Assessment Group concluded that the data are insufficient to establish PCBs as a human carcinogen. Dr. Stephen Safe, a leading adviser to the EPA on the toxicity of PCBs, PCDFs, and dioxins, concluded in 1985 that certain PCB structures have toxic properties qualitatively similar to dioxin, although quantitatively much less potent.

The biodegradation of PCBs by both aerobic (requiring oxygen) and anaerobic (functioning without oxygen) bacteria has been, and continues to be, extensively studied. Most soils contain aerobic bacteria with some level of PCB-degrading ability. Some PCB-degrading bacteria are able to degrade some highly chlorinated PCBs, thereby providing a natural mechanism through which PCBs will be broken down.

Examination of PCB-contaminated samples indicated that the degradation by aerobic bacteria is limited to surface layers (up to 2 inches deep) of soils and of river- and lake-bottom sediments. Examination of sediment samples from the Hudson River below the surface layers, however, showed an unexpected type of degradation of PCBs that is believed to be due to anaerobic bacteria.

There are strong indications that protective measures against PCB contamination are working, in terms of PCB concentration in living organisms measured over a long time frame. For example, New York striped bass and Great Lakes herring gull eggs have increased in numbers. However, many PCBs have a very long half-life and therefore persist in environments, and like DDT, with which they share many similarities, have become distributed in trace quantities—parts per million or trillion—worldwide. Fortunately, as yet there is no known clinical significance that can be attributed to these residues in humans.

PEST CONTROL

Pests are organisms out of place—any unwanted plants, animals, or microorganisms. To the farmer, pests include insects and mites that damage crops; weeds that compete with crops for nutrients and soil moisture; aquatic plants that clog irrigation and drainage ditches; plant diseases caused by fungi, bacteria, nematodes and viruses; rodents that feed on grain and the bark of fruit trees; and birds that eat grain from fields, feedlots and storage.

History offers innumerable examples of the mass destruction of crops by diseases and insects. From 1845 to 1851, a massive infection of potatoes in Ireland by the late blight fungus resulted in the deaths of more than a million persons from starvation and the mass migrations to the U.S.

Many kinds of animal and human disease are caused by organisms transmitted by insects. Horse sleeping sickness and Venezuelan equine encephalitis are both carried by mosquitoes and are also diseases of humans. Lyme disease is transmitted by ticks. Since the first recorded epidemic of the Black Death or bubonic plague in 1347, more than 65 million persons have died from this disease transmitted by the rat flea. The same disease occurs endemically in the western states, frequently wiping out thousands of its host, the gopher, and occasionally infecting humans who camp nearby. In the 19th century, the Panama Canal was abandoned by the French because more than 30,000 laborers died from yellow fever transmitted by mosquitoes. World-wide, the annual death rate from malaria has been reduced from 2.5 million in 1965 to less than 1 million today through the use of insecticides. The number of deaths resulting from all wars is paltry beside the toll taken by insect-borne diseases. Currently there is danger to humans from such diseases as encephalitis, typhus, malaria, yellow fever, African sleeping sickness, and many others.

Plants are the world's main source of food. They compete with about 80,000 plant diseases, 30,000 species of weeds the world over (1,800 cause serious economic loss), 3,000 species of nematodes that attack crop plants (more than 1,000 cause damage), and 800,000 species of insects (10,000 are pests), all of which add to the devastating losses of crops throughout the world. One-third of the world's food crops is destroyed by these pests during growth, harvesting and storage, with losses being much higher in developing countries. These losses alone could remediate world hunger. As for weeds, it has been estimated that more energy is expended on the weeding of crops than on any other single human endeavor.

It is estimated that in the U.S. alone, insects, weeds,

and plant diseases account for losses exceeding $20 billion annually in the agricultural sector. Add to that another $3.0 billion for professional pest control, 60% of which comes from the residential market.

Pest Control Methods

Pest control involving the use of chemical pesticides has been the universal approach for the last 50 years, beginning with the discovery of DDT and other synthetic pesticides in the 1940s. During those decades of intensive pesticide use many other forms of pest control were ignored or forgotten. These include cultural control, host plant resistance, physical control, biological control, legal or regulatory control, and more recently, integrated control or integrated pest management (IPM).

Cultural Control

Cultural methods deliberately alter the routine production practices in ways that are detrimental to the biological success of insects, diseases, weeds and other pests. For instance, cultivating the soil kills weeds, buries disease organisms, and exposes soil insects to adverse weather conditions, birds and other predators. Additionally, deep plowing buries some insects, killing them outright or preventing their emergence.

Crop rotation is effective against insects and diseases that are specific in their range of affected plants, and especially against insects with short migration ranges. Movement of crops to different locations kills insect pests by isolating them from their food source. Crop rotation using different plant families also prevents the transmission of disease pathogens by offering unrelated and usually tolerant species.

Planting or harvesting times can be changed to reduce disease vulnerability or keep insect pests separated from susceptible stages of the host plant. Corn and bean seed damage by seed maggots can be greatly reduced by delaying planting until the soil is warm enough to cause rapid germination. This technique also reduces root and seedling diseases that attack slow growing plants. In many instances using healthy transplants overcomes insect and disease damage more readily than growing plants from seed in the field.

Removal of crop residue and disposal of weeds eliminates disease organism sources and food and shelter for many insect pests including cutworms, webworms, white grubs and spider mites. When crops mature, prompt harvest and plowing the crop residue into the soil composts it and destroys pest reservoirs.

Host-Plant Resistance

Varieties of plants that are resistant to attacks by insects, disease organisms, nematodes, or birds are available in many crop cultivars and can be selected by the conscientious grower. Resistance in plants does not mean immunity to damage, but rather distinguishes plant varieties that exhibit less insect or disease damage when compared to other varieties. Resistant varieties may be distasteful to the pest, may not support disease organisms, or may possess certain physical or chemical properties that repel or discourage insect feeding or egg-laying. Some plants are vigorous enough to support insect or disease organisms with no appreciable damage or alteration in quality or yield.

Physical Control

Handpicking of diseased leaves and fruit, or insects and insect egg masses insures immediate and positive control. These labor-intensive methods are usually more practical for gardens than farms. This is especially effective with diseases such as anthracnose and leaf spots and foliage-eating insects as hornworms, potato beetles, or cabbage caterpillars.

Plant guards or preventive devices are easy to use against insects, and even birds, in the case of fruit trees. In the home garden, a stiff stream of water removes insects without injuring the plants or drenching the soil. Heavy mulching or plastic mulches work well in preventing weed seed germination, thus reducing most forms of cultivation.

Traps have always been popular because of the visible and swift results. Simple homemade traps work for slugs, earwigs, grasshoppers, and moths; commercial traps with special lures, as for Japanese beetles, have long been in use. Light traps, particularly black-light (ultraviolet) traps, are a good insect monitoring tool, but provide essentially no protection for gardens or field crops.

Biological Control

This is the use of parasites, predators or disease pathogens (bacteria, fungi, viruses, and nematodes) to suppress pest populations to levels low enough to avoid economic losses. There are three categories of biological control: (1) introduction of natural enemies not native to the area, and which will have to establish and perpetuate themselves; (2) expanding existing populations of natural enemies by collecting, rearing and releasing additional parasites or predators (inundative releases); and (3) conserving resident beneficial insects by the judicious use of insecticides and maintenance of alternate host insects so that the beneficials can continue to reproduce and be available when needed.

The introduction of a parasite or predator does not guarantee its success as a biological control. However, certain conditions can indicate the potential value of a natural enemy. Its effectiveness is usually dependent on (1) its ability to find or reach the host when host populations are small, (2) its ability to survive under all host-inhabited conditions, (3) its ability to utilize alternate hosts when the primary host supply runs short, (4) a high reproductive capacity and short life cycle (high biological potential), and (5) close synchrony of its life cycle with the host to have the desirable host stage available when needed.

Historically, there are many examples of predators being imported and used against known enemies. The best example is the vedalia ladybeetle imported from Australia in 1888, to feed on the cottony cushion scale on citrus in California. Weeds have also been controlled by insect release. The Tansy ragwort was reduced in importance by the release of the cinnabar moth in the Pacific coast states. The puncture vine (an abomination to bare feet and bicycle tires) in the western states was eventually controlled by the release of a seed weevil and crown weevil in 1963.

Legal Control

Unlike the other forms of control, legal or regulatory control is preventive rather than therapeutic pest control, and in the final analysis usually the most economical. More than half the insect damage in the United States is caused by species of foreign origin that have been introduced unintentionally, e.g., the codling moth and cotton boll weevil. There are also many other foreign insects, invertebrates, plant diseases, and weeds which if introduced may become as destructive as those already established. The risk of introducing them increases with the continuing growth of foreign air commerce and tourism. To prevent the introduction of foreign pests and to prevent the spread of pests already in the U.S., systems of inspection and quarantine have been developed under the control of the U.S. Department of Agriculture and departments of agriculture in several states, particularly Hawaii, Florida, California and Arizona. Some of these regulatory groups have the task of eradicating pests detected with early-warning trapping techniques. For instance, the Mediterranean fruit fly, which appears in California every year or so, requires an elaborate eradication program using traps and malathion-laced baits distributed by aircraft. There are many other pest imports that have not been so successfully removed, e.g., the giant African snail, the Formosan termite, and the Asian cockroach—all introduced into Florida.

Chemical Control

Frequently, non-chemical methods of pest control fail. At these times, when all options are exhausted, the use of chemicals may be the only solution. These chemicals are classed under the broad heading of *pesticides*. The bulk of these would be included under *insecticides* for insect pests, *herbicides* for weeds, and *fungicides* for plant pathogens. Other pesticide groups that do not carry the suffix *-icide* and may not actually kill pests, are defoliants, desiccants, disinfectants, repellents, plant growth regulators, and most recently *biorationals*. However, because they fit practically as well as legally, they are included under the umbrella word *pesticides*.

Currently, with only 2% of our population involved in agricultural production, pesticides have become an integral and indispensable part of agriculture, both here and abroad. They remain the first line of defense in pest control when crop injuries and losses become economic, and they are the only answer to a severe pest outbreak or emergency.

Biorationals are pesticides that are naturally occurring or resemble naturally occurring substances and do not disrupt the environment. They are chemicals that are biologically rational, thus when used for specific pests have no adverse effects on non-target organisms, such as beneficial insects in the case of biorational insecticides. The U.S. EPA identifies biorational pesticides as inherently different from conventional pesticides, having fundamentally different modes of action, and consequently, lower risks of adverse effects from their use.

Insect biorationals include the microbials, antibiotics, insect growth regulators, and pheromones. The microbials include the bacteria *Bacillus popilliae*, which causes "milky disease" of Japanese larvae and other grubworms in sod, and the many varieties of *Bacillus thuringiensis*, used on many crops for caterpillar and beetle larvae control. Also in this group are the nuclear polyhedrosis virus, specific against the cotton bollworm and corn earworm. Fungi have been commercially produced to control the citrus rust mite. Protozoa are sold that attack grasshoppers and crickets, and nematodes specific for control of subterranean termites. Antibiotics include the abamectins, derived from *Streptomyces avermitilus*, the same genus from which the medicinal antibiotic streptomycin is derived.

Insect growth regulators (IGRs) are the ultimate in insecticides. These are compounds that resemble closely or are identical to compounds produced by insects and plants, that alter growth and development in insects. They affect all stages of growth and metamorphosis, reproduction, behavior and diapause, and include the molting and juvenile hormones, and more recently the chitin inhibitors (chitin is the hard outer insect exoskeleton). IGRs are very selective and are harmless to warm-blooded animals.

Insect pheromones are highly specific compounds released by insects in very small quantities, that vaporize and are detected by insects of the same species, sometimes for great distances downwind. Pheromones are probably the most potent physiologically active molecules known. The sex pheromones presently offer the greatest potential for insect control, primarily in trapping. Mass trapping of males prevents mating and suppresses the population sufficiently to reduce if not avoid the need for insecticide applications. More than 150 sex lures for moths, beetles, wasps and others have been identified and synthesized over the last 10 years.

Weeds have been controlled by commercially produced fungi or mycoherbicides. For example, the strangler or milkweed vine in citris groves has been controlled by a mycoherbicide, as has the coffee weed in rice, curly indigo in soybeans, and sicklepod in soybeans and peanuts. Bacterial fungicides or bactofungicides have been developed to control damping-off seedling disease in cotton, crown gall infection on deciduous fruits, nuts, vines and ornamentals, and a third for the control of "take-all," a fungal disease of wheat that attacks the roots, causing dry rot and premature stalk death.

Integrated Pest Management

Pesticides are essential and will remain the first line of defense against pests when damage levels reach economic thresholds. However, to be completely dependent on chemical control and ignore all other methods of pest control leads to serious consequences, such as pest resistance, effects on nontarget organisms, carryover residues in soil, and groundwater contamination.

Long-range, intelligent pest control must address the management and manipulation of pests by using all applicable control methods. This combining of methods into one thoughtful strategy is *integrated pest management* (IPM), the practical manipulation of pests using any or all control methods in a sound ecological manner. IPM programs have been developed for virtually all mono-cropping systems, e.g., cotton, deciduous fruits and nuts, corn, and Christmas trees. The crops that are generally less impacted by IPM programs are vegetables. There are several reasons for this inequity, the most notable being that vegetables normally yield very high income per acre, are labor intensive resulting in high labor and cultural costs to the grower, and depend on cosmetic perfection for primary marketing and final sales appeal.

Our arsenal of pesticide tools is not keeping up with our pest problems, because of pest resistance, persistence, and environmental complications. It is essential that we continue to refine our practice of integrated pest management to preserve their periods of usefulness by using pesticides when and only when needed.

GEORGE W. WARE

For Further Reading: Mary L. Flint, *Pests of the Garden and Small Farm: A Grower's Guide to Using Less Pesticide* (1990); William Olkowski, Sheila Daar, Helga Olkowski, *Common-Sense Pest Control: Least-Toxic Solutions for Your Home, Garden, Pets and Community* (1991); George W. Ware, *Complete Guide to Pest Control—With and Without Chemicals* (1993).

See also Biological Control; Integrated Pest Management; Pesticides.

PESTICIDES

Pesticides is a collective term given to a large and diverse group of chemicals that have been developed to provide effective, economic control of a wide variety of pest organisms that, in any of a large number of ways, compete directly with man for food or fiber or threaten his health and well-being as causative agents or vectors of disease. Major categories of pesticides are named according to the organisms at which they are directed. These include insecticides (insects), fungicides (fungi), herbicides (plants), acaricides or miticides (spiders or mites), rodenticides (rodents), moluscicides (snails), and nematocides (nematodes).

The best-known pesticides are those associated with crop protection. Pesticides have become an essential adjunct to modern agriculture in the U.S. and around the world, and since the late 1940s have helped U.S. farmers reduce labor costs by about 75% and increase the production of food and fiber by about 230%. Many modern farming practices such as the use of new cultivation techniques, the use of large monocultures and new high-yielding crop varieties central to the success of the "green revolution" are made possible by the use of pesticides. While pesticides are not the only method of pest control (there are non-chemical methods available for at least some pests), they are an effective, economically efficient means of controlling the majority of agricultural pests. In the U.S. it has been estimated that, even with the use of pesticides, crop losses amount to about 30%; worldwide, pre- and post-harvest losses may be as high as 45%. Without pesticides U.S. food production would drop by at least a third, and prices would increase dramatically. It has been estimated, for example, that fruit and vegetable prices would increase by nearly 50%, bread and cereal prices by about 30% and vegetable oil prices by about 34%.

During the 1980s pesticide production in the U.S. remained fairly constant at about 1 billion pounds per year worth between $4 billion and $5 billion; world production and use has been estimated at about 4 to 5 billion pounds. In the last few years there appears to have been a slight downward trend in production, possibly resulting from the introduction of more efficacious materials. There are approximately 25,000 pesticide products sold in the U.S. incorporating a total of about 600 active ingredients. About 65% of pesticide products are herbicides, 20–25% insecticides, 10% fungicides, and the remainder other types of materials; in the 1960s the amount of insecticides used was almost double that of herbicides. About 80% of all herbicides used in the U.S. are applied to just three crops, corn (44%), soybeans (25%) and cotton (9%); fruit and vegetables constitute the major uses of fungicides (60% and 24%, respectively). Most pesticides in the U.S. (68–75%) are applied to agricultural lands (cropland and pasture), between 8% and 17% are used privately for pest control around the home and garden and the rest is used for commercial or government purposes.

The manufacture, distribution and use of agricultural chemicals in the U.S. are strictly regulated under the Federal Insecticide, Fungicide and Rodenticide Act (FIFRA) enacted in 1947 and amended in 1972, 1975, 1978, 1980 and 1988. FIFRA, administered by the EPA, requires that any pesticide registered in the U.S. must perform its intended function without causing "unreasonable adverse effect on the environment." The latter phrase is defined as meaning "any unreasonable risk to man or the environment taking into account the economic, social and environmental costs and benefits of the use of the material." FIFRA is a statute that requires the balancing of risks and benefits. While use of the term "unreasonable risk" implies that some risks will be tolerated under FIFRA, it is clearly expected that these will be outweighed by the anticipated benefits derived from the pesticide.

In registering new products pesticide manufacturers have the responsibility of providing the data necessary to demonstrate that a material will not present unreasonable risks to man or the environment. This requires the manufacturer to conduct a comprehensive battery of tests to determine acute and chronic mammalian toxicity, potential adverse effects on other non-target species (birds, fish), environmental fate and transport, etc. The likelihood that the material will leave residues in food crops or might leach into groundwater is also evaluated. The tests required to get a single new pesticide product on the market may be as high as $40 million and the process may take 6–8 years. It is becoming increasingly difficult to find new efficacious materials that meet the new and more stringent environmental- and health-related regulatory criteria. Recent estimates suggest that up to 25,000 new chemicals must be synthesized and tested in order to find a single commercial product.

The use of pesticides is not without its problems. Intense pesticide use often leads to the development of pesticide resistance and the use of non-selective materials may result in the destruction of beneficial species that themselves control other various pests; the result of the latter can lead to unanticipated pest flare-ups. Also, despite the economic advantages of chemical pesticides, there has been much public concern about their potentially adverse impacts on human health and the environment. Humans may be exposed directly during application as well as indirectly through residues in food, water and the environment and there have been fears that such residues may be causing adverse health effects. While there is no evidence that any adverse health effects result from consumer exposure to trace residues of pesticides in the U.S. food supply, a number of accidental poisonings, including deaths, are reported each year. Based on figures from U.S. poison control centers, 85,000 calls relating to pesticides were received in 1985. However, most of these were based more on concern rather than actual illness and only about 24% received any medical attention. Pesticide-related fatalities are currently decreasing in the U.S. and average about 30–40 each year; very few of these result from occupational exposures. The majority of the fatalities involve gross safety violations (improper use) or incompetence (alcohol, illiteracy) and about 20% are suicides; 30% of the fatalities are children under 10 years old. Although no accurate figures exist, it is apparent that pesticide-related illnesses and deaths in many developing countries are more prevalent. In the mid-1970s, the World Health Organization estimated there may be as many as 500,000 pesticide-related illnesses and up to 20,000 deaths per year.

Several persistent pesticides (e.g., chlorinated hydrocarbon insecticides like DDT) have undoubtedly caused serious environmental problems through their ability to move through food chains and bioaccumulate in the top carnivores (raptorial birds, fish). Other materials tend to accumulate in surface water or soil, and still others (e.g., aldicarb, a carbamate insecticide) possess properties that allow them to leach into groundwater. As a result of these and other potential problems, many of the older pesticide chemicals are being replaced by newer, much more effective (some are active at a few grams per acre) materials that are more selective for the target pest and possess physical and chemical characteristics that cause them to be rapidly degraded in the environment.

CHRIS F. WILKINSON

For Further Reading: Scott R. Baker and Chris F. Wilkinson, eds., "The Effect of Pesticides on Human Health," in *Advances in Modern Environmental Toxicology* (1990); Nancy Ragsdale and Ronald Kuhr, eds., *Pesticides: Minimizing the Risks* (1987); Graham J. Turnbull, ed., *Occupational Hazards of Pesticide Use* (1985).

See also Agriculture; Federal Insecticide, Fungicide and Rodenticide Act; Green Revolution; Pest Control.

PETROLEUM

Petroleum is a naturally occurring class of combustible fossil substances. Each type of petroleum differs from other types in composition, density, viscosity, ease of combustion, and even shade of blackness, depending on how it is formed and where it is found. Some types of petroleum are highly fluid liquids while others are so viscous and dense at ordinary temperatures that they can be walked on. What the different types have in common is that all are mixtures of hydrocarbons, compounds of carbon and hydrogen, of many different structures and molecular weights. The different types of petroleum also have in common a high degree of utility in the modern world, yielding many kinds of products: industrial, home heating, and transportation fuels; asphalts, petroleum coke, lubricants, waxes, solvents, propane, butane and petrochemicals; and other useful everyday substances. In the U.S., between 40% and 45% of all energy usage is derived from petroleum.

Although in widespread use today, petroleum has been known for many centuries and has seen limited use even in the distant past. The world *petroleum* itself, from the Greek word *petra* (rock) and the Latin word *oleum* (oil), first appeared about a thousand years ago. Asphalt (or bitumen-like forms of this substance) and "rock-oil" were known and used long before even that time by people in many parts of the world: as components of mortar, leak-proofing for boats and canoes, and waterproofing for baskets; flaming petroleum has been used as a weapon of war. As long as 5,500 years ago asphalt was used as an adhesive in Sumerian ornaments and statuary.

For the most part ancient people obtained their supplies from natural surface leaks, from tar pits, or as weathered, asphalt-like material cast up on beaches from ocean bottom seepages as on the California coast. Widespread drilling of wells for petroleum is a comparatively new activity, though the Chinese drilled for oil 2,200 years ago at depths up to 3,500 feet using bamboo poles and brass drill bits.

The modern era of drilling for crude oil, which made large quantity production possible, began much more recently; the first well was in Pennsylvania in 1859. At first, crude oil was simply distilled to produce such products as lamp oil (replacing increasingly scarce whale oil) and stove oil, in addition to lubricants, medicinal oils, and other products. The gasoline produced in those early days was an undesired byproduct, highly flammable and explosive. When internal combustion engines and the vehicles using them, and the modern assembly line, came into being, the oil industry began to grow rapidly.

Formation and Characteristics of Petroleum

As is still true, the seas of millions of years ago were full of life. Two major atomic components of all living beings are carbon and hydrogen, although many other elements are present as well. Microscopic plants and animals, as they died, sank to the bottom and mixed with the inorganic materials of the ancient seabed, more of which were continually being deposited. The resulting sediments were rich in organic materials. As the sediments deepened, pressure and temperature increased, causing chemical reactions to take place. Crude oil and natural gas under pressure were formed, stored in the pores of sedimentary rocks. The characteristics of the petroleum produced depended on the organic and inorganic components of the sediments and the temperatures and pressures to which the sediments were subjected. Petroleum deposits have been found on every continent, on most major and a number of minor islands, and on many of the continental and island shelves under the seas adjacent to these land masses. The formation of petroleum is a millions-of-years-long process that still continues, but at a rate so minute that petroleum is regarded as a *wasting resource*. Once used, once converted to carbon dioxide and water either by combustion to make energy or by slower but inevitable oxidation processes, it cannot be reconstituted and used again on a time scale of practical value to human beings. That there will someday be a worldwide shortage of petroleum is an undeniable fact, given its rapid rate of usage and the slowness of its formation.

Typical hydrocarbon molecules are made up of carbon (C) and hydrogen (H) atoms that are chemically bonded to each other. Methane is the principal component of natural gas and is the simplest hydrocarbon: one carbon atom bonded to four hydrogen atoms (Figure 1a). Each atom must be bonded to others, the carbon atoms having four bonds to be satisfied and the hydrogen atoms having one.

The molecules of *normal paraffins* (such as normal octane) have carbon skeletons that form a single chain (Figure 1b). Molecules of *isoparaffins*, in contrast, have branched carbon skeletons. Any two or more hydrocarbons with the same numbers of carbon and hydrogen atoms but different arrangements are called isomers of each other. Cycloparaffins (also called naphthenes) are molecules at least part of whose carbon skeletons are in the form of a ring.

Octene, an example of an *olefin*, has eight carbon atoms but is not an isomer of octane because it has two fewer hydrogen atoms. Double or olefinic bonds between carbon atoms are stronger (they take more energy to break completely) than single, normal paraffinic bonds. A molecule with only single bonds is said to be saturated; olefins are unsaturated.

Benzene (Figure 1c) is an example of the *aromatics*, very different in the nature of the chemical bonding from the other molecules already discussed. In the other molecules, the bonds are *covalent*, resulting from a sharing of specific electrons between the bonded at-

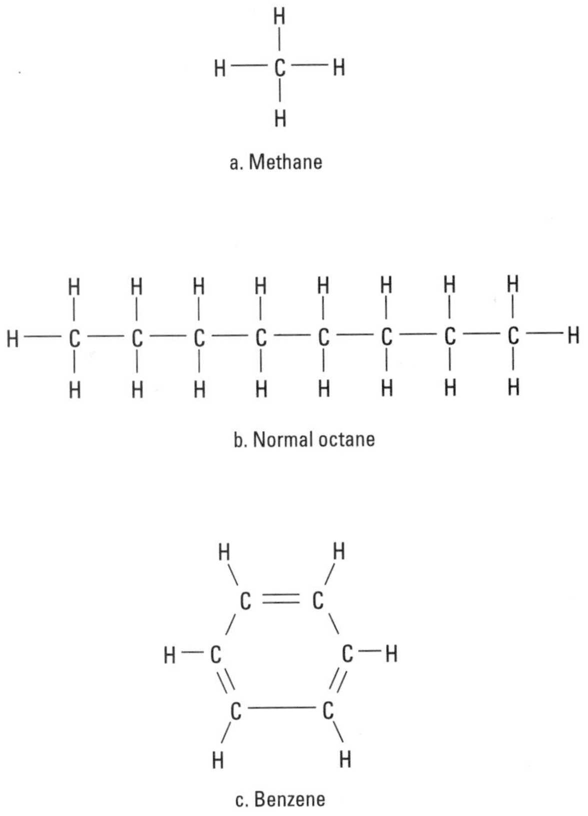

a. Methane

b. Normal octane

c. Benzene

Figure 1. Some typical hydrocarbon molecules which occur in petroleum and its products.

oms. In the aromatics the bonds are *resonant*; the bonding may be thought of as spread out around the ring, each adjacent carbon pair being bound in the same way and as strongly as any other adjacent pair. Resonant bonds are very strong and stable. Individual hydrogen atoms on the benzene ring can be supplanted by carbon atom chains (with their associated hydrogen atoms) to form many different benzene derivatives. Naphthalene is essentially a fusing of two benzene rings, a process that can be continued to form very large polycyclic aromatic molecules.

Since naturally occurring hydrocarbons can have one carbon atom or dozens, many thousands of different hydrocarbons exist in petroleum. The properties of different types of petroleum and its products follow from the properties of the hydrocarbons composing them. The larger the molecule, the higher its specific gravity, freezing point, and boiling point tend to be; these properties are critical in the handling and transportation of crude oils, in their separation into fractions by distillation, and in the performance of products. Structure also plays a role in determining both specific gravity and boiling point, isoparaffins tending

to be more dense and higher boiling than their corresponding normal paraffins, for example. A crude oil and most products do not therefore have a boiling point but, rather, have a range of boiling points, the lighter compounds vaporizing at the lowest temperatures and the heavier ones at increasingly higher temperatures. Characterizing a crude by its "boiling point curve" is very important input to the process of designing crude oil distilling columns; boiling point curves are an important characterization of products as well.

Although modern analytical methods now make it possible to analyze crude oils in great detail, it is still practical to classify them in traditional ways. For example, paraffinic crudes have more paraffins in their makeup, naphthenic crudes more naphthenes, aromatic crudes more aromatics, and so forth. Crudes are also classified in other ways; for example, as waxy, asphaltic, high (or low) in sulfur, vanadium, or other substances, or by their acidity. Crude oils are also characterized by the country, region, or even production field of origin. All of these characterizations are useful in gaining an idea of the values of crude oils and of their suitability for different methods of transportation and refining.

Taken as a whole, hydrocarbons are of low toxicity relative to many other classes of chemicals, but they do have toxic properties that can adversely impact human health and the environment. Benzene is a human carcinogen; isoparaffins can cause kidney damage (at least in animals); many of the saturated components of gasoline and other light fuels and solvents have neurotoxic properties (habitual gasoline sniffers, with their extremely heavy exposures, often suffer serious, irreversible brain damage); and polycyclic aromatics and crude oils themselves, also at heavy, prolonged exposures, are carcinogenic (at least in animals). This kind of knowledge is applied in setting regulations and in taking other precautions for the safe handling of petroleum, its products, and its byproducts.

Finding and Producing Petroleum

Petroleum, under pressure underground, escapes if it can to the surface through pores or fissures, leading to seepages and tar pits. Impervious strata in many cases prevent this escape, which accounts for the existence of underground deposits of petroleum. Petroleum collects readily in the porous strata beneath dome-shaped, impervious strata (*anticlines*), usually with a layer of compressed natural gas on top of it. Sinking a well into such a formation allows the compressed natural gas to drive the petroleum toward the surface where it can be collected. This underground-pressure-driven production is called "primary" production. Where pressure is inadequate, the petroleum must be pumped. As the natural pressure declines or the pumps fail to maintain adequate flow, other ("secondary") means to drive the petroleum out may be taken (water drives, for example).

Most petroleum deposits are deep underground and there are no surface indications. An examination and evaluation of the geology in a region of interest is necessary to determine where there are formations most likely to contain petroleum. This may consist of an examination of preexisting geologic maps, considerations of the surface geology or of fossils, mapping the local gravitational and magnetic fields (which are affected by underground features), and, ultimately and most definitively, of the interpretation of *seismic* data. In this last method sound waves, reflected from underground formations and refracted as they pass through them, are analyzed by very sophisticated computerized mathematical techniques to construct maps, sometimes three-dimensional ones, to locate structures where petroleum might be contained. The sound waves are themselves deliberately generated by such means as explosive charges or vibration devices.

Once likely formations are discovered, an exploration well must be drilled to see if, in fact, hydrocarbons are present. Drill cuttings are examined and the electrical, radioactive, and acoustic properties of the rocks are recorded; all of this information is analyzed to see if a find is likely. Drilling is itself a highly developed technology. With success, further *appraisal* wells are drilled and, if all goes well, actual production wells are drilled and connected to the gathering system. Produced crude oil must have any water and solids removed before it is transported to refineries. Transportation (for crude oils and for their products) is another complex of technologies that must allow for the characteristics of the materials to be transported. Some crude oils, for example, must be maintained hot if they are to be sufficiently fluid to be pumped, transported, and discharged into tanks.

About a third of the world's crude oil comes from under the sea—from shallow depths to depths sometimes in excess of 1,000 feet. Exploration and production under these conditions are far more complicated and difficult than most onshore operations. Specialized drilling ships, diving bells, and production platforms as tall as skyscrapers, anchored to the sea bed with only the topmost parts above sea level and capable of withstanding storms and accommodating hundreds of employees, may be needed for the deepest operations.

Petroleum Products

A refinery may be looked at as a large, complicated, stationary organism whose purpose is to ingest crude oil and natural gas liquids (liquid hydrocarbons ob-tained during the processing of natural gas); divide its feeds into processable portions (*fractions* or *streams*) by distillation; repeatedly convert, treat, separate (by distillation, flash vaporization, solvent extraction or other physical processes) these; and finally blend the resultant streams in the proportions required to form the refinery's products.

In this total process, very little actual "waste" is produced as a percentage of the total feed streams (much of what is produced is treated before being released) and some of the many substances escape into the environment. If all of these materials, products, wastes and other emissions were remixed, the mixture would not resemble the original feeds, so great is the degree to which the components have been chemically transformed. The reason for the transformations is that the different hydrocarbons do not occur in crude oil in the proportions needed to make the necessary quantities of each product. A very extensive and sophisticated understanding of the chemistry of interconverting hydrocarbons has therefore been developed and is applied in refinery conversion processes. Isomerization changes molecules into their more desirable isomers; alkylation combines smaller isoparaffins and olefins to make iso-octanes or isoheptanes (*alkylates*) for blending into gasoline or aviation fuel; cracking (thermal or catalytic) breaks large molecules into smaller ones of higher value in products such as gasoline.

Blending of intermediate refinery streams into the final products is a complicated, computer-optimized operation designed to yield products of stated, reliable performance. Most refinery products are *performance products*, blended not to produce the same chemical mixture but rather a mixture that has stated performance characteristics. In gasoline, such characteristics include antiknock, driveability, and vapor pressure characteristics; two gasolines of the same grade might have identical performance characteristics but differ significantly in their specific chemical compositions. Although there are many ways to achieve the same performance characteristics, classes of chemicals are associated with them. Normal hydrocarbons such as normal octane reduce the smokiness of jet fuel during combustion; iso-octanes and benzene improve the antiknock characteristics of gasoline. Special additives must be blended into some products to give them desired performance characteristics: Additives are blended into gasoline to keep carburetors and spark plugs clean and give better engine performance and into jet fuel to prevent static electricity buildup to avert dangerous explosions.

Petroleum and the Environment

The complex operations of the petroleum industry offer many opportunities for materials to escape into the environment. Even if all goes well—no major fires, explosions, or spills, with their major, immediate consequences, and other emissions strictly controlled—normal operations combined with the certainty of at least some human error means that at least minor amounts of hydrocarbons will be emitted into the environment. Even these, over enough time, can create environmental and human health problems.

Almost any hydrocarbons, released to the atmosphere and reacting with oxygen under the influence of ultraviolet light, produce ozone in the lower atmosphere—a known hazard to human health. Benzene is more soluble than most other hydrocarbons—which are not, as a class, readily soluble—so it is relatively easy to contaminate groundwater from small spills or leaks, over time, of gasoline or other benzene-containing products. The regulation of benzene at levels as low as one part per billion in groundwater is common.

These and many other environmental problems pose major challenges and costs for the oil industry. The key to avoiding both human and environmental damage is avoiding unnecessary exposures and minimizing necessary ones.

PAUL F. DEISLER, JR.

For Further Reading: *Energy for Planet Earth—Readings from Scientific American Magazine* (1991); National Geographic Society, *Earth '88: Changing Geographic Perspective* (1988); Jefferson W. Tester, David O. Wood, and Nancy A. Ferrari, eds., *Energy and the Environment in the 21st Century* (1991).
See also Alaskan Pipeline; Automobile Industry; Cancer; Carbon Cycle; Fuel, Fossil; Methane; Natural Gas; Neurotoxins; Oil Spill; OPEC; Petroleum Industry; Polycyclic Aromatic Hydrocarbons.

PETROLEUM INDUSTRY

The petroleum industry is the worldwide collection of business enterprises, private and state-owned, dedicated to providing petroleum and petroleum products to society. During the 20th century petroleum displaced coal as the primary energy source because of its convenience and relatively low cost. For example, a gallon of gasoline weighing about 8 pounds contains enough energy to propel a 2,000-pound automobile perhaps 20 miles and costs the consumer between one and four dollars, depending on local taxes. The extraordinary utility and availability of petroleum products empowered the spectacular growth of the economies of the industrialized world to their present levels. The petroleum industry has been the enabling mechanism

which made this phenomenal growth possible. As such, it has understandably been under public and government scrutiny as perhaps no other industry has. This intense relationship with the public and governmental policymakers is one of several distinguishing features of the industry which make it truly unique.

Over the years the composition and the dynamics of the industry have changed remarkably. It started in the United States in 1859 with Colonel Drake's discovery well in Titusville, Pennsylvania. From then until about the middle of the 20th century it was largely a collection of U.S. companies, tiny to giant, which had emerged from the ensuing scramble to find and produce petroleum. However, even as early as the turn of the century oil from Russia and Sumatra was important, and sources in the Middle East (Iran), Mexico, and Venezuela would be added to the world's petroleum inventory shortly thereafter. By the 1960s these sources, particularly those in the Middle East, far outweighed those in the U.S.

In the early days, private ownership of oil companies was the norm, and this is still true in the U.S. In Europe state ownership of the industry generally prevailed but the privately held companies continued to operate. In the former Soviet Union and other centrally planned economies, all petroleum resources were state owned. The story in the Middle East has been a changing one: Originally foreign companies with the expertise and resources to find and produce petroleum were granted concessions allowing them to act as though they owned any reserves they might find for the period of the concession. Later the host nations (and their parallels in many other parts of the world) elected to take over the ownership rights, leaving the companies to be operators for which they received a fee. Thus on the worldwide scene the power to set prices and allocate supplies gradually passed from the private companies and U.S. agencies such as the Texas Railroad Commission to the producing countries or their state-owned industrial counterparts. The two worldwide oil crises of the 1970s in which prices soared and supplies were restricted form a visible testimony to this shift in power.

While these crises sparked a traumatic period for the populace and an understandable backlash on the petroleum industry, they had one beneficial effect: Energy conservation began to be taken seriously. For example, automotive fuel efficiency and construction heat loss standards were established in the U.S. As a result petroleum consumption actually fell for a while and growth rates leveled out. The industry itself responded to the social and economic pressures to conserve energy; its own energy utilization indices dropped substantially. Unfortunately, petroleum consumption growth rates are again climbing at the same time that

U.S. production of oil is dropping, so imports are necessarily climbing. This, taken with the perennial unrest in the Middle East, may be setting the stage for another supply upset but this is by no means a certainty. For one thing the U.S. Strategic Petroleum Reserve provides a buffer not present in earlier times.

The question of the size and remaining life of petroleum reserves has always been a matter of debate in the petroleum industry. An industry consensus seems to be building that in the continental U.S., well over half of the petroleum reserves economically recoverable with forseeable technology have been found. The worldwide reserve situation is much less well-defined, but it is likely that half of the ultimate economically recoverable amount has not yet been found. How long the reserves will last is a moot question. It is more a function of the level of consumption permitted in future years than the size of the reserves. The idea held at the time of the oil crises that oil was running out does not have a strong cachet currently and perhaps never did in the industry.

A feature of the petroleum industry from the early days has been an economic driving force to cause the companies, at least those with the resources to accomplish it, to want to link and balance their petroleum producing capability against end-use markets (service stations) under their own control. This drive for integration was commonplace among the larger private companies and still holds to some extent. However, with the shift in ownership of reserves to state-owned entities, two things have happened. One is that petroleum is more of a true commodity traded openly on world financial markets. This means that all end users have equal and largely unlimited access to supplies and pay the same price for them. The other is a thrust by the state-owned companies to gain access to the end-use markets.

Petroleum is a fossil fuel formed by solar-driven biological processes millions of years ago. This means it is a finite, irreplaceable resource. It also means that finding new reserves is likely to be more costly than finding the old ones was. The new ones may be in deep sea or Arctic environments, for example, where access is very costly and strictly controlled. This situation generated a high level of merger and takeover activity; it was cheaper to acquire existing reserves than to find new ones. As a result there are perhaps half as many big oil companies as there were two decades ago. This consolidation plus a drive for cost reduction in the highly competitive environment has reduced industry employment substantially. This downsizing or "restructuring" has become a commonplace yet painful event in the private segment of the worldwide industry.

Perhaps the biggest challenge to the petroleum industry, at least in the U.S., lies in its relationship with the environment. Most end uses of petroleum involve burning it: in automobile engines, jet aircraft, diesel trucks, ships at sea, power plants, residential heating systems, and oil refineries themselves. The combustion products are mainly carbon dioxide and water, but traces of highly undesirable toxic compounds such as sulfur dioxide, nitrogen oxides, and carbon monoxide are emitted. Hydrocarbons or volatile organic compounds also are released; these may be deleterious in their own right and also can contribute to ozone formation, which leads to smog. Previously gasoline also contained tetraethyl lead to control its tendency to knock in automotive engines; gasoline combustion therefore released lead compounds which were also deemed to be a health threat. Starting in 1970 in the U.S., controls began to be placed on these emissions. These have spread to one extent or another throughout the industrialized world. As a result, emissions have dropped substantially. Lead emissions have essentially been eliminated in the U.S. Nonetheless, certain areas of the country (and world) still have less than desirable atmospheric conditions. The U.S. petroleum industry is working with governments and other industries to bring these into attainment with the new standards. Reformulation of gasoline with oxygenates or lower emission hydrocarbons is one approach. The use of cleaner-burning natural gas, a petroleum industry co-product, is another. Eventually electric powered automobiles may return as part of the solution.

The petroleum industry has been under attack for the way its own operations impinge on the environment and the safety and health of its employees. While its safety record has been one of the best of any industry, there still have been fatal explosions and fires. In addition there have been incidents where air, water, or ground pollution occurred as a result of faulty or irresponsible operations. As regulations have tightened, petroleum industry spending on health, safety, and environmental matters has grown to the point that it is a major component of end product costs. Since all companies play by the same government-set rules in a given country, essentially all these costs are inevitably passed on to the consumer who, so far, believes that their value in improving the environment offsets them.

One of the petroleum industry's most visible and emotionally charged environmental incidents, which triggers violent public reaction, is the oil spill. Tremendous quantities of petroleum routinely move about the ocean in tankers. Essentially all reaches its destination safely, but infrequently a spill occurs—far too often in the eyes of the public. Design changes in the tankers and better procedures are being implemented. The petroleum industry has put together worldwide emergency resources to diminish the impacts of these spills. Nevertheless the potential for spills can only be minimized, not eliminated, and the threat will continue so long as oil is shipped by sea.

The petroleum industry position on environmental regulation has for the most part been reactive: "We'll oppose setting the regulation or go along with it, but after it's set, we'll comply." This has been the industry attitude perceived by the public. As a result, the industry's reception in public opinion polls and the press has been dismal. Industry executives have begun to realize that this perceived attitude really hurts the industry. Furthermore and most importantly, they admit that the industry has not fully accepted its responsibilities in our environmentally conscious society. So, they are setting in place an industry-wide compliance effort. This requires each company to take real responsibility for protecting the environment so far as its operations and the use of its products are concerned. Ways to both inform and listen to the public are incorporated. This is perhaps the most positive thing occurring in the U.S. industry's relationship with the environment. It will be interesting to see how effective it is. Similar arrangements are being set up in the chemical industry and in other countries.

The real future of the petroleum industry may hang on the public reaction to a perceived threat to the global climate—the greenhouse effect. This is so named because a number of stratospheric gases such as carbon dioxide, water vapor, and methane, while being transparent to the sun's incoming radiation, absorb the infrared radiation reflected from the earth. Therefore they hold in the heat just as the glass in a greenhouse does. The fear is that major, highly disruptive climate changes may eventually result. The combustion products of petroleum (and coal) contribute many of these gases to the stratosphere, so one way to limit the effect is to control or reduce petroleum combustion. A "carbon tax" on the usage of fossil fuels has been proposed, for example, but it is too early in this debate to predict what will eventually happen. An allied thrust, which potentially affects not only the petroleum industry but all of our industrial civilization, is the thought that to preserve the planet we should operate within an envelope that defines a "sustainable economy." This is one in which the environment is in a long term harmonious equilibrium with human endeavor. Presumably achieving this would mandate restricting growth rates. Again, it is too early to predict how this will turn out but it easily could have a big impact on the petroleum industry.

JAMES F. MATHIS

For Further Reading: Anthony Sampson, *The Seven Sisters: The Great Oil Companies and the World They Shaped*

(1975); Robert Staubaugh and Daniel Yergin, eds., *Energy Future* (1979); Daniel Yergin, *The Prize: The Epic Quest for Oil, Money and Power* (1991).

pH

pH is a measure of the acidity or alkalinity of a solution. It indicates the chemical activity of hydrogen ions in a solution, on a scale ranging from pH 0 (extremely acidic) to pH 14 (extremely alkaline). Pure water has a neutral pH of 7, and the pH of most natural environments ranges from roughly pH 3 to pH 9. pH can be measured by electronic meters and by litmus paper, which indicates pH by changing color.

The pH of living cells and of body fluids must be carefully regulated. For example, the pH of human blood is about 7.4 (slightly alkaline), but if blood pH drops to 7.0, the function of the nervous system is seriously impaired. In the short term, the blood pH is regulated through respiration, which controls the concentration of carbon dioxide, and thus carbonic acid, in the blood. Organisms also adjust the acid-base balance of their blood through uptake and excretion of acid anions, such as chloride, and base cations, such as sodium or calcium.

In agriculture, soil pH affects the growth of many crops. Soil acidity usually ranges between pH 3.5 and pH 8.5. Farmers can add alkaline compounds, such as lime, or acidifying compounds, such as urea, to adjust the pH of their soils to suit their crops.

Each unit of the pH scale represents a tenfold change in acidity. Rainfall with pH of 4 is ten times more acidic than pH 5 rain, and one hundred times more acidic than rainwater with pH of 6. Pure water falling as rain has a pH of 5.6, because it contains dilute carbonic acid from atmospheric carbon dioxide. In regions with several hundred miles of major industrial or population centers, however, pollution can cause acid rain with pH of 4 or below, and extremely acidic fogs with pH between 2 and 3 have been recorded in highly polluted urban environments.

Some lakes and streams are naturally acidic, but others that are poorly neutralized—and in unpolluted environments would have pH values around 6 or 7—can be acidified to pH 5 or below by acid rain. Particularly in very acid-sensitive waters, small changes in acid loading can cause large changes in pH and the rapid disappearance of many species. An important challenge for environmental scientists is predicting the "critical loads" of acid that ecosystems can withstand without biological damage.

JAMES W. KIRCHNER

PHARMACEUTICALS

See Drugs.

PHOSPHATES

See Pollution, Water.

PHOSPHORUS

Phosphorus, one of the six most abundant elements in living matter, is an essential nutrient that must be obtained by organisms from their environment. Plants and animals use phosphorous compounds to store and transfer energy. As part of the nucleic acids DNA and RNA, phosphorous plays a critical role in transmission of genetic information and in synthesis of proteins. Along with calcium, it is also a primary component of bones and teeth.

Phosphorus is rarely found in elemental form in nature; it almost always circulates through the living and nonliving components of an ecosystem in compounds known as phosphates. The primary reservoirs of phosphorus for biological systems and human activities are phosphate rock deposits. Under natural conditions weathering, erosion, and leaching free phosphates from rocks and minerals.

Phosphatic rocks are also mined for the industrial production of fertilizers, beverages (the phosphoric acid in soft drinks), and animal dietary supplements. Phosphorus is used in both matchheads and flame-retardant clothing, and phosphates are used in detergents to enhance the ability of the organic molecules in soap to effectively remove dirt and grease.

In the phosphorus cycle, phosphates move through the living organisms and nonliving components of the land and water. Plants and photosynthetic organisms, including algae, absorb phosphates from the soil and water; animals eat the plants; and the phosphates return to the environment through the excretions of the animals and through the activities of decomposers (bacteria and fungi) from decaying plants and animals. As weathering, leaching, and erosion release the phosphorus in rocks, some of it leaves local terrestrial and aquatic ecosystems. This portion is carried by rivers and streams to oceans, where it becomes part of the sediment. Deep-sea uplifts and sea-floor changes work to restore this phosphorus to the cycle, but this restoration takes place over thousands of years. The sedimentary nature of the process and the relative unavail-ability of phosphates (depending on soil and water conditions) make phosphorus a limiting nutrient in many ecosystems; it is the nutrient most in demand and whose abundance is the critical factor in determining the growth of a biological community.

Phosphate pollution occurs when new sources of phosphorus flood an ecosystem. Runoff from fields containing phosphate fertilizer, sewage containing fecal matter and other wastes, and phosphates from detergents can all increase the amount of phosphorus in an aquatic ecosystem. These sources upset the balance of the phosphorus cycle and create conditions that overwhelmingly favor plant growth, a condition known as eutrophication. Algal blooms and other increases in plant life can reduce the amount of oxygen available to other life forms and rapidly change the character of a body of water. The largest source of phosphate pollution in the U.S. is fertilizer runoff from farms and gardens.

PHOTOSYNTHESIS

Photosynthesis is the synthesis of chemical compounds with the aid of light energy; especially the formation of a variety of organic compounds from carbon dioxide in chlorophyll-containing organisms exposed to light.

Life as we know it would not be possible without an abundance of photosynthetic organisms. Indirectly or directly, photosynthesis supplies nearly all of the chemical energy and organic compounds needed for the structures and dynamics of life. Most of the energy used to drive industrial economies also comes from the combustion of organic residues of photosynthesis preserved in coal, oil, or natural gas. Perhaps most important, the cumulative influence of photosynthesis over billions of years has fundamentally changed the chemical and physical conditions at the surface of planet Earth.

Earth's earliest atmosphere contained little or no oxygen. Significant quantities of this gas probably first appeared in the atmosphere two billion years ago when photosynthetic bacteria began to use electrons taken from water to make reduced organic compounds. As residues of these compounds accumulated in sedimentary rocks, carbon dioxide was depleted from the atmosphere and replaced with oxygen; oxidized forms of iron and sulfur also accumulated. These changes in the chemistry of the atmosphere, soils, and oceans have fundamentally altered the climate and permitted the evolution of aerobic mechanisms of respiration that are essential to all highly organized life forms.

Present-day photosynthetic organisms use very sophisticated mechanisms to capture and store the energy of sunlight. To be used in photosynthesis, a photon of light must first be absorbed by a pigment molecule, bringing electrons of the pigment molecule to an excited state (the energy of the light is used to do electromagnetic work). This excited electronic state may be transferred from molecule to molecule within groups of pigments bound to proteins, forming "antenna complexes" that function to capture light. Other protein complexes known as reaction centers "trap" excited states from associated antenna complexes and use the energy to accomplish thermodynamically unfavorable (uphill) chemical reactions. Reaction centers and antenna complexes in plants are located in membranes of chloroplasts—subcellular structures that support photosynthesis. The processes controlled by a reaction center complex are very fast (similar to the fastest electronic chips used in today's computers) and depend on very precise and complex structures. A Nobel Prize for Chemistry was awarded in 1988 to Johann Deisenhofer, Robert Huber, and Hartmunt Michel for elucidating the structure of a reaction center complex from a photosynthetic bacterium.

The chemical energy, initially derived from the electromagnetic energy of light, is finally stored in chemical bonds. For example, new carbon-carbon and carbon-hydrogen bonds are formed in making sugars from carbon dioxide, and strong hydrogen-oxygen bonds of water are broken in making oxygen gas. The chemical energy stored in the bonds of sugars and oxygen gas can be recovered by respiration at a later time, yielding carbon dioxide and water. Melvin Calvin received a Nobel Prize in Chemistry in 1961 for his work on the mechanisms used in photosynthesis to make sugars from carbon dioxide.

On a global scale most of the photosynthesis is contributed by leaves of higher plants that cloak much of the ice-free land area of the continents, and by phytoplankton, microscopic algae suspended in the upper layers of the oceans. These organisms possess elaborate structures and strongly colored chemical substances, pigments (such as chlorophyll), to absorb sunlight. The light-harvesting features of leaves and phytoplankton alter the color of the Earth's surface as seen from satellites. Analysis of data from these satellites can thus be used to estimate the density of plant cover over the entire globe. Some areas such as tropical rain forests absorb nearly all of the light that could be used for photosynthesis, while desert areas absorb virtually none. Similar contrasts occur in the oceans with highly productive areas near the margins of continents and very poor areas in the central ocean basins. Patterns of productivity on land are driven primarily by rainfall, and those in the oceans by currents that supply nutrients. Averaged over the total ice-free land area 25%–30% of the light that could be used for photosynthesis is absorbed by chlorophyll, while in the oceans only about 5%–7% is absorbed. The oceans, which account for 75% of the total surface area, account for only 40% of the total light captured for photosynthesis. The total quantity of photosynthesis is the product of the quantity of light captured by photosynthetic organisms and the efficiency with which it is used.

The maximum theoretical efficiency of conversion of the energy of red light (the most efficient color for photosynthesis) to energy stored in organic matter is 34%. However, since sunlight contains other colors of light that are used less efficiently, about 10% of its total energy can be used for photosynthesis. Yields of agricultural crops over short intervals under ideal conditions approach 7%, and typical yields over the course of a year are about half of this. The average production for the total land area (almost 0.6 kilogram of combined carbon per square meter per year), is perhaps enough for half a loaf of bread or a pint of gasoline, corresponding to about 1% efficiency. These losses of potential photosynthesis are primarily due to unfavorable climatic conditions (most frequently not enough rainfall) and inadequate soil nutrients (mostly deficits of nitrogen and phosphorus). Much of the most favorable land has already been converted to agriculture, and farming practices are generally designed to mitigate these limitations. Further improvements in agricultural production will require additional inputs of water or other scarce resources, or cultivation of new land or marine environments. It has been estimated that human societies have already preempted some 20%–25% of the total terrestrial photosynthesis. Many scientists are concerned that climate change and degradation of soils could cause photosynthesis and food production of some highly populated areas to decrease in the future. Photosynthesis at the global scale is limited, and this single process places one of the most fundamental limits on the size of the human population that the planet can sustain.

JOSEPH A. BERRY

For Further Reading: Govindjee and W. J. Coleman, "How Plants Make Oxygen," in *Scientific American* (1990); P. S. Nobel, *Physiochemical and Environmental Plant Physiology* (1991); David Walker, *Energy, Plants, and Man* (1992).

See also Algae; Atmosphere; Bacteria; Bioenergy; Carbon Cycle; Electromagnetic Spectrum; Energy; Oxygen Cycle.

PINCHOT, GIFFORD

(1865–1946), American conservationist.

Theodore Roosevelt called Pinchot his conscience on forestry matters; Franklin Delano Roosevelt included him in his braintrust; and John F. Kennedy honored him as the Father of American Conservation. As the nation's first forester and Chief of the U.S. Forest Service, Pinchot overcame stiff Congressional opposition to secure the land base for today's national forests, parks, and wilderness areas.

In 1905, upon taking charge of the federal forest reserves, he renamed them national forests to indicate that, rather than being reserved, they would be used wisely; then he tripled their acreage. Cutting no more in one year than the forests could produce in new growth, Pinchot introduced sustained-yield forestry to the nation. His goal was to show private landowners that they too could harvest trees without stripping the forest and graze livestock without destroying the range. He provided the nation with a practical alternative between no development and land abuse. In 1907, expanding on the principles of forestry, he coined the phrase "the conservation of natural resources" to recognize the interrelationships of forests, water, and minerals and the need for a unified approach to managing them. He became a legend in his day as the man who saved the forests, but at the same time wise-use management earned him the animosity of both developers and preservationists.

The Ballinger-Pinchot Controversy of 1910 epitomized Pinchot's crusade against land sharks. Congressional hearings followed his charge that Interior Secretary Richard Ballinger had approved fraudulent coal claims on a national forest in Alaska. President William Howard Taft ultimately revoked the claims, but, in the process, Pinchot, Ballinger, and Taft lost their federal jobs. The Ballinger-Pinchot Controversy established conservation, once and for all, as a powerful political issue.

Pinchot's split with the preservationists came in the Hetch Hetchy Debate, the grandfather of all environmental debates. In 1913, appearing as a star witness at Congressional hearings, he testified in favor of damming Hetch Hetchy Valley in Yosemite National Park. An advocate of "the greatest good for the greatest number over the long run," Pinchot placed the benefits of the dam—public drinking water and hydropower for San Francisco—above the wilderness value of one remote valley. His friend John Muir led the opposition. The Pinchot-Muir Split that resulted tarnished Pinchot's image as a conservation hero and created opposing camps within the movement. Vestiges of the Pinchot-Muir Split survive today in debates over the best use of old-growth forests and wetlands. Preservation-

ists continue to make a sound case for nature reserves, yet expanding human populations are refocusing attention on wise-use management as a way to perpetuate natural ecosystems in the path of development.

Pinchot's legacy also includes the foresters he recruited and educated. In 1900, when America was in need of colleges to train professional foresters, Pinchot endowed what is now the Yale School of Forestry and Environmental Studies at his alma mater. He served on the faculty, hosted summer sessions at his Pennsylvania estate, and numbered among his students Aldo Leopold, thus providing the wilderness movement with one of its most eloquent voices.

During a half century of public service, Pinchot advised U.S. Presidents from Grover Cleveland to Harry Truman. A confidant of both Roosevelts, Pinchot wrote the conservation plank in Teddy's Bullmoose Progressive platform that FDR later incorporated in the New Deal.

In the 1920s and 1930s, Pinchot combined conservation and politics as a progressive two-term governor of Pennsylvania. His make-work forestry program for the state served as a prototype for the Civilian Conservation Corps, and his Giant Power proposal to distribute low-cost electricity to rural areas prepared the way for the Tennessee Valley Authority. Ahead of his time as a public utilities advocate, Pinchot also supported federal regulation of private timberlands. His campaign for regulation pitted him against private forest industries but, at the same time, motivated many of them to adopt conservative logging practices. Frederick Weyerhaeuser was an early convert.

From the beginning of his career and graduate studies abroad, Pinchot assumed a global stance, and he came to believe that the conservation of natural resources was the only lasting basis for world peace. In 1909 Teddy Roosevelt appointed him chairman of the North American Conservation Conference, the first international meeting held on the subject. He then chaired a World Conservation Conference, scheduled for the Hague but later cancelled by President Taft. Pinchot never abandoned the idea, though, and in 1945 worked at the State Department to update his plans, which FDR carried with him to Yalta. Among Pinchot's last public services was a White House meeting with President Truman on the project. In 1949, three years after Pinchot's death, the United Nations sponsored the Scientific Conference on the Conservation and Utilization of Resources. Held at Lake Success, New York, with Pinchot's widow as a delegate, the conference signaled the start of the global conservation movement.

U.S. sites named to honor Pinchot include Mount Pinchot and Pinchot Pass, 1902, Sequoia-Kings Canyon National Park, California; Gifford Pinchot Redwood in

Muir Woods National Monument, 1945, California; Gifford Pinchot National Forest, 1949, Washington State; Gifford Pinchot State Park, 1961, Pennsylvania; and Pinchot Institute for Conservation Studies, 1963, USDA Forest Service, Milford, Pennsylvania.

BARRY WALDEN WALSH

For Further Reading: Gifford Pinchot, *Breaking New Ground*, reprint with new introduction by George T. Frampton, Jr. (1989); Gifford Pinchot, *Fishing Talk*, reprint of *Just Fishing Talk* with new introduction by Paul Schullery (1993); Barry Walden Walsh, "Natural Cycles," *Wilderness* (Winter, 1989).

PLAINS

See Grasslands.

PLANKTON

The word *plankton* (noun, singular) implies a wandering life style. The plankton is the community of aquatic organisms that have little if any control over their horizontal distribution. They are able to complete their life history with relatively little dependence on the bottom of the body of water they inhabit. The plankton is sometimes described as a floating community of open water (pelagic region), but only some of the organisms are buoyant. In contrast, some animals such as fish can control their horizontal position by oriented swimming: they are called nekton. The community resident on the bottom is the benthos. Some animals divide their activity between the bottom and the open water, operating both as plankton and benthos.

The plankton has all the elements recognized in the generalized concept of community as part of an ecosystem: primary producers (photosynthetic), consumers, parasites, decomposers or transformers (heterotrophs), all submerged in a supporting medium containing the chemical nutritive elements needed by the primary producers. Because of this structural simplicity, the plankton is an excellent source of data to develop concepts of community structure and function. Many general ecological concepts have either been originally based on studies of plankton or refined with data on plankton.

The community is usually described as composed of phytoplankton (plants) and zooplankton (animals), but that terminology is based on an obsolete division of living things into two kingdoms, plants and animals. The phytoplankton is better characterized as composed of photosynthetic organisms. It includes species belonging to several groups of algae and the blue-green bacteria, formerly called cyanophytes or blue-green algae. The phytoplankton might well be designated *photoplankton* because of their dependence on light for photosynthesis. The zooplankton includes organisms generally recognized as animals, both metazoa and protozoa. Heterotrophic (non-photosynthetic) bacteria are bacterioplankton. For some purposes the organisms are classified by size. Several such systems have been proposed. The smallest, less than one micron, are picoplankton; the largest, more than one centimeter, are macroplankton, with several categories in between.

The primary producers are in general much smaller in plankton than in terrestrial communities (algae vs. trees), shorter-lived as individuals, and are smaller than their consumers. The abundance of phytoplankton is under the simultaneous control of nutrient supply and of grazing by zooplankton. Typically in the ocean and large lakes, the populations are largest in spring: the spring bloom.

The freshwater and marine biotas are different. The most common and abundant freshwater phytoplankton belong to the blue-green bacteria and to several groups of algae, most prominent being chrysophytes (including diatoms), cryptophytes, dinoflagellates, and chlorophytes. Blue-greens are likely to dominate in alkaline freshwaters. They differ from algae in having gas vacuoles that permit them to float to the surface where they can form dense aggregations along the downwind shore. Some are able to fix molecular nitrogen. While many species of planktonic algae occur as single cells, most live in groups or colonies. Certain algae have flagella, but most depend on movements of the water.

The freshwater zooplankton is dominated by three groups of metazoa: copepods, cladocera, and rotifers. These animals exhibit several features of behavior and life history that suit them to pelagic existence. Many species of zooplankton migrate vertically, moving down in daytime and coming up during darkness. This pattern minimizes exposure to visual predators but permits access to upper layers richer in food. Under certain conditions there is reverse migration to deep water at night. Some plankton species orient away from dark areas, and swim away from shore, thus reducing loss to littoral fish. Many species have protective spines or other protrusions from the body that interfere with capture or feeding by tactile invertebrate predators. These structures may develop regularly at certain seasons or may develop only in the presence of a substance secreted by a predator.

Most freshwater zooplankton species carry the developing eggs either outside the body or within a brood chamber, preventing them from falling through the water to the bottom. The copepods are limited to biparental reproduction. The cladocera and rotifers have two modes of reproduction. During the reproductive season the population is composed of females reproducing by parthenogenesis, but is then able to switch to biparental reproduction, apparently in response to deterioration of environmental conditions. The fertilized eggs can withstand freezing and drying, and hatch only after a resting period. Many features of structure, color, and life history can be understood in terms of adaptation to the interaction of environmental conditions, food supply, and the character of planktivorous predators.

The marine biota contains more taxonomic groups than that of fresh water, and the plankton therefore has greater diversity. The phytoplankton is often dominated by diatoms and dinoflagellates, with blue-green bacteria much less prominent than in fresh water. Some marine algae (the "seaweeds") are very large, mostly attached to surfaces, but Sargasso weed floats freely.

The zooplankton is very different in the two habitats. Many of the marine forms are large, conspicuous, even spectacular, including jellyfish several meters long. More groups of crustaceans live in the open water in the oceans than in lakes, notably the euphausid and mysids, known as krill, an important food for some species of whales. Larval forms of several groups, including polychaetes and the exclusively marine echinoderms, live a temporary pelagic existence in the sea and form dense swarms in some seasons. Many marine molluscs have free-swimming larvae but few freshwater ones do.

Remembering the area of the globe covered by the ocean and lakes, we can see that the phytoplankton is responsible for a large fraction of the photosynthetic production of the world's organic matter. Plankton is vulnerable to changes in its environment brought about by human activity. One of the most widely known effects of pollution is eutrophication, an increase in the nutrient input to lakes by sewage or other nutrient-rich wastes. Typically such a lake responds by producing large populations of blue-greens which dominate the phytoplankton. Decomposition of large floating masses creates odors and other nuisance conditions, but the main biological effect is change of the food chain relations in the lake. Continued increase in atmospheric carbon dioxide and other possible global changes can be expected to affect all communities including the plankton.

W. T. EDMONDSON

For Further Reading: W. T. Edmondson, "Perspectives in Plankton Research," *Memorie dell'Istituto di Idrobiologia* 47:331–361 (1990); G. E. Hutchinson, *A Treatise on Limnology.* Vol. 2, *Introduction to Lake Biology and the Limno-* *plankton* (1967); J. E. G. Raymont, *Plankton and Productivity in the Oceans* (1980–1983); C. S. Reynolds, *The Ecology of Freshwater Phytoplankton* (1984).

PLANTS, DOMESTICATED

Plant domestication involves the evolutionary process of altering the genetic constitution of a population to suit it to the needs and environments of humans. Plant domestication is responsible, along with the use of fire, for the greatest change in the history of human development. It makes possible the feeding of the billions of people on this planet.

Domesticated plants are not necessary for human survival; hunter-gatherers survived without them for hundreds of millennia and a few persist today. Anatomically modern humans evolved without them. As important as domesticated animals are in most forms of agriculture, a single crop—wheat—produces more than three times the edible dry matter of all meat, milk, and eggs combined. The human diet on a global scale is largely vegetarian by necessity because only domesticated plants can produce the vast tonnage of food required to nourish humanity.

The earliest traces of genuinely domesticated plants so far uncovered are of beans and chile peppers in Guitarrero cave and Pachamachay cave in Peru; squash in Guilá Naquitz cave in Oaxaca, Mexico, and emmer wheat and barley in Jericho and the nearby sites of Gesher, Netiv-Hagdud, and Gilgal in the Jordan Valley. The time range in all three areas is a little before 8000 B.C. Firm evidence of plant domestication from between one and two thousand years later has been found in China, long thought by some to be the primary region of agricultural origins. The earliest traces are represented by only a few seeds or scraps of tissue recognizable as domesticated types, but plant remains become more abundant with time. In the Near East, these tenuous beginnings blossomed in less than a millennium into a full fledged agricultural complex.

Archaeological research has yielded evidence of other independent beginnings in different areas and other times, e.g., North China c. 6500–5000 B.C.; Yangtze Delta c. 5000 B.C.; South China c. 5000 B.C.; Southeast Asia and South Pacific 4000–3000 B.C.; Northeastern Mexico c. 4000 B.C.; Eastern Woodland Complex of lower Ohio–mid-Mississippi watersheds 3500–3000 B.C. Linguistic studies suggest that native agriculture is fairly old (6000–4000 B.C.) in Africa, but actual remains have not yet been found.

Ecology has provided some evidence as to sources of the most important crops. An overwhelming proportion of the major food crops come from regions of long dry seasons. These are generally of two types: Medi-

Table 1
Major food crops

Climate	Foods
Mediterranean	Wheat*, barley*, oats*, rye*, rapeseed*, pea*, chickpea, lentil, olive, almond, pistachio, etc.
Savanna	Maize*, rice*, sorghum*, cassava*, sweet potato*, some millets, bean*, peanut*, yams*, cottonseed (oil)*, etc.
Temperate woodland	Soybean*, grape*, apple*, pear, cherry, etc.
Tropical forest	Cane sugar*, banana and plantain*, orange*, other citrus, mango, taro, other aroids, many fruits and nuts, etc.
Sea coast	Coconut*, beet sugar*, cabbage, tomato
Desert	Date, watermelon
Prairie	Sunflower*, some millets
Tropical highland	Potato*, quinoa, finger millet, enset, minor tubers

*The top 25 crops in EDM.

terranean and tropical savannas. Mediterranean climates have winter rains and summer dry seasons; savannas have summer rains and winter dry seasons. Lands close to the equator tend to have two rainy seasons interrupted by short dry seasons and have contributed fewer major crops.

Estimates of world production of important crops are published yearly by FAO. They are not comparable, however, e.g., rice includes the hull; peanuts include the shell; coconuts are reported as nuts, etc. To make reasonable comparisons, the wastage and moisture content should be subtracted from raw data. The residue is edible dry matter (EDM). (The top 25 crops in EDM are indicated in the table above with an asterisk.) Today, the 10 most productive crops contribute more EDM to the world food supply than all other food crops combined.

Two major adaptations to long dry seasons are important in the evolution of wild species with the potential to become major food crops: annual habit in seed crops and subterranean storage of nutrients in roots and tubers. Dormant seeds and underground roots can survive long periods of drought and the fires that frequently burn vegetation in dry seasons.

Biosystematics has also contributed to our understanding of origins. It has helped to identify wild progenitors and determine their geographical and ecological distributions; experimental hybridization can elucidate genetic and cytogenetic relationships. Fertility of hybrids, chromosome pairing, and karyotype analyses all give clues to routes of domestication. Various electrophoretic techniques have also been employed to compare genetic constitutions. Isozyme and

DNA restriction fragment analyses are among the most powerful tools.

The major food plants have been supplemented and complemented by plant flavorings such as vanilla, chocolate, peppers, spices, and herbs of various kinds; by sweet, flavorful and succulent fruits; and by plants cultivated for their stimulant effect, e.g., coffee, tea, cacao, cola nut, chat, maté, etc.

The wild races of a number of food plants are poisonous and must be detoxified before they are safe to eat. Among the major crops, cassava is the best known. All clones contain at least some cyanogenic glycosides. The so-called sweet types contain less than the bitter ones and simple cooking may be adequate for detoxification. The bitter clones require more elaborate procedures, which were worked out by Indians of tropical America. The procedure is so safe that the early growers actually selected for increase in glycoside content because of improved quality of the cake. Many aroids have toxic loads of oxalic acid; yams contain steroid compounds, and legumes are noted for alkaloids. Even the potato may have toxic levels of solanine in some strains or under some conditions of growth. Selection under domestication usually tends toward reduction in lethal compounds.

On the other hand, toxins of various kinds can be exploited and plants have been domesticated for their production. Modern medicines are generally synthesized at least to some degree, but active principles, e.g., digitalis, cocaine, quinine, caffeine, codeine, ergotine, etc., are extracted from plants. The most popular drug of all, aspirin, is now synthesized, but the salicylic acid on which it is based was originally extracted from willow (*Salix*) bark. Most of the hallucinogenic drugs are collected in the wild, but opium and coca have domesticated races. Both have legitimate medicinal uses as well as narcotic properties.

Among fiber plants, cotton is the most important on the world scene. Three different species were domesticated: a diploid in the Afro-Indian region, and two tetraploids in the Americas—upland cotton from Mesoamerica and long staple Sea Island cotton from South America. The diploid short staple cotton now constitutes less than 2% of world production. Upland cotton of medium staple is by far the most important on a global scale. The Native Americans grew cotton to some extent, but much of their production came from harvesting untended wild and weedy races.

Other fiber plants for cloth, sacking, and cordage include flax (Mediterranean), jute (South Asia), sisal (Mesoamerica), ramie (Southeast Asia), kenaf (Afro-Indian), abaca (Philippines). The main source of natural rubber is produced from trees selected from wild populations in Amazonia. Resins, lac, gums, turpentine, etc., are derived from selected populations of trees.

Bamboos and timber trees have also been domesticated and are different from wild populations.

There is a large class of ornamental domesticates including trees and shrubs, but most especially garden flowers. These seem to be centered in three different regions. In Mesoamerica, the Native Americans exploited zinnia, marigold, salvia, fuchsia, dahlia, cosmos, tuberose, canna, and many others. The Near East and Mediterranean lands developed tulip, iris, poppy, hollyhock, carnation, roses, fox glove, delphinium, oleander, and other popular ornamentals. In China and Japan, the chrysanthemum was almost a sacred flower; other flowers valued in Southeast Asia and the South Pacific included camellia, day lily, tiger lily, primrose, rose, forsythia, flowering cherry, and many orchids.

Forage and fodder plants constitute another class. Although the Romans were well aware of the value of lupine, medicago, and vetch for soil building, most of the forage-crop domesticates are recent developments. A number of species of grasses and legumes have contributed wild material for the domestication process. Highly productive improved pastures and haylands have resulted. Grasses, often related to forage species, have been domesticated for soil conservation, lawns, parks, fairways, and putting surfaces. The process of domestication goes on continually, with additional species and races being brought onto the domestic fold.

JACK R. HARLAN

For Further Reading: C. W. Cowan and P. J. Watson, eds., *The Origins of Agriculture: An International Perspective* (1992); D. R. Harris and G. C. Hillman, eds., *Foraging and Farming: The Evolution of Plant Exploitation* (1989); N. W. Simmonds, ed., *Evolution of Crop Plants*, 2nd ed. (1976).

PLASTICS

Plastics are any of a wide variety of organic polymers, compounds produced by polymerization (joining together basic units—molecules called monomers—in a regular pattern). Plastics are synthetic polymers, but they are similar to natural polymers such as proteins and cellulose.

Preparation and Properties

Rapid technological advances have demanded concurrent advances in materials. In the area of communication, polymeric materials such as liquid crystal polymers have been used as data storage devices. Polycarbonate has infiltrated the homes of all compact disc owners and also hospitals, where it is used for medical

devices and barrier applications. Other plastics are used in automobiles, even for body panels. The many different kinds of plastics have specific properties that make them appropriate for each use, and ultimately dictate the possibility of their reuse.

Two major classes of plastics—thermoplastics and thermosets—are distinguished by their behavior when heated. Thermoplastics soften when heated and harden when cooled; they include high-density polyethylene, polypropylene, polyvinyl chloride, and nylons. Uses for thermoplastics include milk jugs, films, wire and cable coatings, construction pipe, and insulation. Thermosets are more rigid and degrade upon heating; they include phenolics, ureas, and polyurethanes. Applications include electrical connectors, appliances, and automotive parts; they are also used in plywood and varnishes.

The finished product begins as a petroleum byproduct. The feedstock of monomers is synthesized into polymers. Polymerization that proceeds by single additions to the growing chain is called polymerization or addition polymerization. A widely used monomer is ethylene, CH_2CH_2, which has a double bond between the carbon atoms. The double bond is highly unstable and reacts readily. Ethylene is used to manufacture high- and low-density polyethylene (HDPE and LDPE), acrylonitrile butadiene styrene (ABS), and polyvinyl chloride (PVC).

A second type of polymerization is condensation polymerization, in which a small molecule (usually water) is given off when two monomers join. Examples of condensation polymers are nylons and polyesters.

The characteristics of the monomer directly control the physical properties of the polymer. For instance, bulky side groups such as phenyl groups (benzene rings) in polystyrene are unable to pack closely and therefore the polymer has a lower density. Branches along the main chain such as those found in LDPE have the same effect. It is not only the type of branch or group but also their position along the length of the chain that influences the microstructure. Atactic polystyrene, for example, has phenyl groups arranged randomly and cannot be crystallized, whereas isotactic polystyrene is crystallizable since the phenyl groups are only along one side of the chain. The long chains of the polymer crystallize into forms in which the chains fold back and forth regularly in crystalline layers, making the material rigid.

The mechanical properties, specifically tensile strength and impact resistance, are directly correlatable to these microstructural features as well as to the molecular weight. The value of each property has a theoretical maximum (plateau level) that depends on molecular weight, but the actual value of the property depends on the structure of the polymer. For example, the tensile strength of LDPE is one fourth that of HDPE. However, by chlorinating the ethylene to give PVC, the tensile strength increases to six times that of LDPE. Some recently developed polymers have alternating rigid and flexible sections, which impart superior tensile strength while maintaining ease of processing.

Disposal and Recycling

The ideal polymer would cost a minimum amount to make, last as predicted by engineering calculations, and be reusable in some form after being discarded. The most significant effect on the continual increase of plastics is found in landfills. Plastics make up 2.8% by weight of municipal waste in 1970, and 7.3% in 1984; the projection for 2000 is 9.8%. Because plastics are low in density, 7% by weight converts to 30% by volume, a significant problem since the number and availability of landfills is decreasing.

Table 1
Materials composition of municipal waste (Millions of tons)

Material	1970 Tons	1970 %	1984 Tons	1984 %	2000 Tons	2000 %
Paper & Board	37	33.2	49	37.1	65	41.0
Glass	13	11.3	13	9.7	12	7.6
Metal	14	12.2	13	9.6	14	9.0
Plastics	3	2.8	10	7.3	16	9.8
Rubber & Leather	3	2.7	3	2.5	4	2.4
Textiles	2	2.1	3	2.1	4	2.2
Wood	4	3.6	5	3.8	6	3.8
Food Waste	13	11.5	11	8.1	11	6.8
Yard Waste	21	19.0	24	17.9	24	15.4
Misc. Organics	2	1.6	2	1.9	3	2.0
Totals	112	100.0	133	100.0	159	100.0

Source: *Materials Engineering*, February 1988, p. 26

Researchers are investigating ways to reduce the amount of plastic waste in landfills. One possibility that has attracted attention is making plastic that is biodegradable (able to be broken down by microbes and exposure to light and moisture). However, due to the continual addition of waste to landfills, there is little circulation of air and water to enhance degradation; the rate of biodegradation of other landfill wastes such as paper is no faster than that of plastics. Also, the technology to synthesize biodegradable polymers with properties similar to those produced by conventional techniques has not yet been developed. Another method to reduce plastic waste is incineration. Incineration has the advantage of converting waste to energy but the production of acid gases, such as hydrogen chloride, are cause for concern. Effective control of these

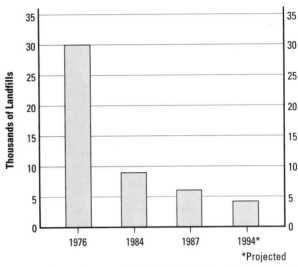

Figure 1. Operating municipal landfills from 1976 to 1994 based on current plastic waste production. Source: Forester 1988 and Environmental Protection Agency 1989.

materials requires removal of the gas before it can form an acid corrosive to the metallic interior.

The two most attractive alternatives to landfills, biodegradation, and incineration are recycling and source reduction. The latter has already taken effect at many major firms. For example, detergents in concentrate form are supplied in HDPE pouches that use only 15% as much HDPE as blow-molded detergent bottles. Despite this obvious advantage and the appeal to the environmentally conscious, one of the greatest problems of both source reduction and recycling is convincing consumers that the products will perform as well as previous products.

Table 2
Economics of plastic recycling

Cost item	Cents/lb
Raw material cost	0
Collection of plastic	5–6
Sorting by major plastic type (i.e., PET, HDPE, water, orange juice and milk bottles, HDPE detergent bottles, HDPE base cups and HDPE bleach bottles)	2–3
Grinding	4–6
Washing and drying	6–10
Recycler cost	17–25
Recycler profit	3–6
Market price for resin flake	20–31
Pelletizing	8
Delivery in HC, EOR[1]	2
Market cost for nonrefortified pellets	30–41

Source: *Hydrocarbon Processing*, May 1990, p. 89

Table 3
Plastics available for recycling from packaging (Millions of pounds)

Type of Resin	1987	1992	1997	2002
LDPE		1,892	3,755	4,990
HDPE	40	1,780	3,526	4,686
PP		516	1,024	1,362
PS		632	1,250	1,663
PET	110	402	796	1,060
PVC		289	568	758
Totals	150	5,511	10,919	14,519

Source: *Materials Engineering*, February 1988, p. 26

Costs must be reduced to make recycled plastic appealing to both the supplier and consumer. Selling costs in October of 1992 for recycled HDPE pellets were 30-37¢/lb which were comparable to virgin materials. The esthetic problems of color and dirt must be eliminated. Alternative markets must be developed since most recycled plastics retain only 70% to 80% of the virgin material's properties. Plastic manufacturers are working with both consumers and recycling firms to eliminate these difficulties, making recycling both cost-efficient and accessible.

JILL MINICK
ABDELSAMIE MOET

See also Carbon; Green Consumerism and Marketing; Incineration; Petroleum; Recycling; Vinyl Chloride; Waste Management, Municipal Solid.

PLUTONIUM

Plutonium is a silvery-white, radioactive metallic element that is produced artificially from uranium in nuclear reactors. Minute quantities of plutonium, which is heavier than uranium, have also been found in some uranium ores. Plutonium was first identified in 1940 as the isotope (a different form of an element with similar properties but different atomic weight) Pu-238 when it was produced by the bombardment of uranium 238 with deuteron (a form of hydrogen). Thirteen isotopes of plutonium are now known, the most stable of which is Pu-244. Plutonium 239 (with a half-life of 24,360 years) is most commonly produced in reactors. Plutonium is considered the most important element derived from uranium fission because of its uses as fuel in some reactors and as a key ingredient in nuclear weapons.

The fissionable isotope Pu-239 is easily produced in large quantities by bombarding uranium 238 with neutrons in breeder reactors (so named because the reactor breeds a heavier element, plutonium, from uranium).

Different reactors produce differing quantities of plutonium, but whatever the reactor type, plutonium production is always inevitable.

Of the three basic fuels for nuclear reactors, including U-235 and U-233, plutonium is the most suitable for fast reactors, usually as the compound plutonium oxide. Used plutonium can be reprocessed and stored for future use or it may undergo thermal recycling, which greatly increases efficiency of uranium fuel.

Small accumulations of plutonium in the body can result in radiation-induced cancer. The limit of absorption by humans without significant injury is about 0.13 micrograms. Because of plutonium's high absorption by bone marrow, it can cause bone lesions similar to radium poisoning, although plutonium is considered much more toxic and its effects much longer lasting. Since it is highly radioactive, extensive and costly safety measures have been employed where plutonium is fabricated and handled. As the use of plutonium as fuel has increased, its environmental distribution and the possibility of harmful effects has become very significant. Some authorities claim that there has been a fourfold increase of plutonium production in the U.S. alone since the 1970s. However, current levels in the environment are considered far from hazardous to human populations.

A much greater threat recognized with the production of plutonium involves its use as an important component in atomic weapons. Only about ten kilograms of plutonium are necessary to make a powerful nuclear weapon. Controlling plutonium production and accounting for how much is produced has become a major international concern. Over one hundred nations are parties to the Nuclear Non-Proliferation Treaty, which is designed to prevent the spread of nuclear weapons.

POLLUTION

Definitions of pollution reflect the perspective of those who are asked to define it. To the economist, pollution is the unwanted portion of production; to the industrialist, it is the stream of matter and energy that can not return a profit; to the environmentalist, it is the effluent from human activity that damages the health of humans and ecosystems; to the regulator, pollution is anything that the law says it is; and to many, including some politicians, pollution is anything that they wish were not around, regardless of whether it is created by human or natural activity. To an optimist, a pollutant is simply a resource that is out of place. Pollution, often described in terms of materials such as chemicals, radioactive elements, solid or biological waste, can also include unwanted quantities of thermal energy, light, and sound.

While pollution is usually thought of as the introduction of toxic or exotic substances into the environment, it may also involve the addition of natural substances in unaccustomed quantities, such as phosphates to water or carbon dioxide to the air. More controversial is how to describe so called "natural pollutants," such as volcanic ash or the organic emissions from trees. It is probably preferable to view these substances as part of the natural background that may influence either human or natural environments.

It has become customary both in description and legal treatment to divide pollution into categories according to type and the medium into which the pollution is released: air, land, or water. When a contaminant such as toxic mercury is released directly, it is referred to as a primary pollutant. But not all pollutants reach environments in this way. For example, the noxious smog component ozone, which damages the lungs, causes eyes to smart, harms plants and destroys materials, is generated largely by chemical reactions between nitrogen oxides from automobile exhaust and the carbon-containing chemicals that vaporize from gas tanks, dry cleaners, and even trees. These secondary pollutants are also regulated under air quality legislation, but determining the best means of control is very difficult both scientifically and politically. Increasingly, it is becoming clear that some of our strategies simply move pollution from one medium to another.

Water Pollution

The major efforts to address pollution have traditionally centered upon those forms that threaten human health. In particular, enormous amounts of money have been spent to separate humans from their own and other biological waste to avoid the spread of infectious disease. During the 19th and early 20th centuries, elaborate sewer systems were constructed in Europe and North America to transport human biological waste from homes and work places to rivers, lakes, and oceans. Before the Clean Water Act of 1970 was passed in the United States it is estimated that half of the nation's population dumped its sewage raw into various receiving waters adding only chlorine to kill the bacteria.

When population numbers were small, "piping pollution away" worked reasonably well, as natural bacteria in streams combined dissolved oxygen with solid and liquid organic waste and converted it to harmless gases. But as the quantities increased, these processes were overwhelmed, the oxygen was used up, and the public began to object to turning water bodies into foul smelling, open sewers that killed wildlife and fish.

Thus began one of the greatest public works programs in history. Sewage treatment facilities were built that first removed solids through primary treatment, and then utilized domesticated bacteria to remove much of the remaining dissolved organic material by converting it to carbon dioxide and water. The treated effluent was then chlorinated and returned to a nearby body of water.

Unfortunately this process did not solve all of the problems. First, by removing the solids, a land disposal problem was created. Little attention was given to the site where disposal occurred, and rainwater often leached organic material and nutrients into the groundwater or in some cases into the water body that was supposed to be protected. The addition of treated sewage effluent to many bodies of water such as Lake Erie was found to produce huge mats of algae that often died and decomposed consuming large amounts of oxygen and producing major fish kills. Unfortunately the sewage treatment process removed very little of the essential nutrients such as nitrogen and phosphorus, which fertilized the growth of aquatic plants.

It has taken a very long time to fully appreciate the basic fact that despite appearances, biological waste does not simply disappear through biodegradation when added to our lakes and streams or even when it is treated—it degrades a wide range of ecosystems. Modern facilities now convert sewage sludge into a productive soil conditioner or fertilizer. Artificial wetlands have also been developed in which aquatic plants and bacteria utilize the nutrients from sewage and release mostly carbon dioxide to the air while producing water and soil.

Air Pollution

Air pollution takes several forms that cause local, regional, and even global problems. Most air pollution results from the burning of either carbon based fuels such as coal, oil, or natural gas, or by burning solid and liquid waste to reduce their volume. Other pollutants evaporate into the air from such diverse sources as chemical manufacturing, paint and industrial solvents, fresh asphalt, and cooking grease from homes and restaurants. Emissions of terpenes and other organic compounds from trees add additional volatile organic compounds (VOCs) to the air.

Carbon monoxide, a primary air pollutant that blocks hemoglobin in the blood from carrying oxygen, is released when fossil fuels or wood are burned without sufficient air. Non-combustible material contained in fuel and incompletely burned carbon soot are released as particulates and reduce visibility, soil clothing and other materials, and can carry cancer-causing chemicals into people's lungs. Finally, nitrogen oxides are produced during combustion from the two dominant components of air. These oxides are transformed chemically and by sunlight into brown smog, common in California, Mexico City, and other sunny regions. Some oxides of nitrogen react with VOCs from both technological and natural sources to produce secondary pollutants, such as ozone. Some nitrogen oxides also are converted to nitric acid and along with sulfur oxides generated from impurities in coal and oil, contribute to acid rain that falls hundreds of miles from their source. In fact, one strategy for keeping sulfur oxide emissions low near their source has been to build tall stacks to dilute and disperse them, helping to turn a local air pollution problem into a regional one.

It is generally assumed that when something is burned, it disappears and leaves only a small amount of ash. Yet if combustion is complete, every carbon atom is converted into a carbon dioxide molecule. This gas is present in only trace amounts in the atmosphere, but the burning of fossil fuels and forests has increased its concentration by 27% since preindustrial times, and it continues to increase at a rate of 0.4% per year. While not usually classified as a pollutant, concern that carbon dioxide increases might raise global temperatures has led to an international treaty that establishes a goal of eventually stabilizing atmospheric levels.

Solid and Chemical Waste Pollution

It is estimated that the average American throws away about four pounds of trash and garbage per day and that industry produces more than one ton of hazardous chemical waste per person per year. Radioactive waste is generated from the electric power sector, medicine, and industry. Traditionally the goal was to make this material disappear. When quantities were small, it was possible to bury trash and garbage, and bacteria would use it as food, converting waste into water and gaseous carbon dioxide and methane. Now the quantities generated are so great and often contaminated with toxic materials that our land system can no longer function as a disposal site.

Attempts to incinerate waste have had mixed success. Since combustion of waste is never complete, a large number of organic compounds may be driven into the atmosphere along with particles of ash and soot that degrade air quality in terms of visibility, human health and ecosystem damage. New, sophisticated incinerators when properly operating, keep air pollutants low and extract useful thermal energy. Some hazardous substances, including toxic metals that don't burn, as well as some organic chemicals, are injected into deep wells. It is argued that this is a safe practice, but a complete knowledge of the geology of the region is never possible, and there have been cases of ground-water contamination, well blowouts, and minor earth tremors. Radioactive waste has proven especially troublesome because of the long lifetime of many dangerous isotopes.

Trying to make solid and liquid waste disappear often results in replacing land contamination with water or air pollution. The trend is towards preventing waste before it is created by using less material or less toxic substances in the first place, or to find a way to recycle or reuse substances now considered to be waste. In many parts of the world, such as the former Soviet Union and in many developing countries, severe damage to human health and environments from toxic and radioactive waste has already occurred and is continuing.

Energy as Pollution

Of all forms of pollution, unwanted energy is clearly a resource out of place. In sheer quantity, the so called "waste heat" from the generation of electricity by thermal, fossil fuel, and nuclear power plants is by far the greatest. Approximately two-thirds of the heat content of the fuel that is used to generate electricity is released into the immediate environment. This is not just a consequence of poor engineering, but is rather a requirement that a large amount of heat must be dumped into a low temperature environment in order to convert some thermal energy into useful work. Waste heat harms environments in two ways: First, by raising the temperature of a river, lake, or even a bay out of the range that can support many of the region's organisms; and second, by decreasing the amount of oxygen dissolved in water, thereby reducing the number of organisms that can live there. Higher temperatures can also hasten the biodegradation of sewage and other natural and waste substances, further lowering the amount of dissolved oxygen. One strategy has been to build huge, expensive cooling towers that dissipate heat by evaporating huge quantities of water. A more productive solution has been to find industrial uses for the waste heat or to use it to supply heat, hot water, and even air conditioning for buildings.

Light pollution of the night sky is a measure of the inefficiency with which light reaches its target: streets, sidewalks, and commercial areas. In some cities, half of the light is wasted to the sky. While all of us are deprived of the beauty of the night sky, one profession (visual astronomy) may be put out of business. The light pollution in southern California is so great that two of the world's largest telescopes are less and less useful in studying the universe. More effective outdoor lighting would not only help to solve the light pollution problem, but would also reduce the amount of electricity needed to light cities.

Finally, noise pollution is an increasing problem in urbanized society. Machines make noise as their parts move and rub against one another, or in the case of vehicle engines as fuel and air combine explosively. Traditionally, the solution has been to try to muffle the sound, but recent design improvements, e.g., in tires, can reduce both noise and energy waste. Use of electric vehicles eliminates engine noise; in fact, some have worried that electric vehicles might be too silent for safe operation.

New Approaches to Dealing with Pollution

As the above examples show, there is no simple way to deal with pollution. Some waste is generated at every stage of technological transformation. Increasingly, however, people are becoming cleverer at figuring out how to use what is now waste as an input to another industrial process. Much waste utilization takes place within production facilities where certain chemicals or solvents are utilized. Currently, the largest component of "recycled" paper never leaves the paper mill but consists of scraps and unused fiber. Industrial systems that analyze the entire product cycle, from resource extraction to product use and disposal, can help close many leakages of materials and energy. This "industrial ecology" approach is gaining ground as the most effective way to prevent pollution before it occurs.

For example, dry cleaners can now purchase a machine that each night redistills the toxic dry cleaning solvent, leaving only a small residue for disposal. Reductions in cleaning solvent use of greater than 99% have been reported. Energy efficiency gains in appliances and in industry have nearly stabilized energy growth over the past 20 years in the United States and have produced actual declines in other industrial countries. Chemical manufacturers have replaced some toxic chemicals with less hazardous ones or have redesigned processes to produce fewer wastes. Safer substitutes have been found for hazardous substances such as lead in gasoline or asbestos in vehicle brakes and steam pipe insulation. While there have been costs associated with these changes, in many cases, companies and utilities have reported that they have been able to provide the same service or product at a lower price. Eliminating aerosol spray cans that used stratospheric ozone depleting chlorofluorocarbons (CFCs) has created new markets for less costly product delivery systems. Reducing unnecessary packaging is another effective way to reduce waste before it is generated. High temperature manufacturing processes, such as steel making and cement manufacturing, may, if well designed, convert hazardous waste into new raw materials for new products. Even as new and better

ways are found to prevent most pollution in future products, it is still necessary to find ways to handle existing wastes and those that remain in an environmentally sound fashion.

Pollution has attracted the attention of many nations which not only address it within their own country, but set global standards as well. During the past 20 years treaties have been put into place to control ocean dumping of oil and other wastes, regulate the trade in hazardous chemical waste, eliminate CFCs by 1996, reduce transboundary air pollution in Europe and North America, and control the pollution of international rivers and regional seas. In 1992 more than 150 nations (a near record) agreed to work towards stabilizing emissions and eventually the concentration of carbon dioxide in the atmosphere.

Although pollution, like death and taxes, will probably always be with us, it is clear that human societies can live materially comfortable lives in a cleaner and more productive total environment by redesigning industrial processes and altering consumption and recycling patterns. Perhaps all nations may adopt the Japanese definition of "waste" which is reserved only for those things that have no conceivable value or use.

WILLIAM R. MOOMAW

For Further Reading: Francis Cairncross, *Costing the Planet* (1992); Barry Commoner, *The Closing Circle* (1971); Robert Socolow, ed., *Industrial Ecology and Global Change* (1993).

POLLUTION, AIR

The study of air pollution is concerned with potentially harmful substances that are released into the air: their sources, dispersion, transport, and effects on people and other parts of ecosystems. Air pollutants range from chemicals that are naturally occurring and enhanced by human activity to other substances that are uniquely the result of modern technology. Air pollutants can be gases, liquid droplets, solid particles, or fibers. The zone of concern ranges from the immediate proximity of the source of the pollutant (the workplace or home) to regionally distributed pollutants to a few instances where the concern is of a global nature.

Sources of Air Pollution

Air pollution can arise from many different sources. Naturally occurring materials in air that may, at sufficient concentration and duration of exposure, cause injury include volcanic ash, products of combustion, radon and its radioactive daughters, silica, molds, spores, pollen, bacteria, and viruses. Some substances, such as

Table 1
National ambient air quality standards

Pollutant	Sensitive population	Health effects	Averaging time	Primary standard*
Particles	Individuals with pre-existing respiratory disease	Changes in mortality in sensitive populations, increase in respiratory symptoms, reduced pulmonary function	24-hour Annual	150 μg/m^3 50 μg/m^3
Sulfur Dioxide	Asthmatics	Increased respiratory symptoms, reduced pulmonary function	24-hour Annual	0.140 ppm 0.030 ppm
Carbon Monoxide	Individuals with heart disease	Aggravation of angina pectoris	8-hour 1-hour	9 ppm 35 ppm
Nitrogen Dioxide	Young children and asthmatics Individuals with pre-existing respiratory disease	Increased pulmonary symptoms, reduced pulmonary function	Annual	0.053 ppm
Ozone	None identified	Increased respiratory symptoms, reduced pulmonary function	1-hour	0.120 ppm
Lead	Fetuses and young children	Neurobehavioral development, impaired heme synthesis	Quarterly	1.5 μg/m^3

*μg/m^3 = micrograms per cubic meter; ppm = parts per million

volatile organics arising from forest and other plants, are not of direct concern but become so as precursors of the important pollutant ozone.

Many natural pollutants are enhanced in concentration by human activities, e.g., a range of materials arising from incomplete combustion. These include carbon dioxide, carbon monoxide, carbonaceous particulate material, volatile organic compounds, oxides of nitrogen, and sulfur compounds. As early as the 13th century, there was a concern for coal smoke and odor in London. Episodes of high levels of air pollution, related principally to meteorological conditions in areas with extensive burning of coal, caused acute deaths and respiratory disease.

The burning of oil and fat contributed to air pollution. Later with the discovery and exploitation of petroleum as a source of kerosene, the problem was exacerbated. This was followed by the use of gasoline and diesel fuels in internal combustion engines. When lead was added to gasoline to enhance the octane rating and performance, airborne lead from vehicle exhaust became an air pollution issue.

Demand for electricity was met by the construction and use of coal or oil-burning power plants, major sources of pollutants. Mining ores exposed miners to silica and asbestos. Other industrial processes also produced pollution, e.g., vinyl chloride exposure in early days of the plastics industry or fluoride fumes from aluminum smelting. The World War II effort to develop nuclear weapons resulted in exposure of uranium miners to high levels of radon and daughter products, followed decades later by an excess of lung cancer deaths.

In the 1980s attention began to focus on indoor air pollutants. These included off-gassing of chemicals, such as formaldehyde from building materials and a host of volatile organics, e.g., benzene and other hydrocarbon compounds—some arising from use of consumer products and measured in trace quantities. Evidence strongly suggests that exposure of non-smokers to environmental tobacco smoke (ETS), the second-hand smoke created by smokers, can increase their risk of respiratory disease and lung cancer.

Dispersion and Transport of Air Pollutants

Some air pollutants are of concern at or near the point of release. Indeed, perhaps the major air pollution issue—cigarette smoking—occurs as a result of individuals intentionally inhaling high concentrations of toxic materials in cigarette smoke over many years. The sidestream smoke and related smoke is the "environmental tobacco smoke" other individuals are exposed to. This is a part of the indoor air pollution exposure that is a prominent part of the exposure of the general population at home or at work. Workplace exposures vary widely as to the kinds and levels of airborne material, source of the pollutant, the material itself (i.e., chemically reactive or hygroscopic [tending to absorb moisture]), the particle size, and ventilation. All of these variables influence the concentration and distribution of the pollutant. Workplace concentrations of some gases can be monitored by chemical analyzers. Particles or fibers can be collected on filters for chemical analysis or microscopy. Exposure concentrations can be modeled using computer models and validated when possible with actual measurements.

The dispersion of air pollutants farther from the source is heavily influenced by how and where the pollutant is emitted, e.g., continuous low level versus accidental releases, multiple stacks versus a few, or the height of the stacks. The nature of the local terrain, meteorology and the chemistry of the released material strongly influence the pattern of regional and, finally, global dispersion and transport. A highly reactive air pollutant, such as butadiene, will not travel far because it will react with other materials and be removed from the air. A chemically stable compound such as carbon monoxide may be transported some distance. The levels of sophistication and apparent reality of the computerized long-range transport models are rapidly improving. However, much more effort is needed to improve their validity for specific applications.

For some air pollutants, concern includes their fate in other media. Deposition on forage or soil may be important, as with fluoride and lead. Resuspension of soil lead in heavily contaminated areas may contribute significantly to the air concentration of lead. In addition, the ingestion of contaminated soil by young children may contribute significantly to their body burden of lead. Deposition of acidic constituents in the air may impact the pH of soil and ground water. These ecological effects of air pollution may be of greater concern than the direct human health consequences of air pollutants. Concern has been expressed for airborne materials serving as a major source of pollutants in large bodies of water such as the Great Lakes. This important issue is being debated, and continued research should contribute to its resolution.

Intake of Air Pollutants by People

Exposure of people to air pollutants is determined by the concentration. Some pollutants directly impact people without being inhaled, e.g., glass fibers in the workplace may deposit on the skin, causing local irritation. Ozone and other oxidant gases may cause eye irritation.

Typically, human health concern relates to air pollutants that are inhaled and deposited in the respiratory tract. The quantity deposited (a measure of dose) is influenced by the quantity in the air, its physical and chemical properties, and the individual's breathing characteristics. If two individuals inhale the same concentration of a pollutant, an exercising individual will take in a larger quantity than a sedentary individual. The individual exercising inhales a larger volume of air and pollutants and draws it deeper into the respiratory tract.

Gases may be deposited or absorbed in all compartments of the respiratory tract, dependent upon the chemical characteristics. For example, a highly reactive water-soluble gas, such as formaldehyde, is deposited largely in the nasal cavity. A less reactive gas, such as ozone, will be absorbed in the nose, but significant quantities will be deposited in the small conducting airway. Relatively non-reactive gases such as oxygen and carbon monoxide reach the alveoli, where they move into the blood of the capillaries. Some gaseous pollutants, such as vinyl chloride, benzene, and butadiene, may be absorbed from the lungs and transported via the blood to liver, bone marrow, or other organs.

The deposition of inhaled particle droplets and fibers and their subsequent fate are much more complicated and are strongly influenced by their size and shape. Fibers have a length that is at least three times the diameter. If particles are very small, less than 0.5 micrometers in diameter, they will be deposited by *diffusion* in the nose or the alveoli. This involves random motion of particles and the probability that a particle will come in contact with the surfaces of the respiratory tract. The deposition of particles larger than 0.5 micrometers in diameter is influenced most heavily by their aerodynamic characteristics. They may deposit by *impaction* when air moves at high velocity around a bend in the conducting airways. Particles may also *sediment* (settle out) when air is relatively still, for example in the alveoli. Fibers deposit much like particles, except they also deposit significantly by *interception*, the process by which a tip of the fiber comes in contact with the airway's surface as the fiber turns a corner.

The proportion of inhaled particles and fibers deposited in the parts of the respiratory tract is a function of their physical and chemical characteristics and the individual's respiration. Large particles will be deposited primarily in the nose. Smaller particles have a high probability of depositing in the airways. Pollen particles a few micrometers in diameter can reach the bronchi and bronchioles, the smaller airways. Particles a micrometer or so in diameter have a high probability of being deposited in the pulmonary compartment.

Several things can happen to fibers or particles once deposited in the respiratory tract. Particles and fibers in the forward part of the nose may be blown out. Those deposited in the back of the nose may move to the mouth and be swallowed or expectorated. Particles or fibers deposited below the larynx may be cleared to the mouth by the action of cilia in the mucous membrane. Most inhaled particles or fibers are quickly phagocytized (eaten) by macrophages. Once within macrophages they also may be carried to the mouth by the mucous blanket lining the airways. This mucous blanket is propelled by the cilia or hair-like structures that are present on a large portion of the cells of the airways. These cilia are constantly beating, carrying the mucous blanket above them. Other phagocytes containing particles or fibers may move into the lung tissue. Some particles may reside in the lungs for a long time, hence the development of coal miners' pneumoconiosis or "black lung" disease in coal miners. Particles of fibers within the lung tissue may dissolve over time and be transported via the lymph or blood to other tissues. Other particles may be transported via lymph channels to lymph nodes, where the material may be retained for a long time. Some fibers—for example, certain kinds of asbestos—may move within the lung to the pleura on the peripheral surfaces, which are lined with mesothelial cells. This is thought to be a mechanism by which mesothelioma, a kind of cancer, is produced.

Health Responses of the Respiratory Tract to Air Pollutants

The response of the respiratory tract to inhaled air pollutants is complicated because of its complex nature and the myriad of air pollutants that can impact it. The respiratory tract contains more than 40 different types of cells that perform, in an integrated manner, many different functions. The complex responses arise as a result of molecular, biochemical, structural, and functional alterations.

In some cases the lungs are simply a portal of entry for a pollutant, e.g., carbon monoxide, which binds to hemoglobin in blood cells resulting in an impact elsewhere. In this case it may be reduced cardiac or nervous tissue function because they are especially dependent upon an adequate supply of oxygen. Other examples of air pollutants that enter via the respiratory tract and have effects elsewhere are: (1) benzene— bone marrow: aplastic anemia and leukemia; (2) carbon tetrachloride—liver: liver cell damage and liver disease; (3) lead—nervous tissue: altered nerve function and impaired learning. Effects can also extend to the developing fetus. It is well recognized that women who smoke have babies that weigh less at birth. The specific causes are not known, but are thought to be an effect of reduced oxygen supply and/or exposure to nicotine.

Depending on the air pollutant, cells may be killed or their function temporarily or permanently altered. One of the most prominent effects of inhaled pollutants is inflammation with attendant reduction in function. If it occurs in the nose, it is called rhinitis; in the bronchi, bronchitis; and in the pulmonary region, e.g., pneumonitis or pneumonia. Rhinitis may be caused by exposure to a number of pollutants, e.g., chlorine gas. Chronic bronchitis, frequently accompanied by increased susceptibility to bacterial or viral infections, is

common in heavy smokers. With continued inflammation there can be destruction of cells and, in some cases, permanent loss of structure and function. If this occurs in the alveolar portions of the lung, the alveolar walls may be destroyed and balloon-like structures result. This is called emphysema. Because the alveolar wall surface available for gas exchange is reduced, the emphysematous individual may have difficulty getting enough oxygen into the bloodstream. Heavy cigarette smoking is a major cause of emphysema, which tends to occur late in life.

Continued injury of cells can give rise to altered cells (dysplasia, bad growth) or to metaplasia (different growth) where one type of cell changes to another type. A common occurrence with high levels of exposure to some air pollutants is for the ciliated cells lining the lower airways to change to squamous epithelium, which does not have cilia and is leather-like. And finally, neoplasia (new growth) or cancer occurs. This involves the uncontrolled growth of cells with encroachment on surrounding tissue and, in some cases, destruction of adjacent normal tissue by compression and invasion. The process by which cancers arise, including those of the respiratory tract, is not yet well understood. However, it is increasingly recognized as a complex, multistep process that includes changes in the genetic constituents of the cells. A number of air pollutants cause lung cancer, based on evidence from epidemiological studies of people and/or results from controlled studies of laboratory animals.

The best-known cause of lung cancer is cigarette smoking. Lung cancer in uranium miners relates to exposure of cells of the airways to alpha particles from deposited daughter products of radon—a radioactive gas from the decay of uranium. The lifetime risk to the most highly exposed uranium miners is on the order of 1 in 10 to 1 in 100. Exposure to radon in homes is also of concern. Asbestos-exposed individuals also are at increased risk for development of lung cancer, as well as for development of mesothelioma.

Studies with laboratory animals also been used to evaluate the cancer-causing potential of various materials. This approach is perhaps the primary one available when a new material has been developed and there has been no exposure of people. Other approaches, for example, using cell cultures, can give an indication of the ability of a new material to cause mutations (changes in the genetic makeup of cells that are indicators of a material's cancer-causing potential). However, cancers can only be observed in humans or other integrated, living organisms. Debate continues over how to quantitatively extrapolate the results of animal studies typically conducted at high levels of exposure to low level exposures likely to be encountered by people. In this case, two extrapolation issues are involved: from high to low levels of exposure, and from laboratory animals to people.

The study of the effects of air pollutants has frequently focused on evaluation of pulmonary function in populations selected for differing levels of exposure. A wide range of techniques had been used. Some studies have been carried out over many years, in some cases over decades. In some heavily exposed occupational populations (e.g., coal miners and stone workers), changes are detected that are exposure-concentration-related and progress with age. In general populations, it is much more difficult to detect effects because of the lower exposures typically encountered.

Regulation of Air Pollutants

In the U.S., air pollutants are regulated by three agencies: the Occupational Health and Safety Administration (OSHA), the Mine Safety and Health Administration (MSHA), and the Environmental Protection Agency (EPA). The first two are concerned with occupational exposures and the EPA primarily with environmental exposures, although it has some authority as regards occupational exposure under the Toxic Substances Control Act.

Many responsible companies (especially the largest) that have workers potentially exposed to air pollutants maintain internal management systems in addition to those mandated by law. Using data developed from multiple sources, including epidemiological and laboratory animal studies, workplace exposure guidance values are set to avoid adverse effects in workers.

In a similar manner, OSHA sets workplace exposure limits and has drawn heavily on threshold limit values (TLV) for a number of compounds set by the American Conference of Governmental Industrial Hygienists (ACGIH). MSHA has also set limits for the exposure of miners to dust. The EPA has the central role in regulating environmental exposures to air pollutants by virtue of its authority under the Clean Air Act. The regulations are best considered as they relate to the so-called criteria pollutants, and hazardous pollutants, sometimes called "air toxics."

The six current criteria pollutants are identified in Table 1. The Clean Air Act specifies the basis for setting limits, specifically stating the need to set standards with an ample margin of safety that are protective of health in sensitive populations, such as asthmatics. For all six pollutants, substantial human data are available either from epidemiological studies of exposed populations or controlled exposure studies with human volunteers, e.g., with carbon monoxide, ozone, and oxides of nitrogen. Data from laboratory animals or studies with isolated cells and tissues provide supporting information. Of the criteria pollutants, ozone has received particular attention because a large portion of the U.S. population lives in areas where the ozone standard is regularly exceeded. The current standard is 120 parts per billion, which when continued for several hours, produces demonstrable effects in young men undergoing moderate exercise. Thus, some people say there is not a "margin of safety" as specified in the Clean Air Act. Others have noted that a change in the standard would have no positive impact because so many areas are already out of compliance and strenuous efforts are already being made to achieve compliance.

In addition to a standard protective of human health, the Clean Air Act also has a provision for setting a standard based on welfare considerations such as damage to crops or buildings. There is currently considerable debate over a standard for ozone based on reduced crop productivity. Rather than a one-hour standard as for human health, a seasonal standard might be more appropriate.

Hazardous air pollutants (sometimes called air toxics) are individual substances (e.g., 1,3-butadiene and formaldehyde) or groups of compounds (e.g., chromium compounds) regulated because of concern for their specific toxicity and their potential carcinogenic effects. Under the Clean Air Act as amended in 1990, the initial phase for control of the hazardous air pollutants is technology-based with a requirement that various source categories (such as the paper industry) use the maximum achievable control technology (MACT) within a specified period of time. In a second phase, the residual risk after installation of MACT will be examined to determine if further controls need to be applied. Implementation of the second phase will require quantitative estimates of risk potency, i.e., what level of exposure will be estimated to yield a lifetime cancer risk of one in a million. Such potency estimates are not available for many of the hazardous air pollutants listed in the Clean Air Act. This emphasizes the need for additional research.

ROGER O. MCCLELLAN, DVM

For Further Reading: American Lung Association, *Health Effects of Ambient Air Pollution* (1992); Roger O. McClellan and Rogene F. Henderson, eds., *Concepts in Inhalation Toxicology* (1989); U.S. Environmental Protection Agency, *What You Can Do to Reduce Air Pollution* (1992).
See also Acid Rain; Asbestos; Ash; Atmosphere; Cancer; Carbon Monoxide; Clean Air Act; Coal; Dose Response Relationship; Electric Utility Industry; Mining Industry; Nuclear Power, History and Technology; Occupational Safety and Health; Ozone Layer; pH; Pollution, Indoor Air; Radon; Risk; Tobacco; Toxicology; Vinyl Chloride; War and Military Activities: Environmental Effects.

POLLUTION, INDOOR AIR

Chronic exposure to air pollutants indoors can produce a variety of health effects and raise the incidence of debilitating chronic disease above background rates. The risks to any given individual are generally small, but when large proportions of the public receive elevated exposure, the overall public health impact can be appreciable.

The indoor air pollutants that are of greatest concern to health are radon and daughters, environmental tobacco smoke (ETS), asbestos, products of unvented combustion, volatile organic compounds, and lead.

Lung cancer can be produced by exposures to some of the indoor pollutants: radon daughters, ETS, asbestos, and some of the volatile organic compounds (VOCs). Yet cancer may represent only a small part of their overall adverse effects on general population health. The effects of ETS on cardiovascular mortality may be ten times as high as its effect on lung cancer mortality; its effects on respiratory disease in children, while difficult to interpret in terms of disease incidence or severity, may represent its greatest public health impact. Some of the same respiratory effects have been associated with the unvented products of combustion. It is still not clear what the contributions of the individual products, e.g., nitrogen dioxide (NO_2), nitrous acid (HONO), and increased humidity, are to the overall effects. In any case, the net effects of VOCs and unvented combustion products are less well established than those of ETS exposure, and they almost certainly have lesser overall impacts on public health.

Asbestos

Asbestos, a naturally occurring group of minerals in fibrous crystalline forms, was hailed as a miracle fiber in the earlier part of the 20th century. The fibers, having excellent mechanical and insulating properties, and being highly resistant to heat and chemical corrosion, were widely used for insulation and fireproofing, and were incorporated into composite materials such as asbestos-cement tiles and pipes, brake and clutch linings, and vinyl-asbestos floor tiles. Unfortunately, asbestos fibers also proved to have unique properties in terms of toxicity, and many workers who inhaled asbestos fibers at relatively high concentrations, generally over many years, developed asbestosis, a diffuse form of lung fibrosis; lung cancer; and/or cancer of the pleural or peritoneal membranes (mesothelioma). These chronic diseases, usually occurring decades after initial occupational exposure, are often progressive and fatal. In recent years, it has been found that maintenance and custodial workers in schools and other public buildings have been exposed to airborne fibers from friable asbestos and damaged asbestos-containing materials that they may disturb during their work activities. Many of them develop pleural plaques, which are markers of exposure, may restrict lung function, and may be precursor lesions for mesothelioma.

Reviews of worker experience and laboratory studies of animals have indicated that cancer risks increase in direct proportion to the extent of exposure. It is therefore prudent to consider that any exposure to airborne asbestos fibers produces some finite added risk of cancer. Worker exposure has long been defined in terms of the number of fibers that are longer than $5\mu m$ (micrometer) per milliliter of air (f/mL), as measured by phase-contrast optical microscopy (PCOM). These "long" fibers are a small minority of all fibers, but they are much more toxic than shorter ones.

Fiber concentrations in schools and public buildings are usually measured by transmission electron microscopy (TEM), which has a much better resolving power than PCOM. Since fibers as thin as $0.05~\mu m$ are seen by TEM, while only fibers thicker than about $0.25~\mu m$ are seen by PCOM, the TEM analyses generally show higher fiber concentrations, even when the counts are restricted to long fibers. Building concentrations are much lower than those that occurred in the industries where disease incidence was high. Historic occupational exposures were often in the range of 5–20 f/mL as measured by PCOM. The permissible exposure limit for workers in the United States was 2 f/mL between 1976 and 1986, when it was lowered to 0.2 f/mL. In 1990, the Occupational Safety and Health Administration (OSHA) proposed reducing the limit further to 0.1 f/mL.

There are few published data on fiber concentrations in background air, in schools, or in other public buildings. In 1991, the Health Effects Institute–Asbestos Research Literature Review Panel gathered all available data on airborne concentrations of asbestos fibers longer than $5~\mu m$. Based on 1,377 air samples obtained in 198 buildings, the mean concentrations were 0.00051, 0.00019, and 0.00020 f/mL in schools, residences, and public and commerical buildings.

In most situations, public concern over asbestos in buildings has focused on risks to general building occupants. However, the data indicate that their exposures are very low. Therefore, the potential risk to more highly exposed custodial and maintenance workers in such buildings should be the primary consideration in determining whether to take remedial action. Asbestos removal activities sometimes cause a considerable and persistent increase in measured levels, and it is impossible to be certain, on the basis of available data, whether removal will, in practice, reduce or increase lifetime exposures of general building occupants.

Unvented Combustion Products

Products of indoor combustion, when not vented to the outside, contribute to pollutant exposures in indoor air. Unvented cooking and heating appliances using natural gas or kerosene as fuels include gas stoves, ovens, and water heaters, as well as gas and kerosene space heaters. All of them generate a variety of nitrogen oxides (NO_x). These include nitric oxide (NO), nitrogen dioxide (NO_2), nitrous acid ($HONO$) and nitric acid (HNO_3). Commercial kerosenes often contain enough sulfur for the combustion effluent to contain sulfur dioxide (SO_2) and sulfuric acid (H_2SO_4) in appreciable concentrations. Other combustion products include water vapor (H_2O), carbon dioxide (CO_2), and carbon monoxide (CO). In some cases the added water vapor can increase humidity to the point that the growth of mold, mildew, and microorganisms is facilitated, creating a confounding factor affecting the interpretation of increased respiratory effects. When burner adjustments are improper, there can be enough generation of CO and products of incomplete combustion to cause fatalities and other serious consequences. When gas is used for cooking and baking, additional releases of water and organic vapors add further complexity to the composition of the indoor air.

A number of studies of populations living in homes with unvented combustion appliances have reported abnormalities in respiratory function and/or symptoms that were greater than those found in control populations. Yet, it is not possible, on the basis of the available data, to draw any definitive conclusions regarding adverse health effects of NO_2 or the other effluents. There have been both positive and negative findings at various levels of NO_2 exposure, with various degrees of precision in measuring actual outdoor exposure levels, and generally with gas stove use as a surrogate measure of indoor exposure. Some results are suggestive that an increase in acute respiratory illness, especially in young children, may be associated with chronic ambient exposure to NO_2 at concentrations generally found in the home. Although the extent of any such effect is small, this result is consistent with the increased bacterial infectivity found in toxicological studies at much higher concentrations.

Lead

Lead poisoning of children due to the ingestion of leaded dust and paint chips from walls and woodwork in deteriorated older buildings continues to be a major public health concern in the United States and in many other countries. A study of data from over 200,000 screening tests of children in Chicago performed between 1976 and 1980 showed that lead paint was a significant predictor of the probability of a child

having lead toxicity. The reduction in leaded gasoline sales during the period of their study reduced blood lead levels and increased the percentage of children whose elevated blood lead concentrations could be attributed to paint lead.

Evidence is accumulating that the effects of low-level lead intoxication in children, as measured in terms of neurobehavioral development and growth, are persistent in later life. Concern has also grown about the potentially toxic effects that lead accumulated in the bones may have on other organs or to an infant during pregnancy and lactation, when calcium in the bones is mobilized for fetal nourishment and milk production.

Elevated blood lead concentration has been associated with increased blood pressure in adults and an increase in left ventricular hypertrophy. The results suggest that a halving of the population mean blood level would reduce myocardial infarctions (heart attacks) by approximately 24,000 events per year, and incidence of all cardiovascular disease by over 100,000.

Volatile Organic Chemicals

A variety of household solvents, cleansers, cosmetics, "air fresheners," paints, and other consumer products release volatile organic chemicals into the indoor air. Other sources include gasoline vapors from attached garages, contaminated soil gases that diffuse through basement walls, and off-gassing from contaminated drinking water. Volatile chemicals can cause odors and a syndrome known as "building related illness" (BRI). Some of these chemicals are also established as carcinogens based on long-term exposures of humans or laboratory animals at much higher concentrations. However, at this time, the exposure conditions associated with BRI and the extent of the cancer risks at realistic indoor concentrations are poorly defined and highly speculative.

MORTON LIPPMANN

For Further Reading: Environmental Protection Agency, *Indoor Air-Reference Bibliography* (1989); Morton Lippmann, ed., *Environmental Toxicants: Human Exposures and Their Health Effects* (1992); Dade W. Moeller, *Environmental Health* (1992).

See also Asbestos; Cancer; Carbon Monoxide; Lead; Occupational Safety and Health; Paint Toxicity; Pesticides; Pollution, Air; Radon; Risk; Sick Building Syndrome; Tobacco.

POLLUTION, LAND

See Pollution.

POLLUTION, NOISE

Noise is generally categorized as unwanted sound. *Noise pollution* is defined in the EPA Report to the President and Congress (1972) as " . . . any sound . . . that may produce an undesired physiological or psychological effect in an individual . . . or group." The U.S. Congress concluded, in the Noise Control Act of 1972, that the policy was to promote an environment for all Americans that is free from noise that jeopardizes their health or welfare.

In an industrialized society, noise is produced by numerous commercial and personal activities. Major sources of commercial noise include trucks, railroad equipment, aircraft, and construction equipment. Personal sources include musical activities, stereo equipment, lawn and garden equipment, home shop tools, firearms, and some children's toys.

Sound is produced by the movement of air molecules caused by some pressure-producing force. The amplitude of the sound depends on the increase in pressure. The tone or pitch of the sound is related to the rate of change of the pressure, and thus the motion of the air molecules, with time. In acoustic terms, the

amplitude of the pressure change is referred to as the sound pressure level (SPL). The rate of pressure change with time is the sound frequency expressed in cycles per second or, more commonly, Hertz (abbreviated Hz).

Sound pressure level is a logarithmic ratio of the pressure amplitude to a reference pressure, expressed in decibels (abbreviated dB). Zero dB represents the threshold of human hearing, the lowest sound pressure a young person with normal hearing can perceive. A 3 dB increase in the SPL represents a doubling of sound energy. A 10 dB increase represents a tenfold increase; 20 dB a 100-fold increase; 30 dB a 1000-fold increase, etc. Therefore, an apparent small numerical increase in SPL represents a large increase in sound energy.

In addition to amplitude and frequency, the duration of a sound is an important parameter of noise pollution. Continuous sounds exhibit little or no amplitude variation with time. Varying sounds exhibit changes in amplitude over time. Impulsive sounds are characterized by very short duration and generally high amplitude. Intermittent sounds are continuous or varying sounds with quiet periods interspersed. Loudness is a subjective descriptor that relates to sound amplitude,

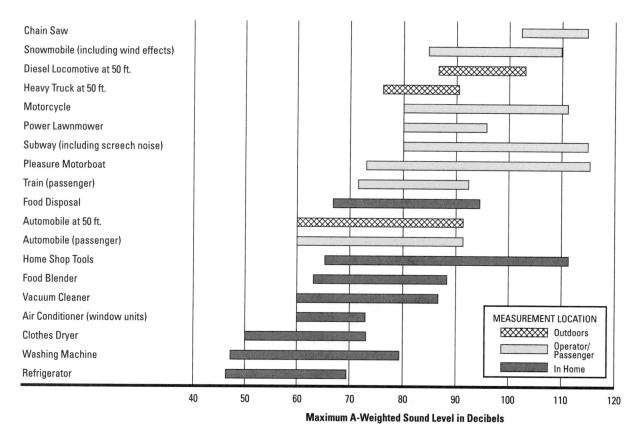

Figure 1. Typical range of common sounds. Source: Environmental Protection Agency Levels Document, reference 3.

intensity, and pressure. The ear perceives a 10 dB increase in SPL as approximately a doubling of loudness.

The SPL is measured with a sound level meter (abbreviated SLM). The SLM consists of a microphone, electronic amplifier, filters and a meter or digital display. The SLM permits measurement of sounds over a very wide dynamic range as well as impulsive sounds. Electronic filters or "weighting networks" allow the measurement of sound (or noise) within a particular frequency spectrum.

Most environmental noise situations change as a function of time. Instantaneous changes in sound field over time are obtained with a strip chart recorder. A single value time averaged sound level is obtained with an integrating SLM. Other sound measuring instruments include personal noise dosimeters and community noise analyzers which can measure, compute, store, and display comprehensive data on a noise field. SLM's contain filter networks that essentially duplicate the frequency response of the ear. This "shaping" of the received sound approximates the equal-loudness response of the ear and is referred to as *"A" weighting*; the resultant SPL is expressed in *dBA*.

Although the adverse effects of noise are not generally considered catastrophic, they are frequently cumulative. Noise can produce loss of hearing and has been found to produce non-auditory physiological and psychological effects.

Exposure to high-intensity sounds for short durations can produce a temporary loss of hearing. Prolonged exposure can result in permanent hearing loss; this is particularly true with noise in the workplace. In 1991 the U.S. Public Health Service reported that of the estimated 21 million Americans with hearing impairments, 10 million suffered hearing loss due to noise exposure.

Physiological changes in the body result from loud sounds that produce an arousal response. Adrenalin is released into the bloodstream; heart rate, blood pressure, and respiration tend to increase; peripheral blood vessels constrict; and muscles tense. On the conscious level we are alerted and prepared to take action. Even though the sound may have no relationship to danger, the body will respond automatically to certain sounds as a warning. These sound-induced physiological changes are considered a health risk in that they are generally associated with stress-related conditions such as high blood pressure, coronary disease, ulcers, colitis, and migraine headaches. Other adverse effects of noise include communication interference, performance interference, and sleep disturbance.

U.S. EPA regulations limit noise emissions from trucks, motorcycles, air compressors, and certain railroad equipment and facilities. Federal Aviation Administration regulations limit noise emissions from aircraft. The Department of Labor and the Department of Interior (Bureau of Mine Safety) have regulations that limit noise exposure in the workplace. The Department of Housing and Urban Development and the Veterans Administration have noise standards for private dwellings whose mortgages they insure. The Department of Defense has noise standards and practices for various military operations.

Some states and political subdivisions have noise ordinances that govern many private and commercial noise-producing products and activities.

Individuals can minimize personal and community noise exposure through the practice of "safe sound." This includes the purchase, use, or demand for quiet products; manufacturers generally respond to consumer demand. Hearing loss can be avoided through use of hearing protectors in high noise environments and reducing sound levels of entertainment devices. Community action groups can petition state and local governments for new or more stringent noise ordinances. Education of young children as to the adverse effects of noise and its control can be the most effective means of reducing present and future noise pollution.

KENNETH E. FEITH

For Further Reading: J. J. Earshen, "Sound Measurement: Instrumentation and Noise Descriptors," in Elliot H. Berger, James C. Ward, and Lawrence H. Royster, *Noise and Hearing Conservation Manual* (1986); Kenneth Kryter, *The Effects of Noise on Man* (1971); Alice H. Suter, *Noise and Its Effects*, Administrative Conference of the United States (1991); U.S. Environmental Protection Agency, "Information on Levels of Environmental Noise Requisite to Protect Public Health and Welfare with an Adequate Margin of Safety," EPA 550/9-74-004 (March 1974).

POLLUTION, THERMAL

The most common form of thermal pollution is the release of heated waste water into natural bodies of water by electric power plants. Other sources of thermal pollution are metal smelters, processing mills, petroleum refineries, and chemical manufacturing plants. In these plants water is used as a coolant for various industrial and manufacturing processes. Drawing water from an ocean, river, lake, or aquifer is the least expensive way to remove heat from an industrial facility. The water runs through a heat exchanger to extract excess heat from production processes and then is returned to its original source. A "thermal plume" extends outward from the point of discharge and can disrupt many processes in aquatic ecosystems, although some organisms flourish in the warmer water.

The electric power industry consumes 75% of all cooling water in the U.S. In fossil fuel plants, 60% of the energy used to produce electricity is lost as waste heat; in nuclear power plants, about 66% is waste heat. U.S. industry uses more than 225 billion liters of water annually for cooling purposes.

Since water temperatures change very slowly from day to day and season to season, aquatic organisms tend to be very sensitive to sudden changes. Also, oxygen solubility in water decreases as temperature increases, making respiration more difficult for most organisms. The spawning behavior of many fish, triggered by seasonal temperature changes, can be impaired. The type of plant growth necessary to sustain some marine ecosystems may become unavailable and thus disrupt entire food chains in an area.

Thermal plumes may appear to be beneficial to aquatic environments, allowing many birds and marine animals to find food and refuge, especially in cold weather. These artificial environments can become traps, however. For example, the manatee, an endangered marine mammal in Florida, is attracted to the abundant food supply in thermal plumes of power plants and is enticed to spend the winter much farther north than it normally would. On several occasions, a midwinter power plant breakdown has exposed a dozen or more of these rare mammals to thermal shocks that became fatal.

There are three alternative methods for releasing cooling water directly back into the source. The simplest is to construct a large, shallow pond that dissipates heat into the atmosphere. The second is to use a cooling tower, in which heated water is sprayed into the air and cooled by evaporation. The disadvantages of both methods are the large quantities of water lost to evaporation and the occasional creation of localized fog. The third and most effective method, a dry tower, which pumps water through tubes and dissipates heat into the air, is the most expensive to construct and operate and reduces the efficiency of the cooling process.

POLLUTION, VISUAL

To pollute is to render impure or unclean, to contaminate. Visual pollution encompasses human-produced elements and conditions that detract from or defile the environment as seen.

Visual pollution received national attention during the environmental movement of the 1960s and 1970s. It varies in scale from litter around a park bench or on a city street or beach, to abandoned derelict buildings and structures, to chaotic commercial strip developments and their associated billboards along highways, to auto exhaust and industrial smoke that reduce the clarity of the atmosphere and impede vision.

A notable example of large-scale visual pollution is the frequently encountered impairment of visibility at Grand Canyon National Park, caused by auto emissions from as far away as southern California and smoke from electric power generating plants in the region. Visitors are deprived of the pleasure and satisfaction of viewing clear, sharp images of the dramatic geological formations for which the Grand Canyon is world famous.

Concern with visual pollution is not a recent phenomenon. It was addressed by local civic improvement associations in the eastern U.S. early in the twentieth century. Programs were initiated and brochures published to stimulate public interest in restraining the proliferation of billboards and reducing or eliminating smoke that was responsible for sooty and discolored buildings and landscapes.

As early as 1905 legal action was initiated, albeit unsuccessfully, against billboards in Passaic, New Jersey. Nearly fifty years later, in 1954, the U.S. Supreme Court ruled that "it is within the power of the legislature to decide that the community should be beautiful as well as healthy." Nevertheless, it was the 1965 White House Conference on Natural Beauty that resulted in a national focus on the issue of visual pollution. Among the topics addressed at the conference were derelict riverfronts in cities, auto junkyards, billboards, and strip development that lined entry roads to towns and cities. Within the succeeding fifteen years, legislation was enacted to beautify highways and control billboards along the interstate highway system, to require the restoration of strip mines, and to address visual blight associated with timber harvest practices in national forests.

The National Environmental Policy Act of 1969 required that all development projects with federal funding be subject to environmental impact assessments, including assessment of visual impacts. The intent was to prevent or minimize visual pollution. Subsequently, a number of states followed the lead of the federal government and adopted similar environmental quality legislation that provided protection from negative visual impacts resulting from development projects and controlling locations and sizes of billboards.

Attention to visual pollution and action to control the many forms it can take have been slow and sporadic during the last ninety years. Review of progress to date suggests that the national leadership manifest in the 1960s and 1970s was important in focusing attention on the issue. Nevertheless, it also suggests that partnerships among the several levels of government can be the most effective in controlling and preventing visual pollution.

ERVIN H. ZUBE

POLLUTION, WATER: CASE STUDIES

Clean water is essential to human health, productive economies, and robust ecosystems. Yet all water is subject to contamination from both natural and man-made sources. Three case studies of different hydrologic systems illustrate the issues that are raised in attempting to implement pollution-control programs. Understanding the physical interconnection between groundwater and surface water is essential to mitigating the degradation of these hydrologic systems.

Danube River, Central Europe

The Danube River flows eastward from Germany's Black Forest across eight countries in western and central Europe to discharge into the Black Sea. The cumulative effects of municipal, industrial, and agricultural discharges and the complexity of attempting to take corrective actions make the Danube susceptible to periodic episodes of impaired use. The river basin occupies 817,000 square kilometers and contains 67 million people. The river is an important water course for navigation and 49 low dams have been built or are planned in order to maintain suitable water depths.

A large part of the drinking water in the basin is supplied from wells in the alluvial gravels of the river's floodplain, although a few cities draw water directly from the river. Irrigation and industrial uses also are important along the length of the river and its tributaries. Approximately 30% of the cultivatable land in Romania is irrigated with water from the Danube River basin, and plans have been made to double this amount. Irrigation return flows carry dissolved salts, nutrients, and pesticides from fields to the Danube's tributaries.

Disposal of municipal and industrial wastewater occurs along most of the 2,850-kilometer length of the river. The assimilative capacity of the river is quite large and several large cities still discharge raw sewage into it. Industrial plants, mainly pulp and paper mills, iron and steel works, chemical and petrochemical plants, and oil refineries discharge wastes into the Danube and its tributaries. The tributaries are generally more polluted with organic matter than the main stream.

The waste water discharges have created numerous problems. While dissolved oxygen is near saturation in many parts of the river, bacterial quality is seriously degraded locally below major pollution sources, such as urban areas. During the 1980s, the algal mass of the river has been increasing in response to the increasing load of nutrients from urban wastes and agricultural activities. In addition to fecal contamination problems, toxic metals are present in some places in both the water and the bottom sediments. While floods flush some of the metals into the Black Sea, the bulk are stored in the bottom sediments. The pesticide lindane was found in over 93% of water samples taken between 1979 and 1982. Aldrin, Dieldrin, and DDT occurred in over 40% of the samples, though at concentrations that never exceeded the World Health Organization standards.

The low dams along the Danube trap sediment, raise water temperatures, and increase the amount of biological activity in the nutrient-rich impoundments behind the dams. The bottom sediments have been enriched in heavy metals. Observations at the Altenworth barrage (dam) in Austria have shown little adverse impact of the dam on river quality. In fact, a slight improvement in quality was observed due to the half- to one-day retention time of the impounded reach, which improved the self-purification capacity, increased sedimentation of organic-rich particulate matter, and raised nitrification rates.

Management of water quality in an international river such as the Danube is very complex. High-quality information about the river and its quality and consistent interpretation of the information requires a high degree of cooperation among the basin states. Accidental spills frequently occur, so an international warning system is indispensable. Those dependent on the river for drinking water need to know of any new or temporary deterioration of water quality. Inventories of pollution sources, the adoption of uniform water quality standards, and integrated pollution control have yet to be instituted.

Pesticide Residues in Groundwater, San Joaquin Valley

The 500-mile-long Central Valley of California is one of the great agricultural regions of the world. The valley, lying between the Sierra Nevada Mountains on the east and the Coast Ranges on the west, is filled with sediments eroded from the surrounding mountains. The San Joaquin Valley, which occupies the southern two-thirds of the Central Valley, provides an example of the impact of irrigated agriculture on groundwater quality. Average annual precipitation in the valley ranges from 5 to 16 inches, most of which is lost to the atmosphere by evaporation and the transpiration of vegetation. Most of the floor of the San Joaquin Valley is irrigated farmland. Therefore, most of the shallow groundwater in the valley is from the infiltration of irrigation water derived from either deep groundwater or imported surface water.

About 10% of the total pesticide use in the United States occurs in the San Joaquin Valley. The leaching of pesticides from the soil surface into the groundwater depends on the amount of excess irrigation water applied to the field, the application patterns of the pesticide, soil texture (coarse-grained soils allow water to move rapidly downward to the groundwater), the organic content of the soil (organic soils absorb some pesticides), the time it takes the pesticide to decompose, and the depth to groundwater. Some 13 pesticides or products of pesticide decomposition products were detected in a recent reconnaissance of the valley's groundwaters.

Although pesticides are widely used throughout the valley, groundwater appears to be susceptible to contamination in only a few areas. This is related in part to the distribution of soil textures. The soils of the western part of the valley and in most of the valley trough are fine-grained and are less susceptible to the leaching of pesticides because of their low permeability. The soils of the eastern part of the valley are coarse-grained and have low organic contents, which permits irrigation water to transport pesticides downward. Pesticide contamination of the groundwater generally occurs in zones where the groundwater is within 100 feet of the land surface. Deeper groundwater is uncontaminated.

The Great Lakes

The Great Lakes contain 18% of the world's fresh surface water. Except for Lake Michigan, the five interconnected lakes are shared with Canada. They drain the southern part of the Province of Ontario and eight states in the U.S. The Great Lakes basin serves 46 million people, most living in the U.S. Because of the economic value of the basin, the U.S. and Canada have established the International Joint Commission of Great Lakes Activities whose mandate is derived from the International Boundary Treaty of 1909. Under the treaty the governments of Canada and the U.S. may refer a problem of mutual concern to the International Joint Commission for investigation and the development of recommendations under the premise that neither country may use the water on its side of the boundary to the detriment of the water, health, and property of the other side. Over the years referrals have resulted in studies of eutrophication in Lake Erie and Ontario and recommendations for controlling the discharge of phosphorus into the lakes. These recommendations, in turn, led to the adoption of the Water Quality Agreement of 1972. Subsequent studies of water quality in Lakes Huron and Superior and of the effects of land use on pollution resulted in a renegotiation of

the Great Lakes Water Quality Agreement in 1978, which included recommendations to address toxic substances and adopt an ecosystems approach to monitoring and managing the water quality. The document was revised in 1987 and remains in effect.

The recovery of the Great Lakes from their degraded condition in the late 1960s, when the press declared Lake Erie "dead or dying" and television news programs showed the Cuyahoga River at Cleveland burning because of gasoline on its surface, is a remarkable environmental success story. Subsequent studies, however, have shown the presence of toxic chemicals that continue to cycle through the Great Lakes ecosystem. Some of the sources of these toxic substances are thousands of miles away. Significant portions of pesticides such as DDT and toxaphene, industrial chemicals such as PCBs, and toxic metals such as mercury and cadmium are entering the Great Lakes through wet and dry precipitation from sources outside the lakes' basin. DDT is of particular interest because its use has been banned in the U.S. for many years. The influx of DDT to the Great Lakes is thought by some scientists to be derived from Central America. Because these substances are persistent and accumulate in the food chain, they pose a threat to the health of wildlife and the inhabitants of the basin who consume fish from the Great Lakes.

Toxic discharges within the basin have their roots in the industrial and agricultural developments following World War II. The Great Lakes attracted large chemical industrial complexes, and agriculture in the region boomed. Contaminants discharged into tributaries found their way into the lakes. Wildlife species at the peak of observed contamination levels during the late 1960s and early 1970s exhibited severe population distress. Prompted by human health concerns, uses of contaminants such as DDT, dieldrin, and PCBs were greatly restricted. Permit systems were instituted to manage discharges and the concentrations of many chemicals declined in sediments and in fish and wildlife tissues. Reductions, however, leveled off in the early 1980s at concentrations that still are of concern to public health authorities.

While concentrations of toxic substances in lake waters are below detection limits in most cases, the toxins in the sediments are of concern because they increase through biomagnification in the food chain. Immigrant bald eagles, for example, after two years of feeding on fish in the Great Lakes experience reproduction problems. Because the eagle is at the top of the food chain, the International Joint Commission has suggested that the eagle be used as an indicator species to measure the health of the Great Lakes ecosystem. The use of concentrations of contaminants in specific wildlife species as an indicator of ecosystem health will guide future efforts to control or eliminate the discharge of toxic substances into the Great Lakes.

As pollution is controlled at its source within the Great Lakes basin, the residual effects of contaminated sediments, waste disposal, the discharge of contaminated groundwater, and the atmospheric deposition and long-distance transport of contaminants from outside the basin will become increasingly important. Atmospheric deposition will have to be approached in a global context. All of these examples show how hydrologic processes, pollution generating activities, and human institutions interact to affect environmental quality.

DAVID W. MOODY

For Further Reading: C. Dale Becker and Duane A. Neitzel, eds., *Water Quality in North American River Systems* (1992); Michel Meybeck, Deborah Chapman, and Richard Helmer, eds., *Global Freshwater Quality, A First Assessment* (1989); Ruth Patrick, *Surface Water Quality: Have the Laws Been Successful?* (1992); U.S. Geological Survey, *National Water Summary 1990–91 — Hydrologic Events and Stream Water Quality* (1993).
See also Dams and Reservoirs; DDT; Dissolved Oxygen; Food Chains; Irrigation; Lakes and Ponds; Law, International Environmental; PCBs; Pesticides; Pollution, Water: Processes; Rivers and Streams; Toxic Metals; Water, Drinking; Watersheds.

POLLUTION, WATER: PROCESSES

Water pollution commonly means degradation of natural water quality, although any alteration of water conditions may be considered as pollution. It occurred in early civilizations, but has increased substantially since the Industrial Revolution owing to (1) increasing human population, (2) greater per capita use for hygienic, manufacturing, and luxury purposes, (3) the use of natural waters for disposal of an increasing array and amount of waste products, and (4) increasing use of natural resources.

Water is the only substance on Earth that occurs naturally in all three states: solid, liquid, and gas. In its many forms it interacts significantly with the Earth's energy balance and geologic processes. The energy balance linkage is because of water's high specific heat (the amount of energy required to raise 1 gram of water from 20°C to 21°C), its almost universal distribution, its high mobility, and its highly variable ability to reflect incoming solar (short wave) radiation.

The geologic processes linkage is because of water's ability to weather (erode) rock and soil. Its molecular structure provides the ability to dissolve both a large variety and a great quantity of substances. Because it expands upon freezing, it breaks up rock and keeps soils porous. It physically erodes and chemically dissolves rock and soil.

Too much or too little of any substance in environments may be unwelcome or even poisonous. If water is made unfit for a certain use, it is said to be *polluted* (for example, too warm for fish). If water becomes toxic, that is, causes disease or death, then it is said to be *contaminated* (for example, carries typhoid fever). Polluted water may not be contaminated; but contaminated water is always polluted. Water may be polluted by nonhuman agents such as beaver, for example, that alter a variety of aquatic conditions by changing the water level and by defecation or urination that pollutes the water so we cannot use it. Thus, pollution must be defined in terms of its use. To be considered *potable*, drinking water must be free of contaminants and contain neither too small nor too great an amount of dissolved substances to suit people's taste. Water for some manufacturing processes must be almost pure, without any dissolved salts, and often with certain specified properties. Anything that alters these desired properties is a pollutant.

How Pollution Is Measured

Water quality standards and pollution are generally measured in terms of concentrations or load.

Concentration is the rate of occurrence of a substance in aqueous solution (weight/volume). The units mg/l (milligrams per liter) or ppm (parts per million) are the same only at standard pressure and temperature, where one cubic centimeter (one thousandth of a liter) of water weighs exactly one gram. A concentration of 150 ppm means that there is 150 pounds of that substance in 1 million pounds (about 120,000 gallons) of water. For all but the most exacting laboratory research and litigation, mg/l and ppm are close enough that either may be used. The total load of a pollutant in a stream is the sum of all the material delivered over a certain period of time (weight/time). Total load may be calculated by multiplying the average flow rate for the period (volume per unit time) times the concentration (weight/volume).

How, when, and where the constituents are sampled is important because concentration varies with hydrologic conditions, especially the amount of water in storage, stream discharge (runoff), and time since the last runoff-causing event. Turbidity or the total suspended solids in the water, for example, is usually highest just prior to peak storm runoff when streams flush the bed, banks, and connected wetlands, lakes, or ponds. Total Dissolved Solids (TDS) are probably lowest just as storm runoff commences when the materials are diluted by relatively "pure" rain or runoff water. Water pollution must be reported in context of the hydrology. To fail to do so renders the water quality characterization incomplete at best, and often erroneous, misleading, or useless.

Water quality analysis can be accomplished in a laboratory approved by the Environmental Protection Agency (EPA) or with EPA-approved field kits that utilize standard methods and properly prepared reagents. Field analyses permit better appraisal of the ecological and hydrologic setting in which the measurements are made.

Properties of Water in the Natural Environment

Natural characteristics of water quality are physical, chemical, or biological as shown in Table 1. The following are the chief characteristics of water, each of which can be subject to pollution.

Table 1
General water quality characteristics

Characteristic	Unit of measurement
Physical	
Reaction	pH units (1–14)
Temperature	Degrees F or C
Transparency	Depth in meters
Total Suspended Solids (TSS)	Light transmission
Color	Platinum-cobalt color units
Odor	Subjective
Chemical	
Total Dissolved Solids (TDS)	Conductivity, μmhos
(the μmho is 1/1000 of the reciprocal of the electrical unit of resistance, the ohm)	
Dissolved Gases	Concentration, mg/l or ppm
Alkalinity	Concentration, mg/l or ppm
Hardness	Concentration, mg/l or ppm
Salinity	Concentration, mg/l or ppm
Biochemical Oxygen Demand	Amount of oxygen
Chemical Oxygen Demand	removed from the water
Nitrogenous Oxygen Demand	over a five-day period
Biological	
Fauna	
Flora	
Indicator species	

Reaction is a measure of acidity, reported as pH. Water that is neutral has a pH of 7.0, while bogs may have very acidic water of 3.5. Naturally alkaline water may have a pH value of 9 or so. Rainwater, which normally contains some carbon dioxide dissolved from the atmosphere, has a pH of 5.6 to 5.7. Acid rain is due to discharge of certain gases such as sulfur dioxide in the atmosphere where it combines with water vapor to make sulfuric acid. The pH of acid rain can be quite low, around 3 or 4, and occurs downwind of major manufacturing industries or urban areas where coal and other hydrocarbons are burned as fuel. Acid rain corrodes objects such as statues and can damage natural environments as well.

In the natural aquatic environment respiration and decomposition of plants and animals occur constantly; aerobically if sufficient oxygen is present, producing carbon dioxide and, therefore, more acid water; anaerobically if oxygen is deficient, producing methane and other low-oxygen compounds and hydrogen sulfide if anaerobic bacteria are present. Photosynthesis increases dissolved oxygen and takes place only during the day; pH is therefore affected by the time of measurement.

Water, unlike almost all other naturally occurring substances, is most dense just above its freezing point, about 4°C; it expands as it gets warmer or colder. This causes ice to float, allowing overwintering of life. *Stratification* is the result of temperature distribution that affects density. Ponds and lakes deeper than about 5 meters usually exhibit stratification during summer and winter: the upper, warmer layer that is stirred by wind and usually rich in oxygen is called the *epilimnion*, the lower layer, which is often deficient in oxygen is the *hypolimnion*, and the transition zone is referred to as the *metalimnion*. During fall and spring, the water body may be more uniform in temperature and, as the advent of winter or summer brings temperature changes, there is an *overturn*. Stratification may also be caused by salinity, total suspended solids, or some combination that may result from pollution. During stratification, the hypolimnion may be polluted while the epilimnion is not. For example, if there is an accumulation of dead organic material at the bottom of a lake, it may combine with and therefore lower the dissolved oxygen. If there is no mixing, the hypolimnion may be polluted because of low oxygen levels whereas the epilimnion is still healthy for fish.

Temperature also affects the amount of gases dissolved in water, regulates certain chemical reactions, and affects the metabolic rate of aquatic organisms. As temperature increases, the amount of a dissolved gas that the water can hold goes down. The temperature of most groundwater is 55°F year-round.

Transparency is a combined measure of color, biological activity, and total dissolved and total suspended solids. A quadrant-divided black-and-white Secchi disk, 20 cm in diameter, is lowered on a length-marked cable until it disappears from view, and then raised until it just reappears; the latter measurement is a measure of transparency. It varies with the wind, season and vegetative growth, faunal activity, decomposition of vegetation, and turbidity.

Turbidity, or total suspended solids (TSS), is evaluated by light transmission, since particulates reduce it.

The major contributor to turbidity is sediment, mostly fine soil material eroded by natural geologic processes and by many land uses. The amount and size of particles in a stream varies with the volume and velocity of flow. After a storm, unprotected soil can erode and the small particles that are washed off are suspended in the water, increasing the turbidity and decreasing transparency until they are deposited.

Color and *odor* of water are often the result of organic material, although certain chemical compounds also produce them, and photosynthetic activity may alter them. Tea-colored water from tree tannin (in hemlock bark, for example) is common. Decomposition of organic substances produces a variety of compounds including methane, which is odorless, and hydrogen sulfide, which gives off a "rotten egg" odor as the water warms. In severely polluted rivers, the production of methane can be so great that the river may, if a spark is provided, actually catch fire. There are color criteria for drinking water; odor is subjectively evaluated.

Dissolved gases include carbon dioxide, oxygen, nitrogen, chlorine, hydrogen sulfide, sulfur dioxide, and ammonia. The first three are naturally available in the atmosphere and dissolve readily in water droplets; the others are often absorbed in water vapor or water bodies from volcanic activity and from decaying organic matter. Readily soluble carbon dioxide makes rainwater a weak solution of carbonic acid, with a high capacity to corrode and weather rock and soil materials.

Aquatic biological activity depends on and affects oxygen dissolved in water. High dissolved oxygen (DO)—caused by photosynthetic activity—is associated with high pH values. As a water body warms, fish are adversely affected because their metabolic rate and demand for oxygen increase while the DO decreases resulting in less oxygen to breathe.

Nitrate—usually the limiting nutrient for plant growth—may appear naturally in rainwater as a result of fixation of free nitrogen gas (N_2) by lightning into nitrate (NO_3), and in surface waters from decomposition of organic matter. Nitrate as a pollutant is commonly washed into streams or lakes from the land if too much fertilizer is used on farm fields or lawns, or from animal manure.

Alkalinity of water measures the amount, form, and distribution of bicarbonate in water. Alkalinity is an indicator of the *buffering capacity* of water, that is, how much of a strong or plentiful base or acid the water can absorb before suffering serious change. Waters that are low in buffering capacity (alkalinity under 10 mg/l) are particularly vulnerable to the effects of acid rain; lakes, ponds, and wetlands on old rock, as in the granitic Adirondack Mountains of New York State, have low alkalinities and buffering capacities.

Total Dissolved Solids (TDS) in water, measured by conductivity, include chemical compounds that occur naturally in the Earth's crust and oceans. Salt spray lofted into the atmosphere along with extra-terrestrial, natural and manmade terrestrial dust provide ready sources of water-soluble materials. Water offers less resistance to the passage of an electric current if there are more salts dissolved. Reasonable quantities of TDS are what make waters potable; too much imparts unpleasant taste or other effects.

Hardness is the quantity of calcium or magnesium carbonate which affects the amount of other materials that may be dissolved. Groundwater flowing from limestone rock strata tends to be high in calcium carbonate, or *hard*; groundwater flowing from igneous or metamorphic rock and surface waters are generally *soft*. Ideally, water for domestic purposes should have a hardness below 100 mg/l if accompanied by low alkalinity, but water that is too soft can be undesirable for some purposes. Very soft water may not be able to dissolve even small amounts of soap; and very hard water may not readily dissolve soaps or detergents either. Hard water may also leave a scale that decreases heating efficiency in hot water heaters, teapots, or clothes irons. Low buffering capacity and low hardness often occur together, but are not necessarily connected.

Of the biological water pollutants, *fecal coliforms* and *streptococci* are of particular importance because their existence indicates presence of mammalian waste products. The coliforms themselves are not necessarily harmful, but their presence is an indicator that other, harmful bacteria may be present and thus water is considered polluted. Fecal coliforms can come from deer, beaver, muskrat, and rabbit, as well as humans.

Concern over water-borne bacteria commenced more than a century ago, when the causes of typhoid, dysentery, and cholera were discovered. The use of chlorine to disinfect water dates from this time, although in the early 1970s it was discovered that in the presence of large amounts of organic matter, chlorine caused formation of carcinogenic chloroforms. Also a major concern is the spore of the *giardia* virus that causes giardosis—a currently widespread, debilitating and sometimes fatal disease.

Study of the natural sources of quality characteristics suggests that water has a natural "signature" that helps identify where it has been and, like the status of blood from laboratory analyses, indicate the health of the "patient"—the stream, lake, or wetland watershed.

Regulation of Water Quality

Investigation and regulation of water quality commenced in the late 1800s with chlorination of public supplies to control water-borne diseases. In 1970 the

Public Health Service water pollution control functions were transferred to the EPA.

Current responsibilities for primary water pollution control by federal agencies include: standard setting and enforcement by the EPA; continuous field investigation of water quality by the Geological Survey; education and technical assistance to landowners by the Soil Conservation Service; land use regulation by the Agricultural Commodity and Conservation Service; pollution cleanup by the Coast Guard and the EPA, and wetlands permits by the Corps of Engineers. State agencies complement EPA responsibilities, in particular, and programs vary dependent upon local circumstances and laws.

Major federal water pollution control legislation from 1899 through the 1960s included the prohibition of dumping in navigable waters without a permit from the Army Corps of Engineers, and gradually expanded federal responsibilities in research, water quality control, and federal contribution to costs of construction of municipal water and wastewater treatment plants. In 1972 Congress passed the Water Pollution Control Amendments, now known as the Clean Water Act (CWA). The CWA was amended in 1977 and 1987, and was scheduled for reauthorization in 1992, now in 1993. Related legislation includes the Coastal Zone Management Act (1972), the Federal Insecticide, Fungicide, and Rodenticide Act (1972), the Toxic Substances Control Act (1976), the Safe Drinking Water Act (1976), and the Farm Security Act (1985).

The CWA introduced: (1) national goals, (2) administration by the EPA, (3) the National Pollutant Discharge Elimination System for new point sources of pollution, (4) Areawide Waste Treatment Plans to control nonpoint sources of pollution, (5) application of standards to effluent rather than receiving waters, (6) inclusion under regulation of more water bodies by applying to "waters of the United States" instead of "interstate" waters, (7) pretreatment of industrial wastes prior to discharge into municipal sewers, and (8) stiff enforcement penalties.

The approximately 132 priority polluting substances under the EPA's authority include dissolved gases, heavy metals, toxic substances, chlorinated hydrocarbons, and various compounds associated with petroleum production and use.

Types of Pollution

Two major types of pollution are *point sources*, where polluted water is discharged from a pipe, and *nonpoint sources*, where the discharge is diffused surface runoff from various land uses, including agriculture, urban development, construction, mining, on-land disposal of solid waste, salt water intrusion, and forestry activ-

ities. Both types may impact surface or ground waters.

Point sources are under fairly good control in the U.S. Most cities and smaller communities have modern or updated waste treatment plants. Control involves spending money to monitor pollution concentrations and loads, treatment of polluted waste waters, and permits for the discharge of polluted (or treated) water into receiving water bodies.

Initial plans to identify and control nonpoint sources are underway throughout the U.S., varying from state to state. Control involves a process that: (1) identifies and locates pollution problem areas, (2) designates critical watershed zones that are vulnerable to pollution and which are the best sites for pollution control efforts, (3) selects appropriate Best Management Practices (BMPs) that may mitigate pollution, (4) conducts a public education campaign, (5) selects and implements the appropriate BMPs, and (6) evaluates and re-evaluates performance, with feedback to step (1), as necessary.

PETER E. BLACK

For Further Reading: Peter E. Black, *Watershed Hydrology* (1991); Environmental Protection Agency, *Quality Criteria for Water* (1986ff.); H. A. Swenson and H. L. Baldwin, *A Primer on Water Quality* (1984).
See also Acid Rain; Chlorination; Clean Water Act; Dissolved Oxygen; Eutrophication; Federal Insecticide, Fungicide, and Rodenticide Act; Fertilizers; Groundwater; Insecticides; Lakes and Ponds; Nitrogen Cycle; Ocean Dumping; Pesticides; Petroleum; pH; Sea and Lake Zones; Toxic Chemicals; Toxic Metals; Toxic Substances Control Act; Waste Treatment; Wastewater Treatment, Municipal; Water; Water Cycle; Water, Drinking; Wolman, Abel.

POLLUTION PREVENTION ACT

See Source Reduction; United States Government.

POLLUTION TRADING

Pollution trading is a market-based approach to reducing overall discharges of pollutants by allowing the sale of rights to emit pollutants. Under the traditional "command and control" regulatory approach, each discharger of air or water pollution is subject to the same limits on pollutants.

Critics pointed out that such a system provides dischargers with no economic incentive to reduce their discharges below the permitted limit. As early as the 1960s, and increasingly in the late 1980s, economists and some environmentalists argued that market incentive systems with tradable rights are superior to, or work well in addition to, traditional command and control regulation in stimulating innovative environmental technologies and attaining overall reductions in pollution discharges.

The theory is that companies that are most efficiently and economically able to reduce their emissions—to zero, if possible—should do so, and receive compensation from companies that cannot economically reduce emissions. The notion of allowing companies to buy the right to pollute may be distasteful to some. However, the alternatives are ultimately even less attractive: without pollution trading, clean plants may have no financial incentive or reward for reducing their emissions, and dirtier industries may have to shut down if they cannot buy the right to pollute during the time it takes to modernize or replace older facilities or plants. Pollution trading thus satisfies both the principal of rewarding the non-polluter and making the polluter pay.

Pollution trading programs have been established by the EPA for rights to produce chlorofluorocarbons, by the Southern California Air Quality Management District for smog-producing air pollutants in Los Angeles, and by the Wisconsin Department of Natural Resources for water pollution. A large-scale program for trading the rights to emit sulfur dioxide was established under the Clean Air Act of 1990 to reduce acid rain. In reported deals involving several million dollars, industrial or utility companies have reduced their emissions below the legally established limits by several thousand tons, and sold the right to emit the extra sulfur dioxide to companies that cannot meet the limits. Similar programs are under consideration by the EPA for water pollution, and have been proposed as a way to provide economic incentives to reduce agricultural drainage, greenhouse gas (carbon dioxide) emissions, and other types of waste.

It has been observed that for a successful pollution trading program, the agency administering the program should have clear legal authority and technical capability for designing, implementing, and monitoring the program. The system must be evasion-proof, and may need a regional emphasis. There should be a wide variation of discharges among regulated firms to create a market of buyers and sellers, and minimal transaction costs.

MICHAEL A. GOLLIN

POLYCHLORINATED BIPHENYLS

See PCBs.

POLYCYCLIC AROMATIC HYDROCARBONS

Polycyclic (or polynuclear) aromatic hydrocarbons (PAH) are compounds that occur naturally in living organisms and as a byproduct of the combustion of fossil fuels; they can also be produced synthetically in laboratories. Although PAHs are essential to life and commerce, some are extremely toxic and carcinogenic.

PAHs differ from most other hydrocarbons (compounds of carbon, hydrogen, and oxygen) in two major ways. First, they are highly complex, consisting of dozens of chemical components. Second, they contain aromatic rings, also called benzene rings, which consist of six carbon atoms joined together in a very stable, ring-shaped group. These groups give certain PAHs their characteristic odors and flavors. For example, the scents associated with flowers and the odors and flavors of fruits and spices all arise from the presence of PAHs.

Complex chemical groupings in PAHs create unique molecular shapes, some of which can interfere with life functions. For example, certain PAHs may interrupt DNA replication or attach themselves to proteins, inhibiting their functioning, and giving rise to mutations or cancers. Although government agencies in the United States have identified many carcinogenic and suspected carcinogenic PAHs, understanding of the processes underlying PAH-associated cancers is still limited.

POPULATION

A very broad-brush overview of the global population situation, past and present, helps to imagine what the future holds in store, assuming continuing global cooperation towards population stabilization in the 21st Century. This overview provides an understanding of the population dynamics at work in the world today, an explanation of the marked variations between and among countries, and an outline of what has been accomplished and what remains to be done. It attempts to place population dynamics within the broader context of national social and economic development, and to explain the implications of different developmental patterns and approaches for population change.

The present high rate of global population growth is neither sustainable nor desirable and future global sustainability and quality of life depend upon the earliest possible stabilization of population growth. (By way of example, if the current rate of growth were continued for a hundred years, the population of the world would be almost 30 billion. If continued for 200 years, it would be more than 160 billion). Certainly, since situations in countries differ, this generalization does not apply equally to all countries. Some countries consider themselves underpopulated and capable of sustaining relatively rapid population growth rates, while others feel that the present rates of population growth place a heavy burden on their ability to achieve improvements in living standards for their people. But, overall it seems reasonable to assert that neither the present rate of population growth nor a global population of 15 billion, three times its present level, are in the general interest.

Definition of the *Demographic Transition*

The demographic transition is a term that describes the movement from high birth rates and high death rates which characterize traditional societies to the low birth rates and low death rates that characterize today's industrialized societies. Classically, the transition begins when death rates start to decline. In this initial stage birth rates remain high, so that the population growth rate actually accelerates. The reasons why death rates decline are not fully understood, but the decline is generally attributed to improvements in food supply, nutrition status, and public hygiene and health.

The first demographic transition occurred at the time of the industrial revolution in Europe and North America, and lasted approximately one hundred years. The decline in death rates was rather gradual as was the subsequent decline in birth rates. Generally, overall population growth rates never exceeded 1.5% per

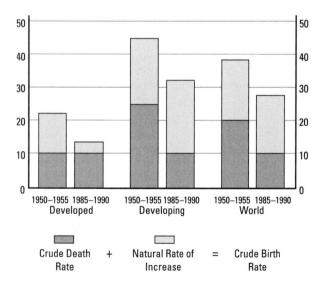

Figure 1. Crude birth and death rates and rates of natural increase in developed and developing countries 1950–1955 and 1985–1990.

year, or a doubling of population over a period of some fifty years. In today's low-income countries, the declines in the death rates have been much steeper than those that occurred when today's industrialized countries were at a similar point along the transition.

Since World War II, the death rates in developing countries have declined by more than 60%, from 24 per 1,000 to 9 per 1,000 (Figure 1). This dramatic decline in mortality, the result of major improvements in the control of communicable diseases and the production and distribution of food, are historically unprecedented. Consequently, in the absence of an immediate commensurate decline in fertility, population growth rates exploded. Countries which had relatively stable population growth rates in a regime of high birth and death rates, suddenly found themselves growing at rates of 3% a year or higher, resulting in doubling times in some cases of less than 20 years. Thus, in the late 1960s, the period of most rapid global population growth, the world as a whole was growing at slightly over 2% per year, implying a doubling of population in the short span of 35 years. During the period 1965–70, the less developed countries experienced a growth rate of over 2.5%, while the more developed countries were growing at somewhat under 1% per year.

By 1990, the growth rate in the less developed countries had dropped to just over 2% while that of the more developed countries had dropped below one half of 1% per year. At the same time, the global population growth rate dropped from over 2% in 1965–1970 to 1.7% in 1985–1990. Clearly, the world has passed the peak of the demographic transition and is moving gradually toward a new equilibrium, called stabilization by demographers, at lower birth and death rates. The question is not whether stabilization will occur, but when.

The different regions of the world are at quite different points along the demographic transition. As has already been noted, the industrialized countries have essentially completed the transition. In fact, some countries in Western Europe are now experiencingencing negative rates of natural population increase because death rates actually exceed birth rates. Austria, Denmark, Germany, and Hungary are in this category. Twenty-five industrialized countries now have below-replacement fertility. In other words, on average, women in these societies are producing fewer children than are required to replace themselves and their spouses. This extraordinarily low level of population growth has become a subject of considerable concern to some European countries which are attempting various approaches to stimulate higher rates of reproduction.

At the other extreme is Africa where death rates remain much higher than the global average and where, in the great majority of countries, birth rates have not declined very much, if at all. Three sub-Saharan African countries, Botswana, Kenya, and Zimbabwe, show clear signs of having started the demographic transition. In all three, birth rates have declined noticeably over the last decade while death rates have stabilized at relatively low levels. But these countries are the exception. Generally speaking, death rates, particularly infant and child mortality rates, are unacceptably high and birth rates remain relatively unchanged from their levels of 15 and 25 years ago. At the other extreme, many of the countries of East and Southeast Asia are near the end of the transition. For East and Southeast Asia as a whole, the crude birth rate has dropped by half, from nearly 44 per thousand to just over 22 per thousand between 1950 and 1990. And the rate of natural increase has dropped from over 2.5% a year in 1965–1970 to just 1.6% in 1990–1995. The countries furthest along the transition are the Republic of Korea, The People's Republic of China, Thailand, Taiwan, Indonesia, Hong Kong, and Singapore.

Much of Latin America is also well along the trajectory toward low birth and death rates (Figure 2). Declines there have been less dramatic than that of East and Southeast Asia because, while birth rates in the early 1950s were nearly identical in the two regions, death rates in Latin America were much lower than those in East and Southeast Asia — approximately ten points less. Thus, Latin America at that time had a considerably higher population growth rate than East and Southeast Asia ever did. However, the rate of natural increase has declined from a peak of 2.7% per year in the early 1950s to just under 2% in the early 1990s. While the decline in population growth rates in Latin America has been

widespread, the countries which experienced the earliest declines were Chile, Argentina, and Uruguay— joined later by Costa Rica, Cuba and Colombia and more recently still by Mexico and Brazil. It is the declines in the Mexican and Brazilian birth rates in the last two decades which account for most of the recent regional decline in population growth rates. Fertility in some of the Andean countries (e.g., Bolivia, Peru) and Central America (e.g., El Salvador, Honduras, Guatemala, Nicaragua) remains quite high.

Mortality continues to decline in the region. The crude death rate in Latin America is now under seven per thousand, the lowest in the world. This is a function of both widespread improvements in the quality of life and of health care in particular, and a youthful population which is itself a result of high fertility in the recent past. It is unlikely that the crude death rate will decline very much further in Latin America; in fact as the population ages, the CDR will increase. Meanwhile, it is likely that fertility will continue to decline along the trajectory established over the last ten to twenty years.

Generally speaking, the rapid fertility decline in East and Southeast Asia is attributed to dramatic improvements in living standards over the last twenty years and vigorous family planning programs in a large majority of the countries. In Latin America the decline is attributed similarly to considerable improvements in living standards and to the spread of family planning services, but in Latin America many of these services were provided by private and commerical rather than government organizations.

The situation in South Asia and the Middle East and

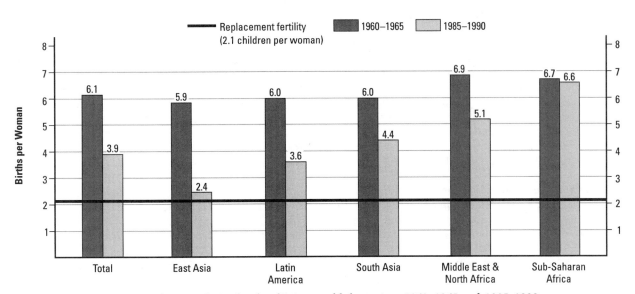

Figure 2. Fertility trends in the developing world, by region, 1960–1965 and 1985–1990.

North Africa is quite different from either the dramatic declines of Southeast and East Asia and Latin America on the one hand, and the virtually unchanged situation with respect to fertility in Africa. South Asia has experienced a very slow but steady decline in birth rates since the 1950s, dropping from a regional average of nearly 45 per thousand in the early 1950s to around 33 per thousand in the early 1990s. However, the crude death rate has declined from around 25 per thousand to under 11 per thousand during the same period so that the population growth rate has only declined from a peak of 2.4% in the 1960s to around 2.25% today. Southern India and Sri Lanka have experienced the most significant declines in fertility, with levels that approach those of some of the most successful countries of Southeast Asia and Latin America. On the other hand, Northern India, Pakistan, and Nepal have experienced very little decline in fertility. Bangladesh, with an extremely vigorous national family planning program, is somewhere between, having achieved significant declines in fertility in the last decade. It is widely believed that further fertility declines in South Asia will depend heavily on the will of the governments of India and Pakistan to expand contraceptive availability to the populations of the states of Northern India and to all of Pakistan.

Until quite recently population growth in North Africa and the Middle East was quite similar to that in sub-Saharan Africa. In the early 1950s, the crude birth rate approached 50 per thousand and the crude death rate was around 25 per thousand, producing a population growth rate of nearly 2.5% per year. Today, while the birth of Sub-Saharan Africa remains in the mid 40s, the crude birth rate in North Africa and the Middle East has fallen into the mid 30s. At the same time, while Africa's death rate has fallen to around 15 per thousand, that in North Africa and the Middle East has declined to less than 9. Recent declines in the birth rate suggest that the Middle East and North Africa are now in the initial stages of demographic transition. The population growth rate actually increased in response to declining mortality between the early 1950s and the early 1980s from 2.5% to 2.8% per year, but it shows signs now of having declined toward the 2.5% level in the last decade, while the rates in Sub-Saharan Africa remain at over 3% per year. Tunisia and Turkey have led the way, followed more recently by Egypt. Differentials among countries in the region are still quite substantial, depending upon the extent to which family planning services have become available. Thus, like South Asia, the rate at which the demographic transition proceeds in this region will depend importantly on the willingness of governments and the ability of the private sector to make family planning services available to the increasing proportions of couples who now appear ready to use them.

Causes of the Transition

Why Death Rates Have Declined

One of the most careful analyses of the causes of mortality decline in the developing countries was carried out by Preston in the late 1980s. He observed that between the 1930s and the 1960s, the developing countries experienced an unprecedented mortality decline, a rate of decline which has slowed since the 1960s. Preston shows that much of the pre-1960s decline can be explained by special public health efforts such as smallpox eradication, efforts to eradicate and control malaria, childhood immunization programs and improvements in personal health practices. Only about 30% of the decline was explained by more general improvements in social and economic conditions—improvements such as increases in income, literacy, and nutrition.

However, over the last thirty years, Preston found that as the decline in mortality slowed, the effects of specific public health interventions also declined. In the more recent period, declines in mortality were much more strongly associated with improvements in general social and economic conditions. Preston's analysis suggests that the extraordinarily high mortality that characterizes the pre-transition developing world was relatively amenable to specific public health interventions but that after a certain point, the residual high mortality that existed was more endemic, responding only to fundamental improvements in living standards and quality of life.

Of course, Preston's analysis was carried out before the full dimensions of the AIDS pandemic began to be revealed. Demographers, as well as epidemiologists, differ dramatically in their assessments of demographic implications of HIV/AIDS in the long run, but it is becoming increasingly clear, at least in Africa, that the AIDS phenomenon will have important implications for mortality (and perhaps for fertility as well) over at least a couple of decades.

Generally speaking, it seems fair to say that in most of the developing world mortality levels will decline less in the future than they have in the past. Even so, crude death rates are projected by the United Nations to decrease by more than 25% in developing countries in coming decades. There will be major declines in Sub-Saharan Africa (more than 50%), Southeastern Asia (30%), and in Southern and Western Asia (about 45% in each area). Crude death rates are now very low in Latin America and East Asia, and cannot be expected to drop significantly in the future. Endemic and expanding exposure to malaria and very incomplete immunization against the major childhood killers mean that African mortality rates could be significantly reduced through effective action on these two fronts alone.

Turning then to the subject of birth rates, we also need to understand why birth rates have declined more rapidly in some places than others. Much has been written about the conditions under which fertility declines in developing countries. For years the debate raged over the question of whether or not fertility could be affected dramatically in the short run through the provision of family planning services. One view assumed a high predisposition on the part of most couples to limit the size of their family, given the opportunity to do so. On the other side were those who argued that fertility would respond only to longer-term changes in social and economic conditions and that family planning programs would have at best a marginal impact on these changes. The 1974 World Population Conference at Bucharest was itself a major battleground for the proponents of these different points of view. The one point on which both schools agreed was that fertility is a function of both general changes in socioeconomic conditions and availability of the means to control childbearing. They differed on the relative importance of these two factors. Berelson very neatly summarized the situation with a simple two-by-two matrix in which he juxtaposed socioeconomic conditions on the one hand and family planning program effort on the other. This simple framework was subsequently elaborated by first Mauldin and Lapham and later Mauldin and Ross (see Table 1) who have carried out analyses over the past decade on the relationship between socioeconomic development, family planning program effort, and fertility decline.

This analytical framework, along with vastly improved data which have come from a series of surveys which began in the early 1970s (the World Fertility Survey followed by Contraceptive Prevalence Studies and most recently the Demographic and Health Surveys) have substantially diminished the intensity of the debate between the proponents of the family planning school and the proponents of the broader development school of fertility decline. What these analyses have shown is that the overall level of development (conventionally defined in terms of income levels, literacy rates, nutrition levels, etc.) set the boundaries within which fertility decline can occur. The lower the level of socioeconomic development, the higher will be fertility and the lower will be the motivation of families to limit the number of children they bear. The higher the general level of development, the higher the motivation of couples to limit family size, and the lower the fertility. However, the Berelson/Mauldin/Lapham/Ross formulation and the vastly improved basic data on factors affecting fertility have made it possible to understand why countries have shown very different demographic patterns while being at similar levels of socioeconomic development.

These analyses have shown that the extent to which countries permit or promote family planning services can have a significant impact above and beyond the general effect of socioeconomic setting. Put more simply, the rate at which fertility *can* decline is heavily influenced by the general level of socioeconomic well being. The rate at which fertility *will* decline is very importantly influenced by the level of effort behind the family planning program. As can be seen in Table 1, the highest fertility declines have occurred in those countries which have both a favorable social setting and a strong program effort (Mauritius, Korea, Taiwan, Mexico, Singapore) while the least decline has occurred in those countries which are poorest and where program effort is weak or non-existent (Sudan, Chad, Mauritania, Somalia, etc.). What the table shows is that both socioeconomic setting and program effort are important factors in explaining fertility decline. The better the social setting, the greater an impact a strong program effort will have on fertility decline. But it also shows that countries with strong socioeconomic settings and weak programs will have fertility decline and that countries with poor socioeconomic performance can still achieve substantial fertility decline through strong program effort. Moreover, Mauldin and Ross have shown that substantial strengthening of program effort in a number of countries over the last fifteen years has led to quite substantial declines in fertility—declines well beyond what changes in their social and economic circumstances alone would have produced. Important analyses by Bongaarts and many others have confirmed through more sophisticated analyses the strong independent effort of family planning programs on fertility decline.

The major regions of the developing world differ quite significantly in the paths they have followed. Nearly all the countries in the upper left hand quadrant of the table are Asian or Latin American, with the Asian countries predominating among those with the strongest program effort and the Latin American countries predominating among those with highest social setting. On the other hand, the countries with the lowest scores on both dimensions are overwhelmingly found in Africa. The South Asian countries tend toward the lower end of the socioeconomic scale but distribute themselves across the program effort scale, from very strong program effort in India to weak or very weak effort in Pakistan and Afghanistan. For example, India with a strong program effort has shown twice the decline in fertility of Pakistan with weak program effort but similar socioeconomic conditions. Likewise, Bangladesh, once part of Pakistan, has

Table 1

Percentage decline in total fertility rate among developing countries 1975–1990, by social setting in 1985 and program effort level in 1989

Social setting	Strong		Moderate		Weak		Very weak or none		Mean
	Country	% decline	Country	% decline	Country	% decline	Country	% decline	
High	South Korea	51	North Korea	46	Brazil	26	Kuwait	43	
	Taiwan	43	Jamaica	44	Jordan	23	Iraq	13	
	Mexico	41	Cuba	34					
	Mauritius	38	Panama	33					
			Colombia	31					
			Costa Rica	24					
			Venezuela	23					
			Lebanon	22					
			Singapore	20					
			Chile	18					
			Trinidad and Tobago	15					
		(43)		(28)		(24)		(28)	31
Upper middle	Thailand	48	Guyana	42	Algeria	29	Libya	10	
	China	39	Peru	33	Turkey	27	Saudi Arabia	2	
	Tunisia	37	Dominican Republic	31	Paraguay	17			
	Sri Lanka	34	Ecuador	29	Syria	14			
	Indonesia	33	Egypt	21	Congo	0			
	Botswana	28	Iran	21					
	El Salvador	21	Philippines	20					
			Malaysia	19					
			Zimbabwe	19					
			Guatemala	13					
		(34)		(25)		(17)		(6)	24
Lower middle	Vietnam	32	Morocco	29	Papua New Guinea	16	Myanmar	26	
	India	18	Honduras	25	Haiti	12	Cambodia	5	
			Kenya	17	Bolivia	8	Liberia	1	
			Pakistan	11	Nigeria	2	Ivory Coast	0	
			Ghana	3	Madagascar	1	Laos	-4	
			Zambia	-2	Zaire	0			
					Lesotho	-1			
					Tanzania	-1			
					Cameroon	-6			
					Central African Republic	-7			
		(25)		(14)		(2)		(6)	8
Low	Bangladesh	22	Nepal	12	Senegal	10	Sudan	5	
					Afghanistan	5	Chad	2	
					Mozambique	3	Somalia	0	
					Rwanda	3	Malawi	-1	
					Benin	0			
					Burkina Faso	2			
					Ethiopia	2			
					Burundi	0			
					Guinea	0			
					Mali	0			
					Mauritania	0			
					Niger	0			
					Sierra Leone	0			
					Togo	0			
					Uganda	-3			
					Guinea-Bissau	-6			
		(22)		(12)		(1)		(1)	2
Mean		**34**		**24**		**5**		**8**	**16**

Note: Means were calculated by unit weights. Mean percentage decline in total fertility rate in each cell is shown in parentheses. Negative entries indicate a rise in the TFR.

achieved more than twice the fertility decline despite being at a substantially lower socioeconomic level than its former Western wing.

Why Birthrates Have Declined More Rapidly in Many Low-Income Countries

Clearly, the availability of modern contraceptive methods has made it possible for low income countries to accelerate the demographic transition far beyond the rate at which it would have occurred in the absence of those methods. The difference between the demographic transition in Thailand and the demographic transition in Britain is not so much explained by the rapidity of socioeconomic development in Thailand since World War II, as by two additional factors: the availability of modern methods of contraception and an explicit public policy aimed at reducing the population growth rate—with the availability of modern methods being probably the more important. This appears to be so because even in countries which did not have explicit policies to reduce population growth rates, birth rates have dropped quite dramatically where modern family planning methods have been generally available through private and commercial channels. From this one can infer that technology has played an even more important role than policies per se, except perhaps in a few countries which have pursued fertility control policies with special vigor (China, and perhaps Indonesia). The introduction in the early 1960s of the oral contraceptive and the intrauterine device revolutionized the ability of individual couples to exercise control over their fertility. The subsequent introduction of simple, low-cost, surgical sterilization, injectable contraceptives, and most recently contraceptive implants has further broadened the range of effective modern methods which are available to family planning programs. While each of these technologies has its drawbacks, taken together they do represent a dramatic change in the capacity of couples to determine both the number and the spacing of their children. Coupled with effective delivery systems, these technologies can be extremely powerful, resulting in fertility transitions that would have been inconceivable fifty years ago.

How Far Along the Path to Replacement Level Fertility Are We Today?

There is an important difference between replacement level fertility and stationary population. A stationary population is one in which there is an equilibrium between births and deaths—in other words zero population growth. Replacement fertility is defined as about 2.1 children per woman during her reproductive lifetime—the number required to replace the parents, allowing for subsequent mortality. The reason the two

are not synonymous is that a stationary population will not be achieved until the number of women entering the reproductive age group is equal to the number leaving that group. In other words, even at replacement fertility, if there are more women entering the reproductive age group than leaving it, there will be more births than deaths in a typical population. This explains what is meant by "population momentum," the fact that for some period after replacement level fertility is achieved, population growth will continue. The magnitude of population momentum is dependent on the age structure and the level of fertility. The developing world today has a very young population; if, magically, replacement fertility could be achieved today, the population would continue to grow for more than six decades and would increase by a little more than three billion.

Looking just at the question of replacement fertility, the developing world has moved halfway from the peak fertility of the 1960s (when women were on average producing six children each) to the replacement level of 2.1. Between the mid-1960s and 1990, the total fertility declined from 6.1 to 3.9. In 1965 the global population was 3.3 billion; today it is 5.4. What will it be 25 and 50 years from today? At what level is it likely that the population of the world will cease to grow, or stabilize? What can be done to keep up, and increase, the pace of fertility decline?

Socioeconomic Factors

Major changes have been wrought in socioeconomic conditions in developing countries during the past two decades (Table 2). Life expectancy at birth has in-

Table 2

Changes in socioeconomic indicators in developing countries, 1965–1970 to 1985–1990

Indicator	1965–1970	1985–1990	Change Amount	Change Percent
Life expectancy	52.7	61.9	9.2	17
Adult literacy	36[1]	61[2]	25	69
Primary plus secondary school enrollment ratios	39[1]	62[2]	23	59
Percent in non-agriculture labor force				
Males	28[2]	40[2]	12	43
Females	8[2]	36[2]	28	350
GNP, per capita	310[2]	635[2]	325	105
Percent users of contraceptives	15	50	35	233
Crude death rate	15	9.8	-5.2	-35
Infant mortality rate	116	78	-38	-33
Total fertility rate	6.01	3.94	-2.07	-34

[1] 1960   [2] 1985

creased by almost nine years to an average of 61 years. Adult literacy has increased from 36% to 61% and primary plus secondary school enrollment ratios have increased by 59%. Female literacy lags behind that of males but it almost doubled from 1960 to 1985, increasing from 28% to 55%. Female school enrollment more than doubled from 27% to 58%. More than 60% of young people of primary and secondary school ages are enrolled as compared with just under 40% in 1960.

Mortality has declined very rapidly since the end of World War II. The crude death rate declined by 38%, from 24 to 15 between 1950–1955 and 1965–1970, and by another 35%, from 15 to 10, between 1965–1970 and 1985–1990. Infant mortality rates, still far too high, have fallen by one-third. If one compares the most recent figures with those of developed countries, one is struck by the continued disparity between the developed and developing worlds. Life expectancy is 12 years lower in developing countries, and infant mortality, at 78 infant deaths per 1,000 live births, is 63 points higher than the rate in developed countries. Despite these differences, however, the changes during the past 25 years have been impressive, and in the desired direction. They have had their own effects upon fertility rates. Socioeconomic conditions in much of the developing world are far from satisfactory, but there have been substantial improvements during the past two decades, and similar changes are likely to occur in the coming decades.

Family Planning Programs

A large number of countries have adopted policies to reduce rates of population growth and/or provide support for family planning programs. We have ratings or scores of a large number of key aspects of family planning programs for three time periods, 1972, 1982, and 1989. The principal finding of these studies is the global improvement of program effort (PE) over the last two decades. The mean PE score more than doubled from 1972 to 1989, rising from 20 to 44 when each country is weighted equally. If the scores are weighted by the number of women of reproductive age (15–49) the average rose from 51 in 1972 to 63 in 1989. The difference in the two sets of scores is brought about by the relatively high scores of large countries such as China, India, Indonesia, Thailand, Colombia, and Mexico throughout the period.

In the 1982–1989 period countries have strengthened each of the four dimensions of program effort. Not every country has participated in these changes— some, in every region, have had no programs at all, and some have begun work only recently. Countries showing greater progress have generally been those with more favorable social and economic infrastruc-

tures, with some notable exceptions. Overall, most countries can substantially further strengthen their efforts.

In 1980 there were about 220 million contraceptive users in developing countries, about 40% of all married women and women in sexual unions in ages 15–49. For convenience we shall refer to these as married women in the reproductive ages, or simply MWRA. By 1990 there were some 380 million contraceptive users, 51% of married women and women in unions. The remarkable and unexpected increases in the prevalence of contraceptive use have closely followed the degree of program strength of family planning programs. Also, the higher the level of contraceptive use the more effective the method mix and the lower the failure rate, which becomes an important fertility determinant where most couples use contraception.

Programs have evolved in terms of their complexity. The original, simple model of family planning based on mere contraceptive supply of selected methods to selected subgroups, is a caricature of most actual programs. We have described the extensive efforts at normative change, undertaken especially by Asian programs, which insist on the benefits of the two-child family, with a pattern of interweaving supply and persuasion by field staff. Persistent, large-scale public campaigns directed to the whole population, to legitimate the small family and to popularize birth control, are common. Attitudinal change has been very extensive, as documented in survey data and in program acceptance patterns showing a steady movement toward contraceptive practice by younger, lower-parity couples.

Sterilization has become the world's leading contraceptive method, and the technology of birth control is far advanced from its 1967 state. The unmarried are included in substantial numbers of programs. Safe abortion is offered as contraceptive backup in some. Different programs have chosen alternative routes to similar objectives, depending upon their circumstances. Some governments have used monetary and other inducements to encourage behavioral change, usually though not invariably of relatively small value. "Incentives," variously defined, have been used even in some African countries. All these modifications to the simple model of the mid-1960s have changed the reality of program action.

The future picture is mixed: the programs themselves are mixed, and socioeconomic progress is not assured everywhere. It is reasonable to expect further strengthening of public efforts, additional increases in contraceptive use, and continuous declines in family size. However, there will be less change in crude birth rates due to unfavorable age distributions, which build in a heavy momentum for continued growth, and even less in population growth rates, due to further mortality declines in some large countries that flow both from reduced age-specific rates and from smaller proportions in the vulnerable years of infancy and early childhood.

Some features of the future are nearly certain: due to the momentum factor, the population base will grow inexorably, since even with the two-child family, young populations have more births than deaths. The force of momentum can hardly be overstated. The age structure is now so unfavorable that even if fertility falls *immediately* to the two-child family, the world's population will still grow by over half (58%), adding another 3.1 billion. That would bring the world's population to 8.5 billion. To this must be added the increment due to the more gradual character of the fertility decline.

Thus programs have grown in numbers, coverage, features, and strength. This is more than could have been foreseen a quarter of a century ago. Yet population stabilization—zero growth—still lies many years in the future. Even China, with its exceptionally vigorous measures, is still above replacement fertility, not to mention zero growth. The positive side is that fertility has fallen halfway to replacement and is still declining, and that both socioeconomic settings and programs may be expected to continue their gains. As we noted above, several developing countries have below replacement fertility and others are approaching that watershed level.

The so-called family planning strategy—a strategy that is centered on efforts to improve the economy, typically without major structural changes, supplemented by a family planning program—probably can lead to eventual stabilization of the world's population at around 10 billion.

Another future feature is the impact of the enlarging base. The United Nations estimates a 22% growth in the number of women of childbearing age in the developing world in this decade, a 28% growth in the number of women in union, and a 49% growth in the number of contraceptive users if the medium projection of the U.N. is to be met. The implications for the mobilization of both public and private resources are clear: if fertility is to be reduced rapidly, governments must continue to strengthen their family planning programs, both donors and governments of developing countries must make more resources available, and the private sector must be encouraged to help make contraceptives available at modest prices.

Costs

The cost of family planning programs in developing countries is estimated by UNFPA to be about $3.5 billion, of which donor countries pay about $1 billion, family planning users about 10% ($350 million), and the remaining three-quarters of costs are borne by the governments of developing countries. By the year 2000 it is anticipated that there will be an increase in the number of contraceptive users in developing countries from 380 million in 1990 to the order of 575 million, and costs will increase at least proportionately, and probably even more. Thus, the costs of family planning programs in the year 2000 are likely to be $5.25 billion or more (an estimate of the World Bank is $8.3 billion, and one by UNFPA is $9.0 billion). The costs only of contraceptive commodities are estimated to increase from about $400 million at the present time to more than $600 million in the year 2000.

Although the magnitude of the increase in the cost of family planning programs is uncertain, it is clear that substantial additional financial resources will be needed. In order to maintain the momentum of increasing contraceptive prevalence firmly established during the decade of the 1980s, it is urgent that donors increase their commitment to assisting developing countries, and that developing countries maintain and increase their commitment to lowering rates of population growth both by strengthening their family planning programs and providing increased resources. At the present time donors contribute less than one percent of foreign assistance to population programs, and only 0.4% of total government expenditures. This is too little, and could and should be at least doubled.

The private sector has played only a minor role in Asia and the Pacific, although serious efforts are now underway in Indonesia to promote "self reliance," working with contraceptive suppliers to keep prices at a modest level. The private sector has played a major role in Latin America, and is likely to become even more important in the decades ahead. The number of contraceptive users in Sub-Saharan Africa is quite small, though growing, and it is likely that most of the growth will be accomplished by expansion of the public sector. Some estimates of how much contraceptive users pay are as high as $500 million per year for services and supplies, and some analysts project that this figure could be doubled by the year 2000.

The governments of developing countries rarely spend more than one percent of their budgets on population programs. While the major increase in funding for population programs must come from the governments of developing countries, it is vital that donors continue to assist, particularly with foreign exchange costs. With the combined efforts of developing countries, donors and the private sector (including non-governmental organizations), it is possible to bring about a stabilization of the world's population at about 10 billion.

STEVEN W. SINDING
W. PARKER MAULDIN

For Further Reading: J. Bongaarts, W. P. Mauldin, and J. F. Phillips, "The Demographic Impact of Family Planning Programs," *Studies in Family Planning* (1990); B. Boulier, "Family Planning Programs and Contraceptive Availability: Their Effects on Contraceptive Use and Fertility," in *The Effects of Family Planning Programs on Fertility in the Developing World*, World Bank (1985); P. Cutright, "The Ingredients of Recent Fertility Decline in Developing Countries," *International Family Planning Perspectives* (1983).

See also Age Composition; AIDS; Carrying Capacity; Contraception; Demographic Transition; Demography; Family Planning Programs; Human Reproduction; Mortality and Life Expectancy; Population in Africa; Population in Asia; Population in Latin America; Population Explosion.

POPULATION IN AFRICA

Africa in the 1990s faces a historic challenge. Never before have so many nations tried to modernize in the face of such extraordinary population growth. The tremendous growth of the population and its increasing concentration in urban areas is contributing to serious environmental deterioration across the continent. Environmental degradation, in turn, makes it even more difficult for these countries to achieve social and economic development.

From the limited data available, the United Nations estimates that Africa has the highest birth and fertility rates in the world in the early 1990s. A common measure used by demographers is the total fertility rate, the average number of lifetime births per woman at current age-specific fertility rates. In 1992, Africa had a fertility rate of 6.1 children per woman. By contrast, Asia had a fertility rate of 3.2 children per woman (3.9 if China is excluded); Latin America, North America, and Europe had fertility rates of 3.4, 2.0, and 1.6 children per woman respectively. While birth rates have been high, and in some cases have even risen over time, death rates have declined gradually in recent decades. As a result, natural increase—birth rates minus death rates—has risen each decade. Africa's natural increase averaged 2.6% in the 1960s, 2.8% in the 1970s, and 3.0% in the 1980s. At the latter rate, the population of the entire continent would double in just 23 years.

The high rate of growth is reflected in population projections for the region. The median or "most likely course of events" projection used by the Population Reference Bureau for its *1992 World Population Data Sheet* assumes substantial decline in the fertility rate over time. It still shows the African population increasing from 654 million persons in 1991 to 1,085 million in 2010 and 1,540 million persons in 2025, or an addition of nearly 900 million persons in 33 years. Beyond this explosive growth, Africa is also urbanizing faster than any other part of the world. While still distinctively rural—only 30% of Africans lived in cities in 1992—the continent has been urbanizing at an extraordinary pace as a result of migration from the countryside and the high rate of natural increase in the urban areas themselves. The United Nations, for example, judges that the urban population of Africa grew from 132 million to 217 million persons over the 1980s, or a rate of 5% per annum. At that rate, the cities would double in size in just 14 years and the pace of growth may be accelerating.

Africa is already experiencing severe environmental degradation. The prospect of hundreds of millions of additional inhabitants and their increasing concentration in the cities is daunting for the African environmental future. Population pressures on the environment can best be seen in five areas: deforestation, soil degradation and loss, water shortages, loss of biodiversity, and increased urban pollution.

Fuelwood shortages and severe deforestation are increasingly characteristic of many countries in Africa. Rapid population growth is influential in at least three ways. First, fuelwood remains the primary source of energy in Africa, especially for rural households. Second, as the rural population grows, the need for new agricultural land also expands, especially when agricultural technologies are not changing. The need for new land is met by cutting the forests. Third, in many cases, the livestock population increases in tandem with the human population. Overgrazing by cattle, sheep, and goats, especially of young trees, also contributes to deforestation. Deforestation is further abetted by destructive commercial logging practices in many countries. This destruction of the forests accelerates soil degradation and results in a loss of diverse plant and animal species.

The pace of deforestation has been rapid. In 1980, the World Bank reports, sub-Saharan Africa contained about 660 million hectares of forests and woodlands (North Africa has few forests). The region lost about 3.3 million hectares per year during the 1980s and the rate of deforestation is accelerating. Only about 91 thousand hectares per annum were replanted, only a small proportion of the amount lost. Alternative fuels are either not available or not affordable. On a global scale, deforestation in Africa is an important contributor to the volume of greenhouse gases, especially carbon dioxide and nitrous oxide, trapped in the atmosphere.

The fuelwood crisis is more acute in certain regions, including the arid and semi-arid areas below the Sahara, the mountainous regions of Central Africa, much of eastern and southeastern Africa, and the savannah zones on both the eastern and western sides of the continent. Nearly half the countries of sub-Saharan Africa will reach a situation in the 1990s where the cutting of forests seriously exceeds the sustainable supply. Many have already done so. Those countries with large tracts of tropical forest remaining, such as Zaire, Cameroon, Gabon, Congo, and the Central African Republic, do not yet face wood shortages.

Soil loss and degradation is a second environmental issue of profound importance. African agricultural production over the past 25 years has increased somewhere around 2% per year, lower than the rate of population growth. Africa is the only major region of the world where per capita food production has been declining over time. One of the important contributing factors to this poor performance is the deterioration of soil, itself partially a consequence of the rapid growth of the population.

Traditional African farming and livestock raising is largely subsistence, based mostly on labor and simple tools. Wood is the predominant source of energy and land tenure practices give users customary rights to the land, although not outright ownership. Farmers are dependent on extensive agriculture in traditional African systems; when old lands are worn out, new lands are brought into cultivation and the old ones are left fallow until they regenerate their productive capabilities. These systems are dependent on low population densities. But with rapidly growing populations, land pressures have increased. The World Bank reports that the amount of arable land per capita dropped from 0.5 hectare in 1965 to 0.3 in 1987, while farming and grazing practices have changed little. The result has been severe soil loss and degradation, as Africans have greatly reduced or eliminated fallow periods, extended cultivation onto marginal and ecologically fragile lands, cut the forests, and overgrazed the land in an effort to cope with the stresses of rapid population growth. This destructive "mining" of soils to meet the immediate needs of a rapidly growing population is occurring across the continent. The World Bank indicates that the soil erosion crisis is most serious in countries which have both high rates of population growth and little arable land remaining and gives the examples of Burundi, Kenya, Mali, Mauritania, parts of Nigeria, Rwanda, and Togo. Few countries have been exempted, however. And in those countries with large tracts of unused, cultivable land remaining, often tropical forests would have to be cut and tropical diseases brought under control for the land to be agriculturally useful.

Desertification—an extreme loss of soil fertility and productive capability—affects about 80% (1.5 billion hectares) of sub-Saharan Africa's formerly productive drylands and rangelands. This problem is particularly acute in Sahel countries and those other countries where the Sahara meets the savanna; the dry areas of southwest and southeast Africa; and nations such as Sudan, Ethiopia, Somalia, and Kenya.

Water availability has also become a critical issue. The World Bank reports examples of actual declines in rainfall, based on analyses by its own staff. Ivory Coast in West Africa has had perhaps the fastest rate of deforestation in Africa. With the cutting of the forests, the dry savannah area has been extended from the northern parts of the country into the middle and beyond. At the same time, evidence suggests an important and increasing decline in mean annual rainfall each year. Ethiopia, northern Nigeria, and Senegal are examples of other areas experiencing similar changes.

One scholarly study projects that by the year 2000 about 250 million Africans will be living in highly water-stressed areas, of whom 150 million will be residing in places with chronic or even absolute water scarcities. By the year 2025, more than a billion Africans will be living under conditions of severe water scarcity. Water scarcity will be an increasingly important issue in virtually all of Africa except the Central African countries with large rain forests.

Africa is renowned for its abundant and diverse wildlife and plant species. Popular imagery of the continent is still inspired by images of the large mammals. The overall diversity of animal and plant life in Africa, while enormously rich, is severely threatened. The rhinoceros and gorilla have dwindled in number; elephant herds have been greatly reduced. Other plant and animal species are disappearing in unknown numbers, but perhaps by the thousand. Not all of the loss is related to population change; poachers destroy rhino, elephant, and gorilla for economic gain. But the biggest contributors to the loss of biodiversity on the African continent are deforestation, desertification, and urban expansion, themselves largely a consequence of the rapid growth of the population. By some reports, Africa has already lost 64% of its original wildlife habitat and the decline continues at an accelerated pace. An unknown but probably large number of animal and plant species are lost forever each year in Africa due to deforestation.

Rapid growth of the cities is also contributing to African environmental problems. Most urban dwellers depend on fuelwood for energy, and some of the most severe deforestation is taking place in rings around the urban areas. With such rapid urban growth, African cities do not have anything approaching an adequate infrastructure to dispose of sewage, garbage, or other pollutants. Water supplies are particularly at risk. Increased air pollution from vehicles and industries is also a product of rapid urbanization. Cars and trucks in poor African cities are generally of such bad quality that there is a constant spillage of pollutants into the atmosphere. In Egypt and elsewhere, invaluable agricultural land is lost as cities expand their geographic boundaries.

The prospects are daunting. Recent trends have already led to serious environmental deterioration in Africa; at the same time, hundreds of millions of additional persons will be inevitably added to the African population in coming decades. Radical changes are necessary. The region needs policies and programs to preserve and safeguard its productive resource base while increasing agricultural production to meet the needs of a rapidly growing population. Africa is also going to have to move towards lower birth and population growth rates, if the continent is to achieve sustainable social and economic development.

THOMAS J. GOLIBER

For Further Reading: Malin Falkenmark, "Rapid Population Growth and Water Scarcity: The Predicament of Tomorrow's Africa," in Kingsley Davis and Mikhail S. Bernstam, eds., *Resources, Environment and Population: Present Knowledge, Future Options* (1991); World Bank, African Region, *The Population, Agriculture and Environment Nexus in Sub-Saharan Africa* (1993); World Bank, *World Development Report 1992: Development and the Environment* (1992).
See also Biodiversity; Carrying Capacity; Deforestation; Desertification; Fuelwood; Population; Soil Conservation; Urbanization; Wastewater Treatment, Municipal.

POPULATION IN ASIA

Asia has probably always been the largest continent with respect to population as well as geographically, and China has probably always been the world's largest country. China grew especially rapidly at two points in its history, once in the 11th century and again in the 18th. The first period of rapid growth was the result of the discovery of fast growing varieties of rice, which permitted multiple cropping in one year; the second period was the result of the diffusion of certain New World crops, particularly corn and peanuts, probably introduced by Portuguese travellers. Population responded rapidly to both of these increases in food supply.

Asia's civilizations are far older than those of Europe, and China in particular regarded Europe as an unimportant peninsula of the Asian continent, with nothing much to offer in trade, and certainly nothing

in cultural exchange. Chinese junks traded in the islands and mainland of Southeast Asia, and in the colonial period Chinese from the more crowded provinces emigrated to work as unskilled laborers in tin mines and plantations. These workers now form minorities in all the countries of the area, and their diligence and abstinence have enabled many to beome rich. Europeans, of whom the best known was Marco Polo, came to China on trading missions along the silk trail that crossed the desert country of Southwest Asia. The major movement of population included the Huns, pastoral horse-riding nomads who attacked China from the north and west; they ultimately moved toward Europe, pushing the Goths ahead of them, and the Goths in turn destroyed the Roman Empire.

Asia contains now, and has contained as far back as statistics can report, the main concentration of population in the world. (See Table 1.) Its growth rate leaped in the 20th century, increasing nearly fourfold between 1900 and 1950, compared to only 25% two centuries earlier, in Asia as in the world as a whole. Asian peoples were an even larger part of the world total in the 18th century, and probably before that as well. The historical memory of Asia reaches back to the time when Europe was a relatively empty peninsula of the Eurasian continent, and the Americas were as yet unknown to it.

Far more than half of the Asian population is in two countries, India and China, followed by Indonesia, Bangladesh, and Pakistan. (See Table 2.) All other Asian countries together contain only about one sixth of its population. Trends in the numbers for individual Asian countries show differentials in the application of birth control.

Population growth in Asia will no doubt slow as economic development progresses. Little can be said about the relation of population to environment that has not become controversial. In fact the controversy is not necessary. We know that poor people destroy the landscape through cutting trees for firewood to cook

Table 1
Population of the world and Asia (in millions)

Year	World	Asia
1750	791	498
1800	978	630
1850	1262	801
1900	1650	925
1950	2350	1393
2000	6122	3698

Estimates by John Durand, as published by the United Nations, *The Determinants and Consequences of Population Trends,* New York, 1973, page 21.

Table 2
Distribution of population in Asia (in millions)

Year	Bangladesh	China	India	Indonesia	Pakistan
1950	42	555	358	80	40
1980	88	996	689	151	86
2000	146	1256	964	211	141
2020	206	1436	1186	262	198

their meals, as in Nepal, or through excessive grazing as in parts of India. From that the conclusion has been drawn that poverty is the worst threat to the environment. Yet when people become rich they damage the environment in other ways: they may not consume firewood, but they use other sources of energy for cooking, most of which damage the ecosphere through emissions of greenhouse gases; they drive automobiles and tractors that do far more damage than the oxen that they displace; they read newspapers that require large consumption of paper; they have far more waste of all kinds, including dangerous chemical and nuclear waste. This applies to all continents, but above all to Asia, the largest in area as well as in population.

Thus people, whether poor or rich, tread heavily on the biosphere. Among measures to keep the damage low, one has to place at the top control of population, plus lower energy consumption, plus greater efficiency in energy consumption. Debate on which of these comes first is not helpful; all three are essential. No one need fear that protection of the environment will prove excessive.

In the dynamic economies of Japan, South Korea, Taiwan, Hong Kong, Singapore, Thailand, Indonesia, and some of the southern provinces of China the demographic transition to low birth and low death rates is the most advanced of all developing countries. Asia's birth rate will continue to decline and probably at a faster rate than the other continents. Still, Asia will continue for the foreseeable future to contain more than half of the world's population.

Improvement of agricultural yields during recent decades has resulted in the extraordinary growth of cities in all of the less developed countries, including those in Asia. Table 3 shows the levels attained by 1990, as estimated by the United Nations.

Some of the world's largest cities are in Asia, including Beijing. Others include Shanghai and Guang-Dong (Canton) in China; Delhi, Calcutta, Madras, and Bombay in India; Jakarta in Indonesia. These have grown considerably beyond the possibilities of employment in the organized modern sector, and a considerable part of the population gets along by providing services not badly wanted, such as collecting scrap and other miscellaneous activities, some legal, some not. Most of

Table 3
Percent urban population

Indonesia	25
Malaysia	38
Thailand	20
South Korea	65
Singapore	100
Hong Kong	92
India	26
China	21
Pakistan	30
Philippines	40

The numbers have to be taken as approximations only, given not only the inevitable inaccuracy of censuses, but more importantly, arbitrariness in defining city boundaries. Estimates are as far as possible counts for the built-up or metropolitan area, which may be much larger than the incorporated city.
Source: The Population Division of the United Nations.

the governments of the continent have tried to keep down the numbers of in-migrants, but even in China they have had only limited success. The cities of Asia have impossibly crowded streets, with traffic tieups that take hours to disentangle, shortage of potable water, and polluted air. China suffers especially from the burning of soft coal, and its city air, in Chungking, for example, must resemble that of the English cities during the industrial revolution.

Japan can trade its manufactures for foodstuffs to support Tokyo (second only to Mexico City in the world with 18.1 million people). Yet the time is now past when cities mostly grew in proportion to the amount of manufactures they attracted, or to the services they provided to a circumscribed hinterland. Calcutta at 11.8 million[1] is in 1990 the 7th largest city in the world; Bombay at 11.2 billion is 9th; Beijing at 10.8 million is 11th; Jakarta at 9.3 million is 14th. Most Asian cities are growing in relation to world urbanization, though still less rapidly than in Latin America.

Education[2] seems to be both a cause and an effect of urbanization. (See Table 4.) Young Asians with a middle-school education or better increasingly find that there is little demand for such schooling in the village, and with good roads and easy public transport nearly universal, they move to the city. Once there, competition for jobs impels many young people to continue their education. College education in Asia has expand-

---

1. City numbers from the United Nations Population Division, *World Urbanization Prospects 1990.*
2. In this analysis and the following table I have relied on UNESCO data as given in its 1986 Statistical Year Book. The UNESCO materials were obtained by a questionnaire addressed to the several countries, so what we have is not always comparable national data.

ed rapidly over the 20 years from 1965 to 1985, although the quality of the education is suspected of having deteriorated in some countries.

Table 4
Percent enrollment of population in cohort

	Secondary 1985	Tertiary 1965	Tertiary 1985
Indonesia	39	1	7
Malaysia	53	2	6
Thailand	30	2	20
South Korea	94	6	32
Singapore	71	10	12
Hong Kong	69	5	13
India	35	5	-
China	39	0	2
Pakistan	17	2	5
Philippines	65	19	38

Education also appears to slow the birth rate. Women who have been to school have fewer children. Whether this is due to greater independence from their husbands or to greater opportunities for them to use their time in the market is not known, nor is the direct effect of men's education in reducing the birth rate.

NATHAN KEYFITZ

For Further Reading: Donald F. Lach, *Asia in the Making of Europe* (1970); Gunnar Myrdal, *Asian Drama: An Enquiry into the Poverty of Nations* (1968); Harry T. Oshima, "Impact of Economic Development on Labor Markets, Education, and Population in Asia," *Ambio* (1992).

POPULATION IN LATIN AMERICA

"Gobernar Es Poblar"

Latin America was settled long before the arrival of Europeans, but the original inhabitants lived in a way that had little permanent impact on the environment. Most had cultural and religious beliefs that reinforced their self-image as integral components of the natural environment.

The Europeans who conquered these lands, decimating the coastal population through disease and driving the survivors deep into the interior, saw themselves as the appointed human masters of their surroundings. The new environment was something to be tamed and exploited, not something they needed to preserve, and

certainly not something of which they were an integral part. Latin America was originally colonized for extractive purposes, as a distant supplier of raw materials and precious metals for Europe. Given the abundance of land, the colonists abandoned the traditional crop rotation practices of Spain and Portugal, and turned to slash-and-burn monoculture. When these practices stripped the soil of its nutrients and led to erosion, the settlers simply moved further inland.

The limiting factor always seemed to be manpower. At first, the colonists pressed the original inhabitants of Latin America into service. When indigenous labor was not enough, they turned to African slaves. After the abolition of slavery, many Latin American governments offered land and free passage to European emigrants looking for opportunities in the New World. It was still not enough. There remained enormous, nearly empty areas in the interior of Latin America, and many governments worried about the integrity of their unsettled border regions. By demonstrating the presence of Portuguese-speaking settlers beyond its original borders, Brazil succeeded in gaining territory from virtually all of its many neighboring countries.

The pronatalist spirit that dominated Latin America was well summarized in the famous phrase *gobernar es poblar* (to govern is to populate).

From Demographic Equilibrium to Explosion

In describing the demographic evolution of Latin America, which includes all countries south of the Rio Grande and in the Caribbean, we must acknowledge that regional averages hide important variations among its component countries. Table 1 illustrates the diversity in terms of population size, birth rate, death rate, and the annual rate of natural increase.

Like most of the developing world, Latin America before 1900 was characterized by very high death rates, with as many as 20% of infants dying before their first birthday. Latin American culture responded to this reality with strong social pressure for even higher birth rates. The result was a near equilibrium between births and deaths, resulting in slow population growth.

During the first half of the 20th century, health conditions gradually improved and death rates declined. The cultural tradition of high birth rates was not easily changed, however, so birth rates remained high. With annual births now exceeding deaths by a significant margin, the population grew by around 1% per year. Latin American and North American (Canadian and U.S.) populations were of similar size during this period, and growing at similar rates.

After World War II, the trend was dramatically ac-

Table 1
Population, fertility and mortality levels, and natural increase: Latin America and the Caribbean (1989 estimates)

Region or country	Population estimate mid-1992 (millions)	Birth rate (per 1,000)	Death rate (per 1,000)	Natural increase (annual, %)
Belize	0.2	37	5	3.1
Costa Rica	3.2	27	4	2.4
El Salvador	5.6	36	8	2.9
Guatemala	9.7	39	7	3.1
Honduras	5.5	40	8	3.2
Mexico	87.7	29	6	2.3
Nicaragua	4.1	38	8	3.1
Panama	2.4	24	5	1.9
Central America	**118**	**31**	**6**	**2.5**
Antigua and Barbuda	0.1	14	6	0.8
Bahamas	0.3	19	5	1.5
Barbados	0.3	16	9	0.7
Cuba	10.8	18	6	1.1
Dominica	0.1	20	7	1.2
Dominican Republic	7.5	30	7	2.3
Grenada	0.1	33	8	2.5
Guadeloupe	0.4	20	6	1.4
Haiti	6.4	45	16	2.9
Jamaica	2.5	25	5	2.0
Martinique	0.4	18	6	1.2
Netherlands Antilles	0.2	19	6	1.2
Puerto Rico	3.7	19	7	1.2
St. Kitts-Nevis	0.04	23	11	1.2
Saint Lucia	0.2	23	6	1.7
St. Vincent and the Grenadines	0.1	23	6	1.6
Trinidad and Tobago	1.3	21	7	1.4
Caribbean	**35**	**26**	**8**	**1.8**
Argentina	33.1	21	8	1.2
Bolivia	7.8	36	10	2.7
Brazil	150.8	26	7	1.9
Chile	13.6	23	6	1.8
Colombia	34.3	26	6	2.0
Ecuador	10.0	31	7	2.4
Guyana	0.8	25	7	1.8
Paraguay	4.5	34	7	2.7
Peru	22.5	31	9	2.2
Suriname	0.4	26	6	2.0
Uruguay	3.1	18	10	0.8
Venezuela	18.9	30	5	2.5
South America	**300**	**26**	**7**	**1.9**
Latin America	**453**	**28**	**7**	**2.1**

Source: Population Reference Bureau, *World Population Data Sheet.*

celerated by the introduction of new medicines and effective public health measures, especially control of malaria, vaccination against infectious diseases, and improved sanitation. Death rates now plunged, touching off a true population explosion. From 1950 to 1970, the annual rate of population growth in Latin America averaged 2.7%, with some countries experiencing growth of 3.5% per year. Such growth rates were unprecedented, and governments proved incapable of generating corresponding increases in education, health services, or employment.

During the 1970s, the new reality of much higher infant survival rates became apparent throughout the region, and created strong financial incentives for couples to control their fertility, especially in urban areas. Meanwhile, new and more effective methods of fertility control had been introduced, including oral contraceptives and intrauterine devices (IUDs). In some countries, governments joined non-profit family planning associations in making these methods available to the poor. Use of contraception spread, and fertility began to decline. Annual rates of population growth dropped from 2.6% during 1965–1970 to 2.3% during 1975–1980. During the 1980s, death rates continued to fall, but fertility fell even more rapidly. Many more governments included family planning in their health programs. By mid-1992, the rate of population growth had fallen to 2.1%.

Survey research shows that approximately 30 million Latin American women would like to use family planning but do not yet have effective access. These women cannot afford family planning through private doctors or pharmacies, and they are outside the reach of existing subsidized programs. Many live "beyond the pavement," either in urban slums where almost no public services penetrate, or in rural areas many miles from a city or town.

The Future in Doubt

While current population trends in Latin America will eventually lead to a new equilibrium, at or near zero annual growth, the extreme demographic imbalance of the 1950s and 1960s has left challenging long-term legacies: (a) At 453 million, the Latin American population is now big enough and sufficiently dispersed to threaten the natural environment of the entire region; (b) with 36% of the population age 15 or younger, the Latin American population still has tremendous "demographic momentum" (far more people entering reproductive age than leaving it), and will probably continue to grow throughout the 21st century. The "medium" United Nations projection predicts that by 2025 the population of Latin America will be 779 million, more than double that of North America, and still growing at 1.1% per year. It is unlikely that Latin America's population will stabilize at anything less than 1 billion.

Meanwhile, Latin American governments are facing other problems that seem even more urgent than population growth and environmental degradation; crushing international debts, chronic economic crises, and political instability. Attempts to respond to these immediate problems have sometimes damaged the environment. The Brazilian government accelerated the destruction of tropical forest by building an unnecessary Transamazonic Highway, and then offering economic incentives to colonists and ranchers who cleared large areas along the highway and tried to grow crops or graze cattle on the depleted Amazonian soil.

The highly unequal pattern of income distribution in Latin America also undermines efforts to reduce unwanted fertility, and to protect the environment. While many poor and uneducated women now practice family planning, many others simply concentrate on the daily struggle to survive. In that daily struggle for food and shelter, there is little time or incentive to think about the long-term health of the environment. If slash-and-burn agriculture is easy, and will put food on the table next month, desperately poor people cannot be expected to worry about the consequences for next year.

Now that environmental consciousness is rising, together with appreciation of the aggravating role played by population growth, Latin Americans are finding that much damage has already been done to their environment, and that the high demographic momentum of their populations will continue to threaten that environment for many decades to come.

HERNÁN SANHUEZA

For Further Reading: International Union for the Scientific Study of Population, *The Peopling of the Americas: Proceedings* (1992); Hernán Sanhueza, *The Population of Latin America* (1984); Benjamin Viel, *The Demographic Explosion: The Latin American Experience* (1976).
See also Contraception.

POPULATION ECOLOGY

The recent rapid increase in human numbers, in conjunction with their many economic activities, exerts a host of adverse impacts upon environments worldwide. In fact, it may well turn out that we have achieved economic advancement in the past at a cost to the future's capacity to supply still more advancement—and even at the more serious cost of an actual decline in human welfare.

Green Revolution agriculture, for example, enabled growth in grain production to keep ahead of growth in human numbers throughout the period 1950 to 1984. However, there appear to have been certain covert costs, such as overloading of cropland soils leading to erosion, depletion of natural nutrients, and salinization. In Pakistan, 32,000 square kilometers of irrigated lands (20%) and in India a vast 200,000 square kilometers (36%) are so salinized that they have lost much of their productivity. Yet these two countries have often been ranked among the prime exponents of Green Revolution agriculture. The world total of salinized lands can be roughly estimated at 600,000 square kilometers (22% of all irrigated lands).

In fact environmental constraints of several sorts are now causing significant cutbacks in food production at a time when population growth often continues with all too little restraint. Soil erosion leads to an annual loss in grain output that is roughly estimated at 9 million metric tons; salinization and waterlogging of irrigated lands, 1 million tons; and a combination of loss of soil organic matter (through burning of livestock manure and crop residues for fuel), shortening of shifting-cultivator cycle, and soil compaction, 2 million tons; total from these forms of land degradation, 12 million tons. Air pollution costs 1 million tons of grain each year, and flooding, acid rain, and increased ultraviolet radiation, another 1 million tons.

The total annual loss from all forms of environmental degradation, 14 million tons of grain, offsets almost half of the world's gains from increased investments in irrigation, fertilizer, and other inputs, 29 million tons. World population growth alone currently requires an additional 28 million tons of grain output each year (not including the demands of economic advancement and enhanced diets). The net gain in grain output is less than 1% per year, whereas population growth is 1.8%.

Generally technology, level of consumption or waste, and numbers of people interact in multiplicative fashion, i.e., they compound each other's impacts on the environment. Thus, both industrialized nations with high degrees of technology and consumption and developing nations with large populations can generate a vast impact on the environment and hence on prospects for sustainable development. Also influencing these three variables are socioeconomic inequities, cultural constraints, government policies, and the international economic order.

To illustrate how these variables interact, suppose that, by exceptional effort, humankind managed to reduce the average per-capita consumption of environmental resources by 5%, and to improve its technologies so that they caused 5% less environmental injury on average, thereby reducing the total environmental impact of humanity by roughly 10%. Unless global population growth were restrained at the same time, it would bring the total impact back to the previous level within less than six years.

One of the greatest challenges to humankind today is the poverty experienced by fully one billion people. Of the poorest fifth of households in developing countries, between 55% and 80% have eight or more members, whereas at the national level the proportion is only 15% to 30%. These people are unusually dependent for their survival upon the soil, water, forests, fisheries, and biotas, and are compelled to misuse and overuse these resources even at cost to their future prospects, thereby entrenching their poverty. In turn, this appears to reinforce their motivation to have large families, thus strengthening the prospect of ever-tightening constraints.

The plight of the world's poorest reflects some of the failures of development in general. They have been bypassed by the usual forms of development, notably Green Revolution agriculture; they cannot afford costly inputs such as high-yielding seeds, fertilizer, irrigation, and farm machinery. They are marginalized in that, lacking economic, political, legal, or social status, they can do little to remedy their plight. All too often this drives them to seek their livelihood in environments that are unsuitable for sustainable agriculture, being too wet, too dry, or too steep, causing deforestation, desertification, and soil erosion on a wide scale.

NORMAN MYERS

For Further Reading: Paul R. Ehrlich and Anne H. Ehrlich, *The Population Explosion* (1990); Nathan Keyfitz, *Population Growth Can Prevent the Development that Would Slow Population Growth* (1990); Michael Lipton, *The Poor and the Poorest: Some Interim Findings* (1985).
See also Acid Rain; Carrying Capacity; Desertification; Erosion, Soil; Floods; Green Revolution; Irrigation; Population; Ultraviolet Radiation.

POPULATION EXPLOSION

The term *population explosion* refers to the unusually rapid world population growth that has taken place in the second half of the 20th century. (Paul Ehrlich popularized the term *population bomb*; René Dubos preferred *population avalanche*). This unprecedented surge in growth resulted from improved health conditions in countries in Africa, Asia, and Latin America, along with increased food production and crop yields. The "explosion" has actually been caused by drops in the death rate, not by a rise in the birth rate.

It had taken all of human history for world population to reach the first billion, by about 1800. The sec-

ond billion arrived much faster, by about 1930. But, as the 20th century progressed, the world's population growth rate continued to rise, until it reached 2% per year in the 1960s. A growth rate of 2% will double world population in just thirty-five years.

The growth after World War II has raised much concern, particularly since over 90% of it takes place in countries of Africa, Asia, and Latin America. In 1992, world population stood at 5.4 billion, with 4.2 billion people in Africa, Asia, and Latin America. About 91 million people are added each year, and the number continues to rise. However, the number added would have been much higher if many countries had not adopted policies to slow the growth rate.

United Nations' projections show the world population growing to 6.3 billion by 2000, and to 8.5 billion by 2025. The U.N.'s projections show the world population ultimately stabilizing at about 11.5 billion, but only if birth rates in the countries of Africa, Asia, and Latin America continue to decrease without interruption.

In 1950, women in developing countries averaged over six children each during their lifetime—three times the level necessary to "replace" population. A *replacement-level birth rate* simply means that each couple in a society replaces itself with two children, not increasing the size of successive generations. In many industrialized nations, such as Europe and Japan, women have fewer than two children each. If that low birth rate is maintained, the populations of those countries will actually decrease.

While it is clear that birth rates are declining in the large majority of the world's countries, the outcome of the population explosion is far from clear. U.N. projections show a range of from about 6 to 21 billion people in the next century, depending on the future birth rate trend. But the actual number of people could, in fact, be even larger.

CARL HAUB

PRAIRIES

See Grasslands.

PRESERVATIONISM

Preservationism is a strand of environmentalism that seeks to preserve the natural environment from the effects (positive or negative) of human intervention. Preservationism is often confused with conservationism, a term that denotes a utilitarian approach to the use of land and other natural resources rather than a desire to see them remain unaltered or not utilized by humans. The debate that took place in the early twentieth century over whether or not to allow the city of San Francisco to dam the Hetch Hetchy River in Yosemite National Park is a classic example of how these two movements are opposed. The preservationists, led by John Muir, argued that humans had no right to alter the natural condition of wilderness for any reason, economic or otherwise. The conservationists, led by Gifford Pinchot and most notably Theodore Roosevelt, felt that San Francisco's need for water took precedence over the preservation of wilderness, and so the river was dammed in 1913. The preservationist-conservationist schism became a permanent feature of American environmentalism.

Preservationists have offered a wide array of arguments, many of them anthropocentric, as to why natural environments should be left unaltered. Saint Francis and John Muir advocated preservationism not only because they believed humans had no right to exploit God's creations and destroy other creatures but also because they believed humans "need" nature for spiritual, moral, and aesthetic reasons. Many people have called for the protection of wildlife simply to avoid the negative effects of environmental degradation on the quality of human life, such as pollution and its contingent health problems. Others fear that the sustainability of the biosphere itself, including the human life it supports, will be threatened unless a part of it remains free of modification.

Aldo Leopold, who wrote extensively on the beauty of various humanized landscapes, was one of the first to advocate wilderness preservation out of the belief that the only way to understand how ecosystems work is to study them in their natural condition, unaltered by human influence. He argued, in what has now become the conventional wisdom among ecologists around the world, that humans need this vital ecological knowledge in order to modify their behavior to meet ecological mandates and limits. Seeing countless benefits to humans, many people have called for the preservation of ancient and tropical forests, as well as other wilderness areas, because of the vast potential for discovering lifesaving drugs and other useful products that remain untapped in their flora and fauna.

PRICE SYSTEM

Price systems are close analogues to ecological systems in that they are highly complex and interdependent. Just as the parts of an ecosystem are intricately linked through the great chain of eating and being eaten, so is a modern industrial economy interconnected through the linkages of prices, production, consumption, and finance.

The notion that "everything is connected to everything else" pervades modern economics. Economies are connected by markets in the production sphere through the inputs and outputs that circulate through the world; they are connected through exchange of goods and services; and they are connected by flows of funds through which some people or nations finance the economic activity of others. It is generally believed that the great macroeconomic crises of the 20th century—the periodic banking panics, the Great Depression of the 1930s, the debt crisis of the 1980s, the breakdown in socialist economies of the 1990s—occurred because the coordinating function of markets failed, not because of a simultaneous burst of individual economic malfunctions.

But complexity does not imply functionality. Is there any reason to believe that the economy functions effectively? Is there any inner logic behind a system of prices and markets? This has been the central theme of economics for over two centuries.

Any observers of an economy will quickly conclude that there is a great deal of interconnection between the different parts: farmer plants and harvests grain, miller turns it into flour, baker turns it into bread, grocer sells it, and cafeteria makes a sandwich for ultimate consumption.

But a deeper look shows that the sequence is in fact circular rather than linear, for the sandwich might be eaten by the original farmer who planted the grain. The circularity of economic life was described and depicted by François Quesnay, a physician in the court of Louis XV and a member of the "Physiocrats," who likened the circulation of goods in an economy to the flow of blood in the body and developed the *Tableau Economique* to describe the way goods and incomes circulate through the economy. The contribution of the *Tableau* was to show the interdependence of economic life. As such, this conception does little to analyze whether the flow of incomes and outputs is efficient or inefficient, or the impact of different conditions on incomes or prices.

Adam Smith's contribution to economics was to show that markets channel individual self-interest to satisfy the general economic interest "as if by an invisible hand." Adam Smith's parable of the invisible hand was no more than assertion until the middle of the 20th century, when developments in mathematical economics allowed economists to understand the efficiency properties of the circular flow of economic life. Say we have millions of farmers, millers, bakers, candlestick makers, and other workers and businesses deciding on their seeding, sowing, and other decisions in an (apparently) completely uncoordinated way.

Firms and consumers make decisions completely on the basis of profits and satisfactions without any intrinsic interest in coordination or in other people's welfare.

How well or badly would this "economic ecosystem" perform? The surprise is that, under the conditions laid out here, the equilibrium is *efficient* in the sense of Vilfredo Pareto. "Pareto efficiency" means that no one can rearrange the firms, the production decisions, or anything else to reduce the amount of labor or other inputs that needed to produce the amount of final output actually produced. The *actual* amount of inputs is the absolute minimum that is needed to produce the bundle of consumption goods.

The key tool of markets is the value (or price) discovery and enforcement mechanism. That is, the ideal competitive market discovers the value of goods to consumers and the costs of production to firms. Prices provide signals to producers about where to expand or contract. For consumers, prices are ration coupons, forcing consumers to allocate their scarce dollar incomes among the available goods and services.

Thus in the ideal market economy, there *is* a coordinating mechanism at work above the level of the individual economic organism, namely prices, which signal the marginal value of goods to consumers and the marginal cost of goods to producers. In a truly competitive market, there is no need for a supercomputer or central planner to try to optimize the entire system, accounting for all the trillions of interactions among the different economic organisms, for the prices are providing the appropriate economic signals. As social engineers, we can allow each little firm and consumer to wander through the economy without our needing to worry about any interactions with other organisms as long as those interactions take place through the marketplace.

The result has a surprising corollary: In our idealized market economy, the usual efficiencies that are staple for physical scientists have no independent claim to virtue. As social engineers dedicated to the pursuit of human satisfactions, we need not fret about thermodynamic constraints or thermal efficiency or the extent to which we are adding to the universe's entropy. These physical constraints will be important only to the extent that they enter into the price of particular resources. Put differently, it is not sensible to minimize the Btu input into a production process; rather, we should minimize the Btus after each Btu is weighted by the appropriate price on that Btu.

This happy and oversimplified view of markets is unrealistically optimistic in a number of respects. What about the growing landfills, the ozone hole, and the threat of climate change? What about the issues concerning equity of the market allocation and growing inequality? Where are the depressions that have thrown millions out of work or the hyperinflations that have destroyed currencies?

These issues are ones of market failures, in which prices are inappropriately set. In these cases, the "invisible hand" of Adam Smith must be supplemented by the "helping hand" of governments. Government corrections are often most effective when they harness markets in the public interest—using taxes, subsidies, and auctions to ensure that individuals have the right incentives to save and invest, produce and consume.

WILLIAM D. NORDHAUS

For Further Reading: Robert Dorfman and Nancy S. Dorfman, eds., *Economics of the Environment: Selected Readings* (1977); Robert L. Heilbroner, *The Worldly Philosophers* (1980); William D. Nordhaus, "The Ecology of Markets," in *Proceedings of the National Academy of Sciences* (February 1992).
See also Economics; Smith, Adam.

PRODUCTIVITY

Economic productivity is the rate at which an economic unit, such as a worker, a factory, or a nation, produces goods and services. It is a measure of the ratio of output to one or more of the inputs, usually labor, used to produce that output. A nation's output is usually measured by its national income (NI) or gross domestic product (GDP), expressed in constant prices. Labor may be measured by employment, total hours worked, or by a sophisticated measure that assigns weights based on earnings to groups who differ in education, demographic characteristics, and other attributes. Alternatively, input may include labor, capital, and land.

Economic productivity has risen since the industrial revolution, which began in the late eighteenth century in England, the early nineteenth century in the United States, and later elsewhere. Many countries began such economic growth only recently and others have yet to begin. Because of this diverse history, output per worker varies enormously among different regions of the world, far more than in earlier centuries. Among advanced countries, however, trends since 1950 are characterized by convergence.

By 1900, the United States surpassed Great Britain in output per worker and widened its advantage until after World War II. In some countries that were far below the leaders in 1900, output per worker grew faster than in the United States from 1900 to 1950. Nevertheless, as late as 1960, worker productivity was about twice as high in the United States as in northwestern Europe and four times as high as in Japan.

From 1948 to 1973 productivity grew at an unprecedented rate throughout the industrialized world, but this was followed by a period of slow growth. In the United States, GDP per hour in private business grew about two percent a year from 1889 to 1948, three percent from 1948 to 1973, and one percent from 1973 to 1991.

Table 1 compares GDP per employed person in thirteen advanced countries. In the time periods from 1950 to 1973 and from 1973 to 1990, the United States' growth rate was below that of any other country shown in the table. The rate dropped sharply in the second period (1973–1990) in all the countries, and the decrease was typically abrupt, starting with the yearly change from 1973 to 1974. The United States remains the industrial country with the highest GDP per person employed, but the margin of difference has been greatly reduced.

Among the long-term determinants of a nation's productivity, the most important have been advances in technological and organizational knowledge of how to produce at low cost. Other major determinants include the education, experience, working hours, and other characteristics of labor; the available capital and land; the efficiency with which resources are allocated among uses; the size of markets; and the quality of management. Crucial in comparisons of unlike countries and eras are institutional arrangements, religious beliefs, and attitudes toward work, saving, and change in general.

Table 1
Real gross domestic product per employed person: comparative levels and growth rates, thirteen countries

Country	Real GDP per employed person (United States = 100)		Growth rate (percent per year)	
	1960	1990	1950–1973	1973–1990
United States	100.0	100.0	2.0	0.5
Canada	77.8	92.3	2.5	1.1
Japan	23.8	76.3	7.5	2.9
Austria	38.7	73.9	4.8	2.1
Belgium	49.9	88.1	3.7	2.1
Denmark	52.3	68.1	3.5[a]	1.3
France	47.0	89.1	4.6	2.3
Germany (West)	48.7	78.6	4.8	1.8
Italy	41.5	87.8	5.7	2.2
Netherlands	56.7	77.1	3.8	0.8
Norway	49.8	79.8	3.5	2.2
Sweden	51.7	67.2	3.4[b]	1.0
United Kingdom	53.9	70.7	2.6	1.4

[a]1955–1973
[b]1960–1973
Sources: U.S. Bureau of Labor Statistics, "Comparative Real Gross Domestic Product, Real GDP Per Capita, and Real GDP Per Employed Person, Fourteen Countries, 1950–1990" (July 1991) and "1960–1990" (January 1992)

Legislation to protect the environment diverts productive resources from producing measured output to producing benefits that, in large part, are not measured. Consequently, such legislation usually reduces measured output. The estimated reduction in the annual growth rate caused by rising pollution abatement costs reached 0.12 percentage point from 1973 to 1982, and then it receded slightly.

EDWARD F. DENISON

PROPERTY RIGHTS

Property ownership is defined by two attributes: the range of uses that society permits and the obligation to use the property responsibly, again as defined by society. For example, no one may operate an automobile without the owner's permission, but the owner is obliged to use it in conformity with traffic and parking laws. The way that society defines property rights, or fails to define them, is at the heart of many environmental problems. For example, water pollution can be a problem if society does not define the quality of water to which downstream users are entitled. If this property right were defined and enforced, upstream polluters could not discharge pollutants beyond a certain level without the permission of downstream users. If society chooses to reserve the quality of water and not allow downstream users to sell water quality, society would be choosing to ban certain trades that might be desired by both polluters and downstream users.

When property rights are poorly defined, individuals tend to overuse a resource because they do not bear the full cost of their action. For example, on the American frontier the bison was almost exterminated because buffalo hunters did not have property rights in live animals, only dead ones. Any hunter who postponed shooting a buffalo did not benefit from such an action because these animals were common property owned by everyone. Another hunter had an equal right to shoot them. Similarly, in the absence of well defined property rights, air and water are treated as common property and are thus overused.

Property rights are not stable. They evolve in response to changes in demand and in the costs of defining rights. For instance, the lack of fencing materials on the western frontier made defining property rights to land prohibitive. However, with increased settlement, land became more valuable, and the invention of barbed wire dramatically lowered the costs of enforcing rights. Likewise, as environmental amenities become more valuable, it will be worthwhile to spend more on enforcement. In addition, new technology has the potential for reducing enforcement costs. For example, it may be possible to "brand" the effluent of every factory with an inert chemical, thus facilitating the use of the court system to assess liability for pollution.

P. J. HILL

PUBLIC HEALTH

Public health is the art and science of preventing illness and injury. The practice of public health requires organizing community efforts for a safe environment, control of infectious diseases, education of individuals in personal hygiene and lifestyle behavior decisions, evaluation and enhancement of systems of medical and nursing services, and advocacy of social policies and social machinery to assure everyone a standard of living adequate for maintenance of good health.

"Health" has been defined by the World Health Organization as "a state of complete well-being, physical, social, and mental, and not merely the absence of disease or infirmity." "Public" represents the necessity to address threats to good health on a community-wide or worldwide basis, not just through individual actions. Public health should not be confused with "public medicine," commonly understood as medical care for the indigent. The principles and aims of public health apply to everyone.

The worldwide eradication of smallpox is an outstanding example of public health in action. The idea of preventing smallpox by vaccination was conceived and put into practice in England by Edward Jenner in 1798 based on his observations of the protection milkmaids gained from exposure to cowpox; that was long before we even imagined viruses. Public health actions need not await full understanding of the underlying science. However, no amount of individual vaccination or treatment of unvaccinated patients would have rid us of smallpox. Epidemiologic studies, reliable reporting systems, organized distribution of vaccine, efficient maintenance of the cold-chain for vaccine, and mass methods for vaccination were required: a global systems approach. Similar commitments have been advocated and are needed for other preventable diseases ranging from measles and polio to childhood lead poisoning.

Public health may be thought of as health care on a population basis, or community basis, complementary to the role of medicine as health care for individuals. Public health stresses prevention and proactive approaches, while medicine provides diagnosis and treatment and generally reacts to sick or injured patients. The health promotion aspects of public health require a broad view, including the important roles of decent housing, effective education, adequate income, protection of human rights, freedom from war and terrorism, and satisfying work in providing the foundation for healthy lives.

The core science of public health is epidemiology. Epidemiology is the systematic, objective study of the natural history of illness and injury in populations and of the factors that cause, predispose to, or ameliorate disease patterns. Epidemiology can be applied to injuries, occupational hazards, environmental sources of illness, the effectiveness of medical care, and all other aspects of public health and medical practice.

Other essential subfields in public health are biostatistics, environmental and occupational health, risk assessment, health administration and health management, health services research and health policy, socio-behavioral research and health education, and biological sciences applied to disease processes.

Biostatistics defines methods of data collection and data analysis crucial to hypothesis-testing in research and practice, including clinical trials of treatments and preventive interventions. Environmental and occupational health embraces toxicology, industrial hygiene and safety, environmental/occupational epidemiology, occupational medicine and nursing services, regulatory policies, and other preventive approaches to protection of the environment and people from physical, chemical, biological, and psychosocial risks. Risk assessment has become a formal process of wide generality, relying on scientific methods of data gathering for hazard identification, well-chosen guidance for characterization of risk in qualitative and quantitative terms, and then policy decisions about communication of risk information and actions to control or reduce exposures and risks.

Health administration and health management have become increasingly important in every society, in order to better utilize human and fiscal resources. Appreciation for the human side of health services, both the health workers and the patients and community served, and understanding of the special aspects of health care and health promotion, have made health administration and health management a career path distinguishable from general business administration or public administration. Evaluation of the effectiveness and cost-effectiveness of health services, especially in terms of health outcomes and the health status of the people served, is the domain of health services research; its findings inform health policy debates at all levels.

Since individual behavior is so important to personal health, sociological and psychological inquiry is a crucial field for health education and health promotion; its complementary inquiry into community and societal actions and motivations is similarly important for health policy. Attention to cultural and racial differences is a central emphasis of research and practice in public health. The HIV/AIDS epidemic and other sexually transmitted diseases reveal the importance of certain behaviors, in this case sexual behaviors and drug abuse. In all groups, poverty is a major factor in ill health. Finally, advances in public health depend upon knowledge of the biology of humans and other living things and improvement of our capability through immunology, toxicology, genetics, nutrition, parasitology, cell biology, biotechnology, and vaccine development to diagnose and prevent important diseases here and abroad.

Organizationally, there are health departments of local and state governments, the U.S. Public Health Service, and the World Health Organization in the leadership of the public side of public health efforts. The six agencies of the U.S. Public Health Service are the Food and Drug Administration, the Centers for Disease Control, the Agency for Toxic Substances and Disease Registry, the National Institutes of Health, the Alcohol/Drug Abuse/Mental Health Services Administration, and the Health Resources and Services Administration. There are also many private sector organizations working for the public's health, including health care professional organizations, disease-oriented organizations such as the American Cancer Society and the American Lung Association, and pharmaceutical, medical device, health education, and publishing companies. Universities and their schools of public health play an essential role in education, research, and local, regional, and national service. There are 26 schools of public health in the United States, allied in a network under the Association of Schools of Public Health and cooperating with state, county, and city health officers' associations, the preventive medicine teachers in medical schools, and the federal agencies.

As described in the 1988 report "The Future of Public Health," three core functions of public health must be addressed by agencies at all levels of government and must be taught by schools of public health:

1. Assessment: to regularly and systematically collect, assemble, analyze, and make available information on the health of the community. These data should include statistics on health status of various subgroups in the population, community health needs, and epidemiologic and other studies of health problems.

2. Policy development: to use scientific knowledge in generating comprehensive public health policies and to lead decision-making processes.

3. Assurance: to encourage actions by private sector or other public sector providers, to require action through regulations, or to provide services directly to assure constituents the services necessary to achieve agreed-upon goals and objectives arising from the assessment and policy development functions. In recent decades, local health departments have had to act as "provider of last resort" to provide family planning, well-baby care, alcohol and drug abuse treatment services, acute and chronic medical care, and mental health services to segments of the population not covered by health insurance or inadequately reached by the private sector.

Cooperation with legal authorities and social agencies is often essential, as in child abuse situations. Similarly, cooperation with EPA and state and local environmental agencies is essential for prevention and control of environmental risks.

Since 1979, the federal government has provided useful leadership in what is now known as "Health Promotion and Disease Prevention" with the publication of the book *Healthy People* and the subsequent blueprints *Health Objectives for the Nation 1990* and *Healthy People 2000*. This initiative has involved thousands of individuals and organizations. Setting numerical goals for improvements in health status, reductions in risk factors, enhancements in professional and public awareness, and access to data within a comprehensive framework has energized the field of health promotion and offers budget makers and citizens a basis for evaluating the progress and the payoff from investment in such efforts. The Healthy People framework embraces:

1. Lifestyle health promotion behavioral decisions: physical activity and fitness, nutrition, smoking, alcohol and other drugs, family planning, mental health, violent and abusive behavior, educational and community-based programs.

2. Community health protection actions: unintentional injuries, occupational safety and health, environmental health, food and drug safety, and oral health (including fluoridation).

3. Preventive services by health professionals: maternal and infant health, heart disease and stroke, cancer, diabetes and chronic disabling conditions, HIV infection, sexually transmitted diseases, immunization and infectious diseases, other clinical preventive services.

4. Surveillance and data systems.

In 1979 priority goals were set for each age group. In 1990 three overarching goals for the year 2000 were added: to increase the span of healthy life (quality-adjusted life years), to reduce disparities in health status among racial groups, and to achieve access to preventive services for all Americans. The underlying principle is that long life without good health is unsatisfactory; prevention of disability and promotion of better health for all are the aims. The framework was fleshed out with 300 objectives designed as quantitative and realistic targets.

Progress can be illustrated with the following targets for each of five major stages of life:

1. For infants, to improve health and reduce infant mortality by 35% to fewer than 9 deaths per 1,000 live births by 1990 from 14 in 1977 (and 29 per 1,000 in 1950). This goal has been achieved, but the U.S. compares very badly with other industrialized nations, ranking 23rd, and the overall rate masks the fact that black babies are twice as likely as white babies to die before their first birthday. In fact, the ratio of black to white infant mortality increased during the 1980s. Low birth weight is the greatest hazard to infant health; about 60% of those who die before age one year are of low birth weight. Risk factors in the mother are lack of prenatal care, smoking, alcohol use, drug use, young age, and low socioeconomic status. The year 2000 goal is no more than 7 per 1,000 live births, with no more than 9 per 1,000 black infants.

2. For children, age 1–14 years, to improve health status, foster optimal childhood development, and reduce deaths by 20% to fewer than 34 per 100,000. That goal was reached by 1985 and progress continues. Nearly half of all childhood deaths are from unintentional injuries, nearly half of those from motor vehicle accidents. Child safety seats were a major contributor to progress. On the negative side, homicides have been increasing. The year 2000 goal is no more than 28 deaths per 100,000 population.

3. For adolescents and young adults, age 15–24, to improve health habits and reduce deaths by at least 20% to fewer than 93 per 100,000. The rate declined from 114 in 1977 to 96 in 1983, but since has increased to 104. The leading causes are motor vehicle accidents (75% of the total), homicides, and suicides, with HIV/AIDS on the rise. The year 2000 goal is no more than 85 deaths per 100,000 population.

4. For adults, age 25–64 years, to reduce deaths by 25% to 400 per 100,000. This goal was achieved in 1990, reflecting favorable changes that began in 1970 in cigarette smoking, control of high blood pressure, and blood cholesterol levels for heart disease and stroke, and lower rates of alcohol use, increased seat belt use, and lower speed limits leading to substantial declines in mortality from motor vehicle accidents. The decline in heart disease mortality has made cancer the number one killer in this age group (with its overall rate essentially unchanged). The year 2000 goal is no more than 340 deaths per 100,000 population.

5. For older adults, age 65 and older, to reduce the average annual number of days of restricted activity due to acute and chronic conditions by 20%, to fewer than 30 days per year. Substantial progress was made, though short of the goal. Falls and fractures are one key preventable cause of restricted activity and bed disability. For year 2000 the goal has been modified to reduce the proportion of older adults who have difficulty performing two or more personal care activities (dressing, eating, washing, moving about).

Among the public health problems to be addressed, we might classify the HIV/AIDS epidemic and the need for financed access to clinical and preventive services for the uninsured as crises; other problems are enduring challenges—including injuries, teen pregnancy, smoking and other substance abuse, high blood pressure, major chronic diseases, water quality, food safety, and exposures to toxic substances—and some are newly emerging, such as Alzheimer-type dementia in our rapidly aging population and the potential nightmare of multiple-drug-resistant tuberculosis, especially in combination with HIV infection and as a threat to health care workers.

Globally there is belated recognition that the greatest threat to public health is the continued rapid growth of the human population and the related problems of poverty and exploitation and degradation of the environment. The key elements of population growth are numbers—the lag between the decrease in death rates and the decrease in fertility and birthrates, the standard of living, and transboundary movements. To have sustainable development and sustainable societies we must seek to stabilize the global population by reducing fertility rates to replacement levels. Current regional upheavals exacerbate the stresses of human society on the environment and on public health, in general.

One of the major public policy challenges for the nineties and beyond is the need for health-care reform. The task involves moderation of cost increases, which have long been far outstripping general inflation, extension of private or public health insurance coverage to all Americans, and assurance of good quality of care. National health care expenditures have risen from $27 billion in 1960 to $74 billion in 1970, $250 billion in 1980, $666 billion in 1990, and an estimated $830 billion in 1992 (13% of GNP). The public health world and citizens at large have an interest in making sure that prevention has a significant role in health care reform. This reform includes funding for public health from both private and public sources, as part of the total health care sector.

GILBERT S. OMENN

For Further Reading: Department of Health and Human Services, *Healthy People 2000* (1991); National Center for Health Statistics, *Health, United States, 1991 and Prevention Profile* (1992); Gilbert S. Omenn, Jonathan E. Fielding, and Lester B. Lave, eds., *Annual Review of Public Health,* volume 13 (1992).

See also Contraception; Health and Disease; Population.

PUBLIC UTILITY COMMISSION

See United States Government.

PUMPING STATIONS

Pumping stations are used to elevate sewage and stormwater or to transfer them to other locations when the gravity flow used to transport them results in excessive depths or carries them to locations inappropriate for processing or disposal. Pumping stations usually consist of a wet well, where incoming flow collects, and a series of pumps that are used to lift the sewage from the wet well. Usually centrifugal pumps are used and additional standby equipment is provided for emergencies and power outages.

Pumping stations are classified according to their capacity, source of energy, type of construction, and function. Small-capacity pumping stations may be predesigned and factory assembled and selected from a catalog. These are often called package pumping stations and are widely used to elevate sewage in a single sewer line. Larger pumping stations, such as those required for stormwater pumping or for large interceptor sanitary sewers, must be designed individually.

General features of a pumping station include a physical facility that contains the pumps, energy source fans and vents to exhaust sewer gases, and appurtenances that allow the removal and repair of equipment. Larger pumping stations used for storm and sanitary sewage may also contain screens for the removal of materials that may damage pumps. Often dehumidifiers are provided to minimize corrosion of equipment, and heaters are also added.

Wet wells are used to store the wastewater before it is pumped. The storage volume depends on the number and type of pumps, the type of wastewater (storm or sanitary), and the fate of the pumped wastewater. Because sewage is stored, volatile gases may be generated, so explosion-proof and spark-proof equipment must be used. Provisions must also be made for cleaning, so a sloping floor and solids-collection system should be installed.

In order to prevent clogging of pumps, screens are usually required. Because most clogging problems result from the buildup of rags, screens capable of removing rags and other materials are used. Most often, screenings are cut into small pieces and fed back into the system.

Large pumping stations contain automatic controls that respond to wet well water levels, provide for sequential operation of pumps, and provide alarm systems for pump failure and the buildup of toxic gases. Flow measurements are also provided. Although the type of pumps used depends on the type of flow and use, nonclogging, centrifugal pumps are preferred.

JOHN T. NOVAK

PURCHASING POWER

Purchasing power shows the degree to which spendable funds have command over the things to be obtained in the market place. To have income, or funds from other sources, for spending on items of living is not enough, by itself, to enable one to obtain the things that are wanted. It all depends on the prices of things that are to be purchased. The funds to be spent now, or saved for future spending, must be jointly considered, together with the prices of things to be bought, in order to determine the purchasing power of funds.

The purchasing power of an income or other source of funds available for spending can be calculated as a ratio of the funds to the average price of things to be bought. Economic statisticians divide monetary value of funds by an index of general prices in order to calculate the real value of the funds. These real values are indicative of purchasing power.

Workers, for example, are not solely interested in the amount of wages they earn; they are more interested in the real value of their wages because these show their purchasing power in the market place. That is why workers generally bargain, at the beginning of a negotiation, for wages that keep up with price changes—mostly increases in modern economic society—in order to preserve the purchasing power of their wages. They also bargain, next, for larger wage gains than simply the amounts that will keep up with the price increases because they want to improve their purchasing power. The main source of the improvement factor comes from productivity gains and other efficiency improvements.

Other kinds of income besides wage income, such as interest income, profit income, or rental income, also have purchasing power. The recipients of these other forms of income regularly watch their purchasing power by comparing their income gains with average price changes in order to be assured that their purchasing power improves as society advances.

In international economics there is a concept of *purchasing power parity*, which compares exchange values of national currencies against one another in relation to the price levels of one country compared to another. For example, purchasing power parity in one form suggests that the exchange value of Country A's currency over Country B's currency would change in the same proportion that Country A's average prices change in relation to Country B's average prices. There is only, in practice, a general tendency of currency exchange rate movements to change in proportion to the movement of relative inflation rates between the two countries, but it might be expected to be a valid prop-osition, on average, over long periods of time. In the short run, other things besides relative inflation rates appear to affect currency exchange rate movements. Also, the concept of purchasing power parity seems to be more realistic if prices or inflation rates are meas-ured by goods and services that are traded internation-ally.

L. R. KLEIN

QUALITY OF LIFE

The phrase "quality of life" refers to (1) a general description of the nature or conditions of the life of an individual or group, and (2) an evaluation of those conditions—the extent to which certain life conditions are desirable, enjoyable, active, important, or satisfactory.

For example, comparing the lives of the 19th-century American coal miners with those of contemporary American coal miners or noting the high-energy consumption of Americans compared with that of the Chinese addresses the first notion of quality of life.

The concept of "well-being" refers to the second meaning of quality of life; it concerns how an individual might evaluate the general goodness or badness of an experience. Because of its greater precision and sharper focus, the well-being concept has been the subject of considerable research, much of which has occurred as a part of the Social Indicators Movement.

The Social Indicators Movement was motivated partly by the need for a broader range of life quality measures than the limited set of demographic, economic, and health indicators that were available in the late 1960s. For example, educational, social, and environmental programs in the U.S. and in developing countries were intended to increase well-being, but their effects often were not reflected by the existing financial or health statistics. In addition, there was a sense that government programs should be guided by better knowledge of current levels of well-being. It was thought, for example, that the urban riots in American cities during the late 1960s and early 1990s could have been avoided if an established social monitoring system had signaled to all segments of society the discontent that induced the disruptions.

The Social Indicators Movement blossomed during the 1970s. Major international organizations, including the United Nations, the United Nations Educational, Scientific, and Cultural Organization (UNESCO), and the Organization for Economic Cooperation and Development (OECD), instituted social indicator programs, as did many individual nations through their census bureaus or ministries of statistics. Also, several groups of university-based scientists began major social research programs to develop and monitor measures of well-being. An international journal, *Social Indicators Research*, a newsletter, *SINET*, and a special

section within the International Sociological Association were established.

The Social Indicators Movement has made substantial progress on several issues: (1) What aspects of well-being should be measured? (2) What are the effective ways to measure those aspects? (3) What is the quality of the obtained measurements? (4) What are the levels of well-being in specific population groups? (5) What are the relationships between well-being and demographic and social variables such as age, gender, race, education, and wealth? (6) How do people come to evaluate their own lives as they do? (7) What changes have occurred in levels of well-being over time, and how do these changes differ for specific groups? In the process of addressing these issues, several conceptual refinements have emerged. One is the distinction between objective versus subjective (perceptual) indicators.

Objective indicators are based on counts or ratings obtained from administrative records or direct enumeration called for by a panel of experts. In contrast, subjective (or perceptual) indicators are based on individuals' own reports of their well-being.

For example, objective indicators relevant to housing include cost and availability of suitable dwellings, amounts of living space, and accessibility to shops, services, and workplaces. Subjective indicators on housing would include people's answers to questions such as "How satisfied are you with your house or apartment?" and "How satisfied are you with your neighborhood?"

Objective and subjective measures of well-being have been found to complement each other. The information provided by the objective indicators is somewhat different from, and does not substantially overlap with, what one learns from the subjective indicators.

Objective Measures of Well-Being

Objective measures of well-being typically come from government agencies and are produced as part of the agency's regular administrative procedure. Often the data are collected to enable the agency to carry out its primary mission and are used as measures of well-being only secondarily.

Part of the challenge posed by these objective measures has been to link the data assembled by government and other agencies to individuals' experienced levels of well-being. Some such linkages are straightforward. For example, it seems obvious that an increase in the number of disability-free work days would indicate that the well-being of workers improved. However, the implications for well-being of an increase in the divorce rate, or in the proportion of families on welfare – to mention just two examples – are not so obvious and have generated debate.

The range of topics that might be assessed by objective measures of well-being is extremely broad. In an attempt to bring conceptual order to the field, in 1973 the Social Indicators Development Program at the OECD assembled a list of social concerns that were common to most of its member countries. The list identifies eight main topics: (1) Health; (2) Individual development through learning; (3) Employment and quality of working life; (4) Time and leisure; (5) Command over goods and services; (6) Physical environment (housing, pollution, etc.); (7) Personal safety and the administration of justice; (8) Social opportunity and participation.

This list has been influential in guiding the development of objective measures of well-being. Most developed countries and some developing ones, as well as some states or provinces and some cities and towns, have issued social indicator reports that present statistics on these and related topics.

A problem with objective measures of well-being has been finding a way to arrive at a general, overall measure. Although one can measure well-being with regard to the eight life concerns listed above (and others), how to combine that information into a single "global" measure has provoked much debate. Various means, including factor analysis, have been tried for combining the objective measures, but no approach has proven theoretically compelling.

One of the most widely used objective measures for comparing life quality in different geographic regions (e.g., countries or states/provinces) has been the Physical Quality of Life Index. This index combines three indicators: life expectancy, the literacy rate, and the infant mortality rate. The obvious relevance to well-being of these indicators, and the fact that such data are collected and published by most governments, have contributed to the popularity of this measure. Although the accuracy of the information on these measures varies, it is widely believed that these data are likely to be more valid than many other government statistics.

Subjective Measures of Well-Being

The primary means of determining which aspects of subjective well-being should be measured has been to ask people to identify what matters to them in life and what changes would significantly improve (or worsen) their lives. Their answers have provided a rich list of life concerns. Then in a survey or poll, individuals are asked how they evaluate these concerns.

These concerns have included all of the eight topics listed above as relevant for objective measures plus others that are psychologically "closer" to individuals. Important to individuals' well-being are how they feel about themselves (e.g., what they are accomplishing, how they handle problems); their marriages, children, and family life; how they spend their spare time; and the respect they receive from other people. Such concerns are rarely assessed using objective measures of well-being but prove to be strong predictors of people's feelings about their life as a whole.

Many different response scales have been used for recording people's assessments of specific life concerns. Examples include scales that range from feelings of "delighted" to "terrible," and from "completely satisfied" to "completely dissatisfied." These scales tend to produce the same general patterns. However, the scales differ in the extent they reflect affective (emotional) or cognitive (rational/intellectual) components, in the precision with which they tap individuals' feelings, and in the ease with which they can be used in different types of surveys. When properly used, these response scales have been shown to produce measures with good validity and reliability.

The issue of how to measure well-being at the global or general level does not arise in the subjective realm. One can simply ask people how they feel about their "life as a whole," or how happy they are "in general." People answer such questions quickly, easily, with substantial reliability, and in ways that link meaningfully to their reports about more specific aspects of their lives.

Well-Being and the Environment

Life quality, including individual well-being, and the natural environment are intimately linked in both causal directions: quality of life affects the environment, and the environment affects life quality. It seems helpful to distinguish between two types of linkages. One type focuses on long-term causal effects involving environmental factors and life quality. This type is most clearly visible over time and at macro levels for units such as whole societies, major geographic regions, or planet Earth. The second type of linkage operates at a more micro level and addresses short-term or immediate cross-sectional relationships (i.e., over individuals) between environmental factors and individuals' well-being.

At the macro level, increasing attention has been devoted to identifying ways in which the quality of life as inferred from life styles influences the global environment. High-energy-consumption living in industrial countries is linked to global warming through generation of large amounts of carbon dioxide, and to depletion of the ozone layer through release of CFCs. In many developing countries in Asia and South America, the rapid rate of population increase and the needs of that burgeoning population for increases in material well-being, decreasing biodiversity through depletion of forests and wetlands, have potentially destructive impacts on the natural environment.

At the micro level, there is little empirical research that actually examines relationships involving the natural environment and the life quality of individual people. Existing studies have principally addressed the impacts of the man-made physical and social environments on the life quality of specific groups of individuals. An exploratory analysis, conducted in 1992 on a large sample of American adults, found virtually no substantial relationships between their subjective well-being and several attitudes and behaviors regarding the natural environment—including concern about pollution, attitudes about population control, enjoyment of outdoor activities, and recycling behaviors. Most individuals apparently did not experience their own life quality being tied directly and immediately to some of the popular concerns about the natural environment.

Of course, the two types of linkages, macro and micro, are interdependent. The long-lasting environmental phenomena have a current and direct impact on individuals' well-being, and individuals' current life styles have a long-term impact on the natural environment. However, scientists are still exploring these linkages between life quality and the environment, and a full understanding of how the macro and micro linkages relate to one another has yet to be attained.

FRANK M. ANDREWS

For Further Reading: Frank M. Andrews and Stephen B. Withey, *Social Indicators of Well-Being: Americans' Perceptions of Life Quality* (1976); Michael Carley, *Social Measurement and Social Indicators: Issues of Policy and Theory* (1981); Richard Doll, "Health and the Environment in the 1990s," in *American Journal of Public Health* (1992).
See also Social and Economic Infrastructure; Standard of Living.

QUALITY STANDARDS

Quality standards are those regulations that dictate the acceptable levels of toxic substances in the environment. The standards are determined by assessing how much of the various pollutants can be discharged into the environment without adversely affecting the desired quality of the environment. The identification of what levels of pollution can be allowed while still maintaining environmental quality can be guided by the specific goals set forth in environmental legislation. One of the goals of the Clean Water Act, for example, is to restore and maintain the chemical, physical, and biological integrity of the nation's waters. Therefore, quality standards must seek to achieve these goals by limiting the maximum amount of each pollutant to a level where it does not diminish the chemical, physical, and biological integrity of the water. One of the goals of the 1990 Clean Air Act amendments, however, quite specifically seeks the reduction of the amount of air pollution by 56 billion pounds a year to achieve quality standards. Quality standards, stating exact limitations on exposure levels for pollutants, are set forth in the act's National Air Quality Standards program.

Quality standards are broadly based on the legislative goals. However, the exact allowable pollution levels are set according to the scientific and technical data available concerning the danger to public health inherent in each pollutant. Regulators must have access to results from epidemiological and animal studies on each pollutant so that they may determine at what level that pollutant begins to diminish quality and harm the natural ecosystem and human health. Without this information, it is hard to know exactly which pollutants, and what levels of them, diminish quality. For example, the Standards and Enforcement provision of the Clean Water Act states that when a pollutant is found to interfere with the attainment or maintenance of water quality, effluent limitations will be placed on that pollutant. Many problems occur during the determination of the effluent limitations, since definitive scientific data on the effects of many pollutants are difficult to obtain.

RADIANT ENERGY

See Energy.

RADIATION, IONIZING

Natural Radioactivity

Ionizing radiation is a natural phenomenon, an inevitable part of our Earth and our universe. Humans have adapted natural sources of radiation for their use and have made other artificial sources for industrial, medical, research and other uses. These developments have taken place in less than a century since humans first became aware of ionizing radiation.

Among the 92 elements occurring naturally on Earth, all those with atomic numbers above bismuth (number 83) have unstable nuclei which are naturally radioactive. In their efforts to become stable, they emit either alpha radiation or beta radiation, often with gamma radiation as well. Some atoms are unable to achieve stability in a single transformation. Instead, they give rise to other unstable atoms and release further emissions. These constitute a chain or series of radioactive nuclides. In addition, some elements with lower atomic numbers, such as potassium and rubidium, have naturally occurring long-lived radioactive isotopes.

Each radioactive nuclide decays with a characteristic half-life ranging from fractions of a second to many millions of years. Those present on earth now, so long after its formation 4.6 billion years ago, must either have a very long half-life (such as uranium 238) or be a member of a radioactive series with a long-lived parent (such as radon 222 whose parent, radium 226, is a member of the U-238 chain). Some other radioactive elements are produced continuously by natural processes. For example, carbon 14 is produced in the atmosphere by the interaction of cosmic ray neutrons with nitrogen. Since the discovery of artificial radioactivity in 1934 and nuclear fission in 1939, humans have been able to make radioactive forms of all the elements. Over a thousand radionuclide species are now known and available for human use. Alpha particles are helium nuclei, positively charged, slow mov-

ing, and of low penetration. They are easily stopped by a few sheets of paper or a thin layer of skin. Beta particles are electrons, negatively charged, fast, light, and easily scattered. They are able to penetrate several millimeters into tissue. Gamma rays are electromagnetic waves with shorter wavelengths than visible light. They have no charge, and are therefore not deflected in a magnetic field. They are very penetrating and will pass through many centimeters of tissue and even right through the human body.

In addition to these natural types of radiation, humans can make others such as x-rays (similar to gamma rays). The machines used in medical diagnosis produce relatively low-energy "soft" x-rays which penetrate the human body to some extent and provide sharp contrast between various tissues and structures such as bone. More energetic "hard" x-rays can pass right through the body or through thick metal slabs, such as a ship's hull. They can be used for detecting flaws in metal.

Radioactivity occurs naturally in a wide variety of circumstances. Uranium and its radioactive products are found in virtually all soils, in water, and in the air. Because of this, all of our food contains traces of these materials and we take in a certain amount of radioactive uranium, thorium, radium, lead, and especially potassium every day. Some of these materials lodge in our tissues, especially in our bones, and come out again only slowly, if at all. These constitute the "body burden" of radioactivity that we all have and accumulate with time. The main components are radium 226, potassium 40, rubidium 87, and some carbon 14. This is a perfectly natural condition for all who inhabit Earth.

The Radiation Dose

The dose of radiation is expressed in sieverts (Sv) or millisieverts (mSv), i.e., thousandths of a sievert, where 1 sievert is the equivalent dose received by tissue in which 1 gray is absorbed. A gray is the absorbed dose corresponding to 10,000 ergs of energy per gram.

From the amount of radioactivity present in our bodies it is possible to estimate the annual dose (or absorbed energy) delivered to the body from natural radiation sources. This amounts to about 0.3 mSv per year from the radionuclides in our food and water. In addition, the body receives radiation from the rocks, soil, and building materials surrounding us. This also amounts to about 0.3 mSv annually, but varies somewhat with location and circumstances.

Cosmic radiation from outer space is yet another source of radiation exposure. Although filtered somewhat by the atmosphere and sometimes diverted by the magnetic field of the Earth, it still delivers to the body about another 0.3 mSv per year at sea level. These three sources total about 1 mSv per year from natural background sources other than radon.

Cosmic radiation increases with altitude and at typical jet airplane altitudes (30,000–40,000 feet) contributes several times the "at sea level" dose, about 0.01 mSv per 1,000 miles or two hours of air travel. A jet flight of 10 hours would add 0.05 mSv to a traveler's dose. A frequent flier, flying 100,000 miles a year or more, adds 1 mSv or so to his or her normal natural background exposure, doubling the contribution from sources other than radon.

Cosmic radiation is a special problem for manned space flight of long duration. The dose rate from galactic cosmic radiation a few thousand miles from Earth is about 500 mSv per year, several hundred times the natural levels on Earth. Consequently, manned missions in space may be limited by radiation considerations.

A special problem of natural radiation exposure on Earth is the result of the fact that one member of the uranium series, radon 222, is a gas. Because uranium is everywhere in soil, the Earth is constantly releasing radon into the atmosphere and this causes exposure of the lungs as people breathe it. This exposure is increased because we live in houses and work in buildings into which radon seeps and is often trapped in concentrations several times higher than in the outside air. The concentration of radon is variable from place to place and house to house. On average, radon contributes an effective dose of about 2 mSv per year, or more than all the other natural background sources put together.

The dose to individuals from all natural sources totals, on the average, about 3 mSv per year, of which 2 mSv per year is from radon. In addition, the average individual will be exposed to a certain amount of radiation from medical procedures (especially later in life) and inadvertently to small levels of radiation from a variety of sources, such as smoke detectors and building materials that contain small amounts of radioactivity. Tiny levels of exposure also result from fallout from nuclear weapons tested years ago, but this contribution is diminishing with time. Under normal conditions, only a very small exposure comes from the nuclear fuel used to provide nuclear power.

Biological Effects of Ionizing Radiation

Ionizing radiation can have harmful health effects if the dose to the exposed individual is large enough. These effects are divided into two types: direct effects and long-term, or "late," effects. Direct effects may be visible such as a reddening of the skin like sunburn, or loss of hair. Or they may be less visible, such as cataracts, reduced sperm count, or effects on bone marrow and blood. Direct effects occur only after high doses and a dose threshold exists below which the effect does not occur. These thresholds are commonly 500 mSv or more. Thus the effects do not occur at doses normally encountered by people, but may occur only in exceptional circumstances such as accidents. These effects are generally not taken into account in normal low dose radiation protection practice.

Long-term effects are believed to arise with a low probability, depending on the dose, but only a long time after the exposure. Thus, if a group of individuals are exposed to ionizing radiation, some of the individuals may suffer the effects many years later. If the dose is low, very few people will be affected; if it is high, more will be. There is no threshold below which there is no effect. The principal effects of this kind are the induction of various types of cancer, which occurs with a frequency of about 5 in 100,000 for 1 mSv, or the induction of hereditary effects in later generations, which has a frequency of about 1 per 100,000 per 1 mSv.

These estimates are based on careful studies of exposed human populations followed subsequently throughout their lives. These populations include people irradiated for therapeutic purposes, the survivors of the atomic bombs dropped on Japan in 1945, certain occupational groups such as radium dial painters and uranium miners, and Russian and British workers exposed in early atomic energy programs. In each case, estimating the dose to which the individuals were exposed is a critical part of the risk assessment.

These assessments have identified the gonads, lungs, stomach, colon, and bone marrow as the principal organs at risk, with the breast, liver, thyroid, bladder, and esophagus as next in line, while skin and bone had small risks of cancer. The assessment of genetic risk is not based on human studies, but on studies of genetic changes induced by radiation in mice. These changes, including those that might lead to multifactorial diseases in man, were estimated to result in a risk of about 1 in 100,000 per mSv.

The natural background sources mentioned above will have a small risk of causing cancer or hereditary effects. Using 3.6 mSv as the average exposure of the average U.S. individual gives a risk of about 2 in 10,000 annually of either a fatal cancer being induced or a severe hereditary effect occurring. The risk is less in older people and greater in younger ones. In any case, these risks are small compared with other risks in life.

In the U.S. workers engaged in the medical and industrial applications of ionizing radiation typically have an average exposure (in addition to normal natural background) of about 2.3 mSv per year. This corresponds to an additional risk of about 1 to 1.5 in 10,000. The risk of accidents in many nominally "safe" industries are similar.

Recommendations on Limits of Exposure

Since the 1920s professional groups such as the International Commission on Radiological Protection (ICRP) and the National Council on Radiation Protection and Measurements (NCRP) in the U.S. have made recommendations on appropriate limits of exposure to radiation from man-made sources. Given that we are all already exposed to natural radioactivity which is variable from place to place, to expect to control additional exposure to zero is not reasonable. But to limit it sensibly is necessary.

Recently, estimates of induced cancer and hereditary effects have been increased by ICRP and NCRP following their appraisals of the most recent human information. ICRP limits exposure to 20 mSv per year averaged over five years, which is equal to a maximum risk of about 12 in 10,000 annually. The average worker is never regularly exposed to this limit, but rather to about 1/10 or 1/20 of it, for a risk of no more than about 1 in 10,000. The NCRP limit is slightly more restrictive at a lifetime limit of age (years) × 10 mSv which in effect means no more than 10 mSv per year later in life. Again the risk is the equivalent of about 1 in 10,000 annually.

For the public, both organizations limit exposure to an additional 1 mSv per year from man-made sources. This carries a risk of about 6 in 100,000 but again very few people ever receive this exposure. Usually, exposure is closer to 1/5 or 1/10 or less, so the risk of additional exposure is in practice about 1 in 100,000. Why shouldn't the limit be even less? The average exposure from natural and medical sources is on the order of 3.6 mSv per year and quite variable. For given individuals it may easily range from 2 to 5 mSv per year. Thus the added 1 mSv (maximum) is well within this range of variation.

These recommendations of the ICRP and NCRP are most often adopted by governments around the world and by agencies in the U.S. (the Nuclear Regulatory Commission, and the EPA, for example) as regulations in order to promote and ensure the safe use of radiation and radioactive materials, and the protection of the public accordingly.

We do not live in a radiation-free world, nor can we. Sensible recommendations about safe practices and limits seem to be the best way for the peoples of the world to derive benefits from radiation, radiation devices, and practices, and at the same time to protect the health of all.

WARREN K. SINCLAIR

For Further Reading: Eric J. Hall, *Radiation and Life* (1976); International Commission on Radiological Protection, *The 1990 Recommendations of ICRP* (1991); National Council on Radiation Protection and Measurements, *Ionizing Radiation Exposure of the Population of the United States* (1987).
See also Electromagnetic Fields; Electromagnetic Spectrum; Radon; Risk.

RADON

Radon is an odorless, colorless, inert gas that has been recognized as a major source of public exposure to ionizing radiation (a ubiquitous carcinogen). Since it is radioactive, uranium-238 decays very slowly to radium which, in turn, decays to radon. The decay chain of radon proceeds through radioactive isotopes of polonium, bismuth, and lead before ending in the formation of a stable isotope. Paradoxically radon itself does not participate in chemical reactions and can be breathed in and out with little if any harm. However, radon's decay products or daughters emit alpha particles that are harmful. These radon daughters will attach themselves to any solid that they encounter—whether a dust particle or the epithelium of the lung—and are thought to be an important cause of lung cancer. Information concerning the high incidence of lung cancer from radon exposure comes from epidemiological studies of radium and uranium miners. The rate of decay is measured in curies (after physicists Marie and Pierre Curie) or becquerels (after physicist Henri Becquerel). Concentrations of radon in the air are measured in picocuries [one-trillionth of a curie] per liter (pCi/l) or becquerels per cubic meter (Bq/m^3). Outdoor air typically contains 0.1 to 0.2 pCi/l (4–7 Bq/m^3) of radon.

Ores of radium and uranium are widely distributed in the Earth's crust, but concentrations may vary considerably among rocks of the same type or even within a single formation. Very high levels of radon-emitting minerals have been found in parts of certain states: e.g., New England, New York, New Jersey, Pennsylvania, the upper Midwest, and the northern Rocky Mountains. This is reflected in the fact that high radon concentrations have been measured in some U.S. drinking water supplies. Some private wells, for example, have actually measured as high as 7,000 pCi/l. The maximum limit proposed by the EPA is 500 pCi/l.

Radon formed outdoors is diluted rapidly and scattered in the air. However, because it is a gas, it passes through rock formations, filters through soils and enters houses through basement floors and foundation cracks. The amount present varies depending on the underlying soil, rock type, and general construction of the house, but it reaches harmful levels when it is concentrated and this is greatest in the basement and lower floors. The emphasis on conserving energy has actually increased the exposure to radon because tightly sealed buildings tend to trap it inside. Fortunately, well-ventilated buildings can dilute the gas to harmless levels. Nevertheless, the EPA estimates that about 8% of the private single-family housing stock in the U.S. exceeds the guidelines of no more than 4 pCi/l. Residential concentrations between 6 and 24 pCi/l have been recorded in different parts of the U.S. Under the authority of a radon protection law passed by Congress in 1988, the EPA established a Radon Division specifically to provide information and technical assistance for people who are concerned about this problem.

The public, in spite of the fact that the EPA task force of experts in 1987 ranked the indoor radon problem a high cancer risk, exhibited much greater concern and apprehension over Chernobyl. This was despite the exposure estimates of up to 5,000 to 20,000 cancer deaths each year from radon in U.S. homes, as compared with the estimate of up to 28,000 total health effects over seventy years due to the Chernobyl nuclear accident.

RAILROADS

See Transportation, Rail.

RANCHING

Ranches are agricultural operations primarily involved in the production of livestock, such as cattle, sheep, goats, or horses. They can be found in the U.S., Canada, Mexico, Venezuela, Argentina, Brazil, Australia, and New Zealand and are generally located in regions that are unsuitable for crop production. Climates in which ranching takes place vary from desert to alpine to tropic. Ranches are sedentary, occupying one place throughout the year, although herds of livestock are moved from pasture to pasture depending upon the availability of feed and water resources.

Modern, sedentary ranching finds its roots in the mid-1700s, when Spanish missionaries and explorers introduced horses, cattle, sheep and goats to the vast, open rangelands of the Western Hemisphere. Local Native Americans were recruited by the Spaniards to tend flocks and herds, and many of these, later known as Vaqueros, became highly skilled at riding horseback and gathering and herding livestock. Generally, these small herds of livestock were allowed to roam wherever grass and water were available.

Ranching changed dramatically in the mid-1800s when settlers in Texas began raising large herds of cattle on the surrounding rangelands. There were few fences at that time, and the prairie was treated as common domain. When the cattle were ready for slaughter, they were driven north in large herds numbering in the thousands to the railheads of Colorado, Kansas, Missouri, Nebraska, and Wyoming where they were shipped to slaughterhouses in Chicago and other locations. The terminal marketplaces or auction barns at these railheads continue today to facilitate the trade of livestock; western towns and cities where these markets are found still are dependent upon the movement of livestock.

The 1874 invention of barbed wire marked the end of the great cattle drives, because settlers were now able to fence out the roving herds of cattle and fence in their own livestock and crops. The ranch cowboy of the late 1800s and early 1900s has left an indelible mark on Western culture. His clothing, boots, music and art can be found in just about any major marketplace in the world. Since the turn of the century thousands of movies and books have been produced and written dealing with cowboy subject matter.

New breeds of cattle introduced from Great Britain in the late 1800s also changed the face of ranching. British cattle were larger and produced more pounds of meat than did their Longhorn counterparts, and ranching became a more efficient and profitable food-production enterprise. It spread north from Texas and California to the ranges of Wyoming, Montana, and Canada in the last two decades of the 19th century, and continues to be one of the major industries of the U.S. today, accounting for close to $50 billion in annual receipts in 1992.

More than 800 million acres of land in the U.S. are used for grazing and pasture. These ranches average between 2,500 and 3,000 acres and raise approximately 250 to 300 head of cows (breeding-age females) or 500 to 600 head of ewes (breeding-age females). Cattle operations significantly outnumber sheep operations.

The largest ranch by area in the U.S. is well over 1 million acres: the largest by numbers of cattle owned exceeds 35,000 head. The stocking rate (the number of cattle or sheep that can be placed on each acre of ground) of each ranch varies from region to region. For instance, in the state of Nevada, the ranch may only support one cow per 100 acres; in parts of Texas where steady and reliable precipitation and good soil allow for abundant grass growth, the stocking rate may be two or three cows per acre.

Generally, ranches in the U.S. are located in the western third of the country (although livestock production can be found in almost all parts), where the land is either too steep, too rugged, too wet, or too dry for the production of crops. Much of the rangeland used for grazing in the western U.S. is owned by the federal government. For example, in Oregon, Washington, California, Colorado, Nevada, Wyoming, Arizona, New Mexico, Utah, Idaho, and Montana, 48% of the land is federally owned. State and local economies are extremely dependent upon the use of and access to these lands.

Agricultural operations located in the eastern two-thirds of the country tend to be more diversified, growing a variety of crops and raising hogs or chicken in addition to cattle or sheep. These operations, because of their diversity, are called farms, and, in total, account for 80% to 90% of the country's beef production. More than half of the nation's sheep spend at least part of the year grazing federal lands in the western U.S.

Because most of the federally owned lands used by ranchers are considered by many to be fragile ecosystems, a public debate over livestock grazing's impact on the environments of the West has raged during the last decade. Opponents of federal lands grazing contend that stream banks (riparian areas) and other sensitive areas in which rare or endangered species live can be severely damaged by cattle or sheep. Ranchers counter that western ecosystems evolved under intense grazing and that livestock grazing is beneficial to healthful and sustained plant growth.

Since the mid-1980s, many ranchers have changed their management practices to address these concerns, adopting long-held principles of pastoralists and nomadic herdsmen in other parts of the world. For example, rather than allowing livestock to graze one large pasture for a long period of time, ranchers are fencing their pastures into smaller parcels so that the grass may be grazed in short, high-intensity periods. Ranchers contend that this approach stimulates new plant growth and allows dried, crusted top soil to be busted up by hooves, which enables better water and oxygen penetration to plant roots.

New breeds of cattle better suited to the western rangelands have been introduced or developed by ranchers to improve range conditions, too. One such breed, called Salers, is ideally suited to the rangelands of the western United States because of its aggressive, restless nature. These cattle spread out across rangelands and pastures and move more often than their docile British breed counterparts, which alleviates pressures on riparian areas and other sensitive ecosystems.

The toughest challenge for ranchers and the ranching industry may be ahead. The industry is confronted with a number of issues, including environmental concerns over federal and private land-use restrictions, endangered species protection, red meat's role in a healthful diet, and animal rights. Much of the challenge lies in the fact that the general public of the U.S. and Canada is far removed from the farm and ranch. As the population base of the U.S. shifts even more from a rural to an urban base, the communication gap between those involved in agriculture and those three or four generations removed from the land will only widen. Survival for ranchers may hinge on their ability to readopt and relearn the old, proven range management techniques of their nomadic predecessors.

ERIC GRANT

For Further Reading: Robert G. Athearn, *The Mythic West in Twentieth Century America* (1986); Robert W. D. Ball and Ed Vebell, *Cowboy Collectibles and Western Memorabilia* (1991); Deed Brown, *Trail Driving Days* (1952).
See also Agriculture; Animal Rights; Animals, Domesticated; Desertification; Domestication; Grasslands; Land Use; Pastoralism; Rangeland.

RANGELAND

Rangelands are those areas of the world that because of physical limitations—low or erratic precipitation, thin or unproductive soils, extreme hot or cold temperatures, rough topography, or poor drainage—are not suitable for cultivated agriculture or intensive forestry. They provide forage for free-ranging wild and domestic animals, and are used for watershed production, recreation, mineral extraction, wood products, and waste disposal. Herbivory, removal of vegetation by animals, is a normal part of rangeland's ecological processes. *Rangeland* is an American term and evokes images of cowboys and western lands. However, the deserts of Africa, the steppes of China, the tundra of Russia, and the marshes of Florida are also rangelands.

Rangelands are managed by ecological principles rather than agronomic science. The profession of *range management* evolved to promote the care and sustainable production of rangelands. It involves the manipulation of herbivory, mainly grazing by wildlife and domestic animals, to produce goods and services for people while maintaining the productive ability of the soil and the integrity of the environment.

Although range management originated in America, both the terms and the concepts of ecological management are accepted worldwide. The Society for Range Management is a professional organization established for the development and promotion of science relating to the management of rangelands. Most of its members are in the U.S., Mexico, and Canada, but it has representation on every continent. The Australian Rangeland Society relates to regional problems, as do newly formed professional groups in Africa, the Arabian Gulf countries, China, and France.

Range and rangelands are central to the identity and development of the U.S. Native Americans depended on bison, deer, antelope, and sheep for high-quality protein. Without native forage for draft animals and milk stock, the first Europeans could not have survived. As the United States developed, "range" became the romantic center of the very culture of the American West.

Uses of Rangelands

The uses of rangelands are determined by both biological and social factors. Goods and services produced are limited by rangeland's ecological potential. The primary products are relatively low yields of herbaceous and shrubby vegetation. The traditional uses of rangelands involve the harvesting of vegetation for direct human use or converting it to animal products with grazing animals.

In developing countries many rangeland products are harvested directly, as was the case everywhere a few centuries ago. Grass seeds, tubers, nuts, and berries are gathered for food. Grass is used as a stabilizer of sun-dried bricks and as thatching for roofs. Woody vegetation and animal dung are major fuel sources. Incense, including frankincense and myrrh, contributes to people's cultural identity.

Most rangeland vegetation cannot be used directly. Instead humans use products from grazing animals. Meat, hides, and hair from wild animals clothed primitive humans. Today in developing countries a goat, sheep, cow, or camel may produce milk and meat for food, dung for fuel, and hides and hair for clothing. Cows, horses, donkeys, and camels provide transportation and draft power for plowing or hauling produce. Energy captured by rangeland plants is tranformed directly to something valuable to the family in the plant-animal-human system. A break in this ecological/cultural system either by drought, as in the Sahel in the 1970s, or civil war, as in Somalia in the 1990s, results in famine, starvation, and human suffering on a scale not readily understood by people of developed countries.

In developed countries, the links between rangeland products and the people who use them are much more-complex. Few people shoot their own elk or butcher

their own lamb from American rangeland. The hide for shoes may originate from a cow in Nebraska, be tanned in Brazil, be sewn into shoes in Italy, and sold in a department store in Boston. Wool from Wyoming sheep and mohair from Texas goats is woven into fabric in Scotland and made into suits in Hong Kong. Few consumers see the relationship between the exotic incense in the specialty store and Ethiopian rangelands or between piñon sticks from Santa Fe and a New Mexico ranch.

In industrial countries, rangeland plants are converted to animal products through livestock. Most often, rangelands are breeding areas for cattle and sheep. The calves or lambs are usually sold as "feeders" and finish their growth in feedlots. Animals may be moved many miles, from the rangeland where they were born, to the feeding area, to the meat processing plant, to the point of ultimate sale. Regardless of the complexity of the system, raising animals is the most important commercial use of rangelands.

Other uses, though not commercial, may be equally valuable. Rangelands are important watersheds. In the vast, open spaces people enjoy solitude, hunting, and other forms of outdoor recreation. Tourism based on the romance of the Old West draws thousands of people into rangeland settings. These lands are important in maintaining biological diversity.

Rangelands' isolation and low human population also make them a target for the disposal of unwanted material from more populated areas. Hazardous wastes, nuclear materials, and other potentially harmful substances are finding their way to rangeland sites.

Extent and Distribution of Rangelands

About half of the world's land surface is rangelands. Found on every continent, rangelands include grasslands, savannah, steppe, shrublands, grazable woodlands, forests, and deserts.

The U.S. has nearly a billion acres, about one sixth of the world's rangeland. Most of this land is in the states west of the 100th meridian, but an increasing amount of pastureland in the southern U.S. is being managed as rangeland.

The grasslands of the Great Plains are primarily rangelands. However, most of what once was tall grass prairie has been converted to cropland, and it is unlikely that it will ever revert to grassland. Much of the short grass region alternates between rangeland and cropland, depending upon economic and political conditions. Extensive cropping of the grasslands between the 100th meridian and the Rocky Mountains led to widespread erosion in the 1930s. Conservation efforts restored them as rangelands; then they were plowed again for the war effort in WWII, and restored again.

These lands are marginal for cropping, but are sustainable as rangelands.

Other grassland ranges of the U.S. include the Intermountain Bunchgrass, Desert Grassland, and California Annual Grasslands. Part of the Intermountain Bunchgrass, the Palouse Prairie of Washington and Idaho, has been converted to farmland.

Most North American rangelands are desert shrublands in both hot and cold climates. The cold desert shrubland lies mainly in the intermountain region between the Rocky Mountains and the Sierra Nevada or Cascade Ranges. The area includes most of Nevada and Utah, and extends into western Colorado and Wyoming, northern Arizona and New Mexico, eastern California, and southern Oregon and Idaho. It is dominated by deeply rooted desert shrubs belonging mostly to either the Compositae or the Chenopodiaceae families.

The hot desert shrub region is represented by the yearlong ranges of the Chihuahuan, Sonoran, and Mojave Deserts. These lands occur in western Texas, New Mexico, Arizona, and California. This region is the most arid in the U.S. and the vegetation, such as the succulent cacti, has adapted accordingly.

Shrub woodland range vegetation includes the California chaparral, the oak brush type, mesquite shrublands, mountain brush type, and piñon-juniper woodlands. Grazable forests include the ponderosa pine forests, aspen forests, and the southern pine types. Other forest and woodland types are grazed by wildlife, with only limited amounts of domestic animal use.

Africa contains about a third of the world's rangelands. These include woodlands, savannah, desert shrub, grasslands, and deserts, which in the past have supported large herds of two dozen or more species of wild ungulates. Except in the southern tip of Africa and in limited areas in east Africa there is little commercial ranching. However, almost all African rangelands are heavily grazed by subsistence herds and flocks of cattle, sheep, goats, donkeys, and camels. Traditionally, pastoral people lived in harmony with their livestock, moving them to adequate forage and living off their products.

It takes the equivalent of about 20 cows to support a family under primitive conditions. As the human population grows, so do livestock numbers and the rangeland becomes overstocked and overgrazed. Today most of the subsistence areas of Africa's rangelands are overgrazed and in poor condition.

Europe and Asia together contain about another third of the world's rangeland. In Europe rangelands are managed for commercial production, while in Asia many rangelands support a subsistence economy similar to those in Africa.

Most of the Australian continent is rangeland, about one tenth the world's total. As in North America, the Australian rangelands were settled by European immigrants who overestimated their potential. Heavy stocking led to widespread deterioration of vegetation and some soil loss. Now, with improved management and new technology, Australian rangelands have improved and are in generally fair condition.

South America has about a sixth of the world's rangelands. Most are managed for commercial purposes, similar to those of North America and Australia. However, because of unstable economic conditions and the general lack of conservation, South American ranges are not, on the average, as well managed as those of North America.

The Changing Condition of Rangelands

Rangelands are constantly changing because of natural ecological factors: fluctuating weather conditions, fire, and animal grazing. Native animal populations thrive during good periods; during droughts they die off, allowing the plant populations to increase. A new cycle of lush vegetation and fat, healthy animals can be followed again by near-barren landscapes and carcasses of starved animals. But as long as the soil resource is not lost, the rangeland will recover.

Range managers have attempted to describe this process using the ecological concept of succession. The model most often used, known as the Clementsian Paradigm, was developed in Nebraska by F. W. Clements. It describes a process of community development from bare soil to a stable condition known as climax. Some took this to mean that a depleted rangeland would restore itself if left alone: Remove grazing animals and the range would return to its original conditions.

Unfortunately, that simple model of ecological succession does not fit most rangelands. Ecologists and land managers working in Australia, Africa, and Asia noticed instead a pattern of alternate stable states. When a rangeland is disturbed, it does not progress smoothly up and down a successional scale, but rather "jumps" or "bounces" into a new state brought about by the interaction of plants, soils, and animals, many of which may not have been present in the original community. This new state is as stable and productive as the original, but different. Even when the disturbance is removed, the rangeland is unlikely ever to return to the original state.

Thus, although measuring deviation from a pristine state is often used as a measure of range health, it actually shows only the relationship of the present range to some historical stage. A committee of the National Academy of Science/National Research Council recently concluded that the deviation from climax method of measuring range condition may be misleading as to the range's true health.

Some rangelands are changed for political reasons. Bombing and missile ranges have altered thousands of square miles of rangelands and made them useless. Landfills, waste dumps, strip mines, roads, airports, and railroads have altered the soil and landscape of vast areas.

Current Issues Involving Rangelands

The grazing of livestock on public rangelands has become a major point of disagreement between environmentalists and ranchers in the western United States. At the extremes, environmental activists proclaim that any grazing is overgrazing while poor ranchers state that there is no overgrazing. The debate is due partly to the confusing data on range condition discussed above, partly to different levels of access to public lands by urban and rural groups, and partly to disagreement whether ranchers are paying the full rental value for the land. There is also much debate over forage allocation between wild and domestic animals.

Access by sport hunters to rangeland is part of a major conflict between recreational users of rangeland and ranchers. Extractive industries such as pumping aquifers to provide water for urban use, oil and gas development, and surface mining may become important to the local economy, but deemed destructive to the long-term health of the rangeland.

The use of rangeland as an environmental "sponge" creates local conflicts. Incorporating sewage sludge from eastern states on western rangelands may increase soil fertility, but raise objections with local people. As landfills in populated areas become full, issues of waste disposal on rangelands will increase.

On a global basis, the most important issue on rangelands is still overgrazing. The worst cases are in developing countries where the human population is growing and the number of grazing animals is directly tied to the food supply of the people.

Sustainable Rangelands

Grazing is a natural, evolutionary process in the development of rangelands. Therefore, the question is not whether grazing of rangelands is a proven, sustainable use, but what kind and level of grazing can be sustained by rangelands. The science of range management, though less than a century old, has developed principles that will allow continued grazing of either wild or domestic animals without destroying the soil or ecological potential.

Extractive uses such as mining, aquifer depletion, or erosion caused by off-road vehicles are not sustainable.

There is little information on whether additive uses, such as waste disposal, can be sustained. Some, such as landfills, solid waste disposal, and bombing ranges, probably alter the site sufficiently to make sustained production unlikely. However, even though the soil itself may be changed, the addition of organic waste could allow a new community to be sustained at another level.

At the present time, the most reasonable sustainable use for rangelands is grazing. Timing, numbers, and kinds of animals grazed are social decisions that can be debated, but the evidence is that populations of either domestic or wild animals can be sustained without destroying the rangeland resource.

THADIS W. BOX

For Further Reading: J. Holechek, R. Peiper, and C. Herbel, *Range Management* (1989); Neil R. Sampson and Dwight Hair, *Natural Resources for the 21st Century* (1990); L. A. Stoddart, A. D. Smith, and T. W. Box, *Range Management* (1975).
See also Animals, Domesticated; Chaparral; Desertification; Ecological Stability; Grasslands; Land Use; Ranching; Succession.

RCRA

See Resource Conservation and Recovery Act.

REACTORS, NUCLEAR

See Nuclear Power: History and Technology.

RECREATION, OUTDOOR

Philosopher Holmes Rolston III has written that "Persons go outdoors for the repair of what happens indoors." The natural environment provides the stage for individuals and groups to re-create individual wholeness as a member of society. Recreation contributes to an individual's growth and development of the human personality. Time spent in recreational activities provides a sense of enjoyment, satisfaction, and character well-being. Experiences that are enjoyable and freely selected foster creativeness and learning and help to provide an emotional balance to life. Recreation is frequently defined as a form of individual leisure experience while leisure is defined as the time left over after work. Leisure occurs in a time period of perceived freedom when activities including recreation are freely chosen.

As this nation developed there was little leisure time. The environment of forests and plains had to be conquered for survival. By the 1870s as the industrial revolution gained momentum, a romantic, aesthetic notion of the remaining wilderness tracts developed and large acreages were conserved for parks. In the industrialized cities the work ethic was gradually reduced and the Victorian image of the city beautiful was promoted in which parks became more common. The work week hours were gradually shortened, thus providing time for leisure and recreation.

Recreational use of public and private land and water areas has increased significantly in the industrialized nations since World War II. The time available for leisure including outdoor activities has been slowly decreasing. More women in the work force, two-income families, an aging population, lower birth rates, later marriaes, and accelerated economic changes are just a few factors that have changed the way people manage time. Long-distance vacations have given way to more short weekend breaks occurring the year around. Weekend use of recreation resources has accelerated creating the need for innovative research and management techniques to protect environments and still provide opportunities for recreation.

Outdoor recreation opportunities on public or private land are frequently classified as intensive (concentrated) or extensive (dispersed) while the natural resource supply is categorized as occurring on land, water, or snow and ice. In the United States approximately ten percent of the 2.2 billion acres of land has been set aside as public land, much of which is accessible for recreation. Outdoor activities close to home are the most popular. Most post-World War II surveys list picnicking; visiting zoos, parks, and fairs; and driving for pleasure as the top three outdoor activities. The President's Commission on Americans Outdoors in 1983 found that in selecting a recreation site, people consider natural beauty, crowding, toilets, and parking. Once at a site, however, surveys indicate that the visit may be of short duration and the environmental encounter somewhat superficial.

Americans spend many hours of leisure time on driving. Automobile use has a greater impact on the urban and rural environment than many other forms of recreation. Roadways and parking areas change drainage patterns, alter the landscape, and may reduce the amount of productive land. Exhaust emissions contribute to long-term atmospheric physical and chemical changes and ultimately affect the ecosystem. Without the automobile, access to recreation sites would be limited, but easy accessibility by automobile may detract from the outdoor experience.

Activities that make direct use of the natural environment have steadily increased in popularity since World War II. In the 1980s, a growing interest in physical wellness stimulated exercise walking, running or jogging, and bicycling, activities that are usually undertaken close to home. Equipment sales, user counts, and surveys by government and private institutions indicate that many high-use activities will increase in popularity in the 1990s. These activities include the use of recreation vehicles on highways, off-road mechanized travel, fishing, hunting, day hiking, climbing, downhill skiing, and cross-country skiing. White water rafting, kayaking, canoeing, and camping on developed sites have also increased in popularity and are projected to continue.

The impact of these outdoor activities on the environment consists of two types: physical/ecological and social/psychological. Resource deterioration occurs where facilities and people are concentrated at camping and picnicking sites, scenic overlooks, popular trails, remote campsites, and trail junctions. Unfortunately, severe environmental damage may occur even with minimal use.

Physical/ecological deterioration depends upon the type of activity, visitor behavior, physical site durability, and season of use. The impact increases from human use (hiking) to animal use (horseback riding) to motorized activities, with off-road vehicle use causing the most damage. Such vehicles cause noise pollution, may affect migration and breeding patterns of wildlife, and in arid climates create impacts requiring long periods for rejuvenation. Most research on the physical/ecological impact of recreational activities has been limited to soil and vegetation. Human trampling compacts soils, reduces water infiltration, and increases runoff and erosion. The effect on vegetative reproduction capability is a reduction in plant vigor, abundance, and the height of the vegetation story. Less is known about water and air deterioration from recreational use, although water and land resources have been abused by litter, destructive behavior, and agricultural and industrial pollution. Concentrated water activities can cause physical and bacteriological changes in water quality. Oil seepage from boats may increase toxicity levels that affect aquatic plants, fish, and wildlife.

A major social/psychological impact is crowding. Absence of crowding is one of the chief factors people consider in seeking outdoor recreation, yet because it is such a subjective concept, it is difficult to predict and plan for. Studies have shown that crowding among groups engaged in similar activities can be assessed, but crowding among dissimilar activity groups is more difficult to appraise. Conflict among groups engaged in different activities is common at recreational sites where the resources and the capacity to provide for multiple activities have been surpassed. Conflicts have been documented between canoes and motorized boats, snowmobiles and cross-country snow skiers, off-road vehicles and hikers, horseback trail riders and hikers, water skiers and boat fishermen, and bicyclers and pedestrians. Scheduling and zoning of the areas involved has solved some of these conflicts. However, managing people and natural resources in recreational environments is still more an art than a science and solutions must often be based on subjective values.

Recreation takes many forms from an aesthetic appreciation of the environment, to philosophic and spiritual experiences, to the exhilaration obtained from a variety of physical activities. If the outdoor environment is to re-create — to be a haven to repair what happens indoors — then both the manager and user must yield to practices that conserve and protect the environment.

CARLTON S. VAN DOREN

For Further Reading: Marion Clawson and Carlton S. Van Doren, eds., *Statistics on Outdoor Recreation* (1984); Robert E. Manning, *Studies in Outdoor Recreation* (1985); President's Commission on Americans Outdoors, *A Literature Review* (1986).

RECYCLING

In the U.S., governments consider recycling as part of an "integrated waste management" hierarchy that includes source reduction (minimizing waste generation at the source), recycling and composting, incineration, and landfilling.

Thousands of municipal and county recycling programs have sprung up in America to collect such materials as newsprint, corrugated cardboard, aluminum cans, steel food cans, certain plastics (milk jugs and soda bottles being the most common), yard wastes, wood wastes, tires, lead-acid automobile batteries, used oil, and major appliances. However, by the early 1990s only about 15% of "post-consumer" solid waste (i.e., materials discarded by consumers from their residences, offices, shops, and institutions) was being recycled.

Recycling involves a complex chain of different activities, and its success is intimately linked to national and international industrial market economies.

Materials first have to be collected from the waste stream. Collection methods and locations vary greatly depending on local circumstances. In some cases, citizens and companies can take separated materials to drop-off centers. Sometimes drop-off centers pay for the materials, especially for high-value commodities such as aluminum or for beverage containers that are

covered by some type of deposit system (such systems exist in 10 states). In other cases, they may separate one or more materials and leave them at the curbside for pickup by a public or private hauler. In yet other cases, various mixtures of materials may be collected together and taken by the haulers to materials recovery facilities, where they are further sorted into different components.

These collected materials are known as "secondary" materials, in contrast to the "virgin" materials produced via mining, timber operations, and the like.

Once secondary materials are collected, they must be prepared or processed for sale. The quality of a particular collected material greatly affects the price that interested buyers are willing to pay. Uncontaminated materials — for example, glass that is free of stones, ceramics, or other material — command prices substantially greater than contaminated materials. To help in evaluating the quality of a material, government and industry organizations have developed numerous quality standards. The actual price of a material depends on additional factors such as its relative abundance and the subsequent cost of transporting it to manufacturers.

Manufacturers then buy secondary materials from processors and use them to make new products. Sometimes materials are used to make the same types of products that they originally came from — for example, some newsprint is recycled into new newsprint. More often, though, the materials are recycled into other types of products — for example, newsprint into tissue paper, plastic milk jugs into fiberfill for carpets and sleeping bags.

The last major link in the chain is the consumers (including those in households, businesses, and government agencies) who purchase and use new products that contain secondary materials. Without the demand for such products, the cycle would not take place.

Unfortunately, in the U.S. there is a lack of manufacturing capacity for some collected secondary materials, which lowers the demand and hence the price for the collected materials. In the late 1980s, prices for old newsprint declined, sometimes to the point that communities had to pay processors and manufacturers to take the materials. Even aluminum, one of the highest-value secondary commodities, experienced a severe downturn in its value in 1991, attributed in part to large amounts of aluminum that former Soviet republics made available on the world market in order to gain much-needed hard currency for their precarious economies.

Developing additional capacity to use secondary materials in manufacturing new products is not easy. Market conditions must be appropriate for companies to even consider new investments, credit must be available for financing projects, and sites have to be found and regulatory permits obtained.

New domestic collection programs can often still find markets for their materials outside national borders. The U.S., for example, exports prodigious quantities of scrap iron and used paper to countries in Asia and elsewhere. Manufacturers in these countries are capable in some instances of paying higher prices for those materials than many U.S. manufacturers. Depending on their location and access to transportation, municipal collection programs may be able to serve this market. However, as the amount of collected materials increases and as other countries increase their own supplies of materials at home, prices are likely to drop for the highest-quality materials (underscoring the need for close attention to the quality of materials in collection programs).

Beyond these considerations, the public is often confused by the terminology of recycling, which has yet to be standardized. Some define *recyclable* to mean that technology exists somewhere for using the collected materials to make new products. This probably is true in most circumstances. However, it does not necessarily mean that the technology actually exists in a location to which the materials can be transported economically, so it does not guarantee that the materials collected in a given area will be recycled. To counter this, others define *recyclable* to mean that the materials will actually be sold in the marketplace and used in manufacturing plants to make new products.

Likewise, *recycled* means different things. To some, it means that the material is used to manufacture the same types of products that it came from; more often, though, it means use in manufacturing any type of product. Moreover, there often is no indication as to whether the recycled materials in a product come from the post-consumer waste stream or from the scrap left over from industrial processes. Industrial scrap has long been recycled by the private sector because it is economically beneficial to do so. Most new government collection programs focus on the post-consumer materials, but manufacturers can use both industrial scrap and post-consumer materials in the same product.

Public attention to recycling originally focused on its role in reducing solid waste problems. The early 1990s brought a broader view of recycling as a feature of national and international economies. Recycling seems to be part of a shift in industrialized economies toward a more resource-conserving economy. It is increasingly viewed as a factor in developing future industries that are more sustainable than our current, predominantly extractive industries that are based on generally (but not always) more energy- and pollution-intensive mining of raw materials.

Despite this shift, environmental laws in other countries and international trade agreements might affect trade in recyclables. The General Agreement on Trade and Tariffs, the major international agreement that governs trade, is beginning to take into account the relationships between environmental regulations and trade—for example, whether regulations designed to protect human health and the environment or to promote certain policies (e.g., increased recycling) can be considered barriers to free trade. Although it has not happened yet, "minimum content" laws—which require that a prescribed percentage of recycled materials be used in new products—passed by some states conceivably could be considered in violation of GATT.

Debate within individual nations on how best to stimulate development of domestic markets for recycling secondary materials has yielded many ideas, some of which have been put into practice. These include: providing more information about how to recycle materials and about products with recycled contents; establishing product labeling programs; mandating that products contain a minimum percentage of recyclable material; mandating that government agencies purchase a specified percentage of recycled products; providing low-interest loans and other financing for the development of new recycling manufacturing capacity; and mandating that manufacturers be responsible for collection and recycling of residuals from their products.

PAUL RELIS
HOWARD LEVENSON

For Further Reading: Community Environmental Council, *Manufacturing with Recyclables: Removing the Barriers* (1993); Environmental Defense Fund, *Recycling & Incineration, Evaluating the Choices* (1990); Howard Levenson and Kathryn D. Wagner, "Japan Manages Waste—Their Way," in *Waste Age* (November 1990), Paul Relis, *The Future of Recycling* (1990).
See also Ecolabeling; Environmental Movement; Law, International Environmental; Materials Recovery Facility; Waste, Municipal Solid; Waste Disposal.

RELIGION

In 1986, in the ancient town of Assisi, Italy, birth place of St. Francis, representatives of five major world religions, many smaller faiths, and a selection of indigenous peoples came together with the world's main conservation and environmental groups to discuss what each faith had to offer to the struggle for the environment. It was a historic meeting, but it was not called by the faiths. It had taken the initiative of HRH

Prince Philip, International President of the World Wide Fund for Nature (WWF), to convene this group.

It was soon discovered that all the world's main religions have powerful teachings on how we should care for the world, albeit coming from very different understandings and leading to different goals. However, the faiths gathered in Assisi in 1986 had to confess that these teachings had remained largely undeveloped and unexplored for many centuries. At the end of the gathering, each of the five major faiths, Buddhism, Christianity, Hinduism, Islam, and Judaism, issued formal statements about their individual faith's attitude and response to the environmental crisis. These brief but authoritative statements were the first time most of the faiths had ever clearly committed themselves to a position on the environment. Since Assisi, the Network on Conservation and Religion has been formed to assist the religions—a total of eight religions have joined the Network, with the Baha'is, Sikhs, and Jains joining after Assisi—in furthering and developing both theoretical and practical concern for the environment, founded upon the particular insights of each given faith.

The involvement of the world's religions and faiths in environmental concerns is a recent phenomenon, but it has become one of the most significant developments in both religious activities and in environmental mobilization of recent years. There is no other major social issue upon which all faiths so readily agree nor upon which they have all undertaken such major plans.

Broadly speaking, the faiths fall into three categories. The first category is that of faiths which are localized or indigenous to a given area. In such societies—for example the Yanomamo of the Amazon, the Australian Aborigines or the tribes of upper Papua New Guinea—the religious outlook is simply the outlook on life. The beliefs of such groups are highly oriented towards an awareness of and relationship with the particular natural environment they inhabit. However, in almost all cases, this has not prevented use and even at times abuse of the environment. At times in recent years, the West has been guilty of projecting wishful thinking and romanticism onto indigenous peoples, and much has been spoken about these groups and their closeness to the environment. They certainly do have a keener sense of the reliance of human life upon the wider framework of nature. Their lifestyles are marked by a certain respect, sometimes reverence, for other forms of life such as animals or plants. But they have also changed or adapted the landscape around them and have been responsible for the extinction of numerous species over periods of ten or twenty thousand years. That they have evolved a working relationship with the rest of nature and that this is

sanctified by myths, legends and rituals which often help in protecting the environment is true. But this should not be romanticized. There is a strong tendency of Western writers to want the "noble savage" to exist and for there to be an environmental primeval utopia. This does not exist, but within traditional indigenous cultures much practical and spiritual wisdom has been gathered over the centuries, which we lose at our peril. In particular, the belief that all aspects of nature are imbued with divine force or spirits has meant that traditional societies have treated the rest of creation with reverence and with awe. This, while not stopping use of the environment, has usually meant that a working balance with the rest of nature has been maintained.

The second group of faiths are those that function on a cyclical basis of time, that is to say, those faiths that believe in some form of reincarnation. Such a belief is found in Buddhism, Hinduism, certain aspects of traditional Chinese religion, Sikhism, and Jainism. Belief in reincarnation offers a view of life that believes that within each living creature there is a spark of the divine, a form of soul, which passes from one life to another. Depending upon the behaviour and resulting karma of the previous life of any given soul, the next reincarnation is determined. Thus someone can have been born as a human being, but after living a life of selfishness and depravity, could then be reborn in an animal form, even as an insect. This gives an understanding of life that values each life for itself and that, as the Bhagavad Gita of Hinduism says, means that the wise man sees the dog, the priest, the elephant as being the same.

Traditionally this belief in reincarnation has led towards a strong environmental dimension within such faiths. Many of these faiths are vegetarian, at least for the monks or priests within the faith. Traditional attitudes of reverence for the land, for trees, and for other living creatures has been a hallmark of these faiths, although this has not prevented development of both agriculture and in certain areas, urban development and industry.

The Jains are a case in point. The monks and nuns of Jain faith (founded ca. 550 B.C. in India) take great pains to ensure that they never wittingly take the life of another creature. They sweep the ground in front of them before walking forward in order to prevent killing by accident any insect. Likewise they strain all water to remove living beings, which would be killed when the water is drunk. Jain laypeople are traditionally forbidden to engage in any trade or profession that involves the killing of any life form. This has led them to specialize in jewels, stonework, or metalwork. In recent years they have awoken to the fact that great environmental damage is done by mining for diamonds, drilling and exploiting oil, and the development of iron foundries. As a result of joining the Network on Conservation and Religion in 1989, Jain laypeople have had to reexamine the meaning of their faith for the trades they have traditionally developed; the discovery of the environmental damage these trades involve has led to a search for cleaner and more environmentally friendly trades.

These teachings or beliefs and their environmental consequences have been innate rather than specifically propounded and as such have been diminished in recent decades by the march of development and the growth of consumerist economies. Nevertheless, in many parts of Southeast Asia, the only remaining areas of forest or sanctuaries for endangered species have been the lands of monasteries and special temples where to cut down or to hunt is forbidden by the sacred nature of the site. This passive environmentalism has now been taken up as a positive contribution to caring for the environment. From Buddhist centers in Thailand to Hindu pilgrimage centers on the Yamuna River in India, believers are developing religious programs for practical conservation, showing the central importance of compassion for the sufferings of all other forms of life, which with us form part of the inter-related web of life.

The third category is that of the linear, "you only live once" faiths such as Judaism, Christianity, Islam, the Baha'is, and to a certain degree the spin-off "faith" of Marxism. In this tradition, one's one life is lived on Earth. The human being is traditionally seen as being the pinnacle of creation and thus of greater importance than other species. This is understood in different ways within these faiths. For instance, Judaism sees humanity as being the most important creature on the planet, but also sees humanity having a responsibility to use in a wise way the riches provided by God in creation. Christians differ on what the role of the human being is, but fall into two major camps: those who argue that we are here to be good stewards of creation, namely to use what is provided, well and to never endanger any life form by overuse; and those who believe that we are meant to be servants of the rest of creation, to be like priests who stand before God on behalf of others. In this case the others are the rest of creation. This idea has a strong sense of humanity being co-creators with God in the furtherance of creation. Islam teaches that we have a very specific role: that of being vice-regents of creation. We have no authority of our own, but we have been given permission to act on behalf of God/Allah, so long as this does not endanger any aspect of creation.

While it is possible to show that all faiths have positive environmental teaching within them, it has also to be noted that until very recently, all faiths had failed to practice what was inherent within their faiths with regard to the environment. In some areas, where traditional lifestyles based on, say, Sri Lankan Buddhism, Southern Indian Hinduism, or Greek Orthodox Christianity had been able to continue, the environment was protected or cared for by the faithful. But a serious breakdown in the implementation of the core beliefs as they affect the environment was and to some extent still is to be found within all faiths.

But there has also been a considerable challenge thrown at many of the linear faiths, especially Judaism and Christianity. In 1967, Lynn White wrote an article about the historical roots of our environmental crisis. In it he noted and commented upon the role of religion, particularly Christianity, in desacralizing nature and presenting the world as a place created for human use and, by inference, abuse. He traced the influence of the Genesis account of creation and the command given to humanity to "dominate the Earth and subdue it" from its Biblical roots right through to the development of consumerist capitalism. He saw the teachings of superiority and the idea of the divine being transcendent rather than immanent in creation as having licensed us to exploit.

The reaction to this article and the charges leveled against Christianity in particular and religion in general had a number of effects. It drove Christians to look inside their own tradition in order to answer these charges. But it also developed a deep suspicion of religion within the emerging environmental movements. Many felt that science and the knowledge of environmental data was the way forward and that the religious approach was at best obscurantist and at worst responsible for the crisis. Today this has changed. Now the religious dimension is taken as being of significance for several reasons. Some, looking for a new world ethic, hope that the religions will teach people to be ethical and thus to save the world. This naive idea ignores the great differences in teachings of the different faiths and also seems to know what it wants the faiths to say. This attitude is now diminishing in favour of recognizing the areas that all belief and value systems have in common, and exploring and allowing to develop those aspects of any given belief system that contribute to a widening understanding and way of relating to the environment.

There has also been the search for new or alternative spiritual/religious understandings of the environment, loosely based on either an old tradition such as paganism or a modern fusion of different ideas drawn from a variety of religions. The sense that the environmental crisis is not just a crisis of resources but a crisis of relationship one with another, with nature and with the divine, deeply affects and influences many activist groups. The quest for some new or renewed spiritual link with creation has led to great interest in indigenous wisdom, shamanism, and various pagan groups as well as to Taoism and Buddhism. For instance, in recent years there has been a growth within Christianity of creation-centered Christianity, which seeks to draw upon older traditions but to place their insights within a Christian framework. This attempt to find and express a new or renewed spiritual relationship with nature has been given added impetus by the development on a religio-spiritual basis of the Gaia hypothesis. The use of the name *Gaia* from the ancient Greek earth deity Gaia and the notion of the planet as a living being has captured the imagination of many and has led to a growth of interest in those beliefs which stress the goddess or mother aspect of the Earth and the sanctity of life itself.

The turning away from faith in science and facts to solve the environmental crisis and the rediscovery of the need for spiritual and moral values has fundamentally altered contemporary environmental work. It remains to be seen whether it will ultimately prove more effective in helping solve the world's environmental problems.

MARTIN PALMER

For Further Reading: O. P. Diviredi, ed., *World Religions and the Environment* (1989); Martin Palmer, et al., *Faith and Nature* (1987); Steven C. Rockefeller and John C. Elder, eds., *Spirit and Nature: Why the Environment Is a Religious Issue* (1992).

RELIGION AND THE ENVIRONMENT IN A NEW ERA

All peoples have a sacred story that enables them to deal with their world: a story of life's origins and humanity's place within the scheme of things. In philosophical terms it is a "cosmology" that articulates the basic assumptions of a culture which become the lens through which its people view the world.

In the Western world, it was the biblical story of Genesis that shaped thinking and directed the development of institutions. Thus Western law, education, economics, and morality together with Western expansionism in all its forms have been influenced by a biblical cosmology that saw God as the distant-transcendent creator and ruler of all creation, and saw humans as the peak of this creation because they were uniquely made in the image of this God. Nature was viewed as inferior, even corrupt, and salvation, therefore, was escape from its snares. In the present world this meant exercising control over nature, including the human body, and, by extension, woman, who was closely identified with the body. Recent events, from

scientific discovery to environmental and economic decline have caused many to question these assumptions and to look for a new story as a way of addressing the roots of the problems that threaten to destroy all life: a new story that would resituate the human in a creative rather than a destructive relationship with the world.

Pierre Teilhard de Chardin, a French priest-scientist (d. 1955), was perhaps the first figure in modern times to attempt to bring an emerging view of reality together with the Christian perspective in order to foster the creation of a new story that would direct human thinking and activity. Teilhard's achievements were many: in the first place he did much to bridge the chasm between humans and the earth by highlighting the sacred dimension of the universe itself. In Christian terms, stretched with a skilled poetic license, he spoke of a Cosmic Christ who was the "Omega Point" of the emergent universe, the creative force that drew all things to itself in a final unity. In this he not only enriched Christian thought but brought science to a new level of significance in the human process.

However, Teilhard has also been criticized for what in effect are the negative aspects of these same insights. Clearly shaped by the humanist influences of his own time, he saw the human as carrying all nature in himself. Nature had no intrinsic value, no integrity of its own but was simply raw material for human development. The role of the human was to release the power within nature as it were for the sake of evolution's ultimate peak—the human. Damage done to the natural world was simply incidental to the advance of the evolutionary order. However, Teilhard died before the modern ecological movement was born, before the impact of human activity became so distressingly obvious, before it became clear that human alienation from the rest of the natural order was the doubtful foundation of modern society.

Thomas Berry, a cultural historian, has brought Teilhard's important contribution into today's context. For Berry, the glory of the human that Teilhard promoted at any cost has become the desolation of the earth, a desolation which in time could become the destiny of the human. What is required, therefore, is a redefinition of the human at species level that will put him/her in a mutually enhancing relationship with the earth. This redefinition will happen through the creation of a "new story" of life's origins and process. Today, just such a common creation story that reveals a world of infinite interdependence is beginning to be told by the scientific community in a way that echoes the visions and sensitivities of artists and poets, naturalists and historians, ecofeminists and process theologians. In a particularly ironic way, it reflects today's new appreciation of the spirituality of the indigenous cultures who have been among the most tragic victims of the old story's aggression toward the natural world.

Like Teilhard, Berry sees the contribution of science as essential to the earth process, but, in the light of today's environmental crises, he draws Teilhard's insights to a new level where the earth community and not the human is the highest expression of the evolutionary process. Berry's thesis is appropriately radical: the human community and the natural world, he asserts, will go into the future as a single sacred community or both perish in a desert of human making. There is one destiny for all—human and earth—for the human, like everything else, is intimately connected and related to the earth. The human is actually the earth—the universe, in fact—in a conscious mode. In the human, the earth comes to reflective self-consciousness, in human intelligence the evolutionary process reaches consciousness. This perspective is an expression of a growing body of scientific thought that today describes the world as unfolding rather than fixed, in process and relational rather than inert and reductionist.

The implications are enormous. In this context, the role of humans changes from being the center of a world where everything is there for human use to being a participant in an interdependent process where everything has its own integrity and rights, and humans have special responsibilities. Here the earth is primary and everything else is derivative. This includes all human institutions which now must be assessed on the basis of whether and how they promote the life of the whole community. The shift is from an aggressive conquest that characterized human activity in the past to intimacy with the world and a deepened appreciation of the dynamism that flows into the human soul through contact with the natural world.

This shift would also be incorporated into the religious world. It is true, of course, that the great religious traditions have many rich insights about how humans should live in and with the world. There is the indigenous world's sensitivity that sees the earth as alive and the Hindu tradition's notion of divine manifestation that similarly sees the world as alive with the presence of many gods. The goal of Shinto practice is life in harmony with nature and with others, based on a proper sense of gratitude while Buddhist teachings state that we can attune ourselves to this harmony by living lives of gratitude and service. Taoist wisdom actually defines the human as the "mind and heart of the universe." In the monotheistic world, Judaism, while it elevates the human to a place of superiority over the rest of creation, nonetheless emphasizes limits, reminding us that our superiority is partial and contains responsibilities. The Christian focus is somewhat similar in that growth in maturity—conversion—comes

through the practice of love and forgiveness of others in daily life. The Moslem emphasis is on the need to rediscover the sacred that is the real foundation of this apparently independent world which is its manifestation. The Baha'i vision stresses the unity of material and spiritual evolution, that humans are part of the world's unfolding.

However, these suggestions of a benign attitude toward the earth have never been the central thrust of religion, nor have they been able to seriously mitigate the impact of Western culture especially as human ability to manipulate the earth has increased. In fact, throughout the religious world today, a strong separatist and fundamental backlash, which would appear to be born of fear and resistance to change, offers its followers the comfort of rigid, structured codes.

But as a new common creation story emerges to form the foundation for all human activity, religion too will be changed. In the light of an emerging cosmology, traditional insights will be expanded to reflect this foundation. Already some of this has begun to happen in the work of Thomas Berry and a growing number of other religious thinkers. Berry speaks of the "macrophase" of religion when insights, born of a particular time and place, will come together to address the new challenges of a global community. Out of this encounter that is both interfaith and interdisciplinary, are emerging "meta-religious" sensitivities that are perhaps the first signs of religion in a new era.

DR. DANIEL MARTIN

For Further Reading: Thomas Berry, *Dream of the Earth* (1989); Thomas Berry and Brian Swimme, *The Universe Story* (1993); Pierre Teilhard de Chardin, *Human Energy* (1969).
See also Buddhism; Christianity; Deep Ecology; Ecology as a Perspective; Gaia Hypothesis; Hinduism; Historical Roots of Our Ecological Crisis; Islam; Judaism; Religion.

RELIGIOUS-ENVIRONMENTAL ALLIANCES

In the growing worldwide movement of religious-environmental alliances, the Global Forum of Spiritual and Parliamentary Leaders on Human Survival is the largest and most multi-representative, in addition to being the most celebrated because of its orientation, its distinguished membership, and its major global conferences.

Created by the Global Committee of Parliamentarians on Population and Development together with the international Temple of Understanding (the oldest interfaith organization in the U.S.), the Global Forum was conceived by Akio Matsumura, formerly of the U.N. Fund for Population Activities, wtih early assistance from Claes Nobel and the organizing efforts of

James P. Morton, Dean of the Cathedral of St. John the Divine.

The challenging idea of finding a way to conjoin the disparate domains of religion and government had long interested Matsumura during his work at the U.N. since the 1970s. In 1985 he felt the time was right for setting up "an encounter between the spiritual and the parliamentary" that "might yield the fresh vision that seemed so necessary and so lacking . . . to connect the practical and pragmatic with a concern for eternal values."

An appeal for a conference on global survival was issued by an initial core group of international legislators and spiritual leaders from Christianity, Islam, Judaism, Buddhism and Hinduism—"to pursue the search for . . . solutions to the critical issues of poverty; political, racial, economic, cultural and religious persecution; unrestrained population growth; the degradation of the environment and the depletion of the earth's resources. . . ."

Toward that end, the first Global Forum conference—the Global Survival Conference—was convened in Oxford, England, in 1988 and was attended by 200 spiritual and secular leaders including the Archbishop of Canterbury, Mother Teresa, the Dalai Lama, Angier Biddle Duke (Chairman of the Global Forum's International Advisory Committee), the High Priest of Togo's Sacred Forest, Cardinal Koenig of Vienna, U.S. Congressmen Rose and Scheuer, astronomer Carl Sagan, Russian physicist Evguenji Velikhov, Sir George Sinclair, Gaia scientist James Lovelock, Kenyan environmental leader Wangari Maathai, cosmonaut Valentina Tereshkova, Native American spiritual leader Chief Oren Lyons and Chairman of the MacArthur Foundation Thornton Bradshaw. It was the first time spiritual and political leaders were able to discuss global survival with renowned experts on the issues.

The unprecedented gathering at Oxford held up a vision to which they would dedicate themselves: that of "a new community . . . with a new consciousness that transcends all barriers of race and religion, ideology and nationality, linking humanity into an extended family . . . committed to preserving all other forms of life and the natural resources of our planet."

World-wide media attention was riveted on the 1990 Moscow Conference—the Global Forum on Environment and Development for Survival—hosted by Mikhail Gorbachev and the first freely elected Supreme Soviet, the U.S.S.R. Academy of Sciences, and the long-suppressed faith communities of the Soviet Union. The event drew 1300 representatives from 83 countries in government and religion, and from science, law, education, journalism, business, industry, and the arts. Keynote speakers and participants were the former Norwegian Prime Minister Gro Harlem

Brundtland; Secretary-General of the United Nations, Javier Pérez de Cuéllar; Worldwatch Institute's Lester Brown; U.S. Senators Al Gore (a member of the Forum's Executive Committee), Claiborne Pell, and Timothy Wirth; Nobel Peace Laureate Elie Wiesel; Administrator for the United Nations Development Programme (UNDP) William H. Draper III; Director-General Federico Mayor of UNESCO; Executive Director James P. Grant of UNICEF; Ambassador of India to the U.S. Dr. Karan Singh; Elder of the Onondaga Nation of the Iroquois Confederacy Audrey Shenandoah; Executive Director of the United Nations Population Fund Nafis Sadik; and astronomer Carl Sagan.

A series of smaller conferences has been organized, bringing the momentum for dialogue and action between the spiritual and parliamentary world to the local, regional, and national levels throughout the world, including Harvard Divinity School, the Aspen Institute, and the first state-level conferences: the Georgia Global Forum and the Minnesota Global Forum. Separate conferences within the sciences, journalism, and the arts have taken place, such as in Shimane, Japan, and the Sundance Ranch International Artists Conference on the role of the arts in communicating global survival issues, co-chaired by actors Robert Redford and Ron Silver, artist Maria Cooper-Janis, and concert pianist Byron Janis, who composed and performed the Global Forum theme, "Alpha Code: the Beginning," at both the Oxford and Moscow Conferences. In addition, the Global Forum sponsored a major conference during UNCED Earth Summit in Rio de Janeiro in 1992 at which former Senator Al Gore and the Dalai Lama delivered major addresses.

An off-shoot of the Global Forum is the International Green Cross headed by Mikhail Gorbachev and devoted to relief of man-made environmental disasters, and to value change for a sustainable future through education and communications. It is in the process of building local chapters at the state level throughout the world, collaborating in consultation with its international board, and cooperating with other groups in the global and local environmental movement.

The International Coordinating Committee on Religion and the Earth (ICCRE) is a unique consultative body of representatives from the world's major spiritual traditions and environmental groups, created by Daniel Martin, a Catholic priest who was formerly Religious Advisor to the United Nations Environment Programme (UNEP), "to foster the relationship between religion and ecology and to address the spiritual dimension of the ecological crisis imperilling the earth.

Its Earth Charter, written for the UNCED Earth Summit in Rio, inspired the only NGO (non-governmental organization) document to reach negotiation stage during the entire UNCED preparatory process. The result of a series of interfaith gatherings around the world, it proclaims principles of interdependence, the value of all life, beauty, humility, and human responsibility. This document came so close to being adopted that ICCRE was urged by U.N. representatives and other NGO groups to continue its international work on the Earth Charter toward possible adoption by the U.N. as a "declaration of principles and bill of rights for all life on earth" for the U.N.'s 50th anniversary in 1995.

Established in 1989 ICCRE evolved out of Martin's experience organizing UNEP's Environmental Sabbath, an annual earth holy day. The establishment of such a network seemed the logical next step to engage the religious community in teaching its constituents to care for the earth and to help in the wider spiritual vision needed for nations striving to institute ecological change: "We feel it is crucial that the religious/spiritual voice be heard. Only fundamental changes in our deepest attitudes and values will stop the degradation of the earth and the exploitation of its peoples, and allow us to live in harmony with the natural order and with one another. The religious world has a special role to play in such a transformation."

In addition to its workshops, conferences and retreats on religion and ecology, ICCRE organizes consultations with leaders in business, education, legislation and the arts, and is helping to develop international conferences as well. At the UNCED Earth Summit, ICCRE organized the celebration of the Day of Commitment and Prayer, the largest interfaith event of its kind, with participation around the world.

ICCRE's Advisory Board and Steering Committee include key leaders from Anglican, B'hai, Buddhist, Catholic, Episcopalian, Hindu, Islamic, Judaic, Lutheran, Methodist religious faiths, as well as Native American Indian and other indigenous peoples. Its representatives also include leaders from the Global Forum of Spiritual & Parliamentary Leaders, Temple of Understanding, World Council of Churches, World Congress of Faiths, World Conference on Religion and Peace, and environmental groups like National Audubon Society and World-wide Fund for Nature International.

The Joint Appeal by Religion and Science for the Environment, a unique organization under the directorship of Paul Gorman, emerged in response to an Open Letter to the Religious Community by 34 internationally prominent scientists, organized by Carl Sagan in 1990. Calling for moral leadership to help deter the grave environmental peril facing all life on earth, the Nobel laureates declared: "Problems of such magnitude must be recognized from the outset as having a religious as well as a scientific dimension. . . . Efforts

to safeguard and cherish the environment need to be infused with a vision of the sacred."

Moved by such an unprecedented plea from scientists, religious leaders from all five continents began a process of dialogues and exchanges between the scientific and religious communities to mobilize major denominations to educate, create programs and networks, integrate an informed environmental perspective into the religious life, and facilitate formal consultations between religious leaders and scientists."

Two major gatherings were organized, the second of which took place in 1992 in Washington, D.C., with over 100 representatives in attendance, including Nobel laureates and heads of denominations. "Our two ancient, sometimes antagonistic traditions now reach out to one another in a common endeavor to preserve the home we share," the participants affirmed in a joint statement. "We do not have to agree on how the natural world was made to be willing to work together to preserve it." The Washington meeting captured the attention of congressional leaders, inviting scientists and religious principals to meet with members of Congress.

The Joint Appeal's "Directory of Environmental Activities and Resources in the North American Religious Community" documents for the first time a comprehensive list of resources and models throughout the country: from a Maryland Catholic parish testifying at state forestry department hearings to a N.Y. Presbyterian congregation regularly monitoring pH in a local creek and lobbying the community to designate it as a critical environmental area. A national hotline number was established by the Joint Appeal office in New York to enable congregations to report on activities or to receive information on how to begin programs and resolve environmental conflicts.

As a result of the interfaith collaborations, four major faith groups and denominations—the U.S. Catholic Conference, the National Council of Churches of Christ, the Consultation on the Environment and Jewish Life and the Evangelical Environment Network (an association of evangelical Christian bodies)—have established an unprecedented partnership, the National Religious Partnership for the Environment, to better reach and educate a membership of over 100 million Americans on environmental justice, individual and corporate responsibility, and deeper moral commitment to help change environmental values.

SUSAN FISHER

For Further Reading: Joint Appeal by Religion and Science for the Environment, *A Directory of Environmental Activities and Resources in the North American Religious Community* (1992); Rev. Dr. Daniel Martin, "A Religious Voice for the Earth," in *Celebrating Earth Holy Days*, Susan J. Clark,

ed., (1992); *Shared Vision*, quarterly newsletter of the Global Forum of Spiritual and Parliamentary Leaders on Human Survival (see esp., Autumn 1987, No. 1; Summer 1988, No. 3; Winter 1988, No. 4; and Vol. 4, 1990, No. 7).
See also Religion.

REMOTE SENSING

The science of remotely sensing the Earth's environments from space was made possible by the ability to launch scientific satellites into orbit around the Earth. Remote sensing is using our senses, or recording instruments, to gather information from a distance.

The space age also brought about the global awareness that fostered the age of the environment through remote sensing. The single photograph of the Earth from Apollo 8 in 1968, for example, has been credited with creating more awareness of the finiteness of our planet than any other image in history. Within a year of its publication the first Earth Day was in planning. Within three years the world was preparing for the 1972 United Nations Conference on the Human Environment.

Remote sensing from a higher vantage point has always been a natural tendency. Humans throughout history intuitively knew the value of the view from above: Primitive hunters and scouts climbed a rock or tree. Balloons developed in the 18th century and primitive cameras in the early 19th century were combined to provide aerial views. The airplane in the 20th century quickly became a reconnaissance vehicle for war, mapping, rescue, and law enforcement.

The worldwide environmental interest during the early 1970s was a perfect opportunity for scientists and engineers to learn if satellites, in addition to so much accurate atmospheric data, could also provide information on the Earth's land and life forms. The objective was to demonstrate the usefulness of the space vantage point for remote sensing of the Earth's surface both on a global scale and on a repetitive basis. This "synoptic" view could theoretically provide more precise global data about geography, geology, forests, crops, urban growth, pollution, ice formation, etc. A Nimbus spacecraft was chosen as the platform for what was named the Earth Resources Technology Satellite, or ERTS.

ERTS was developed to show scientists, environmentalists, economists, and planners what kind of economic, social and political value could accrue from remote sensing. The spacecraft called ERTS-A, weighing 1,965 pounds (893 kilograms), was launched into a 570 miles (912 kilometers) high orbit that took it over both poles. The Delta launch vehicle lifted off on July 23, 1972, carrying the spacecraft and its 485 pounds (220 kilograms) of sensors and communications equipment.

The two primary sets of instruments were a Return Beam Vidicon (RBV) camera, a type of high quality television, and a Multispectral Scanner Subsystem (MSS). This second array was a cluster of four sensors that looked at red and green light and two different bands of infrared. Combinations of information from these provided a powerful tool for interpreting what was seen below. The spacecraft could cover a continuous strip of the Earth's surface 115 miles (184 kilometers) wide. Information was stored on videotape recorders that could transmit the data to Earth when the spacecraft was over a ground station.

ERTS-A, renamed Landsat, produced a wealth of information. Water pollution along the Atlantic coast was found to coalesce rather than disperse as had been thought. Water-bearing rocks were discovered in New York, Illinois, and Nebraska. Two more spacecraft of this design were launched in January 1975 and February 1978. The information they provided was applied to many areas, including agriculture, environmental monitoring, geological mapping and prospecting, marine resources, water resources, meteorology, and land use. A fourth, advanced Landsat with a new spacecraft design and more precise instruments and a short-lived Seasat were also launched. The French SPOT spacecraft and the Soviet Kosmos series spacecraft now provide most of the civilian remote sensing since the Landsats ceased operation.

Environmental information remotely sensed from satellites is usually used as part of a three-tier system. Satellite images from over 500 miles high can usually distinguish only objects larger than about 10 yards (9 meters) across. Aircraft operating at much lower altitudes (from one to eleven miles high) can see less, but with much greater detail. Finally, someone actually on the ground must verify what the satellites and airplanes think they are seeing. This is called "ground truth."

Thousands of pieces of data from ground measurements are still needed to monitor such things as acid rain, water pollution, crop blights, or the extent of deforestation from logging or fires, but the spacecraft offers the best and quickest way to give planners immense quantities of information. In 1972, for example, the whole Mississippi River Valley was flooded. Landsat mapped the entire 2,000-mile affected area in a matter of days—a task impossible without a satellite. Even anthropology has benefited from remote sensing by satellites. The variations in jungle foliage revealed a lost ancient city of a Maya-related culture. Biblical sites have been identified in the desert of the Middle East, and evidence of extensive civilization across northern Africa in areas that are now desert has been mapped.

Using infrared sensors that literally see heat in the dark, polluters have been caught by satellite, wildlife migration patterns have been tracked, urban sprawl has been monitored, and likely places to search for oil have been identified. These "eyes in the sky" have become one of the most powerful tools in the inventory of all those concerned with the condition of the planet.

KERRY M. JOELS

For Further Reading: Craig Brosius et al., *Remote Sensing and the Earth* (1977); Robert Colwell, *Monitoring Earth Resources from Aircraft and Spacecraft* (1971); Norman Short et al., *Mission to Earth: Landsat Views the World* (1976).
See also Earth Day; Electromagnetic Spectrum; Geography; Meteorology; Resource Management; Space Environment.

REPRODUCTION

Reproduction is the process by which offspring are generated, thereby providing for the continuation of a species. Three and a half billion years of reproduction and evolution have resulted in a rich diversity of life forms. The first protozoa and bacteria simply divided in order to reproduce; as multicellular organisms evolved, more complex methods of reproduction developed.

There are two basic types of reproduction: asexual and sexual. Asexual reproduction gives rise to offspring that are genetically identical to the single parent. In one-celled organisms reproduction is usually by fission, in which the parent cell splits into two identical "daughter" cells. Another form of asexual reproduction occurring in some bacteria, fungi, algae, and nonflowering plants is the formation of spores, which develop into new organisms. Budding, which occurs in a variety of organisms including hydra and yeast, is yet another asexual process. Some organisms, like flatworms, earthworms, and sea stars, can regenerate, that is, produce entire individuals from small portions of a single organism. Vegetative propagation occurs in plants to generate whole plants from parts—cut leaves, stems, or roots—of the former plant. Some plants, such as ferns and mosses, have both sexual (gametophyte) and asexual (sporophyte) phases in their life cycles.

Most animals and plants reproduce sexually. Sexual reproduction involves the creation of a new individual through the union of sex cells called gametes, which are usually from two different individuals. Gametes result from meiosis, a type of cell division that produces cells with half the number of chromosomes of the original cell. During fertilization, a male and female gam-

ete unite to form a zygote, the first cell of a new organism with a full chromosome number. Sexual reproduction ensures that each offspring is genetically different from the parents. The union of independent gametes brings about new variation on which natural selection can act in the evolutionary process.

All living things have a great potential to reproduce at rates which, if unchecked, would soon flood their environments. This inherent capacity to reproduce is limited by many forces of nature such as disease; starvation; accidents; natural disasters, such as drought; and other hazards, including predators and parasites. The effect of these and other factors is referred to as environmental resistance—the tendency to maintain populations of organisms at stable levels.

REPRODUCTION, HUMAN

See Human Reproduction.

RESERVOIRS

See Water.

RESOURCE CONSERVATION AND RECOVERY ACT

In 1976, Congress responded to the nation's solid waste crisis by enacting the Resource Conservation and Recovery Act (RCRA). Amending the Solid Waste Disposal Act of 1965 and the Resource Recovery Act of 1970, the RCRA legislation, with its 1984 amendments, directed the Environmental Protection Agency (EPA) to establish a solid waste management program. The legislation was to apply to wastes that were potentially dangerous to human health or the environment: hazardous, nonhazardous, medical, and storage tank wastes. Solid wastes were defined as waste solids, liquids, sludges, and contained gases.

The EPA was directed to identify hazardous wastes according to a developed set of criteria, and then to publish a list of the wastes with the characteristics that make them hazardous. In addition, the EPA was to develop specific laws regulating the existence of hazardous waste from "cradle-to-grave." Safety standards were also to be established for handling hazardous wastes from their origin through all phases of treatment, transport, storage, and disposal. A manifest system was required along with the record-keeping and labeling of hazardous wastes. All operators of hazardous waste treatment, storage, and disposal facilities were required to obtain operating permits.

The act also required the EPA to develop criteria to be followed for creating solid waste disposal sites that protect the environment from toxic substances. Open dumps were to be closed or upgraded within five years. Thereafter, the EPA was to compile a list of all the disposal sites in the United States that did not meet the guidelines. Waste handlers could not obtain a permit unless they complied with the EPA's guidelines for safe disposal. States were given a timetable within which they could assume control of the implementation of solid waste management, issue permits, and develop safety rules, as long as their programs met federal standards and were approved by the EPA. State regulations could be more stringent, but not weaker, than federal regulations.

In promoting resource conservation and recovery, RCRA directed the EPA to conduct research and development on the recovery of glass and plastic waste and on creating alternative uses for solid waste, such as for fuel. It also required government agencies to purchase items composed of recycled material and to use fuel generated from recovered materials. The 1984 amendments reflected Congressional impatience with the EPA's slow pace in implementing RCRA. The amendments set new deadlines for standards on the disposal of specific wastes and the regulation of new waste disposal activity. As of 1992, the statute was awaiting reauthorization by Congress.

Despite its many regulations, RCRA is surprisingly quiet on the issue of municipal solid waste, or garbage, and recycling. Environmentalists hope that the reauthorization of this statute will include a measure that contributes to the solution of the solid waste disposal problem and that adequately promotes the conservation and recovery of the nation's resources: a federal requirement for the recycling of garbage.

RESOURCE MANAGEMENT

Resource management is purposeful action taken by people to mitigate the potential negative effects of using resources beyond their sustainable capacity to yield benefits. Conservation is the wise and careful use, protection, and preservation of resources to prevent their exploitation, destruction, or neglect. Resource management is what people do to achieve the goals of conservation.

A resource is a means of support, a source of wealth, information or expertise for coping with challenges. All living things depend on resources: from the solar energy that fuels the engines of virtually all life on Earth to the foods, fuels, shelter, and spiritual places that support human livelihood and growth. Natural and cultural resources essential for human life include productive soils, clean water, clear air, minerals, energy, and healthy forests and wetlands; as well as the knowledge, wisdom, and technologies that countless human societies have accumulated over the millennia. Many of these resources are renewable, but they are not inexhaustible. The more the natural productivity of resources is drawn on, the greater their capacity to provide future benefits is strained. When people use renewable resources without reinvesting in their productivity and resilience, the capacity to support life declines. Either the quantity or quality of life is affected and sometimes both.

The need for resource management is a simple result of growth in the human population. If population growth relative to the fixed size of the planet is considered, the amount of space and resources available to each human has declined by more than 80% during the past 300 years. In sparsely populated, hunter-gatherer societies the tendency to overuse resources is usually held in check by lack of technology, natural mortality factors, and periodic famine. Resource management in such societies is not essential; nature keeps the balance. In technological societies and primitive societies with a high population density, it is essential.

A prime example of successful resource management is the history of forests and wildlife in the U.S. Forests cover about one-third of the land surface of the world and about one-third of the land surface of the U.S. — about 66% of what existed prior to the advent of agriculture and industry. Since 1700 the global number of humans has grown by 11 times due to the benefits of agriculture, public health, and industry; from an estimated 500 million to about 5.5 billion people. The population of the U.S. has grown by 25 times during this period; from an estimated 10 million prior to when introduced diseases decimated native Americans to about 250 million in the early 1990s. In per capita terms, this means each global citizen had an average of about 12 hectares (30 acres) of forest resource in 1700, while in 1990, each had only 0.75 hectares (1.6 acres). Each American had on average 45 hectares (110 acres) of forested area in the 1700s, whereas each now has on average 1.2 hectares (2.9 acres).

In the latter half of the 19th century, the U.S. population was expanding rapidly with European immigrants moving westward. Forests were cleared on a massive scale for farmland. Increasing urbanization and industrialization created a huge demand for lumber. By 1900 many U.S. forests were exhausted and populations of many wildlife species were depleted almost to the point of extinction. These included the white-tailed deer, which had been extirpated entirely from most eastern states; wild turkey; pronghorn antelope; moose; bighorn sheep; and beaver. Many waterfowl, including swans, wood ducks, Canadian geese, and all manner of plumed wading birds (such as herons, egrets, and ibises) were also dangerously depleted. Several bird species including the passenger pigeon and the Carolina parakeet were lost to extinction.

A policy framework emerged after the conservation movement in the latter half of the 19th century that emphasized protection of forests from wildfire and of wildlife from overharvest, and the management of both forests and wildlife under scientific principles. Specific actions focused on: (1) acquisition of scientific knowledge through research on forest and wildlife culture and management, and its enlightened application by resource professionals, both public and private; (2) promoting and encouraging the protection of forests, regardless of ownership, from wildfire, insects, and disease; (3) encouraging the productive management of private forest lands through tax incentives and technical and financial assistance; (4) adoption and enforcement of strong state and federal wildlife conservation laws; and (5) acquisition and management of public lands for both commodity and amenity uses and values. A key element of the resource management framework was strong cooperation among federal, state and private sector interests to achieve common goals. The result of these policies was a general and dramatic recovery of U.S. forests and wildlife. The total volume of wood growing in U.S. forests is now 25% greater than it was in 1952.

Because people heeded the call for conservation and resource management and took purposeful actions to recover, restore, and sustain the basic resources of soils, waters, forests, and wildlife, the United States in the 1990s enjoys more productive and abundant forest and wildlife resources than at any time during the century, even though the human population more than doubled in size and the material standard of living of Americans increased. But today's challenges are different from yesterday's and a major change is underway in how resource management is practiced.

Resource management is not a perfect solution to sustain ideal conditions. It can and does lead to both desired and undesired outcomes. Management of freshwaters to provide trout for recreational fisheries, for example, while providing many people with fine outdoor experiences, has caused the extirpation of native fishes in many streams and lakes. European forests have been intensively managed for several centuries to provide sustainable supplies of wood. But an undesired result has been the depletion of soil nutrients and wildlife diversity in simplified forests. Project Tiger in India halted the decline in tigers by establishing national parks and wildlife sanctuaries. But in at least one tiger park, elimination of villagers and their grazing livestock may be leading to a decline in tigers that relied on wild animals in the area that thrived on habitat conditions maintained by the villagers. The important point is that resource management is essential to achieve some objectives but it may not always provide only benefits.

By the early 1900s resource management had begun to move beyond a focus on individual resources such as timber, water, game wildlife, and commercial fish, and toward management of entire ecosystems—the places where plants, animals, soils, waters, climate, people, and the processes of life work as an integrated whole. An ecosystem can be a habitat as small as a rotting log or as large as the whole biosphere. What people do or fail to do to protect and conserve resources and ecological systems in their "backyards" affects larger ecosystems of which their backyards are a part. Moreover, ecosystems are not only composed of natural but economic and social elements as well.

Because ecosystems are all interconnected, it does little good to conserve resources in one place only to have human consumption and waste deplete them and damage the global environment in other places. Local action is the key to sound resource management, but global responsibility requires that local actions also be positive on national and global scales. Hence the motto, "Think globally, act locally."

HAL SALWASSER

For Further Reading: Arnold Toynbee, *Mankind and Mother Earth: A Narrative History of the World* (1976); James B. Trefethen, *An American Crusade for Wildlife* (1975); Michael Williams, *Americans and Their Forests: An Historical Geography* (1989).

See also Biomes; Conservation Movement in the U.S.; Ecological Stability; Economics of Renewable and Nonrenewable Resources; Energy; Environmental Movement; Human Ecology; Land Use; Mining Industry; Population; Restoration Ecology; Sustainable Development; Think Globally, Act Locally; Wildlife Refuges.

RESTORATION ECOLOGY

Restoration ecology is the science of returning a damaged ecosystem to a close approximation of its condition prior to disturbance. The important factor is that the ecological damage to the resource is repaired. Merely returning the species that were lost does not constitute ecological restoration. If restoration is fully successful, the system will be self-maintaining and integrated into the larger ecological landscape in which the damaged patch exists.

Natural ecosystems provide a variety of life-supporting services such as climate regulation and rainfall modification, carbon storage, transformation of hazardous wastes, and, of course, the production of oxygen. Even if ecosystems remain untouched, a rapidly rising human population means a precipitous decline in ecosytem services per capita. If people were housed in a building where the number doubled in approximately 50 years, it would soon be quite clear that the quality of life was deteriorating and that the capability to accommodate new people was not infinite. The ultimate solution is to stabilize or even possibly reduce the total human population. During the transition period, however, the need for additional ecosystem services to keep the per capita level constant can be wholly or partially met with restoration ecology.

Although restoration ecology is a newly developing field, a growing body of evidence indicates that dramatic results can be achieved in a relatively short period of time using present technology and without damaging local economic systems. One of the best-known ecosystem recoveries is the tidal Thames River, which flows through London, England, one of the most heavily populated areas of the world. The Thames has suffered stress from human activity for centuries. By the late 1940s, few or no fish could be found in its tidal waters. Roughly 30 years later, 100 species could be observed. The cleanup used waste treatment methods existing in the 1950s. Probably most important, the Thames Water Authority coordinated and integrated the effort so that there was a system-level improvement in water quality. This was accomplished without undue financial hardship upon the municipalities and industries involved and the financial gains from improved amenities on the river (tourism, for example) improved the regional economy markedly. Stench no longer emanated from the river and what had been a blight became an economic asset.

During the 1980s Professor Dan Janzen of the University of Pennsylvania began restoring the Guanacaste dry forest on the western coast of Central America. The guanacaste tree is believed to have occupied 550,000 square kilometers, roughly the size of France. The size of the forest diminished to less than 1% of its area after the Spaniards arrived. The reasons were typical of all developing areas of the world: clearing for agricultural purposes, timber for housing and other construction, use of forest as fuel wood, highway construction, and the like. In 1988 Janzen, with the enthusiastic support of the local people and President Arias of Costa Rica, launched a $11.8-million project in northwestern Costa Rica. The innovative restoration project will allow the dry forest organisms in Santa Rosa National Park to spread throughout the area, and to re-occupy the largely deforested low-quality

pasture and agricultural land on the evergreen-forested slopes of two nearby volcanoes. Simultaneously Janzen intends this project in tropical restoration ecology to have a management focus designed to integrate the proposed park itself, Guanacaste National Park, into Costa Rican local and national society as a major new cultural resource in an area that is agriculturally rich but culturally deprived.

A third successful exploratory restoration project is the Kissimmee River in Florida. In the 1970s the Army Corps of Engineers dredged a straight channel through the curves of the river. Re-creating both the structure and the function of the Kissimmee River ecosystem required: (1) reconstruction of antecedent physical and hydrologic conditions; (2) biological manipulation; and (3) adjustment of some parts, such as the wetlands, so that they are re-flooded annually as they were historically. The periodicity of flooding will be regulated more by natural climatic events than by human-regulated flow. As a consequence, sometimes the flooding will be modest, other times much greater. However, the project is designed to eliminate or markedly minimize flooding that would cause major property damage. In order to ensure that this goal is realized, subsidies and other encouragements to persons wishing to utilize the floodplain for housing and other construction should be eliminated or markedly reduced.

Restoration to precise predisturbance conditions is almost invariably impossible because (1) ecosystems are the result of a sequence of climatic, ecological, and biological events that are unlikely to be repeated in precisely the way they originally occurred, and (2) many of the predisturbance species may have been lost and, even if they are available elsewhere for reintroduction, they may not be identical physiologically to the original ones.

Obviously restoration to predisturbance conditions is presently inconceivable in some areas, such as those now occupied by most of the world's large cities or prime agriculture areas. Present human population size and distribution mitigate against full ecological restoration. However, there is no shortage of opportunities! Restoring hazardous waste sites to ecologically superior condition will include recolonization by organisms eliminated by the hazardous waste and, thus, more sensitive to it than those now present. As a consequence, these more sensitive organisms will become sentinels, providing an early warning of a miscalculation in hazard to human health and indigenous species by dying or responding in some less dramatic way. Thus, if appropriate biological monitoring is carried out on the restored sites (accompanied by chemical and physical monitoring), the possibility of leachates and other hazardous materials being exported into adjacent ecosystems will be reduced, and, if it does occur, evidence should be available in time to take early corrective action.

In the United States in 1988 the Conservation Foundation started promoting a goal of no net loss of wetlands, meaning that, if a wetland is destroyed (to build an airport taxiway or shopping mall, for example) a badly damaged wetland should be restored or a constructed wetland should be established. The constructed wetland should be maintained by natural flooding and, thus, be part of the normal hydrologic cycle where it is constructed. In short, the system would be self-maintaining because of natural hydrologic events. Isolation of a constructed wetland from the natural hydrologic cycle means that the water would have to be artificially introduced at a considerable cost in energy and management.

While restoring any ecological damage is beneficial, restoration is most effective from both a cost and ecological standpoint if carried out at the landscape level. Such large-scale restoration projects require social acceptance if they are to have lasting effects. This means acquainting inhabitants with the ways the restoration will benefit them and enlisting their aid in restoration.

A combination of preservation and restoration policies established in every country in the world could achieve the goal of no net loss of ecosystems of any kind. After a no-net-loss situation has been achieved, that is, when the rate of restoration equals the rate of destruction, the next goal should be restoration of the ecological capital of the planet, which means that the rate of restoration would exceed the rate of destruction by a substantial margin

Preservation has a greater economic benefit because restoration is considerably more costly, and the outcome of restoration is more uncertain. Nevertheless, restoration ecology will always be necessary because of accidents.

JOHN CAIRNS, JR.

For Further Reading: John Cairns, Jr., *Rehabilitating Damaged Ecosystems* (1988); John Cairns, Jr., et al., eds., *Recovery and Restoration of Damaged Ecosystems* (1977); National Research Council, *Restoring Aquatic Ecosystems: Science, Technology, and Public Policy* (1992).

RIPARIAN RIGHTS

Riparian rights refer to those rights of use and ownership of streams and rivers belonging to the owner of the banks or riverbeds. An ancient concept of property law, developed in their current American form as part of the English common law dating from the Colonial period, the concept originally existed to ensure access to river water in an agrarian-based society under a legal doctrine called "natural flow." The "natural flow" doctrine allows all the riparian owners to use the water so long as the flow is not diminished in either quality or quantity. The owner at the end of a river is thus guaranteed, as much as possible, the same ability to use the river as those upstream, and the river remains in its natural state throughout its course.

This scheme of rights, however, cannot meet the demands of an industrial/agricultural society that needs the freedom to dam rivers for hydroelectric power and drinking water and the ability to divert river flow for irrigation. Respect for the natural state of the river flow has evolved into a doctrine of "reasonable use," in which the riparian owners use the water for their own ends while trying to ensure that their use does not interfere with the potential uses of others. Since this framework allows much greater intrusion and alteration of the water flow, disputes between owners frequently end up in the courtroom.

Modern applications of riparian rights doctrines involve an assortment of different issues. More often than not, the issues involve access to the water resource itself. In Virginia attempts by the state to divert water from rural Lake Gaston to urban Virginia Beach have met with numerous court challenges from landowners and environmentalists concerned about the power of a state to usurp the water rights of the local landowners. River projects in Canada implicate the riparian rights of the indigenous peoples, whose objections often go unheeded by the proponents of such hydroelectric efforts as Hydro-Quebec's James Bay. In the Middle East Turkey's damming of the Euphrates River for power and irrigation effectively ceased flow for a month to Syria and Iraq—a side-effect of Turkey's attempts to modernize its economy.

RISK

Risk is intrinsic to life: as occupational risks; risks in the home; recreational risks; transportation risks; risks from the production of energy; risks from the manufacture, use, and disposal of chemicals and consumer products. Risk analysts commonly define *risk* as the possibility of suffering harm from exposure to a risk agent. Risk agents include chemical substances, biological organisms, and radioactive materials. There are three elements to a risk: a hazard (source of risk), a probability of exposure to the hazard, and a probability of harm or adverse consequences from the exposure.

Risk Assessment, Management, and Perception

Techniques for analyzing and estimating risks have evolved from a variety of fields such as medicine, occupational safety, engineering, and environmental impact studies. The passage of laws designed to protect public health and the environment has generated rapid growth in these techniques. From these efforts emerged a process called **risk assessment**: the systematic use of data, assumptions, and expert judgments to analyze the information available to estimate the probability of adverse effects.

Several types of analysis may be conducted in assessing a risk. These include: (1) Source/release assessment methods are used to estimate the amounts, frequencies, and locations of the release of risk agents from specific sources into specific environments. (2) Exposure assessment provides quantitative data on populations or ecosystems that may be exposed to a risk agent, the concentrations of the risk agent, and the duration and other characteristics of exposure. (3) Dose-response assessment provides data on the amounts of a risk agent that may reach the organs or tissues of exposed individuals or populations. It also attempts to estimate the percentage of the exposed populations that might experience harm or injury and relevant characteristics of such populations.

Risk estimates—one of the main outcomes of the risk assessment process—are statistical descriptions of risks based on available information. Risk estimates cover a wide array of phenomena, ranging from the number of excess cancers expected to be caused by the use of a new pesticide to the health or environmental consequences of low probability, high consequence events such as explosions at industrial facilities. In some cases, risk estimates inspire great confidence in their accuracy; in other cases they are not much better than guesses. To help users understand the value of a risk estimate, risk analysts often provide statistical information on the strength of support for the estimate.

Much debate within the risk literature has centered on the distinction between risk assessment and risk management. The basic point of contention is the degree to which risk assessment and risk management are different and should be viewed independently. Decision-makers often find risk assessments most useful in situations where answers are not obvious and information is missing, ambiguous, or uncertain. Risk assessment techniques provide one means for organizing the relevant information and estimating the consequences of different decisions.

Risk management, in contrast, refers primarily to the integration of risk information with information about social, economic, and political values and various response options to determine what actions to take. These actions include, but are not limited to, reducing or eliminating the risk. Critical factors influencing the decision process are the resources and the technical capabilities available. Risk management also includes the design and implementation of policies and strategies that result from this decision process. It involves weighing the risks of alternatives, and weighing tradeoffs between the health and environmental benefits of efforts to reduce risks and the costs to society of using its resources to obtain those benefits. Making such tradeoffs is usually keyed to a consideration of what constitutes an acceptable level of risk. The risk-management process can be highly political. Balancing the costs of risk reduction against the benefits of the activity is a controversial and value-based process. The extent to which risk assessment and risk management can be distinct is still unresolved. The controversy revolves around the degree to which, in practice, the scientific risk assessment is, or should be, kept free from biases or values that typically are part of a management decision. For example, the practice of risk assessment within the federal government has been criticized from both ends of the environmental political spectrum. Proponents of less restrictive environmental regulation have charged that the techniques and assumptions used in risk assessment reflect an unjustified bias in favor of overly protective risk management values. Others have argued precisely the opposite, claiming that risk assessments used to support government decisions reflect techniques and assumptions that understate risks for the purpose of relieving regulatory burdens on industry.

A classic case in this regard was the controversy about Alar—a growth hormone used on apples—that erupted in 1989. In this and many similar cases, the public was exposed to inconsistent and contradictory information about risks. Experts from environmental groups said the risks were high. Experts from industry said the risks were negligible. Aside from technical issues, an understanding of the differences between risk assessment and perception can help sort out these differences.

When scientific experts talk about risk, they generally mean risk as estimated by the basic risk assessment process—that is, the likelihood or probability of specific adverse health or environmental effects. In contrast, there is a common-sense notion of risk, as people perceive it in their everyday lives, that is linked with numerous other social and psychological considerations.

Researchers from the fields of psychology, social psychology, and decision analysis have identified several qualities or dimensions of risk that influence risk perceptions (see Table 1). Some people judge riskiness solely on the basis of the likelihood of the risk actually occurring, while others are primarily concerned about effects or consequences—for example, whom it affects, how widespread the effects may be, and how familiar and dreadful the effects are. Furthermore, perceptions of risks are influenced strongly by issues of choice and control. Risks often seem riskier to people if they have not voluntarily agreed to bear the risks and if they have no control over the source and management of the risks. In addition, people often incorporate into their perceptions of risks consideration of the benefits derived from accepting the risks. Fairness, equity, and the distribution of risks and benefits are also critical factors in risk perception. However, the most important risk perception factors are trust and credibility.

Based on these factors, a small actual risk may be perceived as a large risk, and a large actual risk may be perceived as a small risk. An understanding of risk perception is critical to understanding the dynamics of public debates about risk. For example, in the Alar controversy, public response and media coverage were influenced to a significant degree by the perceptual characteristics of Alar above and beyond whatever the actual risks were. Alar had an exceptionally large number of perceptual characteristics that heighten public perceptions of risk. These included perceptions that accepting Alar was involuntary, unfair, provided few public benefits, posed an imminent risk, affected children, caused a dreaded disease (cancer), could not be controlled by consumers (Alar was a systemic chemical and could not be washed off the apple), and contaminated apples—an important cultural symbol of health, education, motherhood and country (via apple pie), love, success, and even temptation and original sin (via Adam and Eve).

Acceptable Risk

Decisions about risk are made at both the individual and societal level. At an individual level, the choice of whether to accept a risk is primarily a personal decision, such as choosing whether to eat foods known to have trace amounts of residual pesticides. On a societal level, where values conflict and decisions often result in winners and losers, decision-making is often more difficult. Determining the acceptability of a risk for society is in large part a social and political decision, not a scientific one.

Several approaches have been used to help decision-makers choose acceptable levels of risk. Historically, decision-makers have relied upon the judgment of technical experts to choose risk levels that will be considered safe. To allow for uncertainties, margins of safety have been incorporated into many risk-related decisions.

To formalize acceptable risk decisions, as well as to quantify the costs and benefits of alternative policy decisions, decision-makers increasingly have turned to techniques such as cost-benefit analysis, risk-benefit analysis, cost-effectiveness analysis, and decision analysis. Some analysts prefer a comparative and precedent-based approach, guiding acceptable-risk decisions by comparison with other risks, including those that

Table 1
Factors important in risk perception and evaluation

Factor	Conditions associated with increased public concern	Conditions associated with decreased public concern
Catastrophic potential	Fatalities and injuries grouped in time and space	Fatalities and injuries scattered and random
Familiarity	Unfamiliar	Familiar
Understanding	Mechanisms or process not understood	Mechanisms or process understood
Uncertainty	Risks scientifically unknown or uncertain	Risks known to science
Controllability (personal)	Uncontrollable	Controllable
Voluntariness of exposure	Involuntary	Voluntary
Effects on children	Children specifically at risk	Children not specifically at risk
Effects manifestation	Delayed effects	Immediate effects
Effects on future generations	Risk to future generations	No risk to future generations
Victim identity	Identifiable victims	Statistical victims
Dread	Effects dreaded	Effects not dreaded
Trust in institutions	Lack of trust in responsible institutions	Trust in responsible institutions
Media attention	Much media attention	Little media attention
Accident history	Major and sometimes minor accidents	No major or minor accidents
Equity	Inequitable distribution of risks and benefits	Equitable distribution of risks and benefits
Benefits	Unclear benefits	Clear benefits
Reversibility	Effects irreversible	Effects reversible
Personal stake	Individual personally at risk	Individual not personally at risk
Origin	Caused by human actions or failures	Caused by acts of nature or God

people have already chosen to accept, or the risks of natural hazards.

Increasingly, decision-makers have encountered opposition to acceptable-risk decisions based on these techniques, especially when large components of expert judgment are required. In some cases, however, acceptable-risk decisions have been upheld by the courts based on the determination that the decision was reasonable. Indeed, the term "reasonable" – or, more commonly, "unreasonable" – appears in several laws as the primary criterion for making decisions about risks and costs. "Reasonableness" is left undefined in these laws, however, requiring the regulatory agencies or the courts to determine what is a reasonable decision.

Partially in response to the ambiguous nature of the reasonableness criterion and criticisms of formal methods of analysis, decision-makers have increasingly focused their attention on changes in the process by which acceptable risk decisions are made. For example, regulatory agencies have begun to make greater use of negotiation, consensus building, and other strategies designed to broaden outside involvement in acceptable-risk decisions.

In any discussion of risk, key questions include: What level of risk is insignificant? To what level should risks be reduced? Is it possible to establish a risk threshold below which a risk is so small that it can be ignored for all practical purposes?

Although it is appealing to give zero risk as an answer, it is also unrealistic since it would force people to give up many things of great social benefit. Moreover, there are usually risks associated with alternatives or substitutes. Thus, except where prohibited by law, decision-makers and risk managers typically face the task of identifying levels of risk that are greater than zero but that are also "safe enough" in light of other factors. These factors include the costs of risk reduction, the benefits gained from the activity or substance that poses the risk, and the availability of substitutes for that activity or substance.

Even zero-risk goals have proven difficult if not impossible to achieve. The classic case in point is the difficulties experienced by the Food and Drug Administration in implementing the Delaney Amendment to the federal Food, Drug, and Cosmetic Act, which mandates that no substance that has been found to cause cancer in animals or humans shall be added to the food supply.

Many federal environmental laws and regulations explicitly or implicitly recognize that very small levels of risk are not significant. However, the determination of how small those levels should be is often controversial. A *de minimis* risk is a specific level below which risks are so small that they are usually ignored.

(The term is derived from the legal doctrine *de minimis non curat lex*, "the law does not concern itself with trifles.") Proponents of a *de minimis* risk-management principle contend that regulatory agencies should establish *de minimis* levels and regulate only those hazards that pose a risk greater than these levels.

People are willing to tolerate some risk in order to gain the benefits of an activity. People continue to drive automobiles even though the annual risk of dying in an automobile accident in the U.S. is about 1 in 4,000, and the lifetime risk of dying in an automobile accident is about 1 in 65. In comparison, a *de minimis* risk level of 1 in 1 million for other hazards, such as the risk of cancer from exposure to a chemical (an increase of 0.0003% over the current 1 in 3 chance of developing cancer) may for some seem quite reasonable. There is as yet no consensus among risk analysts on what constitutes a *de minimis* level of risk. Many federal agencies consider a risk level of one in a million as a *de minimis* risk level. However, the legality of using *de minimis* risk standards is yet to be decided.

At the state level, laws are often quite specific about what constitutes a "de minimis" risk. For example, California's Safe Drinking Water and Toxic Enforcement Act of 1986 (Proposition 65) states that, for chemicals assessed using appropriate methods:

> . . . the risk level which represents no significant risk shall be one which is calculated to result in one excess case of cancer in an exposed population of 100,000, assuming lifetime exposure at the level in question, except where sound considerations of public health support an alternative level as, for example, where a cleanup and resulting discharge is ordered and supervised by an appropriate government agency or court of competent jurisdiction.

Proponents argue that a *de minimis* level, if widely accepted and adopted by the government, could help agencies decide whether a hazard poses a significant public health risk, help agencies set consistent levels of risk requiring regulatory action, and encourage agencies to focus their attention on truly risky activities and avoid spending scarce agency resources on trivial risks.

Critics of the *de minimis* criterion argue that the use of a *de minimis* risk criterion is problematic because of the difficulties of defining a level of risk that is insignificant. Critics also argue that the probability of experiencing an adverse effect is not the only factor that defines a risk. For example, a risk for which the probability of death is one in a million may be perceived as trivial if only one hundred people are exposed to the hazard, but significant if one million or one billion people are exposed. A level of risk that is in itself insignificant may not be insignificant if it is part of a cumulative burden of risk. Establishing a *de minimis* risk level will be difficult as long as experts and the public perceive risks differently. Moreover, a risk that at one time was deemed by society to be *de minimis* may not be so in the future.

Despite these reservations, focusing on serious risks first and trivial ones last can save lives. One goal of risk management is to reach consensus on a risk ranking or comparative risk system that allows public and private sector organizations to allocate risk-management resources more efficiently, while accounting for the dimensions of risk that people consider important.

Risk Communication

Partly in response to the problems and uncertainties of risk assessment and risk management, a new field of research and practice has emerged known as risk communication. Much of its focus is on overcoming difficulties in communicating information about risks of exposures to environmental risk agents. **Risk communication** can be defined as the exchange of information among interested parties about the nature, magnitude, significance, or control of a risk. Interested parties include government agencies, corporations or industry groups, unions, the media, scientists, professional organizations, special interest groups, communities, and individual citizens.

Why the interest in risk communication? One explanation is increased public interest and concerns about exposures to environmental risk agents. Another is the increased number of hazard communication and environmental right-to-know laws. A third is the flood of media and other reports on risks in the air, water, land, and food. But a fourth explanation underlies the first three—the loss of trust in government and industry as trusted and credible sources of information about environmental risks.

Information about risks can be communicated through a variety of channels, ranging from media reports and warning labels on products to public meetings or hearings involving representatives from government, industry, the media, and the general public. These communication efforts can be a source of difficulty for both risk communicators and for the intended recipients of the information. Industry officials, government officials, and scientists, for example, often express frustration, arguing that laypeople do not accurately perceive and evaluate risk information. Individuals and representatives of citizen groups express equal frustration, perceiving risk communicators and risk assessment experts to be uninterested in their concerns and unwilling to solve seemingly straightforward environmental problems. The media often serve as transmitters and translators of risk information, but they have been criticized for distorting or exaggerating risk information and for emphasizing conflict or drama over scientific facts.

More detailed analyses of risk communication efforts suggest that risk communication problems arise from message problems, source problems, channel problems, and receiver problems. Message problems include uncertainties in risk estimates, and technical language that is unintelligible to laypersons. Source problems include lack of trust and credibility, disagreements among experts, failures to disclose limitations of risk assessments and resulting uncertainties, limited understanding of the priorities of individuals and public groups, and use of legalistic or technical language. Channel problems include selective or biased reporting, premature disclosures of information, and oversimplifications and inaccuracies in interpreting technical risk information. Receiver problems include inaccurate perceptions of levels of risk, overconfidence in one's ability to avoid harm, strong beliefs and opinions that are resistant to change, and a reluctance to make tradeoffs between different types of risks or between risks, costs, and benefits.

Reflecting the broad scope of risk communication research, solutions to these problems have emerged from such diverse fields as psychology, consumer behavior, marketing, advertising, economics, mass communications, linguistics, anthropology, decision science, sociology, political science, health education, behavioral medicine, public health, environmental health, law, and philosophy.

One of the principal conclusions of risk communication research is that three key equations underlie effective risk communication: (1) $P = R$. Perceptions (P) are realities (R); what is perceived as real is real in its consequences. Fundamental to effective risk communication is understanding perceptions, especially those that relate to perceptions of risk. (2) $C = S$. Communication (C) is a skill (S). Effective risk communication is often counter-intuitive and is a product of knowledge, preparation, training, and practice. (3) $G = T + C$. The goal (G) of the risk communication effort is to establish trust (T) and credibility (C). Factors determining perceptions of trust and credibility: perceived caring and empathy; perceived competence and expertise; perceived honesty and openness; and perceived dedication and commitment to public safety and the environment. Communication and education are ineffective if trust and credibility have not been established.

In summary, risk is a dynamic, multi-faceted concept with scientific, economic, social, and political ramifications. To examine, evaluate, manage, and communicate these facets, analysts have developed a broad array of scientific methods and techniques. These methods and techniques are most useful in situations where answers are not obvious and information is ambiguous and uncertain. The findings and conclusions produced by these methods and techniques enhance the ability of scientists and decision-makers to identify, evaluate, control, reduce, and communicate the risks associated with human activities.

VINCENT T. COVELLO

For Further Reading: Vincent Covello and Miley Merkhofer, *Risk Assessment Methods: Approaches for Quantifying Health, Safety, and Environmental Risks* (1993); Vincent Covello, Joshua Menkes, and Jeryl Mumpower, *Risk Evaluation and Management* (1986); National Research Council, *Improving Risk Communication* (1989).
See also Benefit-Cost Analysis; Dose-Response Relationship; Economics of Risk; Exposure Assessment; Risk Assessment; Risk-Benefit Analysis; Risk Communication; Risk Management; Risk Perception.

RISK ASSESSMENT

Risk assessment and risk management, along with risk communication, are the three overlapping components of risk analysis, a broad "umbrella" term often used synonymously with risk assessment. Risk assessment is the attempt to characterize the probability that potentially adverse health or environmental effects will occur from exposures to environmental hazards. It generally provides a characterization of the types of health or environmental effects expected, an estimate of the probability (risk) of occurrence of these effects, an estimate of the number of cases with these effects, and suggested acceptable concentrations of toxicants in air, water, soil, or food. It is thus a key tool in formulating regulatory policy. When used in this context, it provides a formal technique to forecast risks that cannot be measured directly.

Basic concepts of quantitative risk assessment, such as risk and probability theory, were well developed in Europe by the nineteenth century and were used to predict both life expectancy and the chance of contracting specific diseases. In the early twentieth century, the use of pesticides to increase agricultural production brought with it concern about adverse health effects from exposure to these chemicals. Toxicologists began conducting tests on laboratory rodents to find the highest level of exposure the rodents could withstand without showing an observed effect. By the 1960s, public apprehension about environmental hazards had increased dramatically. Rachel Carson's 1962 book, *Silent Spring*, sensitized the American public to the dangers of environmental pollution. In the United States, federal programs for air pollution, water pollution, and pesticide use, and public health programs for drinking water and radiological health were combined into the United States Environmental Protection Agency (EPA) in 1970. During the 1970s, widely publicized incidents at Love Canal, New York, and Times Beach, Missouri, resulted in federal buyouts when resident organizations demonstrated their outrage over existing and potential health problems resulting from chemical contamination. At this time, also, fears about exposure to radiation due to the testing of nuclear weapons increased. The Nuclear Regulatory Commission (NRC) responded to these concerns by developing an ambitious program of research aimed at quantifying the likelihood of environmental releases of radiation from nuclear power plants. The analytical processes developed by this program were collectively termed "probabilistic risk assessment." This led to a formal recognition of risk assessment and risk management and its emergence, with the intertwining of science, policy, and public administration.

Risk management uses information from risk assessment and combines it with social, economic, and political values; cost-benefit analysis; and determinations of acceptable risk to select, design, and implement the most appropriate regulatory actions. Risk management seeks to reduce the potential adverse impacts identified in risk assessment. *Risk communication* refers to the process of explaining risk assessment findings and describing the basis for risk management decisions. Risk communication is extremely important to public perception of risk because of the difficulty in explaining to the public in clear, yet non-patronizing terms, conclusions regarding health or environmental hazards and risk management options which are based partly on facts and partly on subjective judgment. The public is more apt to accept (and be able to have meaningful input to) a risk management decision if they are involved from the onset of the process and they understand the significance or meaning of risks and decisions and the way they, the public, will be affected as a result of actions or policies aimed at managing or controlling risks.

The risk assessment process itself is composed of four basic tasks outlined by the National Academy of Sciences and fully adopted by the EPA (National Academy of Sciences, 1983). These steps include hazard identification, dose-response, exposure assessment, and risk characterization.

The hazard identification step determines whether a substance may pose a risk under given conditions. Hazard identification may be used to determine if a substance is of itself hazardous or it may be used to determine whether constituents found in waste streams or at waste sites pose a hazard. The goal of the hazard identification is to determine which, if any, substances are sources of risk; that is, which are potential agents for inducing adverse responses. This phase of investigation uses a range of quantitative and qualitative information to determine if there is a hazard and what the nature of the hazard is. Assessors may look at data from human exposures to the substance or they may examine a range of short- and long-term tests in animals, *in vitro* cell and tissue culture tests, or structure-activity relationships to determine whether or not a substance poses a hazard. If no evidence exists to support the premise that harmful health or environmental effects could result from exposure to the substance, the risk assessment may end with this step.

Should evidence exist that indicates the substance is a potential agent for adverse effects, the assessor will move to the dose-response investigation, in which the assessor evaluates the relationship between the dose of an agent and the incidence of adverse effect in the population. The incidence of the effect is estimated as a function of exposure, taking into account the amount of material to which the subject is exposed, the time period over which the exposure occurs, and the number of responses seen in the studied population. The dose-response equation does not explicitly account for variables that may affect the response seen in the exposed population (e.g., intensity and pattern of exposure, ages of test subjects, metabolism of material).

To estimate the potency of a compound in inducing a particular adverse response (e.g., cancer), the numbers of responses in the test population are plotted against the amount of compound (dose) administered to the animals. The compound's potency is believed to be roughly equivalent to the slope of the line which best "fits" these data points (this line is referred to as the dose-response curve).

Several assumptions affect the establishment of the dose-response curve for carcinogens. First, there is a fundamental assumption that exposure to any amount of carcinogen or mutagen can induce the genetic changes necessary for cancer to develop. Second, scientists believe that cancer develops in multiple stages and that all cancer begins with some alteration in the animal's genetic material. Thus, to account for the belief that exposure to even one molecule of a cancer-causing agent could cause this genetic change, dose-response curves for carcinogens are always fit from the experimental data points through the origin (i.e., through zero), indicating that there is no "threshold" dose below which the adverse effect is believed not to occur. Further, the characteristics of the curve (i.e., the slope and shape) often must be estimated from a handful of experimental data points by extrapolation from the observable range to the range for which we have no data. Scientists perform this extrapolation by fitting

the data to one of several mathematical models: the probit model, the "one-hit model," the Weibull model, etc. The model most often used by the Environmental Protection Agency in estimating the potency of the carcinogenic model is the linearized, multi-stage model. This model accounts for the non-threshold theory, extrapolates the curve through the "unobservable range," and expresses the carcinogenic response in multiple stages. Clearly, much of the uncertainty in quantitative risk assessment is introduced in the development of the dose-response relationship.

For noncarcinogenic dose-response, scientists must also develop a dose-response curve based on experimental data which relate exposure to response. However, because noncarcinogenic responses generally only occur above some "threshold" dose, these curves do not have to be fit back through zero, and they may not reflect multiple stages of development. Dose-response exercises for noncarcinogens generally result in the establishment of various dose levels which describe these thresholds of effect: acceptable daily intakes, risk reference doses, permissible exposure limits, etc. These dose levels indicate the level of exposure above which adverse effects can be expected to occur. The assumption that data on health effects detected in laboratory animals exposed to high concentrations of a suspect chemical can be extrapolated to predict health effects in human populations exposed to lower concentrations of the same chemical clearly introduces a measure of uncertainty in the potency estimate.

Interspecies variability between humans and other mammals is another source of uncertainty in risk assessment. The EPA uses an uncertainty factor to justify interspecies variability when deriving the dosage of the contaminant that will produce a critical toxic effect. Interspecies extrapolation is questioned by many scientists who protest that it is impossible to predict human response using data from a species that has a much different physiological makeup.

Dose-rate and dose-route extrapolations are applied to account for variability in effect. These, too, are a source of uncertainty in the risk assessment. Dose-rate extrapolation must be used to account for any differences in the timing of exposure. For example, extrapolation from a short-term or acute exposure to laboratory rodents to a long-term or chronic exposure to humans may give rise to uncertainty in actual risks. So too, may extrapolations from oral exposures in the laboratory to other exposure routes, such as dermal exposures, that may occur in the actual population of concern. Critical information about metabolism and distribution of the chemical within the body are often lacking, contributing additional uncertainty.

The third step in the risk assessment process is exposure assessment. It is here that the amount of a harmful substance with which a human comes in contact is described, measured, or estimated. This is accomplished by identifying and evaluating exposure pathways. An exposure pathway consists of five elements: a source of contamination (where contaminant release into the environment originated), environmental media and transport (e.g., groundwater, surface water, air, sediment, biota), a point of exposure (e.g., residence, playground, hand-drawn well), a route of exposure (e.g., ingestion, inhalation, dermal contact, and dermal absorption) and a receptor population (population exposed or potentially exposed to contaminants of concern at point of exposure).

The last and perhaps most influential step in the risk assessment process is risk characterization. Risk characterization uses information from the dose-response and exposure assessments to communicate the nature and magnitude of possible or likely human risk to the risk manager or other audience. The risk characterization requires a complex integration of information gathered during hazard identification (Does the material pose a hazard? If so, what kind of hazard and under what conditions is it likely to be manifested?), the dose-response assessment (What is the magnitude of the relationship between dose and effect? What are the specific adverse impacts associated with specific exposure levels?), and the exposure assessment (Who is exposed? By what means? How often?).

Uncertainty is prevalent throughout the entire risk assessment process and requires considerable scientific judgment and skill to interpret. Because the risk assessment process involves a series of essentially independent steps, the effect of combining the uncertainties is multiplicative. Thus, the uncertainty bounds for the total risk estimate may be quite large, in some cases spanning several orders of magnitude. Further, because each of these independent steps often are based on conservative parameters, the combination of these estimates in the final risk estimate may produce an estimate several orders of magnitude greater than more plausible estimates.

The uncertainty in risk assessment leads to different types of problems. One involves the scientific question of how to arrive at the best estimate (or range of estimates) of risk, taking all of the uncertainty into account. Uncertainties cannot be removed, but can be identified clearly so as to provide risk managers and the public with a clear, accurate and honest picture of the risk estimates generated.

Despite its many limitations, risk assessment is the most rational approach currently available for managing risk. The information it contains must be summarized in a succinct way while identifying the components containing the strongest scientific support available. All of the underlying uncertainties must be incorporated. The assessment should not be reduced to a set of numerical estimates, but must be presented in a way so as to present the complexity of the data and the inferences drawn from them. If the risk assessment is done properly and if the people affected by the problem are made part of the solution, true progress in managing risks can be made.

ROBIN K. WHITE*

RENEE A. SHAW*

For Further Reading: Rachel Carson, *Silent Spring* (1962); John J. Cohrssen and Vincent T. Covello, *Risk Analysis: A Guide to Principles and Methods for Analyzing Health and Environmental Risks* (1989); Richard C. Cothera, Myron A. Mehlman, and William L. Marcus, eds., *Risk Assessment and Risk Management of Industrial and Environmental Chemicals* (1988); W. H. Hallenbeck and K. M. Cunningham, *Quantitative Risk Assessment for Environmental and Occupational Health* (1986); National Academy of Sciences, *Risk Assessment in the Federal Government: Managing the Process* (1983); U.S. Environmental Protection Agency, Office of Emergency and Remedial Response, "Risk Assessment Guidance for Superfund, Vol. 1, Human Health Evaluation Manual (Part A)" (December 1989).

RISK-BENEFIT ANALYSIS

All societies and individuals have recognized exposure to personal risk as a normal aspect of the hazards of living. Presumably, such ordinary risk exposures are accepted as necessary to attain a compensating benefit. When individuals voluntarily take risks for personal needs, pleasure, or profit, they appear to be willing to accept relatively high risk levels in return for rather modest quantifiable benefits. For example, in sports people frequently explore the physical limits of their chosen activity for intangible benefits, with accident records disclosing their risk-taking propensities. The controlling factor appears to be their perception of their individual ability to manage the risk-creating situation. If they believe they can do so, they are likely to take the chance.

The situation changes markedly when individuals no longer believe they can control their risk exposure. In such "involuntary" situations, the risk management is in the hands of a societal entity usually remote operationally from the individual's risk exposure. Technical systems all create such involuntary risk exposures: for example, transportation systems, energy supply systems, public utilities, and food sup-

*Oak Ridge National Laboratory, Oak Ridge, Tennessee 37831-6285; managed by Martin Marietta Energy Systems, Inc., for the U.S. Department of Energy under Contract No. DE-AC05-84OR21400.

ply systems. Whereas in the voluntary case the feedback loop of "control-risk-benefit-balance" is very tightly coupled by the individual, in the involuntary case this loop is usually very weakly coupled and dimly perceived, with its elements usually dispersed geographically, politically, managerially, and in time. Under these circumstances, the individual exposed to an involuntary risk is fearful of the consequences, makes risk aversion a primary goal, and therefore demands a probability for such involuntary risk exposure that is as much as one thousand times smaller than would be acceptable on a voluntary basis.

Inherent in all major technical systems is an implicit choice of an acceptable level of risk for the involuntary exposure of the public. The study of existing technical systems and public acceptance of risk has suggested the relation between risk and benefit in Figure 1 as a historical basis for decisions on "how safe is safe enough?" It illustrates the domain of acceptability as defined by the per capita benefits (in equivalent income on an arbitrary scale) and the acceptable risk (expressed in deaths per unit time of exposure) from the operations of a technical system and exposure to its hazards. The highest reference level of acceptable risks appears to be the normal U.S. death rate from disease (about 1 death/year per 100 people). The lowest reference level is set by the risk of death from natural events—lightning, flood, earthquakes, insect and snake bites, etc. (about 1 death/year per 1,000,000 people).

Between these two bounds, the public is apparently willing to accept involuntary exposure (risks imposed by societal systems and not easily modified by the individual) in relation to the benefits derived from the

operations of such societal systems. Between these bounds society tries to reduce the risks. Events approaching the low levels of risk have historically been treated as "acts of God" by the public—an implicit recognition of their relatively minor impact on our welfare as compared to the cost of removing these risks. Thus, any risk created by a socio-technical system has been "safe enough" if the resultant risk level was below the curve of the figure. If, as is usually the case, a new technical system has a range of uncertainty in its risks, a design target might be set below the curve by an appropriate safety factor of ten or a hundred.

Although the relationship hypothesized in Figure 1 appears reasonable, its quantitative aspects should be considered as illustrative, primarily useful for comparative purposes. The evaluation of the comparative benefits derived from the availability of the goods and services provided by technical systems is complex, difficult, and presently more empirical than analytic. Because most new technical systems initially appear to be replacements for existing systems (for example, nuclear for coal power) public policy is generally concerned with comparative risk levels of alternatives rather than comparative benefits. However, what is usually not foreseen is that new alternative technologies may profoundly change societal values and structures. For example, the automobile became more than a replacement for the horse, although it was originally perceived only as a "horse-less carriage."

To choose is to govern. Thus the comparative risk analysis of alternatives is a key element of national decision-making, but its usefulness depends on the credibility of each risk assessment. The study of risk analysis has as its objective the development of a methodology for the predictive evaluation of future risks. Unfortunately, the literature on "futures" is apt to be mixed with personal value-system assessments and imaginative scenarios of alternatives. For this reason, it is useful to recognize the existence of four different evaluations of future risk, as follows:

1. *Real future risk* as disclosed by the fully matured future circumstances when they develop;
2. *Statistical risk*, as determined by currently available data, typically as measured actuarially for insurance premium purposes;
3. *Projected risk*, as analytically based on system models structured from historical studies;
4. *Perceived risk*, as intuitively seen by individuals.

Air transportation illustrates all the above risks. To a flight insurance company, flying constitutes a statistical risk; to a passenger purchasing the insurance at the airport, a perceived risk. To the Federal Aviation Administration, anticipated changes in air traffic patterns and equipment determine a projected risk. It is hoped

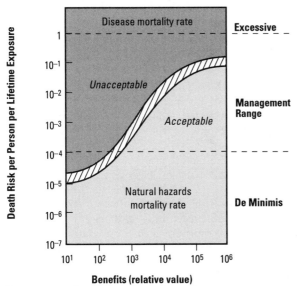

Figure 1. Benefit-risk pattern of involuntary exposure to technical systems.

that the real risks will turn out to be less than projected.

The risk that an analyst projects through careful study of experimental and historical data, through the use of scientific laws, experience with similar systems, or a combination of these, is often not consistent with either an individual or societal perception of risk. Perhaps the single most important factor in perception is manageability or controllability of the risk. An individual feels safer if given some control over the degree of risk from the exposure. Examples of this may be found in transportation: Many individuals perceive more risk from flying than from driving, a perception not borne out by accident statistics. Perception of control is the significant factor that explains why voluntary activities have acceptable levels of risk 100 to 1,000 times higher than those associated with involuntary activities.

The benefit-risk balance that should determine the acceptability of risk is often perceived differently by the individual, the public, and the analyst. In principle, the analyst's assessment should be objective and unbiased by the fears, intangible preferences, and personal values that influence the preference of individuals or the public. Further, the perception of the individual directly exposed to a nearby risk differs from that of a social group removed from the risk but directly benefited by the goods and services. This is commonly evident in the debates on siting refineries or other industrial plants. In this domain of social policy and decision-making, the role of risk analysis is to provide an analytic basis of the social benefits, social costs, and risks for evaluation by the political processes of arriving at a social consensus.

CHAUNCEY STARR

For Further Reading: James J. Bonin and Donald E. Stevenson, *Risk Assessment in Setting National Priorities* (1989); Theodore S. Glickman and Michael Gough, *Reading in Risk* (1990); Chauncey Starr, "Social Benefit versus Technological Risk," *Science* (1969).

RISK COMMUNICATION

In the history of language, "Watch out!" was almost certainly an early development. "Stop worrying" probably came on the scene a little later, as it reflects a less urgent need, but both poles of risk communication—alerting and reassuring—undoubtedly predate written language.

So does the discovery of how difficult risk communication is. If there is a central truth of risk communication, this is it: "Watch out!" and "Stop worrying" are both messages that fail more often than they succeed. The natural state of humankind vis-à-vis risk is apathy; most people are apathetic about most risks, and it is extremely difficult to get them concerned. But when people are concerned about a risk, it is also extremely difficult to calm them down again.

Taking "Watch out!" and "Stop worrying" as the defining goals of risk communication embeds an important and very debatable assumption: that risk communication is essentially a one-way enterprise, with an identifiable audience to be warned or reassured and a source to do the warning or reassuring. For this to be an acceptable assumption, at least three other interconnected assumptions must be accepted as well: that the source knows more about the risk than the audience; that the source has the audience's interests at heart; and that the source's recommendations are grounded in real information, not just in values or preferences. In many risk communication interactions these specifications are not satisfied. A parent warning her children about the risks of marijuana may know less than they do about the drug; a chemical company reassuring neighbors about its effluent may be protecting its own investment more than its neighbors' health; an activist urging shut-down of all nuclear power plants may be motivated more by a preference for a decentralized energy industry than by data on the hazard. To the extent that these things are so, risk communication ought to be multi-directional rather than one-directional, a debate instead of a lecture. And the criteria for "effective risk communication" ought to be things like the openness of the process to all viewpoints and the extent to which values are distinguished from scientific claims, rather than whether the audience's opinions, feelings, and actions come to reflect the source's assessment of the risk.

The judgment that risk communication should be multi-directional is well established in the literature *about* risk communication, but not yet in its practice. Except in the growing area of environmental dispute resolution, which is grounded in the negotiation of competing risk claims, it is considered almost heretical to assert that industry, government, activist groups, and the media (the principal risk communicators) should perhaps talk less and listen more. There is, however, progress on the more modest claim that even one-directional goals are best served by multi-directional means—that is, that it is easier to design effective messages if the source pays attention to what the prospective audience thinks and feels.

Many risk communicators, especially in government, try to avoid the problem by defining their goal in strictly cognitive terms: to explain the risk so that people can make up their own minds how to respond. Though still not multi-directional, this approach is at least respectful of the audience's autonomy. It measures success not by what the audience decides, but by what the audience knows, and whether it believes it knows enough to make a decision. A source that takes knowledge gain as equivalent to making the "right" decision is likely to be misled about the effort's success; knowledge about radon, for example, is virtually uncorrelated with actually doing a home radon test. But often enough knowledge is the real goal. A 1991 California law requires factories to send out a notification letter if they pose a lifetime mortality risk to neighbors of more than ten in a million. Merely letting people know puts pressure on management to get the risk down below the trigger point; the notification letter itself need not aim at provoking *or* deterring neighborhood activism. Informed consent warnings, similarly, can be considered successful whatever the forewarned audience decides.

Whether the process is one-directional or multi-directional, and whether the goal is persuasion or knowledge, risk communicators typically start out with a gap they hope to bridge between their assessment of a particular risk and their audience's assessment. In other words, "Watch out!" and "Stop worrying" are still the archetypes.

Risk-aversion, risk-tolerance, and risk-seeking are often assumed to be enduring traits of character (in individuals and in cultures), but the variations are more impressive than the consistencies. There is no great surprise in encountering a sky-diver who is terrified of spiders. Concern about personal risks (like cholesterol) shows only modest correlations with concern about societal risks (like industrial effluent). When the domain of "risk" is extended even further, the correlations may disappear or even reverse. Quite different groups lead the way in concern about environmental risks (global warming, toxic waste dumps), economic risks (recession, unemployment), and social risks (family values, violent crime). Cultural theories of risk try to make sense of these patterns; one such theory attributes them to distinctions among hierarchical, entrepreneurial, and egalitarian cultural values. Depending on the hazard under discussion, in short, we are all both over- and under-responders to risk.

"Watch Out!"

The most serious health hazards in our lives (smoking, excessive fat in the diet, insufficient exercise, driving without a seatbelt, etc.) are typically characterized by under-response—that is, by apathy rather than panic. This is apparently true even where the list of serious hazards is dominated by war, famine, and infectious diseases instead. Considering how many lives are at stake, the enormous difficulty of warning people gets surprisingly little comment. The new risk communication industry that has emerged since the mid-1980s

is preoccupied far more with reassuring people; those who seek to warn operate under less trendy labels like "health education." Apart from the fact that industry has more money for reassurance than government and activists have for sounding the alarm, there is a more fundamental reason for the distortion: Apathy makes intuitive sense to most people. We are not especially surprised, bewildered, or offended when others fail to take a risk seriously enough.

The dominant models of self-protective behavior assume a rational under-response to the risk and aim at correcting the misunderstandings that undergird that response. That is, they try to convince the audience that the magnitude of the risk is high ("X is a killer"); that the probability of the risk's occurrence and the susceptibility of the audience are high ("X strikes thousands of people each year and is likely to strike you as well"); and that the proposed solution is acceptably effective, easy, and inexpensive ("Here's what you can do about X"). All these propositions are difficult to convey effectively. People tend to be particularly resistant to the idea that they are at risk. For virtually every hazard, most people judge themselves to be less at risk than the average person: less likely to have a heart attack, less likely to get fired, less likely to become addicted to a drug. This unrealistic optimism permeates our response to risk, and we support it by concocting from the available information a rationale for the conviction that the hazard will pass us by, even if it strikes our neighbors and friends. "This means you" is thus a more difficult message to communicate than "many will die."

Several newer models of self-protective behavior postulate that different messages are important at different stages of the process. Information about risk magnitude may be most important in making people aware of risks they have never heard of, while information about personal susceptibility may matter more in the transition from awareness to the decision to act. And deciding to act is by no means the same as acting. As advertisers have long known, what makes the difference between procrastination and action isn't information, but frequent reminders and easy implementation.

In alerting people to risk, social comparison information is often as important as information about the risk itself. Since most people prefer to worry about the same risks as their friends, they are alert and responsive to evidence that a particular hazard is or is not a source of widespread local concern. (The first person in the neighborhood to worry is a coward if the risk turns out to be trivial and a jinx if it turns out to be serious; read Ibsen's *An Enemy of the People*). Messages aimed at building the audience's sense of efficacy may also be effective in motivating action about a risk. Fatalism makes apathy rational; if you are convinced that nothing you can do will help, why bother?

Emotions are also important. Concern, worry, fear, and the like can be products of the cognitive dimensions of risk, but they also exert an independent influence. Even so, many risk communicators forgo appeals to emotion, sometimes out of principled respect for the audience, sometimes out of squeamishness, and sometimes out of a mistaken belief that emotional appeals inevitably backfire. Any appeal can backfire, but the data do not support the widely shared concern that too powerful an emotional appeal, especially a fear appeal, triggers denial and paralysis. Even if the fear-action relationship turns out to be a ∩-shaped curve (that is, even if excessive fear is immobilizing), virtually all efforts to arouse the apathetic are safely on the lefthand side of the curve, where action is directly proportional to the amount of fear the communicator manages to inspire.

"Stop Worrying!"

In essence, people usually underestimate risks because they would rather believe they are safe, free to live their lives without the twin burdens of feeling vulnerable and feeling obliged to do something about it. Why, then, do people sometimes overestimate risks?

A key can be found in the sorts of hazards whose risk we are most inclined to overestimate. What do nuclear power plants, toxic waste dumps, and pesticide residues—to choose three such hazards at random—have in common? In all three cases, the risk is:

- Coerced rather than voluntary. (In home gardens, where the risk is voluntary, pesticides are typically overused.)
- Industrial rather than natural. (Natural deposits of heavy metals generate far less concern than the same materials in a Superfund site.)
- Dreaded rather than not dreaded. (Cancer, radiation, and waste are all powerful stigmata of dread.)
- Unknowable rather than knowable. (The experts endlessly debate the risk, and only the experts can detect where it is.)
- Controlled by others rather than controlled by those at risk. (Think about the difference between driving a car and riding in an airplane.)
- In the hands of untrustworthy rather than trustworthy sources. (Who believes what they are told by the nuclear, waste, and pesticide industries?)
- Managed in ways that are unresponsive rather than responsive. (Think about secrecy vs. openness, courtesy vs. discourtesy, compassion vs. contempt.)

Any risk controversy can be divided into a technical dimension and a nontechnical dimension. The key

technical factors are how much damage is being done to health and environment, and how much mitigation can be achieved at how much cost. The key nontechnical factors are the ones listed above, and others like them. Consider a proposed incinerator. Assume that the incinerator can be operated at minimal risk to health. Assume also that its developers tried to cram the facility down neighborhood throats with minimal dialogue; they are not asking the neighbors' permission, not offering to grant them oversight responsibilities, not proposing to share the benefits. While the experts focus on the technical factors and insist that the risk is small, neighbors focus on the nontechnical factors, find the risk huge, and organize to stop the facility. Is this an over-response? It is if we accept only technical criteria as valid measures of risk. But it may be a proportionate response, even a forbearing response, to the nontechnical side of the risk.

The two dimensions have been given various sets of labels: "hazard" versus "outrage," "technical rationality" versus "cultural rationality," etc. But it is a mistake to see the two as "objective risk" versus "perceived risk" or as "rational risk response" versus "emotional risk response." For many disputed hazards, in fact, the data on voluntariness, dread, control, trust and the like are more solid, more "objective," than the data on technical risk. These nontechnical factors have been studied by social scientists for decades, and their relationship to risk response is well-established. When a risk manager continues to ignore the nontechnical components of the situation, and continues to be surprised by the public's "overreaction," it is worth asking just whose behavior is irrational.

Since people's response to controversial risks doesn't arise from technical judgments in the first place, explaining technical information doesn't help much. When people feel they have been badly treated, they do not *want* to learn that their technical risk is small; instead, they scour the available documentation for ammunition and ignore the rest. It is still necessary to provide the technical information, of course, but the outcome depends far more on the resolution of nontechnical issues. Communication in a risk controversy thus has two core tasks, not one. The task everyone acknowledges is the need to explain that the technical risk is low. The task that tends to be ignored is the need to acknowledge that the nontechnical risk is high and take action to reduce it. When agencies and companies pursue the first task to the exclusion of the second, they don't just fail to make the conflict smaller; they make it bigger.

Of course, not all nontechnical issues can be resolved. Part of the public's response to controversial risks is grounded in characteristics of the hazard itself that are difficult to change—undetectability, say, or dread. Part of the response is grounded in the activities of the mass media and the activist movement, both of which amplify public outrage even though they do not create it. But the part that most deserves attention is the part that results from the behavior of the hazard's proponents. Risk communication guidelines for the proponents of controversial technologies are embarrassingly commonsensical:

- Don't keep secrets. Be honest, forthright, and prompt in providing risk information to affected publics.
- Listen to people's concerns. Don't assume you know what they are, and don't assume it doesn't matter what they are.
- Share power. Set up community advisory boards and other vehicles for giving affected communities increased control over the risk.
- Don't expect to be trusted. Instead of trust, aim at accountability; prepare to be challenged, and be able to prove your claims.
- Acknowledge errors, whether technical or nontechnical. Apologize. Promise to do better. Keep the promise.
- Treat adversaries with respect (even when they are disrespectful). If they force an improvement, give them the credit rather than claiming it yourself.

Advice like this is not difficult to accept in principle. It is, however, difficult to follow in practice. It runs afoul of organizational norms; sources that do not tolerate much internal debate are unlikely to nurture a more open dialogue with the community. It raises "yes, but" objections, from the fear of liability suits to the contention that it is better to let sleeping dogs lie. Perhaps most important, it provokes the unacknowledged bitterness in the hearts of many proponents, who may ultimately prefer losing the controversy to dealing respectfully with a citizenry they consider irrational, irresponsible, and discourteous.

PETER M. SANDMAN

For Further Reading: National Research Council Committee on Risk Perception and Communication, *Improving Risk Communication* (1989); Peter M. Sandman, *Responding to Community Outrage: Strategies for Effective Risk Communication* (1993); Neil D. Weinstein, ed., *Taking Care: Understanding and Encouraging Self-Protective Behavior* (1987).

RISK MANAGEMENT

Risk management can be defined as the "consideration of social, economic, and political factors in the context of a decision to control risk." In a society plagued with varied, often mysterious risks, the job of "risk management" is a critical one. It can be said that virtually all human activities from cradle to grave involve some element of risk and therefore a need for risk management. As a matter of course, risk management decisions are being made constantly by individuals, corporations, and government agencies. These decisions may be casual or rigorous, simple or elaborate, secret or open. The function of the risk manager can be described in several ways. The most obvious is that he or she must assimilate all relevant information and make a decision on whether, or how much, to control a risk through corporate or government action. Or, one could say that the manager's job is to take a risk assessment in which he has confidence and integrate it with the best available sociological, economic, and political information.

Risk management involves a good deal more than a straightforward, categorical decision to eliminate a risk regardless of the situation or cost. Rather, it involves a complex balancing of positive and negative factors. In this process, a risk manager needs to keep three "quantities" in mind. First, the potential for damage or, to put it another way, the benefits of control in terms of damage avoided. Second, the cost—either in terms of industry control costs or in terms of the benefits of a product forgone under regulatory prohibition or limitation. Third, some indication of the reliability of the data on which risk and cost calculations are made. These three "quantities" must be balanced somehow in making risk management decisions.

The Assessment-Management Interface

A crucial point in the risk management process comes at the very start, when the risk manager receives a risk assessment. The interaction between the risk assessor, who may be a scientist or an engineer, and the risk manager can create many types of friction and confusion, and even subvert the risk management process. It is widely felt that, in order to hold public confidence, the two functions of assessment and management should be as clearly defined and cleanly separated as possible. The risk assessor's job is to stick to the scientific approach, steering clear of value judgements (though it is clear that almost every move, even in the laboratory, involves some judgement), and to present the risk manager with reliable and objective information on risk—that is, his or her best judgement of the risk. On the other hand, the risk manager's job is seen as taking the risk assessment more or less at face value and making a policy judgement in a political context. Problems begin when the risk assessor or manager intrudes too much on the function of the other. Certainly a risk manager, if he or she is to have confidence in a policy decision, must feel comfortable that it is based on strong and reliable scientific data and on a sound assessment. He or she needs to feel that the numbers given are good enough to "operationalize"—for example, by ascertaining the cost per case avoided under a system of control.

Risk experts disagree on just how far scientists should go in interpreting their data for the manager. Since there obviously can be no clear-cut line between the two functions, the outcome seems to depend in great degree on the personalities and management philosophies involved. Science can be closely confined to what is replicable and testable. By this definition the boundary of science stops very early—almost at the laboratory stage. But it is possible for a risk assessor, in effect, to intrude into the realm of the risk manager. He can do this by incorporating so many loaded assumptions or value judgements into his scientific or engineering assessment that he predetermines the outcome and, as a practical matter, usurps the function of making a policy decision. In any case, no one has yet developed a systematic way to deal with such choices. Clearly, however, the line between assessment and policy judgement is likely to continue to be fuzzy and wavering. In some ways, in fact, the nature of an assessment may be dictated by expectations about the risk and the management options available. Usually, too, the process for transmitting scientific information from the data base to the point of ultimate decision is itself very complex.

Setting Priorities

One of the most important components of a risk manager's job is deciding which hazards to assess or control first and foremost, and which to ignore. Inevitably some kind of order will be established, even if it is informal. Obviously, it is not possible with limited resources to go after all risks simultaneously. The question, really, is whether it is necessary, possible, or appropriate to engage in some kind of formal or rigorous order of priority. Most experts seem to agree that, while ordering is necessary, it inevitably is a complicated process that defies attempts to adhere to strict rationality or uniformity, given the diverse nature of the risks and the many incongruous elements to be weighed. Many circumstances—the law, the nature of the regulator, public perceptions, external pressures—determine the extent to which a risk manager is able to control the agenda.

Experts in the field seem to agree that any priority setting must take into account a number of factors other than just the nature of the risk itself. These factors include: (1) the degree to which the risk can be controlled; (2) the costs of control; (3) the sociopolitical feasibility or acceptability of control; (4) the countervailing benefits of the substance or product (the other side of the equation); and (5) the degree to which the risk-taking activity is voluntary or involuntary.

Some further observations on establishing priorities can be made. It is often argued that the risks posed by the use of most chemicals are trivial and, in any case, are so swamped by the risks that nature imposes on us (the sun's radiation, aflatoxin, or whatever) or that we impose upon ourselves (smoking, drinking, driving fast, etc.), that there is little to be gained from managing them. By and large, however, risk experts reject the argument that even small environmental risks should be neglected. For one thing, if one risk is simply compared with another, the benefits involved are ignored, and they should be a factor in management decisions. The laissez-faire approach also does not take into account additive, synergistic, or multiple-causation effects. One of the major problems in setting priorities is the frequent discrepancy between science-based risk assessments and the way the public feels about a risk. The public may continue to have its special fears which, though they may not be totally rational from a statistical standpoint, are valid in subjective terms.

Making Risk Management Decisions

Given the reality of the many risks in society and the need to control them, what are the appropriate elements of a risk management decision? One concept that seems to find general favor among risk managers is that the risk management process should be an open expression of the logic and the values that lead to decisions. Since a risk management decision involves balancing various pros and cons, they should be made explicit—and the more explicit the better. This permits all concerned to understand and develop confidence in the process and its results.

In trying to reach a balance, a manager must consider risks, benefits, and costs. But the balancing of factors is often hidden. For example, a decision-maker might not want to be seen as giving too much weight to the costs of control, so he compensates by making allowances that get hidden in the scientific or engineering calculations.

A risk assessment can express the risk factor in quantitative or qualitative terms. The quantified estimate may be allocated among various sources of the risk, like five plants that manufacture a pesticide. Then, the regulatory decision is derived directly by balancing the numerical risk that has been calculated against what it would take to reduce the risk. In a qualitative evaluation of risk, such as that from a cancer-causing agent, one might choose to control it fully with the Best Available Technology (BAT), regardless of other factors, such as costs, and regardless of whether one or 100 lives are likely to be saved.

In some situations a quantitative risk assessment cannot be made, yet a qualitative estimate may suggest that there is a carcinogenic or other serious risk. Or dose-response data may be available without showing any meaningful relationship to humans. If the ranges within which risk might lie can be expressed quantitatively, the quantitative approach can be used. If not, a qualitative approach must be employed. Obviously, there are difficulties with both. It seems unlikely that any expert agreement or formula can be developed, given the complexity, the incompatible measurements, and the conflicting considerations that attend most management decisions.

Benefits and costs also can be expressed numerically for the benefit of the decision-maker, but such quantification can fail to capture some important dimensions. Costs or benefits can be expressed in terms of industry costs (or profits), consumer costs (or savings), jobs created or lost, lives saved or lost, environmental benefits or losses, and so forth. For instance, society, through its risk decisions, spends vastly different amounts of money to save lives, depending on the type of threat or hazard—and these differences indirectly reflect nonquantifiable values and perceptions. For example, should the Federal Aviation Administration allow the San Diego Airport to spend a huge amount of money on an additional landing control system because there has been one accident there every 10 years that killed more than 100 people or should the money be spent on another program that would save more lives?

Many experts feel that the public is gradually becoming more willing to wrestle with the key questions: Should there be more consistency in the amounts spent to deal with risks from different sources? Is it appropriate to establish a value on a life saved or a per-person year saved? If so, what is the upper limit that society can afford? With any risk calculus, there remains the problem of mixed and unquantifiable costs and benefits. The estimate of cost per life saved is only as accurate as the estimate of what the risk is, and that may be quite inaccurate. In the face of these difficulties, risk experts tend to retreat to a sort of middle ground. Simple numerical guides—such as agreeing to spend $450,000 to save a human life but not $452,000—are currently beyond the limits of what risk managers can reasonably deal with. But the experts feel it is wise to set some limits so as not to go forward spending some unconscionable amount. One

option is to express the cost per life saved in terms of a range.

Types and Degrees of Risk Control

A risk manager must decide what type of control to require and what degree of control or level of risk is to be achieved. Major control techniques are: (1) a ban on manufacture or limits on a product's use; or (2) technological control of manufacture or use. As noted, this can be accomplished by applying BAT (Best Available Technology); it can also be done by setting an appropriate level of emissions, or risk, and then permitting a facility or product to be reworked until the control level is achieved. A less common technique, but one that is approved by many risk managers, is to establish a "process" standard so that, for example, a manufacturing operation is enclosed or otherwise designed to control risk.

Another dimension involves improving worker protection and work practices in a plant. Controls used by individual workers can be much less expensive than controls built into a plant and thus are often favored by industry. For example, industry has fought hard for the use of individual respirators to protect against cotton dust in the workplace, instead of employing technology to keep ambient dust levels down. This has been the subject of a bitter controversy. As with all risk management decisions, choices of control techniques typically hinge on considerations of risk (or need), efficacy, and costs of implementation.

Risk managers also must make the crucial choice of how stringent their controls should be. The degree of control often is prescribed by a law or regulation, thus limiting the manager's choices. The strictest form of control is the concept of "zero risk." Because of the extreme difficulty of establishing thresholds at which there is no risk, considering the complex nature of chemicals and testing procedures, it is generally thought of more as an ideal goal or bargaining position rather than a realistic objective.

Between the extremes of zero risk and zero regulation, various gradations of control are available. For example, establishment of a *de minimis* risk level—that is, an upper limit below which the risk is considered too trivial to attempt to reduce, even if the cost is insignificant. Another is the application of safety measures or control technologies "to the extent feasible" (OSHA), to the extent feasible taking costs into consideration (Safe Drinking Water Act), or using the Best Available Technology "economically achievable" (Clean Water Act). One manifestation of this approach is to try to regulate only down to a level at which no industries, or almost none, will be forced out of business. Another is to establish a limit on the costs to be borne per case avoided, per life saved, per year of life saved, or whatever. Still another is either separately, or in conjunction with a technology requirement, to control a risk so as to provide an "ample margin of safety" or similar safeguard, as several laws mandate. (Among these is the Clean Air Act.) Clearly, this and other elements of risk management decision-making are likely to be grounded in the values and philosophy of the manager or of the interests he seeks to serve. In such matters, there is no meeting of the expert minds and it seems unlikely that there will be in the future.

RICE ODELL

For Further Reading: Duane LeVine and Arthur Upton, eds., *Management of Hazardous Agents* (1992); René Dubos Center for Human Environments, *Living with Risk* (1991); René Dubos Center for Human Environments, *Managing Hazardous Materials* (1988).

RISK PERCEPTION

Risk-perception research examines the opinions that people express when they are asked, in a variety of ways, to evaluate hazardous activities, substances, and technologies. Researchers try to discover what people mean when they say that something is or is not "risky," and to determine what factors underlie those perceptions. The basic assumption underlying these efforts is that those who promote and regulate health and safety need to understand the ways in which people think about and respond to risk.

Risk-perception research attempts to aid policy makers by improving communication between them and the lay public, by directing educational efforts, and by predicting public responses to new technologies (such as genetic engineering), events (such as a good safety record or an accident), and new risk management strategies (such as warning labels, regulations, or substitute products).

Important contributions to our current understanding of risk perception have come from geography, sociology, political science, anthropology, and psychology. Geographical research focused originally on understanding human behavior in the face of natural hazards, but it has since been broadened to include technological hazards as well. Sociological research and anthropological studies have shown that perception and acceptance of risk is mediated by social influences transmitted by friends, family, fellow workers, and respected public officials. In some situations, people acting within social groups may downplay certain risks and emphasize others as a means of maintaining and controlling the group.

Psychological research on risk perception has sought to develop a taxonomy for hazards that can be used to understand and predict responses to their risks. The most common approach to this goal has employed the *psychometric paradigm*, which uses psychophysical scaling and multivariate analysis techniques to produce quantitative representations or "cognitive maps" of risk attitudes and perceptions. Within the psychometric paradigm, people make quantitative judgments about the current and desired riskiness of diverse hazards and the desired level of regulation of each. These judgments are then related to judgments about other properties, such as *(1)* the hazard's status on characteristics that have been hypothesized to account for risk perceptions and attitudes (for example, voluntariness, dread, knowledge, controllability), *(2)* the benefits that each hazard provides to society, *(3)* the number of deaths caused by the hazard in an average year, *(4)* the number of deaths caused by the hazard in a disastrous year, and *(5)* the seriousness of each death from a particular hazard relative to a death due to other causes.

Numerous studies carried out within the psychometric paradigm have shown that perceived risk is quantifiable and predictable. Psychometric techniques seem well suited for identifying similarities and differences among groups with regard to risk perception and attitudes (see Table 1). They have also shown that the concept "risk" means different things to different people. When experts judge risk, their responses correlate highly with technical estimates of annual fatalities. Lay people can assess annual fatalities if they are asked to (and produce estimates somewhat like the technical estimates). However, their judgments of "risk" are related more to other hazard characteristics (for example, catastrophic potential, threat to future generations) and, as a result, tend to differ from their own (and experts') estimates of annual fatalities.

Psychometric studies have demonstrated that every hazard has a unique pattern of qualities that appear to be related to its perceived risk. Figure 1 shows the mean profiles across nine characteristic qualities of risk that emerged for nuclear power and medical x-rays in an early study. Nuclear power was judged to have much higher risk than x-rays and to need much greater reduction in risk before it would become "safe enough." As the figure illustrates, nuclear power also had a much more negative profile across the various risk characteristics.

Many of the qualitative risk characteristics that make up a hazard's profile tend to be highly correlated with each other, across a wide range of hazards. Investigation of these interrelationships by means of factor analysis has indicated that the broader domain of characteristics can be condensed to a small set of higher-

Table 1
Ordering of perceived risks for 10 activities and technologies.

The ordering is based on the geometric mean risk ratings within each group. Rank 1 represents the most risky activity or technology.

Activity or technology	League of Women Voters	College students	Active club members	Experts
Nuclear power	1	1	7	10
Motor vehicles	2	5	3	1
Handguns	3	2	1	4
Smoking	4	3	4	2
Motorcycles	5	6	2	6
Alcoholic beverages	6	7	5	3
General (private) aviation	7	10	9	8
Police work	8	8	6	9
Pesticides	9	4	10	7
Surgery	10	9	8	5

order characteristics or factors. Most important is the factor "dread risk." The higher a hazard's score on this factor, the higher its perceived risk, the more people want to see its current risks reduced, and the more they want to see strict regulation employed to achieve the desired reduction in risk. In contrast, experts' perceptions of risk are not closely related to any of the various risk characteristics or factors derived from these characteristics. Instead, experts appear to see

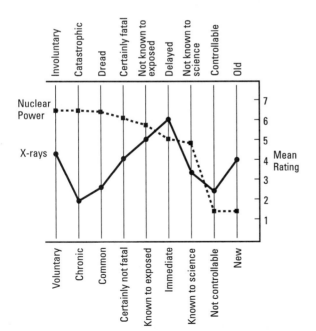

Figure 1. Profiles for nuclear power and x-rays across nine risk characteristics.

riskiness as synonymous with expected annual mortality. As a result, many conflicts about "risk" may result from experts and laypeople having different definitions of the concept. In such cases, expert recitations of "risk statistics" will do little to change people's attitudes and perceptions.

Perhaps the most important message from risk-perception research is that there is wisdom as well as error in public attitudes and perceptions. Lay people sometimes lack certain information about hazards. However, their basic conceptualization of risk reflects legitimate concerns that are typically omitted from expert risk assessments. As a result, risk communication and risk-management efforts are destined to fail unless they are structured as a two-way process. Each side, expert and public, has something valid to contribute. Each side must respect the insights and intelligence of the other.

PAUL SLOVIC

For Further Reading: D. Golding and S. Krimsky, eds., *Theories of Risk* (1992); National Research Council, *Improving Risk Communication* (1989); P. Slovic, "Perception of Risk," *Science*, 236, 280-285 (1987).

RIVERS AND STREAMS

Rivers and streams represent freshwater habitats (about 1% of the water on Earth) that are classified as lotic (from the Latin verb "to wash") for flowing waters as opposed to standing waters such as ponds. There are great differences between streams in temperature, speed and amount of flow, oxygen content, and characteristic plant, animal, and microbial life.

The scientific study of streams and rivers (lotic ecology) started as a "poor sister" of limnology (the study of fresh waters), but saw some excellent early work in North America and Europe at the turn of the century. The field received particular emphasis from fishery biologists in the 1940s and 1950s, and pollution biologists in the 1960s and 1970s. Since the 1970s, lotic ecology has become a sophisticated, multidisciplinary field as rich in theory and paradigms as any ecological discipline. The hallmark of lotic ecology since the 1970s, initiated by H. B. Noel Hynes' (University of Toronto) landmark text on running waters, has been cooperative research between biological stream ecologists, geomorphologists, hydrologists, biogeochemists, and terrestrial plant ecologists.

Since about 1980 the watershed (also referred to as catchment or drainage basin) has generally been recognized as the basic unit of running water ecology. At any point along a stream course, precipitation falling within the watershed and directed by surface flow will follow the channel network. Unique characteristics of watersheds that change with seasons and over longer time spans form a basic pattern to which the stream organisms—microbial, plant, and animal—are adapted. Thus data about the watershed, channel form, annual and long-term patterns of flow and temperature, and the terrestrial vegetation growing along the channels allow for predictions to be made about the expected resident biological communities.

This notion of geomorphic, hydrologic, and vegetative linkages to stream and river biological communities is embodied in the River Continuum Concept. These linkages differ in a predictable fashion along a continuum from the smallest headwaters (first-order streams), through larger streams, mid-sized rivers, and finally to large rivers. The headwater stream sections are the most shaded and thus limited in plant growth, especially in forested areas. In these shaded streams total photosynthesis is less than total respiration so there is a shortage of food that is balanced by imports of organic matter such as vegetation litter from the banks. Broader, less shaded, mid-sized rivers sustain sufficient plant growth and produce excess organic matter that is transported down river. Different invertebrates can serve as indicators of these relationships through their specialized responses to food supplies along the river continuum. For example, animal shredders are associated with leaf litter, scrapers with attached algae, and collectors with fine particulates.

Some examples of important watershed components are area, stream sizes, drainage density, and basin slope (change in elevation from the uppermost headwaters to the point designated as the "mouth" of the watershed).

Lotic organisms in two habitat types—riffles and pools—have adaptations allowing them to maintain position and move about, respire, and acquire nutrients. For example, algae growing in fast water usually have holdfasts or other attachment mechanisms or a hydrodynamically advantageous shape. Many aquatic insects also have special adaptations for attachment in fast water, such as blackfly larvae (Simuliidae), which have a circle of posterior hooks that lock into a layer of silk spun on rocks and logs. From these attachment sites they use filtering fan-like appendages on the head to extract tiny food particles from the passing water. Adaptive body shapes also allow many invertebrate and fish species to avoid the main thrust of the current or to present a hydrodynamically advantageous contour to the flow. In contrast, organisms living in pools must be adapted to counter the continual "rain" of fine sediments being dropped in these habitats. Fish species can remain off the bottom above the suffocating fine sediments. Invertebrates, such as mayfly nymphs, burrow into the fine sediments or have covers to protect delicate gills.

Patterns of annual flow are major features of running waters around which the life cycles of organisms are organized. For example, at temperate latitudes, peak flows usually occur during spring and autumn as runoff from precipitation. At higher latitudes and altitudes, the spring runoff is delayed as snow and ice melt in the early summer. During these generally predictable times of high flow, organisms tend to be in their most resistant aquatic life stage or in a terrestrial stage away from the stream or river. In tropical latitudes the dry and wet seasons divide the annual cycle into two flow regimes requiring different adaptations, such as life stages resistant to desiccation during the dry months. When patterns of high flow are less predictable, life cycles tend to be shorter and not synchronized, increasing the survival probability of a given organism.

Temperature is also a major pattern on which life cycles are organized. The annual range of stream temperature, approximately 0 to 20–30°C in temperate regions and 5–10 to 30–35°C in the tropics, is always less than in the surrounding terrestrial environment. As a consequence, biological activity can proceed in running waters at times when temperature-limited activity in the associated terrestrial habitats is shut down.

Humans have always oriented their activities around streams and rivers, from aboriginal to industrialized times—for drinking water, food fisheries, transportation, waste disposal, recreation, and esthetic inspiration. Given the importance of the geomorphic, hydrologic, and riparian (river bank) influences on stream organisms, the significance of human impacts is clear: bank vegetation removal, reduced (irrigation) or altered (impoundment) flows, and channelization. In addition, changes in organic inputs through pollution alter the balance between aquatic plant growth and total respiration of all the organisms in the stream community. Of course, toxic inputs can drastically alter the microbial, plant, and animal communities of running waters.

KENNETH W. CUMMINS

For Further Reading: H. B. N. Hynes, *The Ecology of Running Waters* (1970); P. J. Boon, P. C. Callow, and G. E. Petts, eds., *River Conservation and Management* (1992); M. Morisawa, *Streams: Their Dynamics and Morphology* (1968).

ROADS AND HIGHWAYS

Streets, roads, and highways permeate our life. They play an important role in providing transportation in both developed and developing economies. They provide channels for transporting people and goods, pathways for pedestrians, and conduits for utilities. Good roads bring society closer together and contribute to its well-being. In countries with high car ownership, such as the United States, roads and streets accommodate most urban and intercity travel. They affect the surrounding environment, of which they form an integral part. They account for one-quarter to one-fifth of all developed land in American cities.

Early Highways

Roads date back to the historic trails used for communication and commerce in antiquity. The early civilizations of the Carthaginians, Chinese, and Incas built extensive road systems, and the streets of Babylon were paved as early as 2000 B.C.

The Romans built the most advanced highways of the ancient world. The first link of the Appian Way—from Rome to Capua—was built about 300 B.C. At its zenith, the Roman road system extended some 53,000 miles. Some Roman roads were built with heavy stone blocks laid over strong foundations of rubble or broken stone. Traces of these roads may be found today, and some form sections of modern highways.

After the fall of the Roman empire in the fifth century, there was little road building in the Western world for some 1,200 years. Although stage coaches were introduced in 1654, the only convenient means of travel between cities was on foot or on horseback. Interest in road building revived in Europe during the late 18th century. The regime of Napoleon in France (1800–1814) gave impetus to road construction for military purposes and led to a national system of highways.

The early American highways in the latter part of the 18th century generally took the form of toll roads or "turnpikes." These included the Philadelphia and Lancaster Turnpike Road (Pennsylvania) and a link between Alexandria, Virginia, and the Blue Ridge Mountains. In 1806, the federal government authorized the construction of the National Pike or Cumberland Road. This route extended some 800 miles from Cumberland, Maryland, to Vandalia, Illinois. Surfaced largely with macadam, it cost about $7 million.

Many roads or "trails" figured prominently in the westward expansion of the United States. These included the Oregon Trail, Sante Fe Trail, and the Overland Trail. However, the emergence of the steam locomotive in the 1830s and the subsequent railroad development led to a hiatus in road building outside of cities during much of the 19th century.

New technologies contributed to road development in two ways: Improved methods of construction made possible hard-surfaced all-weather roads; and the emergence of motor vehicles increased demand for more and better roads.

The early American roads were largely natural—earth, or timber plank roads, while many city streets were paved with cobblestones. In the late 18th century, Trésaquat, a French engineer, advocated using a broken-stone base covered with small stones. At about the same time in England, Telford and McAdam developed similar types of construction. McAdam advocated the use of smaller broken stones both as a base and for the wearing surface—this type of construction is the forerunner of modern macadam bases and pavement. The invention of the power stone crusher and steam roller led to a considerable increase in mileage of broken-stone surfaces. Asphalt was used for paving Pennsylvania Avenue in Washington, D.C., in 1867. Brick pavements of city streets were used since the 1890s. Concrete pavements were introduced about 1893, and the first concrete rural road was built in Wayne County, Michigan, in 1909.

The first automobile was developed by Carl Benz in Mannheim, Germany, during the mid-1880s and appeared in North America around the turn of the century. The first buses operated in Great Britain in 1899, Germany in 1903, and New York City in 1905.

Increased demand for highways emerged around the turn of the century. It came mainly from farmers who wanted better access to markets. The cyclist movement—popular in the 1880s–1890s—also created a demand for smooth roads. The first state aid law was enacted by New Jersey in 1891, and by 1900 six other states had enacted similar legislation. By 1917, every state participated in highway construction, and most states had established a highway agency that was responsible for construction and maintenance of principal state routes.

Growth of the U.S. Highway System

Federal participation in U.S. highway affairs began in 1893 when the Congress established the Office of Local Inquiry. This agency became the Bureau of Public Roads of the Department of Agriculture. In 1939, it became the Public Roads Administration of the Federal Works Administration. In 1949 it was renamed the Bureau of Public Roads and transferred to the Department of Commerce. In 1967, a National Department of Transportation (DOT) was established and the Bureau of Public Roads was made part of the Federal Highway Administration (FHWA) within DOT. It merged with the FHWA in 1970.

The modern era of federal aid for highways began with the passage of the Federal Road Act of 1916. Funds were apportioned to the individual states based on area, population, and rural road mileage. States were required to match funds on a 50–50 basis.

The National System of Interstate Highways was defined in 1944, and its financing was authorized by the Federal Aid Act of 1956. Funds were made available to states on a 90–10 basis.

The U.S. highway legislation of the 1960s recognized the growing transportation needs of urban areas. The 1962 Act required a continuing, comprehensive, and cooperative transportation planning process for urban areas, and required urban highway improvements to be integral parts of soundly based balanced transportation systems. This process introduced social, economic, and environmental considerations into the planning process.

Federal highway legislation in the 1970s reflected the decision to cope with increasing urbanization, environmental impacts of traffic congestion, and fuel shortages. The 1970 legislation allowed highway funds to be used for bus lanes, traffic controls, and parking facilities serving public transport. Environmental considerations included air, noise, and water pollution; displacement of people and activities; and impacts on natural resources, public facilities, and aesthetic values. The 1973 Act allowed Highway Trust Funds to be used for public transport improvements.

The Surface Transportation Assistance Act of 1982 addressed the problems of deteriorating highway infrastructure. The Intermodal Surface Transportation Efficiency Act of 1991 further increased the urban orientation of federal aid. Many funds previously reserved for highways became eligible for public transport or highway projects at the discretion of state and local agencies. The 45,000-mile Interstate system is to become part of a proposed 155,000-mile (±15%) National Highway System, and the primary, secondary, and urban system designations dropped.

Highway Design

Highway design has evolved over the years, reflecting changes in vehicle types, dimensions, and capabilities; operating speeds; and traffic and safety considerations. Travel lanes and shoulders have become wider; side slopes, curves, crowns, and grades gentler; rights-of-way more expansive; and structures and interchanges more plentiful.

The highway cross-section includes travel lanes, auxiliary lanes, shoulders, median islands, and roadsides. The dimensions of each vary with the type of highway. Travel lanes on freeways and main arterial highways are now 12 feet wide. Shoulders are usually 10–12 feet wide on highways with four or more lanes, and 4–8 feet wide on two-lane main rural roads.

The design speed of a highway influences its curvature, superelevation, sight distance, and grade. It is the maximum safe speed attainable over a section of highway when traffic conditions permit design features to govern. It depends upon the type of terrain, density and character of the surrounding environment, function of

the highway, expected traffic volumes, and economic and environmental considerations.

Freeway and arterial roads usually have a design speed of 70 miles per hour in rural areas and 50 miles per hour in urban and suburban settings. A 70 mph design speed usually limits grades to 3 or 4% and curves to radii of 1,600 feet or greater. These controls become relaxed as the design speed is reduced (as in cities or in mountainous terrain).

The higher design standards associated with modern high-type high-speed roads translates into increased land requirements, and in some cases, greater environmental impacts.

The freeway has three essential properties: 1. Opposing streams of traffic are separated by a median divider. 2. Access is limited to interchanges with additional lanes for merging, weaving, and diverging. 3. All intersections are physically separated at different levels.

Most of the early "limited access" roads had their origins in urban aesthetics and park design. The first such road in the United States—the 15-mile Bronx River Parkway—was conceived in 1906 and opened to traffic by 1924. It was followed by the Long Island and Westchester Parkways (1923–1934) and by the New York City Parkway System (1936–1940). These roads were sensitively designed and landscaped.

The initial section of Connecticut's Merritt Parkway, opened in June of 1938, was placed on a 300-foot "park" strip. The Arroyo Seco Freeway in Los Angeles, opened in 1940, was the first freeway to provide acceleration lanes at ramps. The 160-mile Pennsylvania Turnpike, opened in 1940, became the first intercity express highway. The first real freeways in Europe were Germany's Autobahn, built in the 1930s.

Post-war freeways—perhaps best characterized by the New Jersey Turnpike—generally emphasized design and traffic standards and gave environmental impacts secondary concern.

Many modern freeways may be up to 10 to 14 lanes wide. Some provide rail transit lines, busways, or high-occupancy vehicle lanes in the median, and a few plan to incorporate special interchanges for high-occupancy vehicle lanes. Many are sensitively designed in relation to their surrounding environment.

Streets and highways are classified by their importance in the overall road system and by the functions they serve. Roads are normally classified in terms of the proportions of through movement and access they serve. *Freeways* have full control of access and serve only movement. *Arterial* roads and streets normally emphasize movement, but also provide some land access. *Local* streets and roads emphasize land access. Each type has its specific design features, standards, and capacity. A freeway lane can usually carry 1,500 to

1,800 vehicles per hour and an arterial street lane 500 to 600 vehicles per hour, without undue delay.

Traffic engineering and management activities are applied to improve the safety and utility of the road system. Traffic signs, traffic signals, and pavement markings are placed along streets and highways in accord with state and national standards. Other actions include: turn and parking restrictions; one-way streets; reversible lanes and streets; metering of freeway entrance ramps, and motorist information and freeway surveillance systems.

Several states have enacted Highway Access Codes that specify where and how access can be provided to adjacent developments along various classes of roads. The goal of these access management programs is to provide essential access to land development while *protecting* the functional integrity of the arterial road system.

Travel Patterns and Impacts

In 1990, there were 144 million cars and 45 million trucks registered in the United States. These vehicles were operated by 167 million licensed drivers—67% of the nation's 248 million residents. They aggregated more than 2 trillion miles of travel over 3.9 million miles of streets and highways. Highway-based travel accounted for 80% of the domestic intercity passenger miles and 90% of the passenger travel within cities. One third of all intercity freight moved by truck. Highway transportation approximated 16% of the U.S. gross national product. Within cities, one third of the travel took place on freeways. On a global scale, the United States accounted for about a third of the world's motor vehicles.

Freeways and major arterial roads, unless sensitively designed, can preempt large amounts of land, separate communities, and introduce noise and pollutants into areas that the roads traverse. Large parking areas—usually surface parking lots—further contribute to urban and suburban blight.

Conflicts between cities and cars can be resolved in various ways. These include: coordinating land use with transport; emphasizing public transport especially for travel to city centers; changing the type, location, and scale of proposed roads; and creating traffic-free or traffic calming zones.

Planning and policy options for major highways include: cancelling plans for new highways (i.e., not building Boston's Inner Traffic Loop); relocating the highway in a lesser developed area (i.e., relocating I-95 away from Boston's Inner Harbor); placing the roadway below grade, possibly in a tunnel (i.e., placing I-95 in Penn's Landing, Philadelphia, and I-395 under the Mall, Washington, D.C.); achieving multiple use of the highway right-of-way (i.e., building a park over I-5, Seattle); and developing the highway as part of a lineal park (i.e., developing George Washington Memorial Parkway, Washington, D.C.).

Interdisciplinary design teams are sometimes formed when locating and designing new highways in heavily built-up urban areas. These teams normally include urban architects and landscape architects, highway and traffic engineers, urban designers and planners, and sociologists, as well as real estate specialists, lawyers, economists, and educators. Their responsibilities include: developing greater community participation in the design process; evaluating the total environment of the highway corridor; addressing community and social problems as they relate to the highway; suggesting changes in highway design features to make the highway more compatible with its environs; suggesting changes in surrounding land uses to minimize conflicts with the highway; and planning community improvements in conjunction with highway development.

Protecting the Environment

The National Environmental Policy Act of 1969 (NEPA) requires federal agencies to systematically assess the potential impacts of a project on the human environment, and most highway projects involve impacts that require such environmental reviews. This assessment must include possible environmental impacts and unavoidable adverse environmental effects, alternatives to a proposed action or project, local short-term uses of the environment, enhancement of long-term productivity, and irreversible or irretrievable commitments of resources.

Public concerns about air quality led to the Clean Air Act of 1970, which specified auto emission targets and ambient air quality standards. States that could not meet standards were required to prepare traffic control plan components for reducing vehicle miles of travel or reducing emission for vehicle miles. The original 1970 schedule of new-car emission standards has been modified several times.

The South Coast Air Quality Management Plans in response to the provisions of the 1990 Clean Air Act (Regulation X) require all employment sites with 100 or more workers to implement "Trip Reduction Plans." The goals are expressed in "Average Vehicle Ridership"—the number of employees reporting to work between 6 and 10 A.M. and the number of vehicles driven by these employees. Employers are free to choose any combination of travel demand management and transportation control measures to achieve those targets. They can encourage employees to shift to carpools, vanpools, or mass transit (trip reduction); allow employees to "telecommute" or work compressed weeks (trip elimination or trip avoidance); or modify work schedules to encourage off-peak travel (trip shifting). An analysis of 1,110 plans found an actual 2.7% increase of average vehicle occupancy over a one-year period.

Future Highways

Motorization will continue to place greater pressures on existing roads and require new ones throughout the world. More people, occupying more land and driving more cars, will increase demands for road space. The dispersion of residences, work places, and shopping activities will increase the highway orientation of most American urban areas.

Future roads should serve demonstrated social, economic, and mobility needs, with minimum environmental impact. This may entail downsizing roadways in built-up areas, enhancing landscaping and buffer areas, creating transit and pedestrian-friendly road patterns in suburban settings, and applying access control principles to arterial streets and highways.

Intelligent vehicle highway systems (IVHS) and smart highways, now in the formative stage, will become more widespread. They include advanced traffic management systems, advanced driver information systems, automated vehicle control systems; advanced commercial vehicle operations; and advanced public transport systems. Automated highway systems—first proposed at the 1939–1940 New York World's Fair—may become a reality in the 21st century. IVHS may also provide the means for implementing road pricing programs.

A closer alliance between highway transportation and urban development will be needed to improve both mobility and liability. New land use patterns—such as new towns-in-town and suburban new towns—can further reduce the need for roads and transport.

HERBERT S. LEVINSON

For Further Reading: John D. Edwards, ed., *Transportation Planning Handbook* (1992); U.S. Department of Transportation, *Moving America: New Directions—New Opportunities* (1990); Paul H. Wright and Radner J. Paquette, *Highway Engineering* (1987).

ROBERTS, WALTER ORR

(1915–1990), American astro-geophysicist and institution builder. As a graduate student in astronomy, Roberts moved to a high-altitude site in Climax, Colorado, to establish an observing program using the first solar coronagraph in the Western Hemisphere. This instrument, invented by Bernard Lyot in France a few years earlier, allowed the observation of faint light from the sun's atmosphere—light that had previously been obscured by the very intense light from the sun itself.

Roberts's work at Climax employed time-lapse photography to study the evolution of activity on the sun. His application of this technique to larger features of the solar atmosphere resulted in the production of two movies, *Explosions on the Sun* and *Action on the Sun*, which were seen by many scientists around the world and by attendees at his public lectures. The graceful curving motions of material in the solar atmosphere, and the similarity of the paths of motion to drawings of magnetic field lines, contributed to the wide appreciation of the importance of magnetic fields for the behavior of hot ionized gases a decade before the theory of magneto-hydrodynamics was developed. The beauty of these films was also influential in attracting a generation of young people into the study of astronomy, especially solar physics.

The Climax observing station was incorporated in 1946 as the High Altitude Observatory, and later activities were expanded to include a headquarters and research center on the University of Colorado campus in Boulder. The work of the Observatory staff, now including theoretical studies as well as observations, continued to grow and develop into one of the largest astrophysical research and teaching institutions in the country.

In 1960 Roberts was asked to be the first director of the planned National Center for Atmospheric Research (NCAR). He convinced the organizers that the new center should be built on the strong administrative and scientific base of the High Altitude Observatory, and that the new center should be located in Boulder. These conditions were accepted and he was appointed director of NCAR, and he continued an association with this organization, in various roles, for the rest of his life.

While working at Climax, Roberts had become interested in possible connections between solar activity and the weather. Solar weather research was not then a respectable area for scholarly endeavors; it had been compromised by too many claimed correlations of sunspots with stock market performance, wine vintage quality, and specific human diseases. But over the next four decades Roberts persisted in his own studies of the topic and in organizing meetings to survey progress and possibilities. In doing so he was a major factor in restoring sufficient credibility to such studies that today one finds serious measurements and models aimed at elucidating conjectured effects of solar changes on the earth's lower atmosphere. His interest in the weather and climate broadened, especially after assuming the direction of NCAR, and he became a spokesman for the atmospheric sciences and a knowledgeable environmentalist. He was among the first atmospheric scientists to publicize the growing worries about human-induced climate heating and to arrange meetings on past climate changes and their relevance for the future.

In the early 1970s a series of events brought to everyone's attention the relationship between climate and human activities. Unfavorable weather decreased grain production in both the United States and the Soviet Union, and wheat prices soared. Drought in the Sahel region of Africa accompanied famine and deaths in that region. And a major El Niño event along the coast of equatorial South America saw the harvest from the world's largest fishery decrease dramatically. Roberts saw these events as an excellent opportunity to study and delineate the relationship between climate and food production. He therefore raised money for a large study of this period, outlined the nature of the study, found institutions willing to cooperate in the study, and recruited the senior staff for the work.

The result was both a surprise and a hard lesson. Careful examination of the data and events for the critical period showed that the loss in global grain reserves had been quite moderate, but that Soviet leaders had decided to increase meat production in their country and had imported large quantities of feed grain, raising world prices and giving the appearance of a shortage. A severe drought had indeed occurred in the Sahel, but much of the resulting difficulty had originated in the drilling of wells earlier by development agencies, resulting in overgrazing by the additional cattle allowed by the extra water. The drought effects were further exaggerated by growing national restriction on the traditional migration patterns of the nomadic peoples of the region. In Peru temporary reductions in fish catches during El Niño events were well known, but by 1972 harvesting had grown so large that overfishing was about to destroy the fishery for a decade to come—El Niño was simply the last straw. The first book that resulted from this study, authored by study director Rolando Garcia, was appropriately titled *Nature Pleads Not Guilty*. This study was influential in focusing attention on the fact that many tragic events, which we would like to blame on "acts of God," are actually evidence of poor societal organization and foresight.

In every activity Roberts valued above all a positive approach. He had a strong conviction that every problem has a constructive solution and that seeking that solution was where human energies should be placed. In direct opposition to the usual description of scientific research as self-correcting through mutual criticism, he believed that pointing out others' mistakes was counterproductive, that the time would be better spent producing a new piece of work. This deeply held belief in positive action meant that he chose not to expend his personal environmental efforts in activities such as opposing the emission of pollutants. Instead he focused on new forms of agriculture, especially halophyte culture along coasts, on scientific education for the public, and on the exciting opportunities inherent in the requirement to adapt to a different climate regime in the coming decades. Thus his developing environmentalism took forms quite at odds with the growing activism of the 1970s and 1980s.

Professional environmentalists were pleased when Roberts agreed to write an article or give a talk on the possibilities of dramatic climate change, but they were puzzled when he would not join them in proposing controls to limit emissions of climate-changing gases. His approach on these matters did not arise from any naïveté about the Earth and its carrying capacity, but from his lifelong assessment of what was possible and effective in influencing major societal trends, combined with an indefatigable cheerfulness and optimism.

JOHN FIROR

For Further Reading: Paul Ehrlich, Donald Kennedy, Walter O. Roberts, and Carl Sagan, eds., *The Cold and the Dark: The World After Nuclear War* (1984); Walter O. Roberts and Henry Lansford, *The Climate Mandate* (1979); Walter O. Roberts, *A View of Century 21* (1969).
See also Desertification, Natural Disasters.

RODALE, ROBERT

(1930–1990), publisher. Robert Rodale pioneered the scientific investigation of regenerative farming practices, and championed the value of farmers' wisdom and innovation in developing agricultural systems that are productive, profitable, and environmentally sound.

Rodale's concept of regenerative agriculture transcends earlier ideas of organic and sustainable agriculture. Organic farming is commonly defined as production of food or fiber without the use of synthetic fertilizers or pesticides for special consumer markets. Sustainable farming is more broadly defined as using practices that will maintain productivity of the soil and other natural resources indefinitely.

By employing regenerative practices, Rodale maintained that farmers could produce profitable and healthful crops, protect the environment, and improve their resource base even as they use it. His vision was not one of static operations, but thriving farms and rural communities. Farmers could accomplish this best by maximizing the use of their farms' internal resources (such as the sun, water, soil, and natural enemies of pests) and minimizing the use of external inputs (such as chemical fertilizers, pesticides, and weed killers).

For example, a conventional farmer relies almost exclusively on purchased fertilizers to feed crops. A regenerative farmer supplies nutrients by applying manure or compost from on-farm livestock enterprises, and provides additional nitrogen by planting a legume cover crop that captures the sun's energy and fixes free nitrogen from the air over winter. The cover crop also smothers weeds, reducing the need for herbicides, and improves soil structure and health, so succeeding crops' roots can better forage for nutrients and moisture.

Rodale's concepts were not widely accepted at first. In the 1940s his father, J. I. Rodale, founded Rodale Press (publishers of books and magazines such as *Organic Gardening* and *Prevention*) on the premise that nonchemical farming practices produced healthier soil, crops, and people. But these ideas were largely dismissed by the agricultural establishment. After J. I.'s death in 1971 Robert established a 320-acre research facility in Kutztown, Pennsylvania, to prove scientifically that regenerative farming practices are beneficial to farmers and the total environment.

The most influential project conducted at the Rodale Institute Research Center is the ongoing Farming Systems Trial, established in 1981. In a neighboring field that for decades had been farmed using conventional practices, Rodale scientists set up plots comparing chemical and nonchemical crop rotations. In plots where chemicals were abruptly withdrawn, yields and income were lower for the first three years compared to chemical plots. But yields steadily improved, and they continue to equal or exceed those under chemical management. Profits in the nonchemical plots are higher because of lower input costs.

Rodale scientists conducted other pioneering investigations, including the development of perennial grains and breeding programs for new crops such as amaranth, a drought-tolerant grain with superior protein content. The researchers also undertook studies comparing conventional and regenerative management under minimum-tillage systems designed to reduce erosion.

Gradually, Rodale's ideas began to take root within the agricultural establishment. At his urging in 1985 Congress directed the Department of Agriculture (USDA) to begin researching regenerative farming. USDA stationed soil scientists at the Rodale facility, and Rodale scientists forged collaborations with land grant universities.

Rodale long recognized that scientific information did little good unless it was put into practice. So to speed the adoption of regenerative farming practices, the non-profit Rodale Institute established three regional on-farm research networks working with 34 farmers from the Mid-Atlantic states to the Great Plains. The network cooperators test regenerative practices, adopt those best suited for their climate and soils, and hold field days so other farmers can see the practices firsthand. Since 1979 the Rodale Institute has published *The New Farm* magazine, which provides farmers with the technical information they need to make regenerative practices work on their farms.

In the late 1980s Rodale's attention turned increasingly to international agriculture. He saw that the regenerative model that was taking hold in the U.S. would work especially well in developing countries where external inputs were scarce and expensive. The Rodale Institute established farmer networking projects in Tanzania, Senegal, and Guatemala, and Rodale wrote the popular book *Save Three Lives*, detailing how regenerative practices can help prevent famine worldwide.

Even before the breakup of the former Soviet Union, Rodale saw how his ideas could help ease the transition from collective to private farms in that country. He launched *Novii Fermer*, a Russian-language version of *The New Farm* to help the new class of private farmers adopt regenerative practices. In 1990 after finalizing plans for the publication of the first issue of the magazine, Rodale was killed in an automobile accident on the way to the Moscow airport.

Despite the tragic loss, the Rodale Institute continues to carry out the mission begun by Rodale and his father. In 1991 the organization sponsored the first international conference on the soil health–human health connection. At the meeting scientists from around the world recognized the need to develop a "soil health index" to document the gains and losses in soil quality worldwide.

CRAIG CRAMER

For Further Reading: Christopher Shirley, ed., *What Really Happens When You Cut Chemicals* (1993); National Research Council, *Alternative Agriculture* (1989); Robert Rodale, *Save Three Lives—A Plan For Famine Prevention* (1991).
See also Agriculture; Composting; Erosion, Soil; Organic Gardening; Soil Conservation; Sustainable Yield.

ROOSEVELT, THEODORE

(1858–1919), American explorer, conservationist, twenty-sixth President of the United States. In 1916 Theodore Roosevelt wrote in *A Book Lover's Holidays in the Open*: "The extermination of the passenger pigeon means that mankind was just so much poorer; exactly as in the case of the destruction of the cathedral at Rheims. And the chance to see frigate-birds soaring in circles above the storm, or a file of pelicans winging their way across the crimson afterglow of the sunset, or a myriad of terns flashing in the bright light of midday as they hover in a shifting maze above the beach—why the loss is like a gallery of the masterpieces of the artists of old times."

As president, Theodore Roosevelt recognized that Americans were blessed with an enormous natural abundance of beauty, diversity, and wealth of resources. His love of nature, obsession with orderly development, and devotion to the public welfare combined to form a strong sense of duty to long-term interests of the nation and the generations of Americans to follow.

At an early age Roosevelt was fascinated by birds and game. He collected, studied taxidermy, kept journals of his observations, and organized his own small museum; but more importantly, he modeled himself on the naturalists/writers/conservationists of the times. One of those was his uncle, Robert Barnes Roosevelt (R. B. R.) who was an avid sportsman/fisherman who had known excellent trout streams in what is now midtown Manhattan. Even in his era development and commercial overfishing had taken an enormous toll on the fish populations. So, in 1868, R. B. R. set up the New York Fish Commission to regulate the number of fish taken and restock areas to ensure the survival of the fish and the fishing.

Theodore Roosevelt inherited a similar sense of conservation. He organized a group of concerned sportsmen into the Boone and Crockett Club to protect America's big game populations. As President of the Boone and Crockett Club he pressured Congress into passing the 1894 Protection Act, a piece of legislation that kept commercial hunters and timber interests from encroaching on Yellowstone Park.

Protection became one of the hallmarks of the Roosevelt presidency; five new national parks with 40 million acres were established. He created the National Wildlife Refuge designation and set aside 51 areas. Furthermore, he created the National Monuments Act to protect areas of significant natural beauty such as the Grand Canyon, Mount Olympus in Washington State, and Muir Woods in California.

The Roosevelt White House marked a golden age for zoology, exploration, and conservation; naturalists and explorers from all over the world were sought out by the president and encouraged. In his second year of office, he spent several weeks with the poet/naturalist John Burroughs in Yellowstone and went on from there to visit with John Muir in Yosemite. Both of them had tremendous respect for his knowledge of the natural world and for his political accomplishments in protecting wilderness areas.

But Roosevelt brought a broader sense of stewardship to the conservation movement and to government. He felt that conservation of the environment had to go hand in hand with development, that both required sound management practices to assure that nothing was ruined and nothing wasted.

On becoming President in 1901, Roosevelt turned to Gifford Pinchot, head of the Forest Service, and Frederick Newell, his counterpart in the Bureau of Reclamation, to formulate far-reaching policies to assure that the public domain was not jeopardized by private exploitation of minerals, timber, and water rights through the prevailing government leasing system. The public had lost three million acres of government timber land to private enterprise under President McKinley, Roosevelt's predecessor.

In 1902 he signed the Newlands Act, a reclamation act that would create a system of reservoirs and irrigation channels to revitalize arid lands in the Southwest. Roosevelt believed that the states, by allowing streams to pass into private ownership, had not protected the rights of its citizens to the most productive use of the lands that bordered those streams.

Of similar intent was his establishment of an Inland Waterways Commission in 1907. It was authorized to make our waterways more navigable for the stimulation of commerce which was to the greater benefit of the citizens. The Panama Canal was undertaken in the same spirit less than a year later when he authorized the army to begin construction.

Roosevelt was a great believer in effective management and attracted many of the best and brightest of his era into government. Gifford Pinchot had a great influence on him. The Pinchot family started the Yale Forestry School and Gifford was an extremely creative and talented advocate of sound management techniques. He introduced Roosevelt to the European practice of sustained-yield forestry. They were both convinced that the future protection of natural resources lay in the development of renewables.

In 1905 Roosevelt authorized the Forest Service to make arrests for the violation of its regulations. It was clear that the long-term public interests were to be protected over the short-term profitability of timber companies harvesting on public lands.

In 1908 Pinchot and Roosevelt organized a national conference on the environment, inviting federal and state politicians, businessmen, and scientists. In his opening address Roosevelt acknowledged, "Every step of the progess of mankind is marked by discovery and use of natural resources previously unused. Without such progressive knowledge and utilization of natural resources, population could not grow, nor industries multiply, nor the hidden wealth of the earth be developed for the benefit of mankind." He went on to say that the greatness of America can be measured in how we utilize what we have.

When Roosevelt left office, he left a formidable legacy of conservation in government. Much of it resisted the strains of partisan politics and special interest groups. For his part Roosevelt divided his time between writing, exploring, and furthering the cause of the progressive, liberal reform platform.

He explored Africa under the auspices of the Museum of Natural History, and later went to Brazil with several experts to collect specimens and explore the River of Doubt. He contracted jungle fever and nearly didn't make it. The River of Doubt was renamed for Theodore Roosevelt.

He kept records of his observations throughout his life. He is the author of *The Summer Birds of the Adirondacks in Franklin County, The Wilderness Hunter, African Game Trails, Through the Brazilian Wilderness,* and *American Big Game Hunting* with George Bird Grinnell, among others.

JAKE EHLERS

For Further Reading: Paul Russell Cutright, *Theodore Roosevelt: The Making of a Conservationist* (1985); David McCullough, *Mornings on Horseback* (1981); Nathan Miller, *Theodore Roosevelt* (1992).

RURAL COMMUNITIES

Rural communities have been the focus of research for at least a century, and there are almost as many views about what makes a community "rural" as there have been studies, but two approaches predominate. One is to use relative differences—sometimes simply defining rural areas as being those that are "not urban," although more careful definitions focus on relative differences in settlement density, closeness to the land, or most often, population. The second approach is to use absolute differences, most often by identifying specific cutoff populations below which a community is "rural." In the U.S., the two most common cutoff levels reflect figures long used by the Census Bureau. The older one is that rural communities are those having fewer than 2,500 residents—a definition that still includes a majority of the *communities* in the U.S., but that no longer includes a high proportion of the *population*. The increasing scale of society (and the increasing concentration of the U.S. population, into a smaller number of larger urban areas) has tended to increase the importance of a second cutoff population used by the Census Bureau for much of the 20th century, involving communities that are technically "metropolitan"—in general, having 50,000 or more residents and meeting several other criteria. In many studies today, the terms "rural" and "nonmetropolitan" are used almost interchangeably.

In most of the human societies that have ever existed, across most of the history of humanity, the vast majority of the population has lived in relatively small communities. Such arrangements are still common in some of the less-industrialized regions of the world even today. The past two centuries, however, have seen substantial growth in both the percentage and the number of people living in urban areas.

This is true in particular for industrialized countries such as the U.S., where for most of the nation's history, more people have been moving from rural areas to the cities than vice versa. In the 1960s and particularly the 1970s, the rural regions of the U.S. saw an unprecedented population turnaround: For the first time in the nation's history, more people left the cities for the country than vice versa. By the 1980s, however, the so-called turnaround had once again turned around. Preliminary findings from the 1990 census indicate that the majority of rural communities in the U.S. have once again faced a net loss of population.

As might be expected, a number of factors influence the choice of places to live. By far the most important of them have to do with economic strength and weakness. In addition, however, choices are influenced by factors such as the availability of services and the broader range of factors that contribute to community quality of life.

Particularly in the early parts of the 20th century, many of the services now taken for granted in most areas of the U.S.—electricity, telephones, indoor plumbing, paved roads—showed up in urban areas decades before these conveniences became available to most residents of rural areas. While few such urban-rural differences still exist for many of these commonplace conveniences, there still are substantial differences in some of the retail and medical services that are locally available. As a general rule, the more specialized (and the less transportable) the service, the larger the potential service-area population needs to be for that service to be supported. Most small towns have general-purpose or convenience stores, for example, but few have stores for specialized electronic equipment or classical music supplies. Most small towns also have stores that rent or sell at least the more popular videotapes, but very few can support a professional theater company, unless they can draw support from urbanized areas nearby. For medicine, similarly, there are important urban-rural differences in access to general practitioners, but far greater differences in access to more specialized forms of medical care, such as neurosurgery.

Not surprisingly, rural residents tend to express lower levels of satisfaction with most such services than do urban residents. There are a few exceptions, however, as in the case of law enforcement and education, where satisfaction levels tend to be higher in rural areas, even though formal service levels are often lower. In the case of law enforcement, statistics generally bear out the popular perception that rural areas are safer than urban ones, in relative as well as absolute numbers, except in the case of murders, which are actually more prevalent in the most rural areas of the U.S. than in at least the smaller towns. By contrast, rural education programs tend to have not just lower levels of financial support, available facilities, and teacher training, but lower levels of student success, at least as measured by subsequent entrance to or graduation from college and university programs, yet these drawbacks appear to be counterbalanced in the public mind by generally higher levels of classroom safety and lower levels of violence and drug-related activities.

It is particularly in terms of overall judgments, however, that the general tendency in the U.S. is to favor life in rural areas. Many writers over the decades have been hostile or even contemptuous toward rural communities, just as others have detested what they have described as the artificiality of both people and settings in urban areas, but in representative surveys, at least in the U.S., the general public is likely to express greater preference for living in rural communities than in urban ones. The reasons appear to relate to the "less tangible" and non-financial factors involved in quality-of-life judgments, where rural residents express relatively high levels of satisfaction. These aspects of community life tend not to be represented by offices at city hall, but they tend to be no less important for that reason—particularly in rural areas. When discussing the aspects of their lives they find most important, rural residents are much less likely to refer to (formal) community services and cultural activities than to characteristics such as community peacefulness and stability, the ability to enjoy surroundings that are at least somewhat closer to nature, or the fact that "everybody knows [virtually] everybody else."

Still, despite the preference for places that are rural, the reality is that increasing numbers of people have been moving to places that are urban—or more accurately, *sub*urban. Suburban locations appear to reflect the desires of many Americans to live in areas that at least resemble the traditional if romanticized notions about rural life—being relatively quiet, clean, safe, and green—while still maintaining ready accessibility to urban service and employment opportunities. Yet the growth of the suburbs has not been simply a reflection of individual choices; particularly during the latter half of the 20th century, the suburbs have also received huge if often hidden subsidies from governmental policies, including policies that, on the surface, had to do with topics such as road construction, housing, and the creation of special-purpose districts. By the latter decades of the 20th century, the net result was that

there were actually more suburban residents in the U.S. than urban ones.

While there is a tendency to see the remaining "rural" areas of the U.S. as being more or less synonymous with "farm" areas, which most of them once were, this is often no longer true. In the early 1800s, over 80% of all jobs in the United States were in farming, and less than 10% of the U.S. population lived in "urban" areas of 2,500 persons or more. By 1890 the Census Bureau announced the disappearance of an official "frontier"—a continuous string of counties all having populations under two persons per square mile—although much of the territory between the center of the country and the west coast still comes close to meeting that definition today. By 1920 more residents were "urban" (in communities over 2,500) than rural. By the time of the 1990 census, preliminary figures indicated that little more than 2% of the population was still involved in farming, and roughly 75% of the population lived in metropolitan areas. One implication is that, even in nonmetropolitan regions, roughly 11 out of every 12 residents are *not* involved in farming, and even in the most rural of areas, farming is often no longer the most important component of the economic base.

Instead, the rural communities of the U.S. tend to depend on many other kinds of economic activities. Manufacturing, for example, employs a higher percentage of the population in rural areas than in urban ones. In addition, a number of rural areas specialize in tourism or retirement, and some of the larger nonmetropolitan communities perform more diversified roles as regional service centers. The nonagricultural rural activities with the most direct relationship to the environment, however, are those that involve resource extraction (e.g., mining, logging, oil and gas extraction, and fishing)—or increasingly, those that are the virtual opposite of resource extraction, with rural areas being asked to accept materials and activities that urban areas regard as being undesirable or dangerous, ranging from prisons to hazardous waste facilities.

Given the long-term loss of agricultural employment, made more salient by the end of the relative prosperity that marked the turnaround, rural residents and especially rural leaders have shown an increasing level of willingness to accept, or at times seek out, even the kinds of potentially dangerous or environmentally damaging activities being shunned by communities that are in less desperate straits. Ironically, however, at least the preliminary indications are that such efforts may prove to be counterproductive. For most of the history of the U.S., economic prosperity may have been associated with a degradation of environmental quality, but today, several factors appear increasingly to be working against this assumption.

For most of the 19th century, even as farming accounted for ever-decreasing proportions of the U.S. labor force, it was true that the rural extractive industries of mining, logging, and oil and gas development were growing in importance. What is not well known is that this ceased to be true in the early 20th century. Overall, since the 1920 census, the declining economic fortunes of extractive industries have been almost as steep as those for farming. The net result is that, while roughly 90% of all jobs in the U.S. in the early 1800s required that people be located close to raw materials, whether for purposes of agriculture or of raw material extraction, by the late 20th century, the proportion had dropped to well below 7%. Extractive industries do still provide wages that are extremely high relative to the prevailing wages of most rural regions, but today, mining- and logging-dependent counties actually appear to have higher levels of poverty than do agriculture-dependent ones.

Contrary to the usual assumption, moreover, the prices for most natural resource commodities appear *not* to have shown long-term increases, at least to date, in part because of the increased effectiveness of technology and the availability of alternative raw material supplies from less-industrialized countries. The net result has been that many resource-dependent communities have seen the shutting down of local mines, mills, and logging operations, even *before* the raw materials themselves were exhausted. The shutdowns have been due not so much to resource exhaustion as to corporate investment decisions, the availability of richer or more readily accessible resources in other areas, and political factors, such as the ability of resource companies to extract not just raw materials, but concessions—from political leaders, workers, local communities, and the political system. Unless raw-material prices begin a long-term upward trend, there is little reason for optimism about the local economic implications of resource-extraction activities such as mining and logging.

Instead, by the middle of the 20th century, growth was taking place predominantly in sectors of the economy that were information- and skills-intensive, rather than energy- and materials-intensive—e.g., in computers rather than in cars. About 40% of the cost of an automobile, for example, is made up of raw materials, such as iron, steel, and rubber. For the chip that forms the "brain" of a computer, by contrast, raw materials make up roughly 3% of the cost, largely in the form of common materials such as silica or sand, while the majority of the cost is for research and development. Many research and development activities, meanwhile, can take place in rural as well as urban areas—and rural areas with high levels of environmental amenities tend to be more attractive to the relevant specialists than do those with high levels of toxic contamination.

If there are reasons for economic hope for rural areas today, accordingly, those reasons may come from very different activities than those that have traditionally provided the economic base of rural areas. The preference for relatively unspoiled settings remains strong, and may even be growing. Preliminary findings from the 1990 census indicate that, while rural counties across the U.S. experience a return to the net out-migration that characterized earlier decades, there were exceptions in regions having high amenity values, particularly those with mountains, forests, or lakes and seacoasts. The changes are anything but minor; just the real estate changing hands in a tourism-dependent county of western Colorado can exceed the *total* of economic activity in nearby counties having more traditional economic bases of agriculture, forestry, and mining. While there is little evidence that the transition has yet been made except in some of the most prosperous rural communities, it is thus possible that the future economic hope for the rural communities of an industrialized nation such as the U.S. could have less to do with the consumption of natural resources than with their preservation and appreciation.

WILLIAM R. FREUDENBURG
ROBERT GRAMLING

For Further Reading: Cornelia B. Flora and James A. Christenson, eds., *Rural Policies for the 1990s* (1991); William R. Freudenburg, "Addictive Economies: Extractive Industries and Vulnerable Localities in a Changing World Economy," in *Rural Sociology* (1992); Albert E. Luloff and Louis E. Swanson, eds., *American Rural Communities: Trends and Prospects* (1990).

S

SAFE DRINKING WATER ACT

See United States Government.

SAHARA

See Deserts.

SAHEL

See Desertification.

SALMONELLA

Salmonella is a genus of bacteria responsible for many outbreaks of food poisoning that often occur in epidemic proportions. Farm animals such as cows, pigs, and sheep that are infected by the bacteria do not appear to be sick. The bacteria are present in the food derived from infected animals: meat and any products made with a meat base, milk and all dairy products, and eggs. Raw or undercooked eggs in raw-egg dishes and baking account for close to 70% of all cases of salmonella poisoning.

Salmonellae may be spread from person to person on the fingers of anyone handling food containing the bacteria or on contaminated utensils. After recovery from salmonella poisoning, a person may still carry the disease. Food may also be infected by the excretion of cats, dogs, rats, and mice.

Diarrhea accompanied by abdominal cramps, vomiting blood, fever, and occasionally blood in the bowel movements are the main symptoms of this type of food poisoning. Most cases of infection by salmonella bacteria are not very serious and do not require any action beyond avoiding contaminated food. In such cases, the bacteria will be excreted in bowel movements. In more serious cases of salmonella poisoning, the bacteria may spread into the bloodstream, causing inflammation of various organs and excessive loss of body fluids. If untreated, severe cases can cause death by dehydration.

People most susceptible to salmonella poisoning are those who already have compromised immune systems. These include people undergoing cancer treatment, people with AIDS, and pregnant women. Children account for fewer than 1% of those afflicted with salmonella-related food poisoning.

Salmonella poisoning can be prevented in humans by proper food handling and cooking procedures. Most types of salmonellae are killed by heating the infected product to 60°C for 15 minutes. It sometimes is necessary to heat foods to a considerably higher temperature in order to ensure that the heat penetrates fully, and it is important to allow all frozen foods to thaw substantially before being cooked. Because of the ease with which salmonellae can be transferred from person to person and from person to food, it is important to wash hands before and after handling food and utensils.

Current research in New Mexico is aimed at developing an acoustic device capable of detecting eggs contaminated with salmonella bacteria.

SANITATION

See Waste Treatment.

SARA

See United States Government.

SAVANNAHS

See Grasslands.

SCARCITY

Economists say that a good or service is scarce when people want more than is available (at the current price). Thus in the middle of Mexico City, clean air is scarce but pollution is in overabundance. In contrast, in the middle of Montana, clean air isn't scarce; there is virtually no pollution, but the absence of pollution doesn't make it scarce (since no one wants it, even at a zero price). Thus scarcity involves demand (people wanting or desiring something) as well as availability.

Economics is focused on the allocation of scarce resources: how society should manage them to give the greatest satisfaction to people. When there is not enough of a good or service to satisfy everyone who wants it, then some consumers must do with less. One signal of scarcity is high prices. Prices can be used to allocate a scarce good or service, since only those consumers with the money to purchase the good are able to obtain it. When the price of a commodity increases, demand will fall and suppliers will be induced to produce more (if that is possible) thus bringing supply and demand into balance.

Using prices to allocate scarce goods means that richer people are able to get the goods they desire, which may not always be socially desirable. For example, automobiles are rationed by price while medical care for children is free to those who cannot afford to pay. In the OPEC oil embargo of 1973–1974, gasoline was in short supply but its price was prevented from rising for social and political reasons. Long lines formed at gas stations and gas was allocated based on consumers' willingness to wait in line, not their ability to pay.

Clean air in Los Angeles is not available to anyone. To clean the air, California requires control devices on cars and threatens to restrict driving. These regulations are implemented through uniform requirements rather than through prices. For example, each automobile could be charged for the pollutants it emits or the number of cars in use could be restricted and individuals allowed to bid for the right to drive.

Some goods are not produced by humans and so their production cost is zero. However, nature doesn't choose to produce the amount that humans want, e.g., land. Little land is created or destroyed in the U.S.—it is in fixed supply. Thus while it is costless to supply land, there is not enough of it to satisfy demand, e.g., downtown New York City. In such cases, land is properly viewed as scarce.

Natural resources such as oil and gold, or natural wonders such as the Grand Canyon, are also scarce. They are in fixed supply and there is not enough to go around. For many natural resources, there is a cost associated with producing the resource and making it available to consumers. Oil must be pumped to the surface, refined, and transported. While finished gasoline is scarce partly because it is costly to pump, refine, and transport, oil in the ground is also scarce because consumers desire more oil products than would be supplied at a price equal to the cost of pumping, refining, and transporting. But interestingly, some oil in the ground is so expensive to produce that no one wants it; thus even though it is in fixed supply, it is not scarce.

To summarize, scarcity is a concept economists use to characterize goods and services that would be in short supply absent any sort of intervention or allocation mechanism (such as price). Resources may be scarce even though they are costless to produce (e.g., land). How to allocate scarce resources is a major concern of economists.

CHARLES D. KOLSTAD

SCHUMACHER, E. F.

(1911–1977), German-born English economist. Ernst Friedrich Schumacher's book *Small Is Beautiful*, published in 1973, set forth principles of economic development that argued against promoting ever-increasing growth through ever-increasing production in order to enable ever-increasing consumption. It also argued against prevailing economic efforts, especially in poor countries, to achieve economic growth through large-scale, high-technology, labor-saving undertakings.

Instead, Schumacher insisted that economic goals must seek only to satisfy human needs, including contentment, regardless of growth. He advocated "appropriate technology" and conservation of natural resources. He said small was not only beautiful but cheaper and better for most people. Schumacher called for "a new life-style, with new methods of production and new patterns of consumption, a life-style designed for permanence."

In agriculture, he called for production methods that are "biologically sound, build up soil fertility, and produce health, beauty and permanence. Productivity will then look after itself." In industry, he called for "small-scale technology, relatively non-violent technology, 'technology with a human face,' so the people have a chance to enjoy themselves while they are working." He also advocated new forms of common ownership in industry. "We still have to learn how to live peacefully," he wrote, "not only with our fellow men but also with nature and, above all, with those Higher Powers which have made nature and have made us; for, assuredly, we have not come about by accident and certainly have not made ourselves."

Schumacher's *Small Is Beautiful*, sub-titled, "A Study of Economics as Though People Mattered," became a world-wide best-seller. He became a cult figure in the 1970s. Schumacher provided comforting rationalization for taking up modest pursuits outside the mainstream. He had a gift for imagery and a sense of humor. His many quotable allusions and aphorisms were used universally to oppose anything large-scale or complicated.

Schumacher was not, as some might have expected,

an unqualified promulgator of economic theory or applications. Son of Hermann Schumacher, a highly regarded history scholar, E. F. or "Fritz," as he was widely and familiarly known, had been a Rhodes scholar in economics. He was economic adviser to the Allied Control Commission in post–World War II Germany and for 20 years economic adviser to the British Coal Board. He was president of the Soil Association, one of Britain's leading organic farming organizations. In 1966 ("Because the moment comes when you have to take the existential jump from talking to doing," he said), he founded the Intermediate Technology Development Group (ITG). It designed plans and industrial and agricultural equipment to fit the capacities of poor countries.

Schumacher began to fashion his theories in 1955, as an economic adviser to the government of Burma. There he challenged the value of adopting Western economic practices, especially what he regarded as indiscriminate exaltation of limitless growth and the failure to distinguish between renewable and non-renewable resources. Rich people's solutions don't apply to poor people's problems, he said. In Burma, too, he formulated his notions of the attractiveness of a "Buddhist" economy, in which work and leisure were complementary.

After an invitational visit to India in 1961, he recorded his concept of "intermediate technology." A poor country, he declared, could not afford expensive, high technology projects in which the "cost per workplace" was too great for the total benefit accrued. Modern industrial complexes provide too few jobs for large populations in poor countries and most of their products are too expensive for domestic purchase.

When *Small Is Beautiful* first appeared, it attracted little attention. But word of mouth among scholars and environmental activists brought Schumacher invitations to lecture, write articles, and appear on TV. The book was soon translated into 15 languages. A group of his disciples in New York created the Alternative Technology Group, with the same aims as Schumacher's Intermediate Technology Group in London. California established a State Office of Appropriate Technology. Its first action was to build simple solar water heaters to cut the state's energy bill.

Schumacher was not without his critics nor did he and his associates underestimate how hard it might be to apply his proposals. In many less-developed countries, even India, the abandonment of demands for large-scale modern technology was resisted. Some of it had to do with, as a colleague of Schumacher put it, "a kind of technological snobbiness, which regards with disdain anything less than ultramodern."

More significantly perhaps, Schumacher's heirs at the ITG found: "The labor-saving, capital intensive, highly sophisticated technologies, suitable for large-scale production in 'rich' markets, which are commonly used in the rich countries, are very well documented and easily accessible; but technologies applicable on a small scale by (or in) communities with plenty of labor and little capital, lacking technical and organizational sophistication, are, on the whole, poorly documented, difficult to get hold of, and in many cases even non-existent."

Nevertheless, the ITG developed a staff of specialists in agriculture, building materials, energy, transport, small factories, small-scale water supply and food technology. At any given time it operated in 20 poor countries. Schumacher's views began to make increasing sense to the assistance agencies in the developed world. Instead of overly ambitious, expensive, and often wasted high-visibility projects, they began adopting carefully defined programs that did not first require huge power resources and were not dependent upon complicated maintenance.

Schumacher died at age 66 on his way to a conference in Switzerland and a holiday with friends. A mourner told *The Times* of London: "It is hard not to believe that his endless travels and consequent exhaustion helped to hasten his death." So busy had the proponent of "Buddhist economics" been in advocating a healthy mix of work and leisure, he may have worked himself to death.

JACK RAYMOND

For Further Reading: George McRoble, *Small Is Possible* (1981); E. F. Schumacher, *Guide for the Perplexed* (1977); Van B. Weigel, *A Unified Theory of Global Development* (1989). *See also* Appropriate Technology.

SEA AND LAKE ZONES

Marine waters and most freshwater lakes can be divided into distinct spatial zones based on physical characteristics such as water depth and light conditions. In turn, these characteristics lead to distinct biological differences among zones. On a horizontal basis, zones are defined in terms of proximity to the land-water interface (the coastline), reflecting the combined influences of terrestrial inputs and water depth on physical, chemical, and biological properties of the aquatic zone. Life zones in natural waters also are divided vertically based on: (1) light conditions, which limit the depth to which photosynthesis occurs, and (2) thermal structure, which determines water mixing rates and ultimately controls chemical and biological conditions. The major horizontal and vertical zones in lakes and marine waters are well characterized and have widely accepted names.

The land-water interface for many lakes and marine coastal areas is an indistinct boundary of **wetlands**. These ecosystems may be considered a mixture between terrestrial and aquatic environments. Freshwater wetlands are called marshes, bogs, or swamps, depending on the nature of their vegetation. Swamps have woody vegetation (trees and shrubs); marshes have nonwoody, emergent vegetation (rushes, grasses, and reeds); bogs are located in northern climates and have woody shrubs or trees. In temperate marine coastal areas, the most common wetlands are salt (or tidal) marshes, which are dominated by emergent plants like *Spartina* (salt marsh grass). Such marshes commonly are found in estuarine zones, where rivers drain into the oceans. Mangrove swamps are common wetlands in tropical and subtropical coastal areas, but salt marshes also occur in these climatic zones where topographic conditions are suitable. Coastal marshes occur in "low energy" environments, areas of low vertical relief and a broad extent of shallow water so that wave energy is dissipated off shore.

Coastal-zone wetlands are characterized by an abundance of plant and animal life (flora and fauna) and usually are highly productive ecosystems. They serve as the spawning grounds and nursery areas for many species of fish and shellfish and consequently have high ecological value. Coastal-zone wetlands also serve as important buffers between aquatic and terrestrial environments. As such, they are especially vulnerable to impacts of human activity. Because shoreline property is in high demand for residential development, it has a high economic value, and the economic and ecological values of wetlands frequently are in conflict. Strong economic pressures have caused extensive draining and filling of coastal wetlands so that the resulting shorelines could be developed into residential properties. Large expanses of coastline and many lakes have lost their protective wetland buffer altogether as a result of draining and filling of coastal wetlands. In addition, wetlands are often the primary recipients of stormwater runoff from upland urban areas and agricultural land and thus may receive excessive loadings of nutrients, pesticides, heavy metals, and other contaminants.

The shallow open-water region lakeward or seaward of the shoreline is called the **littoral** zone. This zone is also subject to terrestrial influences, especially when wetland buffer zones are absent. The littoral zone often has abundant plant and animal life, great biological diversity, and complicated food webs. Large rooted plants (macrophytes) or mats of algae often grow in the bottom substrate (called the **benthic** zone) of littoral areas and serve as the base of the food web. Littoral

macrophytes provide shelter (habitat) for fish, zooplankton, and other small animals that dwell in the bottom deposits (benthic invertebrates). Because littoral zones are shallow, turbulence from wind-driven waves extends to the bottom and prevents the accumulation of fine-grained mineral and organic sediments. Consequently, littoral bottom deposits tend to be sand, gravel, and rocks with low organic content. The littoral zone in the immediate vicinity of the shoreline sometimes is further subdivided as follows. The **supralittoral** zone is defined as the shoreline region that lies entirely above the water's edge but is affected by waves and sea spray. The **eulittoral** lies between the highest and lowest seasonal water levels (or highest and lowest tidal reaches, that is, the intertidal zone). The terms **sublittoral** and **infralittoral** are sometimes used to denote the balance of the littoral zone (i.e., the portion below the water's edge).

The end of the littoral zone and the start of the **pelagic** zone in a water body is defined by the depth of water at which insufficient light penetrates to the bottom to support the growth of plants. Its location in a lake or ocean area thus depends on water clarity. In turn, this depends on many natural factors and human influences, including runoff of soil and sediment from the land, transport of colored organic matter (aquatic humus) from wetlands, and nutrient status, which affects planktonic production and the standing crop of phytoplankton. The water depth at which the littoral zone ends and the pelagic zone begins may be as little as a few feet in turbid waters or as great as 100–150 feet in highly transparent marine waters and lakes like Baikal (Siberia). The pelagic zone in small lakes typically begins at depths of 10–20 feet. The pelagic zone is divided vertically into the **euphotic** zone (where light is sufficient to allow photosynthesis to exceed respiration) and the **aphotic** zone (where light essentially is absent and photosynthesis does not occur).

Pelagic food webs tend to be simple (more like food chains). Phytoplankton form the base of the chain and convert inorganic nutrients and carbon dioxide into organic matter by the process of photosynthesis (primary production) in near surface waters, where light penetration is sufficiently high. Phytoplankton are consumed (or "grazed") by zooplankton (called herbivores), which are small-bodied animals (microscopic to a few millimeters in size). In turn, the zooplankton are eaten by small fish (planktivores). Large fish (carnivores) prey upon the small fish, and in some systems there are several levels of carnivores. Finally, dead organic matter (detritus) suspended in the water—derived from dead algae and fecal excretions of zooplankton and fish—serves as the food for some zooplankton and fish (detritivores) and is re-mineralized by bacteria and other heterotrophic microbes.

The sediments below the water column in the pelagic zone and the water immediately adjacent to these sediments constitute the **profundal** zone in lakes. The fine-grained sediments here are devoid of plant life, and the fauna living in this zone are referred to as *benthos*. Profundal sediments in lakes often are rich in nutrient elements (nitrogen and phosphorus), metals (iron, manganese), and organic content. Cycling of these elements between the sediments and water column is mediated by microbial processes, and this plays a major role in maintaining the productivity of lake ecosystems.

On a seasonal basis, many lakes develop vertical zones with distinct physical, chemical, and biological characteristics as a result of thermal stratification. For example, in temperate climatic regions, the water column is well-mixed and isothermal from top to bottom in spring (after the ice cover melts) and again in fall (before the onset of ice-cover). Because lake water is heated from the surface by solar radiation, surface water becomes warm faster than bottom water; and because warm water is less dense than cold water, the lighter, warmer surface water tends to float on a bottom layer of colder, denser water. When the temperature difference between the top and bottom is small, only a small amount of wind energy is needed to mix the water. However, as warming continues during spring, top-to-bottom temperature differences may become too great for wind-induced mixing, and three zones of different temperature develop. These zones typically persist in the water column through summer until fall cooling allows the lake to mix completely once more. The zones are: (1) an upper, mixed layer of relatively warm water, called the **epilimnion**; (2) a layer of rapidly decreasing temperature, called the **thermocline** or **metalimnion**; and (3) a bottom layer of cold water, called the **hypolimnion**. The thickness of the epilimnion depends on lake morphometry and climatic conditions. In lakes with small surface areas and wind-protected shorelines, the epilimnion may be only a few meters deep, but in large lakes open to wind action, it may extend 10 meters or more in depth.

Because water in the hypolimnion is trapped by the less dense water of the upper layers, and because most or all of the hypolimnion is below the euphotic zone, the hypolimnion may develop very different chemical characteristics from the epilimnion. In lakes with high biological productivity, organic matter produced in the euphotic zone settles into the hypolimnion, and as it is decomposed by aquatic bacteria, it depletes the oxygen dissolved in the hypolimnion. The loss of oxygen allows a buildup of sulfide and other undesirable chemicals and prevents fish from living in this region. In large deep lakes such as the Great Lakes, the volume of the hypolimnion is sufficient to assimilate all the settling organic matter without causing a serious loss of oxygen. In these cases, the hypolimnion provides a suitable habitat for cold-water fish such as trout.

A similar seasonal pattern of mixing and thermal stratification occurs in the upper few hundred meters of the oceans, but a permanent thermocline below this region in temperate and tropical latitudes results in more or less perpetual isolation of the lower water from the surface. A vast quantity of water occurs below the permanent thermocline, and two life zones are delineated within this region. The upper portion or **bathyl** zone extends from about 100–300 meters to 1,000–3,000 meters, depending on latitude; the large zone below this is called the **abyssal** zone. Both of these zones are aphotic, cold (temperatures near 4°C) and calm (wind and wave-induced turbulence does not extend into these layers). However, the bathyl zone is relatively warmer and richer and more diverse in fauna.

PATRICK L. BREZONIK

For Further Reading: Rhodes W. Fairbridge ed., *The Encyclopedia of Oceanography* (1966); G. Evelyn Hutchinson, *A Treatise on Limnology* (1957–67); Robert G. Wetzel, *Limnology* (1983).
See also Baikal, Lake; Coastlines and Artificial Structures; Food Chains; Lakes and Ponds; Limnology; Oceanography; Water; Wetlands.

SEASONS

In most regions, weather varies with the time of year. Most languages have names for these characteristic seasons. In English, one speaks of winter, spring, summer and autumn (fall). The seasons arise from the Earth's annual journey in its elliptical orbit around the sun, just as day and night depend on its spin around the polar axis.

In middle and high latitudes—poleward of the tropical latitudes of 23½ degrees north and south—the regime is clear and obvious. The Earth's axis of daily rotation is at an angle to the plane of the ecliptic swept out by the annual progression around the sun. This causes variations in the length of day, and in the elevation of the sun at local noon, both of which cause changes in the rate of solar heating. In the northern hemisphere, day is longest and the sun highest at the summer solstice (when at noon the sun is vertically overhead at the Tropic of Cancer), currently on June 21–22; day is shortest at the winter solstice (December 21–22) when the noon sun is vertically above the Tropic of Capricorn. The northern hemisphere has its seasons in opposition to the southern; the northern summer comes simultaneously with the southern winter.

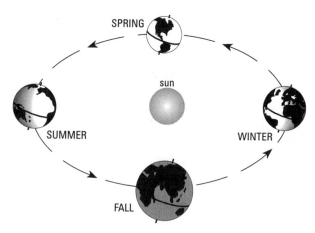

Figure 1. The seasons result from the tilt of the Earth on its axis as it orbits the sun.

At the equinoxes, when the noon sun is vertically above the equator, on September 21–22 and March 21–22, the day is twelve hours long world-wide; but spring in the northern hemisphere corresponds to autumn in the southern.

Between the tropics, these variations are seen as annual swings of the noonday sun from south to north and back. Even at the solstices the noon sun is still high in the sky, and conditions remain warm. There is little variation in length of day, or in the time of dawn and dusk.

This familiar story, known in more or less detail by every human being, conceals a surprising complexity, because the Earth's climate does not respond in a simple way to the varying input of solar radiation. In middle and high latitudes, the thermal response is obvious; winter is cold and summer warm, in varying degrees. But there is nearly always a lag of temperature behind the sun by about 3 to 6 weeks. Thus late July is usually the warmest time of the northern year, the coldest of the southern. A similar lag occurs in daily temperature; the warmest time is about 2 or 3 P.M. These effects are due to heat storage by soil or water.

But there is no simple equivalent story for precipitation (rain, hail or snow). In some climates winter is wet and snowy, and summer dry, as in the Mediterranean countries. In other regions, notably the continental interiors, the reverse is more normal: rain falls mainly in summer. These differences arise from the seasonal variation in solar heating of the Earth, which leads to vast changes in atmospheric circulation, i.e., in the prevailing winds and the disturbances they propel. In general, winds are stronger in the winter hemisphere.

The largest seasonal changes of this second sort affect tropical regions, where complete reversals of the prevailing winds may occur. Such reversing winds—ac-

tually huge currents covering much of Africa, south and east Asia, and northern Australia—are called *monsoons*, as are the seasons they dominate. In sub-Saharan Africa, India, southeast Asia, China and Japan, the southwest monsoon dominates from late June to mid-September, and brings the rainy season to most areas. A reverse, northeasterly monsoon brings the drier weather of December to March or April (though in Australia the seasons are reversed). Many parts of the tropical world have no clear monsoons, but wet and dry seasons are still the usual regime.

Natural ecosystems adapt to these climatic stimuli in striking ways, because the supply of heat and water is crucial to the green plants at the heart of the food web. In many mid-latitude areas, for example, forest vegetation is largely deciduous; trees shed their leaves in the cooler seasons, when photosynthesis is slowed or ceases. A similar strategy is adopted by many tropical forest trees in the dry season. Evergreen forests, typical of the humid tropics, of the entire southern hemisphere, and of the subarctic, also go largely dormant in the cooler or drier seasons. Animal populations adopt many strategies to survive the seasonal changes. Some animals migrate, in the case of birds and large insects for incredible distances, in the search for food or a comfortable place to reproduce. Others hibernate. In the world oceans, the plankton populations multiply and bloom in relation to sea temperature changes, and fish and marine mammal populations, being highly mobile, migrate. One could even define the seasons in terms of ecosystem response, almost as clearly as in terms of varying climate. In the best sense, the word *ecosystem* comprehends the climate, and hence the seasons.

Human beings adapt to the seasons in numerous ways. From earliest times they have varied their activities to meet seasonal needs, and have even evolved small but significant physiological differences between races to protect against extremes. Hunting societies have always been dominated by the seasonal behavior of their prey: for example, the caribou hunters of Canada and Alaska, and the fishermen of the colder oceans. Farmers live by their knowledge of the seasons; when to cultivate, when and what to seed or plant, when to harvest and when to sell, are the elements of the farmers' craft. And in industrial societies there are large interseasonal differences in employment, ease of transportation, market conditions, and raw material and energy costs.

Traditional societies have long absorbed these adaptations into their habits of life. Industrial technology likewise takes the seasons for granted. Sharp anomalies of climate, in which the seasonal pattern is changed (as in the El Niño events on the coasts of Peru and Ecuador), have drastic social and economic consequences. Monsoon rain failures are a recurrent nightmare in India. But in the richer countries, the seasonal changes also offer striking recreational opportunities. Skiing, skating, surfing, sport fishing, and their parallels are direct responses to seasonal changes.

Seasonality—the adjustment of activity to seasonal change—is thus integral to life on this orbiting planet. It has been so throughout human history, and will so remain.

F. KENNETH HARE

For Further Reading: R. G. Barry and R. J. Chorley, *Atmosphere, Weather and Climate* (1976); Bruce Marshall, ed., *The Real World* (1991); National Geographic Society, *Earth '88: Changing Geographic Perspectives* (1988).
See also Atmosphere; Biomes; Climate Change; Geography; Meteorology.

SEAWATER

Water is most often found in nature as seawater, equalling approximately 98% of the world's water supply. Seawater is a concentrated electrolyte solution containing many dissolved salts, which are responsible for its characteristic taste. The saltiness is due to the 3.5% dissolved mineral matter. The most common mineral dissolved in seawater is sodium chloride (table salt), followed by magnesium and sulfur along with all of the common metals. Calcium and silicon are also present; they are important because organisms use them to form shells and skeletons.

It is not just a coincidence that 71% of every human being is salty water and that 71% of Earth's surface is covered by seawater. The salinity of the amniotic fluid that surrounds all people before they are born is similar to the primeval seawater in which all life on Earth originated. All embryos—from protozoa to mammals—develop in their own miniature body of saltwater.

The composition of seawater is affected and regulated by rivers and other sources of fresh water that empty into the oceans. These sources constantly add suspended particles and dissolved substances, such as salts. Because of the size of the oceans and the poor mixing rate, however, the composition of the water remains almost unchanged. Biological and chemical reactions also regulate the composition of seawater, as organisms are constantly using and replacing materials found in the water.

Pollution of freshwater systems that empty into the ocean is altering the composition of seawater. Although the effects are most noticeable at harbors and bays, they are not limited to these areas. Winds, currents, and other mixing processes carry the pollutants to open waters where they affect a much larger area.

Seawater and fresh water have similar physical properties, which are modified by the salinity of seawater. The concentration of salts affects mainly the freezing point and the density of the water. Instead of freezing at 0°C (32°F), water with a salinity of thirty-five parts per thousand freezes at −2°C (28.6°F). The density of fresh water at 4°C (39.2°F) is equal to 1; the density of seawater is slightly higher.

The world's largest seawater salt production facility, operated by Exportadora de Sal S.A. (ESSA), produces seven million metric tons of salt annually from 74,000 acres of seawater evaporation ponds in Baja California. ESSA, Planetary Design Corporation of Tucson, and a Mexican evironmental company (GENESIS) are collaborating on an agricultural project with monumental potential to offset global warming. By pumping the seawater for irrigation instead of evaporation, the project is creating test fields of halophytic (salt-loving) land plants that are literally "greening" desert coasts in Mexico and Saudi Arabia.

SELENIUM

See Pollution, Water: Processes.

SHELTERS: ENVIRONMENTAL CONDITIONING

In the wild, each animal and plant species lives in the type of environment to which it is biologically adapted; this environment is its natural habitat. In contrast, human beings have settled in most parts of the Earth, including many places where physiochemical conditions are unsuited to the biological characteristics of the human species. Even the so-called temperate zones of the Earth are not naturally favorable to human life. Few could survive in them if societies had not developed artifacts that enable them to resist food deficiencies and weather inclemencies. Humans depend on artificial methods of food storage and on the use of clothing, fire, and an immense diversity of shelters. Furthermore, since practically all crops belong to sunloving plant species, large areas of the temperate regions had to be deforested before they could accommodate large human populations.

The cradle of the human species was probably the plateau of East Africa, then as now a subtropical land of alternating rainy and dry seasons, associated with periods of growth and dormancy. Human beings are still biologically adapted to such a subtropical environment. As far back as the Old Stone Age, humans began moving to other parts of the Earth, thereby developing minor biological differences as a result of genetic drift and exposure for many generations to different physical surroundings and ways of life. But despite these changes, all human groups are still fundamentally identical in their physiological requirements and probably also in their psychological drives and emotional needs.

From the physiological point of view, all human beings have remained semitropical animals to such an extent that they cannot truly adapt biologically to either very cold or very hot climates. Even the Eskimos design their clothing and shelters so that their bodies are in a semitropical environment most of the time. People living in hot, dry regions escape the effects of the scorching sun by wrapping themselves in protective clothing, such as the burnoose of the Bedouins, and by living in settlements where the streets are narrow and oriented to provide shade. In human equatorial regions, people have learned to take full advantage of natural breezes. Thus, millennia before the advent of air conditioning, human beings all over the Earth had created microclimates suitable to their semitropical biological nature.

Stone Age people probably first occupied deep caverns and other more or less natural shelters protected by overhanging rocks. Thousands of years ago, however, they began to build shelters of different sizes and shapes with walls, ceilings, and floors of different materials, textures, and colors. Such diversity suggests that many different types of shelter design can be adapted to the universal traits of human nature.

Erik Erikson once wrote that "roundness surrounds you more nicely than squareness," and many Stone Age rooms were indeed round—for example, those built around 50,000 years ago in Moldavia, on the Russian-Rumanian border. But square and rectangular rooms have come to prevail all over the world. Neither climate nor biological necessity can explain the change in the size or shape of shelters and rooms. In the American Southwest, the Navajos lived until recently in round hogans consisting of a single room large enough to house twenty people, while the Pueblo people have long lived in adobe buildings consisting of very small rectangular rooms. Most of these people, however, have moved or are moving into houses conventional in Western culture. Diversity in types of shelters is very striking even on Easter Island, which is only 117 square kilometers. Some of the rooms built there and still occupied two centuries ago were only 2 meters in diameter, whereas others were 100 meters long! Eventually, however, the traditional building practices of Easter Island were replaced with others that favored rectangular rooms of sizes similar to those

in most of the Western world. Everywhere on Earth, cultural and historical forces have thus been at least as influential in design as the natural conditions of the environment.

The type of room most prevalent now in the countries of Western civilization is rectangular with vertical walls that are 3 to 8 meters in length and width, and 2.5 to 5 meters in height. It is probable that these dimensions have been determined by technical aspects of construction and by facility of use by the occupants.

In theory, it should be possible to define a shelter with characteristics that are ideal for *Homo sapiens.* But in practice ethnic groups and individuals differ profoundly, not in what they need but in what they want. Furthermore, social wants are much more variable and therefore less well defined than biological needs, because they reflect historical influences, contemporary social forces, and aspirations for the future. What people want is so largely determined by traditions and expectations that no type of shelter can be universally acceptable even if it satisfies all biological needs.

The design of shelters must therefore be carried out within the constraints imposed by the biological invariants of human nature and must also fit the demands of a particular group or person. Furthermore, it must take into account the fact that individual and social tastes change. It may be useful, nevertheless, to consider some broad principles of building that have long been empirically discovered and widely practiced over a long period of time.

Pretechnological civilizations began very early to rely on thick-walled buildings, perhaps to save on fuel or because fire was difficult to control, and probably also because this type of construction has multiple advantages. Not only is it resistant to storm, fire, and flood, but it has desirable acoustic and thermal properties. During cold weather, thick walls store the heat produced by fireplaces or furnaces and release it slowly at night after heating has been discontinued. During hot weather, thick walls absorb the solar heat, thus retarding the temperature increase in the interior of the house. After sunset they radiate the stored heat into the house and slow down the rate of cooling. Old, well-built adobe houses of Arizona and New Mexico are more comfortable than modern buildings with sophisticated heating and air-conditioning equipment. The same quality of "natural" temperature control through wise design of thick-walled buildings can still be experienced in many ancient houses of Southern Europe.

Massive wall construction became ingrained in most ancient architecture but had to undergo adaptive changes to meet local climatic conditions. In humid tropical regions, for example, glazed windows admitted light but kept out rain; overhanging roofs intercepted the direct sun but did not shut out light; louvered grilles provided for ventilating air but prevented visual intrusions. Other practices naturally developed in very hot, dry countries. Dry heat can be made bearable and even pleasant in shaded courtyards such as those of the Spanish tradition or in buildings where massive walls and relatively small openings insulate the indoor areas.

In most parts of the world, however, thick walls are now considered obsolete because their building is labor-intensive and the control of temperature and humidity is thought to be more easily achieved through modern air-conditioning techniques. It is interesting to note that wherever air conditioning is used, the temperature is generally set at approximately 70–75°F, thus providing evidence for the view that all human beings fundamentally have the same temperature requirements.

Le Corbusier seemingly had biological justification when he asserted that the circulating air of modern buildings should be maintained at a constant temperature around 70°F, but in reality, the biological requirements with regard to temperature are more complex. Even under conditions of health, the body temperature exhibits diurnal and seasonal fluctuations. Temperature and humidity levels that remain constant throughout the day and the year are therefore not likely to create optimal environmental conditions.

Unfortunately, little is known about a possible ideal range of conditions. It can be argued, on the one hand, that air conditioning should be geared as closely as possible to the diurnal and seasonal fluctuations woven into our biological fabric. On the other hand, perfect adjustment to environmental temperature and humidity may not provide sufficient stimulation for the full development of human potentialities. Our bodies and minds possess remarkable mechanisms for adaptation to environmental changes and there are indications that adaptive responses to environmental challenges have creative effects and contribute to a sense of well-being. According to certain geographers and historians, vigorous civilizations are more likely to develop in parts of the world where climatic changes are sufficiently pronounced to cause stimulation but not so violent as to overpower human adaptive mechanisms.

A century ago, the considerations thought to be of major importance in the design of shelters were problems of light, temperature, and the avoidance of "miasmas" (vapors) and bad smells assumed to be responsible for disease. It was furthermore recognized that environmental requirements differ with age and occupation and are not the same in health and in disease. To stimulate the development of human values, however, the environmental control of shelters must go beyond exclusive concern for protection. It should aim also at the production of stimuli capable of favoring the expression of human potentialities. In other words, human beings should live under conditions that keep their adaptive mechanisms in a state of readiness and provide them with the challenges necessary for creativity.

RENÉ DUBOS
(EDITED BY WILLIAM R. EBLEN)

For Further Reading: René Dubos, *So Human an Animal* (1968); James M. Fitch, ed., *Shelters, Models of Native Ingenuity* (1982); James M. Fitch, *American Building: The Environmental Forces That Shape It* (1972).
See also Human Adaptation; Shelters: Social Relevance.

SHELTERS: SOCIAL RELEVANCE

Architectural traditions have commonly led to the neglect of certain spatial experiences and cultural influences that may greatly contribute to physical and mental well-being and development. Nomadic people, who usually live in temporary shelters or tents, tend to group their activities around some external focus—a water hole, a shade tree, a campfire, or the abode of a leader. For them, open space is an important part of their experience. In contrast, people who live in permanent buildings tend to have only restricted visual contact with space, bounded as their daily life is by walls, floors, and ceilings.

The vast majority of people spend a large percentage of their time indoors, yet the effects of such room characteristics as size, height, shape, and color on human comfort, happiness and performance are not well known. In the countries of Western civilization, most dwellings now consist chiefly of a few types of rooms, which, irrespective of esthetic criteria, probably satisfy a few universal human needs. Beyond commonsense requirements, few characteristics of enclosures are essential for biological needs, except perhaps the need for visual contact with open space. Most other criteria deal with values and tastes that are culture-bound and therefore vary with place and time.

The so-called bubble of individual space—the most comfortable distance between persons during social contact—is usually much greater among Anglo-Saxon people than among Mediterranean people. A purely cultural difference of this nature has to be reflected within the shelter in the level of togetherness that is considered objectionable or desirable in a given social situation.

Human behavior is also affected by technological equipment. Until the 20th century, rooms were individually heated by fireplaces or stoves; light was provided at first by candles, and later by lamps using kerosene, acetylene, or gas. Since these techniques were somewhat dangerous, children had to spend their evenings in rooms occupied by adults. The phrase *family circle* probably originated from the grouping of children and parents around the hearth and the lamp. But the situation changed with the introduction of central heating and electric lighting, which were sufficiently convenient and safe so that children could have their own living quarters and indulge in greater freedom of behavior. The trend toward "a room of one's own" is therefore an outgrowth of sociocultural forces rather than the expression of biological imperatives. But it may have social consequences. In the past, the necessity of functioning for several hours with a diversified social group imposed a rather strict social discipline, which may have conditioned the later life of the child.

Furthermore, until recent times, the work of successful planners and builders was guided by ecological wisdom derived from empirical experience. Local constraints made it necessary for them to develop architectural styles suited to the natural conditions of the place where they worked and to the building materials that were readily available. Thus came into existence all over the world types of dwellings wonderfully adapted to local conditions: the snug, well-oriented houses of New England; the breezy plantation houses with broad porches in the South; the thick adobe-walled haciendas in the American Southwest. Everywhere there appeared roofs of different slopes, depending upon the amount of rain or snow in the region. The diversity and beauty of this "architecture without architects" arose empirically, almost subconsciously, through a process of natural selection determined by the environmental constraints and by the availability of local building materials.

Ever since antiquity, there have been many treatises devoted to the theoretical and practical aspects of the plastic arts in general and of architecture in particular. Charles Blanc published in 1876 an influential book called *Grammaire des Arts du Dessein*, in which he advocated that the various styles should aim at expressing the fundamental moods of the human mind—the sublime, the beautiful, the sad, the endearing, the frightening, and so on. According to him, furthermore, each style should be suited to the historical and regional mood of the place. Just as the massiveness of Egyptian temples expressed a religious attitude different from that of Gothic cathedrals, so should every aspect of shelter design convey the social status and lifestyle of the occupants. In his *Grammaire*, Blanc emphasized almost exclusively the technical and artistic aspects of architecture; he barely discussed the effects of shelters on health and comfort, obviously considering these problems of a lower order than those of esthetic appearance and spiritual meaning.

By the end of the 19th century, the traditional philosophy of architecture began to be questioned in England by William Morris, who advocated simpler, unadorned vernacular styles. Morris's ideas were developed in France by Le Corbusier and in Germany by the Bauhaus school of design under the leadership of Walter Gropius. The Bauhaus substituted for conventional styles of architecture a style based on an interior framework of steel, which eliminated the need for supportive walls and accommodated curtain walls of glass. This type of architecture came to be called the International style, because it was based on modern international technology; it was also called functional, because it rejected needless ornamentation. All over the world many buildings, large and small, testify that the austere simplicity of the International functional style can generate beautiful architectural designs that are expressive not of the place where they are located but of the period during which they were built.

The originators of the International functional style expressed their theories in a few arresting formulas. The phrase "form follows function," attributed to the Chicago architect Louis Henri Sullivan (1856–1924), conveys the view that the physical appearance of a building should reveal the function of each of its parts without unnecessary adornment. Mies van der Rohe's "less is more" implies that the very simplicity of a design contributes to its esthetic quality. When Le Corbusier referred to apartment houses and private dwellings as *machines à habiter*, he meant that the design of buildings, whatever their purpose, should primarily aim at efficient function, as is the case for the other machines of modern technology.

The International functional style has now spread all over the world in the form of steel-framed, glass-faced, geometrical buildings that have been assumed to provide rational environments for the pursuit of business, administration, industry, and family life. However, many people, including famous architects, believe that modern buildings are defective from both the human and technical points of view. They are now said to provide poor living and working conditions for their occupants; they generate a monotonous and depressing atmosphere not only for the people who live in them but also for the people who look at them; and they are extremely wasteful of energy.

The present hostility to many aspects of modern architecture and to the various forms of shelters can be explained in large part by the fact that the statement "form follows function" is not as meaningful as it sounds. In its usual architectural meaning, the word *function* is commonly limited to the physical structure and the service systems of the buildings. From the occupants' point of view, however, the functions of a building are chiefly its biological and social uses—to the satisfactions of the mind and the comforts of the body. If all human needs and values were taken into account, the meaning of *functionalism* would be so broad and vague that it could not provide specific guides for design. In practice, form is selected according to the particular kind of function the designer has elected to emphasize, the result being, unfortunately, that most buildings of the so-called functional school do not do justice to the human functions they are expected to serve. The modern rebellion against "functionalism" is largely an effort to reintroduce human and ecological factors into the architectural equation.

Blanc's *Grammaire* symbolized the almost exclusive concern of architects for the aesthetic aspects of their profession. The Bauhaus school emphasized the technological aspects of building. In his *American Building: The Environmental Forces That Shape It* (1972), James Marston Fitch analyzes the relationships between human beings and buildings, including how physiological processes are affected by environmental factors, and how the quality of a building depends not only on visual perceptions but also on the responses of the organism as a whole.

The word *shelter* implies practical considerations for the design, construction, and use of buildings. But important as they are, these considerations fail to take into account the powerful symbolic meanings associated with the words *shelter* and *home*. Since it is difficult to verbalize these symbolic values in precise terms, let alone to express them quantitatively, a large aspect of shelter design will probably remain in the domain of the architect as an artist rather than becoming part of precise architectural science. There is here an analogy with the art of medicine, which, though less well defined than the science of medicine, is nevertheless as important for the welfare of the patient.

In the final analysis, the first and most important steps in design are those that involve choices about human ways of life rather than about technical problems. One does not function the same way in the austere simplicity of a Japanese house as in the sloppy comfort of overstuffed furniture. Before designing shelters, people must ask themselves some basic questions concerning how they want to live, what they want to become, and how they want their societies to evolve.

The sophistication and durability of many structures, both large and small, from ancient civilizations testify to the fact that some human ancestors had practical knowledge of building materials and techniques, as well as a subtle appreciation of the psychological effects of form. The design of shelters has been affected by an immense diversity of social factors—in particular, by fashion.

Style is determined as much by choice, nostalgia, or caprice as by knowledge, resources, or necessity. The coldness and ugliness of shelters and cities is the concrete expression of social ills. If there is reason for hope in the design of better shelters, it is in the increasing awareness that ethical, social, and cultural considerations are as important as materials and techniques in the creation of desirable environments, because these inevitably affect the ways of life.

RENÉ DUBOS
(EDITED BY WILLIAM R. EBLEN)

For Further Reading: René Dubos, *Celebrations of Life* (1981); James Marston Fitch, *American Building: The Environmental Forces That Shape It* (1972); James M. Fitch, ed., *Shelters, Models of Native Ingenuity* (1982).
See also Architecture; Energy-Efficient Buildings; Shelters: Environmental Conditioning.

SICK BUILDING SYNDROME

Sick building syndrome (SBS), sometimes called tight building syndrome, can be described as the expression of a series of vague health or discomfort symptoms in a number of individuals within an office or residence that disappear after the affected people spend time away from the building. Symptoms include eye, skin, or lung irritation; nausea; fatigue; stuffy nose; and others. The condition has arisen primarily as a consequence of the reduction of air movement indoors due to the development of energy-efficient buildings. Other factors include the use of synthetic materials for furniture and possibly the buildup of viable particles (molds) in enclosed environments. However, SBS is not related to just one type of building. Two typical situations conducive to SBS would be the following: (1) structures that totally rely on a central HVAC (heating, ventilation, and air conditioning) system for the movement of the air, and (2) houses that have been sealed with insulation to make them energy-efficient and are designed to be air-tight.

SBS has been hypothesized as a multifactorial problem related to the persistence of stale air and the accumulation of a mixture of inorganic and organic chemicals and biological agents (microorganisms). The levels of all compounds are frequently low with respect to occupational health guidelines or standards. Some compounds, however, are one or more orders of magnitude above the concentrations typically found in the ambient environment. The suggestion has been made, although the hypothesis has not been systematically tested, that when coupled with a poor ventilation in a building, the accumulation of moderate concentrations of a mixture of various contaminants indoors is part of the etiology of the syndrome. A study of sick buildings by the National Institute of Occupational Safety and Health (NIOSH) has pointed out five categories that contribute to SBS: (1) poor ventilation, thermal comfort, and humidity; (2) office equipment; (3) outdoor air pollutants; (4) building materials; and (5) undetermined.

When ascribing the features of sick building syndrome, it is important to consider consequences that go beyond issues of worker sickness and decline in worker efficiency. Instrumentation and equipment used in buildings are also affected by airborne contaminants through deposition on surfaces. The contact and possible reaction of deposited pollutants on electronic surfaces can reduce the lifetime of sensitive components and/or result in malfunctions and costly repairs. Obviously, the problem continues to increase in scope and magnitude by the use of microsilicon and ceramic chips in equipment and the application of high-speed computers and communication systems within all sectors of our society, since these instruments are acutely sensitive to contamination.

In summary, sick building syndrome is a late twentieth-century form of indoor air pollution health effects. It is related to poor ventilation, and multiple pollutant sources, and accumulation of pollutants. These factors require further investigation in various types of buildings.

SIERRA CLUB

See Environmental Organizations; Muir, John.

SLUDGE

See Wastewater Treatment, Municipal.

SMITH, ADAM

(1723–1790), Scottish moral philosopher, parent of modern economic thought. Adam Smith was born in Kirkcaldy, Scotland, and educated at the University of Glasgow and Balliol College, Oxford. His scholarly career began in 1748 with a series of public lectures delivered in Edinburgh. In 1751 he became a professor at the University of Glasgow. He was attracted away from that position in 1763 by an offer of a handsome annuity for life for tutoring the Duke of Buccleuch. From 1764 to 1767 Smith tutored and toured with the Duke.

Financially secure, Smith wrote *The Wealth of Nations*, which was published in 1776. He then took a position as Commissioner of Customs at Edinburgh, which he held until he died in 1790.

In *The Wealth of Nations*, Smith described how a market system can provide a mechanism for the timely and smooth coordination of autonomous individuals' choices in an interdependent world, so as to produce the greatest possible wealth for the nation. This classical liberal vision of an invisible hand guiding the market system to an optimal outcome reflects a central tenet of Smith's larger moral philosophy: The universe is the design of a benevolent Deity. The Deity is to the universe as a watchmaker is to a watch. In both cases the hand of the designer arranges the system and starts it in motion. But in each case the designer's hand is invisible to the observer, who sees only the effect of the effort: the movement of the hands of the watch, or the course of the market economy.

Smith's vision of the elegant and efficient market system required that all the conditions within the system be consistent with the order of the Deity's design. For only when the connecting principles of the design are undistorted can its efficiency, elegance, and beneficence be realized. In the case of the human order, this implies that all autonomous individuals must be acting ethically.

In *The Theory of Moral Sentiments* (1759) Smith described the basis for such moral action. It derives from a proper balance among the sentiments of self-love, beneficence, and justice. The standard by which that balance is measured is the sympathy of an impartial spectator. Actions based on a proper balance are consistent with the intentions of the Deity, and thus contribute to the order of and enhance the well-being of society. To act perfectly ethically a person must have access to the perspective of an impartial spectator and the self-command to maintain the proper balance of sentiments.

Smith recognized, however, that such a perspective and such self-command do not exist in any person. Indeed, in his moral philosophy the key limiting factor in the realization of the greatest possible wealth of the nation lies not in the domain of resources and technology, but in the domain of ethics. He believed that society evolves, and that humanity's ethical and economic advancement go hand in hand.

Smith envisioned society as evolving through stages from hunting/gathering, through pasturage and agricultural stages, to the most advanced stage: a perfectly functioning market society. In such a society all resources would be used most prudently and productively; that is, in a way to generate the greatest sustainable wealth for the nation.

Smith believed that while specific societies might advance and then stagnate or decline, humankind as a whole was on a path that would allow it to approach the ideal. This reflected his faith in the invisible hand's ability not only to guide a well-ordered society, but also to guide humankind toward that well-ordered state. At least he held that sanguine belief until the 1770s, when his exposure to the dynamic nature of powerful distorting commercial interests, a phenomenon he called mercantilism, shook his faith in the inevitability of social improvement.

As a consequence of this new skepticism he made significant revisions to his two major works. In 1783 he published "Additions and Corrections" to *The Wealth of Nations*. The primary focus of these revisions was the mercantilist system and a condemnation of the damage it can do to society. Then in 1790 he made major revisions to *The Theory of Moral Sentiments*. The focus of these efforts is an appeal to private and public virtue.

Gone was his faith that the invisible hand would guide individuals toward the creation of and adherence to a common social ethics. The new passages of *The Theory of Moral Sentiments* are an appeal for active pursuit of a social ethics that emphasizes placing private interest below a commitment to sustaining the well-being of society as a whole for this and succeeding generations: "The wise and virtuous man is at all times willing that his own private interest should be sacrificed to the public interest." According to Smith, the lead in this effort must come from the statesman: "The leader of a successful party . . . may sometimes render to his country a service much more essential and important than the greatest victories and most extensive conquests. He may re-establish and improve the constitution, and from the very doubtful and ambiguous character of a leader of a party, he may assume the greatest and noblest of all characters, that of the reformer and legislator of a great state; and, by the wisdom of his institutions, secure the internal tranquillity and happiness of his fellow citizens for many succeeding generations."

Smith's fully mature voice defines his legacy. He offers us a vision of the allocative efficiency and the productive potential of a well-ordered society of autonomous individuals interacting in the context of a market system. He reminds us that such an ideal requires all individuals to balance their self-love with a commitment to a set of common social values. For, in the absence of such a balance, society is bound not for the efficiency of constructive competition and the greatest wealth for the nation, but for destructive competition and the Hobbesian abyss. He addresses us as individuals, challenging each of us to pursue virtue. He speaks to the leaders of the future challenging

them to assume the noble role of the statesmen working to "secure the internal tranquillity and happiness of fellow citizens for many succeeding generations."

The most significant contribution of Adam Smith's legacy to our understanding of the human condition lies in the holistic approach he took to his analysis of the human condition. As a moral philosopher, Smith viewed the human order as an evolving social, political, and economic system. His work directs our thinking toward the fruitfulness of personal sovereignty in the context of appropriate ethics, toward intergenerational responsibility, and toward the power of education in the broadest sense.

JERRY EVENSKY

For Further Reading: Jerry Evensky, "Adam Smith on the Human Foundation of a Successful Liberal Society," *History of Political Economy* (1994); Jerry Evensky, "The Evolution of Adam Smith's Views on Political Economy," *History of Political Economy* (1989); Adam Smith, *The Glasgow Edition of the Works and Correspondence of Adam Smith* (D. D. Raphael and Andrew Skinner, eds.) (1976–1983).

SMOG–TROPOSPHERIC OZONE

Smog is dirty urban air containing noxious and often aesthetically unpleasant gases and aerosols. (Aerosols are liquid droplets, solid particles, and mixtures of liquids and solids suspended in air.) Historically, smog has been associated with smoke from industrial combustion. The term *smog*, a contraction of *smoke* and *fog*, is thought to have originated in London. Today's smog is a complex brew of many gases (e.g., sulfur oxides, nitrogen oxides, and volatile organic compounds) and aerosols. Tropospheric ozone, located in the first few kilometers above ground, is also a major gaseous component of modern-day smog. Until about 150 years ago the levels of harmful chemicals in the air were quite low and the sources were mainly natural. As the industrial revolution began, the levels of smoke pollution increased due to the increase in fossil fuel combustion. In the early days of air pollution taller stacks were built to better distribute the pollution away from the sources. These local solutions, however, led to regional-scale problems. The pollutants transported away from the sources were converted into oxidants, aerosols, and acids causing health problems, haze, and other environmental insults far away from the sources. More effort is being made to remove pollutants directly from the smokestacks and tailpipes and/or converting to cleaner fuels and technologies.

The use of motor vehicles increased dramatically after World War II, providing another important source of pollution and further complicating the smog mix.

The Los Angeles basin, with its high population of people, cars, and industrial sources, became notorious. In the 1950s Professor Haagen-Smit and colleagues at the California Institute of Technology discovered that the Los Angeles smog, and in particular the key constituent ozone, originated from critical reactions between oxides of nitrogen and volatile organic compounds mainly from motor vehicle exhaust triggered by the presence of sunlight. This discovery was key to the subsequent development of emission control technology to reduce vehicle emissions of volatile organic compounds and oxides of nitrogen. While emission controls have been relatively effective, some difficulties have arisen from malfunctions such as poisoning of control devices by lead from gasoline. The lead problem was a major factor in the decision to remove lead from gasoline. As with the industrial sources, use of cleaner fuel and more efficient and less polluting technologies are the direction of the future. In highly polluted areas such as Los Angeles, a dramatic shift away from fossil fuel combustion is underway.

In addition to the industrial and automotive sources, components of smog are also derived from natural processes associated largely with soils, plants, volcanic eruptions and sea spray. In some areas, such as the southeastern U.S., natural vegetative sources of volatile organic compounds enhance ozone production, thereby complicating the development of air quality management strategies. The severity of smog is influenced by meteorology as well as emissions. When the wind is not blowing, chemicals become trapped and concentrations of the chemicals are higher. When there is more sunlight available and the temperature is higher, chemical reactions are enhanced, producing more noxious gases. When the relative humidity is higher, aerosol formation is enhanced.

Smog can be deadly. Infamous "killer fogs" have been documented for the Meuse Valley (1930), Donora, Pennsylvania (1948), and London (1952 and 1962). In more recent times, severe episodes are a growing problem in many developing areas such as Mexico City and many other cities throughout Asia, Western Europe and South America. Smog also causes chronic severe health effects, including respiratory problems and irritation of eyes and skin. Some studies have shown that stress levels and other adverse behavior patterns are associated with high pollution episodes. Even when the health impacts are not being dramatically felt in an area, the appalling haze associated with smog is a ubiquitous reminder of degraded air quality. In some cities such as Denver, Colorado, the air is so polluted that it appears to be a brown haze on most winter days. In some areas of the western U.S. the pollution haze has become a serious regional effect, contributing to the degradation of scenic vistas in many national parks. Over several decades smog has had other increasingly widespread and serious impacts on the natural environment and human society. Expanding population and industrial growth have made foul air move beyond city limits, cross national boundaries, create international tensions, and present new difficulties for those attempting to control its harmful effects.

PAULETTE MIDDLETON

For Further Reading: William C. Clark and R. E. Munn, eds., *Sustainable Development of the Biosphere* (1986); *1992 Earth Journal* (1991); *Scientific American* special issue: Managing Planet Earth (September 1989).
See also Aerosols; Atmosphere; Automobile Industry; Meteorology; Oxygen Cycle; Pollution, Air.

SMOKING

See Tobacco.

SOCIAL AND ECONOMIC INFRASTRUCTURE

The industrialized countries have undergone rapid transformation over the past two hundred years in the techniques for utilizing water, land, air, minerals, plants, and animals (including human labor) for the production of goods and services that people value. In the process humans have put in place extensive networks of roads, railroads, telecommunications, sewage systems, and electric power grids; public and private structures that serve as schools, hospitals, residences, offices, factories, military installations, and so on; and a wide variety of tools and machines. These constitute a nation's produced capital and are an important part of its wealth. At the same time, there has been substantial public investment in human capital—a relatively educated, disciplined, and healthy work force— which is another important part of a nation's wealth. It is convenient to think of the gifts of nature, the third portion of national wealth, as a stock of natural capital. In the course of normal economic activities, we draw down some stocks of natural capital through the use of raw materials and deteriorate the quality of others, such as clean air and clean water, by the wastes we generate.

Much of the nation's produced capital and human capital have been built up with substantial government planning, regulation, and funding. The rationale for government involvement is that this capital is the durable infrastructure for economic and social services provided to all segments of the society. Government involvement is also justified because the various kinds of networks need to be conceived as relatively integrated systems and implemented on a scale much vaster than that usually undertaken by purely private ventures: public schools and police and fire departments will have clients who are not charged directly for the services received while other services, such as those delivered by public hospitals and transportation systems, are generally subsidized.

The social and economic infrastructure contributes to the overall quality of life. In particular, it promotes a nation's growth and development because it is the basis for providing public services that are indispensable for the production of goods and services in the private sector. Empirical studies have indeed confirmed, at least indirectly, the contribution of the public infrastructure to the private economy by demonstrating systematically positive correlations between government spending in these areas and private income in the economy as a whole. Its importance is perhaps most dramatically confirmed, however, by the difficulties faced by developing countries attempting to industrialize when this infrastructure is not fully in place.

The effective utilization of the social and economic infrastructure requires on-going maintenance and repair and the allocation of adequate funds for regular operations. In fact, a large share of construction projects (private and public) in the industrialized nations involve repair and maintenance while these activities occupy a much smaller, although growing, proportion of construction outlays in developing nations. Nonetheless, there is abundant evidence that even larger outlays are necessary to maintain and operate the economic and social infrastructure in a mature, developed economy than is now committed in the United States. The U.S. clearly faces the challenge of repairing aging bridges, on the one hand, and providing more training and better access to health care to raise the productivity of parts of the labor force, on the other. However, the deterioration in the nation's natural capital is perhaps even more in need of attention because of its possibly irreversible effects.

In the United States, part of the infrastructure is publicly owned and operated and part privately owned or operated. Consequently, its management is governed by a variety of competing motives, incentives, and priorities. These considerations affect decisions about the allocation of resources for expansion, modernization, maintenance, and operations. There are strong pressures toward a much larger private role in these matters in the U.S. today. Proponents argue that the private sector is more efficient in implementation and operations in part because it is more responsive to market demand. While this is often true, only a sub-

stantial government role can assure nearly universal access to vital social services some of which might not otherwise be available at all.

Economists recognize that industrializing economies undergo structural changes as they build up their stocks of produced and human capital while drawing upon their natural capital. The resulting transformation in the relative importance of new and existing economic activities, accompanied and generally instigated by institutional and technological change, is called *development*. A different word, *growth*, is applied to changes in developed economies. Growth implies a change in one dimension only, a rising material standard of living, presumably accompanied by only slow and minor changes to the technological and institutional structure of the economy. Now, as the preservation of natural capital takes a place of importance alongside the accumulation and maintenance of produced capital and human capital, all economies including the rich, industrialized ones will need to place more importance on the kinds of structural changes associated with development and de-emphasize the preoccupation with growth alone.

Social institutions that guide the process of structural change are well developed in the rich, industrialized nations. These include such diverse entities as specialized federal and local government agencies, the banking system, labor unions and trade associations, parent-teacher organizations, and health-care self-help groups. The absence of such institutions presents a serious obstacle to industrialization in many developing countries. In all countries, however, institutions and social and physical infrastructure need to evolve as new challenges arise, notably those associated with demographic changes, new technologies, or the possibility of global climate change. Relatively new institutions, such as the U.S. Environmental Protection Agency, have emerged to guide decision-making especially with regard to resolving existing environmental problems, and this American invention has spread quickly throughout the world. However, many observers are concerned that preserving natural capital and preventing pollution uses resources that would otherwise be devoted to directly improving the social and economic infrastructure and increasing the volume of private production. Whether or not this eventually proves to be true, the time has come for the emergence of new social institutions which take a long-term, strategic perspective toward the mission of balancing social values, efficiency, and environmental concerns.

FAYE DUCHIN

For Further Reading: David Alan Aschauer, *Public Investment and Private Sector Growth* (1990); Michael Barker, ed., *Rebuilding America's Infrastructure* (1984); Alicia Munnell,

"How Does Public Infrastructure Affect Regional Economic Performance?" in *The Third Deficit: The Shortfall in Public Capital Investment*, Boston Federal Reserve Conference Series #34 (1990).

See also Economics; Resource Management.

SOCIAL ECOLOGY

The term *social ecology* is widely and ambiguously used, but has two major meanings. As an academic field of study, it is largely synonymous with *human ecology* as practiced in sociology and other social science disciplines, and focuses on the role of humans in ecosystems.

The second major use of social ecology refers to a strand of contemporary environmental thought and action, epitomized by the work of Murray Bookchin. Social ecologists such as Bookchin view humans and human systems as an integral part of a highly dynamic nature. Rather than seeing humans as an inherently destructive force in nature, social ecologists believe that humans can play a creative and supportive role in ecological protection.

Social ecologists also emphasize that human-environment relations cannot be understood without examining the relationships between humans. Because they see a vital connection between human social problems such as gender, ethnic, and economic inequities and environmental destruction, social ecologists strive to restructure communities, institutions, and government to foster a healthier environment. The impact of social ecology is currently reflected in the "environmental justice" movement concerned with the disproportionate impacts of environmental hazards on the poor and minorities, and efforts to achieve "sustainable development" that recognize the importance of assisting poor nations to achieve economic development as well as environmental protection.

Social ecology has been the subject of criticism, especially by proponents of "deep ecology," for an anthropocentric perspective emphasizing the role of humans in nature. Social ecologists respond to their biocentric critics by arguing that the traditional separation of human and nature is counterproductive to the goal of environmental protection.

SOCIAL INSECTS

The social insects comprise termites, ants, and some species of wasps and bees. All of them are characterized by three traits: an overlap of generations, cooperative brood-care, and most importantly the existence of reproductive and non-reproductive castes. This ev-

olutionary attainment occurred during the Cretaceous period (140–65 million years ago) and appears to have evolved only 13 separate times, once in the line that gave rise to the insect order Isoptera, the termites, and the rest in the order Hymenoptera, the order that contains the ants, wasps, and bees.

Although social insects comprise only 1.5% of the total number of insect species, they are the most abundant and ecologically dominant organisms in many environments. It has been estimated that ants and termites compose one-third of the entire animal biomass of the Amazonian rain forest (8 million ants and 1 million termites per hectare of soil), and they are probably similarly abundant in many other forests and savannahs.

Ants, of which there are some 8,800 extant species, are the leading predators of other insects and small invertebrates in most terrestrial habitats. Leafcutters, which depend for food on a fungus they culture on a substrate of macerated vegetation, are the chief herbivores and most destructive insect pests in Central and South America. Harvester ants are the major seed eaters in the deserts of the southwestern United States. And ants match earthworms in soil moving ability in the temperate zone and far outstrip them in the tropics.

By far the most widely distributed of the social insects, ants live everywhere except above 2,500 meters of elevation on heavily forested mountains in the tropics (where their nests receive insufficient warmth from the sun), in the extreme arctic regions, and on the remotest oceanic islands. Although most ants are ground dwellers, there are also numerous arboreal species. The African and Australian weaver ant builds nests of leaves, held together by silken threads produced by larvae that are manipulated like shuttles by the adult workers. A number of species dwell in specialized cavities in so-called ant-plants, such as the Central American *Azteca* ant that occupies the hollow stems of the *Cecropia* plant, defending the plant from attack by herbivores and obtaining nutrients produced by special glands on the plant. Army ants have dispensed with nest construction altogether, occupying only temporary bivouacs created by interlinking their own bodies. And ants are also frequent occupants of human habitations. "Tramp species" such as the Pharaoh's ant (*Monomorium pharaonis*) are worldwide household pests.

With more than 2,000 species, termites are the second largest group of social insects. Unlike ants, they feed almost exclusively on dead vegetation and rely on symbiotic organisms in their gut to digest the cellulose. This allows an efficiency of assimilation from 54% to 93%. Combined with their great abundance in many habitats, this makes them important components of energy transfer and carbon recycling within ecosystems.

The social bees can be collectively distinguished from wasps as insects that have specialized in collecting pollen rather than insect prey to feed their larvae. The 1,600 species include stingless bees, bumblebees, and the well-known honeybee (*Apis mellifera*). In addition to collecting nectar and pollen for their own use, these insects serve an important role in pollinating flowering plants. Honeybees are especially valuable for this—in addition to the honey and beeswax they yield—and hives of them are widely transported to provide pollination for domestic crops where native pollinators are absent, or have been destroyed through habitat modifications or pesticide use. Researchers estimate that the economic benefit of honeybees to agriculture exceeds the losses caused by all other insect species combined.

Social wasps, with only 800 to 900 species, have the least ecological impact, but paper wasps, hornets, and yellow jackets can be locally abundant. Measurement of yellow jacket abundance in Delaware, for example, has found densities as high as 7.6 colonies per hectare, and foraging workers are frequent pests at picnics where they are attracted to both meats and sweet liquids. The nests of social wasps are almost always elaborate paper structures constructed from chewed wood and vegetable fibers, with the exception of a few species that have reverted to mud. The rapid development of these social wasp colonies allows them to complete a colony cycle in both the extremely short summer seasons of the temperate zone, and the predator-rich tropics where nests are subject to much more frequent attack by ants, their chief enemies.

Large colony sizes and efficient communication systems have enabled all the social insect species to exploit rich and concentrated resources that might otherwise be used only by vertebrates. Social insects also forage over a long season, exploiting a wide variety of food plants and prey. And they have developed ways of storing food—from the honeycombs of bees to the granaries of harvester ants—that allow them to survive harsh or unpredictable environments.

Insect societies evolved a hundred million years before human ones, and are frequently held up as models for human behavior as in the Biblical proverb "Go to the ant thou sluggard, consider her ways and be wise." Kinship plays an important part in determining the social fabric of both societies, but there are major differences as well. Insect societies are intensely impersonal; individual bonds play little or no role. The great size of colonies and the short life of its members make this impossible. For example, an average adult honeybee worker lives but six weeks in a rapidly changing population of 80,000 nestmates. The socialization experiences of developing insects are far less rich than those of human infants, and play is absent. Furthermore, social insects are far more cohesive. Individuals work toward the goal of the colony as a whole, not their own betterment. Finally, though both human and insect societies exhibit divisions of labor, only the insects have strict reproductive specialization, in which neuter castes, containing individuals who were born sterile or have ceased reproduction, serve as workers and soldiers.

ROGER B. SWAIN

For Further Reading: Bert Holldobler and Edward O. Wilson, *The Ants* (1990); Ken Ross and Robert Matthews, eds., *Social Biology of Wasps* (1991); Edward O. Wilson, *The Insect Societies* (1971).
See also Insects and Related Arthropods; Termites.

SOCIOLOGY, ENVIRONMENTAL

Environmental sociology is a new area of inquiry that examines the social dimensions of environmental problems, including the complex interrelations between human societies and their physical environments. It thus represents a major departure from mainstream sociology's focus on the *social* environment.

Sociology emerged as a distinct discipline only in the past century, a unique era of resource abundance, technological progress, and economic growth. As a result sociologists have generally assumed that obtaining sustenance from the physical environment is nonproblematic, at least for industrial societies, and have consequently paid little attention to the environments inhabited by such societies. Indeed, the physical environment was largely ignored, or treated as little more than the stage upon which social life is enacted. In mainstream sociology "the environment" came to mean the social context (e.g., other groups) surrounding the phenomenon under consideration, not its physical setting.

Sociological neglect of the physical environment began to change after our society "discovered" environmental problems in the late 1960s. By the early 1970s a small number of sociologists were studying the emergence of environment as a social problem, especially the nature and activities of the environmental movement, public opinion toward environmental issues, and governmental policy-making. These efforts mainly involved applying traditional sociological perspectives (e.g., social movements theory) to environmental issues, and have been termed the "sociology of environmental issues."

The energy crisis of 1973–74 added a new dimension to sociological interest in environmental topics, highlighting the dependence of industrialized societies on fossil fuels. Increased awareness of potential resource-based limits to growth, along with growing evidence of the seriousness of problems such as air and water pollution, gradually led some sociologists to examine the interrelations between human societies and their physical environments: how societies affect their environments and, in turn, are affected by changing environmental conditions. This concern with societal-environmental interactions represented the arrival of a true "environmental sociology," and by the late 1970s it had become a small but vigorous area of inquiry.

Environmental sociology lost momentum in the 1980s, as the Reagan era deflected societal attention from environmental problems. Nonetheless, the field persisted, and studies of toxic-contaminated communities such as Love Canal attracted considerable attention. In recent years recognition of the reality of human-induced global environmental change, such as climate change and ozone depletion, has sparked renewed interest in environmental sociology. Indeed, global change represents a paradigmatic example of societal-environmental interactions, for the changes are caused by human activities and in turn portend enormous consequences for human societies.

Environmental sociologists currently conduct a wide range of research. Studies of environmentalism and public opinion toward environmental problems (now including risk perceptions) remain prominent, as do studies of energy use and conservation. Other major emphases include analyses of societal reactions to environmental hazards, both natural (e.g., floods and earthquakes) and human-created (e.g., toxic waste sites); the impacts of natural resource development (e.g., off-shore oil drilling, strip mining, and timber harvesting) on local communities; and analyses of environmental protection policies.

Environmental sociology provides insight into the social dimensions inherent in most environmental problems. For example, environmental sociologists emphasize that conditions such as factory smoke may be seen as problematic in one society but not another, or as a sign of economic vitality in one era but as pollution in another. Sociologists thus point to the importance of understanding how conditions come to be recognized as problematic and defined as environmental problems, highlighting the differing roles played by activists, industry, media, and government agencies. In conflict-laden settings such as those surrounding the discovery of local environmental hazards (as occurred, e.g., at Love Canal), special attention is paid to sources of competing definitions of the situation and the tactics used by differing parties to gain official acceptance of their definitions.

Environmental sociologists also examine both the causes and consequences of environmental problems. Their contributions to understanding the causes of environmental degradation can be clarified via the "IPAT equation" made famous by biologists Paul Ehrlich and Barry Commoner in an extended debate over the relative importance of population growth and technological development in generating environmental problems. The equation states that the environmental *impact* (I) of a society is a product of the society's population size (P), the level of technology (T) and the average level of affluence (A), or that $I = P \times A \times T$. While Ehrlich stresses the importance of population, and Commoner that of technology, sociologists emphasize the impact of people's lifestyles on the environment. People in poorer nations like India, for example, consume far fewer resources and produce much less pollution than do residents of affluent nations like the U.S.

Environmental sociologists also emphasize that affluence is an inadequate way of conceptualizing the social dynamics involved in environmental degradation. In particular they note that affluence calls attention to consumption and ignores the production sphere of society. The inherent need for growth leads producers to "create" consumer demand through advertising and planned obsolescence, and the need for profit leads them to produce goods regardless of their environmental consequences. Thus, sociologists argue that the decisions of those in charge of the means of production (e.g., corporate executives) have a much greater impact on the environment than do the choices of individual consumers.

More generally, environmental sociologists point to the complex interconnections between population, technology, and what they prefer to call social organization (rather than level of affluence). Poverty, for example, breeds rapid population growth, and technological advances are stimulated primarily by the economy, making it difficult to single out the effects of either population or technology per se on the environment. Environmental sociologists have been especially critical of simplistic, single-factor explanations of environmental degradation that stress either population or technology as the primary causal factor.

Conversely, environmental sociologists examine the complex set of consequences that can be produced by environmental problems, as exemplified by the potentially vast ramifications of global warming (on agriculture, energy use, migration and settlement patterns, etc.). The current emphasis on understanding the human dimensions — as both cause and consequence — of global environmental change suggests a vital future for environmental sociology. This is reflected by the rapid growth of international interest in the area.

RILEY E. DUNLAP

For Further Reading: Frederick H. Buttel, "New Directions in Environmental Sociology," *Annual Review of Sociology* (1987); Riley E. Dunlap and William R. Catton, Jr., "Environmental Sociology," *Annual Review of Sociology* (1979); William R. Freudenburg and Robert Gramling, "The Emergence of Environmental Sociology," *Sociological Inquiry* (1989).
See also Anthropology; Environmental Movement; Human Ecology; Population Ecology.

SOFT ENERGY PATHS

The term *Soft Energy Path* (SEP), coined by Amory Lovins, denotes a set of technological choices societies could make or are making in the attempt to confront the finite nature of global fossil fuel resources. An SEP consists of three main technical components: (1) using current energy resources efficiently (conservation), (2) securing an increasing amount of energy needs from renewable sources, and (3) using fossil fuels intelligently to meet our energy needs during the transition to a society and economy based on renewable energy sources.

SEPs are often contrasted with Hard Energy Paths (HEPs), which primarily involve capital-intensive energy production. Lovins contends that HEPs are currently being followed. The extraction of more fossil fuels from less conventional sources such as offshore drilling and extraction from shale, as well as a heavier reliance on more abundant sources of coal, natural gas, hydroelectricity, and nuclear energy, are described as HEPs. Although a considerable amount of energy could be obtained from these sources, Lovins portrays these "hard" options as costly, requiring large-scale, capital-intensive projects, and ultimately unsustainable.

Several ordering principles are involved if societies are to follow Soft Energy Paths. First, progress on energy issues would be gauged not by how well a society is able to supply energy to meet increasing needs but rather by how well energy needs and demands can be reduced without sacrificing (and at times even enhancing) the general well-being of the society. Economic growth would continue, but must be contingent upon the efficient use of resources.

Successfully following an SEP is fundamentally based on the use of soft energy technologies. Soft technologies have five main characteristics. First, they are based on renewable flows or cycles of energy (such as solar or hydrologic cycles) rather than on depletable stocks of energy such as fossil fuels or uranium. Sec-

ond, they are much more varied and diverse, reflecting a high degree of flexibility. Third, soft technologies are matched in scale to end-use needs. Fourth, they are often more dispersed, as in electricity generation on site or at the municipal level. Fifth, soft energy technologies are matched in quality to end-use needs.

Lovins has identified four levels (qualities) of energy needed to perform all the tasks of modern life. These are the production of heat not exceeding 100°C, the production of heat exceeding 100°C, liquid-portable energy (fuel), and electricity. Matching the quality of energy to end-use needs reduces wastage associated with producing more energy than is needed for the task involved.

SEPs are usually referred to in the plural because in Lovins' original conception, each country would develop a national plan detailing how it would accomplish the goal of renewability within a fifty-year time frame. Canada and Germany have already done so. The feasibility remains to be demonstrated.

SOIL CONSERVATION

Soil conservation is a method of managing soil use to obtain optimum yields while improving and conserving the soil. Natural processes and some human practices, those primarily associated with the production of food and fiber, can accelerate conditions causing erosion of soil. Soil and water conservation is needed to reduce erosion. Soil conservation has come to mean the whole broad movement of programs to preserve or increase the productive capacity of cropland, forests, and grazing lands.

To control erosion and reduce damage, farmers in some nations have built terraces and used crop rotation for centuries. The destruction of soil in the United States was noted by writers before the Revolutionary War. By 1775 rivers that had once run clear were described as being black with mud. Many references to worn-out land provided evidence that erosion of soil was taking its toll. The folly of exploiting the land caused Thomas Jefferson to urge farmers to adopt techniques that would conserve soil, mainly crop rotation and contour plowing. The newly created colleges of agriculture and the U.S. Department of Agriculture (USDA) later urged care of the soil. Several agriculturalists recognized that ignorance was one of the causes of soil erosion. Family farmers were badly informed about the danger of soil erosion. Land had always been plentiful and many believed that it had no limits. To awaken interest, early conservationists suggested forming agricultural societies, and the wider dispersal of books and farm journals. A bulletin,

Washed Soils: How to Prevent and Reclaim Them, published in 1894, told farmers how to save and use the land they had. Conservationists also pointed out the problems of continued cash-cropping and the necessity for government cooperation in conserving soil.

In 1928, USDA published *Soil Erosion: A National Menace,* a document developed by a new crusader, Hugh Hammond Bennett. He pointed out that the wastage resulting from erosion is "a loss to posterity, and there are indications that our increasing population may feel acutely the evil effects of this scourge of the land, now largely unrestrained." This bulletin stated that the problem faced the entire nation, not just individual farmers. In 1929 the U.S. Congress provided $160,000 to study the causes of soil erosion and methods for its control.

Although conservationists believed that the well-being of farmers and the preservation of the soil were necessary to the well-being of the nation, it required the Great Depression and the Great Plains dust bowl in the 1930s to prompt a national soil conservation program.

In 1933 the world's first comprehensive scientific effort to achieve conservation of soil and water resources on a nationwide scale began as a means of also relieving unemployment with the creation of the Emergency Conservation Work (ECW), later known as the Civilian Conservation Corps (CCC), by President Franklin D. Roosevelt. The Soil Erosion Service (SES) was established in the U.S. Department of the Interior. The SES used CCC enrollees on many reforestation and conservation demonstration projects to curb serious erosion.

In 1935 the Soil Conservation Service (SCS) was established in the USDA (subsuming the SES) to develop and maintain a continuing program of soil and water conservation. Bennett had been the SES Chief and continued as Chief of the SCS. The people working for the SCS represented a variety of disciplines including biology, engineering, economics, research, education, land use planning, and administration. The key field people were trained as soil conservationists.

Less than a year later another program was devised to conserve soil. This action by Congress, later called the Agricultural Conservation Program (ACP), encouraged farmers to voluntarily shift some crop acreage from soil-depleting surplus crops into soil-conserving ones such as legumes, grasses, or trees by making cash payments to landusers carrying out approved soil and water conservation plans.

The legislation was the first of continuing efforts to link conservation and commodity policy. This was strengthened in the Food Security Act of 1985 and the Food, Agriculture, Conservation, and Trade Act of 1990. Both laws recognized the environmental concerns of excessive soil erosion and the impact on water quality, wildlife habitat, and other damage from sediments produced by soil erosion, and established new provisions to deal with soil conservation issues.

From its inception the cardinal principle that guided the field work of the new soil conservation program was that "effective prevention and control of soil erosion and the adequate conservation of rainfall in a field, on a farm or ranch, over a watershed, or on any parcel of land, requires the use and treatment of all the various kinds of land comprising such areas in accordance with the individual needs and adaptabilities (or capabilities) of each different piece of land having any important extent." Bennett promoted the concept that the use and treatment of a given area of land must be determined not only by its physical characteristics, but also wherever possible by such other considerations as available facilities, implements, power, labor, and markets. The landuser's preference as to the type of farming to be followed, financial condition, willingness to try new methods, and persistence, along with the size and location of the farm, were important factors as the individual plan for soil conservation was developed.

To encourage local participation in conservation programs, the president in 1937 sent all state governors *A Standard State Soil Conservation Districts Law.* State laws vary, but generally include provisions for establishing a state conservation agency, a petition and referendum procedure for the creation of individual conservation districts, and a local district governing board with authority to work with landusers. There are nearly 3,000 districts in the U.S., Puerto Rico, the Virgin Islands, Guam, the Northern Mariana Islands, and the District of Columbia. Most coincide with county boundaries as independent local entities. The primary source of technical assistance to the landusers in each district is provided by the USDA-SCS.

As a basis for planning and implementing effective soil and water conservation, soils are evaluated on the basis of soil type, slope, and the degree and kind of erosion, and classified into one of eight degrees of capability. The classes range from I, requiring no special treatment for sustained use, to XIII, land that should never be cultivated under any circumstance.

A formula for estimating the rate of soil erosion, the Universal Soil Loss Equation (USLE) and a recent revision (RUSLE) quantify annual soil loss from water as the product of six factors: runoff erosiveness, soil erodibility, slope length, steepness, cropping and management practices, and supporting conservation practices. Similarly, the Wind Erosion Prediction System (WEPS), a modular computer model based on comparable physical factors, helps determine soil erosion from wind. The term *Soil Loss Tolerance* denotes the maximum level of soil erosion that will permit a high level of crop productivity to be sustained economically over time.

The soil erosion process is complex. However, the reasons for and the extent of soil loss, along with the consequences of excessive erosion and methods of control, are much better understood than at any time in history. It is estimated that despite several decades of soil conservation, as much as five billion tons of soil are moved by water or wind, with about three billion tons of sediment reaching the ponds, rivers, and lakes of the United States each year. Most is from running water causing sheet, rill (small stream), gully, or streambank soil erosion from agricultural land, primarily used for crop production. There are also soil losses from abused range and pasture lands, forest lands, and increasingly during the development of shopping centers, new homes, roads, and other facilities associated with population growth. The Worldwatch Institute estimates 25 billion tons of topsoil loss from cropland each year worldwide.

A ton of soil is about a cubic yard. An inch of topsoil covering an acre weighs about 165 tons. Six inches of topsoil weighs about 1,000 tons. Calculations of soil loss tolerance suggest that as a rule of thumb nature can replenish soil loss of up to five tons per acre per year. However, scientists say that under acceptable farm management systems about one inch of new topsoil will be formed every 100 to 1,000 years. The rate depends on many factors and varies widely in different climates.

By applying the appropriate formulas, conservation planners can recommend production methods for individual fields that limit soil loss to rates that maintain productive potential of the land indefinitely and reduce off-site environmental consequences. Effective erosion control involves reducing the amount of runoff, slowing the runoff or velocity of the wind, dissipating the impact of water and wind on soil by having nonerodible materials absorb those forces, and decreasing wind access to barren soil. There are nearly 100 practices that will control the various types of soil erosion, ranging from a cover of grass or trees or managing of crop residues to reduce runoff, to terracing, windbreaks, shelter belts, and other more costly measures.

Landuser attitudes about profits, government intervention, stewardship of the land, benefits to society, and education are crucial factors in the adoption of erosion control practices. Landusers must be aware of soil erosion problems and the potential results of solving their problems.

Increasingly, the off-site impact of sediment on water quality, wildlife habitat, reservoir capacity, navigation, road maintenance, and other issues will cause more scrutiny of conservation programs. The tradition-al methods of voluntary, incentive-based programs will be complemented by regulation in managing soil resources. The control of excessive soil erosion will eventually require a policy for the nondegradation of soil quality, comparable to the objectives for water quality.

NORMAN A. BERG

For Further Reading: Hugh Hammond Bennett, *Soil Conservation* (1939); R. Burnell Held and Marion Clawson, *Soil Conservation in Perspective* (1965); R. Neil Sampson, *Farmland or Wasteland: A Time to Choose* (1981).
See also Bennett, Hugh H.; Rivers and Streams.

SOIL CONSERVATION SERVICE

See United States Government.

SOLAR AND OTHER RENEWABLE ENERGY SOURCES

Renewable energy sources are those that are replenished by natural processes at rates exceeding their rates of use for human purposes. They are sometimes called "flow-limited" sources, because the rate at which they can be harvested sustainably is limited by the size of the natural replenishment flow. For nonrenewable or "stock-limited" resources, by contrast, natural replenishment is either nonexistent (as for the nuclear fuels) or very slow compared to use rates (as for the fossil fuels).

The most important renewable energy sources are sunlight and certain energy flows and stocks derived from it, including the chemical energy stored by photosynthesis in plant materials ("biomass" energy), the gravitational potential energy stored by the hydrologic cycle, the kinetic energy of the winds, and the thermal energy stored by absorption of sunlight in the surface layer of the oceans. Less important renewable energy sources include ocean waves (which derive their energy from the wind) and tides (the energy of which comes from the rotation of the Earth).

Geothermal energy is heat in the Earth's crust and core, the escape of which through the surface is partly replenished by heat from the decay of naturally occurring radioactive isotopes. It is often treated as a renewable resource in textbooks and tabulations of energy statistics, and, accordingly, it is included here. But the forms in which geothermal energy is being harvested today—isolated underground deposits of steam and hot water—are all too easy to exploit at rates exceeding their rates of replenishment and so are, strictly speaking, nonrenewable.

About 20% of world energy use in 1990 was supplied from renewable sources; the renewable contribution in the U.S. was about 8%. The pattern of U.S. and world energy supply is shown in Table 1. Many official tabulations show smaller renewable contributions than indicated here, because they neglect dispersed uses of renewables by firms and individuals who do not report these uses to any official agency. Hydropower is by far the most important contributor to renewable electricity supply; nearly all of the renewable contribution to nonelectric energy supplies comes from biomass (fuelwood, charcoal, and biomass wastes, as well as liquid and gaseous fuels derived from these and other biomass materials).

The current contributions from renewables bear little relation to the total magnitudes of the available renewable resources, which are shown in Table 2. Sunlight is by far the most abundant renewable energy resource, although so far it has been utilized for civilization's energy supply to a much smaller extent than biomass and hydropower. The attractions of sunlight as an energy source have been offset by its diluteness (requiring large collector areas if large amounts are to be captured) and its intermittency (requiring some form of energy storage or backup supply for energy needs at night and in cloudy weather). These characteristics have tended to make solar energy (and wind, which shares them) more expensive than hydropower and biomass—and, until now, more expensive than fossil fuels.

Comparison of Tables 1 and 2 makes clear that the potential availability of renewable energy sources is far greater than what is being utilized. Practically every inhabited region of the world, moreover, is well endowed with at least some forms of renewable energy resources: The tropics have abundant ocean thermal energy and high potential for growing biomass; the subtropical deserts are the sunniest regions in the world; the temperate zones have good growing conditions, reasonably abundant sunshine, and decent wind and hydropower potential; and the higher latitudes often have wind and hydropower resources in abundance. A wide variety of technologies for capturing and processing renewable energy resources makes it possible to match these resources to the full array of human energy needs (see Table 3).

There is growing interest in utilizing more of this potential. This interest arises in part from concerns about depletion of domestic fossil-fuel resources and excessive dependence on foreign supplies, and in part from the expectation that renewable energy supplies will become cheaper over time while fossil and perhaps nuclear supplies become costlier. Above all, the growing interest in renewables comes from the expectation that they will be less troublesome from the standpoint of environmental impacts and public acceptance than fossil fuels and nuclear energy have been.

That is not to say that the environmental impacts of renewable energy sources are zero. Solar collectors, biomass plantations, and hydroelectric reservoirs can oc-

Table 1
Contributions to U.S. and world energy supply in 1990

	United States		World	
	Electricity generation (3,000 TWh)	Nonelectric energy use (1.8 TWy)	Electricity generation (11,000 TWh)	Nonelectric energy use (9.5 TWy)
Fossil fuels	68 %	95 %	62 %	81 %
Nuclear fission	19 %	negl	17 %	negl
Nonrenewable subtotal	**87 %**	**95 %**	**80 %**	**81 %**
Hydropower	10 %	negl	19 %	negl
Biomass fuels	2 %	5 %	0.9 %	19 %
Geothermal energy	0.5 %	0.1%	0.3 %	0.2 %
Windpower	0.1 %	negl	0.03 %	negl
Solar energy	0.02%	0.1%	0.007%	0.05%
Renewable subtotal	**13 %**	**5 %**	**20 %**	**19 %**
Total	**100 %**	**100 %**	**100 %**	**100 %**

Notes: TWh = terawatt-hour (of electrical energy) = 10^{12} watt-hours = 1 billion kilowatt-hours;
   TWy = terawatt-year (of thermal or chemical energy) = 10^{12} watt-years = 31.5 x 10^{18} joules (31.5 exajoules) = 30 quadrillion British thermal units (30 quads); negl = negligible.
Electrical and nonelectrical energy quantities can be made commensurable using the relation that 1 TWy of chemical energy in fossil or biomass fuels could generate 2,900 TWh of electricity in typical contemporary power plants. On this basis, the sum of electrical and nonelectrical energy use in the United States in 1990 was equivalent to 2.84 TWy or 85 quads of chemical energy, of which 8 percent came from renewables; the corresponding figures for the world were 13.26 TWy or 396 quads of chemical-energy equivalent, 20 percent from renewables. Counting photosynthetic energy devoted to the diets of humans and domesticated animals and to plant fibers used by humans (e.g., in cotton, paper, and lumber)—which customarily are not included in tabulations of civilization's energy use—would add 5–6 TWy to the world total derived here and correspondingly increase the percentage attributable to renewables.

Table 2
Total and potentially harnessable flow-limited energy resources

Resource	Total flow	Plausibly harnessable flow
Sunlight	88,000 TW at Earth's surface, 26,000 TW on land	Converting insolation on 1% of land area at 20% efficiency would yield 52 TW of electricity and/or heat.

Comment: Averaged over 24 hours per day, 365 days per year, and all of the Earth's surface, insolation is 175 watts per horizontal square meter, equivalent to a barrel of oil per square meter per year. Sunniest sites average 250 watts per square meter. High noon on a clear day gives 1,000 watts per square meter. Photovoltaic cells and solar-thermal-electric technologies both have sunlight-to-electricity conversion efficiencies in the range of 5 to 35 percent, depending on sophistication.

Biomass	100 TW global net primary productivity, 65 TW on land	Biomass fuel from 10% of land area at 1% efficiency yields 26 TW of fuel.

Comment: Average efficiency of terrestrial photosynthesis (energy stored in plant material divided by solar input) = 0.25%; efficiency of high-yield agriculture and forestry = 0.5–1.5%.

Ocean heat	22,000 TW absorption of sunlight in oceans	Converting 2% of absorption at 2% efficiency yields 9 TW of electricity.

Comment: Conversion efficiency is limited by the relatively small temperature difference between the heat source (warm ocean surface layer in the tropics at 22–25 degrees C) and the heat sink (deep ocean water at about 4 degrees C).

Hydropower	13 TW potential energy in runoff	Using all feasible sites yields 2–3 TW peak, 1–1.5 TW average (of electricity).

Comment: Potential determined roughly by combining total runoff with average elevation of continents. A flow of 100 cubic meters per second through a drop of 100 meters yields about 100 megawatts of hydropower.

Wind	1,000–2,000 TW driving winds worldwide	Using all cost-effective terrestrial sites may yield 1–2 TW average (electricity).

Comment: The power density in a 20 mph wind is about 440 watts per square meter of area perpendicular to the wind, about half of which is extractable by a good wind turbine.

Notes: TW = terawatt = 10^{12} watts = 1 terawatt-year/year (TWy/y) = 30 quads/y. Civilization's energy use rate in 1990 was about 13 TW.

cupy a lot of land; overharvesting biomass can cause deforestation and soil degradation, and processing and burning biomass fuels can pollute water and air; lining mountain ridges and coastal promontories with windmills may offend aesthetic sensibilities; manufacturing photovoltaic cells can involve substantial quantities of toxic substances. Nonetheless, renewables generate no radioactive wastes and produce greenhouse-gas emissions ranging from zero to modest compared to those of fossil fuels. If concerns about intolerable climate change from greenhouse-gas accumulation continue to grow, this issue may become the dominant driving force promoting replacement of a sizable share of today's fossil-fuel use with renewable energy sources.

Which renewable resources and technologies will play what roles in the future of U.S. and world energy supply will depend on many factors: the state of technological development of the different renewable options; their "fit" with patterns of energy demand in terms of energy form, location, and timing; their monetary costs in relation to each other and to nonrenewable energy options; their environmental and social impacts compared to those of alternatives; and political and institutional factors governing society's willingness and capacity to deploy various energy options.

The subsections that follow summarize, in very abbreviated form, some of the relevant technical, economic, and environmental factors for the main renewable options. Some further issues relating to the magnitude of the renewable-energy contribution to future U.S. and world energy supplies are addressed briefly at the end.

Direct Harnessing of Sunlight

Direct harnessing of sunlight for heating buildings is cost-effective and fairly widely practiced today in the form of "passive-solar" design. This involves the use of building orientation, windows, overhangs, structural elements that absorb and store solar energy, and natural circulation in ways that minimize the need for supplemental heating in winter and for air conditioning in summer. Although these techniques are increasingly used in new construction, the amount of nonsolar energy use they can displace is limited by the rate at which old buildings are replaced by new ones; even in new buildings their use is far from universal, because not all builders know how to use them and not all buyers demand them. As of 1990 about 300,000 passive-solar homes had been built in the U.S.

In "active" solar heating systems for domestic uses, water or another fluid is heated by passage through a

Table 3
Technologies for applying renewable energy to human needs

Energy delivered as:	One-step processes	Multi-step processes
Heat	window as solar collector building as collector flat-plate collector salt-pond collector focusing collector — trough — dish — heliostat geothermal hot water, steam	heat from renewable fuel heat from renewable electricity
Fuel	photosynthesis — waste material — gathered material — energy crops abiotic photochemistry	thermochemical hydrogen production with renewable heat electrolytic hydrogen production with renewable electricity
Mechanical work (motion)	water turbine wind turbine tidal turbine ocean-wave engine	heat engine using renewable heat combustion engine using renewable fuel electric motor using renewable electricity
Electricity	photovoltaic cells	generator driven by renewable motion fuel cell using renewable motion
Light	window as collector fiber optics	light bulbs powered by renewable electricity flames and mantles using renewable fuel

collector (usually a flat plate in configuration, with many variations in details and materials) and then conveyed to the point of application. About 1.5 million such systems had been installed in the U.S. as of 1990, 80% of them to provide domestic hot water, 15% to heat swimming pools, and 5% for space heating. The low usage of active solar collector systems for space heating is the result of unattractive economics compared to passive-solar design or, more simply, to reducing heat losses with insulation and improved windows and then meeting the modest residual heating need with natural gas. Solar water heating has better economics, but it nonetheless is more expensive than water heating with natural gas and electricity in most parts of the U.S. and the world. The use of more sophisticated, higher-temperature solar collectors to supply heat for industrial processes is, likewise, completely feasible technically but not competitive economically in most circumstances, given current fossil-fuel prices.

The two main approaches to generating electricity from sunlight are solar-thermal-electricity conversion (STEC) and photovoltaic cells. STEC systems all use solar heat to drive vapor-cycle heat engines incorporating turbine-generator sets, but variants on this scheme differ widely in the sophistication of the collectors, the temperature at which the heat is collected, and the efficiency with which the heat is converted to electricity.

The simplest, lowest-temperature collectors used in STEC systems are shallow salt ponds with blackened bottoms; the hot brine near the bottom is also the densest, which prevents its rising to the surface where its energy would be lost to the air. Salt-pond collectors are inexpensive and the heat capacity of the water provides "built-in" energy storage, but they generate only modest temperatures and so the vapor cycles to which they are coupled can convert only a small fraction of the heat to electricity. This technology is being tried at pilot-plant scale in Israel.

Focusing collectors used to generate higher temperatures for higher-efficiency STEC systems include parabolic troughs, parabolic dishes, and heliostats (fields of steerable mirrors that focus sunlight on a boiler mounted on a tower). So far the most economical of these for large-scale electricity generation is the parabolic trough; nearly 300 electrical megawatts of STEC capacity using such collectors was operating in the southern California desert as of 1990, by far the largest solar electricity installation in the world at the time. The newest of the California plants was able to generate electricity at a cost of about 10¢ per kilowatt-hour, a figure only moderately higher than fossil-fuel generating costs and competitive with some nuclear plants. More advanced STEC plants may be able to generate electricity for around 6¢ per kilowatt-hour.

Photovoltaic cells, which are able to convert incident sunlight directly to electricity with no moving parts, are the most elegant means of solar electricity generation; they can be built in arrays of any size, are very reliable, and require little maintenance. Such cells can now be manufactured with a wide variety of techniques and materials, and their costs have fallen 100-fold or more from the time when they were used only on satellites and space probes. Still, electricity from photovoltaics cost 25¢ to 35¢ per kilowatt-hour in 1990, 3 to 5 times more than electricity from fossil fuels and nuclear reactors.

Worldwide sales of photovoltaics were about 50 electrical megawatts of capacity per year in the early 1990s, mostly for applications where connecting to an electrical grid is not practical. If the costs of photovoltaic electricity systems continue to fall, they could begin to make a significant contribution to grid-connected electricity supply early in the 21st century.

Other Renewable Energy Sources

Biomass

Plant materials offer a form of solar energy in which the energy-storage problem of intermittent renewable sources is automatically and neatly solved. The chemical energy stored in plants can be extracted at the user's convenience. And although biomass fuels release a greenhouse gas, carbon dioxide (CO_2), to the atmosphere when they are burned, the amount of CO_2 released is equal to the amount extracted from the atmosphere during the growth of the biomass. Thus, as long as biomass fuels are not burned any faster than they are being grown, they make no net contribution to the accumulation of CO_2.

About 75% of the energy derived from biomass fuels worldwide in the early 1990s was used in the less developed countries, most of it in the form of fuelwood, charcoal, crop wastes, and dung burned for cooking and heating water on crude stoves. The resulting indoor smoke constitutes a much bigger total health hazard—measured as a combination of total numbers of people exposed and average concentrations of harmful pollutants—than the outdoor air pollution in all of the world's cities. Improvements in stove design and ventilation offer higher efficiencies and lower pollution. Biomass wastes—from municipal garbage to paper byproducts to feedlot manures—are used for the cogeneration of electricity and process steam in developing and industrialized countries alike, and are being designed to high standards of efficiency and pollution control. The total contribution of energy from biomass wastes eventually will be limited by the magnitudes of the activities that generate wastes suitable for this purpose; studies have indicated that the availability of suitable biomass waste in the U.S. in the early 1990s was sufficient to provide about 12% of energy demand. Biomass "energy crops" can be converted to alcohol and gas fuel; attractive crops for this purpose include grains, sugar crops, fast-growing tree species, and algae. Proven technologies exist for converting such plant materials to alcohol fuels (potential replacements for petroleum-based fuels, especially in the transportation sector) and to clean-burning gas for use in high-efficiency, gas-turbine/combined-cycle electric power plants.

The costs of large-scale production of fuel and electricity from biomass energy crops appear to be close to competitive with the costs of conventional sources of liquid fuels and electricity already, and probably will be fully competitive by the turn of the century. The limits to the amounts of energy ultimately obtainable by civilization in this way will depend on how successfully the environmental impacts of high-yield energy-crop production can be reduced, and on competition with nonenergy uses of biomass materials and high-quality land. Doubling the amount of energy derived from biomass energy today (see Table 1) is almost surely feasible, but whether a quintupling could be managed affordably and substainably is an open question.

Hydropower

As hydropower's 20% share of world electricity supply makes clear, generating electricity from the energy of water flowing downhill through turbines is a well-established technology that is economically competitive with other sources of electricity in at least some circumstances. In the U.S., about half of the ultimately harnessable hydropower potential is already being used; worldwide, about a fifth of the potential is being used. It is unlikely that global use of hydropower will increase even as much as three-fold, however, for reasons of both economics and environmental and social impacts. Economically, the most cost-effective hydroelectric sites in most parts of the world are already being exploited; the remaining sites tend to be farther from the centers of electricity demand or costlier to dam (in relation to the electricity output obtained). With respect to impacts, hydro reservoirs destroy riverine ecosystems and anadromous fish populations, flood fertile land and wildlife habitat, block the flow of renewing nutrients to river deltas, displace human settlements, and pose the hazard of catastrophic loss of human life if a major dam fails due to miscalculation, earthquake, or sabotage. Small-scale hydropower installations spread out the impacts, but except for the risk of dam failure it is not clear that their total impact per kilowatt-hour is smaller.

Wind Power

Wind power at the best sites has monetary costs similar to those of new hydro development—1 to 1.5

times the costs of electricity generation from fossil fuels—but much smaller ecological and social impacts. Some modest environmental hazards are associated with obtaining and processing the materials for windmill construction, but the land occupied by windmills can be used simultaneously for grazing and some other purposes; the threat to birds that may fly into the spinning blades and the visual intrusion of windmills in scenic areas may be significant problems in some regions.

Because of its combination of attractive economics and relatively low environmental impacts, wind power is likely to increase its currently modest contributions to U.S. and world electricity supply in the years immediately ahead. Its ultimate potential, limited by the availability of sufficiently windy sites, could conceivably be comparable to today's electricity demand but most likely considerably smaller than tomorrow's.

Ocean Energy

The thermal energy stored when sunlight warms the surface layer of the ocean represents a very substantial energy resource (see Table 2). This energy can only be harnessed at practical efficiencies in the tropics and in some subtropical regions, however, where the temperature difference between the ocean's surface layer and the cold, deep water below is 20°C or more. Such regions are a long distance from most of the world's centers of population and energy demand, which means that the costs of a global network of energy distribution (in the form of electricity or hydrogen made from electricity) would need to be added to the costs of the ocean thermal-energy conversion (OTEC) technology in order for this energy source to become important at global scale. The costs of OTEC power plants are likely to be quite high, moreover, because the low temperatures and low efficiencies involved mean that very large volumes of water must be handled, and the massive structures for doing this must be strong and corrosion resistant in order to survive prolonged operation in the sea.

The kinetic energy in ocean waves has tempted inventors for centuries, but the challenge of designing structures extensive enough to harness very much of this energy, strong enough to survive the battering of the sea, and yet inexpensive enough to make the energy affordable remains to be met. Even if such a technology were available, the total scale of the global wave-energy resource is such that a contribution equal to more than a few percent of today's world electricity demand is unlikely. Harnessing the energy of the tides, by contrast, using facilities similar to hydroelectric plants, can be done economically at the most suitable sites (a big estuary with a high tidal range and a mouth narrow enough to be dammed). But there are not many

such sites, and the disruption of estuaries would be a high environmental price.

Geothermal Energy

The forms of geothermal energy being exploited today, which altogether account for only a fraction of a percent of world energy supply, are isolated deposits of steam or hot water that become depleted by use within a few decades; nonetheless, the total energy resource represented by these deposits may be in the range of 4,000 terawatt-years, comparable to the world's recoverable coal resources. Tapping the hot rock that is ubiquitous in the Earth's crust offers even larger potential supplies—perhaps in the millions of terawatt-years—but the technology for this remains to be demonstrated and the cost is correspondingly uncertain. Environmentally, today's geothermal technologies entail some water pollution by dissolved salts and toxic metals in geothermal water, as well as worker and public exposure to airborne hydrogen sulfide and radon gases. The environmental impacts of hot-rock geothermal technologies remain to be explored.

The Future of Renewable Energy Supply

The technical potential for expanded use of renewable energy sources is very large. The largest potential contributors over the next several decades are biomass and direct harnessing of sunlight, but wind, geothermal energy, and hydropower could also make significant contributions.

The extent to which this potential becomes reality will depend on many factors. These include the magnitudes of future growth in population and material well-being, the extent to which energy demand growth is alleviated by improvements in energy efficiency, the degree of success that is achieved in reducing the monetary costs of the various renewable-energy technologies, the extent to which the environmental and opportunity costs of extensive land use for high-yield biomass-energy production can be kept within tolerable bounds, and the monetary and environmental costs of the nonrenewable energy options available.

These are of course not just questions of demography, economics, and the "natural" evolution of technologies; they are matters of public perceptions, priorities, and policies. The pace at which renewable-energy technologies improve in efficiency and economics, for example, will depend in large part on the rate at which governments and industries invest money in research and development for this purpose. The prices of nonrenewable options will depend not only on the inherent characteristics of these technologies and on the efforts made to improve them—and, in some cases, on the depletion of the nonrenewable resource bases on which they depend—but also on the extent to which, as a matter of public policy, these options are required to reduce their environmental and social impacts or compensate society for bearing them. Growing concerns about the climate-disruption potential of fossil-fuel use seem especially likely to increase the role of renewables beyond what it would otherwise be.

Various analysts and institutions have developed scenarios indicating the role that renewables can be expected to play under a range of assumptions about the foregoing factors. Under the circumstances most favorable to renewables, their percentage contributions in these scenarios grow to 40% to 50% of a doubled world electricity supply by 2030, and 35% to 40% of a nonelectric energy supply that has increased by about half from its 1990 value. The potential percentage shares of renewables in U.S. electric and nonelectric energy supplies shown in the scenarios are similar. Even larger contributions from renewables—in the range of 60% to 70% of electric and nonelectric energy supplies—would be possible by the midpoint of the 21st century.

JOHN P. HOLDREN

For Further Reading: Thomas B. Johansson, Henry Kelly, Amulya K. N. Reddy, and Robert H. Williams, eds., *Renewable Energy: Sources for Fuels and Electricity* (1993); Nancy Rader, *The Power of the States: A Fifty-State Survey of Renewable Energy* (1990); Carl J. Weinberg and Robert H. Williams, "Energy from the Sun" in *Energy for Planet Earth* (1990).
See also Architecture; Benefit-Cost Analysis; Bioenergy; Dams and Reservoirs; Energy; Energy, Electric; Energy, Geothermal; Energy, Tidal; Energy, Wind; Fuel, Liquid; Fuelwood; Greenhouse Effect; Hydropower; Solar Cell.

SOLAR CELL

Solar cells, which are also referred to as photovoltaic cells, are widely known for their technological simplicity in generating electricity from direct sunlight, and their lack of destructive impact on the environment. These cells employ a phenomenon known as the photoelectric effect, discovered by the German physicist Heinrich Rudolph Hertz (1857–1894), who observed that charged particles are released from certain materials when they absorb radiant energy. Solar cells convert light energy (photons), into electrical energy (electrons). The first solar cells were used by the United States space program in the 1950s. They are now recognized for their acceptability as a safe energy source and for their potential commercial value in generating electricity on a large scale. Their electricity production is limited to the daily solar cycle and the annual season cycle.

Solar cells are typically made of silicon, a semiconducting material in which electrons are unstable and easily released. In addition to being recognized for their simplicity of operation, solar cells are highly regarded as being globally adaptable, modular, and environmentally benign. They release no toxins and are the only power-generating technology that does not rely on mechanical activity. Since the only activity in solar cells is at the atomic level, they require very little maintenance. These systems can also be installed with negligible interference with environmental systems, particularly in desert areas where they are typically located. A much greater benefit to the environment would exist with the wider use of small-scale installations of solar cells, such as on the roofs of residential homes or buildings. However, solar cells used in this manner would require connection to an electric utility network or storage device, such as batteries, for power when the sun is not shining. Such widespread use of solar cells would decrease the peak power demand on large electric utilities.

Solar cells come in several standard designs that typically include crystalline cells, "thick-film" cells, "thin-film" cells, and concentrator cells that use lenses or mirrors to concentrate sunlight onto the cells' surface. All of these cells are capable of an economical 30-year lifetime.

Although solar cells can be installed in a wide range of geographic locations, the greatest potential is in dry equatorial or desert areas. Temperate and populated areas also retain high possibilities as well. Over the continental United States, considering the incidence of sunlight and frequency of meteorological interference, solar cells at the peak of the day have the potential to produce hundreds of watts per square meter, depending on the conversion efficiency of the solar cell, the season, and the geographic location. This means that the globe has the potential to produce billions of watts of electric power from solar energy. (A watt is a unit of power, equal to one joule per second, the rate at which work is done).

United States utilities currently produce approximately 25 megawatts (one million watts) of photovoltaic (photoelectric) energy; the applications are most widely used for irrigation pumps, billboard lighting, pocket calculators, and remote residential systems. Before these cells can become large-scale energy sources, major improvements must be made in capital cost, electricity storage, and conversion efficiency. Typical systems have reduced their prices from six dollars to one dollar per peak watt. New materials and design innovations are making them more affordable. However, even the best systems, at a cost of about 20 cents per kilowatt-hour (kwh), will not compete with other fuels until their costs drop to 12 to 15 cents per kwh. (A kwh is equal to the work done by 1,000 watts acting for one hour.) Nevertheless, with considerable recognition of solar cells' potential energy production and environmental safety the industry is growing about 30% annually, with the long-term potential of supplying 10% to 20% of world energy demands by 2050.

SOLID WASTE MANAGEMENT, MUNICIPAL

See Waste Management, Municipal Solid.

SOURCE REDUCTION

Despite increased reference to source reduction, the concept is imprecise, with different definitions appearing in legislative proposals and the literature on waste. These definitions fall into three categories: (1) source reduction as a set of packaging and product redesign efforts; (2) source reduction as a broader set of waste reduction efforts that include recycling and reductions in energy and nonrenewable resource inputs into packages; and (3) source reduction as measures to reduce hazardous wastes only.

Those offering definitions of source reduction differ over key issues: Is source reduction a pollution control strategy or a resource conservation strategy? Does the concept apply to both hazardous and nonhazardous products and packages? Does source reduction include efforts to increase the recyclability or recycled content of products and packages? Does it include efforts to reduce energy and other resource inputs during the production and transportation of packages?

Because of the ambiguity, other terms have emerged. The Office of Technology Assessment (OTA) has suggested the term *waste prevention*, the EPA has offered the term *pollution prevention*, and others have used the term *waste minimization* or *waste reduction*. These terms imply less focus on reduction of waste at the source (the point of manufacture) and more readily accommodate activities such as recycling and composting.

The solid waste consulting group Franklin Associates advanced the first of the three approaches to defining source reduction in its report *Source Reduction: A Working Definition*. This report describes source reduction as a "collection of activities and actions that, in combination or singularly, lead to a reduction in the quantity and/or toxicity of waste." By this definition,

source reduction is not a technology or process for managing waste, but a set of "front-end" activities that occur during the manufacturing process, resulting in changes in packages, products, or the kinds of materials used; these changes must occur before the product or package is discarded. The EPA has advanced a similar definition of source reduction in its report *The Solid Waste Dilemma*. That report states, "Source reduction is not a technology or process . . . to be applied to the waste stream." OTA concurs with this perspective, claiming that it is "important to make a clear distinction between prevention and management activities to ensure that adequate attention is focused on each." Though the EPA's hierarchy of solid waste management includes source reduction at its apex, the OTA suggests that this is misleading, since source reduction is not a tool available to those managing and disposing of waste, but instead is a set of activities that must be taken by those that manufacture and use products and packaging.

The second approach to defining source reduction is best characterized in a report by the Source Reduction Task Force of the Coalition of Northeastern Governors. That report notes that source reduction includes activities to reduce waste volume, weight, and toxicity, and increase recyclability of products and packaging, or add recycled content to products and packages.

The third definition applies only to hazardous wastes. In the Pollution Prevention Act of 1990, source reduction is defined as activities that: (1) reduce the amounts of hazardous materials entering a waste stream prior to recycling, treatment, or disposal, or (2) reduce the hazards associated with hazardous materials. This narrower concept has been largely eclipsed by definitions that include reductions in the amount and toxicity of all wastes.

All the definitions share an emphasis primarily on activities that prevent or reduce waste "up front." This waste prevention extends to the design and manufacture of products, as well as to consumer behavior that results in generation of less waste.

Source Reduction Techniques

A host of options for reducing waste at the source exist. Franklin Associates identifies a number of source reduction tools that apply to packaging. These include:

(1) Lightweighting of a particular package. For example, in the early 1970s, a plastic milk jug weighed 95 grams; the same size jug by 1990 weighed only 60 grams, as technological improvements allowed manufacturers to reduce the amount of plastic.

(2) Lightweighting by substituting one material for another. A notable example is the replacement of rigid paperboard containers with thin plastic shrink-wrap.

(3) Introduction of bulk packaging. Shipping fresh produce in large containers rather than in pre-packaged, small-serving containers represents a form of bulk packaging that can reduce waste.

(4) Introduction of composite materials to produce more efficient packaging such as the coffee "brick pack." Sixty-five one-pound brick pack containers produce about three pounds of waste. The same number of traditional metal containers produces 20 pounds of waste.

(5) Consumable packaging. For example, in the 1980s some detergent manufacturers introduced dissolvable detergent packets.

(6) Extended packaging life. Reusable, refillable, or reconditioned packaging results in multiple uses of a package before it is ultimately disposed. Refillable milk bottles began to reappear in the late 1980s as manufacturers sought ways of reducing the waste associated with their packages.

(7) Product redesign to reduce packaging requirements for that product. Product concentrates and powders to which one adds liquids, for example, result in less packaging waste.

(8) Elimination of packaging. Manufacturers can sometimes eliminate a plastic overwrap, or a box in which the primary package is contained, resulting in waste reduction.

Organizations that define source reduction more broadly to include recycling and other actions to divert waste from disposal facilities describe additional source reduction techniques. For example, the World Wildlife Fund and the Conservation Foundation, in their report on source reduction, include such actions as backyard composting, use of two-sided copying for written materials, and reductions in junk mail by removing names from mailing lists. Recycling and design for use of recyclable materials also fall under the umbrella of source reduction in its broadest definition.

Source reduction activities are not a recent phenomenon, though the term only regularly appeared in the 1980s. As a result of competition in the marketplace, many packaging manufacturers have reduced the amount of materials (and hence the resulting waste) associated with their packages. In its 1990 study on the U.S. waste stream prepared for the EPA, Franklin Associates reported a decline in the weight of packaging waste from 33% of the waste stream in the early 1980s to 30% in 1990. This decline occurred despite increased consumption and population growth.

Economic pressure to reduce the costs of packaging by reducing materials, energy, and other inputs led to a steady process of source reduction between the 1950s and 1990s. By the late 1980s, increased concern about

environmental impacts of waste disposal resulted in a deliberate push by manufacturers to reduce packaging to address environmental concerns.

In addition to conscious efforts to reduce waste associated with products and packages, manufacturers are developing strategies to reduce waste within their in-house operations, particularly to reduce hazardous wastes and respond to the Pollution Prevention Act of 1990. These strategies include: (1) inventory management to improve storage and handling of materials and to reduce the purchasing of toxic and other materials; (2) equipment modifications to increase recovery and recycling, particularly of toxic materials, and to generate less waste through better operating efficiency; (3) production process changes to substitute nonhazardous for hazardous materials; and (4) recycling and re-use both on-site and off-site, including participation in waste exchanges through which wastes of one company are sold to another company that can productively use them.

Measurement of Source Reduction

Measuring source reduction is both difficult and controversial. The OTA points out two measurement problems: (1) What is the baseline against which improvements are to be gauged? (2) How does one resolve tradeoff issues?

The baseline issue is significant, particularly if source reduction is a legislated goal rather than simply a guideline for action. Since weight reductions in packaging have occurred for several decades, legislative requirements for source reduction that do not take into account these earlier source reductions would, it is argued, discriminate against manufacturers whose products had already undergone significant source reductions.

The tradeoff issue poses equally challenging problems. Some reductions in weight of packaging lead to increases in the volume of waste in the waste stream. Replacement of glass containers with plastic containers, for example, may result in reductions in waste by weight, but increases in waste volume. Or, for those that include use of recyclables as a source reduction technique, increased use of recycled content can also increase the weight and volume of waste when the product is discarded.

Some environmental and technical organizations have challenged the prospects for measuring the relative environmental impacts of different products and packages. These groups point out that efforts to reduce one category of waste might increase other types of waste. For example, increased use of food packaging from the 1950s through the 1990s has been correlated with declines in food waste.

Measuring source reduction, if defined broadly in terms of reductions in environmental impacts associated with a product, confronts the criticisms lodged against all efforts to evaluate the life-cycle environmental impacts of products. How does one weigh the relative impacts of one kind of air pollutant against another, or against water emissions, or habitat degradation? Because all products use a variety of energy and resource inputs and have a variety of air and water emissions, as well as waste, associated with their manufacture, use, and disposal, efforts to compare one product or package with another result in complex assessments of environmental tradeoffs. There is no consensus about how such measurements should be made, or whether they can be made at all.

For this reason, most discussions regarding source reduction and its measurement focus more narrowly on the solid waste attributes of a particular package, product, or activity. That is, measurement of source reduction is discussed primarily in terms of whether reductions in waste by weight (or sometimes volume) result from actions to change products and packages.

Policies and Incentives for Source Reduction

In *The Next Frontier: Solid Waste Source Reduction*, Paul Relis and Karen Hurst identify three sets of options that create incentives or impetus toward source reduction: economic incentives, regulations and product bans, and voluntary targets. A number of these conditions currently exist.

By 1990 the major pressures toward source reduction were economic, particularly among industries that produced toxic or hazardous wastes. These pressures included increased costs associated with disposing of waste, increased financial liabilities for cleaning up hazardous waste disposal sites, and generalized public opposition to hazardous and toxic wastes.

While the immediate impetus toward source reduction of toxic and hazardous wastes was economic in the 1980s, regulations played a key role in increasing disposal costs and therefore encouraging source reduction. For example, the increased costs of disposing of hazardous wastes in the U.S. resulted from two major regulatory measures. The 1984 Hazardous and Solid Waste Disposal Amendments to the Resource Conservation and Recovery Act prohibited land disposal of hazardous wastes. The EPA has continued to increase the minimum treatment standards required in the disposal of hazardous wastes.

For both hazardous and nonhazardous wastes voluntary or legislated source reduction targets also stimulated industry efforts in the 1990s to reduce wastes associated with their manufacturing processes, products, and packaging. The EPA developed its 33/50 voluntary program for industry to reduce the use of toxic materials. Under this program, EPA asked industry groups voluntarily to achieve a 33% reduction in toxic materials use by the end of 1992 and a 50% reduction by the end of 1995, compared with 1988 toxic materials releases.

The most notable source reduction policies aimed at household solid waste have come in the form of economic incentives—household waste disposal charges based on how much waste a household actually disposed of. These fees (based on the number of cans or bags, or the weight of garbage set out for collection) were introduced in over 1,000 cities and local communities by 1993. Such charges appear to encourage households to compost their waste, recycle waste, and change their purchasing habits to buy products and packages that generated less waste. There remains, however, little agreement among waste management officials, environmental and other public officials, and academic researchers about the actual extent of source reduction that takes place in response to this kind of user fee charges for waste collection and disposal.

Despite disagreements about how and why source reduction occurs, there is general agreement that opportunities to reduce both the volume and toxicity of waste exist. This agreement will increasingly make source reduction a conscious part of public- and private-sector waste management practices.

LYNN SCARLETT

For Further Reading: William E. Franklin and Warren A. Bird, *Source Reduction: A Working Definition* (1989); Karen Hurst and Paul Relis, *The Next Frontier: Solid Waste Source Reduction* (1988); Jeanne Wirka, *Wrapped in Plastics: The Environmental Case for Reducing Plastics Packaging* (1988). *See also* Ecolabeling; Hazardous Waste; Life Cycle Analysis; Recycling; Waste, Municipal Solid; Waste Disposal.

SOUTH COAST AIR QUALITY MANAGEMENT DISTRICT

The South Coast Air Quality Management District (AQMD) is the air pollution control agency for the Los Angeles, California, area. Its mission "by law is to achieve and maintain healthful air quality for its residents. This is accomplished through a comprehensive program of planning regulation, compliance assistance, enforcement, monitoring, technology advancement and public education." Through its efforts to alleviate the area's severe smog problem, AQMD has become an international leader in air pollution reduction. Its programs and regulations—the strictest in the U.S.—have served as models for initiatives of the Environmental

Protection Agency (EPA) as well as state and local agencies.

AQMD's jurisdiction covers the counties of Los Angeles, Orange, Riverside, and the nondesert portion of San Bernardino County. Pollution in this 13,350 square mile area — a home to more than 13 million people — is produced by over 9 million vehicles, about 52,000 businesses and industries, and numerous consumer products. Natural conditions exacerbate the problem. Ocean breezes carry pollutants inland where they are trapped in valleys by thermal inversions. Sunshine and warm temperatures cause some pollutants to react, forming more and worse pollution. In 1990 AQMD estimated that some pollutants would need to be reduced by 80% in order to achieve state and federal air quality standards in the district by 2010.

The state of California created AQMD in 1976 through the Lewis Air Quality Management Act. The Act, recognizing the inadequacy of local pollution prevention efforts, mandated AQMD to establish uniform regional plans and programs to attain federal air quality standards by the dates specified in federal law and to attain state standards as quickly as possible. While AQMD is bound by these federal and state standards, it is largely independent in its initiatives to comply with them.

Like other regional air agencies in California, AQMD is legally responsible for controlling stationary sources of air pollution. These sources, which produce 40% of the air pollution, include power plants; gasoline stations; factories; and consumer products, such as house paint and charcoal lighter fluid. To control these sources, AQMD develops and implements an Air Quality Management Plan, reviewed by the California Air Resources Board (CARB), that aims to achieve compliance with state and federal regulations and serves as a blueprint for all future regulations to reduce emissions from specific sources. AQMD evaluates proposals for any new project that causes or controls air pollution, and permits granted generally require use of the best pollution-control technology available. Facilities are inspected periodically to assess compliance. Maximum penalties for offenders are $25,000 and/or a year in jail for each day of violation.

Mobile sources, which account for 60% of the region's air pollution, include primarily cars, trucks, and buses, but also construction equipment, trains, and airplanes. State and federal agencies, such as CARB and EPA, set emissions standards for these sources. Nevertheless, AQMD has various programs to reduce mobile-source pollution. For example, 6,000 work sites employing over two million people participate in a program known as Regulation XV, which requires employers to encourage workers to carpool, use public transportation, or bicycle to work. Citizens can call a toll-free hotline to report violations or to hear information about air pollution conditions. Other programs promote the use of cleaner fuels and vehicles.

Most of AQMD's $110 million operating budget comes from fees paid by businesses for permits. Large polluters pay fees based on the amount of pollution released, providing funding for AQMD and acting as an incentive to reduce emissions. Recognizing that mobile sources are the biggest polluters, AQMD added a surcharge to vehicle registration fees in 1991.

SPACE ENVIRONMENT

Our concept of the environment usually ends in the rarefied upper atmosphere above our home planet. Today, humankind needs to extend that awareness above the atmosphere to space. The increasing use of space for environmental monitoring and communications coupled with future uses for manufacturing, energy production, scientific research and human exploration already raises significant environmental issues.

Barely 150 miles (240 km) overhead, the thousands of civilian and military launches into Low Earth Orbit (LEO) have left tens of thousands of fragments and sections of launch vehicle upper stages, satellite shrouds, spacecraft parts, and other debris. The amount of debris is increasing. Some of this orbiting debris slows by interaction with the thin upper atmosphere and falls into the atmosphere where it burns up. But by the year 2020, as much as ten times more debris of some sizes will be present in the most frequently used orbits and could hit spacecraft with speeds over 30,000 miles per hour. The danger to orbital telescopes, space stations, LEO satellites, and humans will increase the risk of an impact by more than one hundred times over the normal risk caused from natural debris (micro-meteoroids).

The surfaces of the moon and Mars are already subjects of debate about their appropriate use by humans.

The moon, our nearest neighbor in space, is a gigantic mass of rock covered by large areas of lava flows and pounded for hundreds of millions of years by large and microscopic meteors. With virtually no atmosphere, the only "weather" on the moon is the constant solar wind, a stream of radiation particles from the Sun. The temperature on the sunlit side is about 250 degrees Fahrenheit. When an area is unlit, the temperature drops almost instantly to about −150 degrees Fahrenheit. One might ask if a large, airless rock really is an environment at all, much less one that needs protecting. However, work on the moon will require a type of environmental awareness, and some of the activity could have extensive impact on the Earth's environment.

The moon is a promising site for arrays of radio and optical telescopes, which might enable astronomers to detect planets orbiting other stars. The moon would be a good place to stage a practice mission for Mars exploration. It could be used to test the reliability of systems, equipment, and habitats over long periods, and the ability of humans to cope with the high radiation environment and isolation.

The moon, with its gravity one-sixth that of Earth, would be a much cheaper place to get building materials for orbital structures. Lunar material also contains a significantly higher concentration of the isotope helium 3. Helium 3 is seen by fusion physicists as a cleaner, more efficient fuel for fusion reactors. While this development may be far in the future, fusion power has been promoted as a clean limitless source of energy on Earth without burning fossil fuels and increasing temperature, carbon dioxide, and other toxic pollutants. Helium 3 could be produced on the moon and returned to Earth, or used in reactors in space that would beam the power back to Earth, keeping the production facilities completely out of the Earth's environment.

The next planet outward from Earth, Mars is smaller and colder. Coming within forty million miles of Earth, it would take a human crew with an advanced propulsion spacecraft between three months and a year to get there. The earliest optimistic plans for reaching Mars could send humans to the planet about the year 2014.

With long travel times, the impact of humans on the Martian environment will be slow. Mars has long been the subject of speculation about the existence of life. Fuzzy telescopic images from the late 19th and early 20th centuries seemed to show seasonal variations, and visual observations suggested a system of features which came to be called canals. But the arrival of images from the Mariner and Viking exploratory spacecraft showed a cold dead world with temperatures between −225 degrees and 80 degrees Fahrenheit.

There is evidence of water once having flowed on Mars, and there are spectacular rift valleys the length of the United States. Mars was volcanically active, and Olympic Mons, a volcano with a base as large as the state of Texas, is the largest known volcano in the solar system. Mars has a thin atmosphere of carbon dioxide, but 200 mile per hour winds drive global dust storms that might make human habitation difficult.

There is water on Mars, most of which is trapped in the polar ice caps. There may be significant amounts of permafrost in the ground as well. Some estimates say that if the water were properly managed, several hundred thousand people could live on Mars. Mars's 38% of Earth gravity would permit spacesuit-clad humans to operate on the surface, and the Martian surface soil and rocks contain abundant oxygen, which, coupled with groundwater, could make fuel for departing spacecraft and surface activities.

Space exploration expands our definitions of environment. At the same time, the microcosm of spaceflight may be one of our most important laboratories for understanding the limits of human survival here on Earth.

KERRY M. JOELS

For Further Reading: Edward and Linda Ezell, *On Mars: Exploration of the Red Planet 1958–1978* (1984); Duncan Lunan, *New Worlds for Old* (1979); James Oberg, *Mission to Mars* (1982).

SPECIES INTRODUCTION

See Exotic Species.

STANDARD OF LIVING

People consume and use many economic facilities in their daily lives—at home, at work, at play. The standard of living summarizes how well off people are, from an economic point of view, in the implementation of their lifestyles, using such economic facilities. Sometimes standard of living means the actual level of living and sometimes it means a target for a standard that is to be achieved. Normally, people mean level of living when they refer to standard of living.

A principal use of the concept is to compare living conditions among different groups in society or between different countries or areas. In the richest countries, the most advanced industrial countries, it is generally the case that living standards are comparatively high, while in poorer countries, such as developing countries, standards are lower. The same distinctions are made for comparison of different groups or regions within a country.

A standard of living is defined in terms of the total lifestyle, including command over the most necessary items (food, clothing, shelter, fuel) and more luxurious items (recreation, cultural facilities, travel opportunity, educational opportunity, and many conveniences of life). A person or group is said to have a high standard if a total basket or collection of goods and services available is relatively large, i.e. relative to what another person or group has available. Some people may have ample food or shelter but not access to conveniences or other more luxurious facilities. Their standard would be classified lower than that for those who have ample supplies of all or most goods and services.

An entire nation's standard of living is judged on the basis of its total national production or consumption of a comprehensive collection of goods and services. This should include social goods, from the living environment, as well as private goods and should also include such nonmarketable items as climate, scenery, and cultural facilities. Standard of living is increasingly being considered in new dimensions. Attention is being paid to health, education, and infrastructural facilities, in addition to environmental factors such as purity of air and water, and congestion. These are of importance in determining standard of living but are difficult to value; therefore they have, in the past, been neglected but fresh ways of taking them into account are being developed.

In order to measure a country's standard of living in relation to that of other countries, the values of total national production or consumption should be estimated using common prices for all countries being compared. Allowance should also be made for the number of people among whom the goods and services are to be divided; production or consumption *per capita* would be one measure that takes this aspect into account.

The standard of living is high in North America, Western Europe, Japan, and other like places. It is low in sub-Saharan Africa and parts of Asia. These differences are wide and clarify the measuring of the concept. There are many countries, regions, or groups that have standards that are more nearly similar, and it is more difficult to determine who is better or worse off in such closer comparisons.

L. R. KLEIN

STEPPES

See Grasslands.

STOCKHOLM CONFERENCE ON THE ENVIRONMENT

See United Nations Conference on the Human Environment.

STRIP MINING

See Mining Industry.

STRONTIUM-90

Strontium-90 is a radioactive isotope of strontium, an alkali earth-metal with atomic number 38 and atomic weight 87.62. It is present in fallout from nuclear weapons testing and in nuclear energy plants. With a half-life (the time required for its radioactivity to drop to one-half of the original value) of 28 years, it is used for its high-energy beta emission in certain nuclear electric power sources. When strontium-90 is present in nuclear fallout it contaminates the land and water and is absorbed by plants, thus entering the food chain. Low-level radiation exposure often occurs through ingestion of contaminated food and water. Because milk is a good indicator, strontium-90 poisoning is often associated with milk. The isotope is easily absorbed by bones and tissues and may eventually replace the calcium in the body because it is chemically similar to the calcium in bones. Its effect on the bone marrow (where new blood cells are made) varies depending on the level of exposure. It slowly causes atypical blood cells to multiply, thus increasing the incidence of leukemia or blood cancer.

The effects of radioactive fallout that includes strontium-90 were seen in towns where above-ground nuclear weapon testing occurred, such as St. George, Utah. There were two types of exposure to strontium-90: external radiation exposure and ingestion of the radionuclide mix in fallout. The number of reported cases of leukemia among children who grew up there during the testing was two and a half times greater than the number of cases reported among children who grew up there before and after the testing. Significant increases in thyroid cancers and birth defects also followed the testing period. Above-ground nuclear weapon testing was abandoned shortly thereafter.

SUBURBS

In modern American usage the term *suburb* refers to a middle-class residential community located outside—but within commuting distance of—a major metropolis. But suburbs in the more general sense of the word—any settlement just beyond or outside a large city—are as old as cities themselves and have varied widely in different historical periods and cultures. From the middle ages to the 18th century, European suburbs were invariably "suburbs of poverty": shantytowns just outside the city walls to which the poorest inhabitants and the most polluting industries were relegated. Only in late 18th-century England did members of the middle class leave their accustomed townhouses in the urban core for the newly-created "suburb of privilege": a peripheral community made up of large detached houses on landscaped lots.

Even after the middle-class "suburb of privilege" had developed in the 19th century as the predominant form of suburban settlement in both England and the U.S., Continental European cities retained the traditional form of the city with the wealthy and middle class living as close to the center as possible and the working class and poor pushed out to the periphery. This pattern largely persists in Europe today, as well as throughout the Third World. In Latin American, African, and Asian cities the wealthy occupy the urban core, often living in large apartment-houses with modern services. The poor are forced to live in peripheral shantytown "suburbs" that not only lack clean water, electricity, and other services, but require difficult and expensive commutes to the urban center where the bulk of the jobs can still be found.

The Anglo-American middle class "suburb of privilege" has been significant ecologically for two reasons. First, the early middle-class suburbs constituted a small but telling "borderland" between the dense, unhealthy, rapidly expanding industrial cities and the rural areas. In an era of explosive urban growth, these suburbs constituted a critique of overcrowded cities and an alternative tradition of design that embodied healthy densities and a "marriage of town and country."

Second, these early suburbs provided a model of low-density living for the mass suburbia of the 20th century. This mass suburbia, which is now the home of the majority of the American population, is largely free from the egregious overcrowding that plagued the 19th-century city, but mass suburbia has introduced considerable ecological strains of its own. The once-limited suburban borderland between town and country has spread to fill whole regions, overtaking farmland and other vital open space and threatening the survival of the central cities. This low-density suburban "sprawl" has meant near-total dependence on the automobile and high energy consumption.

The modern middle-class residential suburb from which our present sprawl derives was invented by the late 18th-century London bourgeoisie who fled their townhouses in the center of the crowded city and chose to live instead in substantial houses in the picturesque country villages that could then be found no more than three to five miles from the center of town. This new lifestyle was soon copied by the bourgeoisie of the expanding industrial cities of the north of England; by the first half of the 19th century the American middle class was also suburbanizing.

From the first, suburbia was an attempt to combine the health and serenity of a rural village with the prosperity of the rising middle class. Early residents and developers drew from the picturesque tradition of English landscape architecture to create a landscape of "houses in the park." In the 1820s the English master of the picturesque style, John Nash, had created in his Park Village at the edge of Regent's Park, London, a definitive rendering of the suburban tradition in design: substantial detached houses in various period styles laid out along curving roads landscaped with lawns, bushes, and trees.

This suburban design tradition embodied an ecological message of health: low density, detached homes interspersed with natural greenery, quiet as opposed to the clatter of city streets, and the clean air and water that could no longer be found in the urban core. In the new industrial cities of the north of England and the U.S., the middle class suburbs were built just beyond the districts where factories and workers' housing were located, forming a sharp ecological and social contrast.

In the U.S. the English suburban style was domesticated and transformed by three great designers, Andrew Jackson Downing, Catharine Beecher, and Frederick Law Olmsted. Downing popularized the English suburban cottage; Beecher spread the idea of the suburban home with its ample family rooms, verdant grounds, and newly designed kitchen as the best environment for the American housewife and the best refuge for American family values. In his plan for Riverside, Illinois, outside Chicago in 1868, Olmsted created from 1,600 acres of flat prairie a landscape defined by gently winding streets, large houses set back at least thirty feet from the street to provide ample front lawns, and the planting of 35,000 trees and 47,000 shrubs.

Riverside was important not only for its elaborate design but also for its direct rail link to Chicago. Previous suburbs had been served only by private carriages or horse-drawn omnibuses. The second half of the 19th century saw suburbs based on steam rail and later electric trolley links to the central city, yet far enough from the center so that workers could not afford the fares. The houses were clustered within walking distance of the station. The resulting pattern of development consisted of middle class enclaves surrounded by verdant open space.

This pattern changed radically with the invention and popularization of the automobile. The emerging road network allowed suburban development to spread out virtually anywhere on the metropolitan periphery. In the 1920s, suburban developers tried to create a mass suburbia by laying out thousands of subdivisions for lower middle class and working class homebuyers, but this first effort failed because of high costs and the shock of the Great Depression. In the 1930s the New Deal refinanced and reconstructed the moribund suburban housing industry in an attempt to stimulate the economy by making houses more affordable. After World War II all the elements were in place for an explosion of mass suburbanization symbolized by the dramatic success of Levittown, Long Island (1947), where a single developer, Levitt and Sons, constructed over 17,400 homes for 82,000 residents.

These elements were: (1) intensive investment in roadbuilding by the federal and state governments that made vast tracts of farmland at the edge of great cities "convenient" for development; (2) a federally guaranteed home financing system based on savings-and-loan institutions that provided long-term, low-interest mortgages to homebuyers insured by the Federal Housing Administration; (3) loans and other aid for homebuilders that enabled Levitt and other builders to adopt mass production techniques and thus cut costs; (4) a new suburban economy largely based on industrial plants constructed during the war whose workers were prominent among the new suburban homebuyers; (5) a range of federal, state, and local tax advantages that favored suburbanites over city dwellers and enabled suburban municipalities to "tailor" their revenues to the needs of middle class residents.

As suburbia grew, American central cities were hard-pressed by disinvestment and growing social tensions. The massive federal aid to highways was matched by a corresponding neglect of both freight rail and mass transit, eroding the city's traditional advantage as a transportation hub and center of industry. Federal Housing Administration regulations favored suburban projects and virtually banned ("red-lined") federally guaranteed mortgages and improvement loans in large sections of central cities. Finally, the migration of rural blacks from the South to northern cities resulted in racial tensions and "white flight" to the suburbs.

As a result of these positive and negative factors, American suburbs, which housed approximately a quarter of the nation's population in the 1940s, had reached a third of the population by 1960. The 1990 Census shows for the first time an absolute majority living in the suburbs. The growth of suburbs has completely transformed farms and woodlands around major cities. Indeed, suburbia in the post-World War II era consumed ever larger units of land and housing per resident. At Levittown, Long Island, the average house contained 750 square feet of living space, and the houses were spaced to accommodate about 15 people per acre. By 1963 the average suburban house had expanded to 1,450 square feet, and by the late 1980s the average had climbed to 2,000 square feet, built on even more generous lots with densities no more than 10 people per acre. (Houses with 3,000–4,000 square feet at even lower densities are not uncommon.) Ironically,

the American suburban house was growing just when the typical American household was shrinking from an average of 4 members in 1940 to 2.6 today.

Lawns, a design feature of the original English suburban house, had become an integral part of the American concept of a home. Lawn grasses are not native to the North American continent and their cultivation requires 70 million pounds of fertilizers, pesticides and other chemicals annually; the average American homeowner uses a higher concentration of chemicals than the average farmer. Lawn grass requires an inch of water a week, or 10,000 gallons per summer for a 25 by 40 foot lawn. In some areas of the West, two-thirds of the annual water consumption has gone into lawns.

Postwar suburban planning, moreover, has emphasized the separation of housing tracts from commercial or industrial uses. As a result virtually no one in the newer suburbs lives within walking distance of work or shopping, and all significant journeys have to be accomplished by automobile. With population densities low, mass transportation is almost always uneconomical. Thus, the extravagant use of space continues in suburban road and highway design practices that mandate extensive cul-de-sacs in residential areas; wide "collector streets" for through traffic and superhighways that cut swathes of eight lanes or more through the countryside.

Developers have responded to the long, congested commutes "downtown" by moving formerly urban functions to the suburbs; first the shopping mall and then the office park have become familiar features of the suburban landscape. This brought jobs and services closer to the home and temporarily saved time and energy. But the emerging pattern of jobs, services, and shopping dispersed over long highway "growth corridors" has meant that vital functions were dispersed over whole regions, thus ensuring even further suburban dependence on long automobile trips.

Moreover, the space consumed by the new shopping malls and office parks was also enormous. Journalist Joel Garreau has described the largest such mall/office complexes with over 5 million square feet of commercial space as "Edge Cities." Standard parking requirements (mandatory in many suburbs) that for every square foot of shopping or office space the developers must provide 1.5 square feet of parking space have resulted in Edge Cities that dwarf the downtowns they have replaced.

Suburbia in the U.S. is thus no longer a narrow borderland between the city and the countryside, but sprawl that covers whole regions. In his 1961 book *Megalopolis*, devoted to the eastern "corridor" between Boston and Washington, the geographer Jean Gottmann was the first to perceive that formerly separate metropolitan areas were merging into a single "megalopolis" which was neither urban, nor rural, nor suburban in the traditional senses of the words.

Planners and public officials are now attempting to create more balanced and sustainable patterns of suburban development through a number of interrelated initiatives. These include regional or statewide land-use plans that prohibit development on ecologically crucial open space; discourage wasteful, low-density peripheral development by withholding state-subsidized infrastructure (roads, sewer lines, and other utilities); and attempt to re-focus development on neglected core cities where infrastructure is already in place and mass transit is still in operation. Complementing these initiatives are proposals for land-banking, transfer-of-development-rights and tax incentives that help to preserve open space without depriving farmers and other landowners of all financial benefits.

Another important idea is the substitution of "New Towns" for the usual low-density patterns of suburban sprawl. In the United States these ideas derive from the New Deal's Greenbelt program of the 1930s and from two privately developed New Towns of the 1960s: Columbia, Maryland (developed by James Rouse) and Reston, Virginia (developed by Robert Simon). Columbia and Reston introduced two crucial principles which have been widely copied—(1) cluster housing and (2) mixed use.

Cluster housing uses connected townhouses that share common open space to save land, infrastructure, and energy costs. Mixed use means bringing shopping closer to homes and including industrial and commercial sites within the New Town. Despite the careful planning and idealistic aims of their developers, both Columbia and Reston were increasingly constrained by market forces to conform to the dispersed land-use patterns of their time. A government-subsidized New Town program failed in the 1970s.

Other attempts to achieve New Town ecological objectives have recently been initiated by architect-planners such as Andres Duany and Elizabeth Plater-Zyberk (The New Traditionalism) and Peter Calthorpe (The Pedestrian Pocket). Duany, Plater-Zyberk, and Calthorpe wish to return to higher levels of residential density with site plans that recall the American small town. Duany and Plater-Zyberk have been especially adept at utilizing traditional town planning to promote efficient land use and a mixture of classes; Calthorpe has emphasized linking development to mass transit, especially new light-rail lines. Their plans would cut energy consumption—both within homes and for transportation—and preserve open space while staying within the suburban tradition of design.

ROBERT FISHMAN

For Further Reading: Robert Fishman, *Bourgeois Utopias: The Rise and Fall of Suburbia* (1987); Joel Garreau, *Edge City: Life on the New Frontier* (1991); Kenneth T. Jackson, *Crabgrass Frontier: The Suburbanization of the United States* (1985).

See also Architecture; Automobile Industry; Communities, Low and Moderate Income; Green Belts; Land Use; Olmsted, Frederick Law; Transportation; Urban Parks; Urban Planning and Design.

SUCCESSION

Stated in its most general form, succession is a process of community (or ecosystem) change starting from an unoccupied or incompletely occupied site and terminating in a more or less stable condition (climax).

Succession, with its associated endpoint—climax—is the oldest, and one of the most useful concepts of ecology. Succession provides a basis for understanding and predicting how communities and ecosystems change following various kinds of disturbances, and helps to account for some of the variety of communities found in a landscape. However, few ecologists share the same interpretation, and the nature of succession continues to undergo intense debate.

Because succession was originally conceived by plant ecologists, the original principles were derived from observations of changes in plant populations in temperate grassland and forest situations. Later, the concept was applied, with diminishing success and increasing confusion, to deserts, tundra, chaparral, fresh- and saltwater benthic or littoral systems, and even to pelagic systems. Succession seems to be most applicable to communities dominated by "sessile" organisms (fixed in place)—whether land plants or rocky intertidal or coral reef animals—so that there is a definite physical structure composed of these organisms, and an implicit relationship between competition for resources as represented by the occupied space. Succession also seems most useful in situations in which there is a change in species with time, a condition not always met with all of these environments.

Eventually sequential changes in animal populations were noted in some successional situations and the concept was enlarged to include animal populations. Finally the original concept of change in species composition over time with some attendant changes in the abiotic environment was enlarged in the 1950s to include collective properties of ecosystems such as changes in net primary production, accumulation of biomass, alterations in nutrient budgets and cycling rates, and overall biotic diversity. It is mandatory to define the context in discussion of succession: whether it is at the community level—involving species changes—or at the ecosystem level—involving those collective properties.

Succession is divided into two classes: primary succession originating from a newly formed site, and secondary succession originating from disturbance of a formerly occupied site. Newly deglaciated terrain and freshly deposited river bars are examples of sites that would, if left undisturbed, undergo primary succession. Logged forests and plowed grasslands are examples of sites previously modified by biological occupation that would undergo secondary succession.

The differences between primary and secondary succession are profound, so orderly discussion must begin with agreement on successional origins. Primary succession usually involves considerable site alteration through biotic action, such as soil organic matter accumulation, and through abiotic action, such as rock weathering. In primary succession different species typically arrive and colonize the freshly exposed site in an order depending on their dispersal and establishment properties. In addition, it is more likely that earlier occupants facilitate successful invasion by later occupants in primary succession because of the much more pronounced change in environmental conditions brought about by growth and death of earlier occupants.

Secondary succession may not involve much alteration of the site, such as soil development, and initial occupants may exist not only as sprouts or seeds from preexisting occupants but may also be the terminal species of successional development. In this regard, it is useful to differentiate between relay floristics, the serial introduction of species described in the primary succession case, and initial floristics, the introduction of many or all species at the onset of succession as described for the secondary succession case. Facilitation by early occupants for later occupants is also less likely in secondary succession. In fact, mixtures of these characteristics of site alteration, species establishment and facilitation can be found in particular cases of either primary or secondary succession. As is typical of all aspects of succession, these are tendencies rather than strict rules.

A characteristic sequence of changes associated with succession in a particular area is termed a sere. The succession of abandoned agricultural fields on the Piedmont of the southeastern U.S., from weeds to broom sedge to loblolly pine to hardwoods, represents a sere. It is sometimes convenient to divide phases of a sere into more or less definable stages termed seral stages. In actuality, succession is a highly stochastic (random) process because of the accidents of site qualities, dispersal conditions, weather during critical periods of establishment, prior occupation of species, etc. It is difficult to predict exactly how succession will proceed in a particular case, such as if one might wish to assess the impact of pipeline installation or temporary road building, even if the basic patterns are known for a region. It is even less possible to predict successional patterns for specific cases where only general principles can be applied. Probably the best examples of predicting succession are represented by the JABOWA-FORET families of forest growth models, in which local environmental conditions are well-defined and the life history characteristics of participating species are well-known.

The endpoint of succession, termed "climax" by some ecologists, is a point of endless argumentation. In its original concept, climax referred to a more or less stable species configuration characteristic of a climatic region. It was recognized, however, that this stable state would be somewhat different on more congenial sites such as poleward-facing slopes (postclimax), as well as on sites less congenial for the region (preclimax). Early ecologists also recognized that where disturbances such as heavy grazing or frequent fires were more likely, the stable state would be shaped by these high frequency disturbances. For this condition they termed the situation subclimax or disclimax. Clearly this complex of climaxes complicates the identification of stable state, and even the concept of a regional climax. A more practical approach is to avoid these terms altogether, to assume that climax is more of an ideal than a reality, and to expect a more or less stable state to consist of populations that will vary in abundance and importance, depending on position in the complex gradient of environmental factors characterizing any landscape.

Spatial and temporal scale is another important consideration. The succession concept was conceived for spatial scales of more or less uniform sites large enough to accommodate populations of dominating organisms (forest stands, zones of rocky intertidal shoreline). More recently there has been some expansion of the concept to the landscape scale (areas of varying terrain in the order of kilometers) which has produced some interesting insights and new viewpoints on succession and climax. For example, on a larger scale, an unlogged forest is viewed as perhaps consisting of a mosaic of patches of various ages developing from normal small-scale processes such as tree-fall gap formation. Appropriate temporal scale is a time span greater than the life span of dominant species, but less than geomorphological erosion cycles (millions of years). Usually succession is conceived as occurring in shorter times than evolutionary change is likely to occur in the participating species populations.

Application of succession to the collective properties of ecosystems has led to hypothesized trends at the ecosystem level, involving changes in species structure, energy flow, biogeochemical cycles, natural selection and regulation. This has been a very stimulating development, promoting fresh thought about ecosystems, leading to new kinds of studies and measurements, and raising healthy debate. Some trends are self-evident, such as the fact that with succession, biomass must increase. Others require qualification of particular conditions or remain to be tested as to their generality. One of the most difficult of these, the proposition that with succession, system stability increases, depends on the criteria used for defining stability, and on definition of the nature and extent of disturbance. Further research in this area will require incorporation of hierarchy theory and recognition of chaotic behavior at certain scales in ecological systems.

WILLIAM A. REINERS

For Further Reading: Herbert F. Bormann and Gene E. Likens, *Pattern and Process in a Forested Ecosystem* (1979); Robert MacIntosh, "The Relationship Between Succession and the Recovery Process in Ecosystems," in J. Cairns, ed., *The Recovery Process in Damaged Ecosystems* (1980); Robert H. Whittaker, *A Consideration of Climax Theory: The Climax as a Population and Pattern* (1953).
See also Forests, Deciduous.

SUESS, EDUARD

(1831–1914), Austrian geologist and politician. Eduard Suess, the founder of modern global geology, especially modern tectonics, urban and environmental geology, and the "Vienna School" characterized by world-wide paleontological, stratigraphic, paleogeographic, and tectonic studies and syntheses, is considered one of the greatest Earth scientists who ever lived. His work and his "Vienna School" have very strongly influenced the development of geology in the 20th century and created much of its language. Perhaps the most popular orator in the Austrian Parliament in the late 19th century, Suess was one of the most dedicated defenders of rationalism and the role of natural science in politics. A fruitful combination of an unwavering belief in reason inherited from the Enlightenment and an uncompromising love of nature instilled by the Romantics planted in him the conviction that natural science had to play an increasingly important role in approaching the various problems faced by the human society in its struggle for survival.

As a paleontologist, he enthusiastically embraced Darwin's theory of evolution, but criticized it for its insufficient attention to the effects on the organisms of changes in the physical environment. Like the geologist Prince Kropotkin after him and following his Viennese colleagues, the paleontologist Melchior Neu-

mayr and the anatomist Karl von Rokitansky, Suess viewed the struggle for survival as taking place between the organism and its environment, rather than among individuals or species. As a geologist he emphasized the oneness of our present environment and its history and underlined the unity of all natural science. His political work was initially triggered by his concern for the extremely unhealthy relations that existed between his fellow Viennese and their habitat and was always guided by his liberal views and humanist interests. He was one of the first to stress the need for an international body of scientists to exercise influence on the work of governments and, as president of the Austrian Imperial Academy (1894–1911), he initiated the first international cartel of Academies to further that end.

Born in London of Protestant Austrian parents, Suess obtained his formal education in Prague and Vienna, but was compelled to leave the Polytechnical Institute in Vienna without a degree owing to his active involvement in the revolution of 1848. His early interest in paleontology won him an assistant's position in the ancestor of the present Natural History Museum in Vienna, where he worked on a wide range of fossils. Although much of this initial work was paleozoological and stratigraphical in character, Suess showed an early interest in the relations between the organisms and their environment.

Despite his lack of a formal degree, he was appointed the first professor of paleontology in the University of Vienna. In 1861 he also assumed the responsibility for geology for which an institute was formed in 1862, and in 1867, a special chair of geology was created with Suess as the first incumbent. His research in the University initially was a direct continuation of his paleontological and stratigraphical studies. However, his excursions around Vienna aroused an interest in the geology of the Vienna Basin. The appalling typhoid fever epidemics then frequent in Vienna led him to query whether any relation might have existed between the geology of Vienna and the life of the inhabitants of the city. This inquiry inspired Suess's first major book, *The Ground of the City of Vienna*, subtitled "with respect to its structure, composition, and with reference to the lives of its citizens." In this first urban-geology book, Suess showed that the epidemics had been caused by the local groundwater circulating through cemeteries before reaching the wells, and he recommended that a new source of drinking water be found for the city. The book led to his election to the municipal council in 1863, where he became the main promoter of an aqueduct system, 112 kilometers long and still active, to carry unpolluted water from the Alps to Vienna. After its inauguration, the typhoid deaths fell to about less than 10% of their former annual number in the city. He was later elected to the provincial Diet of Lower Austria in 1869, and finally to the Imperial Parliament, where he devoted his energies mostly to improving the standards of education (he fought unsuccessfully against the influence of the church in school education) and placing natural science at the service of his countryman (e.g., the Danube regulation).

Suess worked both as a professor of geology and as a member of Parliament until 1897, when he voluntarily laid down his mandate because of the unwelcome growth of the right-wing reactionary movement in the empire.

His great reputation in the Earth sciences rests on his mammoth work in global tectonics, which he revolutionized by the publication of his epoch-making *The Origin of the Alps.* This book of only 168 pages with no figures was the manifesto of a then novel actualist and indeterminist research program in global tectonics, in which thermal-contraction-driven horizontal motions of parts of the lithosphere—creating the world's major mountain ranges as dominantly asymmetric structures with fore- and hinterlands and subsidences giving rise to its major ocean basins—were used as the major principles to present a synthetic view of the tectonic behavior of the entire planet, in which magmatism was viewed as a consequence and not a cause of tectonism as had been formerly supposed.

Suess's four-volume *magnum opus, The Face of the Earth,* began publication in 1883 and reached completion in 1909. This monumental treatise contains a comprehensive review of world regional geology, interpreted tectonically in terms of the principles he had set forth earlier. But in this book he also expanded his theoretical base and formulated the view that if contraction theory was to be tenable, there should be no primary uplifting movements in the lithosphere. He analyzed in detail the changes in sea level through Earth history and concluded that these changes were due to the movements of the hydrosphere and not to the independent vertical movements of continents. He hinted that changes in the capacity of ocean floors may be their cause, but left the solution of this problem to future generations. Suess was the first to emphasize the universal importance of thrust faulting in mountain building, to recognize that horizontal extension gave rise to rift valley formation, and to acknowledge the importance and common occurrence of strike-slip faults. These, plus his own confession, in his last volume, that thermal contraction alone could not account for the great horizontal mobility of the lithosphere, seem to have paved the way for Wegener's theory of continental drift. Suess's concepts of the former supercontinent Gondwana-Land, and the northern continents Laurentia and Angara-Land, and the former equatorial ocean Tethys were first defined or elaborated in his book.

In the last chapter of *The Face of the Earth,* "Life," Suess re-emphasized the dependence of life on the physical evolution of the environment and named the long-persisting terrestrial abodes of life "the asylums." The last sentence of his *magnum opus* is a monument to environmental awareness: "In face of all these open questions let us rejoice in the sunshine, the starry firmament and all the manifold diversity of the Face of our Earth, which has been produced by these very processes, recognizing, at the same time, to how great a degree life is controlled by the nature of the planet and its fortunes."

A. M. C. ŞENGÖR

For Further Reading: W. H. Hobbs, "Eduard Suess," *Jour. Geol.,* v. 22, pp. 811–817 (1914); E. Wegmann, "Eduard Suess," *Dictionary of Scientific Biography* (ed. C. C. Gillispie), v. 13, pp. 143–148 (1976); A. M. C. Şengör, "Eduard Suess' Relations to the Pre-1950 Schools of Thought in Global Tectonics," *Geol. Rundsch.,* v. 71, pp. 381–420 (1982).

SULFUR

Sulfur is one of the six most abundant elements in living things. It is distributed on Earth in both the free and combined states: in solids such as pure sulfur and metallic sulfide, in gases such as sulfur dioxide and hydrogen sulfide, and in other compounds. Most of these forms cannot be used directly by organisms and are, in fact, toxic to many. Sulfur is found in a number of foods including eggs, cabbage, horseradish, garlic, mustard, onions, and meat. It has an essential role in the structure of proteins; in human beings these proteins are primary components of tissue, cartilage, and tendons. In addition to this biological role, however, sulfur and its compounds are the most widely used chemicals in industrial processes and have been part of human technology for thousands of years. Early human beings used sulfur in their cave drawings; Egyptians bleached their linens with the fumes of burning sulfur as early as 2000 B.C.; and the ancient Greeks used sulfur for its medicinal and fumigating properties.

Sulfur is circulated through the living and nonliving components of the biosphere in the so-called sulfur cycle. Marine algae, cyanobacteria, a photosynthetic species of bacteria, and salt marsh plants release sulfur in a gaseous form as dimethylsulfide, providing a principal source of sulfur in the atmosphere. Sulfur is found in organic sediments such as coal, oil, and peat, and in elemental deposits in soil and rocks. It is also found in seawater and organic matter. Plants take in some sulfur from the atmosphere, but they primarily absorb it through the soil as sulfates. However, since plants can only take up sulfur in the form of compounds, they are dependent upon certain bacteria and fungi that liberate it from organic and inorganic deposits.

Another major source of sulfur is volcanic eruptions. Human activity through industrial processes, however, has come to rival these and other long-term liberators of sulfur into the atmosphere. Sulfur returns to Earth through precipitation as sulfuric acid and is also cycled by the weathering of rock and erosional runoff.

There are many industrial and agricultural uses of sulfur compounds. Its primary use is in the production of fertilizer; sulfuric acid is used to digest phosphatic rocks to produce a variety of different products to improve soil. Sulfur compounds are used in petroleum refining and to produce paper, pulp, plastics, tiles, and highway asphalt, detergents, dyes, pigments, and explosives. The automobile lead-storage battery contains sulfuric acid. Sulfur can be applied directly to soil in order to lower its pH (raise the acidity).

Sulfur as a byproduct of human activity poses environmental hazards. The sulfur dioxide that forms from the burning of fossil fuels is a principle air pollutant that harms plants and contributes to smog. When it is converted to sulfuric acid, it is a component of acid precipitation. When sulfur-containing rocks are exposed during coal mining, some bacteria release the sulfur, allowing sulfuric acid to be formed. Streams become polluted and revegetation of the mining site becomes extremely difficult. Land reclamation efforts in areas of acid mine drainage have been intense; methods include sealing the area to inhibit bacterial activity and creating wetlands that filter out the sulfuric acid.

SULFUR OXIDES

See Pollution, Air; Sulfur.

SUPERCONDUCTIVITY

At very low temperatures, below 20°K (equivalent to −253°C or −423°F) some pure metals lose all resistance to an electric current. This phenomenon, called superconductivity, was discovered in 1911 by Heike Kamerlingh Onnes, a Dutch physicist. Although superconductivity continued to interest physicists and electrical engineers, they worked without a theoretical understanding of the process for almost half a century. Then in 1957 John Bardeen, Leon Cooper, and John Schrieffer published the theory of superconductivity

that bears their names. According to the Bardeen Cooper Schrieffer (BCS) theory, superconductivity is the result of the propensity of electrons in very cold metal to form traveling pairs, which do not scatter off the atoms that make up the material. Superconductors do not resist the flow of strong electric currents, and therefore currents can flow around a loop of superconductive material indefinitely.

The BCS theory suffices for such metals as aluminum and lead, but it could not explain the behavior of a new class of superconductors. In 1986 George Bednorz and Alex Muller at the IBM research laboratory in Zurich, Switzerland, discovered that certain ceramic materials called perovskites become superconductors at temperatures as high as 100°K. The BCS theory cannot account for the phenomenon in these ceramics, which generally consist of copper, oxygen, and rare earth elements. A definitive theory for the ceramic superconductors has not been offered.

Nevertheless, the discovery has generated an intense global effort to exploit the existing superconductors and to develop additional compounds that superconduct at high temperatures. Such materials may turn out to be much more economical to use than conventional superconductors. The expense of cooling a metallic superconductor is far greater than the cost of cooling a ceramic superconductor.

Materials scientists seeking to make useful devices from ceramic superconductors face formidable challenges. The materials are exceedingly brittle, and the strong magnetic fields that they generate can interfere with superconduction in the material. Nevertheless, success would make possible a number of important technological developments. Magnets made from superconductive materials might propel maglev (magnetic levitation) trains, focus beams of particles in a supremely powerful particle accelerator, or hold hot nuclei in a peaceful fusion reactor. Another application would be a device that could store vast amounts of electrical power. Such an innovation could bring renewable energy sources closer to practicality and enhance the efficiency of conventional generation. Exquisitely sensitive detectors of electrical and magnetic fields based on the new superconductors are already in medical use.

SUPERFUND

"Superfund," or the Hazardous Substance Response Trust Fund, first established in 1980, provided $1.6 billion for the immediate cleanup of toxic waste contamination in emergencies. The fund was part of the Comprehensive Environmental Response, Compensation, and Liability Act (CERCLA), which is commonly referred to as the Superfund Act. CERCLA enables the federal government to respond directly to a broad range of environmental dangers and disasters regarding the release of hazardous substances into the environment.

Most of the waste sites containing tons of buried hazardous waste were covered over and abandoned in unlined pits, such as the site at Love Canal. In many instances, the buried chemicals leached out of failed or corroding containers, contaminating groundwater. To prevent or remediate these occurrences, the U.S. Environmental Protection Agency (EPA) is required to identify, supervise, regulate, and provide funds for the cleanup of toxic waste sites, including municipal and industrial landfills, mining waste sites, and federal facilities. Under CERCLA, the EPA has established procedures for conducting site-specific risk assessments to determine whether the hazardous waste poses a threat to human health. These evaluations determine the site's priority and whether a contaminated site requires remedial action, such as removal of the waste for disposal elsewhere, or the closure of the site with an impermeable cap of clay. EPA also addresses liabilities related to chemical pollution and contamination.

The Superfund is supplied by taxes on designated hazardous chemicals and by public funds. However, all money spent from the fund is to be repaid to the federal government by the parties responsible for the hazardous waste release, thereby stimulating the voluntary cleanup by the producer, carrier, or storage companies involved before the sites become subject to EPA enforcement under Superfund.

In 1986, amendments to CERCLA gave people the right to know more about chemicals used by industries within their community. Industries must list the types and quantities of chemicals they release to the air, land, and water. In addition, the data must describe the toxicity, potential human health hazards, first aid, and disposal information for any of the chemicals used at the site that are on the EPA's list of toxic chemicals.

All manufacturing plants that import, process, or produce more than 50,000 pounds per year of any of the EPA's listed chemicals and compounds have to submit completed forms regarding their disposal of these chemicals. The information is coordinated into an annual report, the Toxics Release Inventory (TRI), published by the EPA.

The amendments in 1986 increased available funds for cleanup to $9.6 billion, but more than $100 billion may be needed. The cost for cleaning one site often runs between $21 million and $30 million, according to Clean Sites, Inc., in Alexandria, Virginia.

The EPA is overwhelmed by the responsibility of identifying sites, testing chemicals, and classifying and regulating them. Today, there are over 100 sites on Priorities List that have been cleaned, with nearly 2,000 sites remaining. In an attempt to speed the process, private firms have begun to join the EPA in tasks associated with the cleanup.

SUPPLY

The supply of a product refers to the total quantity that will be brought to market at a specific price. For example, if bakers were offered $0.01 for each standard one-pound loaf of white bread, it is unlikely that any would be supplied, since the cost of the ingredients is greater than the price. If, however, bakers were offered $5 for each loaf, an almost unlimited supply would be offered, since the price exceeds the cost of ingredients, labor, energy, and a reasonable return to the investment in mixers and ovens. For any product at a given time, the higher the price, the greater the supply that will be offered.

An additional consideration is the quality of the product. For example, small loaves would be offered at a lower price than larger ones. In addition, the price must be expected to prevail for some time. If a much higher price were expected tomorrow, producers would hold bread off the market today in order to sell it later at the higher price. Unless a price that covered full cost were expected to last for many years, no one would invest in a bakery.

A competitive market leads to increased efficiency of producers of a standard product and may force less efficient producers out of business. Additionally, a competitive market may lead to increased environmental pollution. For example, if the least expensive method of producing bread involves polluting the environment, and if there is no restriction to this method, bakers would have to choose between polluting the environment or increasing their costs.

Alfred Marshall (1890) and A. C. Pigou (1920) observed that there are unpaid "social costs" from producing goods and services. Unless society finds a way to "internalize" these unpaid costs, industrialized societies will continue to despoil the environment. Little was done to abate environmental discharges until after World War II. The air pollution episodes in London and Donora, the smog problems of Los Angeles and fires on the Cuyahoga River have slowly convinced many people that the unpaid costs need to be taken into account. While the resulting supply prices are greater than when the environmental costs went unpaid, the economic system is more productive and responds better to the demands of individuals as both citizens and consumers.

BRADLEY W. BATEMAN

SURFACE MINING CONTROL AND RECLAMATION ACT

The Surface Mining Control and Reclamation Act (SMCRA), passed in 1977, required that companies reclaim, or restore, landscapes that were left barren by strip-mining. This method was capable of extracting 40% more coal than traditional underground mining. Yet, reclamation was the only way to restore the land to a productive condition.

Prior to 1977 the issue of reclamation was regulated by state governments. The problem inherent in state control was that since reclamation was expensive, states could compete to attract industry by lowering their reclamation standards. In fact, for many years it was cheaper for Europe to import American coal, including the transportation costs, since the European reclamation laws made their own coal expensive. The federal act sought to abolish the economic incentive to leave surface-mined land unreclaimed by mandating a nationwide uniform reclamation standard. The act established federal standards and set up the Office of Surface Mining in the Department of the Interior to publish regulations. Each state creates its own laws conforming with SMCRA. Enforcement of the federal law on a state level allowed for each state to work with its unique conditions while assuring that federal standards were met. The act called for surface miners to meet environmental performance standards in the removal, storage, and redistribution of topsoil; erosion control; and water quality. SMCRA required that the land be restored as much as possible to its original condition and to a use equal to or better than its original use. Prime farmland, steep slopes, and alluvial valley floors in arid and semiarid areas were given special mining and reclamation standards. The act also established a fund, derived from a tax assessed on surface mining, to reclaim abandoned mine sites.

Major amendments were added to SMCRA in 1978, 1980, and 1990. In 1978 SMCRA was amended to create thirteen coal research laboratories at universities in states with abundant coal reserves. In 1980, with James Watt as the Secretary of the Interior, the act was amended to weaken the authority of the Office of Surface Mining by repealing the stipulation that made the office an independent federal regulatory agency. Under the Reagan Administration the office lost federal funding and technical support while the states were given broad regulatory responsibility. The 1990 amendments to SMCRA addressed a major issue by establishing the Abandoned Mine Reclamation Act.

SUSTAINABLE DEVELOPMENT

The concept of sustainable development encapsulates the idea of economic development with due care for the environment. It has received wide currency since the World Commission on Environment and Development (the Brundtland Commission, so called after Prime Minister Gro Harlem Brundtland of Norway, who chaired it) commended it in 1987 as the way to achieve global environmental security. The term itself had been used before the Brundtland Commission endorsed it, notably in the *World Conservation Strategy*, jointly prepared by the World Conservation Union (IUCN), the United Nations Environment Programme (UNEP) and the World Wide Fund for Nature (WWF) in 1980. The *Strategy* was put forward to help achieve sustainable development through conservation, which it defined as "the management of human use of the biosphere so that it may yield the greatest sustainable benefit to present generations while maintaining its potential to meet the needs and aspirations of future generations."

The Brundtland Commission, which presented a detailed case for sustainable development in its report *Our Common Future*, was similarly guided by considerations of intergenerational equity. It described sustainable development as "development that meets the needs of the present without compromising the ability of future generations to meet their own needs."

Several individuals and organizations have offered alternative definitions of sustainable development from the perspective of their disciplines or interests, generally with a view to being more specific. The World Conservation Union and its two partner organizations, in *Caring for the Earth*, their 1991 successor to the *Strategy*, provided this definition: "improving the quality of human life while living within the carrying capacity of supporting ecosystems." While the task of making the definition more precise has attracted interest, the concept itself has been widely endorsed by the United Nations and international agencies like the World Bank, as well as by governments. The Group of Seven leading industrial countries, for instance, in the economic statement issued at their July 1989 summit, said: "In order to achieve sustainable development, we shall ensure the compatiblity of economic growth and development with the protection of the environment." Sustainable development has begun to displace the unqualified maximization of economic growth as the professed aim of economic policy.

The Brundtland Commission's plea for sustainable development reflected mounting concern that human activities, which had increasingly upset the balance between man and nature since the Industrial Revolution, were threatening to deprive the ecosystem of its capacity to sustain human life. The environment was suffering as a result of the extent and character of industrial activity and resource consumption in rich countries as well as of action taken by people trying to eke out an existence in poor countries. Since the Brundtland Commission published its analysis and recommendations, graver dangers arising from ecological disruption caused by human intervention have been revealed. Scientists from many parts of the world warn of potentially calamitous consequences of global warming and a consequent rise in sea levels. Other scientists have reported a sharp deterioration in the stratospheric layer of ozone that shields the Earth's inhabitants from the sun's life-threatening ultraviolet rays. All this further evidence of environmental damage, with very grave consequences for human existence, has massively strengthened the case for making human development conform to the requirements of sustainability.

In popular usage, the term *development* denotes economic and social progress in developing countries rather than in industrial countries. But the prescription of sustainable development is intended to apply to economic activity in all countries without distinction, and both to individual countries and to the world as a whole. The most troubling environmental problems, such as greenhouse warming and the depletion of the ozone layer, have global consequences and have underlined another dimension of global interdependence. Economic activity worldwide must have regard for its impact on the supporting ecosystem. The Brundtland Commission emphasized this when it said that environment and development had to be integrated in all countries, and that the pursuit of sustainable development required changes in the domestic and international policies of all nations.

The advocacy of sustainable development has marked a turning point in the approach to ecological conservation. In the 1960s and 1970s, concern was mainly with preserving endangered species and safeguarding natural amenities, generally from human depredation. It did not pay much attention to the interests of human beings, who were cast in the role of environmental wrong-doers. The new approach, while continuing to press for the conservation of nature, embraces human welfare as a central objective — recognizing that both the quality of development and the neglect of it can result in severe detriment to the natural environment.

Developing Countries

The concept of sustainable development as articulated by the Brundtland Commission gives priority to the satisfaction of people's essential needs — food, clothing, shelter, jobs — particularly for the world's poor. Poverty, besides being an evil in itself, is recognized as a driving force in environmental degradation of many kinds. Such problems as desertification and deforestation are clearly linked to the excessive pressure placed on natural resources by people struggling to survive in conditions of deep poverty. A world in which poverty is endemic remains prone to ecological crisis. In the developing world, sustainable development calls for intensified efforts to speed economic growth oriented to ending poverty. These efforts must be informed by concern to conserve resources such as land, water and forests; to avoid activities that degrade the environment; and to minimize pollution. The prospects of environmental care will be enhanced to the extent that people have a stake in the resources that must sustain them. Wider land ownership and wider involvement of people in planning development projects and in managing resources can help to make development sustainable. Targeting development on the poor and giving them wider options are more likely within a democratic political framework.

An important implication of sustainable development is that world population growth should be braked so that a stable population is reached as quickly as possible. Present demographic trends threaten to lead to a world with many more people than the global ecosystem can support. The number of people added each year to the world total is still rising. The 1990s will add nearly one billion, the highest ever increase within one decade and about as many people as there were in the world in 1800. Population experts at the United Nations expect the world's population to pass 10 billion as early as 2050 and climb for another hundred years to level off around 11.6 billion. Measures to slow population growth are therefore vital to achieving sustainable development.

Population growth is now taking place very largely in developing countries. Of the billion people joining the human family this decade, 95% will be in the developing world. While people in these countries consume far less per head than those in rich countries, growth in their numbers intensifies the pressure on the ecosystem: on land, water, vegetation, and fuelwood. As their incomes rise, consumption-driven demand for other resources, notably commercial energy, will also expand. The historical experience is that families become smaller as countries' living standards improve; in North America and Europe, the average family has fewer children now than earlier in the century. Recent evidence suggests that human or social development — particularly spreading literacy, expanding education, and giving women a better position in society — has a critical role in influencing family size and population growth. Educated women with greater autonomy are more likely to want to—and be able to—determine how many children they have.

Industrial Countries

While in poor countries the lack of development is responsible for the most serious threats to the environment and environmental degradation in turn adds to impoverishment, the nature and level of economic activity in industrial countries—with their high per capita levels of consumption, resource use, and waste—exert a far greater adverse impact on the environment. It is in the common interest of all people that rich countries should themselves follow strategies of sustainable development. The principal implication is that they should modify their economic activities to make them far less burdensome on the biosphere, both in regard to the use of natural resources that cannot be renewed and to the demands made on natural "sinks," whose absorptive capacity—like that of the oceans for carbon dioxide—can be exhausted. These countries must become less intensive in the use of resources, especially of energy from fossil fuels. They must do so to bring down their pressure per head on the environment within prudent limits, while leaving room for developing countries to raise their consumption to reasonable levels without taking the total global impact beyond thresholds of safety.

The developed countries have roughly one quarter of the world's population, but account for the bulk of world consumption, including over three quarters of consumption of many key resources: 80% of commercial energy and of iron and steel, 86% of other metals, 85% of paper. Such disproportionate use cannot continue indefinitely. Nor is it possible for poor nations to lift their consumption—with due care for the environment—if rich countries continue to expand theirs or even maintain their present excessive offtake of materials and energy. If there is to be sustainable development on a global basis, countries that now use more than their fair share of resources or of "sink" capacity should release environmental space for other countries to occupy.

This is well illustrated by the use of fossil fuels. What is critical is not so much their exhaustibility (though that is a concern) but the damage their use causes through the greenhouse effect. The scientific view is that emissions of warming gases, of which carbon dioxide (produced when fossil fuels are burned) is the main one, need to be cut by over 60% just to stabilize their concentration in the atmosphere. At the same time, developing countries need to step up their use of energy if their economies are to expand so that their people may be freed from poverty. While per capita energy consumption in 1989 in the U.S. was 295 gigajoules (the international measure of energy use), it was 23 gigajoules in Brazil and China, 9 in India, and as low as 1 in a dozen African countries such as Ethiopia, Mali, and Tanzania. If all the people in the world are to be able to use 75 gigajoules of energy per head (one-fourth the present U.S. average) and world population doubles to around 11 billion, world energy consumption could be held to three times the present total—but only if industrial countries brought their use down to 75 gigajoules per head. For poor countries to make economic progress, a clear prerequisite is that affluent ones must drastically cut their use of commercial energy in a variety of ways—including, of course, increased energy efficiency and greater use of renewable sources of energy.

Sustainable development therefore requires industrial countries to accelerate efforts to become more efficient in the use of resources and to develop greener technologies. Reducing energy intensity (the amount of energy used to produce one unit of Gross Domestic Product) is of paramount importance. Some progress in improving energy efficiency was made after world oil prices were lifted in the early 1970s, but not in all countries and not after the mid-1980s when the price of oil fell back in real terms. Japan, which imports all its oil, has shown how much can be done in raising energy efficiency; by 1988 it had lowered energy intensity to 0.27 against 0.44 for the U.S., which remains a profligate consumer of energy, aided by low prices.

In the transition to sustainable patterns of economic activity, with reduced dependence on fossil fuels, the transport sector offers scope for action in several directions. These include pricing energy to reflect the environmental cost of fossil fuel use; expanding public transport to reduce the use of private automobiles; incentives to favor the use of vehicles that run more miles per gallon; and intensified research to lower the costs of solar, wind, and other renewable sources of power. Technological innovation is required; as important are public policies to influence consumer behavior and to ensure that technological advances, such as thriftier automobile engines, are widely used.

In an economy that is developing sustainably, reuse and recycling will be extensive. Besides savings on materials such as metals, glass, and paper, there will be reduced energy as well as less pollution and wastes. While reuse—of glass containers, for instance—could offer substantial gains, recycling is reckoned to be capable of yielding savings of 25%–65% in energy. Producing steel from scrap has been reported to need only about one third the energy used in making it from iron ore and to give rise to 85% less air pollution and 75% less water pollution.

Economic Development

The prospect of sustainable development worldwide will be enhanced if there is international action to improve the global arrangements that bear on economic progress in the Third World. Some aspects of these arrangements tend to impede advances by developing nations and make it difficult for them to develop sustainably, following ecologically sound policies. As producers of primary commodities—farm products and minerals—many of these countries suffer because the terms of trade, which reflect the purchasing power of their export earnings, are unstable and generally move against them. The result is that they have to produce and sell more, simply to maintain their purchasing power at the same level. Their efforts to become less dependent on such products and move into basic manufacturing industries—or even to sell their primary commodities in processed or semi-manufactured form—invariably run up against protectionist barriers in the developed countries where they have to find markets. The terms on which they can obtain technology are often too onerous for them.

The debt problem, though past its peak of severity, still burdens many countries; it obliges them to devote too high a share of their export earnings to interest charges and capital repayments, leaving them less for investments in productive or human capital. The global flow of resources, which is normally from rich to poor countries, has for several years taken money out of poor countries. These factors tend to force developing countries to overexploit resources. Preoccupied with efforts to meet the immediate pressure on their external payments, they cannot plan long-term so they neglect sustainability. Most developing countries are unlikely to be able to stick to policies of sustainable development, however much they may wish to do so, unless external conditions are made less obstructive. Their failure, besides having harmful repercussions on their own poeple, could have a wider adverse impact if it involves them, as is only likely, in activities that denude forests, destroy species, or deplete the ozone layer.

To what extent can the world economy expand within the limits of global sustainability? The answer will depend mainly on technological progress. The Brundtland Commission was encouraged by the fact that in the recent past while economic growth had continued, the consumption of raw materials had held steady or declined. It envisaged a five-to-tenfold increase in manufacturing output through a qualitatively different pattern of growth—"producing more with less"—that would raise consumption in poor countries to the level in industrial countries by the time world population

stabilizes. Emerging technologies offered the promise of "high productivity, increased efficiency and decreased pollution" (though stricter controls would be necessary over hazardous products and wastes).

Some environmental economists, while supporting the central Brundtland thrust, have questioned whether the ecosystem can stand expansion of such an order in total output. They think that improving living standards in developing nations will require growth in output to be held back in industrial countries—if the total volume of economic activity is to stay within the limits of sustainability. One way or another, we cannot simply continue on a "business as usual" basis. Survival requires that the human species develops in a sustainable way.

SIR SHRIDATH RAMPHAL

For Further Reading: Lester Brown, Christopher Flavin, and Sandra Postel, *Saving the Planet* (1991); Shridath Ramphal, *Our Country, the Planet* (1992); World Conservation Union, U.N. Environment Programme and World Wide Fund for Nature, *Caring for the Earth* (1991).
See also Greenhouse Effect; Ozone Layer; Population.

SUSTAINABLE YIELD

Sustainable yield is the rate at which humans can harvest economically useful plants, minerals, or other natural resources, while maintaining an ecological balance between resource demand and depletion. Achieving a sustainable rate of resource consumption implies practicing environmentally sound methods of resource extraction, in order to preserve the natural base of material wealth, while equitably meeting the needs of society. Sustainable yield of a product is analogous to overall sustainable development, in that population demands are tempered by the physical limitations of the natural world, and proper stewardship of the land is therefore vital.

Sustainable yield is most often associated with agricultural systems. Since agriculture is by definition an extractive process, whereby soil nutrients are harvested along with crops, intensive management and knowledge of the agroecosystem are required to maintain the integrity of the system. Studies have shown that, for every 7,000 kg of corn harvested, approximately 104 kg of nitrogen, 19 kg of phosphorus, and 22 kg of potassium are removed from the soil. Farmers must therefore address the need to return nutrients to the soil and allow periods of rotation and fallow, so that the land may recover from intensive use and be utilized again. In addition to maintaining soil nutrients, sustainable yields are also based on the judicious expenditure of nonrenewable resources, such as fossil fuel, in the production process. The relative efficiency of human, animal, and petroleum-based energy used to till soil, plant, cultivate, and harvest food crops varies according to the type of crop grown and the amount of calories, or food energy, produced.

An example of a sustainable agricultural system is swidden, or slash-and-burn agriculture. Swidden is a centuries-old farming practice used in the humid tropical regions throughout the world, where large expanses of forest land are systematically burned and cultivated over a period of years. The spatial requirement is dictated by the size of the household and its need to maintain food self-sufficiency, and the relative time period that fragile and poor-quality soil requires to recover from even short-term use. Burning releases scarce nutrients from the biomass to the soils, allowing farmers a brief window of time in which to plant and harvest food and medicinal crops. Afterwards, the area is abandoned and a new one cultivated, until the farmers eventually return to the first area, by which time the soils have recovered, and cultivation may continue. Only when the area of cultivation becomes reduced, due to encroachment by cities and industries in tropical forests, does the slash-and-burn system become unsustainable; this is due to the curtailment of the farmer's land-base, and the reduction of fallow time.

Across the U.S. a number of institutions conduct research on a variety of sustainable agricultural issues, including minimum tillage, organic or low-input systems, fossil-fuel efficiency, and agroforestry. Over the last 50 years U.S. organizations such as The Land Institute, Ecology Action, Rodale Institute, and the Institute for Alternative Agriculture, have introduced sustainable agricultural practices into the mainstream of research. The validity of this work is increasingly acknowledged at the academic and even legislative levels.

For example, the 1960 Multiple Use and Sustained Yield Act formalized, on a national scale, the planned use and management of a historically abused natural resource, the national forests. The act addresses the perennial need to provide a reliable supply of timber to American society, while guaranteeing adequate protection of watersheds and stream flow, and defines "sustainable yield" as "the achievement and maintenance in perpetuity of a high-level annual or regular periodic output of the various renewable resources of the national forests without the impairment of the productivity of the land." Another important achievement for sustainable agriculture was noted in the passage of the 1990 Farm Bill, which established federal and state criteria for organic certification programs, as well as stan-

dards for soil management and fertility, chemical inputs, crop and livestock production, domestic and imported food marketing and distribution, as well as enforcement powers.

To achieve the sustainability of a system, beginning with crop yield, one must recognize the biological dependence of humans on nature, and design short- and long-term usage of their respective ecosystems. As worldwide consumption of natural and often non-renewable resources—particularly energy—becomes more intensive with each generation, the scarcity of these resources threatens to limit the course of modern, highly extractive agricultural practices. In order to change this course, sustainable methods of cultivation may help to reduce our dependency on nonrenewable resources, and help to preserve the environment for future generations.

SYMBIOSIS

Symbiosis is an association of two different kinds of organisms in which each is beneficial to the other and indeed may be esssential to the life of the other. Mutual cooperation is widespread and vital in nature. In the "survival of the fittest," the fittest may be the one that most helps another to survive. As always in nature, there is a continuum from relatively loose associations in which both partners can live apart, through others in which only one partner can live separately, to associations obligate for both. The last are of special interest and far more common and of far greater ecological and economic significance than commonly supposed.

One of the most important is the symbiosis between leguminous plants (such as clover, alfalfa, peas, and beans) and rhizobia, the nitrogen-fixing root nodule bacteria. This association, whereby atmospheric nitrogen is incorporated into compounds useful for plant growth, is responsible for the addition of more nitrogen to agricultural soil than all other fertilizing practices combined. Each year about 100 million tons of atmospheric nitrogen are incorporated into the Earth's surface; 90% of this is accomplished by biological processes, including free-living soil bacteria and, especially, the legume-rhizobia association. Neither the clover plant itself nor the rhizobia alone can utilize nitrogen from the air. The plants will grow, though poorly, without the bacteria if supplied with combined nitrogen as ammonium salts. The rhizobia exist in the soil and can be grown in culture, but do not fix nitrogen. When rhizobia of appropriate strain encounter the root hairs of an appropriate species of leguminous plant, a specific and complex interaction occurs whereby the plant becomes infected. The infection stimulates a response by the plant in which special cells divide and form nodules on the roots. The cells of these nodules are colonized by the bacteria. Each fully formed nodule has outer zones in which the bacteria are rod-shaped, and a large inner zone where the rhizobia are of such changed appearance that they are called bacteroids. This zone has a pink color from the presence of plant hemoglobin. It is in this zone that nitrogen fixation occurs. The hemoglobin formed by the plant as a response to the infection, and enzymes formed by the bacteroids, together carry out the chemical reactions for the combining of gaseous nitrogen into ammonium compounds. Only recently have the details of these reactions been fully understood. The formation of nodules by the host plant is in some ways analogous to a pathological response, and indeed rhizobia are closely related to the pathogenic parasites of plants that cause crown gall, a tumorous growth. The difference is that the rhizobial nodules do not grow in an uncontrolled way but differentiate into special organs. Extensive studies have been done of the genes of both rhizobia and legumes responsible for compatibility between strains of each and for successful nitrogen fixation.

Many species of insects are able to utilize highly specialized diets only because they contain symbiotic microorganisms. This is true of aphids and other plant-sucking bugs and of insects whose food is restricted to the blood of vertebrates throughout their life cycle. The human body louse, for example, has a special organelle, the mycetome, that harbors within its cells a species of large bacteria. These also infect the ovaries of female lice and in this way are transmitted to the next generation. Lice deprived of these bacteria by experimental manipulation cannot survive on a blood diet. Cockroaches are insects renowned for their ability to feed on almost anything, including material having apparently little nutritional value. This facility depends on their having endosymbiotic bacteria restricted to specialized cells found in the fat body. These bacteria also infect the ovaries of the female roaches and in this way the eggs, thereby carrying on the infection to the next generation. There is much evidence to show that for the roaches as for the lice, the symbiotic bacteria supply essential growth factors, as vitamins.

Bacteria play somewhat similar roles in the nutrition of higher animals including people. The importance of the normal intestinal bacterial flora becomes strikingly apparent in the effects of partial sterilization of the alimentary tract produced by certain wide-spectrum antibiotics. These remove some of the normal bacteria, thereby permitting extensive growth of various pathogenic microorganisms, with resulting intestinal upset. Bacteria of the alimentary tract supply adequate amounts of certain essential vitamins such as

biotin that are not produced by the body in their absence. Unlike with insects, transmission of these bacteria is not insured through the egg. Yet it is well established that newborn infants rapidly acquire a characteristic intestinal bacterial flora by contact with other people.

In ruminants, microorganisms of the alimentary tract play an even more important role; they are responsible for the digestion of cellulose. The rumen serves as a fermentation chamber in which anaerobic cellulose-fermenting bacteria act on the chewed cud to produce chiefly short-chain fatty acids. These, absorbed directly through the lining of the rumen, are the main food of the host. This remarkable symbiosis enables ruminants to utilize up to 70% of the fiber (mainly cellulose) of their food plants. Symbiotic cellulose digestion is also found among many species of insects that depend on wood as their principal food. Most famous is the symbiosis between termites and their intestinal protists, first demonstrated by L. R. Cleveland in the 1930s. Wood-feeding termites, and a few species of wood-feeding roaches, have an enlarged region of the hindgut that is packed solid with an assemblage of diverse, highly specialized protists found nowhere else. Many of these ingest particles of the wood chewed and swallowed by the host insect, and digest it to release simpler compounds (probably again fatty acids) that provide food for the host. The termite itself has no enzymes able to break down cellulose.

The utilization of cellulose by the leaf-cutting ants of the genus *Atta* (common in the New World tropics) illustrates another kind of symbiotic association, which involves behavioral adaptations. These ants use comminuted leaves as a substrate on which they rear "fungus gardens." The species of mold forms whitish round bodies, about 0.3 millimeter in diameter, that are the principal food of the ant colony. When a virgin female leaves the colony on her nuptial flight, she carries with her a pellet of hyphae from the fungus garden; when she founds a new colony, she uses this to initiate its fungus garden.

"Cleaning symbioses" are behavioral adaptations common in the oceans. Large fish assemble at certain stations where much smaller fish and shrimps go over them, cleaning away debris and ectoparasites, not only from the outer skin but also inside the gill chambers and even the mouth. Presumably signals keep the large fish from eating the small cleaners.

In the symbiosis between some invertebrate animals and photosynthetic algae, also common among aquatic organisms, the algae supply nutrients to their host and also receive nutrients from their host. There is now much evidence that the chloroplasts of all green plants are derived from a symbiotic association between a photosynthetic prokaryotic organism resembling cyanobacteria and a eukaryotic cell. There is also good evidence supporting the hypothesis that mitochondria, the energy-forming organelles of most modern eukaryotic cells, similarly are derived from symbiosis between a species of aerobic bacteria and a nucleated cell. Such a cell has two genomes, that in its nucleus and that in its mitochondria. The cells of green plants have three genomes, one in the nucleus, one in the mitochondria, and a third in the chloroplasts. Since some proteins of both mitochondria and chloroplasts are encoded by nuclear genes, it is assumed that these genes were originally in the genome of the symbiotic prokaryote and then moved into the nuclear genome of the host cell. There is strong evidence that such movement of DNA sequences from chloroplasts or mitochondria to the nucleus does occur. Hence, symbiosis has played and continues to play a major role in evolution.

WILLIAM TRAGER

For Further Reading: Vernon Ahmadjian and Surindar Paracer, *Symbiosis. An Introduction to Biological Associations* (1986); Paul Buchner, *Endosymbioses of Animals with Plant Microorganisms* (1965); William Trager, *Symbiosis* (1970).
See also Bacteria; Evolution; Genetics; Insects and Related Arthropods; Nitrogen Cycle.

SYNERGISM

Synergism (from the Greek *sunergos,* "working together") has a variety of meanings — chemical, biological, and cultural — that are all related to the outcome of cooperative interaction. In theology it is the doctrine that individual salvation is achieved through a combination of human will and divine grace. In business it is cooperation among subsidiaries or merged parts of a corporation that creates an enhanced combined effect. In physiology it is the harmonious functioning of pairs or groups of muscles that produce coordinated movements or hormones that cooperate, for example, in promoting growth. In pharmacology it is the combined effects of drugs when they reinforce each other in living organisms. In all cases synergism is the interaction of two or more agents or forces so that their combined effect is greater than the sum of their individual effects. Unfortunately, because of synergistic effects in environmental systems, cause-and-effect relationships are difficult to analyze.

At concentrations urban dwellers are usually exposed to, sulfur dioxide (SO_2) alone would not produce the respiratory problems that are known to exist. But SO_2 in combination with suspended particles does contribute to increased bronchitis and emphysema. One explanation is that the deeper penetration and accelerated toxic transformations of the gas pollutant are the result of its attachment to the polluting particles involved. Combinations of herbicides and insecticides can also produce synergistic effects that are complicated by the influence of other factors like soil type. And some synthetic insecticides in combination can produce high toxicity to mammals, for example, when individually the dose-response relationship is not lethal.

Another synergistic effect is that of the insecticide DDT when mixed with oil spills in sea water. DDT is extremely soluble in oil but not in water. As a result marine organisms not usually affected by the lesser concentrations of DDT in the ocean are exposed to much more harmful amounts in oil spills concentrated on the surface. Toxic effects may be additive, enhanced (more than additive) or synergistic (more than enhanced). In some cases toxicity may be decreased instead of increased by the interaction of combinations of chemicals. This opposing effect is called antagonism. Like synergism, antagonism has different but related meanings. These include active resistance (cultural), opposition (biological), or interference (chemical).

Synergism and antagonism of agents producing acute effects are well established; the story is not so clear for chronic effects. Chronic interactions require much difficult testing to produce answers commensurate with the expense. Some cases of synergism, for example, the inhalation of asbestos fibers increasing the chances of lung cancer for cigarette smokers, are relatively easy to recognize. Other synergistic effects continue to be discovered and complicate the process of understanding and analyzing environmental impacts.

T

TAIGA

See Forests, Coniferous.

TECHNOLOGICAL FIX

A technological fix, or countertechnology, is a simplistic technological solution to a complex human problem. Such short-range technological solutions usually create new environmental problems. Superhighways and mammoth underground or multistoried garages constitute almost caricatures of countertechnologies: although they are designed to facilitate traffic they encourage the proliferation of automobiles, thereby increasing congestion and pollution in cities. The association between monocultures and the use of pesticides also illustrates the kinds of problems commonly created by countertechnologies. Monocultures (of almost any crop) make possible the use of farming practices that increase yields and decrease production costs; they also encourage the development of plant pests and therefore require the use of more and more pesticides. Beyond a certain point pesticides become a serious threat to human health. This in turn promotes pest-control measures that become increasingly annoying and costly as the monoculture technology becomes more widespread.

Many medical problems characteristic of our time can similarly be traced to the unpredicted consequences of medical technologies and countertechnologies. A simple example is the concatenation linking cortisone to metabolic and cellular changes that increase susceptibility to infectious disease. Antibacterial drugs may control the bacterial infections evoked into activity by cortisone, but they also open the way for invasion by yeasts and fungi that in turn demand the use of still other drugs. Kidney dialysis, organ transplantation, and the various forms of prostheses are creating their own medical, social, and ethical problems.

Technological fixes are of course needed to alleviate critical situations, but generally they have only temporary usefulness. More lasting solutions must be based on ecological knowledge of the physiochemical and biological factors that maintain the human organism in a viable relationship with the environment. In most cases such knowledge is not available, because certain areas of science have not yet been sufficiently developed or have been neglected altogether. Nature conservation and human adaptation constitute important examples of fields that deserve far greater emphasis from the scientific community.

RENÉ DUBOS
(EDITED BY WILLIAM B. EBLEN)

TECHNOLOGY

Technology is perhaps the most powerful agent impacting the environment. Virtually all human activities are realized through a technological medium or agent. Technology has transformed individual lives by reducing the effort needed to obtain the basic necessaries—food, clothing, shelter—and by opening up a world far beyond these necessaries. At the same time, technology lies at the heart of what many call the dilemma or problems of the modern age.

Poverty and other social breakdowns have not disappeared; technology itself may be partly responsible for the large increases in world population in recent decades. Improved medical care and more secure food supplies have reduced death rates and increased the fraction of live births. Environmental degradation is rampant and continues to grow in spite of stringent regulatory structures in the industrialized parts of the world.

Technology has become so pervasive a part of the everyday context of living that many see it as a fundamental human characteristic. Karl Marx called our species *homo faber* (man the maker) rather than *homo sapiens* (man the thinker).

One of the oldest forms of technology was the taming of fire for human purposes. The *intentional* use of fire, a natural phenomenon that exists independent of human activity, transforms it into a technological phenomenon. The Promethean myth sees this act as a profanation by human beings in appropriating what had been heretofore a substance available only to the Gods. The myth continues by claiming that this new technological artifact would enable human beings to hold sway over other animals and tame the Earth to produce food and shelter.

The great ages of human development carry names that are metaphors for the technological developments that occurred during them. The Stone or Neolithic Age produced tools for hunting, domestic activities, or battle—based on using stones as found lying about or, in its higher level of development, on shaping flint into tool-like forms. Early humans could only pick up and physically transform resources that presented themselves. The Iron Age saw the beginnings of a more modern set of activities based on the chemical conversion of natural substances followed by the shaping of the products into useful forms. The Greek sense of "techne" or craft, the word from which technology springs, was the bringing forth of artifacts out of nature, as if the shaping of an earthen pot unloosed something already existing in the world. The modern view has shifted to see humans as challenging and exploiting nature in the process of creating technology. The balance and control has subtly shifted from nature to humans.

The basic notion of the Promethean myth is still with us; technology brings with it a promise to enrich the human species, protect it from the vagaries of the environment, and lead to individual freedom and the realization of other democratic ideals.

Whether this promise has been realized has been pondered by countless thinkers who suggest that the benefits are not without a dark side. Historically much of the questioning about technology was directed to the reality of the basic social, liberating promise; more recently, attention has expanded to a set of concerns over the very sustainability of a world based on ever-increasing technological development. Technology appears to threaten the essential environmental context that the human species needs for survival.

These critical views of technology see it as both arising out of the social fabric and, at the same time, shaping that fabric. The simple instrumental view pictures technology as morally neutral, simply doing human bidding. If the ends that are chosen or the consequences that crop up in use are ultimately deemed to be antagonistic to human or natural ethical or moral principles, this view argues that it is the fault of the humans that created and used the technology, not the technology itself. This argument is implicit in the ongoing debate about guns. Guns are clearly a technological artifact with a potential for much harm outside of the primarily military usage for which the device was created. The opponents of gun control argue that guns do not kill people; people kill people. Such a limited view of technology ignores interrelationships between technology and behavior.

The social deterministic view of technology suggests that technology embodies the prevailing social, cultural, and political values. Lewis Mumford's perceptive analysis of the history of technology and its cultural consequences pointed to the development of the mechanical clock by monks as a manifestation of the importance of order and timeliness in monastic life. Mumford claims that this particular invention led to a new era in technological developments. In a modern setting, one could argue that the emergence and hegemony of the automobile reflects the American cultural preferences for individual freedom and autonomy. One fundamental theme in Karl Marx's theories is that the dominating class interests of a society shape the technological means of production and, in particular, that the capitalist mode of industrial organization produces systems that enslave and alienate the working class.

As the pace of technological change has quickened and spread to more distant reaches of the globe, the social determinist view seems inadequate. Another perspective, called technological determinism or autonomous technology, attributes a force or life to technology. Technology, in this mode, takes over the human rational process of creating and selecting among means serving specific ends and becomes the primary determinant of individual and social human development. In a mirror-like stance to social determinism, autonomous technology sees the machine as concentrating power in a technocratic elite who possess the economic resources and technical skills to drive the machine towards its inexorable destination. Even warfare, which has contributed historically to critical technological developments such as the transistor and the modern electronics revolution, has become a technological game played out through high technology. The radar screen and smart bomb convert human beings and nature to dim objects.

Beyond the device-like character of technology inherent in each of the above viewpoints, technological determinists go further to claim that technology now shapes the very way that humans think and experience the phenomena of the world. The omnipresence of technological artifacts and our increasing reliance on using them to satisfy all of our needs has created a deep-seated way of seeing the world itself as a technological object. The world-shaping nature of technology has been the subject of attention from the Greek era to the present. In his 1964 book *One-dimensional Man*, Herbert Marcuse argues that our technological way of thinking has reduced the world to a set of artifacts. In the process, societies have become inhumane and authoritarian. Individuals have lost the ability to fully experience life itself. The more we become reliant on larger and larger technological systems, the more we become dependent on social structures that demand elite, technocratic, and fundamentally dominating institutional forms. The world appears always merely as "standing reserve" waiting only to satisfy some material desire.

Opposed to these critical notions is the dominant view in the industrialized nations of the world that the promise of technological progress continues to unfold and, tightly coupled to it, the positive development of the species towards liberation. The same technologies that have the potential to shape and control also can and do enhance personal freedom and escape from drudgery. The awesome power of technologies to bring harm can be kept in tow as in the case of the nuclear bomb ever since its original use in 1945. Modern production technologies have provided affordable and plentiful goods available to virtually everyone in the industrialized and market-driven world.

However, the concerns about the potential mischief that technology could wreak on humans have expanded to include dangers to the natural world. In the very process of serving its human masters or in controlling humankind (depending on which side of the argument one takes), technology has a great potential to impact the environment in a deleterious way. Use of any technology almost inevitably produces side-effects not directed at the main task itself. Cars not only emit air-polluting chemicals, but their manufacture has caused depletion of resources and other pollution of the environment. Some sort of technological object always seems to be the cause of any of the environmental troubles we attribute to our actions.

Society's first response to the discovery that productive and consumptive practices were adversely polluting the environment has been, generally, to call on technology to solve problems that other technologies were creating. The history of environmental management practices during the last two decades or so is almost completely based on this set of means.

Air pollution control systems have extremely high efficiencies, removing up to 99.9999% of the particulate matter from stack gases, but the collected dust now must be put somewhere on the land or in the sea. Air pollution has been converted to a set of land or sea problems to solve. Technological solutions seem almost inevitably to cause or reveal a new set of environmental problems at the same time they work to mitigate or solve the old constellation of concerns. Thus there is a circularity in the role of technology in protecting the environment. As long as technology is seen to be such a solution, the environment is likely to continue to suffer.

More recently, environmental policy is beginning to shift to a different paradigm, that of waste minimization and pollution prevention. The emphasis is on technologies that are fundamentally non-polluting and conservative with respect to natural resource consumption. Clean technologies are being developed and installed. Instead of dumping large quantities of water polluted with copper and other heavy metals from plating shops, new production technologies have almost entirely eliminated the effluent, recovering and recycling the former wastes in the process. Chlorofluorocarbons (CFCs), once thought to be necessary to the manufacture of all sorts of electronic goods, have been eliminated by using other less-polluting solvents and in some cases by altering the fabrication technology so that cleaning is not required at all. Although clean technology cannot completely eliminate pollution and consumption of natural resources, many unsustainable past practices can be managed satisfactorily. At minimum, the old practice of moving society's wastes from one medium to another can be greatly reduced over time.

It is clear that progress, as measured by any set of criteria, requires a continuing critical examination of technology. It is not a matter of which of the many views of technology is the right one, but of being ever aware that technology always has several sides, one of which is opening and freeing, the other closing and diminishing. By being careful of the ways in which we design our products, processes, and large technological systems, human beings should be able to continue on the progressive trajectory that many argue has been the history of humankind.

JOHN R. EHRENFELD

For Further Reading: Albert Borgmann, *Technology and the Character of Everyday Life* (1984); Lewis Mumford, *Technics and Civilization* (1963); Langdon Winner, *Autonomous Technology* (1977).
See also Automobile Industry; Biotechnology, Environmental; CFCs; Source Reduction; Technological Fix.

TECHNOLOGY AND THE ENVIRONMENT

Humans developed tools and technology to enable them to survive in a natural environment which was not always beneficent. While technology enabled humans to survive—and sometimes to control—the natural environment, the overuse or misuse of technology, especially as industrialization takes place, has sometimes had adverse or unanticipated effects on the natural environment. This has led some people to blame technology for many of the ills which today beset the total environment—posing a threat to mankind's future—but there is no reason to believe that technology, if properly used, cannot undo its previous (and oftentimes unforeseen) damages and, in the process, restore the natural environment and also create a better world for the future of mankind.

Most anthropologists believe that technology, i.e., the use of tools, enabled our ancestors to adjust to and survive great changes in the natural environment, such as floods, earthquakes, volcanic explosions, and climatic changes, which produced changes in lake and ocean levels and altered the terrain and paths of rivers. Humans, they point out, are too weak to cope with nature with only their hands and teeth, but tools serve as extensions of human hands and multipliers of human muscle power. That is why many anthropologists define the human species as a tool-making and tool-using (or tool dependant) animal. They call the species *Homo sapiens* (Man the Thinker), but the species might also be described as *Homo faber* (Man the Maker).

As a result of their use of hand and mind to cope with nature and to shape it when necessary so that they could survive, humans and the natural world form part of a dynamic ecological system. According to its Greek root, "ecology" is the study of "houses"; so Xenophon named his study of the management of households, *Oeconomicus* (giving us the word "economics"). While some ecologists limit their scientific research to the natural environment, the fact is that ecology, in accordance with its original meaning, must also include the social environment, the "household of mankind." For humans, with the aid of technology, have created their own physical and social environment, which has thus become part of the larger ecological system.

How did the use of technology enable *Homo sapiens* to survive and then to develop complex civilizations? One of humanity's first great steps in controlling the natural environment was the conquest of fire. Every technical advance requires other techno-cultural developments to make its use effective. At first, food was cooked by holding it over the fire or putting it in the fire. But the range of cooking and nutritional possibilities was enormously increased when suitable containers—pots and pans—could be made, making it possible to boil, stew, fry, and bake foods.

At the same time, fire also made possible the use of metals—as mankind advanced from the Stone Age into the Bronze Age and began using metals for tools, weapons, and ornamentation. But long before using metals, humans had already initiated a whole new technology which changed their lifestyles and gave birth to civilized existence: agriculture. Agriculture meant that humans learned how to cooperate with nature to obtain food—and that required new and special tools, such as hoes to till the soil, sickles to reap the grain, flails to thresh it, and querns (hand-turned mills) to grind it.

No matter where it began, agriculture required—and

made possible—new technological means for changes in life and society. Since people had to remain in one place to plant seeds and tend the crops, they could begin to accumulate goods and to build permanent shelters. At the same time, any surplus of foodstuffs could be exchanged for other goods. This became important when metallic tools, instruments, and adornments became available. The Bronze and Iron Ages could not have occurred without the surplus of food grown by the agriculturists, for the metal deposits were found in uplands, not in the fertile valleys where food could be grown. So there was an exchange of foodstuffs for metal tools, and then for textiles, which became necessary when agriculture largely displaced hunting, so that clothing could no longer be made from the pelts of hunted animals. People grew some crops that could provide threads for textiles, and they domesticated animals, such as sheep, whose wool could be made into textiles.

The invention of agriculture helped make possible important socio-cultural innovations: a state system with a social hierarchy, the beginnings of town life, concentration of economic resources, and strong political control, leading to laws and the other accoutrements of government.

The geographical environment and political changes often determined conditions for technological change. In the Roman Empire, for example, the chief energy source was human muscle power, mostly exerted by slaves. But by the 4th century A.D., the Empire had ceased to expand and the supply of new slaves from conquered peoples dried up. The introduction of Christianity also militated against slavery. To assure a continued supply of manpower on the farms and in the craft industries of the times, people were bound to their occupations.

The conquest of the Roman Empire by the "barbarian" Teutonic tribes, while it meant the downfall of Rome, also transformed the Teutonic invaders and extended the practice of agriculture to northern European areas. But the climatic and physical environmental differences in northern Europe brought about changes in technology—and also socio-political institutions—which in turn fostered still other changes in technology. For example, the Teutons brought new types of grain, such as oats and rye, into the diet. Climatic conditions in northern Europe also led to the adoption of a three-field system of agriculture, with one-third of the cropland being sowed in spring, another one-third in fall, and one-third lying fallow each year in order to recover its fertility for the next planting season. Because domesticated farm animals were allowed to graze on the fallow portions, the result was a "balanced" agriculture of grain and stock-raising, which provided an agricultural surplus for trading with those living in towns who were engaged in production of craft goods.

Increased grain production meant that the old system of grinding grain by hand, using stones on querns, was no longer adequate. So a veritable "power revolution" occurred in the Middle Ages. Although waterwheels had been known to the Romans, their use with heavy grindstones became widespread for grinding the increased quantities of grain. Another source of inorganic power for grinding purposes—the wind—was also brought into the service of humans. Although wind had been used to drive sailing vessels since antiquity, its use as a power source in European windmills dates from about the 12th century, when the device was introduced from North Africa. This medieval "power revolution" also enlarged animal power by improvements in the harness. The rigid horse-collar, introduced from Asia during the early Middle Ages, together with horseshoes and the tandem harness, multiplied the effective pulling power of horses some 3 to 4 times. A horse driving a turnstile grinding machine with the more efficient harness was the equivalent of 10 slaves, while a good waterwheel or windmill provided the work of up to 100 slaves.

Other technical innovations advanced productivity in the Middle Ages, such as improvements in textile making through the introduction of the spinning-wheel and new types of looms. Similarly, the introduction of the lathe and the crank increased the ability of machine tools to produce more sophisticated metallic substances for armor—because the advent of the stirrup had changed the nature of military combat by armored knights. But the introduction of gunpowder from China in the 13th century made human armor obsolete. Indeed, when gunpowder began to be used to propel missiles in the 14th century, it revolutionized warfare, changing the balance of political power throughout the world and bringing about a host of other technical changes.

In the meantime, the improvements in agriculture, machines, and energy devices in the Middle Ages laid the foundations for the renewal of town life. In some cases this had unexpected environmental consequences. For example, growing prosperity meant a growing demand for meat as food, and the townsmen responded by allowing more livestock to graze on the town commons, which then became overgrazed and the local environment and the people suffered thereby. Nevertheless, the quickening of economic exchange, including new objects of trade, and the opening of new areas of commerce, such as the Far East, helped bring the Middle Ages to a close and laid the foundation for the Renaissance and the Age of Exploration, which entirely changed the contours of the natural world as it appeared to Europeans.

The growth of overland trade with Asia—silk from China, products from India through the Arab territories, and spices from the East Indies—spawned an interest in navigation. A series of technical improvements in ships (e.g., the sternpost rudder, skeleton-first construction, and fore-and-aft sailing rigs) made them sturdier and capable of carrying larger loads. These, along with the introduction of the magnetic compass from China in the 13th century, made possible the voyages of exploration—which, with Columbus's "discovery of the New World" and subsequent voyages greatly enlarged the world's dimensions in the eyes of Western civilization.

The quickening of European economic life in the early modern period, aided by the flow of wealth from the New World and by increased trade with the Far East, provoked an energy crisis in Europe by the 16th century. This crisis was caused by a shortage of wood, which throughout the medieval and early modern periods was both the primary material for making things as well as a major source of energy; wood fueled every domestic and industrial fire. (For example, it took almost four wagonloads of wood to evaporate one wagonload of salt from brine.) This shortage of wood forced its price rapidly upward. Thus, from the end of the 16th to the middle of the 17th century, the general price level of goods in England rose three times, but the price of wood went up eightfold. To meet this fuel crisis, the English government imposed conservation and recycling measures. An act of 1593, for example, compelled beer exporters either to return the original barrels or to bring back enough foreign wood suitable for making the same number of barrels as had been exported.

With wood in such short supply, coal began replacing it as a fuel. But coal production was limited by the high cost of pumping water from the mines. The result was a frenzied hunt for cheaper pumping devices. This search was heightened when, during the 18th century, Abraham Darby showed that iron could be smelted with coke, made from coal, instead of charcoal made from wood, which was steadily rising in price. The need for a new prime mover for pumps was finally met by James Watt's invention of the steam engine which became the characteristic power source of the Industrial Revolution. The steam engine transformed life and society in the Western world in the 19th century, and the Industrial Revolution has continued to alter the technology–environment relationship as it took on larger dimensions and turned in new directions during the 20th century.

Industrialization is more than a change from small-scale, handicraft production to large-scale machine production. Indeed, it involves a series of fundamental technological changes in the production and distribution of goods, accompanied by a series of economic, political, social, and cultural changes of the first magnitude.

The mechanization which transformed industrial production from handcraft to mass-production factory work also affected agriculture. The steel-blade ploughshare was invented in 1830 and the reaper in 1834 by Cyrus McCormick; then followed a series of other mechanical devices to assist in seeding, tillage, and harvesting. During the last quarter of the century, mechanical power, through steam tractors (and later automotive tractors), was applied more widely.

The scale of agricultural production was magnified by the opening of vast new areas for cultivation, in the U.S., Canada, Argentina, and Australia. Large-scale farming stimulated mechanization. By allowing the cultivation of land with less manpower, mechanization increased cereal crop production in the U.S. some ten to fourteen times; in harvesting alone, man-hours of labor were reduced to one-sixth the time previously required—and those advances have continued throughout the 20th century. Furthermore, drainage and irrigation in the American high plains, coupled with the technical innovation of barbed wire, allowed for a shift from the open range to crop production.

Commercial fertilizers were increasingly applied, with improved transportation allowing the importation of nitrates and guano from South America. Then the discovery in Germany (1903) of means of "fixing" atmospheric nitrogen enabled greater production and use of fertilizer. Commercial pesticides were also introduced to fight against fungal diseases of plants. And scientific plant- and animal-breeding further contributed to the development of varieties capable of faster growth, higher volumes of production, and adaptation to unfavorable environments.

A host of secondary technological changes, primarily in transportation and processing of foods, allowed for further specialization. Crops could be selected that were particularly suited to specific climatic and soil conditions, and advances in transportation allowed the crops to be widely and cheaply distributed. Railroads could haul produce faster and over greater distances than canal barges or wagons, so that commercial dairying, market gardening, and horticulture could spread out from urban centers. The development of refrigeration allowed for the preservation and transport of perishable food over long distances, thereby providing more varied and nutritious diets to people in the technologically advanced nations.

Technical advances in food processing also helped to industrialize agriculture. Canning, introduced at the beginning of the 19th century, became a common method for processing, preserving, and transporting foods. Pasteurization, derived from the work of Louis Pasteur, helped preserve milk and other products. In certain fields, such as meat-packing, industrial methods converted home industries into mechanized factory production.

While advances in processes for making iron and steel continued throughout the 19th century, new materials and new energy sources came into being as a result of scientific advances. Thus, the application of electricity made possible new means of exploiting mineral resources; electrometallurgy, applied first to plating metals, made other metals (especially aluminum) and alloys cheaper and more readily available. The late 19th century also witnessed the creation of synthetic materials. The marriage of laboratory science to technology proved exceptionally fruitful in bringing forth new material, first to replace older materials or to make them cheaper, and then to produce wholly new, artificial materials which surpassed natural products in meeting specialized needs.

Although Charles MacIntosh had discovered a way to bond rubber to cloth and make raincoats in 1824, it was not until 1839 that Charles Goodyear discovered a process for vulcanizing rubber, so that it could have wider commercial uses. Its first widespread use came as a result of the development of the bicycle, which increased the market for rubber tires. The later development of the automobile made the rubber industry into a major adjunct of transportation technology, while the growing utilization of electricity created still another market for rubber as insulation material. Rubber trees grew first in Brazil, then in countries of equatorial Africa and in Malaysia. But when World War II cut off supplies, synthetic rubber was developed for vehicle tires, while plastics took over as insulating materials. So here is a case where scientific-technological change both made one kind of agricultural production profitable, and then caused its abandonment in favor of more efficient growth.

MELVIN KRANZBERG

For Further Reading: David F. Channell, *The Vital Machine: A Study of Technology and Organic Life* (1991); Frederick Ferre, ed., *Technology and the Environment* (1992); Al Gore, *Earth in the Balance: Ecology and the Human Spirit* (1992). *See also* Agriculture; Industrialization; Technology.

TECHNOLOGY-BASED STANDARDS

See Toxic Chemicals: Management and Regulation.

TECHNOLOGY FORCING

The principle behind technology-forcing regulations is that government may shape the direction of industrial development by restraining technologies that damage the environment while promoting those that do not. A variety of "technology forcing" laws and regulations have formed a sort of eco-industrial policy since the late 1960s. The development of this policy may be seen as an effort to unravel the paradox that industrial growth often produces pollution and environmental destruction—but that advances in technology are critical to protecting and restoring the environment.

A common feature of such regulations is that they permit more flexibility than so-called "action-forcing" regulations. The latter approach stipulates the specific act a regulated company is required to undertake, such as obtaining a permit or preparing an environmental impact statement and adopting the least destructive alternative that is identified.

In the first phase of technology forcing regulation, as exemplified in the National Environmental Policy Act, the general policy was to impose strict controls on new industrial processes and products that may damage the environment. Permitting programs were set up to review and restrict the operation of new plants to ensure adequate pollution control measures were installed.

During the 1970s the Clean Water Act, the Clean Air Act, and the Resource Conservation and Recovery Act were amended to incorporate the technology-forcing approach. These laws included provisions requiring new industrial plants to use the "best available technology" (BAT) or some form of BAT to control the emission of pollutants into water and air, and the disposal of waste on the land.

BAT standards are intended to become more stringent over time—as new technologies are developed, new plants are required to adopt them. They also take into account economic availability and the standard of environmental performance otherwise required by law. In theory, BAT requirements also create a market demand, and hence an incentive, for new pollution control technology.

In practice, BAT requirements have forced massive investment into pollution control technology such as water treatment systems, air scrubbers, and hazardous waste management systems. These requirements are to a large extent responsible for the environmental gains of the 1970s–80s.

However, forcing technology advance through BAT requirements is a flawed approach. For one thing, older plants with more lenient standards have a competitive advantage over new plants subject to BAT standards. Additionally, new plants face delays and uncertainties in the permitting process, creating disincentives for innovation. Finally, when one particular control technology is identified as BAT, there is an incentive for industry to go along rather than to adopt new and different approaches.

Recently, then, there has been a shift away from treating pollution as an inevitable by-product of new plants and industrial development, and toward harnessing ingenuity and efficient industrial practices to the goal of reducing pollution. One important development has been measures that encourage industry to undertake pollution prevention instead of pollution control. Pollution prevention involves replacing raw materials, re-engineering manufacturing processes, and redesigning products to reduce or eliminate the creation of pollutants in all media (air, water, and land). This is generally a much more efficient and effective approach than that required under BAT standards, which typically require pollution control at the end of the pipe or smokestack rather than further up the manufacturing stream.

MICHAEL A. GOLLIN

TEMPERATE RAIN FORESTS

Temperate rain forests are found between the tropics and the polar circles, from 35° to 55° north or south latitudes, where mild, wet winters and warm summers occur. Such areas include the Pacific Coast of North America, Norway, Japan, southern Chile, southern New Zealand, extreme south Australia, and Tasmania. Similar forests probably existed in Europe but have long since disappeared. These forests, the coldest of rain forests, are rare. The annual precipitation exceeds 250 centimeters (100 inches) in some places, predominantly in the form of fog or snow. Conifers are the dominant trees because they require a period of cold weather and can overcome the possibility of a brief growing season by photosynthesizing in mild, cold weather.

Fires do not occur regularly in temperate rain forests. The interval between fires averages 300 to 400 years but may be more than 1,000; the area covered by a fire may exceed 100,000 to 200,000 acres. The multilayered canopies produce heavily filtered light. This low light enables shrubs, ferns, wildflowers, mosses, lichens, and liverworts to thrive. There is a large diversity of these underground plants, many of which are still not well studied. The several different types of epiphytic plants, mainly mosses and leafy lichens, root on branches of trees. Various shelf fungi and mushrooms also live on mature trees. Insect pests are a minor element.

Fallen logs in the temperate rain forest enable seedlings to root where they would otherwise be hindered by the thick growth of understory herbs and shrubs. A few of these new trees straddle the log and appear to rise on stilts after the log has decayed. The rotting logs persist for approximately 300 years. Most of the nutrients that are associated with the foliage and cambium of the tree leach out very early in the log's decomposition, leaving the log mainly to serve as a long-term carbon source for a variety of plants and animals that feed on it.

Wildlife has adapted to the specific niches in old growth and second growth temperate rain forests. Animals such as the Douglas squirrel in the Pacific Northwest have adapted to eat the seeds of the Douglas fir, hemlock, and spruce. Chipmunks, deer mice, and birds eat the fallen seeds. Large mammals such as elk and deer feed on underbrush and young conifers. Browsing by Roosevelt elk has created an unusual thinness of understory vegetation in areas of old growth within the Olympic National Forest in Washington state. Slugs, snails, and salamanders are prominent features of a temperate rain forest.

Earth's greatest temperate rain forest extends nearly 2,000 miles from the northern tip of Alaska's panhandle almost to San Francisco. It is defined as the wettest part of the coastal forest. Although not rich in numbers of tree species, these forests contain many trees over 200 feet tall and 6 or more feet in diameter. Several conifers (western hemlock, western red cedar, Sitka spruce, Douglas fir, coastal redwoods, noble fir, and Pacific silver fir) are among the largest and oldest.

Though not as diverse, temperate rain forest stands surpass tropical rain forests in sheer mass by seven to one and are therefore the densest forests on Earth. The average above-ground biomass of a temperate Douglas fir/western hemlock forest is 870 metric tons per hectare (1 metric ton = 1.1 U.S. tons, 1 hectare = 2.47 acres). This is significantly greater than the above-ground biomass of a tropical rain forest, which is up to about 450 metric tons per hectare in mature stands.

Temperate rain forests provide recreation and aesthetic value to humans but are also valued for their softwood timber, resulting in conflicts over use.

TEMPERATE RIVER BASINS

The basic principles by which streams and rivers organize themselves in drainage basins were discovered by Leonardo da Vinci (1452–1519). One of his pen and ink sketches, now in the Bibliothèque de l'Institute de France, shows branching patterns in tree limbs. With lines drawn through successively more distal and bifurcated portions of the system, Leonardo shows that the cross-sectional area at each level of branching remains constant. This, he explains in a margin of the drawing, is the result of sap conveyed by the branches being proportional to the cross section. In the same way the small tributaries to main trunk stream channels are organized in a hierarchy whose regularity reflects the transport of water off the landscape. Other drawings by Leonardo show canal systems on one side of the page and patterns of plant growth on the other. These, along with his more famous anatomical renderings of branching arteries and lung airways, reveal a profound understanding of relations between the organic and physical worlds. Ironically, even though the term "dendritic" (tree-like) reflects this heritage, the concept of organic wholeness has played little role in the piecemeal engineering that is so commonly done to control and use river networks for human needs.

In the protoscientific world of Leonardo's immediate successors, nature held a special role. Its mysteries allowed for human existence, as in the case of rain clouds providing water to river basins in which people grew crops and sustained their existence. The basins conveyed this water to the sea, but miraculously water was returned to the land. Surely this reflected the wisdom of God.

Modern hydrologists have rediscovered Leonardo's principles of drainage basin composition and explained them mathematically. The quantitative regularity of drainage segments in a hierarchy of tributaries results in downstream channels becoming larger in direct proportion to their increased flow. The channel hierarchy conveys water delivered by precipitation over drainage basins plus the sediment and dissolved load transported by that water. The latter can sometimes be of immense magnitude. The Huang He of northern China transports as much as 2 billion metric tons of sediment each year, so much that the sea into which it delivers this load is called the Yellow Sea.

The sediment of temperate rivers is not in continuous transport, but resides in deposits closely associated with river channels. Alluvial rivers, which have beds and banks of this transported sediment, may have braided patterns in their steeper sections, where coarse sediment is deposited in midchannel bars. Less steep reaches, generally with finer sediments, develop the characteristic smooth curves named for the River Meander in Asia Minor. Alluvial rivers have channels shaped by their most persistent and continuous flows. In humid temperate regions of relatively uniform rainfall and vegetative cover, streamflow is commonly continuous, and channels adjust to the average flood sizes achieved every year or two. In arid regions, by contrast, the channels may be dry for prolonged periods. These ephemeral channels are adjusted to the large floods that occur when relatively rare extreme rainstorms occur in areas lacking dense vegetative cover.

When the floods of alluvial rivers exceed the capacity of channels, they commonly inundate zones of relatively smooth and broad valley floor that parallel the low-flow river path. These surfaces, called floodplains, are constructed through time by the activity of the river. They are the sites of distinctive ecological communities known as the riparian zone. Floodplain plants and soils are adjusted to the rare, but inevitable influx of water and sediment during floods. In this sense, over long time scales, the floodplain is a part of the river.

Human civilization was invented on floodplains. Nomadism and localized horticulture were replaced by organized, large-scale agriculture on the fertile, well-watered soils of the Nile, Tigris, Euphrates, Indus, Chang Jiang (Yangtze), and Huang He (Yellow River) floodplains. Along these rivers people learned to live with cycles of flood and drought. Predicting the cycles came to be a most-honored profession, relying on astronomical observations and mathematics to explain them. Ultimately the science of western civilization arose from these practical beginnings.

The river basin containing the most humanity is the Chang Jiang (Chinese for "Long River"). Nearly as long as the Amazon and Nile, it rises in the Tibetan Plateau and flows 6,300 km to the East China Sea at Shanghai. Its 300 million human inhabitants mostly farm 70 million acres (about 28 million hectares) of highly fertile land, producing 10% of China's grain. In contrast to the Huang He, with its devastating floods, the Chang Jiang has long been considered China's benevolent river. This was not the case for its major northern tributary, the Minjiang, which drains a basin centered at Chengdu, capital of Sichuan Province. The Minjiang was subject to devastating floods until governor Li Bing initiated a diversion project and the Dujiangyan irrigation system. This plan, completed in 250 B.C. by his son Li Erlang, was so enlightened that it continues to function today, and a temple beside the river honors its two noble engineers. More than 100 million people live in this most productive agricultural region of China.

Along the Chang Jiang River the huge human population lives and works in traditional harmony with the river. The famous Three Gorges section, downstream of Chongqing (Chungking), contains numerous holy sites marked by temples and pagodas for Taoist and Confucian traditions. The daily, monthly, and seasonal patterns of peasant life conform to long-recognized cycles of the river and its associated biological systems.

This traditional reverence for rivers by Asian cultures contrasts to their profane treatment in the West. The Colorado River of western North America drains the drier portion of that continent's temperate zone. Since its survey in the 19th century, Colorado River water has been largely viewed according to a philosophy summarized by an engineer who became a U.S. president, Herbert Hoover: "Every drop of water that runs to the sea without serving a commercial use is a public waste."

The only holy sites on the Colorado River are those of Native Americans, and nearly all are now under waters impounded behind a series of massive dams. Those dams were constructed to provide water and electric power to the rapidly growing cities and irrigated agricultural lands of the southwestern U.S. Ironically, however, as the river became fully utilized in the late 20th century, new factors have arisen in the demands placed on it. The free-flowing Colorado in Grand Canyon National Park has become a recreational resource to the region's population. Rafting and fishing require a more natural river condition than allowed by upstream dams. Moreover, legal opinions have restored rights of water use to Native Americans, who now have priority over some water allocated to agriculture and industry.

As the Colorado comes into conflict with values reminiscent of the orient, the Chang Jiang is being transformed to allow modern China to compete with western cultures that progressed by exploiting their river basins among other resources. The Three Gorges is planned to be innundated by one of the world's largest dams, producing electric power but displacing millions from the reservoir site. The cycles of history seem to mimic those of water, with the difference that wisdom seems only associated with one.

VICTOR R. BAKER

For Further Reading: Victor R. Baker, Peter C. Patton, and Craig R. Kochel, eds., *Flood Geomorphology* (1988); John E. Costa and Victor R. Baker, *Surficial Geology* (1981); Luna B. Leopold, M. Gordon Wolman, and John P. Miller, *Fluvial Processes in Geomorphology* (1964).
See also Floods; Rivers and Streams.

TERMITES

Although they resemble ants, wasps, and bees in their social complexity, termites belong to an unrelated order of insects—the *Isoptera* (meaning equal wings)—a group most closely related to cockroaches and mantids. These small to medium-sized insects are characterized by a soft, light-colored body (earning them the sobriquet "white ants"), an abdomen that is broadly attached to the thorax, and antennae that resemble strings of small beads. Most termites are cryptobiotic, rarely appearing in the open but instead foraging, feeding, and nesting within a system of tunnels and galleries in soil or wood or in elaborately constructed nests (termitaria).

Termites occur in a wide range of habitats from rainforests to grasslands, from the ground to the treetops, but in terms of latitude and altitude they are entirely absent from alpine habitats and from latitudes greater than 42–45 degrees. The 2,000–3,000 recognized species are classified into five families of so-called primitive termites, and one advanced family the *Termitidae* which comprises 75% of the existing species.

All termites have a highly developed caste system with which they divide both labor and reproduction. Primary reproductives are fully winged forms that swarm from the parental nest—often in large numbers at certain seasons of the year—and mate. The queen and king then shed their wings, establish a new colony, and live in it together. Supplementary reproductives have rudimentary wings and less-pigmented eyes and can replace the loss of one or both of the resident royal pair. Workers are small, sterile, totally wingless forms—both nymphs and adults—responsible for building, foraging, and feeding both the brood and the reproductives. Soldiers are a sterile defensive caste with greatly enlarged heads and mandibles. Nasute soldiers have a snout-like projection that connects a large gland from which they can squirt a sticky and irritating secretion that immobilizes enemies, chiefly ants. Unlike the all-female castes of ants, termite workers and soldiers include both sexes.

Termites are best known for their consumption of wood. They destroy buildings, furniture, books, fabric, fenceposts, and utility poles. To foil their appetites, elaborate efforts have been taken: from termite shields, to chemical wood preservatives, to whole-building fumigation. Termites, however, are also important decomposers of organic matter, responsible for breaking down dead trees, fallen leaves, grass, and dung. They cannot digest the cellulose they consume directly but rely on symbiotic flagellated protozoans and bacteria in their digestive tracts to break it down. The exchange of anal liquid by termites, which transfers the necessary symbionts to the young, is one of the forces believed to have led to the development of sociality in the group.

Primitive termites, such as the powderpost termite (*Cryptotermes*) that is so destructive to homes in the southern United States, typically nest in the wood on which they are feeding. Advanced species, however, often build elaborate independent nests. The mound of the so-called magnetic termite (*Amitermes meridionalis*) of northwestern Australia, for example, may be twelve feet high and ten feet long but only three feet thick, and is invariably oriented on a north-south axis. This exposes the mound to a maximum amount of so-

lar radiation early and late in the day, and a minimum at noon. The fungus-growing *Macrotermes* mounds of Africa are larger still. Like fungus-cultivating ants of the New World, these termites subsist on a diet of fungus that they raise within their mounds, and have an elaborate system of air channels that regulates the temperature of the nest interior. A single one of these nests may contain 2 million individuals at any one time, and the queen within—as large as a man's thumb—may live for ten years laying 30,000 eggs a day or as many as 10 million in her lifetime.

The upper limit of termite biomass for a given location has been estimated as roughly 15,000 termites per square meter of ground. By any reckoning they are major consumers, whose biomass equals or exceeds that of grazing mammals on the African plains. Termites can seriously deplete the carbon content of tropical soils and are responsibe for redistributing nutrients in the landscape to such an extent that in parts of Africa and Asia crops are deliberately planted atop termite mounds or on land where mounds have been leveled.

The abundance of termites on the planet and the fact that the microorganisms in their guts produce methane have recently led to concern that they might be contributing significantly to global warming. Methane emission is only a minor component of the microbial carbon metabolism of termites but the gas is twenty times more efficient than carbon dioxide at trapping solar radiation. Some estimates have claimed that termites are contributing as much as 40% of the world's annual methane production. However, others have calculated that the figure is only 4% of the annual global emissions, making termites a minor contributor. All such calculations, however, await a more precise knowledge of just how many termites actually exist and how much carbon they consume.

ROGER B. SWAIN

For Further Reading: M. A. K. Khalil, R. A. Rasmussen, J. R. J. French, and J. A. Holt, "The Influence of Termites on Atmospheric Trace Gasses CH$_4$, CO$_2$, CHCl$_3$, N$_2$, CO, H$_2$, and Light Hydrocarbons," *Journal of Geophysical Research* (1990); Kumar Krishna and Francis Weesner, eds., *The Biology of Termites* (1969, 1970); K. E. Lee and T. G. Wood, *Termites and Soils* (1971).

TERRITORIALITY

Territoriality is an instinctive behavior in vertebrate animals that involves occupying and defending a space referred to as a territory. One of the key forces regulating the interrelationships of many types of animals—with each other and with the other factors in their environments—each territorial species has its own unique form of territorial behavior that provides increased survival value for its way of life.

Prime examples of territorial behavior are shown in birds. As soon as resident males arrive from their winter migration, they select specific areas in which nests will be built. H. E. Howard in *Territory in Bird Life* (1920) was the first to call these areas *territories*. The male proceeds to defend the territory, driving out other males of the same species and proclaiming his presence to competitors and prospective mates by singing from prominent perches or through visual performance. The flight-song of the European skylark, the drumming of the grouse, tattoos of woodpeckers, and plumage displays of the peacock are examples of this behavior. Once a female accepts the male, she seems to be more devoted to his territory than she is to him.

Territory size often depends upon the food habits of a species and the abundance of that species in the area. If a robin or catbird obtains food for its young in the immediate vicinity of the nest, it requires a larger nesting territory than a swallow or heron, which will fly great distances to feeding areas. Even in the case of social birds such as swallows, herons, or penguins, who may nearly touch when nesting, there is a definite territory where others are not allowed to trespass. Some birds have separate nesting and feeding territories and may share some neutral feeding areas. For example, robins may divide a garden into several territories with a pair defending each, and yet may have other areas where all feed peaceably.

Territoriality as a form of social behavior represents a respect for property rights and allows animals to coexist next to one another. This behavior has been observed among different mammalian species. For example, monkeys and baboons may show a wide variation in their methods of individual or group territorial defense. Beavers, like many mammals, mark off the boundaries of their territories by scent and also use the tail slap as a sound warning. Even though territorial neighbors may be antagonistic, once territories have been established, overt hostility can decrease. However, neither individuals nor groups often wander past previously established boundaries. Because the effect of territoriality is to establish neighborhoods of families, territorial units of a species population shape a social network. In these social networks, the relations between territorial neighbors are far from perfect, which helps to reinforce the analogy to human societies.

THEOLOGY OF THE EARTH, A

All ancient civilizations have expressed, each in its own way, wonderment at the beauty of the Earth. Aristotle tried to imagine how people who had spent all their lives under luxurious conditions but in caves would respond when given for the first time the chance to behold sky, clouds, and seas. Surely, he writes, "these humans would think that gods exist and that all the marvels of the world are their handicrafts." The visual evidence provided by space travel now gives larger significance to Aristotle's image. Although the Earth is but a tiny island in the midst of vast reaches of alien space, it derives distinction from being a magic garden occupied by myriads of different living things that have prepared the way for self-reflecting human beings.

When humans emerged in their present biological form 150,000-or-so years ago, they must have been fitted for the conditions prevailing around them. Since fitness in the biological sense implies suitable interrelationships between the organism and the total environment, the environment was fit also and ready for human beings when they appeared on Earth. For Walt Whitman, from the point of view of the poet and humanist, the "primal sanities" of nature were the qualities of the Earth that make for a rich human life.

Whitman's primal sanities refer to the conditions under which humans evolved and to which their biological constitution is still adapted. But while human biological nature has remained much the same since the Stone Age, their surroundings and ways of life have changed profoundly. Civilization is often in conflict with primal sanities as evidenced by the present ecological crisis. This conflict accounts for the unfortunate fact that the science of human ecology, which would be concerned with all aspects of the relationships between human beings and the rest of creation, has come to be identified almost exclusively with the problems of disease and alienation resulting from environmental insults. Yet there is much more to human ecology than this one-sided view of the relationships between humans and the external world.

Human beings are still of the Earth, earthy. The Earth is literally our mother, not only because we depend on her for nurture and shelter but even more because the human species has been shaped by her in the womb of evolution. Each person, furthermore, is conditioned by the stimuli they receive from nature during their own existence.

If humans were to colonize the moon or Mars—with abundant supplies of oxygen, water, and food, as well as adequate protection against heat, cold, and radiation—they would not long retain their humanness, because they would be deprived of those stimuli that only Earth can provide. Similarly, we shall progressively lose our humanness even on Earth if we continue to pour filth into the atmosphere, to befoul soil, lakes, and rivers; to disfigure landscapes with junkpiles; to destroy the wild plants and animals that do not contribute to monetary values, and thus to transform the globe into an environment alien to our evolutionary past. The quality of human life is inextricably interwoven with the kinds and variety of stimuli humans receive from the Earth and the life it harbors, because human nature is shaped biologically and mentally by external nature.

Admittedly, certain human populations have functioned successfully and developed worthwhile cultures in forbidding environments, such as the frozen tundras or the Sahara. But even the most desolate parts of the Arctic or the Sahara offer a much wider range of sensations than does the moon. Eskimo life derives exciting drama from ice, snow, and water, from spectacular seasonal changes, and from the migration of caribou and other animals. The nomadic Tuareg have to cope with blinding and burning sand, but they also experience the delights of oases. Being exposed to a variety of environmental stresses and having to function among them is far different from living in a spacesuit or a confining space capsule, however large it may be, in which all aspects of the environment are controlled and extraneous stimuli are almost completely eliminated.

Participation in nature's endless changes provides vital contact with the cosmic forces, which is essential for sanity. In *The Desert Year*, the American drama critic turned naturalist Joseph Wood Krutch pointed out that normal human beings are not likely to fare well in areas lacking visible forms of life. For example, they rarely elect to stay long in the deserts of the American Southwest, as if this kind of scenery, magnificent as it is, were fundamentally alien to mankind.

> Wherever, as in this region of wind-eroded stone, living things are no longer common enough or conspicuous enough to seem more than trivial accidents, man feels something like terror . . . This is a country where the inanimate dominates and in which not only man but the very plants themselves seem intruders. We may look at it as we look at the moon, but we feel rejected. It is neither for us nor for our kind.

Humans seek contact with other living things probably because our own species has evolved in constant association with them and has retained from the evolutionary past a biological need for this association.

Human nature has been so deeply influenced by the conditions under which it evolved that the mind is in some ways like a mirror of the cosmos. Some of the early Church fathers had a vision of this relationship, as illustrated by Origen's exhortation to man: "Thou art a second world in miniature, the sun and the moon are within thee, and also the stars." More than a thousand years later, the British biologist Sir Julian Huxley reformulated Origen's thought in modern terms and enlarged it to include his own concepts of psychosocial evolution:

> The human type became a microcosm which, through its capacity for self-awareness, was able to incorporate increasing amounts of the macrocosm into itself, to organize them in new and richer ways, and then with their aid to exert new and more powerful influences on the macrocosm.

Sir Julian's statement implies two different but complementary attitudes toward the Earth. The fact that human beings incorporate part of the universe in their being provides a scientific basis for the feeling of reverence toward the Earth. But the fact that humans can act on the external world often makes them behave as if they were foreign to the Earth and her master—an attitude that has become almost universal during the past two centuries.

The phrase "conquest of nature" is certainly one of the most objectionable and misleading expressions of Western languages. It reflects the illusion that all natural forces can be entirely controlled, and it expresses the criminal conceit that nature is to be considered primarily as a source of raw materials and energy for human purposes. This view of humanity's relationship to nature is philosophically untenable and destructive. A relationship to the Earth based only on its use for economic enrichment is bound to result not only in its degradation but also in the devaluation of human life. This is a perversion that, if not soon corrected, will become a fatal disease of technological societies.

The gods of early humans were intimately connected with the Earth, and belief in them generated veneration and respect for it. But respect does not imply a passive attitude; early humans obviously manipulated the Earth and used its resources. Primitive religion was always linked with magic, which was an attempt to manage nature and life through the occult influences that were assumed to lurk in the invisible world. There is a fundamental difference between religion and magic. In the words of the anthropologist Bronislaw Malinowski, "religion refers to the fundamental issues of human life, while magic turns round specific concrete and detailed problems." Our salvation depends upon our ability to create a religion of nature and a substitute for magic suited to the needs and knowledge of modern human beings.

The problems of poverty, disease, and environmental decay cannot be solved merely by the use of more and more scientific technology. Technological fixes usually turn out to be procedures that have unpredictable consequences and are often in conflict with natural forces. Indeed, technological magic is not much better than primitive magic in dealing with the fundamental issues of human existence, and in addition, it is much more destructive. In contrast, better knowledge of humanity's relationships to the Earth may enable us to be even more protective of the natural world than were our primitive forebears; informed reason is likely to be a better guide for the management of nature than was superstition or fear. We do know scientifically that the part of the Earth on which we live is not dead material but a complex living organism with which we are interdependent; we also know that we have already used a large percentage of the resources that have accumulated in the course of its past. The supply of natural resources, in fact, presents a situation in which the practical selfish interests of mankind are best served by an ethical attitude.

For most of its geological history, the Earth had no stores of fossil fuels or concentrated mineral ores. These materials, which are the lifeblood of modern technology, accumulated slowly during millions upon millions of years; their supply will not be renewed once they have been exhausted. They must therefore be husbanded with care—for immediate reasons and also for the sake of the future. The natural resources that we now gouge out of the Earth so thoughtlessly and recklessly certainly should not be squandered by a few generations of greedy humans.

From the beginning of time and all over the world, humanity's relationship to nature has transcended the simple direct experience of objective reality. Primitive people are inclined to endow creatures, places, and even objects with mysterious powers; they see gods or goddesses everywhere. Eventually, humans came to believe that the appearances of reality were the local or specialized expressions of a universal force; from belief in gods they moved up to belief in God. Both polytheism and monotheism are losing their ancient power in the modern world, and for this reason it is commonly assumed that the present age is irreligious. But we may instead be moving to a higher level of religion. Science is at present evolving from the description of concrete objects and events to the study of relationships as observed in complex systems. We may be about to recapture an experience of harmony, an intimation of the divine, from our scientific knowledge of the processes through which the Earth became prepared for human life and the mechanisms through which humans relate to the universe as a whole. A truly ecological view of the world has religious overtones.

The Earth came to constitute a home suitable for humans only after it had become a living organism. The sensuous qualities of its blue atmosphere and green mantle are not inherent in its physical nature; they are the creations of the countless microbes, plants, and animals that it has nurtured and that have transformed its drab inanimate matter into a colorful living substance. Humans can exist, function, enjoy the universe, and dream dreams only because the various forms of life have created and continue to maintain the very special environmental conditions that set the Earth apart from other planets and generated its fitness for life—for life in general and for human life in particular.

Humanity is dependent on other living things and like them must be adapted to its surroundings in order to achieve biological and mental health. Human ecology, however, involved more than interdependence and fitness as these are usually conceived. Human beings are influenced not only by the natural forces of their environment but also and probably even more by the social and psychological surroundings they select or create. Indeed, what they become is largely determined by the quality of their experiences. Henry Beston wrote in *The Outermost House*:

> Nature is part of our humanity, and without some awareness and experience of that divine mystery man ceases to be man. When the Pleiades, and the wind in the grass, are no longer a part of the human spirit, a part of very flesh and bone, man becomes, as it were, a kind of cosmic outlaw, having neither the completeness and integrity of the animal nor the birthright of a true humanity.

These words convey one aspect of the ecological attitude that must be cultivated to develop a scientific theology of the Earth.

But there are other aspects, based on the fact that humans are rarely passive witnesses to natural events. They manipulate the world around them and thus set in motion forces that shape their environment, their life, and their civilizations. In this sense, humans make themselves, and the quality of their achievements reflects their visions and aspirations. Human ecology naturally operates within the laws of nature, but it is always influenced by conscious choices and anticipations of the future.

The relationships that link humankind to other living organisms and to the Earth's physical forces thus involve a deep sense of engagement with nature and with all processes central to life. They generate a spirit of sacredness and of overriding ecological wisdom which is so universal and timeless that it was incorporated in most ancient cultures. One can recognize the manifestations of this sacredness and wisdom in many archaic myths and ceremonials, in the rites of preclassical Greeks, in Sung landscape paintings, in the agricultural practices of preindustrial peoples. One can read it in Marcus Aurelius' statement that "all living things are interwoven each with the other; the tie is sacred, and nothing, or next to nothing, is alien to ought else." In our time, the philosophical writings of Alfred North Whitehead have reintroduced in a highly intellectualized form the practical and poetical quality of ecological thought.

Human ecology inevitably considers relationships within systems from the point of view of humanity's privileged place in nature. Placing humanity at the pinnacle of creation seems at first sight incompatible with orthodox ecological teachings. Professional ecologists, indeed, are prone to resent the disturbing influence of human intervention in natural systems. If properly conceived, however, anthropocentrism is an attitude very different from the crude belief that humanity is the only value to be considered in managing the world and that the rest of nature can be thoughtlessly sacrificed to its welfare and whims. An enlightened anthropocentrism acknowledges that, in the long run, the world's good always coincides with humanity's own most meaningful good. Humans can manipulate nature to their own best interests only if they first love her for her own sake.

While the living Earth still nurtures and shapes man, humans, they now possess the power to change it and to determine its fate, thereby determining their own fate. Earth and humanity are thus two complementary components of a system, which might be called cybernetic, since each shapes the other in a continuous act of creation. The Biblical injunction that humans were put in the Garden of Eden "to dress it and to keep it" (Genesis 2:15) is an early warning that we are responsible for our environment. To strive for environmental quality might be considered as an eleventh commandment, concerned of course with the external world, but also encompassing the quality of life. An ethical attitude in the scientific study of nature readily leads to a theology of the Earth.

RENÉ DUBOS
(EDITED BY RUTH A. EBLEN)

For Further Reading: René Dubos, *Beast or Angel? Choices That Make Us Human* (1974); René Dubos, *So Human an Animal* (1968); Gerard Piel and Osborn Segerberg, Jr., ed., *The Word of René Dubos* (1990).

THERMAL ENVIRONMENT

The thermal (heat) environment is a subenvironment of the physical environment in which humans have evolved. According to James Marston Fitch, it is one of three coextensive environments (the others are the at-

mospheric and the luminous), and any manipulation of one influences the others. The thermal environment of an interior space is influenced by the climate of the geographic place where the building is located, by the construction of the surfaces that define the space, by the activities going on within the space, and by the systems that are designed to maintain its comfort.

Energy always flows from a higher level to a lower level. In summer the heat (high) energy will tend to flow from the outdoors to a cooler indoor space; in winter a warmer indoor space will lose heat to the cold outdoors. Hot or cold climatic extremes will therefore make it more difficult to maintain a comfortable thermal environment. Temperature changes between day and night also affect the thermal environment and require consideration and control.

The construction materials making up the six sides of a space (four walls, ceiling, floor) have a certain resistance to the flow of energy both into and out of a space. By choosing the appropriate materials, such as double-glazed windows or well insulated walls, the resistance can be increased and energy flow can be minimized. In the case of windows or glass walls, it is particularly important to provide shading to prevent the sun from contributing to the space cooling requirement.

The comfort of the thermal environment depends on the activity of the occupants. The metabolic level of persons having sedentary occupations is much lower than that of dancers or workers performing manual labor. This means that very active occupants will require different conditions of temperature and humidity; they will also give off more heat that must be removed to obtain comfort conditions.

In addition to people, every watt of electricity that is supplied to a space for lights and equipment will eventually become part of the thermal environment in the form of heat. In calculating comfort requirements for the space, this heat must be added to the cooling load and subtracted from the heating duty of the equipment selected.

Heating for comfort is achieved by radiation or natural convection using heating elements supplied with steam or hot water. Cooling is supplied through forced convection by blowing air over finned tubes cooled by means of water or refrigerant.

Air motion is an important part of a comfortable thermal environment. It must be maintained at a level below which it will become obtrusive. Maintaining the proper humidity has been found to enhance indoor air quality, because moistened mucous membranes are less affected by certain irritants.

A well controlled thermal environment not only contributes to health but has been shown to improve the productivity of the occupants considerably. Building designers and managers must therefore make every effort to achieve it.

PETER FLACK
GEORGE RAINER

THERMODYNAMICS, LAWS OF

Thermodynamics is the study of energy exchanges between a system and its surroundings. Its strength as a discipline lies in the great generality of the first and second laws of thermodynamics and in the insight these laws provide to understanding numerous problems in energy technology and related environmental sciences. Thermodynamics is both subtle and intensely practical. A system can be anything the analyst wishes it to be. For example, in an automobile engine, the system might be the gasoline vapor within the cylinders. The cylinder walls would be the boundaries between the system and its surroundings. The surroundings would include the air, which conducts heat away from the cylinders, and the shafts connected to the pistons, which do the work that makes the car move as they are pushed by the exploding gasoline vapors.

The First Law of Thermodynamics

The first law of thermodynamics (also called the law of conservation of energy) states that energy cannot be created or destroyed; it can be converted from one form to another. This law recognizes only three kinds of energy: work, heat, and internal energy. To understand thermodynamics, one must know precisely what a thermodynamicist means by these terms.

Work is done whenever a force acts through a distance. Thus when gasoline vapor in an automobile cylinder explodes and pushes the piston shaft outward, the system (gasoline vapor) is doing work on its surroundings (the movable piston shaft). Work results in a transfer of energy from the system to its surroundings, and this kind of transfer is accomplished by a mechanical process (motion of matter on a large scale). Note that the surroundings can also do work on the system. For example, an external agent could push the piston inward, thereby compressing the gas. By convention, we call this negative work.

Heat is also a flow of energy between the system and its surroundings. However, in this case there is no large-scale motion of matter. Instead energy flows because there is a temperature difference between the system and its surroundings, and the boundaries between them are thermal conductors. For example, consider a refrigerator. The system could be the fluid (often freon) that vaporizes when it is inside the box to be

cooled, and recondenses in coils outside this box. The surroundings include the box and the air near the coils where the fluid condenses. As the fluid vaporizes inside the box, the temperature of the fluid goes down. It is now at a lower temperature than the air inside the box. Heat flows from air in the box to the fluid through the metal walls of the container that confines the fluid. In its vapor form, the fluid is pumped out of the box and then recondensed. The condensation process causes the temperature of the fluid to rise. When the fluid is at a higher temperature than the air surrounding the condensing coils, heat will flow out of these coils to the surrounding air.

Note that when heat flows, there is no large-scale movement of matter. Kinetic Theory tells us that on a very small scale, molecules are moving and colliding with each other, and these collisions are the mechanism by which energy is transferred. But thermodynamics deals with large-scale motions and energy transfers. We do not need to know anything about molecules to understand and apply thermodynamics.

Heat, then, is an energy transfer that takes place when a thermal conductor connects two regions of different temperatures. Heat is not the word we should use to describe the energy a body possesses because it is at a high temperature (is hot), although it is often used that way in conversations among lay persons—and sometimes even scientists. The energy that is stored by a body—and which can, in part, be measured by its temperature—is called "internal energy." We say "in part" because internal energy also depends on the amount of matter in the body. If a large copper ball and a small copper ball are at equal temperatures, the large ball will have more internal energy. Of course, if two copper balls are equal in size, the one at the higher temperature will have the greater internal energy. The first law of thermodynamics connects these three forms of energy as follows: During any process, the increase of the internal energy of a system equals the heat that flows into the system minus the work done by the system on its surroundings.

The Second Law of Thermodynamics

Some features of heat flow are familiar, even to the most casual observer. Whenever any kind of behavior is universally observed, one can generalize by creating a statement that describes the observation, and this generalization is called a law. Thus the second law of thermodynamics states that heat, on its own accord, flows from regions of high temperatures to regions of low temperatures. Of course, a refrigerator makes heat flow in the opposite direction, but if you unplug it, heat will flow from the warm room into the cold refrigerator. The refrigerator causes heat to flow toward higher temperatures, but there is another effect: net work is done by the compressor on the freon in the refrigerator.

Careful observation will reveal another kind of consistent thermal behavior. A heat engine absorbs heat from a heat source and uses this energy to do work on its surroundings. This is what a heat engine is designed to do, so it would be a boon to mankind if this conversion had associated with it no other effect. However, in the real world, there are always other effects.

Consider a heat engine that operates in a cycle. During part of the cycle, heat is taken in from a heat source such as a steam generator. Gas in the heat engine expands and does work on its surroundings. During another part of the cycle, the gas in the system is recompressed and some heat is rejected into the surroundings at a temperature lower than the heat intake temperature.

Since the initial state and the final state of the system were identical, the entire cycle produced no change in the internal energy of the system. Applying the first law to the entire cycle, we conclude that the net work done is equal to the net heat intake. Analysis of the individual steps reveals that the ratio of the heat intake to the heat output is the same as the ratio of the temperature during heat intake to the temperature during heat output. Since the intake temperature was higher than the exhaust temperature, the net heat intake is positive and the net work done on the surroundings is positive. In fact we can say: Work equals Heat in at intake minus Heat out at exhaust. This equation is a consequence of the first law. What is new here is the conclusion that, barring the existence of heat reservoirs at a temperature of absolute zero, the heat flow into the surroundings during heat exhaust cannot be zero. This conclusion is consistent with all observations of heat engines, independent of the working substance and the kinds of processes that compromise the cycle. This generalization leads to another formulation of the second law of thermodynamics: "It is impossible to construct a heat engine that operates in a cyclic manner that has no other effect than to absorb heat from a source and convert it into work." The other effect, here, is to move some energy from a reservoir at a high temperature to a reservoir at a lower temperature.

Engineers find it useful to define the maximum efficiency of a heat engine as the net work done during one cycle divided by the heat intake from the heat source. (Actual efficiencies will be less than this maximum if some energy is expended in overcoming friction.) Apparently another formulation of the second law would be the simple statement: "No heat engine can be 100% efficient."

Significance of the First and Second Laws

There are important practical implications of the laws of thermodynamics. In a society where the availability of energy at reasonable cost is synonymous with a high standard of living, it is vital to improve the efficiency of heat engines and, more generally, to cease treating energy as an inexhaustible and inexpensive commodity. The second law of thermodynamics tells us that to have high efficiency, a heat engine must achieve a large temperature difference between the heat source and the region where heat is exhausted into the environment. No other refinements in design can make up for failure in this area.

The second law also makes a prediction about the long-range availability of energy that can be used to do work. As noted above, there are two consequences of every cycle of a heat engine: (1) some work is done; (2) some energy initially stored in a reservoir at a high temperature ends up in a reservoir at a lower temperature. Thus the first law tells us that the total energy in the universe is constant, but the second law tells us that as time goes on, this energy is less and less available for doing work.

The concept that energy, in a very real sense, is being constantly degraded, provides us with an arrow of time. Many laws of physics are reversible. The laws of mechanics tell us that for any elastic collision between particles, the reaction can go either way. Light rays that reflect and refract their way through an optical system can be made to retrace their paths exactly. But thermodynamic processes—except those idealized processes we call reversible—always end up making some energy less available for doing work, never more. This one-directional nature of events can be captured elegantly by defining a quantity called entropy that always increases as energy becomes less available for doing work. One can then express the second law by saying "the entropy of the universe either increases or remains constant during every large-scale process that is consistent with the other laws of nature."

ARNOLD STRASSENBURG

For Further Reading: Energy for Planet Earth, Readings from *Scientific American* (1990); W. F. Kenney, *Energy Conservation in the Process Industry* (1984); Mitchell Wilson and the eds. of *Life, Energy* (1967).

THINK GLOBALLY, ACT LOCALLY

The phrase "Think Globally, Act Locally" was originated by René Dubos at the time he served as chairman of the group of experts advising the United Nations Conference on the Human Environment held in

Stockholm in 1972. His purpose was to convey his conviction that, while all environmental problems have global aspects, sweeping statements about them can hardly be converted into action without considering the ecological, economic, and cultural differences that exist from one place to another. "Think Globally, Act Locally" has been used as a motto and trademark for the environmental education services of The René Dubos Center for Human Environments, Inc., since the mid-1970s.

In "The Despairing Optimist" column that he wrote for *The American Scholar* Dubos elaborated on the meaning of the phrase in 1977:

> [Voltaire's] Candide realized at the end of his tumultuous life that the quest for happiness eventually leads to tending one's own garden; similarly each nation will probably discover that the best way to solve its problems is to stick to its own affairs. But while global thinking is no substitute for local action, perpetuating national sovereignty in its present form would be suicidal because global interdependence will be of increasing importance in many crucial aspects of human life and of the physical environment. As we enter the global phase of human evolution, it becomes obvious that we all have two countries, our own and Planet Earth.

In his article entitled "Think Globally, Act Locally," published simultaneously in 1979 in *Newsweek* and *The Wall Street Journal*, Dubos suggested that ecological consciousness should begin at home. Dubos recounted the many occasions when he lectured in colleges where students were deeply concerned about the quality of the environment, but everywhere he went he had two kinds of experience that account for the title of the essay. The students were eager to discuss the global aspects of environmental problems, but they did not seem disturbed by the messiness of their cafeterias and other public rooms. His message to them was that thinking in a global way is a useful intellectual exercise, but no substitute for the care of the place in which one lives. If they really wanted to do something for the health of our planet, the best place to start is in their own community and in its fields, rivers, marshes, coastlines, roads, and streets, as well as with its social problems. In conclusion Dubos said:

> *E pluribus unum.* Skepticism concerning the value of globalization does not imply isolationism. There are good reasons to believe that we can create a World Order, not a World Government, in which natural and social units maintain or recapture their identity, yet interplay with each other through a rich system of communications. My hope is that we shall learn to create for humankind a new kind of unity out of ever-increasing diversity.

Dubos restated the concept in *Celebrations of Life,* published in 1981:

The most valuable achievement of the international conferences was probably to reveal that the best and commonly the only possible way to deal with global problems is not through a global approach but through the search for techniques best suited to the natural, social, and economic conditions peculiar to each locality. Our planet is so diverse, from all points of view, that its problem can be tackled effectively only by dealing with them at the regional level, in their unique physical, climatic, and cultural contexts.

Dr. Spencer H. Beebe, then Vice President and Director of The Nature Conservancy's International Program, said that the phrase "Think Globally, Act Locally" inspired his program to adopt a whole new approach to conservation abroad. "The task of preserving ecologically significant lands in other countries can be done effectively only by local people, applying local resources and knowledge to local problems." "Think Globally, Act Locally," the theme of the "Decade of Environmental Literacy" launched by The René Dubos Center for Human Environments in 1990 in cooperation with The United Nations Environment Programme has become accepted as an environmental principle throughout the world.

RUTH EBLEN

THOREAU, HENRY DAVID

(1817–1862), American naturalist, essayist, social theorist. Thoreau created a motif of individualism, simple living, love of nature, and social protest that have gone hand in hand ever since. He did this by living alone for 26 months in a 10-by-15 foot cabin he built by Walden Pond, growing his own beans, intensely observing the four seasons, yet writing in his voluminous journal as much about people as about nature. Indeed, he spent a night in jail in July, 1846, near the midpoint of his sojourn to protest the institution of slavery and the conduct of the Mexican War.

Thoreau was not the first nature writer; see, for instance, the hymn to the animal kingdom in the Book of Job, 38-41. Nor was he the first "nature lover"; Petrarch is said to have begun the Renaissance by climbing Mount Ventoux in 1336, being the first learned man ever to climb a mountain only for the view. The English writers Gilbert White and Richard Jefferies produced lovely country sketches, but Thoreau salted his essays with a string of lasting aphorisms, such as "the mass of men lead lives of quiet desperation"; "in Wilderness is the preservation of the World;" and "I would rather sit on a pumpkin and have it all to myself, than be crowded on a velvet cushion."

Thoreau was a far better writer and more trenchant commentator than other visionary American conservationists, such as George Perkins Marsh and John Muir, whose books are instructive but not so delightful. Being odder and more nonconformist in his own time somehow fitted him better for later, as when he ridiculed the new panacea of industrialism—"we do not ride on the railroad; it rides upon us"—and, in a period of loud ballyhoo about the forty-niners' gold rush, wrote that wealth was merely a millstone, boasting of building his house for $28. (In his first eight months at Walden he had averaged 27¢ a week for food.) During this crescendo of the westering movement, he burlesqued go-getterism and money-grubbing, and advocated staying at home, traveling "a good deal in Concord," exploring thyself, and subduing "a few cubic feet of flesh," instead.

With his mentor and friend, Ralph Waldo Emerson, Thoreau fathered the personal essay in America. He may be all in all the best literary stylist the U.S. has produced, and he was an early exemplar of New World intellectual independence and originality. But by helping to inspire, through his essay "Civil Disobedience," Gandhi's nonviolent political campaigns, he became a world figure in nonliterary terms also. For environmentalists Thoreau can represent the ultimate in grace and balance, a cosmopolitan concern for humanity as well as wild things, culture as well as ecology, iconoclasm without cynicism, free thinking without irreligion. He is emblematic of a life devoted to generous helpings of solitude—but not antisocial—absorbed in art and philosophy—not just a hectoring activism—and of dissent without fickle utopianism. As frugal as his habits seemed, he was so broad-gauge a writer that some 1990s environmentalists have been critical of him for being less single-minded than themselves. He was not clairvoyant about what the late 20th century would bring, but, like John Muir, he had a sense of place in which each kind of creature—red squirrels, pickerel, woodchucks, blue jays, and humans—should hold an equal citizenship. And if Thoreau seldom sounded the particular notes of modern conservation ideology (*Walden* came out in 1854, forty years before Muir's first book, *The Mountains of California*), he did nonetheless suggest, in an essay called "Huckleberries," the prescient idea that every town might preserve "a primitive forest . . . where a stick should never be cut for fuel . . . but stand and decay for higher uses—a common possession forever."

"I left the woods for as good a reason as I went there. Perhaps," he says in *Walden*, "I had several more lives to live." He was a first-rate land surveyor and pencil-maker, though advocating only "six weeks" of wage-work a year to leave time for more important endeavors. He wrote notable and luminous books on roaming,

Cape Cod and *The Maine Woods*. Preaching self-reliance, and counting raisin bread (which he was rumored to have invented) and spring water enough of a personal luxury, this Harvard graduate lived more abstemiously than most laborers and was an epitome of the American expression of democracy in his century, as well as of low-tech, nonconsumptive style of living that has drawn approving notice in ours. He was a rhapsodist, however, bragging "for humanity," as he put it, and, like Walt Whitman, he was a singer of lists, gleefully attentive to the daily changes outside his door. In an era of frontier fever, Thoreau stuck to his patch of Massachusetts woods and made them shine for posterity more than the Rockies or the High Sierras do in literature, despite the Great West being all the rage. "There is more day to dawn. The sun is but a morning star," he wrote, in one of the capsule gems of transcendentalism, which summarizes why wherever his home had chanced to be, if it had been in the country, he could still have written *Walden*.

EDWARD HOAGLAND

For Further Reading: Walter Harding, *The Days of Henry Thoreau* (1992); Robert D. Richardson, Jr., *Henry Thoreau: A Life of the Mind* (1986); Henry D. Thoreau, *Faith in a Seed* (1993).

THREE MILE ISLAND

For five days, beginning in the early hours of March 28, 1979, the reactor in Unit 2 of the Three Mile Island Nuclear Plant in Harrisburg, Pennsylvania, was in various stages of breakdown—at times within thirty minutes of a complete meltdown—as a result of failure in multiple safety systems and human error. The standard pressurized water reactor was designed by the Babcock and Wilcox company.

At 4:00 A.M. on March 28, 1979, the plant was operating normally, generating approximately 2,700 megawatts of thermal energy. The accident began when an overnight maintenance crew working on the plant's water-purification system inadvertently stopped the main feedwater pumps. These pumps circulate water through the reactor, extracting the reactor-core heat and sending it through steam generators for the turbine-generators.

The reactor-core temperature and pressure began to rise almost instantly, as is normal, causing the pilot operated relief valve to open. In about 15 seconds, the pressure had dropped to the point where the relief valve should have closed, but it malfunctioned, remaining in the open position and creating a leak in the system to the containment building.

Although the operators stopped the fission reaction almost immediately, the need for coolant continued, as the radioactive decay of the fission products continued to heat the reactor core. The pressure dropped to the point where emergency high-pressure injection pumps were turned on to replace the lost liquid, as they were designed to do.

Meanwhile, the closed block-valves to the feedwater pumps were discovered and reopened about eight minutes after the onset of the event. This should have been soon enough to prevent damage to the plant, but the open relief valve was not detected and blocked off until two hours and twenty minutes later, permitting the core coolant to evaporate into the containment shell. As the recorded pressure continued to drop, the operators, unaware of the open relief valve, turned off the emergency core cooling system and released water from the system via the "let down line," which caused the center of the core to be uncovered and to reach its melting point. As the main circulation pumps were turned off, the molten fuel reacted with the water vapor to produce hydrogen, which collected in the containment building, creating a small detonation. Four hours later the emergency core cooling system was activated, but the core had already been damaged by the partial meltdown. No radioactivity of physiologic significance was released from the containment building.

In the immediate aftermath, the Nuclear Regulatory Commission issued new safety regulations and mandated changes in all existing nuclear plants. The Three Mile Island accident galvanized antinuclear sentiment and had an impact in slowing the proliferation of nuclear power in the U.S. Since 1978, no U.S. power company has ordered a new nuclear power plant, and all plants ordered since 1973 have been canceled, both as a result of Three Mile Island, and of a simultaneous drop in the growth of demand for electricity.

TOBACCO

Tobacco, especially the cigarette, is responsible for more than 480,000 deaths in this country and more than 2.5 million deaths around the world each year. There is no safe way to use commercial tobacco products. Because its smoke is nearly always inhaled, distributing toxins widely throughout the body, the cigarette is the most likely to have devastating effects on health. Cigarettes were first mass produced in the 1880s, but wide use began with the introduction of the Camel brand in 1913.

Nicotine, the psychoactive drug in tobacco, is responsible for the enduring appeal tobacco products have for consumers; nicotine-free products have never been commercial successes. This drug modulates mood in subtle ways. Once one develops tolerance to nicotine, suddenly stopping use, even for a few hours, usually produces discomfort, including varying degrees of difficulty concentrating, irritability, anxiety, and depression. These symptoms are relieved by nicotine. Regular use of tobacco usually results in a dependence on or an addiction to nicotine. The consumer finds it difficult to stop using and usually relapses after an attempt to become abstinent. Addiction to tobacco develops more often after a period of experimentation than do addictions to alcohol, heroin, or cocaine after casual use of these drugs.

Fortunes are regularly made selling tobacco products. At Philip Morris, RJR/Nabisco, Loews Corporation, Brown & Williamson, UST, and American Brands, cigarettes are, by far, the most profitable product. Huge sums are earned at the expense of their best customers' health. The tragic dynamic of profits versus health has led to an intensely adversarial relationship between these businesses and the public health community for more than forty years.

Tobacco companies spend over $4 billion annually marketing cigarettes and spitting tobacco in the U.S., making this industry the leading educator about health and disease. Most of this money is spent on promotional items, discount coupons and dealer incentives, although nearly $1 billion is still spent on conventional advertising. Despite federal laws banning broadcast advertising, many hours of cigarette, chewing tobacco, and moist snuff promotions are featured on televised sporting events each year. Marlboro, Camel, and Newport are the leading cigarettes among teenagers, paralleling the marketing efforts for these brands. Skoal Bandit, Skoal, and Copenhagen are pitched to adolescent males in a "graduation strategy" leading from a sweet, low-nicotine product to the high-nicotine brand, Copenhagen. So-called "low-tar" cigarettes are cynically aimed at people who don't want to get lung cancer, even though these brands are not significantly less hazardous.

Less than one fourth of adults in the U.S. smoke. Women will soon have a higher smoking rate than men, mostly because more females are starting to smoke. Smoking has become relatively more common among less educated, lower income groups. Nicotine addiction is a pediatric disease: It nearly always becomes established in adolescence (an age group to whom it is illegal for merchants to sell tobacco products), with new, teenage users replacing those who quit or die. About 70% of adolescents experiment with smoking, and between one third and one half of this group become regular smokers. Among adults, nearly half of all who ever smoked regularly have already quit. People who have depression or alcohol dependence are more likely to smoke and have more difficulty quitting.

Smoking is increasing worldwide. As the market declines in the U.S., multinational tobacco companies are rapidly expanding into Asia and Eastern Europe. Their entry is accompanied by new marketing efforts aimed at women and adolescents. The U.S. government has accommodated these firms by pressing South Korea and Thailand to change public health laws to permit more aggressive advertising of cigarettes. Even though the domestic market is in decline because smoking kills, foreign trade in tobacco products has not been treated as a health issue by the federal government.

Tobacco causes an enormous variety of illness among users and among nonsmokers exposed to tobacco smoke. Overall, someone who smokes regularly for a prolonged period has a 30% risk of dying because of smoking. Heart disease, cancer (especially lung cancer), and chronic lung disease are the major categories of serious illness caused by smoking cigarettes. Cancers of the mouth and throat frequently result from all forms of smoking as well as from the use of chewing tobacco and moist snuff. The fetus and small child suffer from a variety of unnecessary complications if either parent smokes. Nonsmokers experience more heart attacks, lung cancer, and other forms of cancer if they are regularly exposed to environmental tobacco smoke (ETS). ETS results in an estimated 50,000 unnecessary deaths each year in the U.S.

Public health experts recommend a variety of measures, taken in concert, to control the tobacco epidemic. Limits or bans on advertising and promotion of tobacco products, countering the marketing of these products with pro-health advertising, and making warnings and advice about smoking more prominent and detailed on products themselves are all necessary. Tobacco product regulation should be similar to that used for other drugs. Substantially increasing the price of tobacco products by raising excise taxes lowers use, especially among teenagers. Educating merchants about how to avoid selling tobacco products to minors combined with active enforcement of the law further reduces teenage smoking.

ETS causes more illness and death than all other recognized pollution problems combined. ETS should be eliminated where nonsmokers, especially children, are present. Functionally, this means that there should be no tobacco smoke at all in enclosed airspaces used by nonsmokers.

People who smoke should not face discrimination in employment just as people who are addicted to alcohol are protected from being arbitrarily fired unless drinking interferes with job performance or safety.

Addiction to nicotine has an enormous clinical spectrum, from mild and uncomplicated to very severe and complicated by conditions such as depression and alcohol dependence. It is never too early or too late to stop using tobacco. Most who have quit did so without formal assistance. Until recently, treatment was relatively unavailable, but treatment methods have advanced substantially in the last decade. Ineffective treatment methods are common and include stand-alone hypnosis, acupuncture sessions, and the drug lobeline, the active ingredient in BanTron, CigArrest, and NicOBan.

Most smokers want to quit; nearly a third try to do so each year. Nicotine replacement, with patches or gum, suppresses withdrawal symptoms so a patient can feel comfortable while learning how not to smoke. These medications are not a stand-alone treatment, however. Nicotine replacement only provides support for nicotine withdrawal. It is adjunctive to a treatment program for the addiction itself. The treatment program should be chosen based on the severity of the addiction problem and can vary from written materials combined with brief counseling and limited followup to inpatient treatment. As with all addictions, relapse is common and should prompt reevaluation and further treatment. With current knowledge and techniques, no one should feel hopelessly addicted to tobacco.

JOHN SLADE, M.D.

For Further Reading: Peter Taylor, *The Smoke Ring: Tobacco, Money and Multinational Politics* (1985); U.S. Department of Health and Human Services, *Nicotine Addiction: A Report of the Surgeon General* (1988); U.S. Department of Health and Human Services, *Reducing the Health Consequences of Smoking—25 Years of Progress: A Report of the Surgeon General* (1989).

TOXIC CHEMICALS

Many of the diseases and disorders that are attributed to the "environment" are caused by chemical compounds called toxic chemicals or, less formally, "toxics." The environment (air, water, solid waste, etc.) provides the pathways from the source of these toxic chemicals to the human population. Much of environmental health deals ultimately with the toxicology of these chemicals after human exposure. The adverse human health effects of air pollution and smog are largely due to toxic chemicals. Chemicals in the effluents from industrial plants, leachates from toxic waste dumps, and chlorinated hydrocarbons in drinking water are all potentially toxic. Medicines, natural products, and even foods often contain toxic chemicals.

Essentially all compounds can be toxic—can have deleterious health effects—at some dose or with some use. It has been said that the dose makes the poison. Almost any chemical or physical agent may be toxic in certain normal and usual circumstances, but will certainly show toxicity at high enough doses. Such common agents as oxygen are toxic under certain circumstances. Oxygen is necessary for life and is safe, indeed essential at around 20% in inspired air, but at higher concentrations can damage lungs and in newborns, damage vision.

The typical air pollutants such as sulfur oxides and nitrogen oxides are the products of combustion of organic fuels. These, alone or in combination, are toxic to human lungs above some concentration.

Some poisons, however, have no other use than killing something. Pesticides are supposed to kill pests while leaving human beings and other desirable species undamaged. However, DDT, the World War II miracle pesticide, killed mosquitoes very well indeed, and had very low acute toxicity. But it was discovered that, in the long term, it was toxic—it prevented reproduction by causing egg shell thinning in condors and other birds, and has been shown to have toxic effects in humans, including links to breast cancer in women. Certain compounds, such as TCDD (dioxin) and some of its close relatives, are unintended, and unwanted, byproducts of chemical production or combustion and have no known use, but can be highly toxic.

Persistent chemicals such as DDT, dioxin, PCBs, and other chlorinated hydrocarbons are insoluble in water but highly soluble in fat. These compounds cannot easily be excreted from the body. This persistence can lead to long-term toxic effects.

Acute and Chronic Effects

Acute toxic effects occur soon after exposure or administration. Delayed effects occur a long time after exposure or administration. Tinnitus (ringing in the ears) occurs promptly after aspirin ingestion and warns of the onset of more serious toxicity. Laryngitis and cough occur immediately after exposure to mustard gas; permanent damage to the lungs can become apparent years later. Acute exposure to asbestos elicits no immediate reaction but can be followed decades later by cancer—mesothelioma or lung cancer—and asbestosis, a serious lung disease.

In the early 1960s, the occurrence of many cases of phocomelia, flipper-like hands and feet, in newborn babies was a mystery. It was thought likely that something occurring during pregnancy was the cause. The problem was that the something had happened months ago, probably during early pregnancy. Eventually the culprit was identified as the newly introduced sedative thalidomide. Many months elapsed before this was suggested, and many more months before there was agreement that thalidomide was the cause.

The thalidomide episode and the problems with DDT and asbestos led to greatly increased concerns about the long-term or delayed toxic effects of exposure to chemicals. This lead to the greater concern that some common diseases might be caused by polluting chemicals in the environment. The problem was to associate the effect with the cause.

In 1958 there was an outbreak of sudden deaths in Tennessee. It was quickly established that the victims had taken a newly marketed elixir of sulfonamide, and that the elixir had been formulated with a toxic glycol, which was the lethal agent. This process took less than a week. Typically acute effects are easily seen in animal toxicity (safety) testing or rapidly identified after human use. It is relatively easy to suspect newly introduced agents, but much more difficult for agents introduced years earlier.

We are exposed to many chemicals all at the same time. Which air pollutant causes a particular problem: sulfur oxides, nitrogen oxides, or particulate? Research in the toxicity of combinations of agents is complicated and expensive, but many laboratories are beginning the effort. Some early results suggest that the effects of multiple exposures are often additive—the second chemical adds its effect to the first chemical. This becomes worrisome since many water and air pollutants are not identified. It has been estimated that only 10% of the chlorinated hydrocarbons found in some drinking water supplies have been identified.

There are special problems with agents that are toxic for the brain and nervous system. In general, the central nervous system does not regenerate. If neuronal cells die or are destroyed they do not grow back. Fortunately we are endowed at birth with an apparent surplus of nerve cells. The cells die off at a steady rate, but at a rate that would allow for the maintenance of an adequate number of brain cells until old age. However, some agents kill off brain cells. If exposure to such a neurotoxin occurs at an early age, no effect will be seen on brain function at first, but as old age approaches the missing cells will cause earlier senility. The damage caused much earlier may thus not be noticed for many years or decades.

A significant public health problem is the too frequent occurrence of low birth weight (for age) babies. Such babies have a history of frequent diseases and reduced life expectancy. It is likely that exposure to toxic compounds contributes to this problem. In animal studies testing chemicals for teratogenicity—capacity to cause birth defects—reduced birth weights occur at doses of teratogenic chemicals below those at which teratologic affects are seen and which are otherwise nontoxic.

Toxicity Testing

Compounds are tested for toxicity in a number of ways. Experimental animals are administered the agent for a period of time, and then the animals are carefully observed for signs of adverse effects. The animals are studied as if they were patients with an undiagnosed illness. Blood and urine samples are obtained, and if any animals die, they are autopsied. At the end of the study the surviving animals are killed and carefully autopsied. Tissues are taken for histological study. The toxicologist can determine if exposure to the agent had caused any damage to the organs and tissues of the animals. The cause of death of those that died before the end of the experiment can be determined. Both sexes are studied, and many doses are used, varying from no effect levels to doses that cause disease and death. The duration of the study can be short, following a single dose (acute) or long, following muliple doses lasting up to three years (chronic). The time of observation varies from one day to years. Mice, rats, rabbits, and dogs are most frequently used. Until the thalidomide episode in 1962 long-term studies and long observation periods were rare.

A number of specific toxic effects deserve special mention. Chemicals that cause cancer—carcinogens—are a particularly difficult problem. Because there is such a long time between exposure and the resulting effect, human epidemiological studies, at best, can usually show a causative relationship only after 20 or 30 years. This allows a whole generation of human exposure and perhaps, resulting disease. Epidemiological studies are difficult to perform and interpret. Determining exposure, and controlling for the effects of confounding factors such as diet, medicines and other chemical exposures are uncertain. Small-to-moderate effects are usually lost in small sample sizes.

Among the alternatives to epidemiologic studies are structure-activity studies, short-term tests, and long-term animal studies. Structure-activity analysis uses past experience of the relation between chemical structure and toxic effect to predict the effects from a new structure. This works well with a series of closely related chemicals, but is less effective with new structures, which are the most worrisome. Short-term tests—typically tests in simple systems for mutagenicity—have been shown to predict well for some kinds of carcinogens but not for others. Potentially serious carcinogens may therefore, unfortunately, be missed.

Although time-consuming and expensive, long-term laboratory animal tests are the best method available today to identify carcinogens before human exposure.

All known human carcinogens have been shown to be positive in such tests. Many chemicals that have been positive laboratory animals tests, however, have not been shown to be positive in human studies. This is often because of inadequacies in the epidemiological studies. These include small sample size, inadequate determination of exposure, or too short a time for follow-up.

The decades-long latent period between human carcinogen exposure and cancer onset illustrates a major problem faced today in the field of environmental health—the delayed or long-term effects of environmental agents that show no immediate toxic effects. Yet to associate an early exposure with a late toxic effect, the occurrence and extent of that exposure must be known. How many of us can remember what foods we ate, what medicines we took, or what other toxic chemical we may have been exposed to some decades earlier? Records of air pollution levels, pesticide levels in food, occupational exposure levels, drug prescription records, etc., are generally lacking.

For some peristent compounds, such as DDT and dioxin, which remain in the body for long periods of time, new technologies will help. Chemical methods to detect these compounds are becoming more sensitive and it will become easier to relate chemical residue as measured in the body to disease occurrence. For instance, in a study of the carcinogenicity of dioxin it was possible to measure plasma dioxin concentrations in workers and to show that dioxin levels were higher the longer the duration of employment. Significantly more cancer was seen in workers who had worked longer periods of time. Since the cancers were a late effect, the relationship was not clearly seen until 20 years after exposure.

Many carcinogens act by damaging and altering DNA, causing mutations. New techniques can measure tiny amounts of damaged DNA, which often is retained in the body for some time. This yields information of the nature of the exposure. Some of the resulting mutations in DNA activate oncogenes—fragments of DNA that are involved in initiating cancer. These can now be identified in both induced tumors in experimental animals and in human cancers. In animal studies it appears that many carcinogens cause mutations in the DNA of newly initiated cancer cells that are specific for certain chemicals. This can help relate a human cancer to an earlier chemical exposure.

A number of studies since the 1980s have suggested that most of the compounds in common use today either have not been tested at all or have been tested inadequately. Adequate toxicity data exist for only about 10% of pesticides, 18% of drugs, and 5% of food additives.

A recent OTA study looked at the response of the U.S. health regulatory agencies to laboratory animal carcinogenicity data. It found that most known carcinogens for mice and rats were not regulated. Many regulatory gaps were found, and few of the chemicals were banned. This study suggests that, for toxic chemicals at least, we are not over-regulated.

The identification, characterization and, if necessary, control of toxic substances play a key role in environmental health. Much has been done, but much more is needed. Molecular biology is providing new and powerful tools that will permit faster and more certain identification of toxic environmental hazards.

DAVID P. RALL

For Further Reading: Lester Lave and Arthur Upton, eds., *Toxic Chemicals, Health, and the Environment* (1987); William N. Rom, ed., *Environmental and Occupational Medicine* (1992); Arthur Upton and Eden Graber, eds., *Staying Healthy in a Risky Environment* (1993).
See also Dioxins; Exposure Assessment; PCBs; Pesticides; Risk; Toxicology.

TOXIC CHEMICALS: MANAGEMENT AND REGULATION

For the past two decades in the U.S., people have been learning to confront a wide range of environmental problems that pose inherent risks of injury, disease, or death to both humans and the broader ecological network on Earth. Prominent among these problems are toxic chemicals—substances that are capable of inflicting adverse effects on people in the environment in which they live.

Generally, society through its governmental institutions has addressed the issue of toxic chemicals in a two-step approach. First the *risk assessment* question is asked: "Does the presence of this material in the environment pose a risk?" Second, the *risk management* question is asked: "If it does pose a risk, what are we going to do about it?"

The risk assessor draws on the science of toxicology to answer two fundamental questions regarding any material in the environment: (1) "Is this material toxic and, if so, under what conditions?" (2) "How toxic is it?" The risk assessor then couples this toxicological information with the answer to a third question: (3) "Who/what is exposed to this material in the environment, and under what conditions?" to reach a conclusion about what, if any, risk is posed by the material in the environment. A material could be very toxic in the laboratory, but if its exposure is very small (a very toxic agent stored in a secure location where there is no

exposure to the environment), then the risk would be low. Mathematically, Risk = Toxicity × Exposure.

If there is any risk, the responsibility then passes to the risk manager, who must answer two questions: (1) "Is the risk a significant risk?" (2) "If so, what is going to be done about it; how is it going to be managed?"

There are some situations (many of them from years gone by) when the risk manager's task is comparatively simple. For example, in the late 1960s the U.S. had been sensitized to obvious environmental problems—by declarations of pioneers such as Rachel Carson; by news reports that began to focus on the environment; and by simply looking around and seeing conditions with a new vision. For example, belching smokestacks, once tolerated nuisances, became symbols of environmental risks evident everywhere. Fish kills, once seen as simply transient smelly messes, became harbingers of a dark future that would soon fall upon people unless addressed aggressively. Abandoned dumpsites, once decried for driving down real estate values, were now seen as direct short-term and long-term threats to human health and the environment. Medical wastes—particularly pictures of used hypodermic needles found on public beaches—that were once accepted as inevitable byproducts of "the greatest health care system in the world," became a terrifying symbol of society's loss of control over its own wastes.

These problems are characteristic of those that can be effectively addressed by what might be called a "Just do it" approach to risk management. These problems are clearly evident to everyone. The sources of the problems are also clear and are generally present at a demonstrable geographic site (a "point source"); e.g., the smokestack on the other side of town or the outfall pipe from the industrial park upstream of the city's main fishing hole. And the miscreants can't be missed; e.g., the hospital's labelled wastes that show up floating in the harbor and the chemical company whose name is on the drums at the abandoned dumpsite.

Under the "Just do it" approach, the government passes strict "command-and-control" laws (e.g., the Clean Air Act, the Clean Water Act, and the Federal Insecticide, Fungicide, and Rodenticide Act) that tell the parties involved exactly how they must conduct themselves in order to reduce or eliminate the environmental problems associated with their practices. Such an approach leads to regulatory solutions such as registration of pesticides (complete with a wide range of toxicological information) before they can be used, issuance of a permit (with strict limits on practice and pollution) before a plant can operate, and tracking of proscribed wastes "from the cradle to the grave."

The command-and-control approach to dealing with these problems has particular benefits. First, the approach can be targeted at specific "bad actors," thereby satisfying society's righteous indignation—although the blunt tool of regulation often results in some collateral damage being inflicted on unintended targets as well. Second, it is easy to see whether the regulatory action has had an effect; e.g., anyone can see whether the smokestack stops belching. Seeing progress is a reward in itself and helps to maintain public zeal and support for the work at hand.

In many ways the command-and-control approach has served well. For example, the air-blackened cities of the East Coast are more often seen in photos of a generation ago than in the scene outside the window today. The Cuyahoga River in Ohio no longer catches fire as it did 30 years ago, when a match from a careless smoker ignited organic wastes floating on its surface. The public generally recognizes that programs are in place to address many of the visible problems that originally launched the environmental movement.

However, additional, in some ways more insidious, challenges have arisen to compete for time, attention, and resources with the earlier problems. These new problems are generally harder to see, often more difficult to link to a specific culprit, and increasingly rely upon laboratory experiments with animals to demonstrate the case of possible risks to humans. Many of these cases did not yield easily to simple command-and-control remedies, calling instead for more innovative, imaginative approaches to risk management. Such are the new problems posed by toxic chemicals.

At the same time analytical chemists have improved their craft to such a degree that the levels of many pollutants that can be detected in an environmental sample have fallen 1,000-fold and in some cases 1,000,000-fold over the past 15 years. For example, regulatory action against herbicides containing dioxin was abruptly halted in the mid-1970s when most environmental samples showed "non-detectable" (often interpreted as "zero") levels of the chemical. Ten years later, when analytical techniques were capable of detecting levels of dioxin that were 1,000,000-fold lower, regulatory action for this one toxic chemical skyrocketed. This improved capability of analytical chemists became known as the "vanishing zero" phenomenon, pressed the ability of risk assessors to interpret the significance of these low levels of materials in the environment, and challenged the wisdom of risk managers to decide what to do about it.

To give these new challenges a human face, consider William Ruckelshaus, the only person to serve as Administrator of the EPA on two separate occasions. During his first tour in the early 1970s, environmental problems were manifest, and the solutions were equally evident. The main challenge was to muster the political will to "do the right thing." Given the support of an energized public, significant progress was made quickly.

During his second term in the mid-1980s, Ruckelshaus in some ways lamented the demise of the clear targets of the 1970s. In their stead were generally more vague, but potentially more devastating, concerns about cancer and reproductive effects, posed by a wide array of toxic chemicals whose names alone could challenge a Ph.D. chemist—let alone the average member of environmental groups originally formed to address readily discernable problems such as oil spills and fish kills. In the face of this mysterious, unfamiliar threat, an unbalanced fear of these unseen pollutants arose known as "chemophobia." Ruckelshaus remarked at the time that the public had never been healthier or more worried about their health. To get a rational handle on the problems, he adopted the "risk assessment/risk management" paradigm that has guided the Agency into the 1990s.

As if to illustrate the special problems posed by these new threats, shortly after his return to the EPA in 1984 Ruckelshaus was faced with a firestorm of public concern, anxiety, and anger related to a toxic chemical, ethylene dibromide (EDB), which had long been used as a fumigant to reduce losses in stored grain. When administered to laboratory animals at high doses, EDB results in the formation of many different cancerous tumors in a comparatively short time. Since EDB is a gas, however, it was generally considered to pose no significant risk to humans. People thought that the gas would dissipate from the stored grain with time and that any residual EDB would certainly be driven off during the baking of any food products made from the grain. However, in 1984 sensitive analytical methods detected low level residues of EDB in milled grain and even low levels in baked goods made from such grain.

Following the news of this initial discovery, a series of uncoordinated investigations across the country reported low levels of EDB in stored grain, in bread flour, in muffin mixes, and in baked goods of various kinds. Front page stories about "killer muffins" suddenly appeared, accompanied by reports of clearly adverse effects suffered by workers who had been inadvertently exposed to high levels of EDB.

In the face of rising public concern and a perceived lack of action at the federal level, different states leaped into the breech, taking different actions and setting different standards. The evening TV news reports showed state highway patrolmen in one state removing "killer muffin" mixes from grocery store shelves, while authorities in other states issued cautionary or reassuring statements. Order was finally restored after a few weeks when the EPA took action to phase out the use of EDB as a general grain fumigant over a period of three years. In the aftermath of the near public panic, Ruckelshaus observed, "We did the right thing; but we did not do the thing right."

In fact it was hard to demonstrate that the EPA had done the right thing, or anything at all. In contrast to the smokestacks that no longer belched black smoke, the muffin mixes produced muffins that looked and tasted just the same. The cancer rate did not plummet following the removal of the known animal carcinogen from the food supply. Somehow such actions against toxic chemicals—as correct as they might be—lack the gratification that comes from seeing cleaner air in downtown New York City or fish returning to the once depopulated Potomac River in Washington, D.C.

An additional challenge of managing toxic chemicals in the environment arises from the fact that in some cases the "bad actor" responsible for the problem is either unknown or involves essentially everyone. For example, many scientists believe that naturally occurring radon gas seeping through basement walls represents one of the greatest environmental health hazards facing this country. Yet public support to address this issue—without any real culprit—pales in comparison to the manifest outrage of citizens concerned about arguably lower risks posed by more visible, more easily identifiable sources such as the community landfill down the street. The many sources of lead (e.g., the water distribution systems in our communities and in our homes and the lead-based paint in a significant portion of the nation's housing stock) that are associated with excess lead exposures in millions of children represent another example of the diffuse non-point source problems of toxic chemicals.

In order to address these new challenges, risk managers are trying innovative approaches to supplement or replace the command-and-control methods of the past, which may not be effective in dealing with some of the new problems. For example, the Clean Air Act Amendments of 1990 marshal the forces of the marketplace to provide economic incentives to reduce air emissions of sulfur dioxides in the most cost-efficient way.

On a more fundamental level the EPA is aggressively using public education to highlight the benefits of source reduction (pollution prevention)—the philosophy that it is more efficient to prevent pollution in the first place than it is to clean up after it has entered the environment. The EPA is aggressively pursuing cooperative programs aimed at showing citizens and corporations that acting to prevent pollution is to act in their enlightened self-interest. For example, the EPA has enlisted hundreds of companies in replacing most of their incandescent light bulbs with high-efficiency bulbs that use 20% to 30% less electricity. The lower demand for electricity will reduce the companies' elec-

tric bills while eliminating the pollution associated with the generation of the extra power needed to light incandescent bulbs. The electrical utilities are generally strong supporters of the Green Lights program, since it will reduce the capital requirements to construct the additional generating capacity that would otherwise have been needed.

On a related pollution prevention front, the EPA has enlisted hundreds of companies in the "33/50 Program" in which the participants agree that they will reduce the emissions from their plants by 33% (from a 1991 base) by 1993 and by 50% by 1995. While there are no regulatory "teeth" associated with this program, the Agency believes that a cooperative spirit, backed by the public pressure to live up to these pledges, will result in greater pollution reductions in a shorter period of time than could be achieved through the old command-and-control approach of regulations and enforcement with its lengthy, costly legal proceedings.

The issue of toxic chemicals is clearly international in scope. International trade will require a cooperative approach to risk assessments if we are to avoid an expanded replay of the kind of confusion experienced among the states during the EDB episode. Already toxic chemicals are moving through the global environment without regard to international boundaries. For example, recent increases of PCBs in the Great Lakes, following years of steady decline, are thought to be associated with long-range atmospheric transport of the chemicals from other countries. To address these challenges, a new round of creative, innovative risk management will be required.

Perhaps most importantly, the public and their risk managers will have to confront the necessity of setting priorities among the current and emerging environmental problems associated with toxic chemicals. The country simply does not have sufficient resources to address and eliminate all the risks posed by toxic chemicals in our society. And there is no doubt that analytical chemists will continue to find residues of toxic materials in unexpected places, e.g., dioxin in coffee filters or PCBs in human milk. The challenge will be to be smart enough to distinguish the big risks from the smaller risks in order to set an agenda that will lead to the rational protection of public health and the total environment. While there have been promising steps in this direction, the setting of an environmental agenda will remain a constant challenge to everyone.

DONALD G. BARNES

For Further Reading: Rachel Carson, *Silent Spring* (1954); National Research Council (NRC), "Risk Assessment in the Federal Government: Management of the Process," (1983);

USEPA, *Unfinished Business,* Office of Policy Planning and Evaluation (Feb., 1987).

See also Clean Air Act; Clean Water Act; Dioxins; Economic Incentives; Environmental Protection Agency, U.S.; Federal Insecticide, Fungicide, and Rodenticide Act; Law, Environmental; Law, International Environmental; Lead; Medical Waste; Pollution, Air; Radon; Risk; Source Reduction; Toxic Chemicals.

TOXIC METALS

Toxic metals are unique among pollutants in that they are neither created nor destroyed by human activity but are transformed to more or less toxic forms and redistributed in the environment, often to media that enhance human exposure. People may inhale airborne metals or ingest metals that have been deposited on vegetation or water, which eventually leads to their incorporation into the food chain. Solubility in soil is influenced by pH of soil water and chemical form. Acid rain increases the solubility of metals in soil water and thus the uptake by plants. The relative importance of specific pathways to humans differs for each metal. Most metals, even nutritionally essential ones, may be toxic to people from acute exposure if the dose is high enough. Damage to human health from pollution is most often the consequence of long-term, low dose exposure. The metals of most concern at the present time are lead, mercury, cadmium, and arsenic.

Lead is an ancient metal whose toxicity was recognized by Hippocrates. It is obtained from the ore galena, which also contains silver. Lead was first used by the Egyptians for making animal and human figures and later by the Romans for water conduits, eating utensils, and even as a preservative for wine. Toxicity to the general population has been attributed as a factor in the decline of the Roman Empire. Large-scale pollution of the environment with lead began with the industrial revolution and probably peaked just prior to removal of lead as a gasoline additive in North America.

The major sources of non-occupational exposure to lead today are food, water, and air. Children are at increased risk from lead in house dusts, and hand-to-mouth activity or ingestion of dirt and lead-containing paint. The fetus is at risk from lead in maternal blood. Up to 50% of lead ingested by infants and children is absorbed into the blood, whereas adults absorb only about 10% of any lead they ingest. About 25% of inhaled lead is absorbed. Most non-absorbed lead is excreted in urine. A fraction of it is stored in bone. Dietary deficiency of calcium and iron enhance absorption and toxicity.

Blood lead levels are the generally accepted measure of human exposure. Average levels in children in the U.S. have declined from about 18 micrograms per deciliter (μg/dl) in 1976 to below 10 μg/dl or even lower. Human health effects of lead are related to blood lead levels and result in impaired learning and behavioral development in children with levels as low as 10-15 μg/dl. Higher levels of exposure produce more obvious effects, including an unsteady gait, anemia, and kidney disease, and may progress to coma and death. Toxicity may be treated with an agent that chemically binds the lead to enhance its excretion. Because stopping exposure is the most important step in treating lead poisoning, it is important to identify the sources of exposure. Fetuses, infants, and toddlers are at the greatest risk for subtle effects on the nervous system. Current opinion suggests that levels should not exceed 10 μg/dl. Natural levels without exposure to environmental pollution may be as low as 1 μg/dl.

Mercury is peculiar in that it is liquid in its metallic form and readily enters a vapor phase. The history of mercury is linked to the pursuit of alchemy in ancient China, India, and Persia. It was used as a medicinal in many cultures and, until recent years, was a major treatment for syphilis.

Toxicity of mercury is related to its chemical form. The liquid form as used in thermometers is virtually non-toxic but becomes oxidized in contact with body tissues. Inorganic forms such as mercuric oxide and bichloride of mercury are toxic to the stomach and kidneys. The importance of mercury toxicity today concerns two diseases that affect the nervous system and are largely irreversible. Continuous exposure of infants and children to products that contain mercury, such as house paints, teething powders, ointments, and diaper rinses, results in acrodynia or "pink disease," characterized by redness and swelling of the skin of the extremities, irritability, and progressive loss of mobility. Withdrawal of mercury from various household products has reduced the occurrence of this disease.

The second major health concern is exposure to methyl mercury, which has been used in the paper pulp industry and as a fungicide for grain seed. It is also produced naturally by bacteria in water where it is transferred via the aquatic food-chain to fish and eventually to people. This process is enhanced by acidification or acid rain. Pollution of Minamata Bay in Japan resulted in a large number of the fish-consuming population developing Minamata Disease in the 1960s. Toxicity results in impairment of normal brain development in the fetus and infant with loss of vision and hearing. Older people may also lose the ability to walk and may eventually die from the exposure. Limiting the methyl mercury content of commercial fish to 0.5 parts per million (ppm) helps to protect people in the general population from excess exposure. Reducing exposure of pregnant women (mercury content of mother's hair below 10 or 20 ppm) helps prevent toxicity to the fetus. The potential toxicity of the small quantities of mercury vapor released from dental amalgams is not established. On rare occasions it may cause hypersensitivity or allergic contact dermatitis in certain individuals.

Cadmium is a relatively new toxic metal, discovered only about 150 years ago. It is present in zinc-containing ores in trace quantities but has many industrial uses because of its anticorrosive properties; it is used as a component of batteries. People can be exposed through water and food. Cadmium in soil is soluble in water and is taken up by plants, particularly leafy vegetables and grains. Because the metal is not readily excreted, nearly all that is absorbed is retained in liver and kidneys bound to a small protein (metallothionein). Kidney toxicity results when a critical concentration occurs; for this reason, the recommended daily exposure is limited to 70 micrograms per day (μg/day). North Americans currently ingest less than 20 μg/day but a one-pack-a-day smoker doubles his daily intake. Because there is no effective way of removing cadmium from the body, the kidney effects are not reversible. During the 1930s and 1940s a large number of women in Japan developed a crippling bone disease from excess cadmium and dietary deficiencies. Lesser exposures have produced kidney dysfunction in large populations living in polluted areas.

Arsenic is a ubiquitous element with many different chemical forms. The main health risk is cancer of the skin from ingesting arsenic in polluted water and food. People living in areas with naturally high levels in ground water or other polluted sources are at greatest risk. In North America, most people are not exposed above safe levels, but pollution of water supplies may increase the risk. Some forms of edible marine life are rich in arsenic but in forms that are not toxic.

Because of the persistence of toxic metals after release into the environment, prevention of human exposure and subsequent health effects can only be controlled by reduction of initial release into various media: air, water, food, and soil. Federal and local governments have imposed permissible levels for toxic metals through regulatory measures. Nevertheless, reduction of environmental and human exposure can only be achieved by reducing release and by encouragement of recycling practices.

ROBERT A. GOYER, M.D.

For Further Reading: R. A. Goyer, "Toxic Metals," Chapter 19 in M. O. Amdur, J. Doull, and C. D. Klaassen, eds. *Toxi-cology: The Basic Science of Poisons,* 4th ed. (1991); J. O. Nriagu, ed. *Changing Metal Cycles and Human Health* (1984).
See also Lead; Mercury.

TOXICOLOGY

History

Toxicology is the study of the adverse effects of substances on health and on the environment. Ancient alchemists and mystics practiced what would now be considered toxicology when they determined the deleterious effects of various berries, herbs, fungi, minerals, and other natural materials on people. Ancient Greek and Roman writings describe the effects of natural poisons such as hemlock and snake venoms. During the middle ages, alchemists and physicians sought to identify additional poisonous substances that could be used for political gain. By the 16th century, physicians had become aware of the diverse effects of a great variety of chemicals on human health, and the study of such effects was developing into a science.

Paracelsus, a Swiss physician and alchemist of the early 16th century, is frequently characterized as the father of toxicology. Many of his theories have become part of modern toxicology, including his recognition that experimentation is necessary to determine the adverse effects of substances, and the concept that a toxicologic response (adverse effect) can be caused by a single chemical. By the end of the 19th century, the field of toxicology had made major advances. Physicians and chemists identified several chemicals used by primitive societies as arrow poisons (e.g., curare) and studied their modes of action. Autopsy samples were analyzed to determine cause of death when poisoning was suspected. As new medicines were developed, toxicologists studied their adverse (toxic) effects from overdose, allergic response, or unanticipated side effects. By the early 20th century, toxicology included the study of commonly used hazardous inorganic pesticides, such as arsenic-containing substances, copper salts, and sulfates, which resulted in many poisonings in humans and domestic animals.

The rapid growth of the pharmaceutical and chemical industries was accompanied by the introduction of a large number of new chemicals used as medications, detergents, textile fibers, dyes, food additives, pesticides, etc. The manufacture, use, and disposal of these many new chemicals creates infinite opportunity for excessive human and environmental exposure and the accompanying potential for adverse health effects. In the 20th century, laws were passed to control exposure to hazardous chemicals. To determine the toxicity of the many thousands of chemicals being manufactured every year, the field of toxicology rapidly expanded and became a major scientific discipline.

Modern Toxicology

The modern toxicologist is a scientist trained to determine the types of adverse effects associated with exposure to chemical substances. The toxicologist conducts safety assessments, which include the administration of the chemical substance to animals and observation of the animals for possible adverse responses. After evaluating the type of toxicity expressed, the dose levels used, and the duration of exposure necessary to elicit a toxic response, the toxicologist conducts a risk assessment, which predicts the probability of the toxicity occurring in humans.

Since toxicity can manifest itself in many ways, including birth defects, nerve damage, infertility, and cancer, comprehensive testing programs are conducted to determine the safety of existing products and new substances. To address the many forms of toxicologic effects adequately, the field of toxicology is divided into specific specialty areas, each examining a specific type of toxicologic response. Chemicals are tested for toxicity, using laboratory animals. Since no single species is a satisfactory surrogate for humans, several animal species are customarily used in a testing program. The following types of tests may be conducted as part of a full safety assessment program.

In acute tests, rodents receive a single dose of a test chemical by the oral, dermal, and inhalation routes. A high dose is used to simulate accidental human exposure. The treated animals are observed for up to 14 days. Lethality, eye irritation, behavioral changes, and other symptoms are examined by a toxicologist to determine the degree of acute toxicity associated with the chemical.

To understand the potential adverse health effects that may occur from repeated exposure to chemicals, the substances are administered to laboratory animals daily, for up to two years. Usually, four groups of animals are used in each study. One group receives a high dose, the second group receives a moderate dose, and the third group receives a relatively low dose. A fourth group receives no test chemical but is used as an untreated control, representing "normal" conditions for that species. All test animals are observed daily, and any unusual symptoms are noted. Periodically, blood is collected from the animals and examined. At the end of the test, the animals are sacrificed, and many tissues from each animal are examined by a pathologist. This diagnostic pathology is necessary to determine whether repeated exposure to the chemical substance may result in damage to vital internal organs.

To determine whether exposure to chemicals can adversely affect fertility, the reproductive toxicologist administers a test chemical to male and female animals for several weeks and then mates them. Treatment continues through pregnancy and nursing, usually through two generations. The toxicologist determines whether treatment with the chemical affected fertility, pregnancy, lactation, or growth of offspring. In years past, accidental exposure to some chemicals resulted in human infertility. To prevent this in the future, reproductive toxicity testing in advance of human exposure is necessary.

Teratology is the study of birth defects. Some chemicals have no obvious toxicity to children or adults, but do cross the placenta of the pregnant woman and cause birth defects. Since pregnant women are exposed to a large variety of chemicals in consumer products, foods, and the workplace, it is important to test chemicals for their potential to cause birth defects. Pregnant animals receive the test chemical through the critical parts of their gestation period when the limbs and internal organs are developing. Just prior to birth, the fetuses are removed and examined for abnormalities caused by exposure to the test chemical. To help assure the identification of teratogenic chemicals, usually two animal species are used.

Years ago we were told that a "blood-brain barrier" kept almost all foreign chemicals, including toxicants, from entering the central nervous system, thus protecting the brain and spinal cord from hazardous substances. It is now known that many types of chemicals are able to penetrate the blood-brain barrier and affect nervous system function. A variety of tests have been developed that examine the ability of a substance to cause altered behavior, delayed nerve development, inhibition of learning, and physical damage to central and peripheral nerves. These tests provide an assessment of a chemical's potential to cause neurologic disorders.

In recent years, the field of toxicology has expanded to include an assessment of the adverse effects of chemicals on evironmental systems. One important area concerns toxicity to aquatic organisms. Aquatic toxicity assessment has been divided into freshwater and saltwater testing. Toxicity to freshwater organisms is determined by administering the chemical to warm water (bluegill) and cold water (trout) fish, and to aquatic invertebrates such as crayfish and midge. Saltwater organisms that are used include minnow, silverside, shrimp, and clam. As the chemical industry grows and the releases of chemical wastes into waterways continue, it is important to understand the potential impact of a release on living organisms.

Toxicologists and regulators recognize that birds are exposed to a variety of chemicals, including pesticides applied to millions of acres per year and to effluents from manufacturing plants. An assessment of avian toxicity is conducted using mallard ducks and bobwhite quails. General toxicity is evaluated, as well as effects on bird reproduction.

Chemicals that mutate cells are strictly regulated. Since many chemical carcinogens transform normal cells to malignant cells by mutation, mutagenic chemicals are considered suspect carcinogens until they are tested for carcinogenic activity. Tests for mutagenic activity include exposing cells in petri dishes to the test chemical and then examining the cells for signs of mutation. Mutation occurs when the chemical alters cellular DNA. Complex tests have been developed that can measure the interaction of chemicals with DNA inside living cells. These tests can confirm the adverse interaction with DNA and demonstrate the mutation expressed by the altered cells. The field of mutagenicity testing has become important because of the overwhelming evidence that many cancers occur as a result of gene mutation.

New Directions in Toxicology

A new specialty within the field of toxicology is molecular toxicology, which examines the interaction of test chemicals with intracellular macromolecules. This new specialty attempts to identify the mechanism by which toxic substances cause their adverse effects. Once these mechanisms become known and accurate predictions of toxicity can be made by examining a chemical's structure, the needs for animal testing are greatly diminished.

Historically, toxicology has concentrated primarily on the identification of physical damage caused by exposure to toxic substances. Little effort was made to examine for behavioral changes. Now, behavioral toxicology is an emerging specialty with new, predictive tests being used to determine whether chemicals cause behavioral changes or learning disorders with no observable physical damage.

Another emerging specialty is immunotoxicology, the study of the effects of chemicals on the immune system. Because of the complexity of the immune system and the inability to develop predictive tests for impaired immune function, this specialty remained dormant until recently. New tests just developed can predict the effects of chemicals on the various immune system functions. Immunotoxicants can now be identified with some degree of confidence.

The broad field of toxicology continues to develop. Greater emphasis is being given to the development of environmental tests necessary to predict the effects of chemicals on the environment. Although the chemical industry is actively working to decrease emissions from manufacturing facilities, thousands of tons of chemicals are released into the air, soil, and water every year. In addition, chemicals such as fertilizers and pesticides are purposely introduced into the environment. To assure the accurate evaluation of chemical toxicity to humans and their environment, safety assessment methods must continue to improve.

RALPH I. FREUDENTHAL

For Further Reading: Ralph I. Freudenthal and Susan Loy, *What You Need to Live with Chemicals* (1989); Lester B. Lave and Arthur C. Upton, eds., *Toxic Chemicals, Health, and the Environment* (1987); M. Alice Ottoboni, *The Dose Makes the Poison* (1986).

See also Bioassay; Cancer; Carcinogens and Toxins, Natural; Dose-Response Relationship; Exposure Assessment; Neurotoxins; Risk; Toxic Chemicals; Toxic Chemicals: Management and Regulation.

TOXIC SUBSTANCES CONTROL ACT

The Toxic Substances Control Act (TSCA) was passed in 1976. TSCA authorizes the U.S. EPA to acquire information necessary to identify and evaluate hazardous chemicals, and to use that information to regulate the manufacture, use, distribution, and disposal of toxic substances where necessary to protect health or the environment. Prior to passage of TSCA, hazardous chemicals were regulated only at the point where they enter the environment as wastes (e.g., under the Clean Air Act and the Clean Water Act) or in specific types of exposure (e.g., the Occupational Safety and Health Act and the Consumer Product Safety Act). TSCA was intended to plug the gaps in existing legislation by extending regulation to the entire life cycle of toxic substances. However, it has not had the broad effect originally desired.

Section 5 of TSCA requires a manufacturer to notify EPA 90 days before producing a new chemical substance or commencing a new use of a substance that would increase human or environmental exposure. Testing is required to demonstrate that there will be "no unreasonable risk of injury to health or the environment." Regulations for such "premanufacture notification" were promulgated in 1983. They require submission of extensive test data, studies of health and environmental effects, and data on the fate of the substance in the environment.

If the data show there is an unreasonable risk, the EPA administrator may ban or restrict the manufacture or distribution of the substance. If EPA finds the data to be insufficient, it may require the applicant to submit more data. Because the cost of complying is high, applicants often simply withdraw their notification.

Section 8 of TSCA envisioned an inventory of pre-existing chemicals whose use would not require pre-manufacture notification. However, problems with nomenclature and confidentiality have undercut the usefulness of the inventory. Regulations under Section 8 require manufacturers to report production, release, and exposure data and health and safety studies. Any information about a substantial risk of injury to health or the environment must immediately be reported to EPA.

TSCA gives EPA broad enforcement authority. EPA has fined companies millions of dollars for failure to provide premanufacture notification and for delays in reporting safety data. Under Section 6, EPA has banned production and use of polychlorinated biphenyls (a ubiquitous toxin), prohibited use of chlorofluoro-carbons (ozone depleters) as aerosol propellants, and ordered a phase-out of almost all production and use of asbestos (a cause of lung disease). Much of the asbestos ban was rejected on appeal in 1991.

Title II of TSCA, the Asbestos Hazard Emergency Response Act (AHERA) of 1986, required school systems to identify and abate asbestos hazards in schools. Asbestos removal was frequently required. The current thinking is that asbestos is often best left in place, and the actions under AHERA may have been counterproductive.

Section 7 of TSCA gives EPA authority to seek an enforcement order suspending use of a chemical substance that presents "an imminent and unreasonable risk." As of 1991 EPA had not ever used its authority under Section 7.

For many substances, the EPA administrator has discretion to rely on other laws for regulation, or to request another responsible agency (e.g., OSHA) to undertake regulation.

MICHAEL A. GOLLIN

TRADABLE PERMITS

Tradable permits, also known as *marketable* or sometimes *emissions* permits, are the currency of pollution trading programs. Tradable permit programs specify exactly an allowed aggregate pollution level, and then leave the individual pollution sources responsible for allocating the costs and amounts of reduction among themselves in the most economically efficient manner.

Permits represent the right to emit a certain amount of pollutant, which can then be traded or sold among pollution sources. Those sources able to reduce their emissions output have an incentive to do so; those who exceed their allocated limit must purchase permits for the excess, either from the more efficient sources or from the central regulators, at the price the market will bear.

The U.S. has been a leader in experimenting with and implementing tradable permit programs. In 1974 Environmental Protection Agency (EPA) began a program under the Clean Air Act that awarded emissions credits to industries emitting less pollutant than required by law. As the program evolved, these credits were first used to count against over-limit emissions from other sources operated by the owner (known as *netting*) and later traded among other industry members (*offset*) or stored for internal industrial expansion (*banking*). This program saw little inter-firm trading, but some sources estimate that $5 to $12 billion may have been saved over the course of the effort. A tradable permit system was also employed to ease the transition to low-lead gasoline in the early 1980s, and has been used to control effluent into the Dillon Reservoir in California.

The Clean Air Act Amendments of 1990 (CAAA) enacted one of the broadest tradable permit programs to date in an attempt to control sulfur dioxide emissions from utility corporations, and thus reduce environmental damage from acid rain. In contrast to the 1974 program, the CAAA allocates permits for all emissions: A utility must own a permit for every ton of sulfur dioxide emitted, and EPA strictly controls the number of permits available in a given year. Most of the permits will be allocated to sources without charge according to their expected need, but EPA will be retaining a specified number of permits for auction. By steadily reducing the number of permits available, the program hopes to reduce the emissions by 50% by the year 2000.

The CAAA permit program seeks to control emissions through free market principles. If demand for the permits is high and the price increases, emission controls become a more attractive option. If all sources opt to reduce emissions and the price of the permits plummets, making pollution the less expensive alternative, EPA can reduce the permit supply.

Despite their appeal as an alternative to command-and-control regulatory schemes, tradable permit programs are not universally accepted. Some domestic and international environmentalists are concerned about programs that effectively award polluters the "right" to pollute. Furthermore, permits are not necessarily applicable to all situations. While economic cost and benefit may be freely traded, the local effects of pollution are not. For pollutants with a local environmental effect, persons living near those sources who choose permit increases rather than emissions controls bear a disproportionate amount of the environmental burden. Tradable permit systems function at their economic and environmental best for non-specific pollution effects, such as acid rain or effluent into a discrete body of water.

TRAGEDY OF THE COMMONS

The concept of the "Tragedy of the Commons" has come into common parlance after an important article by Garrett Hardin (1968). Hardin asked us to envision a pasture "open to all." Each herder receives substantial benefits from selling animals and a small harm from overgrazing. When the number of animals exceeds the capacity of the pasture, each herder is still motivated to add more and more animals since the herder receives all of the proceeds from the sale of and only a partial share of the cost of over-grazing. Hardin concluded:

> Therein is the tragedy. Each man is locked into a system that compels him to increase his herd without limit—in a world that is limited. Ruin is the destination toward which all men rush, each pursuing his own best interest in a society that believes in the freedom of the commons.

While Hardin phrased his conclusion as immutable, he did see a way out. The major plea of the article is, in fact, for self-conscious development of new institutional arrangements to reduce the freedom of individuals in relationship to common-pool resources. After rejecting an appeal to conscience as an effective means to break out of the short-run logic of the commons, Hardin accepted the establishment of social arrangements that use coercion. Hardin seriously considered, however, only one type of social arrangement. "The only kind of coercion I recommend is mutual coercion, mutually agreed upon by the majority of the people affected." However, he argued that no piece of legislation can spell out all the conditions "under which it is safe to burn trash in the back yard or to run an automobile without smog-control..." Given the need for discretion about detailed application of general rules, Hardin argued that we need to "delegate the details to bureaus" and to use "administrative law."

Hardin pointed to a major flaw in his own preferred alternative—the eternal threat of corruption. "Bureau administrators, trying to evaluate the morality of acts in the total system, are singularly liable to corruption, producing a government by men, not laws." Having identified the problem, Hardin swept it aside:

> We limit possibilities unnecessarily if we suppose that the sentiment of *Quis Custodiet* denies us the administrative law. We should rather retain the phrase as a

perpetual reminder of fearful dangers we cannot avoid. The great challenge facing us now is to invent the corrective feedbacks that are needed to keep custodians honest. We must find ways to legitimate the needed authority of both the custodians and the corrective feedbacks.

Hardin criticized those who argue that if a proposed reform is imperfect, "we presumably should take no action at all, while we wait for a perfect proposal." Hardin argued that the "alternative of the commons is too horrifying to contemplate." It was better, in his judgment, to adopt administrative systems than to see everything go to total ruin. In his view, it was a centralized administrative system or total ruin. The way out was to pass broad, presumably national, legislation to enclose all commons and turn the details of how to regulate natural resource systems over to public bureaus to develop their own administrative law.

Because many open-access resources have indeed resulted in tragic levels of overuse and sometimes destruction, many scholars and public officials have relied upon Hardin's analysis to justify the need for centralized control of all common-pool resources. National legislation has been passed in many countries, and administrative responsibilities for managing natural resources have been turned over to centralized agencies. Unfortunately, the results of many of these efforts have been the opposite of what was hoped. Evidence has now been amassed that central regulation has frequently accelerated resource deterioration complicated by severe problems of corruption and inefficiency. Disastrous effects of nationalizing formerly communal forests, for example, have been well documented for Thailand, Niger, Nepal, and India. Similar problems have occurred in regard to inshore fisheries when national agencies presumed that they had exclusive jurisdiction over all coastal waters.

In the decades after Hardin's article, extensive theoretical and empirical work on the commons has been undertaken and a new academic association—the International Association for the Study of Common Property—has been established and has held several international meetings. We have learned that the type of resource must be separately analyzed from the type of property arrangement. Common-pool resources exist wherever there are natural resources or human-made facilities where excluding users is costly and consuming by some subtracts from the benefits available to others. Many types of property arrangements exist in relationship to these kinds of resources, some of which are quite efficient and help to sustain the continuity of the resource. Hardin incorrectly presumed that most common-pool resources were open-access resources where property rights had not been well-defined.

It is now known that some who jointly use a common-pool resource will:

- expend considerable time and energy devising workable institutions for governing and managing common-pool resources;
- follow costly rules so long as they believe that most of the others affected also follow these rules;
- monitor each other's conformance with these rules; and
- impose sanctions on each other at a cost to themselves.

The likelihood that resource users themselves will develop effective institutions for regulating the use of common-pool resources is increased by the following factors:

- Homogeneous interests (most resource users share similar technologies, skills, and cultural views of the resource).
- Low discount rates (most resource users have secure tenure and plan on using the resource for a long time into the future).
- The cost of communication among individuals is low.
- The costs of reaching binding and enforceable agreements is relatively low.

Large groups have more difficulty governing and managing common-pool resources but usually because size is negatively associated with the above factors. In relatively homogeneous groups in which mechanisms exist for reaching binding agreements about agreed-upon ways of government and management resource use, even quite large numbers of resource users are able to arrive at effective rules to limit the use of their resource. Further, if large groups are composed of smaller groups that focus on specific parts of a larger problem—such as how to regulate water distribution on a branch of an irrigation canal—then these smaller groups can be clustered into ever larger aggregations that then may be able to tackle problems that affect all participants.

Another lesson learned is that any effort to develop new rules for governing and managing complex resources is likely to generate unexpected results and be subject to errors for some time. Thus, all technological and institutional interventions need to be approached as a learning process that helps generate information about errors so that those involved and others can learn from errors rather than continue to make them. There are no panaceas—remedies that work in all circumstances. For every institutional or technological fix there are ways that highly motivated individuals can cheat without being caught, at least for a while. Wholesale solutions imposed on many different resources in a large terrain are more likely to be ineffective than efforts that enhance the institutional environment that encourages responsible self-governance, self-monitoring, and self-enforcement.

ELINOR OSTROM

For Further Reading: Fikret Berkes, David Feeny, Bonnie J. McCay, and James M. Acheson, "The Benefits of the Commons," *Nature* 340:91-93 (1989); Daniel W. Bromley, ed., *Making the Commons Work: Theoretical, Historical, and Contemporary Studies* (1992); Elinor Ostrom, *Governing the Commons: The Evolution of Institutions for Collective Action* (1990).
See also Community; Environmental Organizations; Law, Environmental.

TRANSACTION COSTS

Transaction costs represent the price of reaching agreement between parties or carrying out a transaction. High transaction costs inhibit change, keep the parties from getting full information, and affect the agreement and actions of the parties.

A classic example concerns a polluter and nearby landowners. Assume that the polluter may discharge effluents freely, bearing none of the costs and reaping $90 worth of benefits of his pollution. The nearby landowners bear the $100 costs of that pollution. Market theory suggests that the landowners will pay the polluter $90 to stop polluting (perhaps by purchasing emissions scrubbers). This result, known as Coase's Theorem, gives the polluter the same benefit he had at the start ($90) but effectively saves the landowner $10 compared to the status quo.

The landowners, however, must assemble and initiate the process of negotiation with the polluter. There are costs associated with drawing up an agreement and enforcing it. Furthermore, the costs and benefits to the parties are rarely obvious and must be developed through extensive study. Each of these activities incurs a cost to the landowners and represents the transaction costs. These costs may well exceed the benefit the landowners ultimately seek. Suppose, for example, that the landowners perceive that this process would cost their group $15. It would be cheaper for them to suffer the pollution in silence.

In this example, $15 worth of transaction costs deterred the landowners because the polluter possessed the "right to pollute" at the outset. The analysis works equally well when reversed: If the value to the polluter of venting the effluents is $110, he could afford to pay the landowners their $90 cost and still make a profit. Thus if all the residents possessed a "right to be free from pollutants" and there were zero transaction costs, the polluter would begin production. However, if the transaction costs were $15 he would not set up a factory.

Transaction costs are highly dependent on context. They are lower when the situation is a free market setting: two parties negotiating when each has a range of competing alternatives. In the classic pollution situation, the parties must deal with each other in a situation filled with emotion, ignorance, and mistrust. Organizing and negotiating between multiple members of an affected group increases transaction costs. In the situation above, the polluter must persuade each and every one of the landowners to accept his solution. These costs may be quite high.

Transaction costs deter bargaining parties from behaving "efficiently." The efficient result in the first example is that the landowners buy out or bribe the polluter. Since the transaction costs deter them from seeking this result, the system remains inefficient with costs ($100 to the landowners) exceeding benefits ($90 for the polluter). Government intervention may be able to correct these inefficiencies by lowering transaction costs or by imposing the efficient solution. However, the government might not impose the efficient system for a host of reasons. Government actions also have social costs.

TRANS-ALASKAN PIPELINE

The eight-hundred-mile-long Trans-Alaskan Pipeline extends from Prudhoe Bay above the Arctic Circle to the Port of Valdez in southern Alaska. A massive feat of engineering, the 48-inch pipeline is constructed of half-inch steel, most of it coursing through mountains and over frozen tundra. Administered by the Aleyska Pipeline Service Company, the pipeline transports almost 80 million gallons of crude oil a day and supplies nearly 25% of the United States' domestic oil.

The massive oil field in the North Slope of Prudhoe Bay was discovered on December 26, 1967. However, Prudhoe Bay is inaccessible to ships during most of the year due to the subfreezing weather. The consortium of oil companies that made the discovery resolved this problem by proposing a pipeline to move the crude oil south to the warmer waters of Valdez, where it could be shipped by tankers to the west coast.

Objections arose as soon as plans for the pipeline were announced. Native American groups held that the land was theirs, and contested whether the government had the authority to allow the construction. Environmental groups argued that the plan lacked sufficient consideration of how the pipeline would affect the environment of the tundra. Both groups

brought legal action, delaying the start of construction for over five years.

The environmentalists objected that the pipeline would negatively affect the migratory and feeding habits of Alaskan wildlife. In addition, since the oil was approximately 160° F when it came out of the ground, the permafrost of the surrounding land would melt when the hot oil moved through the pipeline. To prevent melting, refrigerators were put on the supports that held the pipeline up over the permafrost. Another potentially damaging aspect was the expected increase in heavy tanker traffic at Valdez, Alaska, which would significantly increase the chance of a major oil spill along the Alaskan coast.

Several of these worries were realized. In March, 1989 the *Exxon Valdez* supertanker ran aground, dumping ten million gallons of crude oil into the ocean and causing extensive damage to the ecosystem of Prince William Sound. Also, in early 1990 it became clear that many sections of the pipeline had suffered corrosion at a rate much greater than expected, raising fears of "blowout," or pipeline rupture, which would spill oil onto the landscape. Although shutdown of the pipeline was considered, the plan was dropped because of the economic consequences. In 1991, the Aleyska company completed an inspection and declared the pipe was in good condition.

In recent years, production on the North Slope of Prudhoe Bay has begun to decline due to a depletion of the oil reserves. Oil companies are now looking at the possibility of exploiting the Arctic National Wildlife Refuge sixty-five miles east of Prudhoe Bay.

TRANSCENDENTALISM

Transcendentalism is both a short-lived literary period in American culture as well as a cogent philosophy whose basic tenets, diverse as they are, remain important and influential in philosophy and particularly in environmental thinking in the present day. To the extent to which they can be narrowed, the basic precepts of the philosophy of transcendentalism encompass a belief in an underlying unity of creation and a reliance on personal and individual insight as the path to truth, rather than an externally imposed logic or experience. The philosophy is inseparably linked to a conception of a higher spirituality, the existence of a universal being, but in a mystical yet secular way. Men (and women, for the female transcendentalists were ardent feminists) are seen as capable of achieving a godlike potential in this life. Through individual introspection and inspiration, people can achieve a union with the "Oversoul" that encompasses all existence.

The key to this transcendence is a love of and an immersion in nature. Nature provides physical and aesthetic sustenance; it provides the link between physical reality and inner truth. Humans possess a divine dominance over nature, but it is only through nature that universal union can be reached. Henry David Thoreau's experiences at Walden Pond provide the best illustration of this way of life. In its reaction to traditional conceptions of knowledge and society, transcendentalism became a force for social reform, epitomized by Margaret Fuller's early feminist efforts and Samuel Ripley's Brook Farm cooperative community. Other members of the Transcendental Club included A. Bronson Alcott and Walt Whitman. Its foremost progenitor in American literature was Ralph Waldo Emerson.

Influences of transcendentalism on American thought and culture can be detected in many sources. Herman Melville's writings embody some transcendental philosophies, as do those of Emily Dickinson and Robert Frost. Culturally, some scholars see transcendentalist thinking reflected in the rejection of societal constraints and the search for inner freedom that characterized the youth movements of the 1960s and 1970s.

New England transcendentalism traces its origins to the German writings of Kant, whose works in the 1780s argued that some degree of reason and cognition are somehow innate. Imported to the United States by way of France and England and interpreted (some would say misinterpreted) in light of the Unitarian version of Christianity prominent in New England at the time, transcendentalism emerged as an individualized American reaction to the theories of John Locke that taught reliance on logic and experience. Transcendentalism also appears to be somewhat influenced by the religions and philosophies of East Asia, distilled as they came to New England by the long and circuitous trade route from Europe.

TRANSPORTATION

A transportation system enables society to function efficiently by bringing people together to carry out their daily activities—at work, at the marketplace, at school, at play. Throughout history, transportation has determined where and how human beings live. When we relied on our own power, we lived close to one another, with homes and jobs clustered by a navigable waterway for easy movement of people and freight to places beyond a day's walk. For most of human history, cities grew up on strategic waterways. By the turn of the century, the railroads, the streetcar, and rapid transit systems extended the range of travel outward and created small cities, towns, and neighborhoods clustered within walking distances of streetcar stops and rail stations. The larger central city core remained strong with access to it available from many places and directions.

As the 20th century progressed, the availability of the private automobile to many people has altered this traditional settlement pattern. All land has become more or less equally accessible and ripe for development. The preeminent city has become relatively less important. In a similar manner the airplane has brought cities in the nation and around the world closer together, enabling commerce and tourism to thrive. Air travel has largely replaced travel by rail or ship for long distance trips.

The Motor Vehicle and the Highway System

In the U.S. most travel occurs in automobiles. Today, 168 million licensed drivers each drive an average of over 8,000 miles per year. There are 144 million automobiles registered in the nation, driven an average of 10,000 miles each, or 1.4 trillion miles per year. There is one automobile for every 1.7 persons in the United States, compared to only one for every 12 persons in the rest of the world.

In addition to this fleet of automobiles there are 44 million trucks, 4 million motorcycles, and 600,000 buses registered in the U.S. Travel by all of these vehicles totals 2.1 trillion miles annually on almost 4 million miles of road. The road network to carry this enormous traffic load ranges from high-speed, multilane, limited-access highways to local streets.

The unprecedented mobility that the private automobile and the supporting street and highway network has given us has not been without its problems. The emission of pollutants causes health problems, damages agriculture, and corrodes buildings. Hydrocarbons and nitrogen oxides form ozone, a major ingredient of smog. Carbon monoxide is also produced by vehicle traffic.

On average, automobiles log about 10,000 miles per year, consuming a total of 72 billion gallons of fuel, or about 500 gallons per vehicle. Including trucks, buses, and motorcycles, motor vehicles in the United States consume 114 billion gallons of fuel annually, or about 60% of all the petroleum consumed in the U.S. This dependence on an imported, finite resource poses great political, economic, and environmental problems.

Because of the mobility the automobile provides, development can be far-flung and densities can be low, consuming land at a rapid pace. This has caused the loss of valuable farmland, woodlands, and scenic open spaces. For example, in the New York metropolitan region, in the last generation a population growth of only 6% has been accompanied by an increase in land consumption of 60%. Within cities 40% of land is devoted to the movement, maintenance, and storage of motor vehicles.

The cost of motor vehicle accidents is enormous. Two million people are injured each year, and about 46,000 die. Ten times as many people die in motor vehicle accidents as in all other forms of transportation combined. However, in recent years the death rate from motor vehicles has declined by almost 20%, a result of increased use of seat belts, reduced speed limits, and a change in society's attitude toward driving under the influence of alcohol.

Today, many highways are experiencing serious congestion problems, particularly during the commuting rush hours. The result are mounting delays, wasted time and accompanying economic losses. The causes are many: parking is provided free by employers to most workers, often even in large cities, gasoline prices have dropped relative to other consumer costs, trips to work are longer as jobs spread to locations further from where workers live, and job sites are not designed for transit. The result: higher automobile ownership and fewer people using carpools and transit, adding to highway traffic. A variety of actions are now being pursued to make the highway network work better. Transportation system management includes low-cost high-benefit actions, such as left turn lanes and better-timed traffic signals.

Transportation demand management makes use of such action as shifts in commuting times, four-day work weeks, flexible work schedules, strategically located park-and-ride lots, high-occupancy lanes to bypass congestion, preferential parking spaces and lower parking prices for poolers, tax supported transit subsidies by employers for their employees, organized carpools and vanpools, and matching programs by large companies and by government, are among the methods tried. Telecommuting, in which workers spend some days working at home with a computer and a modem or fax machine, is another idea.

Programs that affect the cost to the commuter, such as parking charges, offer the most prospect for shifting behavior, but their popularity is untested. Convincing individuals to do what is best for society when it is not in their short-term interest is very difficult. The incentive to continue to drive alone remains strong because of its freedom, flexibility, convenience, and low cost; single-occupant drivers can leave for and from work when they wish, can run errands during lunch time, or before or after work, have transportation instantly available when needed unexpectedly, and do not need to compromise their daily schedule for anyone else.

In the last few years, Intelligent Vehicle/Highways Systems (IVHS) have been undergoing development. Systems are being tested and developed that can provide drivers with directions from their present location, or inform them quickly about traffic congestion ahead and their options to avoid it. IVHS can be used to communicate with emergency teams to respond quickly to accidents or other traffic-stopping incidents. This is done by placing detectors in the roadway that are connected to a central data bank and are programmed to interpret the traffic situation by the speed and volume detected. Eventually, technology may even develop collision avoidance systems that can enable vehicles to travel at high speeds closer to one another, increasing capacity.

While technology might hold some of the answers to our transportation problems, in the long run, modifying the design of our built spaces, and the patterns of land uses in which we live, work, and play are likely to be necessary to shift the focus from near total dependence on automobile travel to a wider range of travel options, including transit, walking, and bicycling.

Public Transportation

In the U.S. public transportation (transit) annually carries almost 9 billion riders who collectively travel some 41 billion miles. This is only 3% of the 1.4 trillion miles traveled by automobile. In the years before automobile travel dominated, transit operated profitably. In the late 1920s transit carried about 17 billion riders annually, about twice what it does now. Immediately after World War II, the number rose to over 20 billion riders before the dispersal of population to the suburbs, made possible by the automobile, contributed to the automobile's dominance. This led to the decline of transit ridership as private operators discontinued the maintenance and replacement of equipment and facilities, seeing little gain in continued investment in a declining industry. Eventually, most transit systems were taken over by the public sector, and in recent years major programs have been implemented to replace outmoded facilities and vehicles. In some cities the transit network has been expanded, and in others new light and heavy rail transit systems have been constructed.

Transit is needed for places where transit can carry many more people more efficiently than can automobiles. Transit serves denser urban areas where activity, and therefore travel, is concentrated. Without transit services to carry a sizable share of the load these places could not function, much in the way that low-density places could not function without the public investment in highway and street network, and private individuals' investment in automobiles.

Transit is also a public service, providing essential transportation for people who cannot afford or use an automobile. In the U.S. 9% of households do not own an automobile. Even in households that own them, not all members of the household can use them. Those who cannot represent about 20% of the nation's population.

The most widespread means of transit is the bus. Of the 9.1 billion passengers using public transportation annually in the U.S., 63% are on the 60,000 buses operated by almost 2,700 companies. Buses usually operate over fixed routes and on fixed schedules on the street system in traffic with other vehicles. To increase capacity without adding more buses, higher-capacity buses, with either a second deck or an articulated second section trailing behind, have been put in operation. To speed buses and give them a competitive edge, various preferential treatments are used, including exclusive bus lanes on freeways and city streets. The most successful of these is the exclusive bus lane in New Jersey leading to the Lincoln Tunnel and a large bus terminal in Manhattan, where 1,600 buses whisk 60,000 riders in the morning peak period past three lanes of slow-moving automobiles and trucks. In intercity service buses carry 300 million people annually. The most ubiquitous bus service is the school bus, carrying the majority of America's suburban and rural school children to school each day.

Rapid transit networks (also known as heavy rail, subways, or metros) carry most of the other public transportation passengers. These systems are designed to carry large numbers of people in urban corridors. Of the 2.5 billion passengers carried annually by rapid transit in the U.S., approximately 1 billion are on the New York City subway with other systems in Atlanta, Baltimore, Boston, Chicago, Cleveland, Miami, New Jersey (two systems, one connecting to New York and the other to Philadelphia), Philadelphia, San Francisco, and Washington. A rapid transit line in Los Angeles is being opened in stages.

Light rail is a moderate-capacity version of rapid transit. These systems can operate on the street, elevated, in a cut, or in tunnels. They are usually less expensive to build because they are often not fully grade-separated. Light rail systems have gained popularity where the high capacity of rapid transit is not needed. They operate as single cars or in short trains, and often do not have high-level platforms or off-vehicle fare collection. Thirteen systems are now in operation in the U.S., six having opened since 1980. At least 20 other cities in the nation are in various stages of planning light rail lines.

Commuter rail systems perform a specialized function in carrying commuters to large urban centers from relatively long distances, usually from 10 to 60 miles away. Stations are often spaced far apart to speed the long trip. The majority of riders arrive at outlying stations by automobile, where parking is usually available. About 340 million trips per year are made on 15 systems in ten metropolitan areas in the United States.

A number of automated transit systems are now in operation in the U.S. today. Their major advantage is being able to run without an operator on board, saving costs while still providing frequent service. Most operate over short distances within single-purpose but spread-out facilities, such as airports, amusement parks, or university campuses. In recent years automated transit systems have broadened their applications into urban areas, with the opening of systems in Vancouver, British Columbia, and Lille, France.

Another 75 million riders a year use small buses or vans that respond to a call from riders and vary their route in an area to pick up and drop off riders. By doing this they eliminate the walk to and from the transit route, but at the cost of slowing down those on board. These systems serve a useful purpose for people who need to minimize walking and who are not on a pressing schedule. Often these systems are operated by a social service agency for the elderly or disabled or in rural areas.

Intercity Transportation

The intercity bus network, with a fleet of 18,000 buses, carries 300 million passengers 22 billion miles annually. The intercity rail network operated by Amtrak has 24,000 route-miles and 516 stations. Annually, 22 million passengers—half in the Northeast Corridor between Washington and Boston—travel 6 billion miles.

Scheduled airline service in the United States is made up of 60 airlines carrying over 450 million people annually in 5,700 aircraft to and from over 800 airports. In addition, there are 270,000 other aircraft—most small planes used for recreational purposes, in industries such as agriculture, and for business travel—seeking the convenience of local airports.

Because of the high noise levels of aircraft, noise has long been an issue around airports, particularly in residential areas. Locating new airfields farther from cities in less populated areas is one remedy, but that makes the airport less accessible for more of its users. It is also increasingly difficult to find sites where construction is possible. Several cities—Chicago, Boston, Philadelphia, Atlanta, Cleveland, and Washington—have rail access to their airports from the core of the city. Other cities are contemplating such lines. New federal legislation provides an airport passengers user's tax to finance such systems.

As airports and the airspace around them have become more crowded, many states have begun to consider the construction of very high-speed rail lines to connect their major cities and those in adjacent states. Such systems exist in a number of foreign countries, notably France, Japan, and Germany. In the U.S. these systems could either use conventional rail technology or magnetic levitation, which avoids the slowing effects of friction. One way to make better use of the existing technology is the use of "tilt" trains that negotiate curves at higher speeds. The feasibility of these systems rests with finding pairs of cities of sufficient size at distances from one another that are long enough to compete with driving, while short enough to compete with flying.

Freight Transportation

The nation's freight network handles 3.6 billion ton-miles of freight each year, equivalent to about 15 ton-miles per person. The network delivers raw materials to the manufacturer, finished products to the point of sale, fuel to utilities, and food grown to the marketplace. Rail, truck, ships, pipelines, and air share the load, each used where best suited. Air freight is reserved for high value-to-weight ratio, perishable, or urgently-needed items. Goods shipped over water and by rail usually have the opposite characteristics. Pipelines are used for carrying liquid freight such as petroleum products or chemicals. Because of the flexibility of trucks—most freight moved by other modes eventually finds its way onto trucks for local distribution—trucks predominate in many areas, placing a great burden on highway pavement and bridges. To find the most efficient means of shipping, combinations involving trucks and other modes have led to intermodal facilities for transferring goods from one mode to another.

JEFFREY M. ZUPAN

For Further Reading: George E. Gray and Lester Hoel, eds., *Public Transportation* (1992); Wolfgang S. Homburger, Lewis E. Keefer, and William R. McGrath, eds., *Transportation and Traffic Engineering Handbook* (1982); Vukan R. Vuchic, *Urban Public Transportation—Systems and Technology* (1981). *See also* Automobile Industry; Energy; Fuel, Fossil; Pollution, Air.

TRANSPORTATION, COMMUTER

The great majority of American workers drive to work alone in their cars. High fuel prices in the 1970s briefly increased the use of vanpools and carpools. Concern about the environment encourages the use of high-occupancy vehicles. However, the competing pressures of housing prices, business relocation to suburbs, two-worker households, the high level of vehicle ownership in most households, day care needs, and a general desire on the part of most commuters for "independence" and "freedom" combined to drive transit and ridesharing use to a new low by 1990.

The 1990 U.S. Census revealed that 73% of all American workers commuted by driving alone (compared with 64% in 1980); 13.4% used ridesharing (down from 19.7% in 1980); transit use fell to 5.3% (from its 1980 level of 6.2%); bicycling dropped to 0.4% (from 0.5% in 1980); walking slipped to 3.9% (from 5.6% in 1980). Working at home increased, however, from 2.3% in 1980 to 3% in 1990. Average travel time increased in 1990 to 22.4 minutes from 21.7 in 1980.

Several trends run counter to the efforts of public transit and other "high-occupancy vehicle" (HOV) advocates.

Suburban areas continue to attract both residential and business growth, but the balance between jobs and housing is often not adequate within each suburban community. Housing prices often force people to the fringe of suburbs, creating greater commuting distances. Transit systems, with perennially limited resources, have been unable to adapt quickly to the explosion in suburb-to-suburb commute trips that occurred during the 1980s. Suburban land use patterns do not generally lend themselves to easy transit access. Cul-de-sacs and narrower, meandering streets, favored by residents for the purposes of scenery, privacy, and speed control, are seldom conducive to efficient route design or easy bus navigation.

Although ridesharing is feasible for suburbanites, there is a paucity of designated high-occupancy vehicle traffic lanes to provide carpoolers or vanpoolers with an advantage in time or speed. The suburbs, it would seem, were a creation of and for people with automobiles.

Work week changes can make ridesharing difficult, since schedules can change more rapidly under these arrangements. Flextime also increases the number of variations in starting and stopping time at an employer, narrowing the possibility that a prospective carpooler will find suitable partners.

Americans like mobility: The number of vehicle miles traveled in the U.S. increases consistently. The demand for inexpensive, spontaneous mobility conflicts with perceived constraints of transit use and ridesharing.

Fortunately, other trends and developments offer countervailing pressure to advance mobility improvements.

Telecommuting—working from home or from a satellite location, often using a computer, telephone, and modem or facsimile machine to transmit work—is on the increase, with more and more employers, both public and private, experimenting with demonstration projects involving from a dozen to a few thousand employees. Telephone companies, not surprisingly, are taking the lead in promoting telecommuting. The federal government is testing telecommuting in a variety of agencies with its Flexiplace program, so far with promising results.

The Federal Flexible Workplace Pilot Project (Flexiplace), developed for use by the President's Council on Management Improvement, included approximately 700 employees from 13 federal agencies. The program's results were promising, both from a transportation standpoint and from a workplace and personal perspective. 82% of participants indicated reduced rush hour usage of their private vehicles. 35% indicated reduced non-rush hour vehicle usage. 45% indicated sick leave usage was generally lower during their participation in Flexiplace than it had been before. Job performance of the participants was either unchanged or improved. Work-related interpersonal communications were generally unchanged, with significantly more participants reporting improvements than declines. Over half the participants responding to follow-up surveys reported at least some improvement in the quality of their personal life as a result of the Flexiplace program. Quality of work life was also generally better. More than 70% of survey respondents reported reductions in job-related transportation and miscellaneous costs and no change in dependent care costs. However, about a third of respondents reported increased home maintenance costs (probably for utilities) due to their participation in the program. Nearly 80% of the supervisors and 100% of the participants indicated Flexiplace was a desirable work option requiring, at most, minimal refinement.

Intelligent vehicle highway systems (IVHS) are putting technological developments to work in the name of faster, more efficient transportation. The Intermodal Surface Transportation Efficiency Act of 1991 authorized over $600 million for IVHS research and development. IVHS approaches include so-called "smart cars" that would be equipped with on board navigation/congestion avoidance assistance for drivers, and "smart highways" that control the distance between vehicles and the vehicles' speed automatically to allow a greater number of vehicles to fit on the highway yet run at reasonable speeds.

In addition, transit is benefiting from IVHS research. Advanced Transit Information Systems (ATIS) allow improved communication between transit agencies and their buses and between consumers and transit service. Systems are planned that would utilize on-street kiosks to provide transit users the latest real-time schedule and suggested routes.

Clean air concerns and laws have done the most in the last few years to advance the cause of ridesharing and transit use. Since motor vehicles account for about 70% of the ozone produced in most urban areas, reducing the number of single-occupant vehicles in use, particularly during peak morning and evening congestion periods, is seen as one approach to reducing air pollution. The Clean Air Act Amendments of 1990 included a provision requiring states with severe ozone areas to force employers to increase the average vehicle occupancy of the employees' commute trips by 25% over the average in that region. Eight metropolitan areas of the U.S. are affected by this requirement: Baltimore, Chicago, Houston, Los Angeles, Milwaukee, New York, Philadelphia, and San Diego.

Legislation passed in 1992 raised the nontaxable limit on employer-provided transit passes, excluded from employees' taxable income the value of employer subsidies for vanpools and park-and-ride lots, and capped the nontaxable limit on employer-paid parking. Employer-provided transit passes and vanpool subsidies are capped at a nontaxable limit of $60 per month, while tax-free employer-provided parking and park-and-ride lot subsidies are limited to $155 per month.

Traffic congestion may be tackled increasingly by charging people for road use in addition to regulation-based programs. "Congestion pricing" is used successfully in some foreign countries to limit road use and encourage people to ride transit. Such strategies are being looked at very closely in the U.S., as well.

Employers have demonstrated a willingness to encourage employee use of ridesharing, public transit, bicycle commuting, walking, and sometimes telecommuting. Greater sensitivity is emerging toward employees' need for child care and elder care services, resulting in wider use of "guaranteed ride home" programs. Such programs offer employees—typically limited to those who rideshare or use transit—the security of knowing the employer will pay for their taxi ride home in the event of a family emergency or unexpected overtime, thus eliminating the need for bringing their car to work.

The Association for Commuter Transportation (ACT), a private, non-profit organization headquartered in Washington, D.C., was formed in 1979 as the National Association of Van Pool Operators (NAVPO). NAVPO changed its name to ACT in the mid-1980s and merged with the Association of Ridesharing Professionals to form a unified trade and professional organization serving commuter transportation professionals throughout the U.S. ACT's 1,700 members are comprised of employers, government agencies, regional planning organizations, academic institutions, consultants, transportation management associations, real estate developers, and other organizations concerned about commuter mobility. ACT provides transportation professionals practical information, networking opportunities, and an annual conference and exhibition. ACT also educates legislators and policy makers, the media and the general public about key issues affecting mobility.

As the 21st century begins, the desire and need for mobility, of both people and goods, will continue to compete with society's needs for a cleaner environment, sustainable energy supplies, and limited ability to fund new supplies of public infrastructure.

MARK WRIGHT

For Further Reading: American Public Transit Association, monograph series: *Our Vehicle for Conserving Energy, Lifeline in Time of Community Crisis, Sound Investment for 21st Century* (all 1991) and *Rural America's Link, Works for America, Clean Air Alternative* (all 1992); Institute of Transportation Engineers, *Tool Box for Alleviating Traffic Congestion* (1989); Sandra Spence, *National Commuter Transportation Survey: People and Programs* (1990).
See also Automobile Industry; Suburbs; Transportation; Vehicles, Alternative Energy.

TRANSPORTATION, RAIL

Rail transit systems are a family of urban and regional passenger public transportation modes consisting of electrically powered rail vehicles (with steel wheels supported and guided by steel rails) operating on separate rights-of-way. There are exceptions to this definition: some regional rail systems have diesel propulsion, metro systems in some cities (Lyon, Montreal) use rubber tires for support and guidance, and streetcars usually operate on streets in mixed traffic.

Rail transit systems (particularly metros) require very high investment in infrastructure, but they provide high performance in terms of transporting capacity, speed, reliability, comfort and safety. They also have a strong image and permanence, which can be used to influence the shape and character of a city. Due to rail technology and electric traction, rail transit is energy-efficient and environmentally friendly: it is quiet and produces no exhaust along its rights-of-way.

Rail transit encompasses four major modes: streetcars, light rail, rapid transit and regional rail.

Streetcar, trolley or tramway (British). Electrically powered rail vehicles operating in 1–3-car trains, mostly on urban streets. Once the dominant transit mode, streetcars were later replaced in most cities by buses, or upgraded into light rail. Today streetcars operate efficiently in cities that either do not have major street congestion (Eastern Europe), or have introduced good traffic regulation and priorities for transit (Grenoble, Melbourne, Zürich).

Light rail transit (LRT). Electric rail transit system with high-capacity vehicles (usually articulated with capacities of 120–220 spaces) operating in 1–4-car trains, mostly on separated rights-of-way. Being largely independent of street congestion, LRT provides reliable, comfortable service and represents the basic transit network in many medium-sized cities. LRT systems created by upgrading streetcars exist in many West European cities (Cologne, Rotterdam), while new LRT systems have been built in many cities of developed and developing countries, such as Calgary, Manila, Portland, San Diego, and Tunis.

Rapid transit or metro. Electric rail transit system with 4–10-car trains operating on fully controlled rights-of-way. Rapid transit is a complex, high investment system that provides the highest performance (speed, capacity, reliability and safety) of all urban transport modes. Construction of a metro system is a major effort and requires careful planning. The role of the metro system in transportation and its positive impacts on urban form and on quality of life must be weighed against the required investments to establish overall feasibility of their construction. In cities with good coordination of transportation and land use planning (Stockholm, Toronto), metro stations are developed as urban subcenters. Even without strong planning controls, high accessibility of metro corridors and stations fosters intensive activities. Opening of the metros in San Francisco, Washington and Atlanta triggered multi-billion-dollar investments in their station areas.

Regional (commuter) rail. Regional transit service provided by railroad companies or transit agencies, consisting of electric or diesel-powered trains operating on railroad lines. This mode is characterized by very high reliability, speed and riding comfort. Commuter rail services were initially provided as local trains serving suburban areas at "commuted fares" (hence "commuter" designation). In recent decades many cities have upgraded these lines by extensions through central cities and in suburbs, electrification, more frequent, regular services, integrated with local transit through joint stations, fares and information. Serving the fastest growing portions of metropolitan areas and offering park-and-ride, regional rail service is competitive with the automobile and has an increasing role in North American cities.

Special rail systems, such as funiculars (inclines) and cable cars. These vehicles also belong to rail transit in a broader sense; they are used for special topographic conditions, particularly hilly terrain.

Automobile competition and spatial expansion of cities after World War II led to elimination of streetcars from most cities, but it also greatly increased the need for higher-performance rail modes: LRT, metro and regional rail. Metro construction intensified worldwide: in 1955, only 20 cities in the world, including 5 in North America, had metros; in the early 1990s there are 88 world cities with metros, 15 of which are in North America. LRT mode also had a renaissance: since 1978, 24 new LRT systems have been built in developed and developing countries, of which 13 are in North America.

While rail transit is best suited to high-density corridors, it has also been effectively introduced in many low-density, initially auto-oriented cities (Calgary, Sacramento). Highly automated, rail transit has lower operating costs on major lines than buses. Its frequent reliable service, permanence of lines and strong system identity result in much greater ability of rail systems to attract passengers than buses operating on dispersed networks, often with concentration mostly on peak-hour commuter trips.

Due to its high capacity, rail transit uses very limited land area and can support a great variety of urban forms, which automobiles and buses cannot serve without congestion. Moreover, with its high energy efficiency and electric traction, rail transit is by far the most desirable mode of urban passenger transport with respect to energy conservation and environmental protection. Cities with rail transit therefore tend to stimulate pedestrians, human-friendly environments and higher quality of urban life than auto-based cities.

In the long run, cities with extensive rail transit tend to be more open, with stronger social interactions than cities relying on automobile only, which create privacy-oriented living styles. Non-auto owners—low-income groups, teenagers, tourists and so on—have much lower mobility in auto-based cities (Detroit, Houston) than in cities with rail transit, such as New York, Paris, San Francisco or Toronto.

There is considerable criticism of rail transit in the United States because of the high investments it requires. However, the critics mainly focus on short-term, direct monetary costs. Transit planners and city officials tend to be supportive of rail transit where conditions justify it because they evaluate it not only by its immediate effects, but also by its positive impacts on urban form, permanence, environmental features, lower dependence on cheap gasoline and higher quality of urban life. The general public also strongly supports rail projects; in many cities referenda on the introduction of special taxes to fund rail transit construction have been approved by voters (Atlanta, Los Angeles, San Francisco and other cities).

In the future, it is likely that greater attention to energy conservation, improvements of the urban environment and of the quality of life will stimulate continuation of the renaissance of rail transit in many large and medium-sized cities around the world. In the cities of developed countries rail transit is needed to further improve competitiveness of transit with the automobile (Dallas, Los Angeles, Munich, Vancouver); in developing countries, rail transit will be increasingly needed primarily to provide high transporting capacity in rapidly growing cities (Cairo, Manila, Mexico, Sao Paulo).

VUKAN R. VUCHIC

For Further Reading: Boris Pushkarev and Jeffrey Zupan, *Urban Rail in America* (1982); Transportation Research Board, *Special Reports 161, 182, 195, and 221; State-of-the-Art Report 2; and TR Record 1361* (1975–1992); Vukan R. Vuchic, *Urban Public Transportation Systems and Technology* (1981); Vukan R. Vuchic, "Recognizing the Value of Rail Transit," *Transportation Research News 156* (Sept.–Oct. 1991).

TRANSPORTATION AND URBANIZATION

Transportation made possible the great metropolitan areas that account for much of the world's economic and social progress. Cities are fed, factories supplied, and workers delivered to the job. But the task of moving passengers and goods has also degraded the environment and undermined the quality of urban life.

The negative impacts of transportation began with the soot and cinders of the steam railway, the "fly-infested filth" of horse-drawn transport, and the monopolizing of shorelines for highways, railways, and shipping. Later it was the air pollution of the gasoline age, the dominance of traffic and parking, the unwanted intrusion of highways, the barren streetscapes, and the decline of public transport service. In urban areas living and moving have become incompatible.

Since mobility plays such a vital role in daily life, it is fortunate that many of these negative impacts can be overcome. Ways to do this include landscaping the streets and highways, managing how and where the traffic will move, offering attractive public transit options, and using transportation infrastructure to design and redesign urban communities to make them esthetically appealing and less transport-dependent.

Long commutes can be overcome by locating jobs, housing, and services in closer proximity and by using computers and telecommunications as a substitute for travel. Transportation, combined with regional development strategies, can permit more orderly patterns of dispersal. Energy alternatives for transport such as the electric car can reduce the smog and pollution and free the nation from a transport system dependent on foreign supply lines secured at high cost.

Clearing the air will have to be accompanied by clearing the streets. Single-purpose neighborhoods and subdivisions can be replaced by an urban village concept that favors walking and respects the environment. The largely self-sufficient urban village substitutes ease of access for compulsory travel. Experiments include Reston, Virginia; Columbia, Maryland; Las Colinas and the Woodlands in Texas; Irvine, California; and new towns in existing cities such as New York's Battery Park City and Roosevelt Island. Examples outside the U.S. include Jurong in Singapore, Kista near Stockholm, Tema in Ghana, Tama in Japan, and the Industrial City of Curitiba, Brazil.

Many cities are modeling smaller redevelopment projects on the pioneering efforts of these planned communities. Unnecessary streets are closed and converted to other uses. Obsolete rail yards, piers, and terminals are also becoming valuable sites for redevelopment. In Boston the renovated waterfront has replaced derelict piers with restaurants, condominiums, and office buildings. Removal of the city's elevated central artery and directing the traffic underground will eventually release substantial acreage for development and permit easier access to the harbor. Baltimore and Norfolk are among the many other cities making good use of land salvaged from defunct transport facilities.

The plight of inner cities and the environmental disasters of sprawl are prompting many nations to use transportation to move residents out of the congestion of the big city into planned communities in the suburbs or beyond, and simultaneously to redevelop the obsolete older centers.

In Sweden the city of Stockholm has created twenty or more planned suburban communities connected to downtown by rail and highway. The new suburban towns have large retail and service centers, schools, and recreation. They have removed many traffic generators from downtown Stockholm, including schools, shops, and services. This decentralization has permitted an upgrading of the central city, which has included converting streets to pedestrian malls and designating routes for the exclusive use of buses and trams. Automobile traffic in the central area has been curtailed by parking restrictions, higher parking fees, one-way streets, and other methods of traffic management.

Strategies for dispersal in France have been carried out by massive planned suburbs outside Paris, connected to the city by expressways, subways, and a regional rail system. These suburbs have reduced densities and set the stage for renewal in Paris. France is also encouraging dispersal to more distant provincial growth centers by a combination of financial incentives for business relocation and assurances of good intercity transport connections by air, expressway, and high-speed TGV trains.

In Japan public-private partnerships have evolved to help create and manage large-scale urban projects. They play an important part in new towns outside Tokyo, Osaka, and Kyoto. Japan is aiming to accomplish a nationwide dispersal. High-speed "bullet trains" enhance access to and from Tokyo, and a fiber optic network facilitates communications. Within the new towns are amenities rarely seen in other Japanese cities, featuring open space, playing fields, and pedestrian and bicycle greenways completely separated from motor traffic. But as in other countries, not enough jobs have yet been generated, so commuting volumes remain high.

Tokyo provides evidence of the danger of placing excessive reliance on adding more travel capacity to cope with traffic congestion. Tokyo has succeeded in building a transit system capable of moving 26 million people a day, but the effect has been to divert resources from other vital needs, especially housing, sanitation, recreation, and the environment. Commuters spend hours on overcrowded trains to reach affordable housing, so that moving faster has been at the expense of living better. Efforts to solve transportation problems by transportation alone have proven unworkable, and in late 1991 this reality contributed to the decision of Japan's prime minister Kiichi Miyazawa to call for a new focus on the environment aimed at converting Japan to a "quality of life superpower."

Singapore's urban renewal demonstrates the effectiveness of a different focus. In the 1960s this small island republic suffered some of the worst living conditions in Asia. One-third of the population lived in slums and squatter settlements, and unemployment was the rule. For those who lived and worked in town, traffic congestion had become a problem of monumental proportions. The government of Singapore, confronted with spending large amounts of its limited resources on an elevated rapid transit system, decided instead to give priority to housing the poor. The transport solution was to build modern expressways to escape from the congestion and to serve a series of new towns farther out.

To help the process the unemployed were trained and given jobs in construction, carpentry, the manufacture of building materials, and a host of other occupations. With the incomes earned, workers eventually had sufficient funds to rent or buy the apartments that they themselves had built.

Half a million people were housed in a little over two decades, and the quality of the living and working environments was transformed. Meanwhile the old city was redeveloped, bus service modernized and expanded, and auto traffic reduced by charging motorists the economic and social costs of driving downtown in rush hours. As construction boomed and national output soared, Singapore came to be, in Asia, second only to Japan in per capita income, its environment is second to none, and once destitute people are well housed in modern apartments. Helping the poor made Singapore rich.

In the U.S., urban slums and blight are rooted in the fundamental problems of low incomes, unemployment, crime, drugs, and racial tension. There are certainly differences between what can be done in a democracy and under the often dictatorial procedures of Singapore's autocratic government. But the U.S. has many advantages that were not available to Singaporeans, namely ample resources, skills, and space. On balance the U.S. is in a good position, physically and financially, to renovate its urban areas, given a national commitment to do so. Indeed, many of the world's big cities were completely rebuilt following their total destruction by war, including Rotterdam, Seoul, and Hiroshima. The rebuilding was completed in only a few years and in spite of limited resources.

Washington, D.C., and the surrounding capital region are beginning to demonstrate what can be done. An update of the L'Enfant Plan for the capital issued in 1961 called for a series of transportation corridors into Maryland and Virginia along which, at rapid transit stations and highway interchanges, new communities would be built to accommodate growth. Many of the federal jobs downtown would be moved out, and Washington itself would be redeveloped to accommodate its primary role as a center for policy-making, national and international affairs, museums, the arts, tourism, and housing. The plan has been evolving with the redevelopment of the southwest district of the city, the waterfront, Union Station, and Pennsylvania Avenue. At Metro station stops some twenty subcenters are under construction, carrying out the concept of a multi-centered urban region.

In nearby Arlington, Virginia, the county government, in close collaboration with citizen associations, businesses, and developers, has adopted sector plans for the general guidance of growth and redevelopment at stations on the Metro line. Voluntary public-private partnerships have been created, representing all interests and functioning as if they were development corporations. They help to create harmony among builders and between the developers and government providers of infrastructure. The partnership relies mainly on the help of volunteer professionals, but paid staff is supported by an annual grant from the county, matched by membership dues. Special zoning requires a 50-50 division between commercial and residential use of the land, which reduces traffic by increasing the number of people who both live and work in the neigh-borhood. It is anticipated that the mix of land uses, aided by telecommuting, will make possible a new kind of semi-planned urbanization less transport-dependent and more sensitive to the quality of the environment.

The factors that lead to successful interaction between urban development and transportation emerge from this global experience. They are: first, governmental start-up money to help create viable public-private partnerships, and a pooling of federal resources already available to cities to permit a total community-building approach. Second is a broadening of the mission of transportation agencies from simply supplying transportation to solving transportation problems. This fosters close collaboration with urban developers, regional planners, the communication sector, and others who influence transportation requirements.

Third is not depending on mobility to compensate for the lack of convenience in urban planning. Having a variety of jobs and services close to good housing and satisfying environments is part of the answer to many of our transportation troubles, not just supplying more transportation. Fourth is reallocating savings from transport conservation to the needs of the poorest people, the worst areas, and to training and jobs for inner city residents.

Recent advances in moving people, materials, and information allow us to speculate on a new kind of urban future. In high-density areas, redevelopment could take place in discrete areas, green, pedestrian-oriented, and modeled on the college campus. Road and street location would set the stage for enclaves devoted to offices and shopping, with adjacent areas containing housing, play space, convenience shopping, and community gathering places. Internal circulation, depending on volume, could range from automated guideways and people movers to electric cars and mini-buses. Urban traffic problems would be reduced by high-quality public transit systems and the use of video-phones, computers, fax machines, and other aids to the movement of information.

Transportation will support both vertical concentrations typical of New York, Tokyo, and Hong Kong, and a wider regional dispersal in planned communities of varied sizes and styles linked by magnetically levitated transport and automated roadways traversing areas reserved for open space, forest, and farm.

Along with technology and new approaches to urban and regional development, government agencies supplying transportation will be pursuing the new art of transportation problem-solving, which will involve working on transport demand as well as supply. That will mean avoiding unnecessary travel by creating environments that no longer compel escape, by telecommuting instead of commuting, and by shifting travel budgets from unwanted routine urban trips to vacation, recreational, and international travel.

Urban redesign and reconstruction require a vision that takes into account a transformation in the relation of transportation to urbanization. Negative side-effects can give way to positive contributions to the environment and to the human condition, in a new partnership between mobility and livability.

WILFRED OWEN

For Further Reading: Lester R. Brown and Associates, *State of the World* (1992); Richard V. Knight and Gary Gappert, eds., *Cities in a Global Society* (1989); Wilfred Owen, *Transportation and World Development* (1987).
See also Green Belts; Greenways.

TRANSPORTATION TECHNOLOGY

Industrialization in the 19th century changed the wind-water-wood technological complex of earlier eras to one based on coal and iron, and hence meant the birth of new centers of productivity located near deposits of coal and iron. But the new coal and iron technology also made transportation changes necessary and involved the invention of railroads, so transportation was both cause and effect of technological transformation and of new ecological relationships.

For example, urban civilization had been intimately connected with transportation from its very beginnings, for there had to be means of bringing foodstuffs to people who no longer lived in the countryside and who were not primarily engaged in growing food. Thus transport technology became an essential element in determining the location of human settlements and human use of resources.

Waterways provide the easiest—and cheapest—means of transporting goods over long distances, so the great cities of the world have usually been ports—on rivers or seas. Canals also offered economic advantages over horse-and-wagon transport. Before the Erie Canal was built (1817-25) in upper New York State, it took at least 20 days and cost $100 to move a ton of freight from Buffalo to New York City. Once the canal was opened, it took only eight days and cost only $10. Although the canal had been termed "Clinton's Folly" when first broached by Governor Clinton, for it cost New York State some $8,000,000—an enormous sum in those days—it was soon jammed with traffic because of the cheap transportation it provided, and revenue from tolls returned the original investment within a dozen years.

The success of the Erie Canal led to a canal-building mania as other states sought to emulate the New York example. But movement by canal is slow, and canals can be frozen over and blocked by ice during the winter; furthermore, the natural topography does not allow canals to be built everywhere. So, later in the 19th century, railroads, which could be built where canals were impossible, and which possessed speed and dependability advantages over the slow transport of canal barges, became the dominant form of transporting goods over long distances.

Railroad lines helped determine the location of communities. The predecessors of railroads, stagecoaches and horse-drawn wagons, could stop anywhere without lowering their efficiency. But too many intermediate stops were uneconomic in railroad operations, for the efficiency of the steam engine increased with the length of the intervals between stops for fuel, water, or repair. Hence railroad "division points," locations where fuel was stored or engines and crews were changed, soon grew into towns. Thus the location of many midwestern and far western towns in the U.S. was primarily dependent on the efficiency of a technical device, the railroad. Conversely, the loss of transportation facilities through the relocation of a highway or the discontinuance of a railroad line could bring about the decline or death of a town. So in the mid-20th century, when railroads became more efficient by changing from coal to liquid fuels, or by changing to electric power, they no longer required as many intermediate stops—and many "ghost towns" appeared in the American west.

Transportation technology makes cities possible, for the city must have a hinterland to support it with food and raw materials while it supplies manufactured products and services. But other technological needs also play a role in human demography. For example, when coal and iron mining became important to the economy, it was imperative that new industrial centers be concentrated near the coal beds and the connecting valleys. Just as the development of agriculture depended upon the climate and soils in different places, the iron-and-steel industry provides another, albeit quite different, example of how distribution of the earth's natural resources combined with technology to determine the location of industrial centers. For example, in the U.S. a major source of iron ore is the Mesabi range in upper Michigan, but no coal is available nearby, so little ore is smelted there. Why? Because it takes much more coal than ore to extract iron from its ore. Therefore it was cheaper to move the iron ore to the coal than vice versa. Iron ore from the Mesabi range was transported via the Great Lakes to places like Gary, Indiana, Cleveland, Ohio, and Pittsburgh, Pennsylvania, which were located near abundant coal supplies, for smelting and treatment—and those cities became major centers of the iron and steel industry. Similarly, the Lille district in France, the Ruhr in Germany, the Black Country in England, the Allegheny-Great Lakes Region, and the Eastern Coastal Plain in the U.S. became areas of industrial growth and population concentration.

There were also major improvements in the 19th and 20th centuries which greatly altered population patterns and trade throughout the world. The invention of the steamboat made travel and transport over great oceanic distances faster and cheaper. Indeed, the cheap steerage accommodations allowed large numbers of poor people from European countries to migrate to the U.S. and helped create our "melting pot." Not only did transportation and other technologies lead to population concentration, they also led to an enlargement of the city, for developments in urban transportation were as revolutionary as contemporary advances in long-distance transportation. When cities were small, people lived within walking distance of their workplaces—factories, offices, and stores. But the growth in population density began enlarging the size of cities, forcing people to live further away from their work, so there was a demand for new means of transport. In New York City, for example, population growth was so rapid that steam ferries enlarged the metropolitan area to include the New Jersey shore. And wherever railroad lines went through urban areas, a heavy local traffic developed, so by the middle of the 19th century the railroads in areas such as Boston and New York were engaged in the commuting business, enabling people to live fifteen or more miles from their work in the centers of the cities.

The greatest changes in short-distance passenger transportation, however, came with the development of street railways, at first with horse-drawn cars and then electric trolleys. Walking distance no longer set the limits of city growth, and large numbers of workers moved into residential areas which previously had been too far away for them to walk to work. In the period from 1870 to 1930, the urban population of the U.S. increased 696%—and urban areas expanded equally.

This urban expansion was accelerated by the advent of the automobile, which, during the 20th century, enlarged the city to include suburbia and, with the advent of superhighways, even further to exurbia. Although the concentration of automobile exhaust fumes has created smog problems in many major American cities, it should be pointed out that, at the turn of this century, when the automobile first came on the scene, it was extolled as the solution to the pollution, safety, and congestion problems posed by horse-drawn transportation. That was a time when in New York City alone, horses deposited some 2½ million pounds of manure and 60,000 gallons of urine in one day. The automobile promised relief from those prob-

lems—but the large-scale use of the auto brought back pollution, congestion, and safety problems in heightened and altered form.

While the population growth of the central city slowed down as autos and commuter trains enabled people to move to suburban outskirts for more "pastoral" living areas, the central city nevertheless grew—but it grew upwards, as the skyscraper, one of America's most original contributions to architecture, began to dominate downtown areas and business sections in midtowns and urban corridors. Like all technical advances, the skyscraper depended upon a host of other innovations in structural materials and methods, but most notably, the elevator—and it forced the development of heating, lighting, and communication systems of increasing complexity.

Technological developments, especially those involving transportation, have a dynamic impact upon demography and hence business and home locations. And they frequently have effects quite different from those expected. Thus, when America's great interstate highway system was instituted following World War II, there were some who hoped that it would preserve small-town life by putting small towns within easy driving distance of the city. Instead, it spelled the doom of many small towns and simply enlarged the suburban and exurban areas of the major cities. And the interstate highway system also discouraged the use of American railroads for long-distance passenger traffic, making that the province of airlines and motor vehicles.

MELVIN KRANZBERG

For Further Reading: James J. Flink, *The Automobile Age* (1988); Melvin Kranzberg and Timothy A. Hall, eds., *Energy and the Way We Live: A Course-by-Newspaper Reader* (1980); Wilfred Owen, *Transportation and World Development* (1987); John F. Stover, *American Railroads*, Chicago History of American Civilization Series (1961).

TRAVEL AND TOURISM INDUSTRY

The travel and tourism industry (including transportation, accommodation, catering, recreation/cultural and travel service activities) is the world's largest industry accounting in 1994 for more than 10.4% of the world's gross domestic product, one in every nine workers, 10.9% of capital investment and 11.1% of consumer spending worldwide. This industry is an integral part of the life in over 90% of the population of developed countries and contributes core economic activity for developing countries and regions. For many of them it provides access to one of very few prospects of achieving sustainable economic development. Air transport in particular is the lifeblood of international trade, without which global economic development patterns would be radically different.

Three Myths

The World Travel and Tourism Environment Research Center was established in 1991 in Oxford, U.K., to investigate the environmental impacts of the industry and to explore ways in which environmentally compatible growth may be achieved. The Center's second annual review (1993) addresses three tourism myths or misconceptions. Myth 1: Travel and tourism is a nonessential, "mass" activity of affluent people in developed countries. In fact, tourism, as defined by the World Tourism Organization (WTO), is an integral part of the lives of many populations, for business, social, cultural, religious and recreational reasons as well as holidays. In some countries and many regions it is an essential economic activity; it may be the largest earner of foreign currency, providing the essential economic under-pinning of the local population's employment and welfare services. Myth 2: Tourism's major environmental impact is damage to developing countries. In fact, over 80% of the world's international tourism occurs between developed countries, which also generate the bulk of domestic tourism. Package tourism to developing countries is probably under 5% of world travel and tourism. Myth 3: Ecotourism is the only logical, sustainable response to the environmental impacts of travel and tourism. In fact, ecotourism—that is, tourism with the specific motive of enjoying wildlife or undeveloped natural areas—can only make a marginal, though important, contribution because of the limited nature of its market. Without careful management it is no more sustainable than other forms of tourism development, and may cause more problems than it solves.

Global Warming

While the travel and tourism industry does not consume energy on the scale of heavy industry, it does, nonetheless, use considerable amounts of energy for transport systems, heating, air conditioning, swimming pools, laundry, food preparation and so on. Transport for all purposes contributes more than one-fifth of total carbon dioxide emissions and is likely to be the focus of much attention in strategies to reduce warming gases. Tourist transport, like all transport, results in CO_2, SO_2 and NO_x emissions. While all airlines account for an estimated 2% of CO_2 and NO_x, this sector has been the focus of attention because emissions often occur at high altitudes where they may have greater impact than emissions from other forms of transport. Travel and tourism companies also use products containing CFCs for a range of purposes, including refrigeration, air-conditioning and cleaning.

There are a number of initiatives which have been taken by companies to reduce emissions of gases implicated in global warming. The most immediate are simple energy conservation initiatives. Some companies have developed these initiatives into comprehensive energy management programs which make more efficient use of energy company-wide, thus reducing emissions and producing greater savings. Energy programs have also been undertaken in partnership with electric utilities. Many companies are introducing environmental purchasing policies to avoid potentially damaging products.

The airline sector is taking considerable measures to limit emissions. Aircraft manufacturers have achieved huge reductions in noise around airports, and reductions of 80% in unburnt hydrocarbons, 60% in carbon monoxide and 12% in NO_x since the early 1970s. More efficient aircraft handling facilities also significantly reduce aircraft emissions. Improvements in fuel efficiency have been dramatic, with technology consuming less fuel, and estimates are that by 2005 consumption (and thus emission) will decrease by 50% over 1989 levels (due mainly to aerodynamic alteration and structural improvements). Strategies are also being developed by engine manufacturers to limit NO_x emissions.

Depletion of the Ozone Layer

Travel and tourism is affected by ozone depletion in a number of ways. Regulation is already resulting in changes in fire-fighting and refrigeration technology; after 1994 it will become increasingly difficult to repair CFC refrigerators. The transport sector, and specifically airlines, may be increasingly regulated to reduce NO_x. This could raise costs and, ironically, could increase production of gases implicated in global warming. There will be growing pressure to encourage use of alternative forms of transport, such as rail, which could possibly change the nature of the international travel market.

In the long run, holiday-taking habits will change if ozone depletion increases the real or perceived risk of skin cancer. The traditional beach holiday, still the dominant form of leisure tourism, would decline, with consequent effects upon the economies of some destinations. Mountain-based tourism may also be threatened. High altitude activities are more susceptible to the effects of UV-b because it has less opportunity to disperse in the atmosphere, and is therefore more intense. Because of the potential increased risk of health problems, mountain areas might well become a less attractive choice.

Acid Rain

With the potential to erode statues, stonework, buildings, and monuments, and to cause deforestation and damage to freshwater systems, acid rain is of particular concern to the travel and tourism industry. Damage to sites of historic interest or to sites of natural beauty directly affects the industry; there is the danger that if damage, or the perception of damage, to a particular region is great enough, consumers may choose alternative destinations. In Germany, acid rain is perceived as a major problem relating to tourism and the environment. Acidification threatens the Black Forest, the Bavarian Forest and the German Alps, while the cost of maintaining Germany's historic buildings against erosion is estimated at about $100 million a year.

Hayman Great Barrier Reef

A model for tourism is the Hayman Great Barrier Reef Resort in Australia that was redeveloped with the mandate that it was in every possible way to respect the ecology of the reef and the environment in which it was situated. A range of measures have been introduced, including strict management of water use and waste. No effluent is released into the environment from the resort. It has a water desalination plant to serve its own water needs from seawater blended with rainwater collected from the island's roofs. The desalination plant is operated on waste heat from the resort's power station, cutting down on atmospheric pollution and energy use. The gardens are irrigated with treated effluent from the resort's sewage treatment plant, which eliminates the discharge of effluent water into the sea. Dried sewage sludge is used as mulch and compost, as are kitchen and garden wastes. In addition, the resort has its own spill capture equipment to counter any spills of diesel oil, and staff are trained weekly in the use of all equipment. The resort works with environmental conservation groups in the area; it is a recording station for the Australian Bureau of Meteorology, a watch station for coastal surveillance, and a work experience center for marine biology students.

Other Unique Tourist Sites

Ultimately whole segments of tourism are threatened by the disappearance of habitats and species. The World Wide Fund for Nature estimates that of the $55 billion earned by travel and tourism in developing countries in 1988, some $12 billion was due to ecotourism—a rapidly growing sector. In some destinations, such as the Galapagos Islands, the wildlife of the area is the major attraction. However, even where the tourist's primary aim is not to see wildlife, the opportunity to do so once or twice during the trip may influence the choice of destination. The ecotourism

market cannot be sustained without quality environments. The industry itself can implement careful management of existing destinations, but new destinations may be destroyed by other forms of activity. Fortunately, regulations have developed internationally since the 1930s to protect specific areas for their scenic, aesthetic or biological qualities. These agreements eventually included reference to both cultural and natural heritage as well as criteria for environmental impact assessment in sensitive areas. For example, the World Heritage Convention (1972) facilitates the designation "World Heritage Area" for sites of "outstanding universal value." 359 sites are classified under the Convention and many of these (such as the Taj Mahal) are an important part of the tourism product.

Regulation and Self-Regulation

Three primary influences are being brought to bear on the environmental impacts of travel and tourism. These are the international/national regulatory framework, the local or area regulatory framework embraced within area planning procedures and the self-regulatory mechanisms adopted and implemented by commercial organizations. International/national regulation appears to be most effective where there are single issues to tackle. In many instances, regulation protects tourism environments from the ravages of other industries. Planning procedures appear to work best when dealing with area resource issues, but planning procedures do not work when planners seek to control the nature of products, customer segments or the structure of industries. The worldwide shift away from the economic regulation of business provides all the evidence needed that regulators have little effective influence over demand. Self-regulation works best when businesses combine environmental programs for development and operational practices with effective marketing strategies. Large travel and tourism companies, working with national tourism boards and regional authorities, provide the most effective influence over the preferences, activities and attitudes of visitors. Partnerships also provide organizational arrangements for influencing smaller companies within an overall policy framework.

Summary

From fiscal measures to detailed planning controls, more restrictive legislation for travel and tourism seems inevitable. If the industry develops and strengthens its own self-regulatory mechanisms, it will be in a position to influence the development of effective regulation.

Travel and tourism has two great strengths in environmental matters: it has the potential ability to influence customer demand and visitor behavior, using the massive advantage inherent in face to face contact with visitors, and it can give economic value to resources which have no perceived monetary value and security. Without an underlying economic rationale to sustain them, many of these resources, though globally important, are susceptible to damage or destruction, especially for agricultural, extractive or industrial purposes.

GEOFFREY H. LIPMAN

For Further Reading: M. Barrett, *Aircraft Pollution—Environmental Impacts and Future Solutions* (1991); P. Jenner and G. Smith, *The Tourism Industry and the Environment* (1992); World Travel & Tourism Environment Research Center, *Travel & Tourism—Environment and Development* (1993).
See also Acid Rain; Climate Change; Great Barrier Reef; Greenhouse Effect; Landmarks; Ozone Layer; Recreation, Outdoor; Transportation.

TREE FARMING

Tree farming involves cultivation of large plantations of a single tree species and requires intensive management. Although family tree farms do exist, for example, many Christmas tree farms, tree farms are usually part of the reforestation of clearcut areas.

Historically, the demand for wood did not exceed its natural generation until the Romans needed to establish large groves in order to offset their increasing use. Controlled forest management began in the sixteenth century in Germany, where forest harvest and regeneration were balanced to ensure a sustainable yield of timber. Later in the twentieth century, innovative forestry techniques were used to increase the productivity of forests by converting them into agricultural crops. This approach to raising trees requires the maintenance of an "immature" ecosystem that is highly productive but relatively unstable. The short-term production is maximized by cultivation and fertilization, which, if poorly managed, can cause an imbalance of nutrients, increased pollutants, and heightened susceptibility to pest infestation and diseases.

Tree farms are managed to maintain a sustainable yield, under which a timber crop could be harvested indefinitely year after year while counterbalancing the forest's growth. Regeneration of the forest is quickened by planting seedlings raised in nurseries. To improve the seedlings' survival and growth rate, weeds are controlled and the seedlings may be inoculated with certain symbiotic microorganisms.

Tree farms need to balance nutrient demands, minimize herbicides, and improve the biodiversity of forests. Because they usually consist of specific high-yield, fast-growing species, tree farms deplete soil nutrients and discourage biodynamic growth in forests. After soils have been depleted of nutrients, the land slowly becomes unproductive and may become dependent on heavy fertilization. Such monoculture forests may be resistant to particular diseases or pests, but lack diversity to protect them from unfamiliar infestations. Thus tree farms may become dependent on pesticides, which may kill off pest predators such as other insects and birds.

Several alternatives to current tree farming practices exist. One is "sloppy clear cutting," which permits forests to regenerate with less management and more natural protection. Another incorporates agroforestry, which has been used for years in developing nations to balance land use for both timber and crops. One particularly successful tree species, *Leucaena*, serves as livestock fodder, fixes nitrogen to improve soil, and yields fifty tons of wood per hectare on a sustainable basis.

TROPHIC LEVELS

See Food Chains.

TROPICAL DRY FORESTS

Tropical dry forest once occupied about 60% of the forested tropics, but today less than 1% of this highly threatened vegetation remains in what might be termed its original state. These forests once occupied what are now the breadbaskets, cattle pastures, and fiber fields of the tropics. Virtually all of India, most of tropical Australia, southeast Asia, Madagascar, and eastern Africa, and much of Mexico, Mesoamerica, Brazil, Venezuela, Colombia, and Ecuador were once dry tropical forest. Tropical agricultural and hunting humans gravitated to the dry forest tropics throughout the world and, except for those on the fertile soils of rain forest river deltas, developed their major tropical lowland cultures there. Even when there were major upland cultures, they were strongly fueled by croplands in the dry forest at nearby middle to low elevations. Tropical dry forests today are almost entirely understood only as dwindling remnants, or as forest regenerating under mild human impact.

Tropical dry forest is deciduous to semi-evergreen during the dry season, has a canopy with few epiphytes, and ranges from 2 to 40 meters in height in its undisturbed state. In the rainy season it receives 500 to 3500 millimeters of precipitation, and the rain-free dry season is 4 to 8 months long. There is often strong variation in the starting date and the length of the rainy season, and the length and intensity of the short dry season that occurs during the central part of the rainy season. Once the rainy season has started, there is also variation in the continuity of the rains and the amount of rain that falls. Rainy seasons also differ strongly as to whether the rainfall is spread evenly throughout or strongly peaked.

The hottest days and the greatest differences between nocturnal and diurnal temperatures occur during the dry season in tropical dry forests. The weather is usually hottest just before the rainy season begins. In most tropical dry forest areas, there are no natural fires; there is no lightning during the rain-free dry season, and during the rainy season the living forest is too moist to burn. However, in some areas in which a high frequency of lightning strikes occurs at the beginning of the rainy season, flammable grasslands may burn. Large grazing and browsing mammals may exacerbate this situation by breaking up forest canopies, letting in more light. This in turn generates more herbaceous fuel for a fire.

However, the impact of naturally occurring fires on tropical dry forest is largely an academic question, since dry season fires set by humans now burn enormous areas of anthropogenic grasslands. These pastures, or savannas as they are often called, are often rich in species of grass introduced from other continents within the past century. These fires burn into the grass-rich forest margins, but do not cause "forest fires" in the sense of a fire in a northern coniferous forest. Rather, such fires within the forest characteristically occur in the litter and in dead herbaceous vegetation in the understory.

A patch of a hundred square kilometers of tropical dry forest generally contains 50% to 90% as many species of plants and vertebrates, but can contain nearly as many species of invertebrates as are to be found in an equal-sized nearby rain forest at the same elevation, latitude, and substrate of similar geological origin. Dry forests tend to contain large faunal elements that are thought of as primarily arid-land dwellers as well as large faunal elements made up of "rain forest species." While tropical dry forests are rich in complex mutualisms, parasitisms, predatory chains, and other forms of symbiosis, they are not quite as rich in these interactions as are nearby rain forests.

Many species of dry forest insects, birds, and mammals are highly migratory over distances of a few tens of meters to hundreds of kilometers. They seek everything from shady dry riverbeds and single waterholes to distant rain forests and high mountains in which to pass the dry season. Other organisms pass the dry season as dormant seeds, larvae, pupae, or adults. Vertebrate estivation to pass the dry season is, however, almost non-existent. A variety of dry forest plants

remain evergreen during the dry season, by having very deep root systems and by producing drought-resistant leaves. Many species of plants produce both flowers and seeds in the dry season (and grow vegetatively during the rainy season). Since the dry season is dry yet still warm, a variety of vertebrates and insects that feed on these nutrient-rich reproductive parts often reproduce only in the dry season. For them, the rainy season is the harsh season.

Why has tropical dry forest been such a focus of human activity? Because dry forest is easily removed, and kept out by fire, and because dry forest soils are much better for agriculture and ranching than are rain forest soils (except for certain rain forest delta and alluvial soils, and new volcanic rain forest soils). Weeds and insect pests are more easily controlled in dry forest because the dry season reduces their numbers, much as the northern winter gives the northern farmer an advantage. Food storage is easier in the dry season and even in the erratic rainy season of the dry forest, than in the soggy wet tropics. Diseases are less of a problem for the same reason. Livestock are more resistant to diseases in dry forest and can find better and less toxic wild to semi-wild forage in dry forest pastures. Dry forest access roads are more passable for a longer portion of the year, and animals used for transport tend to be healthier in dry forest pastures.

In brief, the reason why we have very little original dry forest to conserve today is because the dry tropics are comparatively friendly. If there were crops that do as well in ordinary rain forest conditions as do the crops of dry forest, there would be virtually no rain forest for the conservationists to argue over today. It is even worth noting that game and edible plants are much more abundant in wild dry forest interiors and edges than they are in rain forest. It is an easy conjecture that hunting and gathering omnivorous terrestrial primates—our ancestors—found dry forest habitats to be much more friendly and resource-rich than were rain forests.

Tropical dry forest plants and animals are very resistant to perturbations of their populations by such events as drought, fragmentation, fire, logging, and hunting. It is not surprising to find that dry forest regenerates on its own throughout the tropics, if there is a seed source nearby and the environmental insults such as hunting, burning, logging, grazing, and herbicides are removed. Dry forest organisms invade abandoned fields and pastures much more readily than do rain forest species for three additional reasons. First, the physical conditions of a wind- and sun-blasted open field or pasture are much more similar to the physical conditions during the worst part of the dry season in intact dry forest than they are to those of the moist, cool, and shady rain forest understory at any time of year. Second, dormant mycorrhizal spores abound in the soils of abandoned dry forest pastures and fields, but are largely absent from rain forest pasture and field soils. Both of these factors make it much easier for the plants of adjacent forest to invade a dry forest early successional site than a rain forest early successional site. Third, dry forest is rich in wind-dispersed seeds (and wind), and dry forest is rich in seed-dispersing animals that are willing to enter large sunny areas. In contrast, rain forest is very poor in wind-dispersed seeds (and wind), and many rain forest seed dispersers are not willing to cross large open sunny areas.

The conservation status of tropical dry forest is precarious at best. Despite the fact that dry forests originally covered more than half of the forested tropics, humanity is in no mood to return large areas of productive cropland and rangeland to wildland dry forest. The only real hope for dry forest wildlands and their organisms lies in intensive management for restoration of a few large areas that contain virtually entire dry forest ecosystems, and adjacent refugia in which to pass the dry season. Such areas are very friendly to ecotourists, be they local or international. Such areas contain hundreds of thousands of species of great importance to the commercial and agricultural prospector for chemicals and genes. Such areas are extremely important for dry season irrigation and wet season flood control. And such areas are repositories of a multitude of species of plants, animals, and microorganisms with potential for food, fiber, timber, fuelwood, drugs, and other uses. In short, the tropical dry forest's only hope for serious survival is if its biodiversity is put to work for society in a non-damaging manner.

DANIEL H. JANZEN

For Further Reading: D. H. Janzen, *Costa Rican Natural History* (1983); D. H. Janzen, "Guanacaste National Park: Tropical Ecological and Biocultural Restoration," in *Rehabilitating Damaged Ecosystems,* Vol. II, J. Cairns, Jr., ed. (1988); P. G. Murphy and A. E. Lugo, "Ecology of Tropical Dry Forest," *Annual Review of Ecology and Systematics* (17:67–88).

TROPICAL RAIN FORESTS

Tropical rain forests lie between the tropic of Cancer (23.5° N. latitude) and the tropic of Capricorn (23.5° S. latitude). Within this region, the type of vegetation depends upon the climate and the landform (Figure 1). Only a small proportion of the vegetation in the tropics, designated R in Figure 1, is strictly rain forest.

Within the tropics, length of dry season varies from most of the year, to no dry season at all. A dry season occurs when the monthly potential evapotranspiration is greater than the monthly rainfall, usually between

100 and 200 millimeters per month. True tropical rain forest occurs only in areas where there is never a water deficit, that is, all months have precipitation greater than potential evapotranspiration. In the popular press however, many tropical forests that have three or four dry months per year are still called rain forests. The forest remains green, but individual trees may shed their leaves.

The structure of forests and the plant types they contain vary within the rain forest region. For example, on well-drained fertile soils, standing stock of tree biomass could be 500 metric tons or more (dry weight) per hectare, while savanna areas could have less than 60 tons per hectare. In cloud forests, epiphytes such as mosses, lichens, and pineapple-like plants called bromeliads make up a large proportion of the biomass.

Tropical rain forests differ from other forests in their primary productivity, that is, the rate at which they fix carbon through photosynthesis, and the rate at which that carbon moves through the food chains and decomposers. Net primary productivity in rain forests on the best soils can be as high as 25 metric tons of biomass per hectare per year. This is about twice the rate for temperate zone forests, in which photosynthesis ceases for half a year or more due to low moisture or cold.

Species diversity, the number of species per unit area, is higher in the tropics than in temperate latitudes. For example, in forests of north temperate regions, there may be five or fewer species of trees per hectare, whereas in tropical forests, tree diversity of 100 species per hectare is not uncommon, and sites with more than 200 species have been recorded.

This high species diversity in tropical forests holds for almost all taxa, including mammals, insects, and fish. There are also more "guilds," that is, functional groups of species.

Many reasons for the great diversity of tropical species have been advanced: Temperate communities are younger due to disruption by glaciation; tropical communities have had more time to evolve. Tropical environments are more heterogeneous, therefore can accommodate more kinds of species. At higher latitudes, organisms are influenced strongly by physical factors; in the tropics, interactions such as competition and mutualism are more important. Finally, because productivity is higher in the tropics, the energy captured can be divided more finely. All of the theories probably contain some truth, but it is unlikely that a single theory can account for all the diversity observed in tropical rain forests.

Rain Forest Fragility

The high rate of productivity and profusion of species led early scientists and explorers to believe that rain forests had a high potential for agricultural production.

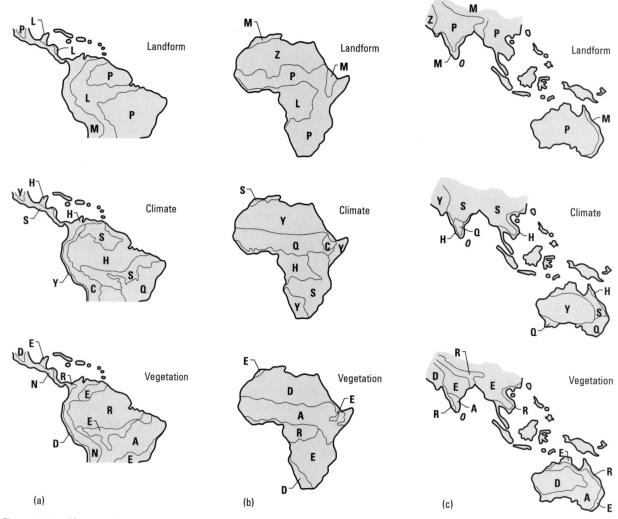

Figure 1. Landforms, climates, and vegetation types in (a) tropical America, (b) Africa, and (c) Southeast Asia and Australia. Relationships on islands are on a scale too small to indicate. Map key: *Landforms*: L, lowland; P, plateau; Z, desert; M, mountain. *Climate*: H, humid; S, seasonal or monsoonal; Q, semi-arid; Y, dry; C, cool. *Vegetation*: R, rain forest; E, seasonal or monsoonal forest; A, savanna; D, desert; N, montane.

As a result, much forest was cleared for pasture, plantation forestry, and agricultural crops. Such development schemes frequently have led to disaster: Pastures did not produce, disease spread through plantations, and crops died for lack of nutrition.

The explanation for the apparent paradox of the high productivity of the natural rain forest and the low potential for conventional development lies in the fragility of the rain forest. Tropical rain forests have a lower resistance to disturbance and lower resilience following it than other forest types due to low soil fertility, mutualistic interactions, and high natural biodiversity.

As soils age, nutrients such as calcium and potassium are slowly removed through the process of weathering. In northern North America, soils are relatively young. They were formed after the retreat of glaciers during the last ice age, approximately 12,000 years ago. In contrast, soils of the lowland tropics are very old. For example, the soils in the central Amazon Basin of Brazil are ten to one hundred million years old. Soils in Central Africa and much of Thailand and the rest of peninsular Southeast Asia are also millions of years old.

Some soils on tropical mountains are younger, but have low fertility for other reasons. For example, volcanic soils common in Central America and islands of the South Pacific have a tendency to rapidly immobilize phosphorus, that is, physically bind it so that it cannot be used by plants. On the steep slopes of the Andes, the soil is often thin and easily eroded. At high

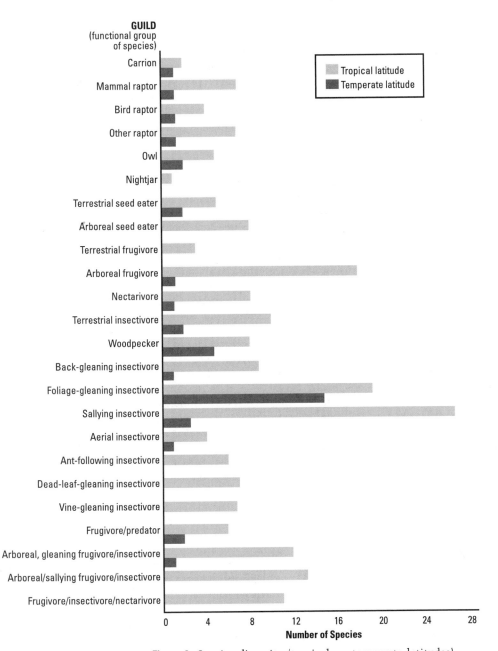

Figure 2. Species diversity (tropical vs. temperate latitudes).

elevations temperatures may drop below freezing every night, resulting in slow decomposition, accumulation of humus, and consequent shortages of nitrogen. Only in mid-elevational valleys are the soils fertile and conditions good for agriculture.

The lush vegetation and apparently vigorously growing trees of the tropical rain forest depend on a system of nutrient recycling in which there is relatively little interchange between the actively cycling nutrients and the underlying mineral soil.

In "open" systems, which are more common in temperate zones, the clay usually has a high exchange capacity, that is, a high ability to hold nutrients and prevent their loss. Roots of trees or crops growing on these soils send down roots three feet or more into the mineral soil, where nutrients exchanged on the clays move into the roots. In contrast, roots in tropical forest ecosystems generally are concentrated near the soil surface, sometimes even forming a mat on top of the soil. Leaf litter and dead wood on the forest floor are attacked by decomposers such as fungi and bacteria, which incorporate the nutrients in their bodies. As these organisms die, the nutrients released are taken up by plant roots, or by mycorrhizal fungi that live in symbiosis with the roots. Because competition for the nutrients is extremely intense, plants with roots and mycorrhizae on the soil surface have an advantage. Nutrients are taken up and recycled by the plants as soon as they are released with very little opportunity for loss. It is through this direct recycling that magnificent tropical forests are able to survive on nutrient-poor soils.

When the forests are cleared, the soil can quickly lose the little fertility it has. Calcium and potassium are rapidly leached into nearby streams. Phosphorus reacts with iron and aluminum in the soil and becomes immobilized. Nitrogen, which is held mainly by organic matter, volatilizes as the organic matter decomposes. As a result, crop productivity decreases rapidly. Fertilizers can be added, but in many regions of the tropics fertilizers are too expensive.

Species Interactions

Mutualistic interactions are more common and more important in the tropics than in temperate regions. For example, there are no obligate ant-plant mutualisms north of 24°, no nectarivorous or frugivorous bats north of 33°, and no orchid bees north of 24°. Extrafloral nectar glands on plants drop off drastically between the northern limits of the neotropics in Mexico, and Texas. Also, within the tropics, mutualistic interactions are more prevalent in the warm, wet evergreen forests than in the cooler and more seasonal habitats.

It is not clear whether high species interdependency is a cause or an effect of high species diversity. Regardless, the high incidence of mutualistic interactions in the tropical rain forest is another reason for its fragility. In temperate forests, for example, many of the trees such as oaks and pines are wind-pollinated. If birds or insects are reduced due to increasing agriculture, the trees still can pollinate successfully. In contrast, most trees in tropical rain forests are pollinated by insects, bats, or birds. If the habitat for a particular pollinator is destroyed by logging or agriculture, survival for the tree that depends upon that pollinator may be impossible.

At high latitudes or in the dry tropics, cold or drought seasons reduce pest populations. As one proceeds toward uniformly moist tropical conditions, the year-round impact of insect herbivores on plants increases substantially.

The high diversity of species within tropical forests is thought to be an adaptation to resist the spread of disease and pests. Where species diversity is high and only a few individuals of each species exist on each hectare, it is more difficult for pest species to spread. Although

this theory is controversial, it has been observed that diseases race through plantations of some tropical trees very rapidly. For example, it is very difficult to plant mahogany in plantations because of the shoot-tip borer, a pest that moves rapidly from tree to tree. In the native forest where individual mahogany trees are scattered, the shoot-tip borer is less of a problem. Monoculture plantations are not well adapted to the environment of the continuously hot and wet tropics.

The combination of high human population growth in the tropics and the fragility of tropical forests is causing what many scientists consider one of the most serious problems in the world today: extinction of species. While there is disagreement on the rate of extinction, all agree that the consequences can be tragic. The losses include genetic material of plants and animals that have the potential to cure many human diseases, including cancer, and to provide genetic resistance to diseases of food crops.

Although there is great need to conserve tropical rain forests and their species, it cannot be done by strict preservation. Too many people already depend upon forest products and services, or upon tropical lands that are used for shifting cultivation, colonization projects, hydroelectric projects, charcoal, pulp, pasture, and agricultural projects.

Yet with care, humans can use the tropical rain forest without destroying it. Humans have lived in tropical rain forests for thousands of years. Indigenous peoples were, and in some regions still are, adapted to the forests of Asia, Africa, and South America, extracting oils and medicinal products from the native species and practicing an agriculture that mimics natural processes such as secondary succession (Table 1). This is the orderly sequence of annuals that are replaced by short-lived perennials, which are replaced in turn by long-lived perennials.

Along river banks in many tropical regions are alluvial deposits that are re-deposited each year. The deposits are high in fertility, and in such locations, more intensive agriculture is possible.

Immigrants to the tropics have also settled in the forests, practicing sustainable agriculture. Sometimes derogatorily called "slash-and-burn agriculture" or "shifting cultivation," this type of agriculture can be destructive when too many farmers try to cultivate too little land, but at low population densities it is not permanently harmful.

Problems arise when large-scale monocultures typical of U.S. Midwestern agriculture are imposed on the rain forest environment. Crop production is sustainable only with massive inputs of fertilizer, pesticides, and herbicides, items beyond the reach of most farmers. Even worse is when tropical rain forests are cut down and converted to pastures for cattle. Because of low fertility,

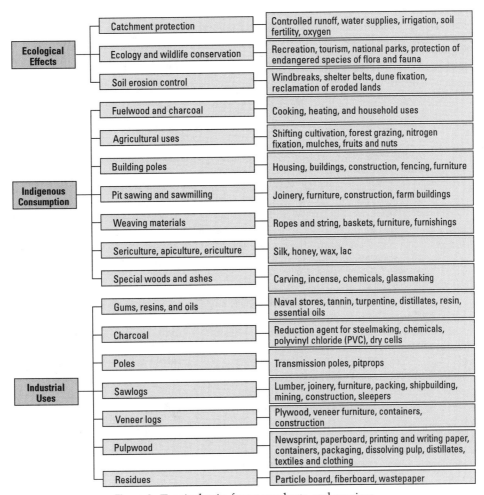

Ecological Effects	Catchment protection	Controlled runoff, water supplies, irrigation, soil fertility, oxygen
	Ecology and wildlife conservation	Recreation, tourism, national parks, protection of endangered species of flora and fauna
	Soil erosion control	Windbreaks, shelter belts, dune fixation, reclamation of eroded lands
Indigenous Consumption	Fuelwood and charcoal	Cooking, heating, and household uses
	Agricultural uses	Shifting cultivation, forest grazing, nitrogen fixation, mulches, fruits and nuts
	Building poles	Housing, buildings, construction, fencing, furniture
	Pit sawing and sawmilling	Joinery, furniture, construction, farm buildings
	Weaving materials	Ropes and string, baskets, furniture, furnishings
	Sericulture, apiculture, ericulture	Silk, honey, wax, lac
	Special woods and ashes	Carving, incense, chemicals, glassmaking
Industrial Uses	Gums, resins, and oils	Naval stores, tannin, turpentine, distillates, resin, essential oils
	Charcoal	Reduction agent for steelmaking, chemicals, polyvinyl chloride (PVC), dry cells
	Poles	Transmission poles, pitprops
	Sawlogs	Lumber, joinery, furniture, packing, shipbuilding, mining, construction, sleepers
	Veneer logs	Plywood, veneer furniture, containers, construction
	Pulpwood	Newsprint, paperboard, printing and writing paper, containers, packaging, dissolving pulp, distillates, textiles and clothing
	Residues	Particle board, fiberboard, wastepaper

Figure 3. Tropical rain forest products and services.

grasses quickly become unpalatable and the pastures are abandoned to worthless secondary scrub.

In recent years, researchers have begun to experiment with native systems of cultivation, and adapt them to the requirements of increasing populations in the tropics. A general term for these agricultural systems is agroforestry, that is, mixing of trees with crops. An example is "alley cropping." In this system, one-meter-wide hedges of fast-growing trees or shrubs, preferably nitrogen-fixing legumes, are planted. Between the hedges, a grain crop is planted in "alleys" about four meters wide. Several times each growing season, the branches of the shrubs are lopped off. The litter forms a mulch that improves the microclimate at the soil surface, improves soil structure, and increases soil fertility. Most important, the litter increases the amount of soil organic matter, a component extremely important in the tropics for maintaining soil fertility. Other advantages of agroforestry systems are that the trees tend to protect the soil from

erosion, and serve as a source of firewood and construction material for local villages.

Tropical rain forests are among the most diverse and highly productive ecosystems of the world—as long as they are left undisturbed. However, when development is imposed upon them, they quickly are destroyed due to poor soils and to loss of complex interactions among species necessary for ecosystem survival.

This does not mean that tropical rain forests should not be used for human benefit. There are ways to manage the forest without destroying it. But such management requires an understanding of how tropical rain forests work, and an incorporation of this understanding into management practices.

CARL F. JORDAN

For Further Reading: Mark Collins (ed.), *The Last Rain Forests. A World Conservation Atlas* (1990); Andrew Mitchell, *Vanishing Paradise. The Tropical Rainforest* (1990); John Terborgh, *Diversity and the Tropical Rain Forest* (1992).

Table 1
Succession of harvestable plants in Bora Indian fields and fallows, Peru.

Stage	Planted harvestable	Spontaneous harvestable
High forest	None	Numerous high forest construction, medicinal, utilitarian, handicraft, and food plants available.
Newly planted field 0–3 mo	All species developing	Dry firewood from unburnt trees for hot fires.
New field 3–9 mo	Corn, rice, cowpeas	Various useful early successional species.
Mature field 9 mo–2 yr	Manioc, some tubers, bananas, cocona (*Solanum sessiliflorum*), and other quick maturing crops	Abandoned edge zone has some useful vines, herbs.
Transitional field 1–4 or 5 yr	Replanted manioc, pineapples, peanuts, coca, guava, caimito (*Chrysophyllum cainito*), uvilla (*Pourouma cecropiifolia*), avocado, cashew, barbasco (*Lonchocarpus nicou*), peppers, tubers; trapped game	Useful medicinals, utilitarian plants within field and on edges. Seedlings of useful trees appear. Abandoned edges yield straight, tall saplings, including *Cecropia* and *Ochroma lagopus*.
Transitional fruit field 4–6 yr	Peach palm (*Bactris gasipaes*), banana, uvilla, caimito, guava, annatto (*Bixa orellana*), coca, some tubers; propagules of pineapple and other crops; hunted and trapped game	Abundant regrowth in field. Many useful soft construction woods and firewoods. Palms appear, including *Astrocaryum*. Many vines; useful understory aroids.
Orchard fallow 6–12 yr	Peach palm, some uvilla, macambo (*Theobroma bicolor*); propagules; hunted game	Useful plants such as above; self-seeding *Inga*. Probably most productive fallow stage.
Forest fallow 12–30 yr	Macambo, umari (*Poraqueiba sericea*), breadfruit, copal (*Dacryodes* sp.)	Self-seeding macambo and umari. High forest successional species appearing. Early successional species in gaps. Some useful hardwoods becoming harvestable, e.g., cumala. Many large palms: huicungo (*Astrocaryum huicungo*), chambira, (*Astrocaryum chambira*), assai, (*Euterpe* sp.), ungurahui (*Jessenia bataauas*).
Old fallow, high forest over 30 yrs	Umari, macambo	Only a few residual planted and managed trees.

W. M. Denevan, C. Padoch, and S. F. Paitan, "Swidden-fallow Agroforestry, Iquitos, Peru." U.S. Dept. of State, *People and the Tropical Forest*, U.S. Man and the Biosphere Program, Washington, D.C. (1987).

See also Agroforestry; Biodiversity; Carbon Cycle; Deforestation; Food Chains; Mineral Cycles; Photosynthesis; Temperate Rain Forests; Tropical Dry Forests.

TROPICAL RIVER BASINS

The characteristics of all rivers largely reflect climate, topography, and vegetation, as well as human modifications within the river basin. Tropical river basins, especially those situated in humid climatic zones, differ from other basins largely because of their climatic attributes. On average, humid tropical regions experience greater annual precipitation than non-tropical areas. In addition, precipitation occurs solely as rain (no snow), and rainfall usually occurs in the form of intense thunderstorms.

The net result of the rainfall attributes of humid tropical areas is that per unit of area, tropical river basins generally have the highest river flows. Of the world's five largest rivers, in terms of river flow, two (Amazon and Zaire) are situated solely within the tropics. The other three (La Plata/Parana, Yangtze, and Ganges/Brahmaputra) receive significant runoff from the tropical areas within their basins. The Nile, the longest river in the world, flows through the Sahara desert and is able to reach the Mediterranean solely due to the large riverflows in its headwaters, which are situated in the humid tropical highland areas of eastern Africa.

As most tropical streams are fed solely by rainwater, rainfall pattern determines the seasonal pattern (regime) of tropical river flows. Basins that are largely equatorial, such as the Amazon and Zaire, have relatively stable flow regimes in spite of their huge discharges. This reflects the minimal seasonal variability of rainfall in the equatorial zone. Tropical river basins that are more poleward than equatorial ones generally are situated in areas that have wet and dry seasons. In these tropical savanna areas, rivers experience large discharges during the rainy season and lower flows during the drier period. For example, the Orinoco and Niger rivers' discharges vary greatly between their high and low flows.

The relation between climate and river flow is

strongly modified by vegetation cover under natural conditions. Vegetation influences stream flows by the amount of moisture it requires for growth, its interception of rainfall, and how it alters the local climate. Broadleaf evergreen trees consume more water than other vegetation types. Grasses and permanent crops such as coffee and bananas are intermediate in their water demands. Annual crops such as corn (maize), beans, and manioc consume the least moisture. In terms of interception, trees capture the highest percentage of rainfall. Permanent crops are intermediate between trees and annual crops. With the rapid deforestation that is occurring in many tropical basins today as well as land clearing for agricultural and pastoral activities, the flow characteristics of many tropical streams reflect the vegetation changes due to human activities.

As land is cleared, the percentage of rainfall entering the river system increases due to lower interception by the vegetation and to lower transpiration. In addition, human activities often compact the soil. This lowers the infiltration of rains entering the soil and greater runoff occurs. The net effect of these human alterations in tropical basins is to increase the overland flow of water within basins. This not only alters the flow characteristics of streams but increases soil erosion. In the areas of dense forest in Amazonia, it is estimated that only 25% of rainfall eventually becomes riverflow. The remaining 75% is evaporated or transpired from the plants or ground. This moisture is then recirculated in the winds throughout the basin. Thus, when these areas are deforested, greater runoff and soil erosion would be expected. However, the trend toward greater runoff is offset to a degree since the moisture not recycled by evapotranspiration results in lower rainfall in the western portions of the Amazon Basin.

Lands within humid tropical stream basins require careful management once disturbed from their natural state. Because of the warm temperatures and moist conditions found in many of them, conditions are almost ideal for bacteria and fungi. These as well as insects and other animal forms that humans consider to be pests are difficult to control by spraying and other conventional methods. The high rainfall, moist ground conditions, and lack of a dormant season minimize the impact of most disease and pest controls crucial in most modern agriculture and livestock systems. When natural continuous groundcover is replaced by annual crops, care must be practiced to minimize erosion. Furthermore, with the exception of young tropical soils, most soils within the tropics are relatively infertile. As a result, lands within tropical river basins generally require fertilizer supplements to sustain permanent agriculture and pastures. With high annual rainfalls, most fertilizers are easily washed out of the soils. All of these problems increase as land clearing occurs on steep slopes.

Because the land and water environments within tropical river basins differ in many fundamental ways from their middle latitude counterparts, they require different strategies if they are not to degrade once their natural state is disturbed. Successful rubber, fruit, rice, and teak production in many tropical river basins illustrates that sustainable development is possible when good management techniques and proper crops are utilized. Discovery of uses for the indigenous plants found in these areas represents another avenue for successful utilization of the generally fragile lands found within tropical river basins.

LAURENCE A. LEWIS

For Further Reading: L. A. Lewis and L. Berry, *African Environments and Resources* (1988); E. Salati, et al., "Amazonia," in B. L. Turner, et al., eds., *The Earth Transformed by Human Action* (1990); J. Tricart, *Landforms of the Humid Tropics, Forests and Savannas* (1972).

TROPOSPHERE

The troposphere is the lowest major layer of the atmosphere. Its boundaries extend from the Earth's surface to an altitude of 11 to 12 miles (18 to 20 kilometers) at the equator and 5 to 6 miles (8 to 10 kilometers) at the poles. Heat as well as latitude affects the height of the troposphere. Temperature decreases at a rate of 18 degrees Fahrenheit per mile (6.5 degrees Centigrade per kilometer) as altitude increases. At the top of the atmosphere, known as the tropopause, the temperature can be as low as −112 degrees Fahrenheit.

The troposphere contains approximately four fifths of the mass of the atmosphere as well as most of its water vapor and carbon dioxide. Both of these gases are integral in the heat balance of the Earth, as they trap much of the infrared radiation that is reradiated from the sun-heated ground.

The troposphere also contains the majority of clouds in the atmosphere. Because the troposphere is the zone of interchange between the Earth's surface and the atmosphere, most of the important weather processes happen there. Evaporation, precipitation (in the form of rain, snow, and so on), and the transport of pollutants account for the interchange between the Earth's surface and the atmosphere.

These weather processes serve as the major route of atmospheric pollution. The troposphere is where acid rain is formed from emissions such as nitrogen oxide and sulfur dioxide. Exhaust from cars and other fossil-fuel-burning sources enters the troposphere. Sunlight reacts with the compounds to produce smog and a high concentration of ozone. Common forms of atmospheric pollution, especially in urban areas, are acid deposition and photochemical smog. The effect of these pollutants remains local as long as they remain below an altitude of about half a mile (about one kilometer). If the contaminants reach higher altitudes, they remain airborne longer, and their effects are spread over an extended area. Although pollutants are able to travel farther and faster at higher altitudes, vertical transport of pollutants is usually slow, since the layers of the atmosphere do not mix easily. Yet, pollutants do get transported rapidly. Models suggest that thunderstorms are one of the major forms of rapid vertical transport. It may take months for pollutants to spread from the Earth's surface to a height of six miles (ten kilometers), yet during a thunderstorm the mixing of the pollutants may take only a few hours to reach the same height. One of the major results of the rapid transport of pollutants is that the effects of acid deposition are lessened, since the acid becomes diluted as it is transported. However, atmospheric transport can cause localized problems to develop into regional or even global problems. Additionally, by supplying the needed chemicals, the spreading of pollutants aids in the production of atmospheric ozone.

TSCA

See Toxic Substances Control Act.

TUBERCULOSIS

Tuberculosis (TB), the deadly "white plague" that has killed more people than any other single contagious disease, has been known throughout history. In the 19th century, TB was associated with the exclusion of air and sunlight from mines, mills, and city slums; today it is found among the homeless, malnourished, and neglected.

Tuberculosis was estimated to be responsible for a fifth of all deaths in the early 19th century. Although TB deaths gradually fell by 50%, this chronic and often incapacitating disease remained the leading cause of death in 1900.

Infection can occur in healthy individuals without producing active disease, as natural resistance or preventive medication inhibits illness. But if disease becomes established, the victim becomes feverish and loses weight (hence the 19th-century name "consumption"). Although the bacillus can infect any part of the body, the lungs are most often involved; coughing and hemorrhage are the most common symptoms.

In the early 20th century, tuberculosis was described as "the disease of the masses" because of its persistent association with the living and working conditions of the poor. With better standards of living and the development of antibiotics after 1940, tuberculosis mortality ceased to be an important measure of public health in industrialized nations. In developing countries, however, many of the social hazards that made TB dangerous continued to prevail. The International Union Against Tuberculosis reports that despite treatment programs, there were 3 million deaths from tuberculosis in 1989. These experts estimate that there are annually at least 8 million new cases of tuberculosis plus another 10 million persons infected with the tubercle bacillus, many of them without the economic and medical resources that could protect them from breaking down into active disease.

Organized efforts to control tuberculosis began with observations of differences in the cirumstances of its victims. Well before they agreed that tuberculosis was a contagious disease, 19th-century physicians and sanitarians gathered statistical evidence to support advice on personal hygiene for those who inherited a constitutional weakness of the lungs, and detailed evidence about the differences between healthy and sickly environments in order to protect the public. Although by today's standards their arguments for cleanliness and fresh air were more moralistic than scientific, they nonetheless identified and directed attention to the significance of environmental factors in the epidemiology of tuberculosis.

Robert Koch identified the tubercle bacillus in 1882. In the same decade, the microorganisms that cause typhoid, diphtheria, and cholera were identified. These discoveries stimulated both enthusiasm and skepticism over whether the new science of bacteriology would make possible the detection and control of contagion.

If the disease was communicable and widespread, some argued, why did some people escape illness even after exposure? Some doctors challenged the idea of contagion by pointing to tuberculosis in infants, which they insisted was inherited, underscoring the essential role of family history. Others believed that TB was communicable only when personal circumstances lowered resistance.

By the early 20th century bacteriological and epidemiological evidence supported the view that the transmission of TB through bacillus-infected sputum could be restricted by personal and domestic cleanliness, the control of dust, and the supply of fresh air at work and in public places. Once infected, the TB victim could regain health but never be freed from the threat of future breakdown. Rest was deemed the most effective prophylaxis, coupled with early discovery of "incipient disease" through the identification of secondary signs and x-ray screening. Sanatorium cure was the best hope for the sick, but the personal and social cost of long term care limited this prospect; institutional policy often excluded the terminally ill. Tragically, those TB patients who were least likely to benefit from medical attention were also most likely to be contagious and least able to care for themselves. An undercurrent of assumption continued to credit inheritance, predisposition, ignorance, and poverty with responsibility for the dramatically uneven distribution of TB.

In the mid-20th century strategies turned to the detection of diseased individuals, who constituted reservoirs of TB as the incidence of disease declined. New aids to strategies for reducing the prevalence of TB came from laboratories: tuberculin for identification of asymptomatic infection; Bacillus Calmette-Guerin (BCG) for short-term immunization of the most susceptible, particularly young children; antibiotics and chemotherapies to boost resistance and to treat infection and disease. With the decline in tuberculosis prevalence after World War II, new cases in wealthy nations were usually the result of the reactivation of unrecognized infection that had occurred years before. With relatively abundant social and medical resources available, children could look forward to many years of adult health because they were freed from the risks of infection in early life.

In 1989 the U.S. Public Health Service announced its "Strategic Plan for the Elimination of Tuberculosis in the United States" targeted for 2010. Less than a year later the fragile balance of microbial, individual, and social factors that control TB was threatened in reports of outbreaks in prisons, among some patients already fighting AIDS, and most seriously among the destitute in cities where public health budgets had been cut.

In the developing world the exposure of children to active disease leads to the infection of virtually everyone by age 20. Despite the potency of 20th-century scientific knowledge and public health authority, tuberculosis appears destined to reinforce its 19th-century identification with the socially vulnerable.

BARBARA GUTMANN ROSENKRANTZ

For Further Reading: Barbara Bates, *Bargaining for Life, A Social History of Tuberculosis, 1876–1938* (1992); Linda Bryder, *Below the Magic Mountain; A Social History of Tuberculosis in Twentieth-Century Britain* (1988); René and Jean Dubos, *The White Plague: Tuberculosis, Man, and Society* (1987).
See also AIDS; Antibiotics; Bacteria; Epidemiology; Health and Disease; Public Health.

TUNDRA

Tundra prevails above the tree line, and the vegetation and climate that distinguishes it occurs in polar latitudes or at high altitudes. Polar latitude regions are called arctic tundra, while similar conditions on upper mountain slopes are alpine tundra. The word *tundra* derives from the Finnish *tunturi* which means "treeless heights." This definition is generally true, although the occasional dwarf tree species may exist, including black spruce or birch, and willows can survive when protected from the wind.

Nearly one third of the Earth's land surface is tundra; most is in the Northern Hemisphere, covering over nine million square miles. The history of the tundra in geological terms is relatively young. The fossil record indicates the tundra biome developed during the early Pleistocene, a mere two million years ago. The earliest specimens of tundra flora and fauna were found in the highlands of central Asia and the Rocky Mountains of the U.S.; they date back to the late Miocene/early Pliocene, around five million years old.

Many areas covered by tundra today were once home to great forests of coniferous and deciduous trees, and the change to smaller plants may have been a result of a major global shift to cooler temperatures during the Miocene. Several advances of glaciers have since occurred, each time disrupting or halting the succession of plants.

The tree line does not coincide neatly with latitudes, but with weather patterns. Trees diminish along the same curves that mark the southernmost extension of arctic air masses.

The northern half of Alaska is rich in tundra, particularly around Prudhoe Bay, Point Barrow, and Eagle Summit. Vast tracts of Canada, parts of Greenland, and northern regions of Scotland, Ireland, Scandinavia, China, and Siberia are also tundra. Tundra also exists in fairly extreme southern latitudes, on Macquarie Island south of Australia, and in the southern Atlantic Ocean, on South Georgia and Signy Islands, between South America and Antarctica. Alpine tundra occurs in such otherwise temperate latitudes as the White Mountains of Arizona, and California's Sierra Nevada range.

Climatic conditions may be harsh, with strong winds, low rainfall, and low temperatures year-round, especially in arctic tundra. The four seasons that we know throughout much of the contiguous U.S. do not pertain with any distinction. The reasons for this are obvious for arctic tundra, because of the higher latitudes and the seasonal effects of the Earth's tilt. In terms of biological activity and climate, it is better to think of the arctic tundra as having two seasons, spring and winter, since temperatures vary from brutal to brisk. Annual highs average around 40°F, with average lows of −25°F (50°C to −32°C). Rainfall in the arctic tundra may actually be as low as in the Mojave Desert.

Temperatures are more forgiving on alpine tundra, and the rainfall is much higher, but drainage is swift. Drainage in arctic tundra is poor, and hydration is a central feature of many tundras in Canada's Northwest Territory, especially around the MacKenzie delta and Victoria Island. Boggy stretches may occur between lakes and rivers; these are also called mires. Along the coastal tundra, cotton grass is plentiful, with rich sedges and mosses. There have been some attempts to order inland tundra vegetation into categories of "wet" and "dry," but these intermingle.

While the long dark days of the arctic limit photosynthesis, tree growth is also constrained by cooler temperatures. Taller plants suffer from the winds that have a profound effect on the temperature, stealing warmth. Even during the warmest months, temperatures just one foot off the ground may be 15°F cooler than at ground level. Short, low plants that hug the Earth maintain the warmth necessary to function.

The complex and extensive roots of trees are also constrained by soil conditions. Just beneath a shallow level of the tundra, the soil may be frozen permanently; this layer of permafrost does not allow roots to reach the larger volume of water that trees require, nor the anchorage that trees need to support and brace their trunks and crowns. The permafrost depth is 2,000 feet in parts of Alaska's North Slope and almost a mile deep in Siberia. The soil layer above this tends to freeze and thaw according to the season, and this is called the active layer, since it provides the basis for biological activity. The active layer may be only a yard, or about one meter, in depth. Permafrost inhibits root growth on arctic tundra, while the denuded mountain slopes that are home to alpine tundra tend to be rock solid as a result of erosion.

Because of the random effects of frost, there is often a mosaic of bare soil or rock dotting arctic vegetation, called spotted tundra. Another, more geometrical formation occurs in dried and repeatedly frozen bogs: clusters of polygons, configurations of cracked soil large enough to be visible from an airplane at 5,000 feet. Large polygons can be over 100 feet across and cover many square miles. Such polygons also occur on dried lake beds in Africa; basically, the shapes result from swelling, then shrinking of soil, a result of hydration, then dehydration.

Another pattern phenomenon that occurs in arctic tundra is waves or wrinkles of topsoil called solifluction. The creeping behavior is similar to a mudslide, only the slide occurs in slow motion, over a long period of time. Solifluction occurs when permafrost prevents the penetration of rain water; the active layer becomes waterlogged, and gently slides down an incline. In their descent, these waves of topsoil collect animal bones and plants, which accumulate at the bottom of the incline. The peaty soil produces well-defined fossils; those discovered include a 17,000-year-old musk ox. Pingos are another formation unique to tundra; these mounds, about thirty feet tall, form when permafrost surrounds soil in a lake. As the water freezes, it cannot expand sideways, so it pushes the soil up.

Arctic soil is generally acidic, very low in nitrogen, and poorly drained. There is very little humus or bacterial activity; consequently, decomposition is slow, and dead leaves and flower petals stay preserved long after they die. In the hummocks that occur inland, the soil is soggy and rich in peat. The more peat that accumulates, the less decomposition occurs, enhanced by the soil's anaerobic conditions, and the cumulative effect of being waterlogged. Once peat begins to accumulate, drainage is further impaired when domed masses of peat build up an impermeable series of dams.

Dominant tundra vegetation includes verdant mosses, lichens, low shrubs, tussocks of grasses and herbs, and cushion plants, which have small, compact leaves and a robust taproot; cushion plants tend to inhabit outwash plains and gravels. The prevailing colors of tundra vegetation are brown and green, but there are heathers of white, mauve, and steel blue. Lichen is prevalent, and provides food for thousands of caribou and reindeer. One common species of lichen is *Cladonia*, also called reindeer moss. When the snows melt, there is an explosion of color among flowering plants, including sunflowers and poppies. Many flowering species in the tundra are dwarfs of their relatives that thrive in friendlier soils and climates; the Lapland rosebay, only four inches tall in the tundra, is a miniature rhododendron related to those that tower over twenty-five feet in the eastern U.S.

Most of the plant species in the tundra are perennials, and all are vascular. Small mounds of soil and vegetation called tussocks are rich in microorganisms, insects, and berry-producing plants, including cranberries. The dung of mammals, especially numerous lemmings, adds nitrogen to the soil, which results in rich pockets of vegetation and color.

Plants and animals have evolved remarkable adaptations to the harsh climate. Polar bears and grizzlies slow down their metabolism during the coldest months, while arctic spiders and insects lie still and frozen. Some insects and tundra fishes have developed an internal antifreeze. The leaves of some plants, including Labrador tea, have hairs or thick surfaces that conserve precious water.

In comparison to the much older tropical forest ecosystems in central Africa and South America, the tundra has larger biomasses and lower productivity. In short, there are fewer species in the tundra, but these few species have greater numbers. Mosquitoes and flies travel in dark swarms, and caribou and snow geese move in herds and flocks of hundreds of thousands. The caribou give birth on the flowering tundra, and rely on lichen and other plants that thrive during a brief summer.

But the biological niches for diet and flexibility of behavior are constrained, and the thin windows for breeding and reproduction are perilous by comparison. For example, birds that nest in temperate or tropical zones can try again if storms destroy their nest of eggs, but birds in the tundra have a very short period of time to replenish their population. The growth period for vegetation may be limited to less than four months of the year, so plants also have a limited time to seed and reproduce. Bad weather conditions in one year can cripple a species for several generations. The evolutionary adaptations for this are impressive, but they are also very young, which makes this biome of tundra particularly sensitive to global climatic change and human interference. For many reasons, tundra is often referred to as fragile terrain.

Throughout the tundra ecosystem, succession of plant species is very slow in comparison to other temperate or tropical environments, and once disturbed, plant life is slow to recover. Disturbances can be natural, from climatic traumas, or as a result of the migration of caribou, which travel in such great numbers as to make trails with their hooves. But increasingly a condition known as thermokarst begins when pressure on the landscape is applied during the summer thaw, when the earth is soggy. The ground ice melts, and the surface cover of vegetation and soil diminishes. Thermokarst results when cross country vehicles are used on the tundra, and in Alaska, state laws forbid off-road driving during the summer thaw.

The arctic tundra invites great migrations of caribou and snow geese, and serves as a permanent home to grizzly and polar bears, moose, musk ox, voles, lemmings, mink, weasel, wolves, wolverines, red and arctic foxes, snowy owls, and rock ptarmigan. The arctic fox and ptarmigan turn white during the winter. The ground squirrel hibernates, while lemmings remain active in their tunnels. Lemmings are considered key to the pyramid of life in the tundra, not only because their large populations add nitrogen to the soil, but they serve as food supply to many predators, including the snowy owls, wolverines, and foxes.

Many of the animals that dominate tundra terrain migrated across the Bering Strait, the same way that early humans did. Between 20,000 and 12,000 years ago, the Ice Age glaciers took up so much moisture

from the sea that a land bridge occurred across the Bering. This invited migration by woolly mammoth, saber-tooth cats, arctic horses, giant ground sloths, yak, and saiga antelope, all now extinct.

Humans have been part of the tundra ecosystem far longer in Eurasia than in the North American arctic. Humans migrated into the upper reaches of Scandinavia following the retreat of glaciers during the Ice Age, and Laplanders who herd reindeer are their descendants. Eskimos continue to live in areas from East Cape, Siberia, around Thule, Greenland to southern Labrador, along the Northwest Passage and in Alaska. The word *Eskimo* derives from the French *Esquimaux*, and the Algonquin *eskipot*, which means "eater of raw flesh." *Inuit* refers to Eskimos of the eastern Canadian arctic; *Yup'ik* to those who live near the Bering Sea, and *Inupiat* to the natives of the North Slope.

Their sleds and kayaks were fashioned from animal skins and wood, their harpoons from bone, their shelters from blocks of snow. Their boats were especially ingenious; thirty-foot-long masted vessels, fitted with a square sail or a single, stepped mast, were fashioned from walrus hide or seal skin, and rowed by women. The men followed in shorter kayaks, free to hunt, and stored their bounty into the "women's boat." This tradition began with the original Dorset tribe, which evolved into the whale-hunting Thule, directly ancestral to the modern Eskimo.

Winter clothing was made from caribou skin; seal skin was used in summer. Slippers inside boots were bird skins turned inside out. Foul weather gear was fashioned from the intestines of seals; polar bear fur went on soles to make a silent approach. Fat from caribou was used as a lubricant for bowstrings, tendon for thread. The horn of a musk ox was used for fish spears; the clear gut of bearded seal for a window, handy since it folded for travel, and did not frost over in cold weather.

Inuit joined the whaling industry in the 19th century, and trapped furs in the 20th century. Since Europeans, Russians, and North Americans made contact, drastic changes occurred in the Eskimos' culture and ecological ways, with the introduction of firearms, steel knives, sugar, flour, canoes, and prefabricated houses. Now, even in remote areas, the traditional hunters who remain fear that their culture will not survive the pressures that have been introduced by the influx of foreign lifestyles. The traditional sled, with bracings of antler horns and a flexible base to accommodate peaks and dips, has virtually disappeared. Few dogsled teams are used anymore; most Eskimos now use snowmobiles, and tow their modern rafts behind. The bonds the Inuit felt with nature are being eroded; in bush settlements like Old Crow in Canada and Alaska's Arctic Village, the traditions of culture and diet associated with the caribou migration have been diminished.

Like hunter-gatherers in Africa, Eskimos relied on a nomadic way of life, moving with the seasons and with the migrations of plentiful food, including caribou, and marine mammals, like seals and narwhals. Their traditional diet was high in fat and protein, essential for survival in cold temperatures, and they obtained considerable vitamin C from raw fish. By the 1970s, their diet had shifted towards carbohydrates to such an extent that young people tend to grow thinner and taller, as opposed to the shorter, stockier stature of their ancestors. Their change in diet, now including sugar, has caused increases in diabetes, dental cavities, gall bladder diseases, and obesity. A shift from breast feeding to bottle feeding has created new health problems among children, and an increased birth rate. The shift from nomadic ways has also had a profound effect on the tundra in Siberia, where over 3 million reindeer have been domesticated, and overgrazing has resulted in some parts of the tundra used as pasture.

Oil exploration on Alaska's North Slope began in 1923 and intensified with the first major strike at Prudhoe Bay in 1968. Construction of a pipeline from Endicott field, east of Prudhoe Bay, to deliver crude to the port of Valdez began in 1974 and was completed in 1977, at a cost of eight billion dollars. The trans-Alaska pipeline delivers 25% of the U.S. domestic oil supply, carrying 60 million tons of crude a year. Mining for gold continues in the Canadian and Siberian tundra, and copper deposits, hydroelectrical energy, and other industrial developments pose major threats to the tundra.

Critical features in the tundra ecosystem, including the thin air at high altitudes, and the low content of nitrogen in the soil, make the tundra important for studies of potential global warming. Changes in the atmospheric and terrestrial carbon budget of the tundra may serve as early warnings of global climatic changes. Global warming could create a lethal cycle in the tundra. As permafrost thawed, it would release huge amounts of ice-locked methane, and as peat bogs decomposed, additional carbon would be released into the atmosphere.

One of the best, comprehensive scientific studies of tundra was conducted between 1966 and 1974, by the International Biological Program (IBP), which drew together scientists from various countries. Their aim was conservation, specifically "the recognition that the rapidly increasing human population called for a better understanding of the environment as a basis for the rational management of natural resources."

DELTA WILLIS

For Further Reading: L. C. Bliss, O. W. Heal, and J. J. Moore, eds., *Tundra Ecosystems: A Comparative Analysis* (1981); Barry Lopez, *Arctic Dreams* (1987); Ann H. Zwinger and Beatrice E. Willard, *Land Above the Trees, A Guide to American Alpine Tundra* (1972).
See also Alaskan Pipeline; Arctic River Basins; Carbon Cycle; Glaciers; Nomadism; Water; Whaling Industry.

TURNER, JOSEPH MALLORD WILLIAM

(1775–1851), British painter in oils and watercolor. Born in Maiden Lane, Covent Garden, the son of a barber, Turner was a precocious artist, exhibiting his first drawings in the window of his father's shop. He entered Royal Academy schools in 1789, worked on architects' drawings, received instruction from Thomas Malton in perspective, learned the current topographical style of Dayes and Hearne with Thomas Girtin, copied drawings at Dr. Monro's house, and traveled extensively through Western Europe. He was a member of the Royal Academy. With Constable, he was one of the leaders of English Romantic landscape painting, whose works covered all aspects of landscape painting, from private revolutionary sketches, to historical landscapes of overwhelming impact.

Turner, one of the great proponents of Romantic landscapes, was interested in the "mutual relations between organisms and their environment" before the first definition of œcology, as it was then called, was proposed in 1873. He was equally well aware of its other definition of "the relationship of people to their surroundings and how they affect one another," as many of his major pictures and studies reveal. His earliest watercolors are records of ruins and picturesque landscapes, not only as seen through the developing interest in antiquarian pursuits, where man's ruins had been overtaken by nature, but also, along with others of his generation, in the way nature could be seen as an overpowering force. In this vein, his paintings of the *Plagues of Egypt* and *Snow Storm, Hannibal Crossing the Alps*, based on personal observations of natural effects, reveal human endeavors to be puny in the face of nature. His shipwreck scenes, notably during the first decade of the nineteenth century, remind us that more people were lost at sea than are now killed on the roads. As a Romantic artist, he painted events ancient and modern, for example, *The Wreck of a Transport Ship*, depicting a contemporary event; depictions of ancient legends, for example, *Ulysses Deriding Polyphemus*; and scenes of elemental power, for example, *Snowstorm, Avalanche and Inundation, a Scene in the Upper Part of Val d'Aouste, Piedmont*. He set great store by observing individual effects of nature closely, when he wrote in the margin of Opie's *Lectures on Painting*:

> He that has that ruling enthusiasm which accompanies abilities cannot look superficially. Every glance is a glance for study. Every look at nature is a refinement upon art. Each tree and blade of grass or flower is not to him the individual tree, grass or flower, but what it is in relation to the whole, its tone, its contrast and its use, and how far practicable: admiring nature by the power and practicability of his Art, and judging of his Art by the perceptions drawn from Nature . . .

His "plein-air" sketches in a boat along the River Thames, ca. 1807, observe trees, skies, and water together, as he advises. They anticipate Constable's campaign of "Skying," as Constable called it, when during 1821–22 he observed the clouds and changes of weather, noted the date and the time of day, the direction of the wind, and the characteristic of the weather. Constable even described the cloud formations in recent scientific terms, such as "cirrus." A modern meteorologist, using contemporary weather records, has found Constable's studies to be totally accurate. Turner was pursuing the same sort of study, around 1807, as well as studying perspective, linear and aerial, to get his facts right.

He was also fishing and hunting. Modern attitudes towards "ecology" might view these activities as reprehensible. Turner was out early in the morning, sketching, and then enjoying himself with rod and gun, particularly on the estates of his friends.

There are many stories of his success as an angler, "although with the worst tackle in the world," as a friend remarked. "Every fish he caught and showed to me, and appealed to me to decide whether the size justified him to keep it for the table, or return it to the river; his hesitation was often almost touching, and he always gave the prisoner at the bar the benefit of the doubt." Turner's painting of *Trout Fishing in the Dee, Corwen Bridge and Cottage* is an embodiment of his life-long interest in rivers, how they flow and how a fisherman could play them. His *Rivers of France* series continues this interest in the relationship of humans to the rivers of Europe. He was equally well aware of the kingfisher's superiority in a rocky stream in Yorkshire. The preliminary sketch is of the stream and its banks (*On the Washburn*). In the finished watercolor Turner has placed a kingfisher on the rocks, in mute deference, as all fishermen know, to that bird's superiority over fly and rod.

He was equally happy shooting birds in the company of his friend and patron, Walter Fawkes, in Yorkshire. Fawkes's brother was an amateur ornithologist, and Turner's *Book of Birds* consists of seventeen exquisite watercolor studies. There are other, equally exact studies of pheasants, grouse, and woodcock. Turner may have shot them all himself. Two other watercolors in the Wallace Collection, London, show similar expeditions. Yet at home, he was known by local children as "Old Blackbirdy," because he would not let them take birds' nests and their eggs from his garden hedges.

He does not seem to have ridden to hounds, but for a famous hunting patron, the Earl of Darlington, he was pleased to show a hunt in full flood (*Raby Castle, the Seat of the Earl of Darlington*). At the same time he was painting his magical view of Dordrecht, *Dort or Dordrecht, or the Dort Packet-Boat from Rotterdam Becalmed*; the sublime and tragic view of the battlefield at *Waterloo*; fifty watercolors of his views on the Rhine as a result of his trip to Europe in 1817; and a large watercolor of Tivoli, which he had not yet actually seen.

He had an equally sharp view of the degradations of the new Industrial Revolution. His views of *Leeds, Yorkshire* and *Dudley, Worcestershire* place these new industrial centers with their factories and smoking chimneys against their natural surroundings.

A number of his major pictures clearly make a comment on the passing of the old, and its replacement by a new order. In his famous *The Fighting "Temeraire," Tugged to her Last Berth to be Broken Up*, an old wooden-walled veteran of Nelson's fleet is being towed by a new steam tug to be dismantled. The scene is set against a phenomenal sunset. With *Rain, Steam, and Speed (The Great Western Railway)*, Turner shows the new steam train crossing the Thames on Isambard Kingdom Brunel's newly engineered railway bridge. These new railway bridges drastically shaped the English landscape, but Turner also includes ancient nymphs on the bank and a hare attempting to outpace the train.

His later views of nature are imbued with a sense of cataclysm, with avalanches in the Alps or a cynical view of recent history, with Napoleon contemplating a rock limpet (*War, the Exile and the Rock Limpet*), *The Deluge: Shade and Darkness, the Evening of the Deluge*, and *Light and Color (Goethe's Theory), the Morning after the Deluge*.

Although these powerful images seem to imply a philosophical view of man against nature, they are equally concerned with visual problems of light and color. At the very end of his life, he produced haunting sky studies at Margate or over the Thames. If the 1873 definition of ecology can be interpreted broadly, there is no doubt that Turner, along with Constable and the other Romantic landscapists and poets, observed nature closely, as part of their concern to awaken great sentiments about man's relation with nature. He would have agreed with his admirer, the artist Samuel

Palmer, that "Landscape is of little value, but as it hints or expresses the haunts and doings of man."

MALCOLM CORMACK

For Further Reading: Martin Butlin and Evelyn Joll, *The Paintings of J. M. W. Turner*, 2nd rev. ed. (1984); John Gage, *J. M. W. Turner. A Wonderful Range of Mind* (1987); Jack Lindsay, *J. M. W. Turner, His Life and Work* (1966; rev. ed. *J. M. W. Turner, The Man and His Art*, 1985).

TYLOR, EDWARD B.

(1832–1917), British anthropologist. The son of a prosperous Quaker brass manufacturer, Tylor entered the family business in 1848 as a clerk, but early in 1855 a chest complaint led him to quit his native London and undertake a leisurely trip to America where it was hoped the warmer climate would restore his health. Having wandered about the United States for the better part of a year, he made his way to Havana, Cuba, in 1856. Here he met, quite by chance, another London Quaker, Henry Christy (1810–1865), whose success in business had allowed him to indulge a passion for prehistoric archaeology. Christy was on his way to study the ancient ruins in the Valley of Mexico, and sensing Tylor's interest in his plans, invited the younger man to accompany him. Under Christy's tutelage, Tylor acquired a practical knowledge of archaeological and anthropological fieldwork. The expedition lasted six months, and after its conclusion Tylor returned to London with his mind firmly set on making anthropology his life's work.

From his field notes and letters he had written to his family while in Mexico, he assembled the text of his first book: *Anahuac, or Mexico and the Mexicans, Ancient & Modern*, published in 1861. While this work was primarily conceived as travelogue, it nevertheless embodies all of the characteristics of his future works, namely, his ability not only to master a multitude of facts and crystallize them into graceful and intelligible prose, but also to develop a logical and erudite amalgam of method and theory. This work was followed four years later by *Researches into the Early History of Mankind and the Development of Civilization*, which immediately established his reputation as a leading anthropologist. Here he outlined his view of human cultural development, which recognized essentially three techno-cultural stages: savagery, barbarism, and civilization. But in addition to being committed to the idea of progressive evolution, he was, as his later works reveal, equally bound to the notion of the psychic unity of the human species. To Tylor progress was inextricably linked with rationalism, and what truly established his fame was the book published in 1871: *Prim-*

itive Culture: Researches into the Development of Mythology, Philosophy, Religion, Language, Art, and Custom, in which these and related ideas were elaborated on more fully.

In tracing the development of human civilized societies from prehistoric times, he viewed the primitive world as a pre-scientific state in which events in the human and natural world were erroneously explained. It was in this context that he surmised religion had had its origin. He coined the term *animism* to characterize what he believed was the earliest form of religious belief. Animism, or the belief in spirit beings, was, according to Tylor, a primitive attempt to explain life and death, as well as other universal human experiences such as dreams. And as a cultural evolutionist he depicted a sequential development from animism to polytheism from which ultimately emerged monotheistic systems.

Another concept developed in *Primitive Culture* was what he called "survivals," by which he meant cultural practices and beliefs that in the passage of time had lost their original function and meaning but continued on into later stages of cultural development in various forms like superstitions. This concept was important to his general thesis since it allowed him to establish an evolutionary relationship between the past and the present—and thereby to demonstrate that modern societies had indeed progressed through earlier, primitive evolutionary stages. In developing this thesis he also introduced into the anthropological vocabulary the term *culture* and a definition that is still valid today: "that complex whole which includes knowledge, belief, art, morals, custom, and any other capabilities and habits acquired by man as a member of society." Prior to Tylor, the term *culture* in its original Latin usage described the process of cultivation or nurture, and more often than not was used in a horticultural context. Although German scholars had since the late 18th century been using terms such as *kulturgeschichte*, they had not used the word *kultur* explicitly in the sense that Tylor had.

The influence of *Primitive Culture* was far-reaching. For example, it led the Cambridge classicist (Sir) James Frazer (1854–1941) to anthropology and subsequently to write his equally influential treatise on the evolution of religion: *The Golden Bough*, first published in 1890.

Tylor's fourth and final book, *Anthropology: An Introduction to the Study of Man and Civilization*, was published in 1881 and represents a summary of both his work and the extent of anthropological knowledge at that time.

Together with his scholarly productions (which provided an agenda for ethnographic research), Tylor's other activities as an organizer and lecturer did much to secure anthropology as a legitimate and independent scientific discipline in Britain. He played an active role in the movement that led the British Association for the Advancement of Science to create in 1884 a separate section for anthropology, with Tylor as its first president. Coinciding with this event was his ensconcement at Oxford University where he was successively appointed Keeper of the University Museum, then reader in anthropology (1884), and finally first professor of anthropology (1896). Without question the present status currently enjoyed by anthropology at Oxford and elsewhere in British academe is due in large part to his pioneering efforts.

During the course of his career Tylor received many honors, including his election as a Fellow of the Royal Society (1871) and a belated Knighthood (1912).

FRANK SPENCER

For Further Reading: R. H. Lowie, *Edward B. Tylor: American Anthropologist* (1917); R. R. Marett, *Tylor* (1936); G. Elliot Smith, "Edward Burnett Tylor," in H. J. Massingham, ed., *The Great Victorians* (1932).

U

ULTRAVIOLET RADIATION

Ultraviolet radiation (UVR) encompasses radiation emissions within a wavelength of 200 to 400 nanometers (nm) (one nm equals one billionth of a meter), falling between x-rays and visible light in the electromagnetic spectrum. An invisible component of sunlight, UVR is also discharged by man-made sources, such as fluorescent lamps, lasers, and welding arcs. There are three UVR wavelength bands. The first two, UVA (400 to 320 nm) and UVB (320 to 290 nm), are the components of sunlight that reach the earth. The third, UVC (290 to 200 nm), includes emissions from the sun that are usually absorbed by the upper atmosphere before reaching the earth.

UVR does not have as much energy as x-rays do. Nevertheless, human exposure to UVR is responsible for a number of acute and chronic health effects, a subject of increasing alarm in recent years with the discovery that levels of exposure are gradually intensifying.

Normally, most UVR from sunlight is absorbed by a layer, or "shield," of ozone molecules in the upper atmosphere. New evidence shows that this shield, which extends some 20 to 50 kilometers above the earth, is thinning at a rapid rate, thereby leading to increased exposure of the earth's surface to UVR. The major factors responsible for the thinning are chlorofluorocarbons, which are used principally in refrigerants, propellants, and solvents, and other man-made chemicals. These substances migrate to the upper atmosphere, where they decompose into free chlorine atoms, which are highly destructive to ozone.

What are the effects of UVR on health? Because UVR cannot penetrate deeply into human tissues, the critical organs for UVR are the eyes and the skin. At very high levels of exposure, such as sunlight reflected from snow or emissions from a high intensity source of artificial UVR, photokeratitis, a painful irritation of the eye, can ensue (for example, "snow blindness" or "welder's flash"). Skin is at risk from common sunburn. Some medications, such as tetracycline, and natural chemicals found in foods, such as lemons and celery, concentrate in skin and can act as photosensitizers, increasing the risk of sunburn.

Of greater concern, however, are the long-term effects of UVR exposure at lower but more prolonged levels. UVR enhances aging of the skin, characterized by thinning, dryness, and wrinkles, and it induces approximately 70% of the 500,000 new skin cancer cases diagnosed each year in the United States. UVR exposure also may contribute to melanoma, a serious type of skin cancer, although this effect is less straightforward than photoaging of the skin and other forms of skin cancer, and it may require bursts of sun exposure and predisposing factors. The development of cataracts, dense opacities of the lens of the eye that can progressively lead to blindness, has also been linked to chronic low-intensity UVR exposure. These harmful UVR effects can be partially mitigated by clothing, barrier lotions and creams, and prescription glasses or sunglasses. As the ozone shield continues to thin, public health advisories on these measures are likely to increase.

It is feared, however, that the full impact of globally increased UVR exposure on human health and welfare may be even greater than suspected. Recent research has suggested UVR can depress the immune system, a disastrous possibility for those regions of the world where infections by parasites, bacteria, and viruses are prevalent. Finally, increased UVR could conceivably alter delicate balances among photosynthetic organisms, with consequent effects on the food chain and the ecosystem.

HOWARD HU

UNCED

See United Nations Conference on Environment and Development.

UNEP

See United Nations Environment Programme.

UNESCO

See United Nations System.

UNIT-BASED PRICING

Unit-based pricing is a concept under which the fees paid for solid waste collection and disposal vary with the amount of waste put out for collection by the consumer. The concept is also known as pay as you throw, volume-based pricing, pay by the bag, or variable rates.

Historically, residential solid waste revenues have been raised through property taxes. Some communities have charged identified users in the form of fixed fees for unlimited collection of solid waste. Unit-based pricing provides for more efficient use of scarce landfill or disposal options and adoption of waste reduction and recycling activities.

These incentive-based fees were adopted in a handful of communities in the U.S. before 1988, including Olympia, Washington, Jefferson City, Missouri, Duluth, Minnesota, and Grand Rapids, Michigan. The rate of adoption of these systems increased significantly by the late 1980s, and exceeded 200 communities in 19 states by 1992. This activity particularly followed widespread landfill closures, rapidly increasing waste management costs, and publication of articles noting the success of these systems in Seattle and other communities.

A number of communities have reported reductions of over 25% in the tonnage of waste delivered to transfer stations and landfills after implementation of unit-based (incentive) pricing and recycling programs. They have been successful in providing strong incentives for residents to participate in waste reduction and recycling efforts. It is also generally perceived to be more equitable for customers to pay in relation to the amount of waste generated. Unit-based pricing also helps communities fund and provide an integrated array of services and reflect the relative costs of waste management options to users.

These incentive-based fees have taken four main forms: variable can rates, pre-paid bag systems, pre-paid tag/sticker-based fees, and weight-based systems. In each case, the principle of the system is to provide higher fees for customers who dispose of more waste, and rewards in the form of lower fees for using less service.

LISA A. SKUMATZ

UNITED NATIONS CONFERENCE ON ENVIRONMENT AND DEVELOPMENT

The United Nations General Assembly, responding to the report of the Brundtland Commission (the UN Commission on Environment and Development), decided in December 1989 to hold the Conference on Environment and Development (UNCED) in Rio de Janeiro the first two weeks of June 1992, coinciding with the 20th anniversary of the first UN Conference on the Human Environment held in Stockholm. It was decided further that nations would be represented at UNCED by their heads of state or government, so that the Conference would be the first ever "Earth Summit."

UNCED was the largest international conference ever held, with 110 nations led by their heads of state or government and 172 nations participating. Resolution 44/228, which established the mandate of the Conference, made it clear that it was to be a conference on "environment and development" and that these must be dealt with on an integrated basis for every issue considered—from climate change to human settlements.

A series of concrete measures resulted from the Conference, including two historic documents. The first was *The Earth Charter:* A declaration of basic principles for the conduct of nations and peoples to ensure the future viability and integrity of the Earth as a hospitable home for humans and other forms of life. The second was *Agenda 21:* An agenda for action establishing the agreed work program of the international community for the period beyond 1992 and into the 21st century in respect of the issues to be addressed by the Conference. This also included means to implement the agenda through: new and additional financial resources; transfer of technology; and strengthening of institutional capacities and processes. Also, there were agreements on specific legal measures, e.g., conventions on global warming and biological diversity, negotiated prior to the Conference and opened for signature by all governments in Rio de Janeiro.

As a follow-up to UNCED, the UN General Assembly in 1992, by resolution 47/191, called for the Economic and Social Council (ECOSOC) to establish the Commission on Sustainable Development to monitor and report on the implementation of Agenda 21 by governments and the international community as well as the activities related to the integration of environmental and developmental goals throughout the United Nations system. ECOSOC formally established the new 53-member Commission at UN Headquarters and its first organizational session was held at UN Headquarters in February 1993, electing as its chairman Ambassador Razall Ismail of Malaysia.

When the global environment first emerged as a concern in the late 1960s and early '70s, it was the industrialized countries that placed it on the international agenda and took the initiative of convening the Stockholm Conference in 1972. Developing countries saw this preoccupation by the rich as a potential constraint on their own economic development. They insisted that the environmental agenda and dialogue be broadened to accommodate their concerns and the issues of poverty, under-development, inequity and natural resources, which are intimately and inextricably bound up with environmental conditions and prospects in these countries.

The Stockholm Conference recognized the essential relationship between environment and development but, since then, little had been done to give practical effect to the integration of environment and development in economic policy and decision-making. While a great deal of progress was made toward environmental improvement in particular instances, the Brundtland Commission made it clear that overall the environment of the planet had deteriorated since 1972 and there had been serious acceleration of such major environmental risks as ozone depletion and global warming.

The primary responsibility for the future of developing countries rests, of course, with them, and their success will depend largely on their own efforts. But they deserve and require an international system that lends strong support to these efforts. This includes substantially increased financial assistance, and much better access to markets, private investment and technology to enable them to build stronger and more diversified economies, to effect the transition to sustainable development and to reduce their vulnerability to changes in the international economy.

Agenda 21 called for "new and additional" annual investment and expenditure of $625 billion 1992 GDP (gross domestic product). $500 billion is to be supplied by the developing countries themselves. The critical balance of $125 billion that is to come from the industrial countries constitutes the catalytic investment required to put the under-employed labor and under-utilized resources of the developing countries to work. That figure comes to no more than the promised 0.7% of the 1992 combined industrial-country GNP (gross national product). The implementation of Agenda 21, under the surveillance of the Commission on Sustainable Development, begins with fulfillment of that undertaking.

Responding to the belief that the sustainable development goals of UNCED could only be achieved through the involvement of all sectors of society, the Preparatory Committee for UNCED (PrepCom) decided, at its first session in August 1990, to accredit all non-governmental organizations (NGOs), whether national or international, which could demonstrate their "competence and relevance," to address the comprehensive UNCED agenda.

In each of the four PrepCom meetings, accredited NGOs made oral and written interventions in the formal proceedings. At times they proposed alternative wording to text under negotiation and had the satisfaction of seeing their work accepted. UNCED was thus able to draw upon a wide range of experience, knowledge and capacity, not only of governmental personnel but of scientists, leaders of business and industry, religious and cultural leaders and such special constituencies as women, youth and indigenous peoples. In addition to the 5,000 members of the formal delegations to UNCED, Rio found itself playing host to more than 20,000 NGO people in attendance at their own Global Forum. A larger number, and wider range, of NGOs actively participated in UNCED than in any previous United Nations Conference, as well as record numbers of media representatives.

UNCED focused largely on the changes that must be made in economic behavior and international relations to ensure global environmental security. Its main task was to move the joint environment and development issues into the center of economic policy and decision-making and provide the basis for the transition to a sustainable way of life on our planet.

MAURICE F. STRONG

For Further Reading: Shridith Ramphal, *Our Country, The Planet* (1992); Stephan Schmidheiny, *Changing Course* (1992); World Commission on Environment and Development, *Our Common Future* (1987).

UNITED NATIONS CONFERENCE ON THE HUMAN ENVIRONMENT

The United Nations Conference on the Human Environment took place in Stockholm, Sweden, 5–16 June 1972. In response to disquieting events occurring worldwide, it brought industrialized and developing nations together for the first time to acknowledge the rights of humanity to a healthy and productive environment.

Economic prosperity following the flood of industrial reconstruction and technological growth after World War II led to an affluent-consumption revolution in the industrialized countries. In developing countries, the rapid penetration of post-war medical technology dramatically reduced death rates and accelerated population growth. The resulting worldwide surge in consumption placed increasing stresses on air, water, and land resources for the production of food, fibres, water, and minerals and for the disposal of waste products, many of them persistently toxic.

These human pressures on the biosphere led to unforeseen negative side-effects. Driven by poverty, overgrazing and overharvesting in developing countries caused woodland and soil degradation and loss, leading to falling water-tables, especially in semi-arid regions, further threatening the availability of wood fuel, food, and water and weakening agricultural export potential. This aggravated episodes of hunger, drought, poverty, deprivation, and ill health, triggering migration to the rapidly growing cities and social turbulence. In the industrialized countries, increasing contamination by potentially harmful chemical residues and other urban-industrial wastes generated by consumer affluence accumulated to the point where food, water, and air were increasingly affected, with costly consequences for human health.

Throughout the 1950s and 1960s, international technical and scientific investigations of these generic issues gradually led to a slow recognition of the environment and development problems familiar today. To the developing countries' issues of soil erosion, desertification, drought, poverty-linked malnutrition, and vulnerability to natural disasters, with all their related problems of disease incidence and mortality, were added the industrialized countries' problems of chemical pollution and its effects on wildlife, forests, fisheries, crops, livestock, and human health. To the anxieties arising over air pollution in cities, oil pollution at sea, poisoning episodes from mercury and other metallic compounds and from organochlorine pesticides, were added the somewhat sinister news that traces of these substances could now be found in the most distant oceans, in the recently formed ice on the tops of remote mountains, and on the polar caps. This gave clear enough hints that barely understood planetary biogeochemical transport processes were at work and brought growing fears that the health of the biosphere was being jeopardized, perhaps threatening future generations. Towards the end of the 1960s these fears were aggravated by a sharpened perception of major regional and global problems such as acid rain in Europe; loss of biodiversity, especially through tropical forest destruction; depletion of the stratospheric ozone shield; and global warming.

As the disquieting news of all these discoveries steadily grew, natural history and conservation groups were stimulated to turn their attention to these problems, and new, non-governmental organizations were formed, especially in industrialized countries. As their causes broadened from local natural history to regional and global issues affecting humanity, their environmental perception was transformed from plant and animal taxonomy and ecology to cover the interrelationships between these subjects and human health, natural resource availability, and ecodevelopment. These new organizations played an important part in the Stockholm process.

It was with developments such as these in mind that in 1968 the UN Economic and Social Council forward-ed a proposal to the UN General Assembly that it should consider holding an environment-development conference. The General Assembly decided to convene such a conference in 1972 and subsequently played an important role in guiding the overall conceptual structure of the Conference. It called for a report on the main problems to be considered, which was drawn up after consultation with member states, UN specialized agencies, and other appropriate intergovernmental and non-governmental organizations.

The report, accepted by the General Assembly a year later, stressed the importance of enabling developing countries to forestall environmental problems by learning from the mistakes of the industrialized countries. The Assembly also accepted the Swedish government's invitation to hold the Conference in Stockholm, and UN Secretary-General Kurt Waldheim was given authority to appoint a small Conference secretariat and a Secretary-General. Maurice F. Strong, organizer of the International Development Research Center in Canada, was subsequently appointed as Secretary-General, with a Preparatory Committee of representatives from 27 Member States to advise him. Four preparatory meetings were held between March 1970 and March 1972 to develop the Conference documentation. Member States were active; 86 submitted national reports on their environmental experiences and concerns. UN agencies and other intergovernmental and non-governmental institutions as well as individual experts also submitted material.

In 1970 the General Assembly stressed the linkages between environment and development. With this in mind, a number of regional seminars and special meetings were held. The Panel of Experts on Development and Environment meeting at Founex, Switzerland, in June 1971, broke new ground in elaborating environmental considerations as an integral part of the development process, a theme taken up by a series of UN Economic Commission seminars in Addis Ababa, Bangkok, Mexico City, and Beirut. The world science community met in Canberra, Australia, from August to September 1971, by special request of Maurice Strong, to bring together the newly formed Special (later Scientific) Committee on Problems of the Environment (SCOPE, part of the International Council of Scientific Unions) and the UN Advisory Committee on the Application of Science and Technology to Development.

During the 3 to 4 years of the preparatory process, governments worldwide were becoming exposed and sensitized to the global aspects of environment and development issues, and international perceptions evolved rapidly. At the start of Conference preparations, the industrialized countries understandably saw their problems as arising from urban-industrial pollution, damaging their quality of life and natural resources, but to their dismay, the developing countries, who had no industry to speak of and hence virtually no industrial pollution, demanded a much wider discussion. Equally understandably, developing countries saw concern with pollution as an issue for the developed countries only, who had created it by their own activities and could well afford to clean it up: They saw their own problems as environmental damage stemming from and aggravating a fundamental poverty-driven deprivation of the basic needs of human life, which threatened not only their future development but even present-day survival.

But as the Conference approached, another question arose. In the light of all the poverty and population-induced degradation and loss of renewable resources in the developing nations and the luxury-consumption-induced pollution in the developed world, would the natural supplying power of the Earth and its cleansing and renewal processes become overwhelmed and eventually fail to support human development within the next 100 years or so? These fears were spelled out in *A Blueprint for Survival* and *Limits to Growth*, which appeared in early 1972. As warnings began to circulate about the limits to the planet's carrying capacity, the developing world started to suspect that the developed world would pressure them to forgo their own development in the interests of environmental preservation. The developing countries argued that rich developed nations, such as the U.S. (with 6% of world population but consuming 35% of world resources), had a moral responsibility to share more equitably by reducing their numbers or per capita consumption.

Although rich countries appealed for environmental restraint to protect biosphere resources, they were rather hostile to discussing their own, often highly wasteful, per capita consumption. Likewise, the poorer nations, while seriously questioning environmental control as "anti-development," were equally hostile to discussing the consumption problems of their own rapidly growing populations.

In this potentially divisive climate, 1200 delegates of 113 governments and an equal number of non-government participants crowded into Stockholm for the two-week environment marathon. Apart from having contributed to the Conference process itself, several non-governmental groups held alternative meetings. These "fringe" activities on occasion produced a level of debate and treatment of development-environment issues equally as useful as the official UN Conference. A major political row had already broken out over East Germany, who was not a member of the UN and hence ineligible to participate. In protest, the Soviet Union and all other members of the Eastern Bloc except Romania and Yugoslavia boycotted the Conference and

thus, unfortunately, did not participate in formulating the 26 principles and 109 recommendations that finally emerged. The Conference was opened by Kurt Waldheim, who stressed that as the rich-poor gap widened, soaring poverty, population, and pollution could not be overcome without diverting current arms expenditures to these problems.

The main work of the Conference was divided into three committees, each with two subcommittees—(1) Planning and Management of Human Settlements for Environmental Quality and Educational, Informational, Social and Cultural Aspects of Environmental Quality; (2) Environmental Aspects of Natural Resources Management and Development and Environment; and (3) Identification and Control of Pollutants of Broad International Significance and International Organizational Implications of Action Proposals. A working group was established, following debate on a resolution by China, amended by Iran, to discuss the Preparatory Committee's draft Declaration on the Human Environment.

In general, the approach to the conference deliberations could be seen as falling into three levels. The first, generic, level was relevant to basic principles, particularly those elaborated by the "Declaration," regarded by many governments as the most important outcome of the Conference. This conceptual approach was expressed in the book *Only One Earth*, written by Barbara Ward and René Dubos, published in ten languages, and distributed at the start of the Conference, which came to be seen by many as embodying the "spirit of Stockholm." The second level was the development of an "Action Plan" of key issues for evaluating environmental states and trends worldwide and managing them for human economic and social development. The third level was the identification of areas suitable for draft conventions or other agreements that could be settled in principle at the Conference.

The Declaration, apart from being an inspirational message, was seen as the first essential step in the development of international environmental law. The Preparatory Committee's draft received 8 days of intensive review and was approved with acclamation by the last plenary of the Conference. After an introductory proclamation about creating a healthy environment for the social and economic development of all countries and affirming that "of all things in the world, people are the most precious," the Declaration sets out 26 principles. These declare that human beings have the fundamental right to freedom and equality in an environment that permits a life with dignity and well-being, and that they have the fundamental responsibility to protect the environment for the future generations by safeguarding renewable and non-renewable natural resources from pollution. Hence, policies of discrimination—apartheid, racial segregation, and colonialism—were condemned.

The remaining principles cover other crucial issues such as the importance of adequate financial, technical, and training assistance and stable commodity prices as prerequisites for accelerating development; the principle that human population growth should be managed when conditions dictate; the need to promote public education in environment and development issues and research and development in their scientific and technological aspects; the need for states to cooperate in developing international environmental law; the principle that when exercising their sovereign right to exploit their own resources, states must not harm others; and the elimination of nuclear weapons.

The output of the three main committees was adopted by the Conference as an "Action Plan." A large number of comprehensive and extremely detailed, sometimes idealistic, proposals were condensed into 109 recommendations, covering every aspect of development and environment, both in actual substance and in all the processes, including research and development; monitoring; education and training; and financial, organizational, and institutional requirements for promoting ecodevelopment. The Action Plan was intended to draw together the knowledge and expertise of national governments and of the entire UN system and to be run as the UN Environment Programme (UNEP). The plan was subsequently endorsed by the General Assembly on 15 December 1972, resolution 2994 (XXVII).

Many other issues received a priority boost which set them on course for later action. Highlighting ocean pollution and whaling at Stockholm subsequently assisted the outcome of the London Dumping Convention and the International Whaling Commission's whaling moratorium. Calls for protection of the world's genetic resources as a reservoir of new genes for agriculture and support of conventions to regulate trade in endangered species (CITES) and to protect world heritage sites (developed by UNESCO) were all acted on. Similarly, the detailed recommendations on human settlements resulted in the formation of the UN Center for Human Settlements.

GORDON GOODMAN

For Further Reading: M. W. Holdgate, M. Kassas, and G. White, eds., *The World Environment 1972–1982* (1982); United Nations Document A/CONF. 48/14/Rev. 1 (a detailed account of the proceedings and conclusions of the Stockholm Conference); Barbara Ward and René Dubos, *Only One Earth* (1972).

UNITED NATIONS DEVELOPMENT PROGRAMME

Created in 1965, the United Nations Development Programme (UNDP) is the largest multilateral source of grant funding for development cooperation in the world. A Governing Council made up of 48 nations is responsible for the use of its funds ($1.5 billion for 1991) that come from annual voluntary contributions of UN Member States. UNDP works with 152 governments through a network in 115 developing countries. In cooperation with over 30 international and regional agencies, the Programme promotes higher standards of living, faster and equitable economic growth, and environmentally sound development. It works extensively with nongovernmental organizations (NGOs) and its projects cover such diverse fields as agriculture, forestry, land reclamation, energy, urban management, education, transportation, communications, public administration, health, housing, trade, and finance. Thus the Programme (a) focuses on the effective management of natural resources, industrial, commercial, and export potentials, and other development assets; (b) stimulates capital investments; (c) trains people in a broad range of professional and vocational skills; (d) transfers appropriate technologies; and (e) fosters economic and social development, emphasizing the needs of the poorest segments of the population.

UNDP has the main coordinating role for development activities undertaken by the entire UN system in developing countries. For example, it has administered special funds such as those entrusted to the Capital Development Fund, the UN Development Fund for Women (UNIFEM), and the UN Volunteers programme. It chaired the interagency steering committee of the International Drinking Water Supply and Sanitation Decade (1981–90). The Programme played a key role in the preparations for the UN Conference on Environment and Development and continues to provide support in transforming UNCED's Agenda 21 into national action plans. With UNEP and the World Bank it co-manages the Global Environment Facility that provides grants to developing countries for reducing global warming and protecting biodiversity, international waterways, and the ozone layer.

UNITED NATIONS EDUCATIONAL, SCIENTIFIC, AND CULTURAL ORGANIZATION

See United Nations System.

UNITED NATIONS ENVIRONMENT PROGRAMME

Environment as a major global issue was first brought formally to the UN system at the UN Conference on the Human Environment in June 1972. The initiative was largely that of Sweden, but the U.S. and other industrialized countries supported the need for the conference and for effective follow-up mechanisms to enlist the capacities of all relevant programs and agencies of the UN system. In addition to adopting a "Stockholm Action Plan," the General Assembly acted on institutional and financial recommendations, leading to the creation of the UN Environment Programme (UNEP), and a voluntary supporting fund.

Environmental matters had long been treated by different parts of the UN system. For example, the World Health Organization (WHO) was concerned with effects of the environment on human health, such as high infant mortality rates due to unsafe drinking water; the World Meteorological Organization (WMO) was concerned with atmospheric conditions, including the early study of greenhouse gases that might affect climate and agriculture and other human activities.

Thus, by the time of the Stockholm Conference many parts of the UN system were already working with governments to help them tackle environmental problems. At Stockholm it was agreed that environmental problems qualified for international attention if the impacts were felt transnationally (pollutant effects outside the borders of the releasing state), whether directly (e.g., down-wind or down-stream flows of pollutants) or indirectly (e.g., trade barriers to protect domestic consumers against unsafe imports). The preparatory process for Stockholm also mobilized significant new or increased contributions from international nongovernmental organizations (INGOs), notably the International Council of Scientific Unions (ICSU) and the International Union for the Conservation of Nature (IUCN).

At Stockholm governments approved 109 recommendations, redistributed into three functional components: the global environmental assessment program ("Earthwatch"), environmental management activities, and supporting measures.

Assessment functions are intended to provide a rational basis for environmental management, also termed management for sustainable development. Most assessment activities, like research and monitoring of environmental parameters and evaluation of resulting data, are carried out by national institutions working together in cooperative international programs to improve the quality, compatibility, and relevance of their findings. UNEP develops these collabo-

rative programs through the Earthwatch Program, in close cooperation with other UN and international organizations. Examples include the Global Environmental Monitoring System (GEMS), the information system (Infoterra), and the International Register of Potentially Toxic Chemicals (IRPTC). Many of the "Earthwatch" programs launched since 1972 have proven their worth, whether with regard to global problems such as measuring changes in mass balance of glaciers as an indicator of climate change or the levels of methyl mercury in regional fisheries.

There is widespread agreement on the need for strengthening the capacity of the UN system to provide "early warning" of major environmental risks, assess these risks, and help states develop cooperative measures to reduce them or mitigate the consequences, such as through better contingency planning. To cope with scientific uncertainty about basic cause-and-effect relationships, continuing assessment has relied on international groups of experts like the UN Scientific Group of Experts on the Effects of Atomic Radiation (UNSCEAR), set up by the Assembly 30 years ago. Another group well-known in environmental circles is the Group of Experts on Scientific Effects of Marine Pollution (GESAMP), appointed by the UN system to provide assessment information. GESAMP assessments have gained increasing credibility, in part because its expert members—many of whom come from government institutions—are expressly working in an expert, noninstructed status.

Respect for state sovereignty precludes a direct management role for the UN system. UN *management* activity, therefore, involves primarily development of policies, practices, and agreements—including, but not limited to, formal treaties—that encourage changes in state practice and, ultimately, human behavior. The role of the UN system is to facilitate the processes by which states negotiate agreements and help administer them, sometimes designing and implementing agreed programs, projects, and activities required to give effect to agreements.

Parallel to the treaty route to effective international management action is the generation of agreed guidelines or recommended practices in various parts of the UN system that are endorsed by agreed declarations. Thus, the 1981 "Geneva Guidelines on Off-Shore Mining" and the 1985 "Montreal Guidelines for the Protection of the Marine Environment against Pollution from Land-Based Sources" have been consulted by states and international organizations in developing international agreements in this field. The UNEP series "Environmental Law—Guidelines and Principles" also includes the agreed guidelines "Weather Modification," "Banned and Severely Restricted Chemicals," and "Environmental Impact Assessment." While not binding, these political agreements can modify national practice, particularly when noncompliance is brought to public attention by nongovernmental organizations (NGOs) and the media.

The UN system also has long experience in generating agreed criteria as a basis for national standard-setting as well as other so-called soft law techniques. Under some treaties expert groups set up by agencies like the International Civil Aviation Organization (ICAO), the International Telecommunications Union (ITU), and WMO and IMO have the power to revise treaty standards in technical annexes which can become effective in the absence of formal objection, i.e. without having to go through the ratification process each time.

UNEP's *support functions* include strengthening human, institutional, and other resources to ensure that *all* key actors—including developing countries—have the means to contribute to agreed actions and to share in the benefits. Supporting measures required for actions in the assessment and management components include education, training, public information, organization of national and international activities, financing, and technical cooperation. Many of these activities are a part of the normal "technical assistance" programs in which UNDP and the UN system have extensive experience. Consideration of "international organizational implications of action proposals" led governments to forward institutional recommendations to the General Assembly, that were adopted in Resolution 2997 (XXVII), "Institutional and Financial Arrangements for International Environmental Cooperation," on 15 December 1972.

Also in 1972, the Assembly, "aware of the urgent need for a permanent institutional arrangement within the United Nations system for the protection and improvement of the environment," established four new mechanisms that together constitute UNEP: (1) a Governing Council of 58 states elected by the Assembly; (2) an environment secretariat headed by an Executive Director elected by the Assembly; (3) an Environment Fund to provide additional financing on a voluntary basis; and (4) an Environment Coordination Board to coordinate environmental work of the UN system.

UNEP's Governing Council promotes international cooperation and provides general policy guidance for direction and coordination of environmental programs throughout the UN system; reviews the world environmental situation and ensures adequate consideration of emerging problems; promotes scientific and other professional research; and reviews the impact and costs of environmental policies on developing countries.

UNEP's Executive Director, elected by governments in the General Assembly on the nomination of the UN Secretary-General, and a small secretariat (based in Nairobi, Kenya) provide substantive support and reports to the Council; coordinate environment programs in the UN system and review and assess their effectiveness; advise intergovernmental bodies of the UN system on environmental programs; secure cooperation from scientific and other professional communities worldwide; provide advisory services for promotion of international cooperation; submit proposals to the Council on medium- and long-range planning for UN programs; and administer the Environment Fund (and keep the problem of additional financial resources for developing country needs under review).

Maurice F. Strong, a Canadian and Secretary-General of the Stockholm Conference, was appointed the first Executive Director of UNEP in 1973; Mostafa Tolba, an Egyptian and Deputy Director of UNEP under Strong, served as Executive Director of UNEP from 1975 through 1992; Elizabeth Dowdeswell, another Canadian, was appointed to serve as Executive Director from 1 January 1993–31 December 1996.

The Environment Fund was established "to enable the Governing Council . . . to fulfill its policy-guidance role for the direction and coordination of environmental activities." To do this the Assembly mandated that the Fund "finance wholly or partly the cost of the new environmental initiatives undertaken within the UN system." Among the "programmes of general interest" for which this fund should be used, the Assembly specified the following:

regional and global monitoring, assessment and data collecting systems, including, as appropriate, costs for national counterparts; the improvement of environmental quality management; environmental research; information exchange and dissemination; public education and training; assistance for national, regional and global environmental institutions; the promotion of environmental research and studies for the development of industrial and other technologies best suited to a policy of economic growth compatible with adequate environmental safeguards; and such other programs as the Governing Council may decide upon.

The combined efforts of UNEP's Governing Council, Executive Director, Environment Fund, and coordination mechanisms have enabled governments to reach agreement much sooner than would otherwise have been possible on issues related to global security and risk management, including specific international accords dealing with stratospheric ozone protection, hazardous wastes, the loss of biological diversity, and, most recently, climate change. There are GEMS activities in 142 countries. IRPTC provides information about more than 80,000 chemicals in use and assists developing nations in establishing their own chemical information systems. Infoterra provides data on environmental problems in 137 countries. More than 120 countries have participated in UNEP's regional seas programmes, resulting in successful negotiations on eight international conventions and fifteen protocols and agreements. Under UNEP auspices the Montreal Protocol (1987) focused on damage to the Earth's ozone layer, and the Basle Convention (1989) dealt with the movement and disposal of harmful industrial wastes. All of these results demonstrate UNEP's strengths in promoting international cooperation and coordination, providing policy guidance within the UN system, and enlisting contributions from scientific and other professional communities to develop knowledge and programs.

PETER S. THACHER

For Further Reading: Mostafa Tolba, *Evolving Environmental Perceptions: From Stockholm to Nairobi* (1988); UNEP, *Earth Matters: Environmental Challenge for the 1980's* (1983); UNEP, *Saving Our Planet: Challenges and Hope* (1992).

UNITED NATIONS POPULATION COMMISSION

Established in 1947, the UN Population Commission is one of the functional commissions of the Economic and Social Council (ECOSOC) and has 27 members. Its mandate is to arrange for studies and advise the Council on (1) the size and structure of populations, and changes and policies which influence them; (2) the interplay of demography with economic and social factors; and (3) other demographic questions on which the UN might seek advice. The Population Division of the Department of Economic and Social Development (DESD) supports the Commission and carries out studies requested by it.

The Commission also monitors population trends and policies and reviews and evaluates progress towards accomplishing the goals of the World Population Plan of Action that was adopted in 1974 (World Population Year) at the World Population Conference held in Bucharest. Likewise, it follows up on recommendations of the 1984 International Conference on Population held in Mexico City.

The 1974 Plan of Action stressed the basic connections between population factors and economic development. The 1984 Conference approved an updated global population strategy—88 recommendations identifying where further action was needed. Principles underlying these recommendations were contained in the Mexico City Declaration of Population and Development.

In 1991 ECOSOC decided to convene the International Conference on Population and Development in Cairo in 1994. The overall theme of the Conference is population, sustained economic growth, and sustainable development, and within that theme there are six groups of issues. These include (a) population growth and demographic structure, (b) population policies and programs, (c) interrelationships between population, development, and the environment, (d) population distribution and migration, (e) population and women, (f) and family planning, health, and family well-being.

Some of the objectives of the Conference are (1) to evaluate the progress made on the World Population Plan of Action; (2) to strengthen awareness of population issues on the international agenda and their linkage to development; (3) to consider the desired action at the global, regional, and national levels, and ways and means of treating population issues in their proper development perspective; (4) to adopt a set of recommendations for the next decade in order to respond to the population and development issues of high priority; and (5) to enhance the mobilization of resources needed—especially in developing countries—for the implementation of the results of the Conference.

Dr. Nafis Sadik, Executive Director of the UN Population Fund (UNFPA), is Secretary-General of the Conference. (A trust fund operating since 1969, UNFPA is the largest internationally funded provider of population assistance to developing countries.) Shunichi Inoue, Director of DESD, is Deputy Secretary-General.

UNITED NATIONS SYSTEM

A framework for international cooperation on a scale unprecedented in human history was put in place with the founding of the United Nations (UN). Officially established when representatives of 50 States adopted and signed the United Nations Charter in San Francisco on 26 June 1945, the UN represents the second time in the twentieth century that industrial powers had established an organization for international cooperation. The League of Nations, organized at the end of the First World War, had failed to prevent the Second.

The UN includes nearly every State on the planet (182 as of July 1993). Member States are legally committed to cooperating in supporting the principles and purposes set out in the Charter. These include maintaining international peace and security, securing justice and human rights, and promoting social progress and friendly relations among all nations.

Now liberated by the demise of the cold war, the United Nations has never before been in such a good

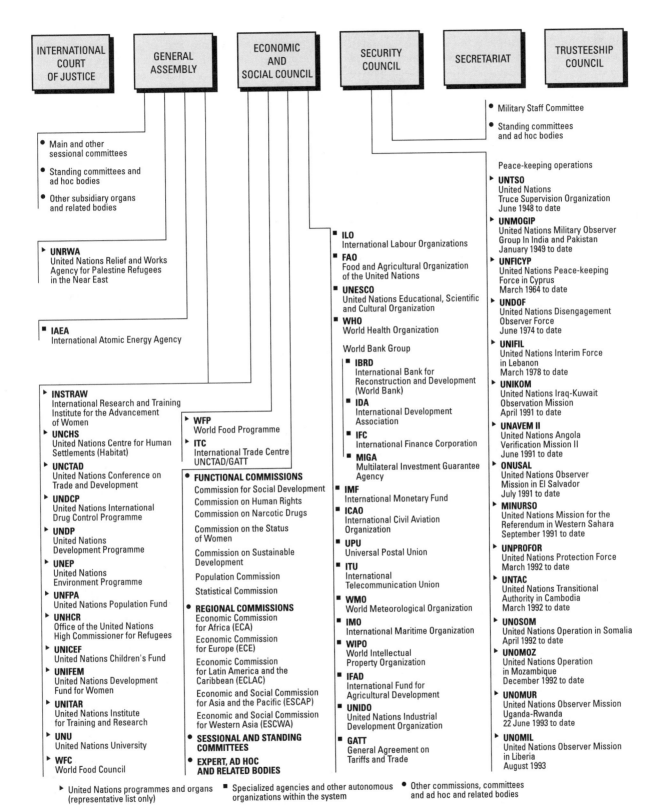

Figure 1. Principal organs of the United Nations. Source: United Nations Department of Public Information, *Basic Facts About the United Nations.*

position to fulfill the purposes for which it was established. It has achieved numerous solid gains in its almost five decades of existence. These include: providing shelter and relief to millions of refugees from war and persecution; being a major catalyst in the evolution of 100 million people from colonial rule to independence and sovereignty; and establishing peacekeeping operations 26 times to contain hostilities and to help resolve conflicts. The UN has expanded and codified international law and wiped smallpox from the face of the planet. Since its founding it has adopted some 70 legal instruments promoting or obligating respect for human rights, thus facilitating an historic change in the popular expectation of freedom throughout the world.

United Nations Day is celebrated each year on October 24th—the day in 1945 on which the Charter was ratified by China, France, the Soviet Union, the United Kingdom, the United States, and a majority of the other signatories.

Under the Charter the official languages of the United Nations are Chinese, English, French, Russian, and Spanish. Arabic has since been added as an official language of the General Assembly, the Security Council, and the Economic and Social Council.

The Charter established six principal organs of the UN:

The General Assembly is the main deliberative organ composed of representatives of all Member States, each with one vote. The Charter may be amended by a vote of two thirds of the Members of the UN, including the five permanent members of the Security Council. Decisions on important questions, such as those on peace and security, also require a two-thirds majority. Decisions on other questions are reached by a simple majority. The General Assembly's regular session begins on the third Tuesday of September each year and usually adjourns in mid-December, except when special sessions are called by the Security Council. At the start of each regular session the General Assembly elects a new President. To ensure equitable geographical representation, the presidency rotates among five groups of States: African, Asian, Eastern European, Latin American, and Western European and other States.

Since issues among nations require negotiation, what is called debate in the General Assembly consists of voluntary declarations by each nation of its position on the issue at hand. After members have had their say, the agenda item is referred to one of the Assembly's seven standing committees. It is in the committees that the real debate, deliberation and, finally, negotiation take place. Although they are smaller bodies, committees are still large enough to be representative of the size and diversity of the UN membership. Committees deal first of all with questions of international

security, and economic and social questions. Other committees take up the more abstract questions, such as human rights, or mundane matters, such as UN finances. The committees return to the General Assembly with draft resolutions which most often carry.

The work called for by resolutions of the General Assembly is accomplished in several ways. Various commissions and committees study and report on issues of the day, such as human rights violations, extension of the arms race to outer space, environment and development. International conferences, for which commissions may have done the preparatory work (e.g. the Brundtland Commission for the UN Conference on Environment and Development held in Rio in June 1992), are held periodically. The associated technical agencies, such as the World Health Organization (WHO) and the Food and Agriculture Organization (FAO), also carry out the resolutions of the General Assembly. It is from the floor of the General Assembly that the questions of environment and development have found their way into public consciousness and given people some comprehension of the scope of environmental issues today.

The Security Council has primary responsibility, under the Charter, for the maintenance of international peace and security. All members of the UN agree to accept and carry out the decisions of the Security Council. The Council has 15 members: five permanent members—China, France, the Russian Federation (with support of 11 member countries of the former USSR), the United Kingdom, and the United States—and 10 members elected by the General Assembly for two-year terms. Each member has one vote and the five permanent members have the right of veto.

While the world has often listened anxiously to the deliberations of the Security Council, developing countries do not accept China as their representative on the Council. The caucus of developing countries—the "Group of 77"—argues for India and Nigeria as fully qualified for permanent membership on the Security Council. Such a change in membership can only be accomplished by amending the UN Charter.

The Economic and Social Council (ECOSOC) was established by the Charter as the principal organ to coordinate the economic and social work of the UN and the 17 specialized agencies and institutions—known as the "United Nations Family" of organizations. ECOSOC has 54 members who serve three-year terms and have one vote. The Council generally holds two month-long sessions each year, one in New York and the other in Geneva. Its year-round work is carried out in its subsidiary bodies—six functional and five regional commissions and committees—which meet at regular intervals and report back to the Council. The functional Commissions include Population and Status of Women. The regional Commissions are for Africa, Europe, Latin America and the Caribbean, Asia, and the Pacific.

Specialized agencies include intergovernmental agencies related to the UN by special agreements, and separate, autonomous organizations which work with the UN and each other through the coordinating machinery of ECOSOC. Examples include FAO, UNESCO, WHO, and the World Bank Group, four institutions with the common objective of helping developing countries.

Under the UN Charter ECOSOC has consultative status with some 900 non-governmental organizations (NGOs) concerned with matters within the Council's competence. NGOs bring expertise and the pressure of public opinion to bear in UN deliberations on the different issues of concern to their diverse membership. The International Council of Scientific Unions (ICSU) conducts studies for ECOSOC; Amnesty International originates many of the complaints of human-rights abuse that ECOSOC carries to the General Assembly. The persuasiveness of numerous NGOs helped convene The Stockholm Conference on the Human Environment that created the UN Environment Programme.

The Trusteeship Council was assigned the task of supervising the administration of Territories placed under the Trusteeship System. The major goals are to promote the advancement of the inhabitants of Trust Territories and their progressive development towards self-government or independence. Their aims have been fulfilled to such an extent that only one of the original 11 Trusteeships remains: Palau, in the Pacific Islands, administered by the United States.

The International Court of Justice is the principal judicial organ of the United Nations. The seat of the Court is at The Hague, Netherlands. Its Statute is an integral part of the UN Charter. All Member States of the UN are automatically parties to its Statute. The Court consists of 15 judges elected by the General Assembly and the Security Council, voting independently. They are elected on the basis of their qualifications, not on the basis of nationality, and care is taken to ensure that the principal legal systems of the world are represented in the Court. No two judges can be citizens of the same State. The judges serve for a term of nine years and may be reelected. They cannot engage in any other occupation during their term of office.

The International Court of Justice created in 1946 is a response to the call in Article I of the Charter for the settlement of disputes between and among nations by peaceful means and in accordance with the law. Its jurisdiction depends upon the willingness of nations to yield to it their sovereignty in the issue at hand. One measure of progress from anarchy to harmony in the international community has been its adjudication of some 60 cases. The Court normally sits in plenary session, but it may also form smaller units, called chambers, if the parties so request. Judgments given by chambers are considered as though they were rendered by the full Court.

The Secretariat, an international staff of civil servants working at United Nations Headquarters in New York and in the field, carries out the diverse day-to-day work of the UN organization. The Secretariat is made up of more than 25,000 men and women from more than 150 countries who have taken an oath to answer to the United Nations alone for their activities and not to seek or receive instructions from any government or outside authority. The Secretariat, headed up by the Secretary-General, services the other organs of the United Nations and administers the programmes and policies laid down by them.

The Secretary-General is described by the United Nations Charter as the "chief administrative officer" of the Organization. He is, of course, much more than that. Equal parts diplomat and activist, conciliator and provocateur, the Secretary-General stands before the world community as the very emblem of the United Nations. The present Secretary-General of the United Nations, and the sixth occupant of the post, is Boutros Boutros-Ghali of Egypt, who took office on 1 January 1992. His predecessors as Secretary-General were: Javier Perez de Cuellar of Peru from 1982 to 1991; Kurt Waldheim of Austria from 1972 to 1981; U Thant of Burma (now Myanmar) from 1961 to 1971; Dag Hammarskjold of Sweden from 1953 until his death in a plane crash in Africa in 1961; and Trygve Lie of Norway from 1945 to 1953.

The Budget of the United Nations is approved by the General Assembly biennially. The budget is submitted initially by the Secretary-General and reviewed by a 16-member expert committee—the Advisory Committee on Administrative and Budgetary Questions. The programmatic aspects are reviewed by the 34-member Committee for Programme and Coordination.

Funds for the regular budget come from the contributions of Member States, who are assessed on a scale specified by the Assembly on the recommendation of the 18-member Committee on Contributions and based on the real capacity of Member States to pay. The Assembly has fixed a maximum of 25% of the budget for any one contributor and a minimum of 0.01%. For example, (1992 figures) the U.S. is assessed the maximum 25%, with the population estimated at 253,887,000, and Japan, with 123,921,000 people, is assessed 12.45%.

The budget approved for appropriations for the 1992–93 biennium was $2.4 billion. With many members chronically failing to pay in full or on time, the overall financial situation of the UN has been extremely precarious for several years.

Outside the regular budget Member States are also assessed in accordance with a modified version of the basic scale, for peacekeeping operations (13 current or recently completed in 1993). Many other UN activities are financed mainly by voluntary contributions outside the regular budget, including the UN Development Programme, the World Food Programme, the Office of the UN High Commissioner for Refugees, the UN Children's Fund, the UN Relief and Works Agency for Palestine Refugees in the Near East, and the UN Population Fund.

RUTH A. EBLEN

For Further Reading: UN Department of Public Information: *Basic Facts About the United Nations* (1992); *Notes for Speakers* (1993); and *Report of the Secretary-General on the Work of the Organization* (1993).

UNITED STATES AGENCY FOR INTERNATIONAL DEVELOPMENT

See USAID.

UNITED STATES GOVERNMENT

As with most government functions in the U.S., environmental responsibility is split among three branches—executive, legislative, and judicial—and three levels of government—federal, state, and local. For example, the legislative branch (Congress) periodically updates the Clean Water Act; the executive branch EPA oversees the law's implementation by state executive agencies; and courts resolve differing interpretations of the law. The Act empowers state governments to set water quality standards and issue permits restricting pollution discharges, subject to federal approval. The federal government provides subsidies for publicly owned sewage treatment works and may act jointly with states to enforce the law. Local governments build and operate municipal sewage treatment works. Nearly all U.S. environmental administration is decentralized in some manner.

The Executive Branch

While all federal entities are bound by law to consider and minimize the environmental impacts of their activities, there are more than 30 departments and of-

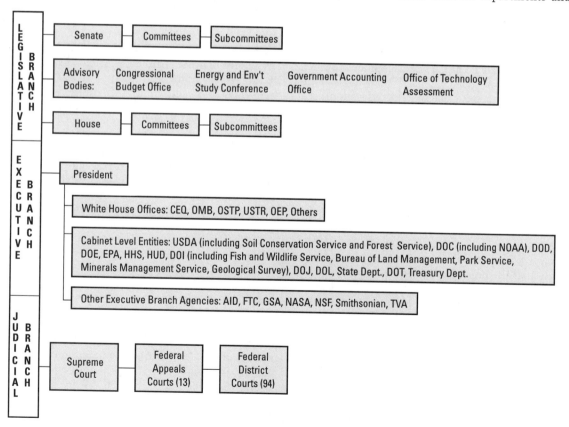

Figure 1. Environment-related institutions of the U.S. government.

fices that have primary responsibilities for various aspects of environmental affairs.

The White House. Because environmental duties are spread throughout the government, presidents traditionally employ White House aides to coordinate and oversee broad policy matters. Although each President organizes the White House differently to suit his own purposes, the offices most likely to influence environmental policy are the following:

Council on Environmental Quality (CEQ). CEQ was established by Congress in the National Environmental Policy Act of 1969 to oversee implementation of broad environmental policy goals throughout the federal government. In particular, CEQ oversees federal agency implementation of regulations that require environmental impact analyses (EIA) for all major proposed actions.

Office on Environmental Policy. This office, created by President Bill Clinton in 1993, is charged with the coordination of environmental policy across agency boundaries.

Office of Management and Budget (OMB). This powerful office assists the President in establishing and enforcing federal budget goals and overseeing the regulatory activities of the executive branch.

Office of Science and Technology Policy (OSTP). OSTP serves as a source of scientific and technical expertise for the President and as a focal point for federal science and technology initiatives.

Office of the U.S. Trade Representative (USTR). USTR develops U.S. trade policies and represents the government in international trade negotiations. Relevant topics include the use of trade sanctions for environmental purposes (e.g., banning importation of tuna caught in nets that also destroy dolphins) and the competitive effects of environmental regulation (e.g., the possibility that strict regulations may cause companies to move abroad).

Other offices. Other White House offices that often are involved in environmental decisions include the Council of Economic Advisors, the National Security Council, and the Office of the Vice President.

Department of Agriculture (USDA). Most federal farm program benefits—commodity price supports, loans, and crop insurance—are contingent on the application of land stewardship practices. USDA programs prevent soil erosion on farmlands, conserve wetlands, enhance wildlife habitat, and promote alternatives to pesticides and fertilizers that can pollute water resources. Within the USDA, the agencies chiefly responsible for implementing conservation requirements are the Agricultural Stabilization and Conservation Service and the Soil Conservation Service.

USDA is also home of the U.S. Forest Service, which manages 191 million acres of land for various uses, including recreation and timber production, and assists state foresters and private nonindustrial landowners with forest stewardship.

Department of Commerce (DOC). DOC plays an indirect role in environmental affairs by setting many economic, technology, and international trade policies.

However, the National Oceanic and Atmospheric Administration (NOAA) within DOC is one of the nation's premier environmental agencies. NOAA explores, monitors, and manages the conservation of the oceans, including living resources such as fisheries and marine mammals. The agency also conducts satellite observations of global climate patterns and research concerning oceans and inland waters, the lower and upper atmosphere, and the space environment.

Department of Defense (DOD). Though DOD is not traditionally considered to have environmental responsibilities, it does manage 25 million acres of public land and thousands of military-industrial facilities. Since the late 1980s, it has devoted increasing attention to compliance with federal environmental regulations, as well as cleanup and restoration of contaminated sites and initiatives to develop cleaner technologies and prevent pollution. It also conducts programs to protect natural and cultural resources on DOD lands.

The department also contains the Army Corps of Engineers, which designs, executes, and manages major civil works projects such as dams for flood control and dredging of harbors and waterways. In so doing, the Corps has inherited the authority, exercised jointly with EPA, to approve or block activities that affect the nation's wetlands.

Department of Energy (DOE). DOE programs and regulations influence the energy efficiency of the U.S. economy as well as the choice and availability of power from petroleum, coal, nuclear, solar, and wind sources. The department's laboratories are world leaders in research and development of advanced energy technologies. Like the DOD, it manages many Cold War era facilities that require long-term, costly environmental cleanup. Through several regional systems dating to the 1930s, it also produces and markets hydroelectric power.

The Federal Energy Regulatory Commission, which is independent but closely affiliated with DOE, regulates the sale and transportation of natural gas, electricity, and oil transported by pipeline. The Nuclear Regulatory Commission, also independent of DOE, regulates and licenses all aspects of civilian nuclear power.

Environmental Protection Agency (EPA). The EPA was established in 1970 when President Nixon consolidated various elements of the Department of Health and Human Services and the Department of the Interior into the new entity. The EPA bears wide-ranging responsibilities to establish and enforce antipollution regulations in the areas of air, water, solid waste, pesticides, radiation, and toxic substances. It also conducts programs of environmental research, education, and awareness at all levels.

In recent years, the agency has shifted its priorities in several ways. Increasingly it has emphasized *prevention* rather than just cleanup of pollution; protection of *ecosystems*, not just parts of the landscape; regulations tailored to the *degree of risk* posed, rather than arbitrary mandates; and *international* coordination, recognizing the global nature of ecosystems and the economy.

Department of Health and Human Services (HHS). HHS performs assessments of public health risks from such hazards as lead in drinking water, airborne fumes, medical wastes, and contaminated sites. It supports research and education in environmental hazards and supports various other institutions, such as the World Health Organization, in conducting similar duties.

Department of Housing and Urban Development (HUD). Although most of its energies are devoted to the nation's housing needs, particularly in low-income communities, HUD assists communities with historic preservation, parks, and long-range design and planning.

Department of the Interior (DOI). DOI history spans more than a century of conservation. Together with USDA, DOE, EPA, and NOAA, it is one of the premier environmental entities in the U.S. Government.

As the steward of more than 500 million acres of land held by the Bureau of Land Management, Fish and Wildlife Service, and National Park Service, DOI perennially is involved in disputes over competing demands for land uses. The department's varied duties include supervising the use of publicly held resources such as reservoirs and mineral deposits (including those found offshore), protecting the nation's wealth of biological resources, and preserving scenic, historic, and recreational places. The U.S. Geological Survey maps and classifies the nation's lands, waters, energy, and other natural resources. The department is also charged with overseeing Native American reservation communities and various island territories.

Department of Justice (DOJ). Working closely with other federal agencies, the Environment and Natural Resources Division of the Department of Justice represents the U.S. Government in litigation involving public lands, natural resources, environmental quality, and wildlife. The fastest growing area of responsibility in recent years has been enforcement of antipollution statutes designed to ensure that polluters, rather than

the public, bear the burden of pollution costs. It also defends the government in suits brought by industry or environmental groups challenging agency decisions and policies.

Department of Labor. As part of its mission to promote the interests of American wage earners, the Department of Labor administers laws guaranteeing workers' rights to safe and healthful working conditions.

Department of State. As it implements the President's foreign policies, including international environmental policies, the State Department engages in continuous consultations with the public, other government entities, and foreign governments; negotiates treaties and accords with other nations; and speaks for the U.S. in the UN and dozens of additional international fora and conferences. The department's Bureau of Oceans and International Environmental and Scientific Affairs (OES) is concerned with such issues as climate change, fisheries, Antarctica, technology cooperation, and population.

Department of Transportation (DOT). DOT administers programs to ensure that the nation's transportation systems, including roads, rail and transit systems, airports, waterways, and pipelines, operate in harmony with the environment. Key issues of concern are air pollution, noise, congestion, vehicle efficiency, oil spills, and disruption of neighborhoods, historic districts, and ecosystems.

Department of the Treasury. The U.S. Treasury plays a major role in setting financial and tax policies that influence environmental protection. For example, the department represents the U.S. in World Bank deliberations on loans for dams, roads, and power systems in developing nations. In recent years, the Treasury has required improved environmental management as a condition of such loans. The department also oversees tax policy, which affects investment decisions, and the U.S. Customs Service, which blocks the importation of endangered animals or products made from them, such as African elephant ivory.

Other Executive Branch Agencies. In addition the following executive branch entities have significant environmental responsibilities:

Agency for International Development (AID). The primary foreign assistance arm of the U.S. Government is AID, which supports environmental programs in 90 countries. Projects include assistance in agriculture, forestry, biodiversity, coastal zone management, family planning, urban and industrial pollution, and water and resources management.

Federal Trade Commission (FTC). FTC is an independent quasi-regulatory agency that prevents monopolies and other unfair or deceptive market practices. In the early 1990s it developed and began to enforce guidelines concerning the environmental claims of advertisers, clarifying the appropriate uses of such terms as "environmentally friendly," "biodegradable," and "recyclable."

General Services Administration (GSA). On behalf of many federal agencies, GSA administers the construction and operation of hundreds of buildings, as well as the procurement and management of supplies and motor vehicles. In so doing, the agency administers an array of programs to reduce waste and fuel use, increase recycling and the use of recycled products, phase out products using CFCs, and improve energy efficiency in federal buildings and vehicles.

National Aeronautics and Space Agency (NASA). NASA's chief environmental contribution is in the area of scientific research. Using space- and ground-based systems to observe and analyze the Earth, the agency's Mission to Planet Earth helps scientists understand such phenomena as global climate change, ozone layer depletion, and distribution of rainfall.

National Science Foundation (NSF). NSF promotes science and engineering through research, educational programs at all levels, international cooperation, and long-range science policy planning. It sponsors research across the full range of environmental issues.

Smithsonian Institution. Among the Smithsonian's many activities to increase the diffusion of knowledge are museums of natural history and ancient cultures, the National Zoo, a Tropical Research Institute, and other programs that enhance understanding of humanity and its interactions with the total environment.

Tennessee Valley Authority (TVA). Established in the 1930s to develop the resources of the Tennessee Valley region, TVA more recently has adopted the goal of becoming a leader in solving environmental problems of national concern. Ongoing activities include electric power generation and conservation, waste reduction, watershed management (especially fertilizer and animal waste control), land management, and environmental education.

The Legislative Branch

The legislative branch of the federal government, called the Congress, consists of two bodies: the Senate, composed of 100 members, two elected at large from each state; and the House of Representatives, composed of 435 members, elected from single-member districts of approximately equal population. Legislative proposals called "bills" may be introduced by any member of either chamber, and are referred to specialized committees and subcommittees for consideration. Committee members may shelve proposed bills or, after holding public hearings and gathering comments from the executive branch and all affected parties, make changes and recommend a vote by the full chamber.

In carrying out its functions, the Congress is assisted by several expert advisory offices:

Congressional Budget Office (CBO). CBO provides the Congress with assessments of the economic impact of the federal budget.

Energy and Environment Study Conference (EESC). EESC is a bipartisan office that provides objective information on energy and environmental issues to members of Congress and their staffs. A related organization, the Energy and Environment Study Institute, conducts informational briefings and research projects.

Government Accounting Office (GAO). GAO is the investigative arm of Congress, often assigned to scrutinize executive branch implementation of Congressional mandates. Its reports may contain recommendations of ways to cut waste and fraud in costly government programs.

Office of Technology Assessment (OTA). OTA helps the Congress to deal with complex scientific and technical issues. It helps resolve competing claims, identifies policy options, maintains contacts with scientists in the executive branch and the nongovernment sectors, and alerts Congress to new developments that could have implications for future policy.

The Judicial Branch

In the U.S., most environmental cases are litigated in the federal courts. Cases usually enter the system in one of 94 federal district courts. Appeals are usually heard first in one of 13 federal appeals courts. The U.S. Supreme Court is the second and final appellate court.

The process of filing civil environmental cases in the federal court system is generally the same as that for other types of civil actions. Complaints brought on behalf of the federal government to enforce environmental laws are filed by the attorney general or designate. Any other person (regardless of citizenship) may file a grievance against a polluter or against the government for failing to enforce a law.

Over 90% of all federal civil cases, including environmental cases, are settled out of court or disposed of without a trial. However, cases that do proceed to trial follow the same legal procedures as in other matters, eventually resulting in sanctions on the polluter and/or relief to the aggrieved party. For example, the courts may order that contaminated sites be cleaned up, that environmental impacts be assessed before a project proceeds, or that the defendant undertake beneficial environmental projects.

Table 1

Congressional committees and their environmental jurisdictions

House of Representatives

Agriculture	production practices, pesticides, soil conservation, national forests, biofuels
Appropriations	spending authority for all federal programs
Banking, Finance, and Urban Affairs	multilateral development bank lending, urban development
Energy and Commerce	national energy policy generally, air pollution, pesticides, drinking water, groundwater, transportation, hazardous and other solid wastes, consumer and regulatory matters
Foreign Affairs	international environmental issues
Government Operations	general oversight of operations of all government programs and agencies
Interior and Insular Affairs	all public lands, irrigation and water use generally, Indian and territorial affairs, minerals and mining, regulation of the domestic nuclear energy industry, including waste disposal
Judiciary	legal aspects of environmental and energy laws
Merchant Marine and Fisheries	oil spills, ocean dumping, coastal zone management, endangered species, fisheries, wildlife refuges, outer continental shelf, National Environmental Policy Act (environmental impact analysis)
Public Works and Transportation	transportation systems, water pollution, toxic waste site cleanup, Army Corps of Engineers, water projects, wetlands, Tennessee Valley Authority
Science, Space and Technology	environment and energy research and development, space-based Earth-observing systems, technology policy
Small Business	energy and environmental issues as they affect small business
Ways and Means	all tax and trade issues

Senate

Agriculture, Nutrition, and Forestry	production practices, pesticides, soil conservation, national forests, biofuels
Appropriations	spending authority for all federal programs
Commerce, Science, and Transportation	interstate commerce and consumer matters generally, coastal zone and oceans management, highway safety, marine fisheries, space programs, technology policy
Energy and Natural Resources	national energy policy generally, energy research and development, all public lands, irrigation and water use generally, Indian and territorial affairs, minerals and mining, regulation of the domestic nuclear energy industry, including waste disposal
Environment and Public Works	NEPA, air and water pollution, hazardous and solid wastes, global environmental issues, endangered species, wetlands, regulation of nuclear power, drinking water, groundwater, ocean dumping, water resource development generally, highways and infrastructure
Finance	all tax and trade issues
Foreign Relations	international environmental issues generally
Government Affairs	general oversight of operations of all government programs and agencies
Judiciary	legal aspects of environmental and energy laws

Criminal cases under federal environmental laws may be brought only by the U.S. attorney general or state attorneys general. A criminal investigation may lead to an arrest or indictment, followed by an arraignment, where the defendant must enter a plea of guilty, not guilty, or no contest. In such cases, as in all other criminal cases, the government must prove to a jury that the defendant is guilty beyond a reasonable doubt. Sanctions in criminal cases may include fines and imprisonment.

DALE CURTIS

For Further Reading: U.S. Government, Council on Environmental Quality, *Environmental Quality: 23rd Annual Report of the Council on Environmental Quality, together with the President's Message to Congress* (1993); U.S. Government, Council on Environmental Quality, *United States of America National Report to the United Nations Conference on Environment and Development (1992)*; U.S. Government, National Archives and Records Administration, Office of the Federal Register, *The United States Government Manual 1993/1994* (1993).

See also Environmental Protection Agency.

URANIUM

Uranium, named by the German chemist Klapproth in 1789 after the newly discovered planet Uranus, is a dense, hard, silvery-white metal. The heaviest naturally occurring element, it is one of the few that are naturally radioactive (that is, its nuclei disintegrate into alpha and beta particles and gamma rays). Uranium is the eighth least plentiful mineral, comprising only about two parts per million of the earth's crust. Nevertheless, uranium deposits are numerous, and if converted to nuclear fuel, would contain many times the energy in all recoverable fossil fuels. One pound of uranium, for example, yields as much energy as three million pounds of coal. Because of this, uranium has assumed a significant place in the production of large-scale commercial energy.

Henri Becquerel discovered uranium's radioactivity in 1896. All the isotopes of uranium (different forms of the element with similar chemical properties but different atomic weights) are radioactive, each with a long half-life. By the end of each half-life, radioactivity of a particle is decreased by half. After a series of half-lives, uranium ultimately decays into lead, which is "stable" and does not decompose. Uranium-238 (with a half-life of 4,510,000,000 years) is the most commonly occurring isotope and comprises over 99% of all uranium. Uranium-235 (713,000,000-year half-life) and uranium-234 (247,000-year half-life) are the next most common types. The isotope U-235 is used most easily for energy production in nuclear power plants. U-238 can be used by conversion in a nuclear reactor to plutonium, which is also a nuclear fuel.

The long half-life of uranium is useful in determining the age of the earth. Uranium ores undisturbed by geological changes since they were formed contain all of the products of the radioactive disintegration of the original uranium. The last of these products, a lead nucleus, is converted from a uranium nucleus after 14 nuclear disruptions. Since lead is stable, the amount of lead present in a rock reveals its age, making it possible to calculate the age of certain rock formations by using the known rate at which lead is formed from uranium.

Uranium's role in energy production began in the 1930s, when the United States pioneered the development of nuclear reactors. Study intensified just prior to World War II, prompted in the United States by the urgent necessity to compete with German efforts to create nuclear weapons and to develop an inexpensive, powerful energy source. Now less significant in nuclear weapons than in nuclear energy, uranium is the source of large percentages of the electricity generation of industrialized nations. The largest users of nuclear energy are the United States, which produced 575 billion kilowatt-hours (kwh) of electricity from nuclear energy in 1990; France, which produced 298 billion kwh; and Japan, which produced 186 billion kwh.

URBAN CHALLENGES AND OPPORTUNITIES

By the year 2000, for the first time in history, the world will be predominantly urban. In 1800 only 3% of the world's population lived in cities; in 1950, it was 29%; and shortly after 2000, it will be over 50%. Furthermore, cities are reaching unprecedented sizes. By the turn of the century, there will be 23 mega-cities with populations exceeding ten million, and eighteen of these will be in developing countries.

Virtually every country has responded to the "urban explosion" by trying to limit the growth of their largest cities. These efforts range from restricting in-migration; to dispersing the would-be migrants (to growth poles, new capitals, smaller cities, or resettlement areas); to stimulating regional and rural development in hopes of equalizing the level of living in the countryside and the city.

Efforts to limit the growth of large cities have had limited success, however. Some, such as rural development, have proven counterproductive, actually hastening out-migration from the countryside. The fundamental reason for the failure of these policies is not only the lack of resources, enforcement mechanisms, or political will, but also the fact that cityward migration benefits the individuals, families, communities of origin, cities, and the nation as a whole.

Not only is there more economic opportunity in the city, but the larger the city the greater the opportunity. Empirical evidence also shows that large cities are more productive, and the largest cities are most productive relative to others in a less developed country. These cities typically account for 80 to 85% of their national GNP. Furthermore, detailed analyses of revenues and national budget expenditures show that funds and resources from central cities are transferred to, and benefit, the rest of their countries.

This is not to say that mega-cities do not have severe problems. In fact, these problems are often so linked with city size and management capacity that in many ways Rio de Janeiro, Bombay, Shanghai, and New York City have more in common with each other than with the smaller cities and towns in their own countries. To begin with, the sheer size of the mega-cities presents a situation for which we have no collective experience. No precedent exists for feeding, sheltering, or transporting so many people in so dense an area, nor for removing their waste products or providing clean drinking water. Urban systems based on human settlements of 100,000 or 250,000 may be able to accommodate urban populations of one million, but begin to break down at four million, and are blatantly unworkable at 10 million. What is needed is a more sophisticated and sensitive management capability than anything we have thus far developed.

What is required is not simply a set of innovative ideas that happen to be transferable from one context to another. More important is the cumulative effect of these ideas in helping to rethink the cities of the future and substitute hope for despair. Given the deeply vested interests in the status quo, how can the political will be found for urban transformation towards a more socially just, ecologically sustainable, politically participatory, and economically viable city?

It is in this light that Andre Gorz' concept of "non-reformist reforms," or "traditional reforms," is particularly helpful. Gorz discusses the struggles between workers and owners, and distinguishes between palliative reforms which are often simple material possessions, and transitional reforms which change the incentive system and the rules of the game. Manuel Castells makes the urban policy analogy clear in his book *The City and the Grassroots*. Some innovations may be intriguing in and of themselves and could help to improve the quality of life if more widely diffused. Others, like the decentralization of budgetary, zoning, land-use, and service delivery decisions to the neighborhood level, or the granting of equity shares to local community residents in large-scale private sector development projects, are primary level changes which generate other innovations and will have profound consequences.

Throughout history cities have been the crucibles of culture and the source of major advances of civilization. The boldness of the quest for deliberate social change and the transformation of urban practices (from the neighborhood level all the way to city, national, and international levels) is at the heart of whether 19th century solutions continue to be projected onto tomorrow's world, or the leap to the 21st century city is finally made.

JANICE E. PERLMAN
ELWOOD HOPKINS

For Further Reading: Manuel Castells, *The City and the Grassroots* (1983); Janice Perlman, "Mega-Cities: Global Urbanization and Innovation," in *Urban Management: Policies and Innovations in Developing Countries* (1993); United Nations, Population Division, *Urban Agglomerations* (1992).

URBAN FORESTRY

See Forests, Urban.

URBAN HEAT ISLAND EFFECT

See Pollution, Thermal.

URBANIZATION

See Communities, Low and Moderate Income; Homelessness: Urban Environment; Housing; Industrialization; Transportation and Urbanization; Urban Challenges and Opportunities; Urban Planning and Design; Urban Renewal; Waterfront Development.

URBAN PARKS

The idea of an open space for public use is ancient in origin. The town square or open space between buildings, the place for gathering or for celebration, is part of the long tradition of public space in European culture. In the U.S. the 17th-century equivalent might be the New England town common or village green. The idea of the park as a designed space maintained at public expense is, however, more recent in origin.

The idea of public parks belongs to the era of Victorian social reform in England during the first quarter of the 19th century. The garden critic and editor of the *Gardener's Magazine*, J. C. Loudon, was one of the first to advocate the idea of public gardens and parks. Loudon, a friend of Jeremy Bentham and a supporter of liberal causes, published in 1822 a design for a greenbelt system of "breathing zones" for London. But Birkenhead park in Liverpool is the first notable example of the public park as we think of them today.

Designed in 1843 by Sir Joseph Paxton, Birkenhead — located in a growing suburb of Liverpool — relied in part on the Picturesque aesthetic developed by Sir Humphry Repton toward the end of the 18th century. This idea implied the creation of scenery that seemed "natural." The American designer Frederick Law Olmsted was impressed by the design when he visited the park in 1852, noting favorably the undulating lawns, the double lakes, winding carriage ways, and paths that led around the park. What most impressed Olmsted was the democratic character of the park; it was enjoyed equally by all classes. "In democratic America, there was nothing to be thought of as comparable with this People's Garden."

The American landscape designer and author of an influential book, *Theory and Practice of Landscape*

Gardening (1841), Andrew Jackson Downing was one of the first to advocate the idea of public parks in this country. It was Olmsted, however, who designed in partnership with Downing's former associate Calvert Vaux the first major public park in this country, Central Park in New York City, in 1856.

Park design in the U.S. until recently has relied primarily on the Picturesque model, formulated in England and pioneered during the Romantic era by Downing and later Olmsted. Like Downing, Olmsted thought of the park as an instrument for democratic social change. His designs for Central Park and subsequent designs for Prospect Park (Brooklyn) and the Boston and Chicago park systems were thought to provide city dwellers with a source of refuge from the harsh realities of 19th-century urban life. Several other mid-19th-century designers, including H. S. W. Cleveland, who designed the Roger Williams Park in Providence in 1876 and later the Minneapolis park system, as well as Olmsted's brilliant protégé Charles Eliot, shared Olmsted's view of the public park as a democratic experiment. Writing of his proposal for a Metropolitan Boston Park Commission in 1893, Eliot declared that "Crowded populations if they would live in health and happiness must have space for air, for light, for exercise, for rest and for the enjoyment of that peaceful beauty of nature . . . so wonderfully refreshing to the tired souls of townspeople."

Olmsted's vision of the picturesque park with its emphasis on pastoral scenery and seeming naturalness remained the guiding aesthetic of American park design well into the 20th century and is still imitated by some designers today. Although his vision is still visibly intact in some cities such as New York, much is required in the face of increasing maintenance costs and collapsing infrastructure to preserve this vision.

Beginning in the 1920s the demand for active recreation prompted park designers to provide space for community sport. By 1934, New York Park Commissioner Robert Moses could boast, for example, of the construction of 51 baseball diamonds and 240 tennis courts as well as the reconstruction of St. James, Crotona, and Macombs Dam parks in the Bronx, Mount Morris, Manhattan Square, and Carl Schurz park within the space of a single year.

Park design after World War II became increasingly associated with the development of regional park systems, linked by freeways, and in the 1950s by massive superhighway construction. Road construction during this period often resulted in the fragmentation of earlier 19th-century parks. Major urban parks suffered from lack of funding in the 1960s and many parks became the scene of political demonstrations and rock concerts. The visible decline of public parks prompted park planners to seek private funding for the rehabilitation of parks throughout the U.S.

The example of Central Park provided a model throughout the U.S. for park restoration. Beginning in 1980 Central Park Administrator Elizabeth Barlow Rogers began to raise private funds for the restoration of the park. The Central Park Conservancy was established as the official financial instrument for restoration of landscape as well as architectural features of the park. By 1987 $2.5 million had been privately raised, resulting in the renovation of the Sheep Meadow, the Bethesda Fountain and Terrace, and the Dairy as well as other historic structures within the park.

Much of the success of the revitalization of Central Park as well as other 19th-century parks is due to the establishment of new programs that interpret the history of these parks and their place in the evolution of park design. Educational programs in urban forestry, botany, and urban ecology have also contributed to public appreciation of the value of parks in densely populated areas throughout the U.S.

The successful Central Park initiative in park restoration has inspired similar efforts in other parts of the U.S., notably Boston, where in 1983 the Massachusetts Association for Olmsted Parks pressed for funding at the state level for the restoration of the Boston park system, the so-called Emerald Necklace. This work is still in progress, in spite of recent financial setbacks. Programs in Buffalo, Seattle, and Chicago have helped to prompt legislation at the national level that would provide funding for the rehabilitation of many other 19th-century parks designed by Olmsted and his contemporaries, Cleveland and Eliot.

While it is clear that a well-defined constituency exists for the restoration of 19th-century parks, much new thinking points in the direction of the design of new parks that are not green nor pastoral in their imagery. Many parks designed within the past two decades respond to the reality of urban life within the context of urban revitalization. These parks suggest a vision of city life that does not seek to escape the reality of urban experience, but rather to embrace it.

Lawrence Halprin is one of several American designers to have pioneered new directions in park design. His Lovejoy Fountain Plaza (Portland, Oregon, 1972), associated with the revitalization of downtown Portland, is one of the icons of contemporary park design. Bold in its conception and spare in its use of materials, the plaza has as its focal point a massive concrete fountain whose image offers the metaphorical equivalent of Oregon's Cascade Mountains. Visitors are encouraged to experience the torrential flow of water, to move through the plaza over stepping stones.

Isamu Noguchi's Dodge Fountain Plaza (Detroit, 1979) provides another bold example of modern park design. In a draft of his presentation to the Detroit Common Council, Noguchi spoke of revealing the special "genius of the place." The Dodge Fountain would be the focal point, which in one of many symbolic images he proposed would take the "form of a bird rising, more American Indian in a way, though it still retains an evocation of the airplane and jet propulsion."

Richard Haag's Gas Works Park (Seattle, 1975) incorporates into the overall scheme the idea of saving industrial relics, in this case an abandoned gasworks. The entire site was restructured and detoxified. The exhaust house became a children's play barn and the boiler house transformed into a picnic shelter.

On the international scene some of the most striking new parks have also been associated with urban revitalization and the restructuring of earlier sites.

In Paris opportunities have arisen to create new parks, gardens, or places. In the wake of the 1970s, technological change and restructuring of industrial sites have released redundant market, rail, or dock land for new uses. Bernard Huet's primary objective for his design of the Place Stalingrad was to restore urban and historical continuity, on the site of Ledoux's 1787 Rotunde de La Villette Customs Toll House and the La Villette basin of the Canal St. Martin. In this case, Huet's handsomely detailed design creates a traffic-free public open space and promenade along the Canal.

Bernard Tschumi's design for Parc de La Villette (Paris, 1982) has prompted many park designers to entirely rethink the idea of park design. Parc de La Villette, on the site of what had been the meatpacking district of Paris, is now a successful park with little reference to traditional approaches to design. The park is conceived, according to Deconstructivist design theory, as a complex multilayered sequence of activities, marked by references to technology, to film and popular culture. Distributed in linear fashion throughout the design are bright red metallic Folies, marking focal points of activity within the park's complex transit system.

The Urban Public Spaces of Barcelona (Spain), completed between 1981 and 1987, represents a large and important body of public works at widely different scales spread throughout a city of 1.7 million people that covers close to 100 square kilometers. The program began officially in December 1980, when the then Mayor Narcis Serra appointed a commission to assess urban issues confronting the city. Projects were undertaken within all ten districts of Barcelona. By 1987, when the city began to focus attention on hosting the Olympic Games, over 100 urban space projects had been completed, beginning with small-scale urban plazas and ending with extensive improvements to the Moll de la Fusta.

Three kinds of specific urban space projects have been completed: plazas, parks, and streets. Some, such as the Plaça Reial and the Parc Güell, designed originally by Gaudí, are renovations of existing urban spaces, whereas many others, such as the Plaça dels Paisos Catalans and the nearby Parc de L'Espanya Industrial, are new improvements within the city.

Some plazas such as the Plaça de la Mercè are relatively small and hard-surfaced. Others, for example the Plaça dels Paisos, are far more extensive, although equally hard-surfaced. Neighborhood parks, such as the Plaça de la Palmera, provide a variety of recreational activities as well as public art works by international artists. Other parks, such as the Parc de Joan Miro on a site formerly occupied by a slaughterhouse, are intended for city-wide use.

The Barcelona urban space project, awarded the Prince of Wales prize in Urban Design in 1990, represents a bold and humane experiment in urban planning and preservation. In announcing its decision the jury recognized the "significant role public spaces play in contemporary cities and the need for renewal within the public realm to further the quality of urban life." The jury also cited the need for public spaces as places for recreation and for "public gathering that mark the civic life of a community." The Barcelona program offers an exemplary model for urban design and park planning in other cities throughout the world, as a visible reminder of the critical role played by civic life within modern urban society.

ELEANOR M. MCPECK

For Further Reading: George Chadwick, *The Park and the Town* (1966); Elizabeth B. Kassler, *Modern Gardens and the Landscape* (1964); Roy Rosenzweig and Elizabeth Blackmar, *The Park and the People: A History of Central Park* (1992). *See also* Humanized Environments; Olmsted, Frederick Law; Parks, U.S. National; Urban Planning and Design; Waterfront Development.

URBAN PLANNING AND DESIGN

People have been trying to plan and design cities since cities began, but the task of shaping growth and change in the modern metropolis is very different from the planning and design of preindustrial cities.

The nature of the city changed relatively little from the beginning of recorded history until about 1600; even in Europe, where modern urbanization began, the pace of change continued to be slow until the latter half of the 18th century. Since then the city has changed so rapidly that planners and designers have had serious difficulty keeping up with development, much less defining it in advance.

Of course, there were always significant variations among cities of different cultures and from different historical periods, but preindustrial technology was similar from culture to culture throughout most of history: Goods were moved by water, or in carts pulled by animals; people walked, rode on animals, rode in boats and carriages. Building materials were wood, brick, and stone—and, in ancient Rome, concrete. Similar technologies meant that cities survived and prospered in similar locations: often at a natural harbor, a ford, or at the confluence of two rivers, in places that could be made healthy, were well supplied with food and water, and could be effectively defended.

Urban planning and design in preindustrial cities sometimes had a mystical component as well as a functional one, with the design of cities embodying ritual meaning, generic functional components, and adaptations to particular circumstances. The principal issues were city location, the layout of streets and fortifications, and, perhaps, the management of development within a religious precinct or around a central market square. Ordinary buildings within the city belonged to recognizable typologies, whose design evolved gradually over centuries. Such buildings were so imbedded in local tradition that their design and placement was implied by the street layout.

The first new technology to modify European city design was the introduction of gunpowder, beginning in the late 1300s. Straight walls of stone that could be broken down by cannon fire were replaced by sloping ramparts that ultimately evolved into triangular bastions. The regular geometry of these fortifications helped create the symmetrical star shapes that were a Renaissance ideal of city form. The rediscovery of Roman and Greek architecture and a new understanding of pictorial perspective also contributed to the reshaping of cities during the Renaissance. Sir Christopher Wren's plan for rebuilding London after the great fire of 1666 summed up major themes of Renaissance design: symmetrical plazas and squares, long straight streets like those built in Rome during the building campaign of Pope Sixtus V in the 1580s, and the vistas and rond-points of French Garden design as exemplified by André Le Nôtre's plans for Vaux and Versailles.

Wren's visionary plan was not adopted; it was faster and less expensive to rebuild London on its old foundations and within existing property lines—with better fire-control measures, such as somewhat wider streets and a code that required brick walls between buildings. This conflict between investing for future improvement and dealing with immediate practical problems is a recurring theme in modern city planning.

The unprotected suburbs were the least desirable location for a walled city, and suburbs were originally the shantytowns of the poor and the place for slaughterhouses and other activities that people of means did not want near their homes. As national frontiers began to replace individual city fortifications, and living behind walls was no longer necessary, the suburb became an attractive alternative to the crowded, noisy, polluted city. Londoners were already living in districts beyond the walls at the time of the great fire, and, by the mid-18th century, London merchants were commuting by carriage from villages several miles from the outskirts of the city. By the mid-19th century, the greater accuracy and range of modern artillery and the power of explosive shells made city walls obsolete, leaving cities more vulnerable but removing limits to their expansion, just at the time when the effects of industrialization set off unprecedented urban growth.

The first effect of the Industrial Revolution on cities was to increase trade, creating bigger dock installations on the waterfront and new districts of elegant houses. The rapid development of these new districts, such as Beacon Hill in Boston, the west end of London, and Edinburgh New Town, permitted designers to create compositions of squares and crescents that emulated earlier development for aristocrats but on a larger scale.

The factories themselves were located near water power and natural resources, usually a long way from established cities. The steam railway and the stationary steam engine brought industry to the cities, beginning in the 1830s. Cities were unprepared for the noise and pollution of railroads and industry, or for the influx of factory workers, so the initial effects of industrial development in cities were disastrous for the environment, and the conditions provided by real-estate speculators to house workers and their families were the worst imaginable.

The first efforts at what was to become modern city planning were attempts to engineer water and sewer services to replace disease-breeding cesspools and wells, and new roads to support increased traffic. There were also attempts to improve the living conditions of the poor through charity housing, and building regulations to control the most egregious exploitation.

The elegant uniform architecture, parks, boulevards, sewers, and other public works created in Second Empire Paris under the direction of Baron Georges Eugène Haussmann became models for cities everywhere. The image of Paris was enormously influential in the U.S., where the City Beautiful Movement that followed the World's Columbian Exposition of 1893 produced park and boulevard plans for many American cities—the most famous being Daniel Burnham's and Edward Bennett's Plan for Chicago, published in 1909.

While these plans for reconstructing and gaining control of industrialized cities were being prepared, others were making plans to develop the suburbs. The railroad had permitted the creation of garden suburbs away from the outskirts of the city, such as Llewelyn Park in what is now West Orange, New Jersey, and Riverside outside Chicago. Model company towns such as Port Sunlight, near Liverpool, were also planned communities related to, but removed from, existing cities. Ebenezer Howard, in a book published in 1898 and still in print, became the theorist of a complete escape from the polluted and congested 19th-century city, advocating a network of self-contained garden cities each served by its own railway station, and separated from each other by stretches of natural and agricultural landscape, or greenbelts. Howard proposed a way of life that combined the advantages of town and country, replacing existing cities altogether.

The beginning of the 20th century also saw the first government-subsidized housing in cities and much stricter building regulations. The first zoning regulations, separating land uses in cities such as Rotterdam and Frankfurt, also date from this time. New York's 1916 zoning ordinance included provisions that took the design principle of Haussmann's Paris and, by changing the scale, used it to shape streets of tall buildings. Buildings could rise to a height that was a function of the width of the street, and then had to set-back behind an imaginary angled plane similar to the slope of a Parisian attic roof.

Around 1910, at about the time that the modern planning profession was created, it looked as if cities could be perfected by replanning their central districts with parks and boulevards along Haussmannesque lines, replacing slums with government-subsidized housing like that built by the London County Council, and surrounding the city with a greenbelt that led to garden suburbs, garden cities, and model company towns. What subverted this vision was the continued presence of slums and industry in the city center, tall buildings located in a far more scattered pattern than the buildings that line Haussmann-style boulevards, and the invention of the automobile, which destroyed Howard's neat assumptions about urban expansion in the country by letting urbanization happen almost anywhere, not just around railway stations.

The New York Regional Plan of 1929 was a heroic attempt to synthesize earlier concepts of subsidized housing, parks, boulevards, and garden suburbs and relate them to tall buildings and new highways. Among the illustrations in the Regional Plan was Clarence Stein's and Henry Wright's design for Radburn, in Fairlawn, New Jersey, which was to have been a garden city on Howard's model, but redesigned to accept the automobile. Unfortunately the depression intervened before more than a fragment of Radburn could be completed.

At the same time in Europe, other theorists, notably the French-Swiss architect who called himself Le Corbusier, were advocating far more sweeping changes in the city: cutting new highways through the congested and outmoded streets and blocks, clearing the old unsanitary neighborhoods and replacing them with apartment towers surrounded by parks, decanting the old business center into modern skyscrapers.

The worldwide economic depression of the 1930s made sweeping changes in existing cities less likely, but made subsidized housing and inventive government regulation more necessary. In the U.S., the adoption of zoning and subdivision ordinances continued, and new highways were constructed especially designed for automobiles and trucks. The New Deal funded public works, experimented with the creation of Greenbelt Towns according to Ebenezer Howard's theories, provided subsidies to cities to build publicly aided housing, and through the Tennessee Valley Authority engaged in regional planning and development on an unprecedented scale. At the same time, private philanthropy was recreating colonial Williamsburg, an amazingly comprehensive attempt at preserving and reconstructing an entire community.

The destruction of cities during World War II had the side effect of clearing the way for experiments in city planning and design along the lines that Le Corbusier and other modernists had advocated. It was remarkable, nevertheless, how many historic central areas of cities were painstakingly reconstructed, even in Communist countries of eastern Europe. However, there were many new districts of modernist housing, and many new highways cut through the urban fabric.

In the end, suburbanization, automobile traffic, and postwar population movements were more devastating to existing cities than the war. In the 1950s cities and their suburbs still retained patterns of development that had been created over the previous century, and, because of the depression and World War II, had been essentially frozen for a generation. Up to the 1950s, while planning and city design had never had more than isolated successes, it was possible to believe that cities could be perfected according to an established model: Haussmann's Paris, Howard's Garden City, the synthesis attempted by the New York Regional Plan of 1929, Le Corbusier's vision of a city of towers set in parkland. In the 1950s television, computers, and intercontinental jet travel were new and their effects were just beginning to be felt. World-wide industrialization and mechanized agriculture were just beginning; the population of the world was a third what it is today.

By the 1990s cities everywhere had been transformed, and the techniques for planning and designing them have had to be transformed as well. Old ideas about what cities should be are not necessarily invalid; but they can no longer be seen as either static or comprehensive. The process of urbanization has now speeded up to the point where city design and planning are a means for managing change, not for codifying an end product. It has also become more and more difficult to separate city planning and design from environmental conservation, historic preservation, economic development, social service delivery, education, and all the other techniques for managing the continued transformation of the modern metropolis. With urbanization and industrialization putting the future of the planet as a hospitable home for people in question, planning and design assume a new urgency.

The changes of the last generation are least apparent in Western Europe, where each piece of land has a history, and attachment to the forms of the past is high. But beneath the historic surface are far more diverse populations, an information-based society, and patterns of urbanization that have adapted historic villages and towns to uses not so different from American suburbia. Planning and design are part of a highly directed economy; many decisions are made centrally, and the government can direct major investments in industry and infrastructure.

At the opposite end of the spectrum is the state of planning and design in developing countries where millions of people may be living in shantytowns and modernist housing projects around a historic preindustrial core and districts of European-style boulevards and garden suburbs. By going from a preindustrial, primarily rural society to the late 20th century in a generation, these cities have no 19th- and early 20th-century infrastructure and industry to rely upon, and there is no precedent for how to manage new development and the integration of their people into a workforce. People may draw water from wells and come home to movies on a VCR; there is little or no public transit, and the automobile is the primary means of transportation, with all its attendant traffic congestion and air pollution. Baron Haussmann's task in transforming Paris is nothing compared to problems confronting planners in Mexico City.

In countries such as the U.S., Canada, and Australia, which took part in the industrial revolution from its beginning and have plenty of undeveloped land, the principal issue is the urbanization of what used to be suburban and rural areas. Modern highway construction opened up huge new areas of land to development; bigger houses on cheaper land (and, in the U.S., subsidies for homeowners) plus a desire to evade problems of the central cities drew people to the new suburbs.

Then stores followed them to suburban downtowns and to shopping centers, then office jobs moved to suburban office parks, and industry to highway locations far from old railway centers. By the mid-1970s more urban development was taking place in what had been suburbs and the rural fringe of the metropolitan area than in the old central cities. New clusters of urban development formed, with shopping centers, office towers, hotels, and apartment buildings, miles from established centers on the edge of open fields.

In the U.S., older cities are fighting for their lives. Some, such as Camden, New Jersey, Gary, Indiana, and East St. Louis, Illinois, have been almost totally bypassed by new investment. Others, such as Boston, Minneapolis, and Portland, Oregon, have done much better, but almost every city has its deteriorated areas, and every metropolitan area has its new edge cities. Planners and designers have evolved sophisticated methodologies for renewing older cities, although they have not been able to do much about intractable problems of joblessness, crime, deteriorating school systems, and crumbling inner-city neighborhoods.

In the new suburban centers, zoning and subdivision based on models devised in the 1930s have proved totally inadequate tools for safeguarding the environment and channeling new investment into livable patterns. This new development is wasteful: Since the 1960s, the amount of urbanized land in the New York metropolitan region has increased by two thirds, without any increase in population.

Statewide planning offers hope of planning for a more sustainable future. Growth limits at the urban fringe, strong measures to safeguard the natural environment, and policies to redirect development toward areas with existing infrastructure offer hope of a better future.

JONATHAN BARNETT

For Further Reading: Peter Hall, *Cities of Tomorrow* (1988); Spiro Kostof, *The City Shaped; The City Assembled* (1992); Kevin Lynch, *A Theory of Good City Form* (1981).
See also Architecture; Automobile Industry; Communities, Low and Moderate Income; Land Use; Population; Suburbs; Transportation and Urbanization; Wastewater Treatment, Municipal.

URBAN RENEWAL

In the decades following World War II many cities in the United States experienced rapid suburbanization accompanied by dramatic deterioration of their centers. Between 1950 and 1970 the proportion of the national population in central cities fell by 5% while suburbs increased their share by 11%. Businesses, industry, retail stores, and middle income residents moved from central cities resulting in rapid disinvestment in the built environment. The demographic composition of central cities changed to include greater proportions of minority and low-income populations, some of whom migrated from rural areas as mechanization and land consolidation reduced the demand for agricultural labor. Retail sales growth in these central cities between 1954 and 1967 was only half of that in surrounding suburbs where new climate-controlled shopping malls with plentiful parking lured consumers.

The causes of this suburbanization and concomitant urban decline were many and their relative importance debatable, but most agree that government policies were major contributors. Insurance on residential mortgages made available through the Federal Housing Administration and the Veterans Administration enabled millions of families to realize the "American Dream" of a single-family home in the suburbs. The Interstate Highway System, initiated by the federal government in 1956, speeded travel between central cities and outer suburbs, facilitating commuting as well as increasing the accessibility of suburban shopping centers. As consumers and employees suburbanized, so did commerce and industry, attracted by cheaper and more open land with lower taxes. Higher crime rates, congestion, pollution, and regulation in cities were deterrents to remaining. New factory technologies such as the horizontal production line had large space requirements most easily met on the urban fringe. Containerization reduced the locational advantages of traditional port and rail centers as interstate freeways and truck transportation made urban peripheries accessible.

Programs to clear slums and build publicly subsidized housing began before the Second World War but escalated after the 1949 Housing and Slum Clearance Act. This led to the construction, in many cities, of high-rise "projects," which perpetuated and intensified racial and income segregation. While the initial inspiration for the design of these tower blocks was Le Corbusier's "city in a garden" concept, the reality of public housing projects in most cities has been one in which public spaces are ill maintained and dangerous, where residents are ghettoized and stigmatized, and where drugs and guns prevail. Although the great majority of legal residents of such projects are women and children, this is far from a nurturing environment.

The federal program dubbed Urban Renewal began with the 1949 Housing Act, later modified by a subsequent Housing Act in 1954, which provided federal funds to local authorities to clear slums. Using eminent domain, local governments acquired land and properties in deteriorated central areas, demolished ex-

isting structures, provided new streets and utilities, and "wrote down" (subsidized) the cost of land. The goal was to attract private investors back to rebuild the cities. Between 1949 and 1974 government spent almost $10 billion on urban renewal projects in about 1,250 cities in the United States.

But in numerous cities although land was cleared, little new investment resulted. The "slums" which were cleared were often minority neighborhoods. Displacement earned Urban Renewal the moniker "Negro Removal." By the late 1960s urban distress, racial polarization, and income inequalities had reached crisis proportions erupting in civil disturbances in cities across the country. Subsequently, the Model Cities program and various anti-poverty programs of the Great Society's War on Poverty sought more progressive strategies for urban renewal, but the tide of decline continued in large parts of many central cities, exacerbated during the 1970s by economic stagnation and recessions, migration of people and capital to the "Sunbelt," and fiscal crises in numerous Northeastern and North Central cities.

Nevertheless, during the 1970s small parts of numerous otherwise distressed cities underwent "revitalization." These included central business district commercial rebuilding, and both gentrification and incumbent upgrading in residential neighborhoods. Conventional explanations for this wave of revitalization include the increase in commuting and home maintenance costs with the oil shortage following the embargo of the early 1970s, demographic changes resulting in smaller households with fewer children, social changes such as the increase in dual-career households, and increases in house prices and mortgage costs. These led to increased demand for smaller, less expensive, more central housing units. Urban neighborhoods with structurally sound but deteriorated housing, especially those with historic or aesthetic appeal, were ripe for gentrification. Government programs such as Urban Development Action Grants provided often large subsidies to commercial projects. Alternative explanations for urban redevelopment lie in the need for capital to preserve the value of investments and debt in central cities and in the "rent gap" of residential neighborhoods in which the land and property are generating less revenue than their potential after gentrification.

Localized urban renewal/revitalization/redevelopment has thus been fostered by both public and private interests in many places, but America's central cities remain depressed. From Boston to Los Angeles, Seattle to Miami, Detroit to Atlanta, many urban residents live in dilapidated housing, attend segregated schools, receive only emergency health care, and experience high rates of unemployment and poverty.

While there have been small successes, Urban Renewal failed overall.

BRIAVEL HOLCOMB

For Further Reading: Susan Fainstein, et al., *Restructuring the City: The Political Economy of Urban Redevelopment* (1983); H. Briavel Holcomb and Robert A. Beauregard, *Revitalizing Cities* (1981); Jon C. Teaford, *The Rough Road to Renaissance: Urban Revitalization in America 1940–1985* (1990).
See also Communities, Low and Moderate Income; Housing.

USAID

The United States Agency for International Development (USAID or AID) is an agency of the United States government that administers most of the nation's nonmilitary foreign aid programs. AID programs are designed to improve the quality of life in poor countries around the world and to help these countries become self-supporting.

The agency generally seeks to support American economic, political, and security interests abroad. AID mainly administers foreign assistance on a bilateral basis, whereby aid is given from one country directly to another. There are two main types of aid programs: Development Assistance and Economic Support Funds. The Development Assistance program focuses on areas such as agriculture, rural development, nutrition, health, population, child survival, AIDS infection control, education and human resource development, private enterprise, environment and energy-related activities, and a special fund for sub-Saharan Africa. AID money and assistance for these programs may be used for various activities, such as building schools, monitoring farm production, and establishing communications systems. The Economic Support Fund program provides resources to maintain political stability while helping countries establish market-based economies. The fund consists of loans and grants that may be used to bolster a nation's economy, to address governmental financial problems, or to improve infrastructure.

AID also works with other departments in the federal government to coordinate various projects. For example, AID and the Department of Agriculture handle the sales and donations of agricultural commodities and food in a "Food for Peace" program. AID also supports nonprofit private organizations that provide relief and technical aid to victims of natural disasters.

USAID was originally established by Congress in 1961, the same year that President Kennedy created the Peace Corps. Both programs were created at a time when concern for developing countries was growing in response to strained relations between capitalist and communist powers. AID was established as part of the State Department under the Foreign Assistance Act of that year. The administrator of AID serves as the principal international development advisor to both the President and the Secretary of State. AID remains within the general jurisdiction of the State Department.

UTILITIES INDUSTRY

See Electric Utility Industry.

UTOPIANISM

Utopianism is the belief that a perfect human society can be created, and that through moral, political, or scientific organization, societal ills such as war, poverty, and disease can be eliminated. The term comes from the Latin prose work *Utopia*, published in 1516 by Sir Thomas More, which pictures an idealized agricultural society. Today, *utopian* is often confused with *Quixotic*, and it is used to describe a set of goals or a course of action that is a practical impossibility. However, for many centuries utopian beliefs were considered not only achievable, but inevitable.

Utopianism had a profound influence on the growth of environmentalist thought as it slowly formed in reaction to the Industrial Revolution. Even today environmental measures as basic as recycling newspapers are in some way attempts to realign human society with the "utopia" of the perfectly integrated natural system. The realization that mankind is only one component in a complex web of natural and cultural relationships, and the sense that through scientific and social planning we can be brought back into harmony with this web, owes much to the inherent optimism of the utopian school of thought.

Early utopian experimenters were concerned with the moral dimension of human character. They attempted to perfect that character by isolating themselves in self-sufficient colonies far away from outside influences. Most of the experimental utopian colonies in America in the 19th century were founded on the frontier, away from the established cities of the east coast. In order to maintain their self-sufficiency, they were based on communal, agricultural lines. However, they did not cut themselves off completely from the outside world. The Oneida community, founded in 1848, was originally a religious commune in which all property was held in common and marriages were arranged to produce the healthiest, most intelligent offspring. The community produced furniture, silverware, and animal traps that were sold to the outside world, and after many years metamorphosed into a thriving business concern.

By the 20th century the focus of utopian thought moved from trying to create moral perfection to the more practical concern of how to set up a functioning ideal society. Ideas from the 19th century were applied on a larger scale, the most striking example being the rise of Communism and the creation of such planned states as the Soviet Union and the People's Republic of China. Echoes of utopianism can be seen in Israel's kibbutz movement and the brief flourishing of agricultural communes in the United States in the late 1960s.

In this century, few actual attempts at an ecological utopia have been made in the mode of the self-sustaining colonies of the 19th century. A large part of present-day environmentalism focuses on the need to take up environmentally safe ways of living, while still maintaining the comforts of industrial society. Ernest Callenbach's *Ecotopia*, published in 1975, depicts a society that maintains a modern social organization while restructuring itself along environmental lines. In *Ecotopia* no nonbiodegradable material is allowed, garbage is completely recycled, organic waste is turned into fertilizer, and energy is generated by solar power and other alternative sources. The growing awareness of ecological concerns—the spreading of recycling programs, the founding of "green" interest groups and political parties, and the importance of the environment in the public debate—all point to a general trend to bring present society closer to an ecologically safe and ideal state.

Utopian has also been used to describe a whole literary genre, including works as diverse as Plato's *Republic*, St. Augustine's *City of God*, and Bellamy's *Looking Backward*. The subgenre of "negative utopias," or "dystopias," includes such works as Huxley's *Brave New World* and Orwell's *1984*.

V

VARIABLE CAN RATES

Historically, residential solid waste revenues have been raised through property taxes or in the form of fixed fees for unlimited collection. However, neither of these methods provides residents with incentives to reduce waste. As the availability of solid waste disposal options has decreased and the costs have increased, communities have begun to explore non-traditional methods for managing solid waste. Under variable can rate systems, the level of payment varies more directly with a measure of the volume of waste disposed.

A variable can system involves having customers select subscription levels based on the normal number of cans of garbage they need to dispose of each week. Their bills are calculated based on the subscribed service level, with higher levels leading to higher bills. The community usually offers subscription levels in standard 30-gallon increments (one can, two cans, etc.). In some communities, smaller can sizes are offered to provide stronger waste reduction incentives (e.g., 10-gallon can in Olympia, Washington, and 19-gallon mini-can in Seattle, Washington). In a limited number of communities, the number of units is recorded by the collectors, and the fee based on the actual number of cans set out for collection. Jurisdictions either allow customers to supply their own cans, or require community-provided or hauler-provided wheeled toters sized to the subscription level (60-gallon for two-can subscribers, etc.). The first known variable can program was implemented in Olympia, Washington, in 1954.

Primary advantages of the system are increased equity between users; improved incentives for recycling and waste reduction; improved utilization of programs; ability to implement the incentives quickly; and impressive tonnage reduction results. The most commonly reported problems with variable can rates include the possibility of increased levels of illegal dumping and burning; concerns about costs to low-income customers, worries about more complicated billing procedures; and greater revenue variability. In addition, the costs of purchasing, maintaining, and delivering different-sized carts to reflect customer changes in subscription levels can be a significant cost to a community. In response to some of these concerns, communities have implemented low income rates, combined the system with other municipal billing systems, and implemented penalty systems for dumping.

However, case study results from communities with variable can systems have been very positive. The success of the program's implementation in Seattle in particular influenced many communities to evaluate and implement similar programs across the country. Seattle's variable can system was implemented in 1981. By 1987, before the implementation of a city-sponsored curbside recycling program, the city reported recycling rates of 24% of its tonnage. Seattle credits the city's variable can program as its most important recycling program, and it is an integral part of the city's plan for achieving an aggressive 60% waste diversion goal. Other communities have reported reductions of over 25% of transfer station or landfill tonnage with the introduction of volume-based rates and recycling programs. These systems have been seen as important contributors in achieving legislatively mandated recycling goals.

LISA A. SKUMATZ

VECTOR-BORNE DISEASES

Vector-borne diseases are those that are transmitted from person to person, vertebrate host to vertebrate host, or vertebrate host to humans by an arthropod which is generally hematophagous (blood sucking) and acts as the vector in the transfer of the pathogen (disease agent) from one host to another.

The relationship between certain arthropods and human disease was recognized by early scholars, but the actual demonstration of transmission is relatively new. Only in the 20th century has the importance of vector-borne diseases been realized. Prior to the advent of modern day insecticides, they were responsible for hundreds of millions of deaths each year. It is estimated that over 25% of Europe's human population died of plague (the Black Death) during the 14th to 17th centuries. Yellow fever caused devastating epidemics with high mortality in the Americas from the 17th through the 19th centuries. Malaria, historically a major cause of disease and death in the world, was brought under control in the 1950s and '60s through organized programs that emphasized a combination of mosquito control, active case-finding and treatment. A combination of complacency among health authorities in endemic countries and the development of insecticide resistance in the mosquito populations and drug resistance in the parasite populations caused a major resurgence of malaria in the 1970s and '80s. In 1992 it was estimated that there are over 100 million cases and 1 million deaths due to malaria each year.

There has been a dramatic resurgence of other vector-borne diseases in the past 15 years. Examples include dengue hemorrhagic fever, yellow fever, Japanese encephalitis, oropouche and Rift Valley fever among the viruses, Lyme disease and plague among the bacteria, leishmaniasis among the protozoa and filariasis among the helminths (parasitic worms). Collectively these and other vector-borne diseases cause more morbidity and mortality than all other diseases combined. In most cases, the resurgence in recent years can be directly linked to human activities that have changed the total environment and allowed these diseases to break out of their natural foci and cause major epidemics in human populations. Two of them, dengue/dengue hemorrhagic fever and Lyme disease, will be used as examples to illustrate the importance of humans in their environments and how these changes impact on the transmission of vector-borne diseases.

Types of Transmission

There are two basic types of vector-borne disease transmission—mechanical and biological. Mechanical transmission consists of a simple transfer of the pathogen on contaminated mouth parts or feet or by regurgitation or defecation. There is no multiplication or developmental change of the disease agent on or in the arthropod during this type of transmission. Examples include a number of enteroviruses, bacteria and protozoa that normally have a fecal-oral transmission cycle. Insects such as houseflies may become contaminated with these pathogens while feeding on feces and transport them directly to the food. The incidence of this type of disease transmission is directly related to environmental sanitation and proper disposal of human waste since the insects involved usually lay their eggs in feces. While advances in public health have controlled this type of vector-borne disease in the U.S., it is still a significant problem in developing countries where human waste disposal is inadequate.

In biological transmission—which is the most important—the pathogen must undergo biological development in the body of the arthropod vector in order to complete its life cycle. Blood-sucking arthropods are the most important vectors because this habit provides the ideal mechanism for transfer of blood-borne pathogens between hosts. There are three basic types

of biological transmission. Propagative transmission occurs when the pathogen ingested with the blood undergoes simple multiplication in the body of the arthropod. Examples include arboviruses and bacteria. Cyclopropagative transmission occurs when the pathogen ingested with the blood undergoes a developmental cycle (changes from one stage to another) as well as multiplication in the arthropod host: for example, malaria. In cyclodevelopmental transmission, the pathogen does not multiply, but does undergo developmental changes. Filariae (types of roundworm) are an example. Time is required for development of the pathogen to the stage where it is infective for the vertebrate host. This period of time is called the extrinsic incubation period (EIP) and is generally 7 to 14 days in duration, depending on the pathogen, the vector and a variety of environmental factors.

Arthropod-borne pathogens are transferred to a new vertebrate host in several ways. The most common and important method is direct injection into the bloodstream or tissues. Examples are the arboviruses and malaria. Other pathogens may gain entrance in a variety of ways. Some viruses and rickettsia are transmitted from the infected female arthropod through the eggs to the offspring (transovarial transmission), after which the pathogen is transmitted to subsequent developmental stages of the arthropod. Venereal transmission of certain viruses occurs when adult male arthropods that have become infected transfer the infective virus to uninfected females in the seminal fluid during copulation. These latter types of transmission have obvious epidemiologic significance for human disease as well as for maintenance of the pathogen in nature.

Factors Influencing Transmission

Successful transmission of vector-borne infectious diseases is dependent upon the interactions between three populations: the pathogen, the arthropod vector and the vertebrate host. Having evolved over thousands of years, these host-pathogen-vector associations are very complex, and are highly dependent on environmental factors.

Because they are cold blooded animals, arthropod population densities, mobility, survival/longevity and certain behavioral characteristics are influenced by climatic and seasonal variables such as temperature, humidity and rainfall. For example, even small variations in temperature influence the rate of virus replication in arthropods and can increase or decrease the EIP by days. Cooler temperatures lengthen periods thus decreasing the probability that the arthropod will survive the EIP and transmit the pathogen to a new host. Moderately increased temperatures may increase the probability of transmission by decreasing the EIP. Rainfall is a major determinant of successful mosquito-borne disease transmission. All mosquitoes (the most important of the arthropod vectors) require water to complete their immature stages of development. Eggs are generally laid on the surface or on substrates that will ultimately be flooded by water. Mosquito population densities, therefore, are closely linked with rainfall and certain agricultural practices that result in water accumulation. The latter include irrigation (ponds and canals) and rice fields. During periods of drought, population densities are generally low and as a result, disease transmission is either absent or minimal. Humidity, temperature and wind have a great influence on the survival of arthropods, especially those that fly. Even small increases in daily mortality can significantly reduce disease transmission.

Transmission is also influenced by biologic factors such as the susceptibility of the arthropod to infection with a pathogen. This has been shown to vary greatly between arthropod species and even between geographic strains of the same species; it is genetically controlled and thus may be expected to change with time as a result of various selection pressures. The competence of the arthropod as a vector is also influenced by biological and behavioral characteristics. The degree of contact the arthropod has with humans and other vertebrate hosts is influenced by host preferences, intrinsic blood-feeding behavior and the population density and behavior of both humans and other vertebrate hosts. Longevity, flight behavior and breeding habits of the arthropod population are influenced by environmental factors such as temperature, humidity, wind and rainfall.

Human infection with arthropod-borne pathogens is greatly influenced by human behavior and selected socioeconomic factors. Most arthropod-borne pathogens are zoonoses (animal diseases) and humans are accidental or incidental hosts which do not contribute to the transmission and maintenance cycles. Many of these diseases are focal in their distribution and exist only in limited geographic areas (foci) that have specific ecologic conditions favorable to the existence of the maintenance cycle. Humans become involved in these disease cycles only when they intrude—most commonly by hunting, woodcutting or by building houses and communities in or near disease foci.

Arthropod-borne disease foci may expand or contract with environmental and seasonal changes. For example, natural disasters often result in increased mosquito densities and increased exposure of the human population because shelter is substandard or lacking. Unplanned and uncontrolled population growth are major determinants of urban epidemics of mosquito-borne disease. Construction of new dams and irrigation projects which impound water frequently lead to increased mosquito population densities, immigration of new vertebrate hosts and, therefore, increased disease. Creation of protected bird or wildlife sanctuaries and wetlands can result in increased risk for surrounding communities.

In temperate regions arthropod transmission of disease is usually seasonal and is directly related to temperature. In the tropics and subtropics transmission generally occurs year-round, but is most frequently correlated with rainfall. Because only slight changes in temperature can influence disease incidence, global warming may become an important determinant. Thus warming of a few degrees may expand temperate, seasonal, transmission zones considerably and expose highly susceptible human populations to new or exotic disease agents.

Dengue/Dengue Hemorrhagic Fever

Dengue fever (DEN) and dengue hemorrhagic fever (DHF) are caused by infection with one of four dengue viruses. These are antigenically related, but do not provide cross-protective immunity. Infection in humans causes a spectrum of illness ranging from mild, nonspecific viral syndrome or flu-like illness, to severe and fatal hemorrhagic disease. All age groups are affected, but the severe hemorrhagic form of the disease occurs more frequently in children under the age of 15 years. The viruses belong to a family which contains approximately 60 other related viruses—the best known being yellow fever.

Dengue viruses are mosquito-borne and have three basic transmission cycles, a forest cycle involving lower primates and forest *Aedes* mosquito species, a rural or semirural cycle involving humans and peridomestic *Aedes* species and an urban cycle involving humans and domestic *Aedes* species. There may be some overlap between these cycles, depending on where they occur and the mosquito species involved. Dengue viruses probably evolved in the rainforests of Asia in mosquitoes or monkeys. Humans likely first became involved while working in the forest and transported the viruses to their villages where a peridomestic Asian mosquito (*Aedes albopictus*) became involved. From there the viruses were transported to urban centers where a highly domesticated and efficient mosquito vector (*Ae. aegypti*) was common. Once established the urban *Ae. aegypti*–human cycle became the primary maintenance cycle for dengue viruses.

Prior to World War II dengue fever was characterized by infrequent epidemics that occurred in tropical regions of the world at 10 to 40 year intervals. Early epidemics usually began in port cities and moved inland because transport of viruses and mosquitoes between countries was usually via ship. During and following World War II environmental conditions in Asia changed. The disruption caused by the war resulted in an expanded distribution and higher densities of *Ae. aegypti*. Following the war urbanization and movement of people within and between geographic regions increased dramatically, thus providing an ideal mechanism for increased movement of dengue viruses between population centers. The result was increased frequency of epidemics and the emergence of DHF in southeast Asia. Within 20 years of the first reported epidemic in the Philippines in 1953–54, the disease had spread throughout the region and had become a leading cause of hospitalization and death among children in most countries of southeast Asia. The geographic distribution of the disease in Asia continued to expand during the 1970s and '80s. In the last decade alone, major epidemics of DHF have occurred in the People's Republic of China (PRC), India, the Maldive Islands and Sri Lanka. In PRC and Taiwan the first dengue transmission in over 35 years was reported, and Singapore experienced the largest epidemic in that country's history following 20 years of a very successful DHF control program.

Following a successful *Ae. aegypti* eradication program in much of Central and South America in the 1950s and '60s, epidemic dengue occurred only sporadically in Caribbean Islands. The eradication program was abandoned in 1970 and over the next decade, *Ae. aegypti* reinvaded most of those countries where it had been eradicated. This reinfestation resulted in increasingly frequent epidemics of dengue fever in the 1970s, followed by epidemic DHF in the 1980s. Major epidemics of dengue fever occurred in several American countries in the 1980s, all of which had been free of dengue transmission for 35 to 130 years. The first major DHF epidemic in the American region occurred in Cuba in 1981 and the second in Venezuela in 1989–90. During the intervening period, smaller outbreaks of DHF occurred in Mexico (1984), Nicaragua (1985), Puerto Rico (1986), El Salvador (1987), Brazil (1990) and Colombia (1991). Since 1981 13 countries have reported laboratory-confirmed cases of DHF in the American region.

Similar patterns of epidemics were observed in Africa and the Pacific during the 1980s. Dengue fever is generally underreported and outbreaks are often reported as other diseases. In 1993 DEN/DHF is the most important arbovirus disease of humans with over 2 billion persons at risk. Each year there are 30 to 40 million cases of dengue fever and hundreds of thousands of cases of DHF.

The factors responsible for the resurgence of epidemic dengue activity worldwide and the emergence and geographic spread of DHF are still not completely understood. It is clear, however, that human activity and environmental changes have played major roles. Un-

planned and uncontrolled urbanization in tropical countries in the past 20 years has resulted in a breakdown in environmental sanitation, water and housing systems. This has led to increased use of water storage containers which are the principal larval habitats of *Ae. aegypti*. In addition, most consumer goods are now packaged in nonbiodegradable plastic, which collects rain water and also makes ideal breeding sites for these highly domesticated mosquitoes. The result has been crowded human populations living in intimate contact with large mosquito populations, thus creating ideal conditions for increased transmission of dengue.

Another important factor contributing to epidemic dengue transmission is commercial air travel, which has increased dramatically in the past 30 years. Air travel by humans provides the ideal mechanism for the transport of dengue and other viruses between population centers. Persons infected with disease pathogens with incubation periods up to 2 weeks transport viruses all over the world on a regular basis. Repeated introduction of viruses into permissive areas with large *Ae. aegypti* populations has resulted in increased frequency of epidemic activity, increased disease incidence and the emergence of severe and fatal dengue hemorrhagic fever.

Lyme Disease

Lyme disease is a multi-stage, multi-system bacterial infection caused by the spirochete *Borrelia burgdorferi*. The acute stage of illness is characterized by fever, flu-like symptoms and a characteristic skin rash. The early stage illness usually responds to antibiotic therapy, but untreated or inadequately treated Lyme disease may cause arthritic, neurologic and cardiac disorders.

Lyme disease is one of the most important new and emerging diseases in the U.S. First described in Connecticut in 1975, the disease has spread rapidly throughout the northeast, upper midwest and northern Pacific states in the intervening 17 years. In 1982 when national surveillance was initiated, only 497 cases were reported from 11 states. In 1992 over 9600 cases were reported from 45 states—a 19-fold increase. Nearly 50,000 cases have been reported in that 11-year period.

The disease is transmitted by ticks, primarily *Ixodes scapularis*. A three-host tick, this species has a rather complicated life cycle that spans 2 years. The adult females, which prefer to feed on large mammals such as deer, usually lay their eggs in the spring. The resulting larvae seek hosts, usually small rodents such as mice, and feed most often between July and September. They drop from their host and molt to nymphs which then are ready to seek a new host in the spring of the next year. Peak feeding of nymphal *I. scapularis* is between

May and July, and although they too prefer small rodents such as mice, they feed on a variety of animals, including humans. Adults, which emerge in late summer and fall, may feed in late fall and early winter and lay eggs the next spring.

The principal reservoir host for the Lyme disease spirochete in the northeastern U.S. is the white-footed mouse, which also serves as the preferred host for larval and nymphal *I. scapularis*, thus ensuring maintenance of the pathogen in nature. Larvae, which feed after the nymphs, become infected by feeding on mice, insuring that the next generation of nymphs and adults are infected. The nymphal tick is the stage that normally transmits the pathogen to humans.

The epidemic of Lyme disease in the U.S. can be directly attributed to changes in human behavior and land use during the last half of the 20th century. The primary risk factor appears to be the presence of white-tailed deer, which are the preferred host. These deer tolerate large tick populations (as many as 500 ticks have been removed from a single deer) and are an ideal mechanism to transport ticks to new areas. It is well documented that during the last century, much of the northeastern U.S. had been cleared of forest for agricultural use. Deer populations in the area had been eliminated to the point where they were rarely seen during the early 1900s. Agriculture became less important during the last half of the 20th century in the U.S., and in the northeast, agricultural land was allowed to reforest. By the 1970s and '80s extensive woodlots in various stages of growth were widespread. This resulted in an explosion of the white-tailed deer population and with it an expanding geographic distribution and population explosion of *I. scapularis*.

Another important factor contributing to the emergence of Lyme disease was the trend, which accelerated in the 1970s, of building residential developments in wooded areas of the northeast with little or no disturbance to the ecology. Millions of people thus live in close contact with deer, mice and ticks, with the transmission cycle of *B. burgdorferi* occurring in their backyards. This is the single most important factor responsible for the current Lyme disease epidemic. It has been demonstrated that tick population densities and thus exposure can be dramatically reduced by reducing or removing the deer population from an area. Until better vegetation and animal management practices are developed in residential areas, Lyme disease will remain a part of the way of life in the U.S.

DUANE J. GUBLER

For Further Reading: Duane J. Gubler, "Insects in Disease Transmission," in G. T. Strickland, ed., *Hunters Tropical Medicine* (1991); Duane J. Gubler, "Dengue," in Thomas P. Monath, ed., *Epidemiology of Arthropod-Borne Viral Disease* (1988); Robert S. Lane, Joseph Piesman, and Willy Burgdorfer, "Lyme Borreliosis: Relation of Its Causative Agent to Its Vectors and Hosts in North America and Europe," in *Annual Review of Entomology* (1991).
See also Bacteria; Epidemiology; Health and Disease; Insects and Related Arthropods; Malaria; Viruses.

VEHICLES, ALTERNATIVE ENERGY

Alternative energy vehicles are vehicles that do not use traditional petroleum fuels. The primary goal in searching for alternative ways to power vehicles is avoiding the pollution caused by the burning of gasoline or diesel fuel. Alternative fuels such as ethanol, methanol, reformulated gasolines, and compressed natural gas reduce the amounts of some exhaust pollutants. These fuels offer a viable economic way to reduce pollution, especially when they are used by large industrial or government vehicle fleets. However, the infrastructure investments needed to adopt alternative fuels on a large-scale basis are not considered to be cost-effective when compared to the economic value of the pollution reduction.

Manufacturers are focusing on battery-powered electric vehicles that can run without polluting. Several models are planned for production in the mid-1990s. Their environmental impact comes from the central power plant that produces the electricity necessary to charge batteries. According to the California Air Resources Board, however, an electric vehicle would reduce total pollutant emissions by up to 97%. It is expected that existing power plants in the United States will be able to accommodate electric vehicles in the early stages, using off-peak capacity. Eventually more power plants could be needed.

Several challenges are being addressed in the drive to make electric vehicles popular. Initially the vehicles will be expensive, although increased production driven by regulations requiring their use will bring costs down. It is difficult to produce an electric car that has the speed, acceleration, and range expected by drivers accustomed to traditional cars. The battery recharge time, which can take hours, must be reduced. In addition, batteries must be replaced about every two years at a cost of about $1,500. However, operating costs in general may be lower than those for an internal combustion engine. Other challenges include battery size, weight, safety, and temperature requirements. Researchers are concentrating on improving the efficiency of the batteries and the vehicle.

The ultimate clean automobile eventually may be powered by the hydrogen fuel cell, which should be available in ten to twenty years. The hydrogen fuel cell uses the electrons from hydrogen to form electricity.

The fuel cells today use hydrocarbon liquids as a source of hydrogen and dissociate them at the cell for the hydrogen supply. The production and distribution of hydrogen may be a major barrier. Efforts are underway to reduce the costs of hydrogen fuel cells and build experimental vehicles.

Legislation passed in California has served as a catalyst in the campaign to develop cleaner vehicles. The California law, which applies to automobile makers who sell over 3,000 vehicles in California, requires that by 1998 2% of all new cars will have to be zero emission vehicles (ZEVs). By 2001, 5% must be ZEVs; by 2003, 10% must be ZEVs. California is a significant portion of the United States market. Similar legislation is being considered by other states as well as by the United States government. As a result, it appears that substantial markets will develop for ZEVs. By 2000 Japan plans to have 200,000 electric vehicles on the road in the European community, where legislation is likely in the near future.

VINYL CHLORIDE

Vinyl chloride (VC), a gas at room temperature, is a synthetic hydrocarbon with a pleasantly sweet odor. It is primarily used for the production of polyvinyl chloride (PVC) in a wide variety of plastic products, including pipes, floor tiles, electrical insulation, and food and beverage packaging. The use of vinyl chloride as an aerosol propellant and as an ingredient in drug and cosmetic products has been discontinued in the U.S. since 1974.

Vinyl chloride is a skin irritant; contact with it in the liquid state can cause frostbite upon evaporation. VC also depresses the central nervous system, and acute exposure can cause intoxication, nausea, and respiratory irritation. High concentrations or prolonged exposure can be fatal.

In addition, vinyl chloride is recognized as a human carcinogen, and chronic exposure at relatively low levels manifests itself in multiple tumor sites including the liver, brain, lungs, and kidneys. Very strong evidence implicates it as the cause of a rare cancer, angiosarcoma of the liver, in workers at plants producing VC or PVC products. Other harmful effects may include nonmalignant liver, bone, nerve, and skin damage; it is also a suspected cause of genetic abnormalities.

In 1989 over 4.5 billion kilograms of vinyl chloride were produced in the U.S., most of which was used in the production of polyvinyl chloride. Emissions from both VC and PVC production plants pose the greatest potential health risk for employees and neighboring populations. An additional concern is that vinyl chloride plants are clustered, and are often located near or adjacent to polyvinyl chloride plants. The Environmental Protection Agency (EPA) estimates that emissions of vinyl chloride into the atmosphere prior to 1975 exceeded 100 million kilograms per year, 90% of which was believed to emanate from polyvinyl chloride production plants. Neither VC nor PVC are biodegradable, and the disposal of PVC products has led to a variety of waste problems. VC is known to be released by sanitary landfills and is a byproduct of the incineration of plastics.

Vinyl chloride has also been found in drinking water, beverages, and food. Particles of VC are known to migrate out of PVC products. Although VC concentrations in public water supplies tested so far are well below the minimal levels associated with reported carcinogenic or other toxic responses, levels in foods and beverages may range from detectable to nine parts per million. Among foods tested, vegetable oils and apple cider contained the highest concentrations of VC.

Other ecological effects of vinyl chloride have only recently come under investigation. Studies have found that VC has an effect on the ripening of fruit and on plant growth, since it is a physiologically active gas. No studies have been performed on vegetation near VC or PVC production sites. There is very little data on the effects of vinyl chloride on freshwater or marine organisms, and no criteria have been determined for the protection of aquatic life. Although VC is highly toxic, it is highly valuable, and hence should be used under strictly controlled conditions.

VIRUSES

The term *virus* is derived from Latin and means "poison." Viruses were first described approximately 100 years ago as agents of disease that could pass through the finest filters but could not grow in cell-free cultures. With the introduction of cell culture for their cultivation in 1949, thousands of viruses were isolated and characterized. Although the initial description is still accurate, more than 60 years passed before their nature was understood.

Description

There are two key aspects to the definition of viruses. First, they require living, metabolizing cells for their multiplication. Unlike other intracellular parasites, viruses are totally dependent on the infected cells for the assembly of their proteins. Viruses do not "grow"—they do not divide but rather are assembled in large numbers by viral or cellular protein using cellular functions. Second, all cells contain DNA (deoxyribonucleic acid), in which genetic information is stored in chromosomes, and RNAs (ribonucleic acids), which may have several functions including the transfer of genetic instructions from chromosomes to the machinery that translates this information to make proteins. Viruses store their genetic information in either RNA or DNA and only one nucleic acid is present in virus particles.

Viruses have been likened to well-protected tapes or disks carrying comprehensive instructions in search of a compatible computer—the cell—for the instructions to be carried out. The medium with the instructions is the viral genome (one or more chromosomes). The virus particle, referred to as the virion, protects the genetic material, and enables the virus to enter the cell.

Virtually all forms of life from bacteria to humans become infected with viruses. Although viruses have been designated by the name of the organism they infect or the resulting disease (e.g., influenza, aleutian mink disease, tobacco mosaic virus), the formal classification is based on the composition and structure of the virion. Viruses with a similar structure form a family, e.g., paramyxovirus family (measles) and rhabdovirus family (rabies). Viruses are divided into DNA or RNA groups. Each group is further divided on the basis of the symmetry of the virion or of an internal component into cuboidal, helical, or asymmetric viruses. The third rung of the classification is whether the virion is enclosed in a membrane designated as the envelope. For example, the orthomyxovirus (influenza virus) family contains RNA genomes bound to proteins and enclosed in an envelope with viral proteins projecting from its surface. The helical symmetry is an attribute of the internal RNA-protein complex and not of the surface envelope. The herpes virus consists of an envelope surrounding a protein shell: the capsid—made up of subunits (capsomeres)—which in turn surrounds the DNA core. The number and arrangement of the capsomeres and the size of the capsid are characteristic for each virus family.

Viral genomes can be either single- or double-stranded. In addition, these genomes can be monopartite in which all viral genes are contained in a single chromosome, or multipartite in which the viral genes are distributed in several chromosomes. DNA viruses contain a monopartite genome and are double- or single-stranded, linear or circular. The orthomyxoviruses, reoviruses, and bunyaviruses contain 8, 10, and 3 RNA chromosomes, respectively.

Viral Infection: the Initial Phases

To multiply, a virus must infect a cell. Susceptibility defines the capacity of a cell or organism to become infected. The host range defines both the kinds of tissue cells and the species in which they can multiply. Whereas some viruses (e.g., St. Louis encephalitis) have a wide host range, others (e.g., papillomaviruses) have an extremely narrow host range.

Infection follows exposure of susceptible cells at the portal of entry. To cause disease or death, the virus must spread from the portal of entry to target tissue (e.g., central nervous system). In many instances (e.g., respiratory infections, genital herpes simplex infections), the target cells are at the portal of entry.

Attachment. To infect a cell, the virus must attach to the cell surface, penetrate into the cell, and become sufficiently uncoated to make its genome accessible to viral or cellular enzymes for the expression of its genes. Attachment involves the binding of the virion to a cell surface receptor. As a rule, each virus species uses a different cell surface receptor.

Penetration. Penetration occurs almost instantaneously after attachment and involves one of three mechanisms: (1) translocation of the virion across the plasma membrane used primarily by nonenveloped viruses, (2) endocytosis of either enveloped or nonenveloped virus particle, or (3) fusion of the cellular membrane with the virion envelope.

Uncoating. Tailed bacterial viruses inject their genomes into the bacterial cell. For viruses infecting eukaryotic (containing a distinct membrane-bound nucleus) cells, uncoating is a general term applied to the events that occur after penetration and that set the stage for the viral genome to express its functions. In most instances the virion disaggregates, alone or with the aid of cellular enzymes and the nucleic acid or a nucleic acid-protein complex is all that remains of the virus particle before expression of viral functions.

Strategies of Viral Replication

General viral pattern of replication. Shortly after infection and for up to several hours thereafter, only small amounts of infectious virus can be detected. This interval, known as the eclipse phase, signals the fact that the viral genomes have been released from their protective coats and are exposed to host or viral machinery necessary for their expression, but that progeny virus production has not yet increased to a detectable level. There follows the maturation phase in which viral nucleic acids are made and assemble with viral proteins into progeny virions. These virus particles accumulate in the cell or in the extracellular environment at exponential rates. After approximately 8 to 72 hours of infection with specific viruses, viral synthesis ceases and the cells are irreversibly injured. Cells infected with other kinds of viruses may continue to synthesize viruses indefinitely. The yields per cell range from more than 100,000 poliovirus particles to several thousand poxvirus particles.

What must a virus do to multiply? In the course of

their evolution, viruses have evolved several different strategies for their multiplication. While no single pattern of replication has prevailed, two concepts are key to the understanding of how viruses multiply.

1. *Viruses must express proteins.* Viruses must express their genes and direct the cell to make viral proteins in order to multiply. The ability of a virus to multiply and the fate of an infected cell hinge on the synthesis and function of virus gene products: the proteins.
2. *Viral proteins have three key functions.* Although viruses differ considerably in the number of genes they contain, all viruses encode a minimum of three sets of functions which are expressed by the proteins they specify. Viral proteins (a) ensure the replication of the viral genomes by their own enzymes or those of the host, (b) package the genome into virus particles (the virions), and (c) alter the function and, frequently, that of the infected cell. The capacity to remain latent—a feature essential for the survival of some viruses in the human population—is an additional function expressed by some viruses.

Pathogenesis Associated with Viral Infections

The pathologic effects of viral diseases result from (a) toxic effect of virus gene products on the metabolism of infected cells, (b) reactions of the host to infected cells expressing virus genes, and (c) modifications of host gene expression by structural or functional interactions with viral genetic material. In many instances, the symptoms and signs of acute viral diseases can be related to the destruction of cells by the infecting virus.

Cell destruction by viral proteins occurs as a consequence of two sets of events. First, viruses contain genes whose products shut off the synthesis of cellular proteins and of other macromolecules. We may speculate that these genes have evolved to preclude the cell from reacting to the infection. Indeed, one response to viral infection and specifically to the presence of double-stranded RNA in cells is interferon, a protein that can block the expression of viral genes and has been used in some instances to treat viral infections.

The second cause of cell destruction is viral proteins that are made for the synthesis of viral nucleic acids and virion assembly but that are highly toxic to the cell. In this category are viral proteins that degrade cellular chromatin and cause infected cells to fuse with infected and uninfected cells.

Host Response to Infection

Infection of a susceptible cell does not automatically insure that viral multiplication will ensue. Infection of susceptible cells may be productive, or abortive. Productive infection occurs in permissive cells and results in infectious progeny. Abortive infection can occur in nonpermissive cells which allow a few, but not all, viral genes to be expressed for reasons that are rarely known, and in both permissive and nonpermissive cells infected with defective viruses, which lack a full complement of functional viral genes. The significance of abortive infection stems from the observation that viruses which normally destroy the permissive cell during productive infection may merely injure, but not destroy, abortively infected, permissive or nonpermissive cells. The consequences of this injury may be the expression of host functions which transform the cell from normal to malignant.

Dissemination of Viruses in the Environment

There are several ways by which viruses are disseminated in the general environment. These are (i) accidental entry of a virus from its natural host to another host (e.g., human contact with field rodents, or a bite by a vector feeding on other species); (ii) vector dependent dissemination of viruses (e.g., equine encephalitis viruses transmitted by mosquitos, transmission of plant viruses by insects); (iii) direct contact of susceptible and infected individuals (e.g., herpes simplex); (iv) transmission through aerosols (influenza), polluted beaches (polioviruses), or contaminated blood (poliovirus, hepatitis B, human immunodeficiency virus).

Epidemics result from rapid and widespread dissemination of a virus in a population. The factors required are a large susceptible population and efficient transmission of the virus. Viruses encounter large susceptible populations in environments in which they were not previously present or as a consequence of mutations that alter their antigenicity. For example, influenza virus epidemics periodically sweep through the human population. The most recent super-epidemic (pandemic), that of 1918, was caused by an influenza virus that carried genes derived from swine influenza. This pandemic killed 20 million people—the same number killed in all of World War I.

The most common cause of epidemics by viruses endemic in the population (e.g., influenza virus) is mutations in antigenic sites of surface proteins that shield the viruses from preexisting immunity. A less common cause of epidemics is mutations that alter the host range of viruses. Mutations are more likely to occur in RNA than in DNA genomes since the former do not have mechanisms to correct base pairs mismatched during synthesis. As a consequence RNA viruses tend to vary much more than DNA viruses.

Not all common viral pathogens are transmitted in

epidemics. Herpesviruses, for example, remain in their hosts for life in a latent state. Periodically, the viruses are reactivated and transmitted by direct contact to susceptible individuals.

The rate of dissemination of viruses in the general environment depends on several factors. Enveloped viruses are not very stable and do not survive drying, exposure to detergents, etc. Nonenveloped viruses (e.g., polioviruses) can be extremely stable. Transmission is very efficient through aerosols and fomites, less efficient by vectors or direct contact between infected and uninfected individuals. The longer the duration of the disease during which viruses are excreted, the higher the probability of virus spread.

The list of viral infections eradicated or largely controlled by vaccines includes smallpox (developed by Edward Jenner), rabies (Louis Pasteur), yellow fever, poliovirus (Jonas Salk and Albert Sabin), measles, mumps, rubella, and hepatitis A and B. These vaccines represent humankind's finest achievements; they have enormously reduced the emotional and economic burden of infectious diseases of children and adults. The cost associated with morbidity and mortality caused by viral infectious disease, however, remains high. Genetic engineering techniques developed in the past decade are being used to develop superior vaccines for the prevention of human (influenza, chicken pox, respiratory virus, immunodeficiency virus) and animal infections.

Treatment of viral diseases lags behind the means to prevent them. Although drugs are available for herpes simplex and human immunodeficiency virus infections, they suppress or retard viral replication rather than rid the body of the invading virus. New technologies that allow development of drugs by design rather than mass screening chemical libraries should speed the development of better drugs.

A commonly articulated theme is that viruses affected evolution in several ways: by introducing foreign genetic material into cells, by inactivating viral genes during integration into cellular genomes, and by exerting selective pressure for survival from viral infections. More often than not, viruses and other infectious agents were the true victors of armed conflicts that shaped human history. Notwithstanding their possible role in biologic evolution, the harm caused by viral infections far outweighs any potential benefit. Concerted efforts have resulted in the elimination of human smallpox—a pathogen whose sweeps through human populations resulted in mortality rates as high as 25%. Efforts to eliminate other pathogenic viruses are a major public health objective.

BERNARD ROIZMAN

For Further Reading: B. N. Fields, D. M. Knipe, R. M. Chanock, M. S. Hirsch, J. L. Melnick, T. P. Monath and B. Roizman (eds.), *Virology* (1990); R. E. F. Matthews, "Classification and Nomenclature of Viruses," in *Intervirology* (1982); W. H. McNeill, *Plagues and Peoples* (1976).
See also DNA; Epidemiology; Evolution; Health and Disease.

VISUAL POLLUTION

See Pollution, Visual.

VOLCANOES

See Natural Disasters.

WALLACE, ALFRED RUSSEL

(1823–1913), English naturalist, evolutionist. Wallace was born in a rural district of Wales, a member of an impoverished middle-class family. He went to grammar school at Hertford but left it at 13, acquiring most of his vast later knowledge by copious reading throughout his life. At the age of 15, he was apprenticed to his surveyor brother William, for whom he worked until the end of 1843. Surveying brought him in daily contact with nature, and led to the development of his insatiable interest in all aspects of natural history. In 1844–45, he taught at the Collegiate School in Leicester, but also pursued various other activities until 1848. During this period, he became friends with another young naturalist, Henry Walter Bates. The two young naturalists, after reading the travel books of Humboldt and Darwin, decided to become explorers in South America and earn their living by selling natural history specimens. But, after reading Chambers's *Vestiges of Creation*, they also planned "to gather facts . . . to solving the problem of the origin of species, a subject on which we had conversed and corresponded much together." And for the next 10 years this problem was always on Wallace's mind. Even though Wallace learned a great deal during the four years he spent in the Amazon valley, he lost all of his collections and nearly his life when his ship caught fire on the return voyage and sank. Undaunted by this experience, Wallace left two years later for the Malay Archipelago, where he stayed from 1854 to 1862, again as a natural history collector. The collections he sent back to England were enormous and contained literally thousands of new species of birds, insects, and other kinds of animals. Always intensely curious about everything he could observe, he acquired a fine knowledge of geology, geography, and native peoples, which he used in numerous of his later publications. Most importantly, in the solitude of his camps he thought deeply about nature, species, and evolution. In 1855, in Sarawak (Borneo), he wrote a landmark paper, "On the Law which Has Regulated the Introduction of New Species." In this essay Wallace not only rejected Lyell's claim that species can vary only within certain limits, but he also offered a solution to the problem of how new species are introduced to replace those that had become extinct. Wallace discovered that it was geog-raphy that provided the solution. This observation permitted him to propose a law: *Every species has come into existence coincident both in space and time with a preexisting closely allied species.* By asserting that the distribution pattern of closely related species gave the solution of the problem of speciation, Wallace had not only essentially solved the problem of speciation, but had also implied the concept of common descent.

Realizing how closely Wallace had come to the solution of the problems that had occupied Darwin for the last 17 years, Lyell urged Darwin to publish his ideas on the origin of species lest he be preempted by Wallace. Darwin set to work in 1856 but his book was less than half completed when what Lyell had feared actually happened.

Even though Wallace now knew the pathway (geographical separation) by which new species arose, he was uncertain for another couple of years as to the mechanism by which speciation was effected, but he thought about it incessantly. One day early in 1858, while he was ill with malaria in the Moluccas and unable to go out into the forest to collect, all the pieces of the puzzle suddenly came together. When he recovered, in a few days he wrote a manuscript in which he articulated his ideas and mailed it to Darwin, asking for his opinion, and if he approved to submit it for publication. This manuscript arrived at Down House in June 1858 and left Darwin dumbfounded, because the theory proposed by Wallace, evolution by natural selection, was remarkably similar to Darwin's own. Darwin was severely ill at the time and his friends Lyell and Hooker presented on July 1, 1858, Wallace's essay together with parts of Darwin's unpublished manuscripts to the Linnean Society of London. Wallace, who was a very generous person, never resented having been made to share the priority with Darwin, and remained his friend. Later in life, however, they drifted apart owing to differences in the interpretation of evolution and for other reasons.

After Wallace had returned from the East Indies, he entered an active literary career that lasted until the end of his life. His two-volume *Geographic Distribution of Animals* (1876) became the standard work in the field of biogeography, one might say for the next 100 years. He discovered the sharp faunal break between the islands on the Sunda Shelf and the more easterly islands of Indonesia, now known as Wallace's Line. In numerous smaller papers he made important contributions to systematics, biogeography, and evolution. He presented convincing evidence for the importance of mimicry among Indo-Malayan butterflies and later supported the occurrence of Mullerian mimicry. His highly readable book *Island Life* (1880) documented numerous aspects of evolution.

Eventually, Wallace's ideas diverged from those of Darwin. For instance, he thought that isolating mechanisms between species could be built up by natural selection, something Darwin considered impossible. Also in most other respects he became a more consistent selectionist than Darwin, eventually rejecting any inheritance of acquired characters and joining Weismann's neo-Darwinism in his book *Darwinism* (1889).

Being a believer in extra-sensory forces, he became an adherent of spiritualism and eventually acknowledged that he could not believe in a purely natural evolution of Man, that somehow God must have been the creator of Man. In his later years, Wallace became a socialist and the promoter of a number of liberal and social schemes. Yet owing to his lovable personality, he continued to be on good terms with his old friends. In spite of all his financial setbacks and various personal losses, Wallace always remained an optimist and dedicated himself to the improvement of the lot of humanity up to the end of his life. He was a truly admirable person. He died at Lodgestone, Dorset, England, on the 7th of November, 1913, having outlived Darwin by more than 30 years.

ERNST MAYR

For Further Reading: Ernst Mayr, *Evolution and the Diversity of Life* (1976); Ernst Mayr, *The Growth of Biological Thought* (1982); G. C. Williams, *Adaptation and Natural Selection: A Critique of Some Current Evolutionary Thought* (1966).
See also Darwin, Charles; Evolution; Genetics; Mendel, Gregor.

WAR AND MILITARY ACTIVITIES: ENVIRONMENTAL EFFECTS

War is always damaging to the natural environment. The often profligate employment of high-explosive munitions against enemy personnel and materiel and the major reliance by ground forces on heavy off-road vehicles can be especially disruptive of local habitats and the creatures that depend upon them. The construction of fortifications, base camps, and lines of communication adds to the environmental disruption. Battle-related activities can also lead to considerable amounts of local air and water pollution. In addition, food, feed, and timber resources are often heavily exploited by armed forces in time of war, both within their theater of operations and beyond.

On land, the pursuit of war often involves the intentional destruction of fields or forest to deny the enemy access to water, food, feed, and construction materials as well as cover or sanctuary. This is accomplished by (1) applying chemical poisons (herbicides); (2) introducing exotic (non-indigenous) living organisms, including microorganisms; (3) incendiary means; and (4) mechanical means.

In one prime example of environmental warfare, forests were devastated by the United States during the Vietnam War by far-reaching aerial and ground application of herbicides, by massed bombing, by the extensive use of large tractors, and—to a lesser extent—by fire. Killing the trees of a forest ecosystem leads to substantial damage to that system's wildlife and to its nutrient budget through soil erosion and nutrient dumping (loss of nutrients in solution). Substantial recovery from such unbalancing of the system takes decades.

Bacteriological warfare agents could do serious long-term damage to ecosystems if disruptive exotic microorganisms were employed and then became locally established.

Water resources can serve as both targets and weapons of war. Marine ecosystems could readily be damaged if offshore oil wells, large tankers, or other appropriately located oil facilities were attacked with a resulting release of large amounts of oil into the water, thereby precluding human use of those ecosystems for many years. In the Persian Gulf War of 1991, Iraq released huge amounts of oil into the Persian Gulf off the Kuwaiti coast, causing severe damage of several years' duration to local marine flora and fauna. Similarly, the destruction of nuclear reactors (whether land-based or on shipboard) would release radioactive contaminants that could for decades prevent human utilization of marine resources in the region of attack.

A number of important rivers flow through more than one country, providing an opportunity for a nation upstream to divert or foul the waters before they reach a downstream nation, a major calamity in an arid region.

The breaching of dams has been a spectacularly successful strategy in past wars, including World War II and the Korean War of 1950–1953. The most devastating example of intentional military flooding occurred during the Second Sino-Japanese War of 1937–1945. In order to stop the Japanese advance, in June 1938 the Chinese dynamited the Huayuankow dike along the Yellow River (Huang He) near Chengchow. Several thousand Japanese soldiers drowned and the Japanese advance into China along this front was halted. In the process, however, the flood waters also ravaged major portions of Henan, Anhui, and Jiangsu Provinces. Several million hectares of farmland were inundated in the process, and the crops and topsoil destroyed. The river was not brought back under control until 1947. The flooding inundated 11 Chinese cities and more than 4000 villages. At least several hundred thousand Chinese (possibly many more) drowned as a result and millions were left homeless. Indeed, this act of environmental warfare appears to have been the most catastrophic single act in all human history in terms of number of human lives claimed.

Nuclear facilities constitute particularly dangerous potential targets. A destroyed nuclear facility could contaminate a large surrounding area with iodine-131, cesium-137, strontium-90, and other radioactive debris—an area that would be measurable in hundreds or thousands of hectares. The most heavily contaminated inner zone would be life-threatening; an intermediate zone of lesser contamination would be health-threatening; and a still greater zone beyond would become agriculturally unusable. Such a radioactively polluted area would defy effective decontamination. It would recover only slowly, over a period of many decades, as has been demonstrated by the Pacific nuclear test islands and other test sites. The aftermath of the Chernobyl accident of April 1986 indicates the disruption to the human environment that could be expected from nuclear contamination.

As to the atmosphere, two sorts of hostile manipulations were used by the United States during the Vietnam War. First, clouds over enemy territory were injected with various chemical substances in a massive attempt to increase rainfall so as to make enemy lines of communication impassable—with only trivial, if any, success. Second, the troposphere over enemy territory was injected with substances designed to render enemy radars inoperable. The results of these efforts have never been made public. During the Persian Gulf War of 1991, Iraq set fire to many hundreds of Kuwaiti oil wells for no immediately apparent military gain. Immense amounts of soot and poisonous fumes rose into the troposphere, with deleterious effects of several years' duration on the health of the local human population as well as on the local biota. Local weather patterns may also have been temporarily influenced by the pollution.

Control over the forces of nature for military purposes has long been a human desire. But nature has an enormous capacity to overcome extensive disruption by progressively reestablishing its original state. As René Dubos pointed out in *The Wooing of Earth:*

The native forest has returned even to the 50,000 acres of the Verdun region of northeast France, where the

French and German armies fought the most destructive and longest battle of World War I. At the end of the battle, practically all the trees had been destroyed; yet the original vegetation is now back, as are birds, rabbits, and deer. It is believed that the forest will have returned to its original state within less than a century after the battle of 1916.

Similar phenomena of restoration can be observed in tropical and subtropical countries. When the Korean War ended in 1953, a demilitarized zone (DMZ) of 2.5 miles width was agreed upon between North and South Korea. The DMZ was then a wasteland pockmarked with bomb craters and shell holes; yet it has now become one of Asia's richest wildlife sanctuaries. Abandoned rice terraces have turned into marshes used by waterfowl; old tank traps are overgrown with weeds and serve as a cover for rabbits; herds of small Asian deer take refuge in the heavy thickets.

ARTHUR H. WESTING

For Further Reading: A. H. Westing, ed., *Environmental Warfare: a Technical, Legal, and Policy Appraisal* (1984); A. H. Westing, ed., *Explosive Remnants of War: Mitigating the Environmental Effects* (1985); A. H. Westing, ed., *Environmental Hazards of War: Releasing Dangerous Forces in an Industrialized World* (1990).

WARD, BARBARA

(1914–1981), writer, economic historian. Barbara Ward was a powerful and eloquent voice for addressing the needs of the poor countries. More than anyone else, she laid the foundations for the concept of sustainable development.

Surveying postwar Europe for *The Economist* magazine, she was struck by the degree of human and physical devastation and the immense cooperative effort needed to repair it. The Marshall Plan, therefore, had an immense influence on her thinking. According to one story, she persuaded Foreign Secretary Ernest Bevin (for whom she had campaigned in the election of 1944) to take the Harvard Commencement address by the Secretary of State as a serious proposition. In recognition of her own tremendous efforts to have the Marshall Plan accepted in both the United States and Europe, she was invited to deliver the Commencement address herself on the tenth anniversary of Marshall's speech.

Ward's own political and religious beliefs led her to a concern for the poor and dispossessed. Trips to India and a long residence in Ghana (where her husband was Nkrumah's development commissioner) convinced her of the need for a worldwide plan to end poverty in the developing countries.

Vietnam took a toll on the momentum that had been built up in favor of foreign aid, so Ward persuaded the World Bank to set up the first of the great independent commissions. Chaired by former Canadian Prime Minister Lester Pearson, it reported in 1970, urging that the Organization for Economic Cooperation and Development nations devote at least 0.7% of their GNP to development assistance. When the momentum again was lost in the 1970s, she persuaded Robert McNamara, a close friend, to establish a similar body under Willy Brandt.

In 1968, Sweden and a number of other developed countries introduced a resolution in the General Assembly calling for the United Nations Conference on the Human Environment. Fearing that the developing countries would play little or no part in a conference which they perceived would deal with the pollution of the rich, the Conference Secretary-General, Maurice Strong, commissioned René Dubos to establish and serve as chairman of an international Committee of Corresponding Consultants that would prepare a world report on the human environment for publication in advance of the Conference. Strong asked Barbara Ward to join with Dubos in writing the report and she drafted the first report for review and comment by the Committee, composed of 152 experts from 58 countries. Ultimately Ward and Dubos co-authored the unofficial report for the Conference on the Human Environment that would become the landmark book *Only One Earth*.

For the first time in a popular book, the environmental problems of the developing world were set out in detail. Joining the familiar list of industrial pollutants were the environmental problems of poverty: soil erosion, deforestation, desertification, the lack of clean water, and the massive problems of rapid urbanization. The book and conference firmly established that environment and development must be reinforcing rather than opposing concepts.

The Stockholm Conference also convinced Barbara Ward of the potential power of non-governmental organizations (NGOs). She was impressed with the commitment of the thousands of NGOs present and with their ability to influence the media. In 1973 she became president of the International Institute for Environmental Affairs. At her suggestion the name was changed to the International Institute for Environment and Development and its headquarters moved to London. IIED was a "bully pulpit" that Ward used to "bite the ankles of governments" as she put it. Before the World Food Conference in 1974, she gathered well-known food experts to try to force governments to face up to the problems of hunger.

In 1976, she published the second of her superb books on the environment and people. *The Home of Man*, produced for the United Nations Habitat Conference in Vancouver, reminded the world that some of the worst environments on earth were to be found in the vast cities of the developing countries where the Dickensian problems of poor housing, lack of sanitation, and foul drinking water were combined with twentieth century pollutants.

Barbara Ward was ambivalent about the ability of technology to help solve the problems of the developing countries. Her third book, *Progress for a Small Planet*, concludes with her customary call for international cooperation: ". . . the only fundamentally unsolved problem in this unsteady interregnum between imperial ages which may be dying and a planetary society which struggles to be born is whether the rich and fortunate are imaginative enough and the resentful and underprivileged poor patient enough to begin to establish a true foundation of better sharing, fuller cooperation, and joint planetary work . . . In short, no problem is insoluble in the creation of a balanced and conserving planet save humanity itself."

Throughout her life, Barbara Ward remained a devout Catholic, but she was a thorn in the side of several Popes. She was active in the Pontifical Commission on Justice and Peace and the first woman to address a Synod of Bishops since the tenth century. This faith helped to sustain her through a long fight against cancer.

In her dying days she produced an introduction to *Down to Earth* by Erik Eckholm. After despairing at the resurgence of a narrow-minded nationalism which was reducing the flow of resources to the developing world, she characteristically concluded: "And it is for the world's poor—the nations of the Third World, and the poor majority within those countries—that a decent environment is even more important than it is for the rich west. The poor are always nearer the margin, and the margins of our global environment are today smaller than they were ten years ago in Stockholm."

DAVID RUNNALLS

For Further Reading: Barbara Ward and René Dubos, *Only One Earth* (1983); Barbara Ward, *The Home of Man* (1976); Barbara Ward, *Progress for a Small Planet* (1979).

WASTE

Debris, garbage, trash, refuse, junk, clutter, offal, rubbish, litter, rejectamenta—waste goes by many names. But in the natural world the idea of waste does not really exist. It is part of the life cycle, substance returned to the physical environment simply in a different form. Although humans are not the only living species to generate waste, they are the only one that passes judgment on it.

Michael Thompson, in his book *Rubbish Theory*, identifies three categories of objects. The first is the "durable" object which increases in value over time, such as a fine piece of furniture or a painting. The second is a "transient" object which decreases in value over time and has a finite lifespan, like a refrigerator or a dishwasher. The third is "rubbish"—that which has "zero and unchanging value" and usually does not disappear, but "continues to exist in a timeless and valueless limbo." To which category an object belongs, however, can depend entirely on how it is perceived.

To some, Ford Motor Company's Edsel was a colossal mistake which rightly found its place in a scrapheap to be melted down and recast. To others, the car with the unique grillwork was a curiosity—or even a work of art—which should be preserved for the ages. Old Coca-Cola bottles, once containers for soda, have been picked out of dumps and hailed as a collector's item—an important symbol of our consumer society. Our culture imparts value to some things and declares others worthless.

To the anthropologist garbage, organic waste, is a highly effective research tool. "Garbologist" William Rathje sifting through the discards in a modern-day landfill draws insights about contemporary societal change in much the same way archaeologists do about ancient civilizations.

For all our technological sophistication, Americans are not that different from those who inhabited most of the world's other great (and ostentatious) civilizations. Our social history fits rather neatly into the broader cycles of rise and decline that other peoples have experienced before us. Over time, grand civilizations seem to have moved from efficient scavenging to conspicuous consumption and then back again to the scavenger's efficiency. It is a common story, usually driven by economic realities.

Anthropologists like Rathje can do much to explain the relationship of humans to their environment and culture by what people throw away. However, to contemporaries of almost all societies—ancient and modern—refuse was merely a problem to confront or endure.

Among the vast array of wastes, refuse—or solid waste—has been one of the most abundant, most unwieldy, and most problematic. Beginning about 10,000 B.C. humans began abandoning the nomadic life to establish more permanent settlements. Nomadic tribes that followed herds of game could simply leave their wastes behind. But in the new towns, villages, and cities such habits could not so easily be tolerated, and methods of dealing effectively with refuse took time to

develop. In ancient Troy wastes were sometimes left on the floors of houses or dumped into the streets. When the stench became unbearable in the home, a fresh supply of dirt or clay was brought in to cover the droppings. In the streets, pigs, dogs, birds, and rodents ate the organic remains. One study estimated that the debris accumulation in Troy amounted to 4.7 feet per century. In some communities, the average elevation rose as much as 13 feet a century.

In Mahenjo-Daro, a city founded in the Indus Valley about 2500 B.C., central planning led to the construction of homes with built-in rubbish chutes and trash bins. The city also invested in a drainage system and a scavenger service. In Heracleopolis, founded in Egypt about 2100 B.C., wastes were collected in elite areas, but dumped primarily in the Nile. About the same time, the homes of the Sea Kings in Crete had bathrooms connected to trunk sewers, and by 1500 B.C. the island had land set aside for disposal of organic materials.

Religion often played a role in enforcing sanitary practices. Mosaic law from about 1600 B.C. obliged Jews to bury their waste far from living quarters. The Talmud required that the streets of Jerusalem be washed daily despite the paucity of water in this arid region.

There was, however, no universal standard of cleanliness in the ancient world. In the 5th century B.C., garbage cluttered the outskirts of Athens threatening the health of the citizens. About 500 B.C., Greeks organized municipal dumps, and the Council of Athens began enforcing an ordinance requiring scavengers to dispose of wastes no less than one mile from the city walls. Athens also issued the first known edict against throwing garbage into the streets, and even established compost pits.

Ancient Mayans also placed their organic waste in dumps and used broken pottery, grinding stones, and cut stones as fill. Records from 2nd century B.C. China reveal "sanitary police" who were charged with removing animal and human carcasses and "traffic police" responsible for street sweeping.

Because of its size and dense population, Rome faced sanitation problems unheard of in Greece and elsewhere. The city was more effective in dealing with water and sewerage than in confronting refuse. While garbage collection and disposal were well organized by the standards of the day, they were not sufficient to meet the city's needs. General collection was restricted to state-sponsored events, and property owners were responsible for cleaning abutting streets—although the laws were not always enforced. The wealthy Romans used slaves to collect and dispose of their wastes, and some independent scavengers collected garbage and excreta for a fee and sold the material as fertilizer. As Rome's power waned, however, the quality of the city's environment deteriorated.

As Western Europe "deurbanized" in the Middle Ages, people were spared the massive waste problems experienced by the highly concentrated cities such as Rome. Despite the crudity of medieval dwellings and living conditions, the eventual rise of new cities was accompanied by greater attention to health practices. Cities began paving and cleaning their streets at the end of the 12th century. Paris, for example, started paving its streets in 1184, but street cleaning at public expense was not introduced until 1609.

The migration of rural peoples to urban places also meant the migration of hogs, geese, ducks, and other animals into the city limits. In 1131 a law was passed prohibiting swine from running loose in the streets of Paris after young King Philip was killed in a riding accident caused by an unattended pig. But animals continued to roam the streets and often served as unofficial scavengers.

Until the late 19th century sanitation systems in the great Islamic cities and in China were the most advanced in the world. Sanitary conditions in Europe were less advanced but improved slowly throughout the Middle Ages and Renaissance. With the onset of the Industrial Revolution in the mid-18th century, however, urban sanitation took a decided turn for the worse, first in England, then on the Continent. As Lewis Mumford stated, "[I]ndustrialism, the main creative force of the 19th century, produced the most degraded urban environment the world had yet seen; for even the quarters of the ruling classes were befouled and overcrowded."

The inability to house the growing population migrating to industrial centers led to serious overcrowding and health problems. In one section of Manchester in 1843 there was one toilet for every 212 people. The pages of Charles Dickens overflow with the graphic images of the wretchedness of life in the early industrial city.

Some scholars have argued that, along with the rise of laissez-faire capitalism, the 19th century also experienced a kind of "municipal socialism," that is, a demand for services provided by the city rather than by the individual. The largest industrial cities in England and elsewhere eventually began to develop rudimentary public works and public health agencies to address the most pressing sanitation needs.

The rise of public health science also was crucial. The 1842 report of the Poor Law Commission, authored by Edwin Chadwick, came to the conclusion that communicable disease was related in some way to filthy environmental conditions. The filth—or miasmic—theory of disease became the most significant force for promoting environmental sanitation until the

turn of the century. At that time it was superseded by the germ theory of disease, which successfully identified bacteria as the culprit in spreading typhoid, smallpox, yellow fever, and other maladies.

While Europe was in the throes of its Industrial Revolution, the U.S. was just emerging as a nation. Many of the European lessons about sanitation were not applied immediately and some would not garner attention until the U.S. went through its own industrial revolution in the mid-19th century.

Preindustrial America was highly decentralized, and the smaller American towns and cities did not face the enormous waste problems of London or Paris. Yet habits of neglect affected these communities too. Casting rubbish and garbage into the streets was done casually and regularly. As late as the 1860s, Washingtonians dumped garbage and slop into alleys and streets, pigs roamed freely, slaughterhouses spewed nauseating fumes, and rats and cockroaches infested most dwellings, including the White House.

By and large, sanitation in preindustrial America was determined by local circumstances. Some city leaders placed a high priority on city cleanliness, applying the principles of environmental sanitation. Others simply ignored the problem. Individuals or private scavengers usually collected refuse. Boards of health were slow in developing, understaffed, and limited in power. The burghers of New Amsterdam had been among the first to pass laws against casting waste into the streets (1657), but the condition of the streets remained the responsibility of the householders. In the late-18th and early-19th centuries, New York established municipal control over several sanitary services, but disputes between state and local governments continued.

The impact of the Industrial Revolution on American cities was no less staggering than its impact on European cities. Like Europe, the U.S. experienced environmental and health crises through overcrowding, poor sanitation, and primitive methods of collection and disposal. On the other end of the scale, the eventual increase in affluence associated with economic expansion in this period meant the production of more goods and thus more waste.

Yet by 1900 as industrial cities began to mature and demand "home rule," dependence on scavengers or refuse-collecting franchises granted through short-term contracts was replaced by city service. Borrowing Chadwick's notion that waste was not simply a nuisance but a health hazard, city leaders argued that environmental sanitation and other public health measures could only be carried out through municipal authority. Not surprisingly, health departments became central agents in directing and monitoring collection and disposal of wastes. With the advent of the germ theory of disease, however, public health officials turned their attention to combating communicable diseases through bacteriological laboratories, inoculation, and immunization. This meant that programs in environmental sanitation fell to public works departments, and refuse unfortunately became more of an engineering problem than a health problem.

Whatever the perception, the scale of waste production was colossal in this period. Between 1900 and 1920, each citizen of the major boroughs of New York produced 160 pounds of garbage, 1,231 pounds of ashes, and 97 pounds of rubbish annually. Franz Schneider, Jr., a research associate at MIT Sanitary Laboratory, calculated in 1912 that if an entire year's refuse of New York City was gathered in one place "the resulting mass would equal in volume a cube about 1/8 of a mile on edge. This surprising volume is over three times that of the great pyramid of Ghizeh, and would accommodate one hundred and forty Washington monuments with ease."

To meet the challenge, many cities relied on large streetcleaning and collection crews. In time, motorized trucks replaced horses and carts in the collection phase. Disposal was carried out in numerous ways—dumping in water and on land, open burning, and filling ravines. By World War I, most of these primitive methods were modified or abandoned. Techniques borrowed from Europe, most notably incineration, became popular as early as the 1880s. Briefly, the Vienna or Merz reduction process found support. In this process organic waste was placed under pressure, resulting in the extraction of oils and other residues which could be used for soap and perfume base or for other purposes. The stench from the plants, however, made the process unpopular and most plants were abandoned.

By World War II, the sanitary landfill—or the systematic layering of earth and waste—became the most widely used disposal method in the U.S. Modern sanitary landfills originated in Great Britain in the 1920s under the term "controlled tipping." American versions were first attempted in the 1930s in New York City and Fresno, California. They promised inexpensive and efficient disposal because of the abundance of cheap land.

The development of a solid-waste management system in the industrial era did not mean that the problem of managing waste, let alone confronting the issue of the generation of waste, was settled. In recent times grave concern has arisen over a possible "garbage crisis" in America—especially a crisis in waste volume and a crisis in landfill space. In reality, however, the U.S. has been facing a range of chronic waste problems for decades that while not necessarily at a crisis state, pose serious dilemmas.

In recent years, the volume of wastes in the U.S. has increased to staggering proportions. Between 1968 and 1988, municipal solid waste alone increased from 140 million to 180 million tons per year. While the amount of waste has levelled off at about 4 pounds per capita per day, population increases result in a steady rise in total discards. The composition of materials in the waste stream also has changed significantly over the years, posing obstacles in collection as well as adding to the amount of toxic materials in landfills. In the 1880s, the most likely wastes to challenge the garbage collectors were coal and wood ash, seasonal food wastes, and horse manure. While these items were not always pleasant to remove, today's waste stream includes a much more diverse and complex array of materials, such as an abundance of paper products, glass, plastics, rubber, metals, leather, yard waste, and a host of household chemicals and cleaners.

Composition of the waste stream depends on several external factors, including patterns of consumption and production, availability and use of new materials and technologies, and the obsolescence of older materials and older technologies. Without careful attention to the types of substances entering the waste stream, and without recognition that composition is always changing, collection and disposal practices can prove to be ineffective and in some cases obsolete. In addition, valuable resources can be lost in the process.

Collection of refuse has long been regarded as cumbersome, labor-intensive, and expensive. Surveys estimate that from 70% to 90% of the total cost of dealing with solid waste in the U.S. goes for collection. Mechanization of collection after World War II through the use of compaction vehicles was an important technical breakthrough. But collection remains largely dependent on human labor, resulting in job-related accidents and labor disputes which compromise the process. Also, as cities spread out further from their core, the question of equitable service for both urban and suburban communities becomes an important issue.

The most crucial problem with waste management is that of disposal, especially the shrinking of landfill sites. Three northeastern states—New Jersey, Pennsylvania, and New York—already export eight million tons of garbage each year, much of it to the Midwest. As early as the 1970s solid-waste professionals and others began to doubt the adequacy of the sanitary landfills to serve the future disposal needs of cities. In some places, identifying any space is a problem, while in others strict siting regulations or the resistance of local citizens poses a major concern.

Aside from siting, landfills have been criticized on the basis of safety. Birds, insects, rodents, and other animals frequent the sites and can carry pathogens back to people. If not properly lined, landfills pose a threat to groundwater and surface water. If not properly constructed, landfills produce combustible methane gas. There is also concern about contamination from various hazardous materials and incinerator ash which find their way into landfills from homes and factories.

By current estimates, 75% to 85% of the waste generated in the U.S. ends up in landfills, while the total number of sites is declining. According to the EPA, half of the nation's approximately 6,000 landfills will be closed by 1995. Since 1978, 14,000 facilities—or 70% of all landfills in the U.S.—were closed. New landfills are not being created fast enough to replace old ones.

Most American cities have depended on landfill disposal for at least fifty years and have not prepared for alternatives. After a brief resurgence of using incinerators after the 1930s, the number of new incinerators began to drop again in the 1960s largely because of the high capital costs and the problems of air pollution associated with burning.

In the 1980s, recycling emerged as an important alternative or a complement to more traditional disposal methods. Propounded in the 1960s as a grassroots method of source reduction and a protest against overconsumption, recycling has developed strong political and social appeal because it puts policymakers and practitioners on the side of conservation of resources as well as giving concerned citizens a personal way to participate in facing the solid waste problem. Before 1980 less than 140 communities in the U.S. had door-to-door recycling collection service. Estimates for the 1990s put that total over 1,000. In 1989 there were more than 10,000 recycling drop-off and buy-back centers in operation and more than 7,000 scrap processors. A major goal for most communities and the nation in general is to increase the recycling rate which stood at 10% in the late 1980s.

In the decades following World War II, the trend toward municipal reponsibility for collection and disposal that had been underway in the late 19th and early 20th centuries began to swing back toward privatization. As cities annexed greater numbers of suburban communities, especially in the South and West, they often contracted with existing private firms rather than expand municipal service. In addition, the move to a more competitive system was accelerated by the change from separate to mixed refuse collection and demands for greater efficiency. With the ban on backyard burning of trash in many cities during the 1970s, trash collecting became mandatory along with garbage collecting. The increased volume of collectible wastes led several cities to contract part of the service while maintaining a portion of their original routes.

Between 1973 and 1982 the number of communities with private collection increased from 339 to 486 and in refuse disposal from 143 to 342.

In the 1960s, recognition that solid waste was part of a national environmental problem quickly led to the assumption that control limited to the local level would be inadequate to deal with a problem that extended beyond city limits. In time solid waste was elevated to the status of "third pollution," alongside air and water pollution. This change in perspective about solid waste as an environmental threat may prove to be as pivotal as the transformation of refuse from a nuisance to a health hazard in the 19th century.

The Solid Waste Disposal Act of 1965 was the first major federal law to recognize solid waste as a national problem, and was followed in the next several years by other laws, often focusing more narrowly on hazardous waste and materials recovery. In the early 1970s, EPA acquired extensive regulatory authority over municipal solid waste, especially in the design and operation of landfills and incinerators. But by mid-decade EPA proposed a drastic cutback in the federal solid waste program and recommended that federal activities be limited to regulating hazardous wastes. When Congress and state and local groups balked at this stance, EPA took a less severe position and by the mid-1980s began once again to play a major role in the solid waste issue.

Around the world, solid waste has become a significant political, social, and environmental issue. As in the U.S., siting new disposal facilities to manage these wastes has become more difficult as urban populations continue to grow rapidly and as potential sites grow scarcer. Although many industrialized nations report an increasing reliance on recycling, sanitary landfilling remains a central component of existing as well as new strategies. Japan, however, has become one of the leaders in the use of incineration technology. Despite the extensive rethinking about collection and disposal issues, efforts at integrated solid waste management largely remain on the drawing board.

Much of what has come to be considered the "garbage crisis" in the world is not the product of immediate past practices or present inaction, but a series of chronic problems interrelated in such a way as to defy a clear solution. What once was considered simply a nuisance—or even more seriously a health hazard—in the past has become a major environmental blight in the 20th century. This change in the perception of "garbage" will no doubt prompt new answers to the age-old questions, What is waste? What is wasteful? What should be thrown away? What should be kept?

MARTIN V. MELOSI

For Further Reading: Joseph S. Carra and Raffaello Cossu, eds., *International Perspectives on Municipal Solid Wastes and Sanitary Landfilling* (1990); Martin V. Melosi, *Garbage in the Cities: Refuse, Reform and the Environment, 1880– 1980* (1980); William Rathje and Cullen Murphy, *Rubbish! The Archeology of Garbage* (1992).

See also Composting; Hazardous Waste; Hazardous Waste, Household; Incineration; Public Health; Recycling; Waste, Municipal Solid; Waste, Radioactive; Waste Disposal; Waste Management, Municipal Solid; Waste Treatment; Wastewater Treatment, Municipal.

WASTE, HAZARDOUS

See Hazardous Waste.

WASTE, MUNICIPAL SOLID

Municipal solid waste (MSW) is defined by the U.S. Environmental Protection Agency (EPA) and others as the wastes generated from residences, commercial establishments, institutions, and to a limited extent, industrial facilities. It is generated by everyone in the course of daily life—at home, at school, traveling, and at work. The use of the term generally implies that the waste generation is not a one-time event but occurs regularly over a period of time. Thus, residential wastes are generated every day and industrial wastes are generated every working day.

Some industrial wastes are hazardous, while some might better be classified as municipal solid waste. Large amounts of industrial wastes, however, are not classified as hazardous by EPA definitions. To illustrate their wide variety, they include wastewater treatment and other sludges; ash from power generation; food processing wastes; wastes from ore processing; off-specification products; wiping rags; trimmings and shavings of various kinds; pulp and paper mill sludges; bark and other wood wastes; used filters; blast furnace slag; electric arc furnace dust; foundry sand; waste cullet (glass); and many others.

It is generally agreed that MSW does not include industrial process wastes, construction and demolition wastes, sludges from municipal sewage treatment and other processes, hazardous wastes, agricultural waste, oil and gas waste, and mining waste, even though some of these wastes may be managed along with MSW—for instance, in a landfill.

The materials and products in MSW can be divided into three broad categories: durable goods, nondurable goods, and containers and packaging. In addition, MSW includes food wastes, yard wastes, and some miscellaneous items such as dirt and bits of stone and concrete. The EPA has estimated that containers and pack-

aging made up the largest tonnage of MSW generated in 1990; nondurable goods were the second largest category. Yard wastes ranked next in size, followed by durable goods and other items such as food wastes.

The durable goods in MSW include large and small appliances, furniture, tires, carpets and rugs, and some other miscellaneous items. By definition, these items have lifetimes of about three years or longer; therefore, there is a time lag of some years between the time they are purchased and the time they are discarded.

Nondurable goods in MSW include newspapers, books, magazines, other types of papers, paper and plastic disposable plates and cups, trash bags, disposable diapers, clothing and footwear, and many other miscellaneous items. These items are generally discarded within a year of the time they are purchased.

The containers and packaging category includes paper and paperboard boxes and wraps (which contribute the most tonnage); glass bottles and jars; steel and aluminum cans; plastic bottles, bags, and wraps; wood pellets; and other miscellaneous items. Containers and packaging are normally assumed to be discarded the same year they are purchased. Finally, MSW includes some items that are not manufactured products, mostly food wastes and yard wastes.

Looked at another way, MSW is made up of the various materials that are used to manufacture products. The estimates published by EPA indicate that in 1990 paper and paperboard were the largest material category, at over 37% of total generation (by weight). Yard wastes were the second largest item, at about 18%. Glass, metals, plastics, wood, and food wastes each ranged between 6% and 9%. The remainder is made up of rubber and leather, textiles, and miscellaneous items, none comprising over 3% of the total.

MARJORIE A. FRANKLIN

WASTE, RADIOACTIVE

Radioactive waste originates predominantly from commercial nuclear energy production and from nuclear weapons programs. Because of its high toxicity, radioactive waste is disposed of either by diluting it and dispersing it into the environment, or by concentrating and containing it until the radioactivity becomes neutralized. Containment poses significant problems because it requires security for very long periods of time in leakproof repositories.

Radioactive waste is a byproduct of all of the stages of the nuclear fuel cycle. It generally falls into one of four categories according to the degree of toxicity: low-level waste (LLW), transuranic (TRU) waste (materials derived from uranium fission), uranium mill tailings, and high-level waste (HLW). Low-level waste includes the wide range of wastes that do not come under the other categories, such as contaminated clothing. Most low-level waste has very low toxicity levels and requires only short-term management, although some forms occasionally do possess substantial radioactivity. Until the 1960s the U.S. dumped low-level wastes at sea. Much low-level waste is now buried in shallow trenches, but there are efforts to correct the shortcomings of this method. Transuranic waste is composed of spent plutonium from weapons programs. Although significantly less toxic than high-level waste, it still requires long-term disposal in geological repositories; plans to do so in the U.S. are under way in the Waste Pilot Project in New Mexico.

Uranium mill tailings, which have a level of toxicity similar to transuranics, are the chemical residues of the refining processes of uranium ore. Until recently, tailing piles were left exposed and untreated. Now, however, they must be contained and covered to prevent leakage.

High-level waste, which includes spent fuel elements from reactors, contains the highest degree of radioactivity (10^{10} Curies and above) and requires the longest safe disposal. One-fourth to one-third of the fuel in typical reactors is replaced each year, since fission products gradually build up in the fuel and prevent the chain reaction from occurring. The most practical safeguard is burial 500 to 1,000 meters beneath the surface. As of 1990, however, none had been buried this deep, primarily because of the great complexities involved in building permanent geological repositories to last millennia. Most high-level waste is stored on reactor sites in water-cooled basins. There are plans for a large (though controversial) repository at Yucca Mountain in Nevada, which should accommodate all high-level wastes produced in the U.S. until 2020.

Major considerations in judging the quality of disposal sites include local topography, seismic and geologic history, chemical properties of surrounding rocks (such as permeability), hydrology and hydrogeology, probability of future natural events, and public attitude. The reprocessing of spent fuel, which involves the extraction of reusable uranium and plutonium, eliminates the need to dispose of the spent fuel and reduces costs by eliminating the mining and enriching steps of preparing uranium ore. In addition, the transport of high-level waste is reduced, lowering the risk of accident or theft by terrorists as material for nuclear weapons. This technology, however, is still being developed and is currently only a small factor in waste management. Many countries still have no comprehensive regulation governing packaging, transportation, and disposal of radioactive wastes.

WASTE DISPOSAL

The amount of solid wastes discarded by a family depends on the size of the home and living area, disposable income, the value of time, and the development level of the society. Waste production is an index of life. Solid waste production is an index of the standard of living.

Americans and Canadians lead the world by discarding about 3.3 lbs. (1.5 kg) of household solid wastes per person per day. Other nations, such as Sweden, Germany, Netherlands, and Japan, have increased their per capita waste discard rate as their standard of living and personal wealth increased over the years.

There are only two options for solid waste disposal: long-term storage systems (landfills), and conversion systems with storage for residues.

There are three types of landfills: above-grade mound-type storage; partially below-grade and above-grade mound-type storage; and below-grade storage.

There are four principal conversion systems. *Recycling* makes use of discarded material in remanufacturing the same material for similar use (e.g., making new glass bottles from discarded ones or making new aluminum cans from discarded beverage containers). *Reuse* makes use of discarded material for remanufacturing new products (e.g., making steel plate from discarded coffee cans or converting old newsprint into facial or toilet tissues). *Composting* is the conversion of easily degradable organic material into a stabilized humic material for improving soil productivity (e.g., conversion of selected food scrap into compost or anaerobic decomposition of organic material to produce methane gas and humus). *Energy recovery* is the extraction of usable energy from combustion of solid wastes (e.g., hot water supply for district heating or production of electricity).

An integrated solid waste management system makes use of a combination of conversion systems to reduce the amount of waste to be disposed of in a landfill. The choice of the conversion system is dependent on a number of factors, such as: land availability; markets for recyclable materials; energy prices; and state laws and regulations.

The U.S. EPA has set a national target for recycling at least 25% of the materials we discard. Some states, such as New Jersey, have set higher targets. In order to meet the national target, most communities will have to set up programs for separate collection of materials.

There are three types of systems for separating recyclable materials. In *curbside collection systems*, residents are provided special containers, usually made out of recycled plastic (often called a blue box). Specific materials for recycling, such as old newspapers, glass bottles, tin and aluminum cans, and sometimes plastic soda bottles (PET or No. 1) and milk jugs (HDPE or No. 2), are placed in the blue box by residents and set out for collection on a weekly or biweekly basis. Garbage collection companies pick up the materials in trucks fitted with special compartments. Workers toss materials into appropriate bins. Bagged grass clippings, leaves, and shrub trimmings are often picked up separately for composting. Curbside collection systems cost about $1.25 to $1.50 per month per residence.

In a variation of the curbside collection system, a specially made blue tag is used. Residents are asked to place recyclable materials, such as glass bottles, tin and aluminum cans, and plastic bottles, in blue bags. A regular garbage truck picks up the blue bags on a scheduled week day, and the various materials are separated in a materials recovery facility (MRF). Blue bag systems can collect only a limited variety of recyclables. While the transportation cost is reduced to some extent, the overall cost may be the same as the blue box system because of the costs of the MRF.

In *drop off center systems*, a series of large containers are set up at selected locations. Each container is marked for collection of a specific material. Materials usually collected are newspapers, magazines, telephone books, tin and aluminum cans, green, brown, and clear glass bottles, plastic bottles, cardboard, household batteries, and waste motor oil. Sometimes containers are also set up for discarded tires and white goods (washers, dryers, etc.). Many drop off centers are open 24 hours every day. Most drop off center programs depend on voluntary participation of the residents. The materials collected from the drop off centers can be shipped for either recycling or reuse without further separation through an MRF. The drop off center costs are lower than curbside collection systems and cost about 31 cents per month.

In some communities, voluntary groups operate "buy back" centers, often with a subsidy from the local governments. Residents are offered cash for specific recyclable materials, such as aluminum, steel, cardboard, and old newspaper. The buy back centers also become trading centers where people can swap discarded materials.

Landfills

Modern landfills are located at sites after the geology, hydrology, and the flora and fauna of the area are studied. Flood plains, wetlands, depth of water table, and the use of aquifers are all factors that affect the design of landfills. All new landfills have one or more layers of clay and synthetic flexible membrane liner (FML) combinations. The FML is made of inert polymers, such as high density polyethylene or polyvinyl chloride. There are many special varieties of FMLs, and the choice is dependent on capability of the FML to sustain its integrity even when exposed to many organic solvents which leach out of the waste material. Municipal solid waste leachates contain organic acids, metals, and many complex organic compounds.

In double lined landfills, a sand layer is placed between liners to provide an easy pathway for leachate to reach the leachate collection system. It serves as a leak detection system as well. A network of heavy duty plastic pipes is placed over the top liner in a bed of gravel. The pipes have perforations to allow the leachate to drip into the pipes for collection. The FML and the pipe network are covered with at least two feet (61 cm) of screened, graded sand for protection. The leachate is collected in special tanks for storage. Either a leachate treatment plant treats the liquid on site, or it is conveyed to larger, wastewater treatment plants for treatment. Untreated leachate can pollute both ground and surface waters.

Landfill Gas

Landfills also produce methane gas and small quantities of volatile organic compounds. Landfill gas is a mixture of methane and carbon dioxide. Landfill gas can migrate through soil layers and accumulate in basements of buildings. If the gas is not quickly diluted or vented, it could cause explosions. Modern landfills have extensive gas extraction systems to recover the gas and prevent migration. The recovered gas is often used in engines for production of electricity or as industrial boiler fuel. When neither use is feasible, the gas is combusted in a candle flare, and the methane is converted into carbon dioxide.

Landfill Cap

When a landfill cell reaches its capacity, it is capped by using either soil or a combination of soils and FML. The purpose of the cap is to prevent infiltration of precipitation. The cap also prevents landfill gas from escaping and channels it into gas recovery wells. Grass or other suitable vegetation is added to improve the aesthetics. Landfills are sometimes used for general recreation purposes, such as parkland, tennis courts, jogging and walking trails, and golfing. Well-landscaped and capped landfills act as desirable open space in community planning.

Composting

Composting is a conversion system which stabilizes the organic fraction of solid wastes, such as vegetable matter, food scraps, lawn clippings, and leaves. Even paper and some cardboard can be composted. Sewage sludge is often used with the organic fraction of solid wastes, and such a system is known as co-composting. There are two distinct composting methods: aerobic and anaerobic. In aerobic composting, oxygen is an essential requirement for biological activity. In anaerobic composting, biological activity takes place in the absence of oxygen. There are different types of bacteria in each composting method.

Aerobic Systems

There are three types of aerobic composting systems. In an *in-vessel system*, the compost is made in enclosed structures using mechanical equipment to mix the material, allow the supply of needed oxygen, and to control moisture. In-vessel systems are expensive to construct and operate. However, they reduce the time needed for stabilizing organic matter. Co-composting requires in-vessel systems for odor control. All in-vessel systems should have large buffer areas to prevent odor complaints or have substantial odor scrubbing or destruction systems. The composting time varies between one and two weeks. The partially composted material is cured (i.e., stored in windrows or piles) for about 30 days and screened to remove material which did not compost. In-vessel systems are the most expensive among the composting systems.

In *active windrow systems* the material to be composted is placed in a long windrow over a piping system which extracts air from the compost pile. The outside air migrates through the pile and supplies oxygen. The extracted air contains odor causing compounds and must be filtered or scrubbed. Periodically the windrow is turned using either special equipment or general construction equipment such as front-end loaders. Wood chips are mixed in as bulking material to allow easy passage of air. It takes over three months to produce a good quality compost. All windrow systems need a good buffer area to prevent odor complaints. The compost is screened to remove large wood pieces and plastic.

In *passive windrow systems* the organic matter is placed in a long windrow and allowed to compost slowly. The windrow is turned on a weekly basis using special equipment or front-end loaders. The turning action exposes the organic material to oxygen. Passive systems require very long periods of time for completing the compost. Wood chips are used as bulking material to allow free passage of air. Passive windrow systems are usually located in agricultural areas with substantial open space. The compost is screened to remove undesirable materials such as large wood chunks and plastics. Over six months are needed to produce a useful product, but passive systems offer low-cost performance.

The quality of the compost is determined by the carbon-nitrogen ratio, levels of heavy metal and organic chemical residue, moisture content, and the amount of nutrients such as nitrogen, phosphorus, and potassium. Composts are generally low in nutrients and are best used for soil conditioning.

Anaerobic Systems

Organic material can also be stabilized using anaerobic bacteria. Such systems are completely enclosed in air-tight tanks to maintain anaerobic conditions. Methane gas is produced as a result of biodegradation. The organic material is pulverized and mixed with the material within the tanks. The process is continuous, and tank contents are mixed using either gas or mechanical paddles. The highly moist material is periodically withdrawn and pressed to remove the fluid which is reused with the incoming new material. Anaerobic systems require careful operation. They are used in Europe to a limited extent. In the U.S. there is only one pilot plant currently in use.

Energy Recovery

Many discarded nonrecyclable and nonreusable materials are combustible because they are made up of plastics, paper, or leather. Examples are carpet backing, used carpets, old shoes, sneakers, wax-coated cardboard, paper-plastic laminates, and soiled paper. Instead of storing such materials in a landfill, it is technically possible to burn them in systems specifically designed for waste combustion. Such systems typically achieve 1800–1900°F (982–1038°C) furnace temperatures without supplemental fuel.

Maintaining and controlling such high temperatures is achieved through computer control systems which adjust the feed rate and the volume of air required. Since the materials are primarily made up of carbon, hydrogen, and nitrogen, the combustion process yields carbon dioxide, carbon monoxide, and nitrogen oxides. However, almost all materials have some other elements in their molecular structure (e.g., sulfur, chlorine, lead, cadmium, mercury, chromium, and copper).

The combustion gases (referred to as flue gases) therefore also contain sulfur oxides, hydrogen chloride, vaporized metals, metallic oxides, chlorides, sulfates, and ash particles comprised of aluminum silicon and iron oxides. Because of the highly excited status of atoms in the combustion zone, new molecules are formed. Dioxins and dibenzo furans are examples of such reconstituted compounds.

The combustion systems consist of "grates" or "rotating kilns." Grates are usually made up of special steel capable of withstanding high temperatures and abrasion. Although the design varies, depending upon the manufacturer, all use reciprocating or rotating action to tumble waste materials, exposing new surfaces for combustion. Combustion air is supplied both from above and below. Usually the air from the waste tipping area is evacuated through preheaters to the grates.

In this manner, the energy conversion plant is maintained under negative pressure, thereby preventing the garbage odor from affecting the neighborhood.

Rotary kilns are long cylinders made from special steel and coated with refractory bricks. The kilns rotate and allow the material inside to tumble, thereby exposing new surfaces to the flame. Special steel tubes on the walls of the boiler or the rotary kiln carry water which is converted to steam by the exchange of heat with the flue gas. The fully combusted material (bottom ash) finally drops into a tank where water is used to quench it.

N. C. VASUKI

For Further Reading: James Abert, *Resource Recovery Guide* (1983); T. H. Christensen, R. Cossu, and R. Stegman, *Sanitary Landfilling Process, Technology and Environmental Impact* (1989); Floyd Hasselriis, *Refuse Derived Fuel Processing* (1984).

WASTE MANAGEMENT, MUNICIPAL SOLID

Municipal solid waste management (MSWM) is a diverse and complex enterprise. It involves organizing, financing, operating, and regulating processes that are responsible for the entire municipal solid waste (MSW) stream. These processes depend on institutional linkages and the integration of a variety of management systems. The partnership among three levels of government (federal, state, and local) and MSWM service contractors provides the infrastructure to serve the public. MSWM is striving for a "cradle-to-grave" approach by integrating source reduction (waste minimization), recycling (waste-to-energy), and composting, combustion (waste-to-energy), and disposal (sanitary landfills).

The current terminology for solid waste and its management is embodied in a series of federal legislative statutes. In 1965 Congress passed the Solid Waste Disposal Act (SWDA) that addressed what was known as trash, rubbish, garbage. These waste materials were assigned a new term: solid waste. SWDA was primarily directed at state planning, research, training, and technical assistance.

By 1970 the federal government had learned a great deal more about trash and Congress passed the Resource Recovery Act (RRA). By then the term *solid waste* had become well established and the concept of management had replaced the single-dimensional view of disposal. Congress expected the federal government as well as local and state governments to not only think landfills, but resource recovery too. In this law,

resource recovery was defined as the recovery of energy or materials from solid waste. Finally, the term *municipal solid waste* had begun to creep into usage, but the definition of this term was elusive.

In 1976 Congress passed the Resource Conservation and Recovery Act (RCRA) and did several things: (1) defined liquid, gaseous, or solid wastes as solid wastes; (2) divided the solid waste stream into municipal solid waste (MSW) and hazardous wastes (HW); and (3) provided the federal government with regulatory authority over hazardous wastes. In implementing RCRA (1976–1984) the EPA focused on the development of a cradle-to-grave (generation–transport–storage–treatment–disposal) regulatory program for hazardous wastes. Little federal attention was directed at MSW. MSWM was left to the state and local governments.

Today we think of MSW as being composed of three solid waste streams: residential (RSW), commercial (CSW), and industrial (non-hazardous) (ISW). RSW is solid waste generated at single and multi-family dwellings. CSW is solid waste generated at businesses and commercial enterprises (non-manufacturing). ISW is non-hazardous solid waste generated at manufacturing industries, but does not include sludges. Other solid waste streams that are frequently managed by an MSWM system include medical wastes, sludges, street sweeping wastes, construction/demolition wastes, and dead animals.

In 1984 RCRA was amended by the Hazardous and Solid Waste Amendments (HSWA), which directed the federal government to issue regulations for MSW landfills. In addition, due to the amendments the federal government, particularly the EPA, expanded its focus from strictly HW to MSW. This led to increased attention on recycling and waste reduction while retaining the two basic methods of incineration and landfilling. These four management methods have been combined as Integrated Municipal Solid Waste Management. These consist of: (1) reduction—the minimization of solid waste generation; (2) recycling—the series of activities, including collection, separation, and processing, by which products or other materials are recovered from the solid waste stream and are used or reused in the form of raw materials in the manufacture of new products or as fuel for producing heat or power by combustion; (3) combustion—the combustion of solid waste for the purpose of volume reduction and energy recovery or volume reduction only; and (4) landfilling—the disposal of solid waste by the sanitary landfilling process.

The management of MSW in the U.S. continues to involve sharing many different roles and responsibilities. Ownership of facilities is by local government and MSWM service contractors. Local governments own approximately 80% of the landfills and collect

50% of the residential solid waste. Most incinerators are either owned by or under contract to local government. In addition, operational roles and responsibilities of local government and MSWM service contractors are usually under local government controls. The regulation of MSWM is the responsibility of state governments and in some instances, certain roles are assigned to the lower levels of government (counties, regions, etc.). Federal regulations exist for landfills and for air emissions from waste-to-energy plants. Regulations for air emissions from landfills are expected. Enforcement continues to be the responsibility of state government.

Local government is defined to mean any incorporated or unincorporated jurisdiction including cities, municipalities, towns, townships, boroughs, districts, special purpose districts, authorities, counties, or similar local government entities that have been established by state, provincial, or local government law for the purposes of serving a designated segment of population within a state or province, or interstate/interprovincial areas.

The perceived role of each level of government in the U.S. may, at times, overlap or be shared. Thus, the role of the federal government, based on appropriate legislation, includes: establishment of national guidelines for proper design, operation, and management of systems and technologies; research and development, assessment, and interpretation of systems and technologies; provision of technical information and assistance to state/provincial and local governments; and initiatives to provide capacity through solid waste reduction and recycling. State governments have a very important role, based on appropriate legislation, that includes: establishment and enforcement of regulations for the proper design, operation, and management of systems and technologies (consistent with federal/national guidelines); provision of technical information and assistance to local governments; and initiatives to provide capacity through solid waste reduction and recycling. Finally, local governments have the major and principal role, based on appropriate legislation, that includes: planning; protection of public health and public interest; responsibility for the management of all MSW generated within, exported out of, or imported into their jurisdiction; compliance with all rules and regulations which address systems and technologies; and the provision of efficient and economic systems either with government owned/operated services, or through contractual services under local government control.

H. LANIER HICKMAN, JR.

For Further Reading: O. P. Kharbanda and E. A. Stallworthy, *Waste Management: Toward a Sustainable Society* (1990);

The René Dubos Center for Human Environments, *Integrating Waste Management* (1991); U.S. EPA, *International Perspectives on Municipal Solid Wastes and Sanitary Landfilling* (1990).

See also Composting; Electric Utility Industry; Environmental Protection Agency, U.S.; Hazardous Waste; Incineration; Law, Environmental; Pollution, Air; Recycling; Resource Conservation and Recovery Act; Source Reduction; United States Government; Waste; Waste, Municipal Solid.

WASTE MINIMIZATION

See Source Reduction.

WASTE-TO-ENERGY PROCESSES

See Electric Utility Industry.

WASTE TREATMENT

The technical processes associated with treatment of municipal solid waste to permit its final disposal or to recover energy and/or materials fall into three broad categories: mechanical, thermal, and biological/biochemical. Decisions about the types of management, or treatment, to employ result from a logical determination of whether municipal solid waste (MSW) is to be managed as waste—that is, treated for disposal—or whether it is viewed as a feedstock for a subsequent process to recover its energy or materials content. These decisions are often in contention in public debate, and cannot be based solely on technical criteria; they must be balanced with local needs, economic viability considerations and must fit within a public policy framework which is directed first to protection of health and the environment, and second, to efficient use of resources.

Treatment for disposal is used to reduce toxicity or to stabilize toxic constituents in the wastes. In MSW these toxics are nearly nonexistent, so treatment for toxicity reduction or stabilization is not needed. Sometimes wastes are treated for the purpose of volume reduction, but since there is usually some product associated with the treatment process (e.g., heat from combustion, gas from anaerobic digestion, soil amendment from composting), the two objectives of environmental protection and efficient use of resources are again interconnected. Each of these treatment processes leaves some amount of residue that has no real utility and must be landfilled.

Mechanical Processes

Mechanical (including manual) processing can be used to separate wastes into various components including metals, glass, paper, plastics, and what is referred to as a refuse-derived fuel (RDF) fraction. Mechanical processing has two basic functions: homogenization and separation. Generally, a mechanical process is a preliminary step to the thermal and biological or biochemical technologies that convert the organics to energy. This processing can be done in the home, commercial establishment, or office, or at central facilities utilizing a combination of manual labor and processing equipment. If the mechanical processing is done to precede a thermal or biological application, it is usually done at a site co-located with that technology, to reduce handling and transportation costs. If the processing is done to prepare materials for marketing as recyclable materials for use in manufacture of new products, it is usually done at a facility referred to as a materials recovery facility, or MRF. The inorganic components (and some paper products, as well) are generally separated and processed so they can be further "beneficiated," or prepared for eventual reuse in making new products.

Specific design decisions about processing must be based on the most up-to-date analysis of the particular MSW stream involved. If the organic fraction is to be used in the recovery of energy through a thermal or biochemical process, it must be as uncontaminated by inorganic materials as possible in order to maximize efficiency in its intended use. Any pre-processing adds cost; therefore, this beneficiation must add value and be efficient and/or cost-effective in order to be justified. If the MSW stream is viewed as a source of material and/or energy resources, these recovered materials must compete in the marketplace with virgin materials. They must be as easy to use, must cost no more, and must perform as well as the materials they replace.

No technological process is 100% efficient; each process will produce a residue incidental to the production of the desired end product. The amount of residue from mechanical processes varies, and is related to the degree of contamination that the end-use technology can tolerate, or to the specific process flow (degree of automation) used, or to the amount of manual labor available (affordable). High levels of residue may be related also to lack of convenient markets for the recovered materials or to failure to meet market specifications; often materials are processed, but cannot be reused and must be sent for disposal. For instance, if polyethylene is being separated for recycling or reuse, those users require that the degree of contamination with polyvinyl chloride (PVC) be zero, and

other plastic resins be limited to less than 5%. This is extremely difficult to achieve with a commingled stream of waste. Manual separation is usually needed, adding to the cost of preparation. Several automatic sorting techniques are in development; their operating principles are based upon x-ray fluorescence for detection of PVC, and optical and spectrophotometric methods for distinguishing between PET (polyethylene terephthalate) and HDPE (high density polyethylene) resins. Markets for clear glass require that contamination with colored glass be limited. Other end uses have specific requirements which must be considered in process design and operation. The possibility that specifications may change over time necessitates that mechanical processing be as flexible as possible. For many systems, this means more reliance on manual sorting.

Selection of processing equipment is related to desired properties of the end product. That is, if refuse-derived fuel is being produced for co-firing with pulverized coal or for combustion in a fluidized bed combustion system, processing is designed to produce a homogenous fuel that can be fed into the combustion system with ease, and it must be as free from contamination as possible. Thus the heterogeneity of as-received refuse must be addressed. "Fluff" RDF can be further densified by pelletizing, briquetting, or extruding, providing a denser fuel that is more easily transported and stored. Generally, RDF processing leaves a residue of approximately 25% by weight of the incoming MSW, which is primarily composed of noncombustible materials.

Mechanical processing occurs in several discrete unit operations that must be carefully integrated, but each process addresses the physical properties of the various components of municipal solid wastes. Magnets (either permanent or electromagnetic) are used to recover ferrous metals; eddy current separators (operating on the principle that metals passing through an electromagnetic field generate an electric "eddy current" in each piece of metal) recover aluminum; froth flotation or optical sorting may be used for glass recovery. Froth flotation depends on the tendency of hydrophobic particles to accumulate at the air/water interface of an aqueous system. A fatty substance added to an aqueous mixture of glass and nonglass particles adsorbs to and separates the glass particles. Air classification and screening systems have been used to separate light and heavy materials. Experience has shown that size reduction equipment (such as hammermills, crushers, grinders, shears, etc.), while useful, must be used after glass has been removed, in order to minimize embedding the glass particles in the other materials, such as paper. Excessive amounts of glass in the fuel contribute to slag buildup in the combustion unit, affecting maintenance and overall efficiency. Excess glass in materials processed for anaerobic digestion or other biological techniques is also a problem.

Three principal types of screening equipment have been used: vibrating flat beds, disk screens, and trommel screens. Their purpose is to achieve effective particle separation—based upon differences in physical size of the components of the incoming stream—at reasonable cost and energy efficiency. Once the incoming material has been mechanically processed it usually must be stored. Unfortunately, storage of materials with high moisture content (like MSW at 40–50%) is difficult. The concentration of cellulose materials, around 40%, also appears to be a limiting consideration. Factors affecting storage dictate that this material be used as quickly as possible.

Thermal Processes

Thermal technology either combusts waste for the recovery of heat energy or converts it to a gaseous, liquid, or solid fuel. Waterwall combustors, the most technically proven systems, employ special grates to burn "as received" waste and recover steam for use in industrial processes or to generate electricity. The use of combustion equipment for volume reduction alone is referred to as incineration. To offset the high capital and operating expense of thermal treatment approaches, recovery and utilization of the heat generated for use as replacements for scarce fossil fuels is preferred. Another essentially thermal process relies upon mechanical pre-processing to produce either pelletized or fluff RDF, which is storable and could be co-fired with another fossil fuel, such as coal, to recover its energy content.

As-received MSW has a heat content of approximately 5,500–6,000 Btu per pound, or nearly half the Btu content of coal. Of course this Btu content can be affected by many factors, including the degree of moisture and specific content of incombustible materials such as glass and metals. The Btu content of MSW has increased over time—due in part to the decrease in food waste (or an increase in use of residential food waste disposals) coupled with a general increase in paper and plastic products and their presence in discards. Generally about 500–600 kilowatt-hours of electricity can be generated by each ton of MSW combusted, considering systems that meet applicable air emissions standards.

Combustion of MSW results in gaseous air emissions and solid emissions in the form of ash residue. To the extent that heavy metals—particularly mercury, cadmium, lead, and chromium, incident to the manufacture or performance of products discarded in MSW—are combusted, these heavy metals will be present in the emissions and ash residue. In ash residue,

the concentration of metals will be higher than in as-received MSW, because the combustion process reduces the volume of materials by approximately 90%. An obvious observation can be made: the lower the metal content in incoming MSW, the less metal in the effluents, either air or ash. Operators of combustion systems have no control or influence over metal content in products, and no absolute knowledge about what their systems will receive for combustion on any given day, thus principal measures to mitigate the impacts of these emissions involve air pollution control devices to trap and treat flue gases or ash after the combustion process.

A number of devices are commonly added to the process train to reduce emissions to the air: electrostatic precipitators and fabric filters, such as baghouses, for particulate matter control; wet or dry scrubbers utilizing alkaline substances such as lime for removal of acid gases such as sulfur dioxide (SO_2), hydrochloric acid (HCl), and nitrogen oxides (NO_x) formed during combustion. Dioxins (chlorinated organic compounds that are not intentionally produced) and other trace organic emissions from any source are a concern. This much is known: These trace organics can be formed as products of incomplete combustion; they might result from constituents within the MSW, or they may be formed in the system from other chemical reactions. Management of temperature and retention time within the combustion chamber is essential to provide good burnout of the MSW. Also management of temperature of the flue gases has been shown to be extremely effective in condensing vaporized pollutants onto particles that are in turn trapped by electrostatic precipitators or baghouses and not permitted to become part of the air emissions.

Management of ash residues from combustion is important. With the addition of acid gas scrubbers and baghouses for air emissions control, residues increase from 10% to about 20% by weight. These residues contain metals that need to be prevented from contributing to leachate formed in landfills. (Leachate results from the percolation of liquids—from rainfall as well as from decomposition of MSW—through the landfill.) Generally ash residues are managed in "monofills" containing only ash and are not mixed with MSW in conventional landfills. In fact, due to the stabilization properties of alkaline scrubber residues, the metals contained in the ash tend not to leach. Leachate test data indicate that the primary constituents are inorganic salts of sodium, potassium, and calcium, which are highly soluble, thus the leachate is quite similar to seawater.

Thermal gasification and pyrolysis systems are also in development but have not reached commercial viability. Products from these processes include higher-value chemicals, gases, liquids, and chars. Pyrolysis is a thermal process that produces combustible gas plus liquids and solids of questionable utility. Another thermal system, known as fluidized bed combustion, could be applied to MSW. This system has the potential for reduced air emissions because limestone is often used as the fluidizing medium within the combustion chamber, thus reducing the need for extensive treatment of flue gases. These systems rely upon effective size reduction of the incoming MSW and removal of as much glass as possible. Again, the pre-processing and materials handling needs limit the viability of fluidized bed combustion systems.

Biological/Biochemical Processes

Biological/biochemical techniques use living organisms to convert the organics into useful energy or chemical forms, or into simple soil conditioners or "composts." All biochemical processes for recovery of resources from MSW share common principles (the need for microorganisms to accomplish desired results, and an environment to support or enhance the activity of those microorganisms). Three critical factors affect the rate and extent of biochemical conversion of MSW or specific organic components of MSW: moisture, temperature, and degree of resistance to bacteriological attack. Composting is probably the most straightforward of these processes but as it is practiced in the U.S., it is still basically an art form. While composting

is widely practiced for certain subsets of the municipal wastestream, especially yard wastes, it can be applied to the organic fraction of municipal solid wastes as a whole. The common limitation of this practice occurs when the compost product is not of sufficient quality to meet market specifications, resulting in materials that must be landfilled. An associated problem is related to the difficulty in preprocessing or separating out the inorganic wastes that are contaminants in the final compost product.

Composting is a low-temperature, partial oxidation process which can be applied to the easily degradable simple sugars, carbohydrates, proteins, and fats in discarded wastes. Composting results in an approximate 50% volume reduction—consuming about half of the organics, which are released primarily as carbon dioxide and water. Composting can be done slowly, in windrows outside; or more quickly, in specially designed enclosed vessels. The choice of approach is based on the specific system needs, the type of wastes being composted, amount of wastes available, time constraints, costs, etc.

Other biological/biochemical processes are in various stages of research, development, and demonstration, but most are not ready for commercial application. Some of these processes begin with hydrolysis, which allows the constituent ions in water to chemically interact with the cellulose materials within solid wastes to form glucose, which, in turn, is used to syn-

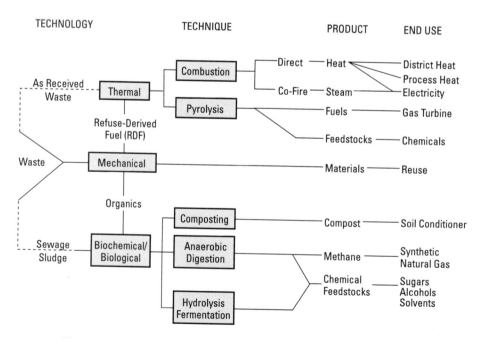

Figure 1. Technological processes in municipal solid waste treatment.

thesize higher-value chemicals, such as ethanol, methanol, and single cell protein. Since paper is a major constituent of MSW, it is a source of cellulose, but this cellulose, because it is mixed with lignin, resists microbial degradation. Lignin restricts the practicality of using processed MSW as chemical feedstocks.

Other biological processing approaches rely upon anaerobic digestion (acting in the absence of oxygen) and result in gaseous fuels. Usually anaerobic digestion of MSW is performed in combination with sewage sludge to take advantage of the moisture content in the sludge. However, since refuse contains many materials that decompose at different rates, the efficacy of anaerobic digestion is directly related to refuse composition, as well as to particle size, which is dependent upon the efficiency of a preliminary mechanical size reduction or homogenization step. Anaerobic digestion yields gases—methane (60–70%), carbon dioxide (29–39%), and trace amounts of hydrogen, hydrogen sulfide, and other gases. Production of methane for electricity generation or direct use is the major objective of anaerobic digestion systems. This gas could also be upgraded to pipeline-quality natural gas or methanol. Residues from anaerobic digestion are relatively high: 50% by weight of the incoming mass. Anaerobic digestion of MSW and sewage sludge is not a commercial technology at this time.

The concern of municipal solid waste managers in developing effective systems for managing MSW is certainty about environmental performance and long term reliability. The technological approaches can be viewed in terms of their current overall technical risk in an increasing hierarchy from, for example, ferrous metal recovery and aluminum separation (lower technical risk) to pyrolysis and enzymatic hydrolysis (higher technical risk). Figure 1 is a schematic representation of the technological processes discussed.

There is no single, simple approach to management or treatment of MSW. Some materials in some areas of the country are more valuable if recovered and reused in manufacture of products; in other areas these same materials are more suitable and more valuable if converted to energy. Whatever approach or combination of approaches is chosen, the technological performance must reliably meet applicable standards to protect public health and the environment.

CHARLOTTE FROLA

For Further Reading: Luis Diaz, George Savage, and Clarence Golueke, *Resource Recovery from Municipal Solid Wastes* (1982); J. Winston Porter, *Trash Facts* (1992); Walter Shaub, *Incineration of Municipal Solid Waste: Scientific and Technical Evaluation of the State of the Art* (1990).
See also Acid Rain; Ash; Bacteria; Biodegradable/Photodegradable; Bioremediation; Composting; Dioxins; Electric Utility Industry; Electromagnetic Fields; Flue Gas Cleanup; Fluidized Bed Combustion; Incineration; Materials Recovery Facility; Methane; Plastics; Recycling; Toxic Metals; Waste; Wastewater Treatment, Municipal.

WASTE TREATMENT, END-OF-PIPE

See Flue Gas Cleanup; Waste Disposal; Waste Treatment.

WASTEWATER SYSTEM, MUNICIPAL

See Bioremediation; Deep Well Injection; Pollution; Wastewater Treatment, Municipal.

WASTEWATER TREATMENT, MUNICIPAL

Municipal wastewater treatment is provided to reduce the impact of municipal sewage on rivers, streams, and lakes. The minimum acceptable level of treatment is called secondary treatment and removes about 90% of the organic matter and particles from sewage. When this level of treatment is inadequate to protect water quality, additional treatment, often called advanced or tertiary treatment, may be required. Current treatment standards in the U.S. are a result of legislation passed by Congress in 1972 during the Nixon administration. Also at that time the Environmental Protection Agency was established to ensure that the treatment requirements would be met.

Secondary treatment involves two stages: (1) settling by way of gravity to remove larger particles and (2) biological treatment to remove small particles and dissolved organic matter. The most commonly used biological treatment is called the activated sludge process and consists of a large reactor in which microorganisms are allowed to consume the organic matter in sewage under well oxygenated conditions. After allowing sufficient time for contact between the microorganisms and sewage (about 6 to 12 hours), the slurry is transferred to a settling basin, where the organisms settle out, and is then returned to the sewage contact basin, where the organisms can consume more organic matter. An alternative biological treatment is the trickling filter process, in which microorganisms are brought into contact with sewage in a reactor that contains plastic surfaces on which the microorganisms grow. The sewage is sprayed over the top of the plastic surfaces where it trickles down, undergoing biological degradation.

When sewage is discharged to lakes and coastal waters, it may promote excessive plant growth, or eutrophication, due to an abundance of nitrogen and phosphorus in the wastewater. For this reason, removal of nitrogen, phosphorus, or both may be required. Phosphorus removal is usually achieved by the addition of iron or aluminum salts, which form chemical solids that can be separated out in a settling basin. Nitrogen is usually removed by a series of complicated biological steps in which it is first converted to NO_3 (nitrate) and then to N_2 (nitrogen gas). Nitrogen gas is not usually available to plants for growth.

Additional removal of both organic matter and particles may be necessary if the stream receiving the treated wastewater discharge is small or if high quality water is required. In these cases more complicated physical or chemical treatment processes must be used. In many instances disinfection of sewage is also required and is usually achieved by the addition of chlorine.

A necessary part of any municipal wastewater treatment scheme is the disposal of solids and microorganisms removed from the sewage (sludge). Sludge dewatering and disposal must be considered part of a complete treatment scheme and frequently involves the application of sludge to farmland, disposal in a landfill, or incineration.

JOHN T. NOVAK

WATER

Earth, unlike the other planets in the solar system, is blessed with an abundant supply of water in its liquid form. Life on planet Earth is inextricably linked to water. It is essential for life and essential for nearly every human endeavor. Water covers 75% of the Earth's surface, is the primary component of living organisms (60% to 70% by weight), and plays key roles in photosynthesis and in regulating the Earth's surface temperatures. The distribution of plant, animal, and human populations is determined primarily by the occurrence of water. The importance of water as the common denominator of life is derived from its unique physical and chemical properties.

Composition and Properties of Water

As compounds go, water's chemical makeup is rather simple. Two hydrogen atoms are covalently bonded (by shared electrons) to one oxygen atom forming an isosceles triangle. The chemist's shorthand for water is H_2O. Water molecules are attracted to each other by hydrogen bonds. This simple arrangement of atoms and the resulting bonding and molecular geometry imparts special chemical and physical properties that make life possible on Earth. In fact, these properties

are so unusual compared to other inorganic compounds that water may be referred to as nature's most fabulous freak! The arrangement of hydrogen and oxygen atoms results in a polar molecule (having an unequal distribution of electrical charges around the compound) and accounts for many of water's unusual properties.

The properties of water can be divided into two classes depending on whether chemical bonds between the H and O atoms are broken or whether only the hydrogen bonds connecting individual H_2O molecules are broken leaving the H_2O molecule intact. Life depends on both classes. An example of a chemical reaction that illustrates the importance of water to life is photosynthesis. Photosynthesis is the process by which green plants use the sun's energy to convert carbon dioxide (CO_2) and water (H_2O) into carbohydrates ($C_6H_{12}O_6$) and oxygen (O_2).

$$6CO_2 + 6H_2O \xrightarrow[\text{chlorophyll}]{\text{light}} C_6H_{12}O_6 + 6O_2$$

Water molecules contribute their hydrogen atoms to form the carbohydrate and their oxygen atoms are liberated. Carbohydrates are further converted into proteins and lipids by plants and ultimately serve as energy sources for animals, including humans. The oxygen liberated from water molecules is the oxygen we breathe. Thus water is a critical part of photosynthesis, the most fundamental and important process making life on Earth possible.

Water is present on Earth in three forms: solid (ice), liquid (water itself), and gaseous (water vapor). Water changes forms over a relatively narrow range of temperatures (0° to 100°C). These different forms result from changes in the strength of the hydrogen bonding between water molecules caused by heating or cooling. The transition of water between its various forms is an example of the second class of reactions in which only the hydrogen bonds between individual H_2O molecules are affected. At 0°C (32°F) water freezes, going from a liquid form to a solid form. At 100°C (212°F) water boils, changing from a liquid to water vapor. When water freezes, it becomes less dense. This is in sharp contrast to most other compounds, which become more dense when they freeze. This causes ice to float. Water has its greatest density at 3.94°C. If this were not the case, there would be no life in the oceans and lakes on Earth. Since ice floats, it forms an insulating cover preventing water bodies from freezing completely. As a result, fish and other aquatic life remain below the ice in a relatively safe and moderate climate.

The physical and chemical dynamics of lakes and

reservoirs are governed by the temperature-density relationship of water. These relatively complex relationships can be summarized by simply noting that cold water is more dense (heavier) than warm water. As the surface water of a lake warms in summer, it becomes less dense (lighter) than the cooler water at the bottom. Like water and oil (liquids of different densities), the surface and bottom water layers of a lake are effectively isolated from each other. Aquatic ecologists refer to this condition as density stratification, and it is of fundamental importance to the functioning of aquatic environments. The distributions of oxygen and nutrients in aquatic environments are strongly influenced by stratification. One unusual application of stratification is the ability of some lakes to support "two-storied fisheries" with warm water species (bass) living in the warm surface waters and cold water species (trout) living below.

Water also has several unusual properties that allow it to control the heat budget of the Earth. First, the specific heat (the amount of heat energy in calories required to raise the temperature of a mass of a substance by 1°C) of liquid water is very high. This high heat capacity means that a gain or loss of a very large amount of heat energy is required to change the temperature of water. The specific heat of water is much higher than that of soil or air, and as a result, water changes temperature slowly relative to the atmosphere and adjacent land masses. This fact accounts for the refreshing coolness of water during the hot summer months and relative warmth of water in the fall and winter. It also accounts for the moderating influences that oceans and large lakes have on the climate of nearby land masses. For example, it is possible to grow fruit orchards near the Great Lakes in Canada and the United States and near the Bay of Fundy in Nova Scotia because of the warming effect of water on the areas around them. Similarly, the climates of the Scandinavian countries are modified by the Atlantic Ocean, giving them a more temperate climate than if they were located inland on the tundra.

Water also has high latent heats of evaporation and fusion. These properties mean that large heat inputs or losses are required to change water from a liquid to either a solid (ice) or a gas. The large amount of heat energy required to evaporate water is responsible for the cooling effect of sweat drying on our skin.

Distribution of Water in the Biosphere

Water covers 75% of the surface of the Earth. The total amount of water on Earth is approximately 1,404,377,000 cubic kilometers (km³). To put this in perspective, one km³ will hold 264,000,000,000 gallons. If all the water on Earth were equally distributed, it would cover the entire surface to a depth of 3 kilo-

meters (1.9 miles). Most of the Earth's water is in the oceans (97.6%). Fresh water makes up less than 3% of water on Earth. Over two thirds of this fresh water is tied up in polar ice caps and glaciers. Fresh water lakes and rivers, so critical for civilization, make up only 0.009% or 127 km³ of the water on Earth. Groundwater, which is also critical for supplying water for people, crops, and industry, makes up 0.28%. Unfortunately, fresh water in lakes and rivers or underground aquifers is not uniformly distributed around the world. Problems of uneven distribution, water pollution, and inequitable access are causing water management problems around the world.

Water is a renewable resource. It is renewed by the hydrologic cycle. Water circulates between the oceans, land masses, and the atmosphere in a predictable manner so that the total amount of water on Earth remains about the same from one year to the next. The sun's energy causes water to evaporate from the land or oceans and to enter the atmosphere as water vapor. The water vapor condenses and is precipitated back to the Earth's surface. If the precipitation falls on land, the water may run off into streams, rivers, lakes, or seas, or may infiltrate and become groundwater. Because of the hydrologic cycle, water is continuously being renewed, redistributed, and purified. The hydrologic cycle is fundamental to the functioning of the Earth. Not only does it recycle water, but it also plays an important role in modifying and regulating Earth's climate. Human threats to the hydrologic cycle include global climate change due to "greenhouse" gases and deforestation.

Importance of Water to All Organisms

Water is essential for all life forms. Even desert animals and plants require water and have evolved physiological and behavioral adaptations to utilize scarce water efficiently. Life itself is thought to have originated in water and evolved to terrestrial environments. The human body, like most other animals, is composed of approximately 65% water by weight. It is the primary component of plant and animal cells. Water provides the medium in which all of life's biochemical reactions occur.

Water is known as the universal solvent. Its ability to dissolve other substances makes it ideal for transporting nutrients to and waste products away from animal and plant cells. Thus water is the primary component of circulatory systems in animals, which supply essential food, oxygen, and hormones. It is also the medium for eliminating waste products of metabolism via the excretory system. In plants, because of its high surface tension and wetting ability, water can transport nutrients and oxygen from the roots to the top leaves of the tallest trees. Evaporation of water from tree leaves or the surface of skin produces a cooling effect that helps maintain the internal temperature of living organisms. At a more global scale, water vapor in the atmosphere traps the sun's heat and keeps the surface of the Earth warm so that life can exist. Without water, life as we know it could not exist.

Humans like all organisms have a basic physiological need for water (approximately 2 liters per day). However, human demands for water have gone far beyond physiological needs. In the United States we now use approximately 5,400 liters (1,400 gallons) per person per day. Much of this water is used to irrigate crops and manufacture goods. In the less developed countries of the world average water use is only 45 liters per person per day.

Water is used for many purposes in today's world and is considered essential to most human endeavors. Water serves as a major means of transportation for materials and goods exchanged in the emerging global economy. Ships ply the waterways and shipping channels of the world oceans carrying raw materials to be refined and return with finished products. Major water resource development projects around the world have developed water supplies for human consumption, irrigation, hydroelectric power generation, industrial use, and recreation.

Development of safe potable water supplies has had a major impact on decreasing water-borne diseases around the world. Unfortunately, many people of the world still do not have safe drinking water. The United Nations estimates that two billion people still do not have adequate water supplies.

Water from underground aquifers and surface supplies irrigates millions of hectares of croplands throughout the world. Irrigation is the largest consumptive use of water worldwide, using approximately 69% of water withdrawn for human use. The benefits of irrigation have been increased crop yields and food supplies. Unfortunately, in some parts of the world irrigation has caused salination of soils, subsidence of land, pollution of surface and ground water, and depletion of underground aquifers (see Desertification).

In developed countries of the world, industrial use of water to produce goods and materials is the second most consumptive use of water. Much of this water is used for cooling. However, some comes in contact with byproducts of manufacturing and becomes contaminated, contributing to water pollution.

Development of hydroelectric dams has provided electricity to many parts of the world. Many reservoirs have also provided recreational and fisheries benefits in addition to electricity. Concomitant with their development has been the loss of productive bottomlands, riparian ecosystems, and free-flowing rivers.

During the twentieth century, water resources developments in many parts of the world have improved human health and economic prosperity by providing safe drinking water, irrigation water for agriculture, and water for industry and manufacturing. However, the growing human population on Earth, combined with increasing demands for consumer goods, is taxing surface and ground water supplies. A major challenge for the twenty-first century will be to manage water resources in ways that prove beneficial to the human environment, without endangering ecological resources.

KENNETH L. DICKSON
ROBERT D. DOYLE

For Further Reading: W. P. Cunningham and B. W. Saigo, *Environmental Science* (1992); B. J. Nebel, *Environmental Science* (1990); R. C. Wetzel, *Limnology* (1990).

WATER, DRINKING

To be suitable for drinking, water should be free of pathogenic, or disease-causing, microorganisms, harmful chemicals, and materials producing a taste or odor. To ensure that drinking water meets the standards for purity, it is usually treated by a series of processes that include disinfection. The modern practice of water treatment was instituted during the period between 1900 and 1930, when first water filtration and then disinfection by chlorination were adopted. Widespread chlorination of water was spurred by the investigations of Abel Wolman, who showed that outbreaks of infectious intestinal diseases could be eliminated by the proper use of chlorine.

Current methods of water treatment depend primarily on the source of water. Drinking water obtained from rivers and streams is in danger of contamination by microorganisms, industrial chemicals, and surface runoff. Treatment of these waters usually involves disinfection to eliminate pathogens, chemical coagulation in which small particles are agglomerated, usually by the addition of alum (aluminum sulfate), and filtration to remove agglomerated particles.

For drinking water produced from lakes and reservoirs, the potential contaminants are similar to those from rivers. However, there is also the likelihood that taste- and odor-causing chemicals produced as a result of algae growing in slow-moving or stagnant water may be present. Removal of taste- and odor-causing chemicals may require the use of activated carbon or the addition of an oxidizing chemical such as potassium permanganate or ozone.

Groundwater is usually more highly mineralized than surface water. Some of the minerals, such as iron and manganese, which impart a rusty stain to water,

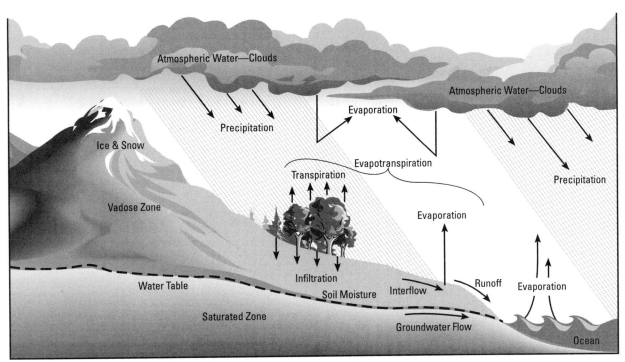

Figure 1. Water cycle.

are removed. Removal is usually accomplished by aeration followed by filtration. In addition, groundwater often has high levels of hardness, causing deposits to occur in hot water heaters. Removal of hardness—or water softening—by addition of lime (CaO) is a common process for groundwater systems.

In the future, regulations regarding drinking water will dictate changes in water treatment that will provide safer drinking water, but increase treatment costs. These changes will include additional monitoring for toxic chemicals, improved removal of trace organic contaminants, and alternatives to the use of chlorine for disinfection, because chlorinated organic compounds are produced when chlorine reacts with organic matter in water.

JOHN T. NOVAK

WATER CYCLE

The water cycle, or hydrologic cycle, is the cycle through which water passes from the ocean through evaporation to the atmosphere, to precipitation, to infiltration and runoff and return to the ocean. The study of water is the purview of two sub-fields of earth science. Meteorology deals with water as it moves through the atmosphere. Hydrology is concerned with water and its movement and distribution at or near the Earth's surface.

The amount of water on Earth is for all practical purposes fixed. There is approximately the same amount of water today as there was during the ice ages and as there will be in the future. An extremely small amount of water, called juvenile water, is produced by chemical reactions in volcanos but this amount is so small that it can be ignored when we speak of water availability and of the water cycle. The total amount of water on Earth is estimated to be 326 million cubic miles (1,360 million cubic kilometers). Most of the world's water is in the oceans, 97.3%; fresh water available to humans in lakes and streams is only 0.009% of the total or 30,000 cubic miles (126,000 cubic kilometers). Groundwater accounts for 0.6% or 2 million cubic miles (8 million cubic kilometers). Much of this groundwater is salty or far too deep for human exploitation. An additional 264 cubic miles (1,100 cubic kilometers) are contained in the world's biota. Not all fresh water can be used by humans because of its location or quality. If you imagine all the water on Earth to be in a 55-gallon oil drum, then the water available for human use is not much more than one teaspoon. Only a small part of the total water amount—102,000 cubic miles (420,000 cubic kilometers) or 0.031%—participates in the water cycle. This cycle is driven by two forces: solar energy evaporates water and lifts it into the atmosphere, and gravity brings it back down as precipitation and moves it along in streams and groundwater flow.

Figure 1 is a simplified diagram of the water cycle. It

has no beginning or end but for the purpose of this description we will begin in the oceans where water evaporates from the surface and is moved along by air masses. As the vapor cools, it condenses and forms visible water droplets that accumulate into clouds. Under proper conditions the tiny droplets grow large enough to fall to earth as precipitation. Depending on temperature such precipitation may occur as rain, sleet, or snow. Some of this precipitation never reaches the ground. It evaporates on the way down and the water vapor rejoins the clouds. If the precipitation occurs over the ocean, the cycle is complete. If it occurs over land, it is either intercepted by the surface of plants (the plant canopy) or falls to the ground. The water that is intercepted and retained by the plant canopy evaporates rapidly and rejoins the atmospheric moisture once precipitation has stopped. Snow may accumulate on the ground and the snowpack formed stores water for a period that may be relatively short, as through a winter season, or long, as in glaciers or ice caps. Once melted, snow and ice rejoin the water cycle. The bulk of the precipitation hits the ground and remains on the surface as puddles, or directly infiltrates at a rate dependent on the type of ground. This infiltration rate varies greatly: rapidly through loose sand but very slowly or not at all through hard-packed clay or paved surfaces. Water in depression storage (on the surface) eventually either infiltrates into the ground or evaporates. If the precipitation rate is greater than the infiltration rate, then the excess water will move along the surface of the ground. This is called surface runoff.

The water that infiltrates the ground enters the unsaturated (vadose) zone of the soil and replenishes the soil moisture. The moisture is then taken up by plants and transpired through the foliage back to the atmosphere. The process of transpiration from plants and evaporation from vegetated land surfaces is called evapotranspiration. Some of the water in the soil moisture zone will become interflow, an underground flow in the unsaturated zone that eventually enters a water course. Most of the water that is not retained in soil moisture, taken up by plants, or discharged as interflow passes into the saturated zone and becomes groundwater. Some groundwater is stored, recharging the aquifers; the rest is moved along by gravity and is eventually discharged into streams, forming their dry weather flow. Some water, estimated at 5% of the groundwater, flows directly into the ocean. Most of the stream flow—the groundwater outflow into the water courses, the interflow, and the surface runoff—eventually ends up in the ocean. On its way to the ocean, water will evaporate from both the surface of the ground and the surface of lakes and streams. Some

groundwater in very deep strata is stored permanently. Such water enters the water cycle only if it is pumped to the surface.

The early Greeks and Romans considered the oceans to be the ultimate source of water. Water was believed to infiltrate from the oceans into the base of mountains where it mysteriously was distilled, rose to the top, and came down in the form of springs. Plato explained the flow of rivers and springs by a system of underground channels in which water constantly surges to and fro. Marcus Vitruvius Pollio (about 100 B.C.) seems to have been the first to recognize the role of the sun as causing water on the surface of the Earth to turn into vapors that rise and form clouds. He indicated that these clouds, colliding with mountains or shocked by storms, caused rainfall as they broke. Leonardo da Vinci (1452–1519) said, "Where there is life there is heat, and where there is vital heat there is movement of vapor. This is proved because one sees that the heat of the element of fire always draws to itself the damp vapors, the thick mists and dense clouds, which are given off by the seas and other lakes and rivers and marshy valleys. And these are often swept away and carried by the winds from one region to another, until at last their density gives them such weight that they fall in thick rain." The founders of modern hydrology were the French physicists Pierre Perroult (1608–1680) and Edmé Mariotte (1620–1684), and the English astronomer Edmund Halley (1656–1742). But even their work did not completely establish belief in the evaporation-condensation-precipitation-infiltration-runoff cycle. The theory that evaporation and condensation were the cause of the existence of water on the land was opposed by many scientists in the 18th and 19th centuries. Even as late as 1921 some people questioned the concept of the water cycle.

The water cycle must also be considered as a part of the larger system of world climate. While the components of the cycle are the same under all climatic conditions, the importance and magnitude of each of the components within the cycle vary greatly. For example, in areas having a hot and dry climate, evaporation directly from precipitation, and from the plant canopy and the surface, are the major components, while infiltration and runoff are rare events. In wet and humid areas, evapotranspiration from plants is a major factor.

It is estimated that 83% of evaporation and 76% of precipitation occurs on the ocean. Practical hydrologists, water managers, and engineers must look at the water cycle on a more localized and limited basis. The technique used to describe the part of the water cycle that occurs over land is called the determination of a water budget or water balance.

The water budget in its simplest form can be described by the equation

I (inflow) − O (outflow) = ΔS (change in storage)

for any defined land area and any defined time period. Inflow is the amount of precipitation and any other water entering the area under study. Outflow is the sum of total runoff (the sum of surface runoff and sub-surface flow) and total evaporation (the sum of evaporation from the ground, water surfaces, and transpiration from plants). Change in storage is the change in the amount of water retained in soil moisture, aquifers, reservoirs, snow and ice, or biologic entities or manufactured products.

A single water project's water budget, such as for a reservoir, would consist of the measurement or estimation of the inflow (surface water and groundwater entering the reservoir and the precipitation falling on the reservoir surface) and the outflow (reservoir releases, infiltration and leakage, and evaporation from the free water surface).

Looking at long time periods, such as a year or more, the change in storage becomes zero for large areas (whole continents or large countries) because the amount of water in all types of storage on large areas is reasonably constant and possible changes in storage are very small when compared to all of the other quantities involved. South America is by far the richest continent in water resources in terms of the depth of precipitation and runoff per unit area. Europe comes next with Asia, North America, Africa, and Australia following. For total water resources Asia is first with South America, North America, Africa, Europe, and Australia following.

The U.S. Water Resources Council estimated the water balance of the United States. Inflow was 4200 billion gallons per day (bgd) or 15.9 cubic kilometers per day (km³/day) from precipitation. Outflow consisted of 995 bgd (3.8 km³/day) discharge to the Atlantic and the Gulf of Mexico, 325 bgd (1.2 km³/day) to the Pacific, 6 bgd (0.023 km³/day) to Canada, and 1.6 bgd (0.006 km³/day) to Mexico. Evaporation was estimated to be 2765 bgd (10.5 km³/day) from surface, plants, and inland water bodies. The remaining 106 bgd (0.4 km³/day) are assumed to be consumed in biologic and manufacturing processes.

The water cycle is an important factor in many natural processes. Water is a solvent for many materials occurring in nature. Thus, the waters falling onto the earth and infiltrating the ground will leach out a number of minerals. The most important are salt, alkali, and limestone. In each case the mineral either precipitates as water evaporates or increases the mineral content of a water "sink," a water body without outflow such as the ocean or some lakes. Examples of the results of this natural process are the salt and alkali flats common in the American Southwest or the heavy mineralized lakes like the Great Salt Lake in Utah or the Dead Sea in Israel. The leaching of lime forms the caves in karst regions and the stalagmites and stalactites in those caves.

The most important geomorphologic impact of the water cycle is the effect of erosion and sedimentation on land forms. Precipitation hitting the ground and water flowing on the ground loosen solid material, which is then carried along by flowing water. Softer rock or looser soil is removed faster, shaping hillsides and valleys accordingly. The amount of water flow and its velocity control the amounts and the size of particles, from fine sands to large rocks, that are moved by the water flow. As the velocity of flow decreases, particles are deposited in order, from large to small sizes. In that way valley bottoms and deltas are formed by sedimentation, and rivers meander, eroding rock and soil in one spot and depositing sediments in others.

Another impact of the water cycle, when combined with a particular temperature regime, is the formation of glaciers. In areas where the annual precipitation in the form of snow exceeds annual melting, glaciers form. The movement of glaciers has a large impact on land forms in some areas, usually forming U-shaped valleys.

The water cycle also plays a major role as a transportation medium for a number of substances. Among them are sulfur compounds (causing acid rain and the acidification of lakes and streams), carbon compounds, and many aerosols (tiny particles and droplets suspended in air).

The water cycle affects the flora and fauna of a region. The amount of precipitation that falls on a given area is the determining factor of the type of vegetation that will grow there. Both the total amount and the distribution of that precipitation over time is important. In turn, the plants that are supported will determine the community of animals involved.

In pre-industrial times human habitation developed at or near water courses because they provided the materials needed for life and also provided avenues of transportation. Areas in which the amounts and availability of precipitation shifted significantly on annual or longer cycles spawned nomadic populations. Very long-term changes in water availability have caused major population shifts and the disappearance of whole civilizations.

Just as the water cycle affects humans, human actions can affect the water cycle. On the regional or local scale they may have considerable effect. Changes in the vegetative cover, such as clear-cutting of forests or allowing pastures to be overgrazed, significantly alter the distribution of rainfall by decreasing evapo-transpiration and increasing surface runoff. Paving areas and constructing buildings decrease infiltration into the ground. The construction of reservoirs increases evaporation from the water surfaces. At local levels heat rising from cities and large industrial installations is likely to decrease rainfall.

Climate change through the greenhouse effect is the only significant impact that humans are likely to make on the global water cycle. Increased temperatures increase evaporation and with it total precipitation. While these global effects are very likely to happen, there are not yet sufficient scientific data to be sure of the amount and the timing of their possible occurrence. Even greater uncertainty exists about the regional and local changes in the water budget caused by climate change. Regions may experience increased or decreased precipitation, different seasonal distribution of precipitation, and changes in the range of extreme events such as hurricanes, severe floods, and droughts.

HARRY E. SCHWARZ

For Further Reading: E. K. Berner and R. A. Berner, *The Global Water Cycle — Geochemistry and Environment* (1987); M. I. Lvovitch, "The Global Water Balance," *Transactions, American Geophysical Union* (1973); John C. Manning, *Principles of Hydrology* (1987).

WATERFRONT DEVELOPMENT

Many of the world's urban centers developed along waterways with natural ports. Historically waterfront cities took advantage of the benefits of fresh water, seafood, strategic views, and trade. As new civilizations and cultures replaced earlier ones, these settlements were rebuilt many times, adapting to new conditions and cultural uses. Similarly American waterfront cities are undergoing a gradual replacement and reconstruction of some of their past features.

There has been a shift from an industrial-based economy to one dominated by support communications, service, and high technology. Declines in shipping and water-dependent industry left large derelict parcels along urban waterfronts. A familiar sight in many American cities has been a long stretch of crumbling piers, unused waterfront rail yards, and empty warehouses. Concurrently competition between downtown central business districts and suburban growth centers (shopping malls and commercial office parks) led to a decline in center cities.

The combination of these changes led to two initiatives for action: (1) assisting water-dependent industries in staying competitive and adapting to market shifts; and (2) reconstructing the derelict waterfront in new forms that help revitalize downtown commerce.

Reconfiguring empty structures has initiated a renaissance of attention and investment in some long-neglected neighborhoods. Post-industrial waterfronts are being redeveloped as housing, parks, pedestrian esplanades, hotels, marinas, restaurants, and commercial space. Many capitalize on the recent public interest in preserving historically interesting details of the old structures, and nearly all include public parks.

Tremendous benefits to people, wildlife, and plant communities occur when a series of green spaces along a waterway are linked. Long strips of open space can be used for bike and running trails and allow a relatively small acreage to serve several communities. Allowing water edges to sustain natural plantings filters storm water from the city on its way to the river, lake, or ocean and provides habitat for urban wildlife. This enhances both water quality and the ecology of the shore.

Getting people to enjoy the waterfront and spend time there is taking many forms. Cities are locating fireboats, ferries, ocean liner docks, and yacht basins where they can add to the bustle and novelty of the active waterfront as well as do their jobs. Thematic restaurants have long been part of urban waterfront, but are now being designed into new complexes and sometimes even subsidized in order to create a desired mix of services and shops.

Water-based transportation is reemerging as an alternative to building more highways. For example, Boston Harbor is now served by several private and semi-private subsidized ferries in order to provide a consistent commuter alternative and encourage public use of the waterfront.

Urban Waterfront Development Projects

Battery Park City in New York is one of the leading examples of mixed-use waterside developments. In phases, new buildings and park spaces are being built on landfill and relieving platforms (massive decks above the water). Combining condominiums, a yacht basin, high rise offices in "The World Financial Center," shops, bookstores, restaurants, and take-out food service, the development is primarily private. However, it was built with considerable public encouragement. Like most new waterfront projects, it incorporates public access and urban design controls, in part based on the concept of the public trust.

Reconnecting downtown Baltimore to the waterfront separated by a major highway, the Inner Harbor Project was an early example of engineering a change in the way the public thought of an older waterfront zone. One of the most important aspects of this project's success was its inclusion of public access along an esplanade edged with shops, restaurants, and informal outdoor activities. It has continued to expand, in phases.

A series of private redevelopments of old finger piers is being built near downtown Boston. It is connected by a publicly planned, privately built esplanade. Getting the public to the water's edge in a safe, attractive, and convenient way has been critical to its success. Without the public component, private housing developments on the waterfront can become isolated luxury space, barricaded "island" communities surrounded by poorer neighborhoods. With the presence of the walkway and services connecting them, Boston's private piers have been developed as a part of the city.

In smaller cities as well, revitalizing the shoreline has been used to attract reinvestment in the adjoining downtown. For example, in Wilmington, N.C., competition with suburban shopping malls had depleted the historic downtown of much of its vitality. However, charming historic buildings, including The Cotton Exchange, attracted non-profit groups to take the lead in promoting investment in the rehabilitation of the historic district. The public sector built a waterfront park and boat launch, which attracted people to the area and helped to stimulate the reconstruction of many older buildings as offices and specialty retail space.

A small number of old piers near Seattle's central business district were reconstructed as public open space and nautical-theme retail and restaurants in the early 1970s. The heart of the development is an aquarium utilizing the historic shell of a former warehouse, attracting many tourists. A recent extension of the harbor open space is a linear park. New aquaria in many cities are part of the trend to replace strictly maritime uses with more popular commercial installations.

Planning for Revitalization

Planning the physical framework to allow cities to adapt to the broad cultural shift to post-industrial societies is complicated and difficult. Waterfronts in many cities have been studied and restudied for years. Often, despite an abundance of studies, little action results.

Cities that have managed to negotiate the revitalization of waterfront areas have typically done so by building coalitions of diverse interest groups, and working in phases, building sequential parts of a large, long-term vision.

Sometimes termed "the new frontier" in urban development, waterfront reconstruction represents an opportunity to both reclaim public access to the water and address some long-standing water quality problems.

MARCHA JOHNSON

For Further Reading: Ann Buttenwieser, *Manhattan Waterbound* (1987); Roy Mann, *Rivers in the City* (1973); Douglas Wrenn, *Urban Waterfront Development* (1983).
See also Coastlines and Artificial Structures; Urban Challenges and Opportunities.

WATER, HARD

The hardness of water is a function of its dissolved mineral content. The type and amount of hardness depends on the geology of the area. As rainfall travels over and through the surface of the earth, minerals are dissolved from the soil and rocks. In general, hard water is found in areas where topsoil is thick and limestone formations are present. Soft water is found where the topsoil is thin and limestone formations are sparse or absent. Surface waters are usually softer than groundwater.

Historically, the hardness of water has been measured in terms of its ability to produce soapsuds. Hardness may be considered the soap-consuming property of water, because suds cannot be produced until the minerals causing hardness have combined with soap. Water that required more soap to form suds was called hard water. Rainwater, on the other hand, is soft and requires little soap to produce suds.

With the advent of synthetic detergents, many of the disadvantages of hard water for household use have been diminished. However, soap is still preferred for some types of laundering and for personal hygiene, and hard water still remains objectionable for these purposes. Besides requiring excessive amounts of soap, hard water clogs pipes and leaves mineral residues in hot water heaters and a ring in the bathtub. Additionally, because it interferes with some industrial processes, it must be controlled for optimum results.

The most popular household method to eliminate hardness employs a process called ion exchange. In this process resins known as zeolites, which usually contain sodium, are used. Zeolites exchange the sodium for calcium and magnesium in water, thereby decreasing its hardness. The harder the water, the more sodium is introduced. However, the addition of sodium in water-softening systems may pose a problem for people who are on sodium-restricted diets. The calcium and magnesium levels found in hard water are not a human health concern.

WILLIAM T. WALLER

WATER HYACINTH

Water hyacinth (*Eichornia crassipes*), a free-floating aquatic plant, is one of the most destructive weeds of the warmer parts of the world. It thrives in both polluted and clean waterways and from drainage ditches to great rivers, and it vastly affects the ecosystems in which it is found. The plant grows in dense masses that choke the waters where it lives, upsetting the ecological balance of available light, nutrients, and space, and killing or driving off other plants and animals. It directly affects human communities by making boat passage impossible, serving as a carrier of disease, and disrupting agriculture and fishing. The plant's vigor, however, is mirrored by the creativity and ingenuity of the methods used to control and utilize it.

Native to the Amazon Basin, water hyacinth was prized in the 19th century for the beauty of its flowers and was introduced throughout the world for cultivation in public gardens and private collections. In every country outside of its native habitat it quickly changed from attractive garden ornament to destructive weed. The plant's history in the United States typifies this pattern. It first arrived in America in 1884, when Japanese delegates to a cotton exposition in Louisiana distributed specimens of the plant as gifts. The plant spread throughout the South within ten years, hindering navigation, diverting water from farm use, and killing off fish by depriving rivers and lakes of oxygen.

Three biological characteristics make the water hyacinth the invasive species that it is: (1) its broad tolerance of environmental change, (2) its physical structure, and (3) its tremendous reproductive ability. The plant is well adapted to all but the faster flowing waterways, to freshwater bodies with up to 15% of the salinity of seawater, and to high levels of pollution and competition. It is distinguished physically by tissue in its swollen leaf stem that enables it to float unattached on the water's surface, aiding its spread. It is also able to root during dry periods. But the rapid growth of the water hyacinth is primarily due to its explosive reproduction by vegetative propagation, or cloning. As the plant grows, it separates into parts that can replicate the whole plant. Estimates of its growth rate describe the area covered by a plant and its offshoots doubling in anywhere from six to twelve days.

Many methods are used to control or make use of this weed. Water hyacinth is used as feed, turned into paper and other products, and collected as a source of fuel. It is even being intentionally grown in some areas for possible use as a pollution filter and wastewater treatment. Systematic harvesting of water hyacinths, after they have absorbed pollutants including nitrates and phosphates, pesticides, and heavy metals, can leave behind cleaner water. Disney World began recycling water in this way in 1980, and San Diego currently processes millions of gallons of sewage daily by this process.

Methods of controlling water hyacinth in agricultural and fishing communities include physical removal of the plants, application of herbicides, and biological control. Many different insects, fungi, and bacteria from the plant's native habitat have been tested as control agents with some success. Biological control methods also have great potential, because the genetic uniformity of the plants due to cloning makes them more susceptible to a population crash.

WATER LAWS

See Clean Water Act; Pollution, Water.

WATER POWER

See Hydropower.

WATERSHEDS

In American usage, a watershed is a unit of the landscape where all water drains toward some common exit point, that is, a drainage area. In European usage, these units are referred to as "catchments." In one sense the term *watershed* is a misnomer because a watershed is commonly defined as the point where water can drain by gravity in more than one direction from that point, or the ridge dividing two drainage areas. Nevertheless, the watershed concept has become ingrained in American usage and is an important theoretical concept for constructing biogeochemical mass balances and for management considerations. Thus the movement of water and what it transports are key ingredients in the determination of a watershed unit, and are what make the approach valuable to resource managers and decision makers.

The watershed concept has been used successfully in evaluating quantitatively the inputs and outputs of water, toxic chemicals, and nutrient chemicals to large drainage basins. For example, the Hubbard Brook Ecosystem Study in the White Mountains of New Hampshire pioneered the use of the watershed-ecosystem approach in evaluating the effects of forestry practices (for example, clearcutting) and atmospheric pollutant loading (for example, acid rain) on landscapes. Such quantitative data are important to decision makers dealing with the environmental problems related to forestry practices and air pollution.

Precipitation falling on watersheds in hilly or mountainous terrain can follow numerous pathways downslope. For example, water can move as overland flow, as shallow subsurface flow, or as groundwater flow. The route that water follows within a watershed-ecosystem can dramatically alter the quantity, quality, location, and timing of fluvial outputs from the drainage area. For example, the residence time for water in groundwater aquifers frequently is long. It may take tens to thousands of years for groundwater to reach discharge points in deep aquifers. Thus, if such an aquifer were to become polluted, it would take a very long time to become decontaminated following elimination of pollutant inputs.

In systems such as forests where infiltration rate of water into soil normally exceeds the rate of precipitation input, overland flow is rare. In watersheds where the biological canopy is sparse or nonexistent, however, overland flow is common during intense storms and can result in high rates of erosion. Such events commonly are referred to as "gully washers" for good reason. Living and dead organic matter play a major role in reducing erosion in a vegetated area by regulating the amount, timing, and erosive effect of flowing water within the watershed-ecosystem. For example, a vegetation canopy and leaf litter on the soil are effective at reducing the erosive impact of falling raindrops, and organic debris dams in stream channels also reduce the export of particulate matter from watersheds.

Surface and shallow subsurface flow of water can be measured quantitatively and relatively easily by means of a gauging weir placed at the mouth of a stream draining a watershed. Deep-water flows are much more difficult to estimate, and usually are determined from differences in hydrologic head and hydrologic conductivity of the porous media in the aquifer.

A critical requirement for the use of the watershed approach is the establishment of boundaries. Boundaries must be established when quantitative, mass balances are desired. Ideally, lateral boundaries of the watershed should be coincident with phreatic (groundwater) divides. For some systems the watershed boundaries may be apparent, for example, a discrete drainage valley. In most landscape units, however, the actual groundwater drainage may be more complicated than would be apparent from the surface topography. In some situations the boundaries may need to be determined at the convenience of the investigator. The usefulness and limitations of the watershed approach are dependent upon the questions being addressed and the actual knowledge about surface and subsurface water movement.

Watershed-ecosystems have linkages both to adjacent and distant watersheds by a variety of inputs and outputs. These inputs and outputs are transported by meteorologic, hydrologic, or biologic vectors. The quantitative measures of these inputs and outputs provide the data for constructing mass balances or budgets of water and other materials.

The United States can be divided into large watersheds or hydrologic regions. The inputs and outputs from individual watersheds within these areas represent their critical linkages with regional as well as global cycles of water and other materials. Entire watershed-ecosystem manipulations such as those done at Hubbard Brook have led to new insights about the linkages between biotic and abiotic factors within landscapes. For example, destruction of forest vegetation, such as would be done in timber harvest, can lead to large increases in losses of drainage water and dissolved nutrients such as nitrate.

GENE E. LIKENS

For Further Reading: F. Herbert Bormann and Gene E. Likens, "Nutrient Cycling," in *Science* (1967); F. Herbert Bormann and Gene E. Likens, *Pattern and Process in a Forested Ecosystem* (1979); Gene E. Likens, *The Ecosystem Approach: Its Use and Abuse* (1992).
See also Aquifer; Forests; Groundwater; Rivers and Streams; Water Table; Wetlands.

WATER TABLE

The water table is the upper water level of a body of groundwater; it is the level below which the ground is completely saturated with water. The water table varies according to the surface topography and subsurface structure. It also fluctuates with seasons and over time, according to precipitation and infiltration rates as well as withdrawal rates. It is normally highest in late winter and early spring. During the summer months, the water table is lower because of evaporation and the intake of water by vegetation. For the sustainability of groundwater resources, water withdrawal must be balanced with recharge. Most recharge occurs naturally from rainfall, but artificial means such as water spreading are sometimes used and can be very important.

Measurement of the water table normally reveals that it slopes toward the nearest stream. Water below the water table moves or percolates from the high points of the water table toward areas where the water table is low. The water moves from high-pressure areas to low-pressure areas. Gravity provides the energy that causes the groundwater below the water table to flow. At this underground level, water is at atmospheric pressure; above the water table, water is at less than atmospheric pressure.

Water wells are the result of penetrating or intersecting the water table. Removing water from a well results in a depression of the immediately surrounding water table. If a well is dug in an area where there is no water table, however, the result is a dry well. A well dug in soil above the water table that is composed of clay and is therefore less permeable would also result in a dry well.

The water table may be used as an indication of the sensitivity or vulnerability of the groundwater to pollution. Areas of land with a higher water table are more vulnerable. At these locations, there is less distance for polluting substances to travel prior to making contact with the groundwater. Hence the ability of the soil to act as a form of cleanser or filter is reduced.

WEATHER MODIFICATION

See Climate Change.

WELLS

A well is a vertical opening made to extract water, oil, brine, or other fluid substance from the ground. Wells provide approximately one-quarter of the water supply used in the United States in the 1990s.

Dug wells are excavations up to approximately 10 m (30 ft) deep, used to reach shallow aquifers. They provide only small quantities of water due to seasonal variation in the height of the water table (the upper water level of a body of ground water). In addition, dug wells are susceptible to pollution. Contamination can occur from nearby feedlots, barnyards, improper disposal of hazardous materials, or septic systems. In South Dakota, a state survey during the 1980s found that over 39% of the state's dug or bored wells were unsafe because of high nitrate levels. Dug wells are still used in some domestic and agricultural areas of the world but are generally no longer recommended for supplying water to the public.

Drilled wells are usually more than 30 m (100 ft) deep and are most commonly used for public water supplies; shallow drilled wells can have the same public health hazards as dug wells. Drilled wells can reach extensive aquifers that tend to yield more water than shallower wells. Deeper drilled wells are more costly than dug or shallow drilled wells and can present a problem in low-income areas.

The various steps required to build a well are well design, which takes into consideration the yield expected from the well; well logging, which involves taking measurements and recording data on the character of the different layers of material unearthed by the drilling and on the quality of water in the permeable zones; well construction, of which there are several methods depending on the type of well needed; and well acceptance and efficiency tests, which are conducted after the well is completed and give information on the yield and efficiency of the well.

The portion of the well above the aquifer is lined with concrete, stone, or a steel casing. A slotted metal screen is attached to the bottom of the casing to allow water, but not debris from the aquifer, to enter the well. Water pumps, which are usually designed to fit inside the well casing, are used in all wells except flowing artesian wells or wells in areas where the water level is high enough so that a suction lift may be used. An artesian system is an aquifer in which the water is confined above and below by impermeable beds. Many artesian systems are under pressure high enough so that water flows naturally from the surface, as in a spring. Flowing artesian wells were important in the early days of ground water development because there was no need for pumping. Because many artesian systems today are overmined and lack sufficient pressure for water to flow naturally, however, pumps must be used to raise water to the surface. Well pumps are placed far below the water level so that they will remain submerged even after pumping begins and the water level drops.

WETLANDS

Wetlands are lands transitional between terrestrial and aquatic systems, where the water table is usually at or near the surface or the land is covered by shallow water. Wetlands must have one or more of the following attributes: (1) at least periodically, the land supports predominantly hydrophytes ("water-loving" vegetation); (2) the substrate (underlying layer) is predominantly undrained hydric (very moist) soil; and (3) the substrate is nonsoil and is saturated with water or covered by shallow water at some time during the growing season of each year.

For years the U.S. Fish and Wildlife Service, an agency of the Interior Department, has been concerned with the classification, inventory, and status of America's wetlands. Alarmed by dramatic reductions in the acreage of wetlands in the 1950s, the service conducted the first nationwide wetlands inventory for purposes other than agriculture conversion. Published in 1956, *Wetlands of the United States*, more commonly known as Circular 39, was a guide to wetlands identification and provided the first benchmark for assessing wetlands losses in the United States. In Circular 39, wetlands are described as "lowlands covered with shallow and sometimes temporary or intermittent waters," including "marshes, swamps, bogs, wet meadows, potholes, sloughs, and river over-flow lands," as well as "shallow lakes and ponds, usually with emergent vegetation as a conspicuous feature."

Flood Conveyance

Wetlands are often referred to as natural sponges. They can slow and retain large amounts of water and, in some instances, absorb flood water and release it slowly. This ability gives some systems extraordinary value as sites for temporary water storage and can be critical in altering flood flows. At times of peak runoff, rivers and streams often overflow their banks into adjacent floodplains. These floodplains retain this overflow and reduce its rate of flow. The result is that peak flows of flood water are reduced and flooding becomes less damaging.

There are numerous case studies demonstrating the important flood conveyance function that wetlands provide. In one instance, the U.S. Army Corps of Engineers elected to preserve wetlands through acquisition rather than construct extensive flood control facilities for a portion of the Charles River near Boston, Massachusetts. The Charles River Natural Valley Storage Project was completed in 1984 and resulted in an annual estimated savings of $17 million in flood damage.

Isolated and other non-riparian wetlands also hold rain and runoff water and contribute to flood control. Wetlands are especially valuable as flood moderators because the water they retain almost never reaches watercourses when they are at flood stage. For example, a study conducted in Wisconsin showed flood flows to be reduced by 80% in basins with wetlands.

Storm Surge Abatement

Wetlands associated with barrier islands, salt marshes, and mangrove swamps act as storm buffers and can weather major storm events without sustaining lasting damage. The low gradient of many shorelines and the capacity of wetland vegetation to absorb and dissipate wave energy combine to counteract storm surges and to prevent shoreline erosion.

Coastal development, which destroys or degrades the wetlands in barrier islands and other critical coastal wetlands, estuarine salt marshes, or lakeshore marshes, is likely to cause costly storm damage. Thus, the assemblage of dunes, marshes, and woody vegetation that make up coastal wetlands are much more important for their natural values than for real estate.

Water Quality—Sediment Control and Nonpoint Pollution

One of the most important values of wetlands is their ability to help maintain and improve the water quality of rivers, lakes, and other bodies of water. They im-prove water quality and control nonpoint pollution in a number of ways that include removing and retaining nutrients, processing chemical and organic wastes, and reducing sediment loads to receiving waters. Since most contaminants adhere to sediment particles, and water entering wetlands is slowed by aquatic vegetation, sediments settle out and wastes are deposited.

The majority of contaminants are not biologically active and are immobilized. In addition, the high rate of biological activity associated with wetlands can result in many wastes being absorbed and incorporated into the plant tissue or given off as a gas. Compounds containing nitrogen are broken down into products usable by plants, and bacteria convert much of the nitrogen gas that escapes into the atmosphere. As these plants die, decompose, and are replaced by new ones, what were once pollutants become part of the detritus that will eventually form the underlying sediment or be exported from the system at the end of the growing season.

Groundwater Recharge and Discharge

In some instances, wetlands play an important part in replenishing or "recharging" groundwater supplies. Surface water bodies connected to groundwater systems can recharge these systems as their waters migrate and percolate into the surrounding aquifer. These wetland recharge sites may serve an important role in maintaining groundwater levels at the local or regional level. For example, Lawrence Swamp, a 2,700-acre wetland in Massachusetts, recharges the underlying shallow aquifer at a rate of eight million gallons per day. This wetland replenishes an area of 16 square miles and provides much of the water supply for the nearby town of Amherst.

Wetlands having value as recharge areas include some prairie potholes, glaciated wetlands of the Northeastern and Midwestern U.S., and southern cypress swamps. These occur where there is an elevated water table and they may contribute to the adjoining, shallow aquifers. Seasonal wetlands in the prairie pothole region of the U.S. and Canada are important to the maintenance of high water tables, which in turn provide water for livestock during droughts and may be vitally important to the long-term water balance of the prairies by providing significant recharge to soil moisture. Not only does enhanced soil moisture recharged by prairie wetlands improve crop production, but emergent wetland plants found in these areas can provide abundant forage for livestock.

Wetlands as Habitats

Wetlands are critical habitats for a variety of plants and animals; even wetlands smaller than one acre support an abundance of life forms. Ducks, geese, and swans are some of the more prominent wildlife species to make use of wetlands, and are important economically. North American waterfowl populations have reached record low levels in recent years, and the U.S. wetlands inventory is also its lowest in recorded history.

Besides waterfowl, a large number of threatened and endangered species rely on wetlands. As of 1991, 43% of the 595 plant and animal species designated as threatened or endangered depend directly or indirectly on wetlands to complete their life cycle successfully. In addition, as many as 700 wetland-dependent or related plants need federal protection.

About 5,000 species of plants, 190 species of amphibians, and 270 species of birds occur in America's wetlands alone. According to the Nature Conservancy, one out of three North American fishes and two out of three of the continent's crayfish are rare or imperiled. Mussels appear even more threatened: One in every ten North American freshwater mussel species has become extinct in this century and 73% of the remaining species are now rare or imperiled. The primary reason for the decline of these barometers of aquatic health appears to be habitat loss and degradation.

Wetlands are—literally—the cradle of the world's seafood industry. Fish and shellfish need estuaries for spawning and nursery grounds, migration, and food production. The annual economic value of estuarine habitats to the U.S. economy is about $14 billion. In the late 1980s commercial landings of estuarine-dependent species alone contributed $5 billion to $6 billion annually. Despite these figures, over half of America's fishery-supporting wetland habitats have been lost, costing the country's fisheries $208 million each year during the mid-1980s.

The percentage of wetland-dependent fish species varies, depending on the region of the country. For example, 98% of all marine species in the Gulf of Mexico spend part of their lives in wetlands and marshes. In the southeastern U.S. this percentage is slightly lower, 94%. Historic commercial fish and shellfish harvests are in steep decline and the primary factor appears to be the loss and degradation of wetlands.

Recreation

Recreational opportunity is another important contribution that wetlands make to the economy. The U.S. Fish and Wildlife Service estimates that, in 1980, 1.9 million hunters of migratory waterfowl spent in excess of $3 million on the sport. Recreational fishing is also an important wetland-dependent activity that generates millions of dollars annually. Finally, the amount of money spent by Americans on other wildlife-related activities associated with wetlands involves billions of dollars each year. For example, the federal government estimates that 55 million people spent almost $10 billion in 1980 observing and photographing waterfowl and other wetland-dependent species of birds, an annual expenditure of almost $200 per person. Wetlands are also important for aesthetic retreats and places of diversity for nature study and so are central to the enjoyment of millions of citizens.

The world's wetlands are being changed. Water that once remained in low spots and provided habitat for wildlife now collects in ditches and tile lines and is rushed to the nearest stream. Millions of acres that once grew cattails, wild rice, and pondweeds now support wheat, corn, houses, factories, airports, and roads. Many millions of acres of good waterfowl habitat have been destroyed and the loss continues.

Despite the recognition of the important functions and values wetlands provide, and despite comparatively strong domestic protection policies, the loss continues at an unacceptably high rate—almost 300,000 acres per year in the U.S. The country has lost 50% of its wetlands, and they now comprise only 5% of its surface area.

JAY D. HAIR

For Further Reading: J. Scott Feierabend and John M. Zelazny, *Status Report on Our Nation's Wetlands* (1987); J. Scott Feierabend, *Endangered Species, Endangered Wetlands: Life on the Edge* (1992); William J. Mitsch and James G. Gosselink, *Wetlands* (1986).

See also Aquifer; Coastlines and Artificial Structures; Floods; Groundwater; Law, Wetlands; Pollution, Water; Waterfront Development; Water Table.

WHALING INDUSTRY

Origins and Early Development

The first humans to eat whale flesh and burn whale oil probably obtained their samples by scavenging, as the bodies of whales that die from natural causes sometimes become stranded on shore or float on the surface of the water due to gaseous bloating. Thus archaeological or ethnographic evidence of people using whale products does not necessarily demonstrate that the whales were actively hunted.

The origins of whaling are obscure, but technologies for chasing, killing, and retrieving whales seem to have been invented independently in a number of discrete localities at different times. Eskimoid peoples probably engaged in communal hunts for the arctic bowhead, *Balaena mysticetus*, white whale or beluga, *Delphinapterus leucas*, and narwhal, *Monodon monoceros*, from as early as the first millennium A.D., using primitive harpoons thrown from skin-covered boats. Settlement sites were often chosen for their proximity to constricted migratory routes or seasonal nearshore aggregations of whales.

Similarly the Aleuts and Indians living along much of the west coast of North America, from Washington State northward, inhabited strategic sites on beaches or headlands, from which they could launch skin kayaks or wooden dugout canoes to intercept migrating gray whales, *Eschrichtius robustus*, and humpback whales, *Megaptera novaeangliae*. At Kodiak and the Aleutian Islands whalers coated their lance tips with aconite, a poison substance derived from the monkshood plant. After darting a whale, they did not pursue it. Rather, they let the animal die from poisoning, then hoped its carcass would drift near enough to their village to be discovered and retrieved.

The Japanese may have harpooned whales from small boats prior to the 10th century. In 1675 they invented a unique method of capturing whales with nets. A whale sighted from shore was surrounded by boats and driven toward specially constructed straw (later, hemp) nets. The entangled animal was harpooned repeatedly until it became weak enough for a sailor to climb onto its body and attach tow ropes.

The Basques of western Europe had developed their own whaling industry by the 11th century. Although they were not the only inventors of whaling techniques, the Basques were responsible for the geographic spread of open-boat, hand-harpoon whaling across the rim of the North Atlantic. The Spanish Basques reached Newfoundland by the 1530s and initiated an intensive cod and whale fishery in the Strait of Belle Isle. Large galleons sailed to the New World annually, loaded with men and provisions. They anchored in harbors and erected processing facilities on shore. Whales caught in nearby waters were delivered to these stations, where blubber and baleen ("whalebone") were removed from the carcasses. Casks of cod and whale oil and bundles of baleen were then loaded onto the galleons for shipment to Europe each fall.

Early whalers were limited to catching whales that could be sighted and followed within a few miles of shore or an ice edge. Catchability was determined less by the animal's size than by its swimming speed, aggressiveness, diving abilities, and buoyancy once dead. Right whales, *Eubalaena* spp., bowhead whales, *Balaena mysticetus*, humpback whales, and gray whales bore the brunt of whaling initially, and they were joined by the sperm whale, *Physeter catodon*, only after the whalers had learned to venture into deep waters far offshore.

Expansion and Decline of Pre-Modern Whale Fisheries

The northern whale fishery, centered on the bowhead,

was a major theater for territorial and mercantile rivalry among European nations for much of three centuries, beginning in 1611. The Dutch dominated until the late 18th century and were thereafter replaced in importance by the British. German, French, Basque, Danish-Norwegian, and American whalers also participated at one time or another in the "Eastern Arctic" (Spitsbergen and Davis Strait) whale fisheries. Bowhead whaling in the "Western Arctic" (Bering Sea northward) was initiated by the Americans in 1848 and continued until about 1915. The stocks of bowheads were sequentially depleted, until by the early 20th century the hunt for them was no longer economical.

New England whalers, having hunted right whales from shore since the early colonial period, inaugurated the southern high-seas whale fishery in the 18th century. A key innovation in the 1760s was the on-board tryworks for processing blubber. It allowed vessels to remain at sea for extended periods and to venture far from home port. Sperm whales were the major targets, but right whales, humpbacks, and North Pacific gray whales were also taken. The oil of sperm whales was especially valued for making candles. At its peak between 1830 and 1850, this American-led, worldwide hunt produced more than 140,000 barrels of sperm oil annually. American vessels undertook some 14,000 voyages and probably killed 300,000 sperm whales between 1804 and 1866. By 1900, few stocks of sperm whales and right whales remained that had not been intensively exploited. All populations of right whales and probably some of sperm whales had been severely depleted.

The pre-modern whaling industry was an important agent of change in aboriginal communities. It caused local or regional game depletion, facilitated trade, and created dependency on external markets. It was also a vehicle for exploration and discovery of new lands and travel routes, and it stimulated technological innovations.

Development and Expansion of Modern Whaling

Modern whaling began in the 1860s as a result of two Norwegian inventions: the deck-mounted harpoon cannon (which shot harpoons with exploding grenade heads) and the engine-driven catcher vessel. These made it possible to add the powerful, fast-swimming balaenopterid whales, including the blue, *Balaenoptera musculus*, and fin, *Balaenoptera physalus*, to the mix of species that could be caught. Catcher boats soon operated from processing plants on shore throughout the northern North Atlantic, first in Norway and later Iceland, Spitsbergen, Newfoundland, the Faroe Islands, and the British Isles. In the 1890s and early

1900s, this "radiation" of Norwegian enterprise reached literally around the world, with whaling stations established on the coasts of Africa, North and South America, the subantarctic islands, and Japan. In 1903 the first floating factory, consisting of a large mother-ship (to process and transport whale products) and one or more fast catcher boats, operated at Spitsbergen. Within only a few years the same concept was applied in many areas, including the Antarctic. In 1926 the introduction of the stern slipway again revolutionized the industry. It was no longer necessary to tow whales ashore or flense them alongside the vessel; they could now be winched on board the mother-ship, where the carcasses were rendered by crews of specialists.

The demand for whale products remained strong throughout the 20th century. Even though previous uses of many of them had become obsolete—spring steel and plastic replaced baleen, and fossil fuels and electricity supplanted whale oil for lighting—new uses arose. Of particular significance was the development of procedures for hydrogenating whale oil, which made it suitable for use as edible oleomargarine. Demand for this product, above all others, drove the frenzied expansion of pelagic whaling in the Antarctic, which eventually involved most western European nations, Japan, the Soviet Union, and the United States. Sperm oil, which is chemically unlike whale oil (obtained only from baleen whales), came to have strategic value as a heat-resistant lubricant for missiles and rockets. In Norway and particularly Japan, whale meat has always provided an additional incentive for catching whales.

Conservation and Protection

The International Convention for the Regulation of Whaling in 1946 marked the start of effective regulation of the industry. Earlier attempts by individual countries or regional blocs had failed to prevent the stepwise collapse of whale stocks. The International Whaling Commission (IWC), a body created by the 1946 convention, was itself flawed in that it lacked enforcement powers and had an internal structure giving more weight to political and economic considerations than to scientific advice. It did, from the start, provide important protection for right, bowhead, and gray whales and, from 1965, for blue and humpback whales. But this protection came only after all of these species were approaching biological extinction.

In the late 1960s and early 1970s the IWC finally responded to the pleas of conservationists by instituting stock-by-stock management, monitoring compliance with conservation measures through an international observer scheme, and strengthening the role of its sci-

entific committee. At its 1982 annual meeting the IWC agreed to ban commercial whaling, but by then most stocks of the commercially valuable species had already been given full protection due to their depleted status. Also, only a handful of countries, most notably Japan, Iceland, and Norway, showed signs of wanting to continue whaling past 1985–86, when the ban was to take effect. Japan continued through the early 1990s to take several hundred minke whales, *Balaenoptera acutorostrata*, in the Antarctic each year as part of a national scientific research program. Iceland, dissatisfied with the commission's reluctance to allow a resumption of whaling on stocks in the North Atlantic not thought to be endangered, withdrew from the IWC in 1992. Norway announced its intention to resume whaling for minke whales in the North Atlantic in 1993, with or without the IWC's approval. The deepening schism between whaling and non-whaling countries led to the establishment in 1992 of a North Atlantic Marine Mammal Commission, a forum for those countries wishing to continue or resume the "rational" exploitation of whales. Whether the formation of similar regional management bodies will erode the effectiveness of and ultimately destroy the IWC remains to be seen.

Whales have become powerful symbols of humanity's past misuse of wild living resources. Their physical grace and beauty, social behavior, acoustic abilities, "friendliness" toward people, and presumed intelligence have fueled strong sentiments against the killing of whales and in favor of their complete, permanent protection. Controversy has thus become centered not only on whether or how to conserve whale stocks for sustained exploitation, but also on the morality of whaling as a human endeavor. Those who view whales as renewable resources are increasingly at odds with those who regard them as special animals that rival humans in intelligence and sensitivity. Industrial whaling as it developed for nearly a millennium, from the Basque rowboats of the 11th century to the Japanese floating factories of the early 1980s, no longer exists.

RANDALL R. REEVES

For Further Reading: Michael F. Tillman and Gregory P. Donovan, eds., *Historical Whaling Records including the Proceedings of the International Workshop on Historical Whaling Records, Sharon, Massachusetts, September 12–16, 1977* (1983); Johann N. Tønnessen and Arne O. Johnsen, *The History of Modern Whaling*, translated from the Norwegian by R. I. Christophersen (1982); Robert L. Webb, *On the Northwest. Commercial Whaling in the Pacific Northwest 1790–1967* (1988).

See also Ecology, Marine; Fishing Industry, U.S. Commercial; Law, International Environmental; Oceanography.

WHO

See United Nations System.

THE WILDERNESS EXPERIENCE

Ecologists define as wilderness any environment that has not been disturbed by human activities, but in the popular mind, the word often has a deep resonance with a feeling of alienation and insecurity. Many people in industrialized societies, for example, use the word *wilderness* to denote huge, anonymous, urban agglomerations that appear to them hostile.

The experience of nature in a native prairie, a desert, a primeval forest, or high mountains not drowned with tourists is qualitatively different from what it is in a well-tended meadow, a wheat field, an olive grove, or even in the high Alps. Humanized environments give people confidence because nature has been reduced to the human scale, but the wilderness in whatever form almost compels them to measure themselves against the cosmos. It makes them realize how insignificant they are as biological creatures and invites them to escape from daily life into the realms of eternity and infinity.

From the beginning of recorded history and even in prehistoric legends, the word *wilderness* has been used to denote barren deserts, deep forests, high mountains, and other inaccessible or harsh environments not suited to human beings, cursed by God, and commonly occupied by foul creatures. Such forms of wilderness evoked a sense of fear for a good biological reason. They are profoundly different from the environmental conditions under which the human species acquired its biological and psychological characteristics during the Stone Age.

The word *wilderness* occurs approximately three hundred times in the Bible, and all its meanings are derogatory. In both the Old and New Testament, the word usually refers to parched lands with extremely low rainfall. These deserts were then as now unsuited to human life, and they were regarded as the abodes of devils and demons. After Jesus was baptized in the Jordan River, he was "led up by the Spirit into the wilderness to be tempted by the devil." The holy men of the Old Testament or of the early Christian era moved into the wilderness when they wanted to find a sanctuary from the sinful world of their times. Thus, while some great events of the Judeo-Christian tradition occurred in the desert, this environment was at best suitable for spiritual catharsis.

Humankind has always struggled against environments to which it could not readily adapt; in partic-

ular, it has shunned the wilderness or has destroyed much of it all over the world. Contrary to what is often stated, this is just as true of Oriental as of Occidental people.

The admiration of wild landscapes expressed in Oriental arts and literature probably reflects not so much the desire to live in them as the intellectual use of them for religious or poetic inspiration. The ancient Chinese, especially the Taoists, tried to recognize in nature the unity and rhythm that they believed to pervade the universe. In Japan the followers of Shinto deified mountains, forests, storms, and torrents and thus professed a religious veneration for these natural phenomena. Such cultural attitudes were celebrated in Chinese and Japanese landscape paintings more than a thousand years before they penetrated Western art, but this does not prove that Oriental people really identified with the wilderness. Paintings of Chinese scholars wandering thoughtfully up a lonely mountain path or meditating in a hut under the rain suggest an intellectual mood rather than life in the wilderness. The Chinese master Kuo Hsi wrote in the eleventh century that the purpose of landscape painting was to use art for making available the qualities of haze, mist, and the haunting spirits of the mountains to human beings who had little if any opportunity to experience these delights of nature. Much of the Chinese land had been grossly deforested and eroded thousands of years before, and the Taoist movement may have been generated in part by this degradation of nature and as a protest against the artificialities of Chinese social life.

In the Christian world, also, there has been a continuous succession of holy men, poets, painters, and scholars who did not live in the wilderness but praised it for its beauty and its ability to inspire noble thoughts or actions. St. Francis of Assisi was not alone among medieval Christians in admiring and loving nature. After Jean Jacques Rousseau, the many Romantic writers, painters, and naturalists of Europe became more than a match for the Chinese poets and scholars of the Sung period. But like the Chinese, they wrote of the wilderness in the comfort of their civilized homes, as intellectuals who preached rather than practiced the nature religion.

In Europe the shift from fear to admiration of the wilderness gained momentum in the eighteenth century. The shift was not brought about by a biological change in human nature but was the consequence of a new social and cultural environment. Fear of the wilderness probably began to decrease as soon as dependable roads gave confidence that safe and comfortable quarters could be reached in case of necessity. There were numerous good roads in western Europe by the time Jean Jacques Rousseau roamed through the Alps and Wordsworth through the Lake District. In the New World access was fairly easy even to the High Sierras when John Muir reached them from San Francisco.

Appreciation of the wilderness began not among country folk who had to make a living in it, but among city dwellers who eventually came to realize that human life had been impoverished by its divorce from nature. People of culture generally wanted to experience the wilderness not for its own sake, but as a form of emotional and intellectual enrichment. In Europe Petrarch is the first person credited with having deliberately searched both mountain and primeval forest for the sheer pleasure of the experience. His account of his ascent of Mount Ventoux in 1336 is the first known written statement of the beauty of the Alps under the snow, but he reproached himself for letting the beauty of the landscape divert his mind from more important pursuits. By the early Romantic period, however, the wilderness came to be seen not only as the place in which to escape from an artificial and corrupt society but also as a place to experience the mysterious and wondrous qualities of nature. The wilderness experience became a fashionable topic of conversation as well as of literature and painting and thus rapidly changed the attitudes of the general public toward nature.

People who express love for the wilderness do not necessarily practice what they preach. In 1871 Ralph Waldo Emerson refused to camp under primitive conditions when he visited John Muir in the Sierras and elected instead to spend the night in a hotel. When Thoreau delivered the lecture with the famous sentence "In Wildness is the Preservation of the World," he was living in Concord, Massachusetts, a very civilized township where the wilderness had been completely tamed. He loved the out-of-doors, but knew little of the real wilderness. His cabin by Walden Pond was only two miles from Concord; woodchucks were the wildest creatures he encountered on his way from the pond to town, where he often went for dinner. In fact, Thoreau acknowledged some disenchantment when he experienced nature in a state approaching real wilderness during his travels in the Maine woods.

Increasingly during recent years, interest in the wilderness and the desire to preserve as much of it as possible have been generated by an understanding of its ecological importance. It has been shown, for example, that the wilderness accounts for some 90% of the energy trapped from the sun by photosynthesis and therefore plays a crucial role in the global energy system. The wilderness, furthermore, is the habitat of countless species of animals, plants, and microbes; destroying it consequently decreases the Earth's bio-

logical diversity. This in turn renders the ecosystem less resistant to climatic and other catastrophes, and less able to support the various animal and plant species. Undisturbed natural environments, including forests, prairies, wetlands, marshes, and even deserts, are the best insurance against the danger inherent in the instability of the simplified ecosystem created by modern agriculture. From a purely anthropocentric point of view, as much wilderness as possible must be saved because it constitutes a depository of genetic types from which people can draw to modify and improve their domesticated animals and plants.

Admiration of the wilderness can thus take different forms. It can lead to direct and prolonged experience of the natural world as in the case of John Muir. For many more people, it derives from a desire to escape from the trials and artificialities of social life, or to find a place where one can engage in a process of self-discovery. Experiencing the wilderness includes not only the love of its spectacles, its sounds, and its smells, but also an intellectual concern for the diversity of its ecological niches.

The human species has now spread to practically all parts of the Earth. In temperate latitudes, although not in tropical or polar regions, they have enslaved much of nature. And it is probably for this reason that they are beginning to worship the wilderness. After having for so long regarded the primeval forest as an abode of evil spirits, they have come to marvel at its eerie light and to realize that the mood of wonder it evokes cannot be duplicated in a garden, an orchard, or a park. The emotional response to the thunderous silence of deep canyons, to the frozen solitude of high mountains, and to the blinding luminosity of the desert is the expression of aspects of humans' fundamental being that are still in resonance with cosmic forces. The experience of wilderness, even though indirect and transient, helps people to be aware of the cosmos from which we emerged, and to maintain some measure of harmonious relation to the rest of creation.

RENÉ DUBOS
(EDITED BY WILLIAM R. EBLEN)

For Further Reading: Michael Frome, *Battle for the Wilderness* (1974); Mary L. Grossman, Shelley Grossman, and John Hamlet, *Our Vanishing Wilderness* (1969); Roderick Nash, *Wilderness and the American Mind* (1967).

WILDFLOWERS

A wildflower is any flowering plant growing in a natural, uncultivated state. The term *wildflower* is used subjectively to distinguish wildflowers from domesticated flowers. However, wildflowers are actually the source of all cultivated garden flowers. For example, there is scarcely a region in the Northern Hemisphere where flowers of the genus *Lilium*—with its nearly 100 wild species—are not present. This plant is so striking that it was brought into gardens wherever it grew wild.

The distinction between the terms *weeds* and *wildflowers* depends upon the purpose or use of the flower with respect to humans. For example, sunflowers are considered weeds when growing in cultivated fields, wildflowers in uncultivated valleys, and crops when grown for their seeds.

Disturbance of the native flora by humans began in prehistoric times and still continues. Native wildflowers have been preserved in national, state (or provincial), and local parks and monuments, particularly in the U.S. and Canada, but vast areas of wildflowers have been lost to cropland, grazing land, or settlements. Disturbance also includes the scattering across the globe of wildflowers once growing on only one continent. Plant "immigrants" become "stowaways" by means of seeds attached to cargoes of merchandise of all sorts. Once within a country the "alien" plants speedily colonize waste places. The common dandelion reached America from Europe shortly after the first colonists and is found in fields, lawns, vacant lots, and roadsides in nearly every civilized country. The bright flowers characteristic of the Hawaiian Islands are nearly all native to other parts of the tropics and subtropics. Other wildflowers are "escapees" from cultivation. Originally imported for use in brewing beer, the common hop that grows wild in thickets and on river banks in the U.S. and Canada is an example.

Many wildflowers grouped in the same family are not alike in size, leaf shape, or flower color but in the number, shape, and arrangement of the different flower parts. Four of the largest families in the world are the rose family (*Rosaceae*), mustard family (*Cruciferae*), pink family (*Caryophyllaceae*), and the composite family (*Compositae*). The composite family is the largest, with more than 20,000 species.

The National Wildflower Research Center, a nonprofit organization in Austin, Texas, was established in 1982 to stimulate and carry out research on the conservation and cultivation of wildflowers and native plants, in cooperation with universities, botanic gardens, aboreta, and other institutions throughout North America. Through these efforts the Center aims to restore and beautify the landscape by reestablishing native plants.

Wildflowers are important for civilization's future, because they constitute the only vast reservoir of genetic material not largely resident in our economically useful plants. This genetic reservoir should be preserved for future use by transferring wildflower genes to domestic plants, thereby creating more valuable crops. Crops produced in this manner may be faster growing, disease—or pesticide—resistant, or sources of new biochemicals, drugs, and other products.

WILDLIFE REFUGES

See National Wildlife Refuge System.

WOLMAN, ABEL

(1892–1989), American civil engineer. Abel Wolman was among the pioneers in expanding the field of sanitary and environmental engineering at the end of the nineteenth and beginning of the twentieth century. His initial work focused on the development of safe drinking water and safe disposal of wastewater. At the turn of the century, drinking water supplies were poor in most places, even in many large cities, and the development of sewerage systems was only emerging as a major environmental effort. Abel Wolman and a chemist, Lynn Enslow, both employees of the Department of Health in the State of Maryland, developed the first controlled system for chlorination of drinking water, transforming the treatment of drinking water supplies throughout the world. Originally developed in the laboratory, then tested in a small municipal system in Maryland, it provided the basis for accurate dosage of chlorine and for monitoring residual chlorine to assure adequate treatment and control of pathogens.

With the adoption of filtration and chlorination, typhoid fever rates in the United States dropped from roughly 36 per 100,000 in 1900 to 3 per 100,000 in 1936. Similar declines were experienced in Europe. As president of the American Water Works Association and of the American Public Health Association, as well as editor of several journals, Wolman was a leader in developing standards of water and wastewater treatment in the U.S. and abroad, and in stimulating efforts to improve the quality of urban and rural environments.

In the 1930s Wolman was chairman of the Water Resources Committee of the National Resources Planning Board. The federal government embarked on major studies of major river basins covering the entire country. Among the first of their kind, they provided a basis for planning potential development on a river basin scale. The Planning Board became the focus of efforts to evaluate the social and economic values, benefits, and costs associated with the development of water resources. These efforts encompassed what later became known as multi-purpose developments involving flood control, water supply, power generation, pollution control, navigation, and recreation. Wolman participated in planning at the federal, state, and local level as the idea of city planning also took hold during this period. A leader in what later became the broad field of environmental engineering, Wolman was instrumental in the development of urban planning and in relating planning to the establishment of environmental standards for land, air, and water.

Following World War II and the development of the atom bomb, the United States embarked on major efforts to develop nuclear energy. Early on it became evident that environmental concerns were only weakly represented in the initial considerations of the potential for nuclear energy. The absence of such concern reflected in part the accelerated effort during wartime to develop an atom bomb and the limited number of individuals and professional fields represented in the emerging peacetime effort. Wolman was appointed to the Reactor Safeguard Committee of the newly formed Atomic Energy Commission. In that role, and through his involvement in the field of public health in the United States, he was instrumental in initiating an effort to assure adequate consideration of health and safety in the development of peacetime uses of atomic energy, particularly for electric power generation. Initially the few individuals familiar with atomic energy believed that, equipped with superior knowledge, they alone could be responsible for the development of the new industry, including protection of the health of the public. Wolman believed that health officials at all levels of government should assume this new responsibility as part of their mission, rather than vesting responsibility in a single agency and the fraternity of experts in the field of atomic energy. He and others succeeded in making a broad political constituency aware of the need to protect those working in the industry and the public at large in considering both site location of power plants and disposal of atomic wastes.

The view that Wolman championed with regard to atomic energy rested upon basic tenets regarding establishment of criteria and standards for environmental protection as well as the structure of government needed to assure their proper application. Wolman drew upon his extensive involvement in the development of standards for water supply and wastewater treatment in arguing the case for adequate safeguards of public health in connection with the development of atomic energy. These issues, including the responsibility of government in protecting public health and the environment, involvement of government and private industry in the development of new and changing technologies, and modes of public involvement in decision making, remain central to environmental policy making today and in the future.

While pursuing work in the United States, Wolman also participated in the development of programs to improve water supply and sanitation in the developing world. As a member of the U.S. delegation to the first assembly of the World Health Organization (WHO), he successfully urged that the field of environmental sanitation and health become a part of the newly created organization. His concern for the value of clean water to people throughout the world is reflected not only by his work with the World and Pan American Health Organizations, but also by his involvement in water and sanitation planning in areas as diverse as Calcutta, India; Sao Paulo, Brazil; Buenos Aires, Argentina; and Israel. These efforts resulted in the development of water systems for urban areas throughout the world. They continue in such programs as the International Decade of Water and Sanitation.

A professor of environmental engineering at Johns Hopkins University throughout most of his life, Abel Wolman's students, both within academia and in the world at large, represent his pervasive influence in each of the activities in which he was a pioneer and leader.

M. GORDON WOLMAN

For Further Reading: Brian Balogh, *Chain Reaction* (1992); G. F. White and D. A. Okun, *Abel Wolman, National Academy of Engineering Memorial Tributes* (1992); Gilbert F. White, ed., *Water, Health and Society: Selected Papers by Abel Wolman* (1969).

WOMEN AND THE ENVIRONMENT

Rachel Carson published *Silent Spring* in the 1960s. That was a turning point. In the 1970s Lois Gibbs protested at Love Canal: another watershed. Norway's Prime Minister, Gro Harlem Brundtland, headed up the Commission on Environment and Development in the 1980s; the resulting Brundtland Commission Report, popularly known as *Our Common Future*, was a milestone document. In June 1992 the Earth Summit in Rio de Janeiro placed women on the global agenda and they are now recognized on the world stage as having a crucial role in environmental management and development.

Throughout history a special relationship between women and the environment has existed. It has primarily grown out of the roles of women in society. The perspective that women bring to the environment is similar to the way women have been acculturated historically: to manage what exists. Whether from the North or the South, each culture has its version of so much allowance, so many hectares, so many babies, so much food. Women have learned to manage the reality of available resources.

In many parts of the world where women are the vast majority of local farmers and land users in the production of food, mostly by their own hand, there exists the dual challenge of survival on a daily basis and doing what is healthy for the land itself. Nature suffers routinely as a result of harmful actions for the necessity of everyday existence. All over the world, the environment also suffers for short-term economic gain. Women and the environment are inextricably linked: by their own nurturing, life-giving natures; by the way women and nature are detrimentally dominated by patriarchal societies; and by the way cultures shape gender roles.

Women who have become icons of environmental leadership are joined by legions of anonymous individuals at the local/community level who contribute, sometimes at great personal risk, to the cause of good stewardship of the environment. Dr. Wangari Matthai represents both. As founder of the grassroots Green Belt Movement in Kenya, she has created a successful tree-planting program that has involved thousands of women, children, and old people in the planting and equally critical caretaking of millions of trees. Her movement is exemplary of the new style of non-governmental organizations that emphasizes local efforts to manage local resources, in this case with modest cash payments to guardians of surviving trees.

Environmental management is integral in the process of attaining the vision of more sustainable development. America's Joan Martin Brown, who heads the United Nations Environment Programme office in Washington, D.C., believes that the cooperative and task-oriented approaches women bring to this challenge are significant. Her own ability to provide language that has helped define the issues (e.g. "environmental bankruptcy," "public awareness," "environmental refugees") and to create forums (e.g. the 1992 Global Assembly of Women and the Environment) has empowered many women to take action, or reach yet higher levels of accomplishment in this field.

One post on the road from the Earth Summit is the Women's World Conference in 1995 in Beijing. Ongoing follow-up on women and the environment by non-governmental organizations based on the progress made in Agenda 21 during the Earth Summit is another. These aspects and more are part of the framework that will enable women to fulfill their significant role in the process of sustainable development.

LIANNE SORKIN FISHER

WORK

Physical work is the transfer of energy performed when a force moves something through a distance. Work is done when water vapor is lifted from land into the air in the water cycle, when a toucan flies through a tropical forest, or when a person drives an automobile. The different kinds of work—such as mechanical, chemical, or electrical—all involve motion. If a barbell with a mass of 10 kilograms (approximately 22 pounds) is lifted one meter (approximately 3.25 feet), work has been done and energy is stored work. The potential energy of the barbell has been increased by lifting it. To lift the barbell, a force is applied equal to its mass (10 kg) times the pull of the earth's gravity on it (9.8 meters per second squared) over a distance of one meter, to yield a net amount of work of 98 joules (or 98 kg·m^2/s^2) in metric system units of energy. The 98 joules of energy are stored in the barbell, and if it is dropped, 98 joules of energy would be released. Potential energy is converted into kinetic energy when the barbell falls. Unfortunately, different forms of energy (stored work) are not equally convertible into useful, applied work.

Most forms of useful energy can be regarded as stored work, where potential energy is converted to kinetic energy and then reconverted to potential work. In whatever form, work comes from a source—according to the law of conservation of energy. Hydroelectric power, for example, is not "free" in a physical sense. It is the result of a complex chain of events that begins with the evaporation of water by the sun. The solar energy stored in the water is released, just as the falling barbell releases its stored energy. When a turbine is placed in a stream of falling water, some of the water's potential energy is tapped when it forces the turbine blades to move. The movement is ultimately converted into electricity. Work from stored energy is also performed when water is pumped up into a tank, when the water flows back down, and when a battery is charged.

Processing work consists of converting initial energy into kinetic energy as matter is moved about and rearranged. An automobile engine performs processing work by moving the automobile. A battery performs processing work by starting an engine. Processing work is performed against the resisting forces, such as friction or air resistance. It is also performed against inertial forces—the resistance of mass to deviate from any straight-line motion.

The availability of stored work is an important consideration that cannot be overemphasized. For example, the total amount of solar energy (stored work) that is stored in photosynthesis is a limiting factor for life on earth. In addition, further limits are established by the patterns of flow of this energy through the earth's ecosystems. Today, an increasing amount of that energy is being diverted to the direct support of humans.

WORLD BANK

The World Bank Group is divided into four sections: the International Bank for Reconstruction and Development (IBRD); the International Development Association (IDA); the International Finance Corporation (IFC); and the Multilateral Investment Guarantee Agency (MIGA).

The IBRD, founded in 1945, is owned by 172 member countries. While it finances lending operations primarily by borrowing from the world's capital markets, the bank also uses money from retained earnings and the flow of repayments on loans. IBRD loans are directed at countries considered "more advanced" in terms of economic and social growth. Its loans involve a 5-year grace period, are repayable over 15 years or less, and charge interest based on the relative cost of borrowing.

The IDA was established in 1960 to provide similar assistance to developing nations, but chiefly to those categorized as "poorer" (membership 146 as of October 1992). IDA's borrowers typically have per capita incomes of less than $765. Balance-of-payment terms are not as heavy as those for IBRD loans. Loans are repayable over 35 to 40 years with a grace period of 10 years. The IFC was created in 1956 to assist economic development in developing countries by promoting private-sector growth and mobilizing domestic and foreign capital for such development. In addition, advisory services and technical assistance are provided to businesses and governments on investment-related matters. The MIGA, begun in 1988, offers those who invest in developing nations guarantees against noncommercial risks and advises developing governments on the design and implementation of policies, programs, and other aspects of foreign investments.

Within the IBRD lies the Environment Department, which sets strategies, conducts research and staff training on environmental issues, oversees environmental assessment procedures, and administers the Global Environment Facility (GEF). In an effort to incorporate environmental interests into actual project work, environmental work has expanded into the operations sector of the World Bank. In addition, some technical and country departments have their own environmental divisions that handle both individual and strategic projects.

The GEF was created in November 1990 to provide grants and low-interest loans to developing countries to help them carry out programs of environmental concerns. The program was established by a coalition of national governments, the World Bank, the United Nations Development Programme (UNDP), and the United Nations Environment Programme (UNEP). It addresses four major areas of environmental concern: reducing and eliminating greenhouse gas emissions, preserving biological diversity and maintaining natural habitats, halting the pollution of international waters, and protecting the ozone layer from further depletion.

Funds provided through the GEF are in addition to those allotted for regular development assistance. The World Bank is responsible for investment projects, as well as administering the facility and chairing governmental meetings. UNEP provides scientific and technological guidance, and UNDP coordinates the financing of technical assistance activities.

The World Bank also administers the Global Environment Trust Fund, the financial mechanism of the GEF, on behalf of the three implementing agencies. While the majority of GEF loans and grants go to recipient governments, resources will be available to intermediaries, nongovernmental organizations (NGOs), or private enterprises (with government agreement). A small grants program provides loans to NGOs for smaller-scale environmental projects. Participating GEF governments have called for an equitable mix of developing and donor country governance of the facility.

WORLD CLIMATE RESEARCH PROGRAM

The World Climate Research Program (WCRP) was established in 1980 for the purpose of determining to what degree climate can be predicted and human activity can influence it. It was developed from the Global Atmospheric Research Program (GARP), which began in the early 1960s, and is jointly sponsored by the International Council of Scientific Unions and the World Meteorological Organization. WCRP focuses on the physical and dynamical aspects of the climate system, while the International Geosphere-Biosphere Program (IGBP) emphasizes the biogeochemical aspects. Together, WCRP and IGBP make up the organizational framework for international scientific research on the problems of climate variability and change. They represent one of the most prodigious and comprehensive scientific undertakings in requiring the highest level of instrumentation (in data collection) and large-scale data management and organization. The Joint Scientific Committee promotes the objectives of WCRP by stimulating research in national agencies, institutions, universities, and other multinational organizations either directly or through international working groups.

WCRP is concerned with the interactions and feedbacks that occur within the entire climate system. Climate variation is of key importance because of the enormous impact it has on every aspect of human activity. The main components of the system include the atmosphere, the oceans, the cryosphere (continental ice, mountain glaciers, surface snow cover, and sea ice), and the land surface.

Each of the three main streams in WCRP's research deals with a different time scale of climate predictability. The first stream seeks to understand the processes that would affect atmospheric circulation on time scales of months to seasons, and to establish the physical basis for improved practical long-range climate prediction. The second stream addresses the interannual variability of both the global atmosphere and tropical oceans. Understanding the dynamic nature of the interactions of the oceans and the atmosphere in the tropics will permit prediction of such phenomena as El Niños. WCRP's program on Tropical Ocean and Global Atmosphere (TOGA) is the focus of that research. TOGA has given substantial attention to understanding the nature of long-term fluctuations of monsoons and their relationships with atmospheric circulation, as well as how the tropical ocean influences monsoonal variability and the predictability of that variation. The third stream concentrates on the reaction of the global climate to anthropogenic forcing factors over a period of several decades to a century. The World Ocean Circulation Experiment was implemented as the focus of the third objective in order to provide detailed models of heat transport by world oceanic circulation and its variability from year to year.

At the core of WCRP's research is the Atmospheric Model Intercomparison Project, which seeks to integrate a variety of climate models based on different parameters such as global budgets of heat, momentum, and moisture; evaluation of surface fluxes, land surface processes, and cloud-radiation interaction; simulations of monsoonal circulations; the treatment of the stratosphere; and stratospheric/tropospheric dynamical interaction. WCRP is also extending its research to investigating model performance in the polar regions, given the demonstrated importance of polar regions' influence on global climate.

The ongoing research of WCRP continuously seeks to further our ability in predicting climate variability and its impacts, and to identify feasible strategies to lessen its potential harmful effects. As recommended during the Second World Climate Conference in 1991 (the first was in 1979), this entails plans to develop a Global Climate Observing System (GCOS) which would include space-based and surface-based observational components, as well as data communications and other infrastructure needed to support operational climate forecasting.

WORLD HEALTH ORGANIZATION

See United Nations System.

WRIGHT, FRANK LLOYD

(1867–1959), American architect. Frank Lloyd Wright is known for spectacular edifices such as New York City's Guggenheim Museum and for his pioneering "prairie" and "Usonian" single family houses of the 1900s and 1930s, which redefined the genre while influencing international design in general. Less known but fundamental to his work was an extraordinary sensitivity to the environment, shaping the very look and operation of his buildings.

In Riverside, Illinois, the 1905–06 F. F. Tomek residence exemplifies specific ways in which Wright used this sensitivity. On the front or southwestern exposure, beneath far overhanging eaves that deflect intense summer sun, twelve contiguous casement windows in the main floor and five in the second admit low winter rays. Facing northeast where eaves are either shallower or altogether absent and casements smaller, the greater wall mass rebuffs prevailing winds. Grilles for three large hot water radiators in the north living room wall are the same size (37″ wide × 44″) as the large windows facing south, whereas grilles for six small radiators in the south wall extending into the dining area correspond in dimension (37″ × 21″) to the six small north facing windows. Wright based the proportions of the living room in part on climate control factors such as window size and radiator capacity. And it is certainly clear that he understood the benefits of passive solar heating: on sunny winter days with temperatures in the low 20s Fahrenheit and the thermostat set at 65 degrees, the furnace remains off until late afternoon.

Three scuttles or vented openings in eave undersurfaces reduce heat build-up in dead air spaces below the roofs whose low pitch retains snow for its insulating value. The eaves also reduce the detrimental impact of elements on upper walls and on windows that may be kept open during gentle rains. These casements are hung so that pairs of two open against each other, from certain angles of vision all but disappearing behind their dividing posts. This increases the already substantial vista, and when windows in the north and south facades are open in tandem there is air movement even on windless days. Two pairs of double doors and two small casements at the east end of the living room work with openings at the west end of the dining room for longitudinal air flow. Since Wright disliked mechanical air conditioning and rarely provided for it even late in his career when it was readily available, he designed his buildings to be self-cooling.

The small alcove at the southeast end of the Tomek living room is an example of Wright's regular use of multi-purpose elements principally for environmental control. The alcove is an intimate spot, very much part of, but psychologically distinct from, the large, more public, space. Book shelves make it a library. With window seat and expanse of plate glass overlooking Longcommon—the aesthetic centerpiece of the Frederick Law Olmsted-designed suburb—it is an observatory. The side seat covers lift for storage. A radiator beneath the window makes the alcove a heat source, a copper water pan (of which there are several throughout the house) under the radiator grille is a humidifier, the picture window and wall sconces are light sources, and flanking casements are ventilators. Most of the functions of this tiny area may be deemed environmental since they regulate indoor climate, establish unique relationships with adjacent interior spaces, and enhance connections with the outside world.

Extending from the alcove's exterior wall is a semi-circular terrace removed from public view by a solid wall under a dramatically cantilevered roof, by native plantings, and by large urns atop side piers. Here is an example of what Wright called "shelter in the open," the state of feeling protected from society and nature while still participating in them. The impulse to control these two kinds of environments explains, for example, architectural features like the long, walled terrace on the south facade of the 1908–09 house in Chicago designed for Frederick C. Robie, who told Wright he wanted to see his neighbors but not be seen by them. Typical among Wright's building objectives was the literal, visual, and symbolic reducing of barriers between inside and out while simultaneously increasing the sense of enclosure. Thus his many publicly visible openings and outdoor extensions of indoor space (like terraces, patios, balconies, entries, or stairs) were screened overhead or peripherally by wall, eave, roof, or planting, the purpose being to domesticate social and natural environments. Exterior worlds gained meaning through design manipulation including such devices as the carefully framed vista, purposefully directed breezes and sun rays, or programmed paths for visitors along complex, always ceremonial, entrance routes.

Other environmentally controlling features were not so subtle. Wherever possible, Wright's buildings were site specific. This could mean, as in the case of Fall-ingwater (1935) in Pennsylvania, his most famous dwelling, that the composition was determined by, and an actual extension of, natural site configurations so that literally, as Wright liked to claim for all his structures, it could not have been erected elsewhere in the same form. Or it could mean that he altered the site to fit particular demands. For example, at the Herbert Jacobs "solar hemicycle" (1948) in Middleton, Wisconsin, soil excavated from outside the front glass wall was packed against the rear wall as a wind deflecting berm, leaving a southerly depression that, when snow-covered, reflected sun as far up inside as a second level rear bedroom deck.

More loosely and more often, site specificity could mean Wright's preference for materials "native" to the area: stone cemented randomly as found in the desert for his own Taliesin West (1937) in Arizona or "field stone" laid up to replicate hillside strata near his Taliesin East (1911) in Wisconsin. He also left materials "natural": stone untrimmed, wood stained and waxed but not painted, glass shaded so as to encourage its reflecting properties. And when, as was often the case, his mostly midwestern construction sites had no memorable features, he simply interpreted the land with long-lined horizontal forms placed close to the ground but, since Midwest terrain can be as rolling as it is flat, rising gently from it as sympathetic comrades.

One of Wright's most lasting contributions was to propose a symbiotic relationship between humanity and the environment, between the built and the natural worlds, a design priority throughout his 72-year practice. His implementations ranged from interpreting terrain to offering passive solar heat, from enhancing reciprocity between indoors and out to providing self-cooling. More than any other American architect to his time, he designed his buildings to work with nature not against it, as he often said. During his long life, few architects followed his lead, and today new technologies and enlightened conventional wisdom make it possible to surpass his achievements. But Wright still personifies the environmentally sensitive attitude at work.

ROBERT TWOMBLY

For Further Reading: Reyner Banham, *The Architecture of the Well-Tempered Environment* (1969); Robert Twombly, *Frank Lloyd Wright: His Life and His Architecture* (1979); Frank Lloyd Wright, *The Natural House* (1954).

XENOBIOTIC

A xenobiotic is an organic or inorganic chemical that is foreign to biological systems. Inorganic elements, such as chromium, nickel, and arsenic, can be toxic and carcinogenic. However, because they are essential to life at very low concentrations, they are not xenobiotic. More toxic elements, such as lead, cadmium, and mercury, have not yet been found to be essential for life and are, therefore, xenobiotic.

In the past hundred years man has created many thousands of organic chemicals that are foreign to the human body. Exposure to these chemicals can produce adverse or toxic effects. These xenobiotics can act directly, or they may be converted into highly reactive forms that bind with crucial large and complex molecules, such as DNA or protein, producing toxic and carcinogenic consequences.

Xenobiotics act at many different sites in complex organisms like humans. Selective toxicity may occur in specific organs because either the xenobiotic is delivered there, or it is metabolized (changed chemically) in that organ to a more reactive form. Examples of toxic organic xenobiotics include carbon tetrachloride, benzene, and organophosphates. Carbon tetrachloride is metabolized to a chemical derivative in the form of a free radical, which breaks down lipids in the liver and destroys the enzymes by which it is metabolized. Benzene is selectively toxic to cells in the bone marrow, where it is metabolized into a toxic derivative. Organophosphate insecticides are selectively toxic to the nervous system, where they inactivate an important enzyme involved in the transmission of impulses between nerve cells.

The toxic action of xenobiotics may also depend on their accumulation in living cells. Although methods of degrading many organic compounds are being developed by genetic engineers, inorganic compounds are more persistent and less easily biodegradable.

MAX COSTA

XERISCAPE

Xeriscape, a recently coined term, refers to a style of water-conserving landscaping currently being promoted in the arid lands of industrialized nations. Derived from the Greek words *xeros*, dry, and *xeron*, dry land, it shares roots with *xeric*, an adjective applied to landscapes low or deficient in moisture available for plant life, and *xerophyte*, plants structurally adapted to such environments, especially by mechanisms that limit transpiration (loss of water vapor) or by which the plant stores water. Xerophytes are native to the world's deserts and other arid lands. They also are the raw materials of xeriscape design. Water conservation is usually enhanced in such "drylandscapes" by the use of drip irrigation systems, which reduce water loss to evaporation, and by water harvesting. The central idea of xeriscaping is to use the relatively thirstiest low-water-use plants, or so-called oasis plantings, near a dwelling place and to gradually move outward in zones of progressively less thirsty plants, until the designed landscape blends with the naturally occurring vegetation.

JOHN M. BANCROFT

Z

ZIGGURATS

Ziggurats are tall, pyramid-shaped structures that first appeared in the valley of the Euphrates and Tigris rivers at the end of the fourth millennium B.C. There is considerable speculation about the purpose of the ziggurats—whether these artificial mounds served as religious structures, as burial places, as sites for astronomical observation, or simply as a means for people to make their mark on the landscape and to dominate their surroundings. They are architectural structures characteristic of the major cities of Mesopotamia (Iraq) from about 2200 to 500 B.C. The name *ziggurat* is from the Babylonian, *sigguratu,* meaning *pinnacle* or *mountain top.*

Ziggurats were constructed on a wide base with each successive level smaller than the one below it, creating a stepped pyramid. Some of the terraces were planted with trees and shrubs, resulting in the appearance of natural rocky cliffs. This practice was perhaps the origin of Semiramis's Hanging Gardens of Babylon of the fifth century B.C., one of the seven wonders of the ancient world. Their average size was 125 feet by 170 feet at the base. The whole construction was solid except for drainage channels and the small temple at the top. Unlike the pyramids of Egypt, which were made of stone, ziggurats were made of clay or clay bricks strengthened with layers of reed and surrounded by a wall of baked bricks. They have been plundered for building materials and flooded and overgrown with reeds from the Euphrates. Consequently, none exists today at its original height of approximately 100 feet.

The Tower of Babel is the most famous ziggurat, and it is generally associated with the ziggurat of the great temple of Marduk in Babylon. The best preserved ziggurat is found at Ur in Sumer. Although ascent to the top of some ziggurats was made by climbing exterior stairways or by a spiral ramp, no means of ascent has been found for almost half of them. Their ornamentation consisted of evenly spaced, high, narrow recesses. The 25 known ziggurats are divided among Sumer, Babylonia, and Assyria.

René Dubos pointed out in *The Wooing of Earth* that the creation of artificial hills has remained as a feature of grand-scale landscape architecture, as seen in the Peking Summer Palace of the 17th century and in the English parks of the 18th century. He suggested that the desire to occupy an elevated position dominating the landscape probably had a biological origin and eventually resulted in an aesthetic experience satisfied by artificial hills, but with no practical utility. As is frequently the case, he concluded, a desire of biological origin eventually evolves into a sociocultural attitude. Thus, from Persepolis, a human settlement of antiquity, to medieval castles and the mansions of 19th century robber barons, physical elevation has come to be associated with social dominance and distinction.

ZONING AND OTHER LAND USE LAWS

Zoning is the most pervasive form of land use regulation in the United States. Zoning predesignates where and what kind of use may exist on a parcel; the minimum and maximum size of a structure; the number of persons who can live in a house; the size of the lot; the limit and location of accessory uses—in short, zoning can substantially restrict an owner's use of their property. Unlike the environmental statutes enacted over the past twenty-five years which are enforced by either federal or state jurisdiction, zoning regulations are enforced at the local governmental level.

At the beginning of the 20th century, land use controls relied on the common law doctrine of nuisance. They were enforced by the use of restrictive covenants placed against individual parcels of land. Rudimentary forms of land use control were enacted by localities to limit the location of noxious uses such as slaughterhouses or stables. But these regulatory schemes were piecemeal and dealt with only a narrow range of land use issues that needed to be confronted by the growing urbanization of the U.S.

Zoning, in contrast, is comprehensive, encompassing not just noxious uses, but all uses of land. Government's power to zone property derives from the police power, the power to promote the general welfare, safety, and public health of its citizens. The first comprehensive zoning ordinance was enacted in New York City in 1916 as an outgrowth of the negative reaction to the construction of an office building at 120 Broadway, a 40-story, 1.4 million square foot, office building that overwhelmed its surrounding environs by substantially reducing light and air. This first ordinance was unique because it established restrictions on all uses that could be built in the city (not just noxious or nuisance uses); limited the height of buildings; and controlled the amount of land upon which a building could be constructed. The statute provided for three separate districts expressed in three different sets of zoning maps: residence, business, and unrestricted districts. Trade and industry were prohibited in residence districts; more noxious trades and businesses were excluded from the second district but were permitted in the unrestricted districts.

By the mid-1920s over 300 municipalities had adopted local zoning ordinances patterned after the New York City statute or the U.S. Department of Commerce Standard Act of 1922.

The village of Euclid, Ohio, a suburb of Cleveland, adopted a comprehensive zoning ordinance, patterned on the New York City code, in 1922. It regulated and restricted the location of trades, industries, apartment houses, two-family houses, single-family houses, lot area, and the size and height of buildings. In 1926 the U.S. Supreme Court upheld the constitutionality of the Euclid Ordinance as a valid exercise of the village's police powers, and a logical, rational, and comprehensive extension of the law of public nuisance. The Court endorsed the fact that the ordinance was comprehensive and regulated all uses, whether they were deemed noxious or not.

Typical "Euclidean" zoning ordinances are cumulative, so that "higher" uses are permitted in areas zoned for "lower" uses such as a business or industrial zone but "lower" uses are not permitted in areas zoned for "higher" uses. Large lot, single-family zones are typically classified as a "higher" use in a Euclidean ordinance. Such dwellings are the only uses permitted in that zoning district. In the early ordinances single-family "higher" uses were permitted in all other zoning districts. Over time, Euclidean zoning ordinances were modified for environmental reasons so that, for example, residential uses were generally excluded from industrial zones.

A Euclidean zoning ordinance is said to be "in accordance with a comprehensive plan," that is, a locality's plan which contains a compilation of that government's standards and goals for future development. This plan may consist of a series of studies, maps, and a zoning text that locates parks, zoning districts, streets, highways, and bridges.

In addition to use restrictions, zoning also controls height, density, and bulk. The ordinance specifies the types, sizes, and heights of structures; where structures can be placed on a building lot; the location of building setbacks so as to permit light and air; and limitations on signage, parking requirements, curb cut locations, minimum lot size, and floor area for residential units. The ordinance is enforced by local officials through the issuance of a building permit. Most jurisdictions have an administrative body, a board of appeals or adjustment, empowered to review the decisions of local zoning officials and, where warranted, grant waivers, if the strict application of the zoning regulation to a specific building lot would cause practical difficulties or undue hardship.

Having acted in the mid-1920s, the Supreme Court did not decide another zoning case for over 50 years. Subsequent cases were occasionally decided by lower federal courts. However, most zoning cases were heard and decided in state courts where zoning law was made. Throughout this period the aims of zoning changed and expanded as localities took zoning initiatives that the Euclid drafters could not have imagined. Zoning has been used to control aesthetics, promote specific uses such as theatres and low-income housing, control population growth, transfer zoning development rights from one building lot to another, preserve historic structures, and require preservation of open space. One of the current issues zoning has addressed resulted from local communities' attempts to control growth.

One such community is the township of Mount Laurel, a township in southern New Jersey consisting of approximately 14,000 acres. From 1950 to 1970 its population quadrupled from approximately 2,800 persons to almost 12,000 persons. Mount Laurel's zoning was primarily Euclidean, separating residential uses from the "lower" zones; over 70% of its land area was zoned for only residential uses. The residential zone permitted only single-family, detached dwelling units; it did not permit mobile homes or attached housing — that is, townhouses or apartments (except for agricultural workers). Minimum lot size controls, together with mandated minimum residential floor area on each building lot, assured that only relatively expensive, single-family homes would be built. However, one portion of Mount Laurel's ordinance permitted the construction of "cluster" units, sometimes known as a "planned unit development" or "PUD." "This scheme differs from the traditional [zoning] in that the type, density and placement of land uses and buildings, instead of being detailed and confined to specific districts by local legislation in advance, is determined by contract, or 'deal,' as to each development between the developer and the municipal administrative authority" (*Southern Burlington County NAACP v. Township of Mount Laurel*, Supreme Court of New Jersey, 1975). While such zones provide for attached, multi-family housing, only persons of medium and upper income were sought as residents.

Each PUD was approved by the township's governing body and was designed to attract "a highly educated and trained population base to support the nearby industrial parks in the township as well as the commercial facilities. The approvals sharply limit the number of apartments having more than one bedroom. Further, they require that the developer must provide in its leases that no school age children shall be permitted to occupy any one-bedroom apartment and that no more than two such children shall reside in any two-bedroom unit" (*Southern Burlington County NAACP v. Township of Mount Laurel*). This requirement and several other low density conditions were contained in a recorded covenant, running with the land. The covenant required that if more than a specific number of children attended the township school system in any school year, the developer was required to pay for the tuition and other school expenses of all excess children.

The New Jersey Supreme Court struck down the Mount Laurel ordinance stating that "over the years Mount Laurel has acted affirmatively to control development and to attract a selective type of growth and that through its zoning ordinances has exhibited eco-nomic discrimination in that the poor have been deprived of adequate housing." In a subsequent decision, known as Mount Laurel II, the Court went further and required that New Jersey's localities affirmatively provide a means of accommodating residents of all economic levels.

Land use regulations entail more and different restrictions than zoning—they include the panoply of environmental statutes regulating wetlands, clean air, water pollution, filling of navigational waters, and historic preservation. These statutes all derive from the same source as zoning—government's inherent power to promote and protect the health, safety, and welfare of its citizens.

MARK A. LEVINE, ESQ.

For Further Reading: Richard F. Babcock and Frederick Bosselman, *Exclusionary Zoning* (1973); Richard F. Babcock, *The Zoning Game* (1967); "The Mount Laurel Symposium," *Seton Hall Law Review* (1984).

Index of Contributors

Subject Index